STATICS STUDY PACK

CHAPTER REVIEWS AND FREE-BODY DAIGRAM WORKBOOK

MECHANICS FOR ENGINEERS

STATICS

THIRTEENTH SI EDITION

R. C. HIBBELER
KAI BENG YAP

SI Conversion by
S. C. Fan

Singapore   London   New York   Toronto   Sydney   Tokyo   Madrid
Mexico City   Munich   Paris   Cape Town   Hong Kong   Montreal

Published in 2012 by
Pearson Education South Asia Pte Ltd
9 North Buona Vista Drive,
#13-01 The Metropolis Tower One
Singapore 138588

Publishing Director: *Mark Cohen*
Acquisitions Editor: *Arunabha Deb*
Project Editor: *Donald Villamero*
Prepress Executive: *Kimberly Yap*

Pearson Education offices in Asia: *Bangkok, Beijing, Ho Chi Minh City, Hong Kong, Jakarta, Kuala Lumpur, Manila, Seoul, Singapore, Taipei, Tokyo*

Authorized adaptation from the United States edition, entitled *Study Pack for Engineering Mechanics: Statics,* Thirteenth Edition, ISBN: 9780132915564 by Hibbeler, Russell C., published by Pearson Prentice Hall, Pearson Education, Inc., Copyright © 2013 by R. C. Hibbeler.

WORLD WIDE adaptation edition published by PEARSON EDUCATION SOUTH ASIA PTE LTD., Copyright © 2013.

Printed in Singapore

12 11 10 9 8 7
16 15

ISBN   978-981-06-9286-5

About the Cover: © iStockphoto.com / James Michael Kruger

www.pearsoned-asia.com

Contents

What's in This Package

The Statics Study Pack was designed to help students improve their study skills. It consists of two study components—a chapter-by-chapter review and a free-body diagram workbook.

- **Chapter-by-Chapter Review and Free-Body Diagram Workbook**—Prepared by Peter Schiavone of the University of Alberta. This resource contains chapter-by-chapter *Statics* review, including key points, equations, and check up questions. The Free-Body Diagram workbook steps students through numerous free-body diagram problems that include full explanations and solutions.

- **Companion Website**—Located at *www.pearsoned-asia.com/hibbeler*, the Companion Website includes the following resources:

  - **Video Solutions**—Complete, step-by-step solution walkthroughs of representative homework problems from each section in the *Statics* text. Developed by Professor Edward Berger of University of Virginia, Video Solutions offer:

    - **Fully worked Solutions**—Showing every step of representative homework problems, to help students make vital connections between concepts.

    - **Self-paced Instruction**—Students can navigate each problem and select, play, rewind, fast-forward, stop, and jump-to sections within each problem's solution.

    - **24/7 Access**—Help whenever students need it, with over 20 hours of helpful review.

To log in to the Companion Website, follow the instructions on the Access for Video Solutions page found in the main book.

Preface

This supplement is divided into two parts. Part I provides a section-by-section, chapter-by-chapter summary of the key concepts, principles and equations from R. C. Hibbeler and Kai Beng Yap's text, *Mechanics for Engineers: Statics,* Thirteenth SI Edition. Part II is a workbook which explains how to draw and use free-body diagrams when solving problems in *Statics*.

Part I: Chapter-by-Chapter Summaries

This part of the supplement provides a section-by-section, chapter-by-chapter summary of the key concepts, principles and equations from R. C. Hibbeler and Kai Beng Yap's text, *Mechanics for Engineers: Statics,* Thirteenth SI Edition. We follow the same section and chapter order as that used in the text and summarize important concepts from each section in easy-to-understand language. We end each chapter summary with a simple set of review questions designed to see if the student has understood the key concepts and chapter objectives.

   This section of the supplement will be useful both as a quick reference guide for important concepts and equations when solving problems in, for example, homework assignments or laboratories and also as a handy review when preparing for any quiz, test, or examination.

Part II: Free-Body Diagram Workbook

   A thorough understanding of how to draw and use a free-body diagram is absolutely essential when solving problems in mechanics.

This workbook consists mainly of a collection of problems intended to give the student practice in drawing and using free-body diagrams when solving problems in *Statics*.

   All the problems are presented as *tutorial* problems with the solution only partially complete. The student is then expected to complete the solution by "filling in the blanks" in the spaces provided. This gives the student the opportunity to *build free-body diagrams in stages* and extract the relevant information from them when formulating equilibrium equations. Earlier problems provide students with partially drawn free-body diagrams and lots of hints to complete the solution. Later problems are more advanced and are designed to challenge the student more. The complete solution to each problem can be found on the back of the page. The problems are chosen from two-dimensional theories of particle and rigid body mechanics. Once the ideas and concepts developed in these problems have been understood and practiced, the student will find that they can be extended in a relatively straightforward manner to accommodate the corresponding three-dimensional theories.

   The workbook begins with a brief primer on free-body diagrams: where they fit into the general procedure of solving problems in mechanics and why they are so important. Next follows a few examples to illustrate ideas and then the workbook problems.

For best results, the student should read the primer and then, beginning with the simpler problems, try to complete and understand the solution to each of the subsequent problems. The student should avoid the temptation to immediately look at the completed solution on the back of the page. This solution should be accessed only as a last resort (after the student has struggled to the point of giving up), or to check the student's own solution after the fact. The idea behind this is very simple: *we learn most when we **do** the thing we are trying to learn*—reading through someone else's solution is not the same as actually working through the problem. In the former, the student gains *information*, in the latter the student gains *knowledge*. For example, how many people learn to swim or drive a car by reading an instruction manual?

Consequently, since the workbook is based on **doing**, the student who persistently solves the problems in the workbook will ultimately gain a thorough, usable knowledge of how to draw and use free-body diagrams.

PETER SCHIAVONE

PART I

Section-by-Section, Chapter-by-Chapter Summaries with Review Questions and Answers

1

General Principles

MAIN GOALS OF THIS CHAPTER:

- To introduce the basic ideas of *Mechanics*.
- To give a concise statement of Newton's laws of motion and gravitation.
- To review the principles for applying the SI system of units.
- To examine standard procedures for performing numerical calculations.
- To outline a general guide for solving problems.

1.1 MECHANICS

Mechanics is that branch of the physical sciences concerned with the behavior of bodies subjected to the action of forces. The subject of mechanics is divided into two parts:

- *statics*—the study of objects in equilibrium (objects either at rest or moving with a constant velocity).
- *dynamics*—the study of objects with accelerated motion.

Although statics can be considered as a special case of dynamics (in which the acceleration is zero), it deserves special treatment since many objects are designed with the intention that they remain in equilibrium.

1.2 FUNDAMENTAL CONCEPTS

BASIC QUANTITIES

- **Length, time, mass, force**

IDEALIZATIONS

Mathematical models or idealizations are used in mechanics to simplify the theory. The more common ones, in order of sophistication, are:

- **Particle**—a *particle* has a mass but a size that can be neglected i.e., the geometry of the body is ignored. A particle is often represented by a *point* in space.
- **Rigid Body**—a *rigid body* has a mass and a size (shape) but it is assumed that any changes in shape can be neglected i.e., the geometry of the body *is* taken into account but any deformations (changes in shape) are ignored. Consequently, the material properties of the body can be ignored. A rigid body is often represented as a collection of particles in which all the particles remain at a fixed distance from each other before and after applying a load.
- **Deformable or Elastic Body**—a deformable body has a mass, a size (shape) and the deformations (changes in shape) of the body are taken into account. Hence the material properties of the body must be considered in describing the behavior of the body.
- **Concentrated Force**—A concentrated force represents the effect of a loading which is assumed to act at a point on a body. This idealization requires that the area over which the load is applied is very small compared to the overall size of the body e.g., contact force between wheel and ground.

NEWTON'S THREE LAWS OF MOTION

Newton's laws apply to the motion of a particle as measured from a nonaccelerating (inertial) reference frame.

- **First Law**—a particle originally at rest or moving in a straight line with constant velocity, will remain in this state provided the particle is not subjected to an unbalanced force.
- **Second Law**—a particle acted upon by an unbalanced force **F** experiences an acceleration **a** that has the same direction as the force and a magnitude directly proportional to the force i.e.

$$\mathbf{F} = m\mathbf{a}$$

- **Third Law**—The mutual forces of action and reaction between two particles are equal, opposite and collinear.

NEWTON'S LAW OF GRAVITATIONAL ATTRACTION

$$F = G\frac{m_1 m_2}{r^2}$$

F = force of gravitation between two particles
G = universal constants of gravitation
m_1, m_2 = mass of each of the two particles
r = distance between the two particles

MASS AND WEIGHT

- **Mass** is a (scalar) property of matter that does not change from one location to another. In other words, mass is an *absolute* quantity
- **Weight** is a *force* (and hence has a magnitude and direction) which refers to the gravitational attraction of the earth on a quantity of mass m. Weight is not an absolute quantity. Its magnitude depends on the elevation at which the mass is located. We write the magnitude of weight as $W = mg$ where g is termed the acceleration due to gravity.

1.3 UNITS OF MEASUREMENT

The four basic quantities *force, mass, length* and *time* are related by Newton's 2nd law. Hence, the units used to define these quantities are not independent i.e., three of the four units are called *base units* (arbitrarily defined) and the fourth unit a *derived unit* (derived from Newton's 2nd law).

SI UNITS (INTERNATIONAL SYSTEM OF UNITS)

- In the SI system, the unit of force, the *newton*, is a derived unit. The meter, second and kilogram are base units.

- One *newton* is equal to a force required to give one kilogram of mass an acceleration of 1 m/s^2.

- In newtons, the weight of a body has magnitude

$$W = mg \text{ where } g = 9.81 \text{ m/s}^2.$$

The following table summarizes the SI units.

Name	Length	Time	Mass	Force
International System (SI)	meter (m)	second (s)	kilogram (kg)	newton* $\left(N = \dfrac{\text{kg·m}}{\text{s}^2} \right)$

Derived Unit

1.4 THE INTERNATIONAL SYSTEM OF UNITS

PREFIXES

When a numerical quantity is either very large or very small, the units used to define its size may be modified by using a prefix. For example:

	Exponential Form	Prefix	SI Symbol
1 000 000 000	10^9	giga	G
1 000 000	10^6	mega	M
1 000	10^3	kilo	k
0.001	10^{-3}	milli	m
0.000 001	10^{-6}	micro	μ
0.000 000 001	10^{-9}	nano	n

RULES FOR USE

You should know the rules for the proper use of the various SI symbols. These are used extensively in engineering practice throughout the world.

1.5 NUMERICAL CALCULATIONS

It is important that the numerical answers to any problem encountered in engineering practice be reported with both justifiable accuracy and appropriate significant figures.

DIMENSIONAL HOMOGENEITY

Each term in any equation used to describe a physical process must be expressed in the same units i.e., the terms must be *dimensionally homogeneous*. Algebraic manipulations of an equation can be checked, in part, by verifying that the equation remains dimensionally homogeneous.

SIGNIFICANT FIGURES

The accuracy of a number is specified by the number of significant figures it contains. A *significant figure* is any digit, including a zero, provided it is not used to specify the location of the decimal point for the number. For example, 0.00546 and 2500 expressed to three significant figures would be $5.46 \times \left(10^{-3}\right)$ and $2.50 \times \left(10^{3}\right)$, respectively (*engineering notation*).

ROUNDING OFF NUMBERS

For numerical calculations, the accuracy of the solution of a problem (generally) can never be better than the accuracy of the problem data. Consequently, a calculated result should always be *rounded off* to an appropriate number of significant figures. To convey appropriate accuracy, there are rules for rounding off numbers. You should know these.

CALCULATIONS

Perform numerical calculations to *several significant figures* and then report the final answer to *three* significant figures.

1.6 GENERAL PROCEDURE FOR ANALYSIS

- The most effective way to learn engineering mechanics is to *solve problems*.
- You must present your work in a logical and orderly manner as follows:
  - Read the problem carefully and try to establish a link between the actual physical situation and the appropriate part of the theory studies.
  - Draw any necessary diagrams and tabulate the problem data.
  - Apply the relevant principles.
  - Solve the necessary equations algebraically, as far as practical, then, making sure they are dimensionally homogeneous, use a consistent set of units and complete the solution numerically. Report the answer with no more significant figures than the accuracy of the given data.
  - Study the answer and see if it *makes sense* physically—in the context of the physical problem.

HELPFUL TIPS AND SUGGESTIONS

- The *language* of engineering mechanics is *mathematics*. Consequently, make sure you review/re-read the necessary mathematical notation/concepts *as they arise* in your mechanics course (trying to review all of the necessary mathematics *at once* is not recommended—there's just too much to digest at one time). You should aim to achieve *fluency* in basic mathematical techniques/notation so that your learning of mechanics is not distracted by trying to remember things which your instructor *assumes* you know e.g., how to solve linear systems of algebraic equations, how to perform basic vector algebra, differentiation and integration etc.

- *Remember* that in solving problems from engineering mechanics you are solving real practical problems and producing real data with physical significance. Thus, you are responsible for making sure your results are correct, consistent and well-presented. Get into the habit of doing this *now* so that it will become second nature by the time you graduate. In the world of professional engineering you have a responsibility to your profession and to the many people that will use the product you will help to design, manufacture or implement.

REVIEW QUESTIONS: TRUE OR FALSE[1]?

1. The subject called *Statics* studies only bodies which are at rest.
2. A *particle* has a mass but negligible shape/size.
3. A rigid body has a mass but negligible shape/size.
4. Newton's three laws of motion can be proved mathematically.
5. Weight is a property of matter that does not change from one location to another.
6. In the *SI* system of units, the newton is a derived unit.
7. When performing numerical calculations, the final answer should be reported to *three* significant figures.
8. In an equation it's permitted to have different terms expressed in different units. This is referred to as dimensionally inhomogeneous.

---

[1] 1. F   2. T   3. F   4. F   5. F   6. T   7. T   8. F

2

Force Vectors

MAIN GOALS OF THIS CHAPTER

In this chapter we define scalars, vectors and vector operations and use them to analyze forces acting on objects. Specifically:

- To show how to add forces and resolve them into components.
- To express force and position in Cartesian vector form.
- To explain how to determine a vector's magnitude and direction.
- To introduce the dot product and use it to find the angle between two vectors or the projection of one vector onto another.

2.1 SCALARS AND VECTORS

Most of the physical quantities in mechanics can be represented by either *scalars* or *vectors*:

- A *scalar* is a real number e.g., mass, time, volume and length are represented by scalars.
- A *vector* has both magnitude and direction e.g., force, velocity and acceleration are vectors.

2.2 VECTOR OPERATIONS

MULTIPLICATION OR DIVISION OF A VECTOR BY A SCALAR

- The product of a vector **A** and a scalar a is a vector $a\mathbf{A}$ with magnitude $|a\mathbf{A}| = |a|\,|\mathbf{A}|$. The direction is the same as that of **A** if a is positive and opposite to that of **A** if a is negative.

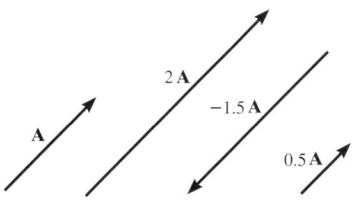

Scalar Multiplication and Division

VECTOR ADDITION

- Two vectors **A** and **B** can be added to form a *resultant* vector $\mathbf{R} = \mathbf{A} + \mathbf{B}$ by using the parallelogram law. If the two vectors are *collinear* (both vectors have the same line of action), the resultant is formed by an algebraic or scalar addition.

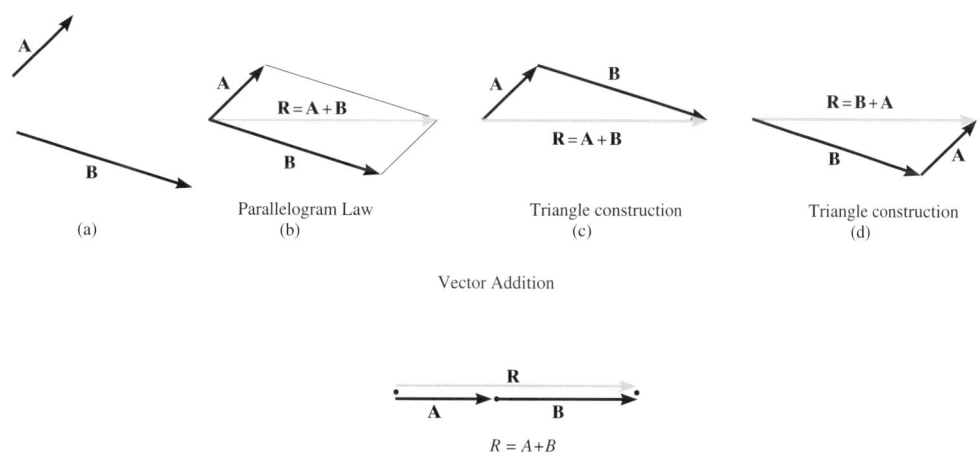

(a)	Parallelogram Law (b)	Triangle construction (c)	Triangle construction (d)

Vector Addition

R

A   B

$R = A + B$

Addition of collinear vectors

RESOLUTION OF A VECTOR

- A vector may be resolved into *components* having known lines of action by using the parallelogram law.

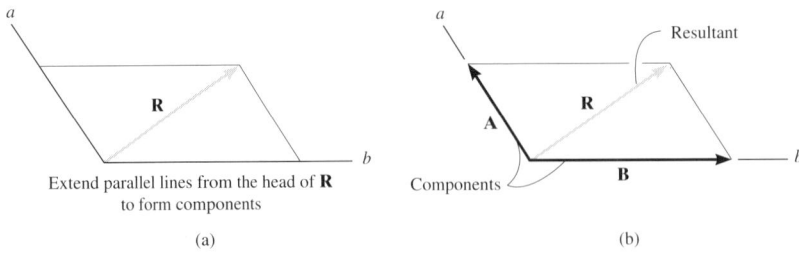

Resolution of a vector

2.3 VECTOR ADDITION OF FORCES

- A force is a *vector* quantity since it has a specified magnitude and direction. Consequently, forces are added together or resolved into components using the rules of vector algebra.

- Two common problems in statics involve either finding the resultant force given its components or resolving a known force into components.

- Often the magnitude of a resultant force can be determined from the law of cosines, while its direction is determined from the law of sines:

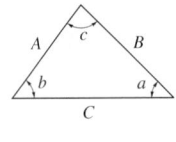

Sine law:
$$\frac{A}{\sin a} = \frac{B}{\sin b} = \frac{C}{\sin c}$$

Cosine law:
$$C = \sqrt{A^2 + B^2 - 2AB \cos c}$$

2.4 ADDITION OF A SYSTEM OF COPLANAR FORCES

In the plane, a force can be resolved into two rectangular components. There are two separate notations for doing this:

- *Scalar Notation*—we write the force **F** as (F_x, F_y) where F_x and F_y are the scalar components of the force **F** in the directions of the positive x- and y-axes, respectively. If F_x and F_y are negative, it means that $|F_x|$ and $|F_y|$ are directed along the *negative* x- and y-axes, respectively.

- *Cartesian Vector Notation*—we write the force **F** as

$$\mathbf{F} = F_x \mathbf{i} + F_y \mathbf{j},$$

where **i** and **j** represent the positive directions of the x- and y-axes, respectively.

COPLANAR FORCE RESULTANTS

- The resultant of several coplanar forces can easily be determined if an x, y-coordinate system is established and the forces are resolved along the axes. For example,

$$\mathbf{F}_1 = F_{1x}\mathbf{i} + F_{1y}\mathbf{j},$$
$$\mathbf{F}_2 = F_{2x}\mathbf{i} + F_{2y}\mathbf{j},$$
$$\mathbf{F}_3 = F_{3x}\mathbf{i} + F_{3y}\mathbf{j},$$

then the resultant is given by

$$\begin{aligned}\mathbf{F}_R &= \mathbf{F}_1 + \mathbf{F}_2 + \mathbf{F}_3 \\ &= \left(F_{1x} + F_{2x} + F_{3x}\right)\mathbf{i} + \left(F_{1y} + F_{2y} + F_{3y}\right)\mathbf{j} \\ &= (F_{Rx})\mathbf{i} + \left(F_{Ry}\right)\mathbf{j}.\end{aligned}$$

- In the general case, the x and y components of the resultant of any number of coplanar forces can be represented symbolically by the algebraic sum of the x and y components of all the forces i.e.

$$F_{Rx} = \sum F_x,$$
$$F_{Ry} = \sum F_y.$$

- The magnitude and direction of the resultant force are given by:

$$|\mathbf{F}_R| = F_R = \sqrt{F_{Rx}^2 + F_{Ry}^2}, \tag{2.0}$$
$$\theta = \tan^{-1}\left|\frac{F_{Ry}}{F_{Rx}}\right|,$$

respectively.

2.5 CARTESIAN VECTORS

- A Cartesian coordinate system is often used to solve problems in three dimensions. The coordinate system is *right-handed* which means that the thumb of the right hand points in the direction of the positive z-axis when the right hand fingers are curled about this axis and directed from the positive x toward the positive y-axis.

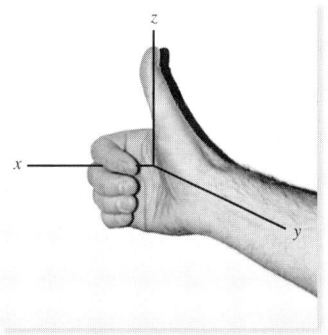

Right-handed coordinate system.

- The unit vector from a vector $\mathbf{A}$ is given by $\frac{\mathbf{A}}{A}$ where $A \neq 0$ is the magnitude of vector $\mathbf{A}$. The unit vector is dimensionless and defines the direction of vector $\mathbf{A}$.
- The positive directions of the x, y, z axes are defined by the Cartesian unit vectors $\mathbf{i}, \mathbf{j}, \mathbf{k}$, respectively. Consequently, any vector $\mathbf{A}$ with scalar components A_x, A_y and A_z can be written in the Cartesian vector form

$$\mathbf{A} = A_x\mathbf{i} + A_y\mathbf{j} + A_z\mathbf{k}. \tag{2.1}$$

- The magnitude of vector $\mathbf{A}$ is given by

$$|\mathbf{A}| = A = \sqrt{A_x^2 + A_y^2 + A_z^2}. \tag{2.2}$$

- The direction of vector $\mathbf{A}$ is defined by the angles α, β and γ measured between the tail of $\mathbf{A}$ and the positive x, y, z axes located at the tail of $\mathbf{A}$.

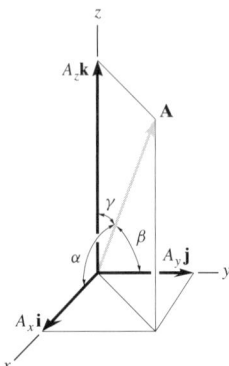

- The angles α, β and γ are found from their *direction cosines*

$$\cos\alpha = \frac{A_x}{A}, \cos\beta = \frac{A_y}{A}, \cos\gamma = \frac{A_z}{A}, \quad \cos^2\alpha + \cos^2\beta + \cos^2\gamma = 1. \tag{2.3}$$

This means that only two of the angles α, β and γ have to be specified—the third can be found from $\cos^2\alpha + \cos^2\beta + \cos^2\gamma = 1$.

2.6 ADDITION AND SUBTRACTION OF CARTESIAN VECTORS

- To find the resultant of a concurrent force system, express each force as a Cartesian vector and add the $\mathbf{i}, \mathbf{j}, \mathbf{k}$ components of all the forces in the system.

2.7 POSITION VECTORS

- The position vector $\mathbf{r}$ is defined as a fixed vector which locates a point in space relative to another point. For example, from the origin of coordinates O, the point in space $P(x, y, z)$ has position vector $\mathbf{r} = x\mathbf{i} + y\mathbf{j} + z\mathbf{k}$.
- More generally, the position vector may be directed from point A to point B in space. In this case, the position vector is again denoted by $\mathbf{r}$ (or sometimes $\mathbf{r}_{AB}$) and is given by

$$\mathbf{r}_{AB} = \mathbf{r}_B - \mathbf{r}_A \tag{2.4}$$

where $\mathbf{r}_B$ and $\mathbf{r}_A$ are the position vectors of A and B from the origin of coordinates O. For example, if $A(x_A, y_A, z_A)$ and $B(x_B, y_B, z_B)$ then

$$\mathbf{r}_A = x_A\mathbf{i} + y_A\mathbf{j} + z_A\mathbf{k}, \quad \mathbf{r}_B = x_B\mathbf{i} + y_B\mathbf{j} + z_B\mathbf{k}, \quad \mathbf{r}_{AB} = (x_B - x_A)\mathbf{i} + (y_B - y_A)\mathbf{j} + (z_B - z_A)\mathbf{k}.$$

2.8 FORCE VECTOR DIRECTED ALONG A LINE

- A force **F** (with magnitude F) acting in the direction of a line represented by a position vector **r** can be written in the form

$$\mathbf{F} = F\left(\frac{\mathbf{r}}{r}\right) = F\mathbf{u}$$

where $\mathbf{u} = \frac{\mathbf{r}}{r}$ is a unit vector representing the direction of the line.

2.9 DOT PRODUCT

- The dot product is used to determine

    - The angle between two vectors.
    - The projection of a vector in a specified direction.

- The dot product of two vectors **A** and **B** is defined as

$$\mathbf{A} \cdot \mathbf{B} = A_x B_x + A_y B_y + A_z B_z = AB\cos\theta, \tag{2.5}$$

where A and B are the magnitudes of **A** and **B**, respectively, and θ is the angle between the tails of **A** and **B**. Consequently,

$$\theta = \cos^{-1}\left(\frac{\mathbf{A} \cdot \mathbf{B}}{AB}\right).$$

- The dot product is commutative ($\mathbf{A} \cdot \mathbf{B} = \mathbf{B} \cdot \mathbf{A}$), and distributive $\mathbf{A} \cdot (\mathbf{B} + \mathbf{D}) = \mathbf{A} \cdot \mathbf{B} + \mathbf{A} \cdot \mathbf{D}$. Also, if $a \in \mathbb{R}$:

$$a(\mathbf{A} \cdot \mathbf{B}) = (a\mathbf{A}) \cdot \mathbf{B} = \mathbf{A} \cdot (a\mathbf{B}) = (\mathbf{A} \cdot \mathbf{B})a. \tag{2.6}$$

- In some engineering applications, you must resolve a vector into components which are parallel and perpendicular (normal) to a given line (direction). The component of vector **A** in the direction specified by the unit vector **u** is given by

$$A_{\parallel} = \mathbf{A} \cdot \mathbf{u} = A\cos\theta. \tag{2.7}$$

This component is also referred to as the scalar projection of **A** onto the line with direction **u** or the component of vector **A** parallel to a line with direction **u**. Clearly, the vector $\mathbf{A}_{\parallel}$ is defined by $\mathbf{A}_{\parallel} = \left(A_{\parallel}\right)\mathbf{u}$.

- Once the parallel component has been determined, we can determine the component of **A** perpendicular (or normal) to a line with direction **u** by

$$A_{\perp} = \sqrt{A^2 - A_{\parallel}^2}. \tag{2.8}$$

- Clearly, in terms of vectors,

$$\mathbf{A} = \mathbf{A}\| + \mathbf{A}\perp.$$

HELPFUL TIPS AND SUGGESTIONS

- Be aware of the differences between vectors and scalars. For example, force, velocity and acceleration are *vectors* while speed, time and distance are *scalars*. If you are asked to find a vector (e.g., a force) you must report *both* magnitude and direction.

- Vector operations are essential in describing the basic principles of mechanics. Make sure you take the time to review basic vector algebra. It doesn't take long but the payoff (in terms of your effectiveness in mechanics) is significant.

REVIEW QUESTIONS

1. How are the scalar components of a vector defined in terms of a Cartesian coordinate system?

2. If you know the scalar components of a vector, how can you determine its magnitude and direction?

3. Suppose you know the coordinates of two points A and B. How do you determine the scalar components of the position vector of point B relative to point A?

4. How do you identify a right-handed coordinate system?

5. What are the direction cosines of a vector? If you know them, how do you determine the components of the vector?

6. What is the definition of the dot product? Is the dot product a vector or a scalar?

7. If the dot product of two vectors is zero, what does that mean?

8. If you know the components of two vectors $\mathbf{A}$ and $\mathbf{B}$, how can you determine their dot product?

9. Simplify
   i. $\mathbf{A} \cdot (\mathbf{B} + \mathbf{D})$
   ii. $(a\mathbf{A}) \cdot \mathbf{B}$ where a is a scalar.

10. How can you use the dot product to determine the components of a vector parallel and perpendicular to a line?

3

Equilibrium of a Particle

MAIN GOALS OF THIS CHAPTER

In this chapter we:

- Introduce the concept of the free-body diagram for an object modelled as a particle (an object with mass but negligible shape/size—henceforth referred to simply as a *particle*).
- Show how to solve particle equilibrium problems using the equations of equilibrium.

3.1 CONDITION FOR THE EQUILIBRIUM OF A PARTICLE

A particle is in *equilibrium* provided it is at rest if originally at rest (*static equilibrium*) or has a constant velocity if originally in motion.

- To maintain equilibrium it is necessary and sufficient that the *resultant force* acting on a particle be equal to zero. In terms of Newton's laws of motion, this is expressed mathematically as:

$$\sum \mathbf{F} = \mathbf{0} \qquad (3.0)$$

where $\sum \mathbf{F}$ is the vector sum of all forces acting on the particle.

3.2 THE FREE-BODY DIAGRAM

To apply the equation of equilibrium (3.0), we must account for all the known and unknown forces ($\sum \mathbf{F}$) which act on the particle. The easiest way to do this is to draw a *free-body diagram*.

- A free-body diagram is simply a sketch which shows the particle 'free' from its surroundings with *all* the forces that act *on* it. There are three main steps:

  ◆ **Draw Outlined Shape.** Imagine the particle to be isolated or cut 'free' from its surroundings by drawing its outlined shape.

  ◆ **Show all Forces.** Indicate on this sketch *all* the forces that act *on the particle*—it may help to carefully trace around the particle's boundary, noting each force acting.

  ◆ **Identify Each Force.** The forces which are known should be labeled with their proper magnitudes and directions. Letters are used to represent the magnitudes and directions of forces that are unknown.

∗ **Connections**—there are two types of connections often encountered in particle equilibrium problems:

(a) *Springs*—The magnitude of force exerted on a linear elastic spring with stiffness k, deformed a distance s measured from its unloaded position is

$$F = ks. \tag{3.1}$$

Here s is determined from the difference in the spring's deformed length and its undeformed length.

(b) *Cables and Pulleys*—Assume cables (or cords) have negligible weight and cannot stretch. Also, a cable can support only tension which always acts in the direction of the cable.

There are several examples and practice problems, as well as much more on drawing free-body diagrams in Part II of this study pack.

3.3 COPLANAR FORCE SYSTEMS

Coplanar force equilibrium problems for a particle can be solved using the following procedure.

1. Free-Body Diagram

- Establish the x, y axes in any suitable orientation.
- Label all the known and unknown force magnitudes and directions on the diagram.
- The sense of a force having an unknown magnitude can be assumed.

2. Equations of Equilibrium

- Resolve each force into its $\mathbf{i}$ (x) and $\mathbf{j}$ (y) components and apply the *scalar equations of equilibrium*

$$\sum F_x = 0, \quad \sum F_y = 0. \tag{3.2}$$

(the algebraic sum of the x and y components of all the forces acting on the particle equal to zero).

- Components are positive if they are directed along a positive axis and negative if they are directed along a negative axis.
- If more than two unknowns exist and the problem involves a spring, apply $F = ks$ to relate the spring force to the deformation s of the spring.
- If the solution yields a negative result, this indicates the sense of the force is the reverse of that shown on the free-body diagram.

3.4 THREE-DIMENSIONAL FORCE SYSTEMS

Three-dimensional force equilibrium problems for a particle can be solved using the following procedure.

1. Free-Body Diagram

- Establish the x, y, z axes in any suitable orientation.
- Label all the known and unknown force magnitudes and directions on the diagram.
- The sense of a force having an unknown magnitude can be assumed.

2. Equations of Equilibrium

- When it's easy to do so, resolve each force into its $\mathbf{i}(x)$, $\mathbf{j}(y)$, and $\mathbf{k}(z)$ components and apply the *scalar equations of equilibrium*

$$\sum F_x = 0, \quad \sum F_y = 0, \quad \sum F_z = 0. \tag{3.3}$$

(the algebraic sum of the x, y and z components of all the forces acting on the particle equal to zero).

- If the three-dimensional geometry appears difficult, then first express each force as a Cartesian vector and substitute these vectors into the *vector equation of equilibrium* (3.0)

$$\sum \mathbf{F} = \mathbf{0}$$

and then set the $\mathbf{i}$, $\mathbf{j}$ and $\mathbf{k}$ components equal to zero.

- If the solution yields a negative result, this indicates the sense of the force is the reverse of that shown on the free-body diagram.

HELPFUL TIPS AND SUGGESTIONS

- Since we must account for *all the forces acting on the (object modelled as a) particle* when applying the equations of equilibrium, the importance of *first* drawing a free-body diagram cannot be over-emphasized.
- **One of the most common mistakes made in writing equilibrium conditions is forgetting to include all of the forces acting.** When drawn carefully, a free-body diagram will make it easier for you to identify *all* the forces acting.
- Use Part II of this supplement to get lots of practice in drawing free-body diagrams and applying the equations of equilibrium for a particle.

REVIEW QUESTIONS

1. What is meant by 'equilibrium of a particle'?
2. What do you know about the sum of the external forces acting on an object modelled as a particle in equilibrium?
3. What are the steps in drawing a free-body diagram?
4. What is a coplanar force system?
5. What is a three-dimensional system of forces?
6. What is the difference between equilibrium of coplanar and three-dimensional force systems?
7. What is the relation between the magnitude of the force exerted on a linear spring and the change in its length?
8. The following is the correct free-body diagram for the ring at E. True or False?

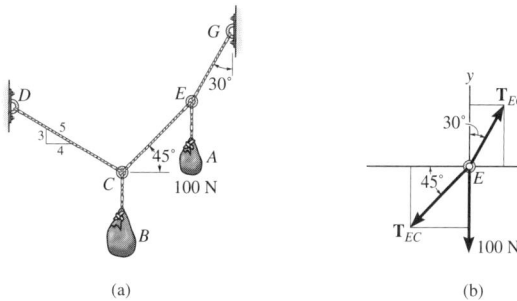

(a)   (b)

4

Force System Resultants

MAIN GOALS OF THIS CHAPTER:

- To discuss the concept of the *moment of a force* and show how to calculate it in two and three dimensions.
- To provide a method for finding the moment of a force about a specified axis.
- To define the moment of a couple.
- To present methods for determining the resultants of nonconcurrent force systems.
- To indicate how to reduce a simple distributed loading to a resultant force having a specified location.

4.1 MOMENT OF A FORCE—SCALAR FORMULATION

- The *moment* $\mathbf{M}_O$ of a force $\mathbf{F}$ about an axis passing through a specific point O provides a measure of the tendency of the force to cause the body to rotate about the axis (sometimes referred to as a *torque*). Clearly the moment is a *vector* and so has *both magnitude and direction*.
- The *magnitude* of the moment is determined from $M_0 = Fd$, where d is the perpendicular or shortest distance from point O to the line of action of the force $\mathbf{F}$.
- Using the right-hand rule, the *direction* (sense) of rotation is indicated by the fingers with the thumb directed along the moment axis or line of action of the moment.

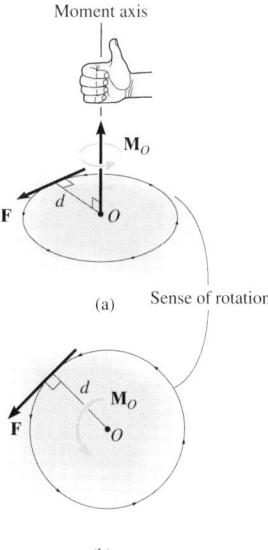

Moment axis

$\mathbf{M}_O$

$\mathbf{F}$ d O

(a)   Sense of rotation

d $\mathbf{M}_O$

$\mathbf{F}$ O

(b)

- If a system of forces lies in the x–y plane, then the moment produced by each force about point O will be directed along the z-axis. The resultant moment $\mathbf{M}_{R_O}$ of the system can be determined by simply adding the moments of all the forces algebraically since all the moment vectors are *collinear* i.e.

$$\curvearrowleft +M_{R_O} = \sum F_d.$$

Here the counterclockwise curl written alongside the equation indicates that the moment of any force will be positive if it is directed along the z-axis, whereas a negative moment is directed along the z-axis.

4.2 CROSS PRODUCT

- The cross product of two vectors $\mathbf{A}$ and $\mathbf{B}$ yields a *vector* $\mathbf{C}$ written $\mathbf{C} = \mathbf{A} \times \mathbf{B}$.

    ◆ The *magnitude* of vector $\mathbf{C}$ is given by $AB \sin \theta$ where θ is the angle between the tails of $\mathbf{A}$ and $\mathbf{B}$.

    ◆ Vector $\mathbf{C}$ has a direction which is perpendicular to the plane containing $\mathbf{A}$ and $\mathbf{B}$ such that $\mathbf{C}$ is specified by the right-hand rule i.e., curling the fingers of the right hand from vector $\mathbf{A}$ (cross) to vector $\mathbf{B}$, the thumb then points in the direction of $\mathbf{C}$.

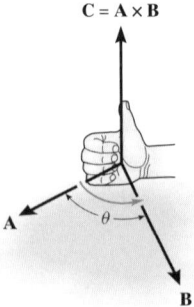

LAWS OF OPERATION

- $\mathbf{A} \times \mathbf{B} \neq \mathbf{B} \times \mathbf{A}$, rather $\mathbf{A} \times \mathbf{B} = -\mathbf{B} \times \mathbf{A}$.
- $a(\mathbf{A} \times \mathbf{B}) = (a\mathbf{A}) \times \mathbf{B} = \mathbf{A} \times (a\mathbf{B}) = (\mathbf{A} \times \mathbf{B})a$.
- $\mathbf{A} \times (\mathbf{B} + \mathbf{D}) = (\mathbf{A} \times \mathbf{B}) + (\mathbf{A} \times \mathbf{D})$.

CARTESIAN VECTOR FORMULATION

- To find the cross product of any two Cartesian vectors $\mathbf{A}$ and $\mathbf{B}$ we use the determinant

$$\mathbf{A} \times \mathbf{B} = \begin{vmatrix} \mathbf{i} & \mathbf{j} & \mathbf{k} \\ A_x & A_y & A_z \\ B_x & B_y & B_z \end{vmatrix}. \tag{4.0}$$

- The following useful results can be obtained by applying the right-hand rule and don't need to be memorized.

$\mathbf{i} \times \mathbf{j} = \mathbf{k}$,	$\mathbf{i} \times \mathbf{k} = -\mathbf{j}$,	$\mathbf{i} \times \mathbf{i} = \mathbf{0}$,
$\mathbf{j} \times \mathbf{k} = \mathbf{i}$,	$\mathbf{j} \times \mathbf{i} = -\mathbf{k}$,	$\mathbf{j} \times \mathbf{j} = \mathbf{0}$,
$\mathbf{k} \times \mathbf{i} = \mathbf{j}$,	$\mathbf{k} \times \mathbf{j} = -\mathbf{i}$,	$\mathbf{k} \times \mathbf{k} = \mathbf{0}$.

4.3 MOMENT OF A FORCE—VECTOR FORMULATION

In three-dimensions it is preferable to use the vector cross product to determine the moment:

- The moment $\mathbf{M}_O$ of a force $\mathbf{F}$ about the moment axis passing through point O and perpendicular to the plane containing O and $\mathbf{F}$ can be represented by

$$\mathbf{M}_O = \mathbf{r} \times \mathbf{F} = \begin{vmatrix} \mathbf{i} & \mathbf{j} & \mathbf{k} \\ r_x & r_y & r_z \\ F_x & F_y & F_z \end{vmatrix}, \tag{4.1}$$

where $\mathbf{r}$ represents a position vector drawn *from O to any point* lying on the line of action of $\mathbf{F}$.

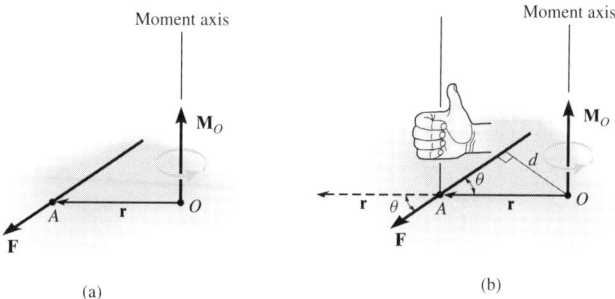

(a)   (b)

PRINCIPLE OF TRANSMISSIBILITY

- Since in (4.1), **r** can extend from O to *any* point on the line of action of **F**, **F** is a *sliding vector* and can act at *any* point along its line of action and create the *same moment* about point O.

RESULTANT MOMENT OF A SYSTEM OF FORCES

- If a body is acted upon by a system of n forces, the resultant moment about O is just the vector sum of the individual moments:

$$\mathbf{M}_{R_O} = \sum_{i=1}^{n}(\mathbf{r}_i \times \mathbf{F}_i).$$

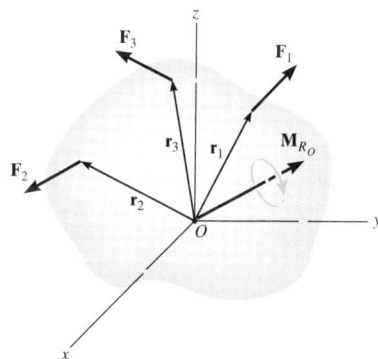

4.4 PRINCIPLE OF MOMENTS

- The *principle of moments* (Varignon's theorem) states that *the moment of a force about a point is equal to the sum of the moments of the force's components about the point.* This is particularly convenient since it is often easier to determine the moments of a force's components rather than the moment of the force itself (e.g., in two dimensions).

4.5 MOMENT OF A FORCE ABOUT A SPECIFIED AXIS

Recall that when the moment of a force is computed about a point, the moment and its axis are *always* perpendicular to the plane containing the force and the moment arm. In some problems it is important to find the *component* of this moment about a *specified axis* that passes through the point.

- In terms of a scalar analysis, the moment of a force **F** about a specified axis can be determined provided the perpendicular distance d_a from both the force line of action and the axis can be determined. Then $M_a = Fd_a$.
- In terms of vector analysis, the component $M_a = \mathbf{u}_a \cdot (\mathbf{r} \times \mathbf{F})$, where $\mathbf{u}_a$ defines the direction of the axis and **r** is directed from *any point* on the axis to *any point* on the line of action of the force. The quantity $\mathbf{u}_a \cdot (\mathbf{r} \times \mathbf{F})$ is called a triple scalar product and can be computed using the determinant

$$\mathbf{u}_a \cdot (\mathbf{r} \times \mathbf{F}) = \begin{vmatrix} u_{a_x} & u_{a_y} & u_{a_z} \\ r_x & r_y & r_z \\ F_x & F_y & F_z \end{vmatrix}$$

Once M_a is determined, we can express $\mathbf{M}_a$ as a Cartesian vector, namely $\mathbf{M}_a = M_a \mathbf{u}_a$.

- If M_a is calculated as a negative scalar then the sense of direction of $\mathbf{M}_a$ is opposite to $\mathbf{u}_a$.

4.6 MOMENT OF A COUPLE

- A couple is defined as two parallel forces that have the same magnitude, opposite directions, and are separated by a perpendicular distance d. Since the resultant force is zero, the only effect of a couple is to produce a rotation in a specified direction.

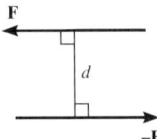

- The moment produced by a couple is called a *couple moment* which is a free vector and, as a result, it causes the same effect of rotation on a body regardless of where the couple moment is applied to the body. Consequently, the couple moment can be computed about *any* point. For convenience, this point is often chosen on the line of action of one of the forces in the couple.
- The couple moment is easily determined from the vector formulation $\mathbf{M} = \mathbf{r} \times \mathbf{F}$ where **r** is directed from *any point* on the line of action of one of the forces to any point on the line of action of the other force **F**.
- A *resultant couple moment* is simply the vector sum of all the couple moments of the system.

4.7 SIMPLIFICATION OF A FORCE AND COUPLE SYSTEM

A force has the effect of both translating and rotating a body and the amount by which it does so depends on where and how the force is applied. It is possible, however, to *replace* a system of forces and couple moments acting on a body with an *equivalent single* resultant force and couple moment acting at a specified point O. Here *equivalent* means that the system and the resultant each produce the same *external effects* of translation and rotation. There are two cases to consider:

- **Point O is on the Line of Action of the Force**—simply slide the force along its line of action to the point O.

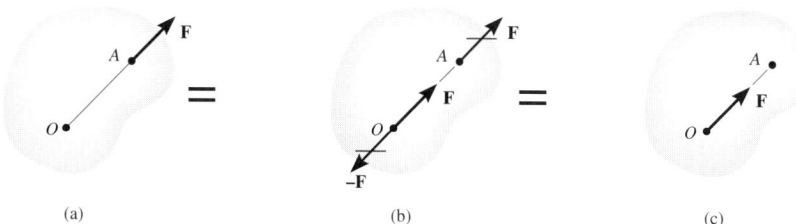

(a)   (b)   (c)

- **Point O is not on the Line of Action of the Force**—move the force to the point O and add a couple moment anywhere to the body.

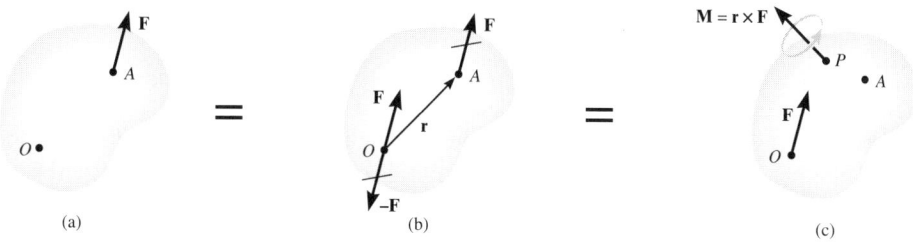

(a)   (b)   (c)

Next, we show how to determine the *equivalent resultants* mentioned above.

- To simplify any force and couple moment system to a resultant force acting at point O and a resultant couple moment we use the following equations

$$F_R = \sum F,$$
$$(M_R)_O = \sum M_C + \sum M_O$$

(4.2)

The first equation states that the resultant force of the system is equivalent to the sum of all the forces. The second equation states that the resultant couple moment of the system is equivalent to the sum of all the couple moments $\sum M_C$ plus the moments $\sum M_O$ about point O of all the forces.

- The following tips may prove useful when applying equations (4.2):

  ◆ Establish the coordinate axes with the origin located at point O and the axes having a selected orientation.

  ◆ If the force system lies in the x–y plane, and any couple moments are perpendicular to this plane i.e. along the z-axis, the equations (4.2) reduce to the scalar equations:

$$(F_R)_x = \sum F_x,$$
$$(F_R)_y = \sum F_y$$
$$(M_R)_O = \sum M_C + \sum M_O.$$

4.8 FURTHER SIMPLIFICATION OF A FORCE AND COUPLE SYSTEM

- In certain special circumstances (when the system of forces is either *concurrent, coplanar or parallel*), the system of forces and couple moments acting on a rigid body reduces at point O to a resultant force $\mathbf{F}_R$ and a resultant couple moment $(\mathbf{M}_R)_O$ which are *perpendicular to one another*. When this occurs, it is possible to further simplify the force and couple moment system by moving $\mathbf{F}_R$ to another point P (on or off the body) so that *no resultant couple moment has to be applied to the body*. That is, *only the force resultant* will have to be applied to the body (at P). The location of point P measured from point O can always be determined provided $\mathbf{F}_R$ and $(\mathbf{M}_R)_O$ are known.
- **Reduction to a Wrench**—A general force and couple moment system acting on a body will reduce to a single resultant force $\mathbf{F}_R$ and a resultant couple moment $(\mathbf{M}_R)_O$ at O which are *not* perpendicular to one another. In this case the force and couple moment system acting on a body can be reduced to a *wrench or screw*: a combination of a *collinear force and couple moment*

4.9 REDUCTION OF A SIMPLE DISTRIBUTED LOADING

In many situations a very large surface area of a body may be subjected to *distributed loadings* such as those caused by wind, fluids or simply the weight of material supported over the body's surface.

- *Distributed loadings* are defined by using a loading function $w = w(x)$ that indicates the intensity of the loading along the length of the member. This intensity is measured in N/m or lb/ft.
- The external effects caused by a coplanar distributed load acting on a body can be represented by a *single resultant force*.
- The *magnitude of the resultant force* is equal to the total *area* under the distributed loading diagram $w = w(x)$.
- The location of the resultant force is given by the fact that it's line of action passes through the *centroid* or geometric center of this area.

HELPFUL TIPS AND SUGGESTIONS

- Be aware of the difference between the *cross product* and the *dot product* of two vectors. The former is a *vector* while the latter is a *scalar*. Also

$$\mathbf{A} \cdot \mathbf{B} = \mathbf{B} \cdot \mathbf{A} \quad \text{BUT} \quad \mathbf{A} \times \mathbf{B} \neq \mathbf{B} \times \mathbf{A} \quad (= -(\mathbf{A} \times \mathbf{B}))$$

- The right-hand rule is essential in the calculation of moments. You should be able to apply this rule quickly and accurately.

REVIEW QUESTIONS

1. What is meant by a *moment*? Is it a vector or a scalar?
2. What's the magnitude of the moment of a force about a point?
3. How do you calculate the sense (direction) of a moment?
4. If the line of action of a force passes through a point P, what do you know about the moment of the force about P?
5. If you know the components of two vectors $\mathbf{A}$ and $\mathbf{B}$, how do you determine their cross product? Is the cross-product a vector or a scalar?
6. If $\mathbf{A} \times \mathbf{B} = \mathbf{0}$, what does this mean?
7. When you use the equation $\mathbf{M}_O = \mathbf{r} \times \mathbf{F}$ to determine the moment of a force $\mathbf{F}$ about O, how do you choose $\mathbf{r}$?
8. If you know the components of the vector $\mathbf{M}_O = \mathbf{r} \times \mathbf{F}$, how can you determine the product of the magnitude of $\mathbf{F}$ and the perpendicular distance d from O to the line of action of $\mathbf{F}$?
9. How do you figure out the sense of the moment of $\mathbf{F}$ about O using the formula $\mathbf{M}_O = \mathbf{r} \times \mathbf{F}$?

10. How would you calculate the moment exerted about a point O by a couple consisting of forces $\mathbf{F}$ and $-\mathbf{F}$?

11. True or false? The moment of a couple about O is different than the moment of the same couple about $P \neq O$?

12. What is meant by an *equivalent system*?

13. How do we *replace* a system of forces and couple moments acting on a body with an *equivalent single* resultant force and couple moment acting at a specified point O? What are the relevant equations?

14. Define what is meant by a *wrench*. When does a force and couple moment system acting on a body reduce to a wrench?

15. How is the resultant force exerted by a coplanar distributed load acting on a body determined from the function $w(x)$?

5

Equilibrium of a Rigid Body

MAIN GOALS OF THIS CHAPTER:

- To develop the equations of equilibrium for a rigid body.
- To introduce the concept of the free-body diagram for a rigid body.
- To show how to solve rigid body equilibrium problems using the equations of equilibrium.

5.1 CONDITIONS FOR RIGID-BODY EQUILIBRIUM

A *rigid body* is the next level of sophistication (after the particle) in the modelling of an object. Basically we 'add' size/shape to the existing model of a particle. Consequently, the main difference between a particle and a rigid body is that a rigid body can support moments. To obtain equations for the equilibrium of a rigid body, therefore, we need to supplement the equations of particle equilibrium with an expression of moment balance.

- The *two equations* of equilibrium for a *rigid body* are

$$\sum \mathbf{F} = \mathbf{0},$$
$$\sum \mathbf{M}_O = \mathbf{0},$$

  where O is an arbitrary point.

EQUILIBRIUM IN TWO DIMENSIONS

5.2 FREE-BODY DIAGRAMS

- No equilibrium problem should be solved without *first* drawing the free-body diagram, so as to account for *all* the forces *and couple moments* that act on the body.
- Part II of this study pack is devoted to the drawing of free-body diagrams including, specifically, *free-body diagrams for rigid body equilibrium in two dimensions*. Study Part II of this study pack making special note of the following important points:

– If a support *prevents translation* of a body in a particular direction, then the support exerts a *force* on the body in that direction.

– If *rotation is prevented*, then the support exerts a *couple moment* on the body.

– Study Table 2.1 in Part II of this supplement (or Table 5-1 of the text).

– Internal forces are never shown on the free-body diagram since they occur in equal but opposite collinear pairs and therefore cancel out.

– The weight of a body is an external force and its effect is shown as a single resultant force acting through the body's center of gravity G.

– *Couple moments* can be placed anywhere on the free-body diagram since they are *free vectors*. Forces can act at any point along their lines of action since they are *sliding vectors*.

5.3 EQUATIONS OF EQUILIBRIUM

• When the body is subjected to a system of forces which all lie in the x–y plane, the forces can be resolved into their x and y components. Consequently, the conditions for equilibrium in two dimensions can be written in scalar form as:

$$\sum F_x = 0, \qquad (5.0)$$
$$\sum F_y = 0,$$
$$\sum M_O = 0,$$

where $\sum M_O$ represents the algebraic sum of the couple moments and moments of all the force components about an axis perpendicular to the xy-plane and passing through an arbitrary point O (on or off the body).

TWO ALTERNATIVE SETS OF EQUILIBRIUM EQUATIONS

•

$$\sum F_a = 0,$$
$$\sum M_A = 0,$$
$$\sum M_B = 0.$$

Here, the only requirement is that a line passing through points A and B is not perpendicular to the a-axis.

•

$$\sum M_A = 0,$$
$$\sum M_B = 0,$$
$$\sum M_C = 0.$$

Here, the only requirement is that points A, B and C do not lie on the same line.

PROCEDURE FOR SOLVING COPLANAR FORCE EQUILIBRIUM PROBLEMS

- **Free-Body Diagram**

  - Establish the x, y coordinate axes in any suitable orientation.
  - Draw an outlined shape of the body.
  - Show all the forces and couple moments acting on the body.
  - Label all the loadings and specify their directions relative to the xy-axes.
  - Indicate the dimensions of the body necessary for computing the moments of forces.

- **Equations of Equilibrium**

  - Apply the moment equation of equilibrium ($\sum M_O = 0$) about a point O that lies at the intersection of the lines of action of two unknown forces. In this way, the moments of these unknowns are zero about O, and a direct solution for the third unknown can be determined.
  - When applying the force equilibrium equations ($\sum F_x = 0$ and $\sum F_y = 0$), orient the x and y axes along lines that will provide the simplest resolution of the forces into their x and y components.
  - If the solution of the equilibrium equations yields a negative scalar for a force or couple moment magnitude, it means that the sense is opposite to that which was assumed on the free-body diagram.

5.4 TWO- AND THREE- FORCE MEMBERS

The solution to some equilibrium problems can be simplified if one is able to recognize members that are subjected to only two or three forces.

- **Two-Force Members**—When a member is subjected to *no couple moments* and forces applied at only two points A and B on a member, the member is called a *two-force member*. In this case, for the member to be in equilibrium, it is necessary that the *resultant* forces at A and B must be *equal, opposite and collinear*. The line of action of both (resultant) forces is known since it always passes through A and B. Hence only the force magnitude (remember *both resultants are equal in magnitude!) needs to be determined or stated.*

- **Three-Force Members**—When a member is subjected to *only three forces*, it is necessary that the forces be *either concurrent or parallel* for the member to be in equilibrium. Once the point of concurrency O (where the lines of action of the forces intersect) is identified, then necessarily $\sum M_O = 0$. If two of the three forces are parallel, the point of concurrency O, is said to be at "infinity" and the third force must be parallel to the other two forces to intersect at this "point."

EQUILIBRIUM IN THREE DIMENSIONS

5.5 FREE-BODY DIAGRAMS

The first step in solving three-dimensional equilibrium problems, as in the case of two dimensions, is to *draw a free-body diagram*. The general procedure for doing this is the same as that outlined for the two-dimensional case in Section 5.2 of the text. However, there are a few subtle differences of which you should be aware:

- It is necessary to be familiar with the different types of reactive forces and couple moments acting at various types of supports and connections when members are viewed in three dimensions. It is important to recognize the symbols used to represent each of these supports and to understand clearly how the forces and couple moments are developed by each support. These are summarized in Table 5-2 of the text. Remember:

  - *As in the two-dimensional case, a force is developed by a support that restricts the translation of the attached member, whereas a couple moment is developed when rotation of the attached member is prevented.*

5.6 EQUATIONS OF EQUILIBRIUM

When the body is subjected to a three-dimensional force system, equilibrium requires that the resultant force and resultant couple moment acting on the body be equal to zero.

- In *vector* form the *two* equilibrium equations are

$$\sum \mathbf{F} = \mathbf{0},$$
$$\sum \mathbf{M}_O = \mathbf{0},$$

where $\sum \mathbf{F}$ is the vector sum of all the external forces acting on the body and $\sum \mathbf{M}_O$ is the sum of the couple moments and the moments of all the forces about any point O (on or off the body).

- Writing

$$\sum \mathbf{F} = \sum F_x \mathbf{i} + \sum F_y \mathbf{j} + \sum F_z \mathbf{k},$$
$$\sum \mathbf{M}_O = \sum M_x \mathbf{i} + \sum M_y \mathbf{j} + \sum M_z \mathbf{k},$$

The *six scalar* equilibrium equations are

$$\sum F_x = 0, \quad \sum F_y = 0, \quad \sum F_z = 0, \tag{5.1}$$
$$\sum M_x = 0, \quad \sum M_y = 0, \quad \sum M_z = 0.$$

5.7 CONSTRAINTS AND STATISTICAL DETERMINANCY

To ensure equilibrium of a rigid body, it is not only necessary to satisfy the equations of equilibrium, but the body must also be properly held or constrained by its supports.

- **Redundant Constraints.** When a body has *redundant supports*, that is, more supports than are necessary to hold it in equilibrium, it becomes *statically indeterminate*. This means that there will be more unknown loadings on the body than equations of equilibrium available for their solution. The additional equations needed to solve indeterminate problems are generally obtained from the *deformation conditions* at the points of support. These equations involve modelling the body not as a rigid body but as a *deformable body* (the next level of sophistication). This is done in courses dealing with "mechanics of materials."

- **Improper Constraints**. In some cases, there may be as many unknown forces on the body as there are equations of equilibrium; however, *instability* of the body may develop because of improper constraining by the supports. When this happens, either the number of available equilibrium equations is reduced by one (making the system *indeterminate*) or we will not be able to satisfy *all* the equilibrium equations. Proper constraining (avoiding instability of a body) requires

  1. The lines of action of the reactive forces do not intersect a common axis **and**
  2. The reactive forces must not all be parallel to one another.

When the minimum number of reactive forces is needed to properly constrain the body in question, the problem will be *statically determinate* and therefore the equations of equilibrium can be used to determine *all* the reactive forces.

HELPFUL TIPS AND SUGGESTIONS

- The first step in solving equilibrium problems is to draw a *free-body diagram*. Don't try to skip this stage no matter how trivial you think it is!
- Make the *free-body diagram* as clear and concise as possible. It will aid your understanding of the problem and it will help you construct the equilibrium equations.

REVIEW QUESTIONS

1. Why is there no moment equilibrium equation for a body modelled as a particle?
2. Write down the six independent scalar equilibrium equations for a rigid body in three dimensions. Adapt these equations to the two-dimensional case explaining why there are now only three independent equations.
3. What does it mean when an object is said to have redundant supports.
4. How do you know if an object is statically indeterminate as a result of redundant supports?
5. How do you avoid instability of a body due to improper constraining?

6

Structural Analysis

MAIN GOALS OF THIS CHAPTER:

- To show how to determine the forces in the members of a truss using the method of joints and the method of sections.
- To analyze the forces acting on the members of frames and machines composed of pin-connected members.

6.1 SIMPLE TRUSSES

A *truss* is a structure composed of slender members joined together at their end points. The members are usually wooden struts or metal bars. The joint connections are usually formed by bolting or welding the ends of the members to a common plate called a *gusset plate*, or by simply passing a large bolt or pin through each of the members.

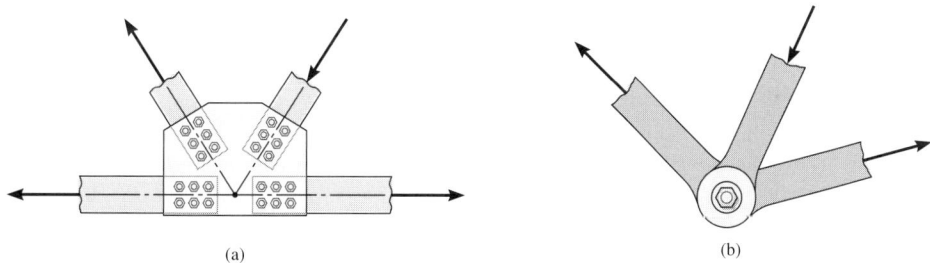

(a)   (b)

- **Planar Trusses** lie in a single plane and are often used to support roofs and bridges.
- **Assumptions for Design.** To design both the members and the connections of a truss, it is first necessary to determine the force developed in each member when the truss is subjected to a given loading. The following assumptions allow us to consider each truss member as a *two-force member* so that the forces at the ends of the member must be directed along the axis of the member.
    - *All loadings are applied at the joints.*
    - *The members are joined together by smooth pins.*

- **Simple Trusses.** To prevent collapse, the form of a truss must be rigid. The simplest form which is rigid or stable is a *triangle*. Consequently, a *simple truss* is constructed by *starting* with a basic triangular element. Additional elements consisting of two members and a joint are added to the triangular element to form a *simple truss*.

6.2 THE METHOD OF JOINTS

In order to analyze or design a truss, we must obtain the force in each of its members. To do this, we consider the *equilibrium of a joint* of the truss. This is the basis for the *method of joints*.

- Since the truss members are all straight two-force members lying in the same plane, the force system acting at each joint is *coplanar and concurrent*. Consequently, moment equilibrium is automatically satisfied at the joint and it is only necessary to satisfy *two independent scalar force equilibrium equations*.

PROCEDURE FOR ANALYZING A (PLANAR) TRUSS USING THE METHOD OF JOINTS

- Draw the free-body diagram of a joint having at least one known force and at most two unknown forces. (If this joint is at one of the supports, it generally will be necessary to know the external reactions at the truss support).
- Establish the sense of an unknown force by either:
  - Assuming that the unknown force is in *tension* and interpreting negative scalar results as members in *compression*. OR
  - *By inspection.*
- Orient the x and y axes such that the forces on the free-body diagram can be easily resolved into their x and y components and then apply the two force equilibrium equations $\sum F_x = 0$ and $\sum F_y = 0$. Solve for the two unknown member forces and verify their correct sense.
- Continue to analyze each of the other joints as above.
- Once the force in a member is found from the analysis of a joint at one of its ends, the result can be used to analyze the forces acting on the joint at its other end. Remember that a member in *compression pushes* on the joint and a member in *tension pulls* on the joint.

6.3 ZERO FORCE MEMBERS

Truss analysis using the method of joints is greatly simplified if one is first able to determine those members which support no loading. These *zero-force members* are used to increase stability of the truss during construction and to provide support if the applied loading is changed.

- Zero-force members of a truss are generally determined by *inspection of each of its joints*. As a general rule:
  If only two members form a truss joint and no external load or support reaction is applied to the joint, the members must be zero force members.
  If three members form a truss joint for which two of the members are collinear, the third member is a zero force member provided no external force or support reaction is applied to the joint.

6.4 THE METHOD OF SECTIONS

The *method of sections* is used to determine the loadings acting within a body. It is based on the principle that *if a body is in equilibrium then any part (section) of the body is also in equilibrium.*

PROCEDURE FOR ANALYZING THE FORCES IN THE MEMBERS OF A TRUSS USING THE METHOD OF SECTIONS

- **Free-Body Diagram**

  - Make a decision as to how to "cut" or section the truss through the members where forces are to be determined.
  - Before isolating the appropriate section, it may first be necessary to determine the truss' *external* reactions. Then three equilibrium equations are available to solve for member forces at the cut section
  - Draw the free-body diagram of that part of the sectioned truss which has the least number of forces acting on it.
  - Establish the sense of an unknown member force by either:

    * Assuming that the unknown member force is in *tension* and interpreting negative scalar results as members in *compression*. OR
    * *By inspection*.

- **Equations of Equilibrium**

  - Moments should be summed about a point that lies at the intersection of the lines of action of two unknown forces, so that the third unknown force is determined directly from the moment equation.
  - If two of the unknown forces are *parallel*, forces may be summed *perpendicular* to the direction of these unknowns to determine *directly* the third unknown force.

6.5 SPACE TRUSSES

A *space truss* consists of members joined together at their ends to form a stable three-dimensional structure. The simplest element of a space truss is a *tetrahedron*, formed by connecting six members together.

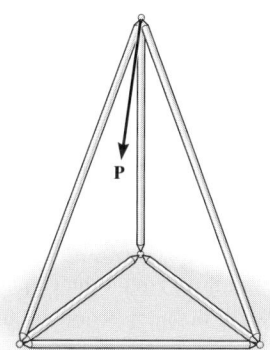

Either the *method of joints* or the *method of sections* can be used to determine the forces developed in the members of a simple space truss:

- **Method of Joints**. If the forces in *all* the members of the truss must be determined, the method of joints is most suitable for the analysis. Solve the three scalar equilibrium equations $\sum F_x = 0$, $\sum F_y = 0$ and $\sum F_z = 0$ at each joint. The solution of many simultaneous equations can be avoided if the force analysis begins at a joint having at least one known force and at most three unknown forces. Use a Cartesian vector analysis if the three-dimensional geometry of the force-system at the joint is hard to visualize.

- **Method of Sections**. If only a *few* member forces are to be determined, the method of sections is most suitable. When an imaginary section is passed through a truss and the truss is separated into two parts, the force system acting on one of the parts must satisfy the six scalar equilibrium equations $\sum F_x = 0$, $\sum F_y = 0$, $\sum F_z = 0$, $\sum M_x = 0$ and $\sum M_y = 0$ and $\sum M_z = 0$. By proper choice of the section and axes for summing forces and moments, many of the unknown member forces in a space truss can be computed *directly* using a single equilibrium equation.

6.6 FRAMES AND MACHINES

Frames and machines are two common types of structures which are often composed of pin-connected *multiforce* members. *Frames* are generally stationary and are used to support loads while *machines* contain moving parts and are designed to transmit and alter the effect of forces. Once the forces at the joints are obtained (see below) it is then possible to *design* the size of the members, connections and supports using the theory of mechanics of materials (deformable bodies) and an appropriate engineering design code.

PROCEDURE FOR DETERMINING THE JOINT
REACTIONS ON FRAMES OR MACHINES
COMPOSED OF MULTIFORCE MEMBERS

- **Free-Body Diagram**

    - Draw the free-body diagram of the entire structure, a portion of the structure, or each of its members. The choice should be made so that it leads to the most direct solution of the problem.
    - When the free-body diagram of a group of members of a structure is drawn, the forces at the connected parts of this group are *internal* forces and are not shown on the free-body diagram of the group.
    - Forces common to two members which are in contact act with equal magnitude but opposite sense on the respective free-body diagrams of the members.
    - Two-force members, regardless of their shape, have equal but opposite collinear forces acting at the ends of the member.
    - In many cases it is possible to tell by inspection the proper sense of the unknown forces acting on a member; however, if this seems difficult, the sense can be assumed.
    - A couple moment is a free vector and can act at any point on the free-body diagram. Also, a force vector is a sliding vector and can act at any point along its line of action.

- **Equations of Equilibrium**

    - Count the number of unknowns and compare it to the total number of equilibrium equations that are available, In two dimensions, there are three equilibrium equations that can be written for each member.
    - Sum moments about a point that lies at the intersection of the lines of action of as many unknown forces as possible
    - If the solution of a force or couple moment magnitude is negative, it means that the sense is the reverse of that shown on the free-body diagrams.

HELPFUL TIPS AND SUGGESTIONS

- The importance of drawing and using a clear and concise free-body diagram cannot be overstated.
- As in most mechanics problems, *practice* is the key. Make sure you read Examples 6-9 through 6-13 in the text and attempt to draw the requested free-body diagrams *yourself.* When doing so, make sure the work is neat and that all the forces and couple moments are properly labelled.

REVIEW QUESTIONS

1. What is a truss?
2. What assumptions allow us to consider a truss member as a *two-force member?*
3. What is the method of joints?
4. How many independent scalar equilibrium equations are available from the free-body diagram of a joint?
5. What is the method of sections?
6. What methods are available to determine the forces developed in the members of a simple space truss?
7. What's the difference between a frame and a machine?
8. When the free-body diagram of a group of members of a structure is drawn, the forces at the connected parts of this group are not shown on the free-body diagram of the group. Why?

7

Internal Forces

MAIN GOALS OF THIS CHAPTER:

- To show how to use the method of sections for determining the internal loadings in a member.
- To generalize this procedure by formulating equations that can be plotted so that they describe the internal shear and moment throughout a member.
- To analyze the forces and study the geometry of cables supporting a load.

7.1 INTERNAL FORCES DEVELOPED IN STRUCTURAL MEMBERS

The design of any structural or mechanical member requires an investigation of **both** the external loads and reactions acting on the member **and** the loading acting *within* the member—in order to be sure *the material can resist this loading*. These internal loadings can be determined using the *method of sections*.

> The idea is to cut an 'imaginary section' through the member so that the internal loadings (of interest) at the section become external on the free-body diagram of the section.

PROCEDURE FOR FINDING THE INTERNAL LOADINGS AT A SPECIFIC LOCATION IN A MEMBER USING THE METHOD OF SECTIONS

- **Support Reactions**

  ◆ Before the member is "cut" or "sectioned," it may first be necessary to determine the member's support reactions, so that the equilibrium equations are used only to solve for the internal loadings when the member is sectioned.

  ◆ If the member is part of a *frame or machine*, the reactions at its connections are determined using the methods outlined in Section 6.6.

- **Free-Body Diagram**
  - ◆ Keep all distributed loadings, couple moments and forces acting on the member in their *exact locations*, then pass an imaginary section through the member, perpendicular to its axis at the point where the internal loading is to be determined.
  - ◆ After the section is made, draw a free-body diagram of the segment that has the least number of loads on it, and indicate the x, y, z components of the force and couple moment resultants at the section.
  - ◆ If the member is subjected to a *coplanar system of forces*, only **N** (normal force), **V** (shear force), and **M** (bending moment) act at the section.

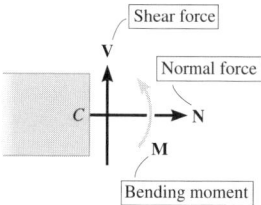

  - ◆ In three dimensions, a general internal force and couple moment resultant will act at the section.

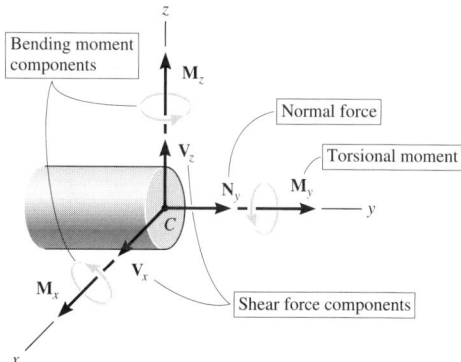

  - ◆ In many cases it may be possible to tell by inspection the proper sense of the unknown loadings; however, if this seems difficult, the sense can be assumed.
- **Equations of Equilibrium**
  - ◆ Moments should be summed at the section about axes passing through the *centroid* or geometric center of the member's cross-sectional area in order to eliminate the unknown normal and shear forces and thereby obtain direct solutions for the moment components.
  - ◆ If the solutions of the equilibrium equations yields a negative scalar, the assumed sense of the quantity is opposite to that shown on the free-body diagram.

7.2 SHEAR AND MOMENT EQUATIONS AND DIAGRAMS

Beams are designed to support loads *perpendicular* to their axes. The actual design of a *beam* requires a detailed knowledge of the *variation* of the internal shear force V and bending moment M acting at *each point* along the axis of the beam. After this, the theory of mechanics of materials is used with an appropriate engineering design code to determine the beam's required cross-sectional area.

The *variations* of V and M as functions of the position x along the beam's axis can be obtained using the method of sections (Section 7.1). However, it is necessary to section the beam at an arbitrary distance x from one end rather than at a specified point. If the results are plotted, the graphical variations of V and M as functions of x are termed the *shear diagram* and *bending moment diagram*, respectively.

These diagrams can be constructed as follows:

- **Support Reactions**

  ◆ Determine all the reactive forces and couple moments acting on the beam and resolve all the forces into components acting perpendicular and parallel to the beam's axis.

- **Shear and Moment Functions**

  ◆ Specify separate coordinates x having an origin at the *beam's left end* and extending to regions of the beam *between* concentrated forces and/or couple moments, or where there is no discontinuity of distributed loading.

  ◆ Section the beam perpendicular to its axis at each distance x and draw the free-body diagram of one of the segments. Be sure **V** and **M** are shown acting in their *positive sense* in accordance with the following sign convention:

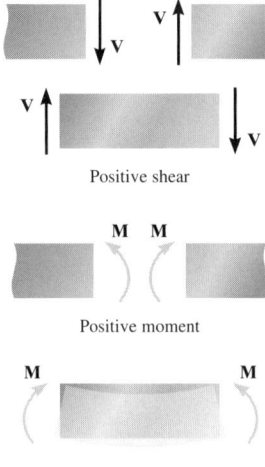

Beam sign convention

  ◆ The shear V is obtained by summing forces perpendicular to the beam's axis.

  ◆ The moment M is obtained by summing moments about the sectioned end of the segment.

- **Shear and Moment Diagrams**

  ◆ Plot the shear diagram (V versus x) and the moment diagram (M versus x). If the computed values of the functions describing V and M are *positive*, the values are plotted above the x-axis, whereas *negative* values are plotted below the x-axis.

  ◆ Generally, it is convenient to plot the shear and bending-moment diagrams directly below the free-body diagram of the beam. See Examples 7-7 and 7-8 in the text.

7.3 RELATIONS BETWEEN DISTRIBUTED LOAD, SHEAR, AND MOMENT

In cases where a beam is subjected to several concentrated forces, couple moments, and distributed loads, the method of constructing the shear and bending moment diagrams may become quite tedious. In this section a simpler method for constructing these diagrams is presented—based on differential relations that exist between the load, shear, and bending moment. The following are the main points:

- The slope of the shear diagram is equal to the negative of the intensity of the distributed loading, where positive distributed loading is downward i.e.

$$\frac{dV}{dx} = -w(x). \tag{7.0}$$

- If a concentrated force acts downward on the beam, the shear will jump downward by the amount of the force.

- The change in the shear ΔV between two points is equal to the *negative of the area* under the distributed-loading curve between the points.

- The slope of the moment diagram is equal to the shear i.e.

$$\frac{dM}{dx} = V. \tag{7.1}$$

- The change in the moment ΔM between two points is equal to the *area* under the shear diagram between the two points.

- If a *clockwise* couple moment acts on the beam, the shear will not be affected, however, the moment diagram will jump *upward* by the amount of the moment.

- Points of zero shear represent points of *maximum or minimum moment* since

$$\frac{dM}{dx} = 0. \tag{7.2}$$

7.4 CABLES

Flexible cables and chains are used in engineering structures for support and to transmit loads from one member to another. In the force analysis of such systems, the weight of the cable itself may be neglected (cable is referred to as '*weightless*') because it is often small compared to the load it carries. In modelling the cable, it is assumed that:

1. The cable is *perfectly flexible* (cable offers no resistance to bending so the tensile force acting in the cable is always *tangent* to the cable at points along its length).

2. The cable is *inextensible* (cable has a constant length before and after load is applied—cable can be treated as a rigid body).

CABLE SUBJECTED TO CONCENTRATED LOADS

- When a cable of negligible weight supports several concentrated loads, the cable takes the form of several straight-line segments, each of which is subjected to a constant tensile force. The equilibrium analysis is performed by writing down a *sufficient number* of equilibrium equations (based on the entire cable or any part thereof) and equations describing the geometry of the cable to solve for all the *unknowns* leading to a description of the tension in (each segment of) the cable. See Example 7-13 in text.

CABLE SUBJECTED TO A DISTRIBUTED LOAD

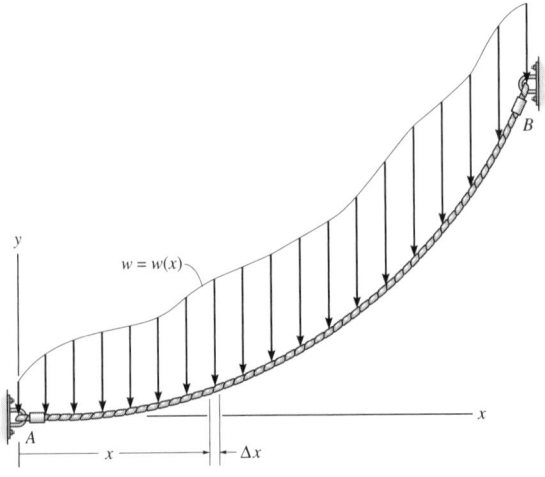

- The equation

$$y = \frac{1}{F_H} \int \left(\int w(x)\, dx \right) dx, \tag{7.3}$$

determines the curve for the cable $y = f(x)$. Here, F_H measures the horizontal component of tensile force at *any point* along the cable and $w(x)$ is the loading function measured in the x-direction. In practice, F_H and the two constants of integration are determined from the boundary conditions for the cable—see Example 7-14 in text.

CABLE SUBJECTED TO ITS OWN WEIGHT

- When the weight of the cable becomes important in the force analysis (e.g., cables used as transmission lines), the loading function along the cable becomes a function of the arc length s rather than the projected length x. The equation of the deflection curve is given by $y = f(x)$ where

$$\frac{dy}{dx} = \frac{1}{F_H} \int w(s)\, ds, \tag{7.4}$$

$$x = \int \frac{ds}{\left\{ 1 + \frac{1}{F_H^2} \left(\int w(s)\, ds \right)^2 \right\}^{\frac{1}{2}}}. \tag{7.5}$$

The two constants of integration from (7.5) are found using the boundary conditions for the cable. First solve (7.5) for $w(s)$ then use (7.4) to get the shape $y = f(x)$ for the cable. See Example 7-15 in text.

HELPFUL TIPS AND SUGGESTIONS

- Use *clear* and concise free-body diagrams.
- As in most mechanics problems, *practice* is the key. Make sure you read Examples 7-1 through 7-15 in the text before you attempt any corresponding problems. These examples will serve as templates with which to solve problems. Draw the requested free-body diagrams *yourself.* When doing so, make sure the work is neat and that all the forces and couple moments are properly labelled.

REVIEW QUESTIONS

1. What are the normal (or axial) force, shear force and bending moment?
2. How are the positive directions (sense) of the shear force **V** and bending moment **M** defined?
3. For a portion of a beam which is subjected only to a distributed load w, how are the shear force and bending moment distributions determined from equations (7.0) and (7.1) ?
4. What does it mean when a cable is assumed to be 'weightless,' 'inextensible' or 'perfectly flexible'?
5. If a cable is subjected to a load that is uniformly distributed along a straight line and its weight is negligible, what mathematical curve describes its shape?

8

Friction

MAIN GOALS OF THIS CHAPTER:

- To introduce the concept of dry friction and show how to analyze the equilibrium of rigid bodies subjected to this force.
- To present specific applications of frictional force analysis on wedges, screws, belts and bearings.
- To investigate the concept of *rolling resistance*.

8.1 CHARACTERISTICS OF DRY FRICTION

As a result of *experiments* that pertain to the foregoing discussion, the following rules which apply to bodies subjected to dry friction may be stated:

- The frictional force acts *tangent* to the contacting surfaces in a direction *opposed to the relative motion* or tendency for motion of one surface against another.
- The maximum static frictional force $\mathbf{F}_s$ that can be developed is independent of the area of contact, provided the normal pressure is not very low nor great enough to severely deform or crush the contacting surfaces of the bodies.
- The maximum static frictional force is generally greater than the kinetic frictional force $\mathbf{F}_k$ for any two surfaces of contact. However, if one of the bodies is moving with a *very low velocity* over the surface of another, F_k becomes approximately equal to F_s i.e. $\mu_s \approx \mu_k$.
- When *slipping* at the surface of contact is *impending (about to occur)*, the maximum static frictional force is proportional to the normal force $\mathbf{N}$, such that $F_s = \mu_s N$.
- When *slipping* at the surface of contact is *occurring*, the kinetic frictional force is proportional to the normal force $\mathbf{N}$, such that $F_k = \mu_k N$.

8.2 PROBLEMS INVOLVING DRY FRICTION

If a rigid body is in equilibrium when it is subjected to a system of forces which includes the effect of friction, the force system must satisfy not only the equations of equilibrium but *also* the laws that govern the frictional forces.

In general, there are three types of mechanics problems involving dry friction. They are classified once the free-body diagrams are drawn and the total number of unknowns are identified and compared with the total number of available equilibrium equations:

- **Equilibrium**—The total number of unknowns is equal to the total number of available equilibrium equations. In this case, once the frictional forces are determined, check that $F \leq \mu_s N$ otherwise slipping will occur and the body will not remain in equilibrium.

- **Impending Motion at all Points**—The total number of unknowns will *equal* the total number of available equilibrium equations plus the total number of available frictional equations or conditional equations for tipping. As a result, several possibilities for motion or impending motion will exist and the problem will involve a determination of the kind of motion which actually occurs.

- **Impending Motion at Some Points**—The total number of unknowns will be *less* than the total number of available equilibrium equations plus the total number of available frictional equations, $F = \mu N$. If motion is *impending* at the points of contact, then $F_s = \mu_s N$ whereas if the body is *slipping*, then $F_k = \mu_k N$.

EQUILIBRIUM VERSUS FRICTIONAL EQUATIONS

When the frictional force $\mathbf{F}$ is an equilibrium force i.e., $F < \mu N$, we can always *assume* the sense of the frictional force $\mathbf{F}$ (since the frictional force *always* acts so as to oppose the relative motion or impede the motion of a body over the contacting surface). The correct sense is determined after solving the equilibrium equations for F. However, in cases where $F = \mu N$ is used, we can no longer assume the sense of $\mathbf{F}$ since the equation $F = \mu N$ relates only the magnitudes of two perpendicular vectors. Consequently, in this case, $\mathbf{F}$ must always be shown acting with its correct sense on the free-body diagram.

PROCEDURE FOR ANALYSIS

- **Free-Body Diagrams**

    ◆ Draw the necessary free-body diagrams and, unless it is stated in the problem that impending motion or slipping occurs, *always* show the frictional forces as unknowns i.e., *do not assume* $F = \mu N$.

    ◆ Determine the number of unknowns and compare this with the number of available equilibrium equations.

    ◆ If there are more unknowns than equilibrium equations, it is necessary to apply the frictional equations at some, if not all, points of contact to obtain the extra equations needed for complete solution.

    ◆ If the equation $F = \mu N$ is to be used, it will be necessary to show $\mathbf{F}$ acting in the proper direction on a free-body diagram.

- **Equations of Equilibrium and Friction**

    ◆ Apply the equations of equilibrium and the necessary frictional equations (or conditional equations if tipping is possible) and solve for the unknowns.

    ◆ If the problem involves a three-dimensional force system such that it becomes difficult to obtain the force components or the necessary moment arms, apply the *vector equations of equilibrium*.

8.3 WEDGES

A *wedge* is a simple machine which is often used to transform an applied force into much larger forces, directed at approximately right angles to the applied force. Also wedges can be used to give small displacements or adjustments to heavy loads. The analysis of problems involving wedges proceeds as above i.e., we draw free-body diagrams of the wedge and any other contacting bodies and formulate the appropriate equilibrium and frictional equations. See Example 8-7 in text.

8.4 FRICTIONAL FORCES ON SCREWS

A *screw* may be thought of simply as an inclined plane or wedge wrapped around a cylinder. In most cases screws are used as fasteners; however, in many applications, they are incorporated to transmit power or motion from one part of the machine to another.

Before proceeding to solve problems involving frictional forces on screws, each of the following cases should be thoroughly understood:

- **Frictional Analysis with Upward Screw Motion.** The moment necessary to cause upward impending motion of the screw is

$$M = Wr\tan(\theta + \phi), \quad \phi = \phi_s = \tan^{-1}\mu_s. \tag{8.0}$$

If ϕ is replaced by $\phi_k = \tan^{-1}\mu_k$, we obtain a smaller value of M necessary to maintain uniform upward motion of the screw.

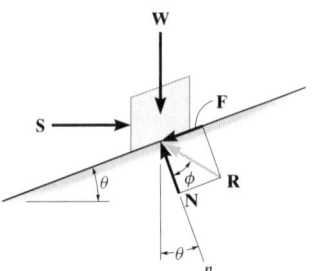

Upward screw motion

- **Frictional Analysis with Downward Screw Motion** (when the surface of the screw is very slippery: $\theta > \phi$). The moment necessary to cause downward impending motion of the screw is

$$M' = Wr\tan(\theta - \phi), \quad \phi = \phi_s. \tag{8.1}$$

If ϕ is replaced by $\phi_k = \tan^{-1}\mu_k$, we obtain a (smaller) value of M necessary to maintain *uniform* downward motion of the screw.

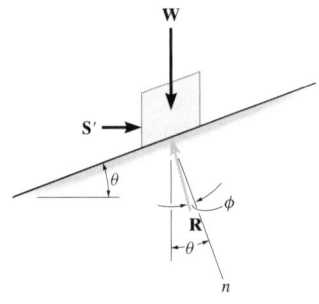

Downward screw motion ($\theta > \phi$)

- **Frictional Analysis with Downward Screw Motion** (when the surface of the screw is very rough: $\theta < \phi$)**.** The moment necessary to cause downward impending motion of the screw is

$$M' = Wr \tan(\phi - \theta), \quad \phi = \phi_s. \tag{8.2}$$

If ϕ is replaced by $\phi_k = \tan^{-1} \mu_k$, we obtain a (smaller) value of M necessary to maintain uniform downward motion of the screw.

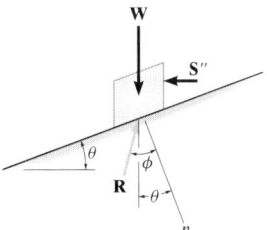

Downward screw motion ($\theta < \phi$)

- **Frictional Analysis with a Self-Locking Screw.** If the moment M (or its effect S) is removed, the screw will remain *self-locking* i.e. it will support the load W by friction forces alone (provided $\phi \geq \theta$).

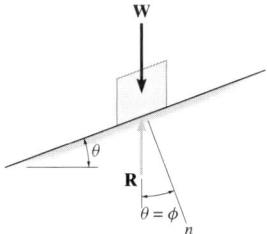

Self-locking screw ($\theta = \phi$)
(on the verge of rotating downward)

8.5 FRICTIONAL FORCES ON FLAT BELTS

Whenever belt drives are designed, it is necessary to dtermine the frictional forces developed between the belt and its contacting surface.

- The tension T_2 in the belt required to pull the belt counterclockwise over the surface and thereby overcome both the frictional forces at the surface of contact and the known tension T_1 (motion or impending motion of belt relative to surface) is:

$$T_2 = T_1 e^{\mu\beta}$$

where μ is the coefficient of static or kinetic friction between the belt and the surface of contact, and β is in radians.

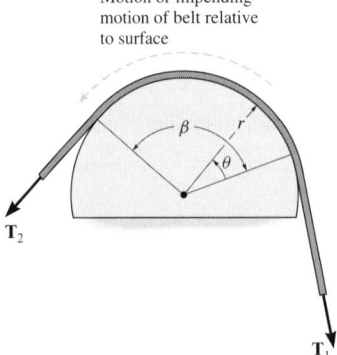

Motion or impending
motion of belt relative
to surface

8.6 FRICTIONAL FORCES ON COLLAR BEARINGS, PIVOT BEARINGS, AND DISKS

Pivot and collar bearings are commonly used in machines to support an axial load on a rotating shaft.

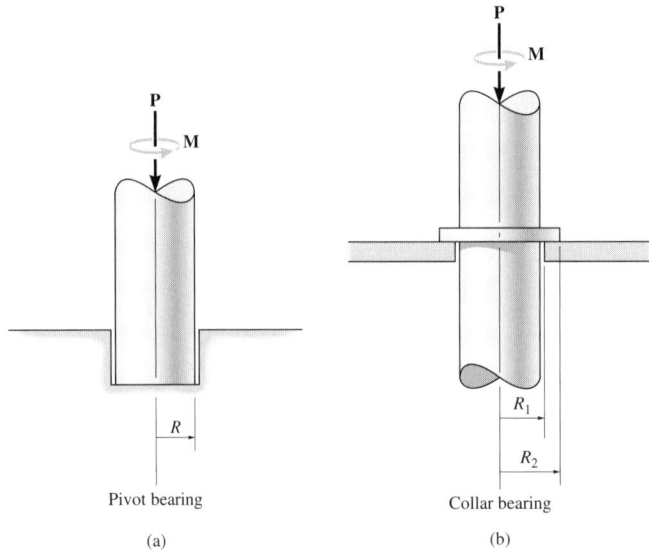

Pivot bearing

(a)

Collar bearing

(b)

- The magnitude of the moment required for impending rotation of the shaft is given by

$$M = \frac{2}{3}\mu_s P \left(\frac{R_2^3 - R_1^3}{R_2^2 - R_1^2} \right).$$

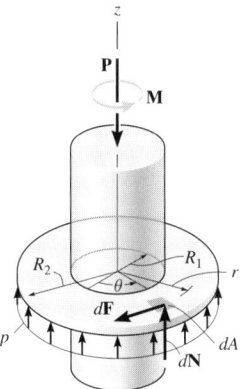

- The frictional moment developed at the end of the shaft, when it is rotating at constant speed can be found by substituting μ_k for μ_s in the expression for M.

8.7 FRICTIONAL FORCES ON JOURNAL BEARINGS

When a shaft or axle is subjected to lateral loads, a *journal bearing* is commonly used for support.

- The moment needed to maintain constant rotation of the shaft is given by

$$M = Rr \sin \phi_k, \tag{8.3}$$

where ϕ_k is the angle of kinetic friction defined by $\tan \phi_k = \mu_k$ and R is the magnitude of the bearing reactive force acting at A.

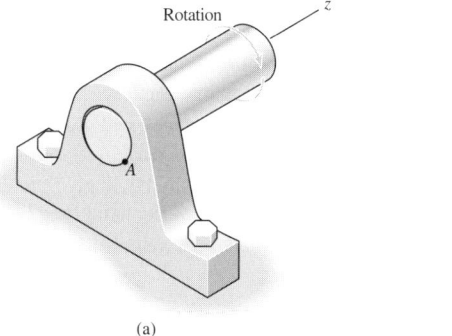

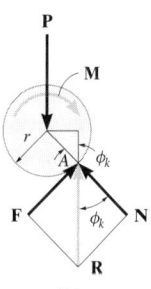

(a)   (b)

8.8 ROLLING RESISTANCE

- The force $\mathbf{P}$ necessary to initiate and maintain rolling (of a cylinder with weight $\mathbf{W}$ and radius r) at constant velocity has magnitude

$$P \approx \frac{Wa}{r}.$$

Here, the *distance a* is referred to as the *coefficient of rolling resistance*.

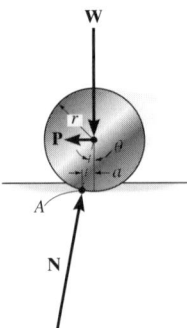

HELPFUL TIPS AND SUGGESTIONS

- Use *clear* and concise free-body diagrams.
- As in most mechanics problems, *practice* is the key. Make sure you read Examples 7-1 through 7-15 in the text before you attempt any corresponding problems. These examples will serve as templates with which to solve problems. Draw the requested free-body diagrams *yourself.* When doing so, make sure the work is neat and that all the forces and couple moments are properly labelled.

REVIEW QUESTIONS

1. If relative slipping of two dry surfaces in contact is impending what can you say about the frictional forces they exert on each other?

2. If two dry surfaces in contact are sliding relative to each other, what can you say about the frictional forces they exert on each other?

3. What are the characteristics of an *equilibrium* problem involving dry friction?

4. What are the characteristics of a problem involving dry friction when motion is impending at all points of contact?

5. What's the first step in solving a mechanics problem involving dry friction?

6. If a screw is subjected to a large axial load, what is the equation which will give the moment necessary to rotate the screw and cause it to move in a direction opposite to the axial load?

7. If a screw is subjected to a large axial load, what is the equation which will give the moment necessary to rotate the screw at *a constant rate* and cause it to move in the direction of the axial load?

8. If the shaft of a journal bearing is subjected to a lateral load with magnitude P, how do you find the moment necessary to maintain constant rotation of the shaft?

9

Center of Gravity and Centroid

MAIN GOALS OF THIS CHAPTER:

- To discuss the concept of the *center of gravity, center of mass*, and the *centroid*.
- To show how to determine the location of the center of gravity and centroid for a system of discrete particles and a body of arbitrary shape.
- To use the theorems of Pappus and Guldinus for finding the area and volume for a surface of revolution.
- To present a method for finding the resultant of a general distributed loading and show how it applies to finding the resultant of a fluid.

9.1 CENTER OF GRAVITY, CENTER OF MASS, AND THE CENTROID OF A BODY

- The *center of gravity G* is a point which locates the *resultant weight of a system of particles*. This coordinates of the center of gravity G of a system of particles is given by

$$\bar{x} = \frac{\sum \tilde{x} W}{\sum W}, \quad \bar{y} = \frac{\sum \tilde{y} W}{\sum W}, \quad \bar{z} = \frac{\sum \tilde{z} W}{\sum W}, \tag{9.0}$$

where $\tilde{x}, \tilde{y}, \tilde{z}$ represent the coordinates of each particle of the system and $\sum W$ is the resultant sum of the weights of all the particles in the system.

- The *center of mass* of a system of particles is obtained by substituting $W = mg$ (assuming g for every particle is constant) into (9.0). Consequently, the center of mass has coordinates

$$\bar{x} = \frac{\sum \tilde{x} m}{\sum m}, \quad \bar{y} = \frac{\sum \tilde{y} m}{\sum m}, \quad \bar{z} = \frac{\sum \tilde{z} m}{\sum m},$$

where $\sum W$ is the resultant sum of the masses of all the particles in the system. Note that the *location* of the center of gravity *coincides* with that of the center of mass. However, the *center of mass is independent of gravity* and so can be used in situations when particles are not under the influence of a gravitational attraction.

- The coordinates of the *center of gravity* G of a body are given by

$$\bar{x} = \frac{\int \tilde{x}\, dW}{\int dW}, \quad \bar{y} = \frac{\int \tilde{y}\, dW}{\int dW}, \quad \bar{z} = \frac{\int \tilde{z}\, dW}{\int dW}, \tag{9.1}$$

where $\tilde{x}, \tilde{y}, \tilde{z}$ represent the coordinates of an arbitrary point in the body.

- The *center of mass* of a rigid body is obtained by substituting $dW = g$ into (9.1) and cancelling g from both the numerators and denominators. This yields (9.1) with W replaced by m.

CENTROID

- The *centroid* C $(\bar{x}, \bar{y}, \bar{z})$ is a point which defines the *geometric center of a body* . This point coincides with the *center of mass* or the *center of gravity* only if the material composing the body is *uniform or homogeneous* (in which case both γ and ρ are constant throughout the body).

- Formulas used to *locate* the center of gravity or the centroid simply represent a balance between the sum of moments of all the parts of the system and the moment of the "resultant" for the system. There are three cases to consider:

  - **Volume.** C $(\bar{x}, \bar{y}, \bar{z})$ is given by

$$\bar{x} = \frac{\int_V \tilde{x}\, dV}{\int_V dV}, \quad \bar{y} = \frac{\int_V \tilde{y}\, dV}{\int_V dV}, \quad \bar{z} = \frac{\int_V \tilde{z}\, dV}{\int_V dV}. \tag{9.2}$$

  - **Area.** C $(\bar{x}, \bar{y}, \bar{z})$ is given by

$$\bar{x} = \frac{\int_A \tilde{x}\, dA}{\int_A dA}, \quad \bar{y} = \frac{\int_A \tilde{y}\, dA}{\int_A dA}, \quad \bar{z} = \frac{\int_A \tilde{z}\, dA}{\int_A dA}. \tag{9.3}$$

  - **Line.** C $(\bar{x}, \bar{y}, \bar{z})$ is given by

$$\bar{x} = \frac{\int_L \tilde{x}\, dL}{\int_L dL}, \quad \bar{y} = \frac{\int_L \tilde{y}\, dL}{\int_L dL}, \quad \bar{z} = \frac{\int_L \tilde{z}\, dL}{\int_L dL}. \tag{9.4}$$

- The centroid will lie on any axis of symmetry of the body. Also, the centroid may be located off the body e.g., in the case of a ring where the centroid is at the center.

9.2 COMPOSITE BODIES

A composite body consists of a series of connected "simpler" shaped bodies.

PROCEDURE FOR ANALYSIS

The location of the center of gravity of a composite body can be determined using the following procedure:

- **Composite Parts**

  - Using a sketch, divide the body or object into a finite number of composite parts that have simpler shapes.

  - If a composite part has a *hole*, then consider the composite part without the hole and consider the hole as an *additional* composite part having *negative* weight or size.

- **Moment Arms**

  - Establish the coordinate axes on the sketch and determine the coordinates $(\tilde{x}, \tilde{y}, \tilde{z})$ of the center of gravity of each composite part.

- **Summations**

  - Determine $\bar{x}$, $\bar{y}$, $\bar{z}$ by applying the center of gravity equations:

    $$\bar{x} = \frac{\sum \tilde{x} W}{\sum W}, \quad \bar{y} = \frac{\sum \tilde{y} W}{\sum W}, \quad \bar{z} = \frac{\sum \tilde{z} W}{\sum W}, \tag{9.5}$$

    where $\sum W$ is the sum of the weights of all the composite parts of the body (total weight of the body).

  - If an object is *symmetrical* about an axis, the centroid of the object lies on this axis.

- **CENTROID FOR A COMPOSITE**—When the (composite) body has *constant* density or specific weight, the center of gravity *coincides* with the centroid of the body which, for lines, areas and volumes, can be found using relations analogous to (9.5) with the W's replaced by L's, A's and V's, respectively [as in (9.2) - (9.4)]. Centroids for common shapes of lines, areas, shells and volumes that often make up a composite body are given in the table on the inside back cover of the text.

9.3 THEOREMS OF PAPUS AND GULDINUS

The following two theorems (of Papus and Guldinus) are used to find the *surface area and volume* of any object of revolution:

- **Surface Area.** The area A of a surface of revolution equals the product of the length of the generating curve and the distance travelled by the centroid of the curve in generating the surface area. That is:

  $$A = \theta \bar{r} L,$$

  where θ is the angle of revolution (radians), $\bar{r}$ is the perpendicular distance from the axis of revolution to the centroid of the generating curve and L is the total length of the generating curve.

- **Volume.** The volume V of a body of revolution equals the product of the generating area and the distance traveled by the centroid of the area in generating the volume. That is:

  $$V = \theta \bar{r} A,$$

  where θ is the angle of revolution (radians), $\bar{r}$ is the perpendicular distance from the axis of revolution to the centroid of the generating area and A is the generating area.

- **Composite Shapes.** These two theorems may also be applied to lines or areas that may be composed of a series of composite parts. In this case, the total surface area or volume generated is the sum of the surface areas of volumes generated by each of the composite parts:

  $$A = \theta \sum \tilde{r} L, \quad V = \theta \sum \tilde{r} A,$$

  where $\tilde{r}$ is the distance from the axis of revolution to the centroid of each composite part (remember that each part undergoes the same angle of revolution θ).

9.4 RESULTANT OF A GENERAL DISTRIBUTED LOADING

In Section 4.9 we discussed the method used to simplify a distributed loading which is uniform along an axis of a rectangular surface. Here, we generalize this method to include surfaces which have an arbitrary shape and are subjected to a variable load distribution.

- **Pressure Distribution over a Surface.** Consider a flat plate subjected to the loading function $p(x, y) \, Pa \, (Pa = 1 \text{N/m}^2)$. The entire loading on the plate can be simplified to a *single resultant force* $\mathbf{F}_R$

- *Magnitude of Resultant Force.*

$$FR = \int V \, dV$$

i.e., total volume under the distributed loading diagram.
- *Location of Resultant Force.* The location $(\bar{x}, \bar{y})$ of $\mathbf{F}_R$ is given by

$$\bar{x} = \frac{\int_V x \, dV}{\int_V dV}, \quad \bar{y} = \frac{\int_V y \, dV}{\int_V dV}. \tag{9.6}$$

In other words,

Line of Action of $\mathbf{F}_R$ passes through the geometric center or centroid of the volume under the distributed loading diagram i.e.

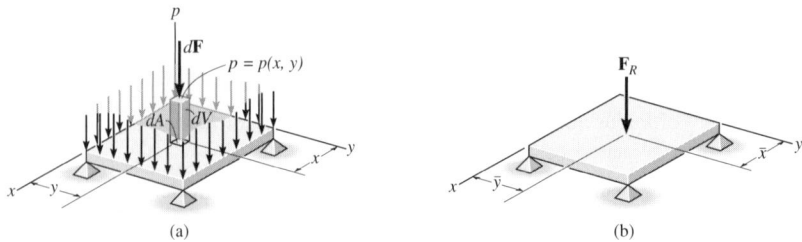

(a) (b)

9.5 FLUID PRESSURE

According to *Pascal's law* a fluid at rest creates a pressure p at a point that is the *same in all* directions. The magnitude of p (force per unit area) is given by

$$p = \gamma z = \rho g z, \tag{9.7}$$

where γ is specific weight, ρ is mass density and z is the depth of the point from the fluid surface. Equation (9.7) is valid only for *incompressible fluids* (in which pressure and temperature variations do not produce any significant density variations).

Using Equation (9.7) and the results of Section 9.4, it is possible to determine the resultant force caused by a liquid pressure distribution and specify its location on the surface of a submerged plate. Three cases are considered:

- **Flat Plate of Constant Width.** The easiest of the three cases. Consider a flat rectangular plate of constant width submerged in a liquid with specific weight γ. The magnitude of the resultant force $\mathbf{F}_R$ resulting from the distribution of pressure over the plate's surface is equal to the trapezoidal volume having an intensity of $p_1 = \gamma z_1$ at depth z_1 and $p_2 = \gamma z_2$ at depth z_2. The line of action of $\mathbf{F}_R$ passes through the *volume's centroid C* [see Equation (9.6)]—this is *not* the centroid of the plate but rather the *center of pressure P of the plate*.
- **Curved Plate of Constant Width.** The calculation of the magnitude of $\mathbf{F}_R$ and its location P is more complicated for a (general) curved plate than a flat plate. There is a simplification, however, when the plate has *constant width*. This method requires separate calculations for the horizontal and vertical *components* of $\mathbf{F}_R$.
- **Flat Plate of Variable Width.** The loading caused by the pressure distribution acting on the surface of a submerged plate having a variable width has resultant $\mathbf{F}_R$ with *magnitude given by the volume described by the plate area as its base and linearly varying pressure distribution as its height.* From Equation (9.6), the centroid of V again defines the point through which $\mathbf{F}_R$ acts i.e., the center of pressure P, which lies on the surface of the plate just below C, has coordinates $P\left(\bar{x}, \bar{y}'\right)$ defined by the equations

$$\bar{x} = \frac{\int V \bar{x} \, dV}{\int V \, dV}, \quad \bar{y}' = \frac{\int V \bar{y}' \, dV}{\int V \, dV}.$$

Note that this point is *not* the centroid of the plate's area.

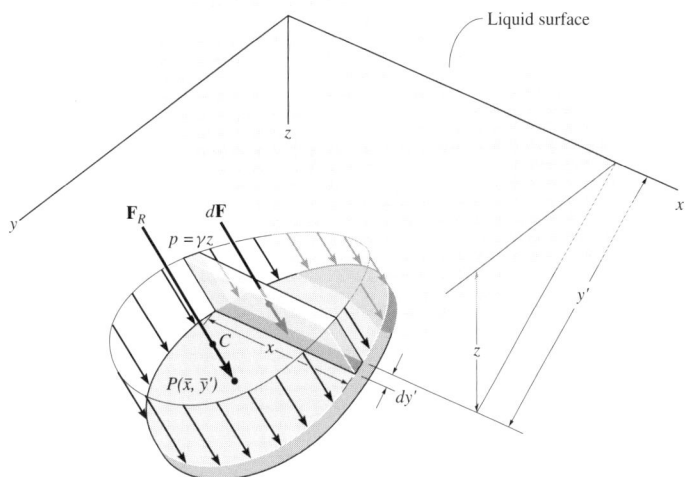

HELPFUL TIPS AND SUGGESTIONS

- Finding the location of the center of mass or centroid involves multiple integration (e.g., over volumes or surfaces). These integrations can often be reduced to single integrations using the procedure outlined at the end of Section 9.1. Study this procedure then read and perform Examples 9-1 through 9-8 of the text.

REVIEW QUESTIONS

1. True or False?
    (a) The location of the center of mass coincides with that of the center of gravity.
    (b) The centroid of a body always coincides with the body's center of mass.
    (c) The centroid of a body is always located on the body in question.
    (d) The formula for the centroid for the surface area of a plate involves integrals over the same surface area.
2. If the total weight of an object is to be represented by a single equivalent force, where must this force act?
3. What's the difference between the center of gravity and the center of mass?
4. How is the specific weight of a body defined?
5. What's the relationship between the mass density ρ and the specific weight γ of a body?
6. What does it mean when we say a body is homogeneous?
7. If an object is homogeneous, what do you know about the position of its center of mass?
8. In the analysis of the location of the center of gravity of a composite body, how do you deal with a hole in the body?
9. What are the theorems of Papus and Guldinus used for?
10. Show that the centroid for the volume of a body coincides with the *center of mass* only if the material composing the body is *uniform or homogeneous*.

10

Moments of Inertia

MAIN GOALS OF THIS CHAPTER:

- To develop a method for determining the moment of inertia for an area.
- To introduce the product of inertia and show how to determine the maximum and minimum moments of inertia of an area.
- To discuss the mass moment of inertia.

10.1 DEFINITION OF MOMENTS OF INERTIA FOR AREAS

- The moments of inertia (second moments) of the area A about the x and y axes are given, respectively, by

$$I_x = \int_A y^2 \, dA,$$ (10.0)

$$I_y = \int_A x^2 \, dA.$$ (10.1)

- The moment of inertia of the area A about the pole O or the z-axis (also known as the *polar moment of inertia*) is

$$J_O = \int_A r^2 \, dA = I_x + I_y$$ (10.2)

where $r^2 = x^2 + y^2$ is the perpendicular distance from the pole (z-axis) to the element dA.

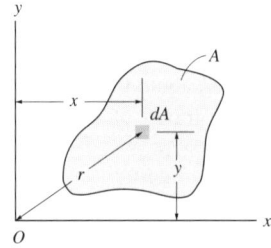

- Clearly I_x, I_y and J_O are always positive and have units of length raised to the fourth power.

- The terminology "moment of inertia" is actually a misnomer in this context—it has been adopted because of the similarity with integrals of the same form related to *mass*.

10.2 PARALLEL AXIS THEOREM FOR AN AREA

- Suppose the centroid of an area is located at $C\left(x', y', z'\right)$. *The moment of inertia of an area A about an axis is equal to the moment of inertia of the area about a parallel axis* **passing through the area's centroid** *plus the product of the area and the square of the perpendicular distance between the axes.* e.g.

$$I_x = \bar{I}_{x'} + Ad_y^2, \quad I_y = \bar{I}_{y'} + Ad_x^2, \quad J_O = \bar{J}_C + Ad^2 \tag{10.3}$$

where $\bar{I}_{x'}$, $\bar{I}_{y'}$ and $\bar{J}_C$ represent moments of inertia of the area about a corresponding parallel axis passing through the area's centroid.

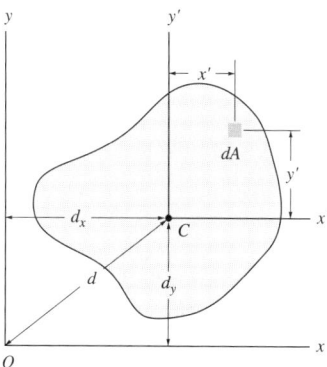

10.3 RADIUS OF GYRATION OF AN AREA

- Provided the areas and moments of inertia are known, the *radii of gyration* of a planar area are determined from the formulas

$$k_x = \sqrt{\frac{I_x}{A}}, \quad k_y = \sqrt{\frac{I_y}{A}}, \quad k_O = \sqrt{\frac{J_O}{A}} \quad \text{(units of length).} \tag{10.4}$$

- It is clear from equations (10.0)–(10.1) that finding moments of inertia requires the evaluation of *area integrals*. If one chooses to describe the area element dA with differential size in two directions (e.g., $dA = dx\, dy$) then a *double integration* must be performed. Most often, however, it is easier to perform only a single integration by *specifying the differential element dA* having a differential size or thickness in only one direction. The procedure is as follows:

PROCEDURE FOR ANALYSIS

- Most often the element dA can be rectangular with a *finite length* and differential width.
- The element should be located so that it intersects the boundary of the area at the arbitrary point (x, y). There are two ways to orient the element dA with respect to the axis about which the moment of inertia is to be determined:
    - ◆ **Case 1**: The *length* of the element can be oriented *parallel* to the axis. In this case, Equations (10.0)–(10.1) can be used *directly* since *all parts* of the element lie at the *same* moment-arm distance from the axis.
    - ◆ **Case 2**: The *length* of the element can be oriented *perpendicular* to the axis. Here neither of Equations (10.0) - (10.1) can be used directly since all parts of the element will not lie at the same moment-arm distance from the axis. Instead, it is necessary to first calculate the moment of inertia of the *element* (separately) and then integrate this result over the area A to obtain the appropriate area moment of inertia.
- See Examples 10-1 to 10-4 in text.

10.4 MOMENTS OF INERTIA FOR COMPOSITE AREAS

Provided the moment of inertia of each of the simpler areas (making up the composite area) is known, or can be determined about a common axis, then the moment of inertia of the composite area equals the *algebraic sum* of the moments of inertia of all its parts.

PROCEDURE FOR ANALYSIS

The moment of inertia of a composite area about a reference axis can be determined using the following procedure:

- **Composite Parts**
    - Using a sketch, divide the area into its composite parts and indicate the perpendicular distance from the centroid of each part to the reference axis.
- **Parallel-Axis Theorem**
    - The moment of inertia of each part should be determined about its centroidal axis, which is parallel to the reference axis—*use the table given on the inside back cover of the text.*
    - If the centroidal axis does not coincide with the reference axis, use the parallel axis theorem to determine the moment of inertia of the part about the reference axis.
- **Summation**
    - The moment of inertia of the entire area about the reference axis is found by summing the results of the composite parts.
    - If a composite part has a "hole," its moment of inertia is found by "subtracting" the moment of inertia for the hole from the moment of inertia of the entire part including the hole.

10.5 PRODUCT OF INERTIA FOR AN AREA

In general, the moment of inertia for an area is different for every axis about which it is computed. In some applications, it is necessary to know the orientation of those axes which give, respectively, the maximum and minimum moments of inertia for the area (see Section 10.6). Essential to this is the idea of a *product of inertia for an area*.

- The product of inertia for the area A is

$$I_{xy} = \int_A xy \, dA$$

- The product of inertia may be negative, positive or zero (unlike moment of inertia). For example I_{xy} will be zero if x or y is an axis of symmetry for the area A.
- The sign of I_{xy} depends on the quadrant where the area A is located. In fact, if the area is rotated from one quadrant to another, the sign of I_{xy} will change.

- **Parallel Axis Theorem for Product of Inertia of an Area** A

$$I_{xy} = \bar{I}_{x'y'} + Ad_xd_y$$

It is important that the *algebraic signs* for d_x and d_y be maintained when applying this result.

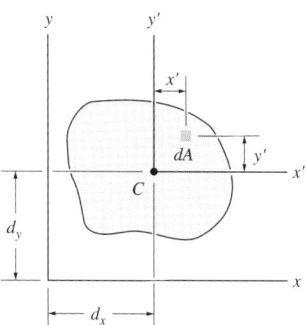

10.6 MOMENTS OF INERTIA FOR AN AREA ABOUT INCLINED AXES

- The moments and products of inertia for an area with respect to a set of inclined (at an angle θ) u and v axes are given by (assuming θ, I_x, I_y and I_{xy} are *known*):

$$
\begin{aligned}
I_u &= \frac{I_x + I_y}{2} + \frac{I_x - I_y}{2}\cos 2\theta - I_{xy}\sin 2\theta, \\
I_v &= \frac{I_x + I_y}{2} - \frac{I_x - I_y}{2}\cos 2\theta + I_{xy}\sin 2\theta, \\
I_{uv} &= \frac{I_x - I_y}{2}\sin 2\theta + I_{xy}\cos 2\theta
\end{aligned}
\tag{10.5}
$$

- The polar moment of inertia about the z-axis passing through the point O is independent of the orientation of the u and v axes i.e.

$$J_O = I_u + I_v = I_x + I_y$$

PRINCIPAL MOMENTS OF INERTIA

- When the angle θ in (10.5) takes the value $\theta = \theta_p$ defined by

$$\tan 2\theta_p = \frac{-I_{xy}}{\left(I_x - I_y\right)/2}, \tag{10.6}$$

the axes u and v are called the *principal axes of the area* since they identify the orientation of the axes u and v about which the moments of inertia I_u and I_v are *maximum or minimum*. In this case, they are called *principal moments of inertia* and are given by

$$I_{\substack{max \\ min}} = \frac{I_x + I_y}{2} \pm \sqrt{\left(\frac{I_x - I_y}{2}\right)^2 + I_{xy}^2}. \tag{10.7}$$

Depending on the sign chosen, this result gives the maximum or minimum moment of inertia for the area.

- The *product of inertia with respect to the principal axes is zero*. Hence any symmetrical axis represents a principal axis of inertia for the area.

10.7 MOHR'S CIRCLE FOR MOMENTS OF INERTIA

Equations 10.5 to 10.7 have a graphical solution which is convenient to use and easy to remember—this solution is called a *Mohr's circle.*

PROCEDURE FOR ANALYSIS

Mohr's circle provides a convenient means for transforming I_x, I_y and I_{xy} into the principal moments of inertia using the following procedure:

- **Determine** I_x, I_y, I_{xy}. Establish the x, y axes for the area, with the origin located at the point P of interest and determine I_x, I_y and I_{xy}.

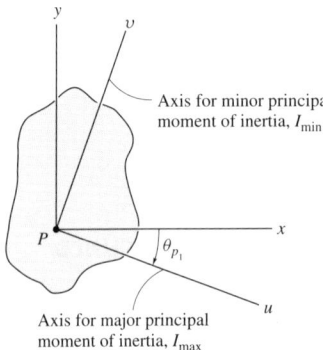

- **Construct the Circle**

  - Construct a rectangular coordinate system such that the abscissa represents the moment of inertia I and the ordinate represents the product of inertia I_{xy}

  - Determine the center of the circle O, which is located at a distance $\dfrac{I_x + I_y}{2}$ from the origin and plot the reference point A having coordinates (I_x, I_{xy}). By definition, I_x is always positive, whereas I_{xy} will be either positive or negative.

  - Connect the reference point A with the center of the circle and determine the distance OA by trigonometry. This distance represents the radius of the circle. Finally draw the circle.

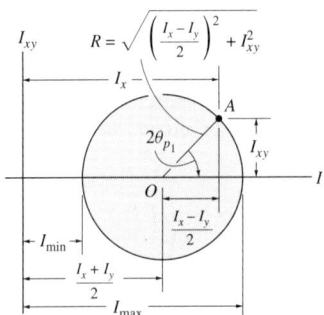

- **Principal Moments of Inertia.** The points where the circle intersects the abscissa give the values of the principal moments of inertia I_{min} and I_{max}. Notice that the *product of inertia will be zero at these points*.
- **Principal Axes.** To find the direction of the major principal axis, determine, by trigonometry, the angle $2\theta_{p_1}$, measured from the radius OA to the positive I-axis. This angle represents twice the angle from the x axis of the area in question to the axis of maximum moment of inertia I_{max}. Both the angle on the circle, $2\theta_{p_1}$ and the angle to the axis on the area, θ_{p_1}, must be measured in the same sense. The axis for minimum moment of inertia I_{min} is perpendicular to the axis for I_{max}.

10.8 MASS MOMENT OF INERTIA

- The mass moment of inertia about the z-axis is given by

$$I = \int_m r^2 \, dm \tag{10.8}$$

where r is the perpendicular distance from the axis to the arbitrary element dm.

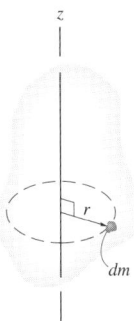

- When the axis passes through the body's mass center G, the moment of inertia is denoted by I_G. The mass moment of inertia is always positive and has units $kg.m^2$ or $slug.ft^2$.
- If the body consists of a material having a variable density $\rho(x, y, z)$, the body's moment of inertia is computed using volume integration as

$$I = \int_V r^2 \rho \, dV.$$

This integral is generally computed as a *triple integral*. The integration process can, however, be simplified to a *single integration* provided the chosen elemental volume has a differential size or thickness in only *one* direction. Shell or disk elements are often used for this purpose.

PARALLEL AXIS THEOREM

- *If the moment of inertia of the body about an axis passing through the body's mass center is known, then the moment of inertia about any other parallel axis can be computed from*

$$I = I_G + md^2 \tag{10.9}$$

where m is the mass of the body and d is the perpendicular distance between the axes.

RADIUS OF GYRATION

- The radius of gyration k has units of length and is related to the mass m and moment of inertia I of the body by

$$I = mk^2 \quad \text{or} \quad k = \sqrt{\frac{I}{m}}$$

COMPOSITE BODIES

- If a body is constructed from a number of simple shapes such as disks, spheres and rods, the moment of inertia of the body about any axis z can be determined by adding algebraically the moments of inertia of all the composite shapes computed about the z-axis. A composite part must be considered as a negative quantity if it has already been included within another part e.g., a "hole" subtracted from a solid plate.

HELPFUL TIPS AND SUGGESTIONS

- Be careful when using the Parallel Axis Theorem (10.9). It is applicable only to the situation when the moment of inertia of the body about an axis *passing through the body's mass center* is known. It *cannot* be applied in the form $I = I_B + md^2$ where B is an arbitrary point.

REVIEW QUESTIONS

1. True or False? The area moments of inertia (10.0)–(10.2) can be negative.
2. It is often the case that the moment of inertia for an area is known about an axis passing through its centroid. What can then be said about the moment of inertia of the same area about a corresponding parallel axis?
3. How are the radii of gyration for a planar area determined?
4. If a composite body has a "hole," how would you find moment of inertia of the body ?
5. Find the moment of inertia of the following composite area about the x-axis.

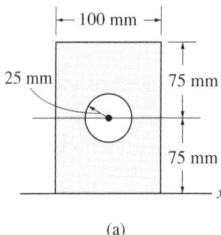

(a)

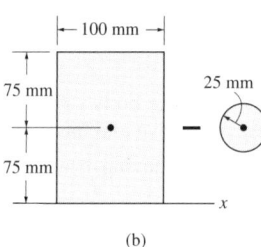

(b)

6. What are the *principal axes of an area* and the *principal moments of inertia*?
7. What are *Mohr's circles* used for?
8. Define the mass moment of inertia and describe how it can be found.

11

Virtual Work

MAIN GOALS OF THIS CHAPTER:

- To introduce the principle of *virtual work* and show how it applies to determining the equilibrium configuration of a series of pin-connected members.
- To establish the potential energy function and use the potential-energy method to investigate the type of equilibrium or stability of a rigid body or configuration.

11.1 DEFINITION OF WORK AND VIRTUAL WORK

WORK OF A FORCE

- A force **F** does work only when it undergoes a displacement in the direction of the force.
- Work is a *scalar quantity* defined by the dot product

$$dU = \mathbf{F} \cdot \mathbf{dr} \qquad (11.0)$$
$$= F \cos \theta \, ds,$$

where dU is the increment of work done when the force **F** is displaced **dr**, θ is the angle between the tails of **dr** and **F**, and ds is the magnitude of **dr**.

- *Positive work* is done when the force and its displacement have the same *sense*. Otherwise *negative work* is done.
- In the SI system, the basic unit of work is a *Joule* (J) ($1\,\text{J} = 1\,\text{N} \cdot \text{m}$). In the *FPS* system work is defined in terms of ft.lb.

WORK OF A COUPLE

- The two forces of a couple do work when the couple *rotates* about an axis perpendicular to the plane of the couple.
- Work done by a couple **M** is a *scalar quantity* defined by the dot product

$$dU = \mathbf{M} \cdot d\boldsymbol{\theta} \qquad (11.1)$$

61

where dU is the increment of work done when the couple **M** is 'displaced' **dθ** (a differential rotation of the body about an axis perpendicular to the plane of the couple).

- The resultant work is *positive* when the sense of **M** is the same as that of **dθ** and *negative* when they have an *opposite sense*. The direction and sense of **dθ** are again defined by the *right-hand rule*. Hence, if movement of the body occurs in the *same plane*, the line of action of **dθ** will be *parallel* to the line of action of **M** and Equation (11.1) becomes

$$dU = Md\theta \tag{11.2}$$

VIRTUAL WORK

- A *virtual* movement (displacement or rotation) is an *imaginary* movement which is *assumed* and *does not actually exist*. A *virtual displacement* is a differential that is given in the positive direction of the position coordinate and is denoted by the symbol δs. Similarly, a *virtual rotation* is denoted by $\delta\theta$.

- The *virtual work* done by a force undergoing a virtual displacement δs is

$$\delta U = F\cos\theta\delta s. \tag{11.3}$$

- The *virtual work* done by a couple undergoing a virtual rotation $\delta\theta$ in the plane of the couple forces is

$$\delta U = M\delta\theta. \tag{11.4}$$

11.2 PRINCIPLE OF VIRTUAL WORK FOR A PARTICLE AND A RIGID BODY

- **Particle.** If a particle undergoes an imaginary or virtual displacement $\delta\mathbf{r}$, then the virtual work (δU) done by the force system must be zero for equilibrium i.e.

$$\delta U = 0. \tag{11.5}$$

In other words we can write *three independent virtual work equations* corresponding to the three equations of equilibrium:

$$\sum Fx\delta x = 0, \quad \sum Fy\delta y = 0, \quad \sum Fz\delta z = 0.$$

- **Rigid Body.** As in the case of a particle we can also write a set of three virtual work equations (11.5) for a rigid body subjected to a coplanar force system (two involving virtual translations in the x and y directions and another a virtual rotation about an axis perpendicular to the $x-y$ plane and passing through an arbitrary point O).

- **NOTE.** As in the case of a particle, no added advantage is gained by solving rigid body equilibrium problems using the principle of virtual work since for each application of the virtual work equation, the virtual displacement or rotation, common to every term, factors out, leaving an equation that could have been obtained in a more direct manner by simply applying the equations of equilibrium.

11.3 PRINCIPLE OF VIRTUAL WORK FOR A SYSTEM OF CONNECTED RIGID BODIES

The method of virtual work is most suitable for solving equilibrium problems that involve a system of several *connected* rigid bodies.

- **Degrees of Freedom.** An $n-degree-of-freedom$ *system* requires n *independent coordinates* q_n to specify the location of all its members.

- **Principle of Virtual Work.** The principle of virtual work for a system of rigid bodies whose connections are *frictionless* may be stated as follows:

  > A system of connected rigid bodies is in equilibrium provided the virtual work done by all the external forces and couples acting on the system is zero for each independent virtual displacement of the system

  Mathematically, we write

  $$\delta U = 0 \tag{11.6}$$

  where δU represents the virtual work of all the external forces and couples acting on the system during any independent virtual displacement.

- This means that for an $n-$ degree-of-freedom system it is possible to write n independent virtual work equations, one for every virtual displacement taken along each of the independent coordinate axes, while the remaining $n - 1$ remaining independent coordinates are held *fixed*.

The following procedure shows how to use the equation of virtual work to solve problems involving a system of frictionless connected rigid bodies having a *single degree of freedom*.

- **Free-Body Diagram**

  - Draw the free-body diagram of the entire system of connected bodies and define the independent coordinate q.
  - Sketch the "deflected position" of the system on the free-body diagram when the system undergoes a positive virtual displacement δq.

- **Virtual Displacements**

  - Indicate position coordinates s_i, measured from a *fixed point* on the free-body diagram to each of the i number of "active" forces and couples i.e., those that do work.
  - Each coordinate axis should be parallel to the line of action of the "active' force to which it is directed, so that the virtual work along the coordinate axis can be calculated.
  - Relate each of the position coordinates s_i to the independent coordinate q; then differentiate these expressions to express the virtual displacements δs_i in terms of δq.

- **Virtual Work Equation**

  - Write the virtual work equation (11.6) for the system assuming that, whether possible or not, all the position coordinates s_i undergo *positive* virtual displacements δs_i.
  - Using the relations for δs_i, express the work of *each* "active" force and couple in the equation in terms of the single independent virtual displacement δq.
  - Factor out the common displacement from all the terms and solve for the unknown force, couple or equilibrium position, q.
  - If the system contains n degrees of freedom, n independent coordinates q_n must be specified. Follow the above procedure and let *only one* of the independent coordinates undergo a virtual displacement while the remaining $n - 1$ coordinates are *held fixed*. In this way, n virtual-work equations can be written, one for each independent coordinate.

- **EXAMPLES.** The above procedure is illustrated in Examples 11-1 through 11-4 in the text. *Note that had these examples been solved using the equations of equilibrium, it would have been necessary to dismember the links and apply three scalar equations to each link. The principle of virtual work, by means of calculus, has eliminated this task so that the answer is obtained directly.*

11.4 CONSERVATIVE FORCES

- If a force **F** is displaced over a path with finite length S the work done by the force is given by the integral

$$U = \int_S dU = \int_S F \cos\theta\, ds.$$

If this integral is *independent of its path* (depends only on the initial and final locations of its path), the force is called a *conservative force*.

- **EXAMPLES OF CONSERVATIVE FORCES:** *Weight, Elastic Springs*.
- **EXAMPLES OF NONCONSERVATIVE FORCES:** *Friction:* the work done by the frictional force *depends on the path*: the longer the path, the greater the work. The work done is dissipated from the body in the form of heat.

11.5 POTENTIAL ENERGY

When a conservative force acts on a body, it gives the body the capacity to do work. This capacity is known as the body's *potential energy* and depends on the location of the body.

- **Gravitational Potential Energy.** Measuring y *positive upward*, the gravitational potential energy of a body's weight **W** is

$$V_g = Wy. \tag{11.7}$$

- **Elastic Potential Energy.** The elastic potential energy V_e that a spring produces on an attached body, when the spring is elongated or compressed from an undeformed position ($s = 0$) to a final position s is

$$V_e = \frac{1}{2}ks^2. \tag{11.8}$$

- **Potential Function.** In the general case, if a body is subjected to both gravitational and elastic forces, the potential energy (function) V of the body can be expressed as the algebraic sum

$$V = V_g + V_e$$

where measurement of V depends on the location of the body with respect to a selected datum in accordance with Equations (11.7) and (11.8).

11.6 POTENTIAL ENERGY CRITERION FOR EQUILIBRIUM

- **System Having One Degree of Freedom (q).** When a frictionless connected system of rigid bodies is in equilibrium, we require that the potential energy (function) V of the body satisfies

$$\frac{dV}{dq} = 0. \tag{11.9}$$

- **System Having n Degrees of Freedom ($q_1, \ldots, q_n$).** When a frictionless connected system of rigid bodies is in equilibrium, we require that the potential energy (function) V of the body satisfies

$$\frac{\partial V}{\partial q_1} = 0, \quad \frac{\partial V}{\partial q_2} = 0, \ldots, \frac{\partial V}{\partial q_n} = 0.$$

In other words, it is possible to *write n independent equations for a system having n degrees of freedom*.

11.7 STABILITY OF EQUILIBRIUM

Once the equilibrium configuration for a body or a system of connected bodies is defined, it is important to investigate the "type" of equilibrium or the stability of the configuration.

- **Types of Equilibrium**

    1. *Stable Equilibrium.* A small displacement of the system causes the system to return to its original position. Potential energy of the system is at a minimum in this case.
    2. *Neutral Equilibrium.* A small displacement of the system causes the system to remain in its displaced state. Potential energy of the system remains constant in this case.
    3. *Unstable Equilibrium.* A small displacement of the system causes the system to move farther away from its original position. Original potential energy of the system is a maximum in this case.

- **System Having One Degree of Freedom (q).** We require that the potential energy (function) V of the body satisfies the following conditions in each case:

    1. *Stable Equilibrium.*

$$\frac{dV}{dq} = 0, \quad \frac{d^2V}{dq^2} > 0. \tag{11.10}$$

    2. *Neutral Equilibrium.*

$$\frac{dV}{dq} = \frac{d^2V}{dq^2} = \frac{d^3V}{dq^3} = \cdots = 0. \tag{11.11}$$

    3. *Unstable Equilibrium.*

$$\frac{dV}{dq} = 0, \quad \frac{d^2V}{dq^2} < 0. \tag{11.12}$$

- **System Having Two Degrees of Freedom** (q_1, q_2). Things become much more complicated as the number of degrees of freedom of the system increases. However, for a system for two degrees of freedom, we can say:

    1. *Equilibrium and Stability* occur at a point $\left(q_{1eq}, q_{2eq}\right)$ when

$$\frac{\partial V}{\partial q_1} = \frac{\partial V}{\partial q_2} = 0,$$

$$[\left(\frac{\partial^2 V}{\partial q_1 \partial q_2}\right)^2 - \left(\frac{\partial^2 V}{\partial q_1^2}\right)\left(\frac{\partial^2 V}{\partial q_2^2}\right)] < 0,$$

$$(\frac{\partial^2 V}{\partial q_1^2} + \frac{\partial^2 V}{\partial q_2^2}) > 0.$$

    2. *Equilibrium and Instability* occur when

$$\frac{\partial V}{\partial q_1} = \frac{\partial V}{\partial q_2} = 0,$$

$$[\left(\frac{\partial^2 V}{\partial q_1 \partial q_2}\right)^2 - \left(\frac{\partial^2 V}{\partial q_1^2}\right)\left(\frac{\partial^2 V}{\partial q_2^2}\right)] < 0,$$

$$\left(\frac{\partial^2 V}{\partial q_1^2} + \frac{\partial^2 V}{\partial q_2^2}\right) < 0.$$

PROCEDURE FOR SOLVING PROBLEMS

Using potential-energy methods, the equilibrium positions and the stability of a body or a system of connected bodies having a *single degree of freedom q* can be obtained using the following procedure.

- **Potential Function**
    - Sketch the system so that it is located at some *arbitrary position* specified by the independent coordinate q.
    - Establish a horizontal *datum* through a *fixed point* and express the *gravitational potential energy* V_g in terms of the weight W of each member and its vertical distance y from the datum, $V_g = Wy$.
    - Express the elastic potential energy V_e of the system in terms of the stretch or compression, s, of any connecting spring and the spring stiffness k, $V_e = \frac{1}{2}ks^2$.
    - Formulate the potential function $V = V_g + V_e$ and express the *position coordinates* y and s in terms of the independent coordinate q.

- **Equilibrium Position**
    - The equilibrium position is determined from Equation (11.9) i.e. $\frac{dV}{dq} = 0$.

- **Stability**
    - Stability at the equilibrium position is determined from Equations (11.10)–(11.12).

REVIEW QUESTIONS

1. What is the work done by a force $\mathbf{F}$ when its point of application is displaced $\mathbf{dr}$?
2. What is the work done by a couple $\mathbf{M}$ when the object on which it acts rotates through an angle $\mathbf{d\theta}$?
3. What does the principle of virtual work say when an object in equilibrium is subjected to a virtual translation or rotation?
4. What is meant by a "conservative force"?
5. What is the potential energy of a body and how is it related to the concept of "conservative force"?
6. What is the potential energy criterion for equilibrium for a frictionless connected system of rigid bodies with one degree of freedom?
7. What does it mean when an equilibrium position of a body is stable or unstable?
8. How do you know when an equilibrium position of a system having one degree of freedom is stable or unstable?

ANSWERS TO REVIEW QUESTIONS

Chapter 1:

  1. F   **2.** T   **3.** F   **4.** F   **5.** F   **6.** T   **7.** T   **8.** F

Chapter 2:

  1. See (2.1)   **2.** See (2.0) (in the plane) or (2.2) and (2.3)   **3.** See (2.4)

  4. See Section 2.5   **5.** See (2.3)   **6.** See (2.5)

  7. Vectors are perpendicular.

  8. See (2.5)   **9.** See (2.6)   **10.** See (2.7) and (2.8)

Chapter 3:

  1. See Section 3.1   **2.** See (3.0)   **3.** See Section 3.2

  4. Lines of action of the forces lie in a plane

  5. Lines of action of the forces lie in three-dimensional space

  6. One more equation—see (3.2) and (3.3)

  7. See (3.1)   **8.** True.

Chapter 4:

  1. See Section 4.1.   **2.** See Section 4.1.   **3.** Right-hand rule.

  4. No moment.   **5.** See (4.0).   **6.** Vectors are parallel.

  7. $\mathbf{r}$ represents a position vector drawn *from O to any point* lying on the line of action of $\mathbf{F}$.

  8. $|\mathbf{M}_O| = |\mathbf{F}|\,|\mathbf{r}|\sin\theta = Fr\sin\theta = Fd$.

  9. Right-hand rule i.e. curling the fingers of the right hand from vector $\mathbf{r}$ (cross) to vector $\mathbf{F}$, the thumb then points in the direction of $\mathbf{M}_O$.

 10. See Section 4.6.   **11.** False.   **12.** See Section 4.7.   **13.** See Section 4.7 and (4.2).

 14. See Section 4.8.   **15.** See Section 4.9.

Chapter 5:

  1. A particle has no size/shape and so cannot support rotation, only translation.

  2. See (5.1) and (5.0).

  3. See Section 5.7: the object has more supports than are necessary to hold it in equilibrium

  4. See Section 5.7: when there are more unknown loadings on the body than equations of equilibrium available for their solution.

  5. Review Section 5.7.

Chapter 6:

1. See Section 6.1: A *truss* is a structure composed of slender members joined together at their end points.
2. See Section 6.1.
3. See Section 6.2: In order to analyze or design a truss, we must obtain the force in each of its members. To do this, we consider the *equilibrium of a joint* of the truss. This is the basis of the method of joints.
4. Two
5. See Section 6.4.
6. See Section 6.5—either the *method of joints* or *method of sections*.
7. See Section 6.6.
8. The forces at the connected parts of the group are *internal* forces and are not shown on the free-body diagram *of the group*.

Chapter 7:

1. See Section 7.1: Normal force **N** acts parallel to the beam's axis. Shear force **V** acts normal to the beam's axis. Bending moment **M** is a couple moment which causes the beam to bend..
2. See Section 7.2. Follow the sign convention shown in the figure.
3. By integration to obtain V and M as functions of x.
4. See Section 7.4. Note that no cable is truly 'weightless,' 'inextensible' or 'perfectly flexible.' These terms are simplifications to aid the modeling.
5. We use equation (7.3) and integrate (noting that w is constant). We obtain

$$y(x) = \frac{1}{F_H}\left(\frac{wx^2}{2} + c_1 x + c_2\right)$$

This is a parabola. If the origin of the x–y coordinate system is chosen so that $y = 0$, $\frac{dy}{dx} = 0$ at $x = 0$, we obtain

$$y(x) = \frac{wx^2}{2F_H}$$

Chapter 8:

1. See Section 8.1: $F_s = \mu_s N$.
2. See Section 8.1: $F_k = \mu_k N$.
3. See Section 8.2: The total number of unknowns is equal to the total number of available equilibrium equations. In this case, once the frictional forces are determined, check that $F \le \mu_s N$ otherwise slipping will occur and the body will not remain in equilibrium.
4. See Section 8.2: The total number of unknowns will *equal* the total number of available equilibrium equations plus the total number of available frictional equations or conditional equations for tipping.
5. Draw a free-body diagram!
6. Equation (8.0).
7. Equation (8.1) or (8.2) with $\phi = \phi_k$.
8. Equation (8.3) with R replaced by P (since the reactive force R is equal in magnitude to the load **P**).

Chapter 9:

1. (i) T   (ii) F (material comprising the body must be homogeneous). (iii) F   (iv) T—see (9.3).
2. See Section 9.1—at the center of gravity (or center of mass).
3. The *location* of the center of gravity *coincides* with that of the center of mass. However, the *center of mass is independent of gravity* and so can be used in situations when particles are not under the influence of a gravitational attraction—see Section 9.1.

4. Specific weight is γ or weight per unit volume.

5. $\gamma = \rho g$—see Section 9.1.

6. A body whose mass density is constant throughout its volume is said to be homogeneous—see Section 9.1.

7. Center of mass coincides with centroid and center of gravity—see Section 9.1.

8. See Section 9.2. If a composite part has a *hole*, then consider the composite part without the hole and consider the hole as an *additional* composite part having *negative* weight or size.

9. To find the *surface area and volume* of any object of revolution—see Section 9.3.

10. From (9.1) with $dW = \rho g\, dV$, the formula for the center of mass of a body is given by

$$\bar{x} = \frac{\int_V \tilde{x}\rho g\, dV}{\int_V \rho g\, dV}, \quad \bar{y} = \frac{\int_V \tilde{y}\rho g\, dV}{\int_V \rho g\, dV}, \quad \bar{z} = \frac{\int_V \tilde{z}\rho g\, dV}{\int_V \rho g\, dV}.$$

If the body is homogeneous ρ is constant. Thus we obtain:

$$\bar{x} = \frac{\int_V \tilde{x}\, dV}{\int_V dV}, \quad \bar{y} = \frac{\int_V \tilde{y}\, dV}{\int_V dV}, \quad \bar{z} = \frac{\int_V \tilde{z}\, dV}{\int_V dV}$$

which is exactly (9.2).

Chapter 10:

1. See Section 10.1: False.

2. See Section 10.2: *It's equal to the moment of inertia about the axis* **passing through the area's centroid** *plus the product of the area and the square of the perpendicular distance between the axes.*.

3. See Section 10.3, Equation (10.4).

4. See Section 10.4: If a composite part has a "hole," its moment of inertia is found by "subtracting" the moment of inertia for the hole from the moment of inertia of the entire part including the hole.

5. The composite area is determined by subtracting the circle from the rectangle. The centroid of each area is located in the figure.

   - **Moment of inertia for Circle:** Using Equation (10.3)

   $$I_x = I_{x'} + Ad_y^2$$
   $$= \frac{1}{4}\pi(25)^4 + \pi(25)^2(75)^2 = 11.4\left(10^6\right)\text{mm}^4$$

   - **Moment of inertia for Rectangle:** Using Equation (10.3)

   $$I_x = I_{x'} + Ad_y^2$$
   $$= \frac{1}{12}100(150)^3 + 100(150)^2(75)^2 = 112.5\left(10^6\right)\text{mm}^4$$

   - **Summation.** The moment of inertia for the composite area is thus

   $$I_x = -11.4\left(10^6\right) + 112.5\left(10^6\right)$$
   $$= 101\left(10^6\right)\text{mm}^4$$

6. See Section 10.6: The *principal axes of the area* identify the orientation of the axes u and v about which the moments of inertia I_u and I_v are *maximum or minimum*. (the *principal moments of inertia*).

7. See Section 10.7: Mohr's circle provides a convenient (graphical) means for transforming I_x, I_y and I_{xy} into the principal moments of inertia.

8. See Equation (10.8). This integral is generally computed as a volume and hence *triple integral* (using the relation $dm = \rho dV$). The integration process can, however, be simplified to a *single integration* provided the chosen elemental volume has a differential size or thickness in only *one* direction. Shell or disk elements are often used for this purpose. The detailed procedure is given in Section 10.8 of the text.

Chapter 11:

1. See (11.0).   **2.**   See (11.1).

4. That the virtual work (δU) done by the force system must be zero—see (11.5).

5. See Section 11.4—that the work done by the force over a finite path is independent of the path itself.

6. See Section 11.5. When a conservative force acts on a body, it gives the body the capacity to do work. This capacity is known as the body's *potential energy* and depends on the location of the body.

7. Equation (11.9).

8. See Section 11.7: *Stable Equilibrium:* a small displacement of the system causes the system to return to its original position. Potential energy of the system is at a minimum in this case; *Unstable Equilibrium.* A small displacement of the system causes the system to move farther away from its original position. Original potential energy of the system is a maximum in this case.

9. Test using Equation (11.10) for stability and Equation (11.12) for instability.

PART II

Free-Body Diagram Workbook

1

Basic Concepts in Statics

Statics is a branch of mechanics that deals with the study of bodies that are at rest (if originally at rest) or move with constant velocity (if originally in motion) that is, bodies which are in (static) equilibrium.

In mechanics, real bodies (e.g. planets, cars, planes, tables, crates, etc) are represented or *modeled* using certain idealizations which simplify application of the relevant theory. In this book we refer to only two such models:

- **Particle.** A *particle* has a mass but a size/shape that can be neglected. For example, the size of an aircraft is insignificant when compared to the size of the earth and therefore the aircraft can be modeled as a particle when studying its three-dimensional motion in space.

- **Rigid Body.** A *rigid body* represents the next level of sophistication after the particle. That is, a rigid body is a collection of particles which has a size/shape but this size/shape cannot change. In other words, when a body is modeled as a rigid body, we assume that any deformations (changes in shape) are relatively small and can be neglected. For example, the actual deformations occurring in most structures and machines are relatively small so that the rigid body assumption is suitable in these cases.

1.1 Equilibrium

Equilibrium of a Particle

A particle is in equilibrium provided it is at rest if originally at rest or has a constant velocity if originally in motion. To maintain equilibrium, it is necessary and sufficient to satisfy Newton's first law of motion which requires the resultant force acting on the particle or rigid body to be zero. In other words

$$\sum \mathbf{F} = \mathbf{0} \tag{1.1}$$

where $\sum \mathbf{F}$ is the vector sum of all the external forces acting on the particle.

Successful application of the equations of equilibrium (1.1) requires a complete specification of all the known and unknown external forces ($\sum \mathbf{F}$) that act on the object. The best way to account for these is to draw the object's *free-body diagram*.

Equilibrium of a Rigid Body

A rigid body will be in equilibrium provided the sum of all the external forces acting on the body is equal to zero and the sum of the external moments taken about a point is equal to zero. In other words

$$\sum \mathbf{F} = \mathbf{0} \tag{1.2}$$

$$\sum \mathbf{M}_O = \mathbf{0} \tag{1.3}$$

where $\sum \mathbf{F}$ is the vector sum of all the external forces acting on the rigid body and $\sum \mathbf{M}_O$ is the sum of the external moments about an arbitrary point O.

Successful application of the equations of equilibrium (1.2) and (1.3) requires a complete specification of all the known and unknown external forces ($\sum \mathbf{F}$) and moments ($\sum \mathbf{M}_O$) that act on the object. The best way to account for these is again to draw the object's *free-body diagram*.

2

Free-Body Diagrams: the Basics

2.1 Free-Body Diagram: Particle

The equilibrium equation (1.1) is used to determine unknown forces acting on an object (modeled as a particle) in equilibrium. The first step in doing this is to draw the *free-body diagram* of the object to identify the external forces acting on it. The object's free-body diagram is simply a sketch of the object *freed* from its surroundings showing *all* the (external) forces that *act* on it. The diagram focuses your attention on the object of interest and helps you identify *all* the external forces acting. For example:

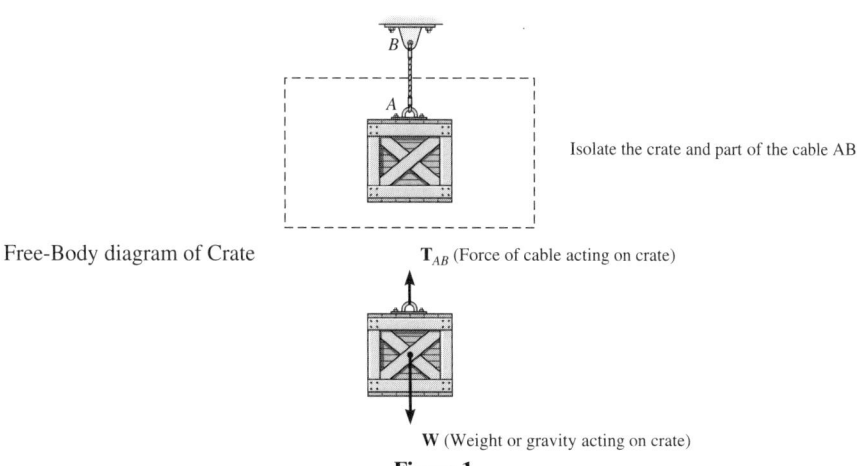

Free-Body diagram of Crate

Isolate the crate and part of the cable AB

$\mathbf{T}_{AB}$ (Force of cable acting on crate)

$\mathbf{W}$ (Weight or gravity acting on crate)

Figure 1

Note that once the crate is *separated* or *freed* from the system, forces which were previously internal to the system become external to the crate. For example, in Figure 1, such a force is the force of the cable *A B acting on the crate*.

Next, we present a formal procedure for drawing free-body diagrams for a particle.

2.1.1 Procedure for Drawing a Free-Body Diagram: Particle

---

1. *Identify the object you wish to isolate.* This choice is often dictated by the particular forces you wish to determine.
2. *Draw the outlined shape of the isolated object.* Imagine the object to be isolated or cut free from the system of which it is a part.
3. *Show all external forces acting on the isolated object.* Indicate on this sketch *all* the external forces that act on the object. These forces can be *active forces*, which tend to set the object in motion, or they can be *reactive forces* which are the result of the constraints or supports that prevent motion. This stage is crucial: it may help to trace around the object's boundary, carefully noting each external force acting on it. Don't forget to include the weight of the object (unless it is being intentionally neglected).
4. *Identify and label each external force acting on the (isolated) object.* The forces that are known should be labeled with their known magnitudes and directions. Use letters to represent the magnitudes and arrows to represent the directions of forces that are unknown.
5. *The direction of a force having an unknown magnitude can be assumed.*

EXAMPLE 2.1

The crate in Figure 2 has a weight of 20 kN. Draw free-body diagrams of the crate, the cord BD and the ring at B. Assume that the cords and the ring at B have negligible mass.

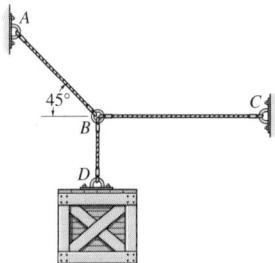

Figure 2

Solution

Free-Body Diagram for the Crate   Imagine the crate to be isolated from its surroundings, then, by inspection, there are only two external forces *acting on the crate*, namely, the weight of 20 kN and the force of the cord BD.

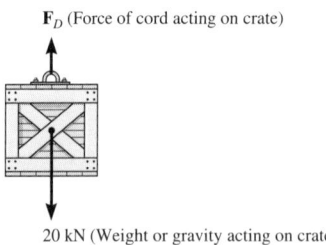

$\mathbf{F}_D$ (Force of cord acting on crate)

20 kN (Weight or gravity acting on crate)

Figure 3

Free-Body Diagram for the Cord BD Imagine the cord to be isolated from its surroundings, then, by inspection, there are only two external forces *acting on the cord*, namely, the force of the crate $\mathbf{F}_D$ and the force $\mathbf{F}_B$ caused by the ring. These forces both tend to pull on the cord so that the cord is in *tension*. Notice that $\mathbf{F}_D$ shown in this free-body diagram (Figure 4) is equal and opposite to that shown in Figure 3 (a consequence of Newton's third law).

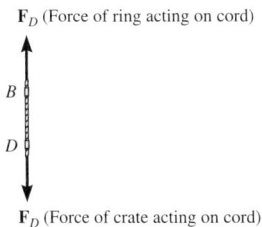

$\mathbf{F}_D$ (Force of ring acting on cord)

$\mathbf{F}_D$ (Force of crate acting on cord)

Figure 4

Free-Body Diagram for the ring at B Imagine the ring to be isolated from its surroundings, then, by inspection, there are actually three external forces acting on the ring, all caused by the attached cords. Notice that $\mathbf{F}_B$ shown in this free-body diagram (Figure 5) is equal and opposite to that shown in Figure 4 (a consequence of Newton's third law).

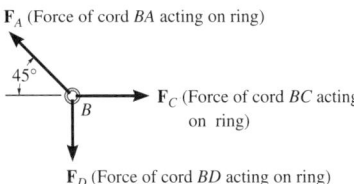

$\mathbf{F}_A$ (Force of cord BA acting on ring)

$\mathbf{F}_C$ (Force of cord BC acting on ring)

$\mathbf{F}_D$ (Force of cord BD acting on ring)

Figure 5

◄

2.1.2 Using the Free-Body Diagram: Equilibrium

The free-body diagram is used to identify the unknown forces acting on the particle when applying the equilibrium equation (1.1) to the particle. The procedure for solving equilibrium problems for a particle once the free-body diagram for the particle is established, is therefore as follows:

---

1. *Establish* the x, y-axes in any suitable orientation.
2. Apply the equilibrium equation (1.1) in component form in each direction:

$$\sum F_x = 0 \text{ and } \sum F_y = 0 \tag{2.1}$$

3. Components are positive if they are directed along a positive axis and negative if they are directed along a negative axis.
4. If more than two unknowns exist and the problem involves a spring, apply $F = ks$ to relate the magnitude of the spring force F to the deformation of the spring s (here, k is the spring constant).
5. If the solution yields a negative result, this indicates the sense of the force is the reverse of that shown/assumed on the free-body diagram.

EXAMPLE 2.2

In Example 2.1, the free-body diagrams established in Figures 3 - 5 give us a 'pictorial representation' of all the information we need to apply the equilibrium equations (2.1) to find the various unknown forces. In fact, taking the positive x-direction to be horizontal ($\rightarrow$ +) and the positive y-direction to be vertical ($\uparrow$ +), the equilibrium equations (2.1) when applied to each of the objects (regarded as particles) are:

For the Crate: $\uparrow + \sum F_y = 0$: $F_D - 20 = 0$ (See Figure 3)

$$F_D = 20 \text{ lb} \tag{2.2}$$

For the Cord BD: $\uparrow + \sum F_y = 0$: $F_B - F_D = 0$ (See Figure 4)

$$F_B = F_D \tag{2.3}$$

For the Ring: $\uparrow + \sum F_y = 0$: $F_A \sin 45 - F_B = 0$ (See Figure 5) $\quad(2.4)$

$\longrightarrow + \sum F_x = 0$: $F_C - F_A \cos 45 = 0$ (See Figure 5) $\quad(2.5)$

Equations (2.2)–(2.5) are now 4 equations which can be solved for the 4 unknowns F_A, F_B, F_C and F_D. That is: $F_B = 20$ lb; $F_D = 20$ lb, $F_A = 28.28$, $F_C = 20$. The directions of each of these forces is shown in the free-body diagrams above (Figures 3–5).   ◀

2.2 Free-Body Diagram: Rigid Body

The equilibrium equations (1.2) and (1.3) are used to determine unknown forces and moments acting *on an object* (modeled as a rigid body) in equilibrium. The first step in doing this is again to draw the *free-body diagram* of the object to identify *all of* the external forces and moments acting on it. The procedure for drawing a free-body diagram in this case is much the same as that for a particle with the main exception that now, because the object has 'size/shape,' it can support also external couple moments and moments of external forces.

2.2.1 Procedure for Drawing a Free-Body Diagram: Rigid Body

---

1. Imagine the body to be isolated or 'cut free' from its constraints and connections and sketch its outlined shape.

2. Identify all the external forces and couple moments that act on the body. Those generally encountered are:

   (a) Applied loadings

   (b) Reactions occurring at the supports or at points of contact with other bodies (See Table 2.1)

   (c) The weight of the body (applied at the body's center of gravity G)

3. The forces and couple moments that are known should be labeled with their proper magnitudes and directions. Letters are used to represent the magnitudes and direction angles of forces and couple moments that are *unknown*. Establish an x, y-coordinate system so that these unknowns e.g. A_x, B_y etc. can be identified. Indicate the dimensions of the body necessary for computing the moments of external forces. In particular, if a force or couple moment has a known line of action but unknown magnitude, the arrowhead which defines the sense of the vector can be assumed. The correctness of the assumed sense will become apparent after solving the equilibrium equations for the unknown magnitude. By definition, the magnitude of a vector is *always positive*, so that if the solution yields a *negative* scalar, the minus *sign* indicates that the vector's sense is *opposite* to that which was originally assumed.

Table 2.1. Supports for Rigid Bodies Subjected to Two-Dimensional Force Systems

	Types of Connection	Reaction	Number of Unknowns
(1)	cable	F	One unknown. The reaction is a tension force which acts away from the member in the direction of the cable.
(2)	weightless link	F or F	One unknown. The reaction is a force which acts along the axis of the link.
(3)	roller	F	One unknown. The reaction is a force which acts perpendicular to the surface at the point of contact.
(4)	roller or pin in confined smooth slot	F or F	One unknown. The reaction is a force which acts perpendicular to the slot.
(5)	rocker	F	One unknown. The reaction is a force which acts perpendicular to the surface at the point of contact.
(6)	smooth contacting surface	F	One unknown. The reaction is a force which acts perpendicular to the surface at the point of contact.
(7)	member pin connected to collar on smooth rod	or F	One unknown. The reaction is a force which acts perpendicular to the rod.
(8)	smooth pin or hinge	F_y, F_x or F, ϕ	Two unknowns. The reactions are two components of force, or the magnitude and direction ϕ of the resultant force. Note that ϕ and θ are not necessarily equal [usually not, unless the rod shown is a link as in (2)].
(9)	member fixed connected to collar on smooth rod	F, M	Two unknowns. The reactions are the couple moment and the force which acts perpendicular to the rod.
(10)	fixed support	F_y, F_x, M or F, ϕ, M	Three unknowns. The reactions are the couple moment and the two force components, or the couple moment and the magnitude and direction ϕ of the resultant force.

2.2.2 *Important Points*

- No equilibrium problem should be solved without first drawing the free-body diagram, so as to account for all the external forces and moments that act on the body.
- If a support *prevents translation* of a body in a particular direction, then the support exerts a force on the body in that direction
- If *rotation is prevented* then the support exerts a couple moment on the body
- Internal forces are never shown on the free-body diagram since they occur in equal but opposite collinear pairs and therefore cancel each other out.
- The weight of a body is an external force and its effect is shown as a single resultant force acting through the body's center of gravity G.
- *Couple moments* can be placed anywhere on the free-body diagram since they are *free vectors*. Forces can act at any point along their lines of action since they are *sliding vectors*.

EXAMPLE 2.3

Draw the free-body diagram of the beam, which is pin-connected at A and rocker-supported at B. Neglect the weight of the beam.

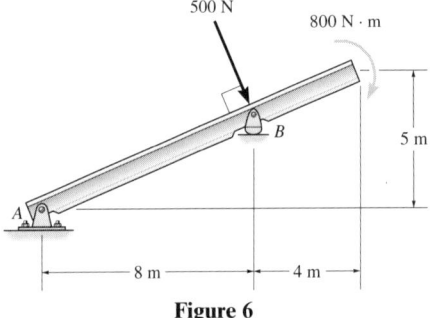

Figure 6

Solution

The free-body diagram of the beam is shown in Figure 7. From Table 2.1, since the support at A is a pin-connection, there are two reactions acting on the beam at A denoted by A_x and A_y. In addition, there is one reaction acting on the beam at the rocker support at B. We denote this reaction by the force $\mathbf{F}$ which acts perpendicular to the surface at B, the point of contact (see Table 2.1). The magnitudes of these vectors are unknown and their sense has been assumed (the correctness of the assumed sense will become apparent after solving the equilibrium equations for the unknown magnitude i.e. if application of the equilibrium equations to the beam yields a negative result for $\mathbf{F}$, this indicates the sense of the force is the reverse of that shown/assumed on the free-body diagram). The weight of the beam has been neglected.   ◀

2.2.3 *Using the Free-Body Diagram: Equilibrium*

The equilibrium equations (1.2) and (1.3) can be written in component form as:

$$\sum F_x = 0, \tag{2.6}$$

$$\sum F_y = 0, \tag{2.7}$$

$$\sum M_O = 0. \tag{2.8}$$

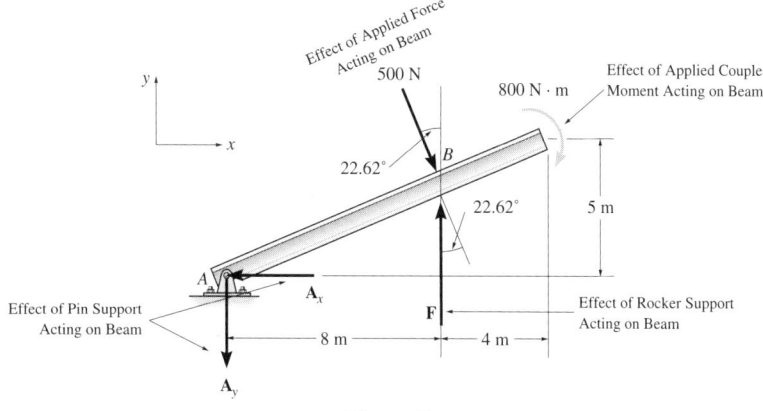

Figure 7

Here, $\sum F_x$ and $\sum F_y$ represent, respectively, the algebraic sums of the x and y components of all the external forces acting on the body and $\sum M_O$ represents the algebraic sum of the couple moments and the moments of all the external force components about an axis perpendicular to the x-y plane and passing through the arbitrary point O, which may lie either on or off the body. The procedure for solving equilibrium problems for a rigid body once the free-body diagram for the body is established, is as follows:

- Apply the moment equation of equilibrium (2.8), about a point (O) that lies at the intersection of the lines of action of two unknown forces. In this way, the moments of these unknowns are zero about O and a direct solution for the third unknown can be determined.
- When applying the force equilibrium equations (2.6) and (2.7), orient the x and y-axes along lines that will provide the simplest resolution of the forces into their x and y components.
- If the solution of the equilibrium equations yields a negative scalar for a force or couple moment magnitude, this indicates that the sense is opposite to that which was assumed on the free-body diagram.

EXAMPLE 2.4

A force of magnitude 150 kN acts on the end of the beam as shown. Find the magnitude and direction of the reaction at pin A and the tension in the cable.

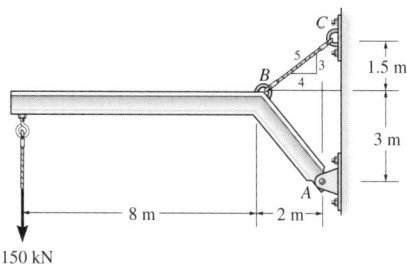

Figure 8

Solution

Free-Body Diagram The first thing to do is to draw the free-body diagram of the beam in order to identify all the external forces and moments acting on the beam.

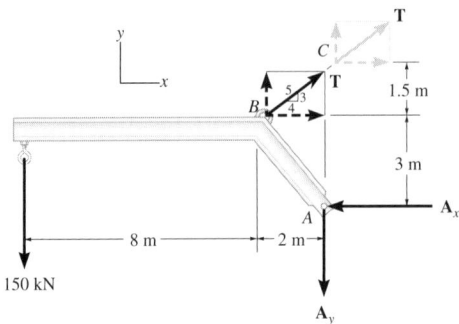

Figure 9

Equations of Equilibrium The free-body diagram of the beam suggests we can sum moments about the point A to eliminate the moment contribution of the reaction forces $\mathbf{A}_x$ and $\mathbf{A}_y$ acting on the beam. This will allow us to obtain a direct solution for the third unknown i.e. the cable tension T. Taking counterclockwise as positive when computing moments, we have:

$$+\circlearrowleft \sum M_A = 0: \; -(3/5)T(2\text{ m}) - (4/5)T(3\text{ m}) + 150\text{ kN}(10\text{ m}) = 0$$
$$-3.6T + 150\text{ kN}(10\text{ m}) = 0$$
$$\underline{T = 416.7\text{ kN}}$$

Ans.

Summing forces to obtain A_x and A_y, using the result for T, we have

$$\longrightarrow +\sum F_x = 0: \; -A_x + (4/5)(416.7\text{ kN}) = 0$$
$$\mathbf{A}_x = 333.3\text{ kN} \longleftarrow$$
$$\uparrow +\sum F_y = 0: \; (3/5)(416.7\text{ kN}) - 150\text{ kN} - A_y = 0$$
$$\mathbf{A}_y = 100\text{ kN} \downarrow$$

Thus, the reaction force $\mathbf{F}_A$ at pin A has magnitude F_A given by:

$$F_A = \sqrt{[(333.3\text{ kN})^2 + (100\text{ kN})^2]} = 348.0\text{ kN}$$

and direction given by

$$\theta = \tan^{-1}[(-100\text{ kN})/(-333.3\text{ kN})] = 196.7°$$

counterclockwise from the positive x-axis or ⟍ 16.7° ⟍ ◀

3

Problems

3.1 Free-Body Diagrams in Particle Equilibrium

Problem 3.1

The sling is used to support a drum having a weight of 4500 N ($\approx$ 450 kg). Draw a free-body diagram for the knot at A. Take $\theta = 20°$.

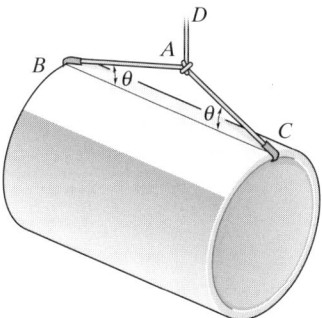

Solution

1. The knot at A has *negligible size* so that it can be modelled as a particle.
2. Imagine the knot at A to be separated or detached from the system.
3. The (detached) knot at A is subjected to three *external* forces. They are caused by:
      i.   **ii.**

      iii.
4. Draw the free-body diagram of the (detached) knot showing all these forces labeled with their magnitudes and directions.

Problem 3.1

The sling is used to support a drum having a weight of 4500 N ($\approx$ 450 kg). Draw a free-body diagram for the knot at A. Take $\theta = 20°$.

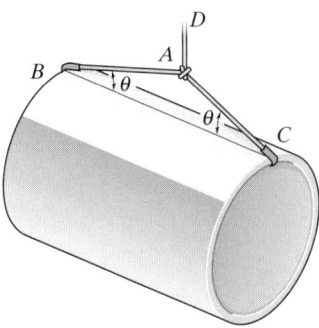

Solution

1. The knot at A has negligible size so that it can be modelled as a particle.
2. Imagine the knot at A to be separated or detached from the system.
3. The (detached) knot at A is subjected to three external forces. They are caused by:
   - **i. CORD** *AB*   **ii. CORD** *AC*
   - **iii. CORD** *AB (weight of drum)*
4. Draw the free-body diagram of the (detached) knot showing all these forces labeled with their magnitudes and directions.

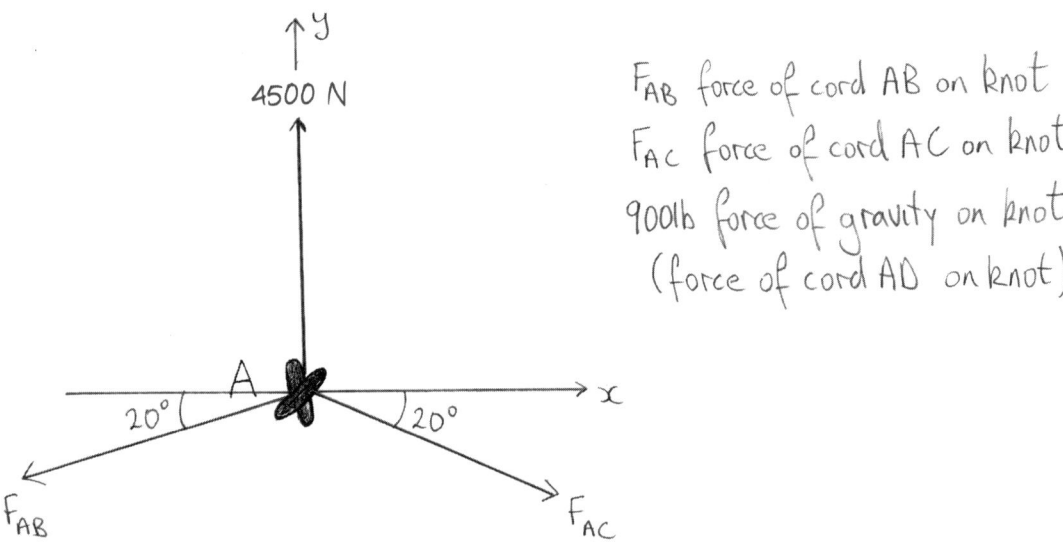

Problem 3.2

The spring ABC has a stiffness of 500 N/m and an unstretched length of 6 m. A horizontal force **F** is applied to the cord which is attached to the *small* pulley B so that the displacement of the pulley from the wall is $d = 1.5$ m. Draw a free-body diagram for the small pulley B.

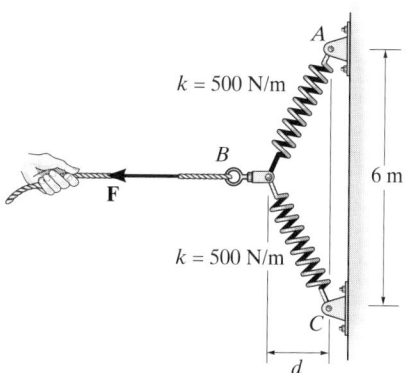

Solution

1. The pulley B has *negligible size* so that it can be modelled as a particle.
2. Imagine the pulley B to be separated or detached from the system.
3. The (detached) pulley B is subjected to three *external* forces. They are caused by:

    i.   **ii.**

    iii.

4. Draw the free-body diagram of the (detached) pulley showing all these forces labeled with their magnitudes and directions. You should also include any other available information, e.g. lengths, angles, etc. — which will help when formulating the equilibrium equations for the pulley.

Problem 3.2

The spring ABC has a stiffness of 500 N/m and an unstretched length of 6 m. A horizontal force **F** is applied to the cord which is attached to the *small* pulley B so that the displacement of the pulley from the wall is $d = 1.5$ m. Draw a free-body diagram for the small pulley B.

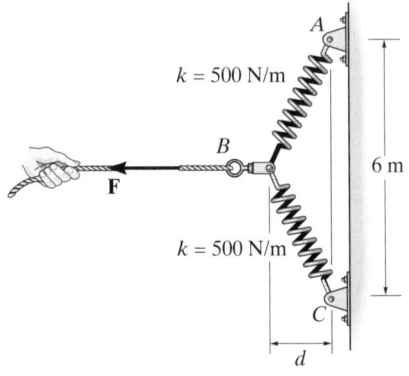

Solution

1. The pulley B has *negligible size* so that it can be modelled as a particle.
2. Imagine the pulley B to be separated or detached from the system.
3. The (detached) pulley B is subjected to three *external forces*. They are caused by:

   i. **Force F**   ii. **Spring** AB

   iii. **Spring** BC

4. Draw the free-body diagram of the (detached) pulley showing all these forces labeled with their magnitudes and directions. You should also include any other available information, e.g. lengths, angles, etc. — which will help when formulating the equilibrium equations for the pulley.

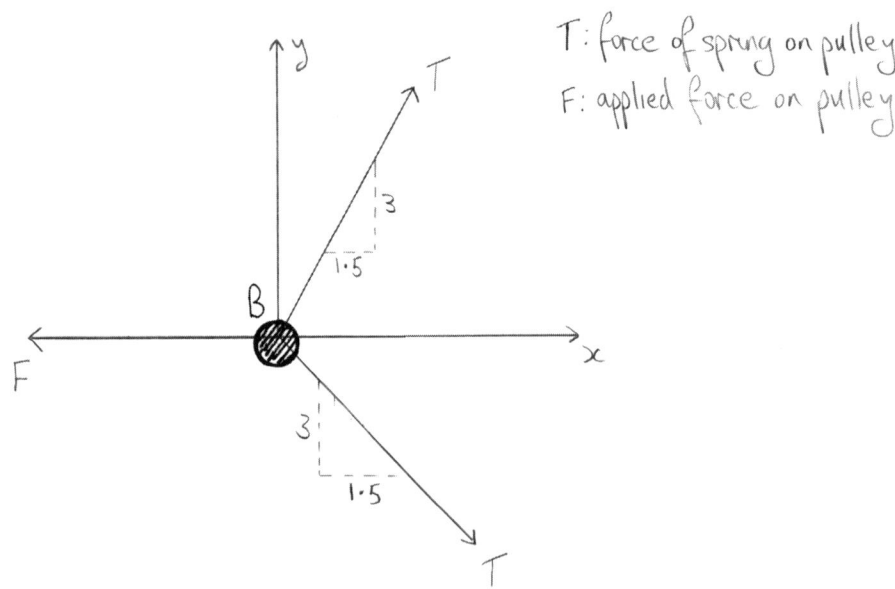

Problem 3.3

The 2-kg block is held in equilibrium by the system of springs. Draw a free-body diagram for the ring at A.

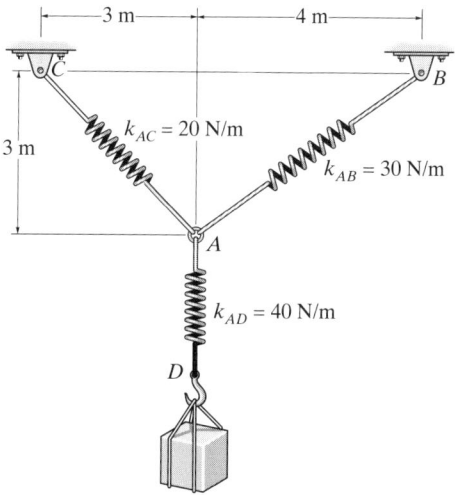

Solution

1. The ring at A has *negligible size* so that it can be modelled as a particle.
2. Imagine the ring at A to be separated or detached from the system.
3. The (detached) ring at A is subjected to three *external* forces. They are caused by:

   i.   **ii.**

   iii.

4. Draw the free-body diagram of the (detached) ring showing all these forces labeled with their magnitudes and directions. You should also include any other available information, e.g. lengths, angles, etc. — which will help when formulating the equilibrium equations for the ring.

Problem 3.3

The 2-kg block is held in equilibrium by the system of springs. Draw a free-body diagram for the ring at A.

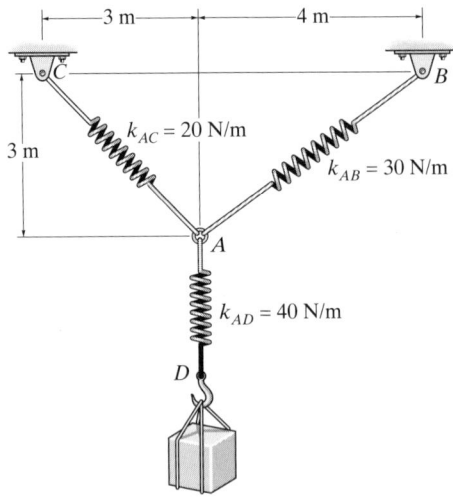

Solution

1. The ring at A has *negligible* size so that it can be modelled as a particle.
2. Imagine the ring at A to be separated or detached from the system.
3. The (detached) ring at A is subjected to three *external* forces. They are caused by:

      i. Spring AD *(weight of block)*   **ii. Spring AC**

      iii. Spring AB

4. Draw the free-body diagram of the (detached) ring showing all these forces labeled with their magnitudes and directions. You should also include any other available information, e.g. lengths, angles, etc. — which will help when formulating the equilibrium equations for the ring.

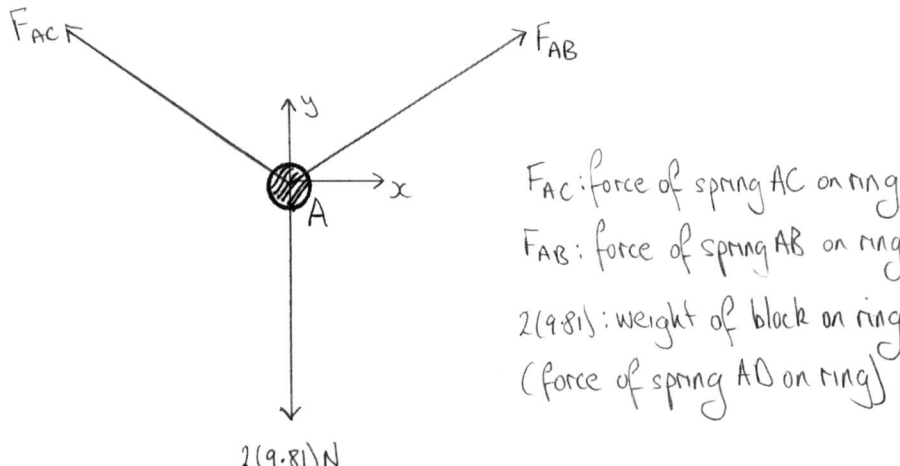

Problem 3.4

The motor at *B* winds up the cord attached to the 30-kg crate with a constant speed. The force in cord *CD* supports the pulley *C* and the angle θ represents the equilibrium state. Draw the free-body diagram of the pulley *C*. *Neglect the size of the pulley.*

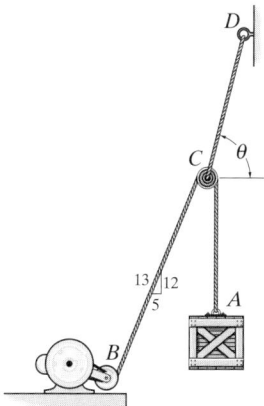

Solution

1. The pulley *C* has *negligible size* so that it can be modelled as a particle.
2. Imagine the pulley *C* to be separated or detached from the system.
3. The (detached) pulley *C* is subjected to three *external* forces. They are caused by:

    i.   **ii.**

    iii.

4. Draw the free-body diagram of the (detached) pulley showing all these forces labeled with their magnitudes and directions. You should also include any other available information, e.g. lengths, angles, etc. — which will help when formulating the equilibrium equations for the pulley.

Problem 3.4

The motor at *B* winds up the cord attached to the 30-kg crate with a constant speed. The force in cord *CD* supports the pulley *C* and the angle θ represents the equilibrium state. Draw the free-body diagram of the pulley *C*. *Neglect the size of the pulley.*

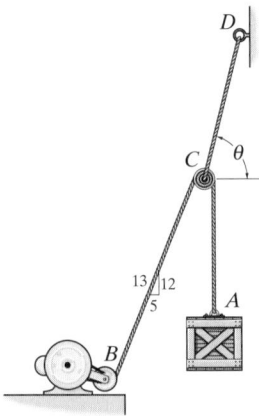

Solution

1. The pulley *C* has *negligible size* so that it can be modelled as a particle.
2. Imagine the pulley *C* to be separated or detached from the system.
3. The (detached) pulley *C* is subjected to three *external* forces. They are caused by:

   i. CORD *CD*   **ii. CORD** *CB*

   iii. CORD *CA (weight of crate)*

4. Draw the free-body diagram of the (detached) pulley showing all these forces labeled with their magnitudes and directions. You should also include any other available information, e.g. lengths, angles, etc. — which will help when formulating the equilibrium equations for the pulley.

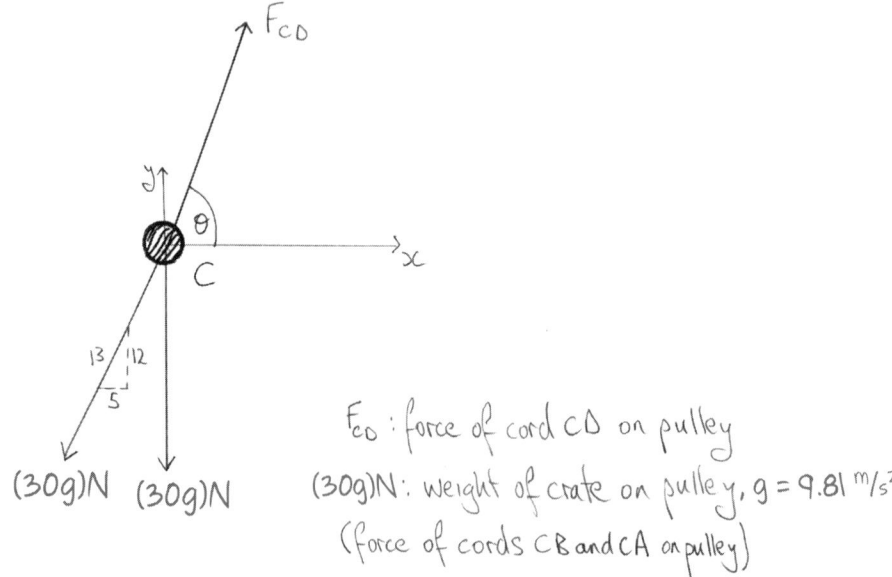

Problem 3.5

The following system is held in equilibrium by the mass supported at A and the angle θ of the connecting cord. Draw the free-body diagram for the connecting knot D.

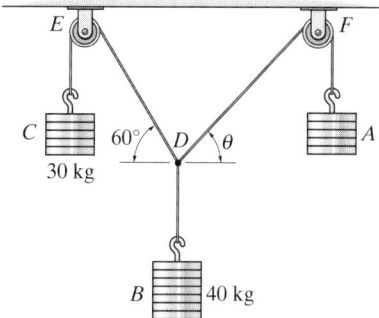

Solution

1. The knot D has *negligible size* so that it can be modelled as a particle.
2. Imagine the knot D to be separated or detached from the system.
3. The (detached) knot D is subjected to three *external* forces. They are caused by:

   i.   **ii.**

   iii.

4. Draw the free-body diagram of the (detached) knot showing all these forces labeled with their magnitudes and directions. You should also include any other available information, e.g. lengths, angles, etc. — which will help when formulating the equilibrium equations for the knot.

Problem 3.5

The following system is held in equilibrium by the mass supported at A and the angle θ of the connecting cord. Draw the free-body diagram for the connecting knot D.

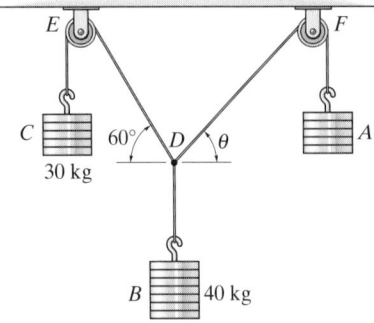

Solution

1. The knot D has *negligible size* so that it can be modelled as a particle.
2. Imagine the knot D to be separated or detached from the system.
3. The (detached) knot D is subjected to three *external* forces. They are caused by:
   i. **CORD** *DE (weight of C)*   ii. **CORD** *DF (weight of A)*
   iii. **CORD** *DB (weight of B)*
4. Draw the free-body diagram of the (detached) knot showing all these forces labeled with their magnitudes and directions. You should also include any other available information, e.g. lengths, angles, etc. — which will help when formulating the equilibrium equations for the knot.

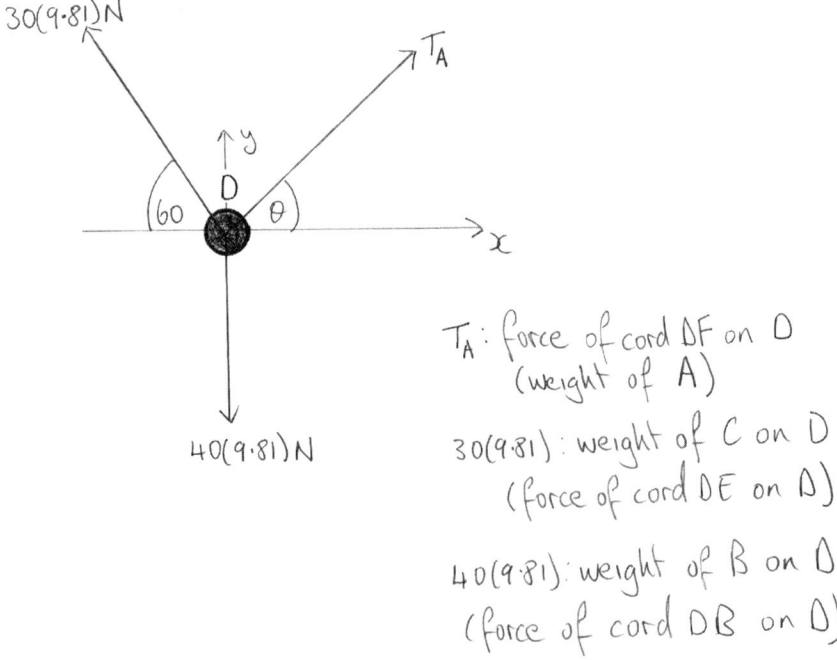

Problem 3.6

The 500-N ($\approx$ 50-kg) crate is hoisted using the ropes AB and AC. Each rope can withstand a maximum tension of 2500 N before it breaks. Rope AB always remains horizontal. Draw the free-body diagram for the ring at A and determine the smallest angle θ to which the crate can be hoisted.

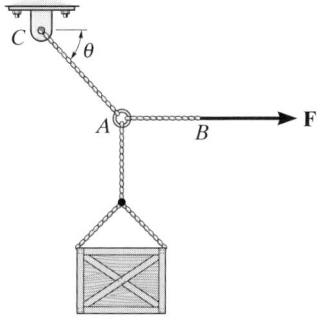

Solution

1. The ring at A has *negligible size* so that it can be modelled as a particle.
2. Imagine the ring at A to be separated or detached from the system.
3. The (detached) ring A is subjected to three *external* forces. They are caused by:

    i.   **ii.**

    iii.

4. Draw the free-body diagram of the (detached) ring showing all these forces labeled with their magnitudes and directions.

5. Establish an xy-axes system on the free-body diagram and write down the equilibrium equations in each of the x and y-directions

$+\uparrow \sum F_y = 0:$

$\xrightarrow{+} \sum F_x = 0:$

6. Solve for the angle θ:

Problem 3.6

The 500-N ($\approx$ 50-kg) crate is hoisted using the ropes AB and AC. Each rope can withstand a maximum tension of 2500 N before it breaks. Rope AB always remains horizontal. Draw the free-body diagram for the ring at A and determine the smallest angle θ to which the crate can be hoisted.

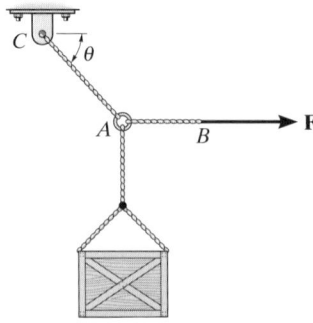

Solution

1. The ring at A has *negligible size* so that it can be modelled as a particle.
2. Imagine the ring at A to be separated or detached from the system.
3. The (detached) ring A is subjected to three *external* forces. They are caused by:

   i. CORD AC **ii. CORD** AB

   iii. CORD AD *(weight of crate)*

4. Draw the free-body diagram of the (detached) ring showing all these forces labeled with their magnitudes and directions.

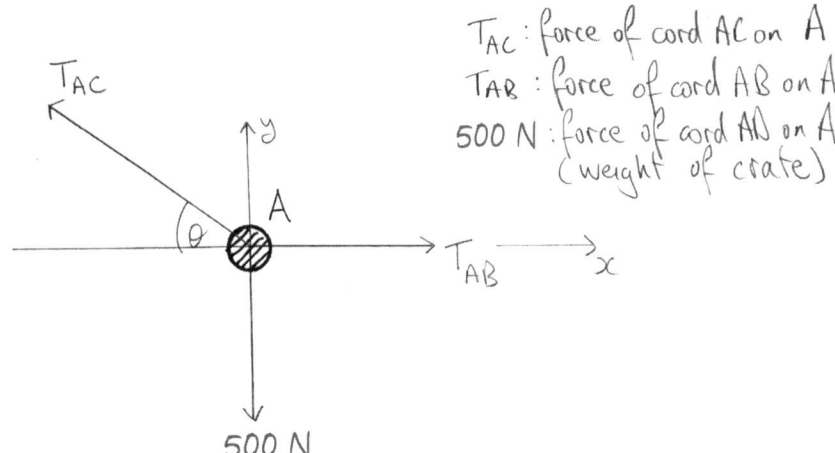

5. Establish an xy-axes system on the free-body diagram and write down the equilibrium equations in each of the x and y-directions

$+\uparrow \sum F_y = 0:\ T_{AC}\sin\theta - 500 = 0$

$\xrightarrow{+} \sum F_x = 0:\ T_{AB} - T_{AC}\cos\theta = 0$

6. Solve for the angle θ:

      Assume $T_{AC} = 2500$ N $\Rightarrow \theta = 11.54°$ and $T_{AB} = 2449.49$ N < 2500 N (O.K!)   **Ans.**

Problem 3.7

The block has a weight of 20 N and is being hoisted at uniform velocity. The system is held in equilibrium at angle θ by the appropriate force in each cord. Draw the free-body diagram for the *small* pulley.

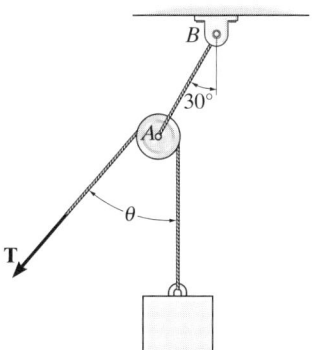

Solution

1. The pulley has *negligible size* so that it can be modelled as a particle.
2. Imagine the pulley to be separated or detached from the system.
3. The (detached) pulley is subjected to three *external* forces. They are caused by:

   i.   **ii.**

   iii.

4. Draw the free-body diagram of the (detached) pulley showing all these forces labeled with their magnitudes and directions. You should also include any other available information, e.g. lengths, angles, etc. — which will help when formulating the equilibrium equations for the knot.

 A

Problem 3.7

The block has a weight of 20 N and is being hoisted at uniform velocity. The system is held in equilibrium at angle θ by the appropriate force in each cord. Draw the free-body diagram for the *small* pulley.

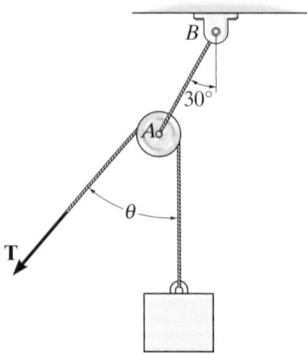

Solution

1. The pulley has *negligible size* so that it can be modelled as a particle.
2. Imagine the pulley to be separated or detached from the system.
3. The (detached) pulley is subjected to three *external* forces. They are caused by:

   i. Cord AB **ii. Force T**

   iii. Weight of block

4. Draw the free-body diagram of the (detached) pulley showing all these forces labeled with their magnitudes and directions. You should also include any other available information, e.g. lengths, angles, etc. — which will help when formulating the equilibrium equations for the knot.

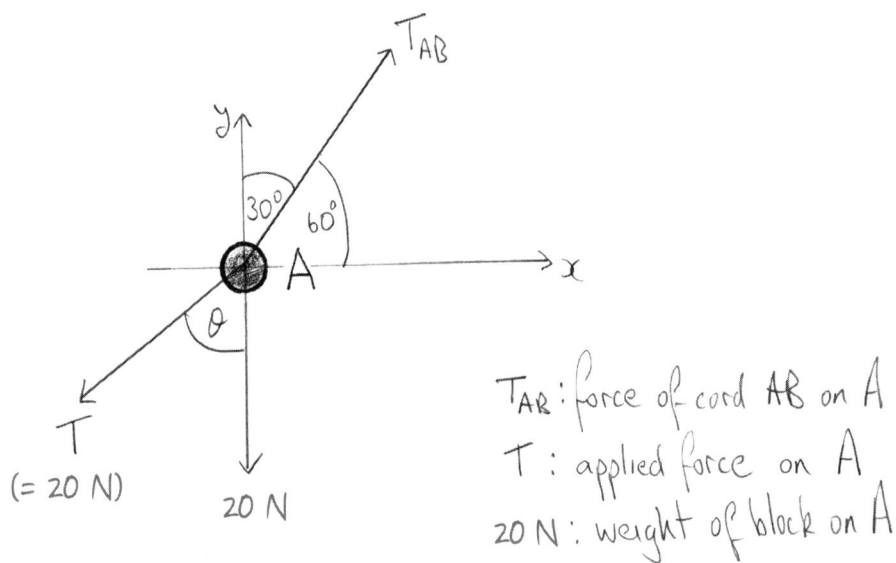

Problem 3.8

Blocks D and F weigh 5 kN ($\approx$ 500 kg) each and block E weighs 8 kN ($\approx$ 800 kg). The system is in equilibrium at a given sag s. Draw the free-body diagram for the connecting ring at A and find s. Neglect the size of the pulleys.

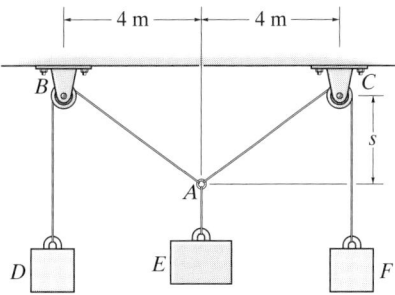

Solution

1. The ring at A has *negligible size* so that it can be modelled as a particle.
2. Imagine the ring to be separated or detached from the system.
3. The (detached) ring is subjected to three *external* forces. They are caused by:

   i.   **ii.**

   iii.

4. Draw the free-body diagram of the (detached) ring showing all these forces labeled with their magnitudes and directions. Include also any other information which may help when formulating the equilibrium equations for the ring.

5. Establish an xy-axes system on the free-body diagram and write down the equilibrium equations in the y-direction only (this is all that is required to solve this problem):

$$+\uparrow \sum F_y = 0:$$

6. Solve for the sag s:

Problem 3.8

Blocks D and F weigh 5 kN ($\approx$ 500 kg) each and block E weighs 8 kN ($\approx$ 800 kg). The system is in equilibrium at a given sag s. Draw the free-body diagram for the connecting ring at A and find s. Neglect the size of the pulleys.

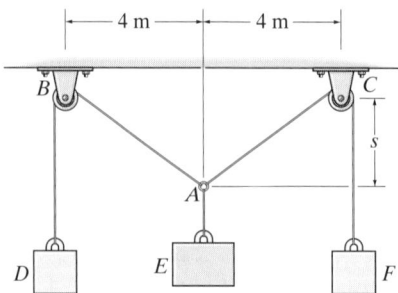

Solution

1. The ring at A has *negligible size* so that it can be modelled as a particle.
2. Imagine the ring to be separated or detached from the system.
3. The (detached) ring is subjected to three *external* forces. They are caused by:

   i. **CORD** AB *(weight of D)*   ii. **CORD** AC *(weight of F)*

   iii. **CORD** AE *(weight of E)*

4. Draw the free-body diagram of the (detached) ring showing all these forces labeled with their magnitudes and directions. Include also any other information which may help when formulating the equilibrium equations for the ring.

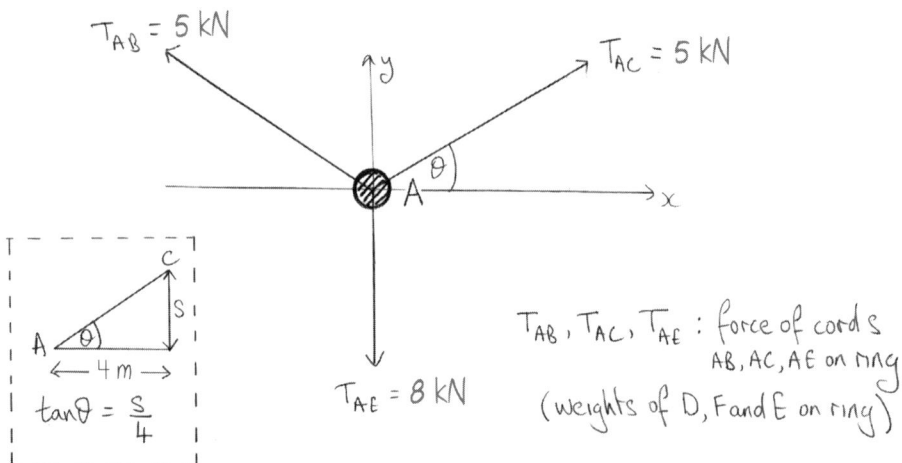

5. Establish an xy-axes system on the free-body diagram and write down the equilibrium equations in the y-direction only (this is all that is required to solve this problem):

   $+\uparrow \sum F_y = 0$: $2(5) \sin\theta - 8 = 0 \Rightarrow \theta = 53.13°$

6. Solve for the sag s:

$$\tan\theta = \frac{s}{4} \Rightarrow s = 4\tan 53.13° = 5.33 \text{ m}$$ **Ans.**

Problem 3.9

A vertical force **P** is applied to the ends of the cord AB and spring AC. The spring has an unstretched length of 2 m and the system is in equilibrium at angle θ. Draw the free-body diagram of the connecting knot at A and write down the equilibrium equations for the knot at A.

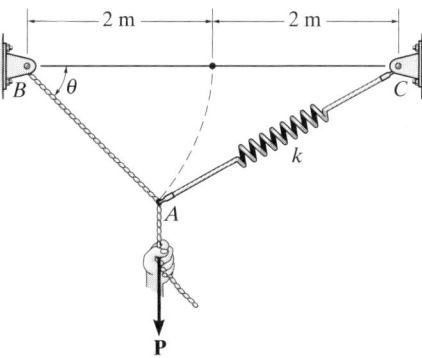

Solution

1. The knot at A has *negligible size* so that it can be modelled as a particle.
2. Imagine the knot at A to be separated or detached from the system.
3. The (detached) knot at A is subjected to three *external* forces. They are caused by:

   i.   **ii.**

   iii.

4. Draw the free-body diagram of the (detached) knot showing all these forces labeled with their magnitudes and directions.

5. Establish an xy-axes system on the free-body diagram and write down the equilibrium equations in each of the x and y-directions

   $+\uparrow \sum F_y = 0:$

   $\xrightarrow{+} \sum F_x = 0:$

Problem 3.9

A vertical force **P** is applied to the ends of the cord AB and spring AC. The spring has an unstretched length of 2 m and the system is in equilibrium at angle θ. Draw the free-body diagram of the connecting knot at A and write down the equilibrium equations for the knot at A.

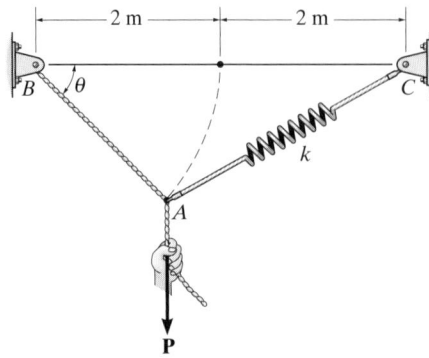

Solution

1. The knot at A has *negligible size* so that it can be modelled as a particle.
2. Imagine the knot at A to be separated or detached from the system.
3. The (detached) knot at A is subjected to three *external* forces. They are caused by:
   i. **CORD** AB ii. **SPRING** AC
   iii. **Force P**
4. Draw the free-body diagram of the (detached) knot showing all these forces labeled with their magnitudes and directions.

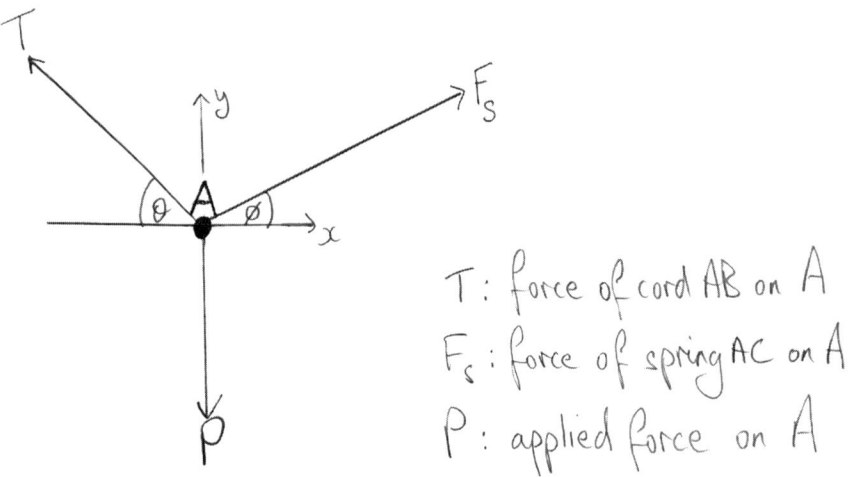

5. Establish an xy-axes system on the free-body diagram and write down the equilibrium equations in each of the x and y-directions

$\xrightarrow{+} \sum F_x = 0: F_s \cos\phi - T\cos\theta = 0$

$+\uparrow \sum F_y = 0: T\cos\theta + F_s \sin\phi - P = 0$

Problem 3.10

The sling *BAC* is used to lift the 10-kN load with constant velocity. By drawing the free-body diagram for the ring at *A*, determine the magnitude of the force in the sling as a function of the angle θ.

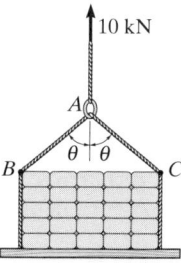

Solution

1. The ring at *A* has *negligible size* so that it can be modelled as a particle.
2. Imagine the ring at *A* to be separated or detached from the system.
3. The (detached) ring at *A* is subjected to three *external* forces. They are caused by:
       i.   **ii.**
       iii.
4. Draw the free-body diagram of the (detached) ring showing all these forces labeled with their magnitudes and directions. Include also any other information which may help when formulating the equilibrium equations for the ring.

5. Establish an *xy*-axes system on the free-body diagram and write down the equilibrium equations in the *y*-direction only (this is all that is required to solve this problem):

$$+\uparrow \sum F_y = 0:$$

6. Solve for the magnitude of the force in the sling:

Problem 3.10

The sling *BAC* is used to lift the 10-kN load with constant velocity. By drawing the free-body diagram for the ring at *A*, determine the magnitude of the force in the sling as a function of the angle *θ*.

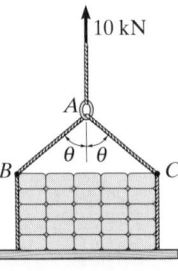

Solution

1. The ring at *A* has *negligible size* so that it can be modelled as a particle.
2. Imagine the ring at *A* to be separated or detached from the system.
3. The (detached) ring at *A* is subjected to three *external* forces. They are caused by:

       i. CORD *AB*   **ii. CORD** *AC*

       iii. CORD *AD (weight of load)*

4. Draw the free-body diagram of the (detached) ring showing all these forces labeled with their magnitudes and directions. Include also any other information which may help when formulating the equilibrium equations for the ring.

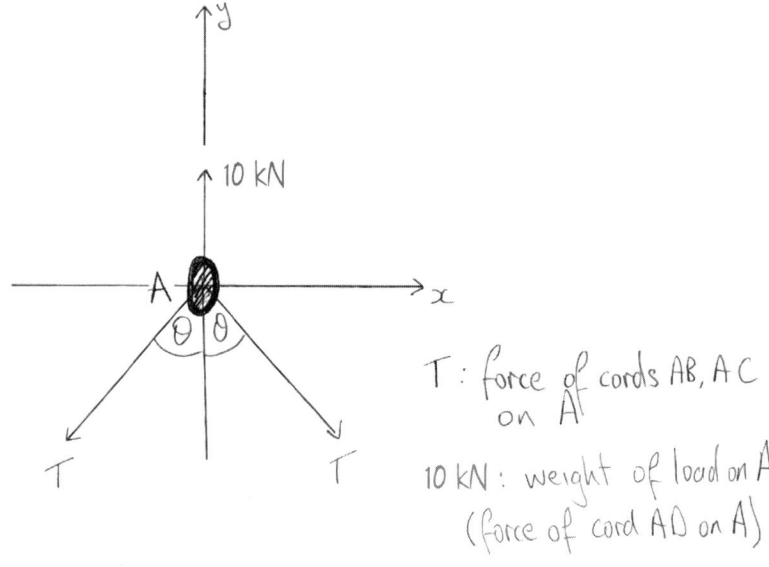

5. Establish an *xy*-axes system on the free-body diagram and write down the equilibrium equations in the *y*-direction only (this is all that is required to solve this problem):

   $$+\uparrow \sum F_y = 0: \ 10 - 2T \cos\theta = 0$$

6. Solve for the magnitude of the force in the sling:

   $$T = \frac{5}{\cos\theta}$$ **Ans.**

Problem 3.11

When y is zero, the springs sustain a force of 300 N. The applied vertical forces **F** and **−F** pull the point A away from B a distance of $y = 1$ m. The cords CAD and CBD are attached to the rings at C and D. Draw the free-body diagrams for point A and ring C.

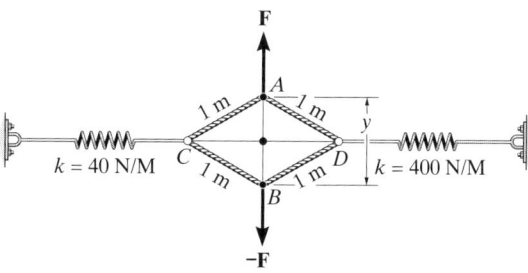

Solution

1. Imagine A and C to be separated or detached from the system.
2. Each of A and C is subjected to three *external* forces. For A, they are caused by:

   i.   **ii.**

   iii.

   For C, they are caused by:

   i.   **ii.**

   iii.

3. Draw the free-body diagrams of A and C showing all these forces labeled with their magnitudes and directions. You should also include any other available information, e.g. lengths, angles, etc. — which will help when formulating the equilibrium equations.

 A

 C

Problem 3.11

When y is zero, the springs sustain a force of 300 N. The applied vertical forces **F** and **−F** pull the point A away from B a distance of $y = 1$ m. The cords CAD and CBD are attached to the rings at C and D. Draw the free-body diagrams for point A and ring C.

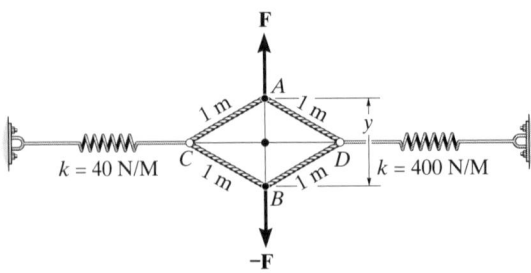

Solution

1. Imagine A and C to be separated or detached from the system.
2. Each of A and C is subjected to three *external* forces. For A, they are caused by:

   i. **CORD** AC ii. **CORD** AD
   iii. **Force F**

   For C, they are caused by:

   i. **CORD** AC ii. **CORD** CB
   iii. **Spring attached at** C

3. Draw the free-body diagrams of A and C showing all these forces labeled with their magnitudes and directions. You should also include any other available information, e.g. lengths, angles, etc. — which will help when formulating the equilibrium equations.

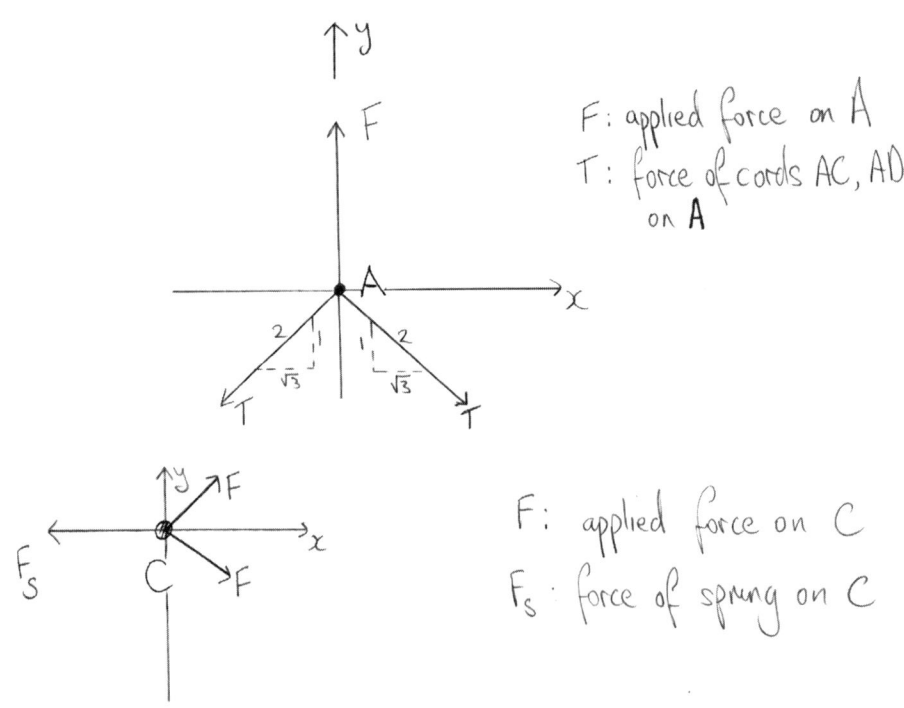

Problem 3.12

By drawing a free-body diagram for the ring at A, determine the maximum weight **W** that can be supported in the position shown if each cable AC and AB can support a maximum tension of 300 N before it fails.

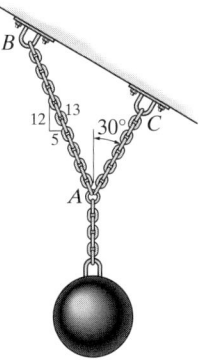

Solution

1. The ring at A has *negligible size* so that it can be modelled as a particle.
2. Imagine the ring at A to be separated or detached from the system.
3. The (detached) ring at A is subjected to three *external* forces. They are caused by:

   i.   **ii.**

   iii.

4. Draw the free-body diagram of the (detached) ring showing all these forces labeled with their magnitudes and directions. Include any other relevant information, e.g. lengths, angles, etc.

5. Establish an xy-axes system on the free-body diagram and write down the equilibrium equations in each of the x and y-directions

   $+\uparrow \sum F_y = 0:$

   $\overset{+}{\rightarrow} \sum F_x = 0:$

6. Set the tension in AB to the maximum of 600 lb and solve for the maximum weight **W**:

Problem 3.12

By drawing a free-body diagram for the ring at A, determine the maximum weight **W** that can be supported in the position shown if each cable AC and AB can support a maximum tension of 3000 N before it fails.

Solution

1. The ring at A has *negligible size* so that it can be modelled as a particle.
2. Imagine the ring at A to be separated or detached from the system.
3. The (detached) ring at A is subjected to three *external* forces. They are caused by:

   i. **CABLE** AB ii. **CABLE** AB

   iii. **Weight of ball**

4. Draw the free-body diagram of the (detached) ring showing all these forces labeled with their magnitudes and directions. Include any other relevant information, e.g. lengths, angles, etc.

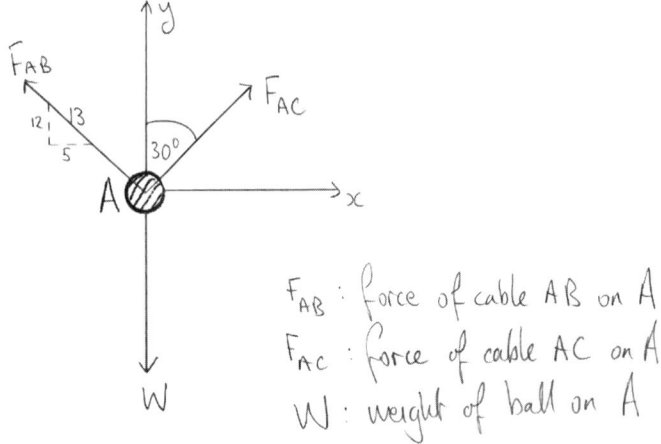

5. Establish an xy-axes system on the free-body diagram and write down the equilibrium equations in each of the x and y-directions

$$\xrightarrow{+} \ \textstyle\sum F_x = 0 \colon \ F_{AB}\left(\frac{5}{13}\right) + F_{AC}\sin 30° = 0$$

$$+\uparrow \ \textstyle\sum F_y = 0 \colon \ F_{AB}\left(\frac{12}{13}\right) + F_{AC}\cos 30° - W = 0$$

6. Set F_{AB}, the tension in AB, to the maximum of 3000 N and solve for the maximum weight **W**:

$$F_{AC} = 2308 \text{ N}(< 3000 \text{ N!!}), \quad \mathbf{W} = 4768 \text{ N} \downarrow \qquad\qquad \textbf{Ans.}$$

Problem 3.13

The cords suspend the two *small* buckets in the equilibrium position shown. Draw the free-body diagrams for each of the points F and C.

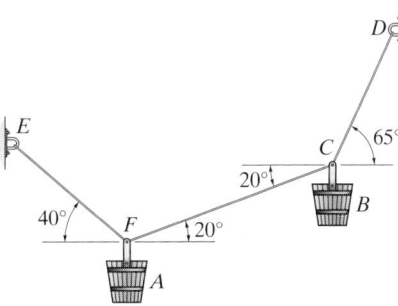

Solution

1. Imagine the points F and C to be separated or detached from the system.

2. Each of F and C is subjected to three *external* forces. For F, they are caused by:

   i.   **ii.**

   iii.

   For C, they are caused by:

   i.   **ii.**

   iii.

3. Draw the free-body diagrams of F and C showing all these forces labeled with their magnitudes and directions. You should also include any other available information, e.g. lengths, angles, etc. — which will help when formulating the equilibrium equations at these points.

Problem 3.13

The cords suspend the two *small* buckets in the equilibrium position shown. Draw the free-body diagrams for each of the points F and C.

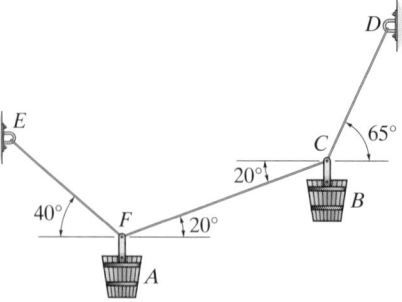

Solution

1. Imagine the points F and C to be separated or detached from the system.

2. Each of F and C is subjected to three *external* forces. For F, they are caused by:

   i. CABLE FE **ii. CABLE** FC

   iii. Weight of A

   For C, they are caused by:

   i. CABLE CF **ii. CABLE** CD

   iii. Weight of B

3. Draw the free-body diagrams of F and C showing all these forces labeled with their magnitudes and directions. You should also include any other available information, e.g. lengths, angles, etc. — which will help when formulating the equilibrium equations at these points.

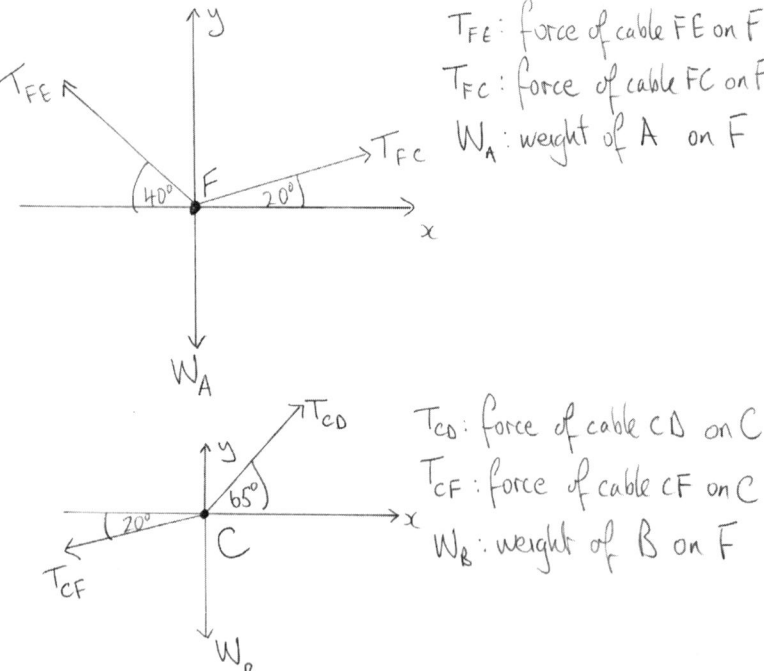

Problem 3.14

The 30-kg pipe is supported at *A* by a system of five cords. Draw the free-body diagrams for the rings at *A* and *B* when the system is in equilibrium.

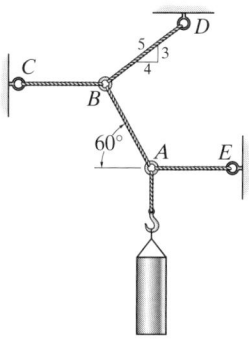

Solution

1. Imagine *A* and *B* to be separated or detached from the system.
2. Each of *A* and *B* is subjected to three *external* forces. For *A*, they are caused by:

   i.   **ii.**

   iii.

   For *B*, they are caused by:

   i.   **ii.**

   iii.

3. Draw the free-body diagrams of *A* and *B* showing all these forces labeled with their magnitudes and directions. You should also include any other available information, e.g. lengths, angles, etc. — which will help when formulating the equilibrium equations.

Problem 3.14

The 30-kg pipe is supported at *A* by a system of five cords. Draw the free-body diagrams for the rings at *A* and *B* when the system is in equilibrium.

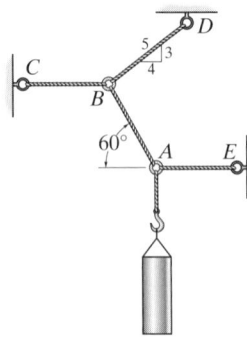

Solution

1. Imagine *A* and *B* to be separated or detached from the system.
2. Each of *A* and *B* is subjected to three *external* forces. For *A*, they are caused by:

      i. CABLE *AB*   **ii. CABLE** *AE*

      iii. Weight of Pipe

   For *B*, they are caused by:

      i. CABLE *BC*   **ii. CABLE** *BD*

      iii. CABLE *BA*

3. Draw the free-body diagrams of *A* and *B* showing all these forces labeled with their magnitudes and directions. You should also include any other available information, e.g. lengths, angles, etc. — which will help when formulating the equilibrium equations.

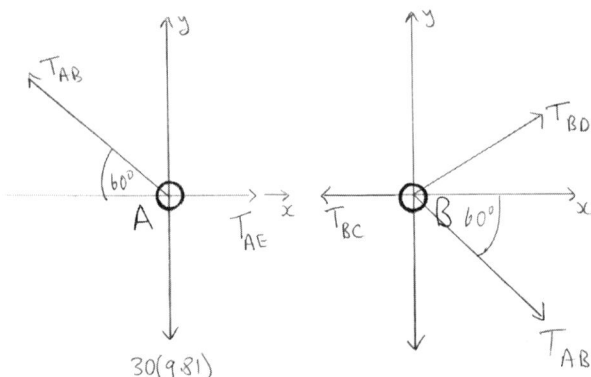

T_{AB}: force of cable AB (BA) on A(B)

T_{AE}: force of cable AE on A

T_{BD}: force of cable BD on B

T_{BC}: force of cable BC on B

30(9.81): weight of pipe on A

Problem 3.15

The cord AB has a length of 5 m and is attached to the end B of the spring having a stiffness $k = 10$ kN/M. The other end of the spring is attached to a roller C so that the spring remains horizontal as it stretches. If a 10-kN ($\approx$ 1 tonne) weight is suspended from B, use the free-body diagram for the ring at B to determine the necessary unstretched length of the spring, so that $\theta = 40°$ for equilibrium.

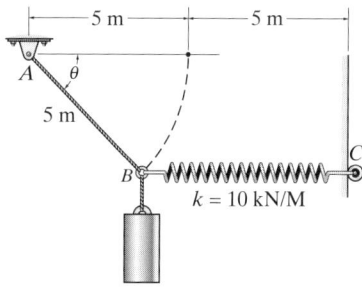

Solution

1. Imagine the ring at B to be separated or detached from the system.
2. The (detached) ring at B is subjected to three *external* forces caused by:

    i.   **ii.**

    iii.

3. Draw the free-body diagram of the (detached) ring showing all these forces labeled with their magnitudes and directions. Include any other relevant information, e.g. lengths, angles, etc.

4. Establish an xy-axes system on the free-body diagram and write down the equilibrium equations in each of the x and y-directions

    $+\uparrow \sum F_y = 0:$

    $\xrightarrow{+} \sum F_x = 0:$

5. Determine the stretch in the spring BC and solve for the necessary unstretched length:

Problem 3.15

The cord AB has a length of 5 m and is attached to the end B of the spring having a stiffness $k = 10$ kN/M. The other end of the spring is attached to a roller C so that the spring remains horizontal as it stretches. If a 10-kN ($\approx$ 1 tonne) weight is suspended from B, use the free-body diagram for the ring at B to determine the necessary unstretched length of the spring, so that $\theta = 40°$ for equilibrium.

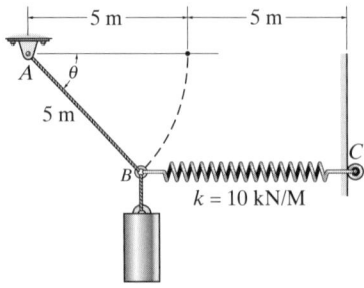

Solution

1. Imagine the ring at B to be separated or detached from the system.

2. The (detached) ring at B is subjected to three *external* forces caused by:

    i. CABLE AB **ii. SPRING** BC

    iii. 10 lb Weight

3. Draw the free-body diagram of the (detached) ring showing all these forces labeled with their magnitudes and directions. Include any other relevant information, e.g. lengths, angles, etc.

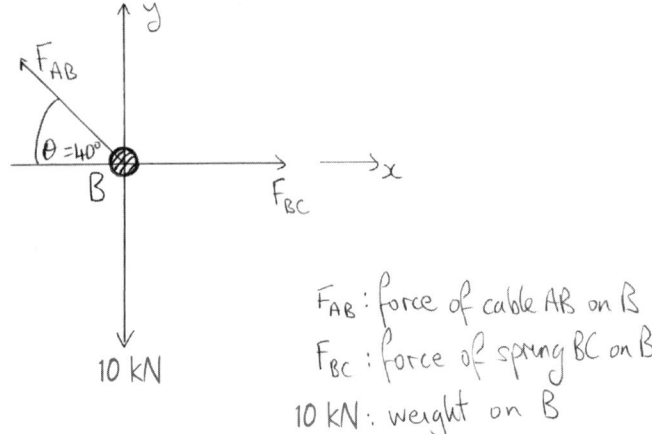

4. Establish an xy-axes system on the free-body diagram and write down the equilibrium equations in each of the x and y-directions

$$+\uparrow \sum F_y = 0:\ F_{AB} \sin 40° - 10 = 0 \Rightarrow F_{AB} = 15.557 \text{ kN}$$

$$\overset{+}{\rightarrow} \sum F_x = 0:\ F_{BC} - F_{AB} \cos 40° = 0 \Rightarrow F_{BC} = 11.918 \text{ kN}$$

5. Determine the stretch in the spring BC and solve for the necessary unstretched length l:

$$F_{BC} = kx \Rightarrow x = \frac{11.918}{10} = 1.1918 \text{ m (stretch in } BC)$$

$$BC = 5 + (5 - 5 \cos 40°) = 6.17 \text{ m}$$

$$l = 6.17 - 1.1918 = 4.98 \text{ m} \hspace{3cm} \textbf{Ans.}$$

3.2 Free-Body Diagrams in the Equilibrium of a Rigid Body

Problem 3.16

Draw the free-body diagram of the 50-kg uniform pipe, which is supported by the smooth contacts at *A* and *B*.

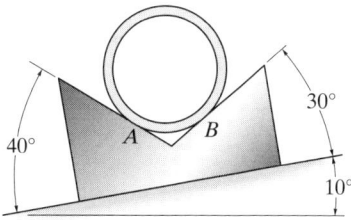

Solution

1. Imagine the pipe to be separated or detached from the system.

2. The supports at *A* and *B* are smooth contacts. Use Table 2.1 (6) to determine the number and types of reactions *acting on the pipe* at *A* and *B*.

3. The pipe is subjected to three *external* forces (don't forget the weight!). They are caused by:

   i.   **ii.**

   iii.

4. Draw the free-body diagram of the (detached) pipe showing all these forces labeled with their magnitudes and directions. *Assume* the sense of the vectors representing the *reactions acting on the pipe* (the correct sense will always emerge from the equilibrium equations for the pipe). Include any other relevant information, e.g. lengths, angles, etc. — which may help when formulating the equilibrium equations (including the moment equation) for the pipe.

Problem 3.16

Draw the free-body diagram of the 50-kg uniform pipe, which is supported by the smooth contacts at A and B.

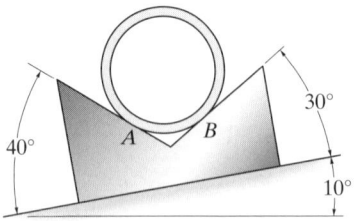

Solution

1. Imagine the pipe to be separated or detached from the system.
2. The supports at A and B are smooth contacts. Use Table 2.1 (6) to determine the number and types of reactions *acting on the pipe* at A and B.
3. The pipe is subjected to three *external* forces (don't forget the weight!). They are caused by:
   i. **The reaction at A**   ii. **The reaction at B**
   iii. **The weight of the pipe**
4. Draw the free-body diagram of the (detached) pipe showing all these forces labeled with their magnitudes and directions. *Assume* the sense of the vectors representing the *reactions acting on the pipe* (the correct sense will always emerge from the equilibrium equations for the pipe). Include any other relevant information, e.g. lengths, angles, etc. — which may help when formulating the equilibrium equations (including the moment equation) for the pipe.

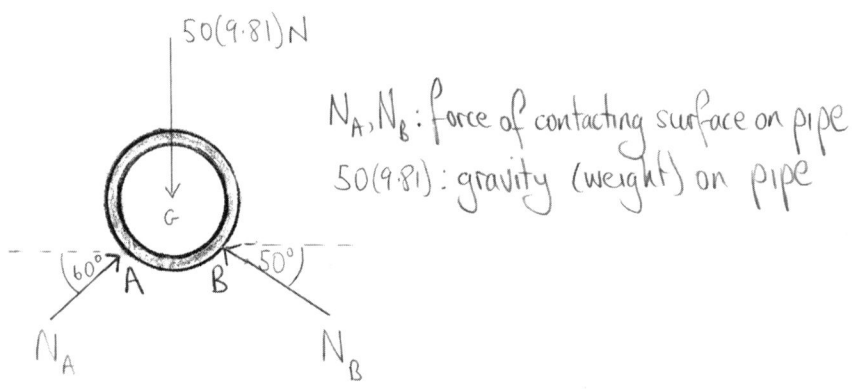

Problem 3.17

Draw the free-body diagram of the hand punch, which is pinned at A and bears down on the smooth surface at B. Neglect the weight of the punch.

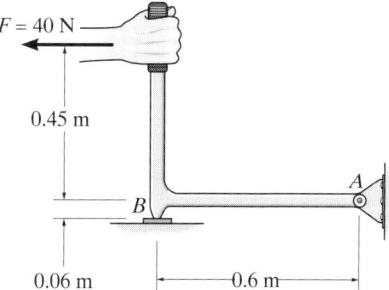

Solution

1. Imagine the hand punch to be separated or detached from the system.
2. The support at B is a smooth contact. The punch is (smoothly) pin-connected at A. Use Table 2.1 (6) and (8) to determine the number and types of reactions *acting on the pipe* at A and B.
3. The punch is subjected to four *external* forces. They are caused by:

   i.   **ii.**

   iii.   **iv.**

4. Draw the free-body diagram of the (detached) punch showing all these forces labeled with their magnitudes and directions. *Assume* the sense of the vectors representing the *reactions acting on the punch* (the correct sense will always emerge from the equilibrium equations for the punch). Include any other relevant information, e.g. lengths, angles, etc. — which may help when formulating the equilibrium equations (including the moment equation) for the punch.

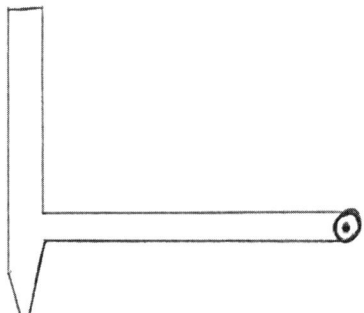

Problem 3.17

Draw the free-body diagram of the hand punch, which is pinned at A and bears down on the smooth surface at B. Neglect the weight of the punch.

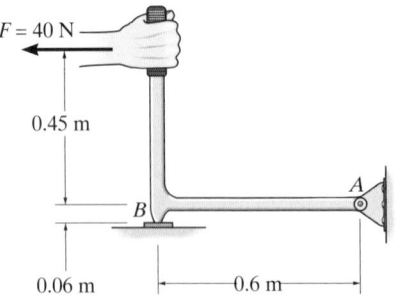

Solution

1. Imagine the hand punch to be separated or detached from the system.
2. The support at B is a smooth contact. The punch is (smoothly) pin-connected at A. Use Table 2.1 (6) and (8) to determine the number and types of reactions *acting on the pipe* at A and B.
3. The punch is subjected to four *external* forces. They are caused by:

   i. The force F   **ii. The reaction at B**

   iii. & iv. The two reactions at A

4. Draw the free-body diagram of the (detached) punch showing all these forces labeled with their magnitudes and directions. *Assume* the sense of the vectors representing the *reactions acting on the punch* (the correct sense will always emerge from the equilibrium equations for the punch). Include any other relevant information, e.g. lengths, angles, etc. — which may help when formulating the equilibrium equations (including the moment equation) for the punch.

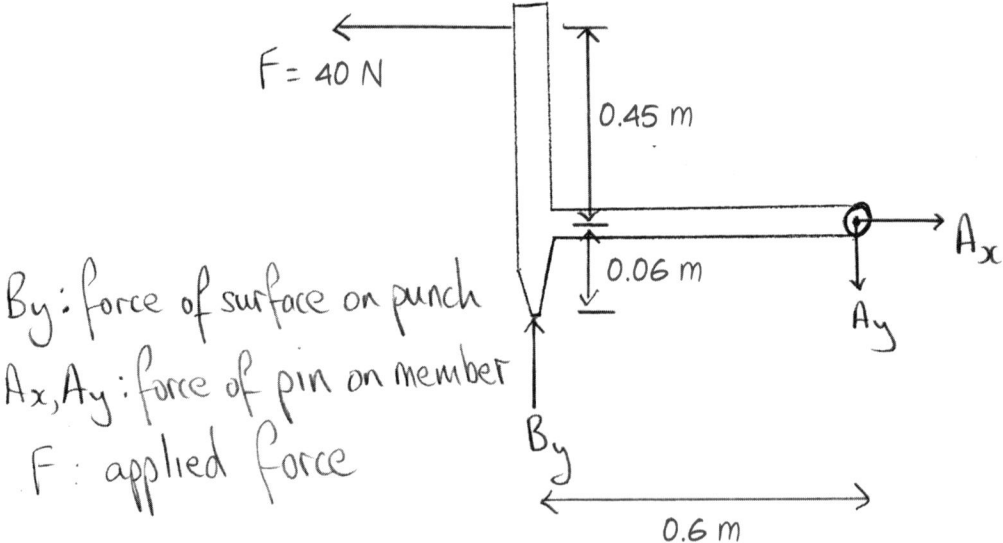

Problem 3.18

Draw the free-body diagram of the jib crane AB, which is pin-connected at A and supported by member (link) BC. Neglect the weight of the crane.

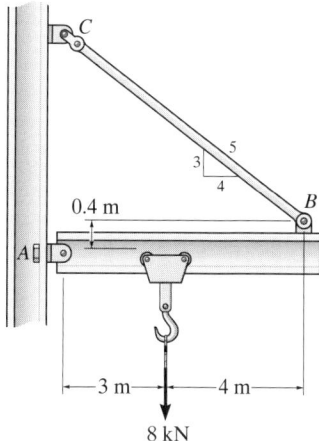

Solution

1. Imagine the jib crane AB to be separated or detached from the system.
2. There is a link support at B and the jib crane is (smoothly) pinned at A. Use Table 2.1 (2) and (8) to determine the number and types of reactions *acting on the jib crane* at A and B.
3. The jib crane is subjected to four *external* forces. They are caused by:

    i.   **ii.**

    iii.   **iv.**

4. Draw the free-body diagram of the (detached) crane showing all these forces labeled with their magnitudes and directions. *Assume* the sense of the vectors representing the *reactions acting on the crane* (the correct sense will always emerge from the equilibrium equations for the crane). Include any other relevant information, e.g. lengths, angles, etc. — which may help when formulating the equilibrium equations (including the moment equation) for the jib crane.

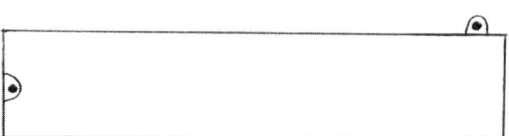

Problem 3.18

Draw the free-body diagram of the jib crane AB, which is pin-connected at A and supported by member (link) BC. Neglect the weight of the crane.

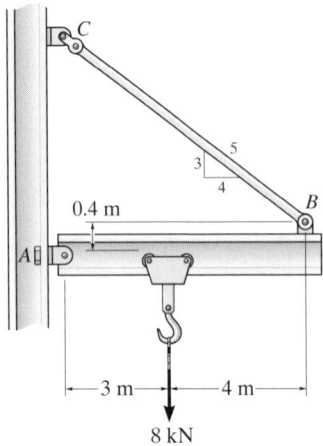

Solution

1. Imagine the jib crane AB to be separated or detached from the system.

2. There is a link support at B and the jib crane is (smoothly) pinned at A. Use Table 2.1 (2) and (8) to determine the number and types of reactions *acting on the jib crane* at A and B.

3. The jib crane is subjected to four *external* forces. They are caused by:

    i. & ii. The reactions at A   **iii. The reaction at B**

    iv. The 8 kN load

4. Draw the free-body diagram of the (detached) crane showing all these forces labeled with their magnitudes and directions. *Assume* the sense of the vectors representing the *reactions acting on the crane* (the correct sense will always emerge from the equilibrium equations for the crane). Include any other relevant information, e.g. lengths, angles, etc. — which may help when formulating the equilibrium equations (including the moment equation) for the jib crane.

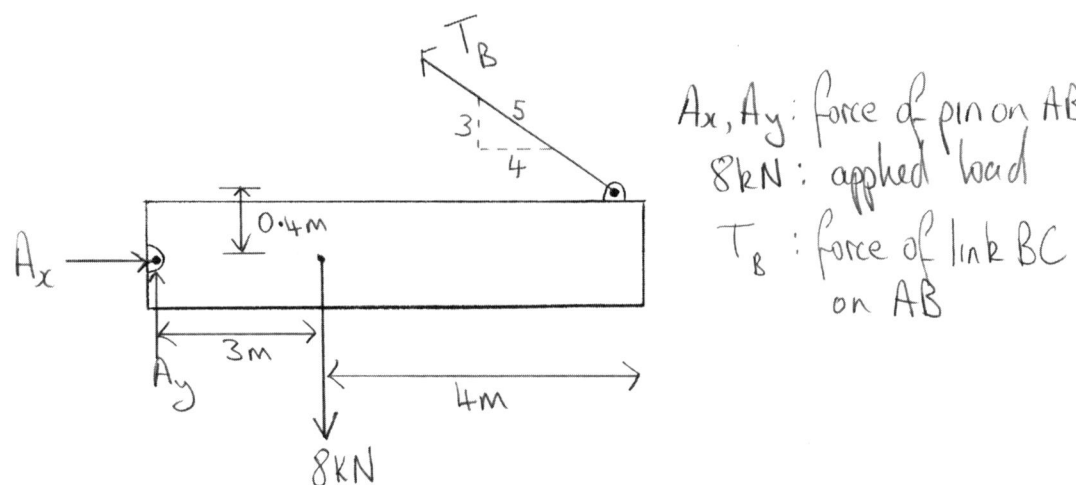

Problem 3.19

Draw the free-body diagram of the dumpster D of the truck, which has a weight of 25 kN ($\approx$ 2.5 tonnes) and a center of gravity at G. It is supported by a pin at A and a pin-connected hydraulic cylinder BC (short link).

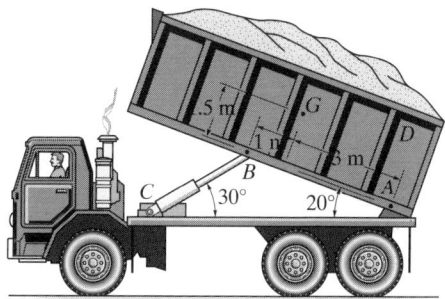

Solution

1. Imagine the dumpster D to be separated or detached from the truck.

2. There is a pin support at A and the dumpster is supported by a short link support at B. Use Table 2.1 (2) and (8) to determine the number and types of reactions *acting on the dumpster* at A and B.

3. The dumpster is subjected to four *external* forces. They are caused by:

    i.   **ii.**

    iii.   **iv.**

4. Draw the free-body diagram of the (detached) dumpster showing all these forces labeled with their magnitudes and directions. *Assume* the sense of the vectors representing the *reactions acting on the dumpster* (the correct sense will always emerge from the equilibrium equations for the dumpster). Include any other relevant information, e.g. lengths, angles, etc. — which may help when formulating the equilibrium equations (including the moment equation) for the dumpster.

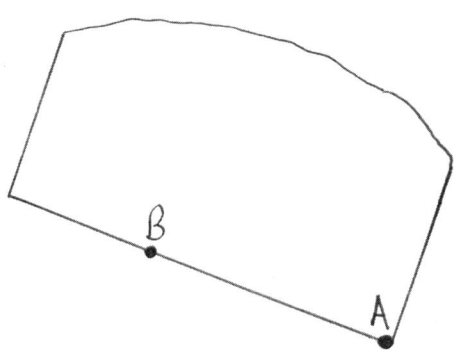

Problem 3.19

Draw the free-body diagram of the dumpster D of the truck, which has a weight of 25 kN ($\approx$ 2.5 tonnes) and a center of gravity at G. It is supported by a pin at A and a pin-connected hydraulic cylinder BC (short link).

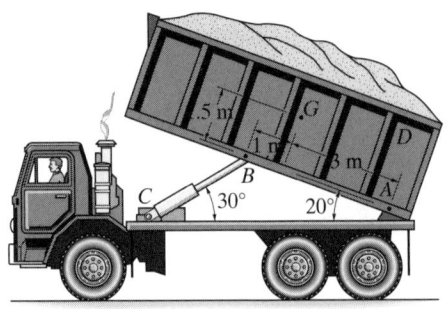

Solution

1. Imagine the dumpster D to be separated or detached from the truck.

2. There is a pin support at A and the dumpster is supported by a short link support at B. Use Table 2.1 (2) and (8) to determine the number and types of reactions *acting on the dumpster* at A and B.

3. The dumpster is subjected to four *external* forces. They are caused by:

   i. & ii. The reactions at A   **iii. The reaction at B**

   iv. The weight of the dumpster

4. Draw the free-body diagram of the (detached) dumpster showing all these forces labeled with their magnitudes and directions. *Assume* the sense of the vectors representing the *reactions acting on the dumpster* (the correct sense will always emerge from the equilibrium equations for the dumpster). Include any other relevant information, e.g. lengths, angles, etc. — which may help when formulating the equilibrium equations (including the moment equation) for the dumpster.

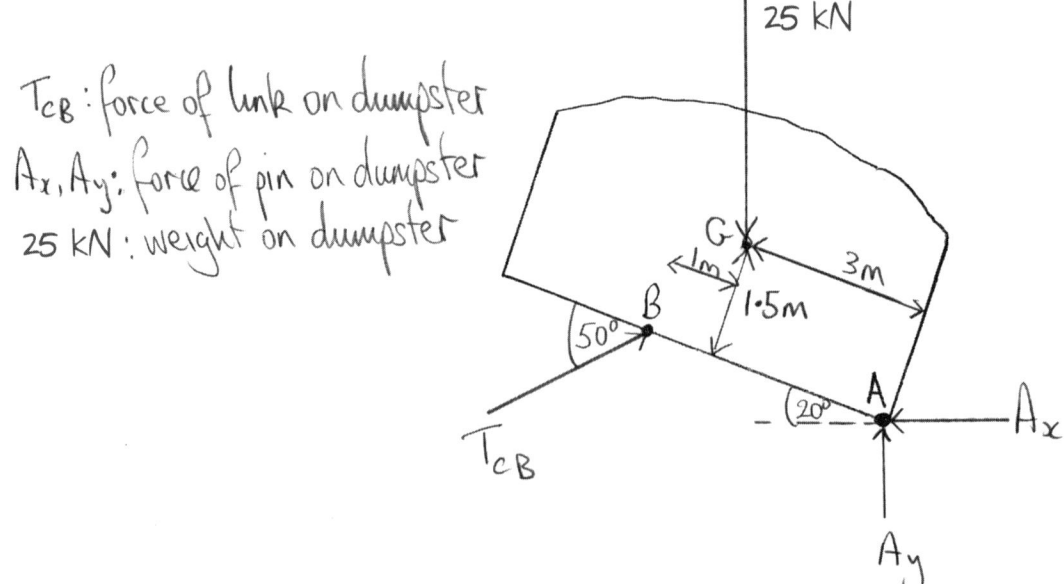

Problem 3.20

Draw the free-body diagram of the link *CAB*, which is pin-connected at *A* and rests on the smooth cam at *B*. Neglect the weight of the link.

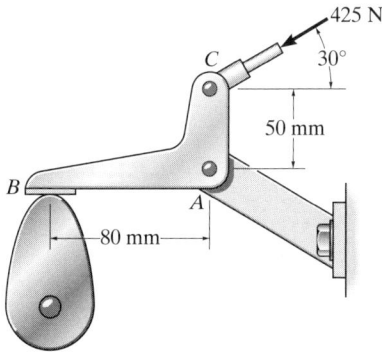

Solution

1. Imagine the link *CAB* to be separated or detached from the mechanism.

2. There is a pin connection at *A* and the link rests on the smooth surface at *B*. Use Table 2.1 (6) and (8) to determine the number and types of reactions acting on the link at *A* and *B*.

3. The link is subjected to four *external* forces. They are caused by:

   i.   **ii.**

   iii.   **iv.**

4. Draw the free-body diagram of the (detached) link showing all these forces labeled with their magnitudes and directions. *Assume* the sense of the vectors representing the *reactions acting on the link* (the correct sense will always emerge from the equilibrium equations for the link). Include any other relevant information, e.g. lengths, angles, etc. — which may help when formulating the equilibrium equations (including the moment equation) for the link.

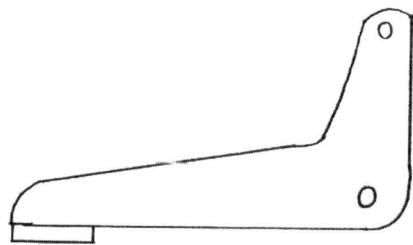

Problem 3.20

Draw the free-body diagram of the link CAB, which is pin-connected at A and rests on the smooth cam at B. Neglect the weight of the link.

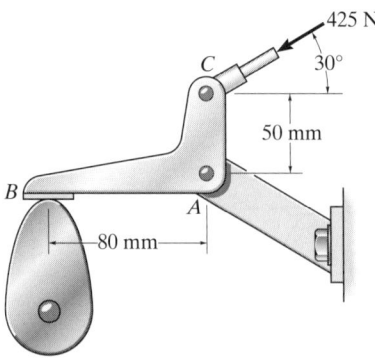

Solution

1. Imagine the link CAB to be separated or detached from the mechanism.

2. There is a pin connection at A and the link rests on the smooth surface at B. Use Table 2.1 (6) and (8) to determine the number and types of reactions *acting on the link* at A and B.

3. The link is subjected to four *external* forces. They are caused by:

    i. & ii. The reactions at A **iii. The reaction at** B

    iv. The 425 N load at C

4. Draw the free-body diagram of the (detached) link showing all these forces labeled with their magnitudes and directions. *Assume* the sense of the vectors representing the *reactions acting on the link* (the correct sense will always emerge from the equilibrium equations for the link). Include any other relevant information, e.g. lengths, angles, etc. — which may help when formulating the equilibrium equations (including the moment equation) for the link.

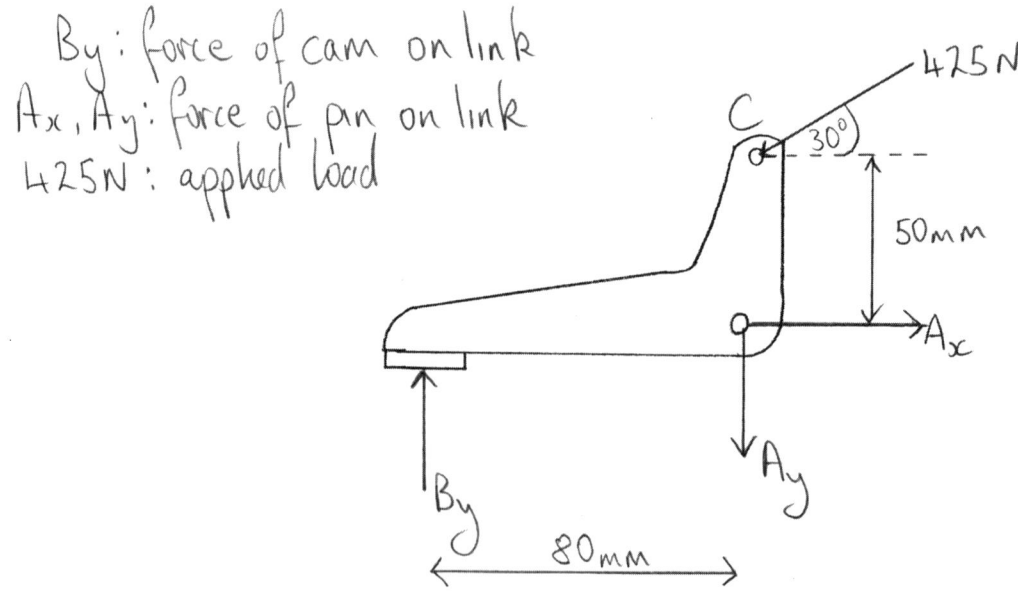

Problem 3.21

Draw the free-body diagram of the uniform pipe which has a mass of 100 kg and a center of mass at G. The supports A, B and C are smooth.

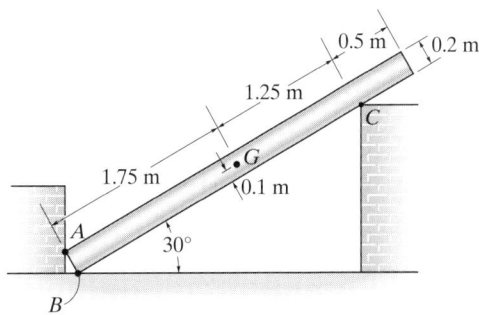

Solution

1. Imagine the pipe to be separated or detached from the system.
2. The pipe rests on smooth surfaces at A, B and C. Use Table 2.1 (6) to determine the number and types of reactions *acting on the pipe* at A, B and C.
3. The pipe is subjected to four *external* forces. They are caused by:

    i.   ii.

    iii.   iv.

4. Draw the free-body diagram of the (detached) pipe showing all these forces labeled with their magnitudes and directions. *Assume* the sense of the vectors representing the *reactions acting on the pipe* (the correct sense will always emerge from the equilibrium equations for the pipe). Include any other relevant information, e.g. lengths, angles, etc. — which may help when formulating the equilibrium equations for the pipe.

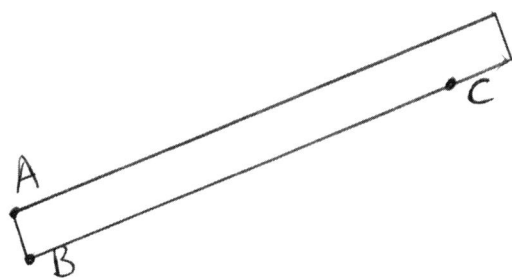

Problem 3.21

Draw the free-body diagram of the uniform pipe which has a mass of 100 kg and a center of mass at G. The supports A, B and C are smooth.

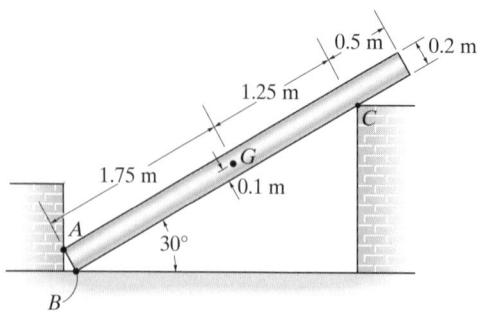

Solution

1. Imagine the pipe to be separated or detached from the system.
2. The pipe rests on smooth surfaces at A, B and C. Use Table 2.1 (6) to determine the number and types of reactions *acting on the pipe* at A, B and C.
3. The pipe is subjected to four *external* forces. They are caused by:

   i. The reaction at A   **iii. The reaction at B**

   iv. The reaction at C   **iv. Pipe's weight**

4. Draw the free-body diagram of the (detached) pipe showing all these forces labeled with their magnitudes and directions. *Assume* the sense of the vectors representing the *reactions acting on the pipe* (the correct sense will always emerge from the equilibrium equations for the pipe). Include any other relevant information, e.g. lengths, angles, etc. — which may help when formulating the equilibrium equations for the pipe.

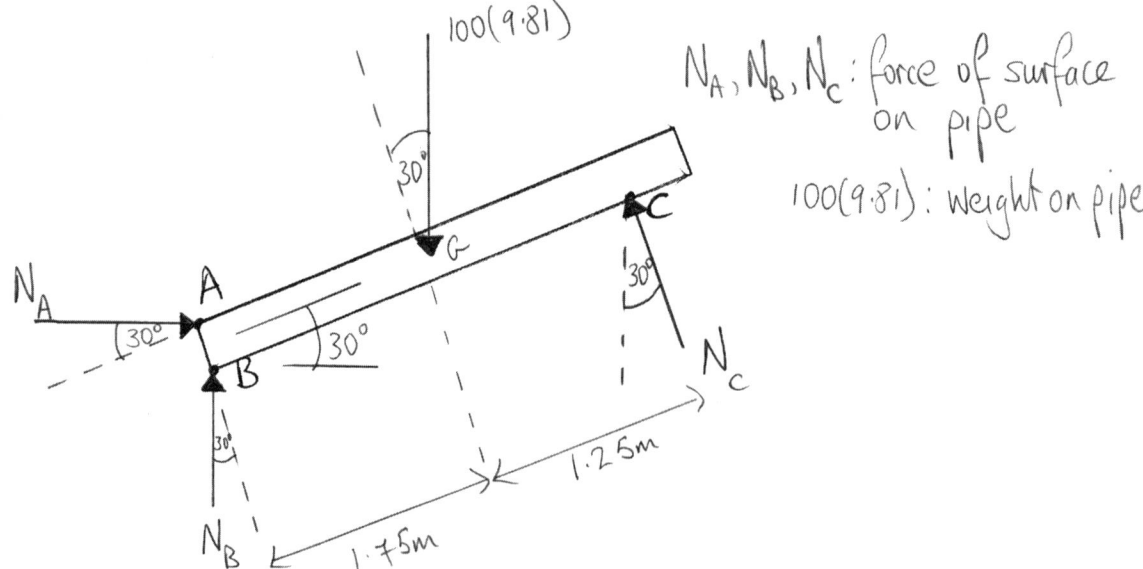

Problem 3.22

Draw the free-body diagram of the beam, which is pin-supported at A and rests on the smooth incline at B. Neglect the weight of the beam.

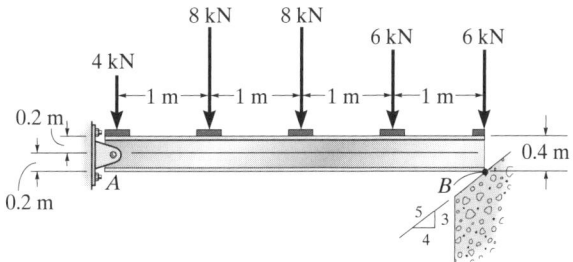

Solution

1. Imagine the beam to be separated or detached from the system.

2. There is a pin connection at A and the beam rests on the smooth (inclined) surface at B. Use Table 2.1 (6) and (8) to determine the number and types of reactions *acting on the beam* at A and B.

3. In addition to the forces shown in the figure, the beam is subjected to three *external* forces. They are caused by:

   i.   **ii.**

   iii.

4. Draw the free-body diagram of the (detached) beam showing all these forces labeled with their magnitudes and directions. *Assume* the sense of the vectors representing the *reactions acting on the beam* (the correct sense will always emerge from the equilibrium equations for the beam). Include any other relevant information, e.g. lengths, angles, etc. — which may help when formulating the equilibrium equations (including the moment equation) for the beam.

Problem 3.22

Draw the free-body diagram of the beam, which is pin-supported at A and rests on the smooth incline at B. Neglect the weight of the beam.

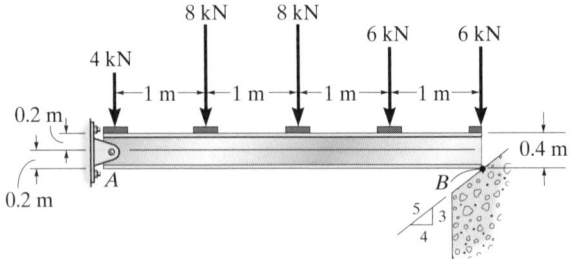

Solution

1. Imagine the beam to be separated or detached from the system.
2. There is a pin connection at A and the beam rests on the smooth (inclined) surface at B. Use Table 2.1 (6) and (8) to determine the number and types of reactions *acting on the beam* at A and B.
3. In addition to the forces shown in the figure, the beam is subjected to three *external* forces. They are caused by:

   i. & ii. The reactions at A   **iii. The reaction at B**

4. Draw the free-body diagram of the (detached) beam showing all these forces labeled with their magnitudes and directions. *Assume* the sense of the vectors representing the *reactions acting on the beam* (the correct sense will always emerge from the equilibrium equations for the beam). Include any other relevant information, e.g. lengths, angles, etc. — which may help when formulating the equilibrium equations (including the moment equation) for the beam.

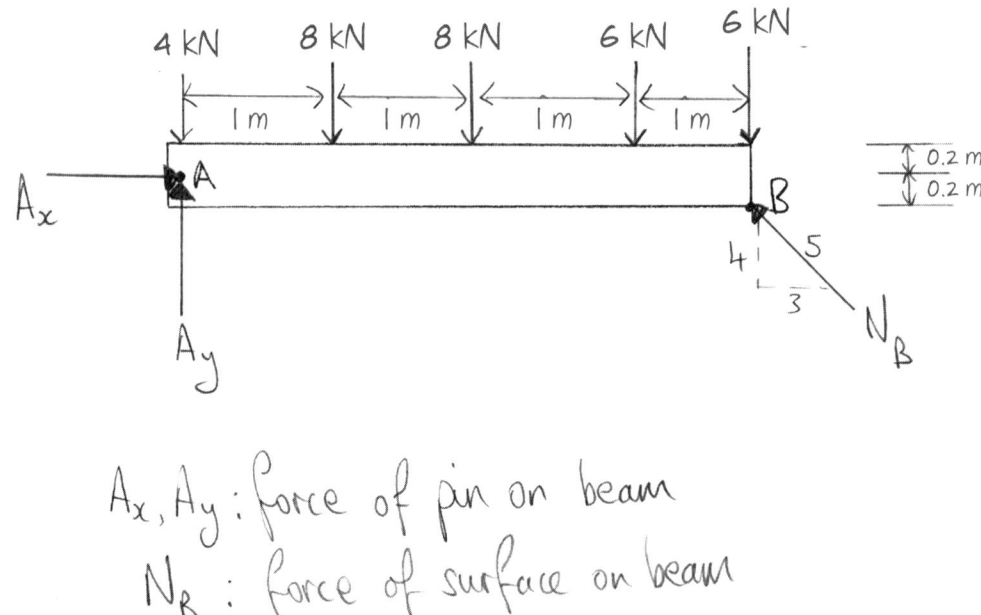

Problem 3.23

Draw the free-body diagram of the member ABC, which is supported by a pin at A and a horizontal short link BD. Neglect the weight of ABC.

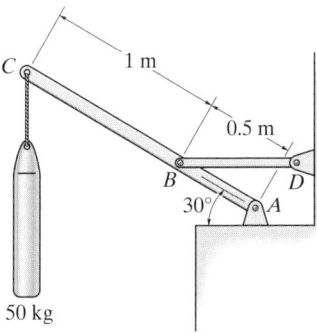

Solution

1. Imagine the member ABC to be separated or detached from the system.

2. There is a pin support at A and the member is supported by a horizontal short link at B. Use Table 2.1 (2) and (8) to determine the number and types of reactions *acting on the member* at A and B.

3. The member is subjected to four *external* forces. They are caused by:

   i.   **ii.**

   iii.   **iv.**

4. Draw the free-body diagram of the (detached) member showing all these forces labeled with their magnitudes and directions. *Assume* the sense of the vectors representing the *reactions acting on the member* (the correct sense will always emerge from the equilibrium equations for the member). Include any other relevant information, e.g. lengths, angles, etc. — which may help when formulating the equilibrium equations (including the moment equation) for the member.

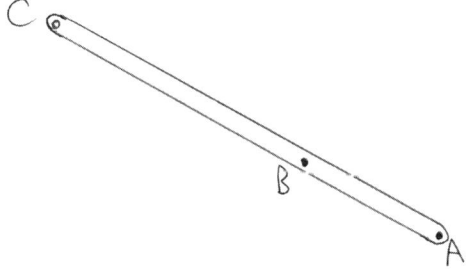

Problem 3.23

Draw the free-body diagram of the member ABC, which is supported by a pin at A and a horizontal short link BD. Neglect the weight of ABC.

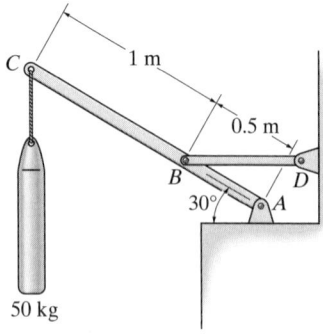

Solution

1. Imagine the member ABC to be separated or detached from the system.

2. There is a pin support at A and the member is supported by a horizontal short link at B. Use Table 2.1 (2) and (8) to determine the number and types of reactions *acting on the member* at A and B.

3. The member is subjected to four *external* forces. They are caused by:

   i. & ii. The reactions at A   **iii. The reaction at B**

   iv. The weight at C

4. Draw the free-body diagram of the (detached) member showing all these forces labeled with their magnitudes and directions. *Assume* the sense of the vectors representing the *reactions acting on the member* (the correct sense will always emerge from the equilibrium equations for the member). Include any other relevant information, e.g. lengths, angles, etc. — which may help when formulating the equilibrium equations (including the moment equation) for the member.

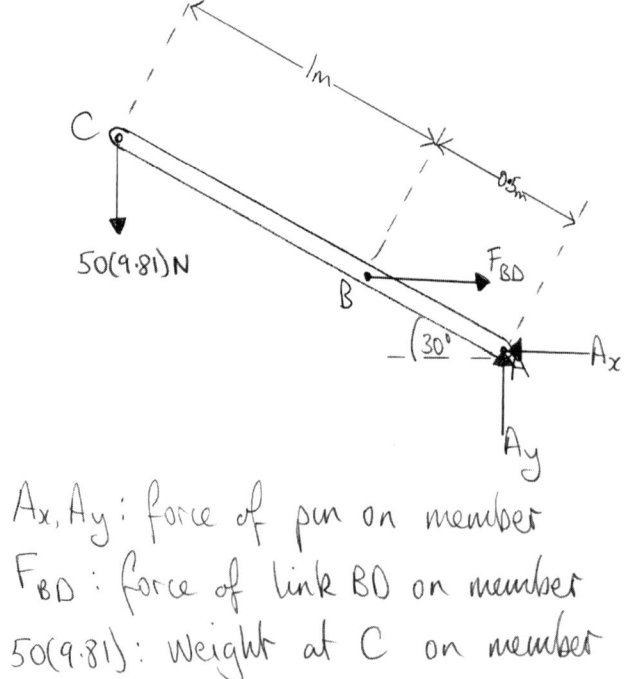

Problem 3.24

Draw the free-body diagram of the beam. The support at B is smooth. Neglect the weight of the beam.

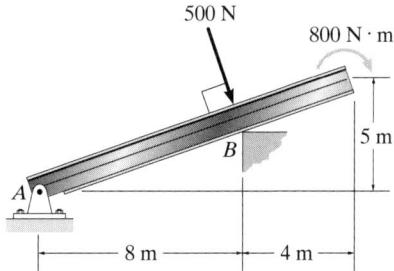

Solution

1. Imagine the beam to be separated or detached from the system.

2. There is a pin support at A and a smooth contact support at B. Use Table 2.1 (6) and (8) to determine the number and types of reactions *acting on the member* at A and B.

3. In addition to those shown in the figure, the member is subjected to three *external* forces. They are caused by:

   i.   **ii.**

   iii.

4. Draw the free-body diagram of the (detached) member showing all these forces and any external applied couple moments labeled with their magnitudes and directions. *Assume* the sense of the vectors representing the *reactions acting on the member* (the correct sense will always emerge from the equilibrium equations for the member). Include any other relevant information, e.g. lengths, angles, etc. — which may help when formulating the equilibrium equations (including the moment equation) for the member.

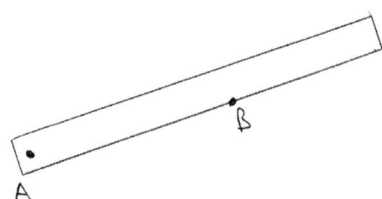

Problem 3.24

Draw the free-body diagram of the beam. The support at B is smooth. Neglect the weight of the beam.

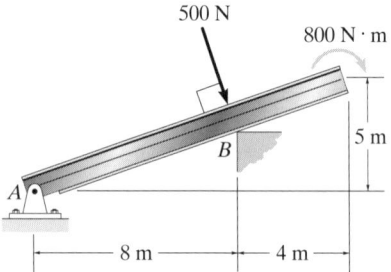

Solution

1. Imagine the beam to be separated or detached from the system.
2. There is a pin support at A and a smooth contact support at B. Use Table 2.1 (6) and (8) to determine the number and types of reactions *acting on the member* at A and B.
3. In addition to those shown in the figure, the member is subjected to three *external* forces. They are caused by:

   i. & ii. The reactions at A   **iii. The reaction at B**

4. Draw the free-body diagram of the (detached) member showing all these forces and any external applied couple moments labeled with their magnitudes and directions. *Assume* the sense of the vectors representing the *reactions acting on the member* (the correct sense will always emerge from the equilibrium equations for the member). Include any other relevant information, e.g. lengths, angles, etc. — which may help when formulating the equilibrium equations (including the moment equation) for the member.

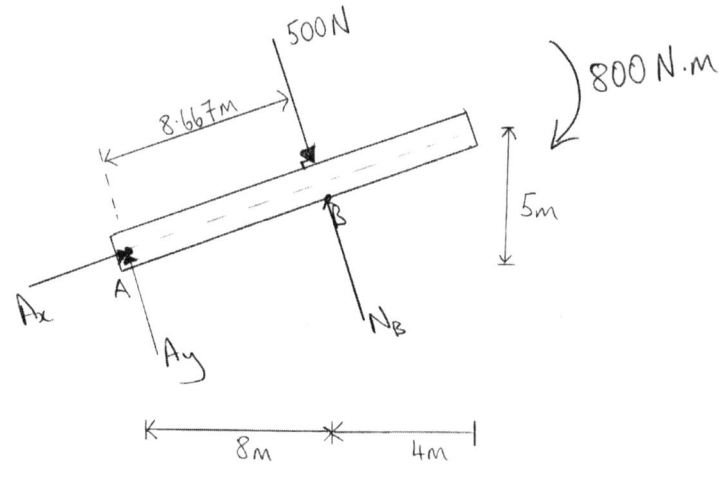

Ax, Ay: effect of pin support on beam
NB: force of rocker support on beam
800N.m: effect of applied couple moment on beam
500N: effect of applied force on beam

Problem 3.25

Draw the free-body diagram of the vehicle, which has a mass of 5 Mg and center of mass at *G*. The tires are free to roll, so rolling resistance can be neglected.

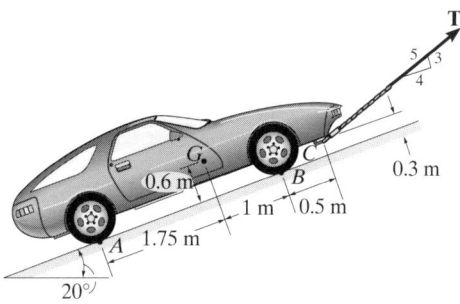

Solution

1. Imagine the vehicle to be separated or detached from the system.
2. There are smooth contacts at *A* and *B*. Use Table 2.1 (6) to determine the number and types of reactions *acting on the vehicle* at *A* and *B*.
3. The vehicle is subjected to four *external* forces. They are caused by:

    i.   ii.

    iii.   iv.

4. Draw the free-body diagram of the (detached) vehicle showing all these forces labeled with their magnitudes and directions. *Assume* the sense of the vectors representing the *reactions acting on the vehicle* (the correct sense will always emerge from the equilibrium equations for the vehicle). Include any other relevant information, e.g. lengths, angles, etc. — which may help when formulating the equilibrium equations (including the moment equation) for the vehicle.

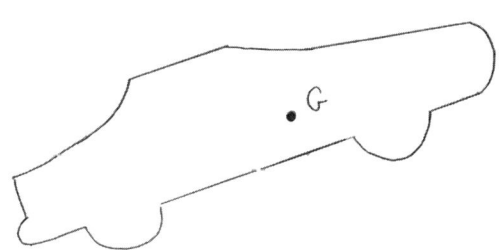

Problem 3.25

Draw the free-body diagram of the vehicle, which has a mass of 5 Mg and center of mass at *G*. The tires are free to roll, so rolling resistance can be neglected.

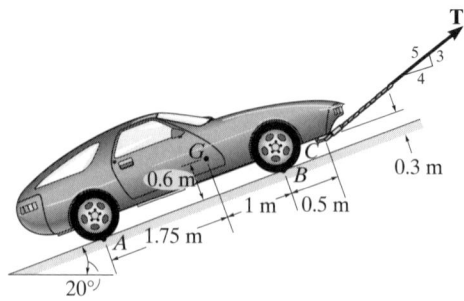

Solution

1. Imagine the vehicle to be separated or detached from the system.

2. There are smooth contacts at *A* and *B*. Use Table 2.1 (6) to determine the number and types of reactions *acting on the vehicle* at *A* and *B*.

3. The vehicle is subjected to four *external* forces. They are caused by:

   i. The reaction at *A*   iii. The reaction at *B*

   iii. Car's weight   iv. Force T

4. Draw the free-body diagram of the (detached) vehicle showing all these forces labeled with their magnitudes and directions. *Assume* the sense of the vectors representing the *reactions acting on the vehicle* (the correct sense will always emerge from the equilibrium equations for the vehicle). Include any other relevant information, e.g. lengths, angles, etc. — which may help when formulating the equilibrium equations (including the moment equation) for the vehicle.

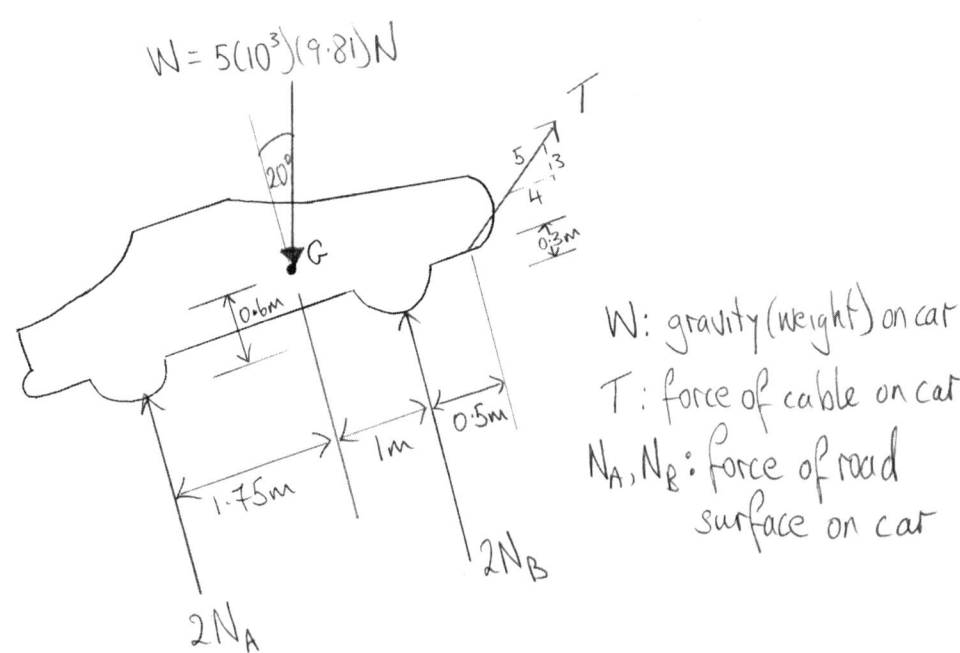

Problem 3.26

Draw a free-body diagram of the crane boom ABC, which has a mass of 45 kg, center of gravity at G, and supports a load of 30 kg. The boom is pin-connected to the frame at B and connected to a vertical chain CD. The chain supporting the load is attached to the boom at A.

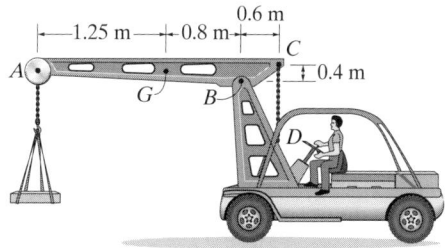

Solution

1. Imagine the boom to be separated or detached from the system.

2. There is a pin connection at B and a vertical chain (cable) support at C. Use Table 2.1 (1) and (8) to determine the number and types of reactions *acting on the boom* at B and C.

3. The boom is subjected to five *external* forces. They are caused by:

   i.   ii.

   iii.   iv.

   v.

4. Draw the free-body diagram of the (detached) boom showing all these forces labeled with their magnitudes and directions. *Assume* the sense of the vectors representing the *reactions acting on the boom*. Include any other relevant information, e.g. lengths, angles, etc. — which may help when formulating the equilibrium equations (including the moment equation) for the boom.

Problem 3.26

Draw a free-body diagram of the crane boom ABC, which has a mass of 45 kg, center of gravity at G, and supports a load of 30 kg. The boom is pin-connected to the frame at B and connected to a vertical chain CD. The chain supporting the load is attached to the boom at A.

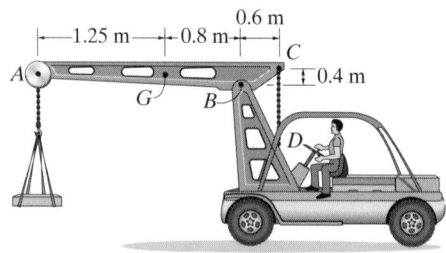

Solution

1. Imagine the boom to be separated or detached from the system.

2. There is a pin connection at B and a vertical chain (cable) support at C. Use Table 2.1 (1) and (8) to determine the number and types of reactions *acting on the boom* at B and C.

3. The boom is subjected to five *external* forces. They are caused by:

     i. & ii. The reactions at B   **iii. The reaction at C**

     iv. Weight of boom   **v. Load at A**

4. Draw the free-body diagram of the (detached) boom showing all these forces labeled with their magnitudes and directions. *Assume* the sense of the vectors representing the *reactions acting on the boom*. Include any other relevant information, e.g. lengths, angles, etc. — which may help when formulating the equilibrium equations (including the moment equation) for the boom.

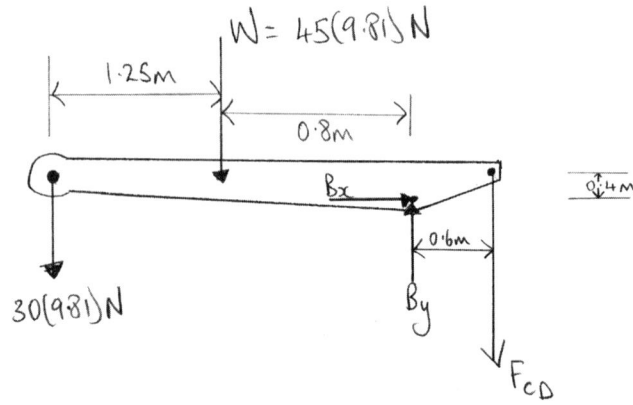

F_{CD} : effect of chain on boom

B_x, B_y : effect of pin on boom

$30(9.81)N$: weight of load (through chain) on boom

W : effect of gravity (weight) on boom

Problem 3.27

Draw a free-body diagram of the beam. Neglect the thickness and weight of the beam.

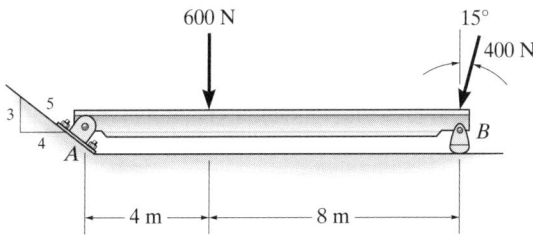

Solution

1. Imagine the beam to be separated or detached from the system.

2. There is a pin connection at A and a rocker support at B. Use Table 2.1 to determine the number and types of reactions *acting on the beam* at A and B.

3. The beam is subjected to five *external* forces. They are caused by:

    i.   **ii.**

    iii.   **iv.**

    v.

4. Draw the free-body diagram of the (detached) beam showing all these forces labeled with their magnitudes and directions. *Assume* the sense of the vectors representing the *reactions acting on the beam*. Include any other relevant information, e.g. lengths, angles, etc. — which may help when formulating the equilibrium equations (including the moment equation) for the beam.

Problem 3.27

Draw a free-body diagram of the beam. Neglect the thickness and weight of the beam.

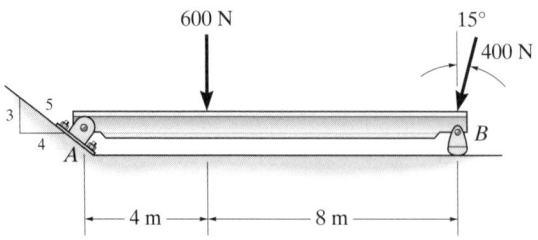

Solution

1. Imagine the beam to be separated or detached from the system.
2. There is a pin connection at A and a rocker support at B. Use Table 2.1 to determine the number and types of reactions *acting on the beam* at A and B.
3. In addition to those shown in the figure, the beam is subjected to three *external* forces. They are caused by:

   i. & ii. The reactions at A   **iii. The reaction at B**

4. Draw the free-body diagram of the (detached) beam showing all these forces labeled with their magnitudes and directions. *Assume* the sense of the vectors representing the *reactions acting on the beam*. Include any other relevant information, e.g. lengths, angles, etc. — which may help when formulating the equilibrium equations (including the moment equation) for the beam.

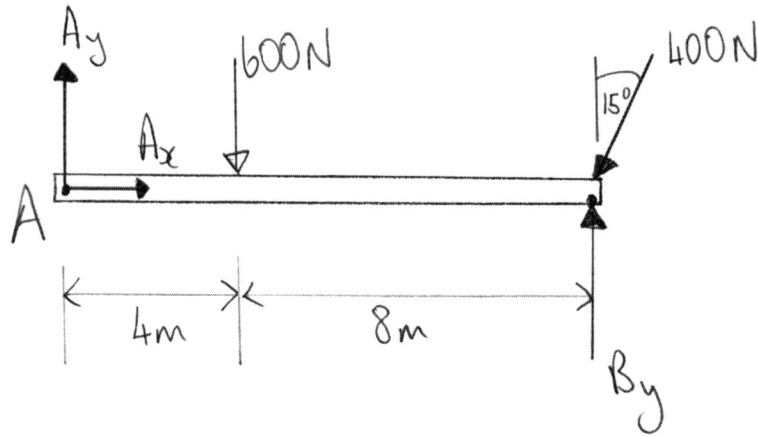

Problem 3.28

The link shown in the figure is pin-connected at A and rests against a smooth support at B. Draw the free-body diagram for link ABC and use it to compute the horizontal and vertical components of reaction at pin A. Neglect the weight of the link.

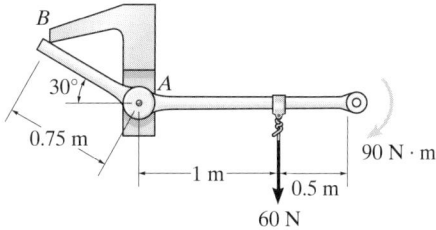

B

30°

0.75 m

A

90 N · m

1 m

0.5 m

60 N

Solution

1. Imagine the link ABC to be separated or detached from the system.

2. There is a pin connection at A and a smooth support at B. Use Table 2.1 to identify the reactions acting *on the link* at A and B.

3. The link is subjected to four *external* forces and one external applied couple moment.

4. Draw the free-body diagram of the (detached) link showing all these forces and couple moments labeled with their magnitudes and directions. *Assume* the sense of the vectors representing the *reactions acting on the link*. Include any other relevant information, e.g. lengths, angles, etc. — which may help when formulating the equilibrium equations (including the moment equation) for the link.

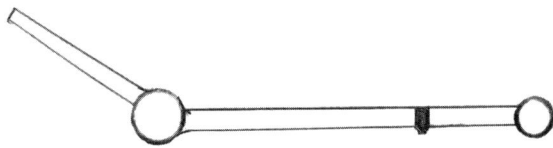

5. Sum moments about A and write down the moment equilibrium equation.

$$\curvearrowleft + \sum M_A = 0:$$

6. Establish an xy-axes system on the free-body diagram and write down the force equilibrium equations in each of the x and y-directions

$$\xrightarrow{+} \sum F_x = 0:$$

$$+ \uparrow \sum F_y = 0:$$

7. Solve the three equations for the required reaction components at pin A:

Problem 3.28

The link shown in the figure is pin-connected at A and rests against a smooth support at B. Draw the free-body diagram for link ABC and use it to compute the horizontal and vertical components of reaction at pin A. Neglect the weight of the link.

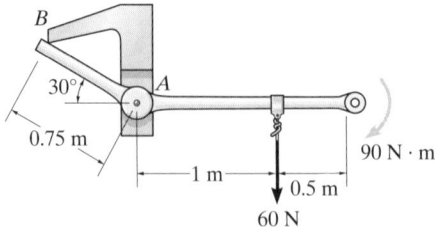

Solution

1. Imagine the link ABC to be separated or detached from the system.

2. There is a pin connection at A and a smooth support at B. Use Table 2.1 to identify the reactions acting on the link at A and B.

3. The link is subjected to four *external* forces and one external applied couple moment.

4. Draw the free-body diagram of the (detached) link showing all these forces and couple moments labeled with their magnitudes and directions. *Assume* the sense of the vectors representing the *reactions acting on the link*. Include any other relevant information, e.g. lengths, angles, etc. — which may help when formulating the equilibrium equations (including the moment equation) for the link.

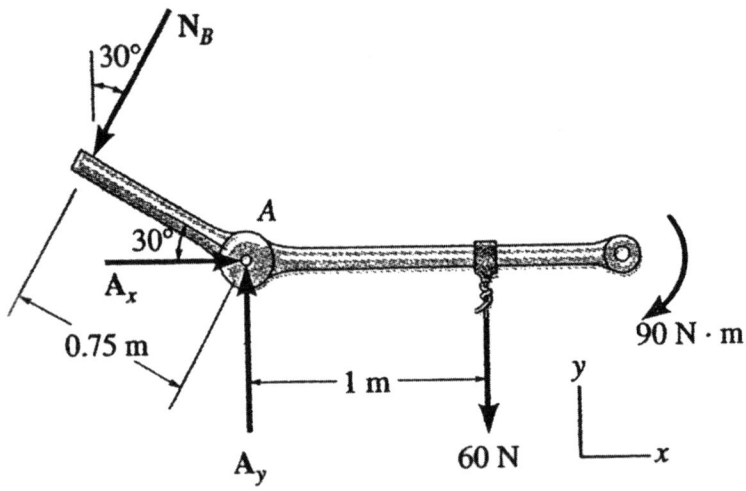

5. Sum moments about A and write down the moment equilibrium equation.

$$\curvearrowleft + \sum M_A = 0: \quad -90N.m - 60N(1m) + N_B(0.75m) = 0$$

6. Establish an xy-axes system on the free-body diagram and write down the force equilibrium equations in each of the x and y-directions

$$\xrightarrow{+} \sum F_x = 0: \quad A_x - N_B \sin 30° N = 0$$

$$+ \uparrow \sum F_y = 0: \quad A_y - N_B \cos 30° N - 60N = 0$$

7. Solve the three equations for the required reaction components at pin A:

$$N_B = 200N, \quad A_x = 100N, \quad A_y = 233N \qquad \textbf{Ans.}$$

Problem 3.29

A force of 150 kN acts on the end of the beam. Using the free-body diagram for the beam, find the magnitude and direction of the reaction at pin A and the tension in the cable. Neglect the weight of the beam.

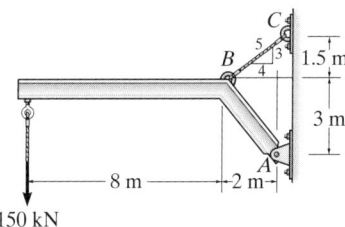

Solution

1. Imagine the beam to be separated or detached from the system.

2. There is a pin connection at A and a cable support at B. Use Table 2.1 to identify the reactions *acting on the beam* at A and B.

3. The beam is subjected to four *external* forces.

4. Draw the free-body diagram of the (detached) beam showing all these forces labeled with their magnitudes and directions. *Assume* the sense of the vectors representing the *reactions acting on the beam*. Include any other relevant information, e.g. lengths, angles, etc. — which may help when formulating the equilibrium equations (including the moment equation) for the beam.

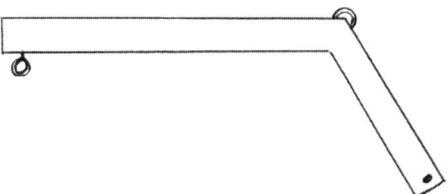

5. Sum moments about A and write down the moment equilibrium equation.

$$\curvearrowleft + \sum M_A = 0:$$

You should obtain the cable tension directly from this equation.

6. Establish an xy-axes system on the free-body diagram and write down the force equilibrium equations in each of the x and y-directions

$$\overset{+}{\rightarrow} \sum F_x = 0:$$

$$+ \uparrow \sum F_y = 0:$$

7. Solve the two equations for the magnitude and direction of the (resultant) force at pin A:

Problem 3.29

A force of 150 kN acts on the end of the beam. Using the free-body diagram for the beam, find the magnitude and direction of the reaction at pin A and the tension in the cable. Neglect the weight of the beam.

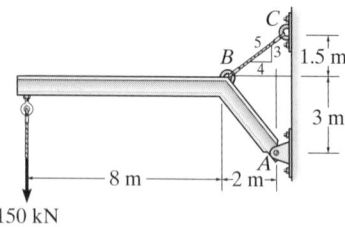

Solution

1. Imagine the beam to be separated or detached from the system.

2. There is a pin connection at A and a cable support at B. Use Table 2.1 to identify the reactions *acting on the beam* at A and B.

3. The beam is subjected to four *external* forces.

4. Draw the free-body diagram of the (detached) beam showing all these forces labeled with their magnitudes and directions. *Assume* the sense of the vectors representing the *reactions acting on the beam*. Include any other relevant information, e.g. lengths, angles, etc. — which may help when formulating the equilibrium equations (including the moment equation) for the beam.

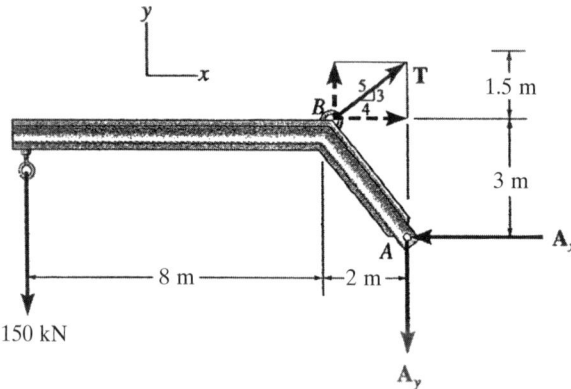

5. Sum moments about A and write down the moment equilibrium equation.

$$\curvearrowright + \sum M_A = 0: \quad -\left(\frac{3}{5}T\right)(2m) - \left(\frac{4}{5}T\right)(3m) + 150 \text{ kN}(10m) = 0$$

You should obtain the cable tension directly from this equation:

$$T = 416.7 \text{ kN.} \hspace{4cm} \textbf{Ans.}$$

6. Establish an xy-axes system on the free-body diagram and write down the force equilibrium equations in each of the x and y-directions

$$\xrightarrow{+} \sum F_x = 0: \quad -A_x + \left(\frac{4}{5}\right)(416.7 \text{ kN}) = 0$$

$$+ \uparrow \sum F_y = 0: \quad \left(\frac{3}{5}\right)416.7 \text{ kN} - 150 \text{ kN} - A_y = 0$$

7. Solve the two equations for the magnitude and direction of the (resultant) force at pin A:

$$A_x = 333.3 \text{ kN} \leftarrow, \quad A_y = 100 \text{ kN} \downarrow$$

Thus, magnitude of force at A is $\sqrt{(333.3)^2 + (100)^2} = 348.0 \text{ kN}$

Direction is $\theta = \tan^{-1} \dfrac{-100}{-333.3} = 196.7°$ $\hspace{1cm}$ 16.7° $\hspace{2cm}$ **Ans.**

Problem 3.30

The oil-drilling rig shown has a mass of 24 Mg and mass center at G. If the rig is pin-connected at its base, use a free-body diagram of the rig to determine the tension in the hoisting cable and the magnitude of the hoisting force at A when the rig is in the position shown.

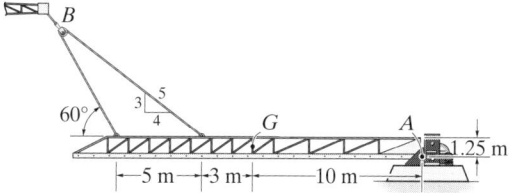

Solution

1. Imagine the rig to be separated or detached from the system.

2. There is a pin connection at A and a cable support at B. Use Table 2.1 to identify the reactions acting on the rig at A and B. Note that since the hoisting cable is continuous and passes over the pulley, the cable is subjected to the same tension T throughout its length.

3. The rig is subjected to five *external* forces.

4. Draw the free-body diagram of the (detached) rig showing all these forces labeled with their magnitudes and directions. *Assume* the sense of the vectors representing the *reactions acting on the rig*. Include any other relevant information, e.g. lengths, angles, etc. — which may help when formulating the equilibrium equations (including the moment equation) for the rig.

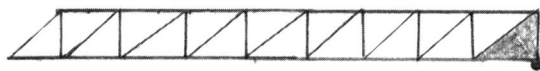

5. Sum moments about A and write down the moment equilibrium equation.

$$\curvearrowleft + \sum M_A = 0:$$

You should obtain the cable tension T directly from this equation:

6. Establish an xy-axes system on the free-body diagram and write down the force equilibrium equations in each of the x and y-directions

$$\xrightarrow{+} \sum F_x = 0:$$

$$+ \uparrow \sum F_y = 0:$$

7. Solve the two equations for the magnitude of the (resultant) force at pin A:

Problem 3.30

The oil-drilling rig shown has a mass of 24 Mg and mass center at G. If the rig is pin-connected at its base, use a free-body diagram of the rig to determine the tension in the hoisting cable and the magnitude of the hoisting force at A when the rig is in the position shown.

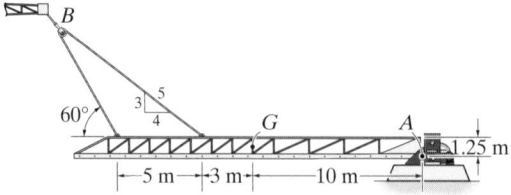

Solution

1. Imagine the rig to be separated or detached from the system.

2. There is a pin connection at A and a cable support at B. Use Table 2.1 to identify the reactions *acting on the rig* at A and B. Note that since the hoisting cable is continuous and passes over the pulley, the cable is subjected to the same tension T throughout its length.

3. The rig is subjected to five *external* forces.

4. Draw the free-body diagram of the (detached) rig showing all these forces labeled with their magnitudes and directions. *Assume* the sense of the vectors representing the reactions *acting on the rig*. Include any other relevant information, e.g. lengths, angles, etc. — which may help when formulating the equilibrium equations (including the moment equation) for the rig.

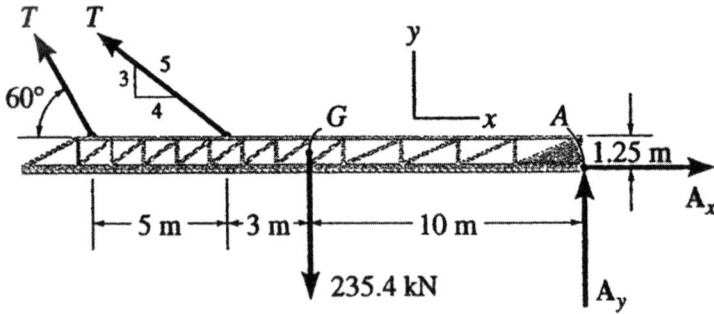

5. Sum moments about A and write down the moment equilibrium equation.

$$\curvearrowright + \sum M_A = 0: \quad (235.4 \text{ kN})(10m) - \left(\frac{3}{5}\right)T(13m) + \left(\frac{4}{5}T\right)(1.25m) - T\sin 60°(18m) + (T\cos 60°)(1.25m) = 0$$

You should obtain the cable tension T directly from this equation: $T = 108.2$ kN

6. Establish an xy-axes system on the free-body diagram and write down the force equilibrium equations in each of the x and y-directions

$$\xrightarrow{+} \sum F_x = 0: \quad A_x - 108.2\left(\frac{4}{5}\right)\text{kN} - 108.2\cos 60° \text{ kN} = 0$$

$$+\uparrow \sum F_y = 0: \quad A_y - 235.4 \text{ kN} + 108.2\left(\frac{3}{5}\right)\text{kN} - 108.2\sin 60° \text{ kN} = 0$$

7. Solve the two equations for the magnitude of the (resultant) force at pin A:

$$A_x = 140.6 \text{ kN}, \quad A_y = 76.8 \text{ kN}$$

Thus, magnitude of force at A is $\sqrt{(140.6)^2 + (76.8)^2} = 160$ kN   **Ans.**

Mechanics for Engineers
S T A T I C S
Thirteenth SI Edition

Access for Video Solutions

Thank you for purchasing a copy of *Mechanics for Engineers: Statics,* Thirteenth SI Edition, by R. C. Hibbeler and Kai Beng Yap. The one-time access code below provides access to the Video Solutions found on the Companion Website.

For students only:

The Video Solutions provide a complete step-by-step walkthroughs of representative homework problems from each chapter. These videos offer:

- *Fully worked Solutions* — Showing every step of representative homework problems, to help students make vital connections between concepts.
- *Self-paced Instruction* — Students can navigate each problem and select, play, rewind, fast-forward, stop, and jump-to sections within each problem's solution.
- *24/7 Access* — Help whenever students need it, with over 20 hours of helpful review.

To access the Video Solutions:

1. Go to www.pearsoned–asia.com/hibbeler.
2. Look for the title *Mechanics for Engineers: Statics,* Thirteenth SI Edition.
3. Click on "Video Solutions" and follow the instructions on the screen.
4. Register the one-time access code below to establish your login name and password.

ISSHES-WHIRR-PLAYS-REDAN-PETRI-LURES

For instructors only:

The Video Solutions can be found on the Instructor Resources webpage on the Companion Website.

To access the Instructor Resources, please follow steps 1 and 2 above and click on "Instructor Resources". Click on the resources (e.g., Video Solutions) you want to access, and you will be prompted to sign in with your login and password. Please proceed if you are already registered and have an existing Instructor Resource Centre access.

If you do not have any access, you may contact your Pearson Education Representative.

IMPORTANT: The access code on this page can only be used once to establish a subscription to the Video Solutions on the *Mechanics for Engineers: Statics,* Thirteenth SI Edition, Companion Website. Each copy of this title sold through your local bookstore is shrink-wrapped to protect the one-time access code. If you have bought a copy of this book that is not shrink-wrapped, please contact your local bookstore.

www.pearsoned-asia.com

MECHANICS FOR ENGINEERS

STATICS

THIRTEENTH SI EDITION

R. C. Hibbeler
Kai Beng Yap

SI Conversion by
S. C. Fan
Nanyang Technological University

Singapore   London   New York   Toronto   Sydney   Tokyo   Madrid
Mexico City   Munich   Paris   Cape Town   Hong Kong   Montreal

Published in 2012 by
Pearson Education South Asia Pte Ltd
9 North Buona Vista Drive,
#13-01 The Metropolis Tower One
Singapore 138588

Publishing Director: *Mark Cohen*
Acquisitions Editor: *Arunabha Deb*
Project Editor: *Donald Villamero*
Prepress Executive: *Kimberly Yap*

Pearson Education offices in Asia: *Bangkok, Beijing, Ho Chi Minh City, Hong Kong, Jakarta, Kuala Lumpur, Manila, Seoul, Singapore, Taipei, Tokyo*

Authorized adaptation from the United States edition, entitled *Engineering Mechanics: Statics,* Thirteenth Edition, ISBN: 9780132915540 by Hibbeler, Russell C., published by Pearson Prentice Hall, Pearson Education, Inc., Copyright © 2013 by R. C. Hibbeler.

WORLD WIDE adaptation edition published by PEARSON EDUCATION SOUTH ASIA PTE LTD., Copyright © 2013.

The authors and publisher of this book have used their best efforts in preparing this book. These efforts include the development, research, and testing of the theories and programs to determine their effectiveness. The authors and publisher shall not be liable in any event for incidental or consequential damages with, or arising out of, the furnishing, performance, or use of these programs.

Printed in Singapore

12 11 10 9 8 7
16 15

ISBN 978-981-06-9260-5

About the Cover: © iStockphoto.com / James Michael Kruger

www.pearsoned-asia.com

To the Student

With the hope that this work will stimulate
an interest in Engineering Mechanics
and provide an acceptable guide to its understanding.

Mechanics for Engineers: Statics, Thirteenth SI Edition, was developed to provide students with a clear and thorough presentation of the theory and applications of engineering mechanics. Renowned for its clarity in explanation and robust problem sets, Hibbeler is currently one of the best-selling course texts in this field.

In this edition, visualization aids such as animations, real-life photos with vectors, and video solutions are provided to enable students to better understand concepts taught and problems posed. The ample problems provided in this text are arranged in increasing level of difficulty to build students' analytical and problem-solving skills, as well as give them the practice they require.

MasteringEngineering, the most technologically advanced and effective online system that offers customized coaching, has also been introduced with this edition to aid in student tutoring. With MasteringEngineering and Hibbeler and Yap's pedagogical approach, this combination will form the definitive authority in the teaching and learning of engineering mechanics that graces lecture halls today.

This edition contains the following elements:

1. Animations that help visualize concepts
2. 24/7 video solution walkthroughs that enable independent revision
3. Wide variety of new problems for practice and problem-solving
4. Realistic diagrams with vectors to demonstrate real-world applications

Animations That Help Visualize Concepts

Animations. On the Companion Website are eight animations identified by the adaptor as fundamental engineering mechanics concepts. The animations, flagged by a film icon, help students visualize the relation between mathematical explanation and real structure, breaking down complicated sequences and showing how free-body diagrams can be derived. These animations lend a graphic component in tutorials and lectures, assisting instructors in demonstrating the teaching of concepts with greater ease and clarity.

571

11.3 PRINCIPLE OF VIRTUAL WORK FOR A SYSTEM OF CONNECTED RIGID BODIES

Each animation is flagged by a film icon.

11.3 Principle of Virtual Work for a System of Connected Rigid Bodies

The method of virtual work is particularly effective for solving equilibrium problems that involve a system of several *connected* rigid bodies, such as the ones shown in Fig. 11–5.

Each of these systems is said to have only one degree of freedom since the arrangement of the links can be completely specified using only one coordinate θ. In other words, with this single coordinate and the length of the members, we can locate the position of the forces **F** and **P**.

In this text, we will only consider the application of the principle of virtual work to systems containing one degree of freedom.* Because they are less complicated, they will serve as a way to approach the solution of more complex problems involving systems with many degrees of freedom. The procedure for solving problems involving a system of frictionless connected rigid bodies follows.

Please refer to the Companion Website for the animation: *Principle of Virtual Work*

Fig. 11–5

Important Points

- A force does work when it moves through a displacement in the direction of the force. A couple moment does work when it moves through a collinear rotation. Specifically, positive work is done when the force or couple moment and its displacement have the same sense of direction.

- The principle of virtual work is generally used to determine the equilibrium configuration for a system of multiple connected members.

- A virtual displacement is imaginary; i.e., it does not really happen. It is a differential displacement that is given in the positive direction of a position coordinate.

- Forces or couple moments that do not virtually displace do no virtual work.

*This method of applying the principle of virtual work is sometimes called the *method of virtual displacements* because a virtual displacement is applied, resulting in the calculation of a real force. Although it is not used here, we can also apply the principle of virtual work as a *method of virtual forces*. This method is often used to apply a virtual force and then determine the displacements of points on deformable bodies. See R. C. Hibbeler, *Mechanics of Materials*, 8th edition, Pearson/Prentice Hall, 2011.

This scissors lift has one degree of freedom. Without the need for dismembering the mechanism, the force in the hydraulic cylinder *AB* required to provide the lift can be determined *directly* by using the principle of virtual work.

11

Instructors can demonstrate the different methods of analysis step-by-step.

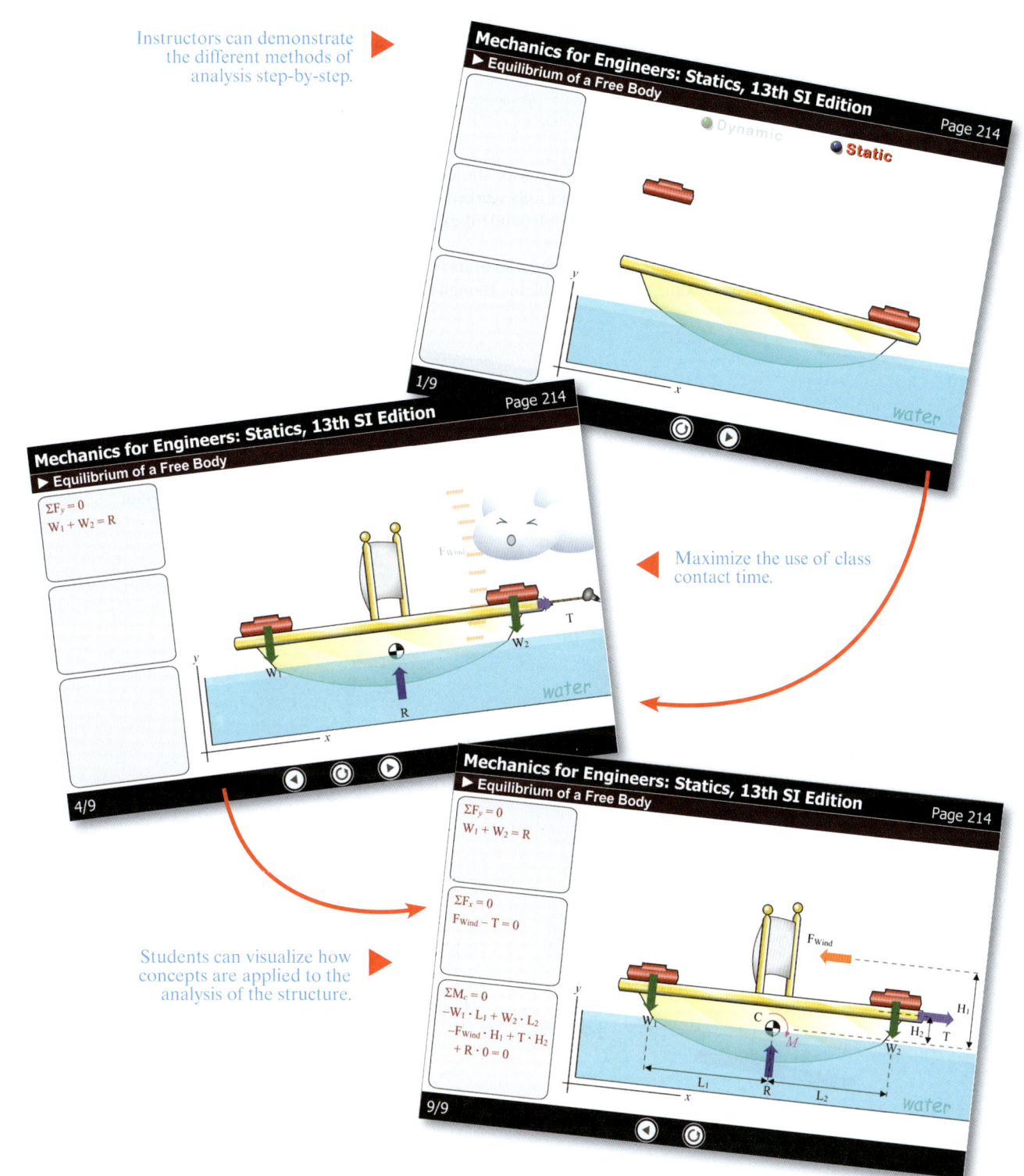

Maximize the use of class contact time.

Students can visualize how concepts are applied to the analysis of the structure.

2 24/7 Video Solution Walkthroughs That Enable Independent Revision

Video Solutions. An invaluable resource in and out of the classroom, these complete solution walkthroughs of representative homework problems from each chapter offer fully worked solutions, self-paced instruction, and 24/7 accessibility. Lecturers and students can harness this resource to gain independent exposure to wide range of examples applying formulas to actual structures.

This book also includes video solutions worked out in Imperial units from the original U.S. edition. This allows students to refine their problem-solving skills by tackling the same problems, though in a different set of units.

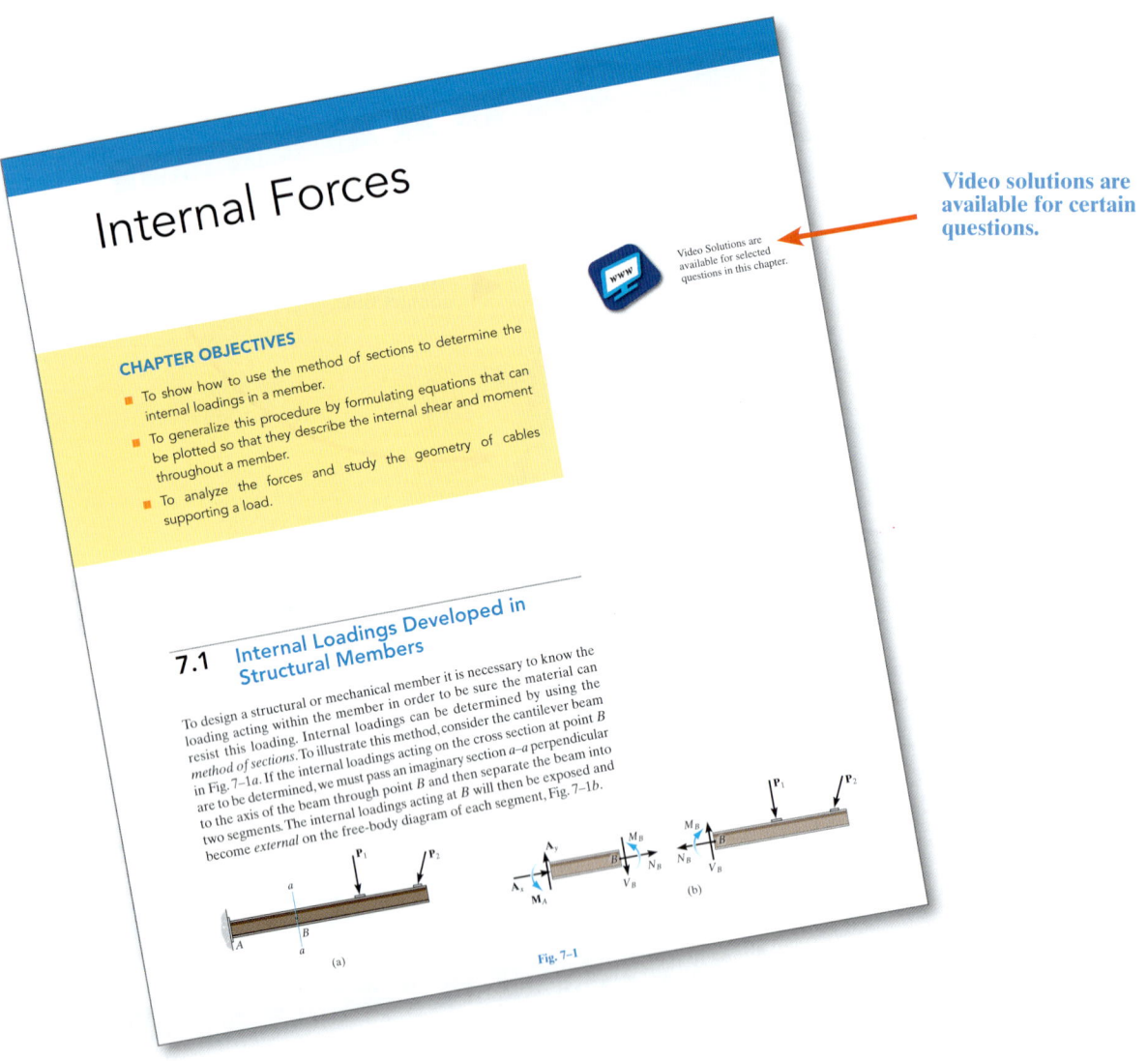

Video solutions are available for certain questions.

Internal Forces

Video Solutions are available for selected questions in this chapter.

CHAPTER OBJECTIVES

■ To show how to use the method of sections to determine the internal loadings in a member.

■ To generalize this procedure by formulating equations that can be plotted so that they describe the internal shear and moment throughout a member.

■ To analyze the forces and study the geometry of cables supporting a load.

7.1 Internal Loadings Developed in Structural Members

To design a structural or mechanical member it is necessary to know the loading acting within the member in order to be sure the material can resist this loading. Internal loadings can be determined by using the *method of sections*. To illustrate this method, consider the cantilever beam in Fig. 7–1a. If the internal loadings acting on the cross section at point B are to be determined, we must pass an imaginary section a–a perpendicular to the axis of the beam through point B and then separate the beam into two segments. The internal loadings acting at B will then be exposed and become *external* on the free-body diagram of each segment, Fig. 7–1b.

Fig. 7–1

Independent video replays of a lecturer's explanation reinforces students' understanding.

PEARSON   ALWAYS LEARNING

VideoSolutions

Video Solution

Chapter 2: Sections 7, 8

Force Directed Along a Line

This problem shows how we can use position vectors to define force vectors corresponding to chains, cables, and ropes. In these problems, we know that the force is directed along the direction of the cable, so we define a position vector in that direction, then use the position vector to derive an expression for the force vector. The usual outcome is that the force vector is expressed as a magnitude (the tension in the cable) times a unit vector in the direction of the cable.

Created by Edward Berger, PhD, University of Virginia

00:00   11:35

PEARSON   ALWAYS LEARNING

VideoSolutions

Force Directed Along a Line

The window is held open by cable AB. Determine the length of the cable and express the 30 N force acting at A along the cable as a Cartesian Vector.

Approach

1. write position vector from A→B, $\bar{r}_{AB}$
2. magnitude of $\bar{r}_{AB}$ (length of cable)
3. $\bar{u}_{AB}$
4. express force as a Cartesian vector

trig

force

B: $(0, 0.15, 0.25)$ m

A: $(0.3\cos30, 0.5, -0.3\sin30)$ m

05:35   11:35

Reduces lecturers' time spent in repetitive explanation of concepts and applications.

PEARSON   ALWAYS LEARNING

VideoSolutions

Force Directed Along a Line

1. position vector A→B

$\bar{r}_A, \bar{r}_B \Rightarrow \bar{r}_{AB} = \bar{r}_B - \bar{r}_A$

$\bar{r}_{AB} = (0.15\bar{j} + 0.25\bar{k}) - (0.26\bar{i} + 0.5\bar{j} - 0.15\bar{k})$

$\bar{r}_B$ $\bar{r}_A$

$\bar{r}_{AB} = (-0.26\bar{i} - 0.35\bar{j} + 0.4\bar{k})$ m

2. magnitude

$|\bar{r}_{AB}| = \sqrt{(0.26)^2 + (0.35)^2 + (0.4)^2}$

$|\bar{r}_{AB}| = 0.59$ m

11:20   11:35

Flexible resource for students where they can learn at a comfortable pace without relying too much on their instuctors.

③ Wide Variety of New Problems for Practice and Problem-Solving

35% New Problems. There are approximately 35% or about 410 new problems in this edition. These new problems relate to applications in many different fields of engineering. Also, a significant increase in algebraic type problems has been added, so that a generalized solution can be obtained.

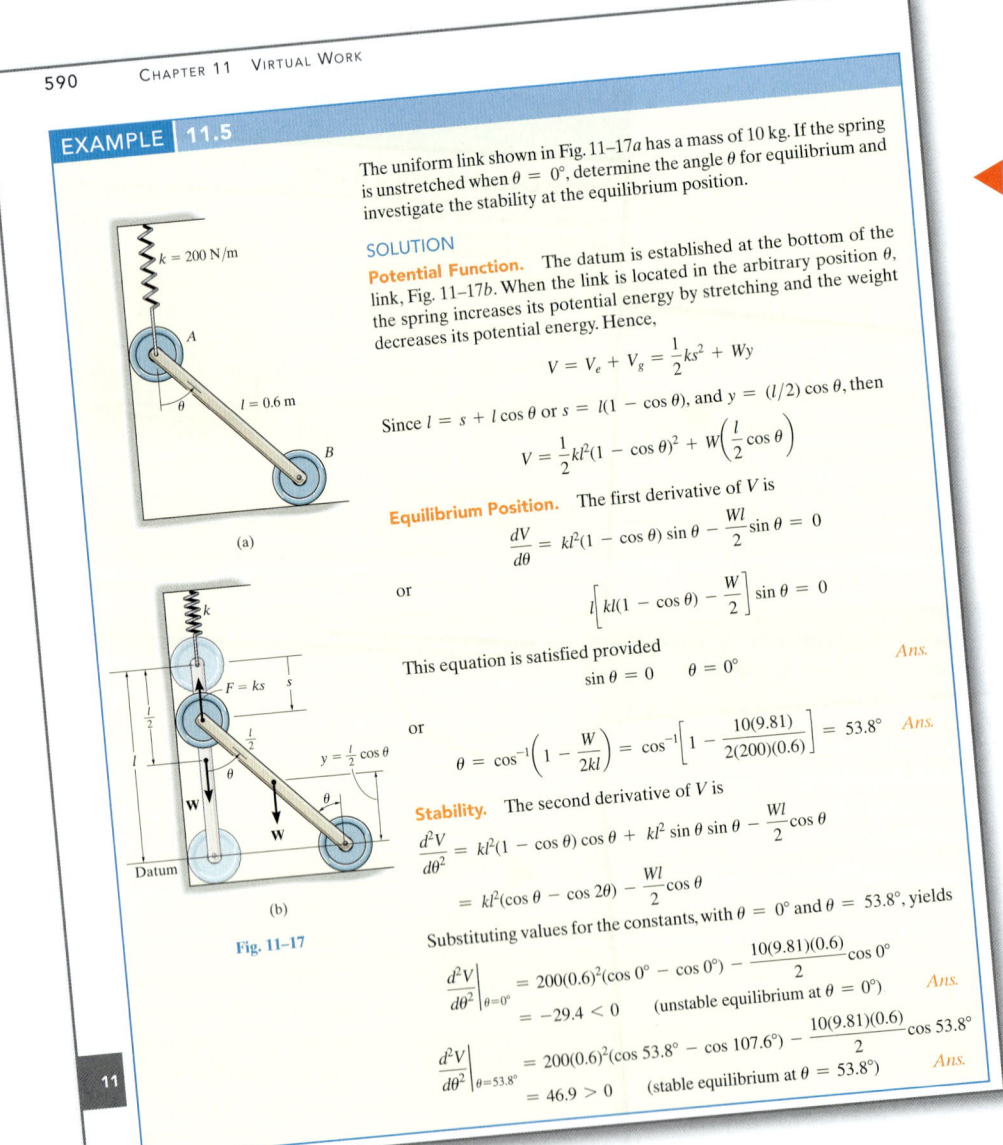

EXAMPLE 11.5

The uniform link shown in Fig. 11–17a has a mass of 10 kg. If the spring is unstretched when $\theta = 0°$, determine the angle θ for equilibrium and investigate the stability at the equilibrium position.

SOLUTION

Potential Function. The datum is established at the bottom of the link, Fig. 11–17b. When the link is located in the arbitrary position θ, the spring increases its potential energy by stretching and the weight decreases its potential energy. Hence,

$$V = V_e + V_g = \frac{1}{2}ks^2 + Wy$$

Since $l = s + l\cos\theta$ or $s = l(1 - \cos\theta)$, and $y = (l/2)\cos\theta$, then

$$V = \frac{1}{2}kl^2(1 - \cos\theta)^2 + W\left(\frac{l}{2}\cos\theta\right)$$

Equilibrium Position. The first derivative of V is

$$\frac{dV}{d\theta} = kl^2(1 - \cos\theta)\sin\theta - \frac{Wl}{2}\sin\theta = 0$$

or

$$l\left[kl(1 - \cos\theta) - \frac{W}{2}\right]\sin\theta = 0$$

This equation is satisfied provided

$$\sin\theta = 0 \qquad \theta = 0° \qquad \textit{Ans.}$$

or

$$\theta = \cos^{-1}\left(1 - \frac{W}{2kl}\right) = \cos^{-1}\left[1 - \frac{10(9.81)}{2(200)(0.6)}\right] = 53.8° \quad \textit{Ans.}$$

Stability. The second derivative of V is

$$\frac{d^2V}{d\theta^2} = kl^2(1 - \cos\theta)\cos\theta + kl^2\sin\theta\sin\theta - \frac{Wl}{2}\cos\theta$$

$$= kl^2(\cos\theta - \cos 2\theta) - \frac{Wl}{2}\cos\theta$$

Substituting values for the constants, with $\theta = 0°$ and $\theta = 53.8°$, yields

$$\left.\frac{d^2V}{d\theta^2}\right|_{\theta=0°} = 200(0.6)^2(\cos 0° - \cos 0°) - \frac{10(9.81)(0.6)}{2}\cos 0°$$

$$= -29.4 < 0 \qquad \text{(unstable equilibrium at } \theta = 0°) \qquad \textit{Ans.}$$

$$\left.\frac{d^2V}{d\theta^2}\right|_{\theta=53.8°} = 200(0.6)^2(\cos 53.8° - \cos 107.6°) - \frac{10(9.81)(0.6)}{2}\cos 53.8°$$

$$= 46.9 > 0 \qquad \text{(stable equilibrium at } \theta = 53.8°) \qquad \textit{Ans.}$$

Fig. 11–17

◄ **New & Revised Example Problems**

Throughout the book, examples have been altered or enhanced in an attempt to help clarify concepts for students. Where appropriate new examples have been added in order to emphasize important concepts that were needed.

Additional Fundamentals Problems ▶

These problem sets serve as extended example problems since their solutions are given in the back of the book. Additional problems have been added, especially in the areas of frames and machines, and in friction.

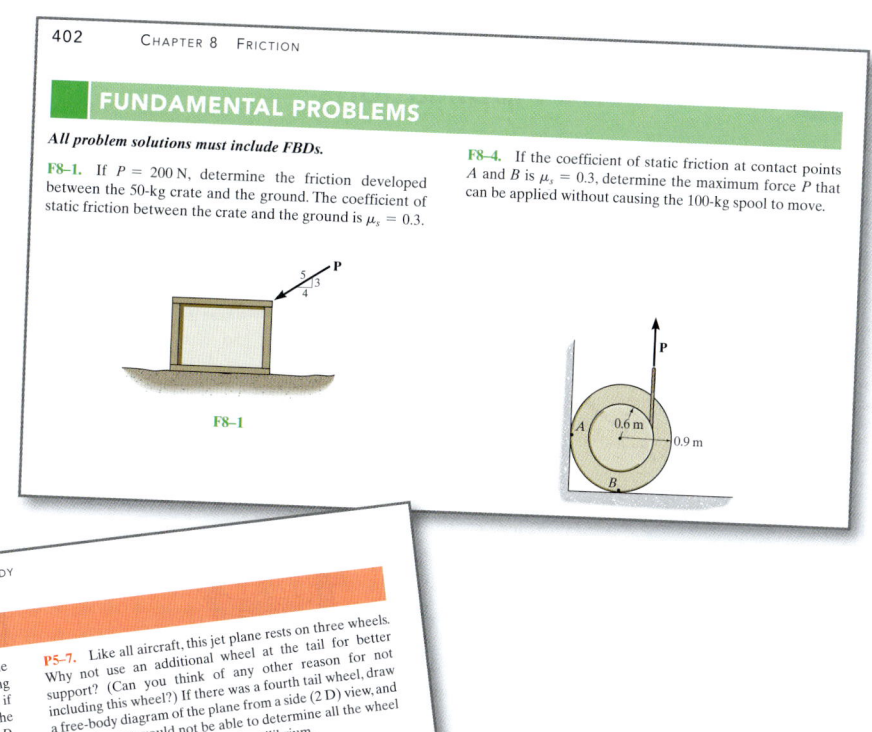

402   CHAPTER 8   FRICTION

FUNDAMENTAL PROBLEMS

All problem solutions must include FBDs.

F8–1. If $P = 200$ N, determine the friction developed between the 50-kg crate and the ground. The coefficient of static friction between the crate and the ground is $\mu_s = 0.3$.

F8–1

F8–4. If the coefficient of static friction at contact points A and B is $\mu_s = 0.3$, determine the maximum force P that can be applied without causing the 100-kg spool to move.

236   CHAPTER 5   EQUILIBRIUM OF A RIGID BODY

CONCEPTUAL PROBLEMS

P5–5. The tie rod is used to support this overhang at the entrance of a building. If it is pin connected to the building wall at A and to the center of the overhang B, determine if the force in the rod will increase, decrease, or remain the same if (a) the support at A is moved to a lower position C, and (b) the support at B is moved to the outer position D. Explain your answer with an equilibrium analysis, using dimensions and loads. Assume the overhang is pin supported from the building wall.

P5–5

P5–7. Like all aircraft, this jet plane rests on three wheels. Why not use an additional wheel at the tail for better support? (Can you think of any other reason for not including this wheel?) If there was a fourth tail wheel, draw a free-body diagram of the plane from a side (2 D) view, and show why one would not be able to determine all the wheel reactions using the equations of equilibrium.

P5–7

P5–6. The man attempts to pull the four wheeler up the incline and onto the trailer. From the position shown, is it more effective to pull on the rope at A, or would it be better to pull on the rope at B? Draw a free-body diagram for each.

P5–8. Where is the best place to arrange most of the logs in the wheelbarrow so that it minimizes the amount of force on the backbone of the person transporting the load? Do an equilibrium analysis to explain your answer.

◀ New Conceptual Problems

The conceptual problems given at the end of many of the problem sets are intended to engage the students in thinking through a real-life situation as depicted in a photo. They can be assigned either as individual or team projects after the students have developed some expertise in the subject matter.

Apart from the Fundamental and Conceptual type problems, other types of problems contained in the book include Free-Body Diagram Problems, which only require drawing the free-body diagram for the specific problems within a problem set; General Analysis and Design Problems, which depict realistic situations encountered in engineering practice; and Computer Problems, which may be solved using a numerical procedure executed on either a desktop computer or a programmable pocket calculator.

The many homework problems in this edition have been placed in two different categories. Problems that are simply indicated by a problem number have an answer and in some cases an additional numerical result given in the back of the book. An asterisk (*) before every fourth problem number indicates a problem without an answer.

4 Realistic Diagrams with Vectors to Demonstrate Real-World Applications

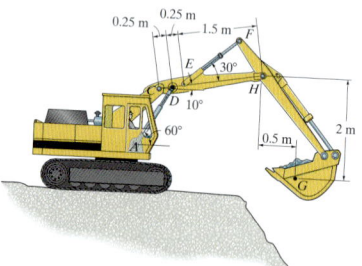
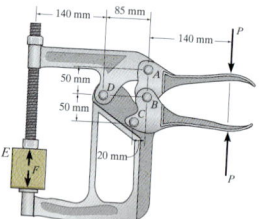

322 CHAPTER 6 STRUCTURAL ANALYSIS

*6–104. Determine the force created in the hydraulic cylinders EF and AD in order to hold the shovel in equilibrium. The shovel load has a mass of 1.25 Mg and a center of gravity at G. All joints are pin connected.

6–106. If P = 75 N, determine the force F that the toggle clamp exerts on the wooden block.

6–107. If the wooden block exerts a force of F = 600 N on the toggle clamp, determine the force P applied to the handle.

Prob. 6–104

Probs. 6–106/107

Illustrations with Vectors

Most of the diagrams throughout the book are in full-color art, and many photorealistic illustrations with vectors have been added. These provide a strong connection to the 3-D nature of engineering. This also helps the student to visualize and be aware of the concepts behind the question.

6–105. The hoist supports the 125-kg engine. Determine the force in member DB and in the hydraulic cylinder H of member FB.

149

4.6 MOMENT OF A COUPLE

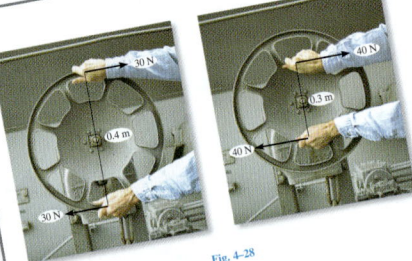

Fig. 4–28

Photographs

The relevance of knowing the subject matter is reflected by the realistic applications depicted by the many photos placed throughout the book. In this edition, 20 new or updated photos are included. These are used to explain how the relevant principles of mechanics apply to real-word situations.

Equivalent Couples. If two couples produce a moment with the same *magnitude and direction*, then these two couples are *equivalent*. For example, the two couples shown in Fig. 4–28 are *equivalent* because each couple moment has a magnitude of $M = 30 \text{ N}(0.4 \text{ m}) = 40 \text{ N}(0.3 \text{ m}) = 12 \text{ N} \cdot \text{m}$, and each is directed into the plane of the page. Notice that larger forces are required in the second case to create the same turning effect because the hands are placed closer together. Also, if the wheel was connected to the shaft at a point other than at its center, then the wheel would still turn when each couple is applied since the 12 N · m couple is a free vector.

Resultant Couple Moment. Since couple moments are vectors, their resultant can be determined by vector addition. For example, consider the couple moments $\mathbf{M}_1$ and $\mathbf{M}_2$ acting on the pipe in Fig. 4–29a. Since each couple moment is a free vector, we can join their tails at any arbitrary point and find the resultant couple moment, $\mathbf{M}_R = \mathbf{M}_1 + \mathbf{M}_2$ as shown in Fig. 4–29b.

If more than two couple moments act on the body, we may generalize this concept and write the vector resultant as

$$\mathbf{M}_R = \Sigma(\mathbf{r} \times \mathbf{F}) \qquad (4\text{–}16)$$

These concepts are illustrated numerically in the examples that follow. In general, problems projected in two dimensions should be solved using a scalar analysis since the moment arms and force components are easy to determine.

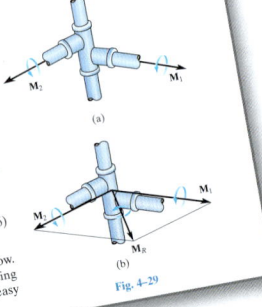

Fig. 4–29

Acknowledgments

The author has endeavored to write this book so that it will appeal to both the student and instructor. Through the years, many people have helped in its development, and I will always be grateful for their valued suggestions and comments. Specifically, I wish to thank all the individuals who have contributed their comments relative to preparing the thirteenth edition of this work, and in particular, O. Barton, Jr., of the U.S. Naval Academy, and K. Cook-Chennault at Rutgers, the State University of New Jersey.

There are a few other people that I also feel deserve particular recognition. These include comments sent to me by H. Kuhlman and G. Benson. A long-time friend and associate, Kai Beng Yap, was of great help to me in preparing and checking problem solutions. A special note of thanks also goes to Kurt Norlin of Laurel Tech Integrated Publishing Services in this regard. During the production process I am thankful for the assistance of Rose Kernan, my production editor for many years, and to my wife, Conny, and daughter, Mary Ann, who have helped with the proofreading and typing needed to prepare the manuscript for publication.

Lastly, many thanks are extended to all my students and to members of the teaching profession who have freely taken the time to e-mail me their suggestions and comments. Since this is too long to mention, it is hoped that those who have given help in this manner will accept this anonymous recognition.

I would greatly appreciate hearing from you if at any time you have any comments, suggestions, or problems related to any matters regarding this edition.

Russell Charles Hibbeler
hibbeler@bellsouth.net

your work...

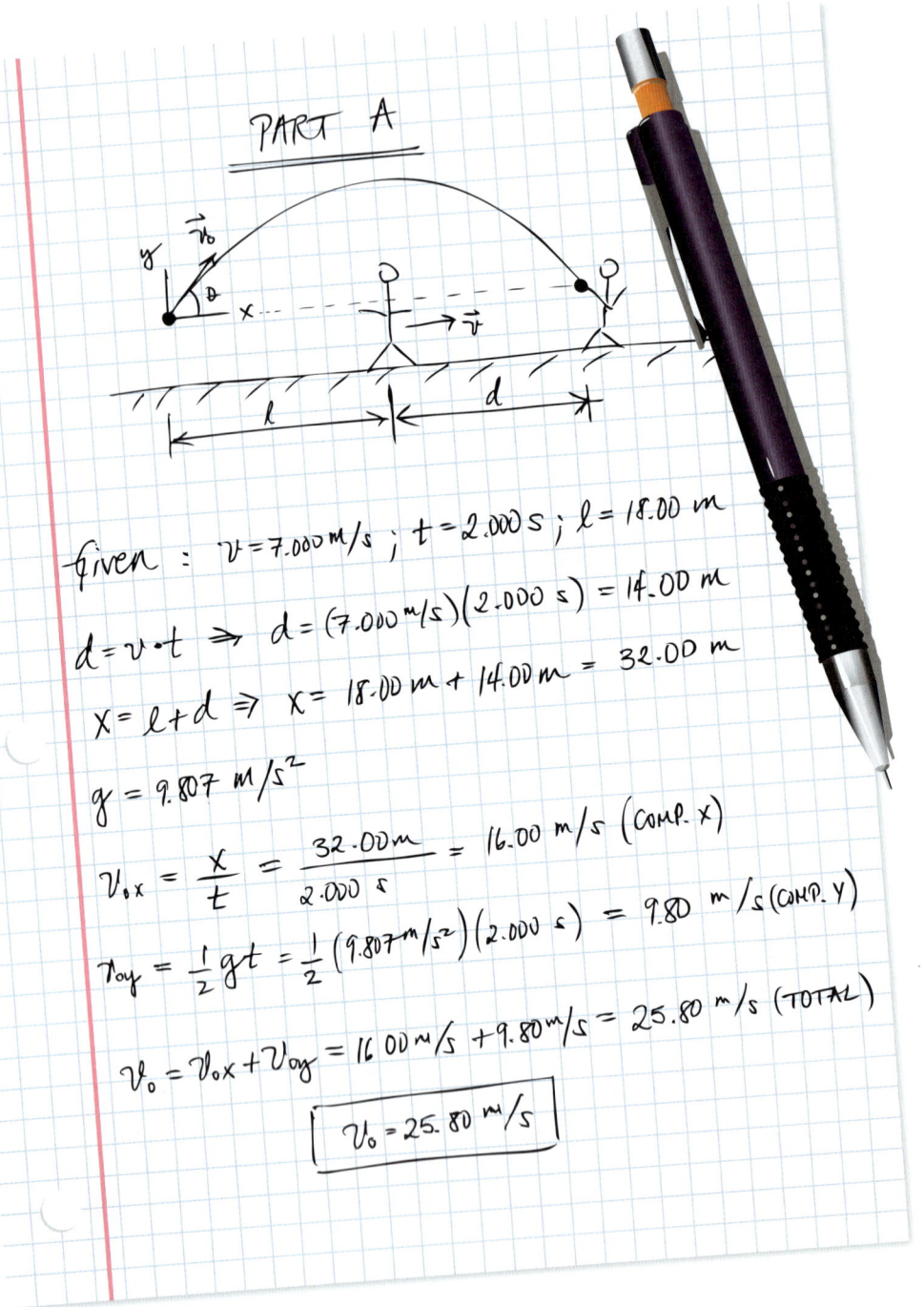

PART A

Given : $v = 7.000$ m/s ; $t = 2.000$ s ; $\ell = 18.00$ m

$d = v \cdot t \Rightarrow d = (7.000 \text{ m/s})(2.000 \text{ s}) = 14.00$ m

$x = \ell + d \Rightarrow x = 18.00 \text{ m} + 14.00 \text{ m} = 32.00$ m

$g = 9.807$ m/s^2

$v_{ox} = \dfrac{x}{t} = \dfrac{32.00 \text{ m}}{2.000 \text{ s}} = 16.00$ m/s (COMP. X)

$v_{oy} = \dfrac{1}{2} g t = \dfrac{1}{2}(9.807 \text{ m/s}^2)(2.000 \text{ s}) = 9.80$ m/s (COMP. Y)

$v_o = v_{ox} + v_{oy} = 16.00 \text{ m/s} + 9.80 \text{ m/s} = 25.80$ m/s (TOTAL)

$\boxed{v_o = 25.80 \text{ m/s}}$

your answer specific feedback

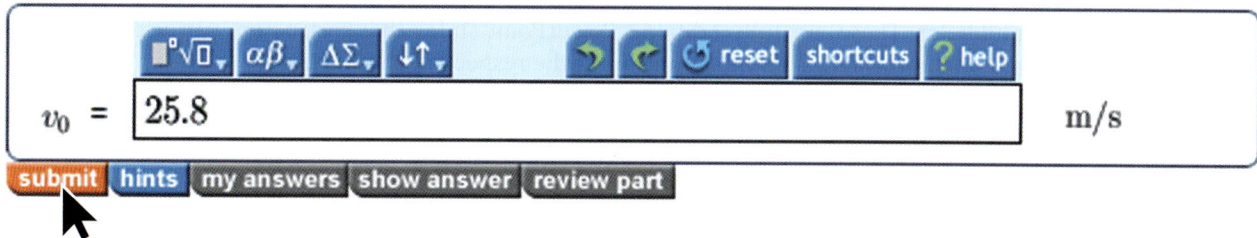

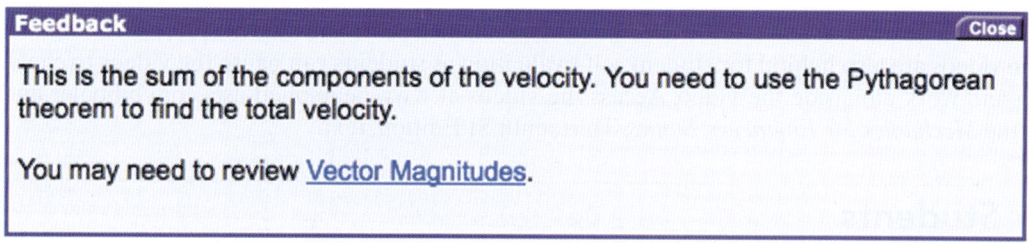

Try Again; 4 attempts remaining

Feedback	Close

This is the sum of the components of the velocity. You need to use the Pythagorean theorem to find the total velocity.

You may need to review Vector Magnitudes.

www.MasteringEngineering.com

Resources for Instructors

- **Instructor's Solutions Manual.** This supplement provides complete solutions supported by problem statements and problem figures. The thirteenth edition manual was revised to improve readability and was checked for accuracy thrice. The Instructor's Solutions Manual is available on the Companion Website: www. pearsoned-asia.com/hibbeler.

- **Teaching Slides.** Visual resources to accompany the text are located on the Companion Website: www. pearsoned-asia.com/hibbeler. These are fully editable PowerPoint slides that instructors can use for frontal teaching. The slides contain information and diagrams from the respective chapters.

- **Image Bank.** The image bank comprises illustrations, diagrams, and photos found in the textbook. Instructors will find this particularly useful in customizing their lesson materials, such as creating worksheets or modifying questions.

- **Video Solutions.** Developed by Professor Edward Berger, University of Virginia, video solutions are located on the Companion Website for the text and offer step-by-step solution walkthroughs of representative homework problems from each section of the text. Make efficient use of class time and office hours by showing students the complete and concise problem-solving approaches that they can access any time and view at their own pace. The videos are designed to be a flexible resource to be used however each instructor and student prefers. A valuable tutorial resource, the videos are also helpful for student self-evaluation as students can pause the videos to check their understanding and work alongside the video. Access the videos at www.pearsoned-asia.com/hibbeler and follow the links for the *Mechanics for Engineers: Statics,* Thirteenth SI Edition, text.

Resources for Students

- **Statics Study Pack.** This supplement contains chapter-by-chapter study materials and a Free-Body Diagram Workbook.

- **Companion Website.** The Companion Website, located at www.pearsoned-asia.com/hibbeler, includes opportunities for practice and review including:

  - **Video Solutions**—Complete, step-by-step solution walkthroughs of representative homework problems from each section. The videos offer:

    - **Fully worked Solutions**—Showing every step of representative homework problems, to help students make vital connections between concepts.

    - **Self-paced Instruction**—Students can navigate each problem and select, play, rewind, fast-forward, stop, and jump-to sections within each problem's solution.

    - **24/7 Access**—Help whenever students need it, with over 20 hours of helpful review.

R. C. Hibbeler graduated from the University of Illinois at Urbana with a BS in Civil Engineering (major in Structures) and an MS in Nuclear Engineering. He obtained his PhD in Theoretical and Applied Mechanics from Northwestern University. His professional experience includes postdoctoral work in reactor safety and analysis at Argonne National Laboratory, and structural work at Chicago Bridge and Iron, as well as Sargent and Lundy in Chicago. He has practiced engineering in Ohio, New York, and Louisiana. He is a Professor at the University of Louisiana, Lafayette, where he teaches courses in civil and mechanical engineering. In the past, he has taught at the University of Illinois in Urbana-Champaign, Youngstown State University, Illinois Institute of Technology, and Union College.

Kai Beng Yap is currently a registered Professional Engineer who works in Malaysia. He has BS and MS degrees in Civil Engineering from the University of Louisiana, Lafayette, Louisiana; and he has done further graduate work at Virginia Polytechnic Institute in Blacksberg, Virginia. His professional experience has involved teaching at the University of Louisiana, and doing engineering consulting work related to structural analysis and design and its associated infrastructure.

About the Adaptor

Sau Cheong Fan is from Nanyang Technological University (NTU), Singapore. He received his PhD from the University of Hong Kong. He is also Deputy Director, Centre for Advanced Numerical Engineering Simulations (CANES) at NTU. His industrial experience includes work and research in bridges, tall buildings, shell structures, jetties, pavements, cable structures, and glass diaphragm walls. He was the adaptor for the 5th and 6th SI editions of Hibbeler's *Mechanics of Materials*, and the 12th SI edition of Hibbeler's *Engineering Mechanics: Statics and Dynamics*.

CONTENTS

5
Equilibrium of a
Rigid Body 199

6
Structural Analysis 263

9
Center of Gravity and Centroid 451

10
Moments of Inertia 515

11
Virtual Work   567

Appendix

Fundamental Problems Partial Solutions and Answers   606

Answers to Selected Problems   624

Index   642

CREDITS

Chapter opening images are credited as follows:

Chapter 1, © Roel Meijer / Alamy
Chapter 2, Atli Mar / Getty Images
Chapter 3, John Foxx / Getty Images
Chapter 4, Huntstock / Getty Images
Chapter 5, © Bettmann / CORBIS
Chapter 6, © Martin Jenkinson / Alamy
Chapter 7, © JG Photography / Alamy
Chapter 8, © INSADCO Photography / Alamy
Chapter 9, © Corbis Premium RF / Alamy
Chapter 10, © RubberBall / Alamy
Chapter 11, © Shaun Flannery / Alamy

All other photos provided by the author.

MECHANICS FOR ENGINEERS

STATICS

THIRTEENTH SI EDITION

Chapter 1

Large cranes such as this one are required to lift extrememly large loads. Their design is based on the basic principles of statics and dynamics, which form the subject matter of engineering mechanics.

General Principles

CHAPTER OBJECTIVES

- To provide an introduction to the basic quantities and idealizations of mechanics.

- To give a statement of Newton's Laws of Motion and Gravitation.

- To review the principles for applying the SI system of units.

- To examine the standard procedures for performing numerical calculations.

- To present a general guide for solving problems.

Video Solutions are available for selected questions in this chapter.

1.1   Mechanics

Mechanics is a branch of the physical sciences that is concerned with the state of rest or motion of bodies that are subjected to the action of forces. In general, this subject can be subdivided into three branches: *rigid-body mechanics, deformable-body mechanics*, and *fluid mechanics*. In this book we will study rigid-body mechanics since it is a basic requirement for the study of the mechanics of deformable bodies and the mechanics of fluids. Furthermore, rigid-body mechanics is essential for the design and analysis of many types of structural members, mechanical components, or electrical devices encountered in engineering.

Rigid-body mechanics is divided into two areas: statics and dynamics. *Statics* deals with the equilibrium of bodies, that is, those that are either at rest or move with a constant velocity; whereas *dynamics* is concerned with the accelerated motion of bodies. We can consider statics as a special case of dynamics, in which the acceleration is zero; however, statics deserves separate treatment in engineering education since many objects are designed with the intention that they remain in equilibrium.

1

Historical Development.   The subject of statics developed very early in history because its principles can be formulated simply from measurements of geometry and force. For example, the writings of Archimedes (287–212 B.C.) deal with the principle of the lever. Studies of the pulley, inclined plane, and wrench are also recorded in ancient writings—at times when the requirements for engineering were limited primarily to building construction.

Since the principles of dynamics depend on an accurate measurement of time, this subject developed much later. Galileo Galilei (1564–1642) was one of the first major contributors to this field. His work consisted of experiments using pendulums and falling bodies. The most significant contributions in dynamics, however, were made by Isaac Newton (1642–1727), who is noted for his formulation of the three fundamental laws of motion and the law of universal gravitational attraction. Shortly after these laws were postulated, important techniques for their application were developed by such notables as Euler, D'Alembert, Lagrange, and others.

1.2   Fundamental Concepts

Before we begin our study of engineering mechanics, it is important to understand the meaning of certain fundamental concepts and principles.

Basic Quantities.   The following four quantities are used throughout mechanics.

Length.   *Length* is used to locate the position of a point in space and thereby describe the size of a physical system. Once a standard unit of length is defined, one can then use it to define distances and geometric properties of a body as multiples of this unit.

Time.   *Time* is conceived as a succession of events. Although the principles of statics are time independent, this quantity plays an important role in the study of dynamics.

Mass.   *Mass* is a measure of a quantity of matter that is used to compare the action of one body with that of another. This property manifests itself as a gravitational attraction between two bodies and provides a measure of the resistance of matter to a change in velocity.

Force.   In general, *force* is considered as a "push" or "pull" exerted by one body on another. This interaction can occur when there is direct contact between the bodies, such as a person pushing on a wall, or it can occur through a distance when the bodies are physically separated. Examples of the latter type include gravitational, electrical, and magnetic forces. In any case, a force is completely characterized by its magnitude, direction, and point of application.

Idealizations.

Models or idealizations are used in mechanics in order to simplify application of the theory. Here we will consider three important idealizations.

Particle. A *particle* has a mass, but a size that can be neglected. For example, the size of the earth is insignificant compared to the size of its orbit, and therefore the earth can be modeled as a particle when studying its orbital motion. When a body is idealized as a particle, the principles of mechanics reduce to a rather simplified form since the geometry of the body *will not be involved* in the analysis of the problem.

Rigid Body. A *rigid body* can be considered as a combination of a large number of particles in which all the particles remain at a fixed distance from one another, both before and after applying a load. This model is important because the body's shape does not change when a load is applied, and so we do not have to consider the type of material from which the body is made. In most cases the actual deformations occurring in structures, machines, mechanisms, and the like are relatively small, and the rigid-body assumption is suitable for analysis.

Concentrated Force. A *concentrated force* represents the effect of a loading which is assumed to act at a point on a body. We can represent a load by a concentrated force, provided the area over which the load is applied is very small compared to the overall size of the body. An example would be the contact force between a wheel and the ground.

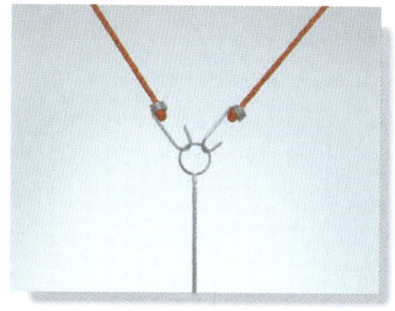

Three forces act on the ring. Since these forces all meet at point, then for any force analysis, we can assume the ring to be represented as a particle.

Steel is a common engineering material that does not deform very much under load. Therefore, we can consider this railroad wheel to be a rigid body acted upon by the concentrated force of the rail.

1

Newton's Three Laws of Motion.
Engineering mechanics is formulated on the basis of Newton's three laws of motion, the validity of which is based on experimental observation. These laws apply to the motion of a particle as measured from a *nonaccelerating* reference frame. They may be briefly stated as follows.

First Law.   A particle originally at rest, or moving in a straight line with constant velocity, tends to remain in this state provided the particle is *not* subjected to an unbalanced force, Fig. 1–1*a*.

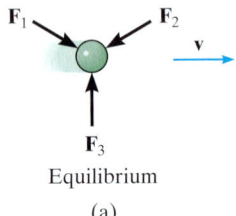

Equilibrium

(a)

Second Law.   A particle acted upon by an *unbalanced force* **F** experiences an acceleration **a** that has the same direction as the force and a magnitude that is directly proportional to the force, Fig. 1–1*b*.*
If **F** is applied to a particle of mass *m*, this law may be expressed mathematically as

$$\mathbf{F} = m\mathbf{a} \tag{1–1}$$

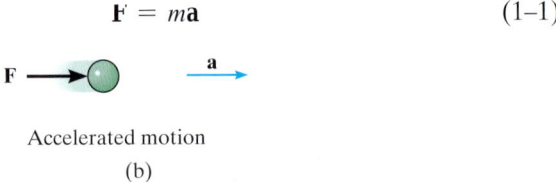

Accelerated motion

(b)

Third Law.   The mutual forces of action and reaction between two particles are equal, opposite, and collinear, Fig. 1–1*c*.

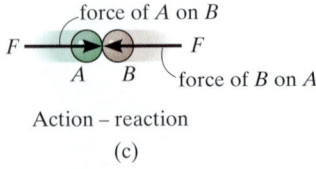

Action – reaction

(c)

Fig. 1–1

---

*Stated another way, the unbalanced force acting on the particle is proportional to the time rate of change of the particle's linear momentum.

Newton's Law of Gravitational Attraction.   Shortly after formulating his three laws of motion, Newton postulated a law governing the gravitational attraction between any two particles. Stated mathematically,

$$F = G\frac{m_1 m_2}{r^2}$$ (1–2)

where

F = force of gravitation between the two particles

G = universal constant of gravitation; according to experimental evidence, $G = 66.73(10^{-12})$ m^3/(kg·s^2)

m_1, m_2 = mass of each of the two particles

r = distance between the two particles

Weight. According to Eq. 1–2, any two particles or bodies have a mutual attractive (gravitational) force acting between them. In the case of a particle located at or near the surface of the earth, however, the only gravitational force having any sizable magnitude is that between the earth and the particle. Consequently, this force, termed the *weight*, will be the only gravitational force considered in our study of mechanics.

From Eq. 1–2, we can develop an approximate expression for finding the weight W of a particle having a mass $m_1 = m$. If we assume the earth to be a nonrotating sphere of constant density and having a mass $m_2 = M_e$, then if r is the distance between the earth's center and the particle, we have

$$W = G\frac{mM_e}{r^2}$$

Letting $g = GM_e/r^2$ yields

$$\boxed{W = mg}$$ (1–3)

The astronaut's weight is diminished, since she is far removed from the gravitational field of the earth.

By comparison with $\mathbf{F} = m\mathbf{a}$, we can see that g is the acceleration due to gravity. Since it depends on r, then the weight of a body is *not* an absolute quantity. Instead, its magnitude is determined from where the measurement was made. For most engineering calculations, however, g is determined at sea level and at a latitude of 45°, which is considered the "standard location."

1.3   Units of Measurement

The four basic quantities—length, time, mass, and force—are not all independent from one another; in fact, they are *related* by Newton's second law of motion, $\mathbf{F} = m\mathbf{a}$. Because of this, the *units* used to measure these quantities cannot *all* be selected arbitrarily. The equality $\mathbf{F} = m\mathbf{a}$ is maintained only if three of the four units, called *base units*, are *defined* and the fourth unit is then *derived* from the equation.

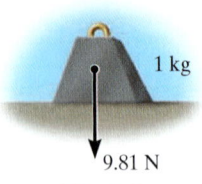

1 kg

9.81 N

Fig. 1–2

SI Units.   The International System of units, abbreviated SI after the French "Système International d'Unités," is a modern version of the metric system which has received worldwide recognition. As shown in Table 1–1, the SI system defines length in meters (m), time in seconds (s), and mass in kilograms (kg). The unit of force, called a newton (N), is *derived* from $\mathbf{F} = m\mathbf{a}$. Thus, 1 newton is equal to a force required to give 1 kilogram of mass an acceleration of 1 m/s^2 ($N = kg \cdot m/s^2$).

If the weight of a body located at the "standard location" is to be determined in newtons, then Eq. 1–3 must be applied. Here measurements give $g = 9.806\ 65 \text{ m/s}^2$; however, for calculations, the value $g = 9.81 \text{ m/s}^2$ will be used. Thus,

$$W = mg \quad (g = 9.81 \text{ m/s}^2) \tag{1–4}$$

Therefore, a body of mass 1 kg has a weight of 9.81 N, a 2-kg body weighs 19.62 N, and so on, Fig. 1–2.

TABLE 1–1 Systems of Units				
Name	Length	Time	Mass	Force
International System of Units	meter	second	kilogram	newton*
SI	m	s	kg	N $\left(\dfrac{kg \cdot m}{s^2}\right)$

*Derived unit.

1.4   The International System of Units

The SI system of units is used extensively in this book since it is intended to become the worldwide standard for measurement. Therefore, we will now present some of the rules for its use and some of its terminology relevant to engineering mechanics.

Prefixes.   When a numerical quantity is either very large or very small, the units used to define its size may be modified by using a prefix. Some of the prefixes used in the SI system are shown in Table 1–2. Each represents a multiple or submultiple of a unit which, if applied successively, moves the decimal point of a numerical quantity to every third place.* For example, $4\ 000\ 000$ N $= 4\ 000$ kN (kilo-newton) $=$ 4 MN (mega-newton), or 0.005 m $= 5$ mm (milli-meter). Notice that the SI system does not include the multiple deca (10) or the submultiple centi (0.01), which form part of the metric system. Except for some volume and area measurements, the use of these prefixes is to be avoided in science and engineering.

TABLE 1–2 Prefixes			
	Exponential Form	Prefix	SI Symbol
Multiple			
1 000 000 000	10^9	giga	G
1 000 000	10^6	mega	M
1 000	10^3	kilo	k
Submultiple			
0.001	10^{-3}	milli	m
0.000 001	10^{-6}	micro	μ
0.000 000 001	10^{-9}	nano	n

*The kilogram is the only base unit that is defined with a prefix.

Rules for Use.   Here are a few of the important rules that describe the proper use of the various SI symbols:

- Quantities defined by several units which are multiples of one another are separated by a *dot* to avoid confusion with prefix notation, as indicated by $N = kg \cdot m/s^2 = kg \cdot m \cdot s^{-2}$. Also, $m \cdot s$ (meter-second), whereas ms (milli-second).

- The exponential power on a unit having a prefix refers to both the unit *and* its prefix. For example, $\mu N^2 = (\mu N)^2 = \mu N \cdot \mu N$. Likewise, mm^2 represents $(mm)^2 = mm \cdot mm$.

- With the exception of the base unit the kilogram, in general avoid the use of a prefix in the denominator of composite units. For example, do not write N/mm, but rather kN/m; also, m/mg should be written as Mm/kg.

- When performing calculations, represent the numbers in terms of their *base or derived units* by converting all prefixes to powers of 10. The final result should then be expressed using a *single prefix*. Also, after calculation, it is best to keep numerical values between 0.1 and 1000; otherwise, a suitable prefix should be chosen. For example,

$$(50 \text{ kN})(60 \text{ nm}) = [50(10^3) \text{ N}][60(10^{-9}) \text{ m}]$$
$$= 3000(10^{-6}) \text{ N} \cdot m = 3(10^{-3}) \text{ N} \cdot m = 3 \text{ mN} \cdot m$$

1.5   Numerical Calculations

Numerical work in engineering practice is most often performed by using handheld calculators and computers. It is important, however, that the answers to any problem be reported with justifiable accuracy using appropriate significant figures. In this section we will discuss these topics together with some other important aspects involved in all engineering calculations.

Computers are often used in engineering for advanced design and analysis.

Dimensional Homogeneity.   The terms of any equation used to describe a physical process must be *dimensionally homogeneous;* that is, each term must be expressed in the same units. Provided this is the case, all the terms of an equation can then be combined if numerical values are substituted for the variables. Consider, for example, the equation $s = vt + \frac{1}{2}at^2$, where, in SI units, s is the position in meters, m, t is time in seconds, s, v is velocity in m/s and a is acceleration in m/s². Regardless of how this equation is evaluated, it maintains its dimensional homogeneity. In the form stated, each of the three terms is expressed in meters $[m, (m/\cancel{s})\cancel{s}, (m/\cancel{s}^2)\cancel{s}^2]$ or solving for a, $a = 2s/t^2 - 2v/t$, the terms are each expressed in units of m/s² [m/s², m/s², (m/s)/s].

Keep in mind that problems in mechanics always involve the solution of dimensionally homogeneous equations, and so this fact can then be used as a partial check for algebraic manipulations of an equation.

Significant Figures.

The number of significant figures contained in any number determines the accuracy of the number. For instance, the number 4981 contains four significant figures. However, if zeros occur at the end of a whole number, it may be unclear as to how many significant figures the number represents. For example, 23 400 might have three (234), four (2340), or five (23 400) significant figures. To avoid these ambiguities, we will use *engineering notation* to report a result. This requires that numbers be rounded off to the appropriate number of significant digits and then expressed in multiples of (10^3), such as (10^3), (10^6), or (10^{-9}). For instance, if 23 400 has five significant figures, it is written as $23.400(10^3)$, but if it has only three significant figures, it is written as $23.4(10^3)$.

If zeros occur at the beginning of a number that is less than one, then the zeros are not significant. For example, 0.008 21 has three significant figures. Using engineering notation, this number is expressed as $8.21(10^{-3})$. Likewise, 0.000 582 can be expressed as $0.582(10^{-3})$ or $582(10^{-6})$.

Rounding Off Numbers.

Rounding off a number is necessary so that the accuracy of the result will be the same as that of the problem data. As a general rule, any numerical figure ending in a number greater than five is rounded up and a number less than five is not rounded up. The rules for rounding off numbers are best illustrated by examples. Suppose the number 3.5587 is to be rounded off to *three* significant figures. Because the fourth digit (8) is *greater than* 5, the third number is rounded up to 3.56. Likewise 0.5896 becomes 0.590 and 9.3866 becomes 9.39. If we round off 1.341 to three significant figures, because the fourth digit (1) is *less than* 5, then we get 1.34. Likewise 0.3762 becomes 0.376 and 9.871 becomes 9.87. There is a special case for any number that ends in a 5. As a general rule, if the digit preceding the 5 is an *even number*, then this digit is *not* rounded up. If the digit preceding the 5 is an *odd number*, then it is rounded up. For example, 75.25 rounded off to three significant digits becomes 75.2, 0.1275 becomes 0.128, and 0.2555 becomes 0.256.

Calculations.

When a sequence of calculations is performed, it is best to store the intermediate results in the calculator. In other words, do not round off calculations until expressing the final result. This procedure maintains precision throughout the series of steps to the final solution. In this text we will generally round off the answers to three significant figures since most of the data in engineering mechanics, such as geometry and loads, may be reliably measured to this accuracy.

When solving problems, do the work as neatly as possible. Being neat will stimulate clear and orderly thinking, and vice versa.

1.6   General Procedure for Analysis

Attending a lecture, reading this book, and studying the example problems helps, but **the most effective way of learning the principles of engineering mechanics is to *solve problems***. To be successful at this, it is important to always present the work in a *logical* and *orderly manner*, as suggested by the following sequence of steps:

- Read the problem carefully and try to correlate the actual physical situation with the theory studied.
- Tabulate the problem data and *draw to a large scale* any necessary diagrams.
- Apply the relevant principles, generally in mathematical form. When writing any equations, be sure they are dimensionally homogeneous.
- Solve the necessary equations, and report the answer with no more than three significant figures.
- Study the answer with technical judgment and common sense to determine whether or not it seems reasonable.

Important Points

- Statics is the study of bodies that are at rest or move with constant velocity.
- A particle has a mass but a size that can be neglected, and a rigid body does not deform under load.
- Concentrated forces are assumed to act at a point on a body.
- Newton's three laws of motion should be memorized.
- Mass is measure of a quantity of matter that does not change from one location to another. Weight refers to the gravitational attraction of the earth on a body or quantity of mass. Its magnitude depends upon the elevation at which the mass is located.
- In the SI system the unit of force, the newton, is a derived unit. The meter, second, and kilogram are base units.
- Prefixes G, M, k, m, μ, and n are used to represent large and small numerical quantities. Their exponential size should be known, along with the rules for using the SI units.
- Perform numerical calculations with several significant figures, and then report the final answer to three significant figures.
- Algebraic manipulations of an equation can be checked in part by verifying that the equation remains dimensionally homogeneous.
- Know the rules for rounding off numbers.

EXAMPLE 1.1

Convert 2 km/h to m/s.

SOLUTION

Since 1 km = 1000 m and 1 h = 3600 s, the factors of conversion are arranged in the following order, so that a cancellation of the units can be applied:

$$2 \text{ km/h} = \frac{2 \cancel{\text{km}}}{\cancel{\text{h}}} \left(\frac{1000 \text{ m}}{\cancel{\text{km}}} \right) \left(\frac{1 \cancel{\text{h}}}{3600 \text{ s}} \right)$$

$$= \frac{2000 \text{ m}}{3600 \text{ s}} = 0.556 \text{ m/s} \qquad Ans.$$

NOTE: Remember to round off the final answer to three significant figures.

EXAMPLE 1.2

Evaluate each of the following and express with SI units having an appropriate prefix: (a) (50 mN)(6 GN), (b) (400 mm)(0.6 MN)^2, (c) 45 MN^3/900 Gg.

SOLUTION

First convert each number to base units, perform the indicated operations, then choose an appropriate prefix.

Part (a)

$$(50 \text{ mN})(6 \text{ GN}) = [50(10^{-3}) \text{ N}][6(10^9) \text{ N}]$$

$$= 300(10^6) \text{ N}^2$$

$$= 300(10^6) \cancel{\text{N}}^2 \left(\frac{1 \text{ kN}}{10^3 \cancel{\text{N}}} \right) \left(\frac{1 \text{ kN}}{10^3 \cancel{\text{N}}} \right)$$

$$= 300 \text{ kN}^2 \qquad Ans.$$

NOTE: Keep in mind the convention kN^2 = (kN)^2 = 10^6 N^2.

1   **EXAMPLE** | **1.2 (Continued)**

Part (b)

$$(400 \text{ mm})(0.6 \text{ MN})^2 = [400(10^{-3}) \text{ m}][0.6(10^6) \text{ N}]^2$$

$$= [400(10^{-3}) \text{ m}][0.36(10^{12}) \text{ N}^2]$$

$$= 144(10^9) \text{ m} \cdot \text{N}^2$$

$$= 144 \text{ Gm} \cdot \text{N}^2 \qquad \qquad \textit{Ans.}$$

We can also write

$$144(10^9) \text{ m} \cdot \text{N}^2 = 144(10^9) \text{ m} \cdot \cancel{\text{N}}^2 \left(\frac{1 \text{ MN}}{10^6 \cancel{\text{N}}} \right) \left(\frac{1 \text{ MN}}{10^6 \cancel{\text{N}}} \right)$$

$$= 0.144 \text{ m} \cdot \text{MN}^2 \qquad \qquad \textit{Ans.}$$

Part (c)

$$\frac{45 \text{ MN}^3}{900 \text{ Gg}} = \frac{45(10^6 \text{ N})^3}{900(10^6) \text{ kg}}$$

$$= 50(10^9) \text{ N}^3/\text{kg}$$

$$= 50(10^9) \cancel{\text{N}}^3 \left(\frac{1 \text{ kN}}{10^3 \cancel{\text{N}}} \right)^3 \frac{1}{\text{kg}}$$

$$= 50 \text{ kN}^3/\text{kg} \qquad \qquad \textit{Ans.}$$

PROBLEMS

1–1. Round off the following numbers to three significant figures: (a) 4.65735 m, (b) 55.578 s, (c) 4555 N, and (d) 2768 kg.

1–2. Represent each of the following combinations of units in the correct SI form: (a) kN$/\mu$s (b) Mg/mN, and (c) MN$/$(kg · ms).

1–3. Represent each of the following combinations of units in the correct SI form using an appropriate prefix: (a) m$/$ms, (b) μkm (c) ks$/$mg, and (d) km · μN.

*1–4. A concrete column has a diameter of 350 mm and a length of 2 m. If the density (mass/volume) of concrete is 2.45 Mg/m^3, determine the weight of the column in newtons.

1–5. Represent each of the following combinations of units in the correct SI form: (a) Mg/ms, (b) N/mm, and (c) mN/(kg · μs).

1–6. Represent each of the following combinations of units in the correct SI form using an appropriate prefix: (a) kN/ms, (b) Mg/kN, and (c) kN/(kg · μs).

1–7. A rocket has a mass $3.529(10^6)$ kg on earth. Specify (a) its mass in SI units, and (b) its weight in SI units. If the rocket is on the moon, where the acceleration due to gravity is $g_m = 1.61$ m/s^2, determine to three significant figures (c) its weight in SI units, and (d) its mass in SI units.

*1–8. Evaluate each of the following to three significant figures and express each answer in SI units using an appropriate prefix: (a) $(0.631$ Mm$)/(8.60$ kg$)^2$, (b) $(35$ mm$)^2$ $(48$ kg$)^3$.

1–9. Determine the mass in ki lograms of an object that has a weight of (a) 50 mN, (b) 250 kN, and (c) 80 MN. Express the answer to three significant figures.

1–10. Evaluate each of the following to three significant figures and express each answer in SI units using an appropriate prefix: (a) $(200$ kN$)^2$, (b) $(0.005$ mm$)^2$, and (c) $(400$ m$)^3$.

1–11. Evaluate each of the following and express with an appropriate prefix: (a) $(430$ kg$)^2$, (b) $(0.002$ mg$)^2$, and (c) $(230$ m$)^3$.

*1–12. Determine the mass of an object that has a weight of (a) 20 mN, (b) 150 kN, (c) 60 MN. Express the answer to three significant figures.

1–13. Represent each of the following with SI units having an appropriate prefix: (a) 8653 ms, (b) 8368 N, (c) 0.893 kg.

1–14. Evaluate $(204$ mm$)(0.00457$ kg$)/(34.6$ N$)$ to three significant figures and express the answer in SI units using an appropriate prefix.

1–15. Two particles have a mass of 8 kg and 12 kg, respectively. If they are 800 mm apart, determine the force of gravity acting between them. Compare this result with the weight of each particle.

*1–16. If a man weighs 690 newtons on earth, specify (a) his mass in kilograms. If the man is on the moon, where the acceleration due to gravity is $g_m = 1.61$ m/s^2, determine (b) his weight in newtons, and (c) his mass in kilograms.

Chapter 2

This electric transmission tower is stabilized by cables that exert forces on the tower at their points of connection. In this chapter we will show how to express these forces as Cartesian vectors, and then determined their resultant.

Force Vectors

Video Solutions are available for selected questions in this chapter.

2.1   Scalars and Vectors

All physical quantities in engineering mechanics are measured using either scalars or vectors.

Scalar.   A *scalar* is any positive or negative physical quantity that can be completely specified by its *magnitude*. Examples of scalar quantities include length, mass, and time.

Vector.   A *vector* is any physical quantity that requires both a *magnitude* and a *direction* for its complete description. Examples of vectors encountered in statics are force, position, and moment. A vector is shown graphically by an arrow. The length of the arrow represents the *magnitude* of the vector, and the angle θ between the vector and a fixed axis defines the *direction of its line of action*. The head or tip of the arrow indicates the *sense of direction* of the vector, Fig. 2–1.

In print, vector quantities are represented by boldface letters such as **A**, and the magnitude of a vector is italicized, *A*. For handwritten work, it is often convenient to denote a vector quantity by simply drawing an arrow above it, $\vec{A}$.

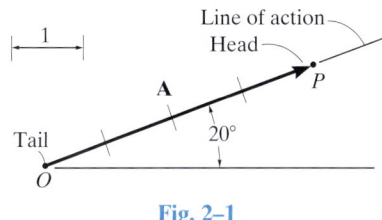

Fig. 2–1

2

A   **2A**   **−A**
                       −0.5**A**

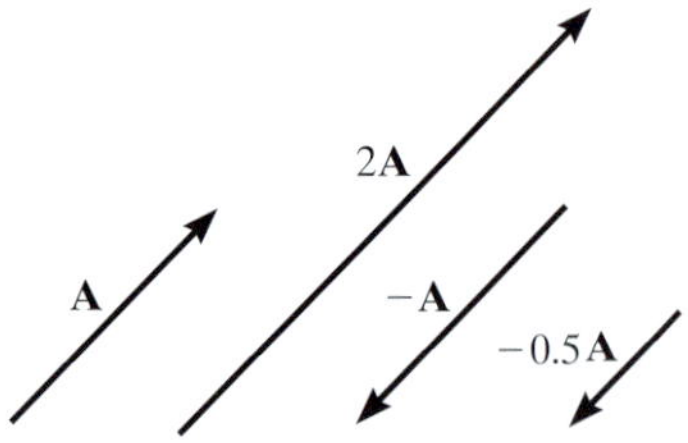

Scalar multiplication and division

Fig. 2–2

2.2   Vector Operations

Multiplication and Division of a Vector by a Scalar.   If a vector is multiplied by a positive scalar, its magnitude is increased by that amount. Multiplying by a negative scalar will also change the directional sense of the vector. Graphic examples of these operations are shown in Fig. 2–2.

Vector Addition.   All vector quantities obey the *parallelogram law of addition*. To illustrate, the two "*component*" vectors **A** and **B** in Fig. 2–3*a* are added to form a "*resultant*" vector **R** = **A** + **B** using the following procedure:

- First join the tails of the components at a point to make them concurrent, Fig. 2–3*b*.
- From the head of **B**, draw a line parallel to **A**. Draw another line from the head of **A** that is parallel to **B**. These two lines intersect at point *P* to form the adjacent sides of a parallelogram.
- The diagonal of this parallelogram that extends to *P* forms **R**, which then represents the resultant vector **R** = **A** + **B**, Fig. 2–3*c*.

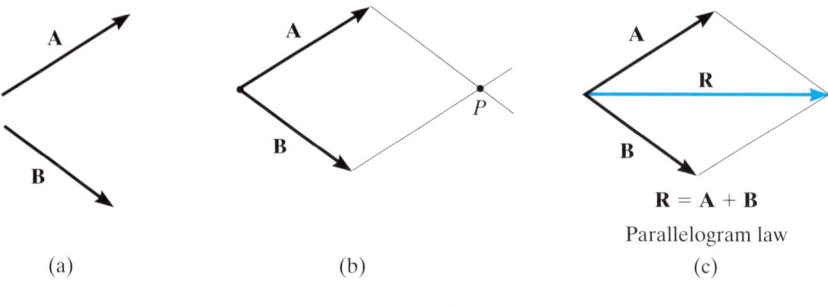

(a)   (b)   (c)

R = **A** + **B**
Parallelogram law

Fig. 2–3

We can also add **B** to **A**, Fig. 2–4*a*, using the *triangle rule*, which is a special case of the parallelogram law, whereby vector **B** is added to vector **A** in a "head-to-tail" fashion, i.e., by connecting the head of **A** to the tail of **B**, Fig. 2–4*b*. The resultant **R** extends from the tail of **A** to the head of **B**. In a similar manner, **R** can also be obtained by adding **A** to **B**, Fig. 2–4*c*. By comparison, it is seen that vector addition is commutative; in other words, the vectors can be added in either order, i.e., **R** = **A** + **B** = **B** + **A**.

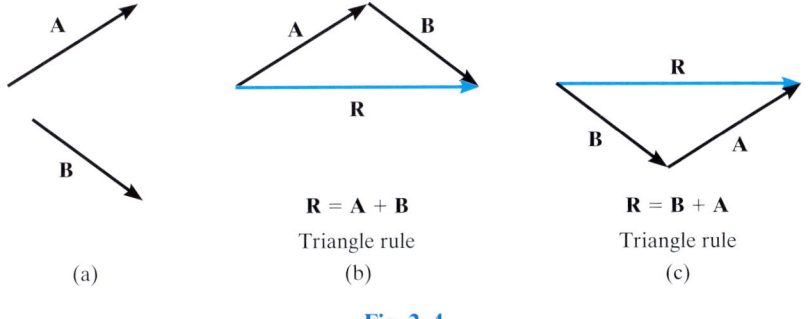

R = A + B
Triangle rule

R = B + A
Triangle rule

(a)   (b)   (c)

Fig. 2–4

As a special case, if the two vectors **A** and **B** are *collinear*, i.e., both have the same line of action, the parallelogram law reduces to an *algebraic* or *scalar addition* $R = A + B$, as shown in Fig. 2–5.

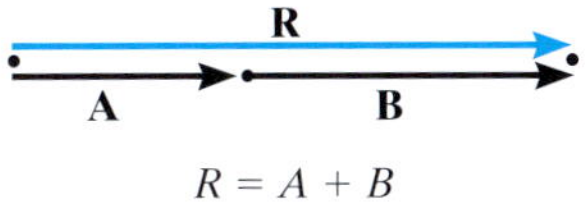

$$R = A + B$$

Addition of collinear vectors

Fig. 2–5

Vector Subtraction.   The resultant of the *difference* between two vectors **A** and **B** of the same type may be expressed as

$$\mathbf{R}' = \mathbf{A} - \mathbf{B} = \mathbf{A} + (-\mathbf{B})$$

This vector sum is shown graphically in Fig. 2–6. Subtraction is therefore defined as a special case of addition, so the rules of vector addition also apply to vector subtraction.

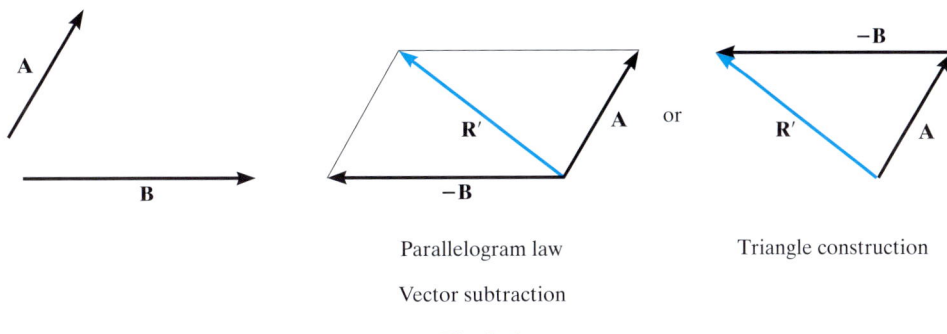

Parallelogram law

Vector subtraction

Triangle construction

Fig. 2–6

2.3   Vector Addition of Forces

Experimental evidence has shown that a force is a vector quantity since it has a specified magnitude, direction, and sense and it adds according to the parallelogram law. Two common problems in statics involve either finding the resultant force, knowing its components, or resolving a known force into two components. We will now describe how each of these problems is solved using the parallelogram law.

The parallelogram law must be used to determine the resultant of the two forces acting on the hook.

Finding a Resultant Force.

The two component forces $\mathbf{F}_1$ and $\mathbf{F}_2$ acting on the pin in Fig. 2–7a can be added together to form the resultant force $\mathbf{F}_R = \mathbf{F}_1 + \mathbf{F}_2$, as shown in Fig. 2–7b. From this construction, or using the triangle rule, Fig. 2–7c, we can apply the law of cosines or the law of sines to the triangle in order to obtain the magnitude of the resultant force and its direction.

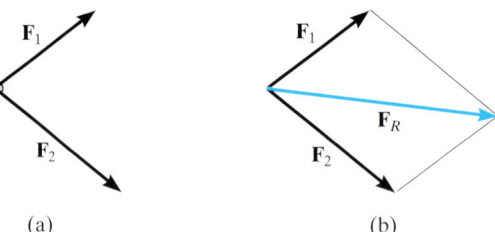

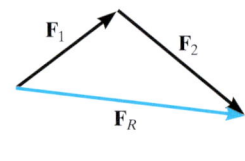

(a)   (b)   (c)

Fig. 2–7

Finding the Components of a Force.

Sometimes it is necessary to resolve a force into two *components* in order to study its pulling or pushing effect in two specific directions. For example, in Fig. 2–8a, $\mathbf{F}$ is to be resolved into two components along the two members, defined by the u and v axes. In order to determine the magnitude of each component, a parallelogram is constructed first, by drawing lines starting from the tip of $\mathbf{F}$, one line parallel to u, and the other line parallel to v. These lines then intersect with the v and u axes, forming a parallelogram. The force components $\mathbf{F}_u$ and $\mathbf{F}_v$ are then established by simply joining the tail of $\mathbf{F}$ to the intersection points on the u and v axes, Fig. 2–8b. This parallelogram can then be reduced to a triangle, which represents the triangle rule, Fig. 2–8c. From this, the law of sines can then be applied to determine the unknown magnitudes of the components.

Using the parallelogram law the supporting force $\mathbf{F}$ can be resolved into components acting along the u and v axes.

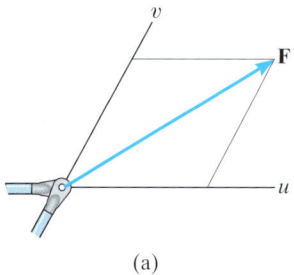

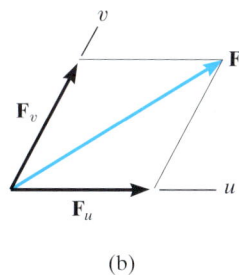

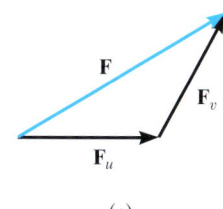

(a)   (b)   (c)

Fig. 2–8

Addition of Several Forces.   If more than two forces are to be added, successive applications of the parallelogram law can be carried out in order to obtain the resultant force. For example, if three forces $\mathbf{F}_1$, $\mathbf{F}_2$, $\mathbf{F}_3$ act at a point O, Fig. 2–9, the resultant of any two of the forces is found, say, $\mathbf{F}_1 + \mathbf{F}_2$—and then this resultant is added to the third force, yielding the resultant of all three forces; i.e., $\mathbf{F}_R = (\mathbf{F}_1 + \mathbf{F}_2) + \mathbf{F}_3$. Using the parallelogram law to add more than two forces, as shown here, often requires extensive geometric and trigonometric calculation to determine the numerical values for the magnitude and direction of the resultant. Instead, problems of this type are easily solved by using the "rectangular-component method," which is explained in Sec. 2.4.

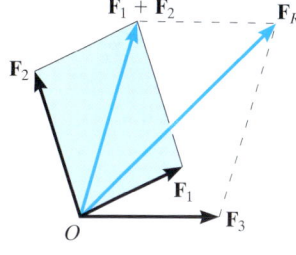

Fig. 2–9

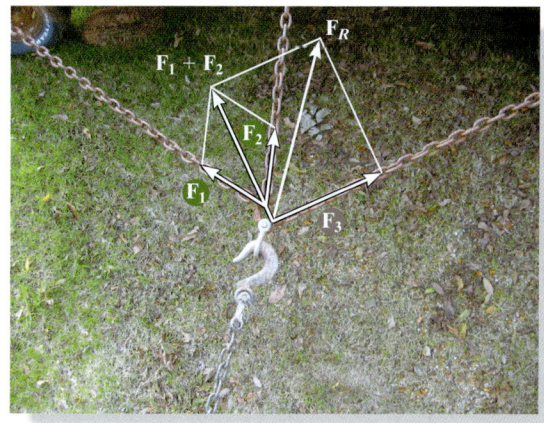

The resultant force $\mathbf{F}_R$ on the hook requires the addition of $\mathbf{F}_1 + \mathbf{F}_2$, then this resultant is added to $\mathbf{F}_3$.

2

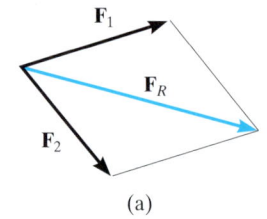

(a)

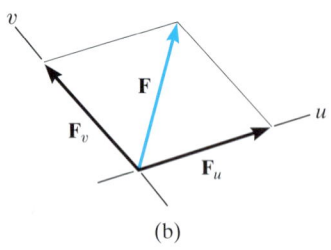

(b)

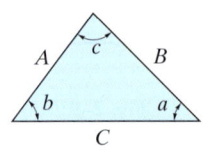

Cosine law:
$$C = \sqrt{A^2 + B^2 - 2AB \cos c}$$
Sine law:
$$\frac{A}{\sin a} = \frac{B}{\sin b} = \frac{C}{\sin c}$$

(c)

Fig. 2–10

Procedure for Analysis

Problems that involve the addition of two forces can be solved as follows:

Parallelogram Law.

- Two "component" forces $\mathbf{F}_1$ and $\mathbf{F}_2$ in Fig. 2–10a add according to the parallelogram law, yielding a *resultant* force $\mathbf{F}_R$ that forms the diagonal of the parallelogram.

- If a force $\mathbf{F}$ is to be resolved into *components* along two axes u and v, Fig. 2–10b, then start at the head of force $\mathbf{F}$ and construct lines parallel to the axes, thereby forming the parallelogram. The sides of the parallelogram represent the components, $\mathbf{F}_u$ and $\mathbf{F}_v$.

- Label all the known and unknown force magnitudes and the angles on the sketch and identify the two unknowns as the magnitude and direction of $\mathbf{F}_R$, or the magnitudes of its components.

Trigonometry.

- Redraw a half portion of the parallelogram to illustrate the triangular head-to-tail addition of the components.

- From this triangle, the magnitude of the resultant force can be determined using the law of cosines, and its direction is determined from the law of sines. The magnitudes of two force components are determined from the law of sines. The formulas are given in Fig. 2–10c.

Important Points

- A scalar is a positive or negative number.

- A vector is a quantity that has a magnitude, direction, and sense.

- Multiplication or division of a vector by a scalar will change the magnitude of the vector. The sense of the vector will change if the scalar is negative.

- As a special case, if the vectors are collinear, the resultant is formed by an algebraic or scalar addition.

EXAMPLE 2.1

The screw eye in Fig. 2–11a is subjected to two forces, $\mathbf{F}_1$ and $\mathbf{F}_2$. Determine the magnitude and direction of the resultant force.

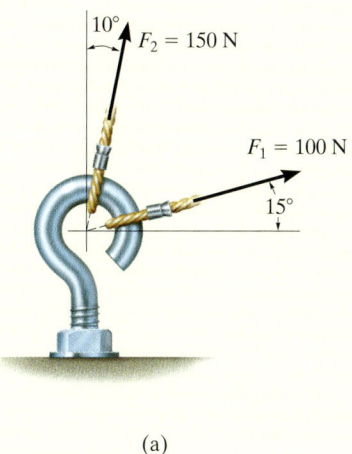

(a)

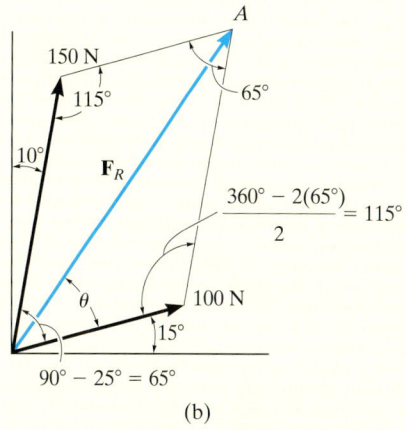

(b)

SOLUTION

Parallelogram Law. The parallelogram is formed by drawing a line from the head of $\mathbf{F}_1$ that is parallel to $\mathbf{F}_2$, and another line from the head of $\mathbf{F}_2$ that is parallel to $\mathbf{F}_1$. The resultant force $\mathbf{F}_R$ extends to where these lines intersect at point A, Fig. 2–11b. The two unknowns are the magnitude of $\mathbf{F}_R$ and the angle θ (theta).

Trigonometry. From the parallelogram, the vector triangle is constructed, Fig. 2–11c. Using the law of cosines

$$F_R = \sqrt{(100\ \text{N})^2 + (150\ \text{N})^2 - 2(100\ \text{N})(150\ \text{N})\cos 115°}$$
$$= \sqrt{10\ 000 + 22\ 500 - 30\ 000(-0.4226)} = 212.6\ \text{N}$$
$$= 213\ \text{N} \qquad\qquad Ans.$$

Applying the law of sines to determine θ,

$$\frac{150\ \text{N}}{\sin\theta} = \frac{212.6\ \text{N}}{\sin 115°} \qquad \sin\theta = \frac{150\ \text{N}}{212.6\ \text{N}}(\sin 115°)$$
$$\theta = 39.8°$$

Thus, the direction ϕ (phi) of $\mathbf{F}_R$, measured from the horizontal, is

$$\phi = 39.8° + 15.0° = 54.8° \qquad\qquad Ans.$$

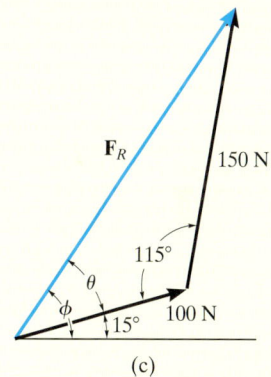

(c)

Fig. 2–11

NOTE: The results seem reasonable, since Fig. 2–11b shows $\mathbf{F}_R$ to have a magnitude larger than its components and a direction that is between them.

EXAMPLE 2.2

Resolve the horizontal 600-N force in Fig. 2–12a into components acting along the u and v axes and determine the magnitudes of these components.

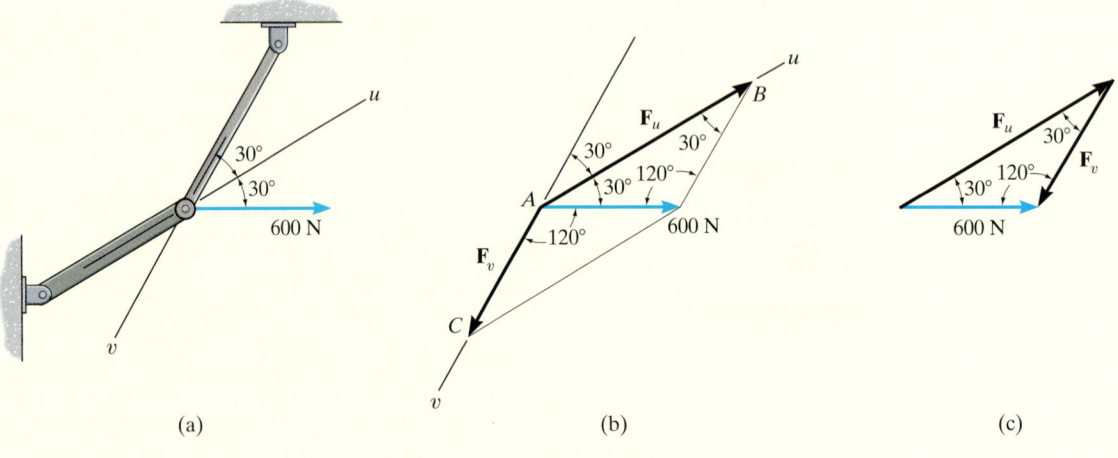

(a)   (b)   (c)

Fig. 2–12

SOLUTION

The parallelogram is constructed by extending a line from the *head* of the 600-N force parallel to the v axis until it intersects the u axis at point B, Fig. 2–12b. The arrow from A to B represents $\mathbf{F}_u$. Similarly, the line extended from the head of the 600-N force drawn parallel to the u axis intersects the v axis at point C, which gives $\mathbf{F}_v$.

The vector addition using the triangle rule is shown in Fig. 2–12c. The two unknowns are the magnitudes of $\mathbf{F}_u$ and $\mathbf{F}_v$. Applying the law of sines,

$$\frac{F_u}{\sin 120°} = \frac{600 \text{ N}}{\sin 30°}$$

$$F_u = 1039 \text{ N} \qquad \qquad \textit{Ans.}$$

$$\frac{F_v}{\sin 30°} = \frac{600 \text{ N}}{\sin 30°}$$

$$F_v = 600 \text{ N} \qquad \qquad \textit{Ans.}$$

NOTE: The result for F_u shows that sometimes a component can have a greater magnitude than the resultant.

EXAMPLE | 2.3

Determine the magnitude of the component force **F** in Fig. 2–13*a* and the magnitude of the resultant force **F**$_R$ if **F**$_R$ is directed along the positive *y* axis.

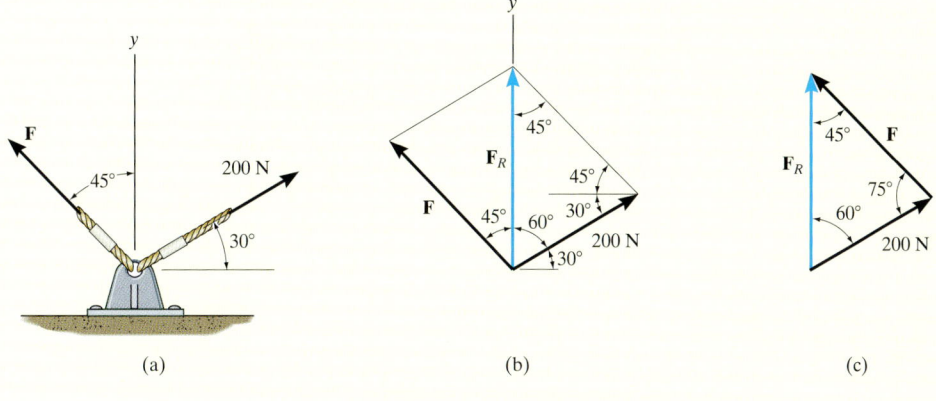

Fig. 2–13

SOLUTION

The parallelogram law of addition is shown in Fig. 2–13*b*, and the triangle rule is shown in Fig. 2–13*c*. The magnitudes of **F**$_R$ and **F** are the two unknowns. They can be determined by applying the law of sines.

$$\frac{F}{\sin 60°} = \frac{200 \text{ N}}{\sin 45°}$$

$$F = 245 \text{ N} \qquad\qquad Ans.$$

$$\frac{F_R}{\sin 75°} = \frac{200 \text{ N}}{\sin 45°}$$

$$F_R = 273 \text{ N} \qquad\qquad Ans.$$

EXAMPLE | 2.4

It is required that the resultant force acting on the eyebolt in Fig. 2–14a be directed along the positive x axis and that $\mathbf{F}_2$ have a *minimum* magnitude. Determine this magnitude, the angle θ, and the corresponding resultant force.

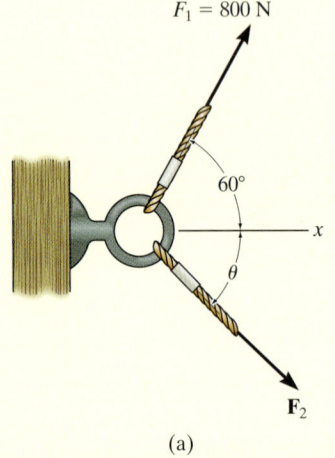

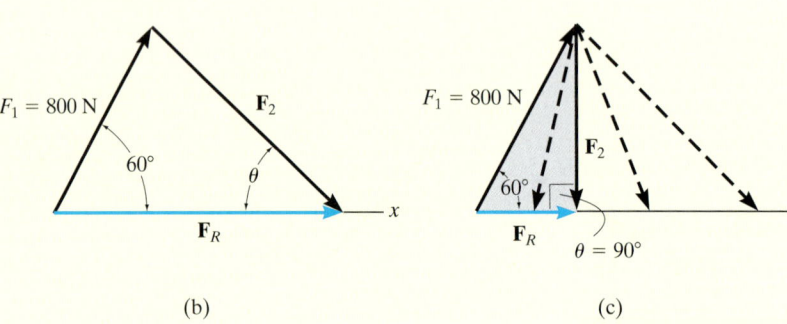

(a)

(b)

(c)

Fig. 2–14

SOLUTION

The triangle rule for $\mathbf{F}_R = \mathbf{F}_1 + \mathbf{F}_2$ is shown in Fig. 2–14b. Since the magnitudes (lengths) of $\mathbf{F}_R$ and $\mathbf{F}_2$ are not specified, then $\mathbf{F}_2$ can actually be any vector that has its head touching the line of action of $\mathbf{F}_R$, Fig. 2–14c. However, as shown, the magnitude of $\mathbf{F}_2$ is a *minimum* or the shortest length when its line of action is *perpendicular* to the line of action of $\mathbf{F}_R$, that is, when

$$\theta = 90° \qquad \qquad \textit{Ans.}$$

Since the vector addition now forms the shaded right triangle, the two unknown magnitudes can be obtained by trigonometry.

$$F_R = (800 \text{ N})\cos 60° = 400 \text{ N} \qquad \textit{Ans.}$$
$$F_2 = (800 \text{ N})\sin 60° = 693 \text{ N} \qquad \textit{Ans.}$$

It is strongly suggested that you test yourself on the solutions to these examples, by covering them over and then trying to draw the parallelogram law, and thinking about how the sine and cosine laws are used to determine the unknowns. Then before solving any of the problems, try and solve some of the Fundamental Problems given on the next page. The solutions and answers to these are given in the back of the book. Doing this throughout the book will help immensely in developing your problem-solving skills.

FUNDAMENTAL PROBLEMS*

F2–1. Determine the magnitude of the resultant force acting on the screw eye and its direction measured clockwise from the x axis.

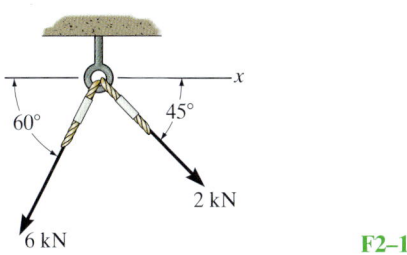

F2–1

F2–2. Two forces act on the hook. Determine the magnitude of the resultant force.

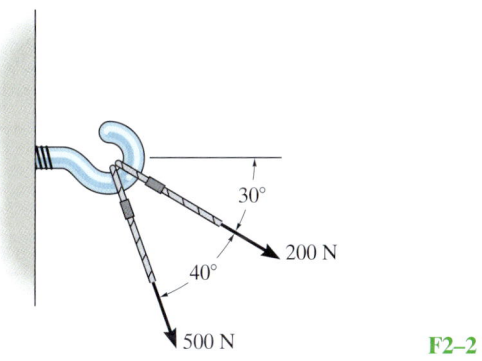

F2–2

F2–3. Determine the magnitude of the resultant force and its direction measured counterclockwise from the positive x axis.

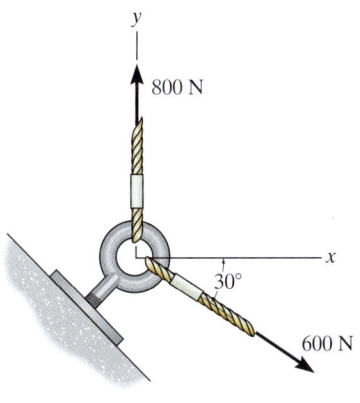

F2–3

F2–4. Resolve the 30-N force into components along the u and v axes, and determine the magnitude of each of these components.

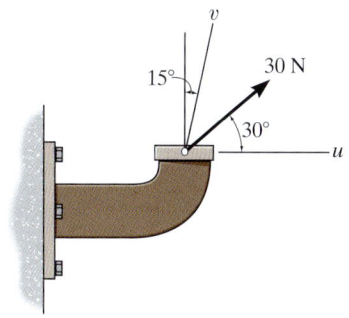

F2–4

F2–5. The force $F = 450$ N acts on the frame. Resolve this force into components acting along members AB and AC, and determine the magnitude of each component.

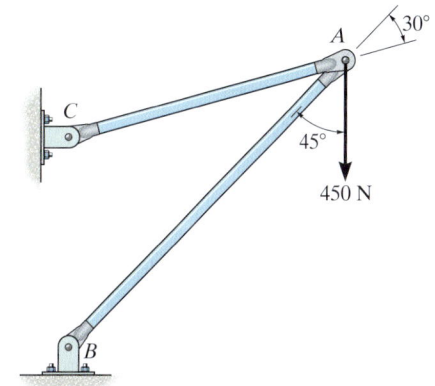

F2–5

F2–6. If force **F** is to have a component along the u axis of $F_u = 6$ kN, determine the magnitude of **F** and the magnitude of its component $\mathbf{F}_v$ along the v axis.

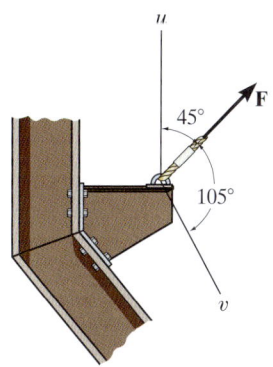

F2–6

* Partial solutions and answers to all Fundamental Problems are given in the back of the book.

PROBLEMS

2–1. Determine the magnitude of the resultant force $\mathbf{F}_R = \mathbf{F}_1 + \mathbf{F}_2$ and its direction, measured counterclockwise from the positive x axis.

*2–4. Determine the magnitude of the resultant force $\mathbf{F}_R = \mathbf{F}_1 + \mathbf{F}_2$ and its direction, measured clockwise from the positive u axis.

2–5. Resolve the force $\mathbf{F}_1$ into components acting along the u and v axes and determine the magnitudes of the components.

2–6. Resolve the force $\mathbf{F}_2$ into components acting along the u and v axes and determine the magnitudes of the components.

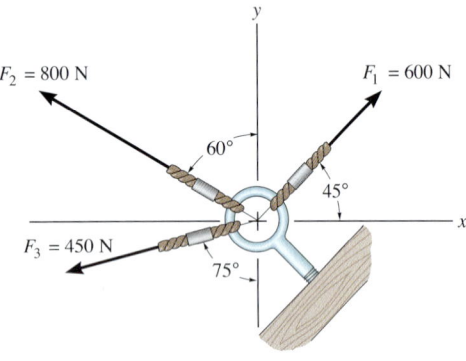

Prob. 2–1

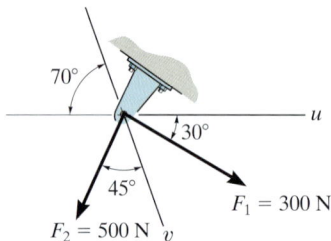

Probs. 2–4/5/6

2–2. If $\theta = 60°$ and $F = 450$ N, determine the magnitude of the resultant force and its direction, measured counterclockwise from the positive x axis.

2–3. If the magnitude of the resultant force is to be 500 N, directed along the positive y axis, determine the magnitude of force $\mathbf{F}$ and its direction θ.

2–7. If $F_B = 2$ kN and the resultant force acts along the positive u axis, determine the magnitude of the resultant force and the angle θ.

*2–8. If the resultant force is required to act along the positive u axis and have a magnitude of 5 kN, determine the required magnitude of $\mathbf{F}_B$ and its direction θ.

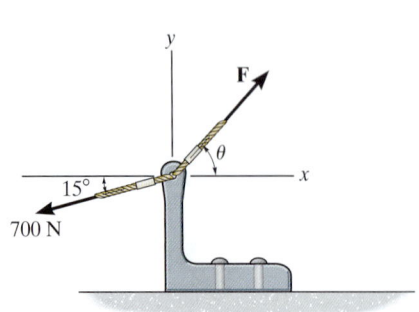

Probs. 2–2/3

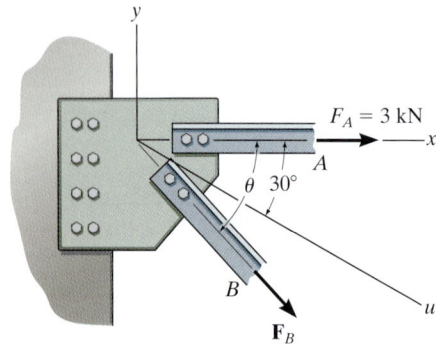

Probs. 2–7/8

2–9. Resolve $\mathbf{F}_1$ into components along the u and v axes and determine the magnitudes of these components.

2–10. Resolve $\mathbf{F}_2$ into components along the u and v axes and determine the magnitudes of these components.

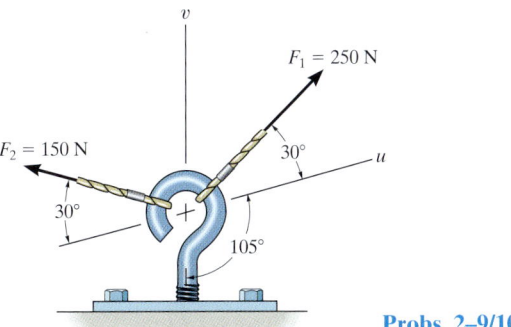

Probs. 2–9/10

2–11. The truck is to be towed using two ropes. Determine the magnitude of forces $\mathbf{F}_A$ and $\mathbf{F}_B$ acting on each rope in order to develop a resultant force of 950 N directed along the positive x axis. Set $\theta = 50°$.

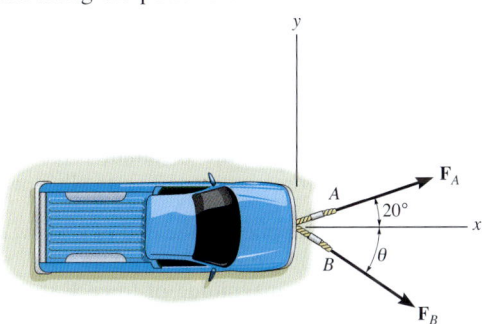

Prob. 2–11

***2–12.** The truck is to be towed using two ropes. If the resultant force is to be 950 N, directed along the positive x axis, determine the magnitudes of forces $\mathbf{F}_A$ and $\mathbf{F}_B$ acting on each rope and the angle of θ of $\mathbf{F}_B$ so that the magnitude of $\mathbf{F}_B$ is a *minimum*. $\mathbf{F}_A$ acts at 20° from the x axis as shown.

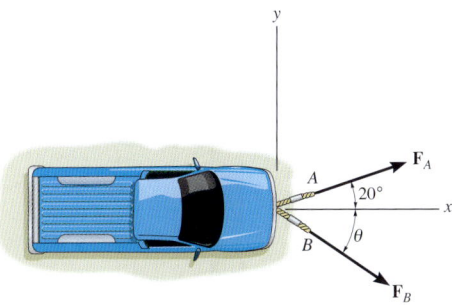

Prob. 2–12

2–13. The device is used for surgical replacement of the knee joint. If the force acting along the leg is 360 N, determine its components along the x and y' axes.

2–14. The device is used for surgical replacement of the knee joint. If the force acting along the leg is 360 N, determine its components along the x' and y axes.

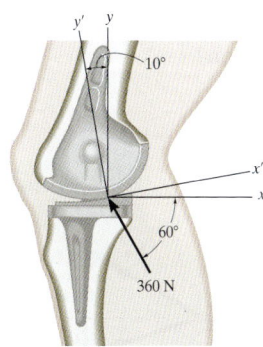

Probs. 2–13/14

2–15. The plate is subjected to the two forces at A and B as shown. If $\theta = 60°$, determine the magnitude of the resultant of these two forces and its direction measured clockwise from the horizontal.

***2–16.** Determine the angle θ for connecting member A to the plate so that the resultant force of $\mathbf{F}_A$ and $\mathbf{F}_B$ is directed horizontally to the right. Also, what is the magnitude of the resultant force?

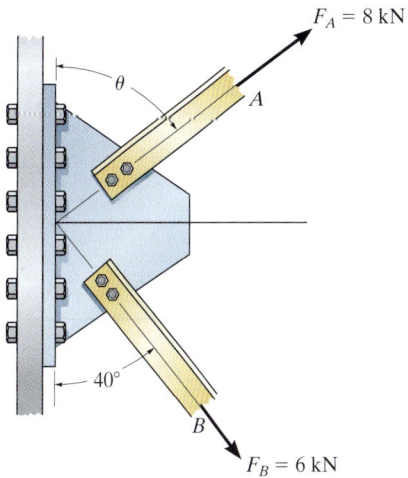

Probs. 2–15/16

2–17. The log is being towed by two tractors A and B. Determine the magnitude of the two towing forces $\mathbf{F}_A$ and $\mathbf{F}_B$ if it is required that the resultant force have a magnitude $F_R = 10$ kN and be directed along the x axis. Set $\theta = 15°$.

2–18. If the resultant $\mathbf{F}_R$ of the two forces acting on the log is to be directed along the positive x axis and have a magnitude of 10 kN, determine the angle θ of the cable, attached to B such that the force $\mathbf{F}_B$ in this cable is minimum. What is the magnitude of the force in each cable for this situation?

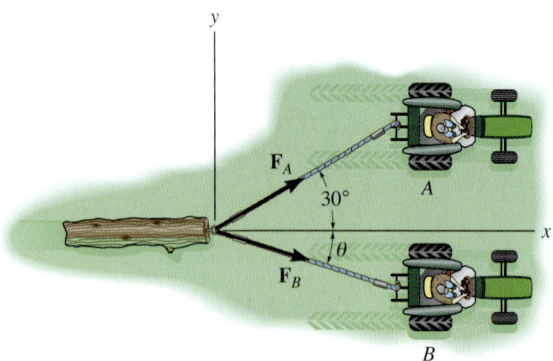

Probs. 2–17/18

2–19. Determine the magnitude and direction of the resultant $\mathbf{F}_R = \mathbf{F}_1 + \mathbf{F}_2 + \mathbf{F}_3$ of the three forces by first finding the resultant $\mathbf{F}' = \mathbf{F}_1 + \mathbf{F}_2$ and then forming $\mathbf{F}_R = \mathbf{F}' + \mathbf{F}_3$.

***2–20.** Determine the magnitude and direction of the resultant $\mathbf{F}_R = \mathbf{F}_1 + \mathbf{F}_2 + \mathbf{F}_3$ of the three forces by first finding the resultant $\mathbf{F}' = \mathbf{F}_2 + \mathbf{F}_3$ and then forming $\mathbf{F}_R = \mathbf{F}' + \mathbf{F}_1$.

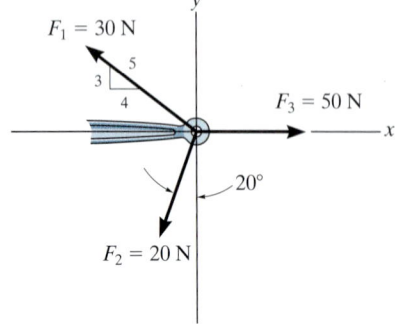

Probs. 2–19/20

2–21. Two forces act on the screw eye. If $F_1 = 400$ N and $F_2 = 600$ N, determine the angle θ ($0° \leq \theta \leq 180°$) between them, so that the resultant force has a magnitude of $F_R = 800$ N.

2–22. Two forces $\mathbf{F}_1$ and $\mathbf{F}_2$ act on the screw eye. If their lines of action are at an angle θ apart and the magnitude of each force is $F_1 = F_2 = F$, determine the magnitude of the resultant force $\mathbf{F}_R$ and the angle between $\mathbf{F}_R$ and $\mathbf{F}_1$.

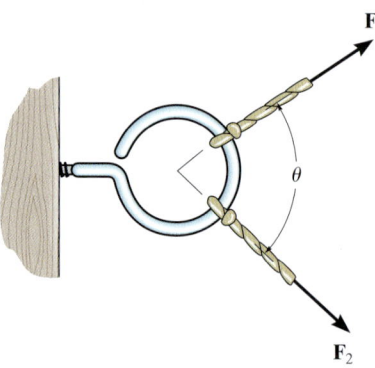

Probs. 2–21/22

2–23. Two forces act on the screw eye. If $F = 600$ N, determine the magnitude of the resultant force and the angle θ if the resultant force is directed vertically upward.

***2–24.** Two forces are applied at the end of a screw eye in order to remove the post. Determine the angle θ ($0° \leq \theta \leq 90°$) and the magnitude of force $\mathbf{F}$ so that the resultant force acting on the post is directed vertically upward and has a magnitude of 750 N.

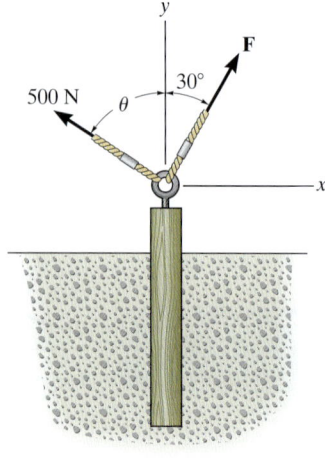

Probs. 2–23/24

2–25. Determine the magnitude and direction of the component force $\mathbf{F}_1$. Express the result in terms of the magnitudes of the component $\mathbf{F}_2$ and resultant $\mathbf{F}_R$ and the angle θ.

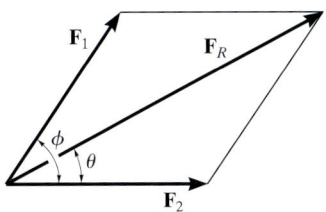

Prob. 2–25

2–26. The beam is to be hoisted using two chains. Determine the magnitudes of forces $\mathbf{F}_A$ and $\mathbf{F}_B$ acting on each chain in order to develop a resultant force of 600 N directed along the positive y axis. Set $\theta = 45°$.

2–27. The beam is to be hoisted using two chains. If the resultant force is to be 600 N directed along the positive y axis, determine the magnitudes of forces $\mathbf{F}_A$ and $\mathbf{F}_B$ acting on each chain and the angle θ of $\mathbf{F}_B$ so that the magnitude of $\mathbf{F}_B$ is a *minimum*. $\mathbf{F}_A$ acts at 30° from the y axis, as shown.

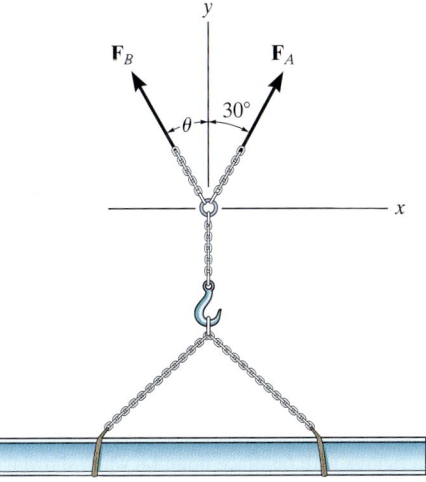

Probs. 2–26/27

*2–28.** If the resultant force of the two tugboats is 3 kN, directed along the positive x axis, determine the required magnitude of force $\mathbf{F}_B$ and its direction θ.

2–29. If $F_B = 3$ kN and $\theta = 45°$, determine the magnitude of the resultant force of the two tugboats and its direction measured clockwise from the positive x axis.

2–30. If the resultant force of the two tugboats is required to be directed towards the positive x axis, and $\mathbf{F}_B$ is to be a minimum, determine the magnitude of $\mathbf{F}_R$ and $\mathbf{F}_B$ and the angle θ.

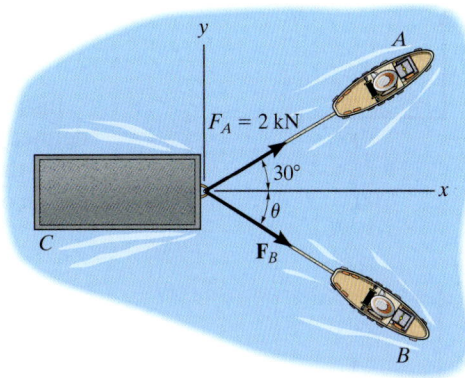

Probs. 2–28/29/30

2–31. If the tension in the cable is 400 N, determine the magnitude and direction of the resultant force acting on the pulley. This angle defines the same angle θ of line AB on the tailboard block.

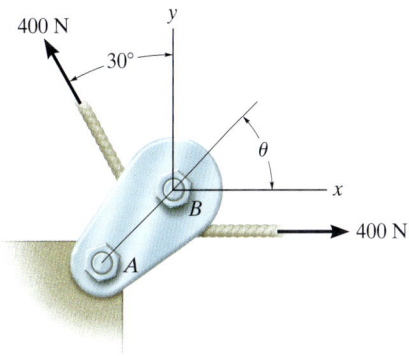

Prob. 2–31

2.4   Addition of a System of Coplanar Forces

When a force is resolved into two components along the x and y axes, the components are then called *rectangular components*. For analytical work we can represent these components in one of two ways, using either scalar notation or Cartesian vector notation.

Scalar Notation.   The rectangular components of force **F** shown in Fig. 2–15a are found using the parallelogram law, so that $\mathbf{F} = \mathbf{F}_x + \mathbf{F}_y$. Because these components form a right triangle, they can be determined from

$$F_x = F \cos \theta \qquad \text{and} \qquad F_y = F \sin \theta$$

Instead of using the angle θ, however, the direction of **F** can also be defined using a small "slope" triangle, as in the example shown in Fig. 2–15b. Since this triangle and the larger shaded triangle are similar, the proportional length of the sides gives

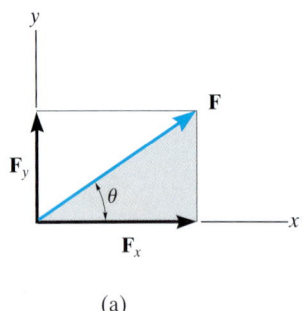

(a)

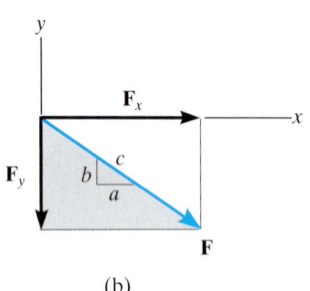

(b)

Fig. 2–15

$$\frac{F_x}{F} = \frac{a}{c}$$

or

$$F_x = F\left(\frac{a}{c}\right)$$

and

$$\frac{F_y}{F} = \frac{b}{c}$$

or

$$F_y = -F\left(\frac{b}{c}\right)$$

Here the y component is a *negative scalar* since $\mathbf{F}_y$ is directed along the negative y axis.

It is important to keep in mind that this positive and negative scalar notation is to be used only for computational purposes, not for graphical representations in figures. Throughout the book, the *head of a vector arrow* in *any figure* indicates the sense of the vector *graphically*; algebraic signs are not used for this purpose. Thus, the vectors in Figs. 2–15a and 2–15b are designated by using boldface (vector) notation.* Whenever italic symbols are written near vector arrows in figures, they indicate the *magnitude* of the vector, which is *always* a *positive* quantity.

*Negative signs are used only in figures with boldface notation when showing equal but opposite pairs of vectors, as in Fig. 2–2.

Cartesian Vector Notation.   It is also possible to represent the x and y components of a force in terms of Cartesian unit vectors $\mathbf{i}$ and $\mathbf{j}$. They are called unit vectors because they have a dimensionless magnitude of 1, and so they can be used to designate the *directions* of the x and y axes, respectively, Fig. 2–16.*

Since the *magnitude* of each component of $\mathbf{F}$ is *always a positive quantity*, which is represented by the (positive) scalars F_x and F_y, then we can express $\mathbf{F}$ as a *Cartesian vector*,

$$\mathbf{F} = F_x\mathbf{i} + F_y\mathbf{j}$$

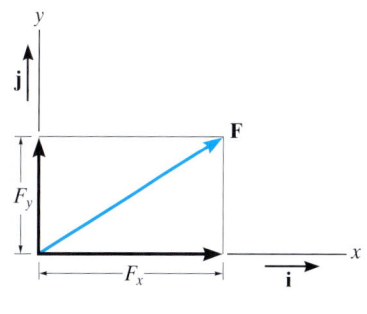

Fig. 2–16

Coplanar Force Resultants.   We can use either of the two methods just described to determine the resultant of several *coplanar forces*. To do this, each force is first resolved into its x and y components, and then the respective components are added using *scalar algebra* since they are collinear. The resultant force is then formed by adding the resultant components using the parallelogram law. For example, consider the three concurrent forces in Fig. 2–17a, which have x and y components shown in Fig. 2–17b. Using *Cartesian vector notation*, each force is first represented as a Cartesian vector, i.e.,

$$\mathbf{F}_1 = F_{1x}\mathbf{i} + F_{1y}\mathbf{j}$$
$$\mathbf{F}_2 = -F_{2x}\mathbf{i} + F_{2y}\mathbf{j}$$
$$\mathbf{F}_3 = F_{3x}\mathbf{i} - F_{3y}\mathbf{j}$$

The vector resultant is therefore

$$\begin{aligned}\mathbf{F}_R &= \mathbf{F}_1 + \mathbf{F}_2 + \mathbf{F}_3 \\ &= F_{1x}\mathbf{i} + F_{1y}\mathbf{j} - F_{2x}\mathbf{i} + F_{2y}\mathbf{j} + F_{3x}\mathbf{i} - F_{3y}\mathbf{j} \\ &= (F_{1x} - F_{2x} + F_{3x})\mathbf{i} + (F_{1y} + F_{2y} - F_{3y})\mathbf{j} \\ &= (F_{Rx})\mathbf{i} + (F_{Ry})\mathbf{j}\end{aligned}$$

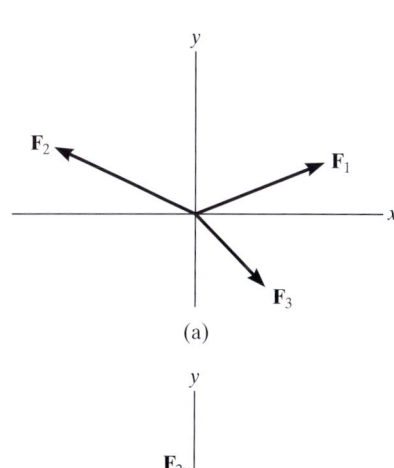

(a)

If *scalar notation* is used, then from Fig. 2–17b, we have

$$(\xrightarrow{+}) \qquad (F_R)_x = F_{1x} - F_{2x} + F_{3x}$$
$$(+\uparrow) \qquad (F_R)_y = F_{1y} + F_{2y} - F_{3y}$$

These are the *same* results as the $\mathbf{i}$ and $\mathbf{j}$ components of $\mathbf{F}_R$ determined above.

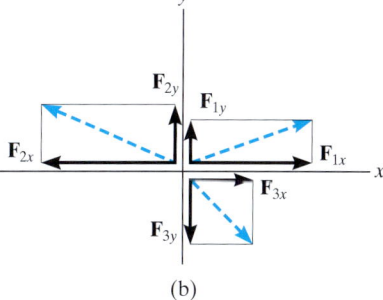

(b)

Fig. 2–17

---

* For handwritten work, unit vectors are usually indicated using a circumflex, e.g., $\hat{i}$ and $\hat{j}$. Also, realize that F_x and F_y in Fig. 2–16 represent the *magnitudes* of the components, which are *always positive scalars*. The directions are defined by $\mathbf{i}$ and $\mathbf{j}$. If instead we used scalar notation, then F_x and F_y could be positive or negative scalars, since they would account for *both* the magnitude and direction of the components.

We can represent the components of the resultant force of any number of coplanar forces symbolically by the algebraic sum of the x and y components of all the forces, i.e.,

$$(F_R)_x = \Sigma F_x$$
$$(F_R)_y = \Sigma F_y$$

(2–1)

Once these components are determined, they may be sketched along the x and y axes with their proper sense of direction, and the resultant force can be determined from vector addition, as shown in Fig. 2–17c. From this sketch, the magnitude of $\mathbf{F}_R$ is then found from the Pythagorean theorem; that is,

$$F_R = \sqrt{(F_R)_x^2 + (F_R)_y^2}$$

Also, the angle θ, which specifies the direction of the resultant force, is determined from trigonometry:

$$\theta = \tan^{-1} \left| \frac{(F_R)_y}{(F_R)_x} \right|$$

The above concepts are illustrated numerically in the examples which follow.

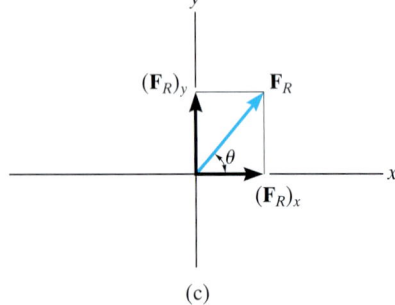

(c)

Fig. 2–17 (cont.)

The resultant force of the three cable forces acting on the post can be determined by adding algebraically the separate x and y components of each cable force. This resultant $\mathbf{F}_R$ produces the *same pulling effect* on the post as all three cables.

Important Points

- The resultant of several coplanar forces can easily be determined if an x, y coordinate system is established and the forces are resolved along the axes.

- The direction of each force is specified by the angle its line of action makes with one of the axes, or by a slope triangle.

- The orientation of the x and y axes is arbitrary, and their positive direction can be specified by the Cartesian unit vectors $\mathbf{i}$ and $\mathbf{j}$.

- The x and y components of the *resultant force* are simply the algebraic addition of the components of all the coplanar forces.

- The magnitude of the resultant force is determined from the Pythagorean theorem, and when the resultant components are sketched on the x and y axes, Fig. 2–17c, the direction θ can be determined from trigonometry.

EXAMPLE 2.5

Determine the x and y components of $\mathbf{F}_1$ and $\mathbf{F}_2$ acting on the boom shown in Fig. 2–18a. Express each force as a Cartesian vector.

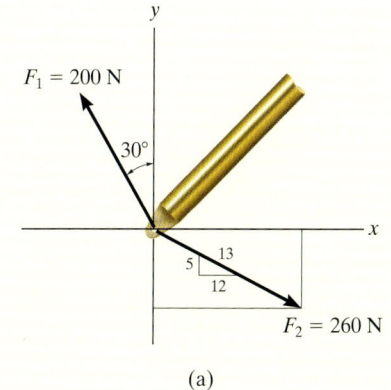

(a)

SOLUTION

Scalar Notation.   By the parallelogram law, $\mathbf{F}_1$ is resolved into x and y components, Fig. 2–18b. Since $\mathbf{F}_{1x}$ acts in the $-x$ direction, and $\mathbf{F}_{1y}$ acts in the $+y$ direction, we have

$$F_{1x} = -200 \sin 30° \text{ N} = -100 \text{ N} = 100 \text{ N} \leftarrow \qquad Ans.$$

$$F_{1y} = 200 \cos 30° \text{ N} = 173 \text{ N} = 173 \text{ N} \uparrow \qquad Ans.$$

The force $\mathbf{F}_2$ is resolved into its x and y components, as shown in Fig. 2–18c. Here the *slope* of the line of action for the force is indicated. From this "slope triangle" we could obtain the angle θ, e.g., $\theta = \tan^{-1}(\frac{5}{12})$, and then proceed to determine the magnitudes of the components in the same manner as for $\mathbf{F}_1$. The easier method, however, consists of using proportional parts of similar triangles, i.e.,

$$\frac{F_{2x}}{260 \text{ N}} = \frac{12}{13} \qquad F_{2x} = 260 \text{ N}\left(\frac{12}{13}\right) = 240 \text{ N}$$

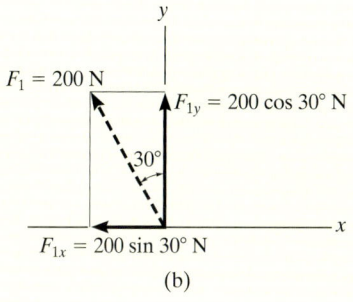

(b)

Similarly,

$$F_{2y} = 260 \text{ N}\left(\frac{5}{13}\right) = 100 \text{ N}$$

Notice how the magnitude of the *horizontal component*, $\mathbf{F}_{2x}$, was obtained by multiplying the force magnitude by the ratio of the *horizontal leg* of the slope triangle divided by the hypotenuse; whereas the magnitude of the *vertical component*, F_{2y}, was obtained by multiplying the force magnitude by the ratio of the *vertical leg* divided by the hypotenuse. Hence, using scalar notation to represent these components, we have

$$F_{2x} = 240 \text{ N} = 240 \text{ N} \rightarrow \qquad Ans.$$

$$F_{2y} = -100 \text{ N} = 100 \text{ N} \downarrow \qquad Ans.$$

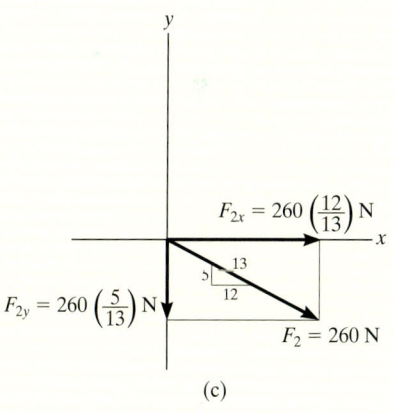

(c)

Fig. 2–18

Cartesian Vector Notation.   Having determined the magnitudes and directions of the components of each force, we can express each force as a Cartesian vector.

$$\mathbf{F}_1 = \{-100\mathbf{i} + 173\mathbf{j}\} \text{ N} \qquad Ans.$$

$$\mathbf{F}_2 = \{240\mathbf{i} - 100\mathbf{j}\} \text{ N} \qquad Ans.$$

2

EXAMPLE | 2.6

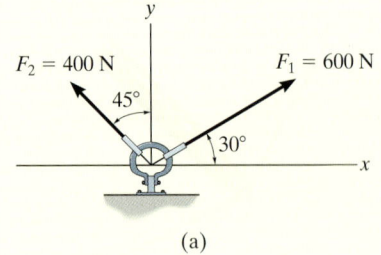

(a)

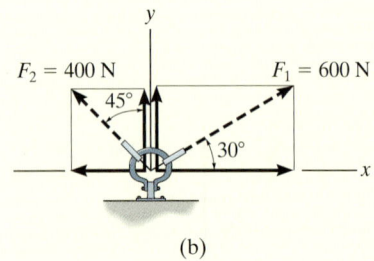

(b)

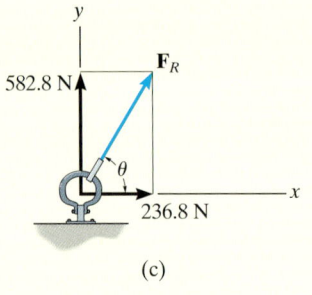

(c)

Fig. 2–19

The link in Fig. 2–19a is subjected to two forces $\mathbf{F}_1$ and $\mathbf{F}_2$. Determine the magnitude and direction of the resultant force.

SOLUTION I

Scalar Notation. First we resolve each force into its x and y components, Fig. 2–19b, then we sum these components algebraically.

$$\xrightarrow{+} (F_R)_x = \Sigma F_x; \qquad (F_R)_x = 600 \cos 30° \text{ N} - 400 \sin 45° \text{ N}$$
$$= 236.8 \text{ N} \rightarrow$$

$$+\uparrow (F_R)_y = \Sigma F_y; \qquad (F_R)_y = 600 \sin 30° \text{ N} + 400 \cos 45° \text{ N}$$
$$= 582.8 \text{ N}\uparrow$$

The resultant force, shown in Fig. 2–18c, has a *magnitude* of

$$F_R = \sqrt{(236.8 \text{ N})^2 + (582.8 \text{ N})^2}$$
$$= 629 \text{ N} \qquad\qquad Ans.$$

From the vector addition,

$$\theta = \tan^{-1}\left(\frac{582.8 \text{ N}}{236.8 \text{ N}}\right) = 67.9° \qquad\qquad Ans.$$

SOLUTION II

Cartesian Vector Notation. From Fig. 2–19b, each force is first expressed as a Cartesian vector.

$$\mathbf{F}_1 = \{600 \cos 30°\mathbf{i} + 600 \sin 30°\mathbf{j}\} \text{ N}$$
$$\mathbf{F}_2 = \{-400 \sin 45°\mathbf{i} + 400 \cos 45°\mathbf{j}\} \text{ N}$$

Then,

$$\mathbf{F}_R = \mathbf{F}_1 + \mathbf{F}_2 = (600 \cos 30° \text{ N} - 400 \sin 45° \text{ N})\mathbf{i}$$
$$+ (600 \sin 30° \text{ N} + 400 \cos 45° \text{ N})\mathbf{j}$$
$$= \{236.8\mathbf{i} + 582.8\mathbf{j}\} \text{ N}$$

The magnitude and direction of $\mathbf{F}_R$ are determined in the same manner as before.

NOTE: Comparing the two methods of solution, notice that the use of scalar notation is more efficient since the components can be found *directly*, without first having to express each force as a Cartesian vector before adding the components. Later, however, we will show that Cartesian vector analysis is very beneficial for solving three-dimensional problems.

EXAMPLE 2.7

The end of the boom O in Fig. 2–20a is subjected to three concurrent and coplanar forces. Determine the magnitude and direction of the resultant force.

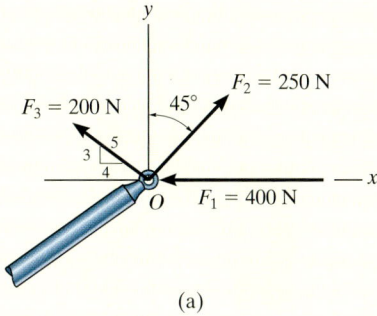

(a)

SOLUTION

Each force is resolved into its x and y components, Fig. 2–20b. Summing the x components, we have

$$\xrightarrow{+}(F_R)_x = \Sigma F_x; \qquad (F_R)_x = -400 \text{ N} + 250 \sin 45° \text{ N} - 200\left(\tfrac{4}{5}\right) \text{ N}$$

$$= -383.2 \text{ N} = 383.2 \text{ N} \leftarrow$$

The negative sign indicates that F_{Rx} acts to the left, i.e., in the negative x direction, as noted by the small arrow. Obviously, this occurs because F_1 and F_3 in Fig. 2–20b contribute a greater pull to the left than F_2 which pulls to the right. Summing the y components yields

$$+\uparrow(F_R)_y = \Sigma F_y; \qquad (F_R)_y = 250 \cos 45° \text{ N} + 200\left(\tfrac{3}{5}\right) \text{ N}$$

$$= 296.8 \text{ N} \uparrow$$

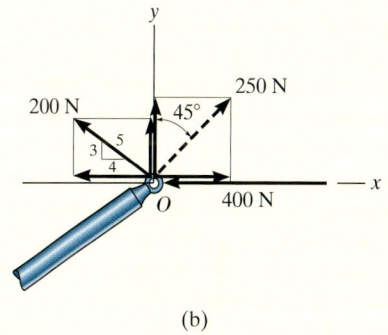

(b)

The resultant force, shown in Fig. 2–20c, has a *magnitude* of

$$F_R = \sqrt{(-383.2 \text{ N})^2 + (296.8 \text{ N})^2}$$

$$= 485 \text{ N} \qquad\qquad Ans.$$

From the vector addition in Fig. 2–20c, the direction angle θ is

$$\theta = \tan^{-1}\left(\frac{296.8}{383.2}\right) = 37.8° \qquad\qquad Ans.$$

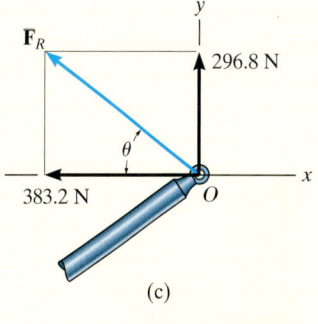

(c)

Fig. 2–20

NOTE: Application of this method is more convenient, compared to using two applications of the parallelogram law, first to add $\mathbf{F}_1$ and $\mathbf{F}_2$ then adding $\mathbf{F}_3$ to this resultant.

FUNDAMENTAL PROBLEMS

F2–7. Resolve each force acting on the post into its x and y components.

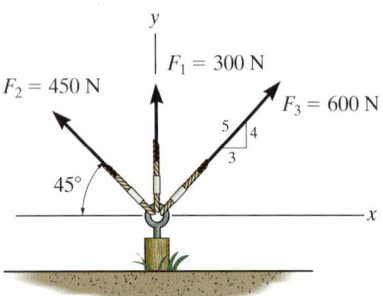

F2–7

F2–8. Determine the magnitude and direction of the resultant force.

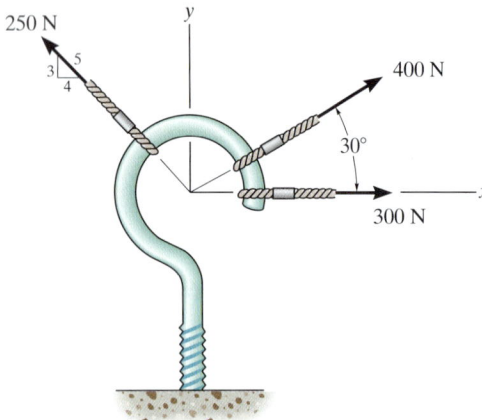

F2–8

F2–9. Determine the magnitude of the resultant force acting on the corbel and its direction θ measured counterclockwise from the x axis.

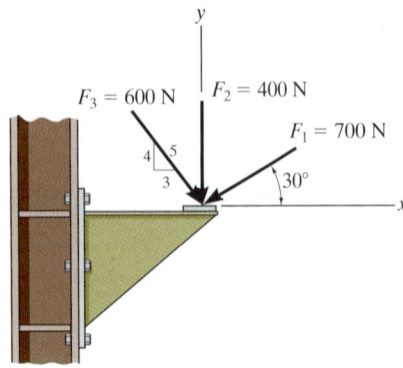

F2–9

F2–10. If the resultant force acting on the bracket is to be 750 N directed along the positive x axis, determine the magnitude of $\mathbf{F}$ and its direction θ.

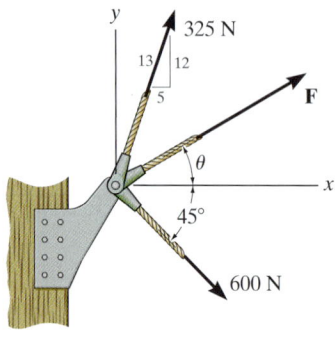

F2–10

F2–11. If the magnitude of the resultant force acting on the bracket is to be 80 N directed along the u axis, determine the magnitude of $\mathbf{F}$ and its direction θ.

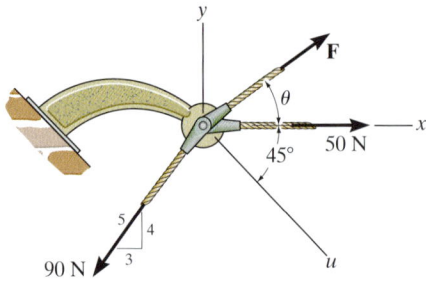

F2–11

F2–12. Determine the magnitude of the resultant force and its direction θ measured counterclockwise from the positive x axis.

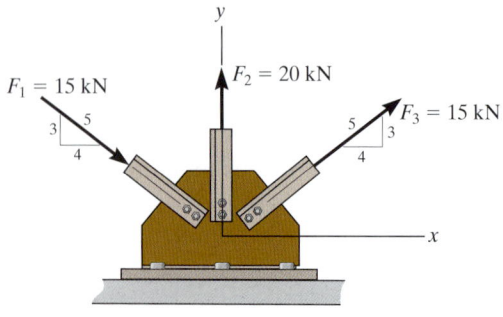

F2–12

PROBLEMS

*2–32. Determine the magnitude of the resultant force and its direction, measured clockwise from the positive x axis.

2–34. Resolve $\mathbf{F}_1$ and $\mathbf{F}_2$ into their x and y components.

2–35. Determine the magnitude of the resultant force and its direction measured counterclockwise from the positive x axis.

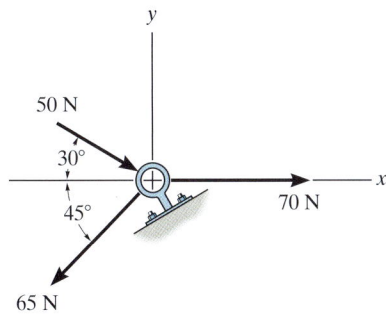

Prob. 2–32

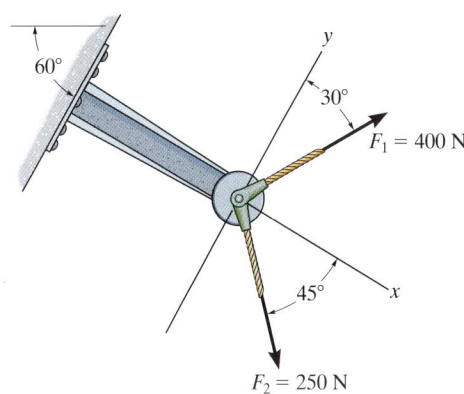

Probs. 2–34/35

2–33. Determine the magnitude of the resultant force and its direction, measured counterclockwise from the positive x axis.

*2–36. Resolve each force acting on the gusset plate into its x and y components, and express each force as a Cartesian vector.

2–37. Determine the magnitude of the resultant force acting on the plate and its direction, measured counter-clockwise from the positive x axis.

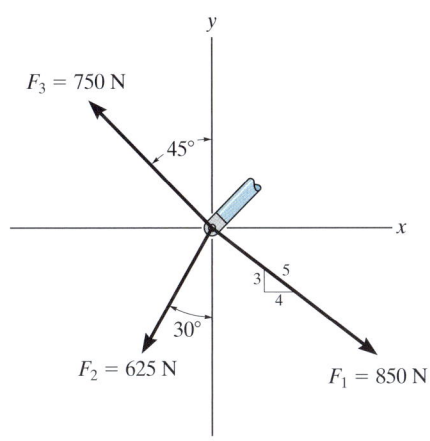

Prob. 2–33

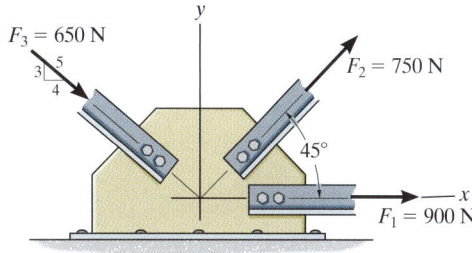

Probs. 2–36/37

2–38. Determine the magnitude of force **F** so that the resultant **F**$_R$ of the three forces is as small as possible.

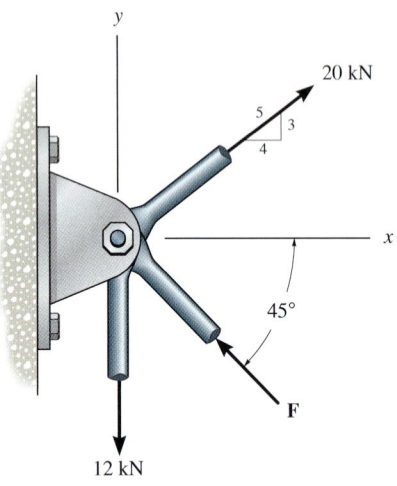

Prob. 2–38

2–41. Determine the magnitude of the resultant force and its direction, measured counterclockwise from the positive x axis.

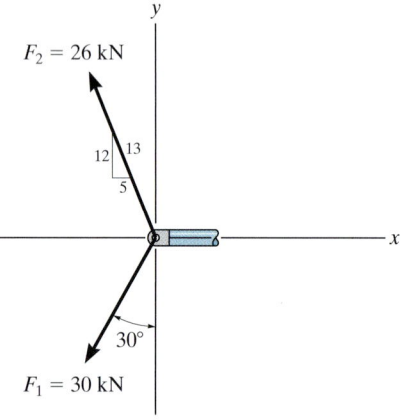

Prob. 2–41

2–39. Resolve each force acting on the support into its x and y components, and express each force as a Cartesian vector.

*2–40. Determine the magnitude of the resultant force and its direction θ, measured counterclockwise from the positive x axis.

2–42. Determine the magnitude and orientation θ of **F**$_B$ so that the resultant force is directed along the positive y axis and has a magnitude of 1500 N.

2–43. Determine the magnitude and orientation, measured counterclockwise from the positive y axis, of the resultant force acting on the bracket, if $F_B = 600$ N and $\theta = 20°$.

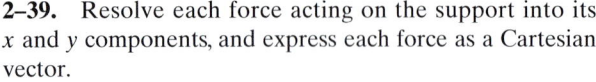

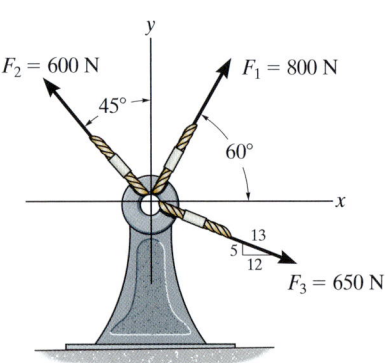

Probs. 2–39/40

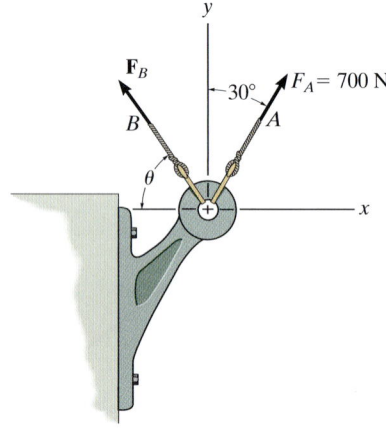

Probs. 2–42/43

*2–44. The magnitude of the resultant force acting on the bracket is to be 400 N. Determine the magnitude of $\mathbf{F}_1$ if $\phi = 30°$.

2–45. If the resultant force acting on the bracket is to be directed along the positive u axis, and the magnitude of $\mathbf{F}_1$ is required to be *minimum*, determine the magnitudes of the resultant force and $\mathbf{F}_1$.

2–46. If the magnitude of the resultant force acting on the bracket is 600 N, directed along the positive u axis, determine the magnitude of $\mathbf{F}$ and its direction ϕ.

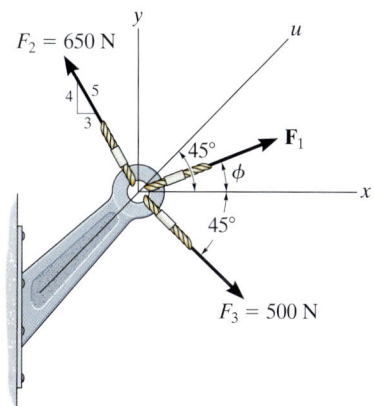

Probs. 2–44/45/46

2–47. Determine the magnitude and direction θ of the resultant force $\mathbf{F}_R$. Express the result in terms of the magnitudes of the components $\mathbf{F}_1$ and $\mathbf{F}_2$ and the angle ϕ.

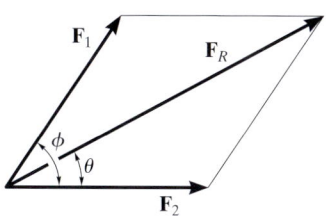

Prob. 2–47

*2–48. If $F_1 = 600$ N and $\phi = 30°$, determine the magnitude of the resultant force acting on the eyebolt and its direction, measured clockwise from the positive x axis.

2–49. If the magnitude of the resultant force acting on the eyebolt is 600 N and its direction measured clockwise from the positive x axis is $\theta = 30°$, determine the magnitude of $\mathbf{F}_1$ and the angle ϕ.

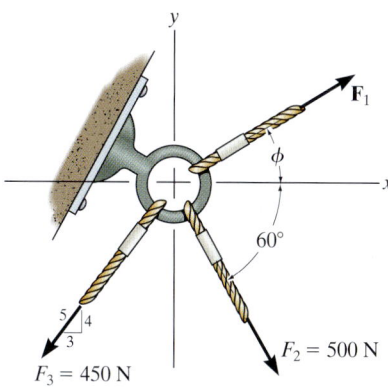

Probs. 2–48/49

2–50. Determine the magnitude of $\mathbf{F}_1$ and its direction θ so that the resultant force is directed vertically upward and has a magnitude of 800 N.

2–51. Determine the magnitude and direction measured counterclockwise from the positive x axis of the resultant force of the three forces acting on the ring A. Take $F_1 = 500$ N and $\theta = 20°$.

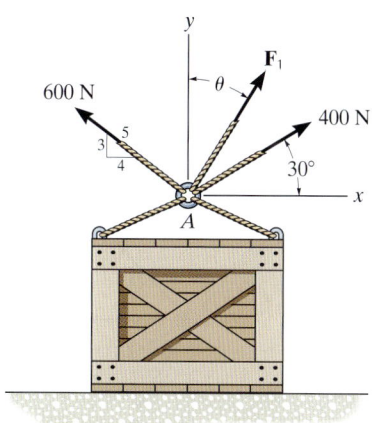

Probs. 2–50/51

*2–52.   Determine the magnitude of force **F** so that the resultant $\mathbf{F}_R$ of the three forces is as small as possible. What is the minimum magnitude of $\mathbf{F}_R$?

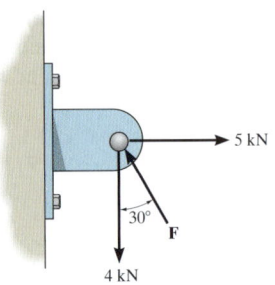

Prob. 2–52

2–53.   Determine the magnitude of force **F** so that the resultant force of the three forces is as small as possible. What is the magnitude of the resultant force?

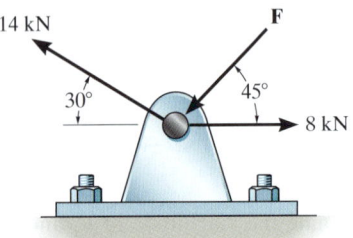

Prob. 2–53

2–54.   Three forces act on the bracket. Determine the magnitude and direction θ of $\mathbf{F}_1$ so that the resultant force is directed along the positive x' axis and has a magnitude of 1 kN.

2–55.   If $F_1 = 300$ N and $\theta = 20°$, determine the magnitude and direction, measured counterclockwise from the x' axis, of the resultant force of the three forces acting on the bracket.

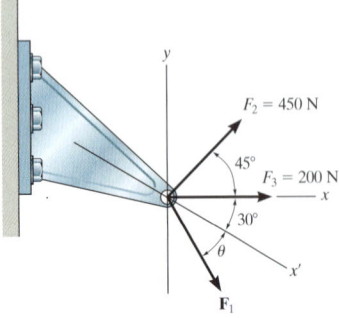

Probs. 2–54/55

*2–56.   Determine the magnitude and direction θ of $\mathbf{F}_A$ so that the resultant force is directed along the positive x axis and has a magnitude of 1250 N.

2–57.   Determine the magnitude and direction, measured counterclockwise from the positive x axis, of the resultant force acting on the ring at O, if $F_A = 750$ N and $\theta = 45°$.

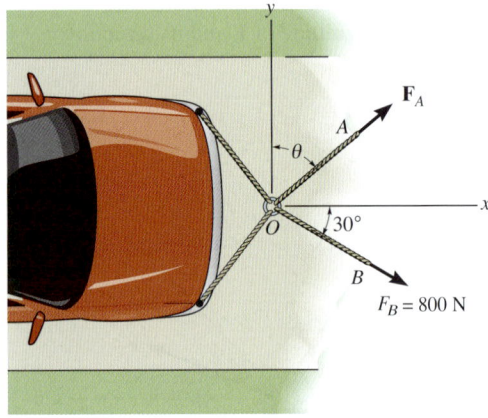

Probs. 2–56/57

2–58.   If the magnitude of the resultant force acting on the bracket is to be 450 N directed along the positive u axis, determine the magnitude of $\mathbf{F}_1$ and its direction ϕ.

2–59.   If the resultant force acting on the bracket is required to be a minimum, determine the magnitude of $\mathbf{F}_1$ and the resultant force. Set $\phi = 30°$.

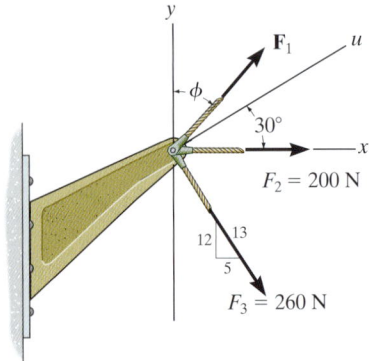

Probs. 2–58/59

2.5   Cartesian Vectors

The operations of vector algebra, when applied to solving problems in *three dimensions*, are greatly simplified if the vectors are first represented in Cartesian vector form. In this section we will present a general method for doing this; then in the next section we will use this method for finding the resultant force of a system of concurrent forces.

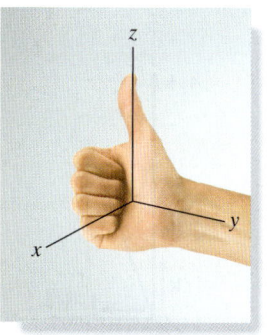

Fig. 2–21

Right-Handed Coordinate System. We will use a right-handed coordinate system to develop the theory of vector algebra that follows. A rectangular coordinate system is said to be *right-handed* if the thumb of the right hand points in the direction of the positive z axis when the right-hand fingers are curled about this axis and directed from the positive x towards the positive y axis, Fig. 2–21.

Rectangular Components of a Vector. A vector $\mathbf{A}$ may have one, two, or three rectangular components along the x, y, z coordinate axes, depending on how the vector is oriented relative to the axes. In general, though, when $\mathbf{A}$ is directed within an octant of the x, y, z frame, Fig. 2–22, then by two successive applications of the parallelogram law, we may resolve the vector into components as $\mathbf{A} = \mathbf{A}' + \mathbf{A}_z$ and then $\mathbf{A}' = \mathbf{A}_x + \mathbf{A}_y$. Combining these equations, to eliminate $\mathbf{A}'$, $\mathbf{A}$ is represented by the vector sum of its *three* rectangular components,

$$\mathbf{A} = \mathbf{A}_x + \mathbf{A}_y + \mathbf{A}_z \qquad (2\text{–}2)$$

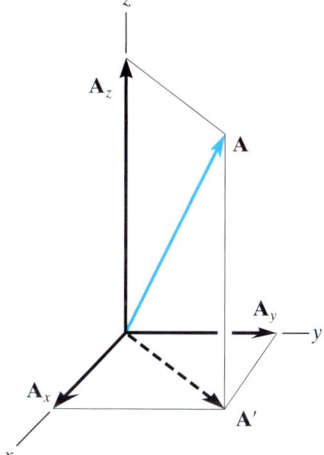

Fig. 2–22

Cartesian Unit Vectors. In three dimensions, the set of Cartesian unit vectors, $\mathbf{i}, \mathbf{j}, \mathbf{k}$, is used to designate the directions of the x, y, z axes, respectively. As stated in Sec. 2.4, the *sense* (or arrowhead) of these vectors will be represented analytically by a plus or minus sign, depending on whether they are directed along the positive or negative $x, y,$ or z axes. The positive Cartesian unit vectors are shown in Fig. 2–23.

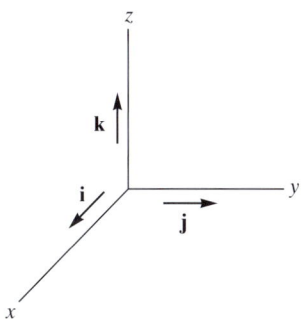

Fig. 2–23

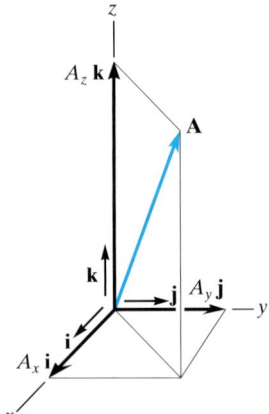

Fig. 2–24

Cartesian Vector Representation.

Since the three components of **A** in Eq. 2–2 act in the positive **i**, **j**, and **k** directions, Fig. 2–24, we can write **A** in Cartesian vector form as

$$\mathbf{A} = A_x\mathbf{i} + A_y\mathbf{j} + A_z\mathbf{k} \tag{2–3}$$

There is a distinct advantage to writing vectors in this manner. Separating the *magnitude* and *direction* of each *component vector* will simplify the operations of vector algebra, particularly in three dimensions.

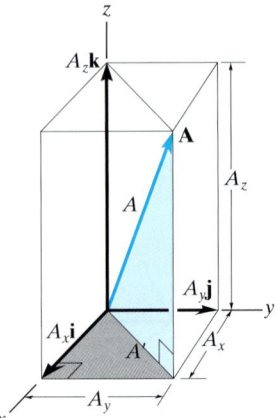

Fig. 2–25

Magnitude of a Cartesian Vector.

It is always possible to obtain the magnitude of **A** provided it is expressed in Cartesian vector form. As shown in Fig. 2–25, from the blue right triangle, $A = \sqrt{A'^2 + A_z^2}$, and from the gray right triangle, $A' = \sqrt{A_x^2 + A_y^2}$. Combining these equations to eliminate A' yields

$$A = \sqrt{A_x^2 + A_y^2 + A_z^2} \tag{2–4}$$

Hence, the magnitude of **A** *is equal to the positive square root of the sum of the squares of its components.*

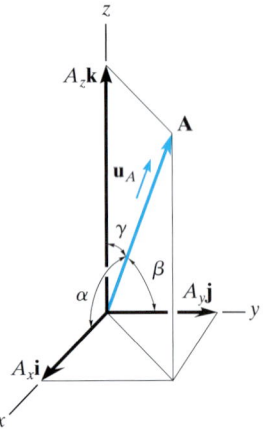

Fig. 2–26

Direction of a Cartesian Vector.

We will define the *direction* of **A** by the *coordinate direction angles* α (alpha), β (beta), and γ (gamma), measured between the *tail* of **A** and the *positive x, y, z* axes provided they are located at the tail of **A**, Fig. 2–26. Note that regardless of where **A** is directed, each of these angles will be between 0° and 180°.

To determine α, β, and γ, consider the projection of **A** onto the *x, y, z* axes, Fig. 2–27. Referring to the blue colored right triangles shown in each figure, we have

$$\cos \alpha = \frac{A_x}{A} \quad \cos \beta = \frac{A_y}{A} \quad \cos \gamma = \frac{A_z}{A} \tag{2–5}$$

These numbers are known as the *direction cosines* of **A**. Once they have been obtained, the coordinate direction angles α, β, γ can then be determined from the inverse cosines.

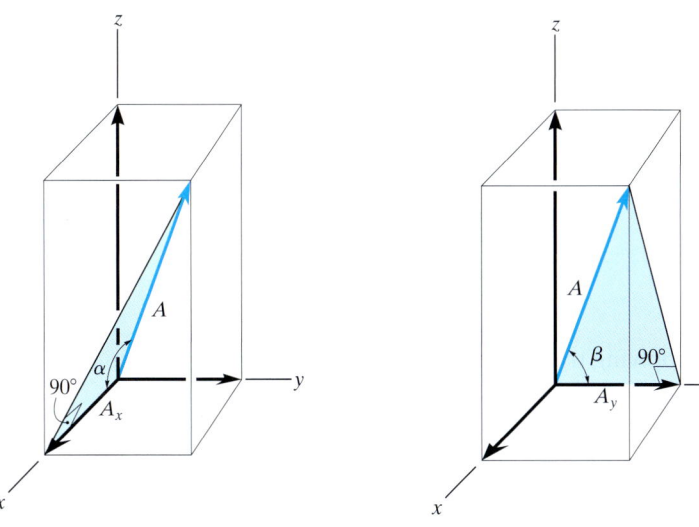
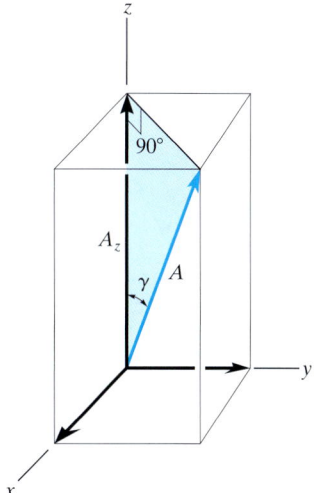

Fig. 2–27

An easy way of obtaining these direction cosines is to form a unit vector $\mathbf{u}_A$ in the direction of A, Fig. 2–26. If $\mathbf{A}$ is expressed in Cartesian vector form, $\mathbf{A} = A_x\mathbf{i} + A_y\mathbf{j} + A_z\mathbf{k}$, then $\mathbf{u}_A$ will have a magnitude of one and be dimensionless provided $\mathbf{A}$ is divided by its magnitude, i.e.,

$$\mathbf{u}_A = \frac{\mathbf{A}}{A} = \frac{A_x}{A}\mathbf{i} + \frac{A_y}{A}\mathbf{j} + \frac{A_z}{A}\mathbf{k} \qquad (2\text{--}6)$$

where $A = \sqrt{A_x^2 + A_y^2 + A_z^2}$. By comparison with Eqs. 2–5, it is seen that *the* $\mathbf{i}, \mathbf{j}, \mathbf{k}$ *components* of $\mathbf{u}_A$ *represent the direction cosines of* $\mathbf{A}$, i.e.,

$$\mathbf{u}_A = \cos\alpha\,\mathbf{i} + \cos\beta\,\mathbf{j} + \cos\gamma\,\mathbf{k} \qquad (2\text{--}7)$$

Since the magnitude of a vector is equal to the positive square root of the sum of the squares of the magnitudes of its components, and $\mathbf{u}_A$ has a magnitude of one, then from the above equation an important relation among the direction cosines can be formulated as

$$\cos^2\alpha + \cos^2\beta + \cos^2\gamma = 1 \qquad (2\text{--}8)$$

Here we can see that if only *two* of the coordinate angles are known, the third angle can be found using this equation.

Finally, if the magnitude and coordinate direction angles of $\mathbf{A}$ are known, then $\mathbf{A}$ may be expressed in Cartesian vector form as

$$\begin{aligned}
\mathbf{A} &= A\mathbf{u}_A \\
&= A\cos\alpha\,\mathbf{i} + A\cos\beta\,\mathbf{j} + A\cos\gamma\,\mathbf{k} \\
&= A_x\mathbf{i} + A_y\mathbf{j} + A_z\mathbf{k}
\end{aligned} \qquad (2\text{--}9)$$

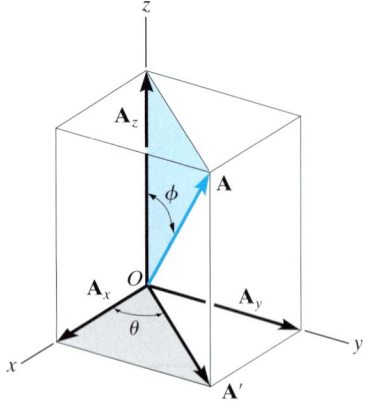

Fig. 2–28

Sometimes, the direction of **A** can be specified using two angles, θ and ϕ (phi), such as shown in Fig. 2–28. The components of **A** can then be determined by applying trigonometry first to the blue right triangle, which yields

$$A_z = A \cos \phi$$

and

$$A' = A \sin \phi$$

Now applying trigonometry to the gray shaded right triangle,

$$A_x = A' \cos \theta = A \sin \phi \cos \theta$$

$$A_y = A' \sin \theta = A \sin \phi \sin \theta$$

Therefore **A** written in Cartesian vector form becomes

$$\mathbf{A} = A \sin \phi \cos \theta\, \mathbf{i} + A \sin \phi \sin \theta\, \mathbf{j} + A \cos \phi\, \mathbf{k}$$

You should not memorize this equation, rather it is important to understand how the components were determined using trigonometry.

2.6   Addition of Cartesian Vectors

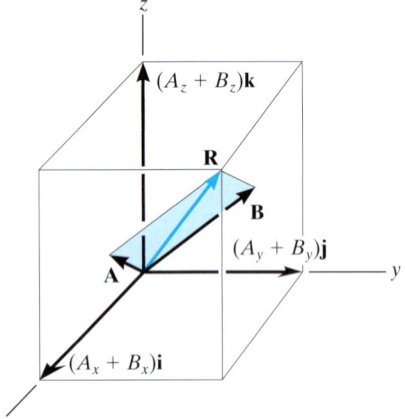

Fig. 2–29

The addition (or subtraction) of two or more vectors is greatly simplified if the vectors are expressed in terms of their Cartesian components. For example, if $\mathbf{A} = A_x\mathbf{i} + A_y\mathbf{j} + A_z\mathbf{k}$ and $\mathbf{B} = B_x\mathbf{i} + B_y\mathbf{j} + B_z\mathbf{k}$, Fig. 2–29, then the resultant vector, **R**, has components which are the scalar sums of the **i**, **j**, **k** components of **A** and **B**, i.e.,

$$\mathbf{R} = \mathbf{A} + \mathbf{B} = (A_x + B_x)\mathbf{i} + (A_y + B_y)\mathbf{j} + (A_z + B_z)\mathbf{k}$$

If this is generalized and applied to a system of several concurrent forces, then the force resultant is the vector sum of all the forces in the system and can be written as

$$\boxed{\mathbf{F}_R = \Sigma\mathbf{F} = \Sigma F_x\mathbf{i} + \Sigma F_y\mathbf{j} + \Sigma F_z\mathbf{k}} \qquad (2\text{–}10)$$

Here ΣF_x, ΣF_y, and ΣF_z represent the algebraic sums of the respective x, y, z or **i**, **j**, **k** components of each force in the system.

Important Points

- Cartesian vector analysis is often used to solve problems in three dimensions.

- The positive directions of the x, y, z axes are defined by the Cartesian unit vectors $\mathbf{i}, \mathbf{j}, \mathbf{k}$, respectively.

- The *magnitude* of a Cartesian vector is $A = \sqrt{A_x^2 + A_y^2 + A_z^2}$.

- The *direction* of a Cartesian vector is specified using coordinate direction angles α, β, γ which the tail of the vector makes with the positive x, y, z axes, respectively. The components of the unit vector $\mathbf{u}_A = \mathbf{A}/A$ represent the direction cosines of α, β, γ. Only two of the angles α, β, γ have to be specified. The third angle is determined from the relationship $\cos^2 \alpha + \cos^2 \beta + \cos^2 \gamma = 1$.

- Sometimes the direction of a vector is defined using the two angles θ and ϕ as in Fig. 2–28. In this case the vector components are obtained by vector resolution using trigonometry.

- To find the *resultant* of a concurrent force system, express each force as a Cartesian vector and add the $\mathbf{i}, \mathbf{j}, \mathbf{k}$ components of all the forces in the system.

EXAMPLE | 2.8

Express the force $\mathbf{F}$ shown in Fig. 2–30 as a Cartesian vector.

SOLUTION
Since only two coordinate direction angles are specified, the third angle α must be determined from Eq. 2–8; i.e.,

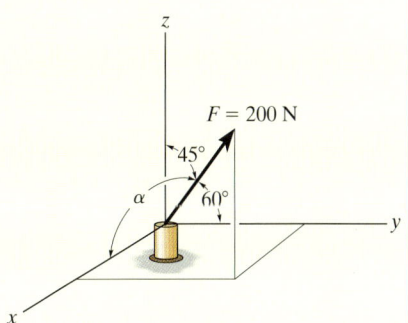

Fig. 2–30

$$\cos^2 \alpha + \cos^2 \beta + \cos^2 \gamma = 1$$
$$\cos^2 \alpha + \cos^2 60° + \cos^2 45° = 1$$
$$\cos \alpha = \sqrt{1 - (0.5)^2 - (0.707)^2} = \pm 0.5$$

Hence, two possibilities exist, namely,

$$\alpha = \cos^{-1}(0.5) = 60° \quad \text{or} \quad \alpha = \cos^{-1}(-0.5) = 120°$$

By inspection it is necessary that $\alpha = 60°$, since $\mathbf{F}_x$ must be in the $+x$ direction.
   Using Eq. 2–9, with $F = 200$ N, we have

$$\mathbf{F} = F \cos \alpha \, \mathbf{i} + F \cos \beta \, \mathbf{j} + F \cos \gamma \, \mathbf{k}$$
$$= (200 \cos 60° \, \text{N})\mathbf{i} + (200 \cos 60° \, \text{N})\mathbf{j} + (200 \cos 45° \, \text{N})\mathbf{k}$$
$$= \{100.0\mathbf{i} + 100.0\mathbf{j} + 141.4\mathbf{k}\} \ \text{N} \qquad \textit{Ans.}$$

Show that indeed the magnitude of $F = 200$ N.

EXAMPLE   2.9

Determine the magnitude and the coordinate direction angles of the resultant force acting on the ring in Fig. 2–31a.

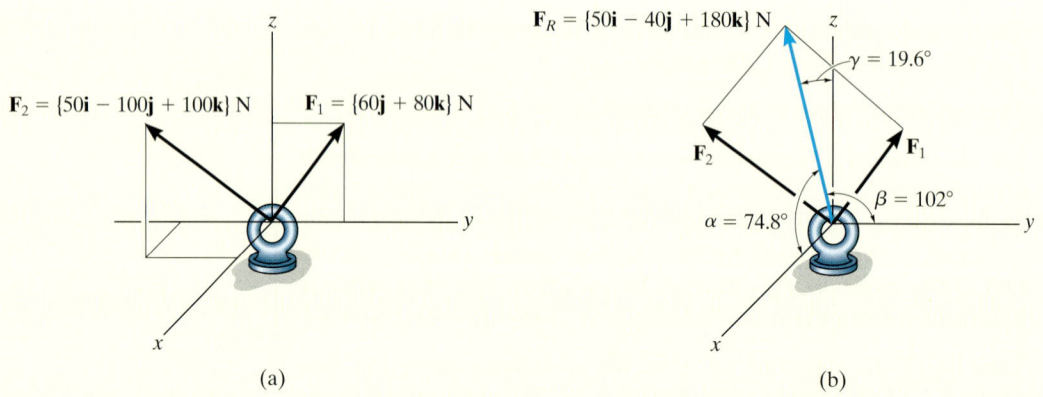

(a)   (b)

Fig. 2–31

SOLUTION

Since each force is represented in Cartesian vector form, the resultant force, shown in Fig. 2–31b, is

$$\mathbf{F}_R = \Sigma\mathbf{F} = \mathbf{F}_1 + \mathbf{F}_2 = \{60\mathbf{j} + 80\mathbf{k}\}\,\text{N} + \{50\mathbf{i} - 100\mathbf{j} + 100\mathbf{k}\}\,\text{N}$$
$$= \{50\mathbf{i} - 40\mathbf{j} + 180\mathbf{k}\}\,\text{N}$$

The magnitude of $\mathbf{F}_R$ is

$$F_R = \sqrt{(50\,\text{N})^2 + (-40\,\text{N})^2 + (180\,\text{N})^2} = 191.0\,\text{N}$$
$$= 191\,\text{N} \qquad\qquad\qquad\qquad Ans.$$

The coordinate direction angles α, β, γ are determined from the components of the unit vector acting in the direction of $\mathbf{F}_R$.

$$\mathbf{u}_{F_R} = \frac{\mathbf{F}_R}{F_R} = \frac{50}{191.0}\mathbf{i} - \frac{40}{191.0}\mathbf{j} + \frac{180}{191.0}\mathbf{k}$$
$$= 0.2617\mathbf{i} - 0.2094\mathbf{j} + 0.9422\mathbf{k}$$

so that

$$\cos\alpha = 0.2617 \qquad \alpha = 74.8° \qquad\qquad Ans.$$
$$\cos\beta = -0.2094 \qquad \beta = 102° \qquad\qquad Ans.$$
$$\cos\gamma = 0.9422 \qquad \gamma = 19.6° \qquad\qquad Ans.$$

These angles are shown in Fig. 2–31b.

NOTE: In particular, notice that $\beta > 90°$ since the $\mathbf{j}$ component of $\mathbf{u}_{F_R}$ is negative. This seems reasonable considering how $\mathbf{F}_1$ and $\mathbf{F}_2$ add according to the parallelogram law.

EXAMPLE | 2.10

Express the force **F** shown in Fig. 2–32*a* as a Cartesian vector.

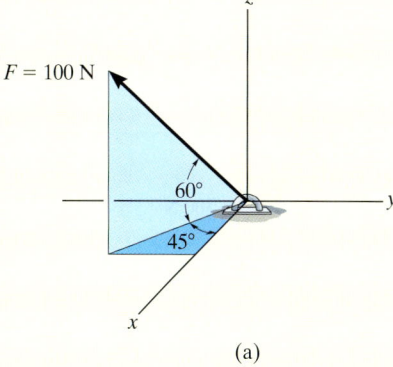

$F = 100$ N

60°

45°

(a)

SOLUTION

The angles of 60° and 45° defining the direction of **F** are *not* coordinate direction angles. Two successive applications of the parallelogram law are needed to resolve **F** into its *x*, *y*, *z* components. First **F** = **F′** + **F**$_z$, then **F′** = **F**$_x$ + **F**$_y$, Fig. 2–32*b*. By trigonometry, the magnitudes of the components are

$$F_z = 100 \sin 60° \text{ N} = 86.6 \text{ N}$$

$$F' = 100 \cos 60° \text{ N} = 50 \text{ N}$$

$$F_x = F' \cos 45° = 50 \cos 45° \text{ N} = 35.4 \text{ N}$$

$$F_y = F' \sin 45° = 50 \sin 45° \text{ N} = 35.4 \text{ N}$$

Realizing that **F**$_y$ has a direction defined by –**j**, we have

$$\mathbf{F} = \{35.4\mathbf{i} - 35.4\mathbf{j} + 86.6\mathbf{k}\} \text{ N} \qquad \textit{Ans.}$$

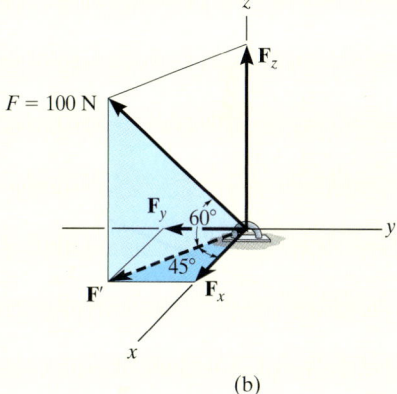

$F = 100$ N

F_z

F_y 60°

45°

F′ F_x

(b)

To show that the magnitude of this vector is indeed 100 N, apply Eq. 2–4,

$$F = \sqrt{F_x^2 + F_y^2 + F_z^2}$$

$$= \sqrt{(35.4)^2 + (35.4)^2 + (86.6)^2} = 100 \text{ N}$$

If needed, the coordinate direction angles of **F** can be determined from the components of the unit vector acting in the direction of **F**. Hence,

$$\mathbf{u} = \frac{\mathbf{F}}{F} = \frac{F_x}{F}\mathbf{i} + \frac{F_y}{F}\mathbf{j} + \frac{F_z}{F}\mathbf{k}$$

$$= \frac{35.4}{100}\mathbf{i} - \frac{35.4}{100}\mathbf{j} + \frac{86.6}{100}\mathbf{k}$$

$$= 0.354\mathbf{i} - 0.354\mathbf{j} + 0.866\mathbf{k}$$

so that

$$\alpha = \cos^{-1}(0.354) = 69.3°$$

$$\beta = \cos^{-1}(-0.354) = 111°$$

$$\gamma = \cos^{-1}(0.866) = 30.0°$$

These results are shown in Fig. 2–32*c*.

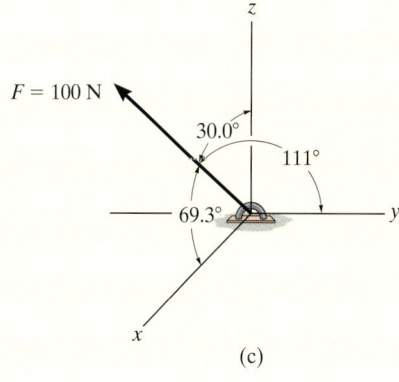

$F = 100$ N

30.0°

111°

69.3°

(c)

Fig. 2–32

EXAMPLE | 2.11

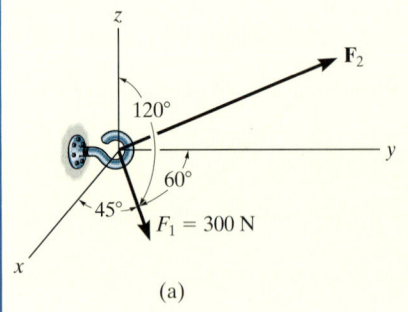

(a)

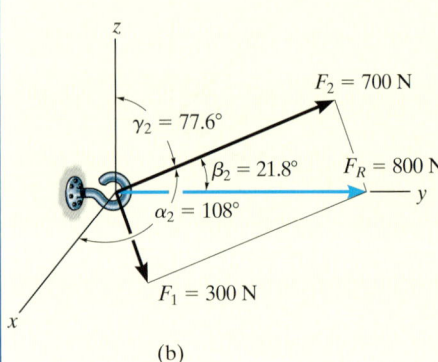

(b)

Fig. 2–33

Two forces act on the hook shown in Fig. 2–33a. Specify the magnitude of $\mathbf{F}_2$ and its coordinate direction angles so that the resultant force $\mathbf{F}_R$ acts along the positive y axis and has a magnitude of 800 N.

SOLUTION

To solve this problem, the resultant force $\mathbf{F}_R$ and its two components, $\mathbf{F}_1$ and $\mathbf{F}_2$, will each be expressed in Cartesian vector form. Then, as shown in Fig. 2–33b, it is necessary that $\mathbf{F}_R = \mathbf{F}_1 + \mathbf{F}_2$.

Applying Eq. 2–9,

$$\mathbf{F}_1 = F_1 \cos \alpha_1 \mathbf{i} + F_1 \cos \beta_1 \mathbf{j} + F_1 \cos \gamma_1 \mathbf{k}$$

$$= 300 \cos 45° \, \mathbf{i} + 300 \cos 60° \, \mathbf{j} + 300 \cos 120° \, \mathbf{k}$$

$$= \{212.1\mathbf{i} + 150\mathbf{j} - 150\mathbf{k}\} \, \text{N}$$

$$\mathbf{F}_2 = F_{2x}\mathbf{i} + F_{2y}\mathbf{j} + F_{2z}\mathbf{k}$$

Since $\mathbf{F}_R$ has a magnitude of 800 N and acts in the $+\mathbf{j}$ direction,

$$\mathbf{F}_R = (800 \text{ N})(+\mathbf{j}) = \{800\mathbf{j}\} \, \text{N}$$

We require

$$\mathbf{F}_R = \mathbf{F}_1 + \mathbf{F}_2$$

$$800\mathbf{j} = 212.1\mathbf{i} + 150\mathbf{j} - 150\mathbf{k} + F_{2x}\mathbf{i} + F_{2y}\mathbf{j} + F_{2z}\mathbf{k}$$

$$800\mathbf{j} = (212.1 + F_{2x})\mathbf{i} + (150 + F_{2y})\mathbf{j} + (-150 + F_{2z})\mathbf{k}$$

To satisfy this equation the $\mathbf{i}, \mathbf{j}, \mathbf{k}$ components of $\mathbf{F}_R$ must be equal to the corresponding $\mathbf{i}, \mathbf{j}, \mathbf{k}$ components of $(\mathbf{F}_1 + \mathbf{F}_2)$. Hence,

$$0 = 212.1 + F_{2x} \qquad F_{2x} = -212.1 \text{ N}$$

$$800 = 150 + F_{2y} \qquad F_{2y} = 650 \text{ N}$$

$$0 = -150 + F_{2z} \qquad F_{2z} = 150 \text{ N}$$

The magnitude of $\mathbf{F}_2$ is thus

$$F_2 = \sqrt{(-212.1 \text{ N})^2 + (650 \text{ N})^2 + (150 \text{ N})^2}$$

$$= 700 \text{ N} \hspace{3cm} \textit{Ans.}$$

We can use Eq. 2–9 to determine $\alpha_2, \beta_2, \gamma_2$.

$$\cos \alpha_2 = \frac{-212.1}{700}; \qquad \alpha_2 = 108° \hspace{2cm} \textit{Ans.}$$

$$\cos \beta_2 = \frac{650}{700}; \qquad \beta_2 = 21.8° \hspace{2cm} \textit{Ans.}$$

$$\cos \gamma_2 = \frac{150}{700}; \qquad \gamma_2 = 77.6° \hspace{2cm} \textit{Ans.}$$

These results are shown in Fig. 2–33b.

FUNDAMENTAL PROBLEMS

F2–13. Determine the coordinate direction angles of the force.

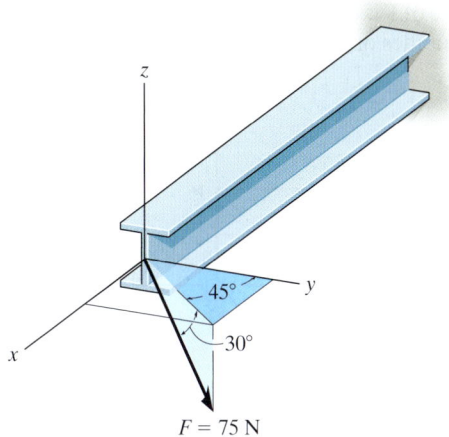

F2–13

F2–14. Express the force as a Cartesian vector.

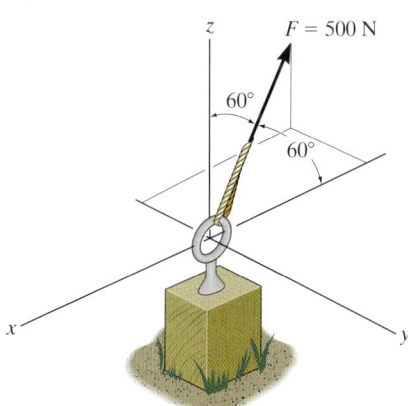

F2–14

F2–15. Express the force as a Cartesian vector.

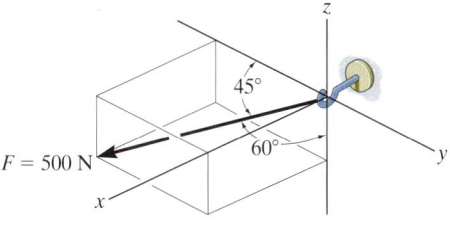

F2–15

F2–16. Express the force as a Cartesian vector.

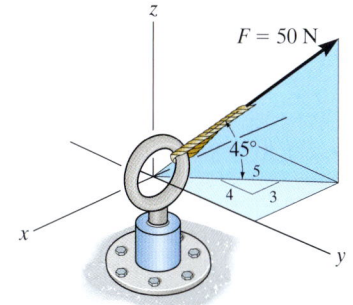

F2–16

F2–17. Express the force as a Cartesian vector.

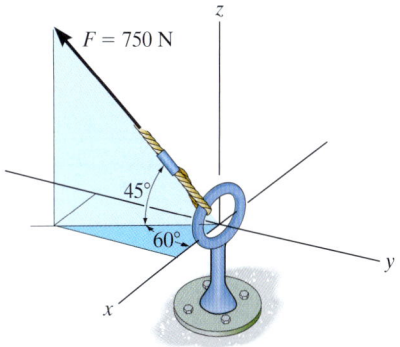

F2–17

F2–18. Determine the resultant force acting on the hook.

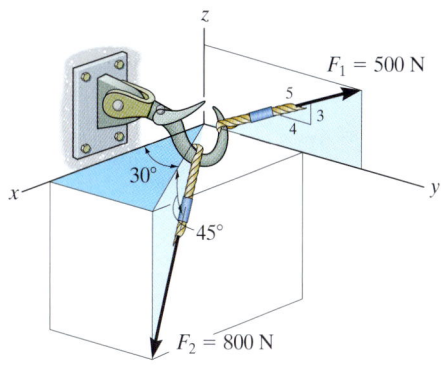

F2–18

PROBLEMS

*2–60. The stock mounted on the lathe is subjected to a force of 60 N. Determine the coordinate direction angle β and express the force as a Cartesian vector.

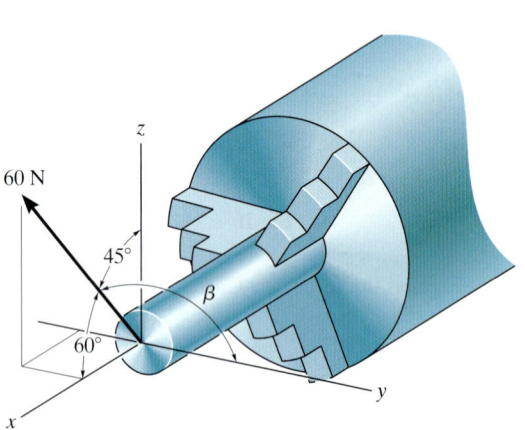

Prob. 2–60

2–61. Determine the coordinate angle γ for F_2 and then express each force acting on the bracket as a Cartesian vector.

2–62. Determine the magnitude and coordinate direction angles of the resultant force acting on the bracket.

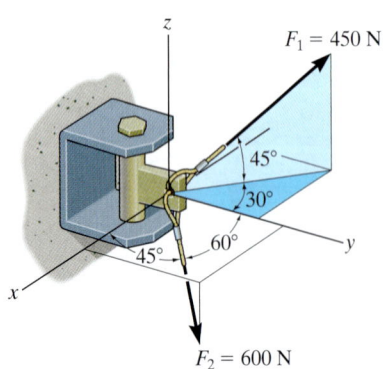

$F_1 = 450$ N

$F_2 = 600$ N

Probs. 2–61/62

2–63. The bolt is subjected to the force F, which has components acting along the x, y, z axes as shown. If the magnitude of F is 80 N, and $\alpha = 60°$ and $\gamma = 45°$, determine the magnitudes of its components.

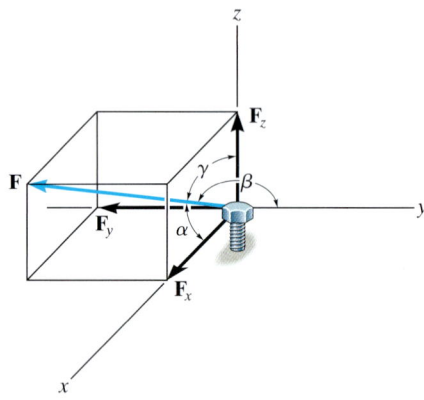

Prob. 2–63

*2–64. Determine the magnitude and coordinate direction angles of the force F acting on the stake.

2–65. Solve Prob. 2–64 if $\theta = 50°$.

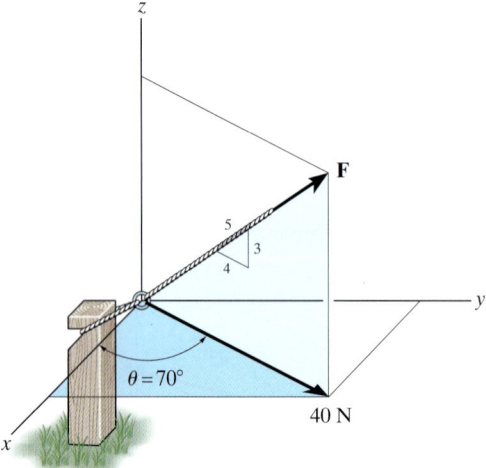

$\theta = 70°$

40 N

Probs. 2–64/65

2–66. Determine the magnitude and coordinate direction angles of $\mathbf{F}_1 = \{60\mathbf{i} - 50\mathbf{j} + 40\mathbf{k}\}\,\text{N}$ and $\mathbf{F}_2 = \{-40\mathbf{i} - 85\mathbf{j} + 30\mathbf{k}\}\,\text{N}$. Sketch each force on an x, y, z reference.

2–67. Express each force in Cartesian vector form.

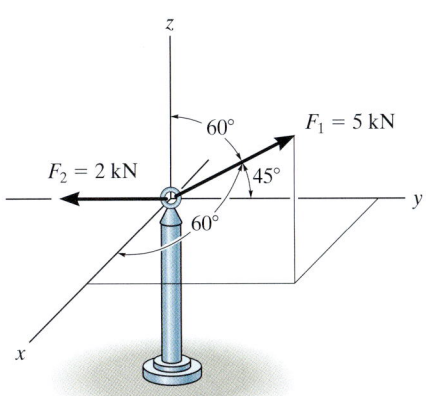

Prob. 2–67

*2–68. Express each force as a Cartesian vector.

2–69. Determine the magnitude and coordinate direction angles of the resultant force acting on the hook.

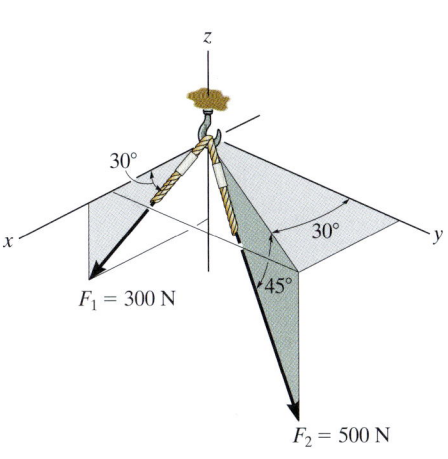

Probs. 2–68/69

2–70. Determine the magnitude and coordinate direction angles of the resultant force and sketch this vector on the coordinate system.

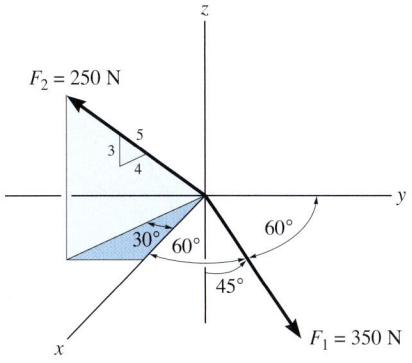

Prob. 2–70

2–71. If the resultant force acting on the bracket is directed along the positive y axis, determine the magnitude of the resultant force and the coordinate direction angles of **F** so that $\beta < 90°$.

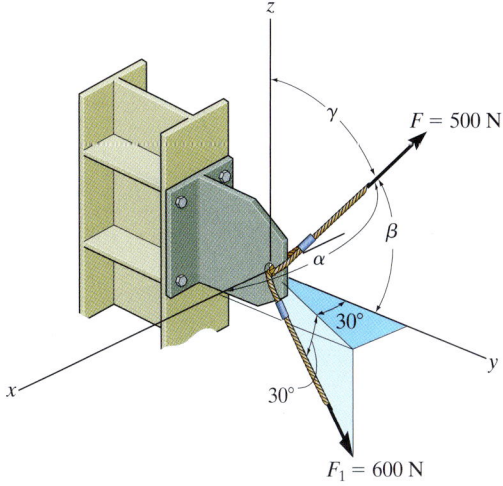

Prob. 2–71

*2–72. Specify the magnitude F_3 and directions α_3, β_3, and γ_3 of $\mathbf{F}_3$ so that the force of the three forces is $\mathbf{F}_R = \{9\mathbf{j}\}$ kN.

2–75. Determine the coordinate direction angles of force $\mathbf{F}_1$.

*2–76. Determine the magnitude and coordinate direction angles of the resultant force acting on the eyebolt.

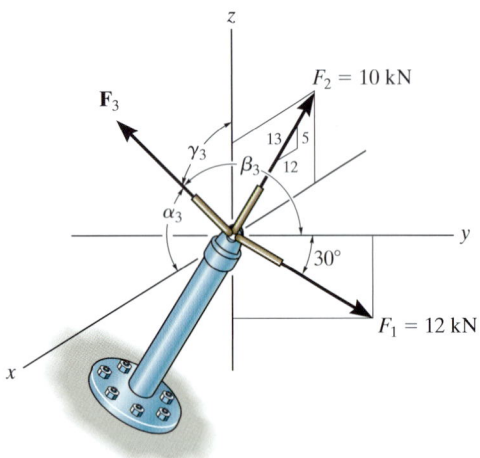

Prob. 2–72

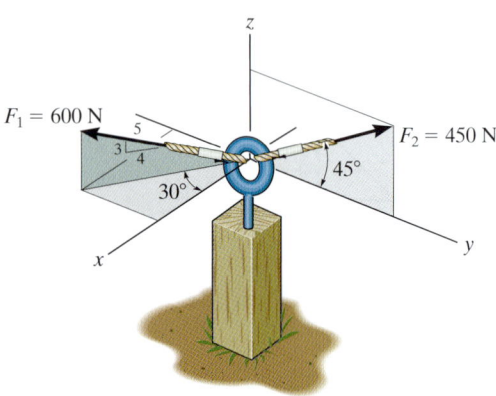

Probs. 2–75/76

2–73. Determine the magnitude and coordinate direction angles of $\mathbf{F}_3$ so that the resultant of the three forces acts along the positive y axis and has a magnitude of 600 N.

2–74. Determine the magnitude and coordinate direction angles of $\mathbf{F}_3$ so that the resultant of the three forces is zero.

2–77. The cables attached to the screw eye are subjected to the three forces shown. Express each force in Cartesian vector form and determine the magnitude and coordinate direction angles of the resultant force.

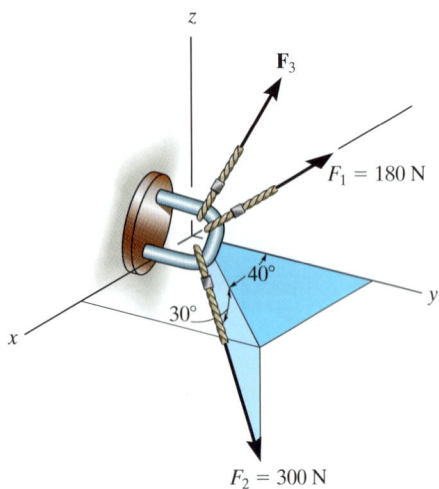

Probs. 2–73/74

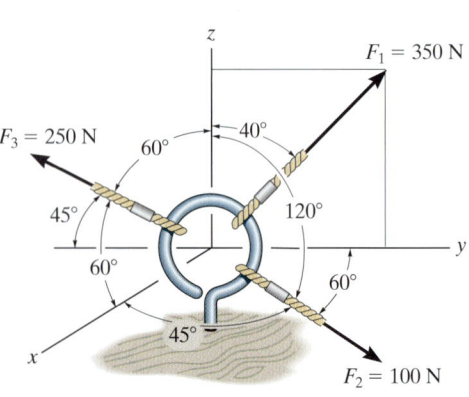

Prob. 2–77

2–78. Three forces act on the ring. If the resultant force $\mathbf{F}_R$ has a magnitude and direction as shown, determine the magnitude and the coordinate direction angles of force $\mathbf{F}_3$.

2–79. Determine the coordinate direction angles of $\mathbf{F}_1$ and $\mathbf{F}_R$.

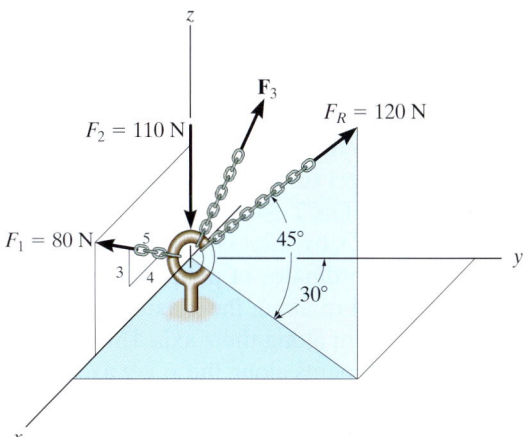

Probs. 2–78/79

*2–80.** The mast is subjected to the three forces shown. Determine the coordinate direction angles $\alpha_1, \beta_1, \gamma_1$ of $\mathbf{F}_1$ so that the resultant force acting on the mast is $\mathbf{F}_R = \{350\mathbf{i}\}$ N.

2–81. The mast is subjected to the three forces shown. Determine the coordinate direction angles $\alpha_1, \beta_1, \gamma_1$ of $\mathbf{F}_1$ so that the resultant force acting on the mast is zero.

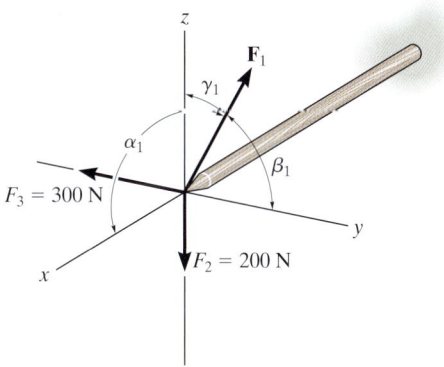

Probs. 2–80/81

2–82. Determine the magnitude and coordinate direction angles of $\mathbf{F}_2$ so that the resultant of the two forces acts along the positive x axis and has a magnitude of 500 N.

2–83. Determine the magnitude and coordinate direction angles of $\mathbf{F}_2$ so that the resultant of the two forces is zero.

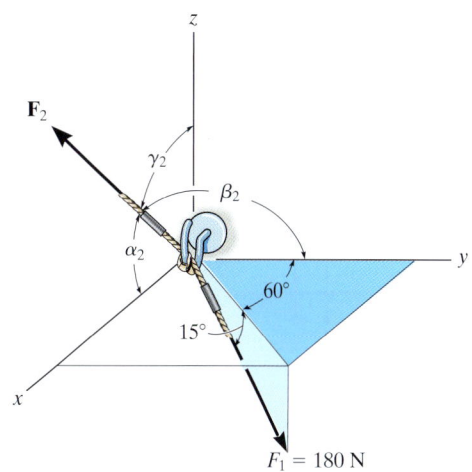

Probs. 2–82/83

*2–84.** The pole is subjected to the force $\mathbf{F}$, which has components acting along the x, y, z axes as shown. If the magnitude of $\mathbf{F}$ is 3 kN, $\beta = 30°$, and $\gamma = 75°$, determine the magnitudes of its three components.

2–85. The pole is subjected to the force $\mathbf{F}$ which has components $F_x = 1.5$ kN and $F_z = 1.25$ kN. If $\beta = 75°$, determine the magnitudes of $\mathbf{F}$ and $\mathbf{F}_y$.

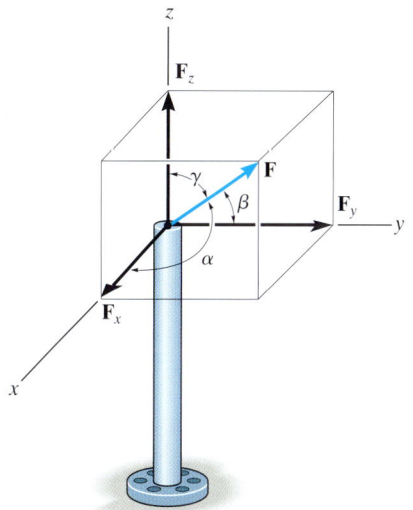

Probs. 2–84/85

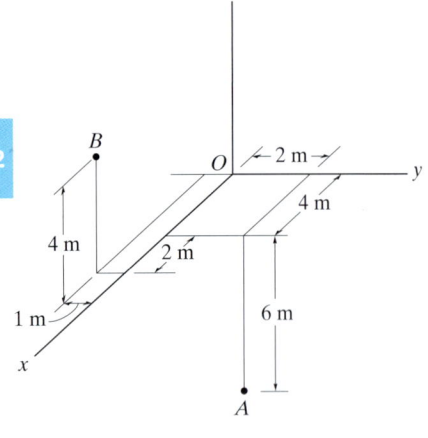

Fig. 2–34

2.7   Position Vectors

In this section we will introduce the concept of a position vector. It will be shown that this vector is of importance in formulating a Cartesian force vector directed between two points in space.

x, y, z Coordinates.   Throughout the book we will use a *right-handed* coordinate system to reference the location of points in space. We will also use the convention followed in many technical books, which requires the positive z axis to be directed *upward* (the zenith direction) so that it measures the height of an object or the altitude of a point. The x, y axes then lie in the horizontal plane, Fig. 2–34. Points in space are located relative to the origin of coordinates, O, by successive measurements along the x, y, z axes. For example, the coordinates of point A are obtained by starting at O and measuring $x_A = +4$ m along the x axis, then $y_A = +2$ m along the y axis, and finally $z_A = -6$ m along the z axis. Thus, $A(4$ m$, 2$ m$, -6$ m$)$. In a similar manner, measurements along the x, y, z axes from O to B yield the coordinates of B, i.e., $B(6$ m$, -1$ m$, 4$ m$)$.

Position Vector.   A *position vector* $\mathbf{r}$ is defined as a fixed vector which locates a point in space relative to another point. For example, if $\mathbf{r}$ extends from the origin of coordinates, O, to point $P(x, y, z)$, Fig. 2–35a, then $\mathbf{r}$ can be expressed in Cartesian vector form as

$$\mathbf{r} = x\mathbf{i} + y\mathbf{j} + z\mathbf{k}$$

Note how the head-to-tail vector addition of the three components yields vector $\mathbf{r}$, Fig. 2–35b. Starting at the origin O, one "travels" x in the $+\mathbf{i}$ direction, then y in the $+\mathbf{j}$ direction, and finally z in the $+\mathbf{k}$ direction to arrive at point $P(x, y, z)$.

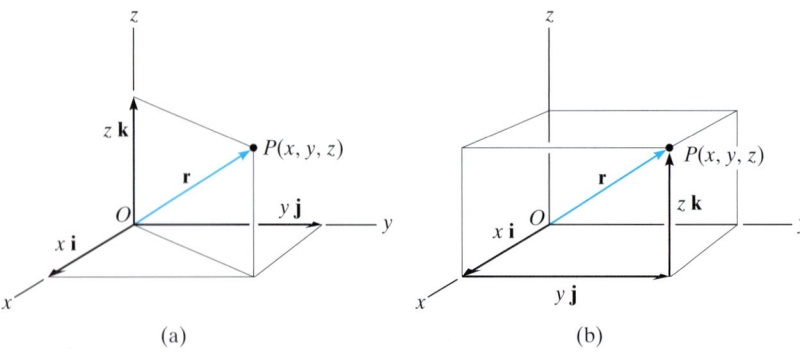

Fig. 2–35

2

In the more general case, the position vector may be directed from point A to point B in space, Fig. 2–36a. This vector is also designated by the symbol $\mathbf{r}$. As a matter of convention, we will *sometimes* refer to this vector with *two subscripts* to indicate from and to the point where it is directed. Thus, $\mathbf{r}$ can also be designated as $\mathbf{r}_{AB}$. Also, note that $\mathbf{r}_A$ and $\mathbf{r}_B$ in Fig. 2–36a are referenced with only one subscript since they extend from the origin of coordinates.

From Fig. 2–36a, by the head-to-tail vector addition, using the triangle rule, we require

$$\mathbf{r}_A + \mathbf{r} = \mathbf{r}_B$$

Solving for $\mathbf{r}$ and expressing $\mathbf{r}_A$ and $\mathbf{r}_B$ in Cartesian vector form yields

$$\mathbf{r} = \mathbf{r}_B - \mathbf{r}_A = (x_B\mathbf{i} + y_B\mathbf{j} + z_B\mathbf{k}) - (x_A\mathbf{i} + y_A\mathbf{j} + z_A\mathbf{k})$$

or

$$\boxed{\mathbf{r} = (x_B - x_A)\mathbf{i} + (y_B - y_A)\mathbf{j} + (z_B - z_A)\mathbf{k}} \qquad (2\text{–}11)$$

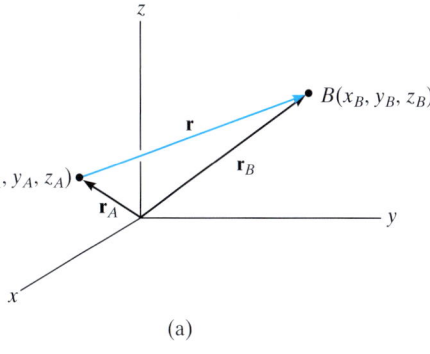

(a)

Thus, the $\mathbf{i}, \mathbf{j}, \mathbf{k}$ *components of the position vector* $\mathbf{r}$ *may be formed by taking the coordinates of the tail of the vector* $A(x_A, y_A, z_A)$ *and subtracting them from the corresponding coordinates of the head* $B(x_B, y_B, z_B)$. We can also form these components *directly*, Fig. 2–36b, by starting at A and moving through a distance of $(x_B - x_A)$ along the positive x axis ($+\mathbf{i}$), then $(y_B - y_A)$ along the positive y axis ($+\mathbf{j}$), and finally $(z_B - z_A)$ along the positive z axis ($+\mathbf{k}$) to get to B.

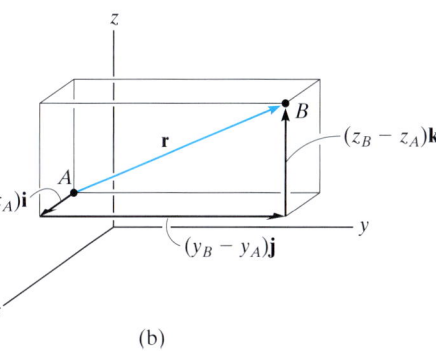

(b)

Fig. 2–36

If an x, y, z coordinate system is established, then the coordinates of two points A and B on the cable can be determined. From this the position vector $\mathbf{r}$ acting along the cable can be formulated. Its magnitude represents the distance from A to B, and its unit vector, $\mathbf{u} = \mathbf{r}/r$, gives the direction defined by α, β, γ.

EXAMPLE 2.12

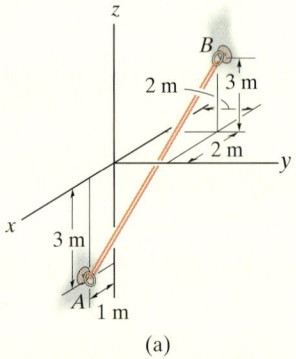

(a)

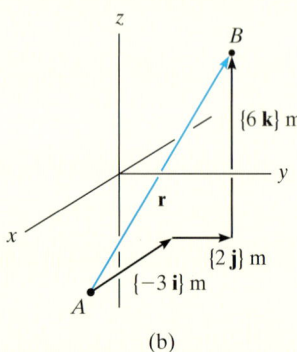

(b)

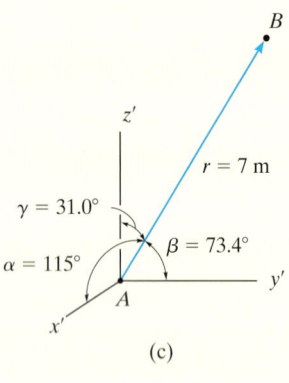

(c)

Fig. 2–37

An elastic rubber band is attached to points A and B as shown in Fig. 2–37*a*. Determine its length and its direction measured from A toward B.

SOLUTION

We first establish a position vector from A to B, Fig. 2–37*b*. In accordance with Eq. 2–11, the coordinates of the tail $A(1\ m, 0, -3\ m)$ are subtracted from the coordinates of the head $B(-2\ m, 2\ m, 3\ m)$, which yields

$$\mathbf{r} = [-2\ m - 1\ m]\mathbf{i} + [2\ m - 0]\mathbf{j} + [3\ m - (-3\ m)]\mathbf{k}$$
$$= \{-3\mathbf{i} + 2\mathbf{j} + 6\mathbf{k}\}\ m$$

These components of $\mathbf{r}$ can also be determined *directly* by realizing that they represent the direction and distance one must travel along each axis in order to move from A to B, i.e., along the x axis $\{-3\mathbf{i}\}$ m, along the y axis $\{2\mathbf{j}\}$ m, and finally along the z axis $\{6\mathbf{k}\}$ m.

The length of the rubber band is therefore

$$r = \sqrt{(-3\ m)^2 + (2\ m)^2 + (6\ m)^2} = 7\ m \qquad \textit{Ans.}$$

Formulating a unit vector in the direction of $\mathbf{r}$, we have

$$\mathbf{u} = \frac{\mathbf{r}}{r} = -\frac{3}{7}\mathbf{i} + \frac{2}{7}\mathbf{j} + \frac{6}{7}\mathbf{k}$$

The components of this unit vector give the coordinate direction angles

$$\alpha = \cos^{-1}\left(-\frac{3}{7}\right) = 115° \qquad \textit{Ans.}$$

$$\beta = \cos^{-1}\left(\frac{2}{7}\right) = 73.4° \qquad \textit{Ans.}$$

$$\gamma = \cos^{-1}\left(\frac{6}{7}\right) = 31.0° \qquad \textit{Ans.}$$

NOTE: These angles are measured from the *positive axes* of a localized coordinate system placed at the tail of $\mathbf{r}$, as shown in Fig. 2–37*c*.

2.8 Force Vector Directed Along a Line

Quite often in three-dimensional statics problems, the direction of a force is specified by two points through which its line of action passes. Such a situation is shown in Fig. 2–38, where the force **F** is directed along the cord AB. We can formulate **F** as a Cartesian vector by realizing that it has the *same direction* and *sense* as the position vector **r** directed from point A to point B on the cord. This common direction is specified by the *unit vector* $\mathbf{u} = \mathbf{r}/r$. Hence,

$$\mathbf{F} = F\mathbf{u} = F\left(\frac{\mathbf{r}}{r}\right) = F\left(\frac{(x_B - x_A)\mathbf{i} + (y_B - y_A)\mathbf{j} + (z_B - z_A)\mathbf{k}}{\sqrt{(x_B - x_A)^2 + (y_B - y_A)^2 + (z_B - z_A)^2}}\right)$$

Although we have represented **F** symbolically in Fig. 2–38, note that it has *units of force*, unlike **r**, which has units of length.

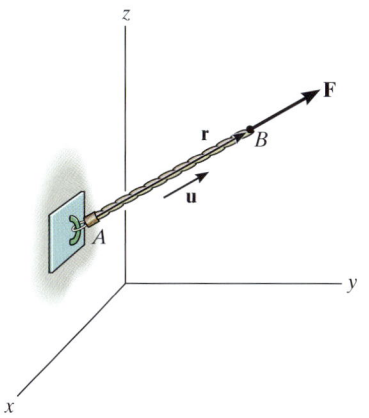

Fig. 2–38

The force **F** acting along the rope can be represented as a Cartesian vector by establishing x, y, z axes and first forming a position vector **r** along the length of the rope. Then the corresponding unit vector $\mathbf{u} = \mathbf{r}/r$ that defines the direction of both the rope and the force can be determined. Finally, the magnitude of the force is combined with its direction, $\mathbf{F} = F\mathbf{u}$.

Important Points

- A position vector locates one point in space relative to another point.

- The easiest way to formulate the components of a position vector is to determine the distance and direction that must be traveled along the x, y, z directions—going from the tail to the head of the vector.

- A force **F** acting in the direction of a position vector **r** can be represented in Cartesian form if the unit vector **u** of the position vector is determined and it is multiplied by the magnitude of the force, i.e., $\mathbf{F} = F\mathbf{u} = F(\mathbf{r}/r)$.

EXAMPLE | 2.13

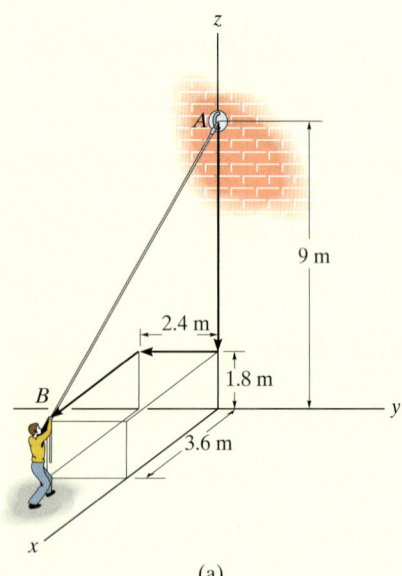

(a)

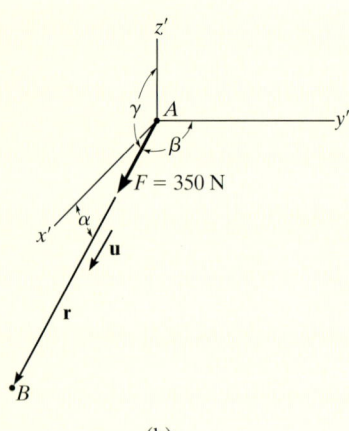

(b)

Fig. 2–39

The man shown in Fig. 2–39a pulls on the cord with a force of 350 N. Represent this force acting on the support A as a Cartesian vector and determine its direction.

SOLUTION

Force **F** is shown in Fig. 2–39b. The *direction* of this vector, **u**, is determined from the position vector **r**, which extends from A to B. Rather than using the coordinates of the end points of the cord, **r** can be determined *directly* by noting in Fig. 2–39a that one must travel from A {−7.2**k**} m, then {−2.4**j**} m, and finally {3.6**i**} m to get to B. Thus,

$$\mathbf{r} = \{3.6\mathbf{i} - 2.4\mathbf{j} - 7.2\mathbf{k}\} \text{ m}$$

The magnitude of **r**, which represents the *length* of cord AB, is

$$r = \sqrt{(3.6 \text{ m})^2 + (-2.4 \text{ m})^2 + (-7.2 \text{ m})^2} = 8.4 \text{ m}$$

Forming the unit vector that defines the direction and sense of both **r** and **F**, we have

$$\mathbf{u} = \frac{\mathbf{r}}{r} = \frac{3.6}{8.4}\mathbf{i} - \frac{2.4}{8.4}\mathbf{j} - \frac{7.2}{8.4}\mathbf{k}$$

Since **F** has a *magnitude* of 350 N and a *direction* specified by **u**, then

$$\mathbf{F} = F\mathbf{u} = 350 \text{ N}\left(\frac{3.6}{8.4}\mathbf{i} - \frac{2.4}{8.4}\mathbf{j} - \frac{7.2}{8.4}\mathbf{k}\right)$$

$$= \{150\mathbf{i} - 100\mathbf{j} - 300\mathbf{k}\} \text{ N} \qquad \textit{Ans.}$$

The coordinate direction angles are measured between **r** (or **F**) and the *positive axes* of a localized coordinate system with origin placed at A, Fig. 2–39b. From the components of the unit vector:

$$\alpha = \cos^{-1}\left(\frac{3.6}{8.4}\right) = 64.6° \qquad \textit{Ans.}$$

$$\beta = \cos^{-1}\left(\frac{-2.4}{8.4}\right) = 107° \qquad \textit{Ans.}$$

$$\gamma = \cos^{-1}\left(\frac{-7.2}{8.4}\right) = 149° \qquad \textit{Ans.}$$

NOTE: These results make sense when compared with the angles identified in Fig. 2–39b.

EXAMPLE   2.14

The force in Fig. 2–40a acts on the hook. Express it as a Cartesian vector.

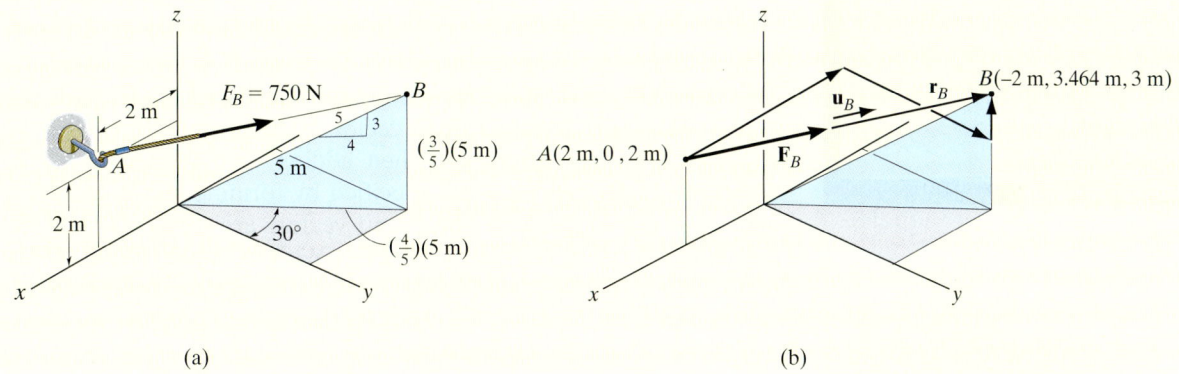

Fig. 2–40

SOLUTION

As shown in Fig. 2–40b, the coordinates for points A and B are

$$A(2 \text{ m}, 0, 2 \text{ m})$$

and

$$B\left[-\left(\frac{4}{5}\right)5 \sin 30° \text{ m}, \left(\frac{4}{5}\right)5 \cos 30° \text{ m}, \left(\frac{3}{5}\right)5 \text{ m} \right]$$

or

$$B(-2 \text{ m}, 3.464 \text{ m}, 3 \text{ m})$$

Therefore, to go from A to B, one must travel $\{-4\mathbf{i}\}$ m, then $\{3.464\mathbf{j}\}$ m, and finally $\{1\mathbf{k}\}$ m. Thus,

$$\mathbf{u}_B = \left(\frac{\mathbf{r}_B}{r_B}\right) = \frac{\{-4\mathbf{i} + 3.464\mathbf{j} + 1\mathbf{k}\} \text{ m}}{\sqrt{(-4 \text{ m})^2 + (3.464 \text{ m})^2 + (1 \text{ m})^2}}$$

$$= -0.7428\mathbf{i} + 0.6433\mathbf{j} + 0.1857\mathbf{k}$$

Force $\mathbf{F}_B$ expressed as a Cartesian vector becomes

$$\mathbf{F}_B = F_B\mathbf{u}_B = (750 \text{ N})(-0.74281\mathbf{i} + 0.6433\mathbf{j} + 0.1857\mathbf{k})$$

$$= \{-557\mathbf{i} + 482\mathbf{j} + 139\mathbf{k}\} \text{ N} \qquad \text{\textit{Ans.}}$$

EXAMPLE   2.15

The roof is supported by cables as shown in the photo. If the cables exert forces $F_{AB} = 100$ N and $F_{AC} = 120$ N on the wall hook at A as shown in Fig. 2–41a, determine the resultant force acting at A. Express the result as a Cartesian vector.

SOLUTION

The resultant force $\mathbf{F}_R$ is shown graphically in Fig. 2–41b. We can express this force as a Cartesian vector by first formulating $\mathbf{F}_{AB}$ and $\mathbf{F}_{AC}$ as Cartesian vectors and then adding their components. The directions of $\mathbf{F}_{AB}$ and $\mathbf{F}_{AC}$ are specified by forming unit vectors $\mathbf{u}_{AB}$ and $\mathbf{u}_{AC}$ along the cables. These unit vectors are obtained from the associated position vectors $\mathbf{r}_{AB}$ and $\mathbf{r}_{AC}$. With reference to Fig. 2–41a, to go from A to B, we must travel $\{-4\mathbf{k}\}$ m, and then $\{4\mathbf{i}\}$ m. Thus,

$$\mathbf{r}_{AB} = \{4\mathbf{i} - 4\mathbf{k}\} \text{ m}$$

$$r_{AB} = \sqrt{(4 \text{ m})^2 + (-4 \text{ m})^2} = 5.66 \text{ m}$$

$$\mathbf{F}_{AB} = F_{AB}\left(\frac{\mathbf{r}_{AB}}{r_{AB}}\right) = (100 \text{ N})\left(\frac{4}{5.66}\mathbf{i} - \frac{4}{5.66}\mathbf{k}\right)$$

$$\mathbf{F}_{AB} = \{70.7\mathbf{i} - 70.7\mathbf{k}\} \text{ N}$$

To go from A to C, we must travel $\{-4\mathbf{k}\}$ m , then $\{2\mathbf{j}\}$ m, and finally $\{4\mathbf{i}\}$. Thus,

$$\mathbf{r}_{AC} = \{4\mathbf{i} + 2\mathbf{j} - 4\mathbf{k}\} \text{ m}$$

$$r_{AC} = \sqrt{(4 \text{ m})^2 + (2 \text{ m})^2 + (-4 \text{ m})^2} = 6 \text{ m}$$

$$\mathbf{F}_{AC} = F_{AC}\left(\frac{\mathbf{r}_{AC}}{r_{AC}}\right) = (120 \text{ N})\left(\frac{4}{6}\mathbf{i} + \frac{2}{6}\mathbf{j} - \frac{4}{6}\mathbf{k}\right)$$

$$= \{80\mathbf{i} + 40\mathbf{j} - 80\mathbf{k}\} \text{ N}$$

The resultant force is therefore

$$\mathbf{F}_R = \mathbf{F}_{AB} + \mathbf{F}_{AC} = \{70.7\mathbf{i} - 70.7\mathbf{k}\} \text{ N} + \{80\mathbf{i} + 40\mathbf{j} - 80\mathbf{k}\} \text{ N}$$

$$= \{151\mathbf{i} + 40\mathbf{j} - 151\mathbf{k}\} \text{ N} \qquad \textit{Ans.}$$

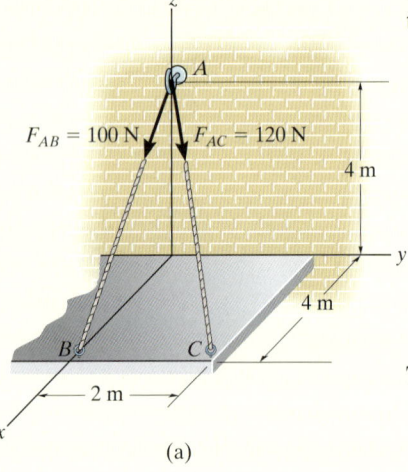

(a)

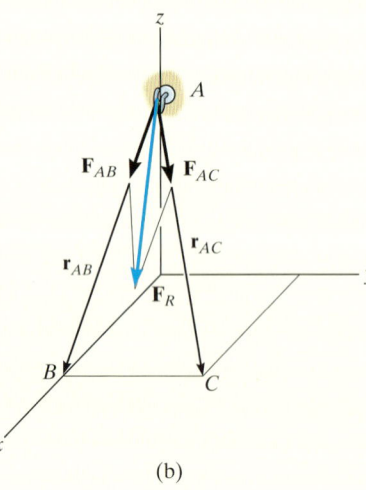

(b)

Fig. 2–41

FUNDAMENTAL PROBLEMS

F2–19.   Express the position vector $\mathbf{r}_{AB}$ in Cartesian vector form, then determine its magnitude and coordinate direction angles.

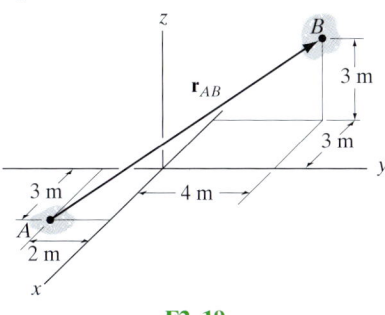

F2–19

F2–20.   Determine the length of the rod and the position vector directed from A to B. What is the angle θ?

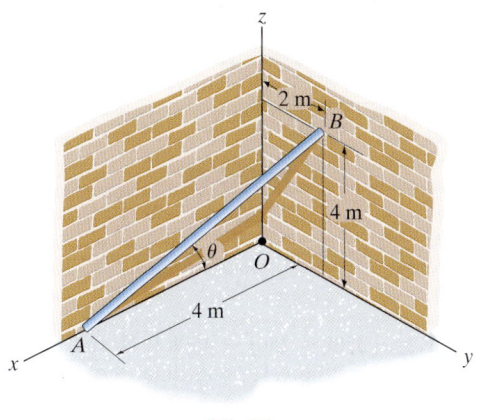

F2–20

F2–21.   Express the force as a Cartesian vector.

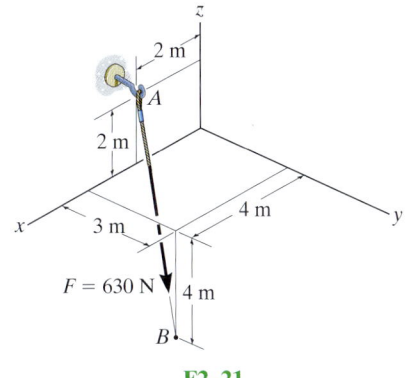

F2–21

F2–22.   Express the force as a Cartesian vector.

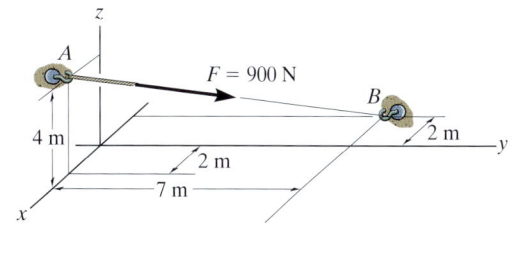

F2–22

F2–23.   Determine the magnitude of the resultant force at A.

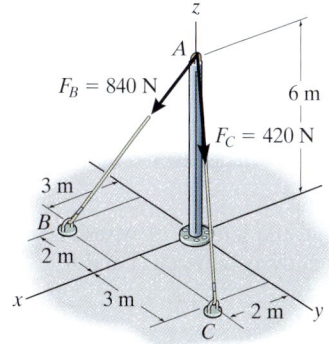

F2–23

F2–24.   Determine the resultant force at A.

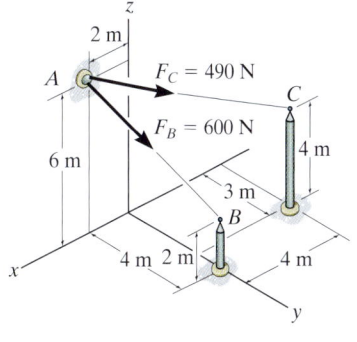

F2–24

PROBLEMS

2–86.   Determine the length of the crankshaft AB by first formulating a Cartesian position vector from A to B and then determining its magnitude.

*2–88.   Determine the length of member AB of the truss by first establishing a Cartesian position vector from A to B and then determining its magnitude.

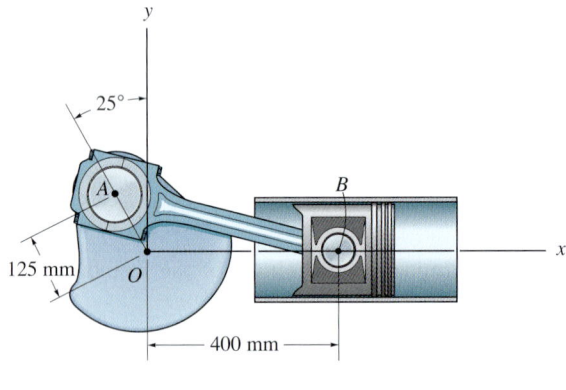

Prob. 2–86

2–87.   Determine the lengths of wires AD, BD, and CD. The ring at D is midway between A and B.

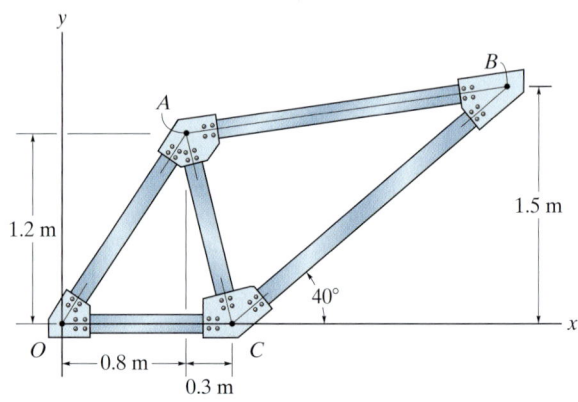

Prob. 2–88

2–89.   If $\mathbf{F} = \{350\mathbf{i} - 250\mathbf{j} - 450\mathbf{k}\}$ N and cable AB is 9 m long, determine the x, y, z coordinates of point A.

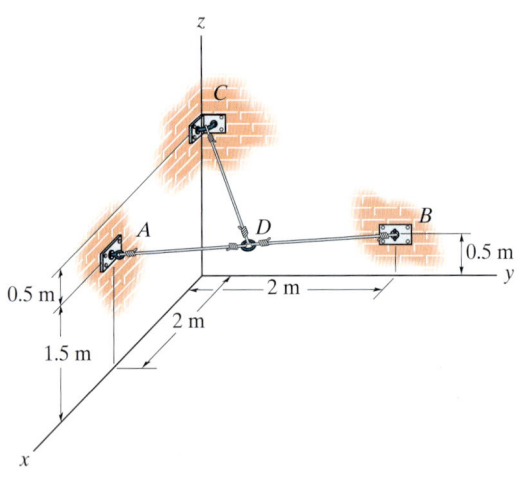

Prob. 2–87

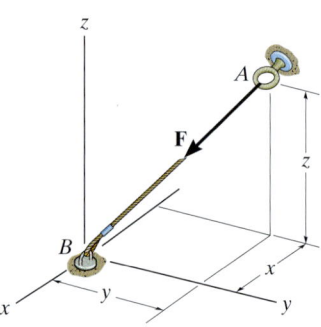

Prob. 2–89

2–90. Express $\mathbf{F}_B$ and $\mathbf{F}_C$ in Cartesian vector form.

2–91. Determine the magnitude and coordinate direction angles of the resultant force acting at A.

2–94. The tower is held in place by three cables. If the force of each cable acting on the tower is shown, determine the magnitude and coordinate direction angles α, β, γ of the resultant force. Take $x = 15$ m, $y = 20$ m.

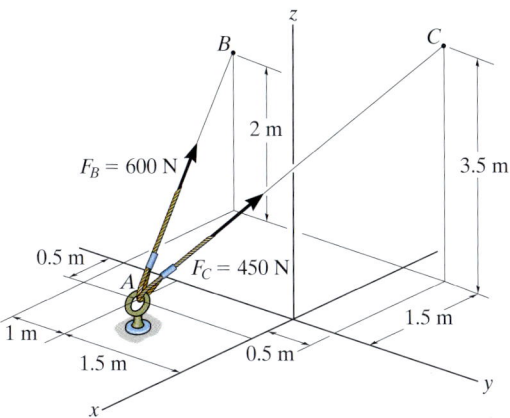

Probs. 2–90/91

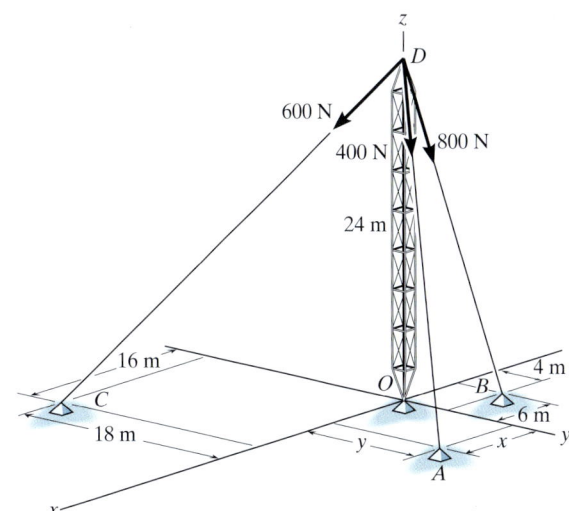

Prob. 2–94

*2–92. If $F_B = 560$ N and $F_C = 700$ N, determine the magnitude and coordinate direction angles of the resultant force acting on the flag pole.

2–93. If $F_B = 700$ N, and $F_C = 560$ N, determine the magnitude and coordinate direction angles of the resultant force acting on the flag pole.

2–95. At a given instant, the position of a plane at A and a train at B are measured relative to a radar antenna at O. Determine the distance d between A and B at this instant. To solve the problem, formulate a position vector, directed from A to B, and then determine its magnitude.

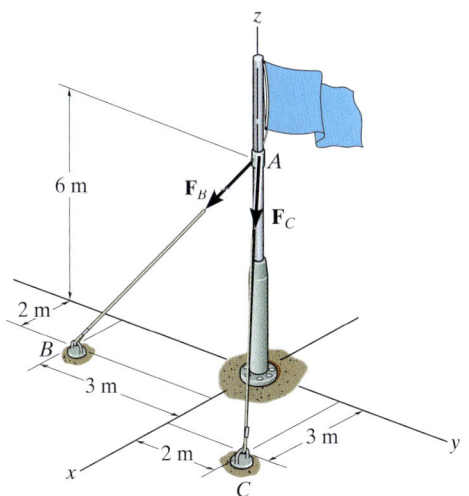

Probs. 2–92/93

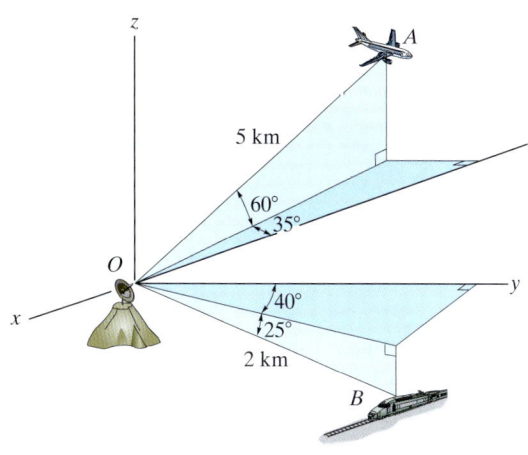

Prob. 2–95

2

*2–96. Two cables are used to secure the overhang boom in position and support the 1500-N load. If the resultant force is directed along the boom from point A towards O, determine the magnitudes of the resultant force and forces $\mathbf{F}_B$ and $\mathbf{F}_C$. Set $x = 3$ m and $z = 2$ m.

2–97. Two cables are used to secure the overhang boom in position and support the 1500-N load. If the resultant force is directed along the boom from point A towards O, determine the values of x and z for the coordinates of point C and the magnitude of the resultant force. Set $F_B = 1610$ N and $F_C = 2400$ N.

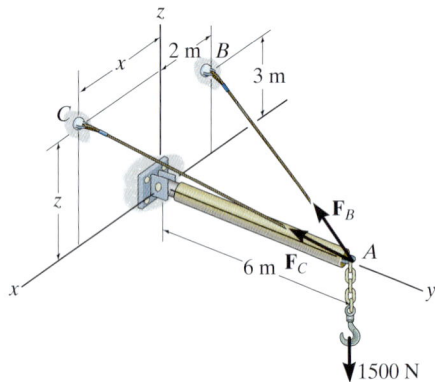

Probs. 2–96/97

2–98. The door is held opened by means of two chains. If the tension in AB and CD is $F_A = 300$ N and $F_C = 250$ N, respectively, express each of these forces in Cartesian vector form.

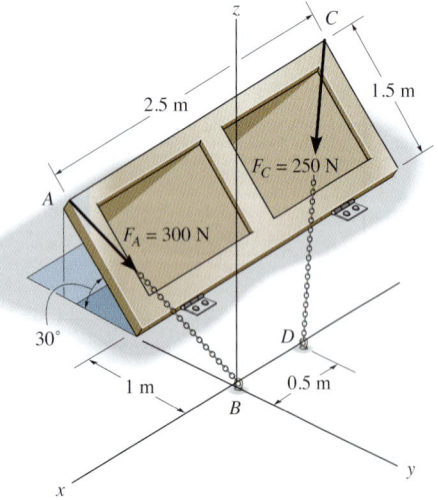

Prob. 2–98

2–99. Determine the magnitude and coordinate direction angles of the resultant force acting at point A.

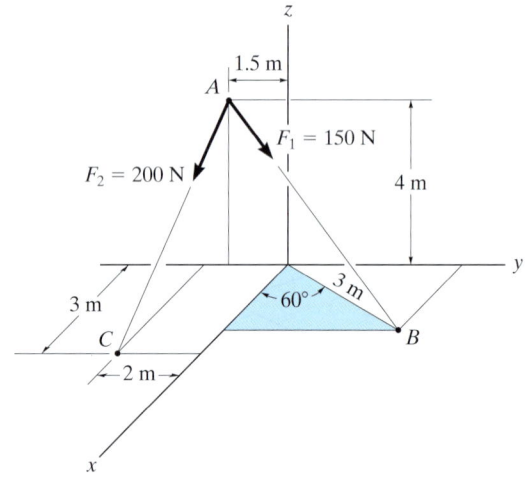

Prob. 2–99

*2–100. The guy wires are used to support the telephone pole. Represent the force in each wire in Cartesian vector form. Neglect the diameter of the pole.

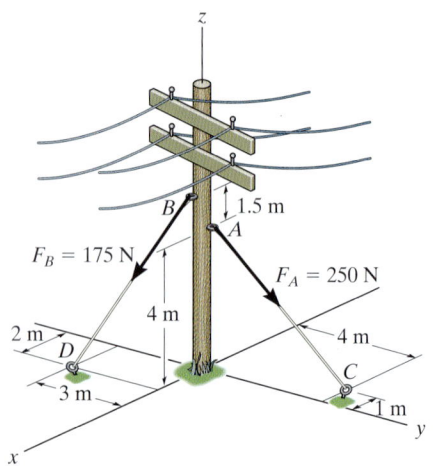

Prob. 2–100

2–101. The force acting on the man, caused by his pulling on the anchor cord, is $\mathbf{F} = \{40\mathbf{i} + 20\mathbf{j} - 50\mathbf{k}\}$ N. If the length of the cord is 25 m, determine the coordinates $A(x, y, -z)$ of the anchor.

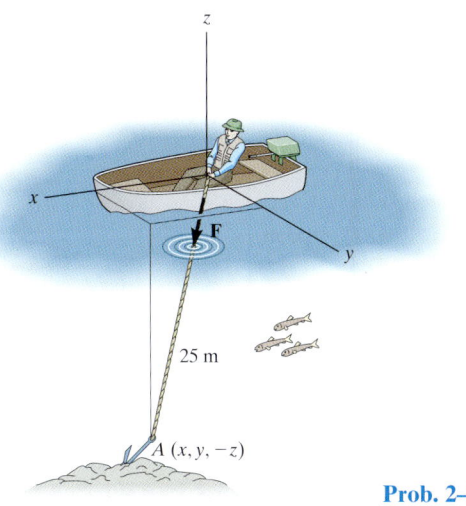

Prob. 2–101

2–102. Each of the four forces acting at E has a magnitude of 28 kN. Express each force as a Cartesian vector and determine the resultant force.

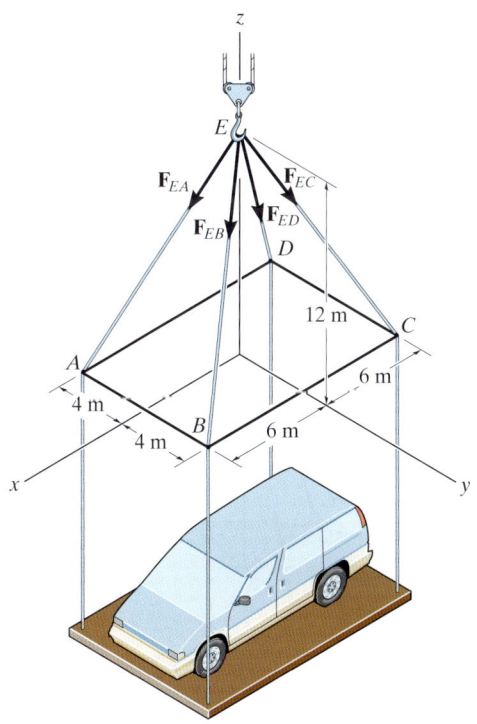

Prob. 2–102

2–103. The cord exerts a force of $\mathbf{F} = \{12\mathbf{i} + 9\mathbf{j} - 8\mathbf{k}\}$ kN on the hook. If the cord is 4 m long, determine the location x, y of the point of attachment B, and the height z of the hook.

*2–104. The cord exerts a force of $F = 30$ kN on the hook. If the cord is 4 m long, $z = 2$ m, and the x component of the force is $F_x = 25$ kN, determine the location x, y of the point of attachment B of the cord to the ground.

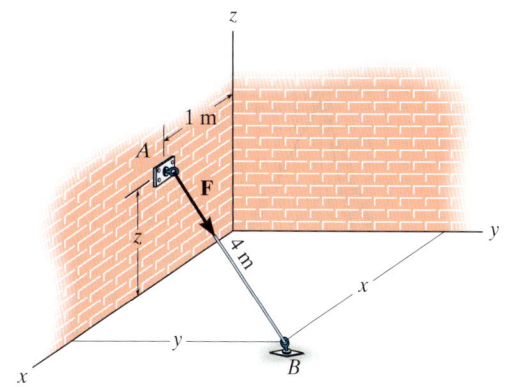

Probs. 2–103/104

2–105. The tower is held in place by three cables. If the force of each cable acting on the tower is shown, determine the magnitude and coordinate direction angles α, β, γ of the resultant force. Take $x = 20$ m, $y = 15$ m.

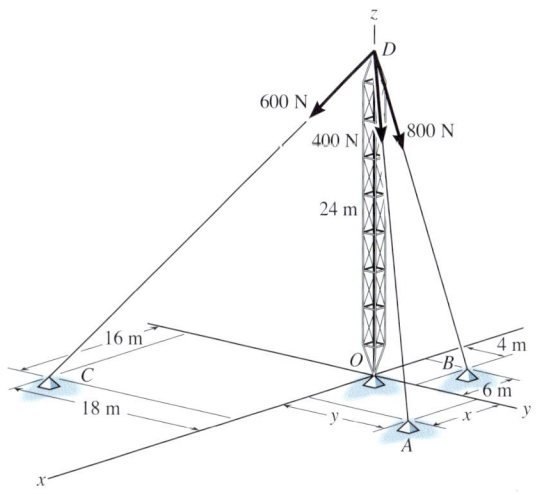

Prob. 2–105

2–106.   The chandelier is supported by three chains which are concurrent at point O. If the force in each chain has a magnitude of 300 N, express each force as a Cartesian vector and determine the magnitude and coordinate direction angles of the resultant force.

2–107.   The chandelier is supported by three chains which are concurrent at point O. If the resultant force at O has a magnitude of 650 N and is directed along the negative z axis, determine the force in each chain assuming $F_A = F_B = F_C$.

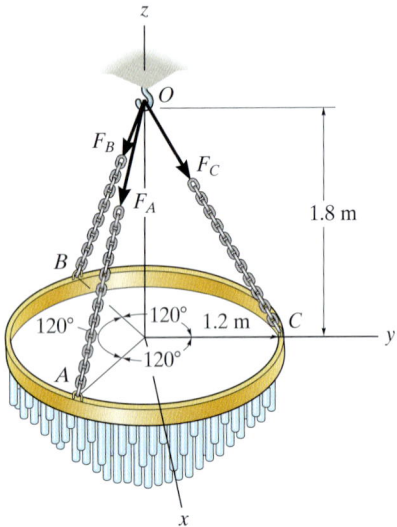

Probs. 2–106/107

*2–108.   Determine the magnitude and coordinate direction angles of the resultant force. Set $F_B = 630$ N, $F_C = 520$ N and $F_D = 750$ N, and $x = 3$ m and $z = 3.5$ m.

2–109.   If the magnitude of the resultant force is 1300 N and acts along the axis of the strut, directed from point A towards O, determine the magnitudes of the three forces acting on the strut. Set $x = 0$ and $z = 5.5$ m.

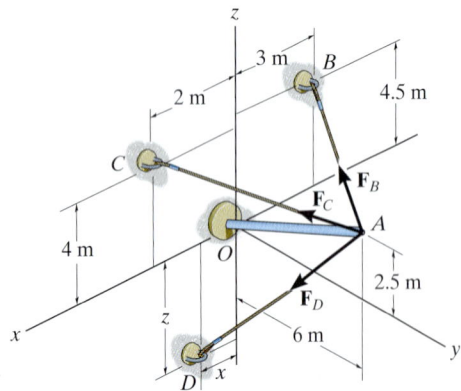

Probs. 2–108/109

2–110.   The positions of point A on the building and point B on the antenna have been measured relative to the electronic distance meter (EDM) at O. Determine the distance between A and B. *Hint:* Formulate a position vector directed from A to B; then determine its magnitude.

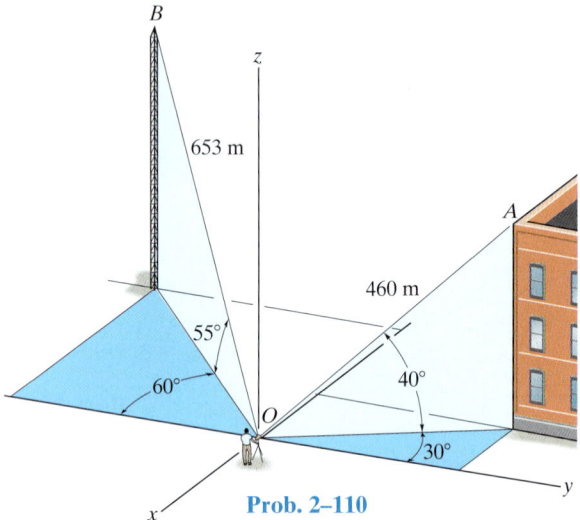

Prob. 2–110

2–111.   The cylindrical plate is subjected to the three cable forces which are concurrent at point D. Express each force which the cables exert on the plate as a Cartesian vector, and determine the magnitude and coordinate direction angles of the resultant force.

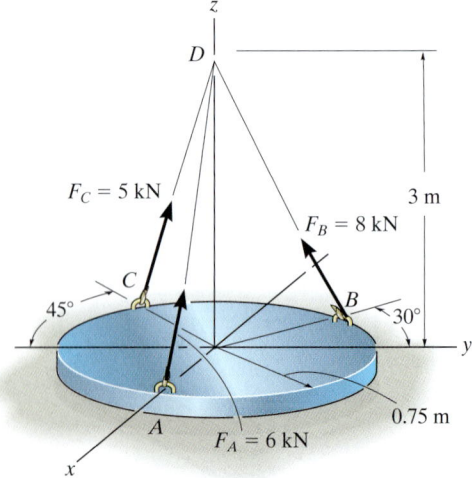

Prob. 2–111

2.9   Dot Product

Occasionally in statics one has to find the angle between two lines or the components of a force parallel and perpendicular to a line. In two dimensions, these problems can readily be solved by trigonometry since the geometry is easy to visualize. In three dimensions, however, this is often difficult, and consequently vector methods should be employed for the solution. The dot product, which defines a particular method for "multiplying" two vectors, is used to solve the above-mentioned problems.

The *dot product* of vectors **A** and **B**, written **A** · **B**, and read "**A** dot **B**" is defined as the product of the magnitudes of **A** and **B** and the cosine of the angle θ between their tails, Fig. 2–42. Expressed in equation form,

$$\boxed{\mathbf{A} \cdot \mathbf{B} = AB \cos \theta} \qquad (2\text{–}12)$$

where $0° \leq \theta \leq 180°$. The dot product is often referred to as the *scalar product* of vectors since the result is a *scalar* and not a vector.

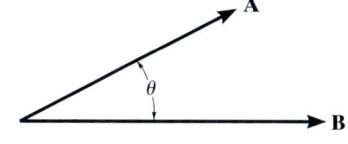

Fig. 2–42

Laws of Operation.

1. Commutative law: $\mathbf{A} \cdot \mathbf{B} = \mathbf{B} \cdot \mathbf{A}$

2. Multiplication by a scalar: $a(\mathbf{A} \cdot \mathbf{B}) = (a\mathbf{A}) \cdot \mathbf{B} = \mathbf{A} \cdot (a\mathbf{B})$

3. Distributive law: $\mathbf{A} \cdot (\mathbf{B} + \mathbf{D}) = (\mathbf{A} \cdot \mathbf{B}) + (\mathbf{A} \cdot \mathbf{D})$

It is easy to prove the first and second laws by using Eq. 2–12. The proof of the distributive law is left as an exercise (see Prob. 2–112).

Cartesian Vector Formulation.   Equation 2–12 must be used to find the dot product for any two Cartesian unit vectors. For example, $\mathbf{i} \cdot \mathbf{i} = (1)(1) \cos 0° = 1$ and $\mathbf{i} \cdot \mathbf{j} = (1)(1) \cos 90° = 0$. If we want to find the dot product of two general vectors **A** and **B** that are expressed in Cartesian vector form, then we have

$$
\begin{aligned}
\mathbf{A} \cdot \mathbf{B} &= (A_x\mathbf{i} + A_y\mathbf{j} + A_z\mathbf{k}) \cdot (B_x\mathbf{i} + B_y\mathbf{j} + B_z\mathbf{k}) \\
&= A_xB_x(\mathbf{i} \cdot \mathbf{i}) + A_xB_y(\mathbf{i} \cdot \mathbf{j}) + A_xB_z(\mathbf{i} \cdot \mathbf{k}) \\
&\quad + A_yB_x(\mathbf{j} \cdot \mathbf{i}) + (A_yB_y(\mathbf{j} \cdot \mathbf{j}) + A_yB_z(\mathbf{j} \cdot \mathbf{k}) \\
&\quad + A_zB_x(\mathbf{k} \cdot \mathbf{i}) + A_zB_y(\mathbf{k} \cdot \mathbf{j}) + A_zB_z(\mathbf{k} \cdot \mathbf{k})
\end{aligned}
$$

Carrying out the dot-product operations, the final result becomes

$$\boxed{\mathbf{A} \cdot \mathbf{B} = A_xB_x + A_yB_y + A_zB_z} \qquad (2\text{–}13)$$

Thus, to determine the dot product of two Cartesian vectors, multiply their corresponding x, y, z components and sum these products algebraically. Note that the result will be either a positive or negative *scalar*.

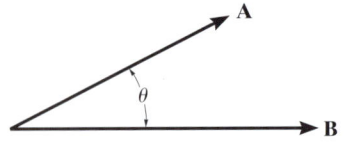

Fig. 2–42 (Repeated)

The angle θ between the rope and the beam can be determined by formulating unit vectors along the beam and rope and then using the dot product $\mathbf{u}_b \cdot \mathbf{u}_r = (1)(1) \cos \theta$.

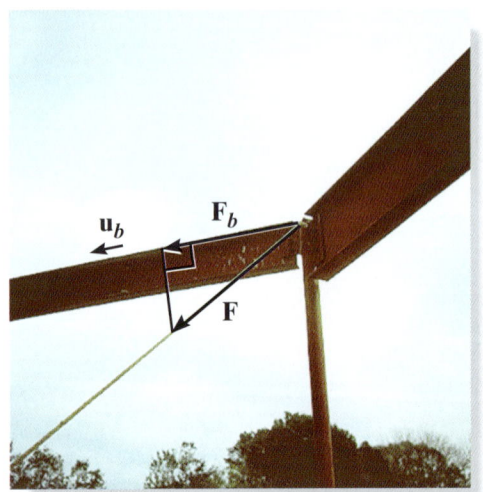

The projection of the cable force $\mathbf{F}$ along the beam can be determined by first finding the unit vector $\mathbf{u}_b$ that defines this direction. Then apply the dot product, $F_b = \mathbf{F} \cdot \mathbf{u}_b$.

Applications.   The dot product has two important applications in mechanics.

- ***The angle formed between two vectors or intersecting lines.***   The angle θ between the tails of vectors **A** and **B** in Fig. 2–42 can be determined from Eq. 2–12 and written as

$$\theta = \cos^{-1}\left(\frac{\mathbf{A} \cdot \mathbf{B}}{AB}\right) \quad 0° \leq \theta \leq 180°$$

Here $\mathbf{A} \cdot \mathbf{B}$ is found from Eq. 2–13. In particular, notice that if $\mathbf{A} \cdot \mathbf{B} = 0, \theta = \cos^{-1} 0 = 90°$ so that **A** will be *perpendicular* to **B**.

- ***The components of a vector parallel and perpendicular to a line.***   The component of vector **A** parallel to or collinear with the line *aa* in Fig. 2–43 is defined by A_a where $A_a = A \cos \theta$. This component is sometimes referred to as the *projection* of **A** onto the line, since a *right angle* is formed in the construction. If the *direction* of the line is specified by the unit vector $\mathbf{u}_a$, then since $u_a = 1$, we can determine the magnitude of A_a directly from the dot product (Eq. 2–12); i.e.,

$$A_a = A \cos \theta = \mathbf{A} \cdot \mathbf{u}_a$$

Hence, the scalar projection of **A** *along a line is determined from the dot product of* **A** *and the unit vector* $\mathbf{u}_a$ *which defines the direction of the line.* Notice that if this result is positive, then $\mathbf{A}_a$ has a directional sense which is the same as $\mathbf{u}_a$, whereas if A_a is a negative scalar, then $\mathbf{A}_a$ has the opposite sense of direction to $\mathbf{u}_a$.

The component $\mathbf{A}_a$ represented as a *vector* is therefore

$$\mathbf{A}_a = A_a \mathbf{u}_a$$

The component of **A** that is *perpendicular* to line *aa* can also be obtained, Fig. 2–43. Since $\mathbf{A} = \mathbf{A}_a + \mathbf{A}_\perp$, then $\mathbf{A}_\perp = \mathbf{A} - \mathbf{A}_a$. There are two possible ways of obtaining $A_\perp$. One way would be to determine θ from the dot product, $\theta = \cos^{-1}(\mathbf{A} \cdot \mathbf{u}_A / A)$, then $A_\perp = A \sin \theta$. Alternatively, if A_a is known, then by Pythagorean's theorem we can also write $A_\perp = \sqrt{A^2 - A_a^2}$.

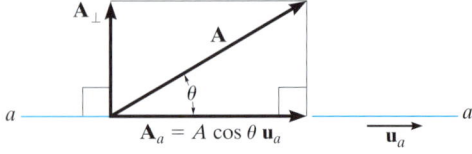

Fig. 2–43

2

Important Points

- The dot product is used to determine the angle between two vectors or the projection of a vector in a specified direction.

- If vectors **A** and **B** are expressed in Cartesian vector form, the dot product is determined by multiplying the respective x, y, z scalar components and algebraically adding the results, i.e., $\mathbf{A} \cdot \mathbf{B} = A_x B_x + A_y B_y + A_z B_z$.

- From the definition of the dot product, the angle formed between the tails of vectors **A** and **B** is $\theta = \cos^{-1}(\mathbf{A} \cdot \mathbf{B}/AB)$.

- The magnitude of the projection of vector **A** along a line aa whose direction is specified by $\mathbf{u}_a$ is determined from the dot product $A_a = \mathbf{A} \cdot \mathbf{u}_a$.

EXAMPLE 2.16

Determine the magnitudes of the projection of the force **F** in Fig. 2–44 onto the u and v axes.

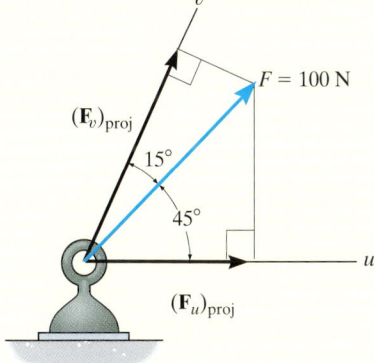

Fig. 2–44

SOLUTION

Projections of Force.   The graphical representation of the *projections* is shown in Fig. 2–44. From this figure, the magnitudes of the projections of **F** onto the u and v axes can be obtained by trigonometry:

$$(F_u)_{\text{proj}} = (100\text{ N})\cos 45° = 70.7\text{ N} \qquad \textit{Ans.}$$
$$(F_v)_{\text{proj}} = (100\text{ N})\cos 15° = 96.6\text{ N} \qquad \textit{Ans.}$$

NOTE: These projections are not equal to the magnitudes of the components of force **F** along the u and v axes found from the parallelogram law. They will only be equal if the u and v axes are *perpendicular* to one another.

EXAMPLE | 2.17

The frame shown in Fig. 2–45a is subjected to a horizontal force $\mathbf{F} = \{300\mathbf{j}\}$. Determine the magnitude of the components of this force parallel and perpendicular to member AB.

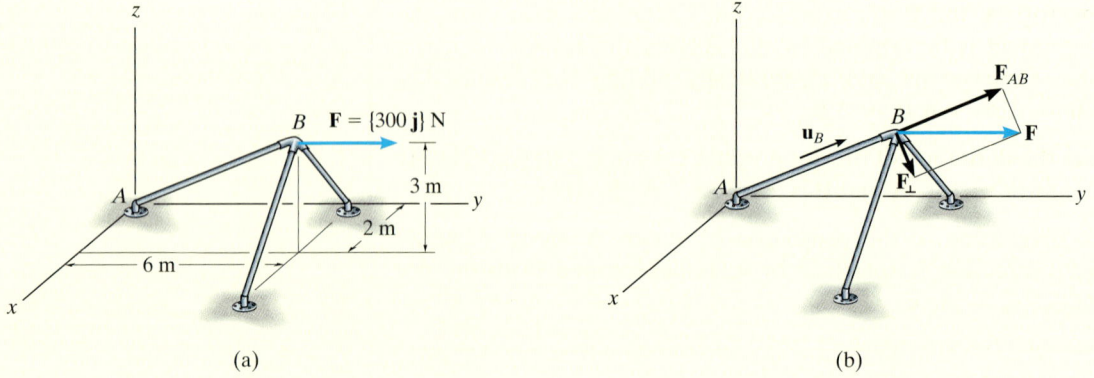

Fig. 2–45

SOLUTION

The magnitude of the component of $\mathbf{F}$ along AB is equal to the dot product of $\mathbf{F}$ and the unit vector $\mathbf{u}_B$, which defines the direction of AB, Fig. 2–45b. Since

$$\mathbf{u}_B = \frac{\mathbf{r}_B}{r_B} = \frac{2\mathbf{i} + 6\mathbf{j} + 3\mathbf{k}}{\sqrt{(2)^2 + (6)^2 + (3)^2}} = 0.286\mathbf{i} + 0.857\mathbf{j} + 0.429\mathbf{k}$$

then

$$F_{AB} = F\cos\theta = \mathbf{F} \cdot \mathbf{u}_B = (300\mathbf{j}) \cdot (0.286\mathbf{i} + 0.857\mathbf{j} + 0.429\mathbf{k})$$
$$= (0)(0.286) + (300)(0.857) + (0)(0.429)$$
$$= 257.1 \text{ N} \qquad\qquad Ans.$$

Since the result is a positive scalar, $\mathbf{F}_{AB}$ has the same sense of direction as $\mathbf{u}_B$, Fig. 2–45b.

Expressing $\mathbf{F}_{AB}$ in Cartesian vector form, we have

$$\mathbf{F}_{AB} = F_{AB}\mathbf{u}_B = (257.1 \text{ N})(0.286\mathbf{i} + 0.857\mathbf{j} + 0.429\mathbf{k})$$
$$= \{73.5\mathbf{i} + 220\mathbf{j} + 110\mathbf{k}\} \text{ N} \qquad\qquad Ans.$$

The perpendicular component, Fig. 2–45b, is therefore

$$\mathbf{F}_{\perp} = \mathbf{F} - \mathbf{F}_{AB} = 300\mathbf{j} - (73.5\mathbf{i} + 220\mathbf{j} + 110\mathbf{k})$$
$$= \{-73.5\mathbf{i} + 79.6\mathbf{j} - 110\mathbf{k}\} \text{ N}$$

Its magnitude can be determined either from this vector or by using the Pythagorean theorem, Fig. 2–45b:

$$F_{\perp} = \sqrt{F^2 - F_{AB}^2} = \sqrt{(300 \text{ N})^2 - (257.1 \text{ N})^2}$$
$$= 155 \text{ N} \qquad\qquad Ans.$$

EXAMPLE | 2.18

The pipe in Fig. 2–46a is subjected to the force of $F = 80$ N. Determine the angle θ between $\mathbf{F}$ and the pipe segment BA and the projection of $\mathbf{F}$ along this segment.

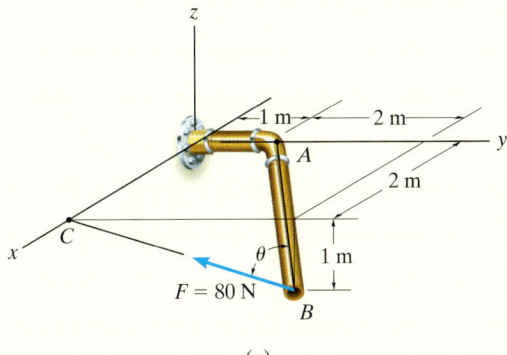

(a)

SOLUTION

Angle θ.   First we will establish position vectors from B to A and B to C; Fig. 2–46b. Then we will determine the angle θ between the tails of these two vectors.

$$\mathbf{r}_{BA} = \{-2\mathbf{i} - 2\mathbf{j} + 1\mathbf{k}\} \text{ ft}, \quad r_{BA} = 3 \text{ m}$$
$$\mathbf{r}_{BC} = \{-3\mathbf{j} + 1\mathbf{k}\} \text{ ft}, \quad r_{BC} = \sqrt{10} \text{ m}$$

Thus,

$$\cos \theta = \frac{\mathbf{r}_{BA} \cdot \mathbf{r}_{BC}}{r_{BA} r_{BC}} = \frac{(-2)(0) + (-2)(-3) + (1)(1)}{3\sqrt{10}} = 0.7379$$

$$\theta = 42.5° \qquad \qquad \textit{Ans.}$$

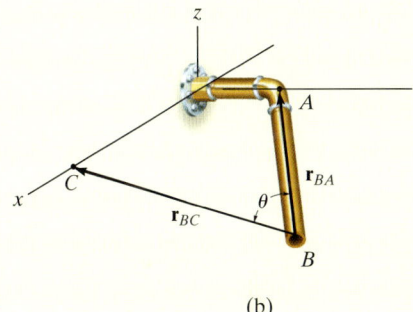

(b)

Components of F.   The component of $\mathbf{F}$ along BA is shown in Fig. 2–46c. We must first formulate the unit vector along BA and force $\mathbf{F}$ as Cartesian vectors.

$$\mathbf{u}_{BA} = \frac{\mathbf{r}_{BA}}{r_{DA}} = \frac{(-2\mathbf{i} - 2\mathbf{j} + 1\mathbf{k})}{3} = -\frac{2}{3}\mathbf{i} - \frac{2}{3}\mathbf{j} + \frac{1}{3}\mathbf{k}$$

$$\mathbf{F} = 80 \text{ N}\left(\frac{\mathbf{r}_{BC}}{r_{BC}}\right) = 80\left(\frac{-3\mathbf{j} + 1\mathbf{k}}{\sqrt{10}}\right) = -75.89\mathbf{j} + 25.30\mathbf{k}$$

Thus,

$$F_{BA} = \mathbf{F} \cdot \mathbf{u}_{BA} = (-75.89\mathbf{j} + 25.30\mathbf{k}) \cdot \left(-\frac{2}{3}\mathbf{i} - \frac{2}{3}\mathbf{j} + \frac{1}{3}\mathbf{k}\right)$$

$$= 0\left(-\frac{2}{3}\right) + (-75.89)\left(-\frac{2}{3}\right) + (25.30)\left(\frac{1}{3}\right)$$

$$= 59.0 \text{ N} \qquad \qquad \textit{Ans.}$$

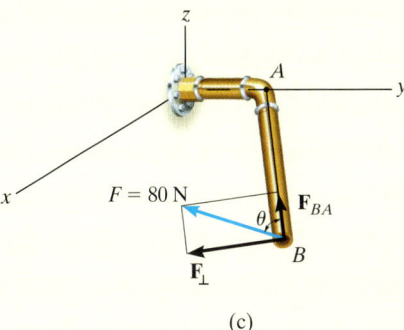

(c)

Fig. 2–46

NOTE: Since θ has been calculated, then also, $F_{BA} = F \cos \theta = 80 \text{ N} \cos 42.5° = 59.0$ N.

FUNDAMENTAL PROBLEMS

F2–25. Determine the angle θ between the force and the line AO.

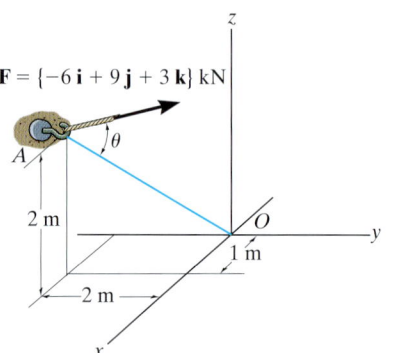

$F = \{-6\,\mathbf{i} + 9\,\mathbf{j} + 3\,\mathbf{k}\}$ kN

2 m

1 m

2 m

F2–25

F2–26. Determine the angle θ between the force and the line AB.

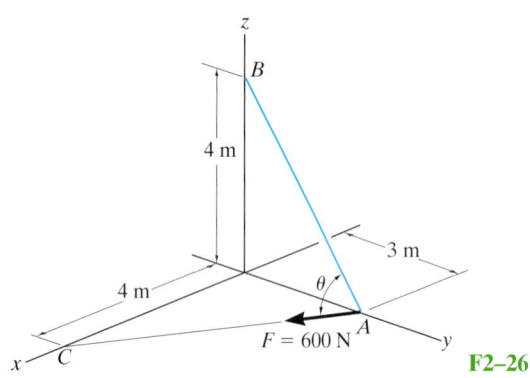

B

4 m

3 m

4 m

C

$F = 600$ N

F2–26

F2–27. Determine the angle θ between the force and the line OA.

F2–28. Determine the component of projection of the force along the line OA.

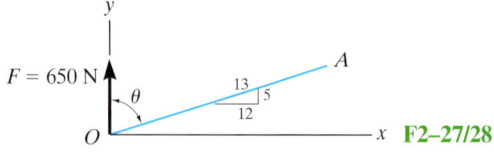

$F = 650$ N

13

12

5

A

F2–27/28

F2–29. Find the magnitude of the projected component of the force along the pipe AO.

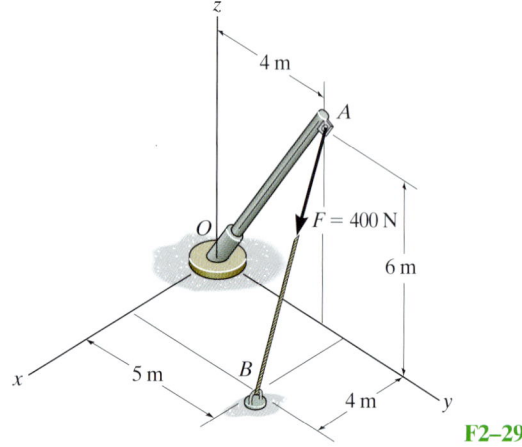

4 m

A

$F = 400$ N

6 m

5 m

B

4 m

F2–29

F2–30. Determine the components of the force acting parallel and perpendicular to the axis of the pole.

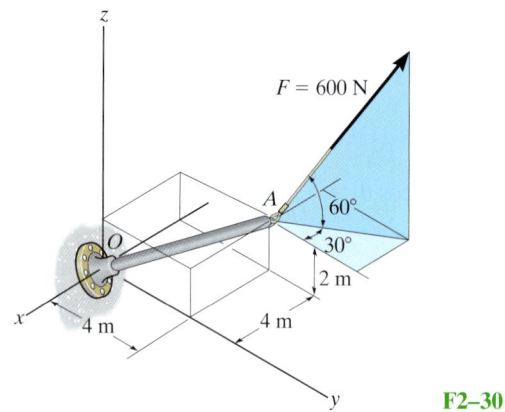

$F = 600$ N

A

60°

30°

2 m

O

4 m

4 m

F2–30

F2–31. Determine the magnitudes of the components of force $F = 56$ N acting along and perpendicular to line AO.

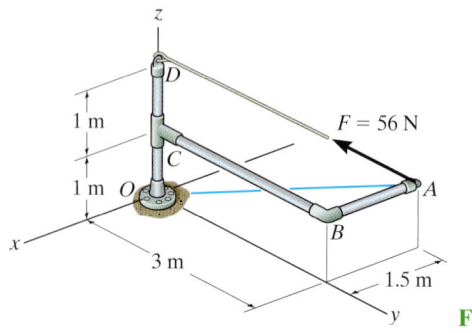

D

1 m

C

1 m   O

$F = 56$ N

A

B

3 m

1.5 m

F2–31

PROBLEMS

*2–112. Given the three vectors **A**, **B**, and **D**, show that $\mathbf{A} \cdot (\mathbf{B} + \mathbf{D}) = (\mathbf{A} \cdot \mathbf{B}) + (\mathbf{A} \cdot \mathbf{D})$.

2–113. Determine the angle θ between the edges of the sheet-metal bracket.

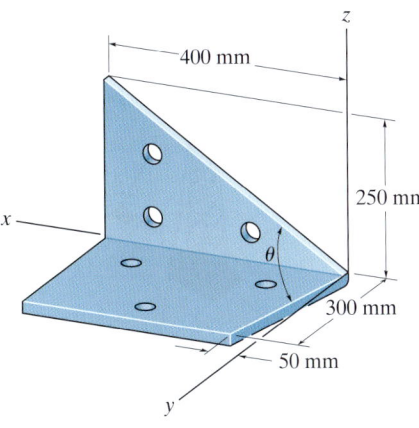

Prob. 2–113

2–114. Determine the angle θ between the sides of the triangular plate.

2–115. Determine the length of side BC of the triangular plate. Solve the problem by finding the magnitude of $\mathbf{r}_{BC}$; then check the result by first finding θ, r_{AB}, and r_{AC} and then use the cosine law.

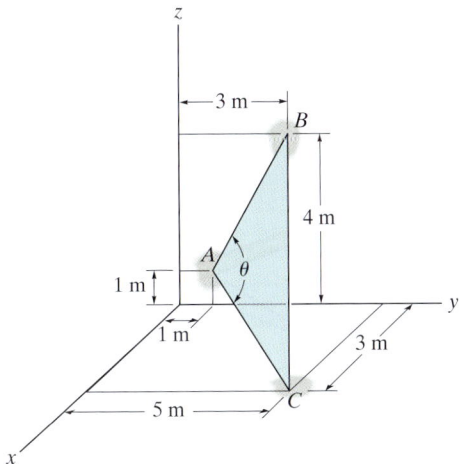

Probs. 2–114/115

*2–116. Determine the angle θ between the tails of the two vectors.

2–117. Determine the magnitude of the projected component of $\mathbf{r}_1$ along $\mathbf{r}_2$, and the projection of $\mathbf{r}_2$ along $\mathbf{r}_1$.

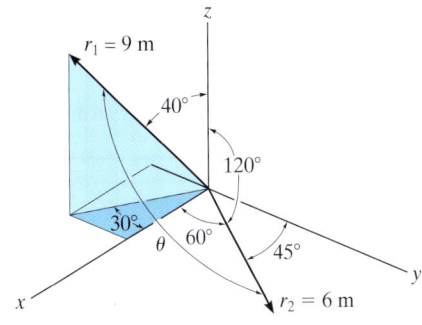

Probs. 2–116/117

2–118. Determine the projection of the force **F** along the pole.

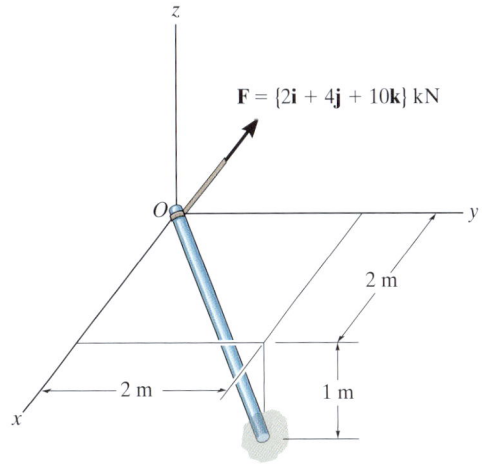

Prob. 2–118

2–119. Determine the angle θ between the two cords.

2–122. Determine the magnitude of the projected component of force $\mathbf{F}_{AB}$ acting along the z axis.

2–123. Determine the magnitude of the projected component of force $\mathbf{F}_{AC}$ acting along the z axis.

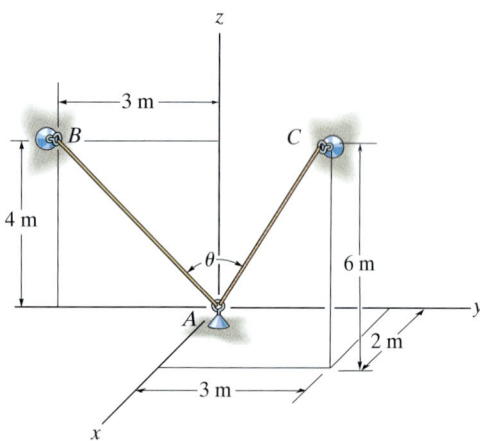

Prob. 2–119

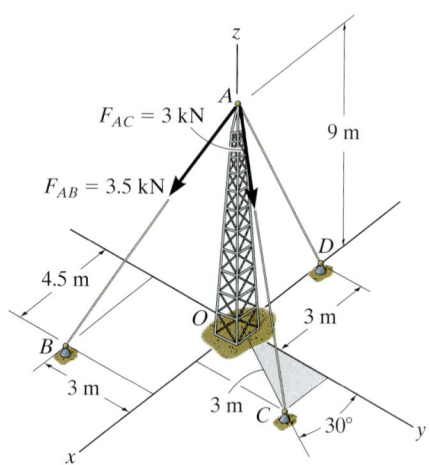

Probs. 2–122/123

2–120. Two forces act on the hook. Determine the angle θ between them. Also, what are the projections of $\mathbf{F}_1$ and $\mathbf{F}_2$ along the y axis?

2–121. Two forces act on the hook. Determine the magnitude of the projection of $\mathbf{F}_2$ along $\mathbf{F}_1$.

2–124. Determine the projection of force $F = 400$ N acting along line AC of the pipe assembly. Express the result as a Cartesian vector.

2–125. Determine the magnitudes of the components of force $F = 400$ N acting parallel and perpendicular to segment BC of the pipe assembly.

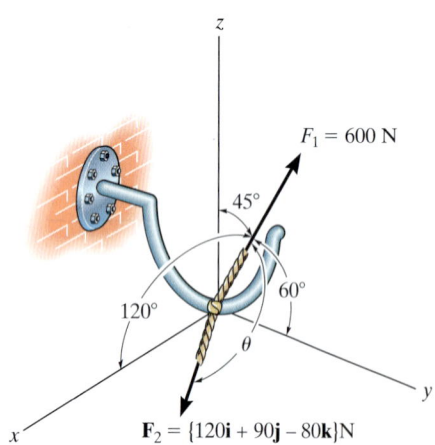

Probs. 2–120/121

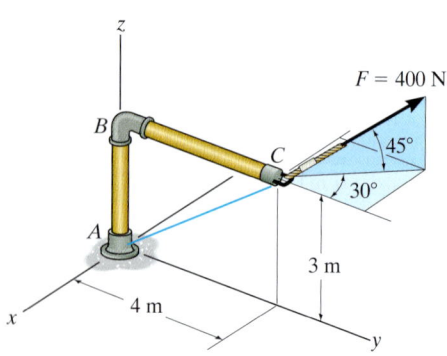

Probs. 2–124/125

2–126. Cable *OA* is used to support column *OB*. Determine the angle θ it makes with beam *OC*.

2–127. Cable *OA* is used to support column *OB*. Determine the angle ϕ it makes with beam *OD*.

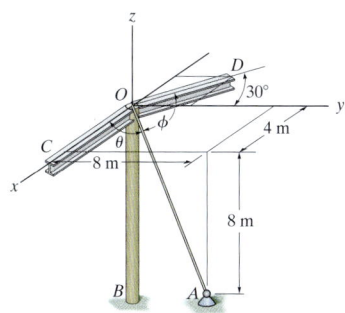

Probs. 2–126/127

*2–128. Determine the angles θ and ϕ between the axis *OA* of the pole and each cable, *AB* and *AC*.

2–129. The two cables exert the forces shown on the pole. Determine the magnitude of the projected component of each force acting along the axis *OA* of the pole.

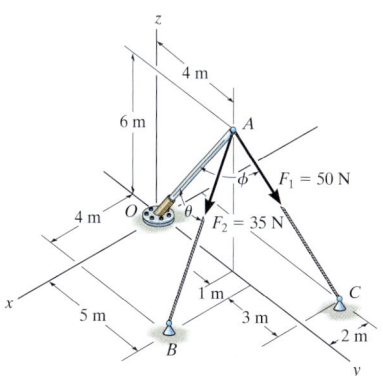

Probs. 2–128/129

2–130. Determine the angle θ between the pipe segments *BA* and *BC*.

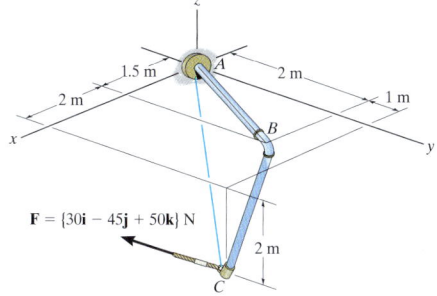

Prob. 2–130

2–131. Determine the angles θ and ϕ made between the axes *OA* of the flag pole and *AB* and *AC*, respectively, of each cable.

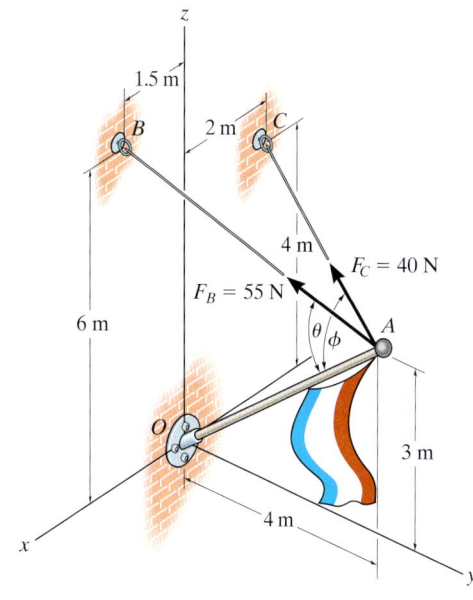

Prob. 2–131

*2–132. The cables each exert a force of 400 N on the post. Determine the magnitude of the projected component of $\mathbf{F}_1$ along the line of action of $\mathbf{F}_2$.

2–133. Determine the angle θ between the two cables attached to the post.

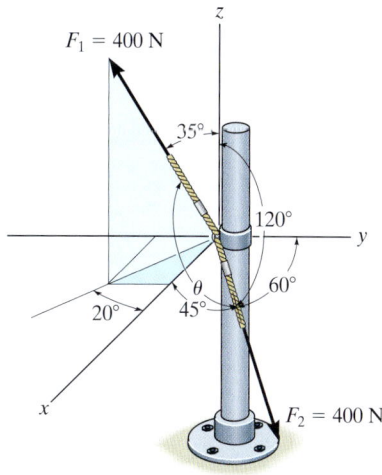

Probs. 2–132/133

2–134. A force of $F = 80$ N is applied to the handle of the wrench. Determine the angle θ between the tail of the force and the handle AB.

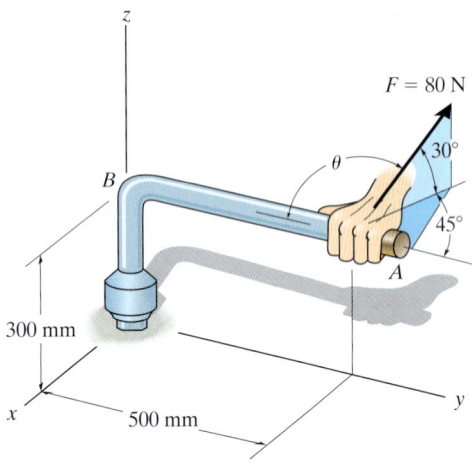

Prob. 2–134

2–135. Determine the projected component of the 80-N force acting along the axis AB of the pipe.

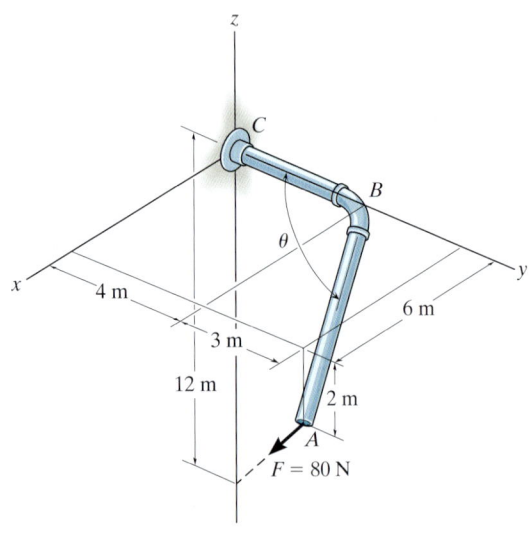

Prob. 2–135

*2–136.** Determine the components of **F** that act along rod AC and perpendicular to it. Point B is located at the midpoint of the rod.

2–137. Determine the components of **F** that act along rod AC and perpendicular to it. Point B is located 3 m along the rod from end C.

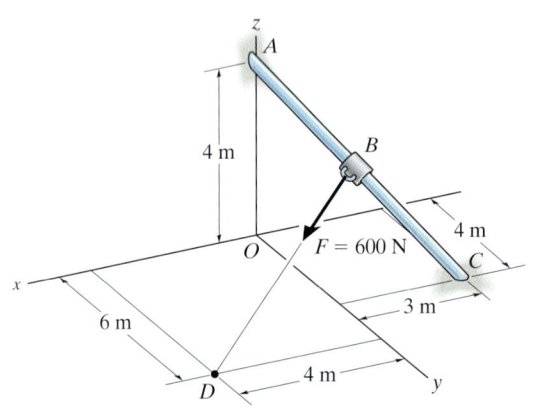

Probs. 2–136/137

2–138. Determine the magnitudes of the projected components of the force $F = 300$ N acting along the x and y axes.

2–139. Determine the magnitude of the projected component of the force $F = 300$ N acting along line OA.

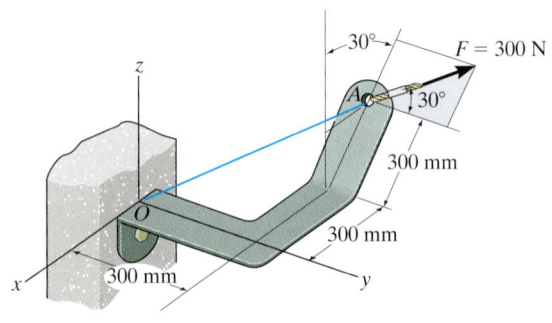

Probs. 2–138/139

CHAPTER REVIEW

A scalar is a positive or negative number; e.g., mass and temperature. A vector has a magnitude and direction, where the arrowhead represents the sense of the vector.		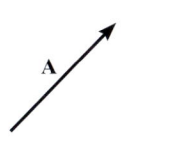
Multiplication or division of a vector by a scalar will change only the magnitude of the vector. If the scalar is negative, the sense of the vector will change so that it acts in the opposite sense.		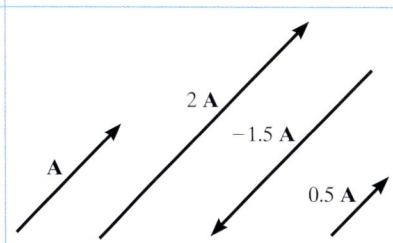
If vectors are collinear, the resultant is simply the algebraic or scalar addition.	$R = A + B$	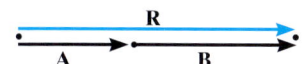
Parallelogram Law Two forces add according to the parallelogram law. The *components* form the sides of the parallelogram and the *resultant* is the diagonal. To find the components of a force along any two axes, extend lines from the head of the force, parallel to the axes, to form the components. To obtain the components of the resultant, show how the forces add by tip-to-tail using the triangle rule, and then use the law of cosines and the law of sines to calculate their values.	$F_R = \sqrt{F_1^2 + F_2^2 - 2F_1F_2\cos\theta_R}$ $\dfrac{F_1}{\sin\theta_1} = \dfrac{F_2}{\sin\theta_2} = \dfrac{F_R}{\sin\theta_R}$	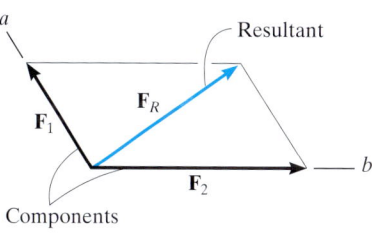 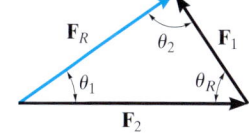

Rectangular Components: Two Dimensions

Vectors $\mathbf{F}_x$ and $\mathbf{F}_y$ are rectangular components of $\mathbf{F}$.

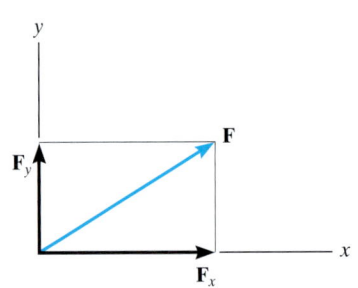

The resultant force is determined from the algebraic sum of its components.

$$(F_R)_x = \Sigma F_x$$
$$(F_R)_y = \Sigma F_y$$
$$F_R = \sqrt{(F_R)_x^2 + (F_R)_y^2}$$
$$\theta = \tan^{-1}\left|\frac{(F_R)_y}{(F_R)_x}\right|$$

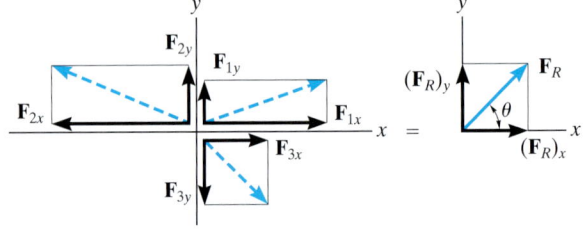

Cartesian Vectors

The unit vector $\mathbf{u}$ has a length of 1, no units, and it points in the direction of the vector $\mathbf{F}$.

$$\mathbf{u} = \frac{\mathbf{F}}{F}$$

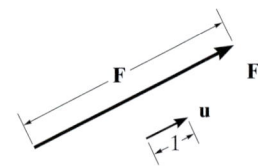

A force can be resolved into its Cartesian components along the x, y, z axes so that $\mathbf{F} = F_x\mathbf{i} + F_y\mathbf{j} + F_z\mathbf{k}$.

The magnitude of $\mathbf{F}$ is determined from the positive square root of the sum of the squares of its components.

$$F = \sqrt{F_x^2 + F_y^2 + F_z^2}$$

The coordinate direction angles α, β, γ are determined by formulating a unit vector in the direction of $\mathbf{F}$. The x, y, z components of $\mathbf{u}$ represent $\cos\alpha, \cos\beta, \cos\gamma$.

$$\mathbf{u} = \frac{\mathbf{F}}{F} = \frac{F_x}{F}\mathbf{i} + \frac{F_y}{F}\mathbf{j} + \frac{F_z}{F}\mathbf{k}$$
$$\mathbf{u} = \cos\alpha\,\mathbf{i} + \cos\beta\,\mathbf{j} + \cos\gamma\,\mathbf{k}$$

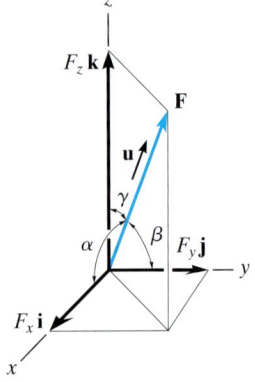

The coordinate direction angles are related so that only two of the three angles are independent of one another.

$$\cos^2 \alpha + \cos^2 \beta + \cos^2 \gamma = 1$$

To find the resultant of a concurrent force system, express each force as a Cartesian vector and add the $\mathbf{i}, \mathbf{j}, \mathbf{k}$ components of all the forces in the system.

$$\mathbf{F}_R = \Sigma \mathbf{F} = \Sigma F_x \mathbf{i} + \Sigma F_y \mathbf{j} + \Sigma F_z \mathbf{k}$$

Position and Force Vectors

A position vector locates one point in space relative to another. The easiest way to formulate the components of a position vector is to determine the distance and direction that one must travel along the x, y, and z directions—going from the tail to the head of the vector.

$$\mathbf{r} = (x_B - x_A)\mathbf{i}$$
$$+ (y_B - y_A)\mathbf{j}$$
$$+ (z_B - z_A)\mathbf{k}$$

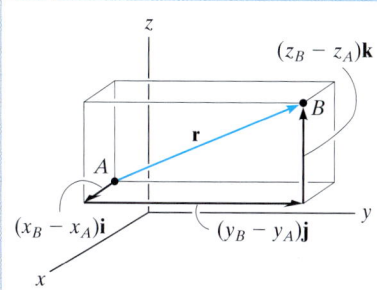

If the line of action of a force passes through points A and B, then the force acts in the same direction as the position vector $\mathbf{r}$, which is defined by the unit vector $\mathbf{u}$. The force can then be expressed as a Cartesian vector.

$$\mathbf{F} = F\mathbf{u} = F\left(\dfrac{\mathbf{r}}{r}\right)$$

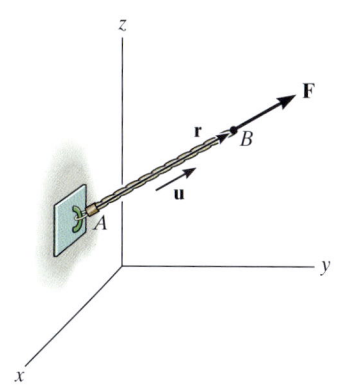

Dot Product

The dot product between two vectors $\mathbf{A}$ and $\mathbf{B}$ yields a scalar. If $\mathbf{A}$ and $\mathbf{B}$ are expressed in Cartesian vector form, then the dot product is the sum of the products of their x, y, and z components.

$$\mathbf{A} \cdot \mathbf{B} = AB \cos \theta$$
$$= A_x B_x + A_y B_y + A_z B_z$$

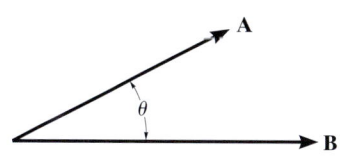

The dot product can be used to determine the angle between $\mathbf{A}$ and $\mathbf{B}$.

$$\theta = \cos^{-1}\left(\dfrac{\mathbf{A} \cdot \mathbf{B}}{AB}\right)$$

The dot product is also used to determine the projected component of a vector $\mathbf{A}$ onto an axis aa defined by its unit vector $\mathbf{u}_a$.

$$\mathbf{A}_a = A \cos \theta \, \mathbf{u}_a = (\mathbf{A} \cdot \mathbf{u}_a)\mathbf{u}_a$$

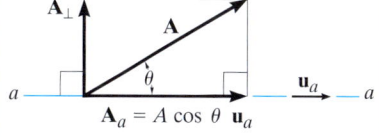

REVIEW PROBLEMS

*2–140. Determine the length of the conneting rod AB by first formulating a Cartesian position vector from A to B and then determining its magnitude.

2–143. Resolve the 250-N force into components acting along the u and v axes and determine the magnitudes of these components.

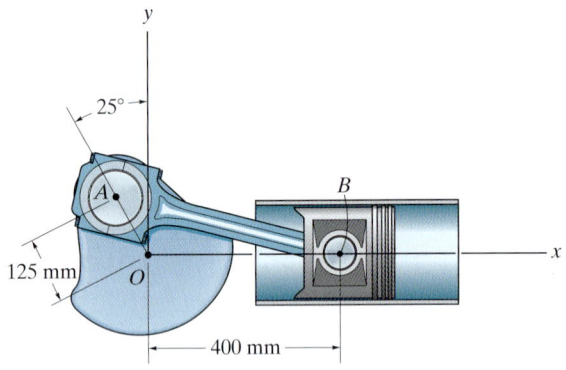

Prob. 2–140

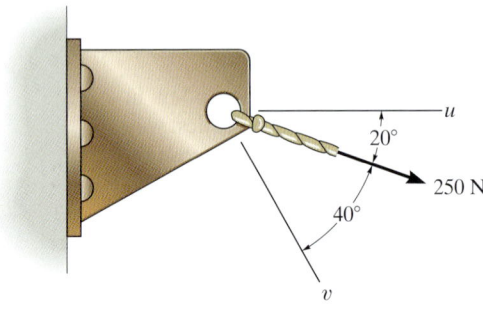

Prob. 2–143

2–141. Determine the x and y components of $\mathbf{F}_1$ and $\mathbf{F}_2$.

2–142. Determine the magnitude of the resultant force and its direction, measured counterclockwise from the positive x axis.

*2–144. Express $\mathbf{F}_1$ and $\mathbf{F}_2$ as Cartesian vectors.

2–145. Determine the magnitude of the resultant force and its direction measured counterclockwise from the positive x axis.

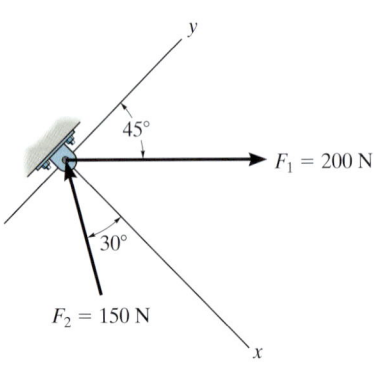

Probs. 2–141/142

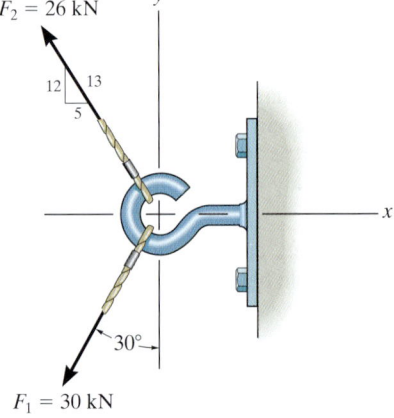

Probs. 2–144/145

2–146. Cable *AB* exerts a force of 80 N on the end of the 3-m-long boom *OA*. Determine the magnitude of the projection of this force along the boom.

*2–148. If $\theta = 60°$ and $F = 20$ kN, determine the magnitude of the resultant force and its direction measured clockwise from the positive *x* axis.

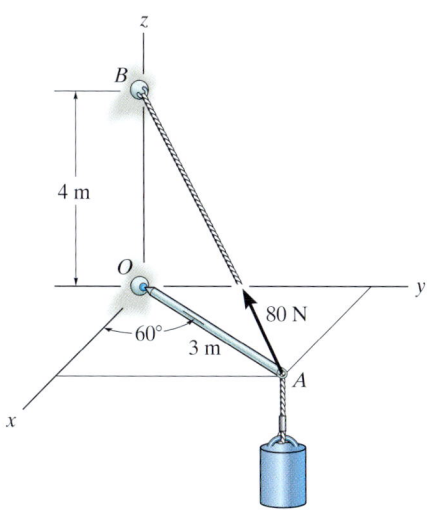

Prob. 2–146

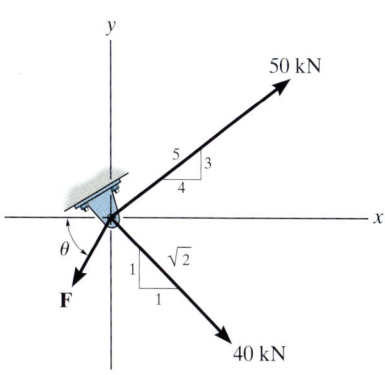

Prob. 2–148

2–147. Determine the magnitude and direction of the resultant $\mathbf{F}_R = \mathbf{F}_1 + \mathbf{F}_2 + \mathbf{F}_3$ of the three forces by first finding the resultant $\mathbf{F}' = \mathbf{F}_1 + \mathbf{F}_3$ and then forming $\mathbf{F}_R = \mathbf{F}' + \mathbf{F}_2$. Specify its direction measured counterclockwise from the positive *x* axis.

2–149. Determine the design angle θ ($0° \le \theta \le 90°$) for strut *AB* so that the 400-N horizontal force has a component of 500 N directed from *A* towards *C*. What is the component of force acting along member *AB*? Take $\phi = 40°$.

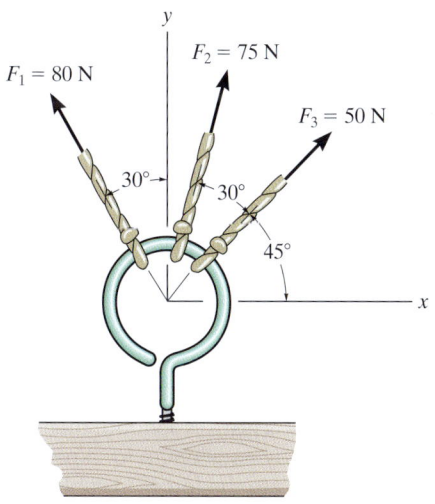

Prob. 2–147

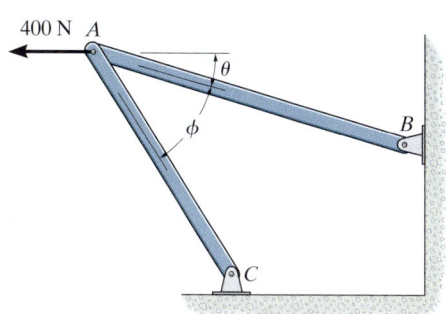

Prob. 2–149

Chapter 3

When this load is lifted at constant velocity, or is just suspended, then it is in a state of equilibrium. In this chapter we will study equilibrium for a particle and show how these ideas can be used to calculate the forces in cables used to hold suspended loads.

Equilibrium of a Particle

CHAPTER OBJECTIVES

■ To introduce the concept of the free-body diagram for a particle.

■ To show how to solve particle equilibrium problems using the equations of equilibrium.

3.1 Condition for the Equilibrium of a Particle

A particle is said to be in *equilibrium* if it remains at rest if originally at rest, or has a constant velocity if originally in motion. Most often, however, the term "equilibrium" or, more specifically, "static equilibrium" is used to describe an object at rest. To maintain equilibrium, it is *necessary* to satisfy Newton's first law of motion, which requires the *resultant force* acting on a particle to be equal to *zero*. This condition may be stated mathematically as

$$\Sigma \mathbf{F} = \mathbf{0} \qquad (3\text{–}1)$$

where $\Sigma \mathbf{F}$ is the vector *sum of all the forces* acting on the particle.

Not only is Eq. 3–1 a necessary condition for equilibrium, it is also a *sufficient* condition. This follows from Newton's second law of motion, which can be written as $\Sigma \mathbf{F} = m\mathbf{a}$. Since the force system satisfies Eq. 3–1, then $m\mathbf{a} = \mathbf{0}$, and therefore the particle's acceleration $\mathbf{a} = \mathbf{0}$. Consequently, the particle indeed moves with constant velocity or remains at rest.

3

3.2 The Free-Body Diagram

To apply the equation of equilibrium, we must account for *all* the known and unknown forces ($\Sigma \mathbf{F}$) which act *on* the particle. The best way to do this is to think of the particle as isolated and "free" from its surroundings. A drawing that shows the particle with *all* the forces that act on it is called a *free-body diagram (FBD)*.

Before presenting a formal procedure as to how to draw a free-body diagram, we will first consider two types of connections often encountered in particle equilibrium problems.

Springs. If a *linearly elastic spring* (or cord) of undeformed length l_0 is used to support a particle, the length of the spring will change in direct proportion to the force $\mathbf{F}$ acting on it, Fig. 3–1. A characteristic that defines the "elasticity" of a spring is the *spring constant* or *stiffness k*.

The magnitude of force exerted on a linearly elastic spring which has a stiffness k and is deformed (elongated or compressed) a distance $s = l - l_0$, measured from its *unloaded* position, is

$$F = ks \tag{3–2}$$

If s is positive, causing an elongation, then $\mathbf{F}$ must pull on the spring; whereas if s is negative, causing a shortening, then $\mathbf{F}$ must push on it. For example, if the spring in Fig. 3–1 has an unstretched length of 0.8 m and a stiffness $k = 500\ \text{N/m}$ and it is stretched to a length of 1 m, so that $s = l - l_0 = 1\ \text{m} - 0.8\ \text{m} = 0.2\ \text{m}$, then a force $F = ks = 500\ \text{N/m}(0.2\ \text{m}) = 100\ \text{N}$ is needed.

Cables and Pulleys. Unless otherwise stated throughout this book, except in Sec. 7.4, all cables (or cords) will be assumed to have negligible weight and they cannot stretch. Also, a cable can support *only* a tension or "pulling" force, and this force always acts in the direction of the cable. In Chapter 5, it will be shown that the tension force developed in a continuous cable which passes over a frictionless pulley must have a *constant* magnitude to keep the cable in equilibrium. Hence, for any angle θ, shown in Fig. 3–2, the cable is subjected to a constant tension T throughout its length.

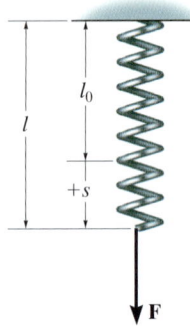

Fig. 3–1

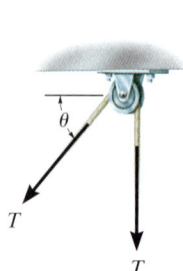

Cable is in tension

Fig. 3–2

Procedure for Drawing a Free-Body Diagram

Since we must account for *all the forces acting on the particle* when applying the equations of equilibrium, the importance of first drawing a free-body diagram cannot be overemphasized. To construct a free-body diagram, the following three steps are necessary.

Draw Outlined Shape.

Imagine the particle to be *isolated* or cut "free" from its surroundings by drawing its outlined shape.

Show All Forces.

Indicate on this sketch *all* the forces that act *on the particle*. These forces can be *active forces*, which tend to set the particle in motion, or they can be *reactive forces* which are the result of the constraints or supports that tend to prevent motion. To account for all these forces, it may be helpful to trace around the particle's boundary, carefully noting each force acting on it.

Identify Each Force.

The forces that are *known* should be labeled with their proper magnitudes and directions. Letters are used to represent the magnitudes and directions of forces that are unknown.

The bucket is held in equilibrium by the cable, and instinctively we know that the force in the cable must equal the weight of the bucket. By drawing a free-body diagram of the bucket we can understand why this is so. This diagram shows that there are only two forces *acting on the bucket*, namely, its weight **W** and the force **T** of the cable. For equilibrium, the resultant of these forces must be equal to zero, and so $T = W$.

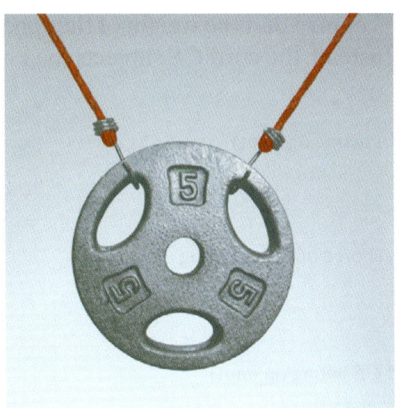
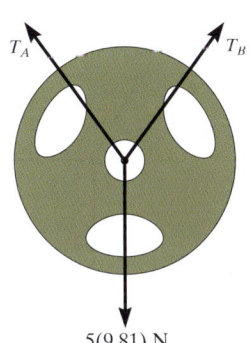

5(9.81) N

The 5-kg plate is suspended by two straps A and B. To find the force in each strap we should consider the free-body diagram of the plate. As noted, the three forces acting on it form a concurrent force system.

EXAMPLE 3.1

The sphere in Fig. 3–3a has a mass of 6 kg and is supported as shown. Draw a free-body diagram of the sphere, the cord CE, and the knot at C.

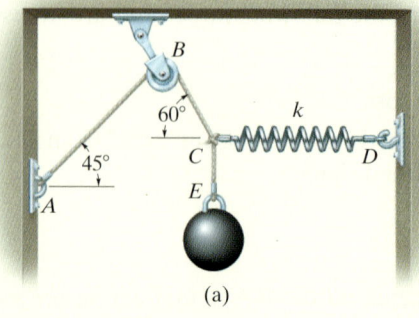

(a)

$\mathbf{F}_{CE}$ (Force of cord CE acting on sphere)

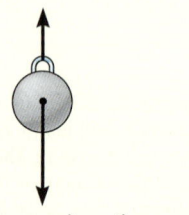

58.9 N (Weight or gravity acting on sphere)

(b)

SOLUTION

Sphere. By inspection, there are only two forces acting on the sphere, namely, its weight, 6 kg (9.81 m/s^2) = 58.9 N, and the force of cord CE. The free-body diagram is shown in Fig. 3–3b.

Cord CE. When the cord CE is isolated from its surroundings, its free-body diagram shows only two forces acting on it, namely, the force of the sphere and the force of the knot, Fig. 3–3c. Notice that $\mathbf{F}_{CE}$ shown here is equal but opposite to that shown in Fig. 3–3b, a consequence of Newton's third law of action–reaction. Also, $\mathbf{F}_{CE}$ and $\mathbf{F}_{EC}$ pull on the cord and keep it in tension so that it doesn't collapse. For equilibrium, $F_{CE} = F_{EC}$.

Knot. The knot at C is subjected to three forces, Fig. 3–3d. They are caused by the cords CBA and CE and the spring CD. As required, the free-body diagram shows all these forces labeled with their magnitudes and directions. It is important to recognize that the weight of the sphere does not directly act on the knot. Instead, the cord CE subjects the knot to this force.

$\mathbf{F}_{EC}$ (Force of knot acting on cord CE)

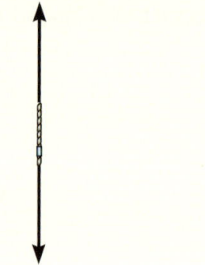

$\mathbf{F}_{CE}$ (Force of sphere acting on cord CE)

(c)

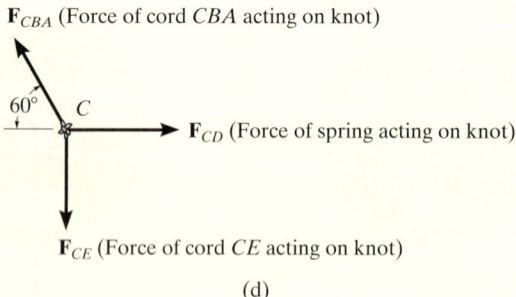

(d)

Fig. 3–3

3.3 Coplanar Force Systems

If a particle is subjected to a system of coplanar forces that lie in the x–y plane, as in Fig. 3–4, then each force can be resolved into its $\mathbf{i}$ and $\mathbf{j}$ components. For equilibrium, these forces must sum to produce a zero force resultant, i.e.,

$$\Sigma \mathbf{F} = \mathbf{0}$$
$$\Sigma F_x \mathbf{i} + \Sigma F_y \mathbf{j} = \mathbf{0}$$

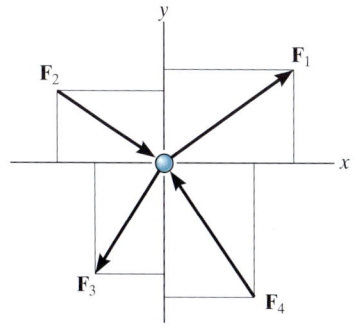

Fig. 3–4

For this vector equation to be satisfied, the resultant force's x and y components must both be equal to zero. Hence,

$$\begin{aligned} \Sigma F_x &= 0 \\ \Sigma F_y &= 0 \end{aligned} \qquad (3\text{–}3)$$

These two equations can be solved for at most two unknowns, generally represented as angles and magnitudes of forces shown on the particle's free-body diagram.

When applying each of the two equations of equilibrium, we must account for the sense of direction of any component by using an *algebraic sign* which corresponds to the arrowhead direction of the component along the x or y axis. It is important to note that if a force has an *unknown magnitude*, then the arrowhead sense of the force on the free-body diagram can be *assumed*. Then if the *solution* yields a *negative scalar*, this indicates that the sense of the force is opposite to that which was assumed.

For example, consider the free-body diagram of the particle subjected to the two forces shown in Fig. 3–5. Here it is *assumed* that the *unknown force* $\mathbf{F}$ acts to the right to maintain equilibrium. Applying the equation of equilibrium along the x axis, we have

$$\xrightarrow{+} \Sigma F_x = 0; \qquad\qquad +F + 10\,\text{N} = 0$$

Both terms are "positive" since both forces act in the positive x direction. When this equation is solved, $F = -10\,\text{N}$. Here the *negative sign* indicates that $\mathbf{F}$ must act to the left to hold the particle in equilibrium, Fig. 3–5. Notice that if the $+x$ axis in Fig. 3–5 were directed to the left, both terms in the above equation would be negative, but again, after solving, $F = -10\,\text{N}$, indicating that $\mathbf{F}$ would have to be directed to the left.

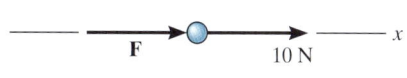

Fig. 3–5

Procedure for Analysis

Coplanar force equilibrium problems for a particle can be solved using the following procedure.

Free-Body Diagram.

- Establish the x, y axes in any suitable orientation.

- Label all the known and unknown force magnitudes and directions on the diagram.

- The sense of a force having an unknown magnitude can be assumed.

Equations of Equilibrium.

- Apply the equations of equilibrium, $\Sigma F_x = 0$ and $\Sigma F_y = 0$.

- Components are positive if they are directed along a positive axis, and negative if they are directed along a negative axis.

- If more than two unknowns exist and the problem involves a spring, apply $F = ks$ to relate the spring force to the deformation s of the spring.

- Since the magnitude of a force is always a positive quantity, then if the solution for a force yields a negative result, this indicates that its sense is the reverse of that shown on the free-body diagram.

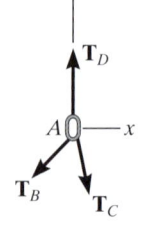

The chains exert three forces on the ring at A, as shown on its free-body diagram. The ring will not move, or will move with constant velocity, provided the summation of these forces along the x and along the y axis equals zero. If one of the three forces is known, the magnitudes of the other two forces can be obtained from the two equations of equilibrium.

EXAMPLE 3.2

Determine the tension in cables BA and BC necessary to support the 60-kg cylinder in Fig. 3–6a.

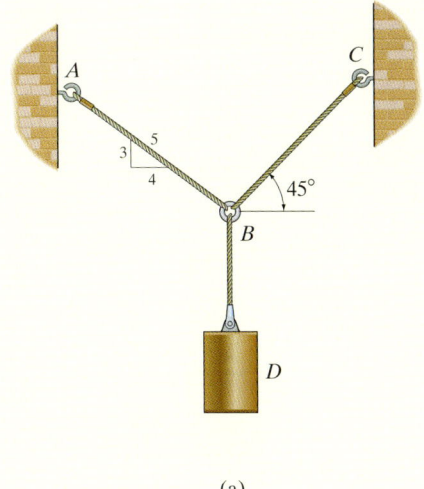

(a)

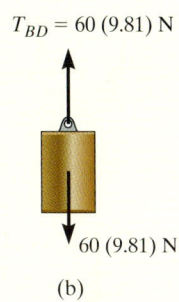

$T_{BD} = 60\,(9.81)$ N

$60\,(9.81)$ N

(b)

SOLUTION

Free-Body Diagram. Due to equilibrium, the weight of the cylinder causes the tension in cable BD to be $T_{BD} = 60(9.81)$ N, Fig. 3–6b. The forces in cables BA and BC can be determined by investigating the equilibrium of ring B. Its free-body diagram is shown in Fig. 3–6c. The magnitudes of $\mathbf{T}_A$ and $\mathbf{T}_C$ are unknown, but their directions are known.

Equations of Equilibrium. Applying the equations of equilibrium along the x and y axes, we have

$$\xrightarrow{+}\ \Sigma F_x = 0; \qquad T_C \cos 45° - \left(\tfrac{4}{5}\right)T_A = 0 \qquad (1)$$

$$+\uparrow \Sigma F_y = 0; \quad T_C \sin 45° + \left(\tfrac{3}{5}\right)T_A - 60(9.81)\ \text{N} = 0 \qquad (2)$$

Equation (1) can be written as $T_A = 0.8839 T_C$. Substituting this into Eq. (2) yields

$$T_C \sin 45° + \left(\tfrac{3}{5}\right)(0.8839 T_C) - 60(9.81)\ \text{N} = 0$$

so that

$$T_C = 475.66\ \text{N} = 476\ \text{N} \qquad\qquad Ans.$$

Substituting this result into either Eq. (1) or Eq. (2), we get

$$T_A = 420\ \text{N} \qquad\qquad Ans.$$

NOTE: The accuracy of these results, of course, depends on the accuracy of the data, i.e., measurements of geometry and loads. For most engineering work involving a problem such as this, the data as measured to three significant figures would be sufficient.

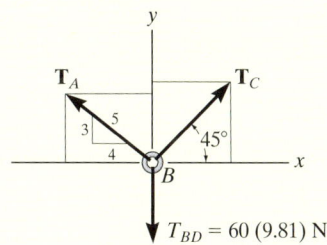

(c)

Fig. 3–6

$T_{BD} = 60\,(9.81)$ N

EXAMPLE 3.3

The 200-kg crate in Fig. 3–7a is suspended using the ropes AB and AC. Each rope can withstand a maximum force of 10 kN before it breaks. If AB always remains horizontal, determine the smallest angle θ to which the crate can be suspended before one of the ropes breaks.

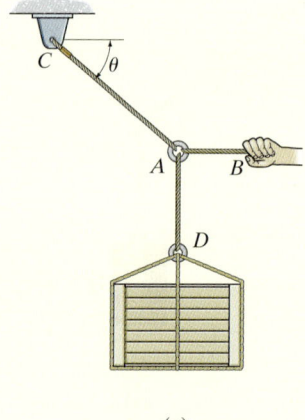

(a)

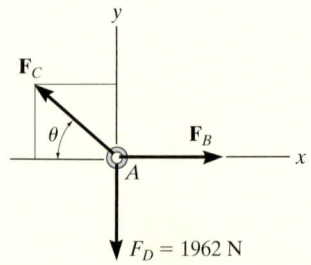

$F_D = 1962$ N

(b)

Fig. 3–7

SOLUTION

Free-Body Diagram. We will study the equilibrium of ring A. There are three forces acting on it, Fig. 3–7b. The magnitude of $\mathbf{F}_D$ is equal to the weight of the crate, i.e., $F_D = 200\,(9.81)$ N $= 1962$ N < 10 kN.

Equations of Equilibrium. Applying the equations of equilibrium along the x and y axes,

$$\xrightarrow{+} \Sigma F_x = 0; \qquad -F_C \cos\theta + F_B = 0; \quad F_C = \frac{F_B}{\cos\theta} \qquad (1)$$

$$+\uparrow \Sigma F_y = 0; \qquad F_C \sin\theta - 1962\text{ N} = 0 \qquad (2)$$

From Eq. (1), F_C is always greater than F_B since $\cos\theta \le 1$. Therefore, rope AC will reach the maximum tensile force of 10 kN *before* rope AB. Substituting $F_C = 10$ kN into Eq. (2), we get

$$[10(10^3)\,\text{N}]\sin\theta - 1962\text{ N} = 0$$

$$\theta = \sin^{-1}(0.1962) = 11.31° = 11.3° \qquad \textit{Ans.}$$

The force developed in rope AB can be obtained by substituting the values for θ and F_C into Eq. (1).

$$10(10^3)\,\text{N} = \frac{F_B}{\cos 11.31°}$$

$$F_B = 9.81\text{ kN}$$

EXAMPLE 3.4

Determine the required length of cord AC in Fig. 3–8a so that the 8-kg lamp can be suspended in the position shown. The *undeformed* length of spring AB is $l'_{AB} = 0.4$ m, and the spring has a stiffness of $k_{AB} = 300$ N/m.

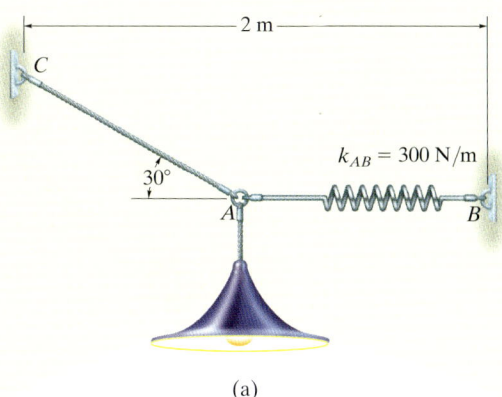

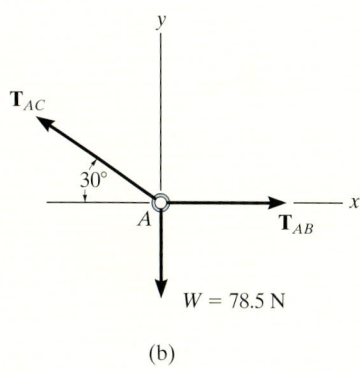

(a)

(b)

Fig. 3–8

SOLUTION

If the force in spring AB is known, the stretch of the spring can be found using $F = ks$. From the problem geometry, it is then possible to calculate the required length of AC.

Free-Body Diagram.   The lamp has a weight $W = 8(9.81) = 78.5$ N and so the free-body diagram of the ring at A is shown in Fig. 3–8b.

Equations of Equilibrium.   Using the x, y axes,

$$\xrightarrow{+} \Sigma F_x = 0; \qquad T_{AB} - T_{AC} \cos 30° = 0$$
$$+\uparrow \Sigma F_y = 0; \qquad T_{AC} \sin 30° - 78.5 \text{ N} = 0$$

Solving, we obtain

$$T_{AC} = 157.0 \text{ N}$$
$$T_{AB} = 135.9 \text{ N}$$

The stretch of spring AB is therefore

$$T_{AB} = k_{AB}s_{AB}; \qquad 135.9 \text{ N} = 300 \text{ N/m}(s_{AB})$$
$$s_{AB} = 0.453 \text{ m}$$

so the stretched length is

$$l_{AB} = l'_{AB} + s_{AB}$$
$$l_{AB} = 0.4 \text{ m} + 0.453 \text{ m} = 0.853 \text{ m}$$

The horizontal distance from C to B, Fig. 3–8a, requires

$$2 \text{ m} = l_{AC} \cos 30° + 0.853 \text{ m}$$
$$l_{AC} = 1.32 \text{ m} \qquad\qquad\qquad \textit{Ans.}$$

FUNDAMENTAL PROBLEMS

All problem solutions must include an FBD.

F3–1.   The crate has a weight of 550 N (≈55 kg). Determine the force in each supporting cable.

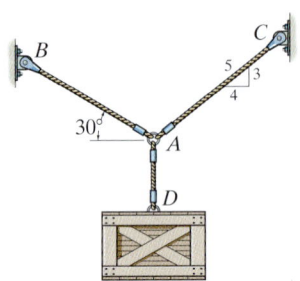

F3–1

F3–2.   The beam has a weight of 3.5 kN (≈350 kg). Determine the shortest cable ABC that can be used to lift it if the maximum force the cable can sustain is 7.5 kN.

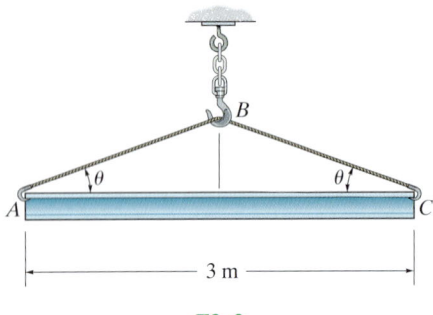

F3–2

F3–3.   If the 5-kg block is suspended from the pulley B and the sag of the cord is $d = 0.15$ m, determine the force in cord ABC. Neglect the size of the pulley.

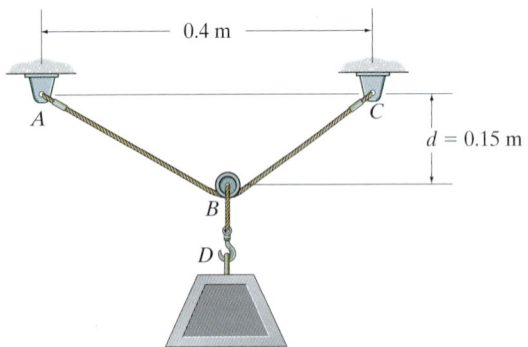

F3–3

F3–4.   The block has a mass of 5 kg and rests on the smooth plane. Determine the unstretched length of the spring.

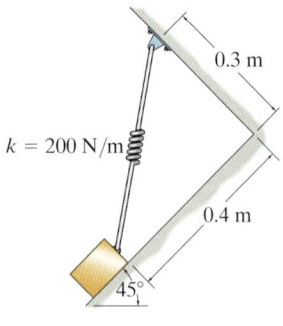

F3–4

F3–5.   If the mass of cylinder C is 40 kg, determine the mass of cylinder A in order to hold the assembly in the position shown.

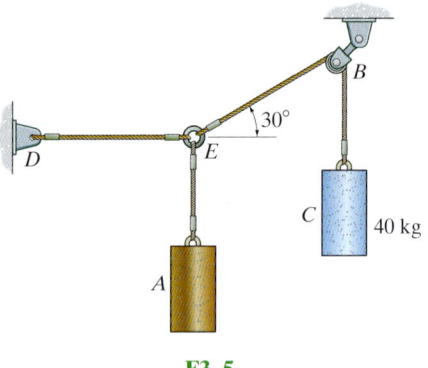

F3–5

F3–6.   Determine the tension in cables AB, BC, and CD, necessary to support the 10-kg and 15-kg traffic lights at B and C, respectively. Also, find the angle θ.

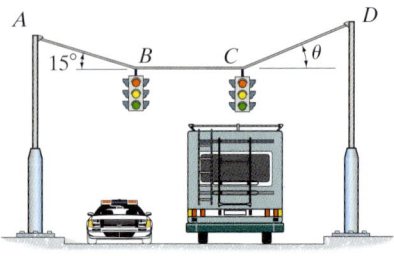

F3–6

PROBLEMS

All problem solutions must include an FBD.

3–1. The members of a truss are pin connected at joint O. Determine the magnitudes of $\mathbf{F}_1$ and $\mathbf{F}_2$ for equilibrium. Set $\theta = 60°$.

3–2. The members of a truss are pin connected at joint O. Determine the magnitude of $\mathbf{F}_1$ and its angle θ for equilibrium. Set $F_2 = 6$ kN.

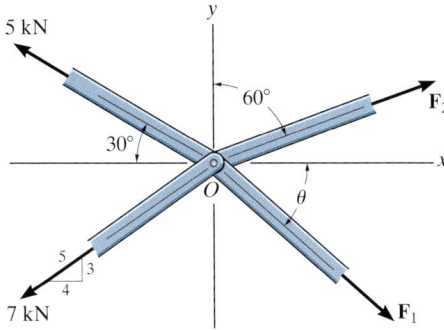

Probs. 3–1/2

3–3. The lift sling is used to hoist a container having a mass of 500 kg. Determine the force in each of the cables AB and AC as a function of θ. If the maximum tension allowed in each cable is 3.5 kN, determine the shortest lengths of cables AB and AC that can be used for the lift. The center of gravity of the container is located at G.

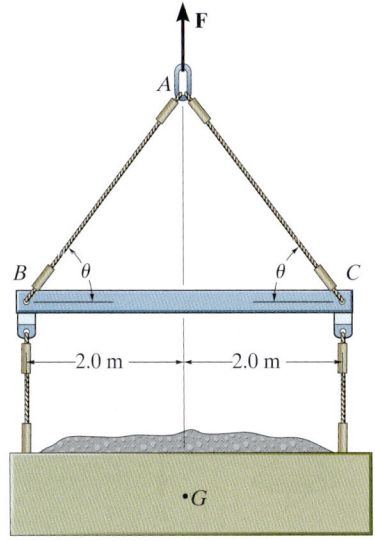

Prob. 3–3

*3–4. The towing pendant AB is subjected to the force of 50 kN exerted by a tugboat. Determine the force in each of the bridles, BC and BD, if the ship is moving forward with constant velocity.

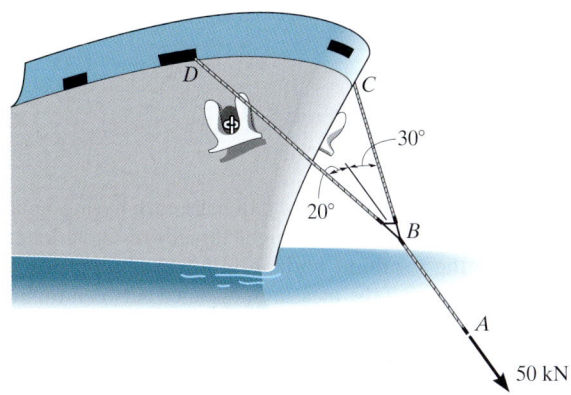

Prob. 3–4

3–5. The members of a truss are connected to the gusset plate. If the forces are concurrent at point O, determine the magnitudes of $\mathbf{F}$ and $\mathbf{T}$ for equilibrium. Take $\theta = 30°$.

3–6. The gusset plate is subjected to the forces of four members. Determine the force in member B and its proper orientation θ for equilibrium. The forces are concurrent at point O. Take $F = 12$ kN

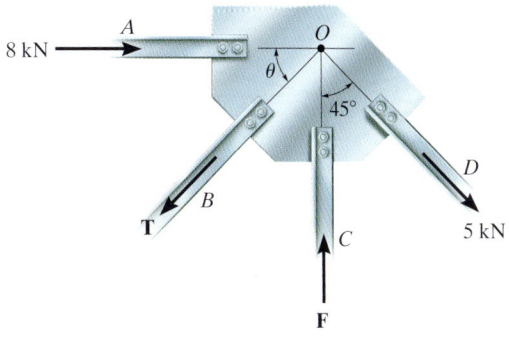

Probs. 3–5/6

3–7. The device shown is used to straighten the frames of wrecked autos. Determine the tension of each segment of the chain, i.e., AB and BC, if the force which the hydraulic cylinder DB exerts on point B is 3.50 kN, as shown.

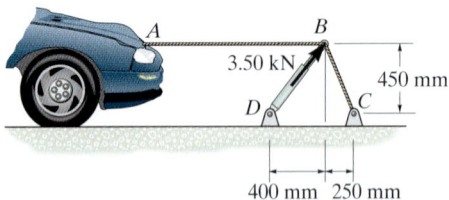

Prob. 3–7

*3–8.** Two electrically charged pith balls, each having a mass of 0.15 g, are suspended from light threads of equal length. Determine the resultant horizontal force of repulsion, F, acting on each ball if the measured distance between them is $r = 200$ mm.

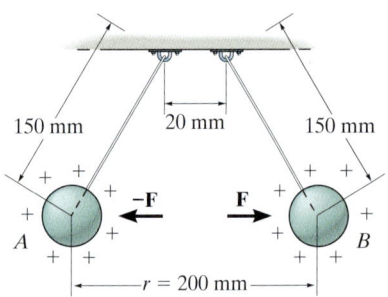

Prob. 3–8

3–9. Determine the force in cables AB and AC necessary to support the 12-kg traffic light.

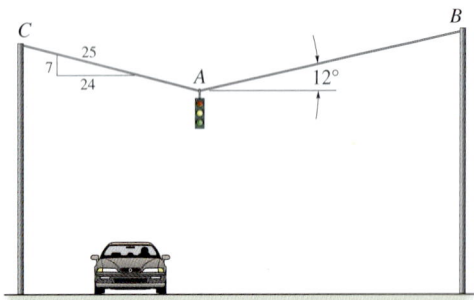

Prob. 3–9

3–10. Determine the tension developed in wires CA and CB required for equilibrium of the 10-kg cylinder. Take $\theta = 40°$.

3–11. If cable CB is subjected to a tension that is twice that of cable CA, determine the angle θ for equilibrium of the 10-kg cylinder. Also, what are the tensions in wires CA and CB?

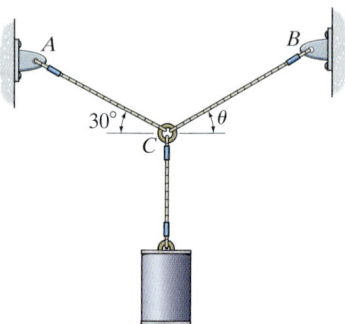

Probs. 3–10/11

*3–12.** Determine the force in each cable and the force F needed to hold the 4-kg lamp in the position shown. *Hint:* First analyze the equilibrium at B; then, using the result for the force in BC, analyze the equilibrium at C.

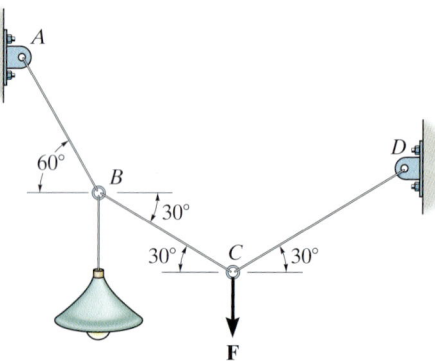

Prob. 3–12

3–13. The springs *AB* and *BC* each have a stiffness of 600 N/m and an unstretched length of 1.5 m. Determine the horizontal force **F** applied to the cord which is attached to the *small* pulley *B* so that the displacement of the pulley from the wall is *d* = 0.75 m.

3–14. The springs *AB* and *BC* each have a stiffness of 600 N/m and an unstretched length of 1.5 m. Determine the displacement *d* of the cord from the wall when a force *F* = 200 N is applied to the cord.

*3–16.** Two spheres *A* and *B* have an equal mass and are electrostatically charged such that the repulsive force acting between them has a magnitude of 20 mN and is directed along line *AB*. Determine the angle *θ*, the tension in cords *AC* and *BC*, and the mass *m* of each sphere.

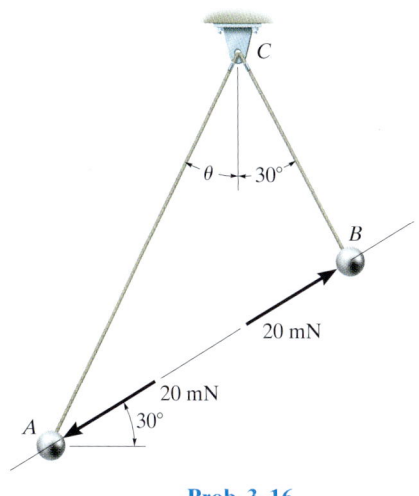

Prob. 3–16

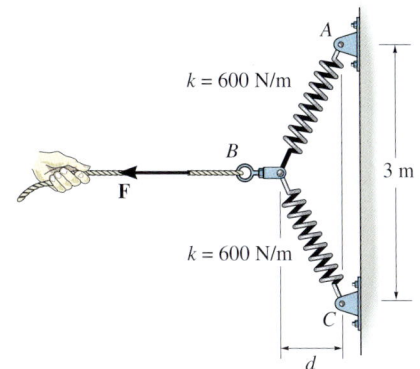

Probs. 3–13/14

3–15. The spring has a stiffness of *k* = 800 N/m and an unstretched length of 200 mm. Determine the force in cables *BC* and *BD* when the spring is held in the position shown.

3–17. Determine the mass of each of the two cylinders if they cause a sag of *s* = 0.5 m when suspended from the rings at *A* and *B*. Note that *s* = 0 when the cylinders are removed.

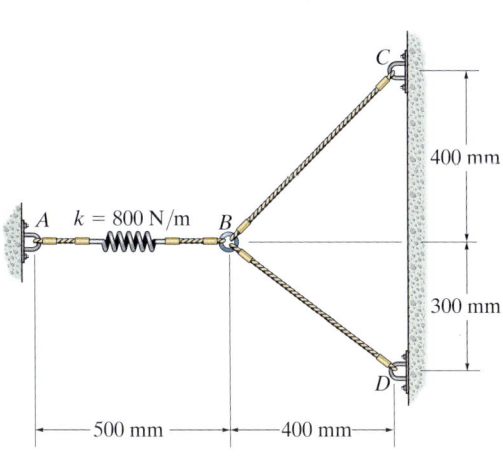

Prob. 3–15

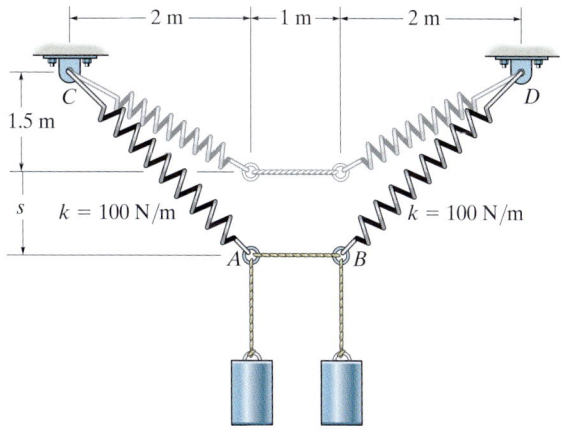

Prob. 3–17

3–18. Determine the stretch in each spring for equilibrium of the 2-kg block. The springs are shown in the equilibrium position.

3–19. The unstretched length of spring AB is 3 m. If the block is held in the equilibrium position shown, determine the mass of the block at D.

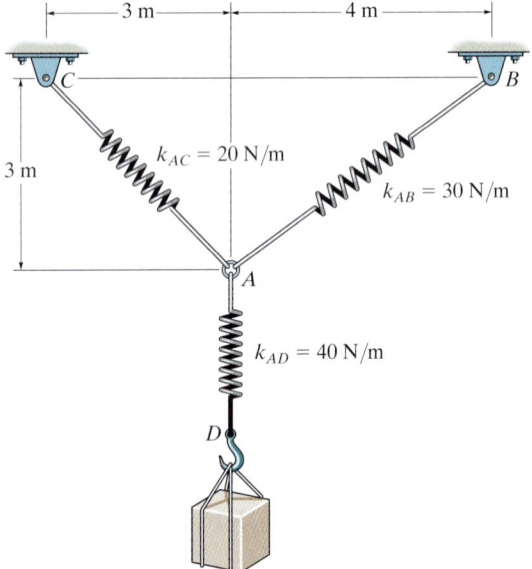

Probs. 3–18/19

*3–20. The springs BA and BC each have a stiffness of 500 N/m and an unstretched length of 3 m. Determine the horizontal force **F** applied to the cord which is attached to the *small* ring B so that the displacement of the ring from the wall is $d = 1.5$ m.

3–21. The springs BA and BC each have a stiffness of 500 N/m and an unstretched length of 3 m. Determine the displacement d of the cord from the wall when a force $F = 175$ N is applied to the cord.

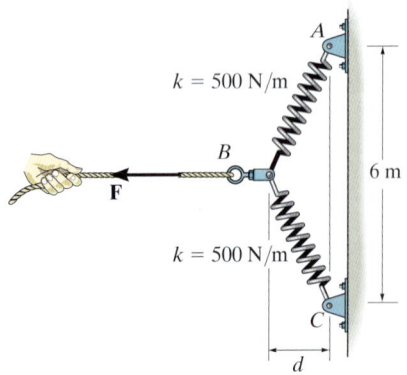

Probs. 3–20/21

3–22. The wire forms a loop and passes over the small pulleys at $A, B, C,$ and D. If its end is subjected to a force of $P = 50$ N, determine the force in the wire and the magnitude of the resultant force that the wire exerts on each of the pulleys.

3–23. The wire forms a loop and passes over the small pulleys at $A, B, C,$ and D. If the maximum *resultant force* that the wire can exert on each pulley is 120 N, determine the greatest force P that can be applied to the wire as shown.

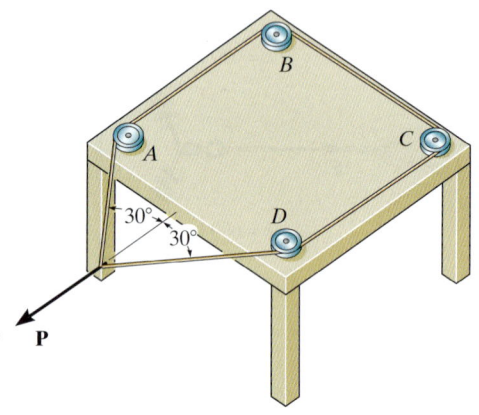

Probs. 3–22/23

*3–24. Determine the force in each cord for equilibrium of the 200-kg crate. Cord BC remains horizontal due to the roller at C, and AB has a length of 1.5 m. Set $y = 0.75$ m.

3–25. If the 1.5-m-long cord AB can withstand a maximum force of 3500 N, determine the force in cord BC and the distance y so that the 200-kg crate can be supported.

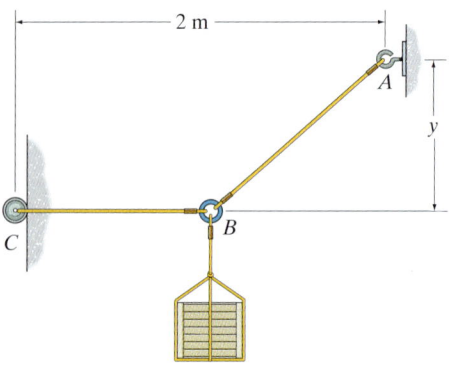

Probs. 3–24/25

3–26. The members of a truss are pin-connected at joint O. Determine the magnitudes of $\mathbf{F}_1$ and $\mathbf{F}_2$ for equilibrium. Set $\theta = 60°$.

3–27. The members of a truss are pin-connected at joint O. Determine the magnitude of $\mathbf{F}_1$ and its angle θ for equilibrium. Set $F_2 = 6$ kN.

3–30. A 4-kg sphere rests on the smooth parabolic surface. Determine the normal force it exerts on the surface and the mass m_B of block B needed to hold it in the equilibrium position shown.

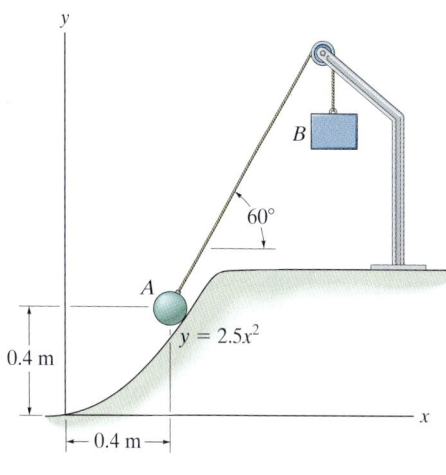

Prob. 3–30

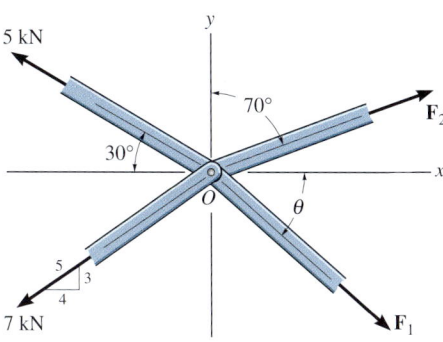

Probs. 3–26/27

3–28. Determine the tension developed in each cord required for equilibrium of the 20-kg lamp.

3–29. Determine the maximum mass of the lamp that the cord system can support so that no single cord develops a tension exceeding 400 N.

3–31. Determine the tension developed in each wire used to support the 50-kg chandelier.

3–32. If the tension developed in each of the four wires is not allowed to exceed 600 N, determine the maximum mass of the chandelier that can be supported.

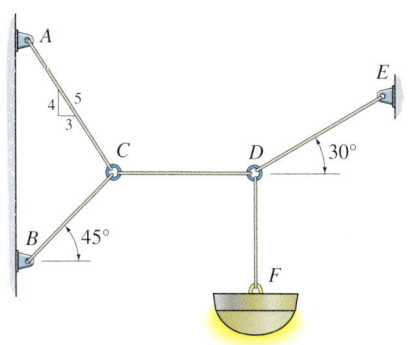

Probs. 3–28/29

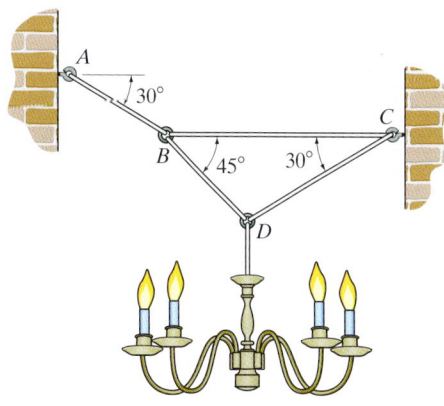

Probs. 3–31/32

3–33. The pail and its contents have a mass of 60 kg. If the cable is 15 m long, determine the distance y of the pulley for equilibrium. Neglect the size of the pulley at A.

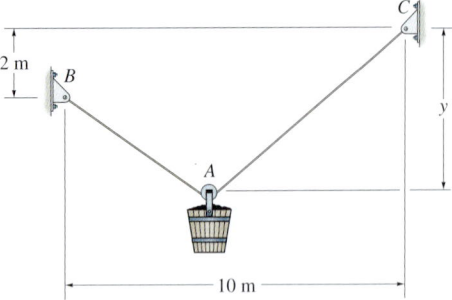

Prob. 3–33

3–34. Two electrically charged pith balls, each having a mass of 0.2 g, are suspended from light threads of equal length. Determine the resultant horizontal force of repulsion, F, acting on each ball if the measured distance between them is $r = 200$ mm.

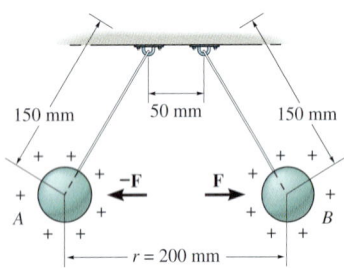

Prob. 3–34

3–35. Cable ABC has a length of 5 m. Determine the position x and the tension developed in ABC required for equilibrium of the 100-kg sack. Neglect the size of the pulley at B.

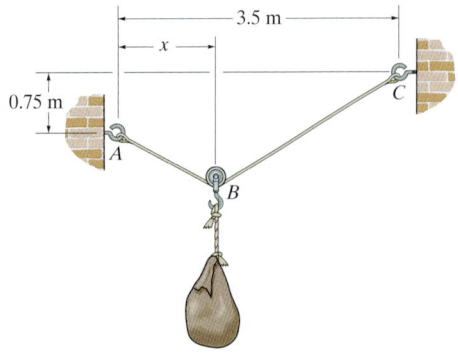

Prob. 3–35

*3–36. A scale is constructed using the 10-kg mass, the 2-kg pan P, and the pulley and cord arrangement. Cord BCA is 2 m long. If $s = 0.75$ m, determine the mass D in the pan. Neglect the size of the pulley.

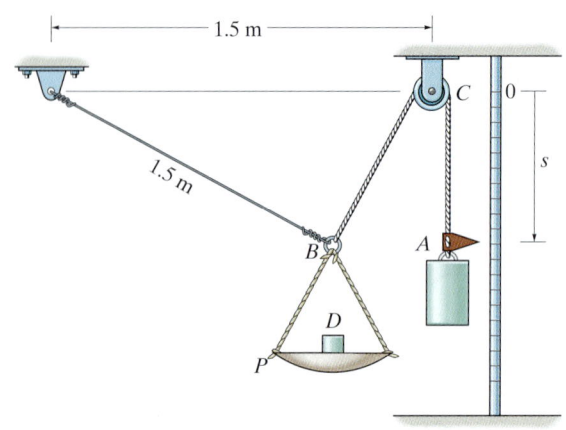

Prob. 3–36

3–37. Determine the magnitude and direction θ of the equilibrium force F_{AB} exerted along link AB by the tractive apparatus shown. The suspended mass is 10 kg. Neglect the size of the pulley at A.

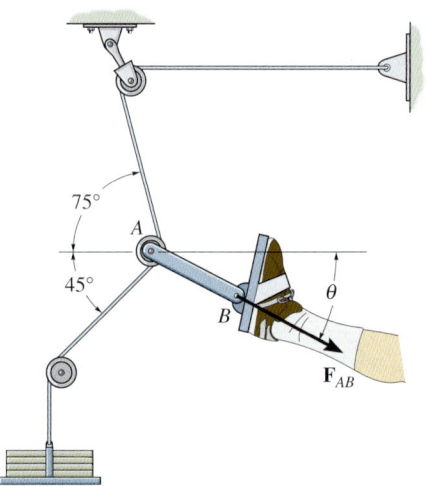

Prob. 3–37

3–38. The lift sling is used to hoist a container having a mass of 500 kg. Determine the force in each of the cables AB and AC as a function of θ. If the maximum tension allowed in each cable is 5 kN, determine the shortest lengths of cables AB and AC that can be used for the lift. The center of gravity of the container is located at G.

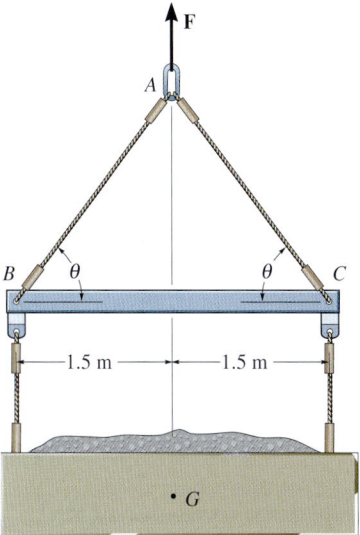

Prob. 3–38

3–39. The 30-kg pipe is supported at A by a system of five cords. Determine the force in each cord for equilibrium.

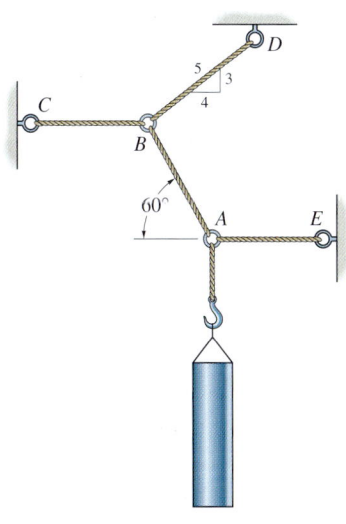

Prob. 3–39

*3–40. The load has a mass of 15 kg and is lifted by the pulley system shown. Determine the force **F** in the cord as a function of the angle θ. Plot the function of force F versus the angle θ for $0 \le \theta \le 90°$.

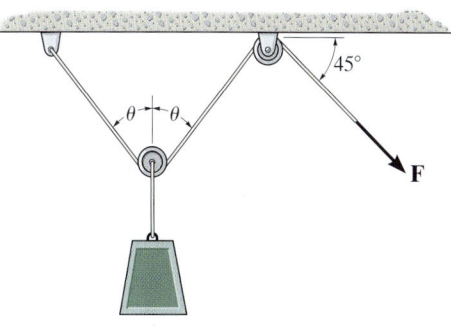

Prob. 3–40

3–41. Determine the forces in cables AC and AB needed to hold the 20-kg ball D in equlibrium. Take $F = 300$ N and $d = 1$ m.

3–42. The ball D has a mass of 20 kg. If a force of $F = 100$ N is applied horizontally to the ring at A, determine the dimension d if the force in cable AC is zero.

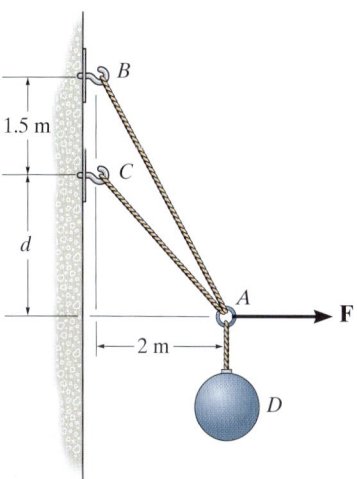

Probs. 3–41/42

CONCEPTUAL PROBLEMS

P3–1. The concrete wall panel is hoisted into position using the two cables *AB* and *AC* of equal length. Establish appropriate dimensions and use an equilibrium analysis to show that the longer the cables the less the force in each cable.

P3–1

P3–2. The hoisting cables *BA* and *BC* each have a length of 6 m. If the maximum tension that can be supported by each cable is 4.5 kN, determine the maximum distance *AC* between them in order to lift the uniform 600-kg truss with constant velocity.

P3–2

P3–3. The device *DB* is used to pull on the chain *ABC* to hold a door closed on the bin. If the angle between *AB* and *BC* is 30°, determine the angle between *DB* and *BC* for equilibrium.

P3–3

P3–4. Chain *AB* is 1-m long and chain *AC* is 1.2-m long. If the distance *BC* is 1.5 m, and *AB* can support a maximum force of 2 kN, whereas *AC* can support a maximum force of 0.8 kN, determine the largest vertical force *F* that can be applied to the link at *A*.

P3–4

3.4   Three-Dimensional Force Systems

In Section 3.1 we stated that the necessary and sufficient condition for particle equilibrium is

$$\Sigma \mathbf{F} = \mathbf{0} \qquad (3\text{--}4)$$

In the case of a three-dimensional force system, as in Fig. 3–9, we can resolve the forces into their respective $\mathbf{i}, \mathbf{j}, \mathbf{k}$ components, so that $\Sigma F_x \mathbf{i} + \Sigma F_y \mathbf{j} + \Sigma F_z \mathbf{k} = \mathbf{0}$. To satisfy this equation we require

$$
\boxed{
\begin{aligned}
\Sigma F_x &= 0 \\
\Sigma F_y &= 0 \\
\Sigma F_y &= 0
\end{aligned}
} \qquad (3\text{--}5)
$$

These three equations state that the *algebraic sum* of the components of all the forces acting on the particle along each of the coordinate axes must be zero. Using them we can solve for at most three unknowns, generally represented as coordinate direction angles or magnitudes of forces shown on the particle's free-body diagram.

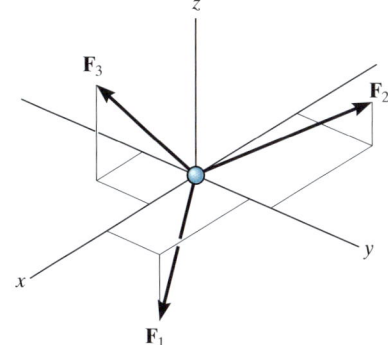

Fig. 3–9

Procedure for Analysis

Three-dimensional force equilibrium problems for a particle can be solved using the following procedure.

Free-Body Diagram.

• Establish the x, y, z axes in any suitable orientation.

• Label all the known and unknown force magnitudes and directions on the diagram.

• The sense of a force having an unknown magnitude can be assumed.

Equations of Equilibrium.

• Use the scalar equations of equilibrium, $\Sigma F_x = 0$, $\Sigma F_y = 0$, $\Sigma F_z = 0$, in cases where it is easy to resolve each force into its x, y, z components.

• If the three-dimensional geometry appears difficult, then first express each force on the free-body diagram as a Cartesian vector, substitute these vectors into $\Sigma \mathbf{F} = \mathbf{0}$, and then set the $\mathbf{i}, \mathbf{j}, \mathbf{k}$ components equal to zero.

• If the solution for a force yields a negative result, this indicates that its sense is the reverse of that shown on the free-body diagram.

The joint at A is subjected to the force from the support as well as forces from each of the three chains. If the tire and any load on it have a weight W, then the force at the support will be W, and the three scalar equations of equilibrium can be applied to the free-body diagram of the joint in order to determine the chain forces, $\mathbf{F}_B$, $\mathbf{F}_C$, and $\mathbf{F}_D$.

EXAMPLE 3.5

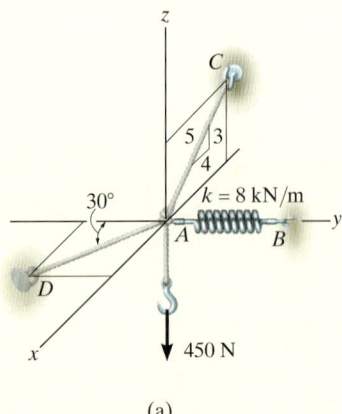

(a)

Fig. 3–10

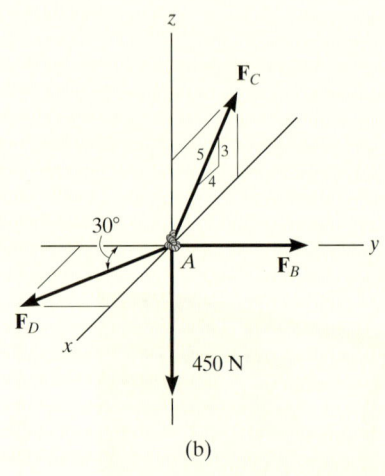

(b)

A 450-N load is suspended from the hook shown in Fig. 3–10 a. If the load is supported by two cables and a spring having a stiffness $k = 8$ kN/m, determine the force in the cables and the stretch of the spring for equilibrium. Cable AD lies in the x–y plane and cable AC lies in the x–z plane.

SOLUTION

The stretch of the spring can be determined once the force in the spring is determined.

Free-Body Diagram. The connection at A is chosen for the equilibrium analysis since the cable forces are concurrent at this point. The free-body diagram is shown in Fig. 3–10 b.

Equations of Equilibrium. By inspection, each force can easily be resolved into its x, y, z components, and therefore the three scalar equations of equilibrium can be used. Considering components directed along each positive axis as "positive," we have

$$\Sigma F_x = 0; \qquad F_D \sin 30° - \left(\tfrac{4}{5}\right) F_C = 0 \qquad (1)$$

$$\Sigma F_y = 0; \qquad -F_D \cos 30° + F_B = 0 \qquad (2)$$

$$\Sigma F_z = 0; \qquad \left(\tfrac{3}{5}\right) F_C - 450 \text{ N} = 0 \qquad (3)$$

Solving Eq. (3) for F_C, then Eq. (1) for F_D, and finally Eq. (2) for F_B, yields

$$F_C = 750 \text{ N} = 0.75 \text{ kN} \qquad \textit{Ans.}$$

$$F_D = 1200 \text{ N} = 1.2 \text{ kN} \qquad \textit{Ans.}$$

$$F_B = 1039.2 \text{ N} = 1.04 \text{ kN} \qquad \textit{Ans.}$$

The stretch of the spring is therefore

$$F_B = k s_{AB}$$

$$1.0392 \text{ kN} = (8 \text{ kN/m})(s_{AB})$$

$$s_{AB} = 0.130 \text{ m} \qquad \textit{Ans.}$$

NOTE: Since the results for all the cable forces are positive, each cable is in tension; that is, it pulls on point A as expected, Fig. 3–10 b.

EXAMPLE 3.6

The 10-kg lamp in Fig. 3–11a is suspended from the three equal-length cords. Determine its smallest vertical distance s from the ceiling if the force developed in any cord is not allowed to exceed 50 N.

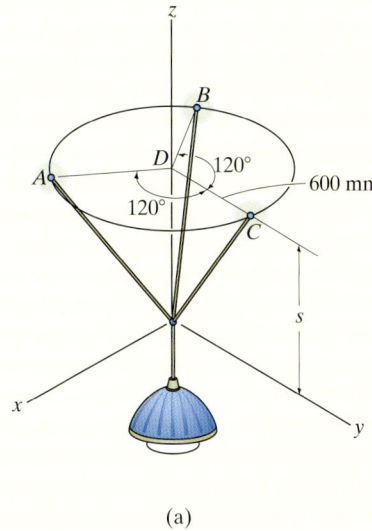

(a)

Fig. 3–11

(b)

SOLUTION

Free-Body Diagram.   Due to symmetry, Fig. 3–11b, the distance $DA = DB = DC = 600$ mm. It follows that from $\Sigma F_x = 0$ and $\Sigma F_y = 0$, the tension T in each cord will be the same. Also, the angle between each cord and the z axis is γ.

Equation of Equilibrium.   Applying the equilibrium equation along the z axis, with $T = 50$ N, we have

$$\Sigma F_z = 0; \qquad 3[(50 \text{ N}) \cos \gamma] - 10(9.81) \text{ N} = 0$$

$$\gamma = \cos^{-1} \frac{98.1}{150} = 49.16°$$

From the shaded triangle shown in Fig. 3–11b,

$$\tan 49.16° = \frac{600 \text{ mm}}{s}$$

$$s = 519 \text{ mm} \qquad \qquad Ans.$$

EXAMPLE 3.7

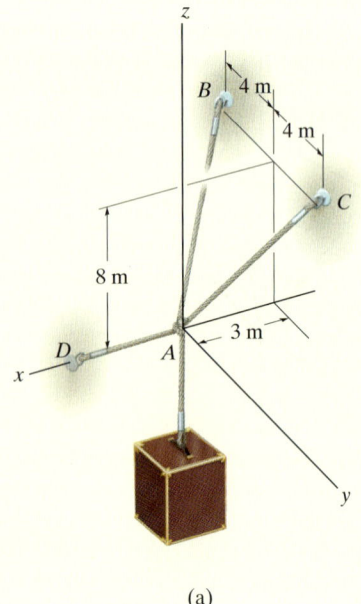

(a)

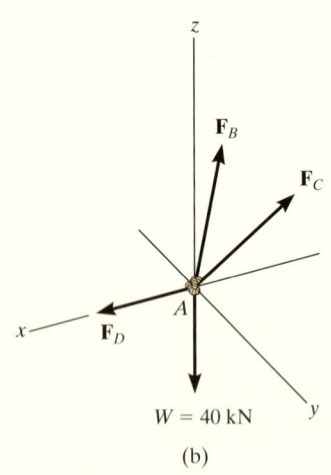

$W = 40$ kN

(b)

Fig. 3–12

Determine the force in each cable used to support the 40-kN($\approx$ 4-tonne) crate shown in Fig. 3–12a.

SOLUTION

Free-Body Diagram.   As shown in Fig. 3–12b, the free-body diagram of point A is considered in order to "expose" the three unknown forces in the cables.

Equations of Equilibrium.   First we will express each force in Cartesian vector form. Since the coordinates of points B and C are $B(-3$ m, -4 m, 8 m) and $C(-3$ m, 4 m, 8 m), we have

$$\mathbf{F}_B = F_B\left[\frac{-3\mathbf{i} - 4\mathbf{j} + 8\mathbf{k}}{\sqrt{(-3)^2 + (-4)^2 + (8)^2}}\right]$$

$$= -0.318F_B\mathbf{i} - 0.424F_B\mathbf{j} + 0.848F_B\mathbf{k}$$

$$\mathbf{F}_C = F_C\left[\frac{-3\mathbf{i} + 4\mathbf{j} + 8\mathbf{k}}{\sqrt{(-3)^2 + (4)^2 + (8)^2}}\right]$$

$$= -0.318F_C\mathbf{i} + 0.424F_C\mathbf{j} + 0.848F_C\mathbf{k}$$

$$\mathbf{F}_D = F_D\mathbf{i}$$

$$\mathbf{W} = \{-40\mathbf{k}\} \text{ kN}$$

Equilibrium requires

$$\Sigma\mathbf{F} = \mathbf{0}; \qquad\qquad \mathbf{F}_B + \mathbf{F}_C + \mathbf{F}_D + \mathbf{W} = \mathbf{0}$$

$$-0.318F_B\mathbf{i} - 0.424F_B\mathbf{j} + 0.848F_B\mathbf{k}$$

$$-0.318F_C\mathbf{i} + 0.424F_C\mathbf{j} + 0.848F_C\mathbf{k} + F_D\mathbf{i} - 40\mathbf{k} = \mathbf{0}$$

Equating the respective $\mathbf{i}, \mathbf{j}, \mathbf{k}$ components to zero yields

$$\Sigma F_x = 0; \qquad -0.318F_B - 0.318F_C + F_D = 0 \qquad\qquad (1)$$

$$\Sigma F_y = 0; \qquad\qquad -0.424F_B + 0.424F_C = 0 \qquad\qquad (2)$$

$$\Sigma F_z = 0; \qquad\qquad 0.848F_B + 0.848F_C - 40 = 0 \qquad\qquad (3)$$

Equation (2) states that $F_B = F_C$. Thus, solving Eq. (3) for F_B and F_C and substituting the result into Eq. (1) to obtain F_D, we have

$$F_B = F_C = 23.6 \text{ kN} \qquad\qquad\qquad Ans.$$

$$F_D = 15.0 \text{ kN} \qquad\qquad\qquad Ans.$$

EXAMPLE 3.8

Determine the tension in each cord used to support the 100-kg crate shown in Fig. 3–13a.

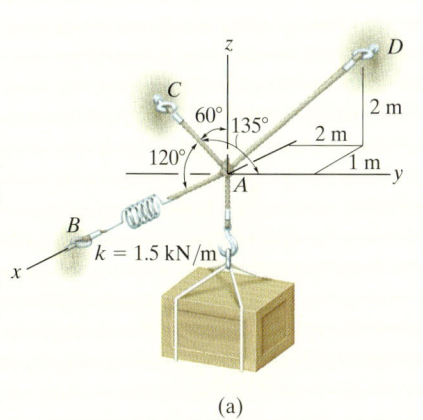

(a)

SOLUTION

Free-Body Diagram.   The force in each of the cords can be determined by investigating the equilibrium of point A. The free-body diagram is shown in Fig. 3–13b. The weight of the crate is $W = 100(9.81) = 981$ N.

Equations of Equilibrium.   Each force on the free-body diagram is first expressed in Cartesian vector form. Using Eq. 2–9 for $\mathbf{F}_C$ and noting point $D(-1\text{ m}, 2\text{ m}, 2\text{ m})$ for $\mathbf{F}_D$, we have

$$\mathbf{F}_B = F_B\mathbf{i}$$

$$\mathbf{F}_C = F_C \cos 120°\mathbf{i} + F_C \cos 135°\mathbf{j} + F_C \cos 60°\mathbf{k}$$

$$= -0.5F_C\mathbf{i} - 0.707F_C\mathbf{j} + 0.5F_C\mathbf{k}$$

$$\mathbf{F}_D = F_D\left[\frac{-1\mathbf{i} + 2\mathbf{j} + 2\mathbf{k}}{\sqrt{(-1)^2 + (2)^2 + (2)^2}}\right]$$

$$= -0.333F_D\mathbf{i} + 0.667F_D\mathbf{j} + 0.667F_D\mathbf{k}$$

$$\mathbf{W} = \{-981\mathbf{k}\}\text{ N}$$

Equilibrium requires

$$\Sigma\mathbf{F} = \mathbf{0}; \qquad \mathbf{F}_B + \mathbf{F}_C + \mathbf{F}_D + \mathbf{W} = \mathbf{0}$$

$$F_B\mathbf{i} - 0.5F_C\mathbf{i} - 0.707F_C\mathbf{j} + 0.5F_C\mathbf{k}$$

$$-0.333F_D\mathbf{i} + 0.667F_D\mathbf{j} + 0.667F_D\mathbf{k} - 981\mathbf{k} = \mathbf{0}$$

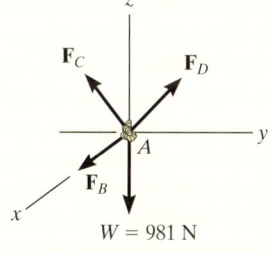

(b)

Fig. 3–13

Equating the respective $\mathbf{i}, \mathbf{j}, \mathbf{k}$ components to zero,

$\Sigma F_x = 0;$	$F_B - 0.5F_C - 0.333F_D = 0$	(1)
$\Sigma F_y = 0;$	$-0.707F_C + 0.667F_D = 0$	(2)
$\Sigma F_z = 0;$	$0.5F_C + 0.667F_D - 981 = 0$	(3)

Solving Eq. (2) for F_D in terms of F_C and substituting this into Eq. (3) yields F_C. F_D is then determined from Eq. (2). Finally, substituting the results into Eq. (1) gives F_B. Hence,

$$F_C = 813\text{ N} \qquad \textit{Ans.}$$

$$F_D = 862\text{ N} \qquad \textit{Ans.}$$

$$F_B = 694\text{ N} \qquad \textit{Ans.}$$

FUNDAMENTAL PROBLEMS

All problem solutions must include an FBD.

F3–7. Determine the magnitude of forces $\mathbf{F}_1$, $\mathbf{F}_2$, $\mathbf{F}_3$, so that the particle is held in equilibrium.

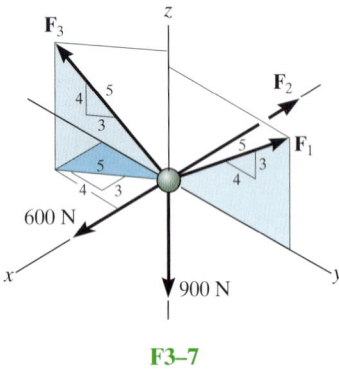

F3–7

F3–8. Determine the tension developed in cables AB, AC, and AD.

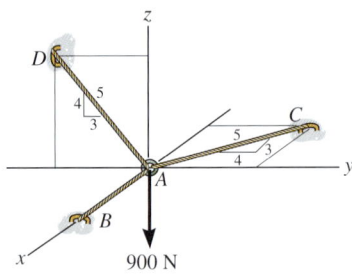

F3–8

F3–9. Determine the tension developed in cables AB, AC, and AD.

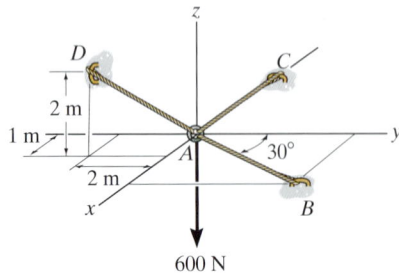

F3–9

F3–10. Determine the tension developed in cables AB, AC, and AD.

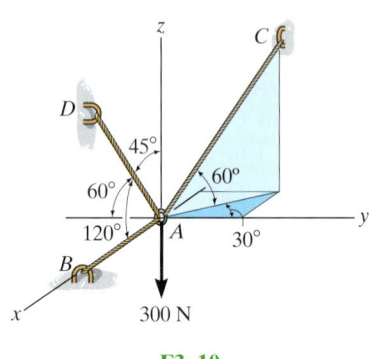

F3–10

F3–11. The 150-N ($\approx$15-kg) crate is supported by cables AB, AC, and AD. Determine the tension in these wires.

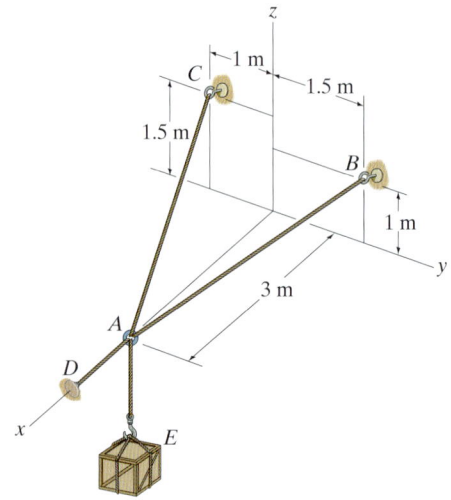

F3–11

PROBLEMS

All problem solutions must include an FBD.

3–43. Determine the magnitude and direction of the force **P** required to keep the concurrent force system in equilibrium.

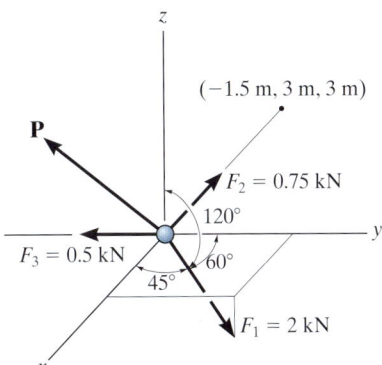

Prob. 3–43

3–45. Determine the magnitudes of F_1, F_2, and F_3 for equilibrium of the particle.

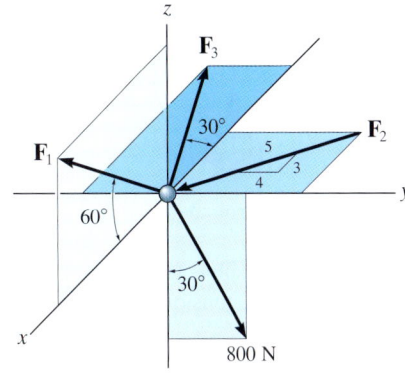

Prob. 3–45

*3–44.** If cable AB is subjected to a tension of 700 N, determine the tension in cables AC and AD and the magnitude of the vertical force **F**.

3–46. The bucket has a weight of 100 N ($\approx$ 10 kg). Determine the tension developed in each cord for equilibrium.

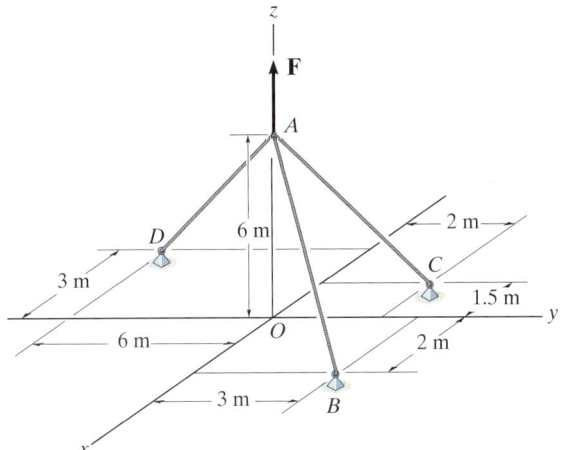

Prob. 3–44

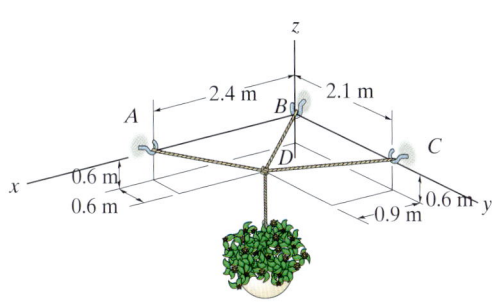

Prob. 3–46

3–47. Determine the stretch in each of the two springs required to hold the 20-kg crate in the equilibrium position shown. Each spring has an unstretched length of 2 m and a stiffness of $k = 300 \text{ N/m}$.

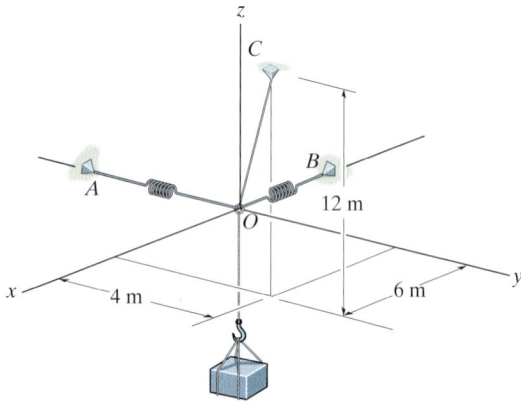

Prob. 3–47

***3–48.** If the balloon is subjected to a net uplift force of $F = 800$ N, determine the tension developed in ropes AB, AC, AD.

3–49. If each one of the ropes will break when it is subjected to a tensile force of 450 N, determine the maximum uplift force **F** the balloon can have before one of the ropes breaks.

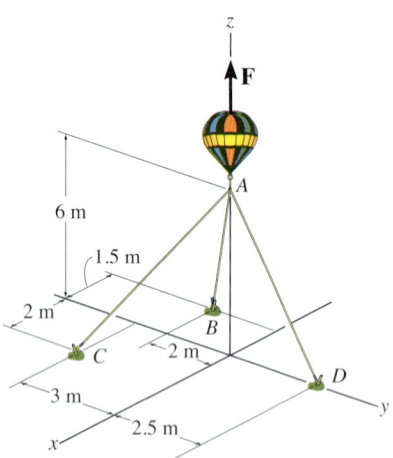

Probs. 3–48/49

3–50. The lamp has a mass of 15 kg and is supported by a pole AO and cables AB and AC. If the force in the pole acts along its axis, determine the forces in AO, AB, and AC for equilibrium.

3–51. Cables AB and AC can sustain a maximum tension of 500 N, and the pole can support a maximum compression of 300 N. Determine the maximum weight of the lamp that can be supported in the position shown. The force in the pole acts along the axis of the pole.

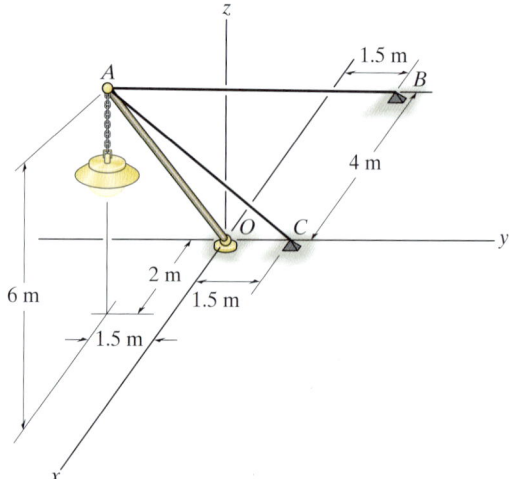

Probs. 3–50/51

***3–52.** The 50-kg pot is supported from A by the three cables. Determine the force acting in each cable for equilibrium. Take $d = 2.5$ m.

3–53. Determine the height d of cable AB so that the force in cables AD and AC is one-half as great as the force in cable AB. What is the force in each cable for this case? The flower pot has a mass of 50 kg.

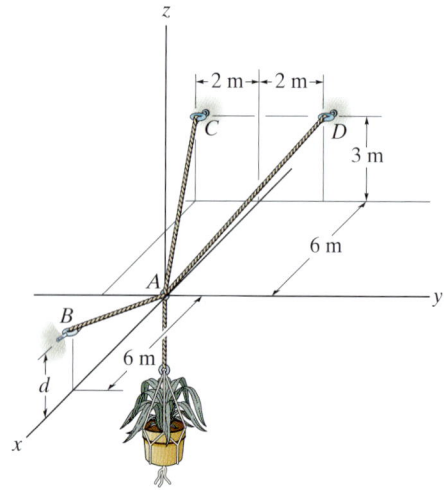

Probs. 3–52/53

3–54. If the mass of the flowerpot is 50 kg, determine the tension developed in each wire for equilibrium. Set $x = 1.5$ m and $z = 2$ m.

3–55. If the mass of the flowerpot is 50 kg, determine the tension developed in each wire for equilibrium. Set $x = 2$ m and $z = 1.5$ m.

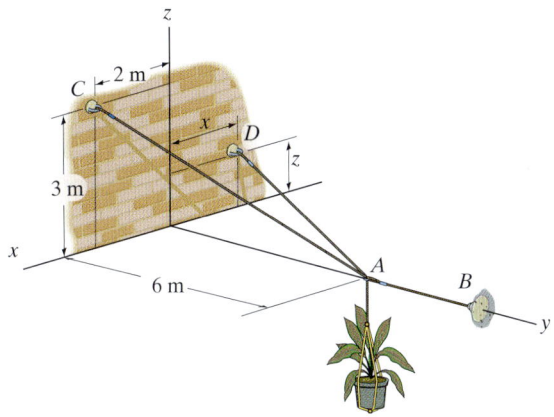

Probs. 3–54/55

*3–56.** The 25-kg flowerpot is supported at A by the three cords. Determine the force acting in each cord for equilibrium.

3–57. If each cord can sustain a maximum tension of 50 N before it fails, determine the greatest weight of the flowerpot the cords can support.

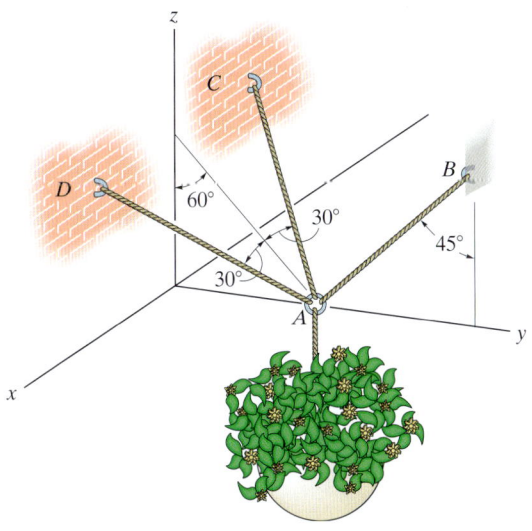

Probs. 3–56/57

3–58. Determine the force in each cable needed to support the 17.5-kN (≈ 1750-kg) platform. Set $d = 0.6$ m.

3–59. Determine the force in each cable needed to support the 17.5-kN (≈ 1750-kg) platform. Set $d = 1.2$ m.

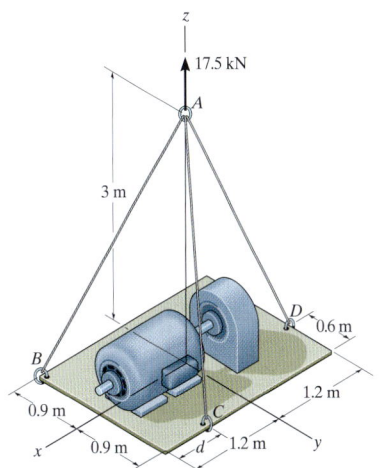

Probs. 3–58/59

*3–60.** A small peg P rests on a spring that is contained inside the smooth pipe. When the spring is compressed so that $s = 0.15$ m, the spring exerts an upward force of 60 N on the peg. Determine the point of attachment $A(x, y, 0)$ of cord PA so that the tension in cords PB and PC equals 30 N and 50 N, respectively.

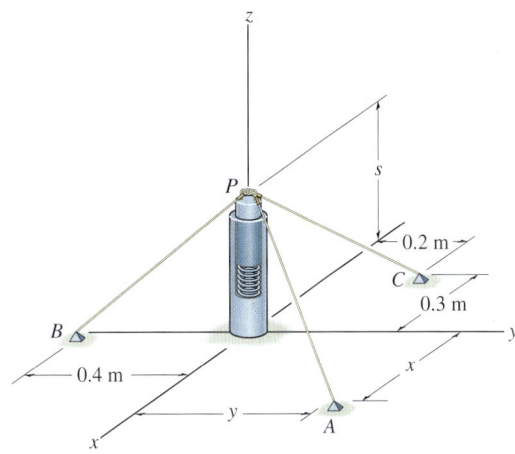

Prob. 3–60

3–61. Determine the tension developed in cables AB, AC, and AD required for equilibrium of the 75-kg cylinder.

3–62. If each cable can withstand a maximum tension of 1000 N, determine the largest mass of the cylinder for equilibrium.

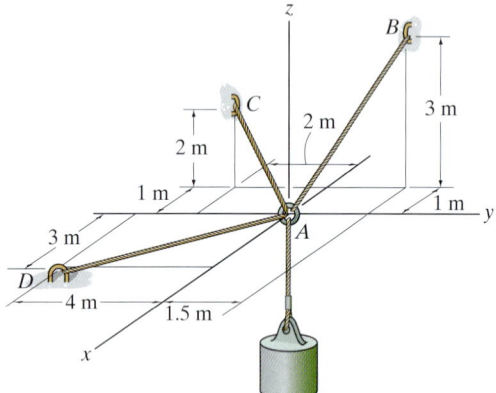

Probs. 3–61/62

3–63. The thin ring can be adjusted vertically between three equally long cables from which the 100-kg chandelier is suspended. If the ring remains in the horizontal plane and $z = 600$ mm, determine the tension in each cable.

*3–64. The thin ring can be adjusted vertically between three equally long cables from which the 100-kg chandelier is suspended. If the ring remains in the horizontal plane and the tension in each cable is not allowed to exceed 1 kN, determine the smallest allowable distance z required for equilibrium.

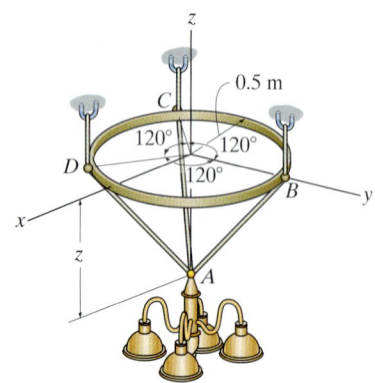

Probs. 3–63/64

3–65. The ends of the three cables are attached to a ring at A and to the edge of a uniform 150-kg plate. Determine the tension in each of the cables for equilibrium.

3–66. The ends of the three cables are attached to a ring at A and to the edge of the uniform plate. Determine the largest mass the plate can have if each cable can support a maximum tension of 15 kN.

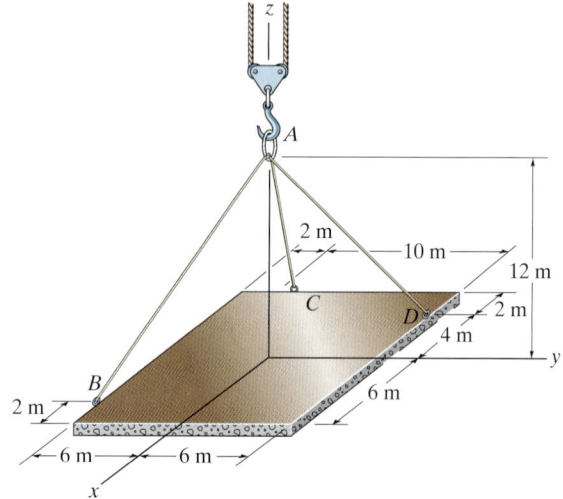

Probs. 3–65/66

3–67. The shear leg derrick is used to haul the 200-kg net of fish onto the dock. Determine the compressive force along each of the legs AB and CB and the tension in the winch cable DB. Assume the force in each leg acts along its axis.

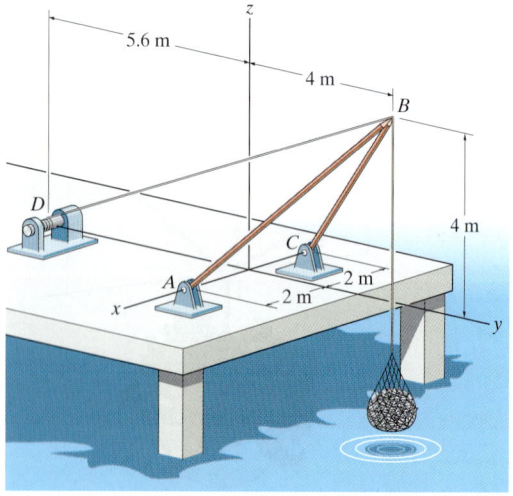

Prob. 3–67

CHAPTER REVIEW

Particle Equilibrium

When a particle is at rest or moves with constant velocity, it is in equilibrium. This requires that all the forces acting on the particle form a zero resultant force.

In order to account for all the forces that act on a particle, it is necessary to draw its free-body diagram. This diagram is an outlined shape of the particle that shows all the forces listed with their known or unknown magnitudes and directions.

$$\mathbf{F}_R = \Sigma\mathbf{F} = \mathbf{0}$$

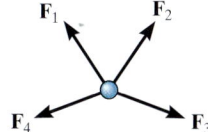

Two Dimensions

The two scalar equations of force equilibrium can be applied with reference to an established x, y coordinate system.

The tensile force developed in a *continuous cable* that passes over a frictionless pulley must have a *constant* magnitude throughout the cable to keep the cable in equilibrium.

If the problem involves a linearly elastic spring, then the stretch or compression s of the spring can be related to the force applied to it.

$$\Sigma F_x = 0$$
$$\Sigma F_y = 0$$

$$F = ks$$

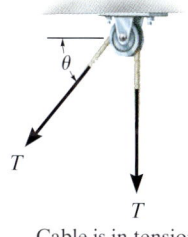

Cable is in tension

Three Dimensions

If the three-dimensional geometry is difficult to visualize, then the equilibrium equation should be applied using a Cartesian vector analysis. This requires first expressing each force on the free-body diagram as a Cartesian vector. When the forces are summed and set equal to zero, then the **i**, **j**, and **k** components are also zero.

$$\Sigma\mathbf{F} = \mathbf{0}$$

$$\Sigma F_x = 0$$
$$\Sigma F_y = 0$$
$$\Sigma F_z = 0$$

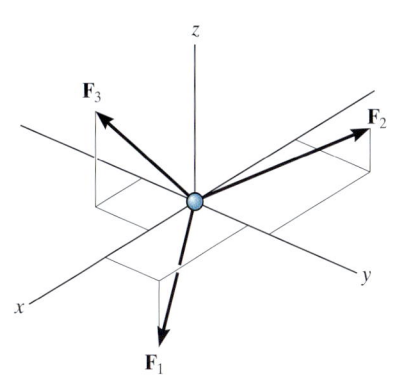

REVIEW PROBLEMS

*3–68. The boom supports a bucket and contents, which have a total mass of 300 kg. Determine the forces developed in struts AD and AE and the tension in cable AB for equilibrium. The force in each strut acts along its axis.

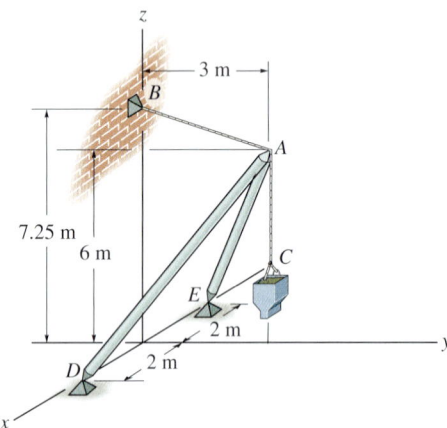

Prob. 3–68

3–69. The members of a truss are pin connected at joint O. Determine the magnitude of $\mathbf{F}_1$ and its angle θ for equilibrium. Set $F_2 = 8$ kN.

3–70. The members of a truss are pin connected at joint O. Determine the magnitudes of $\mathbf{F}_1$ and $\mathbf{F}_2$ for equilibrium. Set $\theta = 50°$.

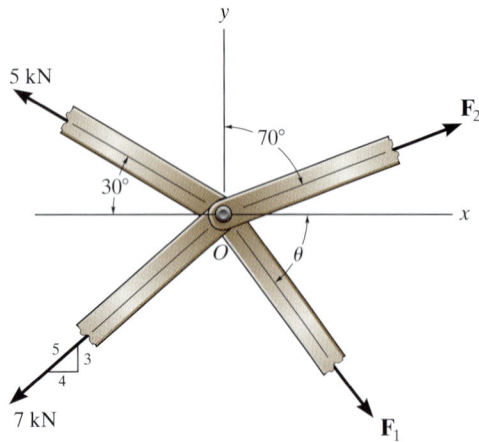

Probs. 3–69/70

3–71. Romeo tries to reach Juliet by climbing with constant velocity up a rope which is knotted at point A. Any of the three segments of the rope can sustain a maximum force of 2 kN before it breaks. Determine if Romeo, who has a mass of 65 kg, can climb the rope, and if so, can he along with Juliet, who has a mass of 60 kg, climb down with constant velocity?

Prob. 3–71

*3–72. Determine the magnitudes of forces $\mathbf{F}_1$, $\mathbf{F}_2$, and $\mathbf{F}_3$ necessary to hold the force $\mathbf{F} = \{-9\mathbf{i} - 8\mathbf{j} - 5\mathbf{k}\}$ kN in equilibrium.

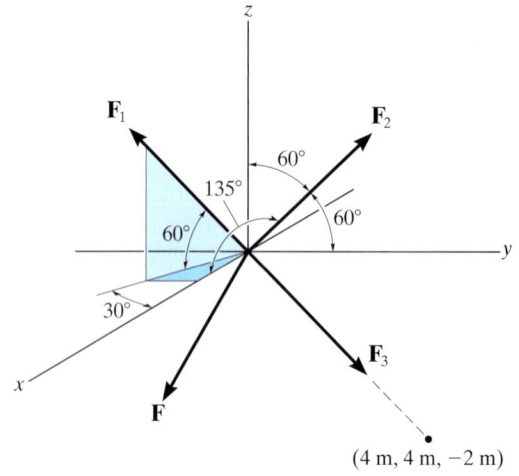

Prob. 3–72

3–73. The three cables are used to support the 800-N (≈ 80-kg) lamp. Determine the force developed in each cable for equilibrium.

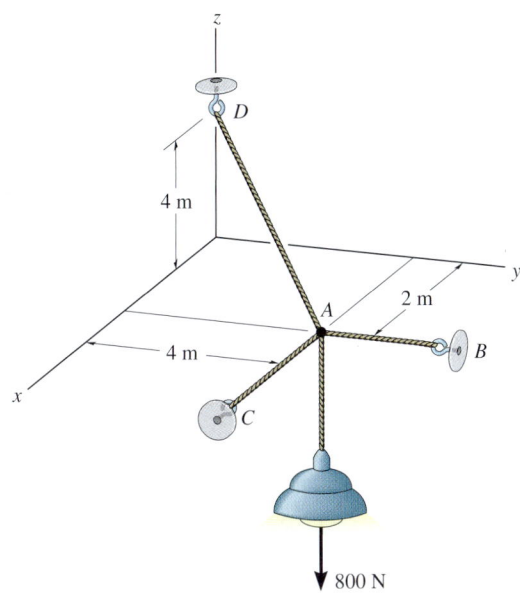

Prob. 3–73

3–74. The ring of negligible size is subjected to a vertical force of 1000 N. Determine the longest length l of cord AC such that the tension acting in AC is 800 N. Also, what is the force acting in cord AB? *Hint:* Use the equilibrium condition to determine the required angle θ for attachment, then determine l using trigonometry applied to $\triangle ABC$.

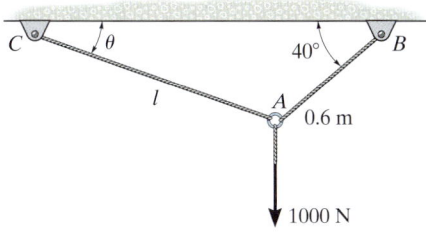

Prob. 3–74

3–75. The "scale" consists of a known weight W which is suspended at A from a cord of total length L. Determine the weight w at B if A is at a distance y for equilibrium. Neglect the sizes and weights of the pulleys.

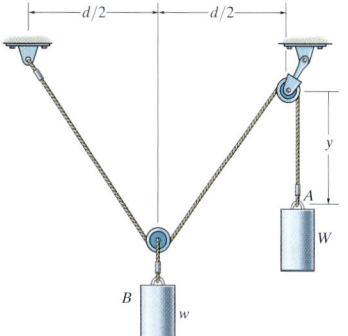

Prob. 3–75

*3–76. The uniform 50-kg crate is suspended by using a 2-m-long cord that is attached to the sides of the crate and passes over the small pulley at O. If the cord can be attached at either points A and B, or C and D, determine which attachment produces the least amount of tension in the cord and specify the cord tension in this case.

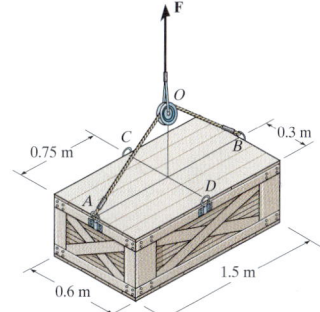

Prob. 3–76

3–77. Determine the magnitudes of $\mathbf{F}_1$, $\mathbf{F}_2$, and $\mathbf{F}_3$ for equilibrium of the particle.

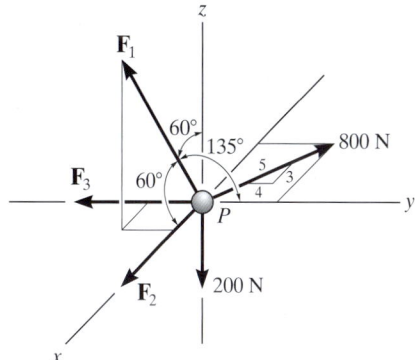

Prob. 3–77

Chapter 4

Forces applied to wrenches and wheels will produce rotation or a tendency for rotation. This effect is called a moment, and in this chapter we will study how to determine the moment of a system of forces and calculate their resultants.

Force System Resultants

CHAPTER OBJECTIVES

- To discuss the concept of the moment of a force and show how to calculate it in two and three dimensions.

- To provide a method for finding the moment of a force about a specified axis.

- To define the moment of a couple.

- To present methods for determining the resultants of nonconcurrent force systems.

- To indicate how to reduce a simple distributed loading to a resultant force having a specified location.

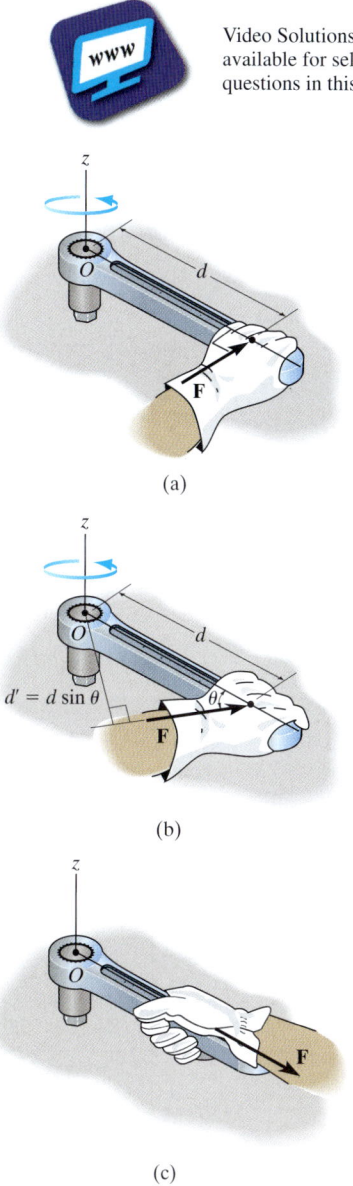

(a)

(b)

(c)

Fig. 4–1

4.1 Moment of a Force— Scalar Formulation

When a force is applied to a body it will produce a tendency for the body to rotate about a point that is not on the line of action of the force. This tendency to rotate is sometimes called a *torque*, but most often it is called the moment of a force or simply the *moment*. For example, consider a wrench used to unscrew the bolt in Fig. 4–1a. If a force is applied to the handle of the wrench it will tend to turn the bolt about point O (or the z axis). The magnitude of the moment is directly proportional to the magnitude of **F** and the perpendicular distance or *moment arm d*. The larger the force or the longer the moment arm, the greater the moment or turning effect. Note that if the force **F** is applied at an angle $\theta \neq 90°$, Fig. 4–1b, then it will be more difficult to turn the bolt since the moment arm $d' = d \sin \theta$ will be smaller than d. If **F** is applied along the wrench, Fig. 4–1c, its moment arm will be zero since the line of action of **F** will intersect point O (the z axis). As a result, the moment of **F** about O is also zero and no turning can occur.

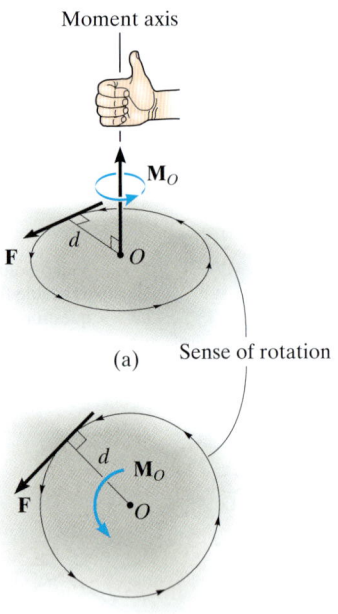

(a)   Sense of rotation

(b)

Fig. 4–2

We can generalize the above discussion and consider the force **F** and point O which lie in the shaded plane as shown in Fig. 4–2a. The moment $\mathbf{M}_O$ about point O, or about an axis passing through O and perpendicular to the plane, is a *vector quantity* since it has a specified magnitude and direction.

Magnitude.   The magnitude of $\mathbf{M}_O$ is

$$M_O = Fd \tag{4–1}$$

where d is the *moment arm* or *perpendicular distance* from the axis at point O to the line of action of the force. Units of moment magnitude consist of force times distance, e.g., $N \cdot m$ or $lb \cdot ft$.

Direction.   The direction of $\mathbf{M}_O$ is defined by its *moment axis*, which is perpendicular to the plane that contains the force **F** and its moment arm d. The right-hand rule is used to establish the sense of direction of $\mathbf{M}_O$. According to this rule, the natural curl of the fingers of the right hand, as they are drawn towards the palm, represent the rotation, or if no movement is possible, there is a tendency for rotation caused by the moment. As this action is performed, the thumb of the right hand will give the directional sense of $\mathbf{M}_O$, Fig. 4–2a. Notice that the moment vector is represented three-dimensionally by a curl around an arrow. In two dimensions this vector is represented only by the curl as in Fig. 4–2b. Since in this case the moment will tend to cause a counterclockwise rotation, the moment vector is actually directed out of the page.

Resultant Moment.   For two-dimensional problems, where all the forces lie within the x–y plane, Fig. 4–3, the resultant moment $(\mathbf{M}_R)_o$ about point O (the z axis) can be determined by *finding the algebraic sum* of the moments caused by all the forces in the system. As a convention, we will generally consider *positive moments* as *counterclockwise* since they are directed along the positive z axis (out of the page). *Clockwise moments* will be *negative*. Doing this, the directional sense of each moment can be represented by a *plus or minus* sign. Using this sign convention, the resultant moment in Fig. 4–3 is therefore

$$\zeta + (M_R)_o = \Sigma Fd; \qquad (M_R)_o = F_1 d_1 - F_2 d_2 + F_3 d_3$$

If the numerical result of this sum is a positive scalar, $(\mathbf{M}_R)_o$ will be a counterclockwise moment (out of the page); and if the result is negative, $(\mathbf{M}_R)_o$ will be a clockwise moment (into the page).

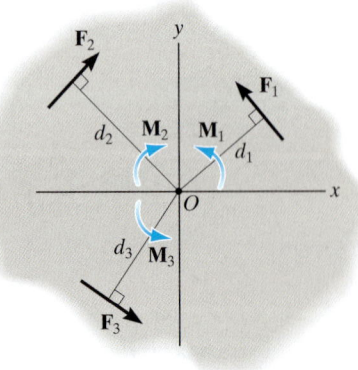

Fig. 4–3

EXAMPLE 4.1

For each case illustrated in Fig. 4–4, determine the moment of the force about point O.

SOLUTION (SCALAR ANALYSIS)

The line of action of each force is extended as a dashed line in order to establish the moment arm d. Also illustrated is the tendency of rotation of the member as caused by the force. Furthermore, the orbit of the force about O is shown as a colored curl. Thus,

Fig. 4–4a $M_O = (100 \text{ N})(2 \text{ m}) = 200 \text{ N} \cdot \text{m}$ ↻   *Ans.*

Fig. 4–4b $M_O = (50 \text{ N})(0.75 \text{ m}) = 37.5 \text{ N} \cdot \text{m}$ ↻   *Ans.*

Fig. 4–4c $M_O = (40 \text{ kN})(4 \text{ m} + 2 \cos 30° \text{ m}) = 229 \text{ kN} \cdot \text{m}$ ↻   *Ans.*

Fig. 4–4d $M_O = (60 \text{ kN})(1 \sin 45° \text{m}) = 42.4 \text{ kN} \cdot \text{m}$ ↺   *Ans.*

Fig. 4–4e $M_O = (7 \text{ kN})(4 \text{ m} - 1 \text{ m}) = 21.0 \text{ kN} \cdot \text{m}$ ↺   *Ans.*

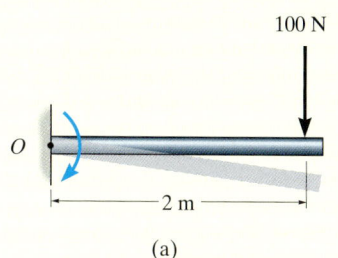

(a)

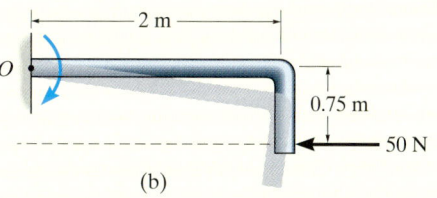

(b)

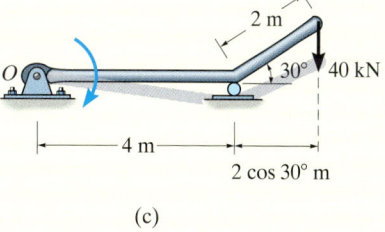

(c)

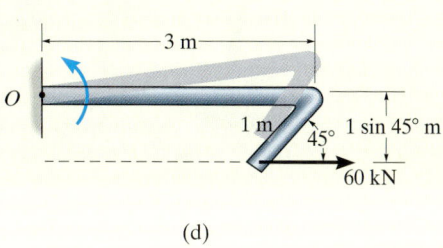

(d)

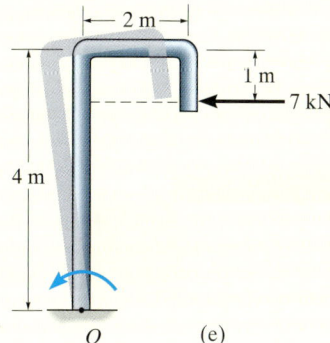

(e)

Fig. 4–4

EXAMPLE | 4.2

Determine the resultant moment of the four forces acting on the rod shown in Fig. 4–5 about point O.

SOLUTION

Assuming that positive moments act in the $+\mathbf{k}$ direction, i.e., counterclockwise, we have

$$\zeta + (M_R)_o = \Sigma Fd;$$

$$(M_R)_o = -50\ \text{N}(2\ \text{m}) + 60\ \text{N}(0) + 20\ \text{N}(3\ \sin 30°\ \text{m})$$

$$-40\ \text{N}(4\ \text{m} + 3\ \cos 30°\ \text{m})$$

$$(M_R)_o = -334\ \text{N} \cdot \text{m} = 334\ \text{N} \cdot \text{m} \,\zeta \qquad \textit{Ans.}$$

Fig. 4–5

For this calculation, note how the moment-arm distances for the 20-N and 40-N forces are established from the extended (dashed) lines of action of each of these forces.

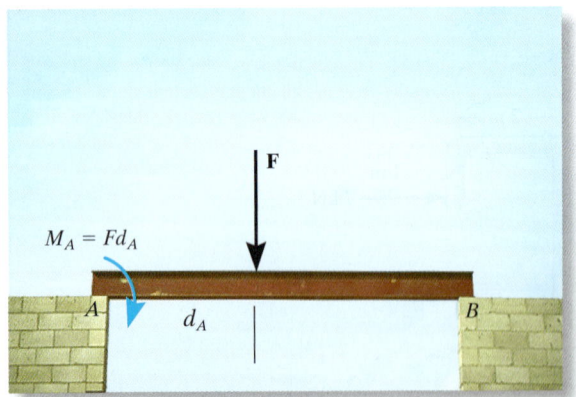

As illustrated by the example problems, the moment of a force does not always cause a rotation. For example, the force **F** tends to rotate the beam clockwise about its support at A with a moment $M_A = Fd_A$. The actual rotation would occur if the support at B were removed.

The ability to remove the nail will require the moment of $\mathbf{F}_H$ about point O to be larger than the moment of the force $\mathbf{F}_N$ about O that is needed to pull the nail out.

4.2   Cross Product

The moment of a force will be formulated using Cartesian vectors in the next section. Before doing this, however, it is first necessary to expand our knowledge of vector algebra and introduce the cross-product method of vector multiplication.

The *cross product* of two vectors **A** and **B** yields the vector **C**, which is written

$$\mathbf{C} = \mathbf{A} \times \mathbf{B} \qquad (4\text{–}2)$$

and is read "**C** equals **A** cross **B**."

Magnitude.   The *magnitude* of **C** is defined as the product of the magnitudes of **A** and **B** and the sine of the angle θ between their tails ($0° \leq \theta \leq 180°$). Thus, $C = AB \sin \theta$.

Direction.   Vector **C** has a *direction* that is perpendicular to the plane containing **A** and **B** such that **C** is specified by the right-hand rule; i.e., curling the fingers of the right hand from vector **A** (cross) to vector **B**, the thumb points in the direction of **C**, as shown in Fig. 4–6.

Knowing both the magnitude and direction of **C**, we can write

$$\mathbf{C} = \mathbf{A} \times \mathbf{B} = (AB \sin \theta)\mathbf{u}_C \qquad (4\text{–}3)$$

where the scalar $AB \sin \theta$ defines the *magnitude* of **C** and the unit vector $\mathbf{u}_C$ defines the *direction* of **C**. The terms of Eq. 4–3 are illustrated graphically in Fig. 4–6.

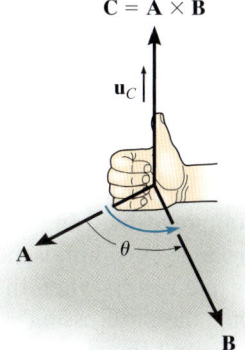

$\mathbf{C} = \mathbf{A} \times \mathbf{B}$

$\mathbf{u}_C$

A

θ

B

Fig. 4–6

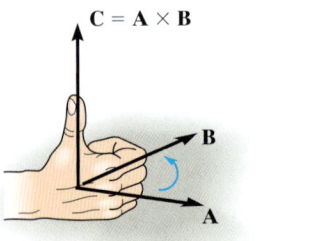

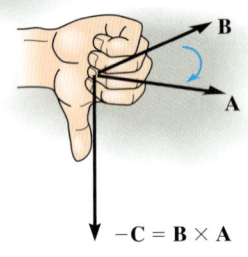

Fig. 4–7

Laws of Operation.

- The commutative law is *not* valid; i.e., $\mathbf{A} \times \mathbf{B} \neq \mathbf{B} \times \mathbf{A}$. Rather,

$$\mathbf{A} \times \mathbf{B} = -\mathbf{B} \times \mathbf{A}$$

This is shown in Fig. 4–7 by using the right-hand rule. The cross product $\mathbf{B} \times \mathbf{A}$ yields a vector that has the same magnitude but acts in the opposite direction to $\mathbf{C}$; i.e., $\mathbf{B} \times \mathbf{A} = -\mathbf{C}$.

- If the cross product is multiplied by a scalar a, it obeys the associative law;

$$a(\mathbf{A} \times \mathbf{B}) = (a\mathbf{A}) \times \mathbf{B} = \mathbf{A} \times (a\mathbf{B}) = (\mathbf{A} \times \mathbf{B})a$$

This property is easily shown since the magnitude of the resultant vector ($|a|AB \sin \theta$) and its direction are the same in each case.

- The vector cross product also obeys the distributive law of addition,

$$\mathbf{A} \times (\mathbf{B} + \mathbf{D}) = (\mathbf{A} \times \mathbf{B}) + (\mathbf{A} \times \mathbf{D})$$

- The proof of this identity is left as an exercise (see Prob. 4–1). It is important to note that *proper order* of the cross products must be maintained, since they are not commutative.

Cartesian Vector Formulation.

Equation 4–3 may be used to find the cross product of any pair of Cartesian unit vectors. For example, to find $\mathbf{i} \times \mathbf{j}$, the magnitude of the resultant vector is $(i)(j)(\sin 90°) = (1)(1)(1) = 1$, and its direction is determined using the right-hand rule. As shown in Fig. 4–8, the resultant vector points in the $+\mathbf{k}$ direction. Thus, $\mathbf{i} \times \mathbf{j} = (1)\mathbf{k}$. In a similar manner,

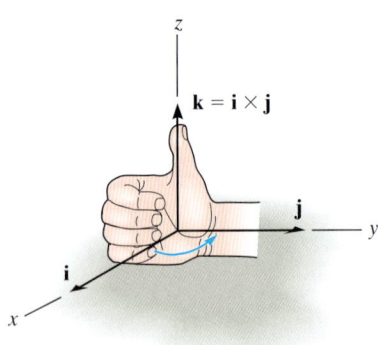

Fig. 4–8

$$\begin{array}{lll} \mathbf{i} \times \mathbf{j} = \mathbf{k} & \mathbf{i} \times \mathbf{k} = -\mathbf{j} & \mathbf{i} \times \mathbf{i} = 0 \\ \mathbf{j} \times \mathbf{k} = \mathbf{i} & \mathbf{j} \times \mathbf{i} = -\mathbf{k} & \mathbf{j} \times \mathbf{j} = 0 \\ \mathbf{k} \times \mathbf{i} = \mathbf{j} & \mathbf{k} \times \mathbf{j} = -\mathbf{i} & \mathbf{k} \times \mathbf{k} = 0 \end{array}$$

These results should *not* be memorized; rather, it should be clearly understood how each is obtained by using the right-hand rule and the definition of the cross product. A simple scheme shown in Fig. 4–9 is helpful for obtaining the same results when the need arises. If the circle is constructed as shown, then "crossing" two unit vectors in a *counterclockwise* fashion around the circle yields the *positive* third unit vector; e.g., $\mathbf{k} \times \mathbf{i} = \mathbf{j}$. "Crossing" *clockwise*, a *negative* unit vector is obtained; e.g., $\mathbf{i} \times \mathbf{k} = -\mathbf{j}$.

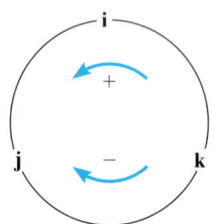

Fig. 4–9

Let us now consider the cross product of two general vectors **A** and **B** which are expressed in Cartesian vector form. We have

$$\mathbf{A} \times \mathbf{B} = (A_x\mathbf{i} + A_y\mathbf{j} + A_z\mathbf{k}) \times (B_x\mathbf{i} + B_y\mathbf{j} + B_z\mathbf{k})$$
$$= A_xB_x(\mathbf{i} \times \mathbf{i}) + A_xB_y(\mathbf{i} \times \mathbf{j}) + A_xB_z(\mathbf{i} \times \mathbf{k})$$
$$+ A_yB_x(\mathbf{j} \times \mathbf{i}) + A_yB_y(\mathbf{j} \times \mathbf{j}) + A_yB_z(\mathbf{j} \times \mathbf{k})$$
$$+ A_zB_x(\mathbf{k} \times \mathbf{i}) + A_zB_y(\mathbf{k} \times \mathbf{j}) + A_zB_z(\mathbf{k} \times \mathbf{k})$$

Carrying out the cross-product operations and combining terms yields

$$\mathbf{A} \times \mathbf{B} = (A_yB_z - A_zB_y)\mathbf{i} - (A_xB_z - A_zB_x)\mathbf{j} + (A_xB_y - A_yB_x)\mathbf{k} \qquad (4\text{--}4)$$

This equation may also be written in a more compact determinant form as

$$\mathbf{A} \times \mathbf{B} = \begin{vmatrix} \mathbf{i} & \mathbf{j} & \mathbf{k} \\ A_x & A_y & A_z \\ B_x & B_y & B_z \end{vmatrix} \qquad (4\text{--}5)$$

Thus, to find the cross product of any two Cartesian vectors **A** and **B**, it is necessary to expand a determinant whose first row of elements consists of the unit vectors **i**, **j**, and **k** and whose second and third rows represent the x, y, z components of the two vectors **A** and **B**, respectively.*

*A determinant having three rows and three columns can be expanded using three minors, each of which is multiplied by one of the three terms in the first row. There are four elements in each minor, for example,

$$\begin{vmatrix} A_{11} & A_{12} \\ A_{21} & A_{22} \end{vmatrix}$$

By *definition*, this determinant notation represents the terms $(A_{11}A_{22} - A_{12}A_{21})$, which is simply the product of the two elements intersected by the arrow slanting downward to the right $(A_{11}A_{22})$ *minus* the product of the two elements intersected by the arrow slanting downward to the left $(A_{12}A_{21})$. For a 3×3 determinant, such as Eq. 4–5, the three minors can be generated in accordance with the following scheme:

For element **i**: $\begin{vmatrix} \mathbf{i} & \mathbf{j} & \mathbf{k} \\ A_x & A_y & A_z \\ B_x & B_y & B_z \end{vmatrix} = \mathbf{i}(A_yB_z - A_zB_y)$

Remember the negative sign

For element **j**: $\begin{vmatrix} \mathbf{i} & \mathbf{j} & \mathbf{k} \\ A_x & A_y & A_z \\ B_x & B_y & B_z \end{vmatrix} = -\mathbf{j}(A_xB_z - A_zB_x)$

For element **k**: $\begin{vmatrix} \mathbf{i} & \mathbf{j} & \mathbf{k} \\ A_x & A_y & A_z \\ B_x & B_y & B_z \end{vmatrix} = \mathbf{k}(A_xB_y - A_yB_x)$

Adding the results and noting that the **j** element *must include the minus sign* yields the expanded form of **A** $\times$ **B** given by Eq. 4–4.

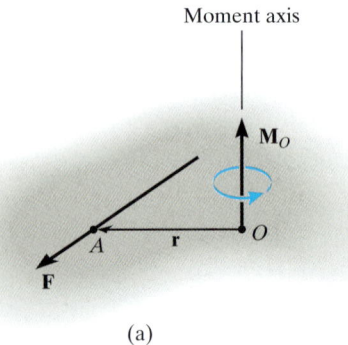

Moment axis

(a)

Fig. 4–10

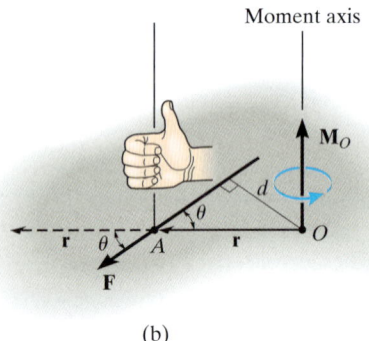

Moment axis

(b)

4.3   Moment of a Force—Vector Formulation

The moment of a force **F** about point O, or actually about the moment axis passing through O and perpendicular to the plane containing O and **F**, Fig. 4–10a, can be expressed using the vector cross product, namely,

$$\mathbf{M}_O = \mathbf{r} \times \mathbf{F} \tag{4–6}$$

Here **r** represents a position vector directed *from O* to *any point* on the line of action of **F**. We will now show that indeed the moment $\mathbf{M}_O$, when determined by this cross product, has the proper magnitude and direction.

Magnitude.   The magnitude of the cross product is defined from Eq. 4–3 as $M_O = rF \sin \theta$, where the angle θ is measured between the *tails* of **r** and **F**. To establish this angle, **r** must be treated as a sliding vector so that θ can be constructed properly, Fig. 4–10b. Since the moment arm $d = r \sin\theta$, then

$$M_O = rF \sin \theta = F(r \sin \theta) = Fd$$

which agrees with Eq. 4–1.

Direction.   The direction and sense of $\mathbf{M}_O$ in Eq. 4–6 are determined by the right-hand rule as it applies to the cross product. Thus, sliding **r** to the dashed position and curling the right-hand fingers from **r** toward **F**, "**r** cross **F**," the thumb is directed upward or perpendicular to the plane containing **r** and **F** and this is in the *same direction* as $\mathbf{M}_O$, the moment of the force about point O, Fig. 4–10b. Note that the "curl" of the fingers, like the curl around the moment vector, indicates the sense of rotation caused by the force. Since the cross product does not obey the commutative law, the order of $\mathbf{r} \times \mathbf{F}$ must be maintained to produce the correct sense of direction for $\mathbf{M}_O$.

Principle of Transmissibility.   The cross product operation is often used in three dimensions since the perpendicular distance or moment arm from point O to the line of action of the force is not needed. In other words, we can use any position vector **r** measured from point O to any point on the line of action of the force **F**, Fig. 4–11. Thus,

$$\mathbf{M}_O = \mathbf{r}_1 \times \mathbf{F} = \mathbf{r}_2 \times \mathbf{F} = \mathbf{r}_3 \times \mathbf{F}$$

Since **F** can be applied at any point along its line of action and still create this *same moment* about point O, then **F** can be considered a *sliding vector*. This property is called the *principle of transmissibility* of a force.

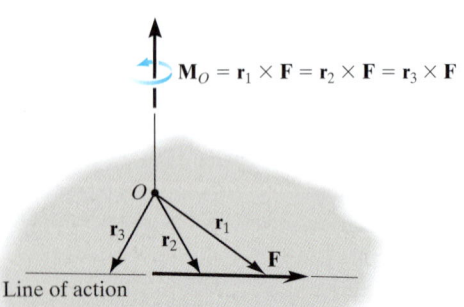

Line of action

Fig. 4–11

Cartesian Vector Formulation.

If we establish x, y, z coordinate axes, then the position vector $\mathbf{r}$ and force $\mathbf{F}$ can be expressed as Cartesian vectors, Fig. 4–12a. Applying Eq. 4–5 we have

$$\mathbf{M}_O = \mathbf{r} \times \mathbf{F} = \begin{vmatrix} \mathbf{i} & \mathbf{j} & \mathbf{k} \\ r_x & r_y & r_z \\ F_x & F_y & F_z \end{vmatrix} \qquad (4\text{–}7)$$

where

r_x, r_y, r_z represent the x, y, z components of the position vector drawn from point O to *any point* on the line of action of the force

F_x, F_y, F_z represent the x, y, z components of the force vector

If the determinant is expanded, then like Eq. 4–4 we have

$$\mathbf{M}_O = (r_y F_z - r_z F_y)\mathbf{i} - (r_x F_z - r_z F_x)\mathbf{j} + (r_x F_y - r_y F_x)\mathbf{k} \qquad (4\text{–}8)$$

The physical meaning of these three moment components becomes evident by studying Fig. 4–12b. For example, the $\mathbf{i}$ component of $\mathbf{M}_O$ can be determined from the moments of $\mathbf{F}_x$, $\mathbf{F}_y$, and $\mathbf{F}_z$ about the x axis. The component $\mathbf{F}_x$ does *not* create a moment or tendency to cause turning about the x axis since this force is *parallel* to the x axis. The line of action of $\mathbf{F}_y$ passes through point B, and so the magnitude of the moment of $\mathbf{F}_y$ about point A on the x axis is $r_z F_y$. By the right-hand rule this component acts in the *negative* $\mathbf{i}$ direction. Likewise, $\mathbf{F}_z$ passes through point C and so it contributes a moment component of $r_y F_z \mathbf{i}$ about the x axis. Thus, $(M_O)_x = (r_y F_z - r_z F_y)$ as shown in Eq. 4–8. As an exercise, establish the $\mathbf{j}$ and $\mathbf{k}$ components of $\mathbf{M}_O$ in this manner and show that indeed the expanded form of the determinant, Eq. 4–8, represents the moment of $\mathbf{F}$ about point O. Once $\mathbf{M}_O$ is determined, realize that it will always be *perpendicular* to the shaded plane containing vectors $\mathbf{r}$ and $\mathbf{F}$, Fig. 4–12a.

Resultant Moment of a System of Forces.

If a body is acted upon by a system of forces, Fig. 4–13, the resultant moment of the forces about point O can be determined by vector addition of the moment of each force. This resultant can be written symbolically as

$$(\mathbf{M}_R)_O = \Sigma(\mathbf{r} \times \mathbf{F}) \qquad (4\text{–}9)$$

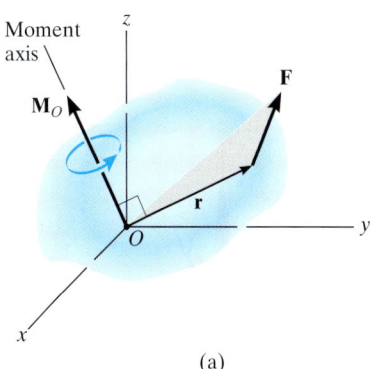

(a)

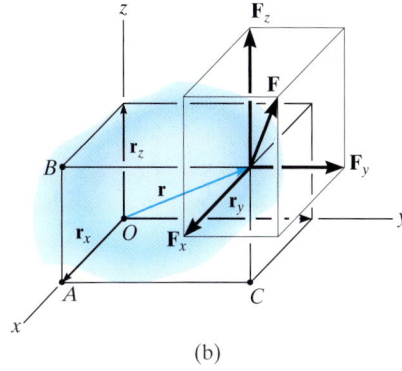

(b)

Fig. 4–12

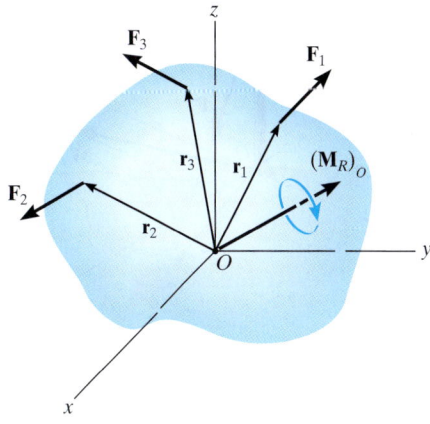

Fig. 4–13

EXAMPLE | 4.3

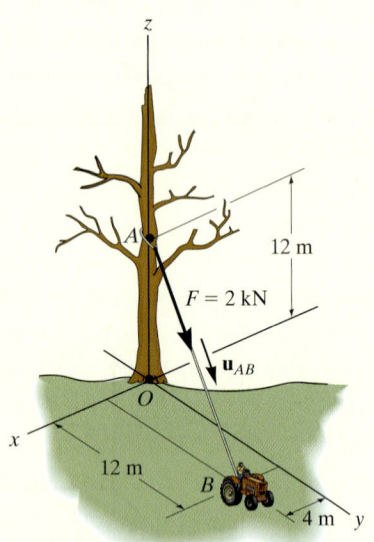

(a)

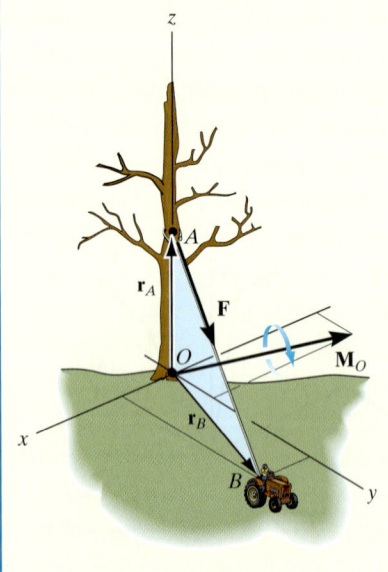

(b)

Fig. 4–14

Determine the moment produced by the force **F** in Fig. 4–14a about point O. Express the result as a Cartesian vector.

SOLUTION

As shown in Fig. 4–14b, either $\mathbf{r}_A$ or $\mathbf{r}_B$ can be used to determine the moment about point O. These position vectors are

$$\mathbf{r}_A = \{12\mathbf{k}\} \text{ m} \quad \text{and} \quad \mathbf{r}_B = \{4\mathbf{i} + 12\mathbf{j}\} \text{ m}$$

Force **F** expressed as a Cartesian vector is

$$\mathbf{F} = F\mathbf{u}_{AB} = 2 \text{ kN} \left[\frac{\{4\mathbf{i} + 12\mathbf{j} - 12\mathbf{k}\} \text{ m}}{\sqrt{(4 \text{ m})^2 + (12 \text{ m})^2 + (-12 \text{ m})^2}} \right]$$

$$= \{0.4588\mathbf{i} + 1.376\mathbf{j} - 1.376\mathbf{k}\} \text{ kN}$$

Thus

$$\mathbf{M}_O = \mathbf{r}_A \times \mathbf{F} = \begin{vmatrix} \mathbf{i} & \mathbf{j} & \mathbf{k} \\ 0 & 0 & 12 \\ 0.4588 & 1.376 & -1.376 \end{vmatrix}$$

$$= [0(-1.376) - 12(1.376)]\mathbf{i} - [0(-1.376) - 12(0.4588)]\mathbf{j}$$
$$+ [0(1.376) - 0(0.4588)]\mathbf{k}$$

$$= \{-16.5\mathbf{i} + 5.51\mathbf{j}\} \text{ kN} \cdot \text{m} \qquad \textit{Ans.}$$

or

$$\mathbf{M}_O = \mathbf{r}_B \times \mathbf{F} = \begin{vmatrix} \mathbf{i} & \mathbf{j} & \mathbf{k} \\ 4 & 12 & 0 \\ 0.4588 & 1.376 & -1.376 \end{vmatrix}$$

$$= [12(-1.376) - 0(1.376)]\mathbf{i} - [4(-1.376) - 0(0.4588)]\mathbf{j}$$
$$+ [4(1.376) - 12(0.4588)]\mathbf{k}$$

$$= \{-16.5\mathbf{i} + 5.51\mathbf{j}\} \text{ kN} \cdot \text{m} \qquad \textit{Ans.}$$

NOTE: As shown in Fig. 4–14b, $\mathbf{M}_O$ acts perpendicular to the plane that contains **F**, $\mathbf{r}_A$, and $\mathbf{r}_B$. Had this problem been worked using $M_O = Fd$, notice the difficulty that would arise in obtaining the moment arm d.

EXAMPLE 4.4

Two forces act on the rod shown in Fig. 4–15a. Determine the resultant moment they create about the flange at O. Express the result as a Cartesian vector.

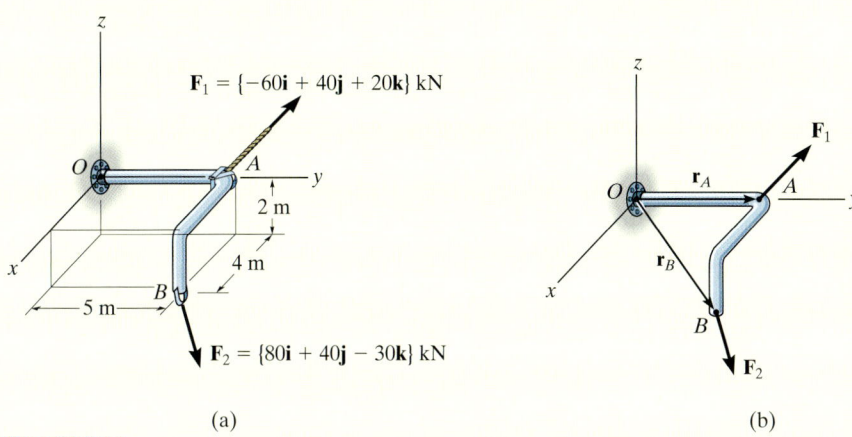

(a)   (b)

SOLUTION

Position vectors are directed from point O to each force as shown in Fig. 4–15b. These vectors are

$$\mathbf{r}_A = \{5\mathbf{j}\} \text{ m}$$

$$\mathbf{r}_B = \{4\mathbf{i} + 5\mathbf{j} - 2\mathbf{k}\} \text{ m}$$

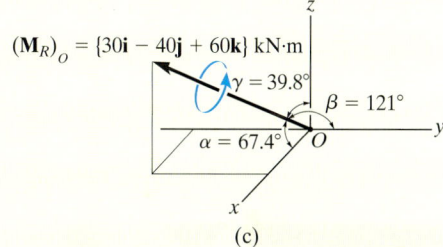

The resultant moment about O is therefore

$$(\mathbf{M}_R)_O = \Sigma(\mathbf{r} \times \mathbf{F})$$

$$= \mathbf{r}_A \times \mathbf{F}_1 + \mathbf{r}_B \times \mathbf{F}_2$$

$$= \begin{vmatrix} \mathbf{i} & \mathbf{j} & \mathbf{k} \\ 0 & 5 & 0 \\ -60 & 40 & 20 \end{vmatrix} + \begin{vmatrix} \mathbf{i} & \mathbf{j} & \mathbf{k} \\ 4 & 5 & -2 \\ 80 & 40 & -30 \end{vmatrix}$$

$$= [5(20) - 0(40)]\mathbf{i} - [0]\mathbf{j} + [0(40) - (5)(-60)]\mathbf{k}$$

$$\quad + [5(-30) - (-2)(40)]\mathbf{i} - [4(-30) - (-2)(80)]\mathbf{j} + [4(40) - 5(80)]\mathbf{k}$$

$$= \{30\mathbf{i} - 40\mathbf{j} + 60\mathbf{k}\} \text{ kN} \cdot \text{m}$$

Ans.

NOTE: This result is shown in Fig. 4–15c. The coordinate direction angles were determined from the unit vector for $(\mathbf{M}_R)_O$. Realize that the two forces tend to cause the rod to rotate about the moment axis in the manner shown by the curl indicated on the moment vector.

Fig. 4–15

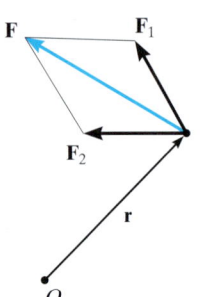

Fig. 4–16

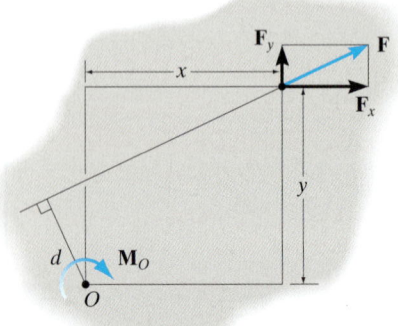

Fig. 4–17

4.4   Principle of Moments

A concept often used in mechanics is the *principle of moments*, which is sometimes referred to as *Varignon's theorem* since it was originally developed by the French mathematician Varignon (1654–1722). It states that *the moment of a force about a point is equal to the sum of the moments of the components of the force about the point.* This theorem can be proven easily using the vector cross product since the cross product obeys the *distributive law*. For example, consider the moments of the force $\mathbf{F}$ and two of its components about point O. Fig. 4–16. Since $\mathbf{F} = \mathbf{F}_1 + \mathbf{F}_2$ we have

$$\mathbf{M}_O = \mathbf{r} \times \mathbf{F} = \mathbf{r} \times (\mathbf{F}_1 + \mathbf{F}_2) = \mathbf{r} \times \mathbf{F}_1 + \mathbf{r} \times \mathbf{F}_2$$

For two-dimensional problems, Fig. 4–17, we can use the principle of moments by resolving the force into its rectangular components and then determine the moment using a scalar analysis. Thus,

$$M_O = F_x y - F_y x$$

This method is generally easier than finding the same moment using $M_O = Fd$.

Important Points

- The moment of a force creates the tendency of a body to turn about an axis passing through a specific point O.

- Using the right-hand rule, the sense of rotation is indicated by the curl of the fingers, and the thumb is directed along the moment axis, or line of action of the moment.

- The magnitude of the moment is determined from $M_O = Fd$, where d is called the moment arm, which represents the perpendicular or shortest distance from point O to the line of action of the force.

- In three dimensions the vector cross product is used to determine the moment, i.e., $\mathbf{M}_O = \mathbf{r} \times \mathbf{F}$. Remember that $\mathbf{r}$ is directed *from* point O *to any point* on the line of action of $\mathbf{F}$.

- The principle of moments states that the moment of a force about a point is equal to the sum of the moments of the force's components about the point. This is a very convenient method to use in two dimensions.

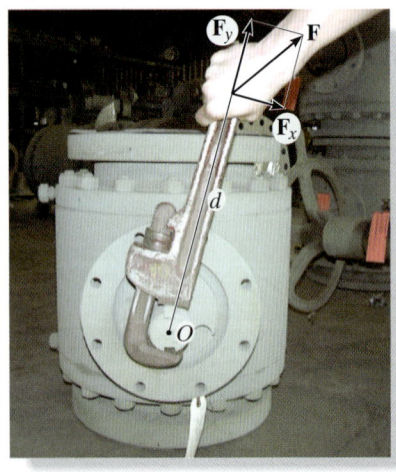

The moment of the applied force $\mathbf{F}$ about point O is easy to determine if we use the principle of moments. It is simply $M_O = F_x d$.

EXAMPLE 4.5

Determine the moment of the force in Fig. 4–18a about point O.

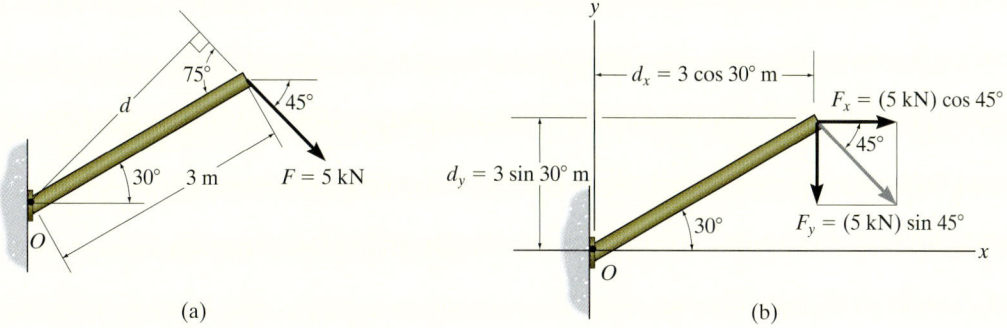

(a)   (b)

SOLUTION I
The moment arm d in Fig. 4–18a can be found from trigonometry.

$$d = (3 \text{ m}) \sin 75° = 2.898 \text{ m}$$

Thus,

$$M_O = Fd = (5 \text{ kN})(2.898 \text{ m}) = 14.5 \text{ kN} \cdot \text{m} \quad\quad \textit{Ans.}$$

Since the force tends to rotate or orbit clockwise about point O, the moment is directed into the page.

SOLUTION II
The x and y components of the force are indicated in Fig. 4–18b. Considering counterclockwise moments as positive, and applying the principle of moments, we have

$$\zeta + M_O = -F_x d_y - F_y d_x$$
$$= -(5 \cos 45° \text{ kN})(3 \sin 30° \text{ m}) - (5 \sin 45° \text{ kN})(3 \cos 30° \text{ m})$$
$$= -14.5 \text{ kN} \cdot \text{m} = 14.5 \text{ kN} \cdot \text{m} \quad\quad \textit{Ans.}$$

SOLUTION III
The x and y axes can be set parallel and perpendicular to the rod's axis as shown in Fig. 4–18c. Here $\mathbf{F}_x$ produces no moment about point O since its line of action passes through this point. Therefore,

$$\zeta + M_O = -F_y d_x$$
$$= -(5 \sin 75° \text{ kN})(3 \text{ m})$$
$$= -14.5 \text{ kN} \cdot \text{m} = 14.5 \text{ kN} \cdot \text{m} \quad\quad \textit{Ans.}$$

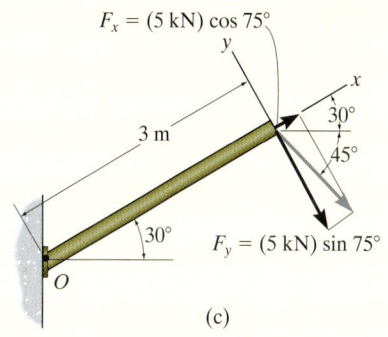

(c)

Fig. 4–18

EXAMPLE 4.6

Force **F** acts at the end of the angle bracket in Fig. 4–19a. Determine the moment of the force about point O.

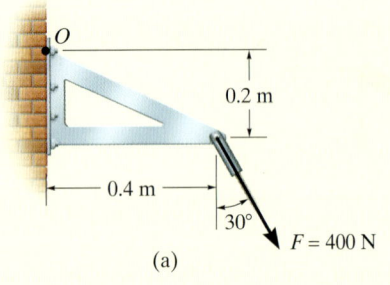

(a)

SOLUTION I (SCALAR ANALYSIS)

The force is resolved into its x and y components, Fig. 4–19b, then

$$\zeta + M_O = 400 \sin 30° \text{ N}(0.2 \text{ m}) - 400 \cos 30° \text{ N}(0.4 \text{ m})$$

$$= -98.6 \text{ N} \cdot \text{m} = 98.6 \text{ N} \cdot \text{m} \; \supset$$

or

$$M_O = \{-98.6\mathbf{k}\} \text{ N} \cdot \text{m} \qquad\qquad Ans.$$

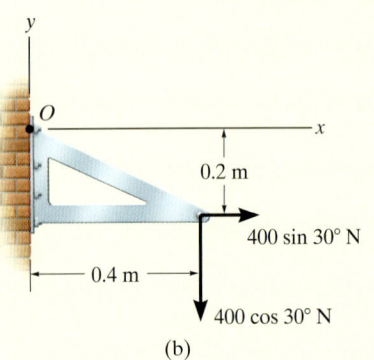

(b)

SOLUTION II (VECTOR ANALYSIS)

Using a Cartesian vector approach, the force and position vectors, Fig. 4–19c, are

$$\mathbf{r} = \{0.4\mathbf{i} - 0.2\mathbf{j}\} \text{ m}$$

$$\mathbf{F} = \{400 \sin 30°\mathbf{i} - 400 \cos 30°\mathbf{j}\} \text{ N}$$

$$= \{200.0\mathbf{i} - 346.4\mathbf{j}\} \text{ N}$$

The moment is therefore

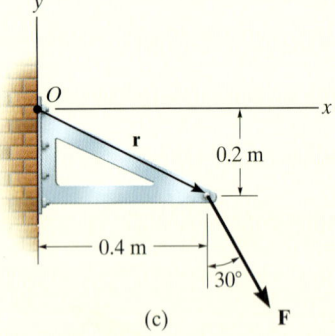

(c)

Fig. 4–19

$$\mathbf{M}_O = \mathbf{r} \times \mathbf{F} = \begin{vmatrix} \mathbf{i} & \mathbf{j} & \mathbf{k} \\ 0.4 & -0.2 & 0 \\ 200.0 & -346.4 & 0 \end{vmatrix}$$

$$= 0\mathbf{i} - 0\mathbf{j} + [0.4(-346.4) - (-0.2)(200.0)]\mathbf{k}$$

$$= \{-98.6\mathbf{k}\} \text{ N} \cdot \text{m} \qquad\qquad Ans.$$

NOTE: It is seen that the scalar analysis (Solution I) provides a more *convenient method* for analysis than Solution II since the direction of the moment and the moment arm for each component force are easy to establish. Hence, this method is generally recommended for solving problems displayed in two dimensions, whereas a Cartesian vector analysis is generally recommended only for solving three-dimensional problems.

FUNDAMENTAL PROBLEMS

F4–1. Determine the moment of the force about point O.

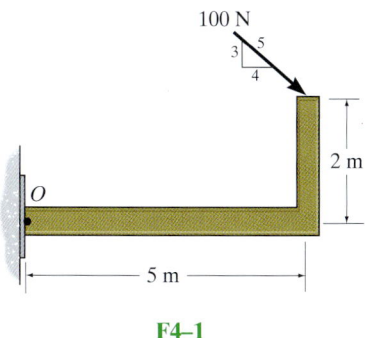

100 N

3 5
4

2 m

O

5 m

F4–1

F4–2. Determine the moment of the force about point O.

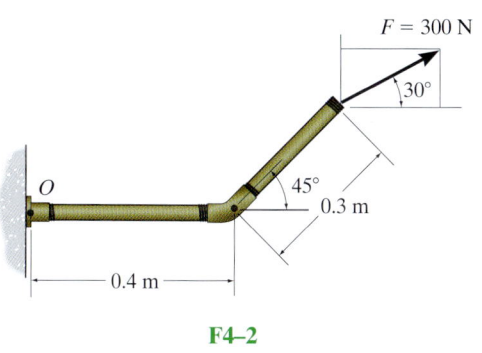

$F = 300$ N

30°

O

45°

0.3 m

0.4 m

F4–2

F4–3. Determine the moment of the force about point O.

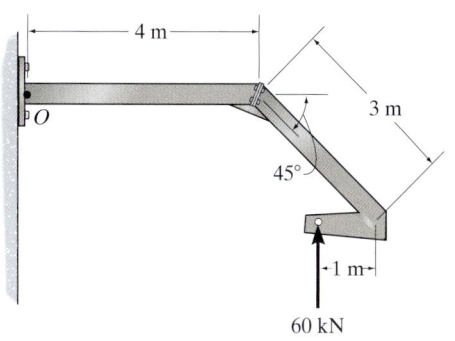

4 m

3 m

45°

1 m

60 kN

O

F4–3

F4–4. Determine the moment of the force about point O. Neglect the thickness of the member.

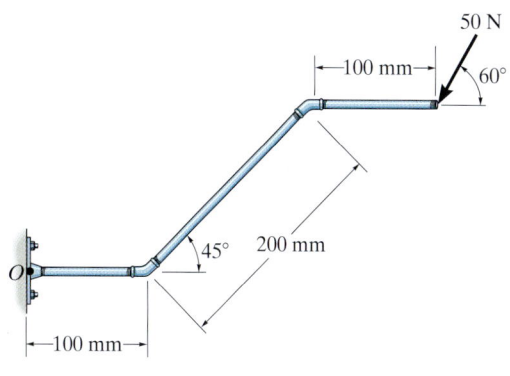

50 N

100 mm

60°

45°

200 mm

O

100 mm

F4–4

F4–5. Determine the moment of the force about point O.

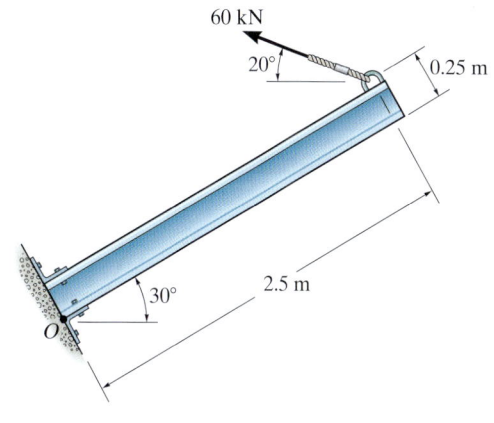

60 kN

20°

0.25 m

30°

2.5 m

O

F4–5

F4–6. Determine the moment of the force about point O.

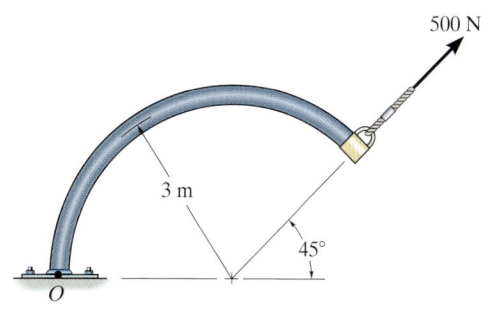

500 N

3 m

45°

O

F4–6

F4–7. Determine the resultant moment produced by the forces about point O.

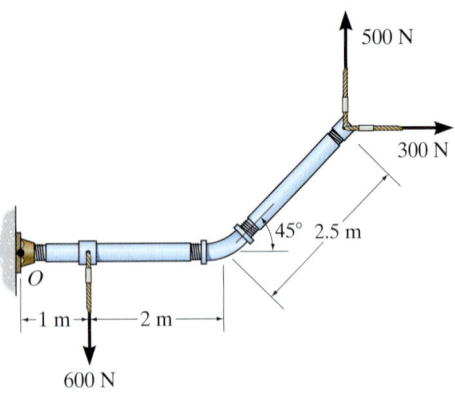

F4–7

F4–8. Determine the resultant moment produced by the forces about point O.

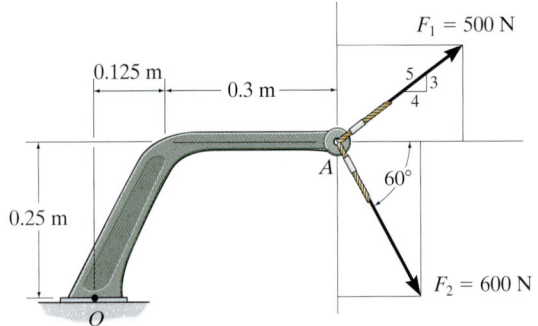

F4–8

F4–9. Determine the resultant moment produced by the forces about point O.

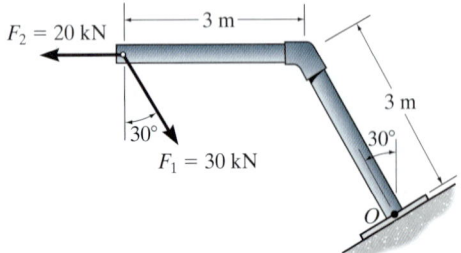

F4–9

F4–10. Determine the moment of force $\mathbf{F}$ about point O. Express the result as a Cartesian vector.

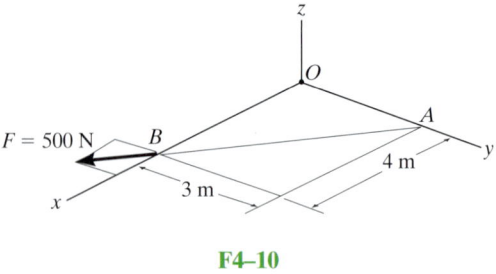

F4–10

F4–11. Determine the moment of force $\mathbf{F}$ about point O. Express the result as a Cartesian vector.

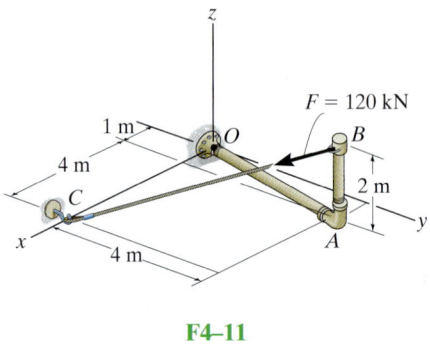

F4–11

F4–12. If $\mathbf{F}_1 = \{100\mathbf{i} - 120\mathbf{j} + 75\mathbf{k}\}$ kN and $\mathbf{F}_2 = \{-200\mathbf{i} + 250\mathbf{j} + 100\mathbf{k}\}$ kN, determine the resultant moment produced by these forces about point O. Express the result as a Cartesian vector.

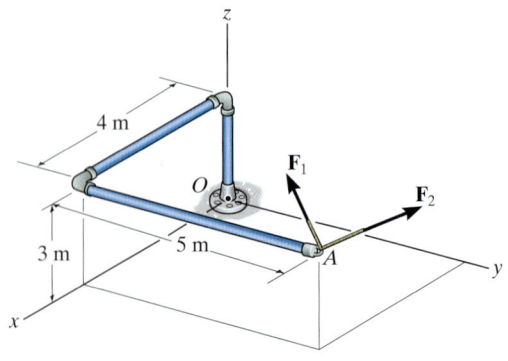

F4–12

PROBLEMS

4–1. If **A**, **B**, and **D** are given vectors, prove the distributive law for the vector cross product, i.e., **A** × (**B** + **D**) = (**A** × **B**) + (**A** × **D**).

4–2. Prove the triple scalar product identity **A** · (**B** × **C**) = (**A** × **B**) · **C**.

4–3. Given the three nonzero vectors **A**, **B**, and **C**, show that if **A** · (**B** × **C**) = 0, the three vectors *must* lie in the same plane.

***4–4.** Determine the magnitude and directional sense of the moment of the force at *A* about point *O*.

4–5. Determine the magnitude and directional sense of the moment of the force at *A* about point *P*.

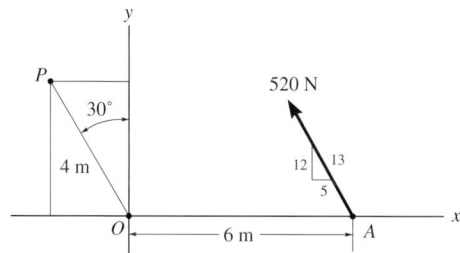

Probs. 4–4/5

4–6. The crane can be adjusted for any angle $0° \le \theta \le 90°$ and any extension $0 \le x \le 5$ m. For a suspended mass of 120 kg, determine the moment developed at *A* as a function of *x* and *θ*. What values of both *x* and *θ* develop the maximum possible moment at *A*? Compute this moment. Neglect the size of the pulley at *B*.

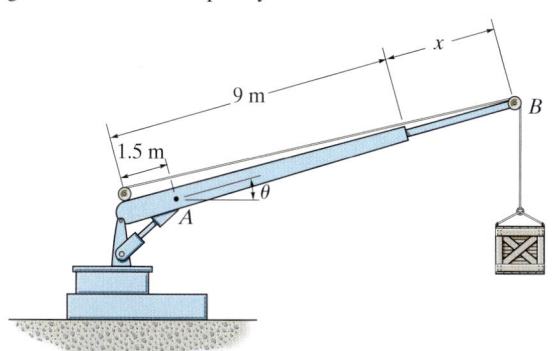

Prob. 4–6

4–7. Determine the moment of each of the three forces about point *A*.

***4–8.** Determine the moment of each of the three forces about point *B*.

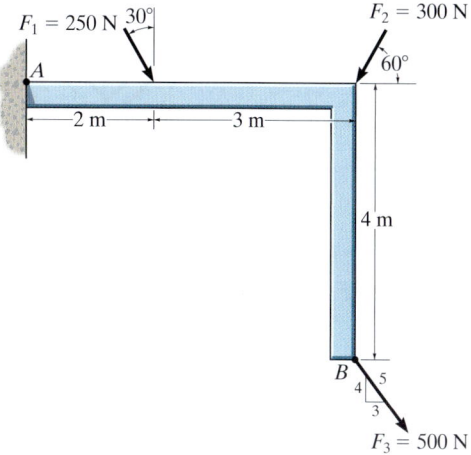

Probs. 4–7/8

4–9. Determine the magnitude and directional sense of the resultant moment of the forces about point *O*.

4–10. Determine the magnitude and directional sense of the resultant moment of the forces about point *P*.

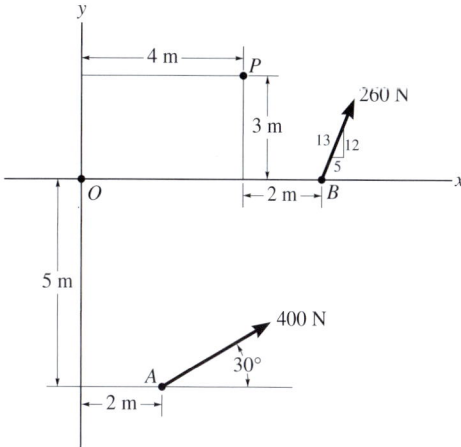

Probs. 4–9/10

4–11. The railway crossing gate consists of the 100-kg gate arm having a center of mass at G_a and the 250-kg counterweight having a center of mass at G_W. Determine the magnitude and directional sense of the resultant moment produced by the weights about point A.

*4–12. The railway crossing gate consists of the 100-kg gate arm having a center of mass at G_a and the 250-kg counterweight having a center of mass at G_W. Determine the magnitude and directional sense of the resultant moment produced by the weights about point B.

4–15. The Achilles tendon force of $F_t = 650$ N is mobilized when the man tries to stand on his toes. As this is done, each of his feet is subjected to a reactive force of $N_f = 400$ N. Determine the resultant moment of $\mathbf{F}_t$ and $\mathbf{N}_f$ about the ankle joint A.

*4–16. The Achilles tendon force $\mathbf{F}_t$ is mobilized when the man tries to stand on his toes. As this is done, each of his feet is subjected to a reactive force of $N_t = 400$ N. If the resultant moment produced by forces $\mathbf{F}_t$ and $\mathbf{N}_t$ about the ankle joint A is required to be zero, determine the magnitude of $\mathbf{F}_t$.

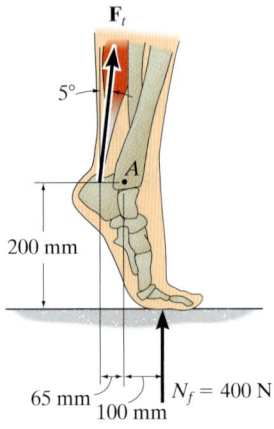

Probs. 4–15/16

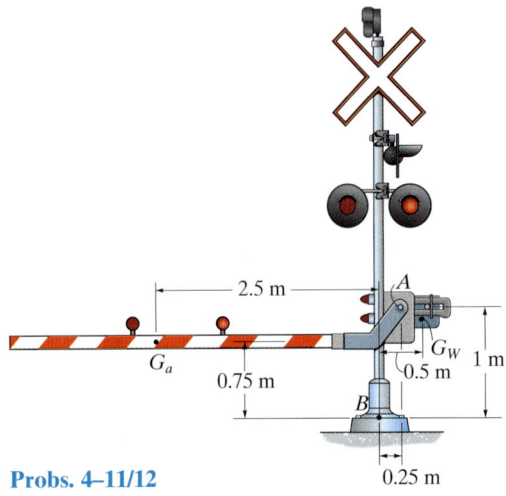

Probs. 4–11/12

4–13. The 70-N force acts on the end of the pipe at B. Determine (a) the moment of this force about point A, and (b) the magnitude and direction of a horizontal force, applied at C, which produces the same moment. Take $\theta = 60°$.

4–14. The 70-N force acts on the end of the pipe at B. Determine the angles θ ($0° \le \theta \le 180°$) of the force that will produce maximum and minimum moments about point A. What are the magnitudes of these moments?

4–17. The total hip replacement is subjected to a force of $F = 120$ N. Determine the moment of this force about the neck at A and the stem at B.

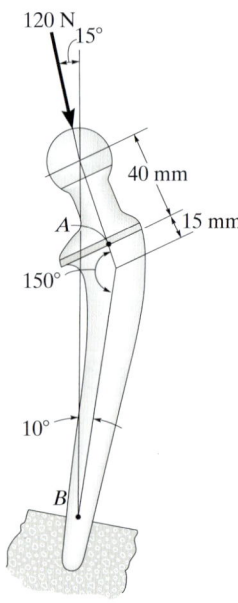

Prob. 4–17

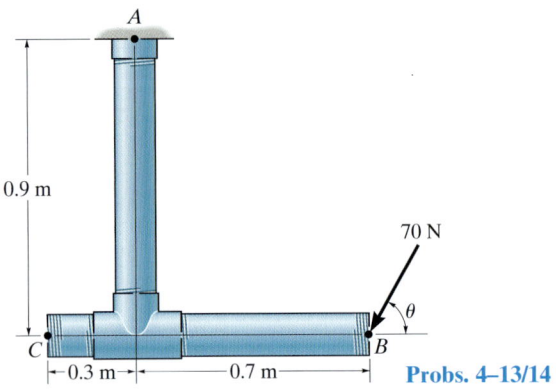

Probs. 4–13/14

4–18. The tower crane is used to hoist the 2-Mg load upward at constant velocity. The 1.5-Mg jib BD, 0.5-Mg jib BC, and 6-Mg counterweight C have centers of mass at G_1, G_2, and G_3, respectively. Determine the resultant moment produced by the load and the weights of the tower crane jibs about point A and about point B.

4–19. The tower crane is used to hoist a 2-Mg load upward at constant velocity. The 1.5-Mg jib BD and 0.5-Mg jib BC have centers of mass at G_1 and G_2, respectively. Determine the required mass of the counterweight C so that the resultant moment produced by the load and the weight of the tower crane jibs about point A is zero. The center of mass for the counterweight is located at G_3.

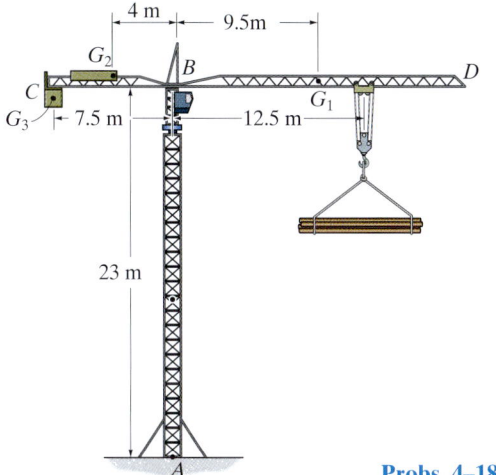

Probs. 4–18/19

*4–20. Determine the magnitude of the force $\mathbf{F}$ that should be applied at the end of the lever such that this force creates a clockwise moment of 15 N·m about point O when $\theta = 30°$.

4–21. If the force $F = 100$ N, determine the angle θ $(0 \le \theta \le 90°)$ so that the force develops a clockwise moment about point O of 20 N·m.

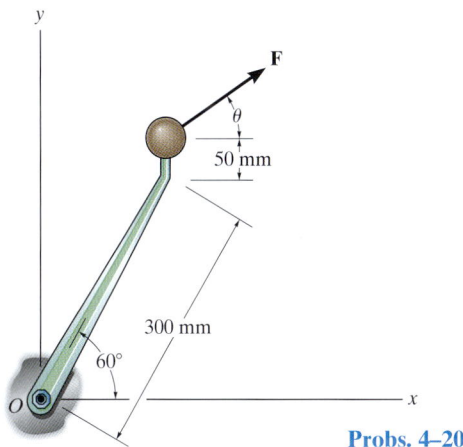

Probs. 4–20/21

4–22. The tool at A is used to hold a power lawnmower blade stationary while the nut is being loosened with the wrench. If a force of 50 N is applied to the wrench at B in the direction shown, determine the moment it creates about the nut at C. What is the magnitude of force $\mathbf{F}$ at A so that it creates the opposite moment about C?

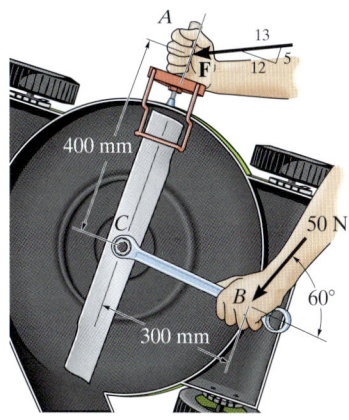

Prob. 4–22

4–23. The towline exerts a force of $P = 4$ kN at the end of the 20-m-long crane boom. If $\theta = 30°$, determine the placement x of the hook at A so that this force creates a maximum moment about point O. What is this moment?

*4–24. The towline exerts a force of $P = 4$ kN at the end of the 20-m-long crane boom. If $x = 25$ m, determine the position θ of the boom so that this force creates a maximum moment about point O. What is this moment?

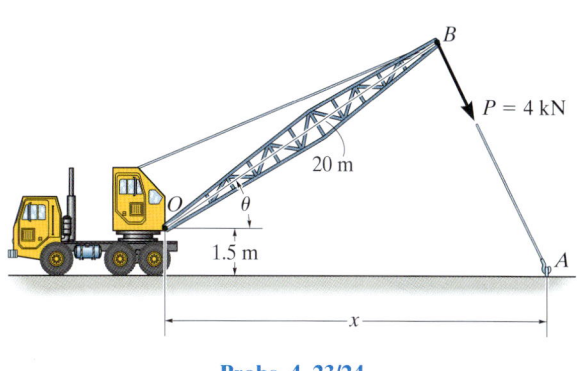

Probs. 4–23/24

4–25. Determine the resultant moment of the forces about point A. Solve the problem first by considering each force as a whole, and then by using the principle of moments. Take $F_1 = 250$ N, $F_2 = 300$ N, $F_3 = 500$ N.

4–26. If the resultant moment about point A is 4800 N · m clockwise, determine the magnitude of $\mathbf{F}_3$ if $F_1 = 300$ N and $F_2 = 400$ N.

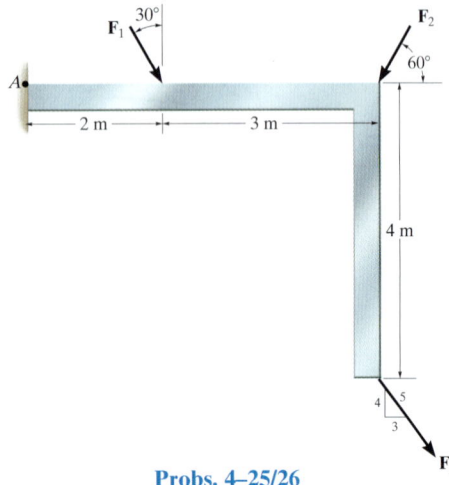

Probs. 4–25/26

4–27. The connected bar BC is used to increase the lever arm of the crescent wrench as shown. If the applied force is $F = 200$ N and $d = 300$ mm, determine the moment produced by this force about the bolt at A.

*4–28. The connected bar BC is used to increase the lever arm of the crescent wrench as shown. If a clockwise moment of $M_A = 120$ N · m is needed to tighten the bolt at A and the force $F = 200$ N, determine the required extension d in order to develop this moment.

4–29. The connected bar BC is used to increase the lever arm of the crescent wrench as shown. If a clockwise moment of $M_A = 120$ N · m is needed to tighten the nut at A and the extension $d = 300$ mm, determine the required force $\mathbf{F}$ in order to develop this moment.

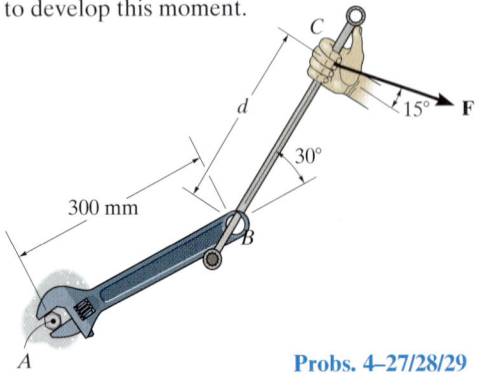

Probs. 4–27/28/29

4–30. A force $\mathbf{F}$ having a magnitude of $F = 100$ N acts along the diagonal of the parallelepiped. Determine the moment of $\mathbf{F}$ about point A, using $\mathbf{M}_A = \mathbf{r}_B \times \mathbf{F}$ and $\mathbf{M}_A = \mathbf{r}_C \times \mathbf{F}$.

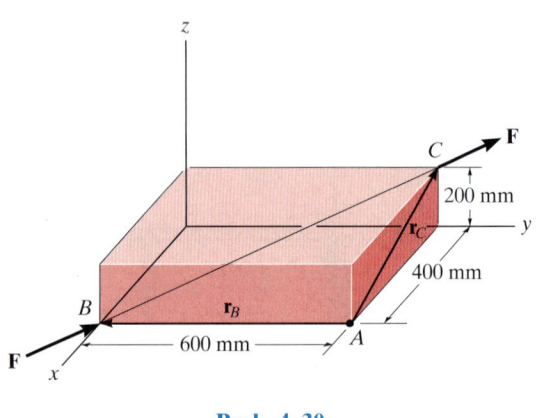

Prob. 4–30

4–31. The force $\mathbf{F} = \{600\mathbf{i} + 300\mathbf{j} - 600\mathbf{k}\}$ N acts at the end of the beam. Determine the moment of the force about point A.

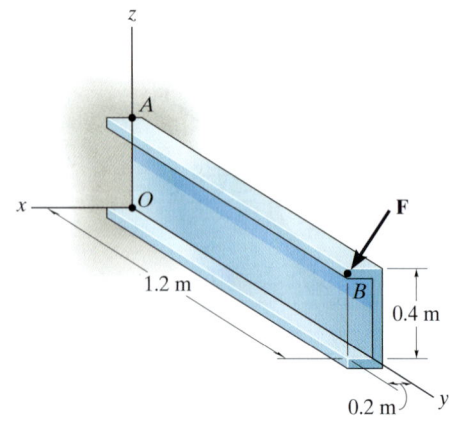

Prob. 4–31

*4–32. Determine the moment produced by force $\mathbf{F}_B$ about point O. Express the result as a Cartesian vector.

4–33. Determine the moment produced by force $\mathbf{F}_C$ about point O. Express the result as a Cartesian vector.

4–34. Determine the resultant moment produced by force $\mathbf{F}_B$ and $\mathbf{F}_C$ about point O. Express the result as a Cartesian vector.

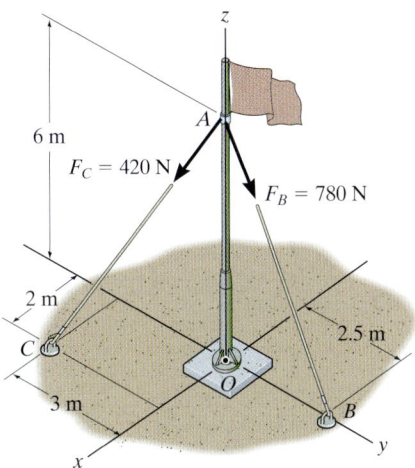

Probs. 4–32/33/34

4–35. Using a ring collar the 75-N force can act in the vertical plane at various angles θ. Determine the magnitude of the moment it produces about point A, plot the result of M (ordinate) versus θ (abscissa) for $0° \le \theta \le 180°$, and specify the angles that give the maximum and minimum moment.

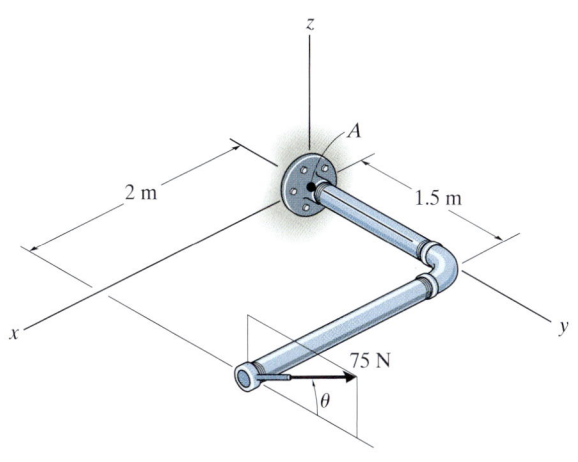

Prob. 4–35

*4–36. The curved rod lies in the x–y plane and has a radius of 3 m. If a force of $F = 80$ N acts at its end as shown, determine the moment of this force about point O.

4–37. The curved rod lies in the x–y plane and has a radius of 3 m. If a force of $F = 80$ N acts at its end as shown, determine the moment of this force about point B.

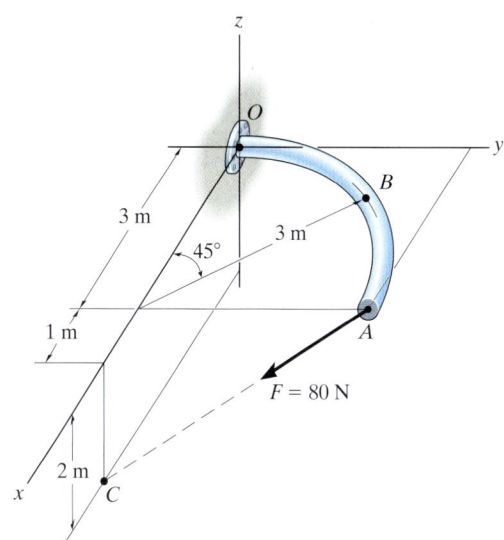

Probs. 4–36/37

4–38. Force $\mathbf{F}$ acts perpendicular to the inclined plane. Determine the moment produced by $\mathbf{F}$ about point A. Express the result as a Cartesian vector.

4–39. Solve Prob. 4–38 if $F = 500$ N.

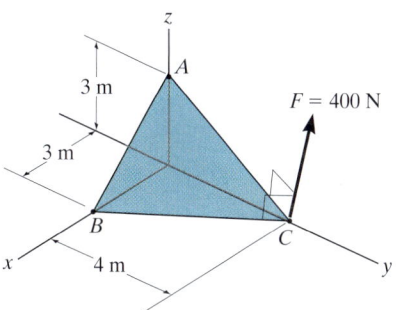

Probs. 4–38/39

*4–40. The pipe assembly is subjected to the 80-N force. Determine the moment of this force about point A.

4–41. The pipe assembly is subjected to the 80-N force. Determine the moment of this force about point B.

4–43. Determine the moment of the force **F** at A about point O. Express the result as a Cartesian vector.

*4–44. Determine the moment of the force **F** at A about point P. Express the result as a Cartesian vector.

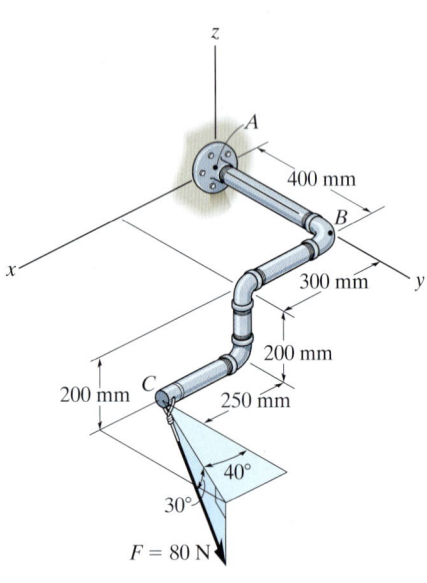

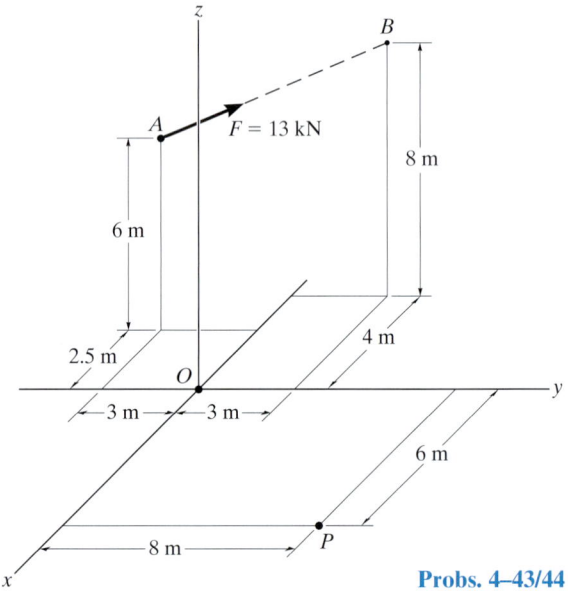

Probs. 4–43/44

Probs. 4–40/41

4–45. A force of $\mathbf{F} = \{6\mathbf{i} - 2\mathbf{j} + 1\mathbf{k}\}$ kN produces a moment of $\mathbf{M}_O = \{4\mathbf{i} + 5\mathbf{j} - 14\mathbf{k}\}$ kN·m about the origin of coordinates, point O. If the force acts at a point having an x coordinate of $x = 1$ m, determine the y and z coordinates.

4–42. Strut AB of the 1-m-diameter hatch door exerts a force of 450 N on point B. Determine the moment of this force about point O.

4–46. The force $\mathbf{F} = \{6\mathbf{i} + 8\mathbf{j} + 10\mathbf{k}\}$ N creates a moment about point O of $\mathbf{M}_O = \{-14\mathbf{i} + 8\mathbf{j} + 2\mathbf{k}\}$ N·m. If the force passes through a point having an x coordinate of 1 m, determine the y and z coordinates of the point. Also, realizing that $M_O = Fd$, determine the perpendicular distance d from point O to the line of action of **F**.

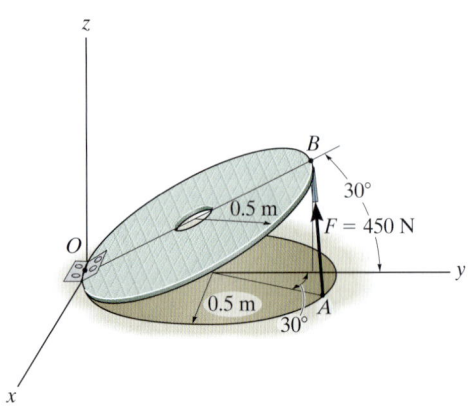

Prob. 4–42

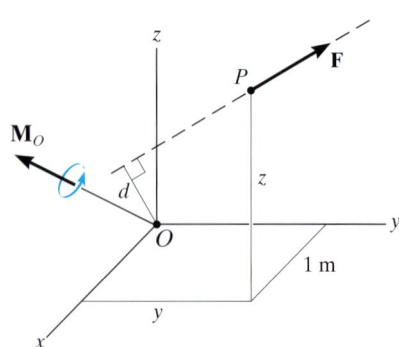

Probs. 4–45/46

4.5   Moment of a Force about a Specified Axis

Sometimes, the moment produced by a force about a *specified axis* must be determined. For example, suppose the lug nut at O on the car tire in Fig. 4–20a needs to be loosened. The force applied to the wrench will create a tendency for the wrench and the nut to rotate about the *moment axis* passing through O; however, the nut can only rotate about the y axis. Therefore, to determine the turning effect, only the y component of the moment is needed, and the total moment produced is not important. To determine this component, we can use either a scalar or vector analysis.

Scalar Analysis.   To use a scalar analysis in the case of the lug nut in Fig. 4–20a, the moment arm perpendicular distance from the axis to the line of action of the force is $d_y = d \cos \theta$. Thus, the moment of $\mathbf{F}$ about the y axis is $M_y = F\, d_y = F(d \cos \theta)$. According to the right-hand rule, $\mathbf{M}_y$ is directed along the positive y axis as shown in the figure. In general, for any axis a, the moment is

$$M_a = F d_a \qquad\qquad (4\text{–}10)$$

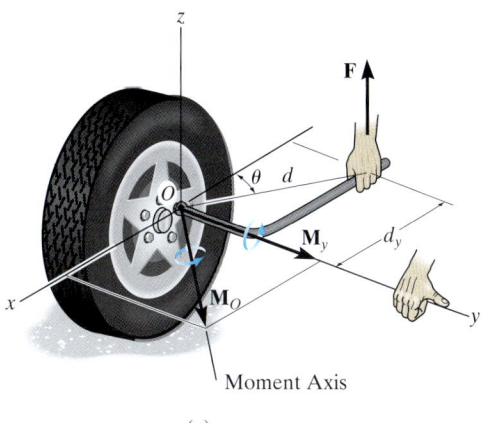

(a)

Fig. 4-20

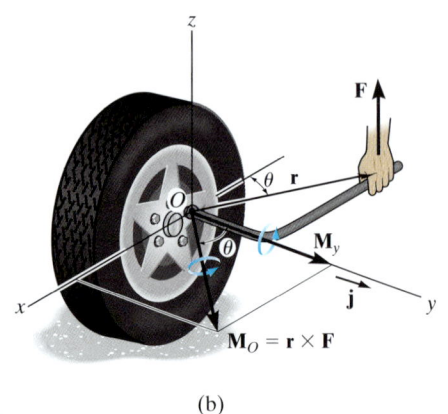

(b)

Fig. 4–20 (cont.)

4

Vector Analysis.   To find the moment of force **F** in Fig. 4–20b about the y axis using a vector analysis, we must first determine the moment of the force about *any point O* on the y axis by applying Eq. 4–7, $\mathbf{M}_O = \mathbf{r} \times \mathbf{F}$. The component $\mathbf{M}_y$ along the y axis is the *projection* of $\mathbf{M}_O$ onto the y axis. It can be found using the *dot product* discussed in Chapter 2, so that $M_y = \mathbf{j} \cdot \mathbf{M}_O = \mathbf{j} \cdot (\mathbf{r} \times \mathbf{F})$, where **j** is the unit vector for the y axis.

We can generalize this approach by letting $\mathbf{u}_a$ be the unit vector that specifies the direction of the a axis shown in Fig. 4–21. Then the moment of **F** about a point O on the axis is $\mathbf{M}_O = \mathbf{r} \times \mathbf{F}$, and the projection of this moment onto the a axis is $M_a = \mathbf{u}_a \cdot (\mathbf{r} \times \mathbf{F})$. This combination is referred to as the *scalar triple product*. If the vectors are written in Cartesian form, we have

$$M_a = [u_{a_x}\mathbf{i} + u_{a_y}\mathbf{j} + u_{a_z}\mathbf{k}] \cdot \begin{vmatrix} \mathbf{i} & \mathbf{j} & \mathbf{k} \\ r_x & r_y & r_z \\ F_x & F_y & F_z \end{vmatrix}$$

$$= u_{a_x}(r_yF_z - r_zF_y) - u_{a_y}(r_xF_z - r_zF_x) + u_{a_z}(r_xF_y - r_yF_x)$$

This result can also be written in the form of a determinant, making it easier to memorize.*

$$M_a = \mathbf{u}_a \cdot (\mathbf{r} \times \mathbf{F}) = \begin{vmatrix} u_{a_x} & u_{a_y} & u_{a_z} \\ r_x & r_y & r_z \\ F_x & F_y & F_z \end{vmatrix} \qquad (4\text{–}11)$$

where

$u_{a_x}, u_{a_y}, u_{a_z}$	represent the x, y, z components of the unit vector defining the direction of the a axis
r_x, r_y, r_z	represent the x, y, z components of the position vector extended from *any point O* on the a axis to *any point A* on the line of action of the force
F_x, F_y, F_z	represent the x, y, z components of the force vector.

When M_a is evaluated from Eq. 4–11, it will yield a positive or negative scalar. The sign of this scalar indicates the sense of direction of $\mathbf{M}_a$ along the a axis. If it is positive, then $\mathbf{M}_a$ will have the same sense as $\mathbf{u}_a$, whereas if it is negative, then $\mathbf{M}_a$ will act opposite to $\mathbf{u}_a$.

Once M_a is determined, we can then express $\mathbf{M}_a$ as a Cartesian vector, namely,

$$\mathbf{M}_a = M_a\mathbf{u}_a \qquad (4\text{–}12)$$

The examples which follow illustrate numerical applications of the above concepts.

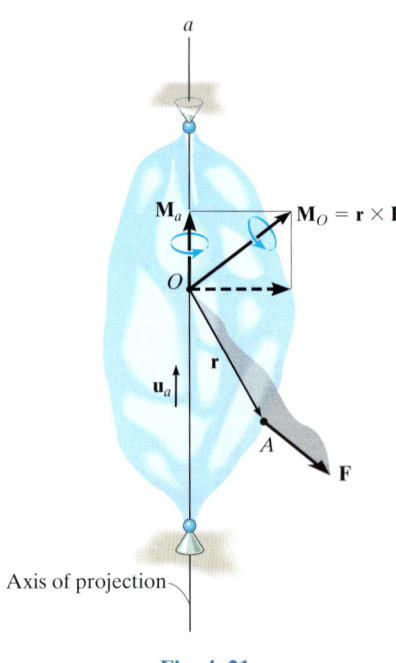

Fig. 4–21

*Take a minute to expand this determinant, to show that it will yield the above result.

Important Points

- The moment of a force about a specified axis can be determined provided the perpendicular distance d_a from the force line of action to the axis can be determined. $M_a = Fd_a$.

- If vector analysis is used, $M_a = \mathbf{u}_a \cdot (\mathbf{r} \times \mathbf{F})$, where $\mathbf{u}_a$ defines the direction of the axis and $\mathbf{r}$ is extended from *any point* on the axis to *any point* on the line of action of the force.

- If M_a is calculated as a negative scalar, then the sense of direction of $\mathbf{M}_a$ is opposite to $\mathbf{u}_a$.

- The moment $\mathbf{M}_a$ expressed as a Cartesian vector is determined from $\mathbf{M}_a = M_a \mathbf{u}_a$.

4

EXAMPLE 4.7

Determine the resultant moment of the three forces in Fig. 4–22 about the x axis, the y axis, and the z axis.

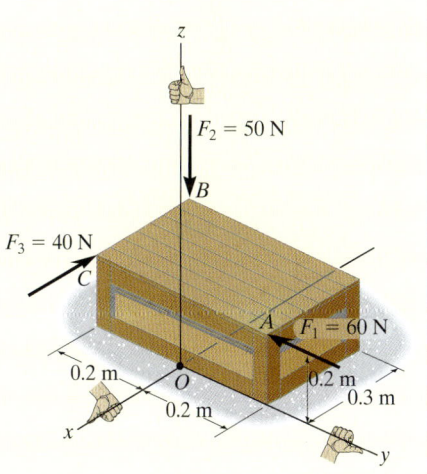

Fig. 4–22

SOLUTION
A force that is *parallel* to a coordinate axis or has a line of action that passes through the axis does *not* produce any moment or tendency for turning about that axis. Therefore, defining the positive direction of the moment of a force according to the right-hand rule, as shown in the figure, we have

$$M_x = (60 \text{ N})(0.2 \text{ m}) + (50 \text{ N})(0.2 \text{ m}) + 0 = 22 \text{ N} \cdot \text{m} \quad Ans.$$

$$M_y = 0 - (50 \text{ N})(0.3 \text{ m}) - (40 \text{ N})(0.2 \text{ m}) = -23 \text{ N} \cdot \text{m} \quad Ans.$$

$$M_z = 0 + 0 - (40 \text{ N})(0.2 \text{ m}) = -8 \text{ N} \cdot \text{m} \quad Ans.$$

The negative signs indicate that $\mathbf{M}_y$ and $\mathbf{M}_z$ act in the $-y$ and $-z$ directions, respectively.

EXAMPLE 4.8

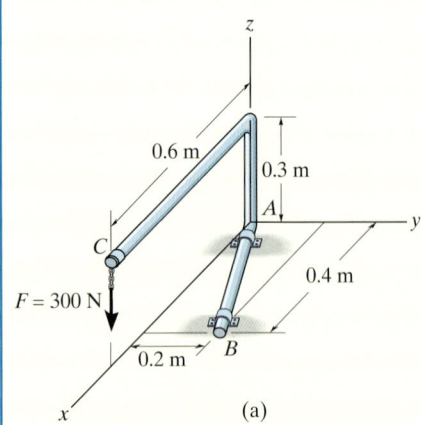

(a)

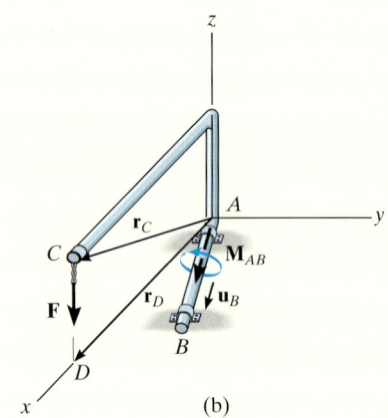

(b)

Fig. 4–23

Determine the moment $\mathbf{M}_{AB}$ produced by the force $\mathbf{F}$ in Fig. 4–23a, which tends to rotate the rod about the AB axis.

SOLUTION

A vector analysis using $M_{AB} = \mathbf{u}_B \cdot (\mathbf{r} \times \mathbf{F})$ will be considered for the solution rather than trying to find the moment arm or perpendicular distance from the line of action of $\mathbf{F}$ to the AB axis. Each of the terms in the equation will now be identified.

Unit vector $\mathbf{u}_B$ defines the direction of the AB axis of the rod, Fig. 4–23b, where

$$\mathbf{u}_B = \frac{\mathbf{r}_B}{r_B} = \frac{\{0.4\mathbf{i} + 0.2\mathbf{j}\}\ \text{m}}{\sqrt{(0.4\ \text{m})^2 + (0.2\ \text{m})^2}} = 0.8944\mathbf{i} + 0.4472\mathbf{j}$$

Vector $\mathbf{r}$ is directed from *any point* on the AB axis to *any point* on the line of action of the force. For example, position vectors $\mathbf{r}_C$ and $\mathbf{r}_D$ are suitable, Fig. 4–23b. (Although not shown, $\mathbf{r}_{BC}$ or $\mathbf{r}_{BD}$ can also be used.) For simplicity, we choose $\mathbf{r}_D$, where

$$\mathbf{r}_D = \{0.6\mathbf{i}\}\ \text{m}$$

The force is

$$\mathbf{F} = \{-300\mathbf{k}\}\ \text{N}$$

Substituting these vectors into the determinant form and expanding, we have

$$M_{AB} = \mathbf{u}_B \cdot (\mathbf{r}_D \times \mathbf{F}) = \begin{vmatrix} 0.8944 & 0.4472 & 0 \\ 0.6 & 0 & 0 \\ 0 & 0 & -300 \end{vmatrix}$$

$$= 0.8944[0(-300) - 0(0)] - 0.4472[0.6(-300) - 0(0)] + 0[0.6(0) - 0(0)]$$

$$= 80.50\ \text{N} \cdot \text{m}$$

This positive result indicates that the sense of $\mathbf{M}_{AB}$ is in the same direction as $\mathbf{u}_B$.

Expressing $\mathbf{M}_{AB}$ in Fig. 4-23b as a Cartesian vector yields

$$\mathbf{M}_{AB} = M_{AB}\mathbf{u}_B = (80.50\ \text{N} \cdot \text{m})(0.8944\mathbf{i} + 0.4472\mathbf{j})$$

$$= \{72.0\mathbf{i} + 36.0\mathbf{j}\}\ \text{N} \cdot \text{m} \qquad \textit{Ans.}$$

NOTE: If axis AB is defined using a unit vector directed from B toward A, then in the above formulation $-\mathbf{u}_B$ would have to be used. This would lead to $M_{AB} = -80.50\ \text{N} \cdot \text{m}$. Consequently, $\mathbf{M}_{AB} = M_{AB}(-\mathbf{u}_B)$, and the same result would be obtained.

EXAMPLE 4.9

Determine the magnitude of the moment of force **F** about segment *OA* of the pipe assembly in Fig. 4–24a.

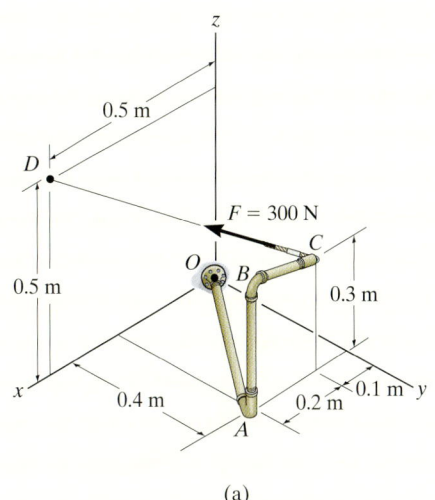

(a)

SOLUTION

The moment of **F** about the *OA* axis is determined from $M_{OA} = \mathbf{u}_{OA} \cdot (\mathbf{r} \times \mathbf{F})$, where **r** is a position vector extending from any point on the *OA* axis to any point on the line of action of **F**. As indicated in Fig. 4–24b, either $\mathbf{r}_{OD}, \mathbf{r}_{OC}, \mathbf{r}_{AD}$, or $\mathbf{r}_{AC}$ can be used; however, $\mathbf{r}_{OD}$ will be considered since it will simplify the calculation.

The unit vector $\mathbf{u}_{OA}$, which specifies the direction of the *OA* axis, is

$$\mathbf{u}_{OA} = \frac{\mathbf{r}_{OA}}{r_{OA}} = \frac{\{0.3\mathbf{i} + 0.4\mathbf{j}\}\text{ m}}{\sqrt{(0.3\text{ m})^2 + (0.4\text{ m})^2}} = 0.6\mathbf{i} + 0.8\mathbf{j}$$

and the position vector $\mathbf{r}_{OD}$ is

$$\mathbf{r}_{OD} = \{0.5\mathbf{i} + 0.5\mathbf{k}\}\text{ m}$$

The force **F** expressed as a Cartesian vector is

$$\mathbf{F} = F\left(\frac{\mathbf{r}_{CD}}{r_{CD}}\right)$$

$$= (300\text{ N})\left[\frac{\{0.4\mathbf{i} - 0.4\mathbf{j} + 0.2\mathbf{k}\}\text{ m}}{\sqrt{(0.4\text{ m})^2 + (-0.4\text{ m})^2 + (0.2\text{ m})^2}}\right]$$

$$= \{200\mathbf{i} - 200\mathbf{j} + 100\mathbf{k}\}\text{ N}$$

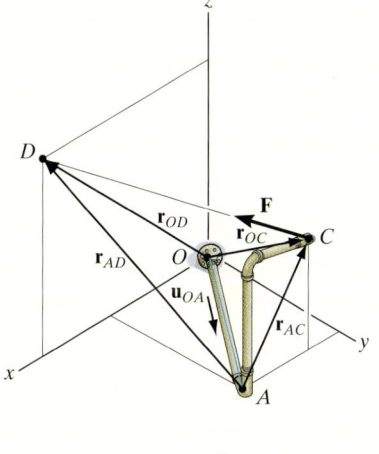

(b)

Fig. 4–24

Therefore,

$$M_{OA} = \mathbf{u}_{OA} \cdot (\mathbf{r}_{OD} \times \mathbf{F})$$

$$= \begin{vmatrix} 0.6 & 0.8 & 0 \\ 0.5 & 0 & 0.5 \\ 200 & -200 & 100 \end{vmatrix}$$

$$= 0.6[0(100) - (0.5)(-200)] - 0.8[0.5(100) - (0.5)(200)] + 0$$

$$= 100\text{ N} \cdot \text{m} \qquad\qquad\qquad Ans.$$

FUNDAMENTAL PROBLEMS

F4–13. Determine the magnitude of the moment of the force $\mathbf{F} = \{300\mathbf{i} - 200\mathbf{j} + 150\mathbf{k}\}$ N about the x axis.

F4–14. Determine the magnitude of the moment of the force $\mathbf{F} = \{300\mathbf{i} - 200\mathbf{j} + 150\mathbf{k}\}$ N about the OA axis.

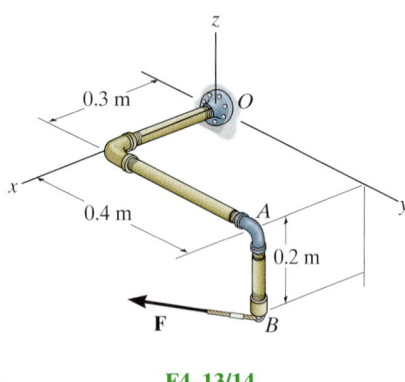

F4–13/14

F4–15. Determine the magnitude of the moment of the 200-N force about the x axis. Solve the problem using both a scalar and a vector analysis.

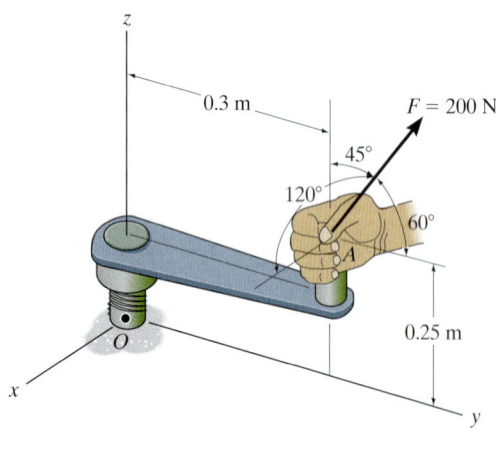

F4–15

F4–16. Determine the magnitude of the moment of the force about the y axis.

$$\mathbf{F} = \{30\mathbf{i} - 20\mathbf{j} + 50\mathbf{k}\} \text{ N}$$

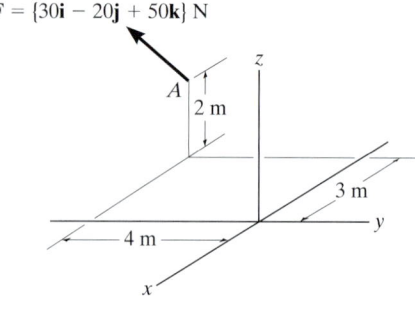

F4–16

F4–17. Determine the moment of the force $\mathbf{F} = \{50\mathbf{i} - 40\mathbf{j} + 20\mathbf{k}\}$ kN about the AB axis. Express the result as a Cartesian vector.

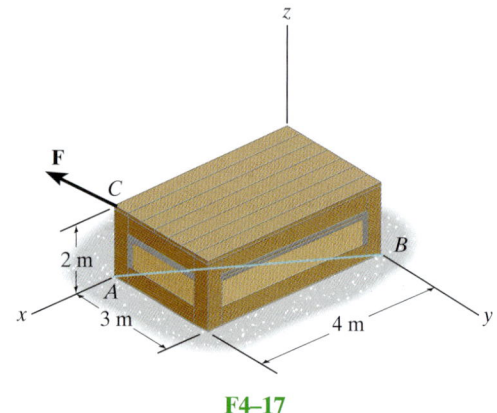

F4–17

F4–18. Determine the moment of force $\mathbf{F}$ about the x, the y, and the z axes. Solve the problem using both a scalar and a vector analysis.

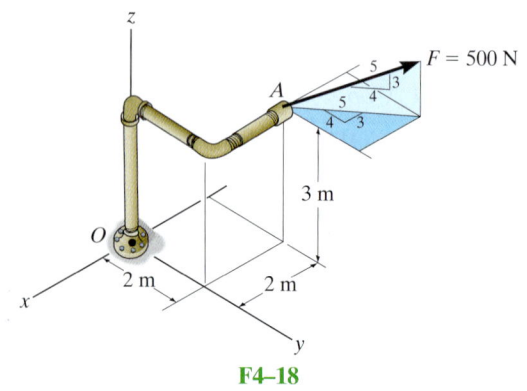

F4–18

PROBLEMS

4–47. Determine the magnitude of the moment of each of the three forces about the axis AB. Solve the problem (a) using a Cartesian vector approach and (b) using a scalar approach.

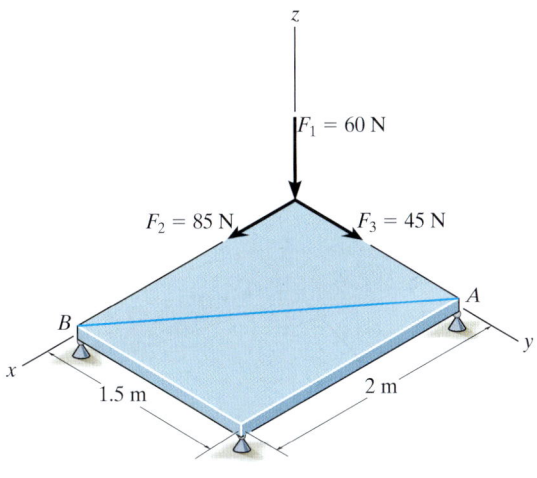

Prob. 4–47

*4–48. Determine the moment produced by force **F** about the diagonal AF of the rectangular block. Express the result as a Cartesian vector.

4–49. Determine the moment produced by force **F** about the diagonal OD of the rectangular block. Express the result as a Cartesian vector.

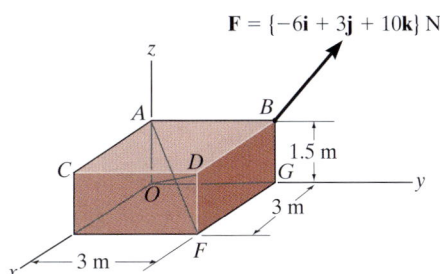

Probs. 4–48/49

4–50. Determine the magnitude of the moment of the force $\mathbf{F} = \{50\mathbf{i} - 20\mathbf{j} - 80\mathbf{k}\}$ N about the base line CA of the tripod.

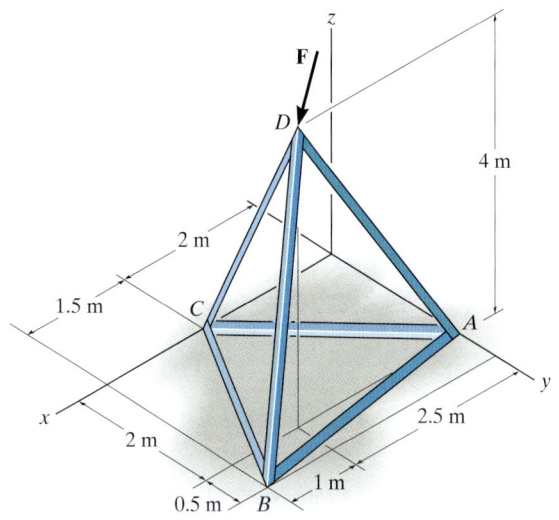

Prob. 4–50

4–51. Determine the magnitude of the moment produced by the force of $F = 200$ N about the hinged axis (the x axis) of the door.

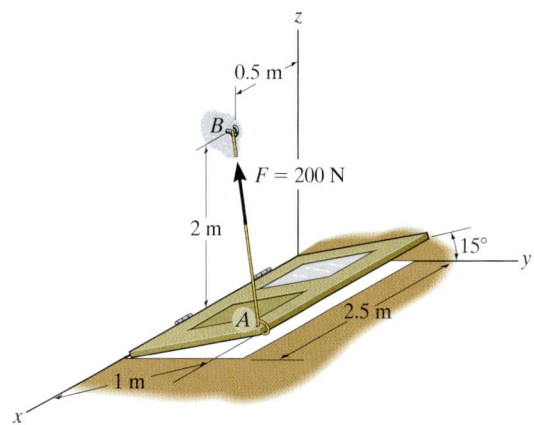

Prob. 4–51

*4–52. The lug nut on the wheel of the automobile is to be removed using the wrench and applying the vertical force of $F = 30$ N at A. Determine if this force is adequate, provided 14 N · m of torque about the x axis is initially required to turn the nut. If the 30-N force can be applied at A in any other direction, will it be possible to turn the nut?

4–53. Solve Prob. 4–52 if the cheater pipe AB is slipped over the handle of the wrench and the 30-N force can be applied at any point and in any direction on the assembly.

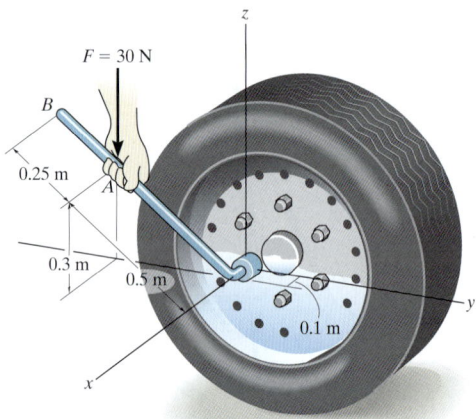

Probs. 4–52/53

4–54. The board is used to hold the end of a four-way lug wrench in the position shown when the man applies a force of $F = 100$ N. Determine the magnitude of the moment produced by this force about the x axis. Force F lies in a vertical plane.

4–55. The board is used to hold the end of a four-way lug wrench in position. If a torque of 30 N · m about the x axis is required to tighten the nut, determine the required magnitude of the force F that the man's foot must apply on the end of the wrench in order to turn it. Force F lies in a vertical plane.

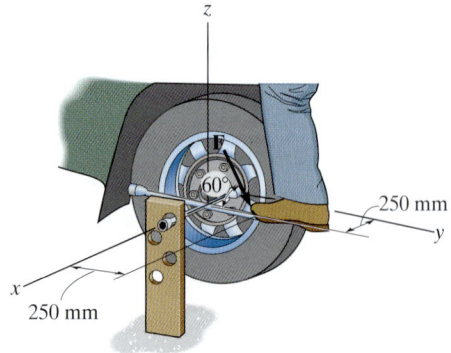

Probs. 4–54/55

*4–56. The cutting tool on the lathe exerts a force F on the shaft as shown. Determine the moment of this force about the y axis of the shaft.

4–57. The cutting tool on the lathe exerts a force F on the shaft as shown. Determine the moment of this force about the x and z axes.

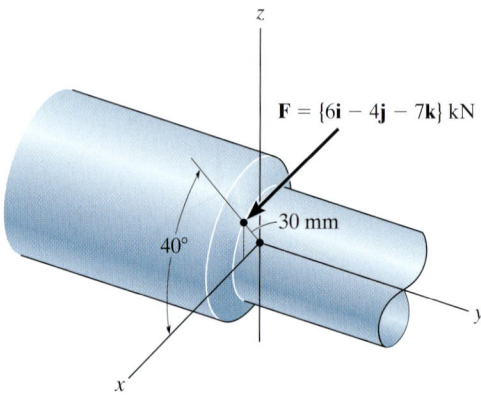

Probs. 4–56/57

4–58. If $F = 450$ N, determine the magnitude of the moment produced by this force about the x axis.

4–59. The friction at sleeve A can provide a maximum resisting moment of 125 N · m about the x axis. Determine the largest magnitude of force F that can be applied to the bracket so that the bracket will not turn.

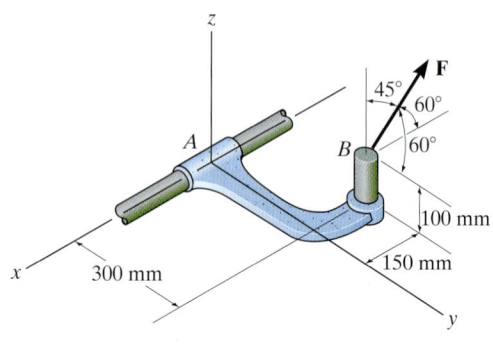

Probs. 4–58/59

*4–60. A vertical force of $F = 60$ N is applied to the handle of the pipe wrench. Determine the moment that this force exerts along the axis AB (x axis) of the pipe assembly. Both the wrench and pipe assembly ABC lie in the x–y plane. *Suggestion:* Use a scalar analysis.

4–61. Determine the magnitude of the vertical force $\mathbf{F}$ acting on the handle of the wrench so that this force produces a component of moment along the AB axis (x axis) of the pipe assembly of $(M_A)_x = \{-5\mathbf{i}\}$ N·m. Both the pipe assembly ABC and the wrench lie in the x–y plane. *Suggestion:* Use a scalar analysis.

*4–64. The wrench A is used to hold the pipe in a stationary position while wrench B is used to tighten the elbow fitting. If $F_B = 150$ N, determine the magnitude of the moment produced by this force about the y axis. Also, what is the magnitude of force $\mathbf{F}_A$ in order to counteract this moment?

4–65. The wrench A is used to hold the pipe in a stationary position while wrench B is used to tighten the elbow fitting. Determine the magnitude of force F_B in order to develop a torque of 50 N·m about the y axis. Also, what is the required magnitude of force $\mathbf{F}_A$ in order to counteract this moment?

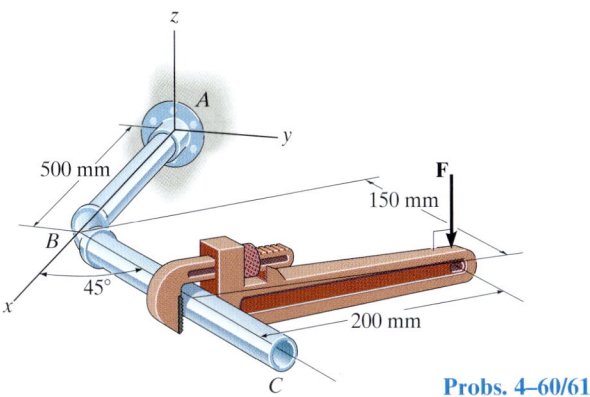

Probs. 4–60/61

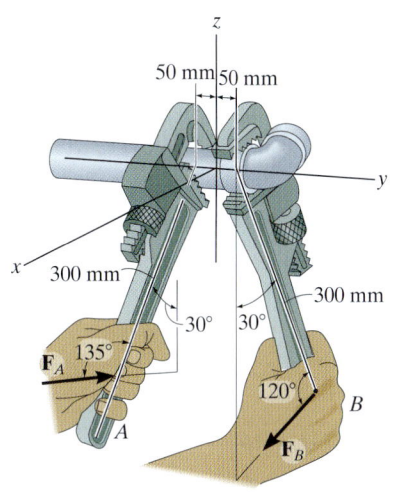

Probs. 4–64/65

4–62. Determine the magnitude of the moments of the force $\mathbf{F}$ about the $x, y,$ and z axes. Solve the problem (a) using a Cartesian vector approach and (b) using a scalar approach.

4–63. Determine the moment of the force $\mathbf{F}$ about an axis extending between A and C. Express the result as a Cartesian vector.

4–66. The wooden shaft is held in a lathe. The cutting tool exerts a force $\mathbf{F}$ on the shaft in the direction shown. Determine the moment of this force about the x axis of the shaft. Express the result as a Cartesian vector. The distance OA is 25 mm.

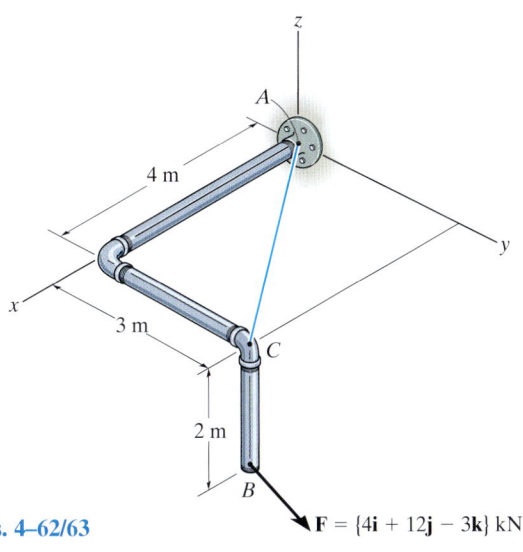

Probs. 4–62/63

$\mathbf{F} = \{4\mathbf{i} + 12\mathbf{j} - 3\mathbf{k}\}$ kN

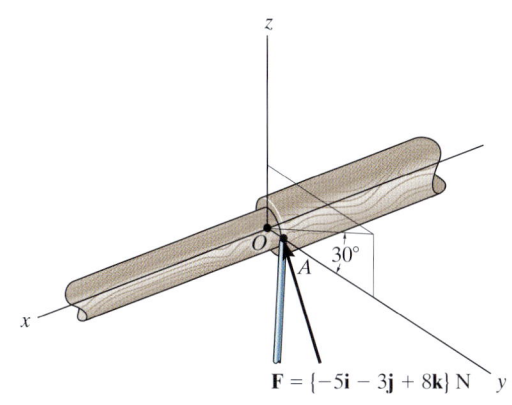

$\mathbf{F} = \{-5\mathbf{i} - 3\mathbf{j} + 8\mathbf{k}\}$ N

Prob. 4–66

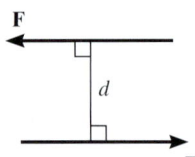

Fig. 4–25

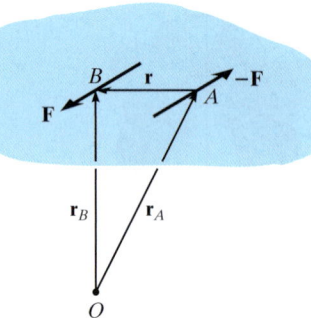

Fig. 4–26

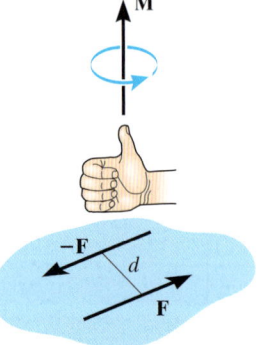

Fig. 4–27

4.6   Moment of a Couple

A *couple* is defined as two parallel forces that have the same magnitude, but opposite directions, and are separated by a perpendicular distance d, Fig. 4–25. Since the resultant force is zero, the only effect of a couple is to produce an actual rotation, or if no movement is possible, there is a tendency of rotation in a specified direction. For example, imagine that you are driving a car with both hands on the steering wheel and you are making a turn. One hand will push up on the wheel while the other hand pulls down, which causes the steering wheel to rotate.

The moment produced by a couple is called a *couple moment*. We can determine its value by finding the sum of the moments of both couple forces about *any* arbitrary point. For example, in Fig. 4–26, position vectors $\mathbf{r}_A$ and $\mathbf{r}_B$ are directed from point O to points A and B lying on the line of action of $-\mathbf{F}$ and $\mathbf{F}$. The couple moment determined about O is therefore

$$\mathbf{M} = \mathbf{r}_B \times \mathbf{F} + \mathbf{r}_A \times -\mathbf{F} = (\mathbf{r}_B - \mathbf{r}_A) \times \mathbf{F}$$

However $\mathbf{r}_B = \mathbf{r}_A + \mathbf{r}$ or $\mathbf{r} = \mathbf{r}_B - \mathbf{r}_A$, so that

$$\mathbf{M} = \mathbf{r} \times \mathbf{F} \qquad (4\text{–}13)$$

This result indicates that a couple moment is a *free vector*, i.e., it can act at *any point* since $\mathbf{M}$ depends *only* upon the position vector $\mathbf{r}$ directed *between* the forces and *not* the position vectors $\mathbf{r}_A$ and $\mathbf{r}_B$, directed from the arbitrary point O to the forces. This concept is unlike the moment of a force, which requires a definite point (or axis) about which moments are determined.

Scalar Formulation.   The moment of a couple, $\mathbf{M}$, Fig. 4–27, is defined as having a *magnitude* of

$$\boxed{M = Fd} \qquad (4\text{–}14)$$

where F is the magnitude of one of the forces and d is the perpendicular distance or moment arm between the forces. The *direction* and sense of the couple moment are determined by the right-hand rule, where the thumb indicates this direction when the fingers are curled with the sense of rotation caused by the couple forces. In all cases, $\mathbf{M}$ will act perpendicular to the plane containing these forces.

Vector Formulation.   The moment of a couple can also be expressed by the vector cross product using Eq. 4–13, i.e.,

$$\boxed{\mathbf{M} = \mathbf{r} \times \mathbf{F}} \qquad (4\text{–}15)$$

Application of this equation is easily remembered if one thinks of taking the moments of both forces about a point lying on the line of action of one of the forces. For example, if moments are taken about point A in Fig. 4–26, the moment of $-\mathbf{F}$ is *zero* about this point, and the moment of $\mathbf{F}$ is defined from Eq. 4–15. Therefore, in the formulation $\mathbf{r}$ is crossed with the force $\mathbf{F}$ to which it is directed.

Fig. 4–28

Equivalent Couples.

If two couples produce a moment with the *same magnitude and direction*, then these two couples are *equivalent*. For example, the two couples shown in Fig. 4–28 are *equivalent* because each couple moment has a magnitude of $M = 30\ N(0.4\ m) = 40\ N(0.3\ m) = 12\ N \cdot m$, and each is directed into the plane of the page. Notice that larger forces are required in the second case to create the same turning effect because the hands are placed closer together. Also, if the wheel was connected to the shaft at a point other than at its center, then the wheel would still turn when each couple is applied since the $12\ N \cdot m$ couple is a free vector.

Resultant Couple Moment.

Since couple moments are vectors, their resultant can be determined by vector addition. For example, consider the couple moments $\mathbf{M}_1$ and $\mathbf{M}_2$ acting on the pipe in Fig. 4–29a. Since each couple moment is a free vector, we can join their tails at any arbitrary point and find the resultant couple moment, $\mathbf{M}_R = \mathbf{M}_1 + \mathbf{M}_2$ as shown in Fig. 4–29b.

If more than two couple moments act on the body, we may generalize this concept and write the vector resultant as

$$\mathbf{M}_R = \Sigma(\mathbf{r} \times \mathbf{F}) \qquad (4\text{–}16)$$

These concepts are illustrated numerically in the examples that follow. In general, problems projected in two dimensions should be solved using a scalar analysis since the moment arms and force components are easy to determine.

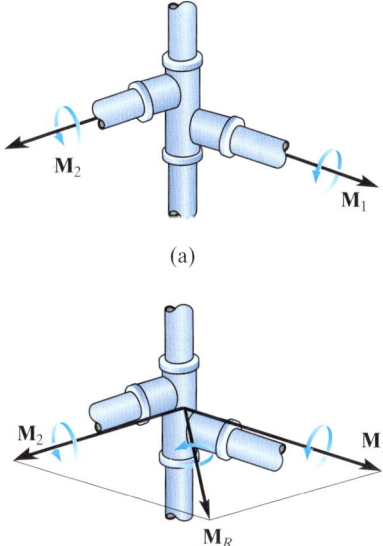

(a)

(b)

Fig. 4–29

Steering wheels on vehicles have been made smaller than on older vehicles because power steering does not require the driver to apply a large couple moment to the rim of the wheel.

Important Points

- A couple moment is produced by two noncollinear forces that are equal in magnitude but opposite in direction. Its effect is to produce pure rotation, or tendency for rotation in a specified direction.

- A couple moment is a free vector, and as a result it causes the same rotational effect on a body regardless of where the couple moment is applied to the body.

- The moment of the two couple forces can be determined about *any point*. For convenience, this point is often chosen on the line of action of one of the forces in order to eliminate the moment of this force about the point.

- In three dimensions the couple moment is often determined using the vector formulation, $\mathbf{M} = \mathbf{r} \times \mathbf{F}$, where $\mathbf{r}$ is directed from *any point* on the line of action of one of the forces to *any point* on the line of action of the other force $\mathbf{F}$.

- A resultant couple moment is simply the vector sum of all the couple moments of the system.

EXAMPLE | 4.10

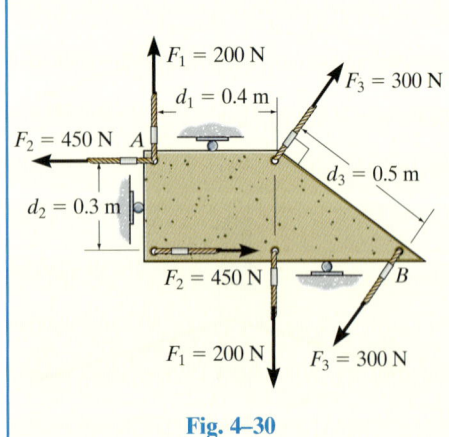

Fig. 4–30

Determine the resultant couple moment of the three couples acting on the plate in Fig. 4–30.

SOLUTION
As shown the perpendicular distances between each pair of couple forces are $d_1 = 0.4$ m, $d_2 = 0.3$ m, and $d_3 = 0.5$ m. Considering counterclockwise couple moments as positive, we have

$$\zeta + M_R = \Sigma M; \quad M_R = -F_1 d_1 + F_2 d_2 - F_3 d_3$$

$$= -(200 \text{ N})(0.4 \text{ m}) + (450 \text{ N})(0.3 \text{ m}) - (300 \text{ N})(0.5 \text{ m})$$

$$= -95 \text{ N} \cdot \text{m} = 95 \text{ N} \cdot \text{m} \; \zeta \qquad \textit{Ans.}$$

The negative sign indicates that $\mathbf{M}_R$ has a clockwise rotational sense.

EXAMPLE | 4.11

Determine the magnitude and direction of the couple moment acting on the gear in Fig. 4–31a.

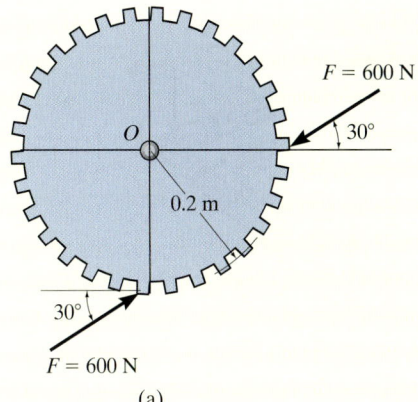

(a)

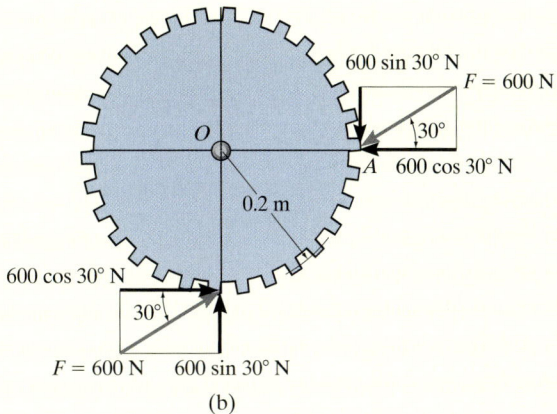

(b)

SOLUTION

The easiest solution requires resolving each force into its components as shown in Fig. 4–31b. The couple moment can be determined by summing the moments of these force components about any point, for example, the center O of the gear or point A. If we consider counterclockwise moments as positive, we have

$\zeta + M = \Sigma M_O$; $M = (600 \cos 30° \text{ N})(0.2 \text{ m}) - (600 \sin 30° \text{ N})(0.2 \text{ m})$

$\qquad\qquad = 43.9 \text{ N} \cdot \text{m} \,\circlearrowright$ *Ans.*

or

$\zeta + M = \Sigma M_A$; $M = (600 \cos 30° \text{ N})(0.2 \text{ m}) - (600 \sin 30° \text{ N})(0.2 \text{ m})$

$\qquad\qquad = 43.9 \text{ N} \cdot \text{m} \,\circlearrowright$ *Ans.*

This positive result indicates that **M** has a counterclockwise rotational sense, so it is directed outward, perpendicular to the page.

NOTE: The same result can also be obtained using $M = Fd$, where d is the perpendicular distance between the lines of action of the couple forces, Fig. 4–31c. However, the computation for d is more involved. Realize that the couple moment is a free vector and can act at any point on the gear and produce the same turning effect about point O.

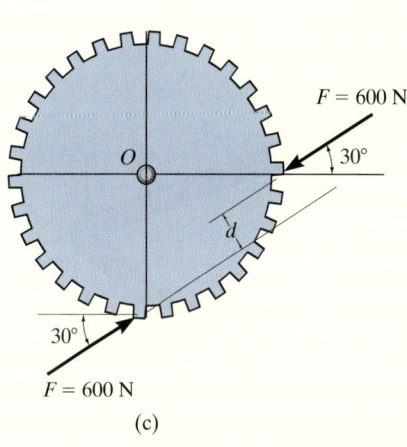

(c)

Fig. 4–31

EXAMPLE | 4.12

Determine the couple moment acting on the pipe shown in Fig. 4–32a. Segment AB is directed 30° below the x–y plane.

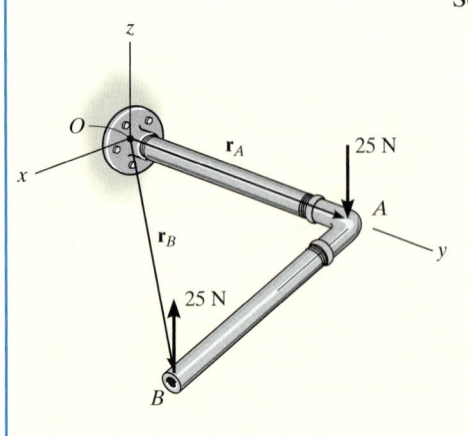

(b)

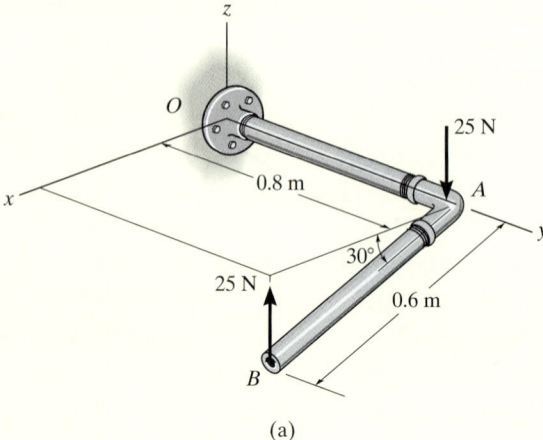

(a)

SOLUTION I *(VECTOR ANALYSIS)*

The moment of the two couple forces can be found about *any point*. If point O is considered, Fig. 4–32b, we have

$$\mathbf{M} = \mathbf{r}_A \times (-25\mathbf{k}) + \mathbf{r}_B \times (25\mathbf{k})$$
$$= (0.8\mathbf{j}) \times (-25\mathbf{k}) + (0.6\cos 30°\mathbf{i} + 0.8\mathbf{j} - 0.6\sin 30°\mathbf{k}) \times (25\mathbf{k})$$
$$= -20\mathbf{i} - 13.0\mathbf{j} + 20\mathbf{i}$$
$$= \{-13.0\mathbf{j}\}\ \text{N}\cdot\text{m} \qquad \textit{Ans.}$$

It is *easier* to take moments of the couple forces about a point lying on the line of action of one of the forces, e.g., point A, Fig. 4–32c. In this case the moment of the force at A is zero, so that

$$\mathbf{M} = \mathbf{r}_{AB} \times (25\mathbf{k})$$
$$= (0.6\cos 30°\mathbf{i} - 0.6\sin 30°\mathbf{k}) \times (25\mathbf{k})$$
$$= \{-13.0\mathbf{j}\}\ \text{N}\cdot\text{m} \qquad \textit{Ans.}$$

(c)

SOLUTION II *(SCALAR ANALYSIS)*

Although this problem is shown in three dimensions, the geometry is simple enough to use the scalar equation $M = Fd$. The perpendicular distance between the lines of action of the couple forces is $d = 0.6\cos 30° = 0.5196$ m, Fig. 4–32 d. Hence, taking moments of the forces about either point A or point B yields

$$M = Fd = 25\ \text{N}\ (0.5196\ \text{m}) = 13.0\ \text{N}\cdot\text{m}$$

Applying the right-hand rule, $\mathbf{M}$ acts in the $-\mathbf{j}$ direction. Thus,

$$\mathbf{M} = \{-13.0\mathbf{j}\}\ \text{N}\cdot\text{m} \qquad \textit{Ans.}$$

(d)

Fig. 4–32

EXAMPLE 4.13

Replace the two couples acting on the pipe column in Fig. 4–33a by a resultant couple moment.

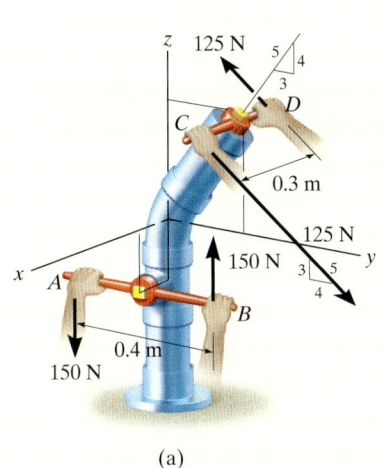

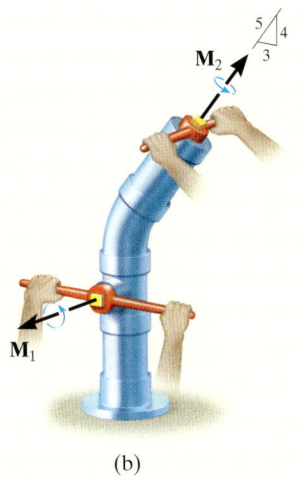

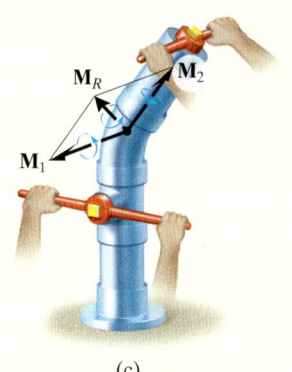

(a)   (b)   (c)

Fig. 4–33

SOLUTION (VECTOR ANALYSIS)

The couple moment $\mathbf{M}_1$, developed by the forces at A and B, can easily be determined from a scalar formulation.

$$M_1 = Fd = 150 \text{ N}(0.4 \text{ m}) = 60 \text{ N} \cdot \text{m}$$

By the right-hand rule, $\mathbf{M}_1$ acts in the $+\mathbf{i}$ direction, Fig. 4–33b. Hence,

$$\mathbf{M}_1 = \{60\mathbf{i}\} \text{ N} \cdot \text{m}$$

Vector analysis will be used to determine $\mathbf{M}_2$, caused by forces at C and D. If moments are calculated about point D, Fig. 4–33a, $\mathbf{M}_2 = \mathbf{r}_{DC} \times \mathbf{F}_C$, then

$$\mathbf{M}_2 = \mathbf{r}_{DC} \times \mathbf{F}_C = (0.3\mathbf{i}) \times \left[125\left(\tfrac{4}{5}\right)\mathbf{j} - 125\left(\tfrac{3}{5}\right)\mathbf{k}\right]$$

$$= (0.3\mathbf{i}) \times [100\mathbf{j} - 75\mathbf{k}] = 30(\mathbf{i} \times \mathbf{j}) - 22.5(\mathbf{i} \times \mathbf{k})$$

$$= \{22.5\mathbf{j} + 30\mathbf{k}\} \text{ N} \cdot \text{m}$$

Since $\mathbf{M}_1$ and $\mathbf{M}_2$ are free vectors, they may be moved to some arbitrary point and added vectorially, Fig. 4–33c. The resultant couple moment becomes

$$\mathbf{M}_R = \mathbf{M}_1 + \mathbf{M}_2 = \{60\mathbf{i} + 22.5\mathbf{j} + 30\mathbf{k}\} \text{ N} \cdot \text{m} \qquad \textit{Ans.}$$

FUNDAMENTAL PROBLEMS

F4–19. Determine the resultant couple moment acting on the beam.

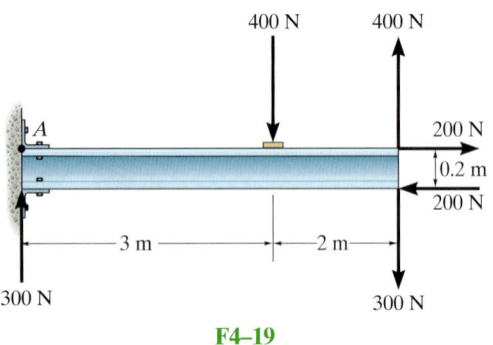

F4–19

F4–20. Determine the resultant couple moment acting on the triangular plate.

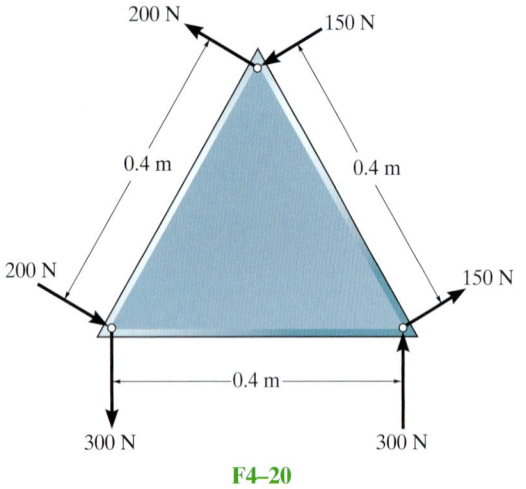

F4–20

F4–21. Determine the magnitude of **F** so that the resultant couple moment acting on the beam is 1.5 kN · m clockwise.

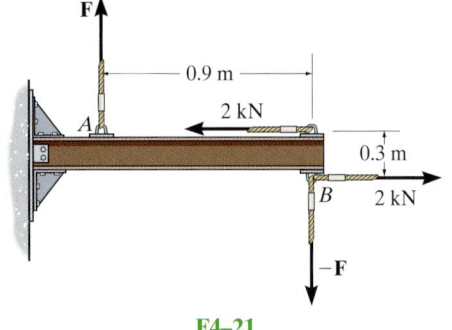

F4–21

F4–22. Determine the couple moment acting on the beam.

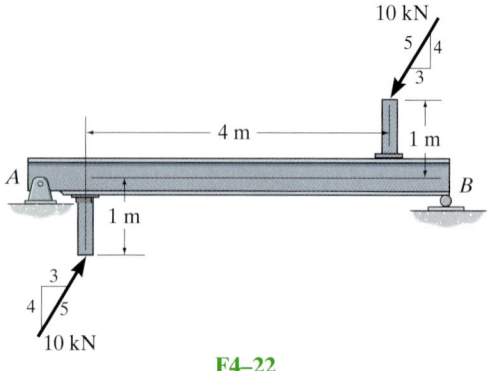

F4–22

F4–23. Determine the resultant couple moment acting on the pipe assembly.

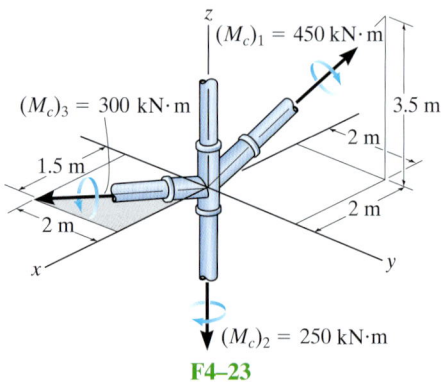

F4–23

F4–24. Determine the couple moment acting on the pipe assembly and express the result as a Cartesian vector.

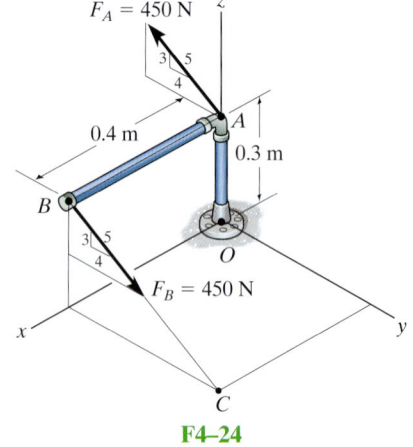

F4–24

PROBLEMS

4–67. A twist of 4 N·m is applied to the handle of the screwdriver. Resolve this couple moment into a pair of couple forces **F** exerted on the handle and **P** exerted on the blade.

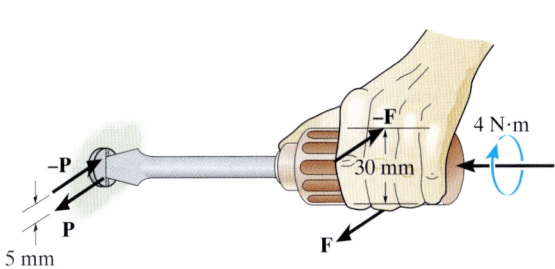

Prob. 4–67

*4–68.** The ends of the triangular plate are subjected to three couples. Determine the plate dimension d so that the resultant couple is 350 N·m clockwise.

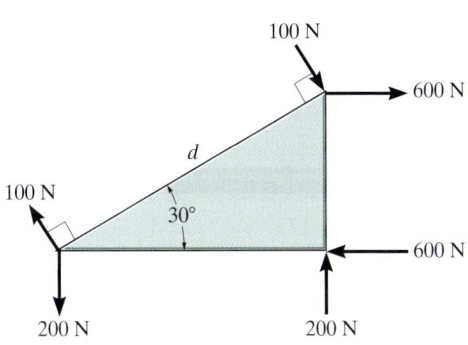

Prob. 4–68

4–69. The caster wheel is subjected to the two couples. Determine the forces **F** that the bearings create on the shaft so that the resultant couple moment on the caster is zero.

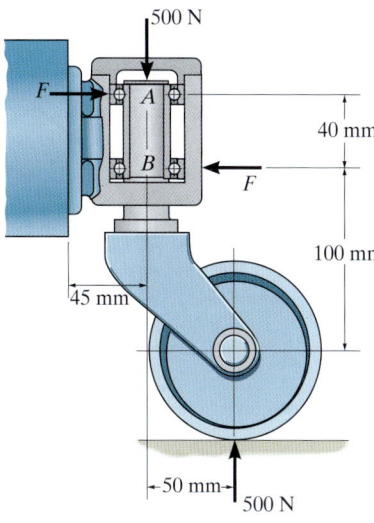

Prob. 4–69

4–70. If $\theta = 30°$, determine the magnitude of force **F** so that the resultant couple moment is 100 N·m, clockwise. The disk has a radius of 300 mm.

4–71. If $F = 200$ N, determine the required angle θ so that the resultant couple moment is zero. The disk has a radius of 300 mm.

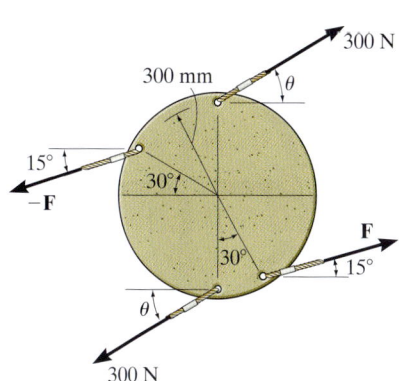

Probs. 4–70/71

*4–72. Friction on the concrete surface creates a couple moment of $M_O = 100$ N·m on the blades of the trowel. Determine the magnitude of the couple forces so that the resultant couple moment on the trowel is zero. The forces lie in the horizontal plane and act perpendicular to the handle of the trowel.

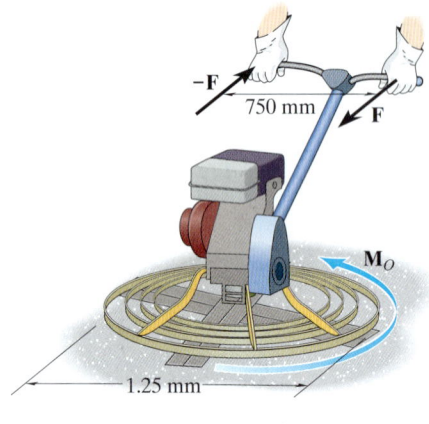

750 mm

−F

F

M_O

1.25 mm

Prob. 4–72

4–73. The man tries to open the valve by applying the couple forces of $F = 75$ N to the wheel. Determine the couple moment produced.

4–74. If the valve can be opened with a couple moment of 25 N·m, determine the required magnitude of each couple force which must be applied to the wheel.

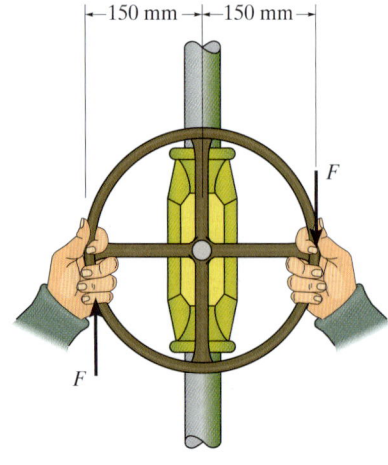

←150 mm→←150 mm→

F

F

Probs. 4–73/74

4–75. A device called a rolamite is used in various ways to replace slipping motion with rolling motion. If the belt, which wraps between the rollers, is subjected to a tension of 15 N, determine the reactive forces N of the top and bottom plates on the rollers so that the resultant couple acting on the rollers is equal to zero.

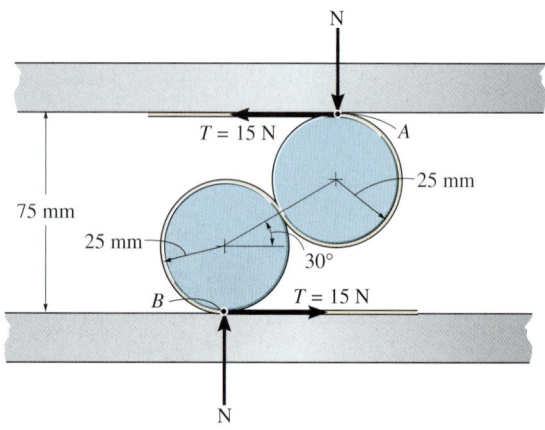

N

$T = 15$ N

A

25 mm

75 mm

25 mm

30°

B

$T = 15$ N

N

Prob. 4–75

*4–76. Three couple moments act on the pipe assembly. Determine the magnitude of M_3 and the bend angle θ so that the resultant couple moment is zero.

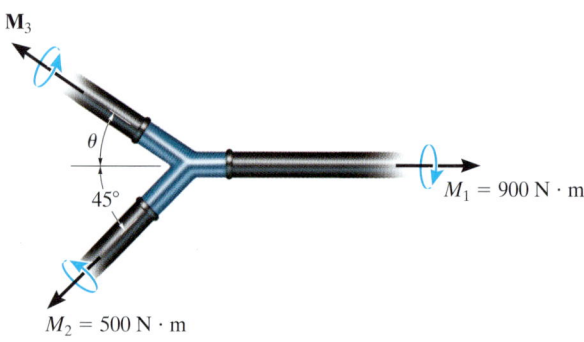

M_3

θ

45°

$M_1 = 900$ N·m

$M_2 = 500$ N·m

Prob. 4–76

4–77. The cord passing over the two small pegs *A* and *B* of the square board is subjected to a tension of 100 N. Determine the required tension *P* acting on the cord that passes over pegs *C* and *D* so that the resultant couple produced by the two couples is 15 N · m acting clockwise. Take *θ* = 15°.

4–78. The cord passing over the two small pegs *A* and *B* of the board is subjected to a tension of 100 N. Determine the *minimum* tension *P* and the orientation *θ* of the cord passing over pegs *C* and *D*, so that the resultant couple moment produced by the two cords is 20 N · m, clockwise.

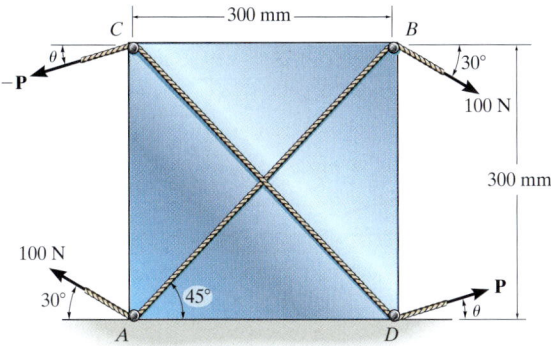

Probs. 4–77/78

4–79. Express the moment of the couple acting on the pipe in Cartesian vector form. What is the magnitude of the couple moment? Take *F* = 125 N.

*****4–80.** If the couple moment acting on the pipe has a magnitude of 300 N · m, determine the magnitude *F* of the forces applied to the wrenches.

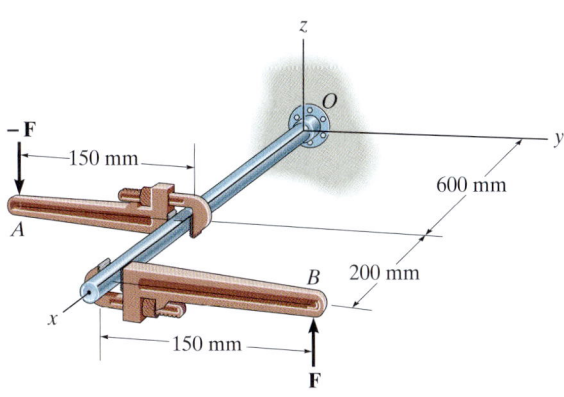

Probs. 4–79/80

4–81. Two couples act on the cantilever beam. If *F* = 6 kN, determine the resultant couple moment.

4–82. Determine the required magnitude of force **F**, if the resultant couple moment on the beam is to be zero.

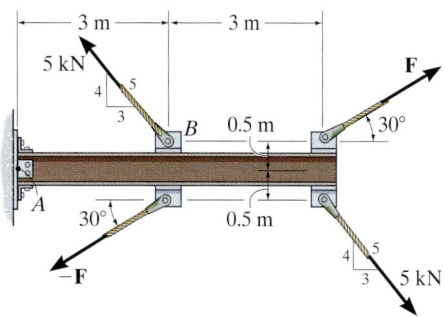

Probs. 4–81/82

4–83. Express the moment of the couple acting on the pipe assembly in Cartesian vector form. Solve the problem (a) using Eq. 4–13, and (b) summing the moment of each force about point *O*. Take **F** = {25**k**} N.

*****4–84.** If the couple moment acting on the pipe has a magnitude of 400 N · m, determine the magnitude *F* of the vertical force applied to each wrench.

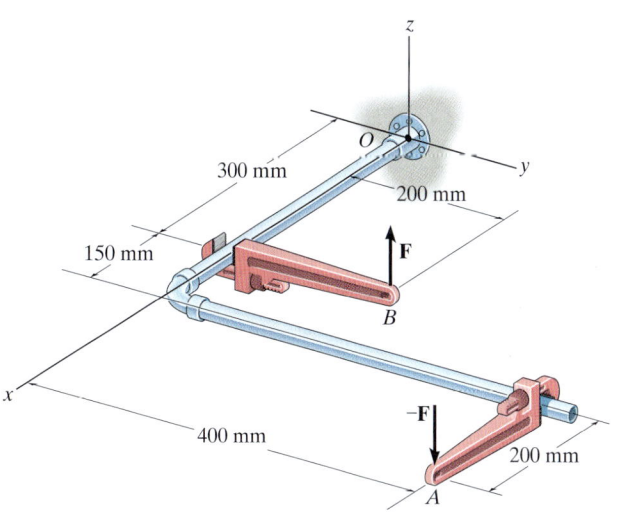

Probs. 4–83/84

4–85. The gear reducer is subjected to the couple moments shown. Determine the resultant couple moment and specify its magnitude and coordinate direction angles.

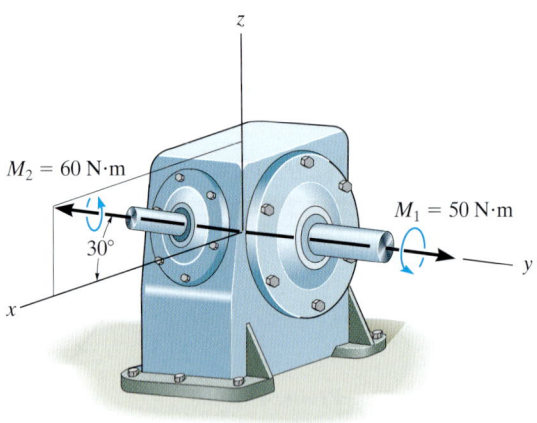

Prob. 4–85

4–86. The meshed gears are subjected to the couple moments shown. Determine the magnitude of the resultant couple moment and specify its coordinate direction angles.

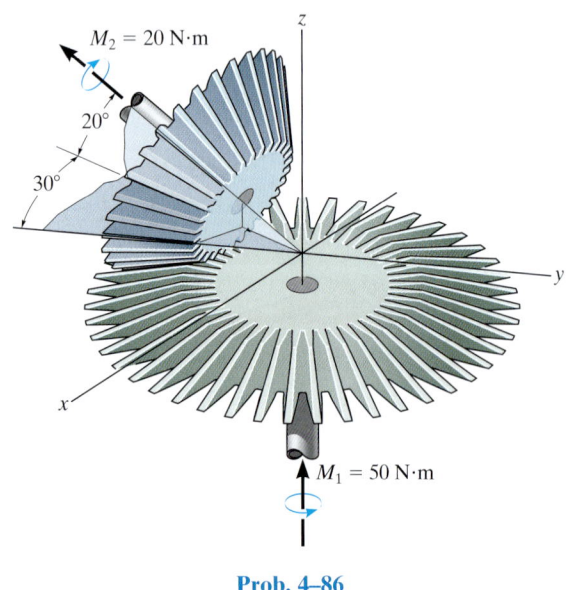

Prob. 4–86

4–87. If the resultant couple of the two couples acting on the fire hydrant is $\mathbf{M}_R = \{-15\mathbf{i} + 30\mathbf{j}\}$ N·m, determine the force magnitude P.

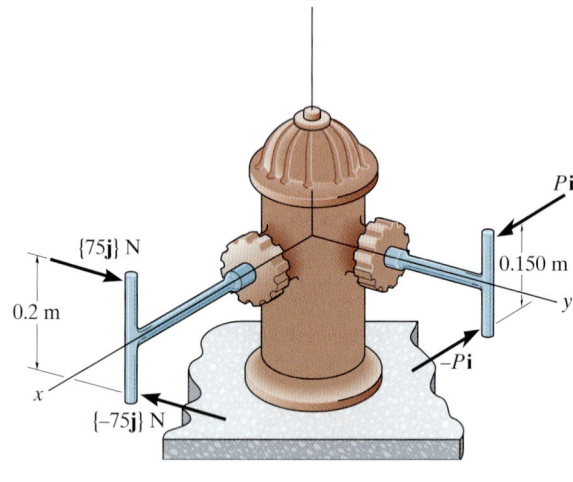

Prob. 4–87

*4–88. A couple acts on each of the handles of the minidual valve. Determine the magnitude and coordinate direction angles of the resultant couple moment.

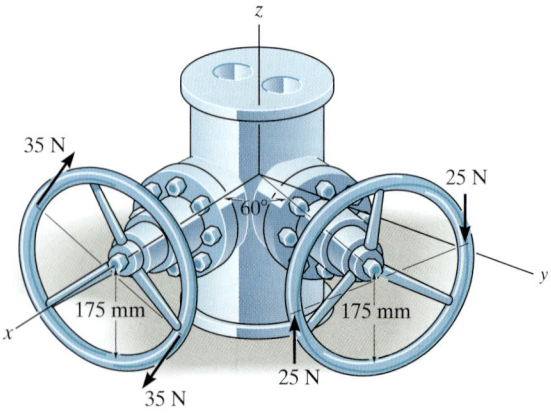

Prob. 4–88

4–89. Determine the resultant couple moment of the two couples that act on the pipe assembly. The distance from A to B is $d = 400$ mm. Express the result as a Cartesian vector.

4–90. Determine the distance d between A and B so that the resultant couple moment has a magnitude of $M_R = 20$ N · m.

4–93. If $F = \{100\mathbf{k}\}$ N, determine the couple moment that acts on the assembly. Express the result as a Cartesian vector. Member BA lies in the x–y plane.

4–94. If the magnitude of the resultant couple moment is 15 N · m, determine the magnitude F of the forces applied to the wrenches.

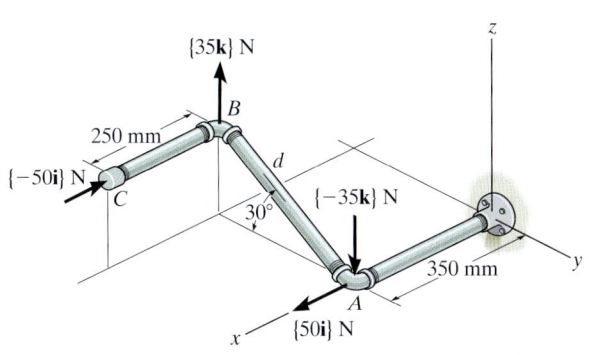

Probs. 4–89/90

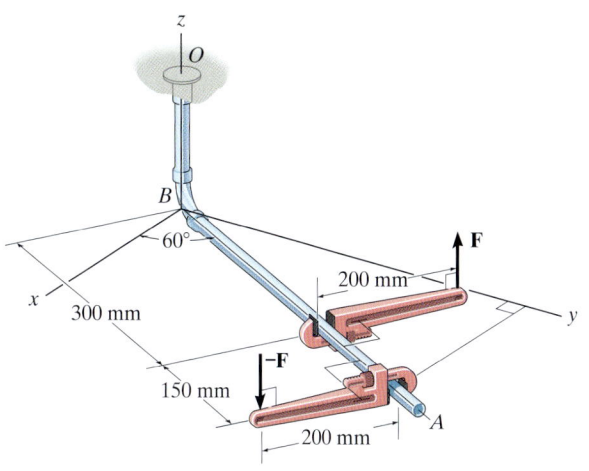

Probs. 4–93/94

4–91. If $F = 80$ N, determine the magnitude and coordinate direction angles of the couple moment. The pipe assembly lies in the x–y plane.

*__4–92.__ If the magnitude of the couple moment acting on the pipe assembly is 50 N · m, determine the magnitude of the couple forces applied to each wrench. The pipe assembly lies in the x–y plane.

4–95. If $F_1 = 100$ N, $F_2 = 120$ N and $F_3 = 80$ N, determine the magnitude and coordinate direction angles of the resultant couple moment.

*__4–96.__ Determine the required magnitude of $\mathbf{F}_1$, $\mathbf{F}_2$ and $\mathbf{F}_3$ so that the resultant couple moment is $(\mathbf{M}_c)_R = [50\,\mathbf{i} - 45\,\mathbf{j} - 20\,\mathbf{k}]$ N · m.

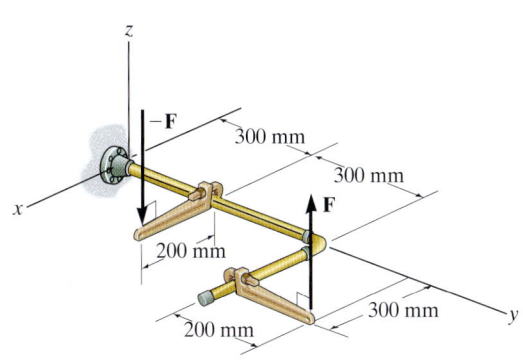

Probs. 4–91/92

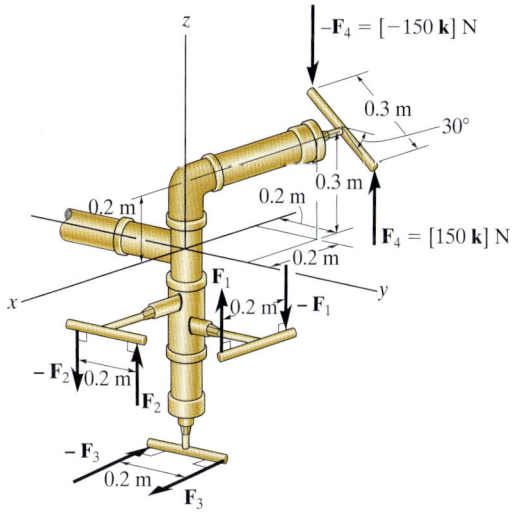

Probs. 4–95/96

4.7   Simplification of a Force and Couple System

Sometimes it is convenient to reduce a system of forces and couple moments acting on a body to a simpler form by replacing it with an *equivalent system*, consisting of a single resultant force acting at a specific point and a resultant couple moment. A system is equivalent if the *external effects* it produces on a body are the same as those caused by the original force and couple moment system. In this context, the external effects of a system refer to the *translating and rotating motion* of the body if the body is free to move, or it refers to the *reactive forces* at the supports if the body is held fixed.

For example, consider holding the stick in Fig. 4–34a, which is subjected to the force **F** at point A. If we attach a pair of equal but opposite forces **F** and −**F** at point B, which is *on the line of action* of **F**, Fig. 4–34b, we observe that −**F** at B and **F** at A will cancel each other, leaving only **F** at B, Fig. 4–34c. Force **F** has now been moved from A to B without modifying its *external effects* on the stick; i.e., the reaction at the grip remains the same. This demonstrates the *principle of transmissibility*, which states that a force acting on a body (stick) is a *sliding vector* since it can be applied at any point along its line of action.

We can also use the above procedure to move a force to a point that is *not* on the line of action of the force. If **F** is applied perpendicular to the stick, as in Fig. 4–35a, then we can attach a pair of equal but opposite forces **F** and −**F** to B, Fig. 4–35b. Force **F** is now applied at B, and the other two forces, **F** at A and −**F** at B, form a couple that produces the couple moment $M = Fd$, Fig. 4–35c. Therefore, the force **F** can be moved from A to B provided a couple moment **M** is added to maintain an equivalent system. This couple moment is determined by taking the moment of **F** about B. Since **M** is actually a *free vector*, it can act at any point on the stick. In both cases the systems are equivalent, which causes a downward force **F** and clockwise couple moment $M = Fd$ to be felt at the grip.

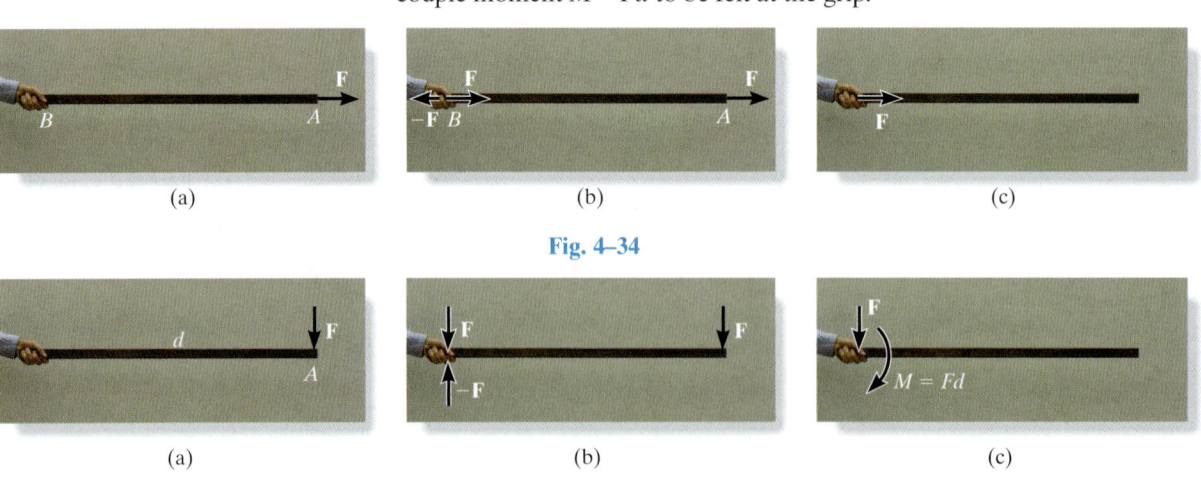

(a)   (b)   (c)

Fig. 4–34

(a)   (b)   (c)

Fig. 4–35

System of Forces and Couple Moments.

Using the above method, a system of several forces and couple moments acting on a body can be reduced to an equivalent single resultant force acting at a point O and a resultant couple moment. For example, in Fig. 4–36a, O is not on the line of action of $\mathbf{F}_1$, and so this force can be moved to point O provided a couple moment $(\mathbf{M}_O)_1 = \mathbf{r}_1 \times \mathbf{F}$ is added to the body. Similarly, the couple moment $(\mathbf{M}_O)_2 = \mathbf{r}_2 \times \mathbf{F}_2$ should be added to the body when we move $\mathbf{F}_2$ to point O. Finally, since the couple moment $\mathbf{M}$ is a free vector, it can just be moved to point O. By doing this, we obtain the equivalent system shown in Fig. 4–36b, which produces the same external effects (support reactions) on the body as that of the force and couple system shown in Fig. 4–36a. If we sum the forces and couple moments, we obtain the resultant force $\mathbf{F}_R = \mathbf{F}_1 + \mathbf{F}_2$ and the resultant couple moment $(\mathbf{M}_R)_O = \mathbf{M} + (\mathbf{M}_O)_1 + (\mathbf{M}_O)_2$, Fig. 4–36c.

Notice that $\mathbf{F}_R$ is independent of the location of point O since it is simply a summation of the forces. However, $(\mathbf{M}_R)_O$ depends upon this location since the moments $\mathbf{M}_1$ and $\mathbf{M}_2$ are determined using the position vectors $\mathbf{r}_1$ and $\mathbf{r}_2$, which extend from O to each force. Also note that $(\mathbf{M}_R)_O$ is a free vector and can act at *any point* on the body, although point O is generally chosen as its point of application.

We can generalize the above method of reducing a force and couple system to an equivalent resultant force $\mathbf{F}_R$ acting at point O and a resultant couple moment $(\mathbf{M}_R)_O$ by using the following two equations.

$$\mathbf{F}_R = \Sigma \mathbf{F}$$
$$(\mathbf{M}_R)_O = \Sigma \mathbf{M}_O + \Sigma \mathbf{M}$$

(4–17)

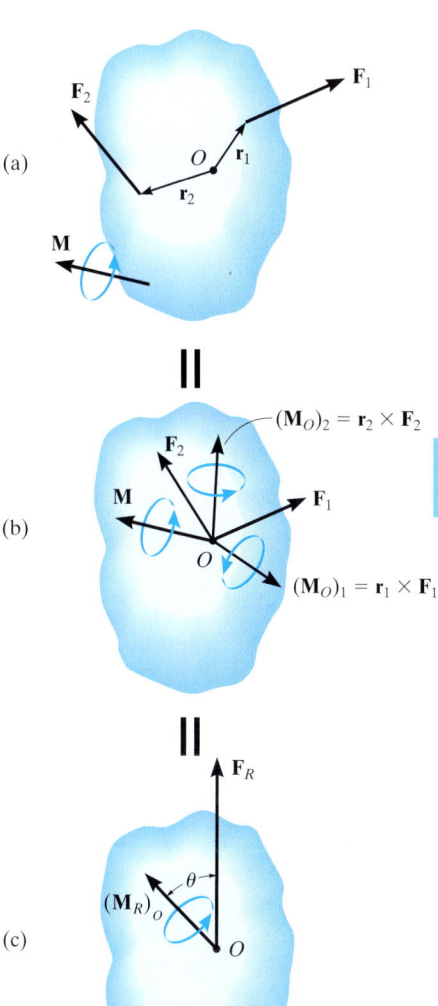

Fig. 4–36

The first equation states that the resultant force of the system is equivalent to the sum of all the forces; and the second equation states that the resultant couple moment of the system is equivalent to the sum of all the couple moments $\Sigma \mathbf{M}$ plus the moments of all the forces $\Sigma \mathbf{M}_O$ about point O. If the force system lies in the x–y plane and any couple moments are perpendicular to this plane, then the above equations reduce to the following three scalar equations.

$$(F_R)_x = \Sigma F_x$$
$$(F_R)_y = \Sigma F_y$$
$$(M_R)_O = \Sigma M_O + \Sigma M$$

(4–18)

Here the resultant force is determined from the vector sum of its two components $(F_R)_x$ and $(F_R)_y$.

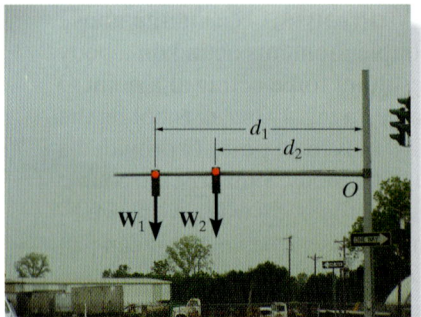 

The weights of these traffic lights can be replaced by their equivalent resultant force $W_R = W_1 + W_2$ and a couple moment $(M_R)_O = W_1 d_1 + W_2 d_2$ at the support, O. In both cases the support must provide the same resistance to translation and rotation in order to keep the member in the horizontal position.

Procedure for Analysis

The following points should be kept in mind when simplifying a force and couple moment system to an equivalent resultant force and couple system.

- Establish the coordinate axes with the origin located at point O and the axes having a selected orientation.

Force Summation.

- If the force system is *coplanar*, resolve each force into its x and y components. If a component is directed along the positive x or y axis, it represents a positive scalar; whereas if it is directed along the negative x or y axis, it is a negative scalar.

- In three dimensions, represent each force as a Cartesian vector before summing the forces.

Moment Summation.

- When determining the moments of a *coplanar* force system about point O, it is generally advantageous to use the principle of moments, i.e., determine the moments of the components of each force, rather than the moment of the force itself.

- In three dimensions use the vector cross product to determine the moment of each force about point O. Here the position vectors extend from O to any point on the line of action of each force.

EXAMPLE 4.14

Replace the force and couple system shown in Fig. 4–37a by an equivalent resultant force and couple moment acting at point O.

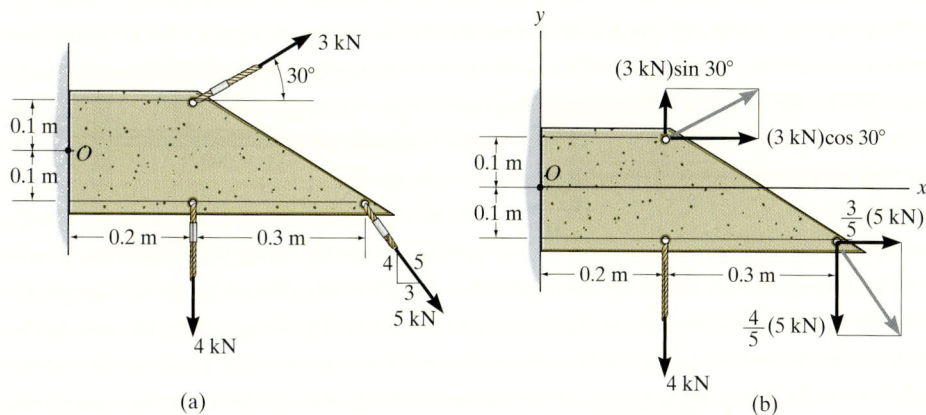

(a)   (b)

SOLUTION

Force Summation.   The 3 kN and 5 kN forces are resolved into their x and y components as shown in Fig. 4–37b. We have

$\xrightarrow{+} (F_R)_x = \Sigma F_x;$ $(F_R)_x = (3 \text{ kN})\cos 30° + \left(\frac{3}{5}\right)(5 \text{ kN}) = 5.598 \text{ kN} \rightarrow$

$+\uparrow (F_R)_y = \Sigma F_y;$ $(F_R)_y = (3 \text{ kN})\sin 30° - \left(\frac{4}{5}\right)(5 \text{ kN}) - 4 \text{ kN} = -6.50 \text{ kN} = 6.50 \text{ kN}\downarrow$

Using the Pythagorean theorem, Fig. 4–37c, the magnitude of $\mathbf{F}_R$ is

$F_R = \sqrt{(F_R)_x^2 + (F_R)_y^2} = \sqrt{(5.598 \text{ kN})^2 + (6.50 \text{ kN})^2} = 8.58 \text{ kN}$ *Ans.*

Its direction θ is

$\theta = \tan^{-1}\left(\frac{(F_R)_y}{(F_R)_x}\right) = \tan^{-1}\left(\frac{6.50 \text{ kN}}{5.598 \text{ kN}}\right) = 49.3°$ *Ans.*

Moment Summation.   The moments of 3 kN and 5 kN about point O will be determined using their x and y components. Referring to Fig. 4–37b, we have

$\zeta + (M_R)_O = \Sigma M_O;$

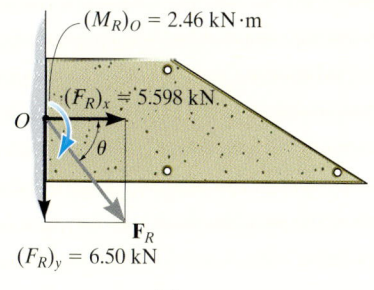

$(M_R)_O = (3 \text{ kN})\sin 30°(0.2 \text{ m}) - (3 \text{ kN})\cos 30°(0.1 \text{ m}) + \left(\frac{3}{5}\right)(5 \text{ kN})(0.1 \text{ m})$
$\qquad\qquad\quad - \left(\frac{4}{5}\right)(5 \text{ kN})(0.5 \text{ m}) - (4 \text{ kN})(0.2 \text{ m})$

$\qquad = -2.46 \text{ kN} \cdot \text{m} = 2.46 \text{ kN} \cdot \text{m} \circlearrowright$ *Ans.*

This clockwise moment is shown in Fig. 4–37c.

(c)

NOTE: Realize that the resultant force and couple moment in Fig. 4–37c will produce the same external effects or reactions at the supports as those produced by the force system, Fig. 4–37a.

Fig. 4–37

EXAMPLE | 4.15

Replace the force and couple system acting on the member in Fig. 4–38a by an equivalent resultant force and couple moment acting at point O.

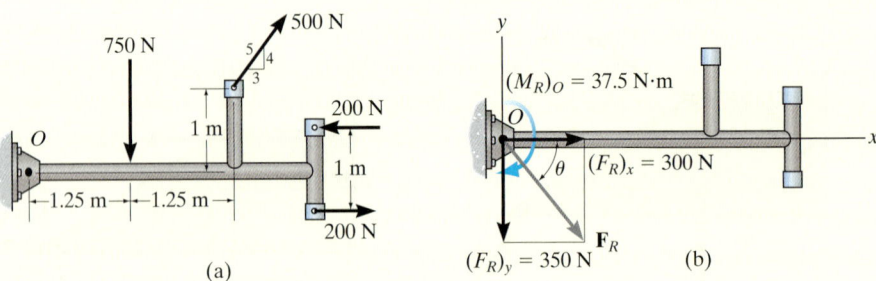

Fig. 4–38

SOLUTION

Force Summation. Since the couple forces of 200 N are equal but opposite, they produce a zero resultant force, and so it is not necessary to consider them in the force summation. The 500-N force is resolved into its x and y components, thus,

$$\xrightarrow{+} (F_R)_x = \Sigma F_x; \quad (F_R)_x = \left(\tfrac{3}{5}\right)(500 \text{ N}) = 300 \text{ N} \rightarrow$$

$$+\uparrow (F_R)_y = \Sigma F_y; \quad (F_R)_y = (500 \text{ N})\left(\tfrac{4}{5}\right) - 750 \text{ N} = -350 \text{ N} = 350 \text{ N}\downarrow$$

From Fig. 4–15b, the magnitude of $\mathbf{F}_R$ is

$$F_R = \sqrt{(F_R)_x^2 + (F_R)_y^2}$$

$$= \sqrt{(300 \text{ N})^2 + (350 \text{ N})^2} = 461 \text{ N} \qquad Ans.$$

And the angle θ is

$$\theta = \tan^{-1}\left(\frac{(F_R)_y}{(F_R)_x}\right) = \tan^{-1}\left(\frac{350 \text{ N}}{300 \text{ N}}\right) = 49.4° \qquad Ans.$$

Moment Summation. Since the couple moment is a free vector, it can act at any point on the member. Referring to Fig. 4–38a, we have

$$\zeta + (M_R)_O = \Sigma M_O + \Sigma M$$

$$(M_R)_O = (500 \text{ N})\left(\tfrac{4}{5}\right)(2.5 \text{ m}) - (500 \text{ N})\left(\tfrac{3}{5}\right)(1 \text{ m})$$

$$- (750 \text{ N})(1.25 \text{ m}) + 200 \text{ N} \cdot \text{m}$$

$$= -37.5 \text{ N} \cdot \text{m} = 37.5 \text{ N} \cdot \text{m} \; \text{⟲} \qquad Ans.$$

This clockwise moment is shown in Fig. 4–38b.

EXAMPLE 4.16

The structural member is subjected to a couple moment **M** and forces **F**₁ and **F**₂ in Fig. 4–39a. Replace this system by an equivalent resultant force and couple moment acting at its base, point *O*.

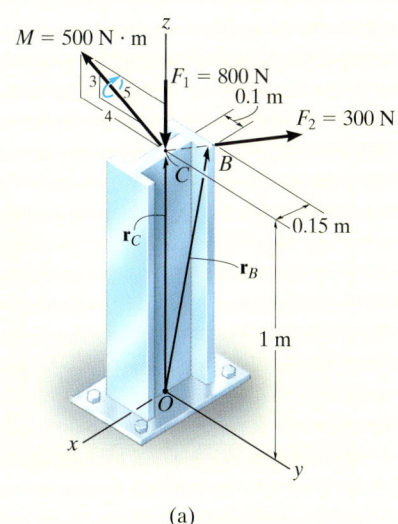

(a)

SOLUTION (*VECTOR ANALYSIS*)

The three-dimensional aspects of the problem can be simplified by using a Cartesian vector analysis. Expressing the forces and couple moment as Cartesian vectors, we have

$$\mathbf{F}_1 = \{-800\mathbf{k}\} \text{ N}$$

$$\mathbf{F}_2 = (300 \text{ N})\mathbf{u}_{CB}$$

$$= (300 \text{ N})\left(\frac{\mathbf{r}_{CB}}{r_{CB}}\right)$$

$$= 300 \text{ N}\left[\frac{\{-0.15\mathbf{i} + 0.1\mathbf{j}\} \text{ m}}{\sqrt{(-0.15 \text{ m})^2 + (0.1 \text{ m})^2}}\right] = \{-249.6\mathbf{i} + 166.4\mathbf{j}\} \text{ N}$$

$$\mathbf{M} = -500\left(\tfrac{4}{5}\right)\mathbf{j} + 500\left(\tfrac{3}{5}\right)\mathbf{k} = \{-400\mathbf{j} + 300\mathbf{k}\} \text{ N} \cdot \text{m}$$

Force Summation.

$$\mathbf{F}_R = \Sigma\mathbf{F}; \qquad \mathbf{F}_R = \mathbf{F}_1 + \mathbf{F}_2 = -800\mathbf{k} - 249.6\mathbf{i} + 166.4\mathbf{j}$$

$$= \{-250\mathbf{i} + 166\mathbf{j} - 800\mathbf{k}\} \text{ N} \qquad\qquad \textit{Ans.}$$

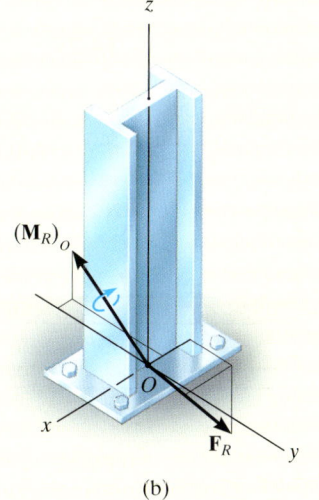

(b)

Fig. 4–39

Moment Summation.

$$(\mathbf{M}_R)_o = \Sigma\mathbf{M} + \Sigma\mathbf{M}_O$$

$$(\mathbf{M}_R)_o = \mathbf{M} + \mathbf{r}_C \times \mathbf{F}_1 + \mathbf{r}_B \times \mathbf{F}_2$$

$$(\mathbf{M}_R)_o = (-400\mathbf{j} + 300\mathbf{k}) + (1\mathbf{k}) \times (-800\mathbf{k}) + \begin{vmatrix} \mathbf{i} & \mathbf{j} & \mathbf{k} \\ -0.15 & 0.1 & 1 \\ -249.6 & 166.4 & 0 \end{vmatrix}$$

$$= (-400\mathbf{j} + 300\mathbf{k}) + (\mathbf{0}) + (-166.4\mathbf{i} - 249.6\mathbf{j})$$

$$= \{-166\mathbf{i} - 650\mathbf{j} + 300\mathbf{k}\} \text{ N} \cdot \text{m} \qquad\qquad \textit{Ans.}$$

The results are shown in Fig. 4–39b.

FUNDAMENTAL PROBLEMS

F4–25. Replace the loading system by an equivalent resultant force and couple moment acting at point A.

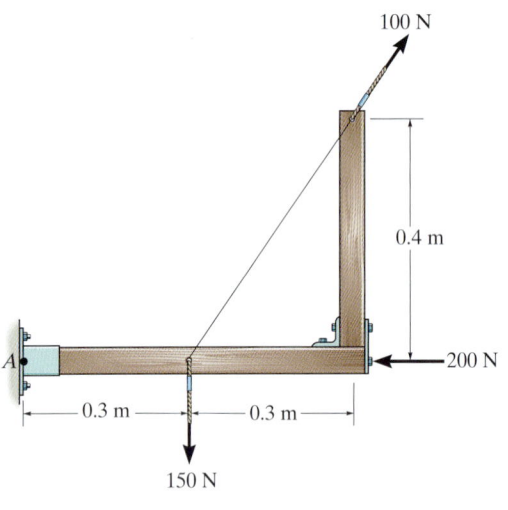

F4–25

F4–26. Replace the loading system by an equivalent resultant force and couple moment acting at point A.

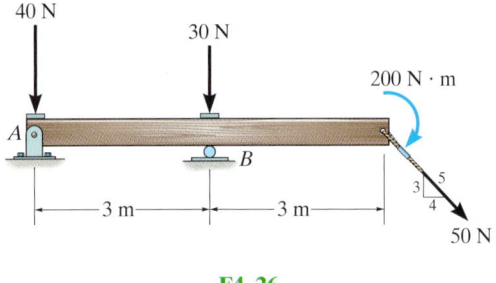

F4–26

F4–27. Replace the loading system by an equivalent resultant force and couple moment acting at point A.

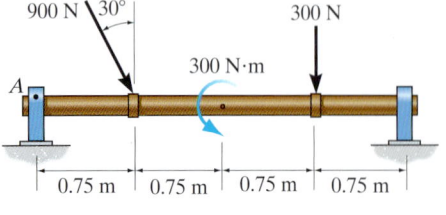

F4–27

F4–28. Replace the loading system by an equivalent resultant force and couple moment acting at point A.

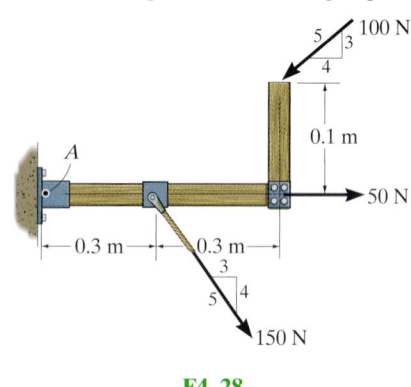

F4–28

F4–29. Replace the loading system by an equivalent resultant force and couple moment acting at point O.

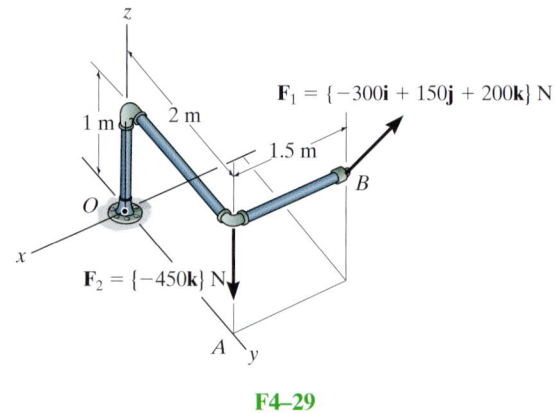

F4–29

F4–30. Replace the loading system by an equivalent resultant force and couple moment acting at point O.

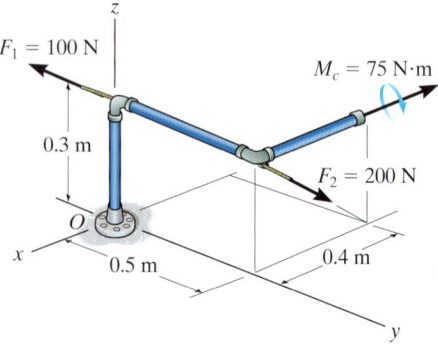

F4–30

PROBLEMS

4–97. Replace the force and couple system by an equivalent force and couple moment at point *O*.

4–98. Replace the force and couple system by an equivalent force and couple moment at point *P*.

4–101. Replace the loading system acting on the beam by an equivalent resultant force and couple moment at point *O*.

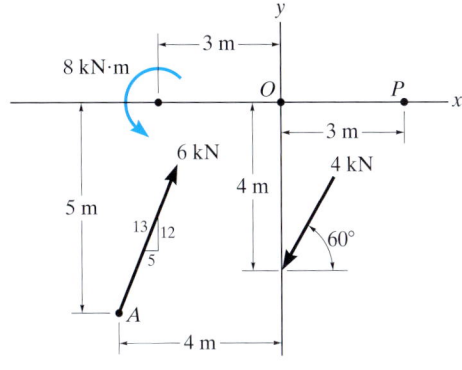

Probs. 4–97/98

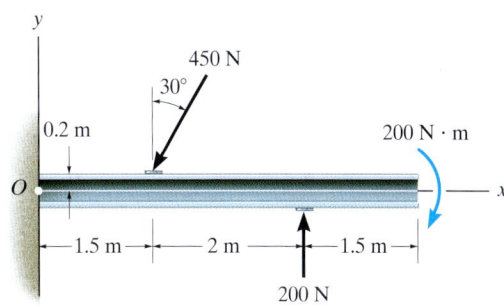

Prob. 4–101

4–99. Replace the force system acting on the beam by an equivalent force and couple moment at point *A*. Ignore the depth of the beam.

*4–100.** Replace the force system acting on the beam by an equivalent force and couple moment at point *B*. Ignore the depth of the beam.

4–102. Replace the loading acting on the beam by a single resultant force. Specify where the force acts, measured from end *A*.

4–103. Replace the loading acting on the beam by a single resultant force. Specify where the force acts, measured from *B*.

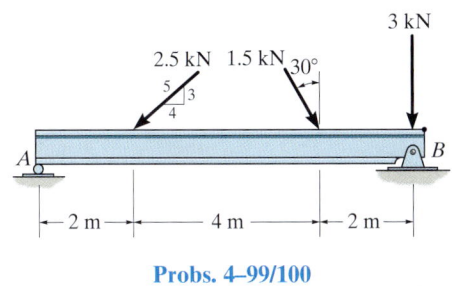
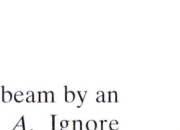

Probs. 4–99/100

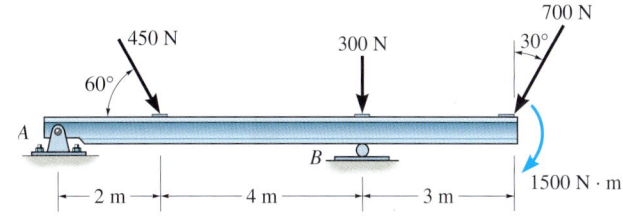

Probs. 4–102/103

*4–104. The system of four forces acts on the roof truss. Determine the equivalent resultant force and specify its location along AB, measured from point A.

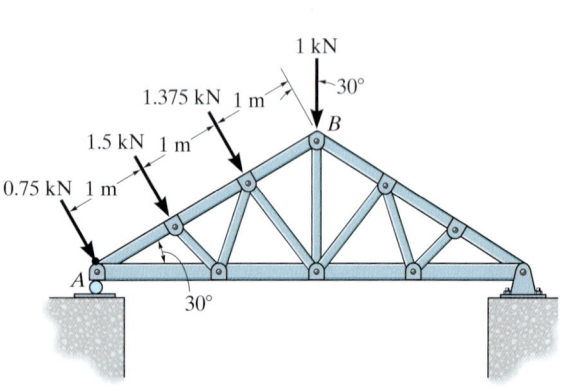

Prob. 4–104

4–105. Replace the force system acting on the frame by a resultant force and couple moment at point A.

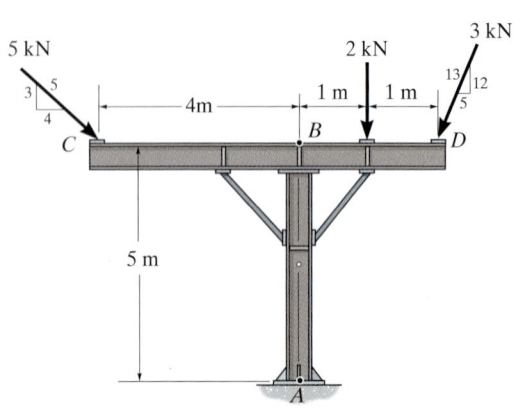

Prob. 4–105

4–106. Replace the force system acting on the bracket by a resultant force and couple moment at point A.

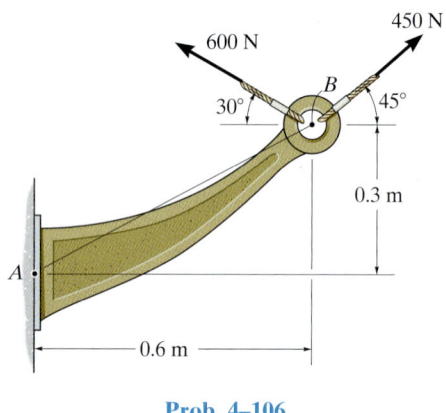

Prob. 4–106

4–107. A biomechanical model of the lumbar region of the human trunk is shown. The forces acting in the four muscle groups consist of $F_R = 35$ N for the rectus, $F_O = 45$ N for the oblique, $F_L = 23$ N for the lumbar latissimus dorsi, and $F_E = 32$ N for the erector spinae. These loadings are symmetric with respect to the y–z plane. Replace this system of parallel forces by an equivalent force and couple moment acting at the spine, point O. Express the results in Cartesian vector form.

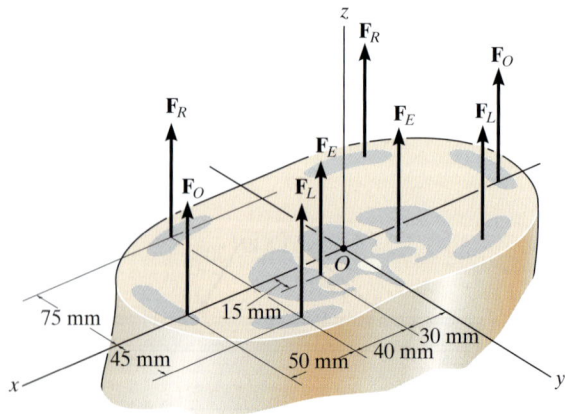

Prob. 4–107

*4–108.   Replace the two forces acting on the post by a resultant force and couple moment at point O. Express the results in Cartesian vector form.

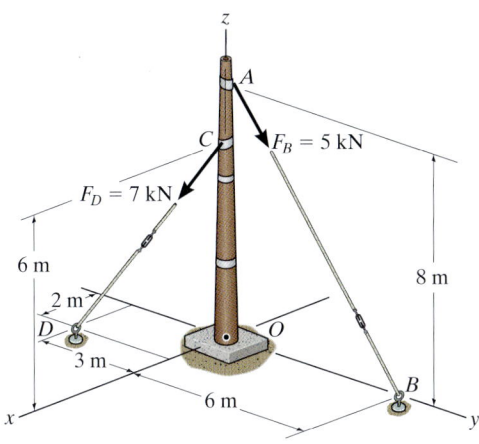

Prob. 4–108

4–109.   Replace the force system by an equivalent force and couple moment at point A.

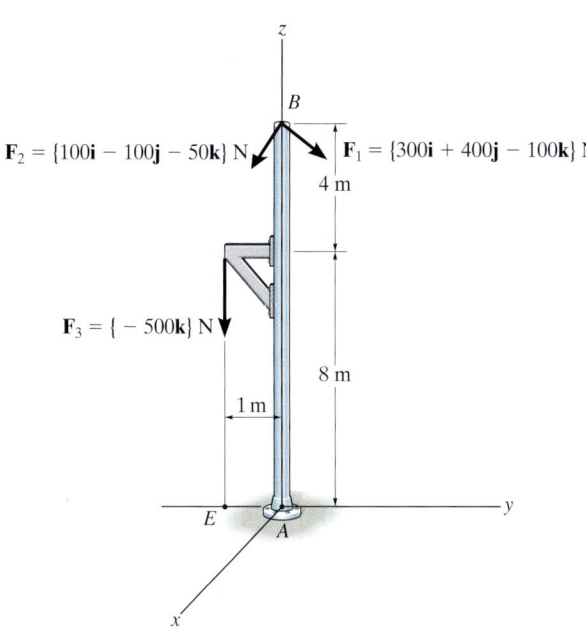

Prob. 4–109

4–110.   The belt passing over the pulley is subjected to forces $\mathbf{F}_1$ and $\mathbf{F}_2$, each having a magnitude of 40 N. $\mathbf{F}_1$ acts in the $-\mathbf{k}$ direction. Replace these forces by an equivalent force and couple moment at point A. Express the result in Cartesian vector form. Set $\theta = 0°$ so that $\mathbf{F}_2$ acts in the $-\mathbf{j}$ direction.

4–111.   The belt passing over the pulley is subjected to two forces $\mathbf{F}_1$ and $\mathbf{F}_2$, each having a magnitude of 40 N. $\mathbf{F}_1$ acts in the $-\mathbf{k}$ direction. Replace these forces by an equivalent force and couple moment at point A. Express the result in Cartesian vector form. Take $\theta = 45°$.

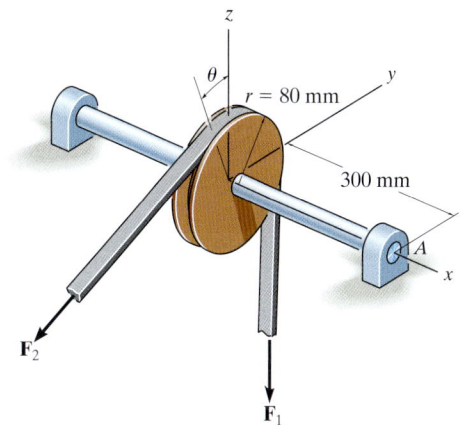

Probs. 4–110/111

*4–112.   Handle forces $\mathbf{F}_1$ and $\mathbf{F}_2$ are applied to the electric drill. Replace this force system by an equivalent resultant force and couple moment acting at point O. Express the results in Cartesian vector from.

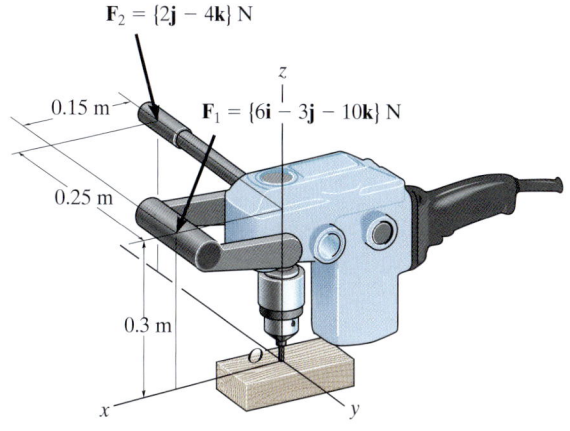

Prob. 4–112

4.8   Further Simplification of a Force and Couple System

In the preceding section, we developed a way to reduce a force and couple moment system acting on a rigid body into an equivalent resultant force $\mathbf{F}_R$ acting at a specific point O and a resultant couple moment $(\mathbf{M}_R)_O$. The force system can be further reduced to an equivalent single resultant force provided the lines of action of $\mathbf{F}_R$ and $(\mathbf{M}_R)_O$ are *perpendicular* to each other. Because of this condition, only concurrent, coplanar, and parallel force systems can be further simplified.

Concurrent Force System.   Since a *concurrent force system* is one in which the lines of action of all the forces intersect at a common point O, Fig. 4–40a, then the force system produces no moment about this point. As a result, the equivalent system can be represented by a single resultant force $\mathbf{F}_R = \Sigma\mathbf{F}$ acting at O, Fig. 4–40b.

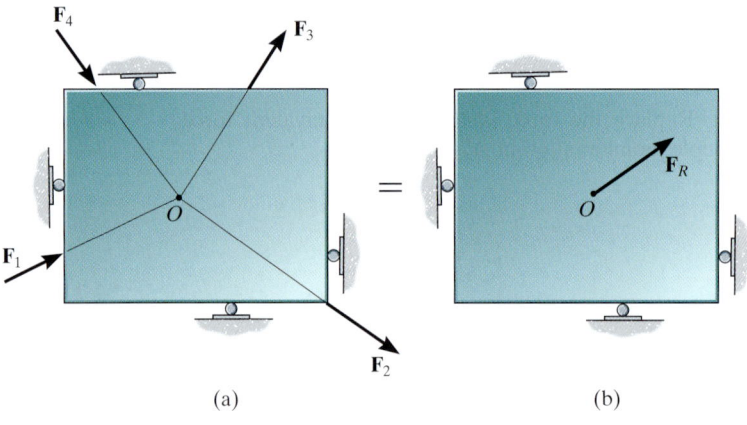

(a)   (b)

Fig. 4–40

Coplanar Force System.   In the case of a *coplanar force system*, the lines of action of all the forces lie in the same plane, Fig. 4–41a, and so the resultant force $\mathbf{F}_R = \Sigma\mathbf{F}$ of this system also lies in this plane. Furthermore, the moment of each of the forces about any point O is directed perpendicular to this plane. Thus, the resultant moment $(\mathbf{M}_R)_O$ and resultant force $\mathbf{F}_R$ will be *mutually perpendicular*, Fig. 4–41b. The resultant moment can be replaced by moving the resultant force $\mathbf{F}_R$ a perpendicular or moment arm distance d away from point O such that $\mathbf{F}_R$ produces the *same moment* $(\mathbf{M}_R)_O$ about point O, Fig. 4–41c. This distance d can be determined from the scalar equation $(M_R)_O = F_R d = \Sigma M_O$ or $d = (M_R)_O / F_R$.

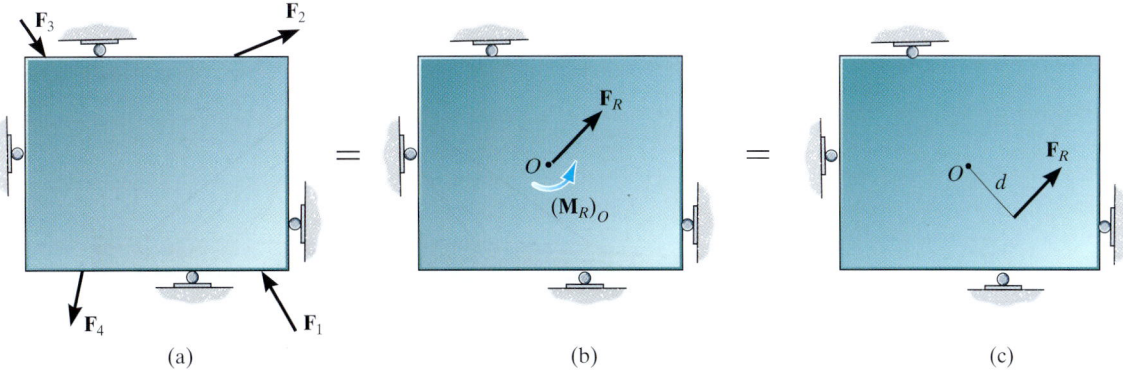

Fig. 4–41

Parallel Force System.

The *parallel force system* shown in Fig. 4–42*a* consists of forces that are all parallel to the *z* axis. Thus, the resultant force $\mathbf{F}_R = \Sigma\mathbf{F}$ at point O must also be parallel to this axis, Fig. 4–42*b*. The moment produced by each force lies in the plane of the plate, and so the resultant couple moment, $(\mathbf{M}_R)_O$, will also lie in this plane, along the moment axis *a* since $\mathbf{F}_R$ and $(\mathbf{M}_R)_O$ are mutually perpendicular. As a result, the force system can be further reduced to an equivalent single resultant force $\mathbf{F}_R$, acting through point P located on the perpendicular *b* axis, Fig. 4–42*c*. The distance *d* along this axis from point O requires $(M_R)_O = F_R d = \Sigma M_O$ or $d = \Sigma M_O / F_R$.

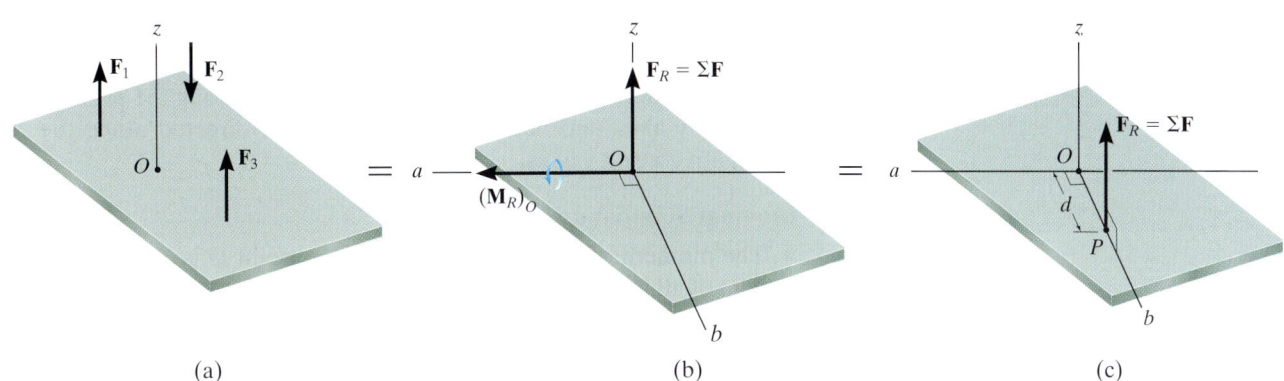

Fig. 4–42

The four cable forces are all concurrent at point O on this bridge tower. Consequently they produce no resultant moment there, only a resultant force $\mathbf{F}_R$. Note that the designers have positioned the cables so that $\mathbf{F}_R$ is directed *along* the bridge tower directly to the support, so that it does not cause any bending of the tower.

Procedure for Analysis

The technique used to reduce a coplanar or parallel force system to a single resultant force follows a similar procedure outlined in the previous section.

- Establish the x, y, z, axes and locate the resultant force $\mathbf{F}_R$ an arbitrary distance away from the origin of the coordinates.

Force Summation.

- The resultant force is equal to the sum of all the forces in the system.

- For a coplanar force system, resolve each force into its x and y components. Positive components are directed along the positive x and y axes, and negative components are directed along the negative x and y axes.

Moment Summation.

- The moment of the resultant force about point O is equal to the sum of all the couple moments in the system plus the moments of all the forces in the system about O.

- This moment condition is used to find the location of the resultant force from point O.

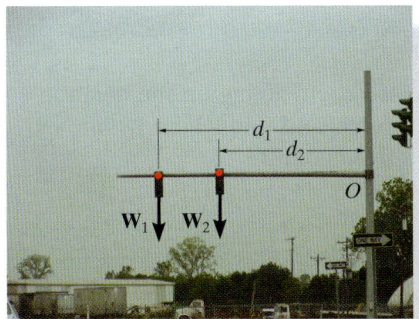

Here the weights of the traffic lights are replaced by their resultant force $W_R = W_1 + W_2$ which acts at a distance $d = (W_1 d_1 + W_2 d_2)/W_R$ from O. Both systems are equivalent.

Reduction to a Wrench.

In general, a three-dimensional force and couple moment system will have an equivalent resultant force $\mathbf{F}_R$ acting at point O and a resultant couple moment $(\mathbf{M}_R)_O$ that are *not perpendicular* to one another, as shown in Fig. 4–43a. Although a force system such as this cannot be further reduced to an equivalent single resultant force, the resultant couple moment $(\mathbf{M}_R)_O$ can be resolved into components parallel and perpendicular to the line of action of $\mathbf{F}_R$, Fig. 4–43a. The perpendicular component $\mathbf{M}_\perp$ can be replaced if we move $\mathbf{F}_R$ to point P, a distance d from point O along the b axis, Fig. 4–43b. As seen, this axis is perpendicular to both the a axis and the line of action of $\mathbf{F}_R$. The location of P can be determined from $d = M_\perp/F_R$. Finally, because $\mathbf{M}_\parallel$ is a free vector, it can be moved to point P, Fig. 4–43c. This combination of a resultant force $\mathbf{F}_R$ and collinear couple moment $\mathbf{M}_\parallel$ will tend to translate and rotate the body about its axis and is referred to as a *wrench* or *screw*. A wrench is the simplest system that can represent any general force and couple moment system acting on a body.

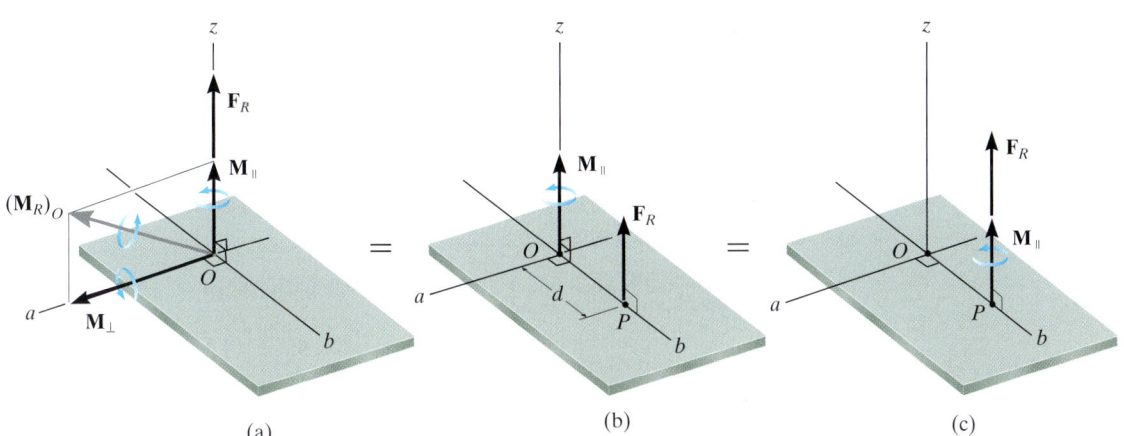

Fig. 4–43

EXAMPLE 4.17

Replace the force and couple moment system acting on the beam in Fig. 4–44a by an equivalent resultant force, and find where its line of action intersects the beam, measured from point O.

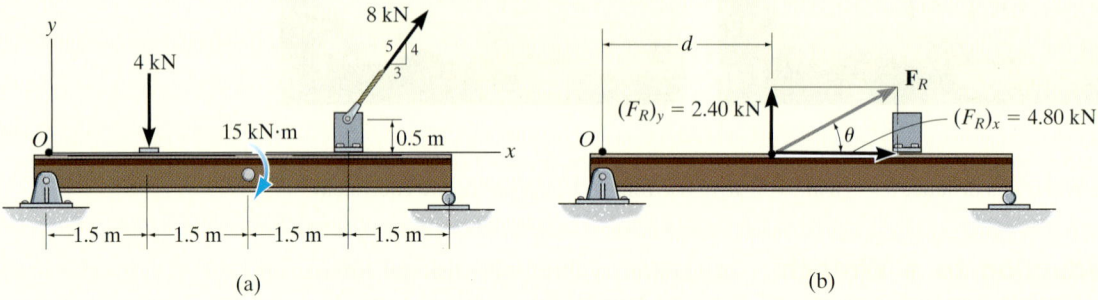

(a)   (b)

Fig. 4–44

SOLUTION

Force Summation.   Summing the force components,

$$\xrightarrow{+} (F_R)_x = \Sigma F_x; \qquad (F_R)_x = 8 \text{ kN}\left(\tfrac{3}{5}\right) = 4.80 \text{ kN} \rightarrow$$

$$+\uparrow (F_R)_y = \Sigma F_y; \qquad (F_R)_y = -4 \text{ kN} + 8 \text{ kN}\left(\tfrac{4}{5}\right) = 2.40 \text{ kN}\uparrow$$

From Fig. 4–44b, the magnitude of $\mathbf{F}_R$ is

$$F_R = \sqrt{(4.80 \text{ kN})^2 + (2.40 \text{ kN})^2} = 5.37 \text{ kN} \qquad \textit{Ans.}$$

The angle θ is

$$\theta = \tan^{-1}\left(\frac{2.40 \text{ kN}}{4.80 \text{ kN}}\right) = 26.6° \qquad \textit{Ans.}$$

Moment Summation.   We must equate the moment of $\mathbf{F}_R$ about point O in Fig. 4–44b to the sum of the moments of the force and couple moment system about point O in Fig. 4–44a. Since the line of action of $(\mathbf{F}_R)_x$ acts through point O, *only* $(\mathbf{F}_R)_y$ *produces a moment* about this point. Thus,

$$\zeta + (M_R)_O = \Sigma M_O; \qquad 2.40 \text{ kN}(d) = -(4 \text{ kN})(1.5 \text{ m}) - 15 \text{ kN} \cdot \text{m}$$

$$-\left[8 \text{ kN}\left(\tfrac{3}{5}\right)\right](0.5 \text{ m}) + \left[8 \text{ kN}\left(\tfrac{4}{5}\right)\right](4.5 \text{ m})$$

$$d = 2.25 \text{ m} \qquad \textit{Ans.}$$

EXAMPLE | 4.18

The jib crane shown in Fig. 4–45a is subjected to three coplanar forces. Replace this loading by an equivalent resultant force and specify where the resultant's line of action intersects the column AB and boom BC.

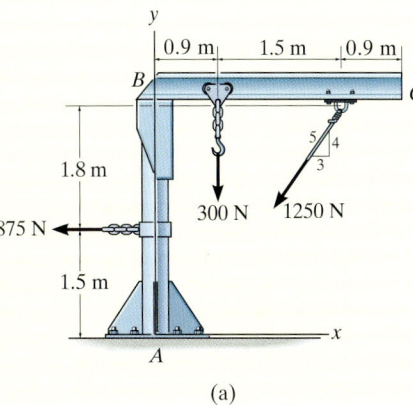

(a)

SOLUTION

Force Summation. Resolving the 1250-N force into x and y components and summing the force components yields

$$\xrightarrow{+}(F_R)_x = \Sigma F_x; \quad (F_R)_x = -1250 \text{ N}\left(\tfrac{3}{5}\right) - 875 \text{ N} = -1625 \text{ N} = 1625 \text{ N}\leftarrow$$

$$+\uparrow(F_R)_y = \Sigma F_y; \quad (F_R)_y = -1250 \text{ N}\left(\tfrac{4}{5}\right) - 300 \text{ N} = -1300 \text{ N} = 1300 \text{ N}\downarrow$$

As shown by the vector addition in Fig. 4–45b,

$$F_R = \sqrt{(1625 \text{ N})^2 + (1300 \text{ N})^2} = 2081 \text{ N} \qquad Ans.$$

$$\theta = \tan^{-1}\left(\frac{1300 \text{ N}}{1625 \text{ N}}\right) = 38.7° \, \nearrow \qquad Ans.$$

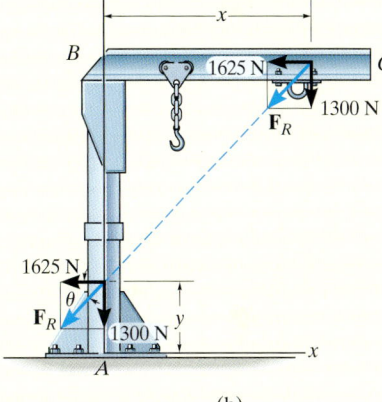

(b)

Fig. 4–45

Moment Summation. Moments will be summed about point A. Assuming the line of action of $\mathbf{F}_R$ intersects AB at a distance y from A, Fig. 4–45b, we have

$$\zeta + (M_R)_A = \Sigma M_A; \qquad 1625 \text{ N } (y) + 1300 \text{ N } (0)$$

$$= 875 \text{ N } (1.5 \text{ m}) - 300 \text{ N } (0.9 \text{ m}) + 1250 \text{ N}\left(\tfrac{3}{5}\right)(3.3 \text{ m}) - 1250 \text{ N}\left(\tfrac{4}{5}\right)(2.4 \text{ m})$$

$$y = 0.688 \text{ m} \qquad Ans.$$

By the principle of transmissibility, $\mathbf{F}_R$ can be placed at a distance x where it intersects BC, Fig. 4–45b. In this case we have

$$\zeta + (M_R)_A = \Sigma M_A; \qquad 1625 \text{ N } (3.3 \text{ m}) - 1300 \text{ N } (x)$$

$$= 875 \text{ N } (1.5 \text{ m}) - 300 \text{ N } (0.9 \text{ m}) + 1250 \text{ N}\left(\tfrac{3}{5}\right)(3.3 \text{ m}) - 1250 \text{ N}\left(\tfrac{4}{5}\right)(2.4 \text{ m})$$

$$x = 3.27 \text{ m} \qquad Ans.$$

EXAMPLE | **4.19**

The slab in Fig. 4–46a is subjected to four parallel forces. Determine the magnitude and direction of a resultant force equivalent to the given force system, and locate its point of application on the slab.

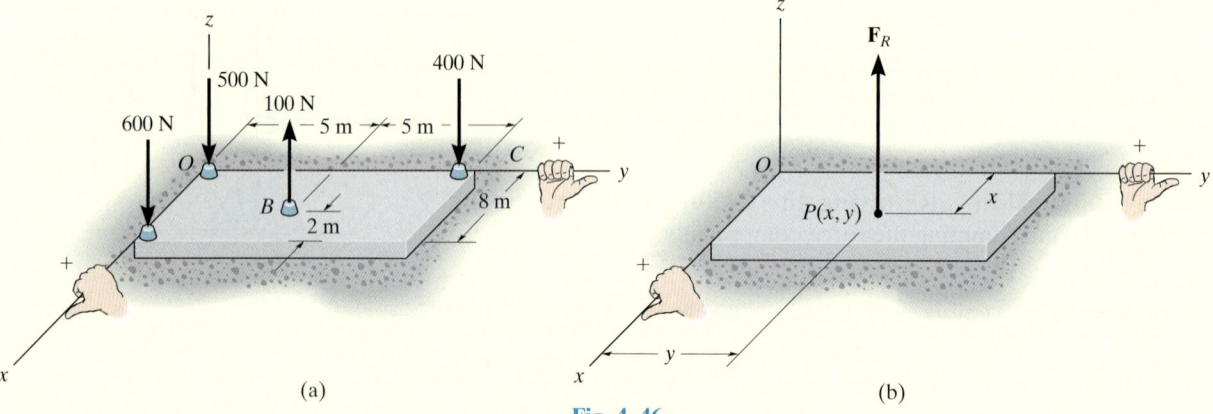

Fig. 4–46

SOLUTION (*SCALAR ANALYSIS*)

Force Summation.   From Fig. 4–46a, the resultant force is

$$+\uparrow F_R = \Sigma F; \qquad F_R = -600\text{ N} + 100\text{ N} - 400\text{ N} - 500\text{ N}$$
$$= -1400\text{ N} = 1400\text{ N}\downarrow \qquad Ans.$$

Moment Summation.   We require the moment about the x axis of the resultant force, Fig. 4–46b, to be equal to the sum of the moments about the x axis of all the forces in the system, Fig. 4–46a. The moment arms are determined from the y coordinates, since these coordinates represent the *perpendicular distances* from the x axis to the lines of action of the forces. Using the right-hand rule, we have

$$(M_R)_x = \Sigma M_x;$$

$$-(1400\text{ N})y = 600\text{ N}(0) + 100\text{ N}(5\text{ m}) - 400\text{ N}(10\text{ m}) + 500\text{ N}(0)$$
$$-1400y = -3500 \qquad y = 2.50\text{ m} \qquad Ans.$$

In a similar manner, a moment equation can be written about the y axis using moment arms defined by the x coordinates of each force.

$$(M_R)_y = \Sigma M_y;$$

$$(1400\text{ N})x = 600\text{ N}(8\text{ m}) - 100\text{ N}(6\text{ m}) + 400\text{ N}(0) + 500\text{ N}(0)$$
$$1400x = 4200$$
$$x = 3\text{ m} \qquad Ans.$$

NOTE: A force of $F_R = 1400$ N placed at point $P(3.00$ m, 2.50 m) on the slab, Fig. 4–46b, is therefore equivalent to the parallel force system acting on the slab in Fig. 4–46a.

EXAMPLE 4.20

Replace the force system in Fig. 4–47a by an equivalent resultant force and specify its point of application on the pedestal.

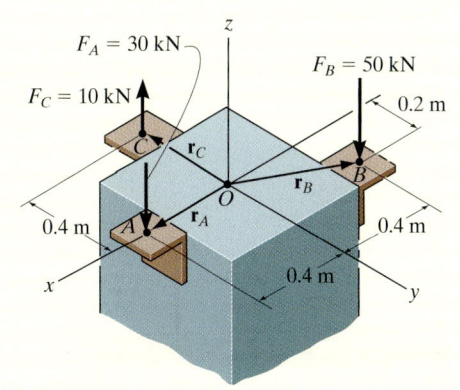

(a)

SOLUTION

Force Summation.   Here we will demonstrate a vector analysis. Summing forces,

$$\mathbf{F}_R = \Sigma \mathbf{F}; \quad \mathbf{F}_R = \mathbf{F}_A + \mathbf{F}_B + \mathbf{F}_C$$

$$= \{-30\mathbf{k}\}\ \text{kN} + \{-50\mathbf{k}\}\ \text{kN} + \{10\mathbf{k}\}\ \text{kN}$$

$$= \{-70\mathbf{k}\}\ \text{kN} \qquad\qquad Ans.$$

Location. Moments will be summed about point O. The resultant force $\mathbf{F}_R$ is assumed to act through point P $(x, y, 0)$, Fig. 4–47b. Thus

$$(\mathbf{M}_R)_O = \Sigma \mathbf{M}_O;$$

$$\mathbf{r}_P \times \mathbf{F}_R = (\mathbf{r}_A \times \mathbf{F}_A) + (\mathbf{r}_B \times \mathbf{F}_B) + (\mathbf{r}_C \times \mathbf{F}_C)$$

$$(x\mathbf{i} + y\mathbf{j}) \times (-70\mathbf{k}) = [(0.4\mathbf{i}) \times (-30\mathbf{k})]$$

$$+ [(-0.4\mathbf{i} + 0.2\mathbf{j}) \times (-50\mathbf{k})] + [(-0.4\mathbf{j}) \times (10\mathbf{k})]$$

$$-70x(\mathbf{i} \times \mathbf{k}) - 70y(\mathbf{j} \times \mathbf{k}) = -12(\mathbf{i} \times \mathbf{k}) + 20(\mathbf{i} \times \mathbf{k})$$

$$- 10(\mathbf{j} \times \mathbf{k}) - 4(\mathbf{j} \times \mathbf{k})$$

$$70x\mathbf{j} - 70y\mathbf{i} = 12\mathbf{j} - 20\mathbf{j} - 10\mathbf{i} - 4\mathbf{i}$$

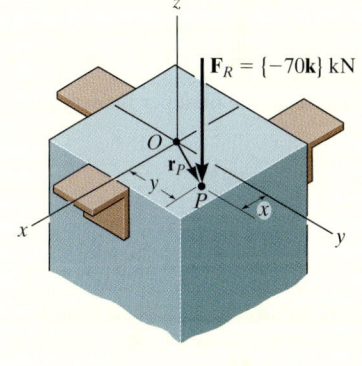

(b)

Fig. 4–47

Equating the $\mathbf{i}$ and $\mathbf{j}$ components,

$$-70y = -14 \qquad\qquad (1)$$

$$y = 0.2\ \text{m} \qquad\qquad Ans.$$

$$70x = -8 \qquad\qquad (2)$$

$$x = -0.114\ \text{m} \qquad\qquad Ans.$$

The negative sign indicates that the x coordinate of point P is negative.

NOTE: It is also possible to establish Eq. 1 and 2 directly by summing moments about the x and y axes. Using the right-hand rule, we have

$$(M_R)_x = \Sigma M_x; \qquad -70y = -10\ \text{kN}(0.4\ \text{m}) - 50\ \text{kN}(0.2\ \text{m})$$

$$(M_R)_y = \Sigma M_y; \qquad 70x = 30\ \text{kN}(0.4\ \text{m}) - 50\ \text{kN}(0.4\ \text{m})$$

FUNDAMENTAL PROBLEMS

F4–31. Replace the loading system by an equivalent resultant force and specify where the resultant's line of action intersects the beam measured from O.

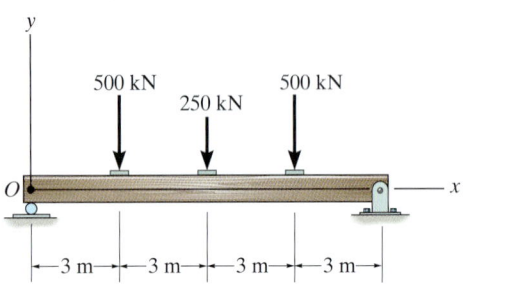

F4–31

F4–32. Replace the loading system by an equivalent resultant force and specify where the resultant's line of action intersects the member measured from A.

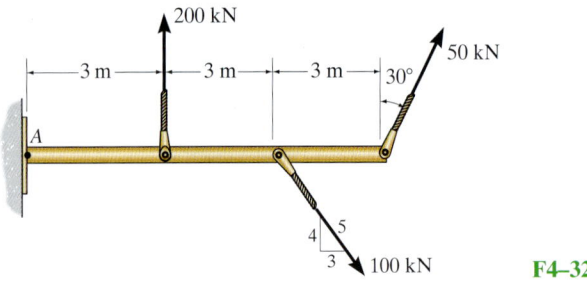

F4–32

F4–33. Replace the loading system by an equivalent resultant force and specify where the resultant's line of action intersects the horizontal segment of the member measured from A.

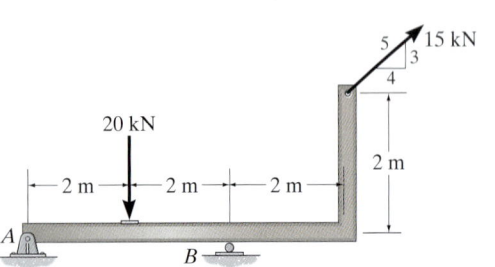

F4–33

F4–34. Replace the loading system by an equivalent resultant force and specify where the resultant's line of action intersects the member AB measured from A.

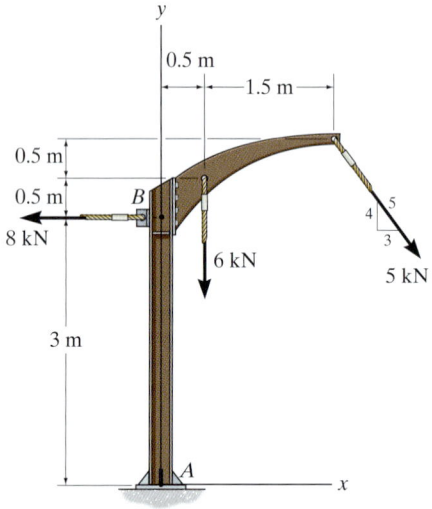

F4–34

F4–35. Replace the loading shown by an equivalent single resultant force and specify the x and y coordinates of its line of action.

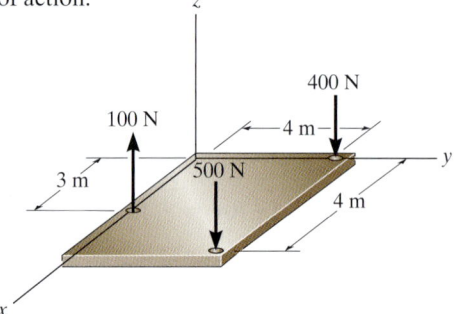

F4–35

F4–36. Replace the loading shown by an equivalent single resultant force and specify the x and y coordinates of its line of action.

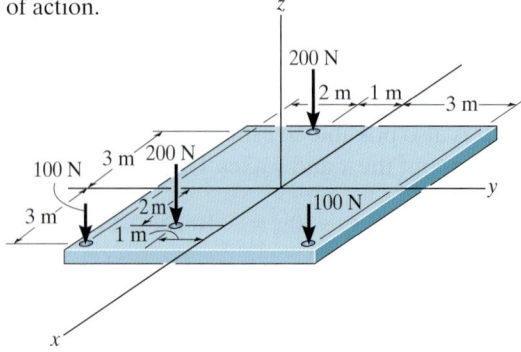

F4–36

PROBLEMS

4–113. Replace the force at *A* by an equivalent force and couple moment at point *O*.

4–114. Replace the force at *A* by an equivalent force and couple moment at point *P*.

4–117. Replace the loading acting on the beam by a single resultant force. Specify where the force acts, measured from end *A*.

4–118. Replace the loading acting on the beam by a single resultant force. Specify where the force acts, measured from *B*.

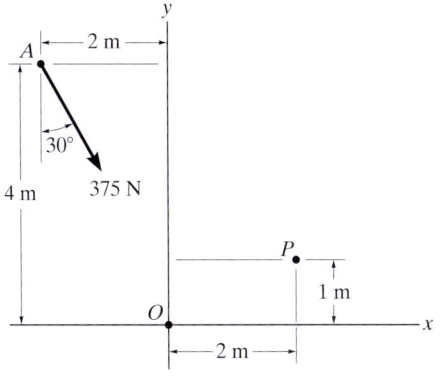

Probs. 4–113/114

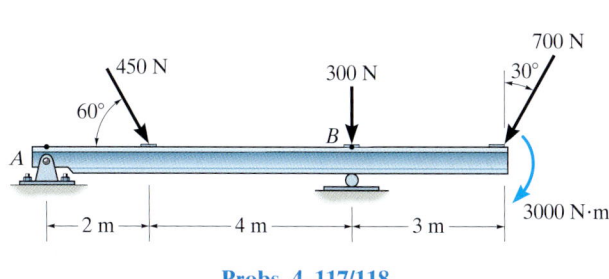

Probs. 4–117/118

4–115. Determine the magnitude and direction θ of force **F** and its placement *d* on the beam so that the loading system is equivalent to a resultant force of 12 kN acting vertically downward at point *A* and a clockwise couple moment of 50 kN · m.

*4–116. Determine the magnitude and direction θ of force **F** and its placement *d* on the beam so that the loading system is equivalent to a resultant force of 10 kN acting vertically downward at point *A* and a clockwise couple moment of 45 kN · m.

4–119. The system of parallel forces acts on the top of the *Warren truss*. Determine the equivalent resultant force of the system and specify its location measured from point *A*.

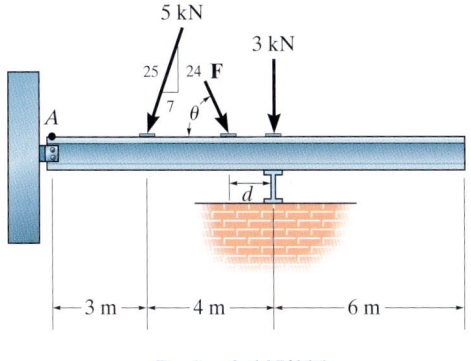

Probs. 4–115/116

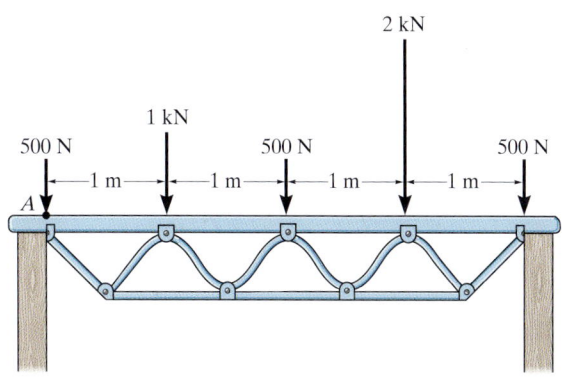

Prob. 4–119

*4–120. Replace the loading on the frame by a single resultant force. Specify where its line of action intersects member AB, measured from A.

4–121. Replace the loading on the frame by a single resultant force. Specify where its line of action intersects member CD, measured from end C.

*4–124. Replace the force system acting on the post by a resultant force, and specify where its line of action intersects the post AB measured from point A.

4–125. Replace the force system acting on the post by a resultant force, and specify where its line of action intersects the post AB measured from point B.

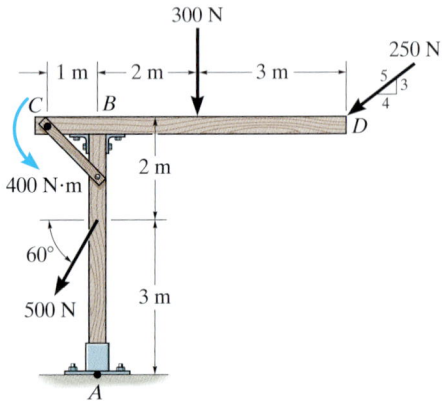

Probs. 4–120/121

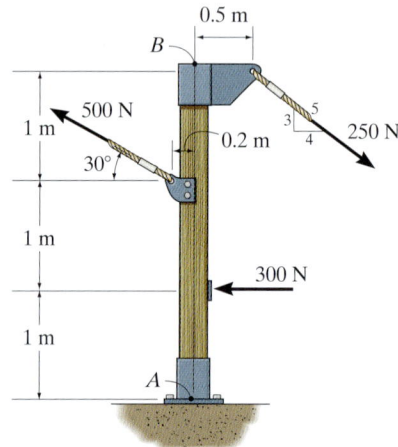

Probs. 4–124/125

4–122. Replace the force and couple system by an equivalent force and couple moment at point O.

4–123. Replace the force and couple system by an equivalent force and couple moment at point P.

4–126. Replace the force and couple moment system acting on the overhang beam by a resultant force, and specify its location along AB measured from point A.

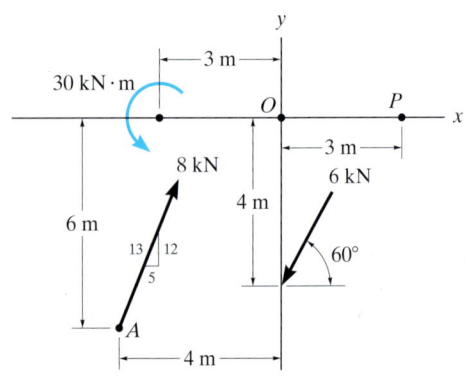

Probs. 4–122/123

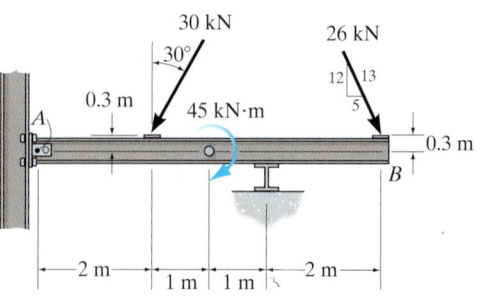

Prob. 4–126

4–127. The tube supports the four parallel forces. Determine the magnitudes of forces $\mathbf{F}_C$ and $\mathbf{F}_D$ acting at C and D so that the equivalent resultant force of the force system acts through the midpoint O of the tube.

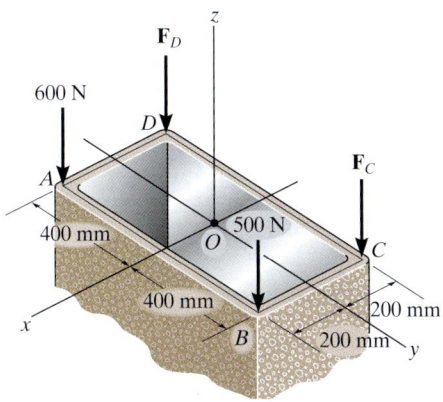

Prob. 4–127

*4–128.** Replace the force and couple-moment system by an equivalent resultant force and couple moment at point P. Express the results in Cartesian vector form.

4–129. Replace the force and couple-moment system by an equivalent resultant force and couple moment at point Q. Express the results in Cartesian vector form.

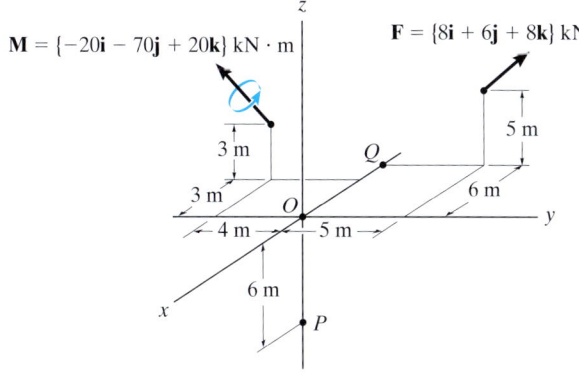

Probs. 4–128/129

4–130. The building slab is subjected to four parallel column loadings. Determine the equivalent resultant force and specify its location (x, y) on the slab. Take $F_1 = 30$ kN, $F_2 = 40$ kN.

4–131. The building slab is subjected to four parallel column loadings. Determine the equivalent resultant force and specify its location (x, y) on the slab. Take $F_1 = 20$ kN, $F_2 = 50$ kN.

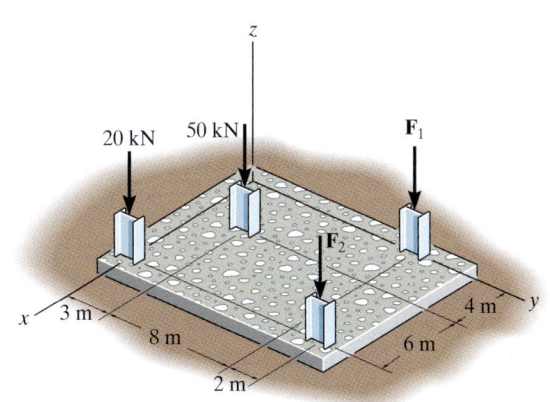

Probs. 4–130/131

*4–132.** If $F_A = 40$ kN and $F_B = 35$ kN, determine the magnitude of the resultant force and specify the location of its point of application (x, y) on the slab.

4–133. If the resultant force is required to act at the center of the slab, determine the magnitude of the column loadings $\mathbf{F}_A$ and $\mathbf{F}_B$ and the magnitude of the resultant force.

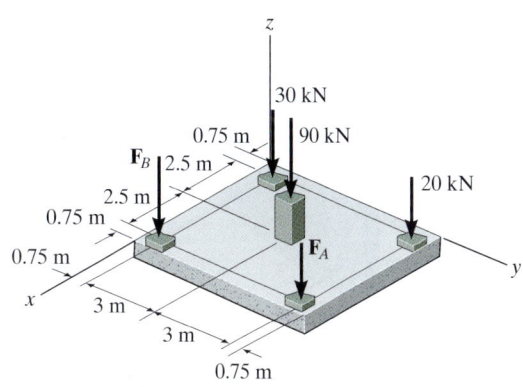

Probs. 4–132/133

4–134. Replace the two wrenches and the force, acting on the pipe assembly, by an equivalent resultant force and couple moment at point *O*.

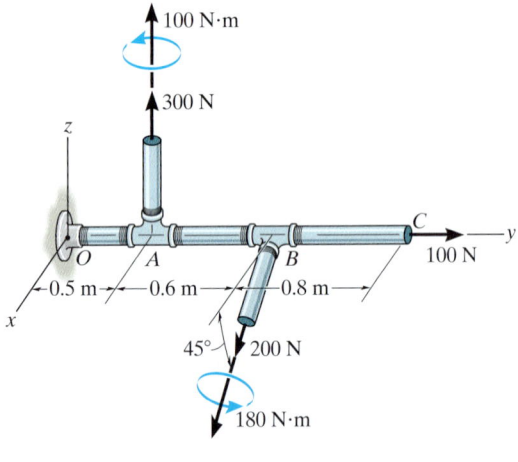

Prob. 4–134

*4–136.** The pipe assembly is subjected to the action of a wrench at *B* and a couple at *A*. Determine the magnitude *F* of the couple forces so that the system can be simplified to a wrench acting at point *C*.

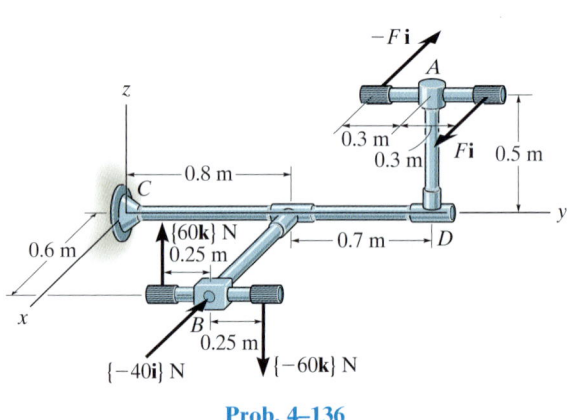

Prob. 4–136

4–135. Replace the parallel force system acting on the plate by a resultant force and specify its location on the *x*–*z* plane.

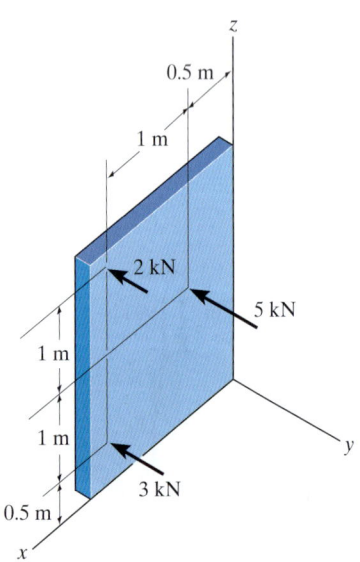

Prob. 4–135

4–137. Replace the three forces acting on the plate by a wrench. Specify the magnitude of the force and couple moment for the wrench and the point *P*(*x*, *y*) where its line of action intersects the plate.

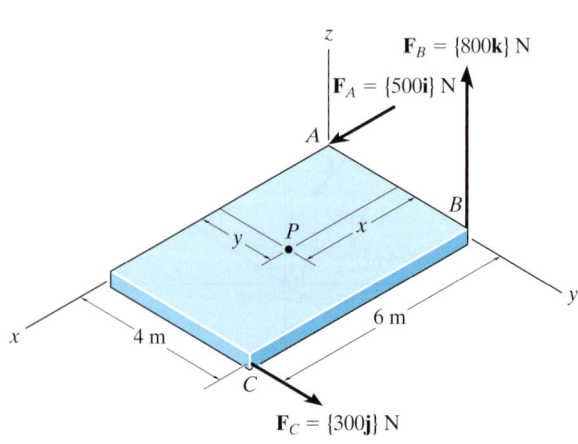

Prob. 4–137

4.9   Reduction of a Simple Distributed Loading

Sometimes, a body may be subjected to a loading that is distributed over its surface. For example, the pressure of the wind on the face of a sign, the pressure of water within a tank, or the weight of sand on the floor of a storage container, are all *distributed loadings*. The pressure exerted at each point on the surface indicates the intensity of the loading. It is measured using pascals Pa (or N/m^2) in SI units.

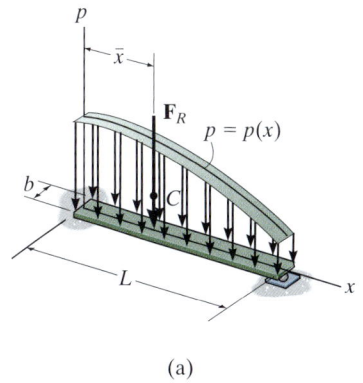

(a)

Loading Along a Single Axis.

The most common type of distributed loading encountered in engineering practice can be represented along a single axis.* For example, consider the beam (or plate) in Fig. 4–48*a* that has a constant width and is subjected to a pressure loading that varies only along the *x* axis. This loading can be described by the function $p = p(x) \ N/m^2$. It contains only one variable *x*, and for this reason, we can also represent it as a *coplanar distributed load*. To do so, we multiply the loading function by the width *b* m of the beam, so that $w(x) = p(x)b \ N/m$, Fig. 4–48*b*. Using the methods of Sec. 4.8, we can replace this coplanar parallel force system with a single equivalent resultant force $\mathbf{F}_R$ acting at a specific location on the beam, Fig. 4–48*c*.

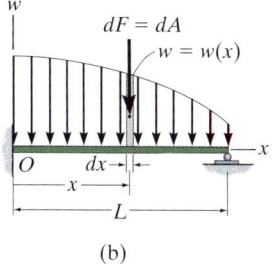

(b)

Magnitude of Resultant Force.

From Eq. 4–17 ($F_R = \Sigma F$), the magnitude of $\mathbf{F}_R$ is equivalent to the sum of all the forces in the system. In this case integration must be used since there is an infinite number of parallel forces $d\mathbf{F}$ acting on the beam, Fig. 4–48*b*. Since $d\mathbf{F}$ is acting on an element of length dx, and $w(x)$ is a force per unit length, then $dF = w(x) \ dx = dA$. In other words, the magnitude of $d\mathbf{F}$ is determined from the colored differential *area* dA under the loading curve. For the entire length *L*,

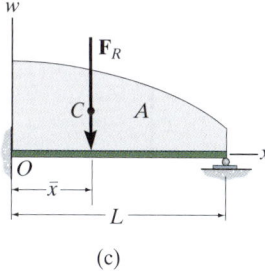

(c)

Fig. 4–48

$$+\downarrow F_R = \Sigma F; \qquad \boxed{F_R = \int_L w(x) \, dx = \int_A dA = A} \qquad (4\text{–}19)$$

Therefore, the magnitude of the resultant force is equal to the area A under the loading diagram, Fig. 4–48*c*.

---

*The more general case of a surface loading acting on a body is considered in Sec. 9.5.

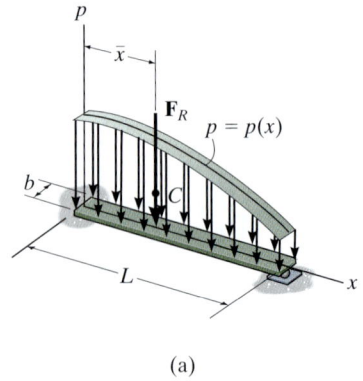

(a)

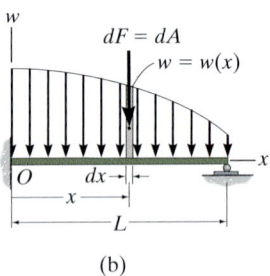

(b)

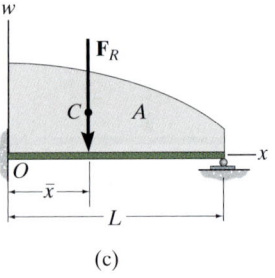

(c)

Fig. 4–48 (Repeated)

Each beam that supports this stack of lumber is subjected to a uniform loading of w_0. The resultant force is therefore equal to the area under the rectangular loading diagram. It acts through the centroid or geometric center of this area.

Location of Resultant Force. Applying Eq. 4–17 ($M_{R_O} = \Sigma M_O$), the location $\bar{x}$ of the line of action of $\mathbf{F}_R$ can be determined by equating the moments of the force resultant and the parallel force distribution about point O (the y axis). Since $d\mathbf{F}$ produces a moment of $x\, dF = xw(x)\, dx$ about O, Fig. 4–48b, then for the entire length, Fig. 4–48c,

$$\zeta + (M_R)_O = \Sigma M_O; \qquad\qquad -\bar{x}F_R = -\int_L xw(x)\, dx$$

Solving for $\bar{x}$, using Eq. 4–19, we have

$$\bar{x} = \frac{\displaystyle\int_L xw(x)\, dx}{\displaystyle\int_L w(x)\, dx} = \frac{\displaystyle\int_A x\, dA}{\displaystyle\int_A dA} \qquad (4\text{–}20)$$

This coordinate $\bar{x}$, locates the geometric center or *centroid* of the *area* under the distributed loading. *In other words, the resultant force has a line of action which passes through the centroid C (geometric center) of the area under the loading diagram*, Fig. 4–48c. Detailed treatment of the integration techniques for finding the location of the centroid for areas is given in Chapter 9. In many cases, however, the distributed-loading diagram is in the shape of a rectangle, triangle, or some other simple geometric form. The centroid location for such common shapes does not have to be determined from the above equation but can be obtained directly from the tabulation given on the inside back cover.

Once $\bar{x}$ is determined, $\mathbf{F}_R$ by symmetry passes through point $(\bar{x}, 0)$ on the surface of the beam, Fig. 4–48a. Therefore, in this case the resultant force has a *magnitude equal to the volume under the loading curve $p = p(x)$ and a line of action which passes through the centroid (geometric center) of this volume*.

Important Points

- Coplanar distributed loadings are defined by using a loading function $w = w(x)$ that indicates the intensity of the loading along the length of a member. This intensity is measured in N/m or lb/ft.

- The external effects caused by a coplanar distributed load acting on a body can be represented by a single resultant force.

- This resultant force is equivalent to the *area* under the loading diagram, and has a line of action that passes through the *centroid* or geometric center of this area.

EXAMPLE 4.21

Determine the magnitude and location of the equivalent resultant force acting on the shaft in Fig. 4–49a.

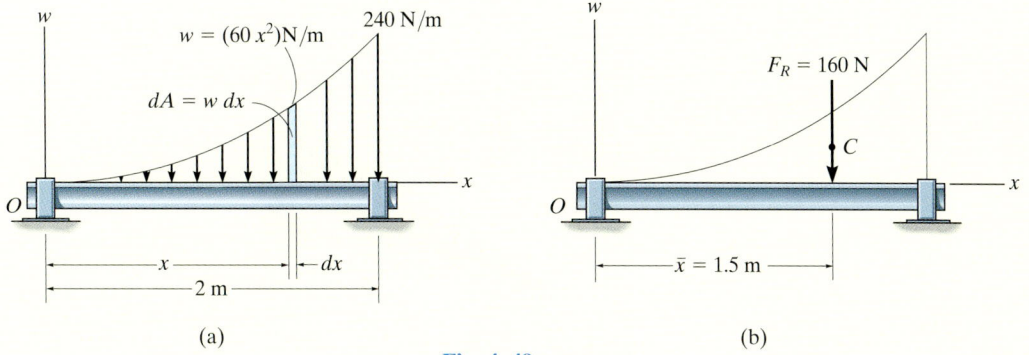

(a)   (b)

Fig. 4–49

SOLUTION

Since $w = w(x)$ is given, this problem will be solved by integration.

The differential element has an area $dA = w\,dx = 60x^2\,dx$. Applying Eq. 4–19,

$+\downarrow F_R = \Sigma F;$

$$F_R = \int_A dA = \int_0^{2\,m} 60x^2\,dx = 60\left(\frac{x^3}{3}\right)\Big|_0^{2\,m} = 60\left(\frac{2^3}{3} - \frac{0^3}{3}\right)$$

$$= 160\ N \hspace{5cm} Ans.$$

The location $\bar{x}$ of $\mathbf{F}_R$ measured from O, Fig. 4–49b, is determined from Eq. 4–20.

$$\bar{x} = \frac{\int_A x\,dA}{\int_A dA} = \frac{\int_0^{2\,m} x(60x^2)\,dx}{160\ N} = \frac{60\left(\frac{x^4}{4}\right)\Big|_0^{2\,m}}{160\ N} = \frac{60\left(\frac{2^4}{4} - \frac{0^4}{4}\right)}{160\ N}$$

$$= 1.5\ m \hspace{5cm} Ans.$$

NOTE: These results can be checked by using the table on the inside back cover, where it is shown that formula for an exparabolic area of length a, height b, and shape shown in Fig. 4–49a, we have

$$A = \frac{ab}{3} = \frac{2\ m(240\ N/m)}{3} = 160\ N \text{ and } \bar{x} = \frac{3}{4}a = \frac{3}{4}(2\ m) = 1.5\ m$$

4

EXAMPLE 4.22

A distributed loading of $p = (800x)$ Pa acts over the top surface of the beam shown in Fig. 4–50a. Determine the magnitude and location of the equivalent resultant force.

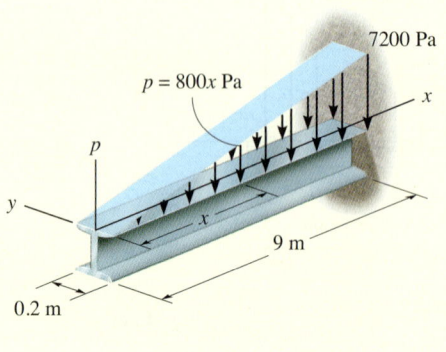

(a)

SOLUTION

Since the loading intensity is uniform along the width of the beam (the y axis), the loading can be viewed in two dimensions as shown in Fig. 4–50b. Here

$$w = (800x \text{ N/m}^2)(0.2 \text{ m})$$
$$= (160x) \text{ N/m}$$

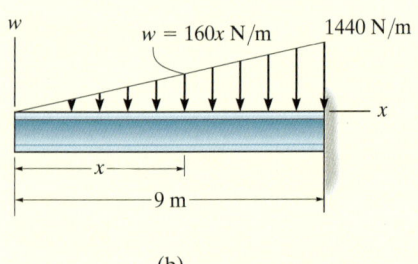

(b)

At $x = 9$ m, note that $w = 1440$ N/m. Although we may again apply Eqs. 4–19 and 4–20 as in the previous example, it is simpler to use the table on the inside back cover.

The magnitude of the resultant force is equivalent to the area of the triangle.

$$F_R = \tfrac{1}{2}(9 \text{ m})(1440 \text{ N/m}) = 6480 \text{ N} = 6.48 \text{ kN} \qquad Ans.$$

The line of action of $\mathbf{F}_R$ passes through the *centroid C* of this triangle. Hence,

$$\bar{x} = 9 \text{ m} - \tfrac{1}{3}(9 \text{ m}) = 6 \text{ m} \qquad Ans.$$

The results are shown in Fig. 4–50c.

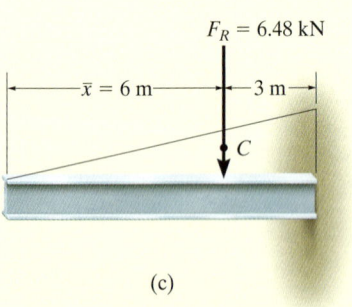

(c)

Fig. 4–50

NOTE: We may also view the resultant $\mathbf{F}_R$ as *acting* through the *centroid* of the *volume* of the loading diagram $p = p(x)$ in Fig. 4–50a. Hence $\mathbf{F}_R$ intersects the x–y plane at the point (6 m, 0). Furthermore, the magnitude of $\mathbf{F}_R$ is equal to the volume under the loading diagram; i.e.,

$$F_R = V = \tfrac{1}{2}(7200 \text{ N/m}^2)(9 \text{ m})(0.2 \text{ m}) = 6.48 \text{ kN} \qquad Ans.$$

EXAMPLE 4.23

The granular material exerts the distributed loading on the beam as shown in Fig. 4–51a. Determine the magnitude and location of the equivalent resultant of this load.

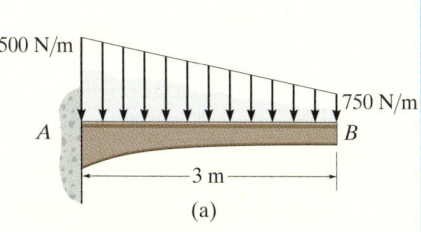

(a)

SOLUTION

The area of the loading diagram is a *trapezoid*, and therefore the solution can be obtained directly from the area and centroid formulas for a trapezoid listed on the inside back cover. Since these formulas are not easily remembered, instead we will solve this problem by using "composite" areas. Here we will divide the trapezoidal loading into a rectangular and triangular loading as shown in Fig. 4–51b. The magnitude of the force represented by each of these loadings is equal to its associated *area*,

$$F_1 = \tfrac{1}{2}(3\text{ m})(750\text{ N/m}) = 1125\text{ N}$$

$$F_2 = (3\text{ m})(750\text{ N/m}) = 2250\text{ N}$$

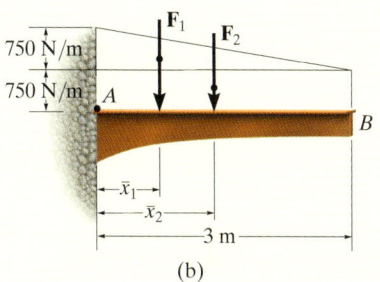

(b)

The lines of action of these parallel forces act through the respective *centroids* of their associated areas and therefore intersect the beam at

$$\bar{x}_1 = \tfrac{1}{3}(3\text{ m}) = 1\text{ m}$$

$$\bar{x}_2 = \tfrac{1}{2}(3\text{ m}) = 1.5\text{ m}$$

The two parallel forces $\mathbf{F}_1$ and $\mathbf{F}_2$ can be reduced to a single resultant $\mathbf{F}_R$. The magnitude of $\mathbf{F}_R$ is

$$+\downarrow F_R = \Sigma F; \qquad F_R = 1125 + 2250 = 3375\text{ N} \qquad Ans.$$

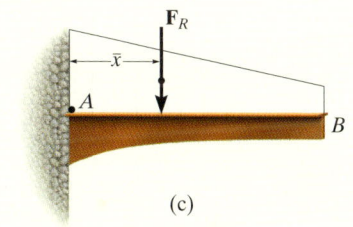

(c)

We can find the location of $\mathbf{F}_R$ with reference to point A, Fig. 4–51b and 4–51c. We require

$$\zeta + (M_R)_A = \Sigma M_A; \quad \bar{x}(3375) = 1(1125) + 1.5(2250)$$

$$\bar{x} = 1.333\text{ m} \qquad Ans.$$

NOTE: The trapezoidal area in Fig. 4–51a can also be divided into two triangular areas as shown in Fig. 4–51d. In this case

$$F_3 = \tfrac{1}{2}(3\text{ m})(1500\text{ N/m}) = 2250\text{ N}$$

$$F_4 = \tfrac{1}{2}(3\text{ m})(750\text{ N/m}) = 1125\text{ N}$$

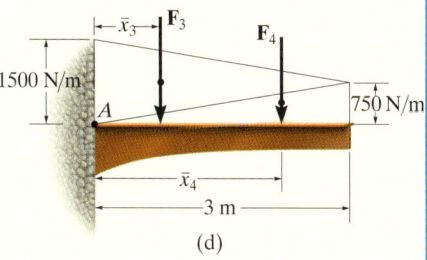

(d)

Fig. 4–51

and

$$\bar{x}_3 = \tfrac{1}{3}(3\text{ m}) = 1\text{ m}$$

$$\bar{x}_4 = 3\text{ m} - \tfrac{1}{3}(3\text{ m}) = 2\text{ m}$$

Using these results, show that again $F_R = 3375\text{ N}$ and $\bar{x} = 1.333\text{ m}$

FUNDAMENTAL PROBLEMS

F4–37. Determine the resultant force and specify where it acts on the beam measured from A.

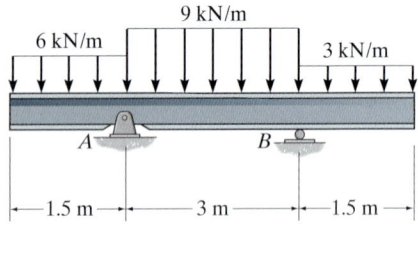

9 kN/m
6 kN/m
3 kN/m

A B

1.5 m   3 m   1.5 m

F4–37

F4–38. Determine the resultant force and specify where it acts on the beam measured from A.

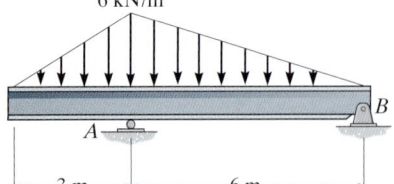

150 kN/m

A B

3 m   4 m

F4–38

F4–39. Determine the resultant force and specify where it acts on the beam measured from A.

6 kN/m

A B

3 m   6 m

F4–39

F4–40. Determine the resultant force and specify where it acts on the beam measured from A.

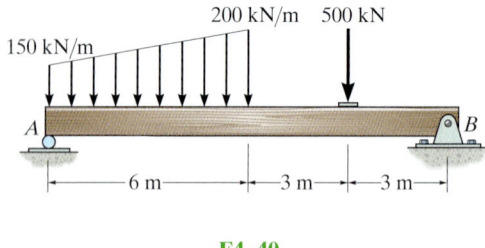

200 kN/m   500 kN
150 kN/m

A B

6 m   3 m   3 m

F4–40

F4–41. Determine the resultant force and specify where it acts on the beam measured from A.

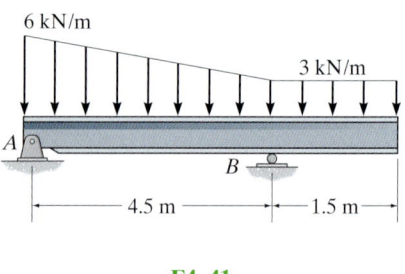

6 kN/m
3 kN/m

A B

4.5 m   1.5 m

F4–41

F4–42. Determine the resultant force and specify where it acts on the beam measured from A.

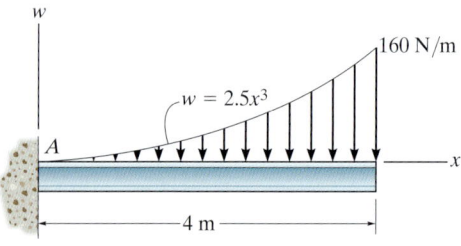

w 160 N/m

$w = 2.5x^3$

A x

4 m

F4–42

PROBLEMS

4–138. The loading on the bookshelf is distributed as shown. Determine the magnitude of the equivalent resultant location, measured from point O.

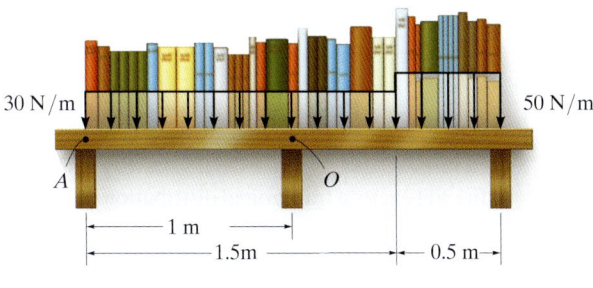

Prob. 4–138

4–139. Replace the distributed loading with an equivalent resultant force, and specify its location on the beam measured from point O.

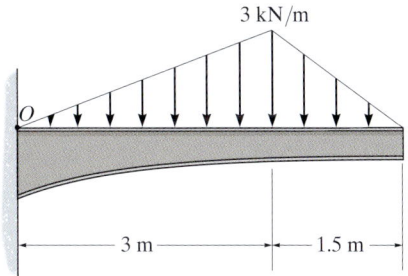

Prob. 4–139

***4–140.** Replace the loading by an equivalent force and couple moment acting at point O.

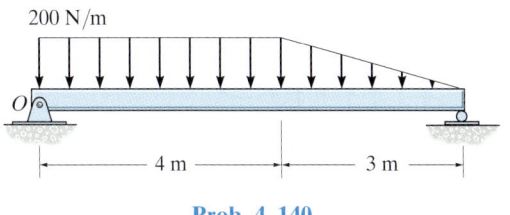

Prob. 4–140

4–141. Replace the loading by an equivalent resultant force and couple moment acting at point A.

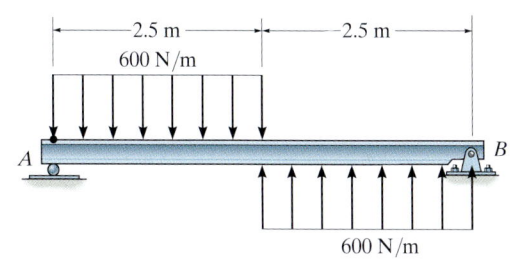

Prob. 4–141

4–142. Replace the loading by an equivalent force and couple moment acting at point O.

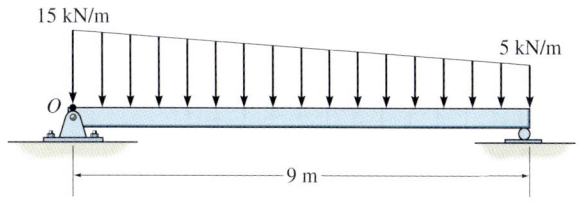

Prob. 4–142

4–143. The masonry support creates the loading distribution acting on the end of the beam. Simplify this load to a single resultant force and specify its location measured from point O.

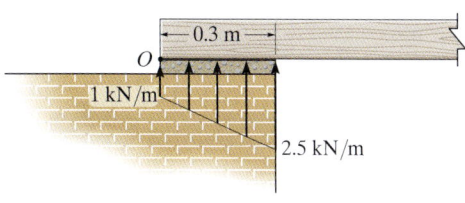

Prob. 4–143

*4–144. Replace the distributed loading by an equivalent resultant force and specify its location, measured from point A.

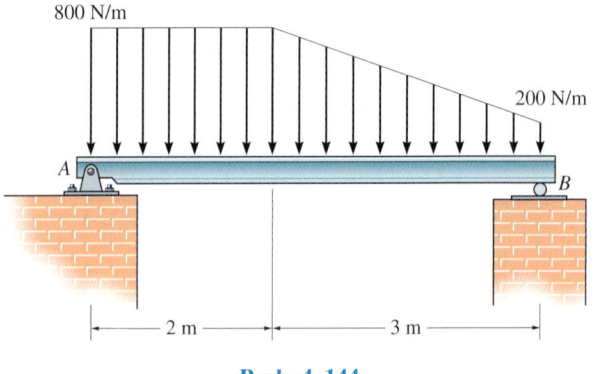

800 N/m

200 N/m

A

B

2 m   3 m

Prob. 4–144

4–145. Replace the distributed loading by an equivalent resultant force, and specify its location on the beam, measured from the pin at C.

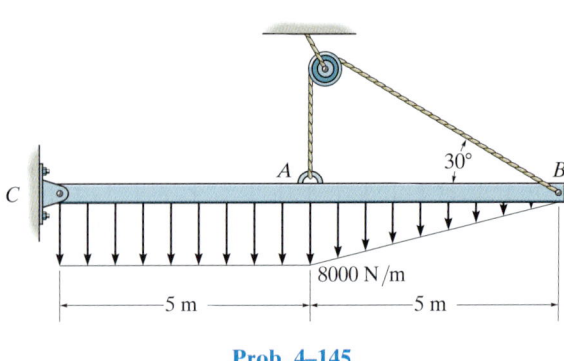

A 30° B

C

8000 N/m

5 m   5 m

Prob. 4–145

4–146. Replace the distributed loading with an equivalent resultant force, and specify its location on the beam measured from point A.

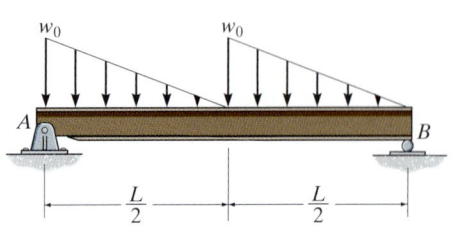

w_0 w_0

A B

$\frac{L}{2}$ $\frac{L}{2}$

Prob. 4–146

4–147. The beam supports the distributed load caused by the sandbags. Determine the resultant force on the beam and specify its location measured from point A.

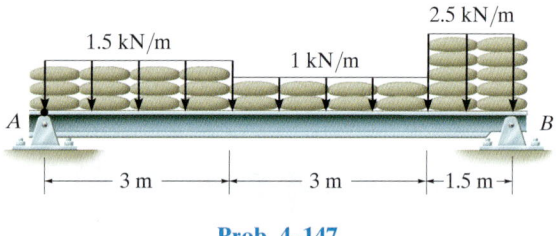

2.5 kN/m

1.5 kN/m   1 kN/m

A B

3 m   3 m   1.5 m

Prob. 4–147

*4–148. If the soil exerts a trapezoidal distribution of load on the bottom of the footing, determine the intensities w_1 and w_2 of this distribution needed to support the column loadings.

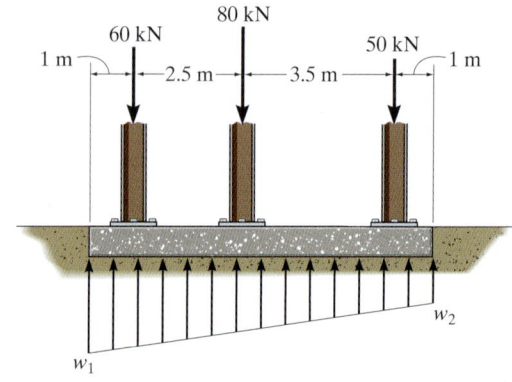

60 kN   80 kN   50 kN

1 m   2.5 m   3.5 m   1 m

w_1 w_2

Prob. 4–148

4–149. Determine the length b of the triangular load and its position a on the beam such that the equivalent resultant force is zero and the resultant couple moment is 8 kN·m clockwise.

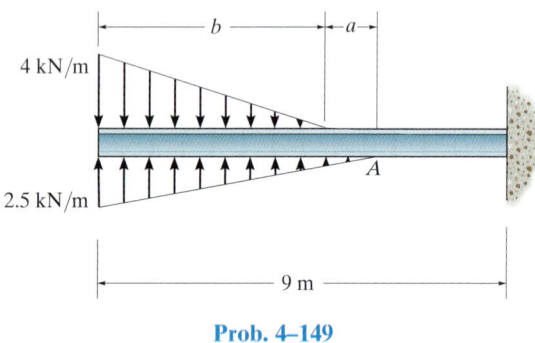

b a

4 kN/m

2.5 kN/m A

9 m

Prob. 4–149

4–150. Replace the loading by an equivalent force and couple moment acting at point O.

4–151. Replace the loading by a single resultant force, and specify the location of the force measured from point O.

4–154. Replace the distributed loading by an equivalent resultant force and specify where its line of action intersects member AB, measured from A.

4–155. Replace the distributed loading by an equivalent resultant force and specify where its line of action intersects member BC, measured from C.

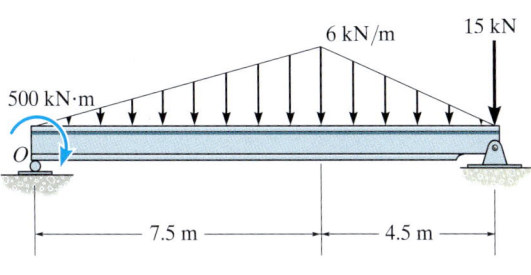

Probs. 4–150/151

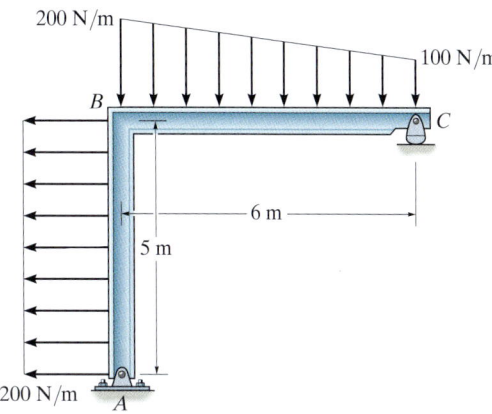

Probs. 4–154/155

***4–152.** Replace the loading by an equivalent resultant force and couple moment at point A.

4–153. Replace the loading by an equivalent resultant force and couple moment acting at point B.

***4–156.** Replace the distributed loading with an equivalent resultant force, and specify its location on the beam measured from point A.

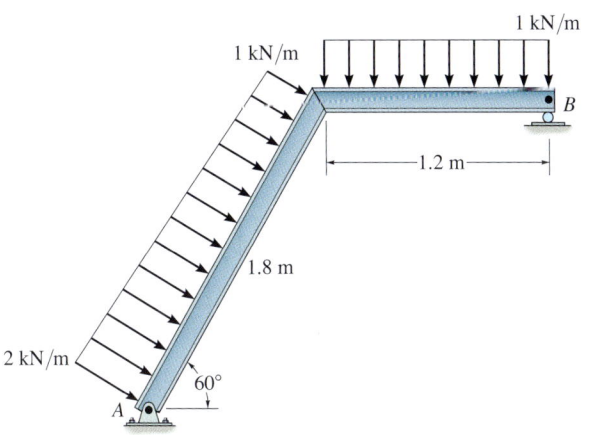

Probs. 4–152/153

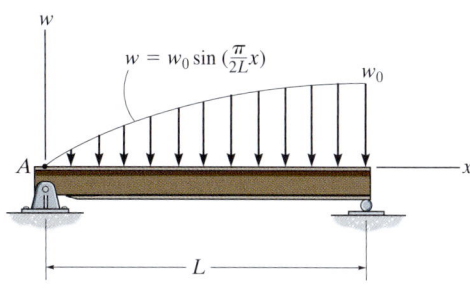

Prob. 4–156

4–157. Replace the distributed loading with an equivalent resultant force, and specify its location on the beam measured from point A.

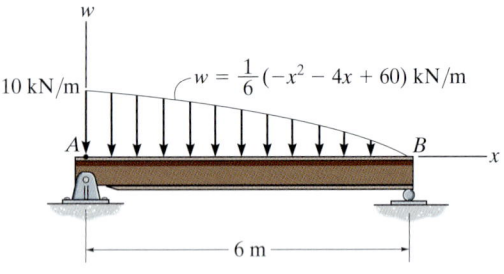

$w = \frac{1}{6}(-x^2 - 4x + 60) \text{ kN/m}$

10 kN/m

6 m

Prob. 4–157

4–158. Determine the equivalent resultant force and couple moment at point O.

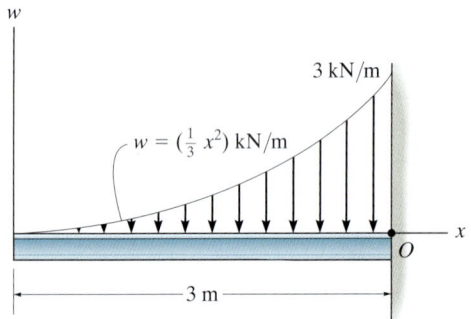

3 kN/m

$w = (\frac{1}{3}x^2) \text{ kN/m}$

3 m

Prob. 4–158

4–159. Wet concrete exerts a pressure distribution along the wall of the form. Determine the resultant force of this distribution and specify the height h where the bracing strut should be placed so that it lies through the line of action of the resultant force. The wall has a width of 5 m.

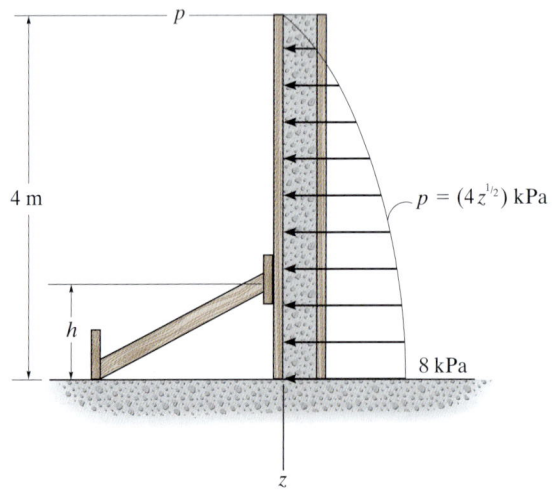

4 m

$p = (4z^{1/2}) \text{ kPa}$

h

8 kPa

Prob. 4–159

*4–160. Replace the loading by an equivalent force and couple moment acting at point O.

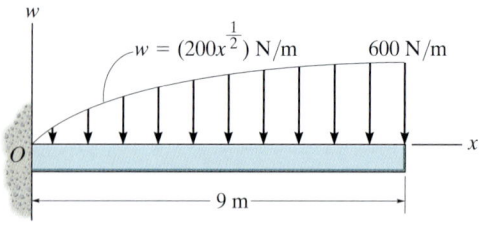

$w = (200x^{\frac{1}{2}}) \text{ N/m}$ 600 N/m

9 m

Prob. 4–160

4–161. Determine the equivalent resultant force of the distributed loading and its location, measured from point A. Evaluate the integrals using Simpson's rule.

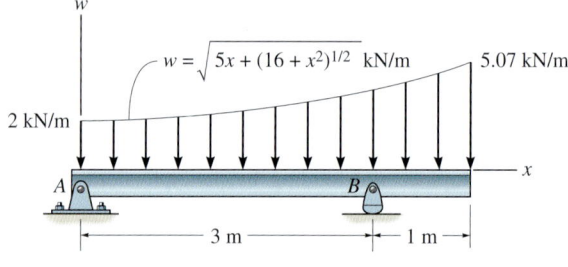

$w = \sqrt{5x + (16 + x^2)^{1/2}} \text{ kN/m}$ 5.07 kN/m

2 kN/m

3 m   1 m

Prob. 4–161

4–162. Determine the equivalent resultant force of the distributed loading and its location, measured from point A. Evaluate the integrals using a numerical method.

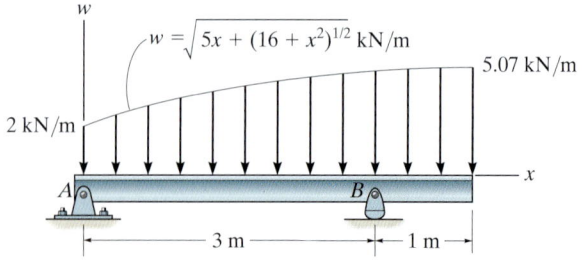

$w = \sqrt{5x + (16 + x^2)^{1/2}} \text{ kN/m}$

5.07 kN/m

2 kN/m

3 m   1 m

Prob. 4–162

CHAPTER REVIEW

Moment of Force—Scalar Definition

A force produces a turning effect or moment about a point O that does not lie on its line of action. In scalar form, the moment *magnitude* is the product of the force and the moment arm or perpendicular distance from point O to the line of action of the force.

$$M_O = Fd$$

The *direction* of the moment is defined using the right-hand rule. $\mathbf{M}_O$ always acts along an axis perpendicular to the plane containing $\mathbf{F}$ and d, and passes through the point O.

Rather than finding d, it is normally easier to resolve the force into its x and y components, determine the moment of each component about the point, and then sum the results. This is called the principle of moments.

$$M_O = Fd = F_x y - F_y x$$

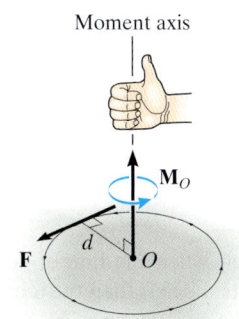

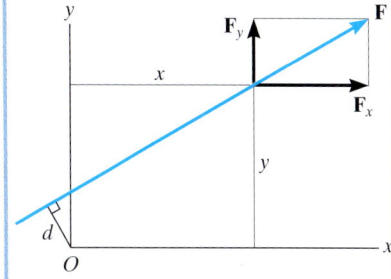

Moment of a Force—Vector Definition

Since three-dimensional geometry is generally more difficult to visualize, the vector cross product should be used to determine the moment. Here $\mathbf{M}_O = \mathbf{r} \times \mathbf{F}$, where $\mathbf{r}$ is a position vector that extends from point O to any point A, B, or C on the line of action of $\mathbf{F}$.

$$\mathbf{M}_O = \mathbf{r}_A \times \mathbf{F} = \mathbf{r}_B \times \mathbf{F} = \mathbf{r}_C \times \mathbf{F}$$

If the position vector $\mathbf{r}$ and force $\mathbf{F}$ are expressed as Cartesian vectors, then the cross product results from the expansion of a determinant.

$$\mathbf{M}_O = \mathbf{r} \times \mathbf{F} = \begin{vmatrix} \mathbf{i} & \mathbf{j} & \mathbf{k} \\ r_x & r_y & r_z \\ F_x & F_y & F_z \end{vmatrix}$$

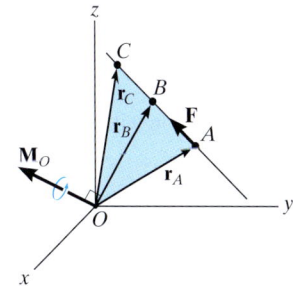

Moment about an Axis

If the moment of a force **F** is to be determined about an arbitrary axis a, then for a scalar solution the moment arm, or shortest distance d_a from the line of action of the force to the axis must be used. This distance is perpendicular to both the axis and the force line of action.

Note that when the line of action of **F** intersects the axis then the moment of **F** about the axis is zero. Also, when the line of action of **F** is parallel to the axis, the moment of **F** about the axis is zero.

In three dimensions, the scalar triple product should be used. Here $\mathbf{u}_a$ is the unit vector that specifies the direction of the axis, and **r** is a position vector that is directed from any point on the axis to any point on the line of action of the force. If M_a is calculated as a negative scalar, then the sense of direction of $\mathbf{M}_a$ is opposite to $\mathbf{u}_a$.

$$M_a = \mathbf{u}_a \cdot (\mathbf{r} \times \mathbf{F}) = \begin{vmatrix} u_{a_x} & u_{a_y} & u_{a_z} \\ r_x & r_y & r_z \\ F_x & F_y & F_z \end{vmatrix}$$

$$M_a = Fd_a$$

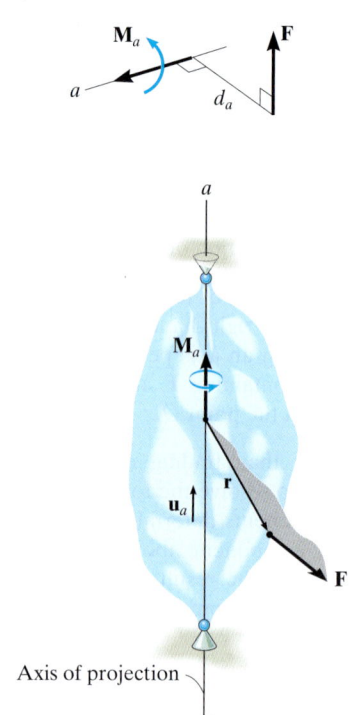

Axis of projection

Couple Moment

A couple consists of two equal but opposite forces that act a perpendicular distance d apart. Couples tend to produce a rotation without translation.

$$M = Fd$$

The magnitude of the couple moment is $M = Fd$, and its direction is established using the right-hand rule.

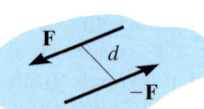

If the vector cross product is used to determine the moment of a couple, then **r** extends from any point on the line of action of one of the forces to any point on the line of action of the other force **F** that is used in the cross product.

$$\mathbf{M} = \mathbf{r} \times \mathbf{F}$$

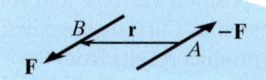

Simplification of a Force and Couple System

Any system of forces and couples can be reduced to a single resultant force and resultant couple moment acting at a point. The resultant force is the sum of all the forces in the system, $\mathbf{F}_R = \Sigma\mathbf{F}$, and the resultant couple moment is equal to the sum of all the moments of the forces about the point and couple moments. $\mathbf{M}_{R_O} = \Sigma\mathbf{M}_O + \Sigma\mathbf{M}$.

Further simplification to a single resultant force is possible provided the force system is concurrent, coplanar, or parallel. To find the location of the resultant force from a point, it is necessary to equate the moment of the resultant force about the point to the moment of the forces and couples in the system about the same point.

If the resultant force and couple moment at a point are not perpendicular to one another, then this system can be reduced to a wrench, which consists of the resultant force and collinear couple moment.

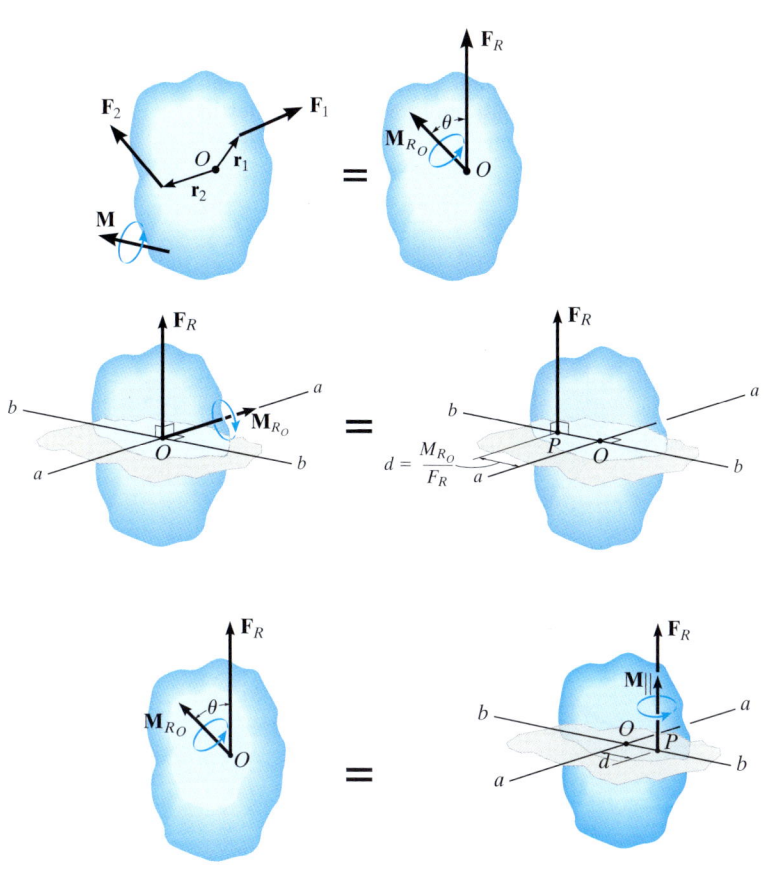

Coplanar Distributed Loading

A simple distributed loading can be represented by its resultant force, which is equivalent to the *area* under the loading curve. This resultant has a line of action that passes through the *centroid* or geometric center of the area or volume under the loading diagram.

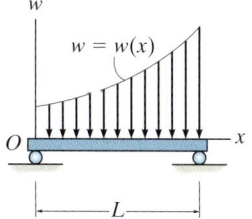

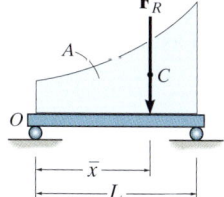

REVIEW PROBLEMS

4–163. Determine the resultant couple moment of the two couples that act on the assembly. Member OB lies in the x–z plane.

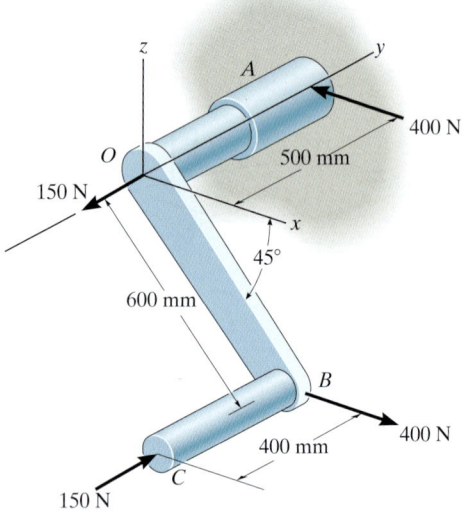

Prob. 4–163

4–166. The forces and couple moments that are exerted on the toe and heel plates of a snow ski are $F_t = \{-50\mathbf{i} + 80\mathbf{j} - 158\mathbf{k}\}$ N, $M_t = \{-6\mathbf{i} + 4\mathbf{j} + 2\mathbf{k}\}$ N·m, and $F_h = \{-20\mathbf{i} + 60\mathbf{j} - 250\mathbf{k}\}$ N, $M_h = \{-20\mathbf{i} + 8\mathbf{j} + 3\mathbf{k}\}$ N·m, respectively. Replace this system by an equivalent force and couple moment acting at point P. Express the results in Cartesian vector form.

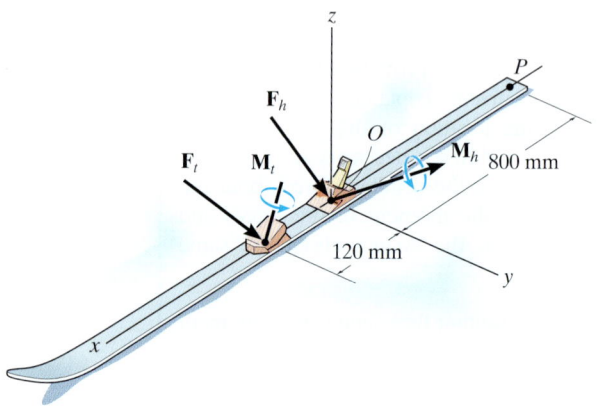

Prob. 4–166

*4–164. The horizontal 30-N force acts on the handle of the wrench. What is the magnitude of the moment of this force about the z axis?

4–165. The horizontal 30-N force acts on the handle of the wrench. Determine the moment of this force about point O. Specify the coordinate direction angles α, β, γ of the moment axis.

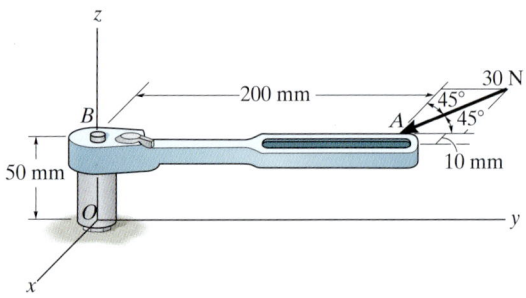

Probs. 4–164/165

4–167. A force of 80 N acts on the handle of the paper cutter at A. Determine the moment created by this force about the hinge at O, if $\theta = 60°$. At what angle θ should the force be applied so that the moment it creates about point O is a maximum (clockwise)? What is this maximum moment?

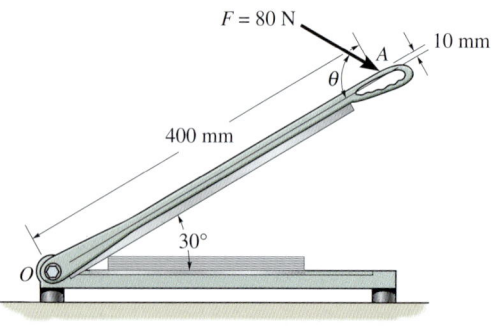

Prob. 4–167

*4–168. Determine the moment of the force at *A* about point *O*. Express the result as a Cartesian vector.

4–169. Determine the moment of the force at *A* about point *P*. Express the result as a Cartesian vector.

*4–172. The ends of the triangular plate are subjected to three couples. Determine the magnitude of the force **F** so that the resultant couple moment is 400 N · m clockwise.

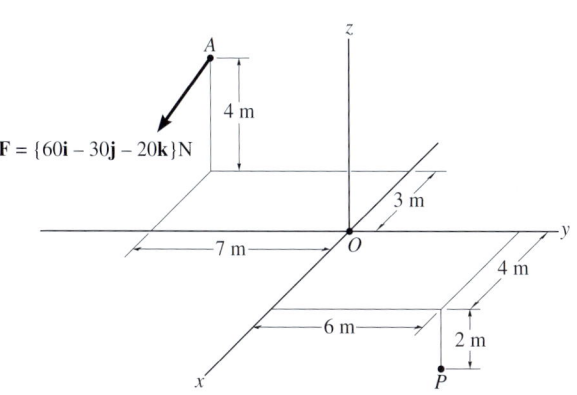

Probs. 4–168/169

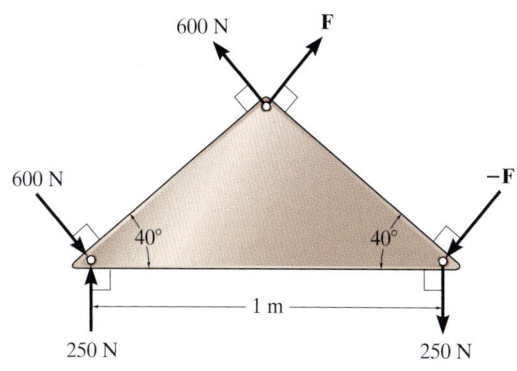

Prob. 4–172

4–170. Determine the moment of the force $\mathbf{F}_c$ about the door hinge at *A*. Express the result as a Cartesian vector.

4–171. Determine the magnitude of the moment of the force $\mathbf{F}_c$ about the hinged axis *aa* of the door.

4–173. The tool is used to shut off gas valves that are difficult to access. If the force **F** is applied to the handle, determine the component of the moment created about the *z* axis of the valve.

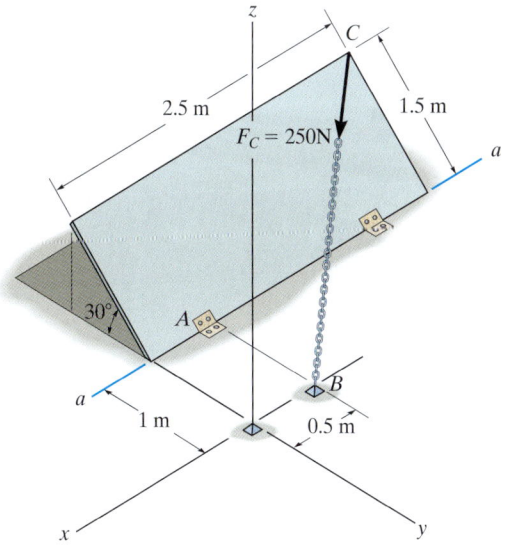

Probs. 4–170/171

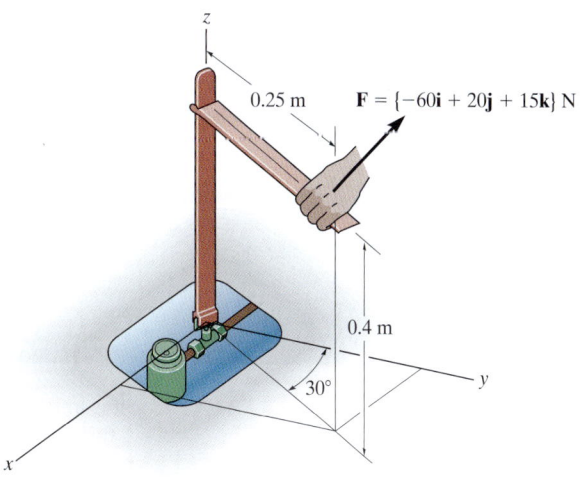

Prob. 4–173

Chapter 5

It is important to be able to determine the forces in the cables used to support this submarine to insure that they do not fail. In this chapter we will study how to apply equilibrium methods to determine the forces acting on the supports of a rigid body such as this.

Equilibrium of a Rigid Body

Video Solutions are available for selected questions in this chapter.

5.1 Conditions for Rigid-Body Equilibrium

In this section, we will develop both the necessary and sufficient conditions for the equilibrium of the rigid body in Fig. 5–1a. As shown, this body is subjected to an external force and couple moment system that is the result of the effects of gravitational, electrical, magnetic, or contact forces caused by adjacent bodies. The internal forces caused by interactions between particles within the body are not shown in this figure because these forces occur in equal but opposite collinear pairs and hence will cancel out, a consequence of Newton's third law.

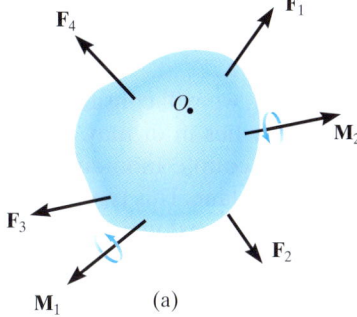

(a)

Fig. 5–1

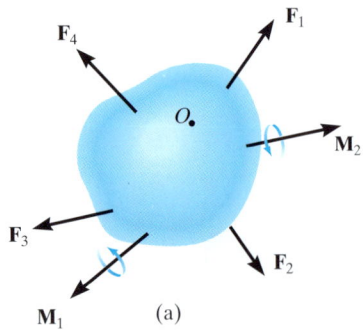

(a)

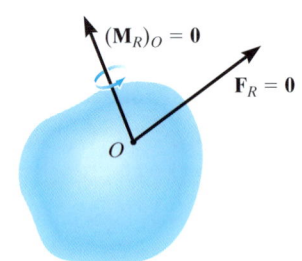

(b)

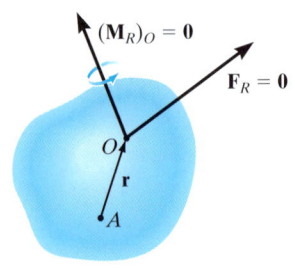

(c)

Fig. 5–1

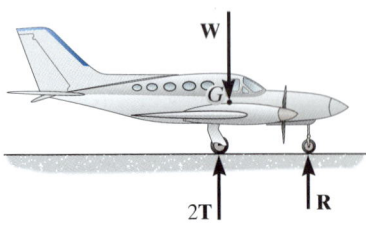

Fig. 5–2

Using the methods of the previous chapter, the force and couple moment system acting on a body can be reduced to an equivalent resultant force and resultant couple moment at any arbitrary point O on or off the body, Fig. 5–1b. If this resultant force and couple moment are both equal to zero, then the body is said to be in *equilibrium*. Mathematically, the equilibrium of a body is expressed as

$$\mathbf{F}_R = \Sigma\mathbf{F} = \mathbf{0}$$
$$(\mathbf{M}_R)_O = \Sigma\mathbf{M}_O = \mathbf{0}$$

(5–1)

The first of these equations states that the sum of the forces acting on the body is equal to *zero*. The second equation states that the sum of the moments of all the forces in the system about point O, added to all the couple moments, is equal to *zero*. These two equations are not only necessary for equilibrium, they are also sufficient. To show this, consider summing moments about some other point, such as point A in Fig. 5–1c. We require

$$\Sigma\mathbf{M}_A = \mathbf{r}\times\mathbf{F}_R + (\mathbf{M}_R)_O = \mathbf{0}$$

Since $\mathbf{r} \neq \mathbf{0}$, this equation is satisfied if Eqs. 5–1 are satisfied, namely $\mathbf{F}_R = \mathbf{0}$ and $(\mathbf{M}_R)_O = \mathbf{0}$.

When applying the equations of equilibrium, we will assume that the body remains rigid. In reality, however, all bodies deform when subjected to loads. Although this is the case, most engineering materials such as steel and concrete are very rigid and so their deformation is usually very small. Therefore, when applying the equations of equilibrium, we can generally assume that the body will remain *rigid* and *not deform* under the applied load without introducing any significant error. This way the direction of the applied forces and their moment arms with respect to a fixed reference remain the same both before and after the body is loaded.

EQUILIBRIUM IN TWO DIMENSIONS

In the first part of the chapter, we will consider the case where the force system acting on a rigid body lies in or may be projected onto a *single* plane and, furthermore, any couple moments acting on the body are directed perpendicular to this plane. This type of force and couple system is often referred to as a two-dimensional or *coplanar* force system. For example, the airplane in Fig. 5–2 has a plane of symmetry through its center axis, and so the loads acting on the airplane are symmetrical with respect to this plane. Thus, each of the two wing tires will support the same load $\mathbf{T}$, which is represented on the side (two-dimensional) view of the plane as 2$\mathbf{T}$.

5.2   Free-Body Diagrams

Successful application of the equations of equilibrium requires a complete specification of *all* the known and unknown external forces that act *on* the body. The best way to account for these forces is to draw a free-body diagram. This diagram is a sketch of the outlined shape of the body, which represents it as being *isolated* or "free" from its surroundings, i.e., a "free body." On this sketch it is necessary to show *all* the forces and couple moments that the surroundings exert *on the body* so that these effects can be accounted for when the equations of equilibrium are applied. *A thorough understanding of how to draw a free-body diagram is of primary importance for solving problems in mechanics.*

Support Reactions.   Before presenting a formal procedure as to how to draw a free-body diagram, we will first consider the various types of reactions that occur at supports and points of contact between bodies subjected to coplanar force systems. As a general rule,

- If a support prevents the translation of a body in a given direction, then a force is developed on the body in that direction.

- If rotation is prevented, a couple moment is exerted on the body.

For example, let us consider three ways in which a horizontal member, such as a beam, is supported at its end. One method consists of a *roller* or cylinder, Fig. 5–3*a*. Since this support only prevents the beam from *translating* in the vertical direction, the roller will only exert a *force* on the beam in this direction, Fig. 5–3*b*.

The beam can be supported in a more restrictive manner by using a *pin*, Fig. 5–3*c*. The pin passes through a hole in the beam and two leaves which are fixed to the ground. Here the pin can prevent *translation* of the beam in *any direction* ϕ, Fig. 5–3*d*, and so the pin must exert a *force* **F** on the beam in this direction. For purposes of analysis, it is generally easier to represent this resultant force **F** by its two rectangular components $\mathbf{F}_x$ and $\mathbf{F}_y$, Fig. 5–3*e*. If F_x and F_y are known, then F and ϕ can be calculated.

The most restrictive way to support the beam would be to use a *fixed support* as shown in Fig. 5–3*f*. This support will prevent both *translation and rotation* of the beam. To do this a *force and couple moment* must be developed on the beam at its point of connection, Fig. 5–3*g*. As in the case of the pin, the force is usually represented by its rectangular components $\mathbf{F}_x$ and $\mathbf{F}_y$.

Table 5–1 lists other common types of supports for bodies subjected to coplanar force systems. (In all cases the angle θ is assumed to be known.) Carefully study each of the symbols used to represent these supports and the types of reactions they exert on their contacting members.

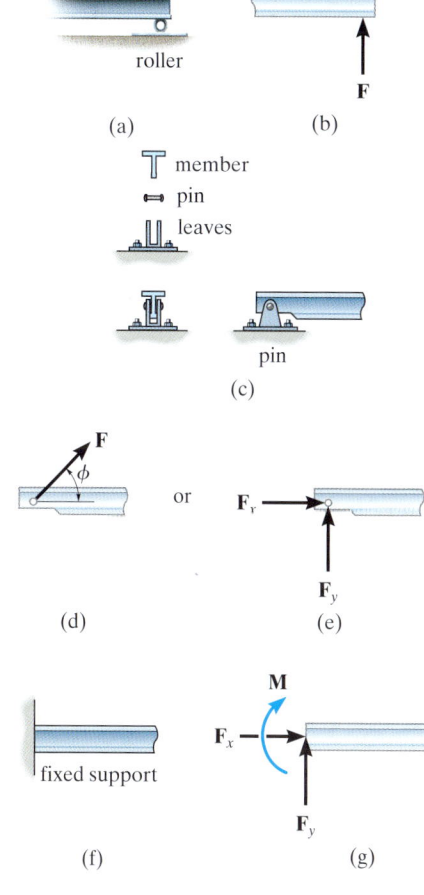

Fig. 5–3

TABLE 5–1 Supports for Rigid Bodies Subjected to Two-Dimensional Force Systems

Types of Connection	Reaction	Number of Unknowns
(1) cable		One unknown. The reaction is a tension force which acts away from the member in the direction of the cable.
(2) weightless link	or	One unknown. The reaction is a force which acts along the axis of the link.
(3) roller		One unknown. The reaction is a force which acts perpendicular to the surface at the point of contact.
(4) rocker		One unknown. The reaction is a force which acts perpendicular to the surface at the point of contact.
(5) smooth contacting surface		One unknown. The reaction is a force which acts perpendicular to the surface at the point of contact.
(6) roller or pin in confined smooth slot	or	One unknown. The reaction is a force which acts perpendicular to the slot.
(7) member pin connected to collar on smooth rod	or	One unknown. The reaction is a force which acts perpendicular to the rod.

continued

TABLE 5–1 Continued

Types of Connection	Reaction	Number of Unknowns
(8) 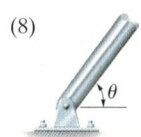 smooth pin or hinge	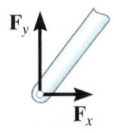 or 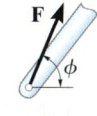	Two unknowns. The reactions are two components of force, or the magnitude and direction ϕ of the resultant force. Note that ϕ and θ are not necessarily equal [usually not, unless the rod shown is a link as in (2)].
(9) 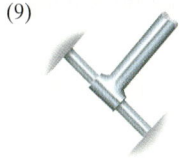 member fixed connected to collar on smooth rod	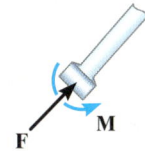	Two unknowns. The reactions are the couple moment and the force which acts perpendicular to the rod.
(10)  fixed support	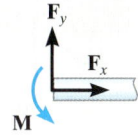 or 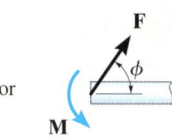	Three unknowns. The reactions are the couple moment and the two force components, or the couple moment and the magnitude and direction ϕ of the resultant force.

5

Typical examples of actual supports are shown in the following sequence of photos. The numbers refer to the connection types in Table 5–1.

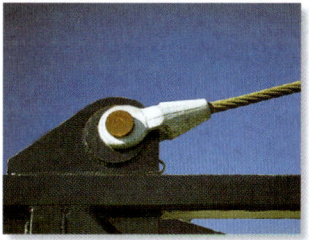

The cable exerts a force on the bracket in the direction of the cable. (1)

The rocker support for this bridge girder allows horizontal movement so the bridge is free to expand and contract due to a change in temperature. (4)

This concrete girder rests on the ledge that is assumed to act as a smooth contacting surface. (5)

This utility building is pin supported at the top of the column. (8)

The floor beams of this building are welded together and thus form fixed connections. (10)

Internal Forces.   As stated in Sec. 5.1, the internal forces that act between adjacent particles in a body always occur in collinear pairs such that they have the same magnitude and act in opposite directions (Newton's third law). Since these forces cancel each other, they will not create an *external effect* on the body. It is for this reason that the internal forces should not be included on the free-body diagram if the entire body is to be considered. For example, the engine shown in Fig. 5–4a has a free-body diagram shown in Fig. 5–4b. The internal forces between all its connected parts, such as the screws and bolts, will cancel out because they form equal and opposite collinear pairs. Only the external forces $\mathbf{T}_1$ and $\mathbf{T}_2$, exerted by the chains and the engine weight $\mathbf{W}$, are shown on the free-body diagram.

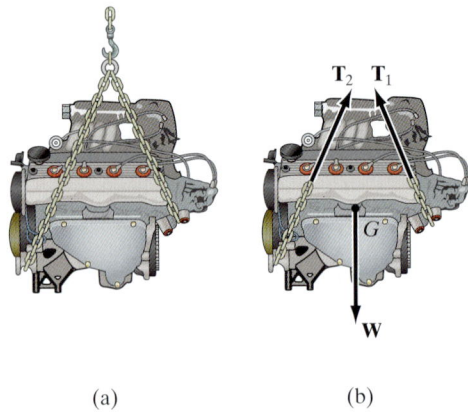

(a)   (b)

Fig. 5–4

Weight and the Center of Gravity.   When a body is within a gravitational field, then each of its particles has a specified weight. It was shown in Sec. 4.8 that such a system of forces can be reduced to a single resultant force acting through a specified point. We refer to this force resultant as the *weight* $\mathbf{W}$ of the body and to the location of its point of application as the *center of gravity*. The methods used for its determination will be developed in Chapter 9.

   In the examples and problems that follow, if the weight of the body is important for the analysis, this force will be reported in the problem statement. Also, when the body is *uniform* or made from the same material, the center of gravity will be located at the body's *geometric center* or *centroid*; however, if the body consists of a nonuniform distribution of material, or has an unusual shape, then the location of its center of gravity G will be given.

Idealized Models.   When an engineer performs a force analysis of any object, he or she considers a corresponding analytical or idealized model that gives results that approximate as closely as possible the actual situation. To do this, careful choices have to be made so that selection of the type of supports, the material behavior, and the object's dimensions can be justified. This way one can feel confident that any design or analysis will yield results which can be trusted. In complex cases this process may require developing several different models of the object that must be analyzed. In any case, this selection process requires both skill and experience.

The following two cases illustrate what is required to develop a proper model. In Fig. 5–5a, the steel beam is to be used to support the three roof joists of a building. For a force analysis it is reasonable to assume the material (steel) is rigid since only very small deflections will occur when the beam is loaded. A bolted connection at A will allow for any slight rotation that occurs here when the load is applied, and so a *pin* can be considered for this support. At B a *roller* can be considered since this support offers no resistance to horizontal movement. Building code is used to specify the roof loading A so that the joist loads **F** can be calculated. These forces will be larger than any actual loading on the beam since they account for extreme loading cases and for dynamic or vibrational effects. Finally, the weight of the beam is generally neglected when it is small compared to the load the beam supports. The idealized model of the beam is therefore shown with average dimensions a, b, c, and d in Fig. 5–5b.

As a second case, consider the lift boom in Fig. 5–6a. By inspection, it is supported by a pin at A and by the hydraulic cylinder BC, which can be approximated as a weightless link. The material can be assumed rigid, and with its density known, the weight of the boom and the location of its center of gravity G are determined. When a design loading **P** is specified, the idealized model shown in Fig. 5–6b can be used for a force analysis. Average dimensions (not shown) are used to specify the location of the loads and the supports.

Idealized models of specific objects will be given in some of the examples throughout the text. It should be realized, however, that each case represents the reduction of a practical situation using simplifying assumptions like the ones illustrated here.

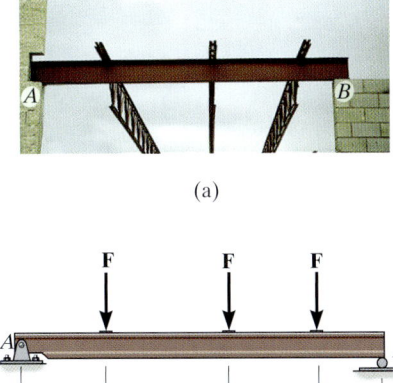

(a)

(b)

Fig. 5–5

5

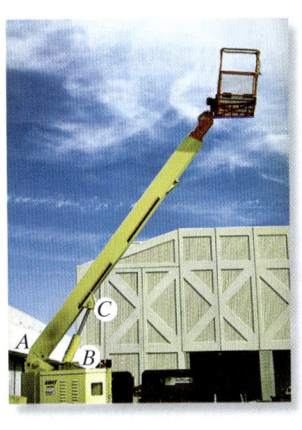

(a)

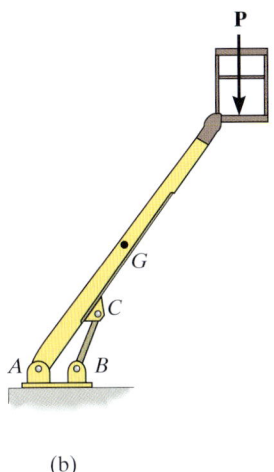

(b)

Fig. 5–6

Procedure for Analysis

To construct a free-body diagram for a rigid body or any group of bodies considered as a single system, the following steps should be performed:

Draw Outlined Shape.

Imagine the body to be *isolated* or cut "free" from its constraints and connections and draw (sketch) its outlined shape.

Show All Forces and Couple Moments.

Identify all the known and unknown *external forces* and couple moments that *act on the body*. Those generally encountered are due to (1) applied loadings, (2) reactions occurring at the supports or at points of contact with other bodies (see Table 5–1), and (3) the weight of the body. To account for all these effects, it may help to trace over the boundary, carefully noting each force or couple moment acting on it.

Identify Each Loading and Give Dimensions.

The forces and couple moments that are known should be labeled with their proper magnitudes and directions. Letters are used to represent the magnitudes and direction angles of forces and couple moments that are unknown. Establish an *x, y* coordinate system so that these unknowns, A_x, A_y, etc., can be identified. Finally, indicate the dimensions of the body necessary for calculating the moments of forces.

Important Points

- No equilibrium problem should be solved without *first drawing the free-body diagram*, so as to account for all the forces and couple moments that act on the body.

- If a support *prevents translation* of a body in a particular direction, then the support exerts a *force* on the body in that direction.

- If *rotation is prevented*, then the support exerts a *couple moment* on the body.

- Study Table 5–1.

- Internal forces are never shown on the free-body diagram since they occur in equal but opposite collinear pairs and therefore cancel out.

- The weight of a body is an external force, and its effect is represented by a single resultant force acting through the body's center of gravity *G*.

- *Couple moments* can be placed anywhere on the free-body diagram since they are *free vectors. Forces* can act at any point along their lines of action since they are *sliding vectors*.

EXAMPLE 5.1

Draw the free-body diagram of the uniform beam shown in Fig. 5–7a. The beam has a mass of 100 kg.

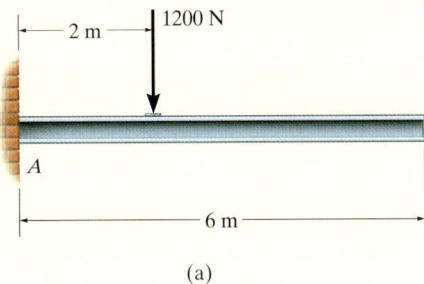

(a)

5

SOLUTION

The free-body diagram of the beam is shown in Fig. 5–7b. Since the support at A is fixed, the wall exerts three reactions *on the beam*, denoted as $\mathbf{A}_x$, $\mathbf{A}_y$, and $\mathbf{M}_A$. The magnitudes of these reactions are *unknown*, and their sense has been *assumed*. The weight of the beam, $W = 100(9.81)\ \text{N} = 981\ \text{N}$, acts through the beam's center of gravity G, which is 3 m from A since the beam is uniform.

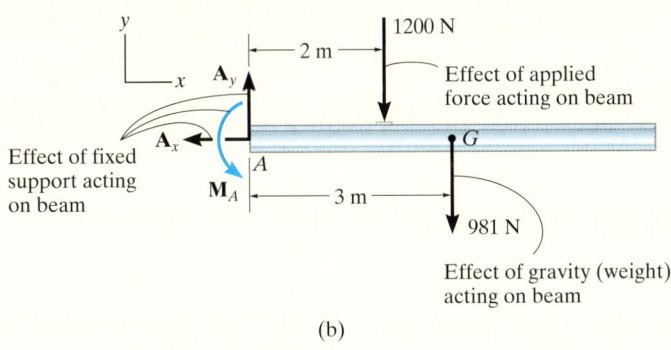

(b)

Fig. 5–7

EXAMPLE 5.2

Draw the free-body diagram of the foot lever shown in Fig. 5–8*a*. The operator applies a vertical force to the pedal so that the spring is stretched 36 mm and the force on the link at *B* is 100 N.

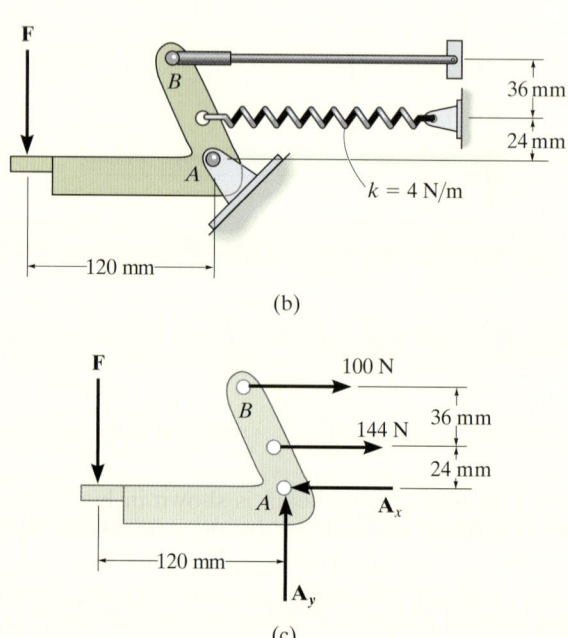

(b)

(c)

(a)

Fig. 5–8

SOLUTION

By inspection of the photo the lever is loosely bolted to the frame at *A* and so this bolt acts as a pin. (See (8) in Table 5–1.) Although not shown here the link at *B* is pinned at both ends and so it is like (2) in Table 5-1. After making the proper measurements, the idealized model of the lever is shown in Fig. 5–8*b*. From this, the free-body diagram is shown in Fig. 5–8*c*. The pin at *A* exerts force components A_x and A_y on the lever. The link exerts a force of 100 N, acting in the direction of the link. In addition the spring also exerts a horizontal force on the lever. If the stiffness is measured and found to be $k = 4 \text{ N/mm}$, then since the stretch $s = 36$ mm, using Eq. 3–2, $F_s = ks = 4 \text{ N/mm} (36 \text{ mm}) = 144$ N. Finally, the operator's shoe applies a vertical force of **F** on the pedal. The dimensions of the lever are also shown on the free-body diagram, since this information will be useful when calculating the moments of the forces. As usual, the senses of the unknown forces at *A* have been assumed. The correct senses will become apparent after solving the equilibrium equations.

EXAMPLE 5.3

Two smooth pipes, each having a mass of 300 kg, are supported by the forked tines of the tractor in Fig. 5–9a. Draw the free-body diagrams for each pipe and both pipes together.

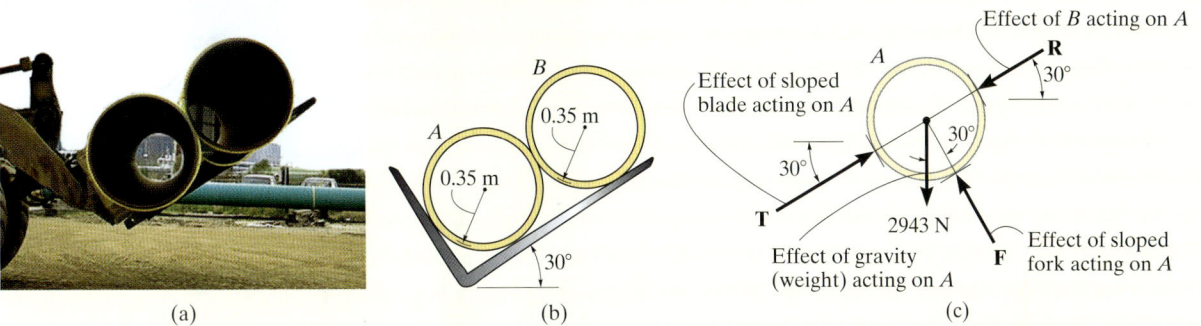

(a)   (b)   (c)

Effect of *B* acting on *A*

A

Effect of sloped blade acting on *A*

T

2943 N

Effect of gravity (weight) acting on *A*

R 30°

F Effect of sloped fork acting on *A*

SOLUTION

The idealized model from which we must draw the free-body diagrams is shown in Fig. 5–9b. Here the pipes are identified, the dimensions have been added, and the physical situation reduced to its simplest form.

The free-body diagram for pipe *A* is shown in Fig. 5–9c. Its weight is $W = 300(9.81)$ N $= 2943$ N. Assuming all contacting surfaces are *smooth*, the reactive forces **T**, **F**, **R** act in a direction *normal* to the tangent at their surfaces of contact.

The free-body diagram of pipe *B* is shown in Fig. 5–9d. Can you identify each of the three forces acting *on this pipe*? In particular, note that **R**, representing the force of *A* on *B*, Fig. 5–9d, is equal and opposite to **R** representing the force of *B* on *A*, Fig. 5–9c. This is a consequence of Newton's third law of motion.

The free-body diagram of both pipes combined ("system") is shown in Fig. 5–9e. Here the contact force **R**, which acts between *A* and *B*, is considered as an *internal* force and hence is not shown on the free-body diagram. That is, it represents a pair of equal but opposite collinear forces which cancel each other.

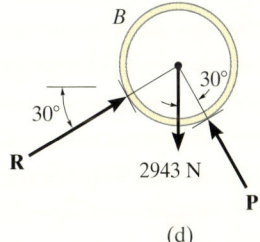

(d)

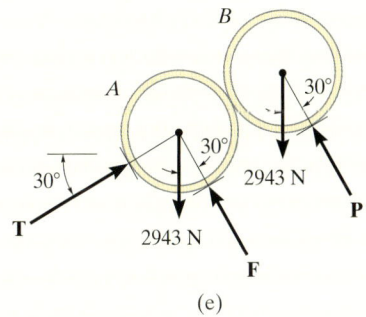

(e)

Fig. 5–9

EXAMPLE 5.4

Draw the free-body diagram of the unloaded platform that is suspended off the edge of the oil rig shown in Fig. 5–10a. The platform has a mass of 200 kg.

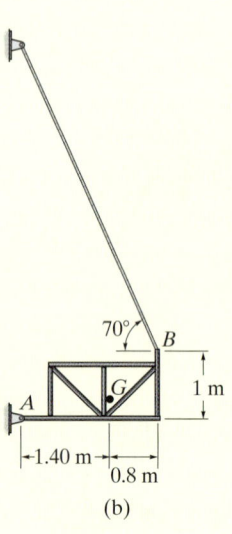

(b)

(a)

Fig. 5–10

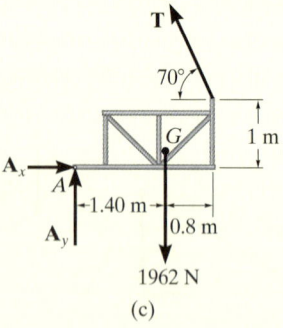

(c)

SOLUTION

The idealized model of the platform will be considered in two dimensions because by observation the loading and the dimensions are all symmetrical about a vertical plane passing through its center, Fig. 5–10b. The connection at A is considered to be a pin, and the cable supports the platform at B. The direction of the cable and average dimensions of the platform are listed, and the center of gravity G has been determined. It is from this model that we have drawn the free-body diagram shown in Fig. 5–10c. The platform's weight is $200(9.81) = 1962$ N. The force components $\mathbf{A}_x$ and $\mathbf{A}_y$ along with the cable force $\mathbf{T}$ represent the reactions that *both* pins and *both* cables exert on the platform, Fig. 5–10a. As a result, half their magnitudes are developed on each side of the platform.

PROBLEMS

5–1.   Draw the free-body diagram of the dumpster D of the truck, which has a mass of 2.5 Mg and a center of gravity at G. It is supported by a pin at A and a pin-connected hydraulic cylinder BC (short link). Explain the significance of each force on the diagram. (See Fig. 5–7b.)

5–3.   Draw the free-body diagram of the beam which supports the 80-kg load and is supported by the pin at A and a cable which wraps around the pulley at D. Explain the significance of each force on the diagram. (See Fig. 5–7b.)

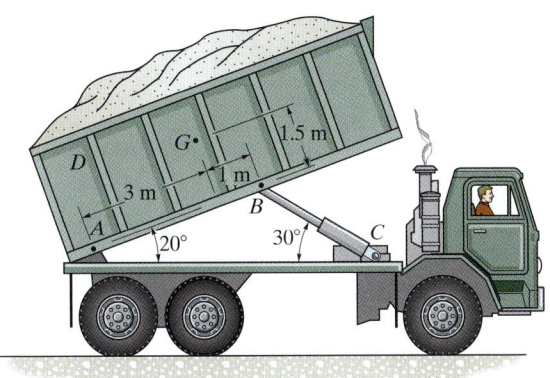

Prob. 5–1

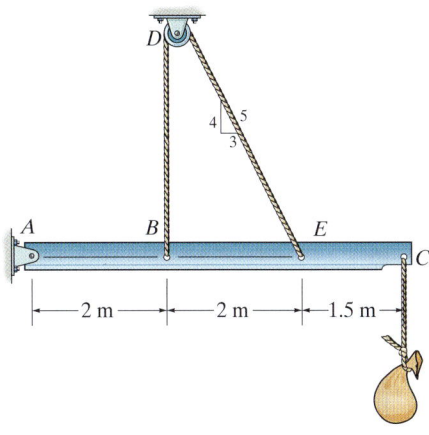

Prob. 5–3

5–2.   Draw the free-body diagram of member ABC which is supported by a smooth collar at A, rocker at B, and short link CD. Explain the significance of each force acting on the diagram. (See Fig. 5–7b.)

*5–4.   Draw the free-body diagram of the hand punch, which is pinned at A and bears down on the smooth surface at B.

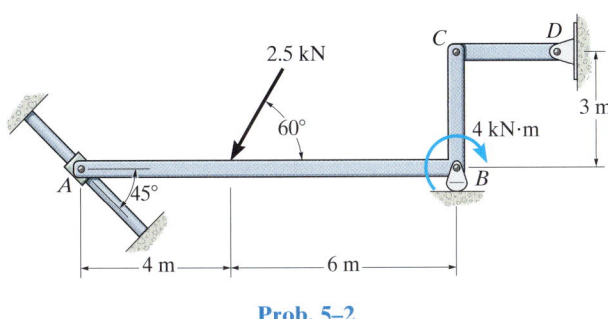

Prob. 5–2

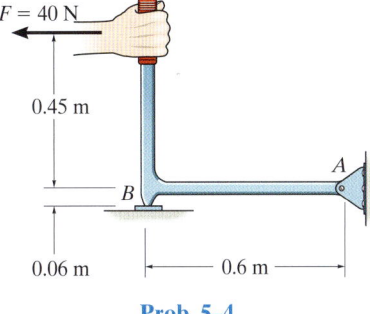

Prob. 5–4

5–5. Draw the free-body diagram of the uniform bar, which has a mass of 100 kg and a center of mass at G. The supports A, B, and C are smooth.

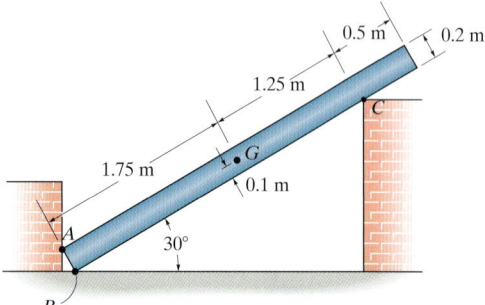

Prob. 5–5

5–6. Draw the free-body diagram of the jib crane AB, which is pin-connected at A and supported by member (link) BC.

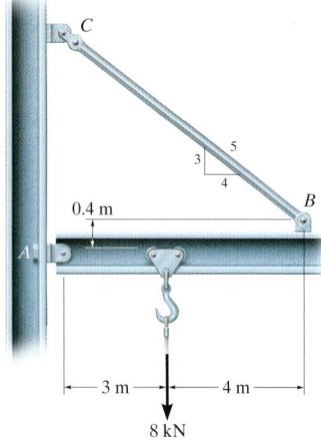

Prob. 5–6

5–7. Draw the free-body diagram of the beam, which is pin connected at A and rocker-supported at B.

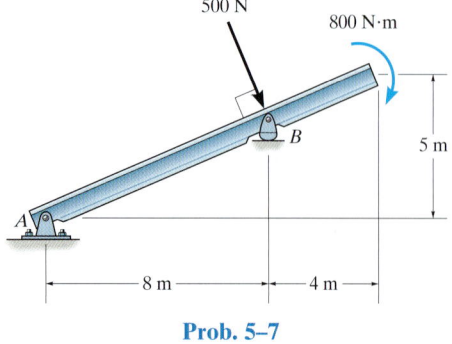

Prob. 5–7

*5–8. Draw the free-body diagram of the truss that is supported by the cable AB and pin C. Explain the significance of each force acting on the diagram. (See Fig. 5–7b.)

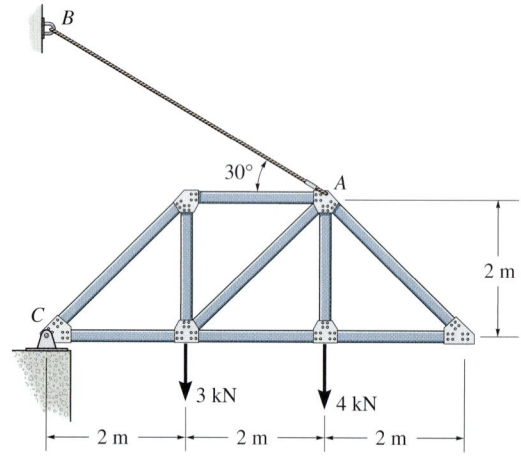

Prob. 5–8

5–9. Draw the free-body diagram of the jib crane AB, which is pin connected at A and supported by member (link) BC.

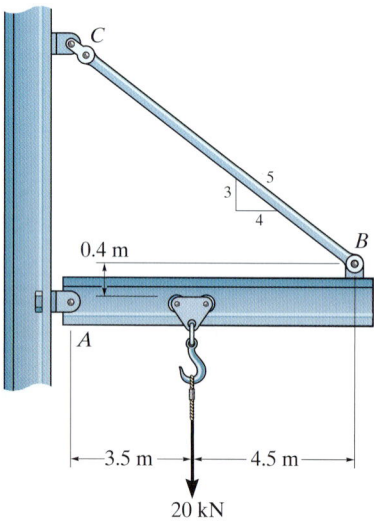

Prob. 5–9

CONCEPTUAL PROBLEMS

P5–1. Draw the free-body diagram of the uniform trash bucket which has a significant weight. It is pinned at *A* and rests against the smooth horizontal member at *B*. Show your result in side view. Label any necessary dimensions.

P5–1

P5–2. Draw the free-body diagram of the outrigger *ABC* used to support a backhoe. The pin *B* is connected to the hydraulic cylinder, which can be considered a short link (two-force member), the bearing shoe at *A* is smooth, and the outrigger is pinned to the frame at *C*.

P5–2

P5–3. Draw the free-body diagram of the wing on the passenger plane. The weights of the engine and wing are significant. The tires at *B* are smooth.

P5–3

P5–4. Draw the free-body diagrams of the wheel and member *ABC* used as part of the landing gear on a jet plane. The hydraulic cylinder *AD* acts as a two-force member, and there is a pin connection at *B*.

P5–4

Please refer to the Companion Website for the animation: *Equilibrium of a Free Body*

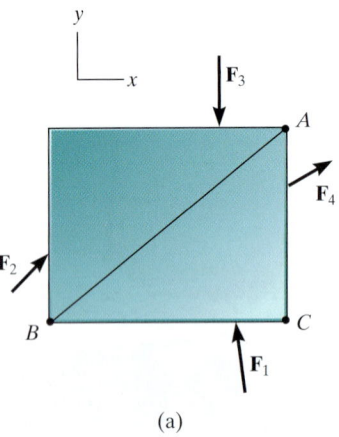

(a)

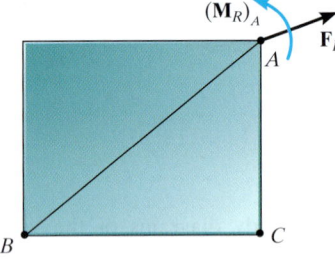

(b)

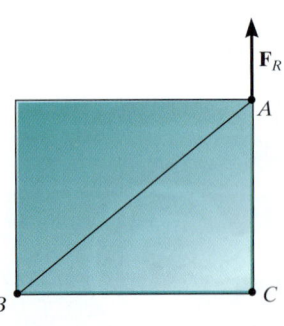

(c)

Fig. 5–11

5.3   Equations of Equilibrium

In Sec. 5.1 we developed the two equations which are both necessary and sufficient for the equilibrium of a rigid body, namely, $\Sigma\mathbf{F} = \mathbf{0}$ and $\Sigma\mathbf{M}_O = \mathbf{0}$. When the body is subjected to a system of forces, which all lie in the x–y plane, then the forces can be resolved into their x and y components. Consequently, the conditions for equilibrium in two dimensions are

$$\begin{aligned} \Sigma F_x &= 0 \\ \Sigma F_y &= 0 \\ \Sigma M_O &= 0 \end{aligned} \qquad (5\text{–}2)$$

Here ΣF_x and ΣF_y represent, respectively, the algebraic sums of the x and y components of all the forces acting on the body, and ΣM_O represents the algebraic sum of the couple moments and the moments of all the force components about the z axis, which is perpendicular to the x–y plane and passes through the arbitrary point O.

Alternative Sets of Equilibrium Equations.

Although Eqs. 5–2 are *most often* used for solving coplanar equilibrium problems, two *alternative* sets of three independent equilibrium equations may also be used. One such set is

$$\begin{aligned} \Sigma F_x &= 0 \\ \Sigma M_A &= 0 \\ \Sigma M_B &= 0 \end{aligned} \qquad (5\text{–}3)$$

When using these equations it is required that a line passing through points A and B is *not parallel* to the y axis. To prove that Eqs. 5–3 provide the *conditions* for equilibrium, consider the free-body diagram of the plate shown in Fig. 5–11a. Using the methods of Sec. 4.7, all the forces on the free-body diagram may be replaced by an equivalent resultant force $\mathbf{F}_R = \Sigma\mathbf{F}$, acting at point A, and a resultant couple moment $\left(\mathbf{M}_R\right)_A = \Sigma\mathbf{M}_A$, Fig. 5–11b. If $\Sigma M_A = 0$ is satisfied, it is necessary that $\left(\mathbf{M}_R\right)_A = \mathbf{0}$. Furthermore, in order that $\mathbf{F}_R$ satisfy $\Sigma F_x = 0$, it must have *no component* along the x axis, and therefore $\mathbf{F}_R$ must be parallel to the y axis, Fig. 5–11c. Finally, if it is required that $\Sigma M_B = 0$, where B does not lie on the line of action of $\mathbf{F}_R$, then $\mathbf{F}_R = \mathbf{0}$. Since Eqs. 5–3 show that both of these resultants are zero, indeed the body in Fig. 5–11a must be in equilibrium.

A second alternative set of equilibrium equations is

$$\Sigma M_A = 0$$
$$\Sigma M_B = 0 \qquad\qquad (5\text{--}4)$$
$$\Sigma M_C = 0$$

Here it is necessary that points A, B, and C do not lie on the same line. To prove that these equations, when satisfied, ensure equilibrium, consider again the free-body diagram in Fig. 5–11b. If $\Sigma M_A = 0$ is to be satisfied, then $(\mathbf{M}_R)_A = \mathbf{0}$. $\Sigma M_C = 0$ is satisfied if the line of action of $\mathbf{F}_R$ passes through point C as shown in Fig. 5–11c. Finally, if we require $\Sigma M_B = 0$, it is necessary that $\mathbf{F}_R = \mathbf{0}$, and so the plate in Fig. 5–11a must then be in equilibrium.

Procedure for Analysis

Coplanar force equilibrium problems for a rigid body can be solved using the following procedure.

Free-Body Diagram.

- Establish the x, y coordinate axes in any suitable orientation.

- Draw an outlined shape of the body.

- Show all the forces and couple moments acting on the body.

- Label all the loadings and specify their directions relative to the x or y axis. The sense of a force or couple moment having an *unknown* magnitude but known line of action can be *assumed*.

- Indicate the dimensions of the body necessary for computing the moments of forces.

Equations of Equilibrium.

- Apply the moment equation of equilibrium, $\Sigma M_O = 0$, about a point (O) that lies at the intersection of the lines of action of two unknown forces. In this way, the moments of these unknowns are zero about O, and a *direct solution* for the third unknown can be determined.

- When applying the force equilibrium equations, $\Sigma F_x = 0$ and $\Sigma F_y = 0$, orient the x and y axes along lines that will provide the simplest resolution of the forces into their x and y components.

- If the solution of the equilibrium equations yields a negative scalar for a force or couple moment magnitude, this indicates that the sense is opposite to that which was assumed on the free-body diagram.

EXAMPLE 5.5

Determine the horizontal and vertical components of reaction on the beam caused by the pin at B and the rocker at A as shown in Fig. 5–12a. Neglect the weight of the beam.

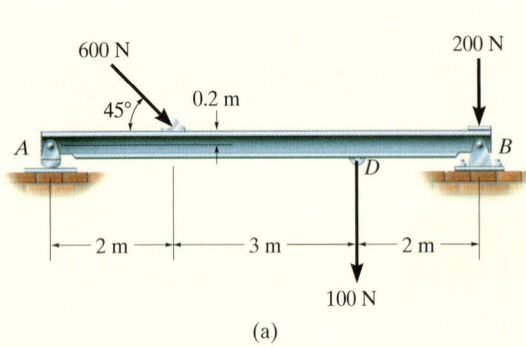

(a)

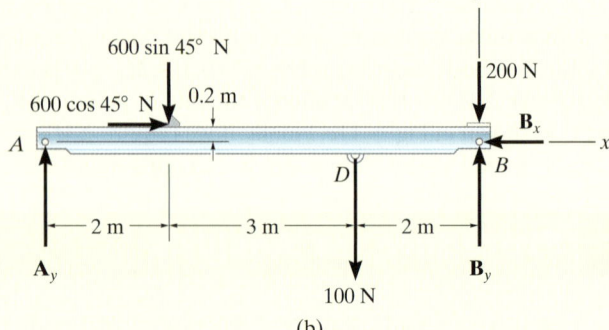

(b)

Fig. 5–12

(c)

SOLUTION

Free-Body Diagram. Identify each of the forces shown on the free-body diagram of the beam, Fig. 5–12b. (See Example 5.1.) For simplicity, the 600-N force is represented by its x and y components as shown in Fig. 5–12b.

Equations of Equilibrium. Summing forces in the x direction yields

$$\xrightarrow{+} \Sigma F_x = 0; \qquad 600 \cos 45° \text{ N} - B_x = 0$$

$$B_x = 424 \text{ N} \qquad\qquad Ans.$$

A direct solution for A_y can be obtained by applying the moment equation $\Sigma M_B = 0$ about point B.

$$\zeta + \Sigma M_B = 0; \qquad 100 \text{ N}(2 \text{ m}) + (600 \sin 45° \text{ N})(5 \text{ m})$$

$$- (600 \cos 45° \text{ N})(0.2 \text{ m}) - A_y(7 \text{ m}) = 0$$

$$A_y = 319 \text{ N} \qquad\qquad Ans.$$

Summing forces in the y direction, using this result, gives

$$+\uparrow \Sigma F_y = 0; \qquad 319 \text{ N} - 600 \sin 45° \text{ N} - 100 \text{ N} - 200 \text{ N} + B_y = 0$$

$$B_y = 405 \text{ N} \qquad\qquad Ans.$$

NOTE: Remember, the support forces in Fig. 5–12b are the result of pins that *act on the beam*. The opposite forces act on the pins. For example, Fig. 5–12c shows the equilibrium of the pin at A and the rocker.

EXAMPLE 5.6

The cord shown in Fig. 5–13a supports a force of 500 N and wraps over the frictionless pulley. Determine the tension in the cord at C and the horizontal and vertical components of reaction at pin A.

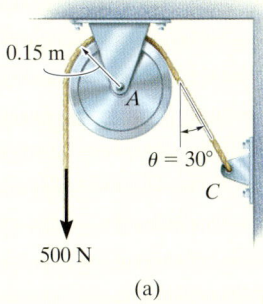

(a)

Fig. 5–13

SOLUTION

Free-Body Diagrams. The free-body diagrams of the cord and pulley are shown in Fig. 5–13b. Note that the principle of action, equal but opposite reaction must be carefully observed when drawing each of these diagrams: the cord exerts an unknown load distribution p on the pulley at the contact surface, whereas the pulley exerts an equal but opposite effect on the cord. For the solution, however, it is simpler to *combine* the free-body diagrams of the pulley and this portion of the cord, so that the distributed load becomes *internal* to this "system" and is therefore eliminated from the analysis, Fig. 5–13c.

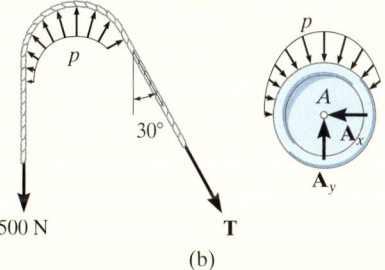

(b)

Equations of Equilibrium. Summing moments about point A to eliminate $\mathbf{A}_x$ and $\mathbf{A}_y$, Fig. 5–13c, we have

$$\zeta + \Sigma M_A = 0; \qquad 500 \text{ N} (0.15 \text{ m}) - T(0.15 \text{ m}) = 0$$
$$T = 500 \text{ N} \qquad\qquad \textit{Ans.}$$

Using this result,

$$\xrightarrow{+} \Sigma F_x = 0; \qquad -A_x + 500 \sin 30° \text{ N} = 0$$
$$A_x = 250 \text{ N} \qquad\qquad \textit{Ans.}$$

$$+\uparrow \Sigma F_y = 0; \qquad A_y - 500 \text{ N} - 500 \cos 30° \text{ N} = 0$$
$$A_y = 933 \text{ N} \qquad\qquad \textit{Ans.}$$

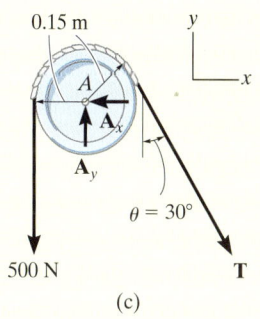

(c)

NOTE: From the moment equation, it is seen that the tension remains *constant* as the cord passes over the pulley. (This of course is true for *any* angle θ at which the cord is directed and for *any radius r* of the pulley.)

EXAMPLE 5.7

The member shown in Fig. 5–14a is pin connected at A and rests against a smooth support at B. Determine the horizontal and vertical components of reaction at the pin A.

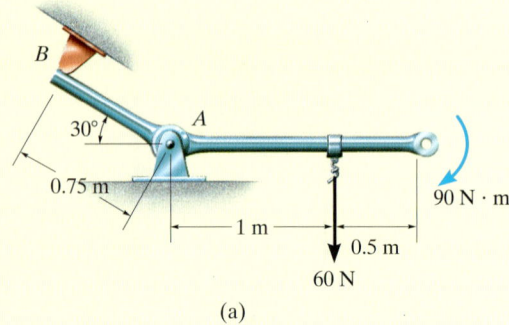

(a)

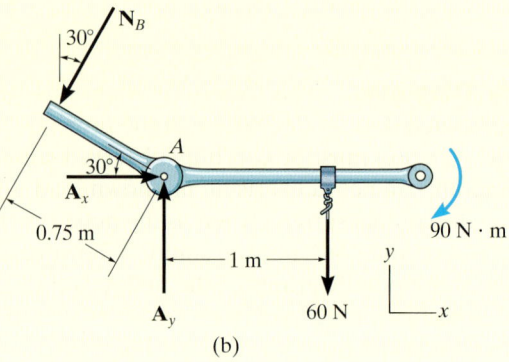

(b)

Fig. 5–14

SOLUTION

Free-Body Diagram. As shown in Fig. 5–14b, the reaction N_B is perpendicular to the member at B. Also, horizontal and vertical components of reaction are represented at A.

Equations of Equilibrium. Summing moments about A, we obtain a direct solution for N_B,

$$\zeta + \Sigma M_A = 0; \quad -90 \text{ N} \cdot \text{m} - 60 \text{ N}(1 \text{ m}) + N_B(0.75 \text{ m}) = 0$$

$$N_B = 200 \text{ N}$$

Using this result,

$$\xrightarrow{+} \Sigma F_x = 0; \qquad\qquad A_x - 200 \sin 30° \text{ N} = 0$$

$$A_x = 100 \text{ N} \qquad\qquad\qquad Ans.$$

$$+\uparrow \Sigma F_y = 0; \qquad\qquad A_y - 200 \cos 30° \text{ N} - 60 \text{ N} = 0$$

$$A_y = 233 \text{ N} \qquad\qquad\qquad Ans.$$

EXAMPLE 5.8

The box wrench in Fig. 5–15a is used to tighten the bolt at A. If the wrench does not turn when the load is applied to the handle, determine the torque or moment applied to the bolt and the force of the wrench on the bolt.

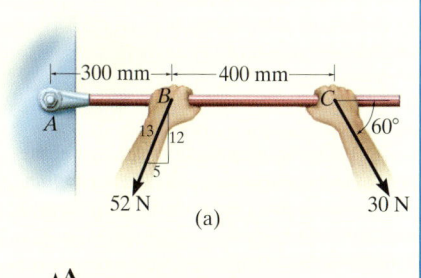

(a)

SOLUTION

Free-Body Diagram. The free-body diagram for the wrench is shown in Fig. 5–15b. Since the bolt acts as a "fixed support," it exerts force components A_x and A_y and a moment M_A on the wrench at A.

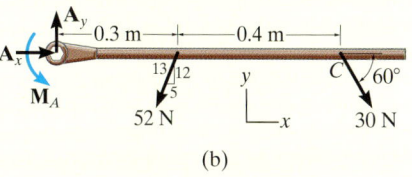

(b)

Fig. 5–15

Equations of Equilibrium.

$$\xrightarrow{+} \Sigma F_x = 0; \qquad A_x - 52\left(\tfrac{5}{13}\right) \text{N} + 30 \cos 60° \text{ N} = 0$$

$$A_x = 5.00 \text{ N} \qquad \textit{Ans.}$$

$$+\uparrow \Sigma F_y = 0; \qquad A_y - 52\left(\tfrac{12}{13}\right) \text{N} - 30 \sin 60° \text{ N} = 0$$

$$A_y = 74.0 \text{ N} \qquad \textit{Ans.}$$

$$\curvearrowleft +\Sigma M_A = 0; \quad M_A - \left[52\left(\tfrac{12}{13}\right)\text{N}\right](0.3 \text{ m}) - (30 \sin 60° \text{ N})(0.7 \text{ m}) = 0$$

$$M_A = 32.6 \text{ N} \cdot \text{m} \qquad \textit{Ans.}$$

Note that $\mathbf{M}_A$ must be *included* in this moment summation. This couple moment is a free vector and represents the twisting resistance of the bolt on the wrench. By Newton's third law, the wrench exerts an equal but opposite moment or torque on the bolt. Furthermore, the resultant force on the wrench is

$$F_A = \sqrt{(5.00)^2 + (74.0)^2} = 74.1 \text{ N} \qquad \textit{Ans.}$$

NOTE: Although only *three* independent equilibrium equations can be written for a rigid body, it is a good practice to *check* the calculations using a fourth equilibrium equation. For example, the above computations may be verified in part by summing moments about point C:

$$\curvearrowleft +\Sigma M_C = 0; \quad \left[52\left(\tfrac{12}{13}\right)\text{N}\right](0.4 \text{ m}) + 32.6 \text{ N} \cdot \text{m} - 74.0 \text{ N}(0.7 \text{ m}) = 0$$

$$19.2 \text{ N} \cdot \text{m} + 32.6 \text{ N} \cdot \text{m} - 51.8 \text{ N} \cdot \text{m} = 0$$

EXAMPLE 5.9

Please refer to the Companion Website for the animation: *Free-Body Diagram for a Beam on Slanting Support*

Determine the horizontal and vertical components of reaction on the member at the pin A, and the normal reaction at the roller B in Fig. 5–16a.

SOLUTION

Free-Body Diagram. The free-body diagram is shown in Fig. 5–16b. The pin at A exerts two components of reaction on the member, $\mathbf{A}_x$ and $\mathbf{A}_y$.

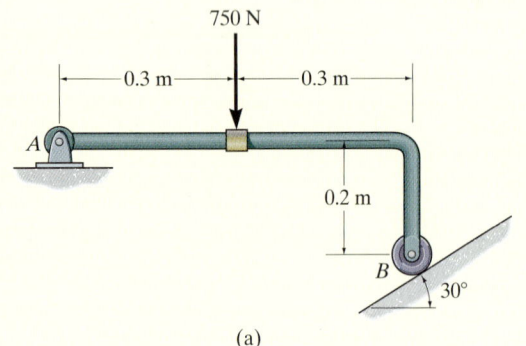

(a)

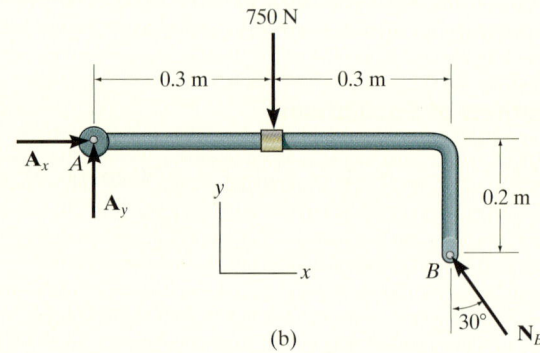

(b)

Fig. 5–16

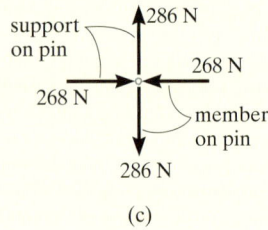

(c)

Equations of Equilibrium. The reaction N_B can be obtained *directly* by summing moments about point A, since $\mathbf{A}_x$ and $\mathbf{A}_y$ produce no moment about A.

$$\zeta + \Sigma M_A = 0;$$

$$[N_B \cos 30°](0.6 \text{ m}) - [N_B \sin 30°](0.2 \text{ m}) - 750 \text{ N}(0.3 \text{ m}) = 0$$

$$N_B = 536.2 \text{ N} = 536 \text{ N} \qquad \textit{Ans.}$$

Using this result,

$$\xrightarrow{+} \Sigma F_x = 0; \qquad A_x - (536.2 \text{ N}) \sin 30° = 0$$

$$A_x = 268 \text{ N} \qquad \textit{Ans.}$$

$$+\uparrow \Sigma F_y = 0; \qquad A_y + (536.2 \text{ N}) \cos 30° - 750 \text{ N} = 0$$

$$A_y = 286 \text{ N} \qquad \textit{Ans.}$$

Details of the equilibrium of the pin at A are shown in Fig. 5–16c.

EXAMPLE 5.10

The uniform smooth rod shown in Fig. 5–17a is subjected to a force and couple moment. If the rod is supported at A by a smooth wall and at B and C either at the top or bottom by rollers, determine the reactions at these supports. Neglect the weight of the rod.

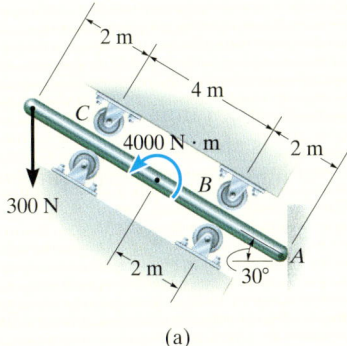

(a)

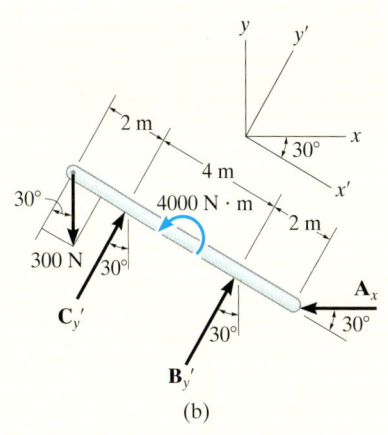

(b)

Fig. 5–17

SOLUTION

Free-Body Diagram.   As shown in Fig. 5–17b, all the support reactions act normal to the surfaces of contact since these surfaces are smooth. The reactions at B and C are shown acting in the positive y' direction. This assumes that only the rollers located on the bottom of the rod are used for support.

Equations of Equilibrium.   Using the x, y coordinate system in Fig. 5–17b, we have

$$\xrightarrow{+} \Sigma F_x = 0; \qquad C_{y'} \sin 30° + B_{y'} \sin 30° - A_x = 0 \qquad (1)$$

$$+\uparrow \Sigma F_y = 0; \qquad -300\ \text{N} + C_{y'} \cos 30° + B_{y'} \cos 30° = 0 \qquad (2)$$

$$\zeta + \Sigma M_A = 0; \qquad -B_{y'}(2\ \text{m}) + 4000\ \text{N} \cdot \text{m} - C_{y'}(6\ \text{m})$$

$$+ (300 \cos 30°\ \text{N})(8\ \text{m}) = 0 \qquad (3)$$

When writing the moment equation, it should be noted that the line of action of the force component $300 \sin 30°$ N passes through point A, and therefore this force is not included in the moment equation.
  Solving Eqs. 2 and 3 simultaneously, we obtain

$$B_{y'} = -1000.0\ \text{N} = -1\ \text{kN} \qquad \textit{Ans.}$$

$$C_{y'} = 1346.4\ \text{N} = 1.35\ \text{kN} \qquad \textit{Ans.}$$

Since $B_{y'}$ is a negative scalar, the sense of $\mathbf{B}_{y'}$ is opposite to that shown on the free-body diagram in Fig. 5–17b. Therefore, the top roller at B serves as the support rather than the bottom one. Retaining the negative sign for $B_{y'}$ (Why?) and substituting the results into Eq. 1, we obtain

$$1346.4 \sin 30°\ \text{N} + (-1000.0 \sin 30°\ \text{N}) - A_x = 0$$

$$A_x = 173\ \text{N} \qquad \textit{Ans.}$$

5

EXAMPLE | **5.11**

(a)

The uniform truck ramp shown in Fig. 5–18a has a weight of 2000 N and is pinned to the body of the truck at each side and held in the position shown by the two side cables. Determine the tension in the cables.

SOLUTION

The idealized model of the ramp, which indicates all necessary dimensions and supports, is shown in Fig. 5–18b. Here the center of gravity is located at the midpoint since the ramp is considered to be uniform.

Free-Body Diagram.   Working from the idealized model, the ramp's free-body diagram is shown in Fig. 5–18c.

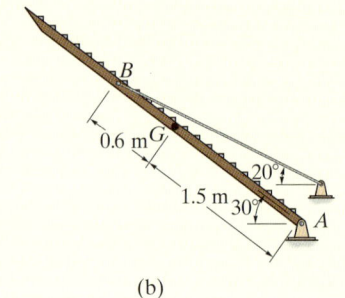

(b)

Equations of Equilibrium.   Summing moments about point A will yield a direct solution for the cable tension. Using the principle of moments, there are several ways of determining the moment of $\mathbf{T}$ about A. If we use x and y components, with $\mathbf{T}$ applied at B, we have

$$\zeta + \Sigma M_A = 0; \qquad -T \cos 20°(2.1 \sin 30° \text{ m}) + T \sin 20°(2.1 \cos 30° \text{ m})$$

$$+ 2000 \text{ N} (1.5 \cos 30° \text{ m}) = 0$$

$$T = 7124.6 \text{ N}$$

We can also determine the moment of $\mathbf{T}$ about A by resolving it into components along and perpendicular to the ramp at B. Then the moment of the component along the ramp will be zero about A, so that

$$\zeta + \Sigma M_A = 0; \qquad -T \sin 10°(2.1 \text{ m}) + 2000 \text{ N} (1.5 \cos 30° \text{ m}) = 0$$

$$T = 7124.6 \text{ N}$$

Since there are two cables supporting the ramp,

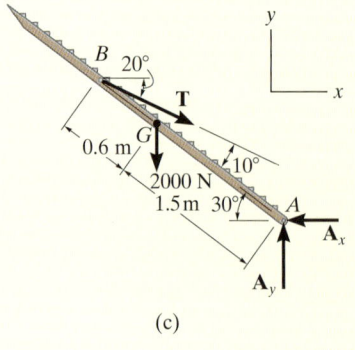

(c)

Fig. 5–18

$$T' = \frac{T}{2} = 3562.3 \text{ N} = 3.56 \text{ kN} \qquad \textit{Ans.}$$

NOTE: As an exercise, show that $A_x = 6695$ N and $A_y = 4437$ N.

EXAMPLE 5.12

Determine the support reactions on the member in Fig. 5–19a. The collar at A is fixed to the member and can slide vertically along the vertical shaft.

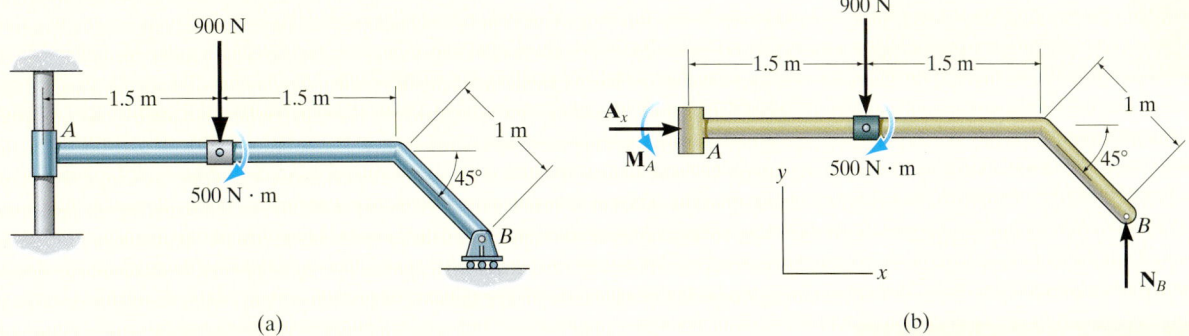

(a)   (b)

Fig. 5–19

5

SOLUTION

Free-Body Diagram.   The free-body diagram of the member is shown in Fig. 5–19b. The collar exerts a horizontal force $\mathbf{A}_x$ and moment $\mathbf{M}_A$ on the member. The reaction $\mathbf{N}_B$ of the roller on the member is vertical.

Equations of Equilibrium.   The forces A_x and N_B can be determined directly from the force equations of equilibrium.

$$\xrightarrow{+} \Sigma F_x = 0; \qquad\qquad A_x = 0 \qquad\qquad\qquad Ans.$$

$$+\uparrow \Sigma F_y = 0; \qquad\qquad N_B - 900\ \text{N} = 0$$

$$N_B = 900\ \text{N} \qquad\qquad\qquad Ans.$$

The moment M_A can be determined by summing moments either about point A or point B.

$$\zeta + \Sigma M_A = 0;$$

$$M_A - 900\ \text{N}(1.5\ \text{m}) - 500\ \text{N} \cdot \text{m} + 900\ \text{N}\,[3\ \text{m} + (1\ \text{m})\cos 45°] = 0$$

$$M_A = -1486\ \text{N} \cdot \text{m} = 1.49\ \text{kN} \cdot \text{m} \,\zeta \qquad\qquad Ans.$$

or

$$\zeta + \Sigma M_B = 0; \quad M_A + 900\ \text{N}\,[1.5\ \text{m} + (1\ \text{m})\cos 45°] - 500\ \text{N} \cdot \text{m} = 0$$

$$M_A = -1486\ \text{N} \cdot \text{m} = 1.49\ \text{kN} \cdot \text{m} \,\zeta \qquad\qquad Ans.$$

The negative sign indicates that $\mathbf{M}_A$ has the opposite sense of rotation to that shown in the free-body diagram.

The hydraulic cylinder *AB* is a typical example of a two-force member since it is pin connected at its ends and, provided its weight is neglected, only the pin forces act on this member.

The link used for this railroad car brake is a three-force member. Since the force **F**$_B$ in the tie rod at *B* and **F**$_C$ from the link at *C* are parallel, then for equilibrium the resultant force **F**$_A$ at the pin *A* must also be parallel with these two forces.

The boom and bucket on this lift is a three-force member, provided its weight is neglected. Here the lines of action of the weight of the worker, **W**, and the force of the two-force member (hydraulic cylinder) at *B*, **F**$_B$, intersect at *O*. For moment equilibrium, the resultant force at the pin *A*, **F**$_A$, must also be directed towards *O*.

5.4   Two- and Three-Force Members

The solutions to some equilibrium problems can be simplified by recognizing members that are subjected to only two or three forces.

Two-Force Members.   As the name implies, a *two-force member* has forces applied at only two points on the member. An example of a two-force member is shown in Fig. 5–20*a*. To satisfy force equilibrium, **F**$_A$ and **F**$_B$ must be equal in magnitude, $F_A = F_B = F$, but opposite in direction ($\Sigma \mathbf{F} = \mathbf{0}$), Fig. 5–20*b*. Furthermore, moment equilibrium requires that **F**$_A$ and **F**$_B$ share the same line of action, which can only happen if they are directed along the line joining points *A* and *B* ($\Sigma \mathbf{M}_A = \mathbf{0}$ or $\Sigma \mathbf{M}_B = \mathbf{0}$), Fig. 5–20*c*. Therefore, for any two-force member to be in equilibrium, the two forces acting on the member *must have the same magnitude, act in opposite directions, and have the same line of action, directed along the line joining the two points where these forces act.*

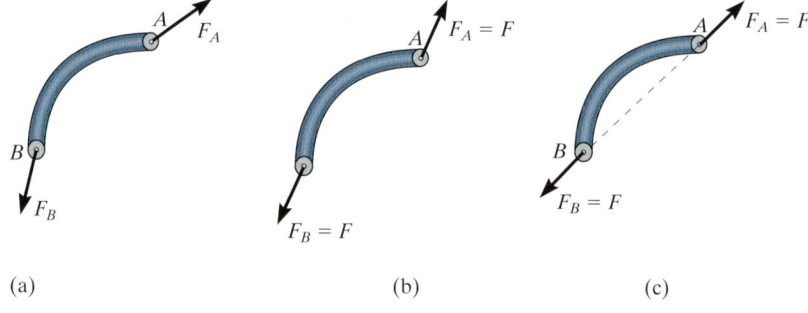

(a)   (b)   (c)

Two-force member

Fig. 5–20

Three-Force Members.   If a member is subjected to only *three forces*, it is called a *three-force member*. Moment equilibrium can be satisfied only if the three forces form a *concurrent* or *parallel* force system. To illustrate, consider the member subjected to the three forces **F**$_1$, **F**$_2$, and **F**$_3$, shown in Fig. 5–21*a*. If the lines of action of **F**$_1$ and **F**$_2$ intersect at point *O*, then the line of action of **F**$_3$ must *also* pass through point *O* so that the forces satisfy $\Sigma \mathbf{M}_O = \mathbf{0}$. As a special case, if the three forces are all parallel, Fig. 5–21*b*, the location of the point of intersection, *O*, will approach infinity.

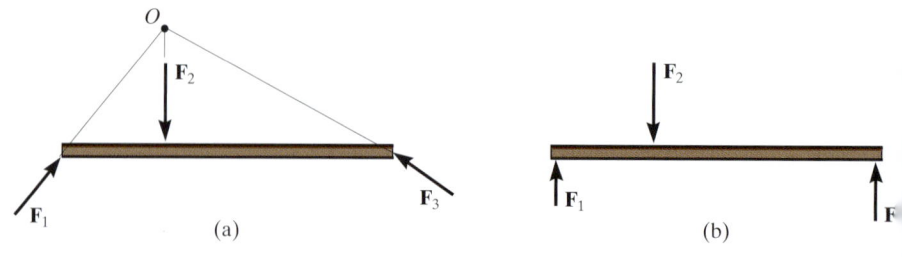

(a)   (b)

Three-force member

Fig. 5–21

EXAMPLE 5.13

The lever *ABC* is pin supported at *A* and connected to a short link *BD* as shown in Fig. 5–22*a*. If the weight of the members is negligible, determine the force of the pin on the lever at *A*.

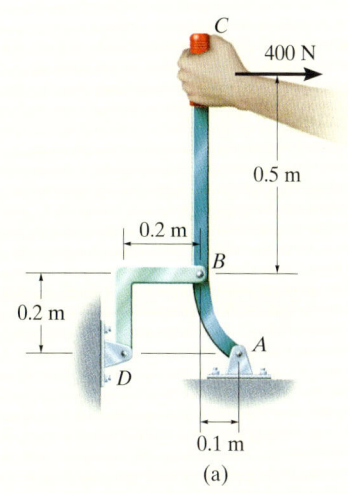

(a)

SOLUTION

Free-Body Diagrams. As shown in Fig. 5–22*b*, the short link *BD* is a *two-force member*, so the *resultant forces* at pins *D* and *B* must be equal, opposite, and collinear. Although the magnitude of the force is unknown, the line of action is known since it passes through *B* and *D*.

Lever *ABC* is a *three-force member*, and therefore, in order to satisfy moment equilibrium, the three nonparallel forces acting on it must be concurrent at *O*, Fig. 5–22*c*. In particular, note that the force **F** on the lever at *B* is equal but opposite to the force **F** acting at *B* on the link. Why? The distance *CO* must be 0.5 m since the lines of action of **F** and the 400-N force are known.

Equations of Equilibrium. By requiring the force system to be concurrent at *O*, since $\Sigma M_O = 0$, the angle θ which defines the line of action of $\mathbf{F}_A$ can be determined from trigonometry,

$$\theta = \tan^{-1}\left(\frac{0.7}{0.4}\right) = 60.3°$$

Using the *x, y* axes and applying the force equilibrium equations,

$$\xrightarrow{+} \Sigma F_x = 0; \quad F_A \cos 60.3° - F \cos 45° + 400 \text{ N} = 0$$

$$+\uparrow \Sigma F_y = 0; \quad F_A \sin 60.3° - F \sin 45° = 0$$

Solving, we get

$$F_A = 1.07 \text{ kN} \qquad\qquad Ans.$$

$$F = 1.32 \text{ kN}$$

NOTE: We can also solve this problem by representing the force at *A* by its two components $\mathbf{A}_x$ and $\mathbf{A}_y$ and applying $\Sigma M_A = 0$, $\Sigma F_x = 0$, $\Sigma F_y = 0$ to the lever. Once A_x and A_y are determined, we can get F_A and θ.

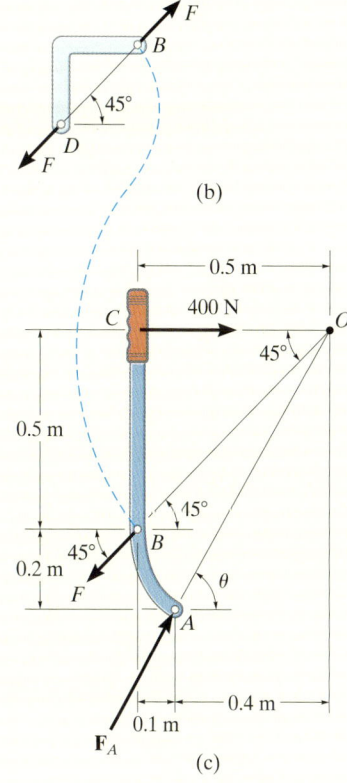

(b)

(c)

Fig. 5–22

FUNDAMENTAL PROBLEMS

All problem solutions must include an FBD.

F5–1. Determine the horizontal and vertical components of reaction at the supports. Neglect the thickness of the beam.

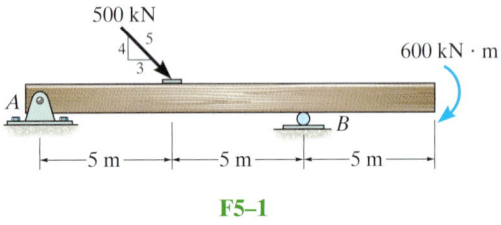

F5–1

F5–2. Determine the horizontal and vertical components of reaction at the pin A and the reaction on the beam at C.

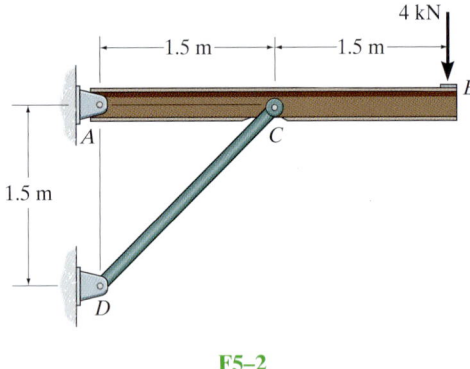

F5–2

F5–3. The truss is supported by a pin at A and a roller at B. Determine the support reactions.

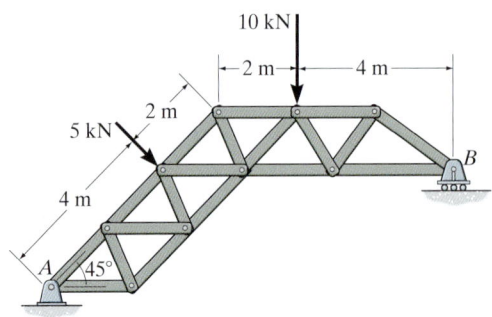

F5–3

F5–4. Determine the components of reaction at the fixed support A. Neglect the thickness of the beam.

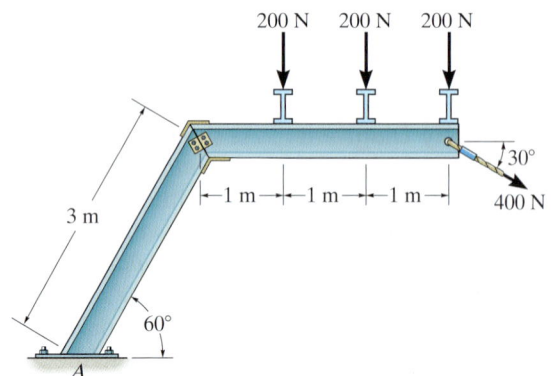

F5–4

F5–5. The 25-kg bar has a center of mass at G. If it is supported by a smooth peg at C, a roller at A, and cord AB, determine the reactions at these supports.

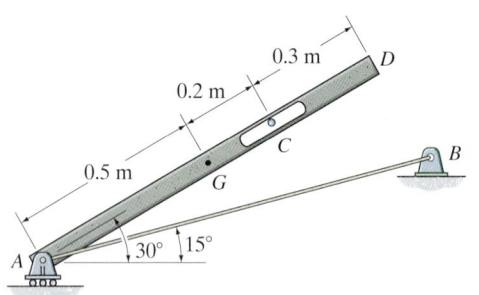

F5–5

F5–6. Determine the reactions at the smooth contact points A, B, and C on the bar.

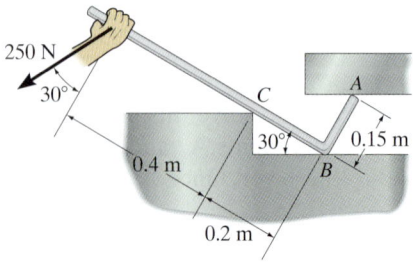

F5–6

PROBLEMS

All problem solutions must include an FBD.

5–10.   Determine the horizontal and vertical components of reaction at the pin A and the reaction of the rocker B on the beam.

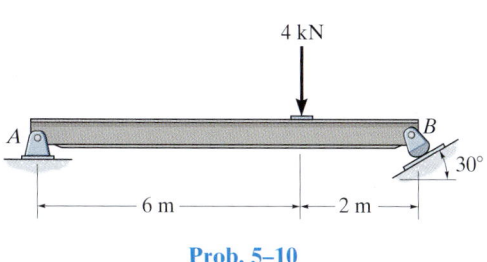

Prob. 5–10

5–11.   Determine the magnitude of the reactions on the beam at A and B. Neglect the thickness of the beam.

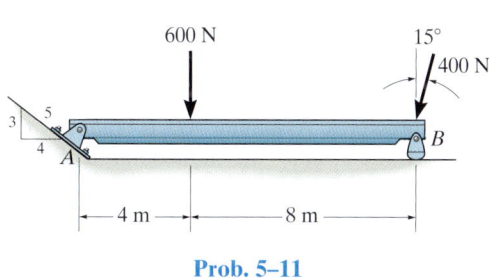

Prob. 5–11

*****5–12.**   Determine the components of the support reactions at the fixed support A on the cantilevered beam.

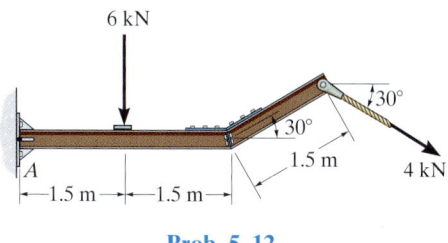

Prob. 5–12

5–13.   The 75-kg gate has a center of mass located at G. If A supports only a horizontal force and B can be assumed as a pin, determine the components of reaction at these supports.

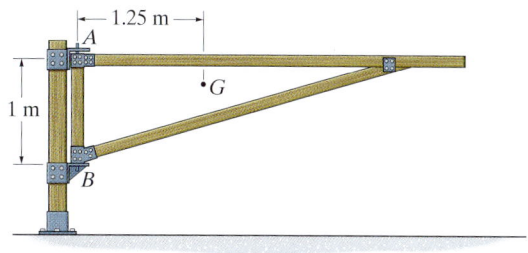

Prob. 5–13

5–14.   The overhanging beam is supported by a pin at A and the two-force strut BC. Determine the horizontal and vertical components of reaction at A and the reaction at B on the beam.

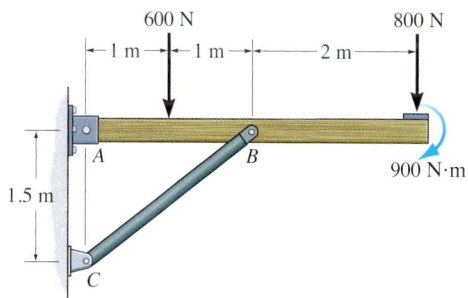

Prob. 5–14

5–15.   Determine the reactions at the pins A and B. The spring has an unstretched length of 80 mm.

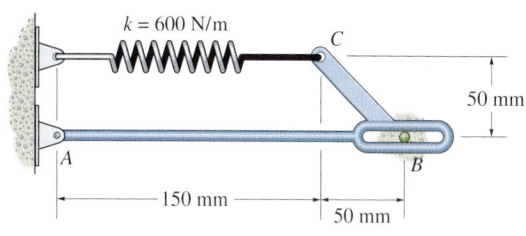

Prob. 5–15

5

*5–16. Determine the components of reaction at the supports A and B on the rod.

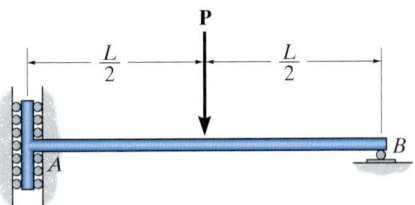

Prob. 5–16

5–17. If the wheelbarrow and its contents have a mass of 60 kg and center of mass at G, determine the magnitude of the resultant force which the man must exert on *each* of the two handles in order to hold the wheelbarrow in equilibrium.

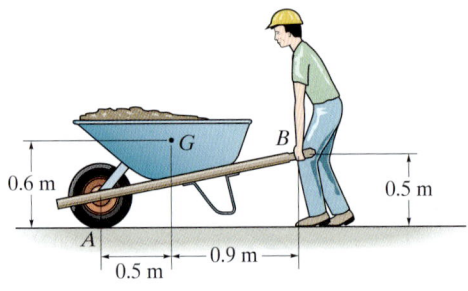

Prob. 5–17

5–18. Determine the reactions at the roller A and pin B.

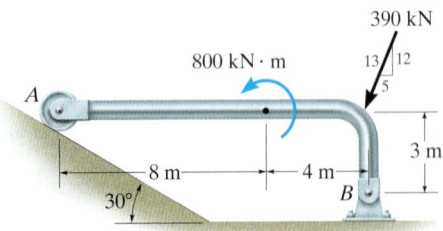

Prob. 5–18

5–19. The shelf supports the electric motor which has a mass of 15 kg and mass center at G_m. The platform upon which it rests has a mass of 4 kg and mass center at G_p. Assuming that a single bolt B holds the shelf up and the bracket bears against the smooth wall at A, determine this normal force at A and the horizontal and vertical components of reaction of the bolt on the bracket.

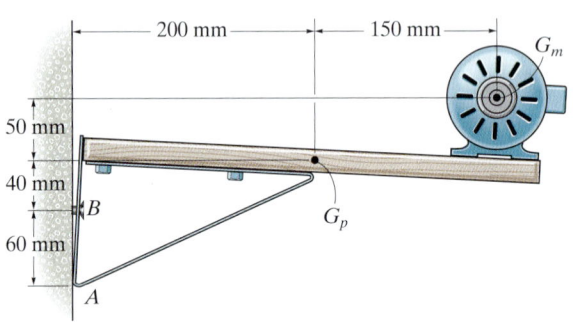

Prob. 5–19

*5–20. The bulk head AD is subjected to both water and soil-backfill pressures. Assuming AD is "pinned" to the ground at A, determine the horizontal and vertical reactions there and also the required tension in the ground anchor BC necessary for equilibrium. The bulk head has a mass of 800 kg.

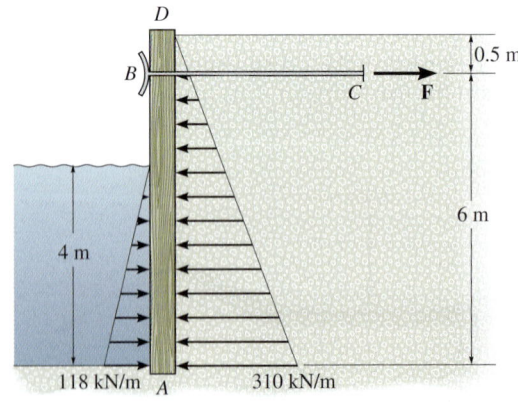

Prob. 5–20

5–21. A cantilever beam, having an extended length of 3 m, is subjected to a vertical force of 500 N. Assuming that the wall resists this load with linearly varying distributed loads over the 0.15-m length of the beam portion inside the wall, detemine the intensities w_1 and w_2 for equilibrium.

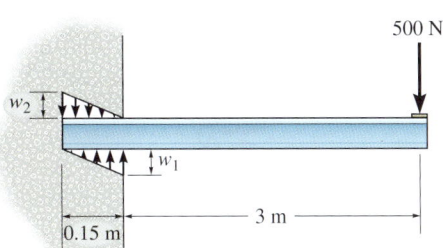

Prob. 5–21

5–22. Spring CD remains in the horizontal position at all times due to the roller at D. If the spring is unstretched when $\theta = 0°$ and the bracket achieves its equilibrium position when $\theta = 30°$, determine the stiffness k of the spring and the horizontal and vertical components of reaction at pin A.

5–23. Spring CD remains in the horizontal position at all times due to the roller at D. If the spring is unstretched when $\theta = 0°$ and the stiffness is $k = 1.5$ kN/m, determine the smallest angle θ for equilibrium and the horizontal and vertical components of reaction at pin A.

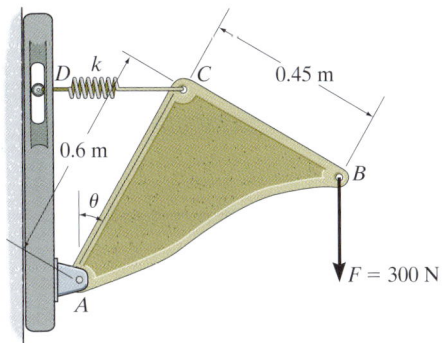

Probs. 5–22/23

5–24. The platform assembly has a weight of 1000 N (≈ 100 kg) and center of gravity at G_1. If it is intended to support a maximum load of 1600 N placed, at point G_2 determine the smallest counterweight W that should be placed at B in order to prevent the platform from tipping over.

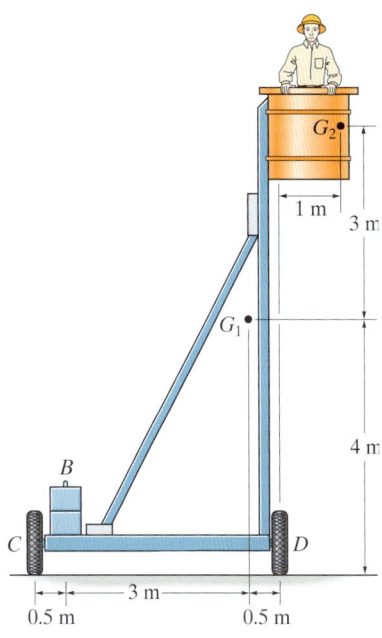

Prob. 5–24

5–25. Determine the reactions on the bent rod which is supported by a smooth surface at B and by a collar at A, which is fixed to the rod and is free to slide over the fixed inclined rod.

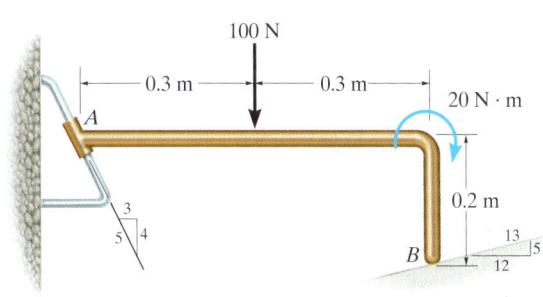

Prob. 5–25

5–26. The toggle switch consists of a cocking lever that is pinned to a fixed frame at A and held in place by the spring which has an unstretched length of 200 mm. Determine the magnitude of the resultant force at A and the normal force on the peg at B when the lever is in the position shown.

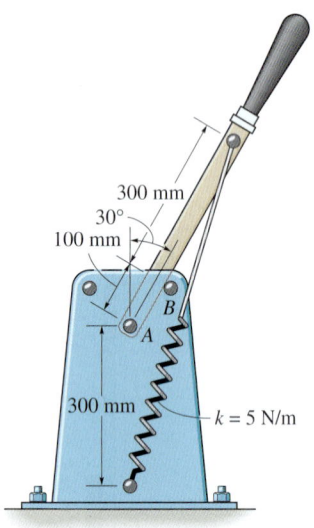

Prob. 5–26

5–27. The sports car has a mass of 1.5 Mg and mass center at G. If the front two springs each have a stiffness of $k_A = 58$ kN/m and the rear two springs each have a stiffness of $k_B = 65$ kN/m, determine their compression when the car is parked on the 30° incline. Also, what friction force $\mathbf{F}_B$ must be applied to each of the rear wheels to hold the car in equilibrium? *Hint:* First determine the normal force at A and B, then determine the compression in the springs.

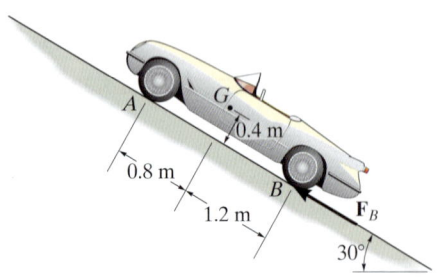

Prob. 5–27

*****5–28.** Determine the magnitude and direction θ of the minimum force P needed to pull the 50-kg roller over the smooth step.

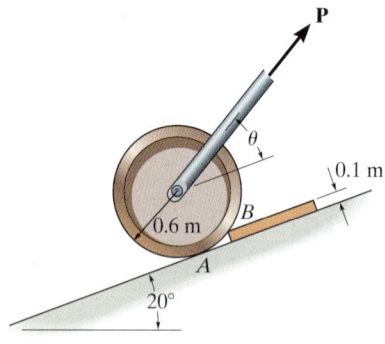

Prob. 5–28

5–29. The horizontal beam is supported by springs at its ends. Each spring has a stiffness of $k = 5$ kN/m and is originally unstretched when the beam is in the horizontal position. Determine the angle of tilt of the beam if a load of 800 N is applied at point C as shown.

5–30. The horizontal beam is supported by springs at its ends. If the stiffness of the spring at A is $k_A = 5$ kN/m, determine the required stiffness of the spring at B so that if the beam is loaded with the 800 N it remains in the horizontal position. The springs are originally constructed so that the beam is in the horizontal position when it is unloaded.

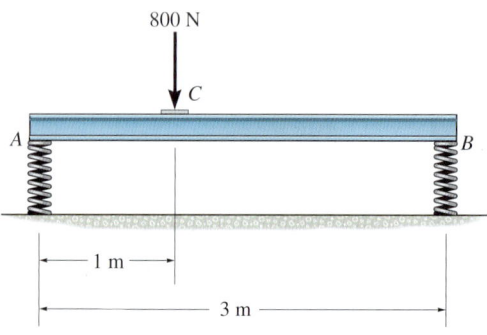

Probs. 5–29/30

5–31. The jib crane is supported by a pin at C and rod AB. If the load has a mass of 2 Mg with its center of mass located at G, determine the horizontal and vertical components of reaction at the pin C and the force developed in rod AB on the crane when $x = 5$ m.

*5–32. The jib crane is supported by a pin at C and rod AB. The rod can withstand a maximum tension of 40 kN. If the load has a mass of 2 Mg, with its center of mass located at G, determine its maximum allowable distance x and the corresponding horizontal and vertical components of reaction at C.

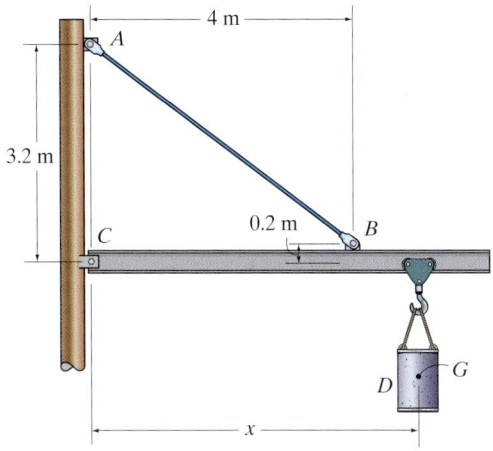

Probs. 5–31/32

5–33. The woman exercises on the rowing machine. If she exerts a holding force of $F = 200$ N on handle ABC, determine the horizontal and vertical components of reaction at pin C and the force developed along the hydraulic cylinder BD on the handle.

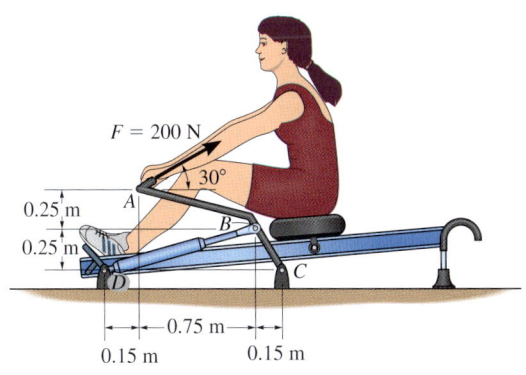

Prob. 5–33

5–34. A uniform glass rod having a length L is placed in the smooth hemispherical bowl having a radius r. Determine the angle of inclination θ for equilibrium.

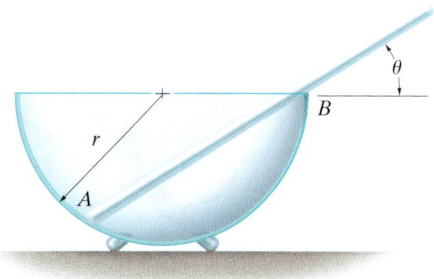

Prob. 5–34

5–35. The toggle switch consists of a cocking lever that is pinned to a fixed frame at A and held in place by the spring which has an unstretched length of 250 mm. Determine the magnitude of the resultant force at A and the normal force on the peg at B when the lever is in the position shown.

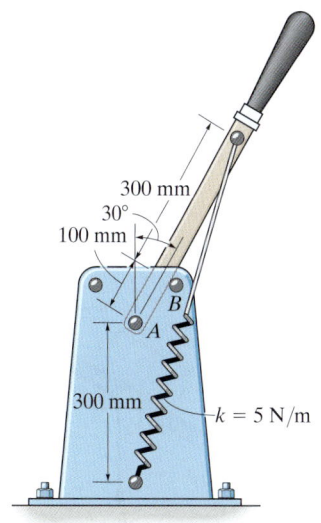

Prob. 5–35

*5–36. The 1.4-Mg drainpipe is held in the tines of the fork lift. Determine the normal forces at A and B as functions of the blade angle θ and plot the results of force (ordinate) versus θ (abscissa) for $0 \le \theta \le 90°$.

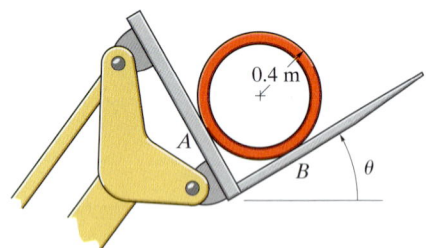

Prob. 5–36

5–37. The boom supports the two vertical loads. Neglect the size of the collars at D and B and the thickness of the boom, and compute the horizontal and vertical components of force at the pin A and the force in cable CB. Set $F_1 = 800$ N and $F_2 = 350$ N.

5–38. The boom is intended to support two vertical loads, $\mathbf{F}_1$ and $\mathbf{F}_2$. If the cable CB can sustain a maximum load of 1500 N before it fails, determine the critical loads if $F_1 = 2F_2$. Also, what is the magnitude of the maximum reaction at pin A?

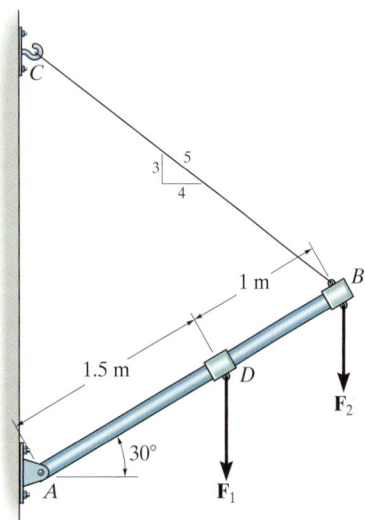

Probs. 5–37/38

5–39. The airstroke actuator at D is used to apply a force of $F = 200$ N on the member at B. Determine the horizontal and vertical components of reaction at the pin A and the force of the smooth shaft at C on the member.

*5–40. The airstroke actuator at D is used to apply a force of $\mathbf{F}$ on the member at B. The normal reaction of the smooth shaft at C on the member is 300 N. Determine the magnitude of $\mathbf{F}$ and the horizontal and vertical components of reaction at pin A.

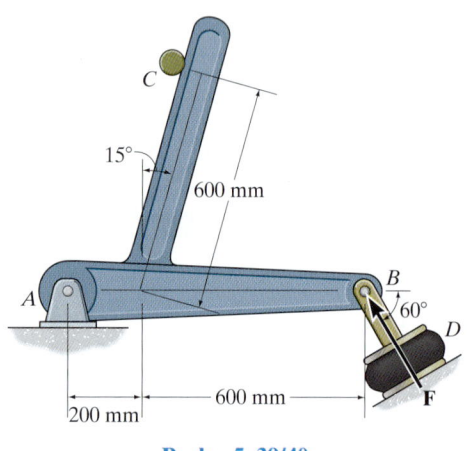

Probs. 5–39/40

5–41. The uniform beam has a weight W and length $2l$ and is supported by a pin at A and a cable BC. Determine the horizontal and vertical components of reaction at A and the tension in the cable necessary to hold the beam in the position shown.

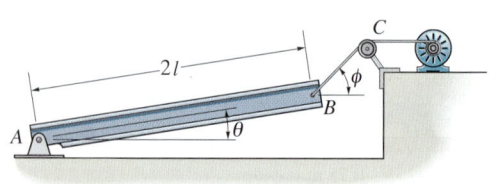

Prob. 5–41

5–42. The rigid metal strip of negligible weight is used as part of an electromagnetic switch. If the stiffness of the springs at A and B is $k = 5$ N/m, and the strip is originally horizontal when the springs are unstretched, determine the smallest force needed to close the contact gap at C.

5–43. The rigid metal strip of negligible weight is used as part of an electromagnetic switch. Determine the maximum stiffness k of the springs at A and B so that the contact at C closes when the vertical force developed there is 0.5 N. Originally the strip is horizontal as shown.

5–45. The device is used to hold an elevator door open. If the spring has a stiffness of $k = 40$ N/m and it is compressed 0.2 m, determine the horizontal and vertical components of reaction at the pin A and the resultant force at the wheel bearing B.

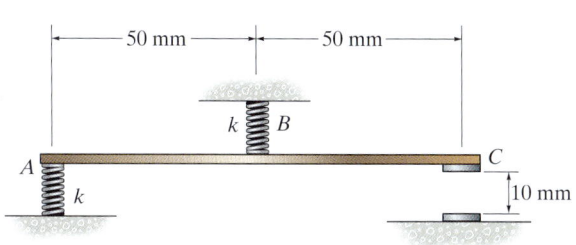

Probs. 5–42/43

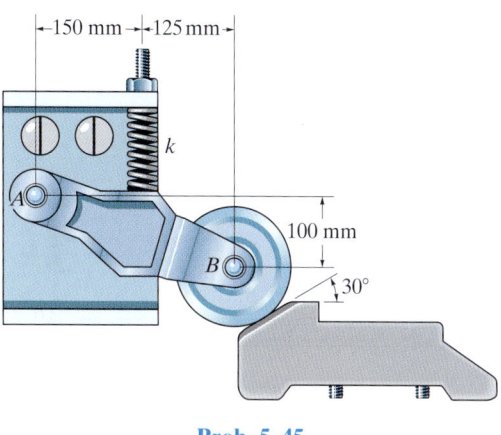

Prob. 5–45

***5–44.** The upper portion of the crane boom consists of the jib AB, which is supported by the pin at A, the guy line BC, and the backstay CD, each cable being separately attached to the mast at C. If the 5-kN load is supported by the hoist line, which passes over the pulley at B, determine the magnitude of the resultant force the pin exerts on the jib at A for equilibrium, the tension in the guy line BC, and the tension T in the hoist line. Neglect the weight of the jib. The pulley at B has a radius of 0.1 m.

5–46. Three uniform books, each having a weight W and length a, are stacked as shown. Determine the maximum distance d that the top book can extend out from the bottom one so the stack does not topple over.

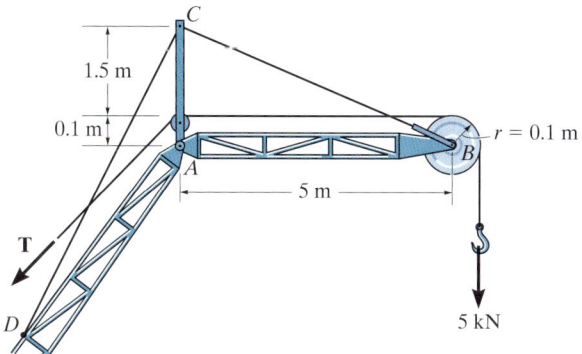

Prob. 5–44

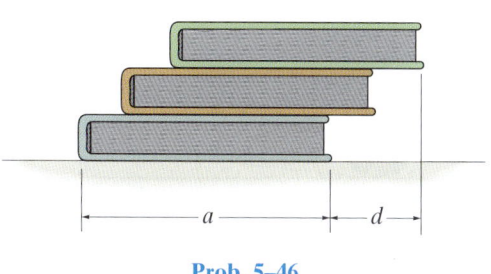

Prob. 5–46

5–47. The horizontal beam is supported by springs at its ends. Each spring has a stiffness of $k = 5\,\text{kN/m}$ and is originally unstretched when the beam is in the horizontal position. Determine the angle of tilt of the beam if a load of 800 N is applied at point C as shown.

*5–48. The horizontal beam is supported by springs at its ends. If the stiffness of the spring at A is $k_A = 5\,\text{kN/m}$, determine the required stiffness of the spring at B so that if the beam is loaded with the 800-N force it remains in the horizontal position. The springs are originally constructed so that the beam is in the horizontal position when it is unloaded.

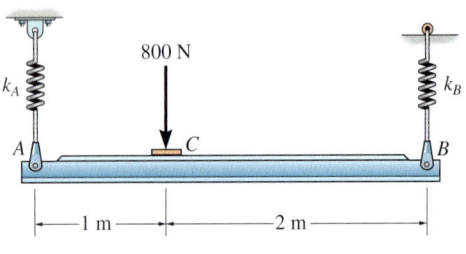

Probs. 5–47/48

5–49. The wheelbarrow and its contents have a mass of $m = 60\,\text{kg}$ with a center of mass at G. Determine the normal reaction on the tire and the vertical force on each hand to hold it at $\theta = 30°$. Take $a = 0.3\,\text{m}$, $b = 0.45\,\text{m}$, $c = 0.75\,\text{m}$ and $d = 0.1\,\text{m}$.

5–50. The wheelbarrow and its contents have a mass m and center of mass at G. Determine the greatest angle of tilt θ without causing the wheelbarrow to tip over.

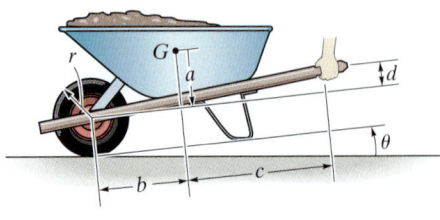

Probs. 5–49/50

5–51. The rigid beam of negligible weight is supported horizontally by two springs and a pin. If the springs are uncompressed when the load is removed, determine the force in each spring when the load $\mathbf{P}$ is applied. Also, compute the vertical deflection of end C. Assume the spring stiffness k is large enough so that only small deflections occur. *Hint:* The beam rotates about A so the deflections in the springs can be related.

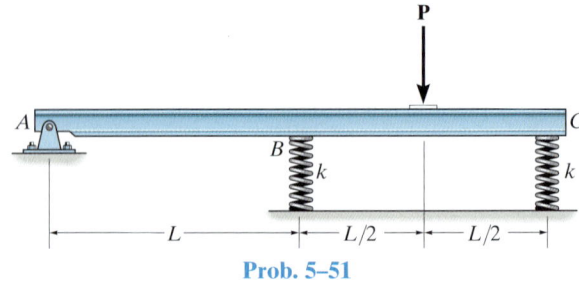

Prob. 5–51

*5–52. A boy stands out at the end of the diving board, which is supported by two springs A and B, each having a stiffness of $k = 15\,\text{kN/m}$. In the position shown the board is horizontal. If the boy has a mass of 40 kg, determine the angle of tilt which the board makes with the horizontal after he jumps off. Neglect the weight of the board and assume it is rigid.

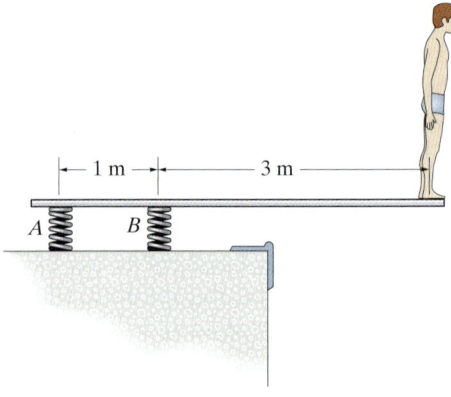

Prob. 5–52

5–53. The uniform beam has a weight W and length l and is supported by a pin at A and a cable BC. Determine the horizontal and vertical components of reaction at A and the tension in the cable necessary to hold the beam in the position shown.

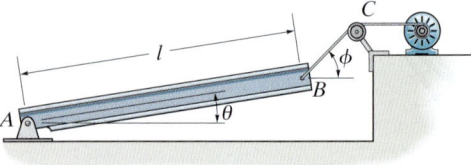

Prob. 5–53

5–54. Determine the distance d for placement of the load **P** for equilibrium of the smooth bar in the position θ as shown. Neglect the weight of the bar.

5–55. If $d = 1$ m, and $\theta = 30°$, determine the normal reaction at the smooth supports and the required distance a for the placement of the roller if $P = 600$ N. Neglect the weight of the bar.

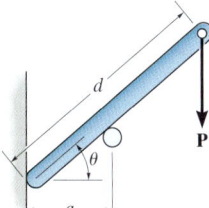

Probs. 5–54/55

***5–56.** The disk B has a mass of 20 kg and is supported on the smooth cylindrical surface by a spring having a stiffness of $k = 400$ N/m and unstretched length of $l_0 = 1$ m. The spring remains in the horizontal position since its end A is attached to the small roller guide which has negligible weight. Determine the angle θ for equilibrium of the roller.

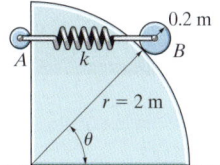

Prob. 5–56

5–57. The winch cable on a tow truck is subjected to a force of $T = 6$ kN when the cable is directed at $\theta = 60°$. Determine the magnitudes of the total brake frictional force **F** for the rear set of wheels B and the total normal forces at *both* front wheels A and both rear wheels B for equilibrium. The truck has a total mass of 4 Mg and mass center at G.

5–58. Determine the minimum cable force T and critical angle θ which will cause the tow truck to start tipping, i.e., for the normal reaction at A to be zero. Assume that the truck is braked and will not slip at B. The truck has a total mass of 4 Mg and mass center at G.x

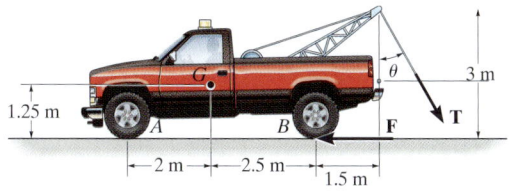

Probs. 5–57/58

5–59. The thin rod of length l is supported by the smooth tube. Determine the distance a needed for equilibrium if the applied load is **P**.

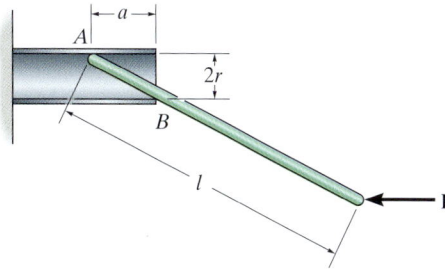

Prob. 5–59

***5–60.** The 30-N uniform rod has a length of $l = 1$ m. If $s = 1.5$ m, determine the distance h of placement at the end A along the smooth wall for equilibrium.

5–61. The uniform rod has a length l and weight W. It is supported at one end A by a smooth wall and the other end by a cord of length s which is attached to the wall as shown. Determine the placement h for equilibrium.

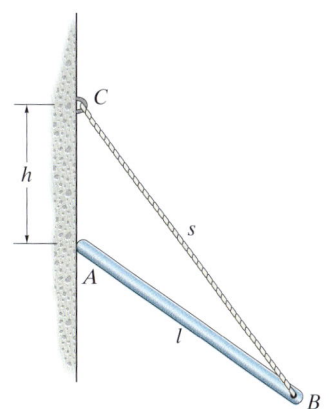

Probs. 5–60/61

CONCEPTUAL PROBLEMS

P5–5. The tie rod is used to support this overhang at the entrance of a building. If it is pin connected to the building wall at A and to the center of the overhang B, determine if the force in the rod will increase, decrease, or remain the same if (a) the support at A is moved to a lower position D, and (b) the support at B is moved to the outer position C. Explain your answer with an equilibrium analysis, using dimensions and loads. Assume the overhang is pin supported from the building wall.

P5–5

P5–6. The man attempts to pull the four wheeler up the incline and onto the trailer. From the position shown, is it more effective to pull on the rope at A, or would it be better to pull on the rope at B? Draw a free-body diagram for each case, and do an equilibrium analysis to explain your answer. Use appropriate numerical values to do your calculations.

P5–6

P5–7. Like all aircraft, this jet plane rests on three wheels. Why not use an additional wheel at the tail for better support? (Can you think of any other reason for not including this wheel?) If there was a fourth tail wheel, draw a free-body diagram of the plane from a side (2 D) view, and show why one would not be able to determine all the wheel reactions using the equations of equilibrium.

P5–7

P5–8. Where is the best place to arrange most of the logs in the wheelbarrow so that it minimizes the amount of force on the backbone of the person transporting the load? Do an equilibrium analysis to explain your answer.

P5–8

EQUILIBRIUM IN THREE DIMENSIONS

5.5   Free-Body Diagrams

The first step in solving three-dimensional equilibrium problems, as in the case of two dimensions, is to draw a free-body diagram. Before we can do this, however, it is first necessary to discuss the types of reactions that can occur at the supports.

Support Reactions.   The reactive forces and couple moments acting at various types of supports and connections, when the members are viewed in three dimensions, are listed in Table 5–2. It is important to recognize the symbols used to represent each of these supports and to understand clearly how the forces and couple moments are developed. As in the two-dimensional case:

- A force is developed by a support that restricts the translation of its attached member.

- A couple moment is developed when rotation of the attached member is prevented.

For example, in Table 5–2, item (4), the ball-and-socket joint prevents any translation of the connecting member; therefore, a force must act on the member at the point of connection. This force has three components having unknown magnitudes, F_x, F_y, F_z. Provided these components are known, one can obtain the magnitude of force, $F = \sqrt{F_x^2 + F_y^2 + F_z^2}$, and the force's orientation defined by its coordinate direction angles α, β, γ, Eqs. 2–5.* Since the connecting member is allowed to rotate freely about *any* axis, no couple moment is resisted by a ball-and-socket joint.

It should be noted that the *single* bearing supports in items (5) and (7), the *single* pin (8), and the *single* hinge (9) are shown to resist both force and couple-moment components. If, however, these supports are used in conjunction with *other* bearings, pins, or hinges to hold a rigid body in equilibrium and the supports are *properly aligned* when connected to the body, then the *force reactions* at these supports *alone* are adequate for supporting the body. In other words, the couple moments become redundant and are not shown on the free-body diagram. The reason for this should become clear after studying the examples which follow.

---

* The three unknowns may also be represented as an unknown force magnitude F and two unknown coordinate direction angles. The third direction angle is obtained using the identity $\cos^2 \alpha + \cos^2 \beta + \cos^2 \gamma = 1$, Eq. 2–8.

TABLE 5–2 Supports for Rigid Bodies Subjected to Three-Dimensional Force Systems

Types of Connection	Reaction	Number of Unknowns
(1)  cable		One unknown. The reaction is a force which acts away from the member in the known direction of the cable.
(2)  smooth surface support		One unknown. The reaction is a force which acts perpendicular to the surface at the point of contact.
(3)  roller		One unknown. The reaction is a force which acts perpendicular to the surface at the point of contact.
(4)  ball and socket	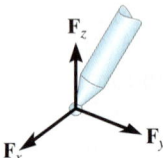	Three unknowns. The reactions are three rectangular force components.
(5)  single journal bearing	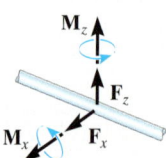	Four unknowns. The reactions are two force and two couple-moment components which act perpendicular to the shaft. Note: The couple moments are *generally not applied* if the body is supported elsewhere. See the examples.

continued

TABLE 5–2 Continued		
Types of Connection	Reaction	Number of Unknowns
(6) single journal bearing with square shaft	M_z F_z M_y M_x F_x	Five unknowns. The reactions are two force and three couple-moment components. *Note*: The couple moments *are generally not applied* if the body is supported elsewhere. See the examples.
(7) single thrust bearing	M_z F_y F_z M_x F_x	Five unknowns. The reactions are three force and two couple-moment components. *Note*: The couple moments *are generally not applied* if the body is supported elsewhere. See the examples.
(8) single smooth pin	M_z F_z F_y M_y F_x	Five unknowns. The reactions are three force and two couple-moment components. *Note*: The couple moments *are generally not applied* if the body is supported elsewhere. See the examples.
(9) single hinge	M_z F_z F_y F_x M_x	Five unknowns. The reactions are three force and two couple-moment components. *Note*: The couple moments *are generally not applied* if the body is supported elsewhere. See the examples.
(10) fixed support	M_z F_z F_x M_x F_y M_y	Six unknowns. The reactions are three force and three couple-moment components.

5

Typical examples of actual supports that are referenced to Table 5–2 are shown in the following sequence of photos.

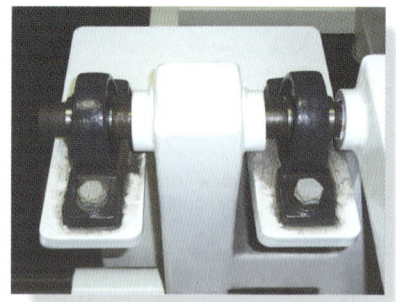

The journal bearings support the ends of the shaft. (5)

This ball-and-socket joint provides a connection for the housing of an earth grader to its frame. (4)

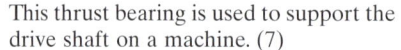

This thrust bearing is used to support the drive shaft on a machine. (7)

This pin is used to support the end of the strut used on a tractor. (8)

Free-Body Diagrams.

The general procedure for establishing the free-body diagram of a rigid body has been outlined in Sec. 5.2. Essentially it requires first "isolating" the body by drawing its outlined shape. This is followed by a careful *labeling* of *all* the forces and couple moments with reference to an established *x, y, z* coordinate system. As a general rule, it is suggested to show the unknown components of reaction as acting on the free-body diagram in the *positive sense*. In this way, if any negative values are obtained, they will indicate that the components act in the negative coordinate directions.

EXAMPLE 5.14

Consider the two rods and plate, along with their associated free-body diagrams, shown in Fig. 5–23. The *x*, *y*, *z* axes are established on the diagram and the unknown reaction components are indicated in the *positive sense*. The weight is neglected.

SOLUTION

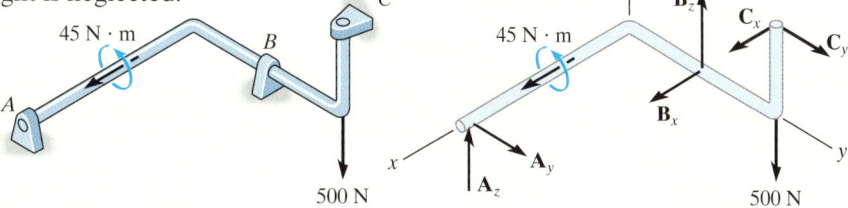

Properly aligned journal bearings at *A*, *B*, *C*.

The force reactions developed by the bearings are *sufficient* for equilibrium since they prevent the shaft from rotating about each of the coordinate axes. No couple moments at each bearing are developed.

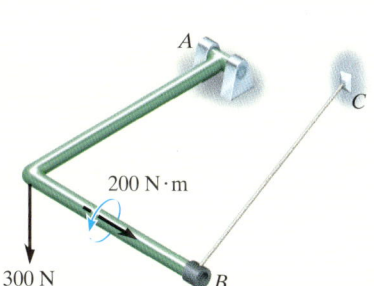

Pin at *A* and cable *BC*.

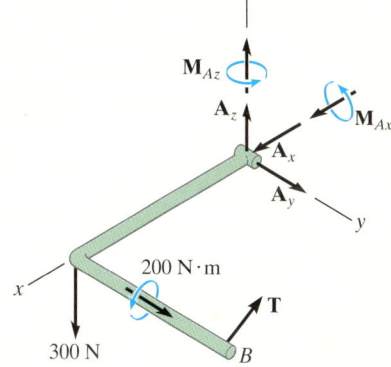

Moment components are developed by the pin on the rod to prevent rotation about the *x* and *z* axes.

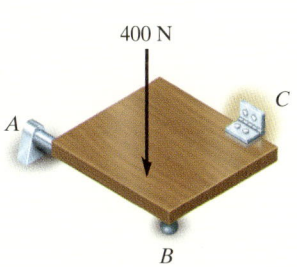

Properly aligned journal bearing at *A* and hinge at *C*. Roller at *B*.

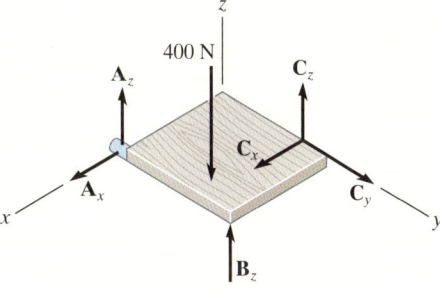

Only force reactions are developed by the bearing and hinge on the plate to prevent rotation about each coordinate axis. No moments are developed at the hinge.

Fig. 5–23

5.6   Equations of Equilibrium

As stated in Sec. 5.1, the conditions for equilibrium of a rigid body subjected to a three-dimensional force system require that both the *resultant* force and *resultant* couple moment acting on the body be equal to *zero*.

Vector Equations of Equilibrium.   The two conditions for equilibrium of a rigid body may be expressed mathematically in vector form as

$$\Sigma \mathbf{F} = \mathbf{0}$$
$$\Sigma \mathbf{M}_O = \mathbf{0} \tag{5–5}$$

where $\Sigma \mathbf{F}$ is the vector sum of all the external forces acting on the body and $\Sigma \mathbf{M}_O$ is the sum of the couple moments and the moments of all the forces about any point O located either on or off the body.

Scalar Equations of Equilibrium.   If all the external forces and couple moments are expressed in Cartesian vector form and substituted into Eqs. 5–5, we have

$$\Sigma \mathbf{F} = \Sigma F_x \mathbf{i} + \Sigma F_y \mathbf{j} + \Sigma F_z \mathbf{k} = \mathbf{0}$$
$$\Sigma \mathbf{M}_O = \Sigma M_x \mathbf{i} + \Sigma M_y \mathbf{j} + \Sigma M_z \mathbf{k} = \mathbf{0}$$

Since the $\mathbf{i}, \mathbf{j}$, and $\mathbf{k}$ components are independent from one another, the above equations are satisfied provided

$$\Sigma F_x = 0$$
$$\Sigma F_y = 0 \tag{5–6a}$$
$$\Sigma F_z = 0$$

and

$$\Sigma M_x = 0$$
$$\Sigma M_y = 0 \tag{5–6b}$$
$$\Sigma M_z = 0$$

These *six scalar equilibrium equations* may be used to solve for at most six unknowns shown on the free-body diagram. Equations 5–6a require the sum of the external force components acting in the x, y, and z directions to be zero, and Eqs. 5–6b require the sum of the moment components about the x, y, and z axes to be zero.

5.7   Constraints and Statical Determinacy

To ensure the equilibrium of a rigid body, it is not only necessary to satisfy the equations of equilibrium, but the body must also be properly held or constrained by its supports. Some bodies may have more supports than are necessary for equilibrium, whereas others may not have enough or the supports may be arranged in a particular manner that could cause the body to move. Each of these cases will now be discussed.

Redundant Constraints.   When a body has redundant supports, that is, more supports than are necessary to hold it in equilibrium, it becomes statically indeterminate. *Statically indeterminate* means that there will be more unknown loadings on the body than equations of equilibrium available for their solution. For example, the beam in Fig. 5–24a and the pipe assembly in Fig. 5–24b, shown together with their free-body diagrams, are both statically indeterminate because of additional (or redundant) support reactions. For the beam there are five unknowns, M_A, A_x, A_y, B_y, and C_y, for which only three equilibrium equations can be written ($\Sigma F_x = 0$, $\Sigma F_y = 0$, and $\Sigma M_O = 0$, Eq. 5–2). The pipe assembly has eight unknowns, for which only six equilibrium equations can be written, Eqs. 5–6.

The additional equations needed to solve statically indeterminate problems of the type shown in Fig. 5–24 are generally obtained from the deformation conditions at the points of support. These equations involve the physical properties of the body which are studied in subjects dealing with the mechanics of deformation, such as "mechanics of materials."*

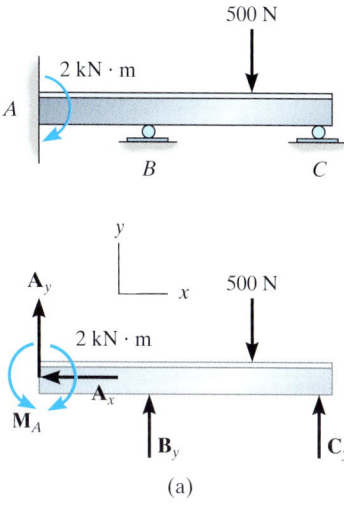

(a)

Fig. 5–24

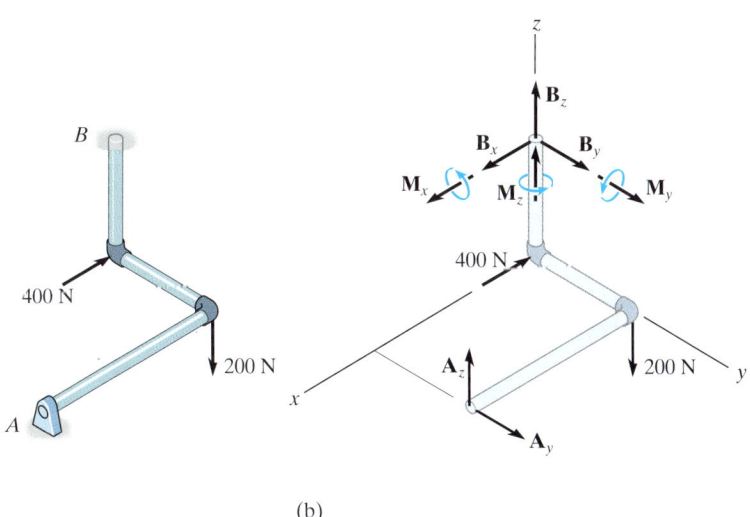

(b)

---

* See R. C. Hibbeler, *Mechanics of Materials*, 8th edition, Pearson Education/Prentice Hall, Inc.

Improper Constraints.   Having the same number of unknown reactive forces as available equations of equilibrium does not always guarantee that a body will be stable when subjected to a particular loading. For example, the pin support at A and the roller support at B for the beam in Fig. 5–25a are placed in such a way that the lines of action of the reactive forces are *concurrent* at point A. Consequently, the applied loading **P** will cause the beam to rotate slightly about A, and so the beam is improperly constrained, $\Sigma M_A \neq 0$.

In three dimensions, a body will be improperly constrained if the lines of action of all the reactive forces intersect a common axis. For example, the reactive forces at the ball-and-socket supports at A and B in Fig. 5–25b all intersect the axis passing through A and B. Since the moments of these forces about A and B are all zero, then the loading **P** will rotate the member about the AB axis, $\Sigma M_{AB} \neq 0$.

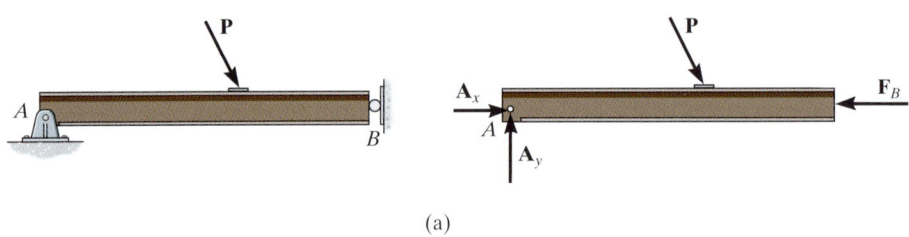

(a)

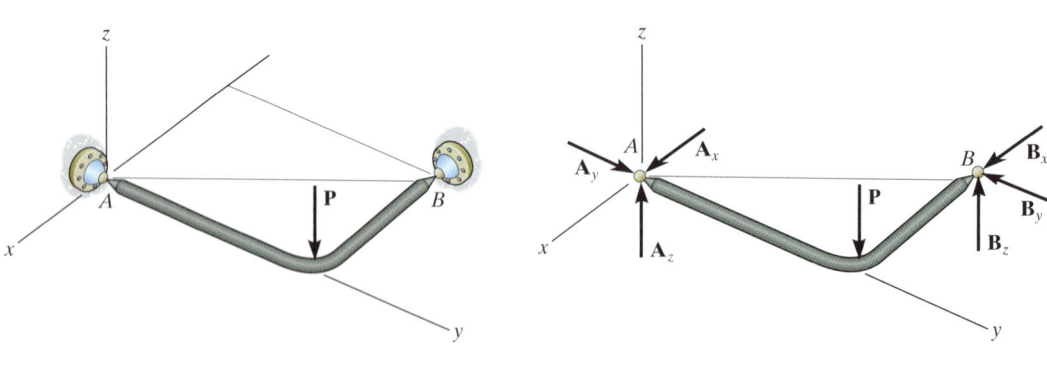

(b)

Fig. 5–25

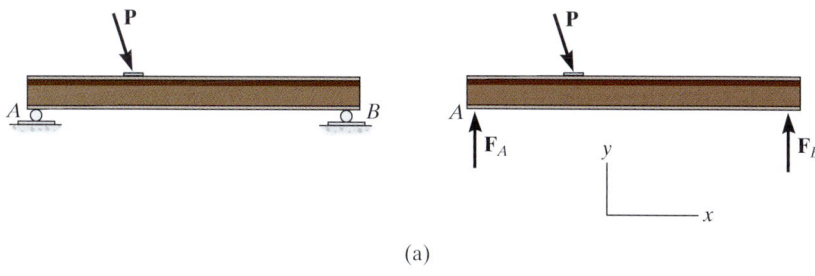

(a)

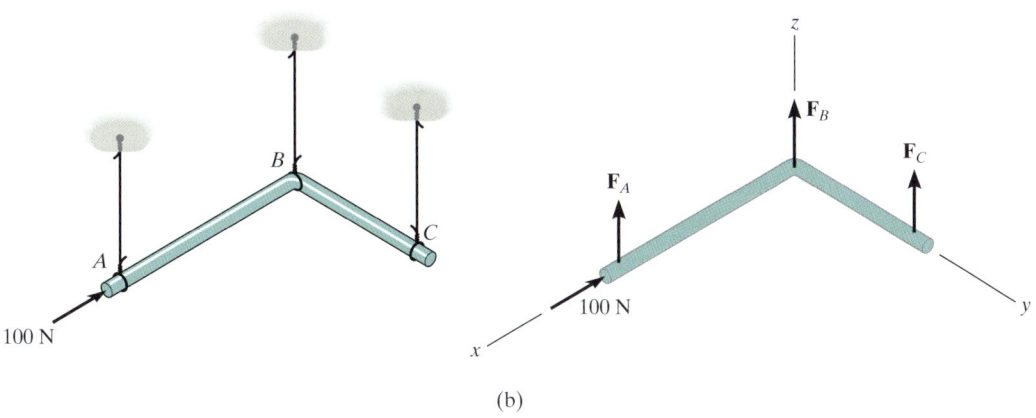

(b)

Fig. 5–26

Another way in which improper constraining leads to instability occurs when the *reactive forces* are all *parallel*. Two- and three-dimensional examples of this are shown in Fig. 5–26. In both cases, the summation of forces along the *x* axis will not equal zero.

In some cases, a body may have *fewer* reactive forces than equations of equilibrium that must be satisfied. The body then becomes only *partially constrained*. For example, consider member *AB* in Fig. 5–27a with its corresponding free-body diagram in Fig. 5–27b. Here $\Sigma F_y = 0$ will not be satisfied for the loading conditions and therefore equilibrium will not be maintained.

To summarize these points, a body is considered *improperly constrained* if all the reactive forces intersect at a common point or pass through a common axis, or if all the reactive forces are parallel. In engineering practice, these situations should be avoided at all times since they will cause an unstable condition.

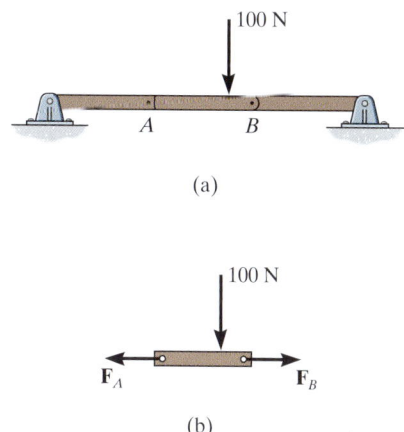

(a)

(b)

Fig. 5–27

Important Points

- Always draw the free-body diagram first when solving any equilibrium problem.
- If a support *prevents translation* of a body in a specific direction, then the support exerts a *force* on the body in that direction.
- If a support *prevents rotation about an axis*, then the support exerts a *couple moment* on the body about the axis.
- If a body is subjected to more unknown reactions than available equations of equilibrium, then the problem is *statically indeterminate*.
- A stable body requires that the lines of action of the reactive forces do not intersect a common axis and are not parallel to one another.

Procedure for Analysis

Three-dimensional equilibrium problems for a rigid body can be solved using the following procedure.

Free-Body Diagram.

- Draw an outlined shape of the body.
- Show all the forces and couple moments acting on the body.
- Establish the origin of the *x, y, z* axes at a convenient point and orient the axes so that they are parallel to as many of the external forces and moments as possible.
- Label all the loadings and specify their directions. In general, show all the *unknown* components having a *positive sense* along the *x, y, z* axes.
- Indicate the dimensions of the body necessary for computing the moments of forces.

Equations of Equilibrium.

- If the *x, y, z* force and moment components seem easy to determine, then apply the six scalar equations of equilibrium; otherwise use the vector equations.
- It is not necessary that the set of axes chosen for force summation coincide with the set of axes chosen for moment summation. Actually, an axis in any arbitrary direction may be chosen for summing forces and moments.
- Choose the direction of an axis for moment summation such that it intersects the lines of action of as many unknown forces as possible. Realize that the moments of forces passing through points on this axis and the moments of forces which are parallel to the axis will then be zero.
- If the solution of the equilibrium equations yields a negative scalar for a force or couple moment magnitude, it indicates that the sense is opposite to that assumed on the free-body diagram.

EXAMPLE 5.15

The homogeneous plate shown in Fig. 5–28a has a mass of 100 kg and is subjected to a force and couple moment along its edges. If it is supported in the horizontal plane by a roller at A, a ball-and-socket joint at B, and a cord at C, determine the components of reaction at these supports.

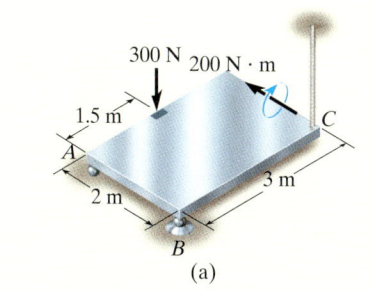

(a)

SOLUTION (SCALAR ANALYSIS)

Free-Body Diagram. There are five unknown reactions acting on the plate, as shown in Fig. 5–28b. Each of these reactions is assumed to act in a positive coordinate direction.

Equations of Equilibrium. Since the three-dimensional geometry is rather simple, a *scalar analysis* provides a *direct solution* to this problem. A force summation along each axis yields

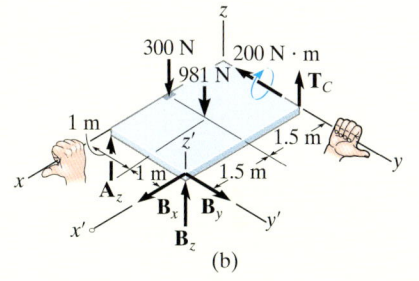

(b)

Fig. 5–28

$$\Sigma F_x = 0; \qquad B_x = 0 \qquad\qquad\qquad Ans.$$

$$\Sigma F_y = 0; \qquad B_y = 0 \qquad\qquad\qquad Ans.$$

$$\Sigma F_z = 0; \qquad A_z + B_z + T_C - 300\ \text{N} - 981\ \text{N} = 0 \qquad (1)$$

Recall that the moment of a force about an axis is equal to the product of the force magnitude and the perpendicular distance (moment arm) from the line of action of the force to the axis. Also, forces that are parallel to an axis or pass through it create no moment about the axis. Hence, summing moments about the positive x and y axes, we have

$$\Sigma M_x = 0; \qquad T_C(2\ \text{m}) - 981\ \text{N}(1\ \text{m}) + B_z(2\ \text{m}) = 0 \qquad (2)$$

$$\Sigma M_y = 0; \qquad 300\ \text{N}(1.5\ \text{m}) + 981\ \text{N}(1.5\ \text{m}) - B_z(3\ \text{m}) - A_z(3\ \text{m})$$
$$- 200\ \text{N} \cdot \text{m} = 0 \qquad (3)$$

The components of the force at B can be eliminated if moments are summed about the x' and y' axes. We obtain

$$\Sigma M_{x'} = 0; \qquad 981\ \text{N}(1\ \text{m}) + 300\ \text{N}(2\ \text{m}) - A_z(2\ \text{m}) = 0 \qquad (4)$$

$$\Sigma M_{y'} = 0; \qquad -300\ \text{N}(1.5\ \text{m}) - 981\ \text{N}(1.5\ \text{m}) - 200\ \text{N} \cdot \text{m}$$
$$+ T_C(3\ \text{m}) = 0 \qquad (5)$$

Solving Eqs. 1 through 3 or the more convenient Eqs. 1, 4, and 5 yields

$$A_z = 790\ \text{N} \quad B_z = -217\ \text{N} \quad T_C = 707\ \text{N} \qquad Ans.$$

The negative sign indicates that **B**$_z$ acts downward.

NOTE: The solution of this problem does not require a summation of moments about the z axis. The plate is partially constrained since the supports cannot prevent it from turning about the z axis if a force is applied to it in the x–y plane.

EXAMPLE 5.16

Determine the components of reaction that the ball-and-socket joint at A, the smooth journal bearing at B, and the roller support at C exert on the rod assembly in Fig. 5–29a.

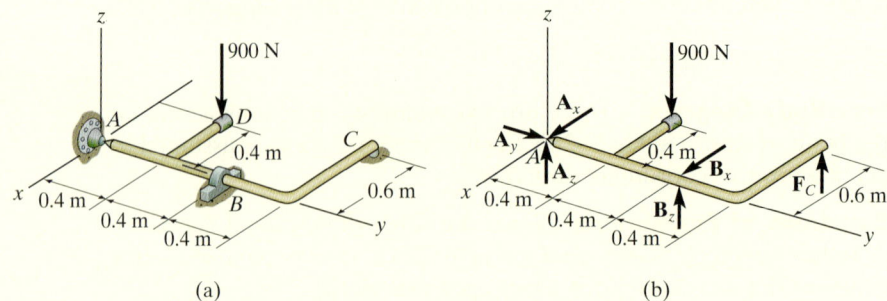

(a)   (b)

Fig. 5–29

SOLUTION

Free-Body Diagram. As shown on the free-body diagram, Fig. 5–29b, the reactive forces of the supports will prevent the assembly from rotating about each coordinate axis, and so the journal bearing at B only exerts reactive forces on the member. No couple moments are required.

Equations of Equilibrium. A direct solution for A_y can be obtained by summing forces along the y axis.

$$\Sigma F_y = 0; \qquad A_y = 0 \qquad\qquad\qquad\qquad Ans.$$

The force F_C can be determined directly by summing moments about the y axis.

$$\Sigma M_y = 0; \qquad F_C(0.6 \text{ m}) - 900 \text{ N}(0.4 \text{ m}) = 0$$
$$F_C = 600 \text{ N} \qquad\qquad\qquad Ans.$$

Using this result, B_z can be determined by summing moments about the x axis.

$$\Sigma M_x = 0; \qquad B_z(0.8 \text{ m}) + 600 \text{ N}(1.2 \text{ m}) - 900 \text{ N}(0.4 \text{ m}) = 0$$
$$B_z = -450 \text{ N} \qquad\qquad\qquad Ans.$$

The negative sign indicates that $\mathbf{B}_z$ acts downward. The force B_x can be found by summing moments about the z axis.

$$\Sigma M_z = 0; \qquad -B_x(0.8 \text{ m}) = 0 \quad B_x = 0 \qquad\qquad Ans.$$

Thus,

$$\Sigma F_x = 0; \qquad A_x + 0 = 0 \qquad A_x = 0 \qquad\qquad Ans.$$

Finally, using the results of B_z and F_C.

$$\Sigma F_z = 0; \qquad A_z + (-450 \text{ N}) + 600 \text{ N} - 900 \text{ N} = 0$$
$$A_z = 750 \text{ N} \qquad\qquad\qquad Ans.$$

EXAMPLE | 5.17

The boom is used to support the 375-N ($\approx$37.5-kg) flowerpot in Fig. 5–30a. Determine the tension developed in wires AB and AC.

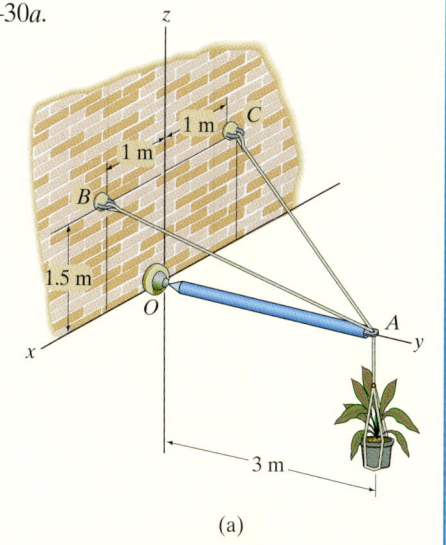

SOLUTION

Free-Body Diagram. The free-body diagram of the boom is shown in Fig. 5–30b.

Equations of Equilibrium. We will use a vector analysis.

$$\mathbf{F}_{AB} = F_{AB}\left(\frac{\mathbf{r}_{AB}}{r_{AB}}\right) = F_{AB}\left(\frac{\{1\mathbf{i} - 3\mathbf{j} + 1.5\mathbf{k}\}\text{ m}}{\sqrt{(1\text{ m})^2 + (-3\text{ m})^2 + (1.5\text{ m})^2}}\right)$$

$$= \tfrac{2}{7}F_{AB}\mathbf{i} - \tfrac{6}{7}F_{AB}\mathbf{j} + \tfrac{3}{7}F_{AB}\mathbf{k}$$

$$\mathbf{F}_{AC} = F_{AC}\left(\frac{\mathbf{r}_{AC}}{r_{AC}}\right) = F_{AC}\left(\frac{\{-1\mathbf{i} - 3\mathbf{j} + 1.5\mathbf{k}\}\text{ m}}{\sqrt{(-1\text{ m})^2 + (-3\text{ m})^2 + (1.5\text{ m})^2}}\right)$$

$$= -\tfrac{2}{7}F_{AC}\mathbf{i} - \tfrac{6}{7}F_{AC}\mathbf{j} + \tfrac{3}{7}F_{AC}\mathbf{k}$$

Fig. 5–30

We can eliminate the force reaction at O by writing the moment equation of equilibrium about point O.

$$\Sigma \mathbf{M}_O = \mathbf{0}; \qquad \mathbf{r}_A \times (\mathbf{F}_{AB} + \mathbf{F}_{AC} + \mathbf{W}) = \mathbf{0}$$

$$(6\mathbf{j}) \times \left[\left(\tfrac{2}{7}F_{AB}\mathbf{i} - \tfrac{6}{7}F_{AB}\mathbf{j} + \tfrac{3}{7}F_{AB}\mathbf{k}\right) + \left(-\tfrac{2}{7}F_{AC}\mathbf{i} - \tfrac{6}{7}F_{AC}\mathbf{j} + \tfrac{3}{7}F_{AC}\mathbf{k}\right) + (-375\mathbf{k})\right] = \mathbf{0}$$

$$\left(\tfrac{18}{7}F_{AB} + \tfrac{18}{7}F_{AC} - 2250\right)\mathbf{i} + \left(-\tfrac{12}{7}F_{AB} + \tfrac{12}{7}F_{AC}\right)\mathbf{k} = \mathbf{0}$$

$$\Sigma M_x = 0; \qquad \tfrac{18}{7}F_{AB} + \tfrac{18}{7}F_{AC} - 2250 = 0 \qquad (1)$$

$$\Sigma M_y = 0; \qquad\qquad 0 = 0$$

$$\Sigma M_z = 0; \qquad -\tfrac{12}{7}F_{AB} + \tfrac{12}{7}F_{AC} = 0 \qquad (2)$$

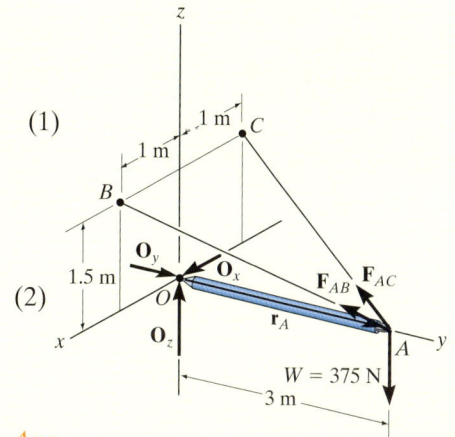

Solving Eqs. (1) and (2) simultaneously,

$$F_{AB} = F_{AC} = 437.5\text{ N} \qquad\qquad \textit{Ans.}$$

(b)

EXAMPLE 5.18

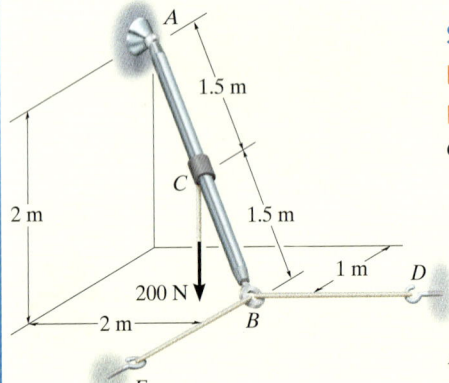

(a)

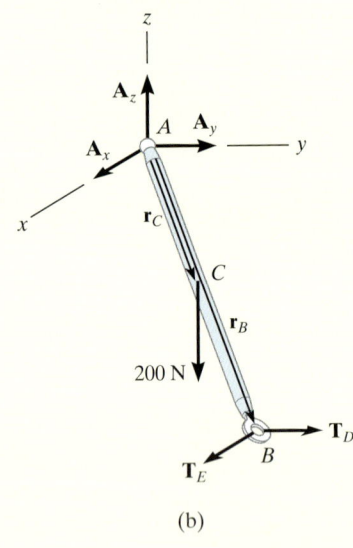

(b)

Fig. 5–31

Rod AB shown in Fig. 5–31a is subjected to the 200-N force. Determine the reactions at the ball-and-socket joint A and the tension in the cables BD and BE. The collar at C is fixed to the rod.

SOLUTION (*VECTOR ANALYSIS*)

Free-Body Diagram.   Fig. 5–31b.

Equations of Equilibrium.   Representing each force on the free-body diagram in Cartesian vector form, we have

$$\mathbf{F}_A = A_x\mathbf{i} + A_y\mathbf{j} + A_z\mathbf{k}$$
$$\mathbf{T}_E = T_E\mathbf{i}$$
$$\mathbf{T}_D = T_D\mathbf{j}$$
$$\mathbf{F} = \{-200\mathbf{k}\}\text{ N}$$

Applying the force equation of equilibrium.

$\Sigma\mathbf{F} = \mathbf{0};$ $\qquad\qquad$ $\mathbf{F}_A + \mathbf{T}_E + \mathbf{T}_D + \mathbf{F} = \mathbf{0}$

$$(A_x + T_E)\mathbf{i} + (A_y + T_D)\mathbf{j} + (A_z - 200)\mathbf{k} = \mathbf{0}$$

$\Sigma F_x = 0;$ $\qquad\qquad$ $A_x + T_E = 0$ $\qquad\qquad$ (1)
$\Sigma F_y = 0;$ $\qquad\qquad$ $A_y + T_D = 0$ $\qquad\qquad$ (2)
$\Sigma F_z = 0;$ $\qquad\qquad$ $A_z - 200 = 0$ $\qquad\qquad$ (3)

Summing moments about point A yields

$\Sigma\mathbf{M}_A = \mathbf{0};$ $\qquad$ $\mathbf{r}_C \times \mathbf{F} + \mathbf{r}_B \times (\mathbf{T}_E + \mathbf{T}_D) = \mathbf{0}$

Since $\mathbf{r}_C = \frac{1}{2}\mathbf{r}_B$, then

$$(0.5\mathbf{i} + 1\mathbf{j} - 1\mathbf{k}) \times (-200\mathbf{k}) + (1\mathbf{i} + 2\mathbf{j} - 2\mathbf{k}) \times (T_E\mathbf{i} + T_D\mathbf{j}) = \mathbf{0}$$

Expanding and rearranging terms gives

$$(2T_D - 200)\mathbf{i} + (-2T_E + 100)\mathbf{j} + (T_D - 2T_E)\mathbf{k} = \mathbf{0}$$

$\Sigma M_x = 0;$ $\qquad\qquad$ $2T_D - 200 = 0$ $\qquad\qquad$ (4)
$\Sigma M_y = 0;$ $\qquad\qquad$ $-2T_E + 100 = 0$ $\qquad\qquad$ (5)
$\Sigma M_z = 0;$ $\qquad\qquad$ $T_D - 2T_E = 0$ $\qquad\qquad$ (6)

Solving Eqs. 1 through 5, we get

$T_D = 100\text{ N}$ $\qquad\qquad$ *Ans.*
$T_E = 50\text{ N}$ $\qquad\qquad$ *Ans.*
$A_x = -50\text{ N}$ $\qquad\qquad$ *Ans.*
$A_y = -100\text{ N}$ $\qquad\qquad$ *Ans.*
$A_z = 200\text{ N}$ $\qquad\qquad$ *Ans.*

NOTE: The negative sign indicates that $\mathbf{A}_x$ and $\mathbf{A}_y$ have a sense which is opposite to that shown on the free-body diagram, Fig. 5–31b.

EXAMPLE 5.19

The bent rod in Fig. 5–32a is supported at A by a journal bearing, at D by a ball-and-socket joint, and at B by means of cable BC. Using only *one equilibrium equation*, obtain a direct solution for the tension in cable BC. The bearing at A is capable of exerting force components only in the z and y directions since it is properly aligned on the shaft. In other words, no couple moments are required at this support.

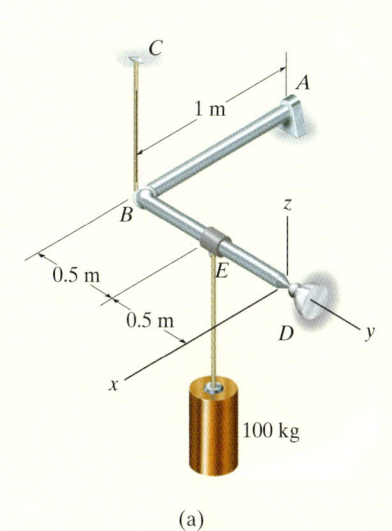

(a)

5

SOLUTION (*VECTOR ANALYSIS*)

Free-Body Diagram.   As shown in Fig. 5–32b, there are six unknowns.

Equations of Equilibrium.   The cable tension T_B may be obtained *directly* by summing moments about an axis that passes through points D and A. Why? The direction of this axis is defined by the unit vector **u**, where

$$\mathbf{u} = \frac{\mathbf{r}_{DA}}{r_{DA}} = -\frac{1}{\sqrt{2}}\mathbf{i} - \frac{1}{\sqrt{2}}\mathbf{j}$$

$$= -0.7071\mathbf{i} - 0.7071\mathbf{j}$$

Hence, the sum of the moments about this axis is zero provided

$$\Sigma M_{DA} = \mathbf{u} \cdot \Sigma(\mathbf{r} \times \mathbf{F}) = 0$$

Here **r** represents a position vector drawn from *any point* on the axis DA to any point on the line of action of force **F** (see Eq. 4–11). With reference to Fig. 5–32b, we can therefore write

$$\mathbf{u} \cdot (\mathbf{r}_B \times \mathbf{T}_B + \mathbf{r}_E \times \mathbf{W}) = \mathbf{0}$$

$$(-0.7071\mathbf{i} - 0.7071\mathbf{j}) \cdot \left[(-1\mathbf{j}) \times (T_B\mathbf{k})\right.$$

$$\left. + (-0.5\mathbf{j}) \times (-981\mathbf{k})\right] = \mathbf{0}$$

$$(-0.7071\mathbf{i} - 0.7071\mathbf{j}) \cdot [(-T_B + 490.5)\mathbf{i}] = \mathbf{0}$$

$$-0.7071(-T_B + 490.5) + 0 + 0 = 0$$

$$T_B = 490.5 \text{ N} \qquad Ans.$$

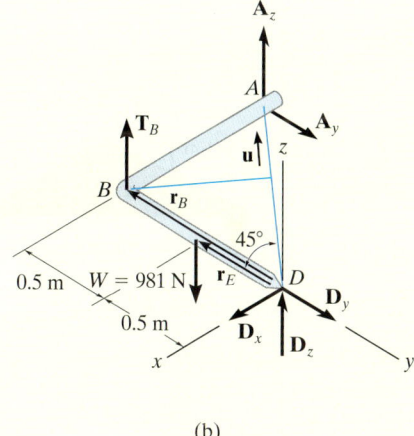

(b)

Fig. 5–32

Since the moment arms from the axis to $\mathbf{T}_B$ and **W** are easy to obtain, we can also determine this result using a scalar analysis. As shown in Fig. 5–32b,

$$\Sigma M_{DA} = 0; \quad T_B(1 \text{ m sin } 45°) - 981 \text{ N}(0.5 \text{ m sin } 45°) = 0$$

$$T_B = 490.5 \text{ N} \qquad Ans.$$

FUNDAMENTAL PROBLEMS

All problem solutions must include an FBD.

F5–7. The uniform plate has a weight of 50 kN. Determine the tension in each of the supporting cables.

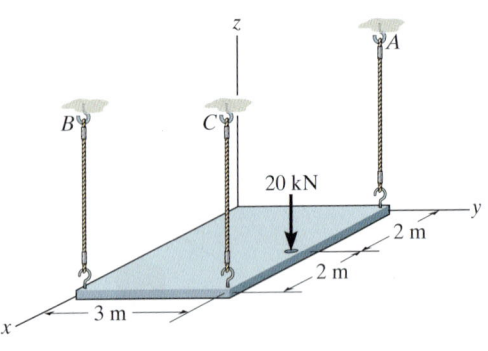

F5–7

F5–8. Determine the reactions at the roller support A, the ball-and-socket joint D, and the tension in cable BC for the plate.

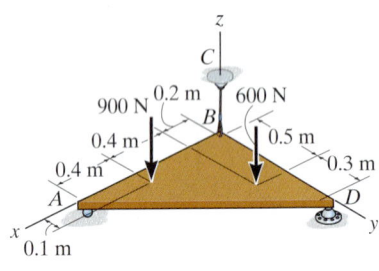

F5–8

F5–9. The rod is supported by smooth journal bearings at A, B, and C and is subjected to the two forces. Determine the reactions at these supports.

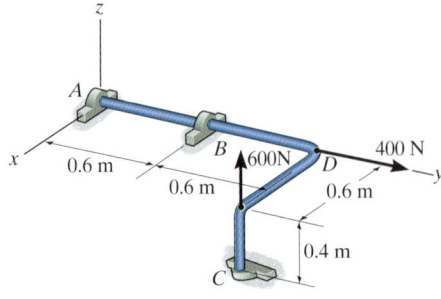

F5–9

F5–10. Determine the support reactions at the smooth journal bearings A, B, and C of the pipe assembly.

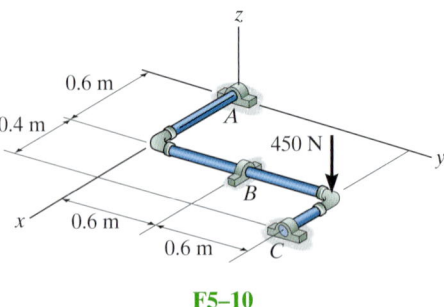

F5–10

F5–11. Determine the force developed in the short link BD, and the tension in the cords CE and CF, and the reactions of the ball-and-socket joint A on the block.

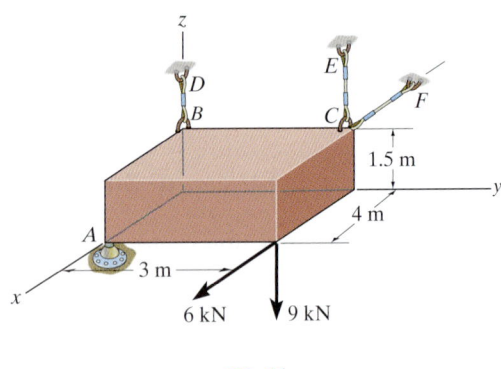

F5–11

F5–12. Determine the components of reaction that the thrust bearing A and cable BC exert on the bar.

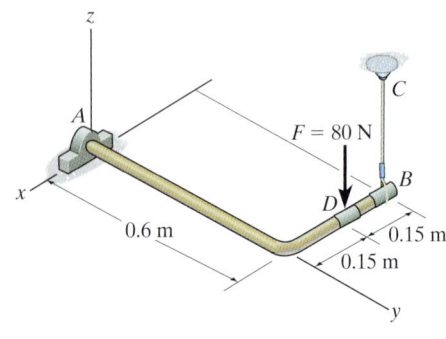

F5–12

PROBLEMS

All problem solutions must include an FBD.

5–62. The uniform load has a mass of 600 kg and is lifted using a uniform 30-kg strongback beam *BAC* and four ropes as shown. Determine the tension in each rope and the force that must be applied at *A*.

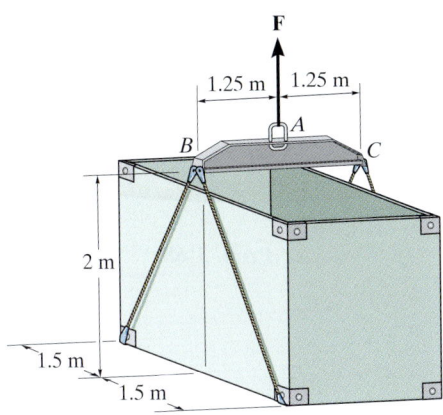

Prob. 5–62

5–63. Due to an unequal distribution of fuel in the wing tanks, the centers of gravity for the airplane fuselage *A* and wings *B* and *C* are located as shown. If these components have weights $W_A = 225$ kN, $W_B = 40$ kN, and $W_C = 30$ kN, determine the normal reactions of the wheels *D*, *E*, and *F* on the ground.

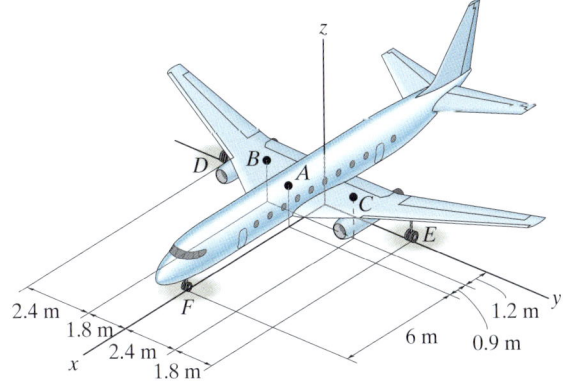

Prob. 5–63

5–64. The wing of the jet aircraft is subjected to a thrust of $T = 8$ kN from its engine and the resultant lift force $L = 45$ kN. If the mass of the wing is 2.1 Mg and the mass center is at *G*, determine the *x, y, z* components of reaction where the wing is fixed to the fuselage *A*.

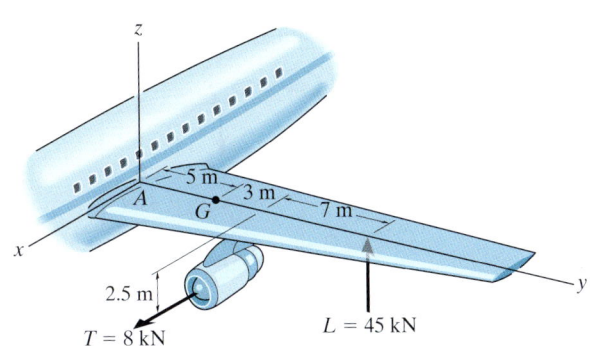

Prob. 5–64

5–65. The cart supports the uniform crate having a mass of 85 kg. Determine the vertical reactions on the three casters at *A, B,* and *C*. The caster at *B* is not shown. Neglect the mass of the cart and assume the casters at *A* and *B* are at the rear corner of the cart.

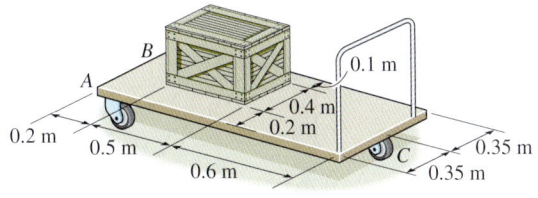

Prob. 5–65

5–66. The pipe assembly supports the vertical loads shown. Determine the components of reaction at the ball-and-socket joint A and the tension in the supporting cables BC and BD.

*5–68. Determine the force components acting on the ball-and-socket at A, the reaction at the roller B and the tension on the cord CD needed for equilibrium of the quarter circular plate.

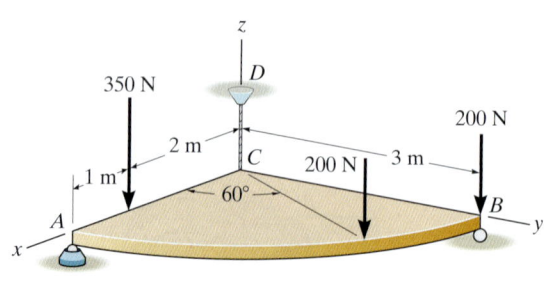

Prob. 5–68

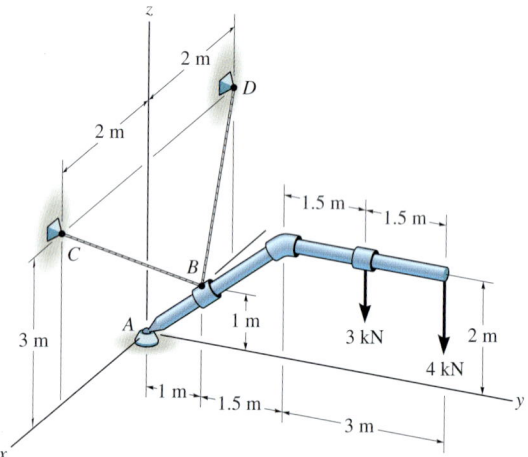

Prob. 5–66

5–67. The boom is supported by a ball-and-socket joint at A and a guy wire at B. If the 5-kN loads lie in a plane which is parallel to the x–y plane, determine the x, y, z components of reaction at A and the tension in the cable at B.

5–69. Determine the components of reaction acting at the smooth journal bearings A, B, and C.

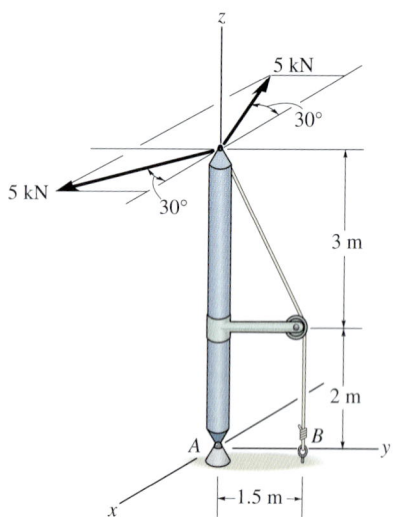

Prob. 5–67

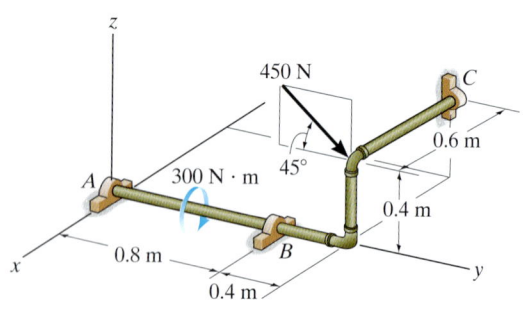

Prob. 5–69

5

5–70. The circular plate has a weight W and center of gravity at its center. If it is supported by three vertical cords tied to its edge, determine the largest distance d from the center to where any vertical force **P** can be applied so as not to cause the force in any one of the cables to become zero.

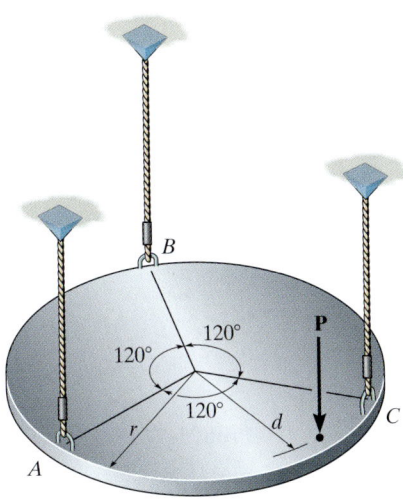

Prob. 5–70

5–71. Determine the support reactions at the smooth collar A and the normal reaction at the roller support B.

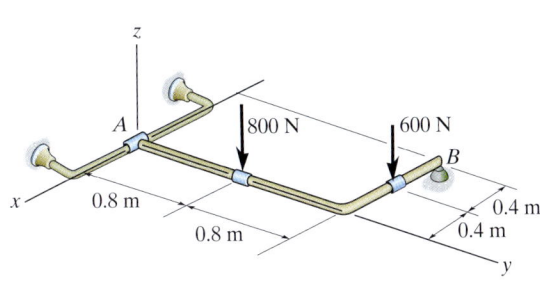

Prob. 5–71

*5–72.** The pole is subjected to the two forces shown. Determine the components of reaction of A assuming it to be a ball-and-socket joint. Also, compute the tension in each of the guy wires, BC and ED.

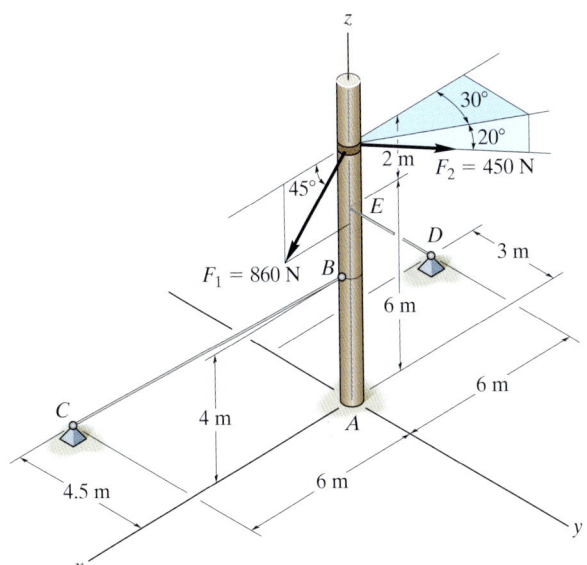

Prob. 5–72

5–73. If $P = 6$ kN, $x = 0.75$ m, and $y = 1$ m, determine the tension developed in cables AB, CD, and EF. Neglect the weight of the plate.

5–74. Determine the location x and y of the point of application of force **P** so that the tension developed in cables AB, CD, and EF is the same. Neglect the weight of the plate.

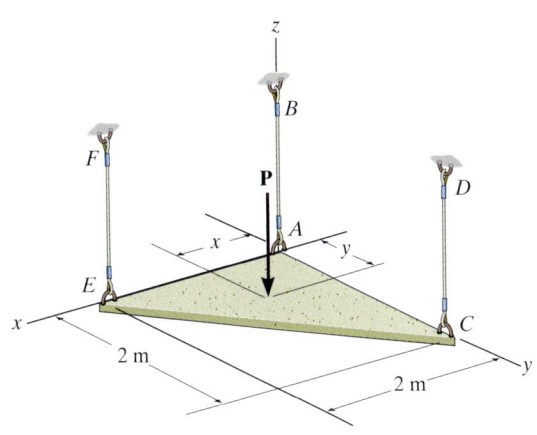

Probs. 5–73/74

5–75. If the pulleys are fixed to the shaft, determine the magnitude of tension **T** and the x, y, z components of reaction at the smooth thrust bearing A and smooth journal bearing B.

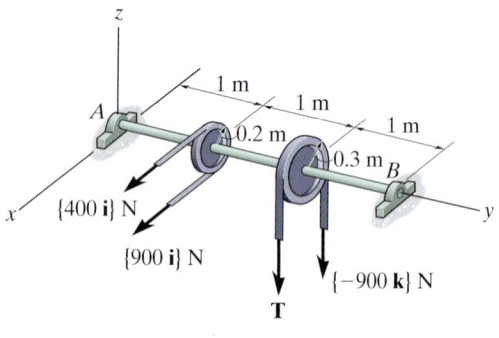

Prob. 5–75

5–77. A uniform square table having a weight W and sides a is supported by three vertical legs. Determine the smallest vertical force **P** that can be applied to its top that will cause it to tip over.

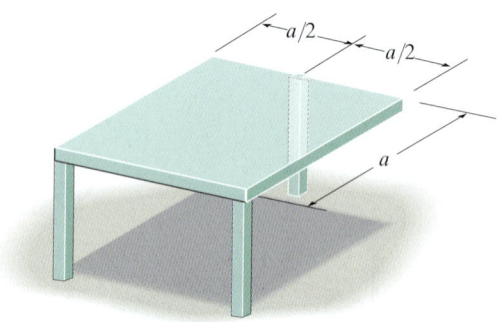

Prob. 5–77

*5–76. The 2-kg ball rests between the 45° grooves A and B of the 10° incline and against a vertical wall at C. If all three surfaces of contact are smooth, determine the reactions of the surfaces on the ball. *Hint:* Use the x, y, z axes, with origin at the center of the ball, and the z axis inclined as shown.

5–78. The shaft is supported by three smooth journal bearings at A, B, and C. Determine the components of reaction at these bearings.

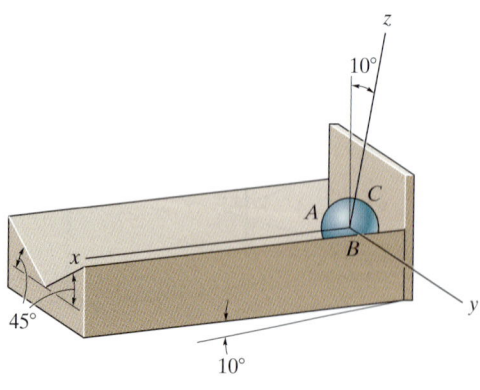

Prob. 5–76

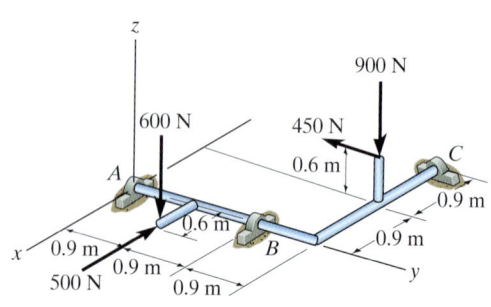

Prob. 5–78

5–79.   The platform has a mass of 3 Mg and center of mass located at G. If it is lifted with constant velocity using the three cables, determine the force in each of the cables.

*5–80.   The platform has a mass of 2 Mg and center of mass located at G. If it is lifted using the three cables, determine the force in each of the cables. Solve for each force by using a single moment equation of equilibrium.

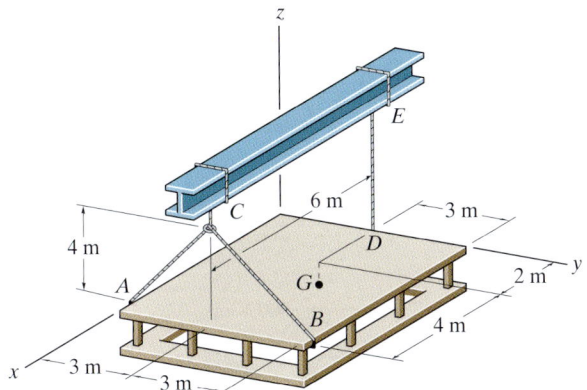

Probs. 5–79/80

5–81.   The sign has a mass of 100 kg with center of mass at G. Determine the x, y, z components of reaction at the ball-and-socket joint A and the tension in wires BC and BD.

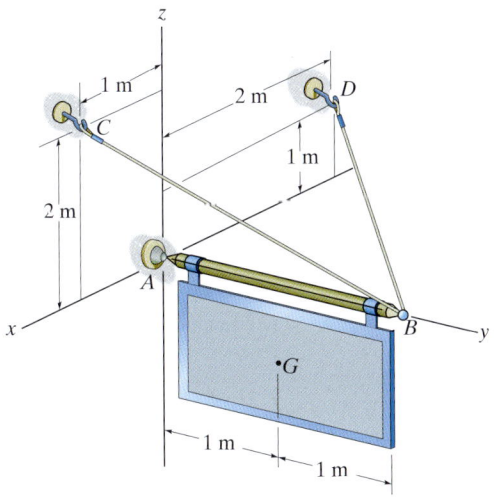

Prob. 5–81

5–82.   Determine the tension in cables BD and CD and the x, y, z components of reaction at the ball-and-socket joint at A.

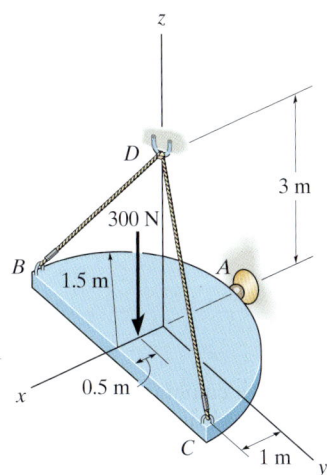

Prob. 5–82

5–83.   Both pulleys are fixed to the shaft and as the shaft turns with constant angular velocity, the power of pulley A is transmitted to pulley B. Determine the horizontal tension **T** in the belt on pulley B and the x, y, z components of reaction at the journal bearing C and thrust bearing D if $\theta = 0°$. The bearings are in proper alignment and exert only force reactions on the shaft.

*5–84.   Both pulleys are fixed to the shaft and as the shaft turns with constant angular velocity, the power of pulley A is transmitted to pulley B. Determine the horizontal tension **T** in the belt on pulley B and the x, y, z components of reaction at the journal bearing C and thrust bearing D if $\theta = 45°$. The bearings are in proper alignment and exert only force reactions on the shaft.

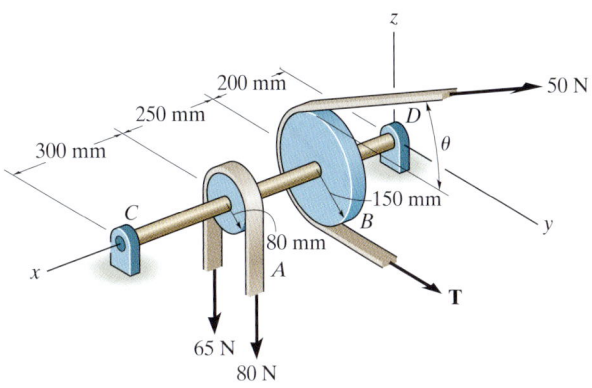

Probs. 5–83/84

CHAPTER REVIEW

Equilibrium

A body in equilibrium is at rest or can translate with constant velocity.

$$\Sigma F = 0$$
$$\Sigma M = 0$$

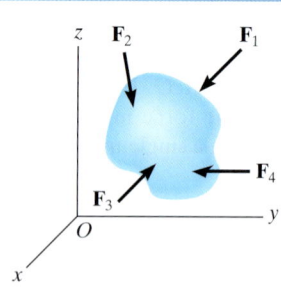

Two Dimensions

Before analyzing the equilibrium of a body, it is first necessary to draw its free-body diagram. This is an outlined shape of the body, which shows all the forces and couple moments that act on it.

Couple moments can be placed anywhere on a free-body diagram since they are free vectors. Forces can act at any point along their line of action since they are sliding vectors.

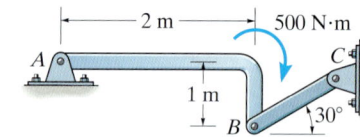

Angles used to resolve forces, and dimensions used to take moments of the forces, should also be shown on the free-body diagram.

Some common types of supports and their reactions are shown below in two dimensions.

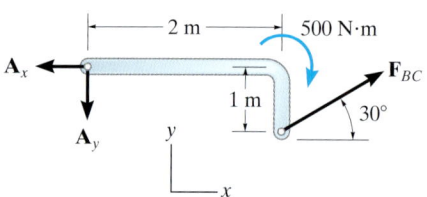

Remember that a support will exert a force on the body in a particular direction if it prevents translation of the body in that direction, and it will exert a couple moment on the body if it prevents rotation.

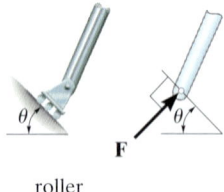

roller

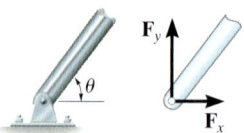

smooth pin or hinge

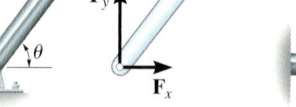

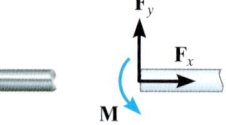

fixed support

The three scalar equations of equilibrium can be applied when solving problems in two dimensions, since the geometry is easy to visualize.

$$\Sigma F_x = 0$$
$$\Sigma F_y = 0$$
$$\Sigma M_O = 0$$

For the most direct solution, try to sum forces along an axis that will eliminate as many unknown forces as possible. Sum moments about a point A that passes through the line of action of as many unknown forces as possible.

$\Sigma F_x = 0;$

$A_x - P_2 = 0 \quad A_x = P_2$

$\Sigma M_A = 0;$

$P_2 d_2 + B_y d_B - P_1 d_1 = 0$

$B_y = \dfrac{P_1 d_1 - P_2 d_2}{d_B}$

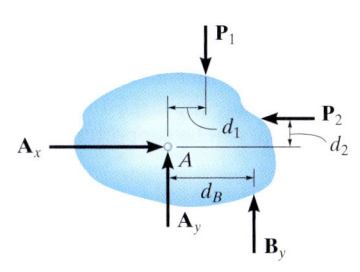

Three Dimensions

Some common types of supports and their reactions are shown here in three dimensions.

 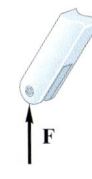   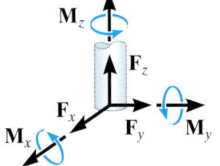

roller   ball and socket   fixed support

In three dimensions, it is often advantageous to use a Cartesian vector analysis when applying the equations of equilibrium. To do this, first express each known and unknown force and couple moment shown on the free-body diagram as a Cartesian vector. Then set the force summation equal to zero. Take moments about a point O that lies on the line of action of as many unknown force components as possible. From point O direct position vectors to each force, and then use the cross product to determine the moment of each force.

The six scalar equations of equilibrium are established by setting the respective $\mathbf{i}$, $\mathbf{j}$, and $\mathbf{k}$ components of these force and moment summations equal to zero.

$$\Sigma \mathbf{F} = \mathbf{0}$$
$$\Sigma \mathbf{M}_O = \mathbf{0}$$

$\Sigma F_x = 0 \qquad \Sigma M_x = 0$
$\Sigma F_y = 0 \qquad \Sigma M_y = 0$
$\Sigma F_z = 0 \qquad \Sigma M_z = 0$

Determinacy and Stability

If a body is supported by a minimum number of constraints to ensure equilibrium, then it is statically determinate. If it has more constraints than required, then it is statically indeterminate.

To properly constrain the body, the reactions must not all be parallel to one another or concurrent.

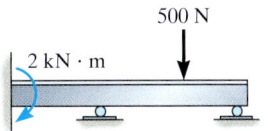

Statically indeterminate, five reactions, three equilibrium equations

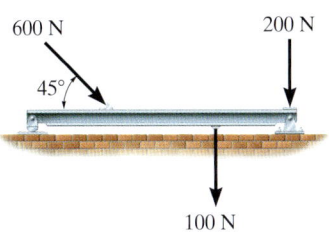

Proper constraint, statically determinate

REVIEW PROBLEMS

5–85. If the roller at B can sustain a maximum load of 3 kN, determine the largest magnitude of each of the three forces $\mathbf{F}$ that can be supported by the truss.

5–87. The symmetrical shelf is subjected to a uniform load of 4 kPa. Support is provided by a bolt (or pin) located at each end A and A' and by the symmetrical brace arms, which bear against the smooth wall on both sides at B and B'. Determine the force resisted by each bolt at the wall and the normal force at B for equilibrium.

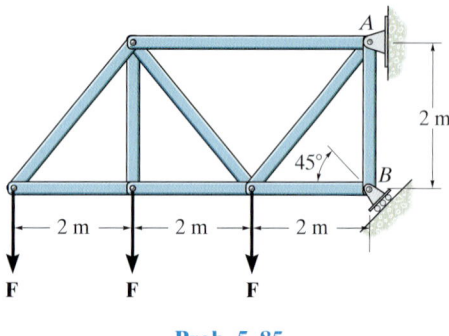

Prob. 5–85

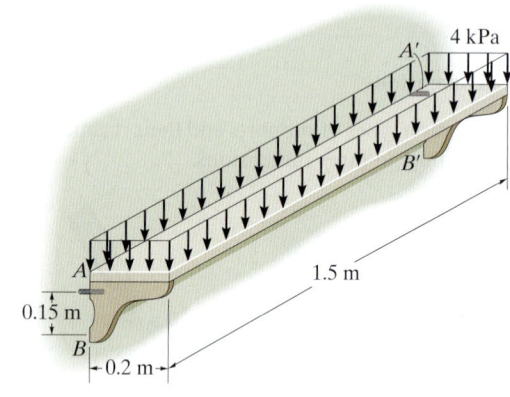

Prob. 5–87

5–86. Determine the normal reaction at the roller A and horizontal and vertical components at pin B for equilibrium of the member.

*5–88. Determine the x and z components of reaction at the journal bearing A and the tension in cords BC and BD necessary for equilibrium of the rod.

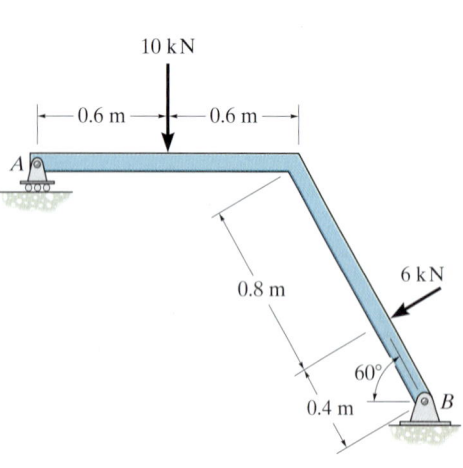

Prob. 5–86

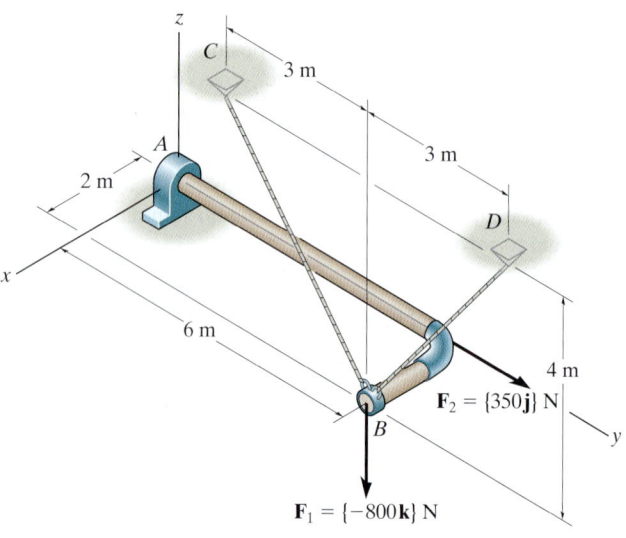

Prob. 5–88

5–89. The uniform rod of length L and weight W is supported on the smooth planes. Determine its position θ for equilibrium. Neglect the thickness of the rod.

5–91. Determine the x, y, z components of reaction at the fixed wall A. The 150-N force is parallel to the z axis and the 200-N force is parallel to the y axis.

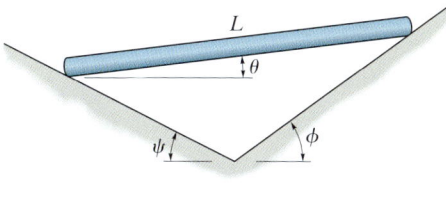

Prob. 5–89

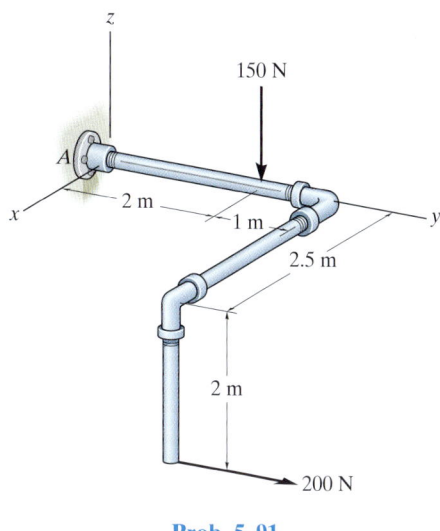

Prob. 5–91

5–90. Determine the horizontal and vertical components of force at the pin A and the reaction at the rocker B of the curved beam.

*5–92. Determine the reactions at the supports A and B for equilibrium of the beam.

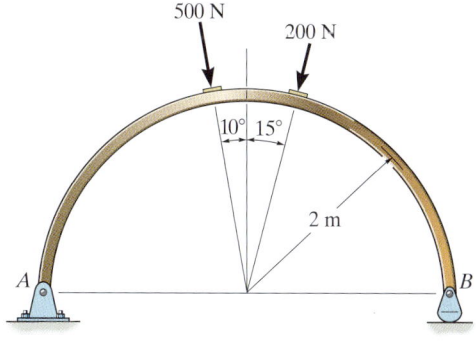

Prob. 5–90

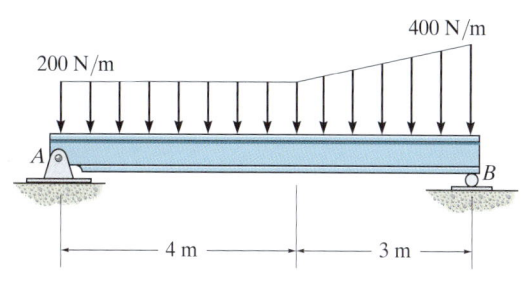

Prob. 5–92

Chapter 6

In order to design the many parts of this boom assembly it is required that we know the forces that they must support. In this chapter we will show how to analyze such structures using the equations of equilibrium.

Structural Analysis

Video Solutions are available for selected questions in this chapter.

6.1 Simple Trusses

A *truss* is a structure composed of slender members joined together at their end points. The members commonly used in construction consist of wooden struts or metal bars. In particular, *planar* trusses lie in a single plane and are often used to support roofs and bridges. The truss shown in Fig. 6–1a is an example of a typical roof-supporting truss. In this figure, the roof load is transmitted to the truss *at the joints* by means of a series of *purlins*. Since this loading acts in the same plane as the truss, Fig. 6–1b, the analysis of the forces developed in the truss members will be two-dimensional.

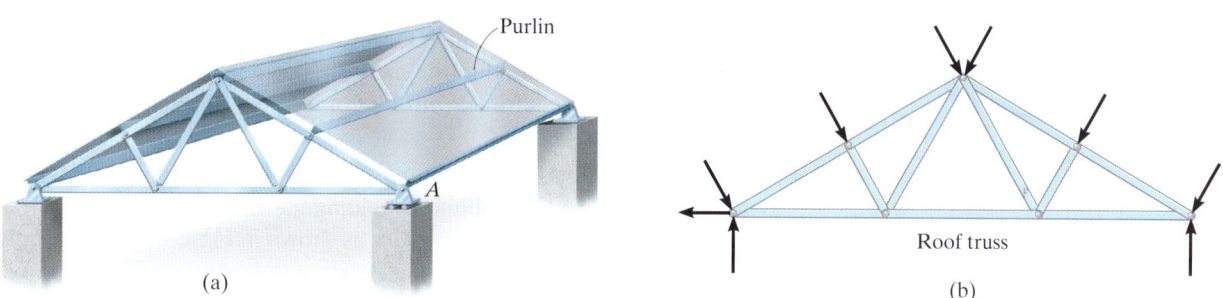

Purlin

A

(a)

Roof truss

(b)

Fig. 6–1

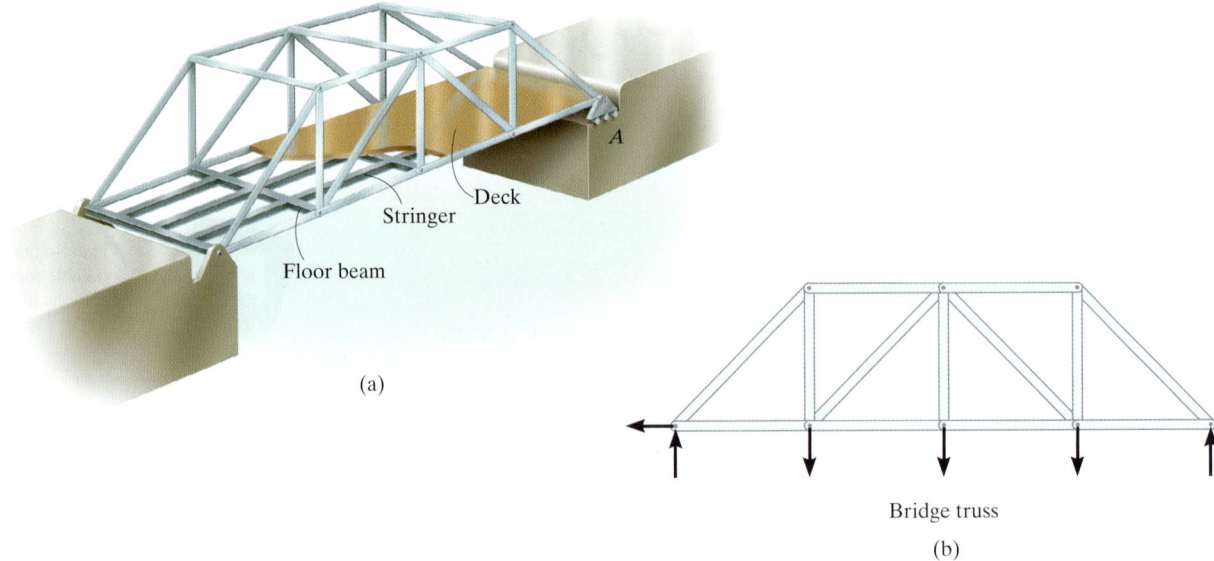

(a)

Stringer
Deck
Floor beam
A

Bridge truss

(b)

Fig. 6–2

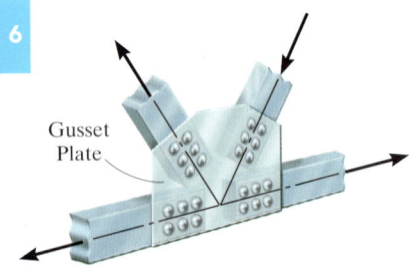

Gusset
Plate

(a)

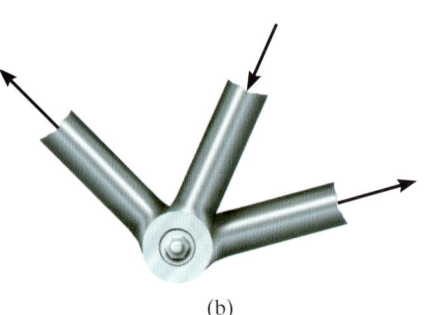

(b)

Fig. 6–3

In the case of a bridge, such as shown in Fig. 6–2a, the load on the deck is first transmitted to *stringers*, then to *floor beams*, and finally to the *joints* of the two supporting side trusses. Like the roof truss, the bridge truss loading is also coplanar, Fig. 6–2b.

When bridge or roof trusses extend over large distances, a rocker or roller is commonly used for supporting one end, for example, joint *A* in Figs. 6–1a and 6–2a. This type of support allows freedom for expansion or contraction of the members due to a change in temperature or application of loads.

Assumptions for Design. To design both the members and the connections of a truss, it is necessary first to determine the *force* developed in each member when the truss is subjected to a given loading. To do this we will make two important assumptions:

• **All loadings are applied at the joints.** In most situations, such as for bridge and roof trusses, this assumption is true. Frequently the weight of the members is neglected because the force supported by each member is usually much larger than its weight. However, if the weight is to be included in the analysis, it is generally satisfactory to apply it as a vertical force, with half of its magnitude applied at each end of the member.

• **The members are joined together by smooth pins.** The joint connections are usually formed by bolting or welding the ends of the members to a common plate, called a *gusset plate*, as shown in Fig. 6–3a, or by simply passing a large bolt or pin through each of the members, Fig. 6–3b. We can assume these connections act as pins provided the center lines of the joining members are *concurrent*, as in Fig. 6–3.

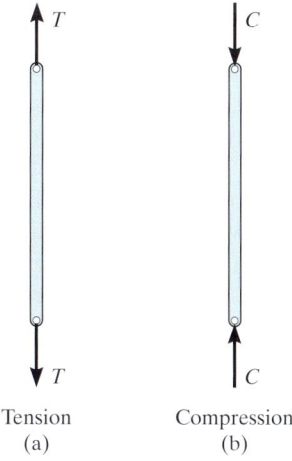

Tension
(a)

Compression
(b)

Fig. 6–4

Because of these two assumptions, *each truss member will act as a two-force member*, and therefore the force acting at each end of the member will be directed along the axis of the member. If the force tends to *elongate* the member, it is a *tensile force* (T), Fig. 6–4a; whereas if it tends to *shorten* the member, it is a *compressive force* (C), Fig. 6–4b. In the actual design of a truss it is important to state whether the nature of the force is tensile or compressive. Often, compression members must be made *thicker* than tension members because of the buckling or column effect that occurs when a member is in compression.

Simple Truss. If three members are pin connected at their ends, they form a *triangular truss* that will be *rigid*, Fig. 6–5. Attaching two more members and connecting these members to a new joint D forms a larger truss, Fig. 6–6. This procedure can be repeated as many times as desired to form an even larger truss. If a truss can be constructed by expanding the basic triangular truss in this way, it is called a *simple truss*.

The use of metal gusset plates in the construction of these Warren trusses is clearly evident.

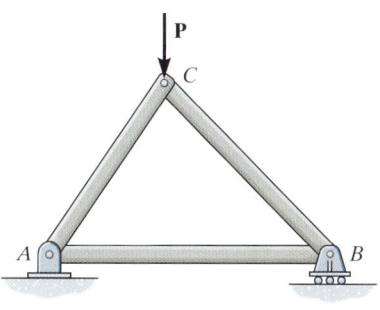

Fig. 6–5

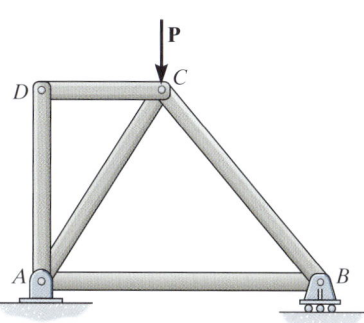

Fig. 6–6

Please refer to the Companion Website
for the animation: *Free-Body Diagram
for an A-Frame*

6.2   The Method of Joints

In order to analyze or design a truss, it is necessary to determine the force in each of its members. One way to do this is to use the method of joints. This method is based on the fact that if the entire truss is in equilibrium, then each of its joints is also in equilibrium. Therefore, if the free-body diagram of each joint is drawn, the force equilibrium equations can then be used to obtain the member forces acting on each joint. Since the members of a *plane truss* are straight two-force members lying in a single plane, each joint is subjected to a force system that is *coplanar and concurrent*. As a result, only $\Sigma F_x = 0$ and $\Sigma F_y = 0$ need to be satisfied for equilibrium.

For example, consider the pin at joint B of the truss in Fig. 6–7a. Three forces act on the pin, namely, the 500-N force and the forces exerted by members BA and BC. The free-body diagram of the pin is shown in Fig. 6–7b. Here, $\mathbf{F}_{BA}$ is "pulling" on the pin, which means that member BA is in *tension;* whereas $\mathbf{F}_{BC}$ is "pushing" on the pin, and consequently member BC is in *compression*. These effects are clearly demonstrated by isolating the joint with small segments of the member connected to the pin, Fig. 6–7c. The pushing or pulling on these small segments indicates the effect of the member being either in compression or tension.

When using the method of joints, always start at a joint having at least one known force and at most two unknown forces, as in Fig. 6–7b. In this way, application of $\Sigma F_x = 0$ and $\Sigma F_y = 0$ yields two algebraic equations which can be solved for the two unknowns. When applying these equations, the correct sense of an unknown member force can be determined using one of two possible methods.

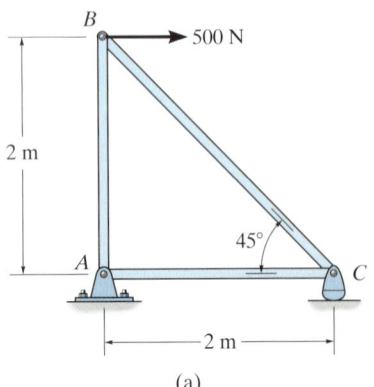

(a)

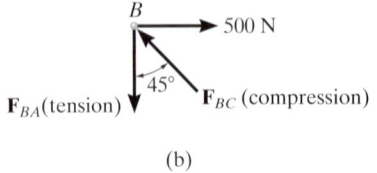

(b)

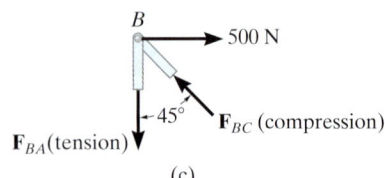

(c)

Fig. 6–7

- The *correct* sense of direction of an unknown member force can, in many cases, be determined "by inspection." For example, $\mathbf{F}_{BC}$ in Fig. 6–7b must push on the pin (compression) since its horizontal component, $F_{BC} \sin 45°$, must balance the 500-N force ($\Sigma F_x = 0$). Likewise, $\mathbf{F}_{BA}$ is a tensile force since it balances the vertical component, $F_{BC} \cos 45°$ ($\Sigma F_y = 0$). In more complicated cases, the sense of an unknown member force can be *assumed*; then, after applying the equilibrium equations, the assumed sense can be verified from the numerical results. A *positive* answer indicates that the sense is *correct*, whereas a *negative* answer indicates that the sense shown on the free-body diagram must be *reversed*.

- *Always assume* the *unknown member forces* acting on the joint's free-body diagram to be in *tension*; i.e., the forces "pull" on the pin. If this is done, then numerical solution of the equilibrium equations will yield *positive scalars for members in tension and negative scalars for members in compression*. Once an unknown member force is found, use its *correct* magnitude and sense (T or C) on subsequent joint free-body diagrams.

The forces in the members of this simple roof truss can be determined using the method of joints.

Procedure for Analysis

The following procedure provides a means for analyzing a truss using the method of joints.

- Draw the free-body diagram of a joint having at least one known force and at most two unknown forces. (If this joint is at one of the supports, then it may be necessary first to calculate the external reactions at the support.)

- Use one of the two methods described above for establishing the sense of an unknown force.

- Orient the x and y axes such that the forces on the free-body diagram can be easily resolved into their x and y components and then apply the two force equilibrium equations $\Sigma F_x = 0$ and $\Sigma F_y = 0$. Solve for the two unknown member forces and verify their correct sense.

- Using the calculated results, continue to analyze each of the other joints. Remember that a member in *compression* "pushes" on the joint and a member in *tension* "pulls" on the joint. Also, be sure to choose a joint having at most two unknowns and at least one known force.

6

EXAMPLE 6.1

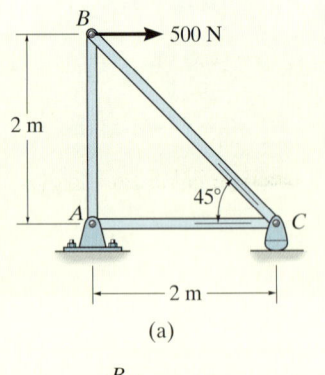

(a)

Determine the force in each member of the truss shown in Fig. 6–8a and indicate whether the members are in tension or compression.

SOLUTION

Since we should have no more than two unknown forces at the joint and at least one known force acting there, we will begin our analysis at joint B.

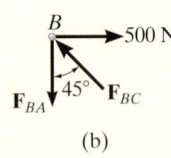

(b)

Joint B. The free-body diagram of the joint at B is shown in Fig. 6–8b. Applying the equations of equilibrium, we have

$$\xrightarrow{+} \Sigma F_x = 0; \qquad 500\ \text{N} - F_{BC} \sin 45° = 0 \qquad F_{BC} = 707.1\ \text{N (C)} \quad \textit{Ans.}$$
$$+\uparrow \Sigma F_y = 0; \qquad F_{BC} \cos 45° - F_{BA} = 0 \qquad F_{BA} = 500\ \text{N (T)} \quad \textit{Ans.}$$

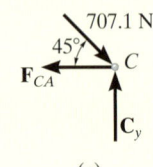

(c)

Since the force in member BC has been calculated, we can proceed to analyze joint C to determine the force in member CA and the support reaction at the rocker.

Joint C. From the free-body diagram of joint C, Fig. 6–8c, we have

$$\xrightarrow{+} \Sigma F_x = 0; \quad -F_{CA} + 707.1 \cos 45°\ \text{N} = 0 \quad F_{CA} = 500\ \text{N (T)} \quad \textit{Ans.}$$
$$+\uparrow \Sigma F_y = 0; \qquad C_y - 707.1 \sin 45°\ \text{N} = 0 \qquad C_y = 500\ \text{N} \quad \textit{Ans.}$$

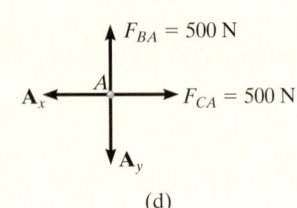

(d)

Joint A. Although it is not necessary, we can determine the components of the support reactions at joint A using the results of F_{CA} and F_{BA}. From the free-body diagram, Fig. 6–8d, we have

$$\xrightarrow{+} \Sigma F_x = 0; \qquad 500\ \text{N} - A_x = 0 \qquad A_x = 500\ \text{N}$$
$$+\uparrow \Sigma F_y = 0; \qquad 500\ \text{N} - A_y = 0 \qquad A_y = 500\ \text{N}$$

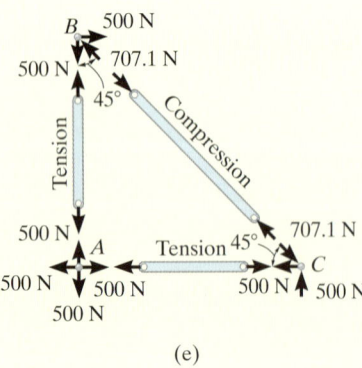

(e)

Fig. 6–8

NOTE: The results of the analysis are summarized in Fig. 6–8e. Note that the free-body diagram of each joint (or pin) shows the effects of all the connected members and external forces applied to the joint, whereas the free-body diagram of each member shows only the effects of the end joints on the member.

EXAMPLE | 6.2

Determine the forces acting in all the members of the truss shown in Fig. 6–9a.

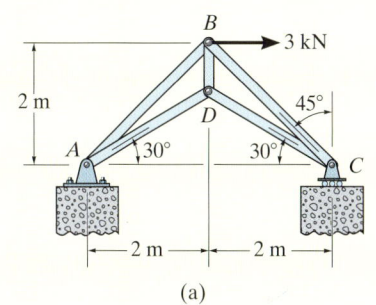

(a)

SOLUTION

By inspection, there are more than two unknowns at each joint. Consequently, the support reactions on the truss must first be determined. Show that they have been correctly calculated on the free-body diagram in Fig. 6–9b. We can now begin the analysis at joint C. Why?

Joint C. From the free-body diagram, Fig. 6–9c,

$$\xrightarrow{+} \Sigma F_x = 0; \qquad -F_{CD}\cos 30° + F_{CB}\sin 45° = 0$$
$$+\uparrow \Sigma F_y = 0; \quad 1.5\ \text{kN} + F_{CD}\sin 30° - F_{CB}\cos 45° = 0$$

These two equations must be solved *simultaneously* for each of the two unknowns. Note, however, that a *direct solution* for one of the unknown forces may be obtained by applying a force summation along an axis that is *perpendicular* to the direction of the other unknown force. For example, summing forces along the y' axis, which is perpendicular to the direction of $\mathbf{F}_{CD}$, Fig. 6–9d, yields a *direct solution* for F_{CB}.

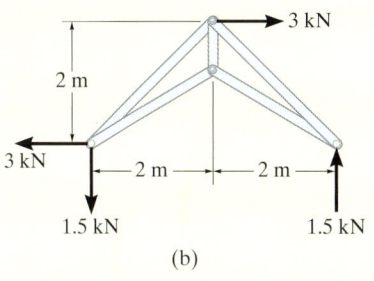

(b)

$$+\nearrow\Sigma F_{y'} = 0; \qquad 1.5\cos 30°\ \text{kN} - F_{CB}\sin 15° = 0$$
$$F_{CB} = 5.019\ \text{kN} = 5.02\ \text{kN (C)} \qquad \textit{Ans.}$$

Then,

$$+\searrow\Sigma F_{x'} = 0;$$
$$-F_{CD} + 5.019\cos 15° - 1.5\sin 30° = 0;\ F_{CD} = 4.10\ \text{kN (T)} \quad \textit{Ans.}$$

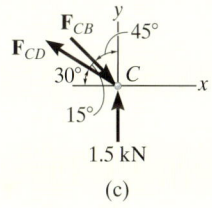

(c)

Joint D. We can now proceed to analyze joint D. The free-body diagram is shown in Fig. 6–9e.

$$\xrightarrow{+} \Sigma F_x = 0; \qquad -F_{DA}\cos 30° + 4.10\cos 30°\ \text{kN} = 0$$
$$F_{DA} = 4.10\ \text{kN (T)} \qquad \textit{Ans.}$$
$$+\uparrow \Sigma F_y = 0; \qquad F_{DB} - 2(4.10\sin 30°\ \text{kN}) = 0$$
$$F_{DB} = 4.10\ \text{kN (T)} \qquad \textit{Ans.}$$

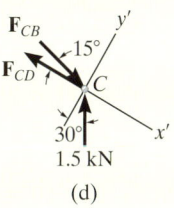

(d)

NOTE: The force in the last member, BA, can be obtained from joint B or joint A. As an exercise, draw the free-body diagram of joint B, sum the forces in the horizontal direction, and show that $F_{BA} = 0.776\ \text{kN (C)}$.

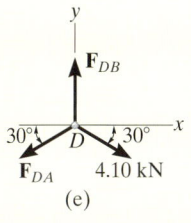

(e)

Fig. 6–9

EXAMPLE 6.3

Determine the force in each member of the truss shown in Fig. 6–10a. Indicate whether the members are in tension or compression.

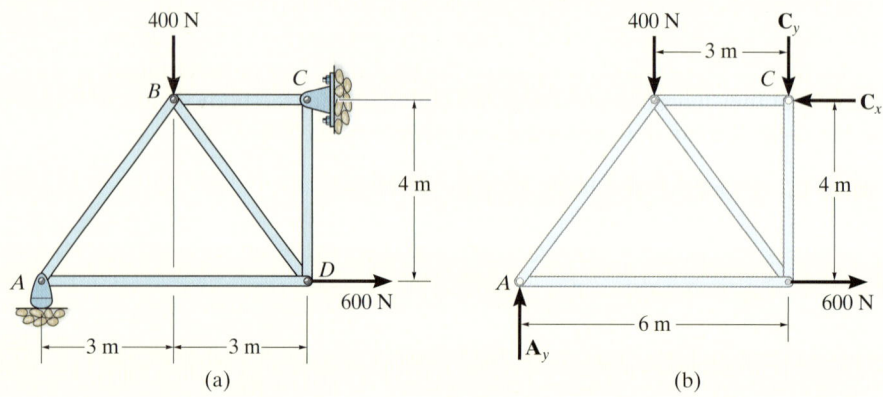

Fig. 6–10

SOLUTION

Support Reactions. No joint can be analyzed until the support reactions are determined, because each joint has at least three unknown forces acting on it. A free-body diagram of the entire truss is given in Fig. 6–10b. Applying the equations of equilibrium, we have

$$\xrightarrow{+} \Sigma F_x = 0; \qquad 600\text{ N} - C_x = 0 \qquad\qquad C_x = 600\text{ N}$$

$$\zeta + \Sigma M_C = 0; \qquad -A_y(6\text{ m}) + 400\text{ N}(3\text{ m}) + 600\text{ N}(4\text{ m}) = 0$$

$$A_y = 600\text{ N}$$

$$+\uparrow \Sigma F_y = 0; \qquad 600\text{ N} - 400\text{ N} - C_y = 0 \qquad C_y = 200\text{ N}$$

The analysis can now start at either joint A or C. The choice is arbitrary since there are one known and two unknown member forces acting on the pin at each of these joints.

Joint A. (Fig. 6–10c). As shown on the free-body diagram, $\mathbf{F}_{AB}$ is assumed to be compressive and $\mathbf{F}_{AD}$ is tensile. Applying the equations of equilibrium, we have

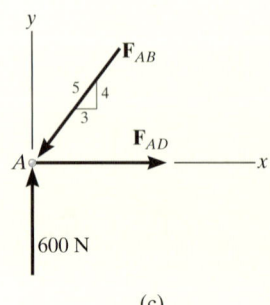

(c)

$$+\uparrow \Sigma F_y = 0; \qquad 600\text{ N} - \tfrac{4}{5}F_{AB} = 0 \qquad F_{AB} = 750\text{ N}\ \ (C) \qquad Ans.$$

$$\xrightarrow{+} \Sigma F_x = 0; \qquad F_{AD} - \tfrac{3}{5}(750\text{ N}) = 0 \qquad F_{AD} = 450\text{ N}\ \ (T) \qquad Ans.$$

Joint D. (Fig. 6–10d). Using the result for F_{AD} and summing forces in the horizontal direction, Fig. 6–10d, we have

$$\xrightarrow{+} \Sigma F_x = 0; \qquad -450 \text{ N} + \tfrac{3}{5}F_{DB} + 600 \text{ N} = 0 \qquad F_{DB} = -250 \text{ N}$$

The negative sign indicates that $\mathbf{F}_{DB}$ acts in the *opposite sense* to that shown in Fig. 6–10d.* Hence,

$$F_{DB} = 250 \text{ N (T)} \qquad\qquad \textit{Ans.}$$

To determine $\mathbf{F}_{DC}$, we can either correct the sense of $\mathbf{F}_{DB}$ on the free-body diagram, and then apply $\Sigma F_y = 0$, or apply this equation and retain the negative sign for F_{DB}, i.e.,

$$+\uparrow \Sigma F_y = 0; \quad -F_{DC} - \tfrac{4}{5}(-250 \text{ N}) = 0 \qquad F_{DC} = 200 \text{ N} \quad (C) \quad \textit{Ans.}$$

Joint C. (Fig. 6–10e).

$$\xrightarrow{+} \Sigma F_x = 0; \qquad F_{CB} - 600 \text{ N} = 0 \qquad\qquad F_{CB} = 600 \text{ N} \quad (C) \quad \textit{Ans.}$$
$$+\uparrow \Sigma F_y = 0; \qquad\qquad 200 \text{ N} - 200 \text{ N} \equiv 0 \quad (check)$$

NOTE: The analysis is summarized in Fig. 6–10f, which shows the free-body diagram for each joint and member.

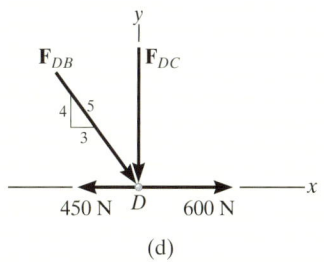

(d)

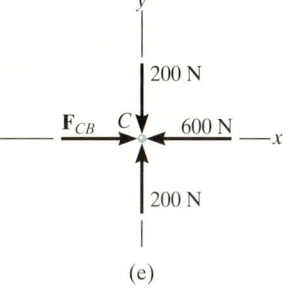

(e)

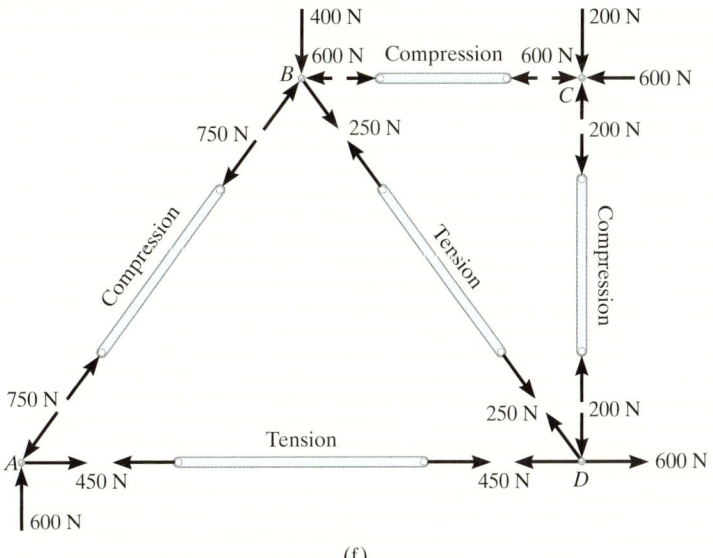

(f)

Fig. 6–10 (cont.)

*The proper sense could have been determined by inspection, prior to applying $\Sigma F_x = 0$.

6.3 Zero-Force Members

Truss analysis using the method of joints is greatly simplified if we can first identify those members which support *no loading*. These *zero-force members* are used to increase the stability of the truss during construction and to provide added support if the loading is changed.

The zero-force members of a truss can generally be found *by inspection* of each of the joints. For example, consider the truss shown in Fig. 6–11*a*. If a free-body diagram of the pin at joint *A* is drawn, Fig. 6–11*b*, it is seen that members *AB* and *AF* are zero-force members. (We could not have come to this conclusion if we had considered the free-body diagrams of joints *F* or *B* simply because there are five unknowns at each of these joints.) In a similar manner, consider the free-body diagram of joint *D*, Fig. 6–11*c*. Here again it is seen that *DC* and *DE* are zero-force members. From these observations, we can conclude that *if only two non-collinear members form a truss joint and no external load or support reaction is applied to the joint, the two members must be zero-force members.* The load on the truss in Fig. 6–11*a* is therefore supported by only five members as shown in Fig. 6–11*d*.

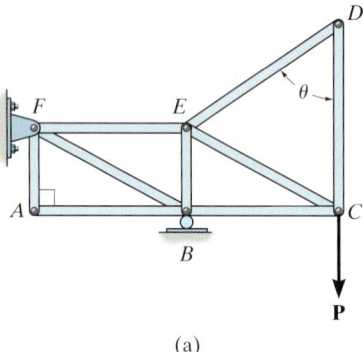

(a)

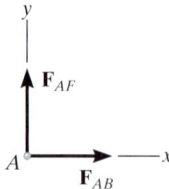

$$\xrightarrow{+} \Sigma F_x = 0; \ F_{AB} = 0$$
$$+\uparrow \Sigma F_y = 0; \ F_{AF} = 0$$

(b)

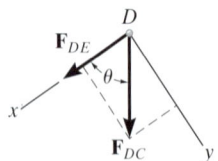

$$+\searrow \Sigma F_y = 0; F_{DC} \sin \theta = 0; \ \ F_{DC} = 0 \text{ since } \sin \theta \neq 0$$
$$+\swarrow \Sigma F_x = 0; F_{DE} + 0 = 0; \ \ F_{DE} = 0$$

(c)

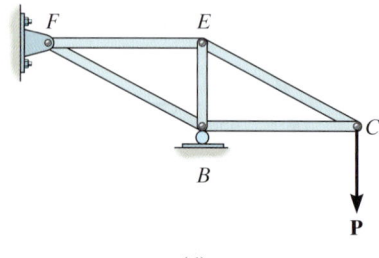

(d)

Fig. 6–11

Now consider the truss shown in Fig. 6–12a. The free-body diagram of the pin at joint D is shown in Fig. 6–12b. By orienting the y axis along members DC and DE and the x axis along member DA, it is seen that DA is a zero-force member. Note that this is also the case for member CA, Fig. 6–12c. In general then, *if three members form a truss joint for which two of the members are collinear, the third member is a zero-force member provided no external force or support reaction is applied to the joint*. The truss shown in Fig. 6–12d is therefore suitable for supporting the load **P**.

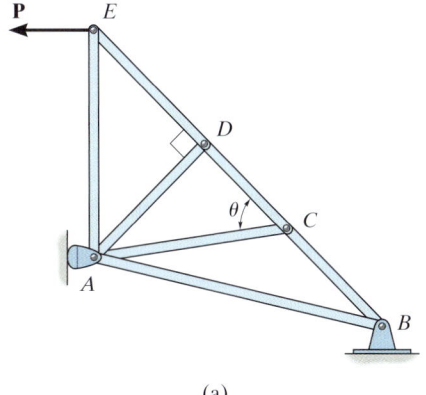

(a)

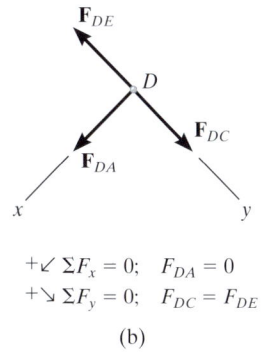

$$+\swarrow \Sigma F_x = 0; \quad F_{DA} = 0$$
$$+\searrow \Sigma F_y = 0; \quad F_{DC} = F_{DE}$$

(b)

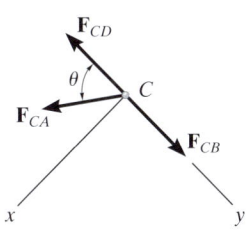

$$+\swarrow \Sigma F_x = 0; \quad F_{CA}\sin\theta = 0; \quad F_{CA} = 0 \text{ since } \sin\theta \neq 0;$$
$$+\searrow \Sigma F_y = 0; \quad F_{CB} = F_{CD}$$

(c)

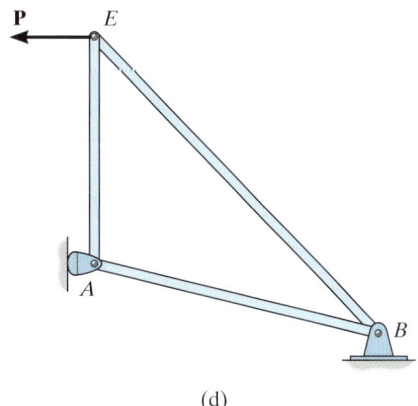

(d)

Fig. 6–12

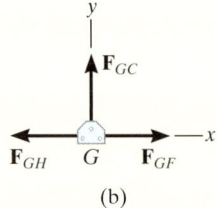

(b)

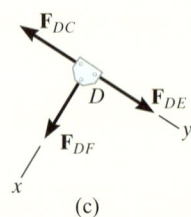

(c)

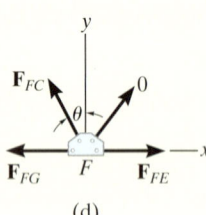

(d)

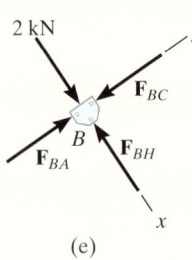

(e)

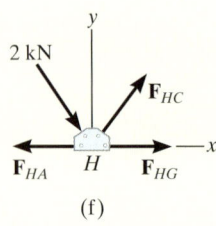

(f)

EXAMPLE | 6.4

Using the method of joints, determine all the zero-force members of the *Fink roof truss* shown in Fig. 6–13a. Assume all joints are pin connected.

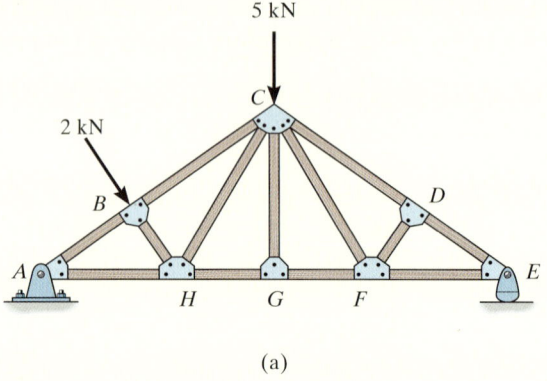

(a)

Fig. 6–13

SOLUTION

Look for joint geometries that have three members for which two are collinear. We have

Joint G.   (Fig. 6–13b).

$$+\uparrow \Sigma F_y = 0; \qquad\qquad F_{GC} = 0 \qquad\qquad Ans.$$

Realize that we could not conclude that GC is a zero-force member by considering joint C, where there are five unknowns. The fact that GC is a zero-force member means that the 5-kN load at C must be supported by members CB, CH, CF, and CD.

Joint D.   (Fig. 6–13c).

$$+\swarrow \Sigma F_x = 0; \qquad\qquad F_{DF} = 0 \qquad\qquad Ans.$$

Joint F.   (Fig. 6–13d).

$$+\uparrow \Sigma F_y = 0; \quad F_{FC}\cos\theta = 0 \quad \text{Since } \theta \neq 90°, \quad F_{FC} = 0 \quad Ans.$$

NOTE: If joint B is analyzed, Fig. 6–13e,

$$+\searrow \Sigma F_x = 0; \qquad\qquad 2\text{ kN} - F_{BH} = 0 \quad F_{BH} = 2\text{ kN} \quad (C)$$

Also, F_{HC} must satisfy $\Sigma F_y = 0$, Fig. 6–13f, and therefore HC is *not* a zero-force member.

FUNDAMENTAL PROBLEMS

All problem solutions must include FBDs.

F6–1.   Determine the force in each member of the truss. State if the members are in tension or compression.

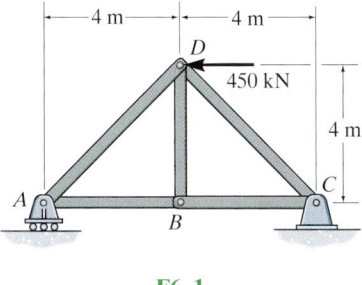

F6–1

F6–2.   Determine the force in each member of the truss. State if the members are in tension or compression.

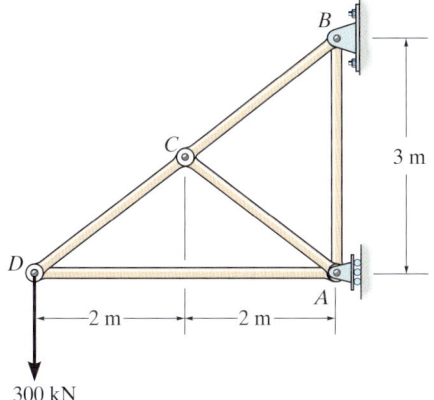

F6–2

F6 3.   Determine the force in members *AE* and *DC*. State if the members are in tension or compression.

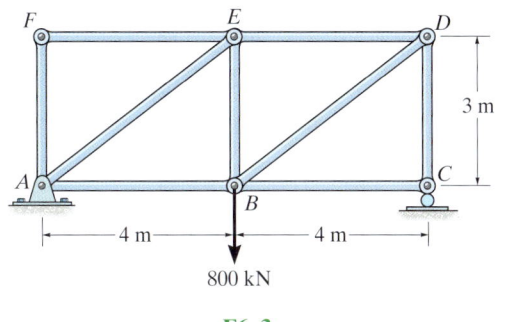

F6–3

F6–4.   Determine the greatest load *P* that can be applied to the truss so that none of the members are subjected to a force exceeding either 2 kN in tension or 1.5 kN in compression.

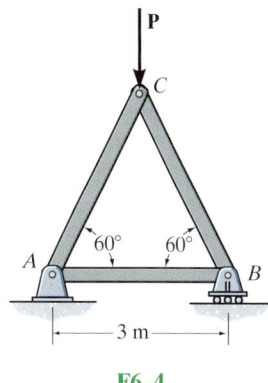

F6–4

F6–5.   Identify the zero-force members in the truss.

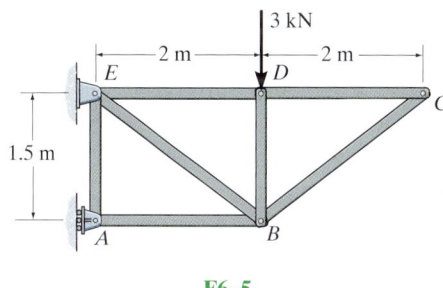

F6–5

F6–6.   Determine the force in each member of the truss. State if the members are in tension or compression.

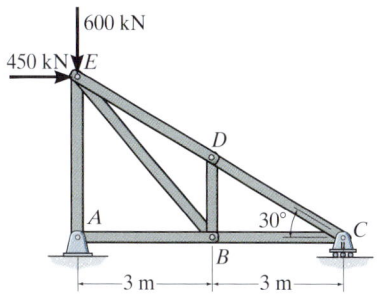

F6–6

PROBLEMS

All problem solutions must include FBDs.

6–1. The truss, used to support a balcony, is subjected to the loading shown. Approximate each joint as a pin and determine the force in each member. State whether the members are in tension or compression. Set $P_1 = 60$ kN, $P_2 = 40$ kN.

6–2. The truss, used to support a balcony, is subjected to the loading shown. Approximate each joint as a pin and determine the force in each member. State whether the members are in tension or compression. Set $P_1 = 80$ kN, $P_2 = 0$.

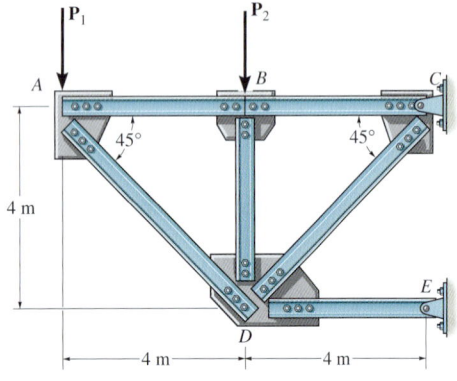

Probs. 6–1/2

6–3. Determine the force in each member of the truss, and state if the members are in tension or compression. Set $\theta = 0°$.

*6–4. Determine the force in each member of the truss, and state if the members are in tension or compression. Set $\theta = 30°$.

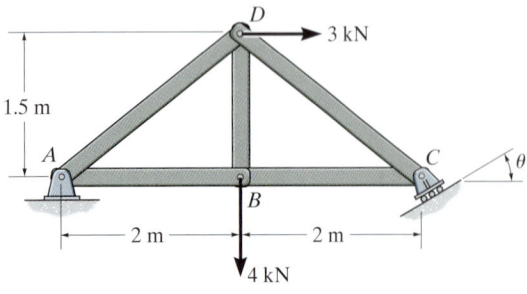

Probs. 6–3/4

6–5. Determine the force in each member of the truss, and state if the members are in tension or compression.

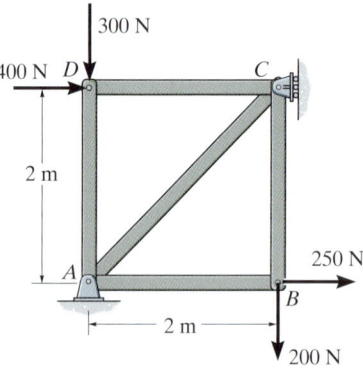

Prob. 6–5

6–6. Determine the force in each member of the truss, and state if the members are in tension or compression.

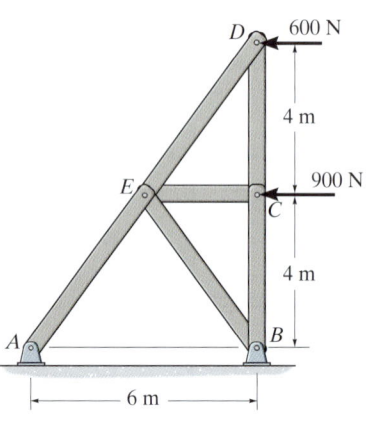

Prob. 6–6

6–7. Determine the force in each member of the *Pratt truss*, and state if the members are in tension or compression.

6–9. Determine the force in each member of the truss and state if the members are in tension or compression. *Hint:* The vertical component of force at *C* must equal zero. Why?

6–10. Each member of the truss is uniform and has a mass of 8 kg/m. Remove the external loads of 6 kN and 8 kN and determine the approximate force in each member due to the weight of the truss. State if the members are in tension or compression. Solve the problem by *assuming* the weight of each member can be represented as a vertical force, half of which is applied at each end of the member.

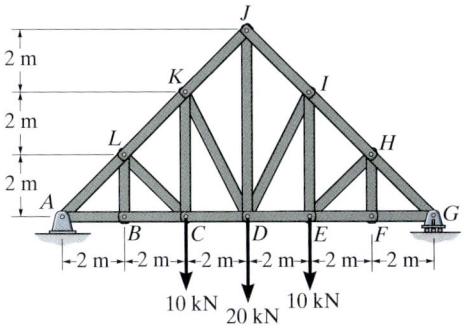

Prob. 6–7

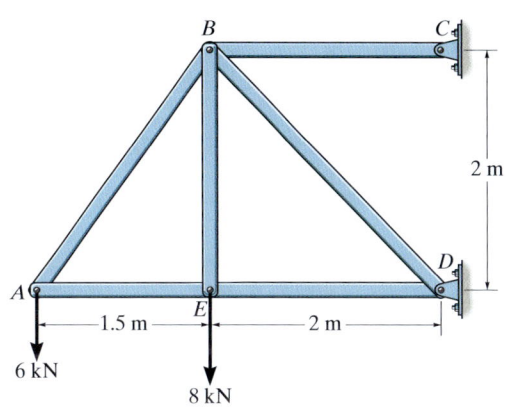

Probs. 6–9/10

*6–8. Determine the force in each member of the truss and state if the members are in tension or compression. *Hint:* The horizontal force component at *A* must be zero. Why?

6–11. Determine the force in each member of the truss and state if the members are in tension or compression.

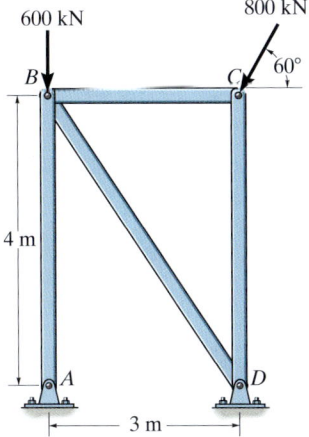

Prob. 6–8

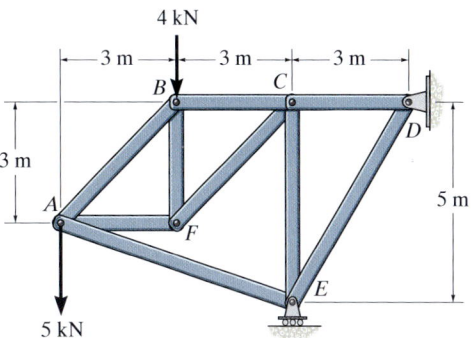

Prob. 6–11

*6–12. Determine the force in each member of the truss and state if the members are in tension or compression. Set $P_1 = 10$ kN, $P_2 = 15$ kN.

6–13. Determine the force in each member of the truss and state if the members are in tension or compression. Set $P_1 = 0, P_2 = 20$ kN.

*6–16. Determine the force in each member of the truss. State whether the members are in tension or compression. Set $P = 8$ kN.

6–17. If the maximum force that any member can support is 8 kN in tension and 6 kN in compression, determine the maximum force P that can be supported at joint D.

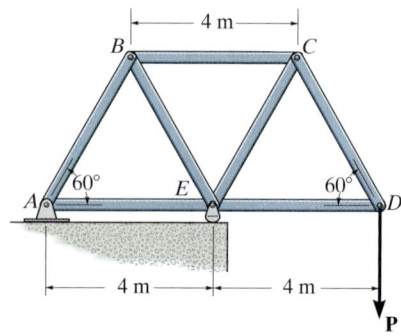

Probs. 6–16/17

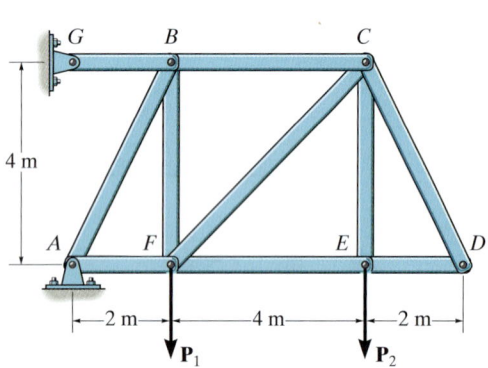

Probs. 6–12/13

6–14. Determine the force in each member of the truss and state if the members are in tension or compression. Set $P_1 = 2$ kN and $P_2 = 1.5$ kN.

6–15. Determine the force in each member of the truss and state if the members are in tension or compression. Set $P_1 = P_2 = 4$ kN.

6–18. Determine the force in each member of the truss and state if the members are in tension or compression. *Hint:* The resultant force at the pin E acts along member ED. Why?

6–19. Each member of the truss is uniform and has a mass of 8 kg/m. Remove the external loads of 3 kN and 2 kN and determine the approximate force in each member due to the weight of the truss. State if the members are in tension or compression. Solve the problem by *assuming* the weight of each member can be represented as a vertical force, half of which is applied at each end of the member.

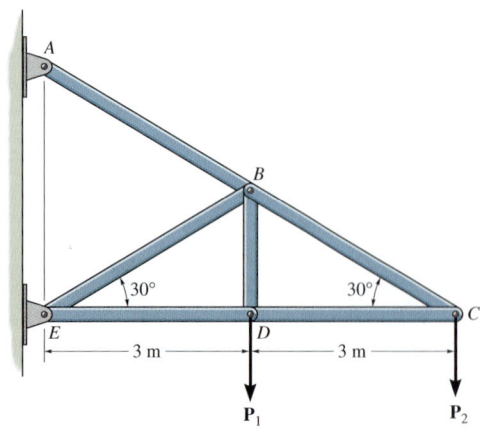

Probs. 6–14/15

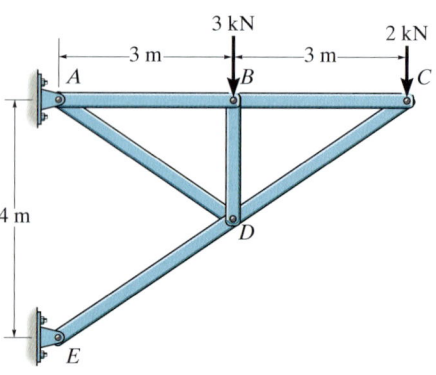

Probs. 6–18/19

*6–20. Determine the force in each member of the truss in terms of the load P, and indicate whether the members are in tension or compression.

6–21. If the maximum force that any member can support is 4 kN in tension and 3 kN in compression, determine the maximum force P that can be supported at point B. Take $d = 1$ m.

6–23. Determine the force in each member of the truss in terms of the load P and state if the members are in tension or compression.

*6–24. Each member of the truss is uniform and has a weight W. Remove the external forces **P** and determine the approximate force in each member due to the weight of the truss. State if the members are in tension or compression. Solve the problem by *assuming* the weight of each member can be represented as a vertical force, half of which is applied at each end of the member.

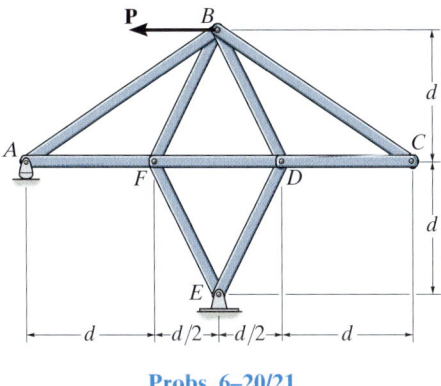

Probs. 6–20/21

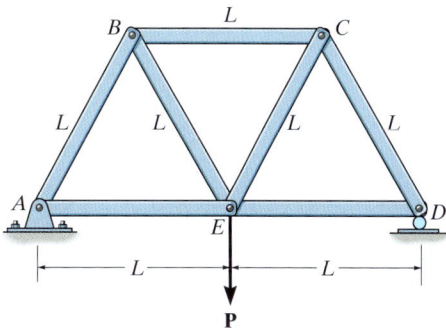

Probs. 6–23/24

6–25. Determine the force in each member of the truss in terms of the external loading and state if the members are in tension or compression.

6–26. The maximum allowable tensile force in the members of the truss is $(F_t)_{max} = 2$ kN, and the maximum allowable compressive force is $(F_c)_{max} = 1.2$ kN. Determine the maximum magnitude P of the two loads that can be applied to the truss. Take $L = 2$ m and $\theta = 30°$.

6–22. Determine the force in each member of the double scissors truss in terms of the load P and state if the members are in tension or compression.

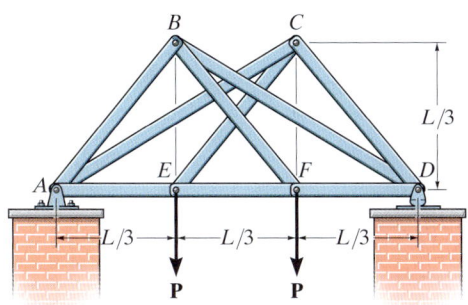

Prob. 6–22

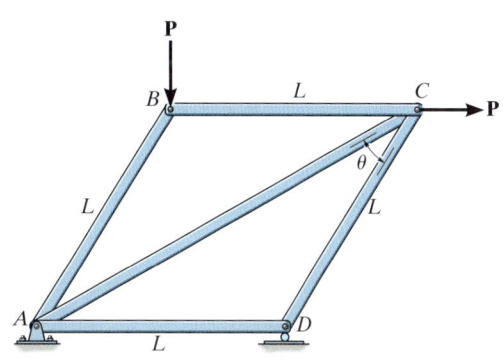

Probs. 6–25/26

6.4   The Method of Sections

When we need to find the force in only a few members of a truss, we can analyze the truss using the *method of sections*. It is based on the principle that if the truss is in equilibrium then any segment of the truss is also in equilibrium. For example, consider the two truss members shown on the left in Fig. 6–14. If the forces within the members are to be determined, then an imaginary section, indicated by the blue line, can be used to cut each member into two parts and thereby "expose" each internal force as "external" to the free-body diagrams shown on the right. Clearly, it can be seen that equilibrium requires that the member in tension (T) be subjected to a "pull," whereas the member in compression (C) is subjected to a "push."

The method of sections can also be used to "cut" or section the members of an entire truss. If the section passes through the truss and the free-body diagram of either of its two parts is drawn, we can then apply the equations of equilibrium to that part to determine the member forces at the "cut section." Since only *three* independent equilibrium equations ($\Sigma F_x = 0$, $\Sigma F_y = 0$, $\Sigma M_O = 0$) can be applied to the free-body diagram of any segment, then we should try to select a section that, in general, passes through not more than *three* members in which the forces are unknown. For example, consider the truss in Fig. 6–15a. If the forces in members BC, GC, and GF are to be determined, then section aa would be appropriate. The free-body diagrams of the two segments are shown in Figs. 6–15b and 6–15c. Note that the line of action of each member force is specified from the *geometry* of the truss, since the force in a member is along its axis. Also, the member forces acting on one part of the truss are equal but opposite to those acting on the other part—Newton's third law. Members BC and GC are assumed to be in *tension* since they are subjected to a "pull," whereas GF in *compression* since it is subjected to a "push."

The three unknown member forces $\mathbf{F}_{BC}$, $\mathbf{F}_{GC}$, and $\mathbf{F}_{GF}$ can be obtained by applying the three equilibrium equations to the free-body diagram in Fig. 6–15b. If, however, the free-body diagram in Fig. 6–15c is considered, the three support reactions $\mathbf{D}_x$, $\mathbf{D}_y$ and $\mathbf{E}_x$ will have to be known, because only three equations of equilibrium are available. (This, of course, is done in the usual manner by considering a free-body diagram of the *entire truss*.)

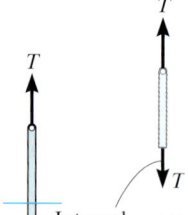

Tension

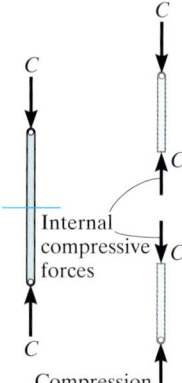

Compression

Fig. 6–14

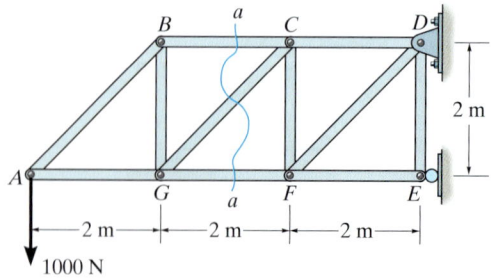

Fig. 6–15

When applying the equilibrium equations, we should carefully consider ways of writing the equations so as to yield a *direct solution* for each of the unknowns, rather than having to solve simultaneous equations. For example, using the truss segment in Fig. 6–15b and summing moments about C would yield a direct solution for $\mathbf{F}_{GF}$ since $\mathbf{F}_{BC}$ and $\mathbf{F}_{GC}$ create zero moment about C. Likewise, $\mathbf{F}_{BC}$ can be directly obtained by summing moments about G. Finally, $\mathbf{F}_{GC}$ can be found directly from a force summation in the vertical direction since $\mathbf{F}_{GF}$ and $\mathbf{F}_{BC}$ have no vertical components. This ability to *determine directly* the force in a particular truss member is one of the main advantages of using the method of sections.*

As in the method of joints, there are two ways in which we can determine the correct sense of an unknown member force:

The forces in selected members of this Pratt truss can readily be determined using the method of sections.

- The correct sense of an unknown member force can in many cases be determined "by inspection." For example, $\mathbf{F}_{BC}$ is a tensile force as represented in Fig. 6–15b since moment equilibrium about G requires that $\mathbf{F}_{BC}$ create a moment opposite to that of the 1000-N force. Also, $\mathbf{F}_{GC}$ is tensile since its vertical component must balance the 1000-N force which acts downward. In more complicated cases, the sense of an unknown member force may be *assumed*. If the solution yields a *negative* scalar, it indicates that the force's sense is *opposite* to that shown on the free-body diagram.

- *Always assume* that the unknown member forces at the cut section are *tensile* forces, i.e., "pulling" on the member. By doing this, the numerical solution of the equilibrium equations will yield *positive scalars for members in tension and negative scalars for members in compression.*

---

*Notice that if the method of joints were used to determine, say, the force in member GC, it would be necessary to analyze joints A, B, and G in sequence.

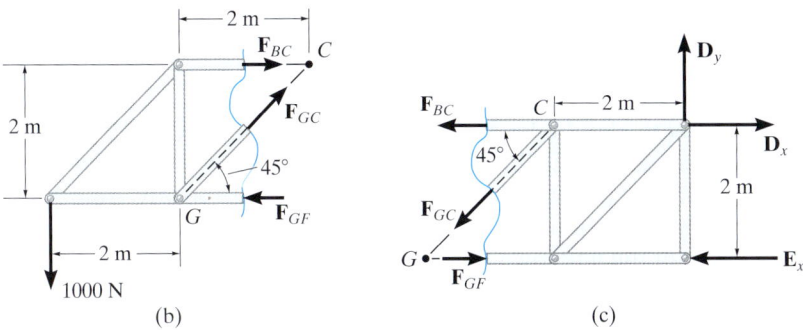

(b)   (c)

Fig. 6–15 (cont.)

Simple trusses are often used in the construction of large cranes in order to reduce the weight of the boom and tower.

Procedure for Analysis

The forces in the members of a truss may be determined by the method of sections using the following procedure.

Free-Body Diagram.

- Make a decision on how to "cut" or section the truss through the members where forces are to be determined.
- Before isolating the appropriate section, it may first be necessary to determine the truss's support reactions. If this is done then the three equilibrium equations will be available to solve for member forces at the section.
- Draw the free-body diagram of that segment of the sectioned truss which has the least number of forces acting on it.
- Use one of the two methods described above for establishing the sense of the unknown member forces.

Equations of Equilibrium.

- Moments should be summed about a point that lies at the intersection of the lines of action of two unknown forces, so that the third unknown force can be determined directly from the moment equation.
- If two of the unknown forces are *parallel*, forces may be summed *perpendicular* to the direction of these unknowns to determine *directly* the third unknown force.

EXAMPLE 6.5

Determine the force in members GE, GC, and BC of the truss shown in Fig. 6–16a. Indicate whether the members are in tension or compression.

SOLUTION

Section *aa* in Fig. 6–16a has been chosen since it cuts through the *three* members whose forces are to be determined. In order to use the method of sections, however, it is *first* necessary to determine the external reactions at A or D. Why? A free-body diagram of the entire truss is shown in Fig. 6–16b. Applying the equations of equilibrium, we have

$$\xrightarrow{+} \Sigma F_x = 0; \qquad 400 \text{ N} - A_x = 0 \qquad A_x = 400 \text{ N}$$

$$\zeta + \Sigma M_A = 0; \qquad -1200 \text{ N}(8 \text{ m}) - 400 \text{ N}(3 \text{ m}) + D_y(12 \text{ m}) = 0$$

$$D_y = 900 \text{ N}$$

$$+\uparrow \Sigma F_y = 0; \qquad A_y - 1200 \text{ N} + 900 \text{ N} = 0 \qquad A_y = 300 \text{ N}$$

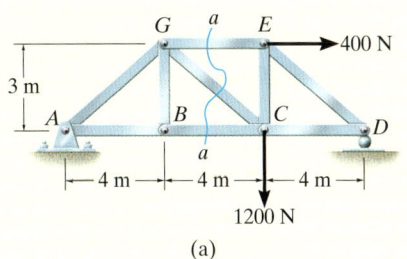

(a)

Free-Body Diagram.   For the analysis the free-body diagram of the left portion of the sectioned truss will be used, since it involves the least number of forces, Fig. 6–16c.

Equations of Equilibrium.   Summing moments about point G eliminates $\mathbf{F}_{GE}$ and $\mathbf{F}_{GC}$ and yields a direct solution for F_{BC}.

$$\zeta + \Sigma M_G = 0; \quad -300 \text{ N}(4 \text{ m}) - 400 \text{ N}(3 \text{ m}) + F_{BC}(3 \text{ m}) = 0$$

$$F_{BC} = 800 \text{ N} \quad (T) \qquad\qquad Ans.$$

In the same manner, by summing moments about point C we obtain a direct solution for F_{GE}.

$$\zeta + \Sigma M_C = 0; \quad -300 \text{ N}(8 \text{ m}) + F_{GE}(3 \text{ m}) = 0$$

$$F_{GE} = 800 \text{ N} \quad (C) \qquad\qquad Ans.$$

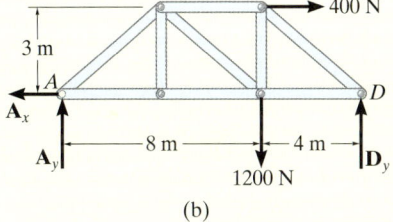

(b)

Since $\mathbf{F}_{BC}$ and $\mathbf{F}_{GE}$ have no vertical components, summing forces in the y direction directly yields F_{GC}, i.e.,

$$+\uparrow \Sigma F_y = 0; \qquad 300 \text{ N} - \tfrac{3}{5}F_{GC} = 0$$

$$F_{GC} = 500 \text{ N} \quad (T) \qquad\qquad Ans.$$

NOTE: Here it is possible to tell, by inspection, the proper direction for each unknown member force. For example, $\Sigma M_C = 0$ requires $\mathbf{F}_{GE}$ to be *compressive* because it must balance the moment of the 300-N force about C.

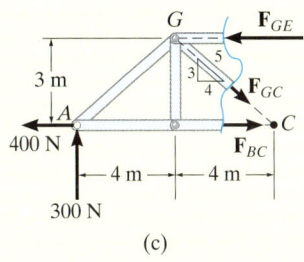

(c)

Fig. 6–16

EXAMPLE 6.6

Determine the force in member CF of the truss shown in Fig. 6–17a. Indicate whether the member is in tension or compression. Assume each member is pin connected.

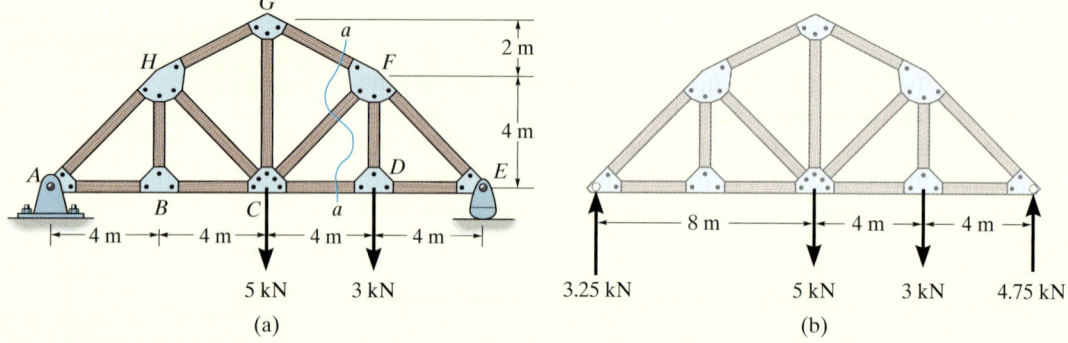

(a)   (b)

Fig. 6–17

SOLUTION

Free-Body Diagram. Section aa in Fig. 6–17a will be used since this section will "expose" the internal force in member CF as "external" on the free-body diagram of either the right or left portion of the truss. It is first necessary, however, to determine the support reactions on either the left or right side. Verify the results shown on the free-body diagram in Fig. 6–17b.

The free-body diagram of the right portion of the truss, which is the easiest to analyze, is shown in Fig. 6–17c. There are three unknowns, F_{FG}, F_{CF}, and F_{CD}.

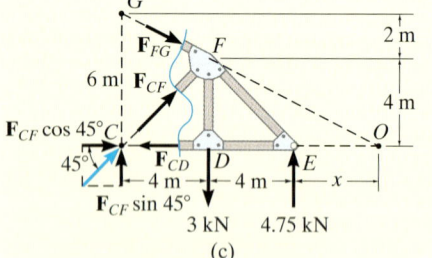

Equations of Equilibrium. We will apply the moment equation about point O in order to eliminate the two unknowns F_{FG} and F_{CD}. The location of point O measured from E can be determined from proportional triangles, i.e., $4/(4 + x) = 6/(8 + x)$, $x = 4$ m. Or, stated in another manner, the slope of member GF has a drop of 2 m to a horizontal distance of 4 m. Since FD is 4 m, Fig. 6–17c, then from D to O the distance must be 8 m.

An easy way to determine the moment of $\mathbf{F}_{CF}$ about point O is to use the principle of transmissibility and slide $\mathbf{F}_{CF}$ to point C, and then resolve $\mathbf{F}_{CF}$ into its two rectangular components. We have

$$\zeta + \Sigma M_O = 0;$$

$$-F_{CF} \sin 45°(12 \text{ m}) + (3 \text{ kN})(8 \text{ m}) - (4.75 \text{ kN})(4 \text{ m}) = 0$$

$$F_{CF} = 0.589 \text{ kN} \quad (C) \qquad\qquad\qquad Ans.$$

EXAMPLE 6.7

Determine the force in member *EB* of the roof truss shown in Fig. 6–18a. Indicate whether the member is in tension or compression.

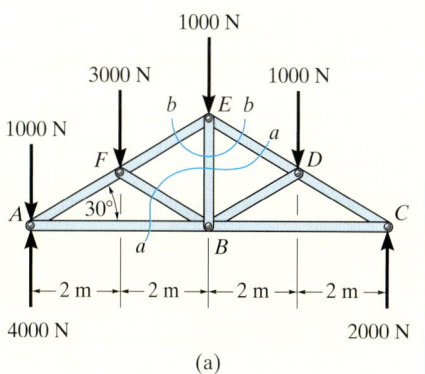

(a)

SOLUTION

Free-Body Diagrams. By the method of sections, any imaginary section that cuts through *EB*, Fig. 6–18a, will also have to cut through three other members for which the forces are unknown. For example, section *aa* cuts through *ED, EB, FB*, and *AB*. If a free-body diagram of the left side of this section is considered, Fig. 6–18b, it is possible to obtain $\mathbf{F}_{ED}$ by summing moments about *B* to eliminate the other three unknowns; however, $\mathbf{F}_{EB}$ cannot be determined from the remaining two equilibrium equations. One possible way of obtaining $\mathbf{F}_{EB}$ is first to determine $\mathbf{F}_{ED}$ from section *aa*, then use this result on section *bb*, Fig. 6–18a, which is shown in Fig. 6–18c. Here the force system is concurrent and our sectioned free-body diagram is the same as the free-body diagram for the joint at *E*.

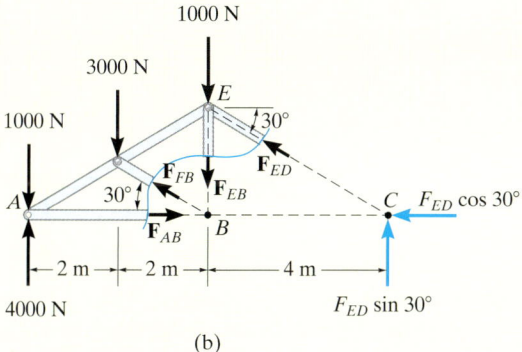

(b)

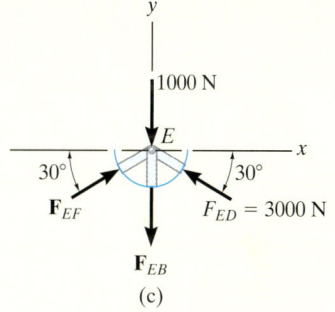

(c)

Fig. 6–18

Equations of Equilibrium. In order to determine the moment of $\mathbf{F}_{ED}$ about point *B*, Fig. 6–18b, we will use the principle of transmissibility and slide the force to point *C* and then resolve it into its rectangular components as shown. Therefore,

$$\zeta + \Sigma M_B = 0; \quad 1000 \text{ N}(4 \text{ m}) + 3000 \text{ N}(2 \text{ m}) - 4000 \text{ N}(4 \text{ m})$$
$$+ F_{ED} \sin 30°(4 \text{ m}) = 0$$
$$F_{ED} = 3000 \text{ N} \quad (\text{C})$$

Considering now the free-body diagram of section *bb*, Fig. 6–18c, we have

$$\xrightarrow{+} \Sigma F_x = 0; \quad F_{EF} \cos 30° - 3000 \cos 30° \text{ N} = 0$$
$$F_{EF} = 3000 \text{ N} \quad (\text{C})$$
$$+\uparrow \Sigma F_y = 0; \quad 2(3000 \sin 30° \text{ N}) - 1000 \text{ N} - F_{EB} = 0$$
$$F_{EB} = 2000 \text{ N} \quad (\text{T}) \qquad \qquad \textit{Ans.}$$

FUNDAMENTAL PROBLEMS

All problem solutions must include FBDs.

F6–7. Determine the force in members *BC*, *CF*, and *FE*. State if the members are in tension or compression.

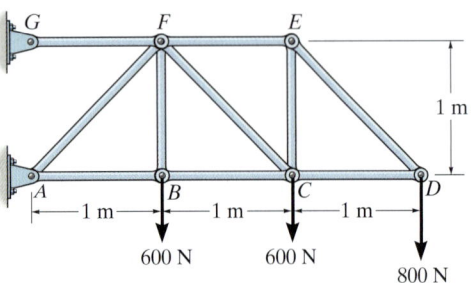

F6–7

F6–8. Determine the force in members *LK*, *KC*, and *CD* of the Pratt truss. State if the members are in tension or compression.

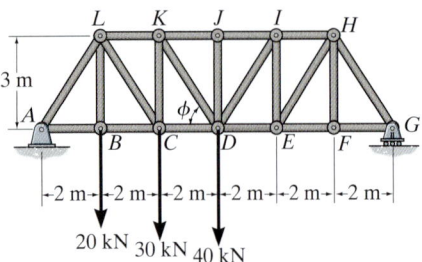

F6–8

F6–9. Determine the force in members *KJ*, *KD*, and *CD* of the Pratt truss. State if the members are in tension or compression.

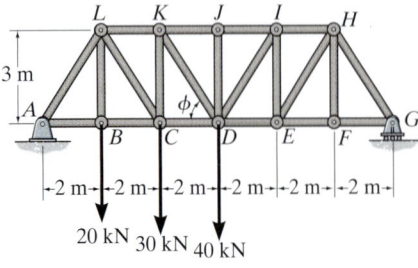

F6–9

F6–10. Determine the force in members *EF*, *CF*, and *BC* of the truss. State if the members are in tension or compression.

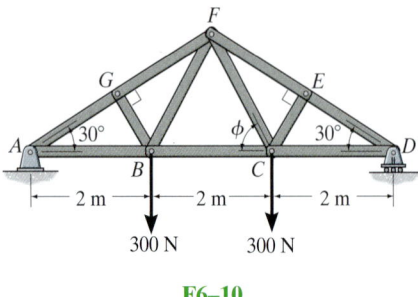

F6–10

F6–11. Determine the force in members *GF*, *GD*, and *CD* of the truss. State if the members are in tension or compression.

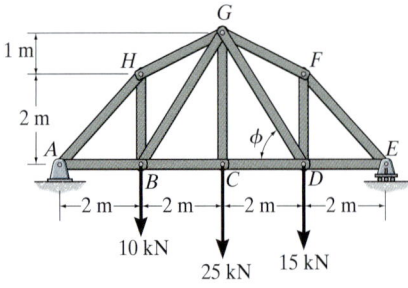

F6–11

F6–12. Determine the force in members *DC*, *HI*, and *JI* of the truss. State if the members are in tension or compression. *Suggestion:* Use the sections shown.

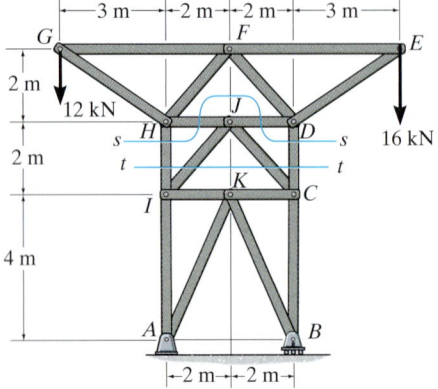

F6–12

PROBLEMS

All problem solutions must include FBDs.

6–27. Determine the force in members *BC*, *CG*, and *GF* of the *Warren truss*. Indicate if the members are in tension or compression.

*6–28. Determine the force in members *CD*, *CF*, and *FG* of the *Warren truss*. Indicate if the members are in tension or compression.

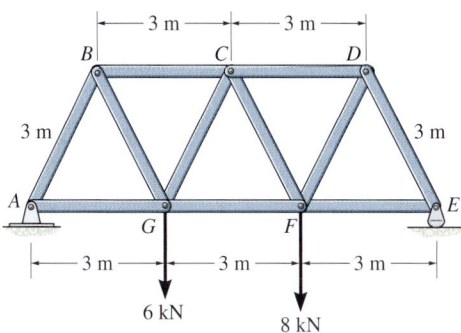

Probs. 6–27/28

6–29. The *Pratt bridge truss* is subjected to the loading shown. Determine the force in members *LD*, *LK*, *CD*, and *KD*, and state if the members are in tension or compression.

6–30. The *Pratt bridge truss* is subjected to the loading shown. Determine the force in members *JI*, *JE*, and *DE*, and state if the members are in tension or compression.

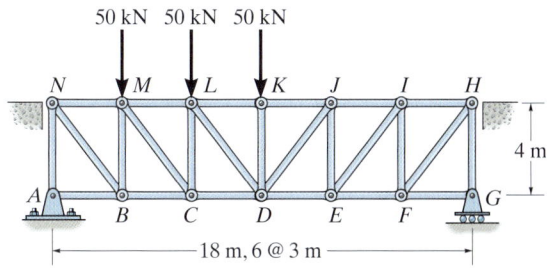

Probs. 6–29/30

6–31. Determine the force in members *BE*, *EF*, and *CB*, and state if the members are in tension or compression.

*6–32. Determine the force in members *BF*, *BG*, and *AB*, and state if the members are in tension or compression.

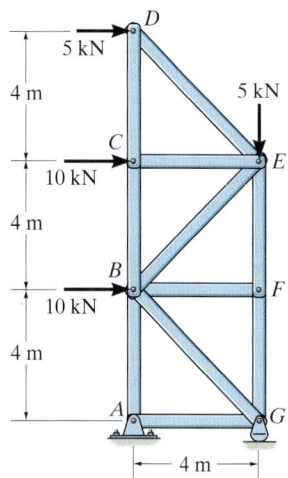

Probs. 6–31/32

6–33. Determine the force developed in members *BC* and *CH* of the roof truss and state if the members are in tension or compression.

6–34. Determine the force in members *CD* and *GF* of the truss and state if the members are in tension or compression. Also indicate all zero-force members.

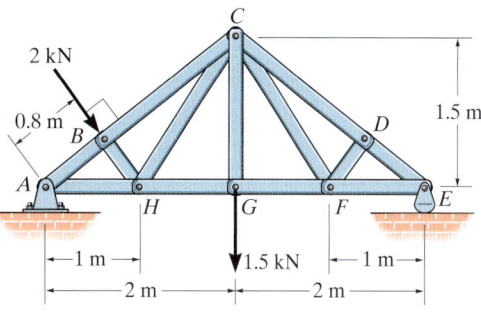

Probs. 6–33/34

6

6–35. Determine the force in members *BC, HC,* and *HG.* After the truss is sectioned use a single equation of equilibrium for the calculation of each force. State if these members are in tension or compression.

*6–36. Determine the force in members *CD, CF,* and *CG* and state if these members are in tension or compression.

6–39. Determine the force in members *IC* and *CG* of the truss and state if these members are in tension or compression. Also, indicate all zero-force members.

*6–40. Determine the force in members *JE* and *GF* of the truss and state if these members are in tension or compression. Also, indicate all zero-force members.

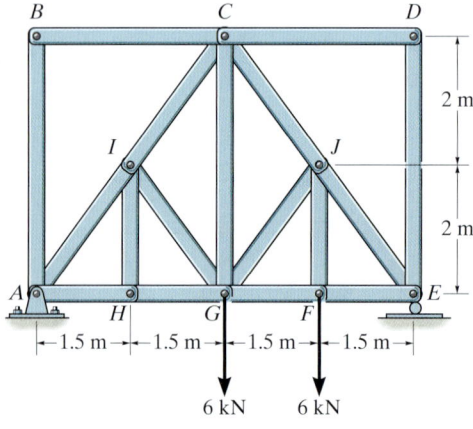

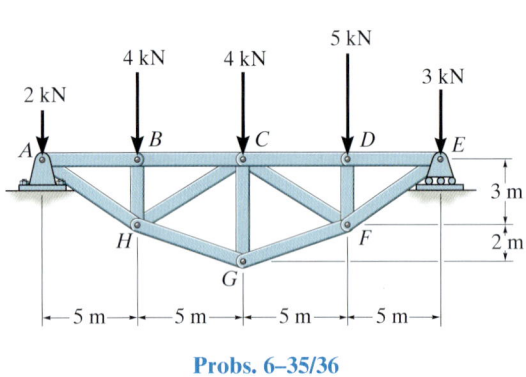

Probs. 6–35/36

Probs. 6–39/40

6–37. Determine the force in members *KJ, NJ, ND,* and *CD* of the *K truss.* Indicate if the members are in tension or compression. *Hint:* Use sections *aa* and *bb.*

6–38. Determine the force in members *JI* and *DE* of the *K truss.* Indicate if the members are in tension or compression.

6–41. Determine the force in members *FG, GC* and *CB* of the truss used to support the sign, and state if the members are in tension or compression.

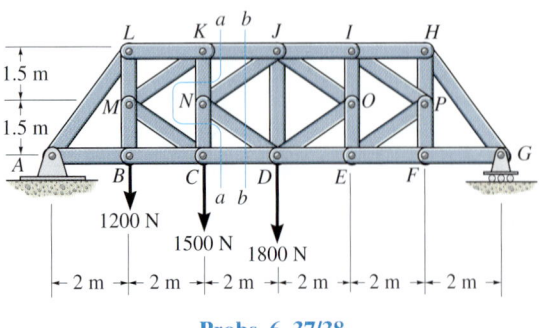

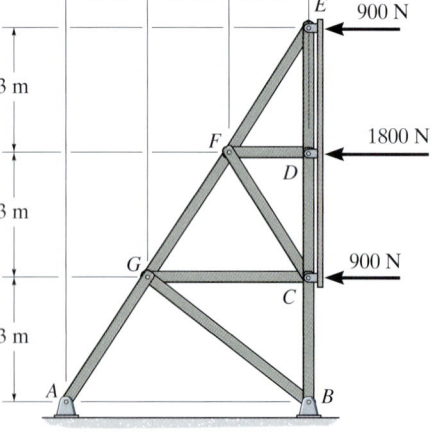

Probs. 6–37/38

Prob. 6–41

6–42. Determine the force in members *DE*, *DL*, and *ML* of the roof truss and state if the members are in tension or compression.

6–43. Determine the force in members *EF* and *EL* of the roof truss and state if the members are in tension or compression.

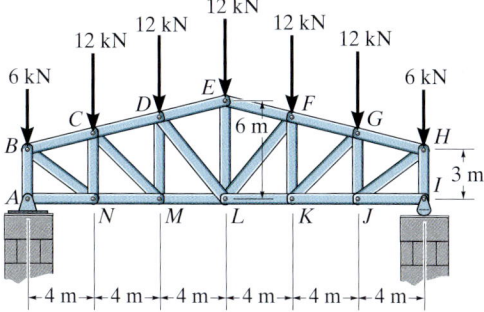

Probs. 6–42/43

*6–44. The skewed truss carries the load shown. Determine the force in members *CB*, *BE*, and *EF* and state if these members are in tension or compression. Assume that all joints are pinned.

6–45. The skewed truss carries the load shown. Determine the force in members *AB*, *BF*, and *EF* and state if these members are in tension or compression. Assume that all joints are pinned.

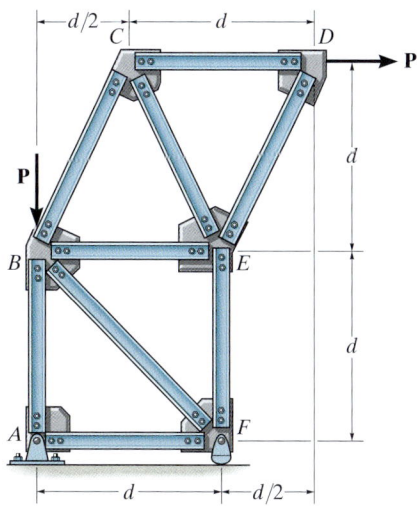

Probs. 6–44/45

6–46. Determine the force in members *CD* and *CM* of the *Baltimore bridge truss* and state if the members are in tension or compression. Also, indicate all zero-force members.

6–47. Determine the force in members *EF*, *EP*, and *LK* of the *Baltimore bridge truss* and state if the members are in tension or compression. Also, indicate all zero-force members.

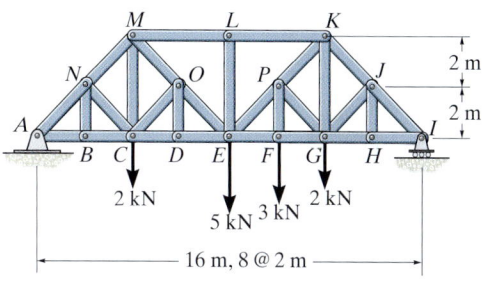

Probs. 6–46/47

*6–48. The truss supports the vertical load of 600 N. If *L* = 2 m, determine the force on members *HG* and *HB* of the truss and state if the members are in tension or compression.

6–49. The truss supports the vertical load of 600 N. Determine the force in members *BC*, *BG*, and *HG* as the dimension *L* varies. Plot the results of *F* (ordinate with tension as positive) versus *L* (abscissa) for $0 \leq L \leq 3$ m.

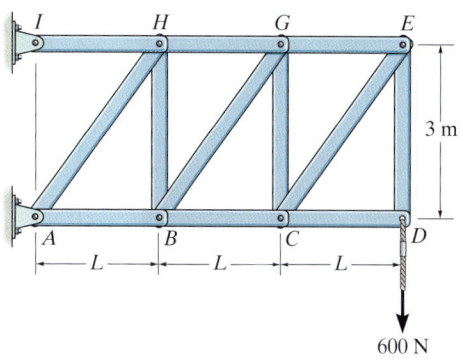

Probs. 6–48/49

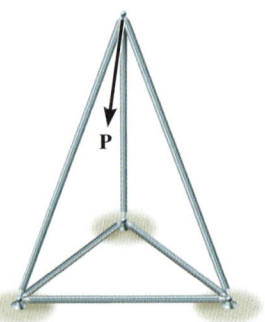

Fig. 6–19

Typical roof-supporting space truss. Notice the use of ball-and-socket joints for the connections.

For economic reasons, large electrical transmission towers are often constructed using space trusses.

*6.5   Space Trusses

A *space truss* consists of members joined together at their ends to form a stable three-dimensional structure. The simplest form of a space truss is a *tetrahedron*, formed by connecting six members together, as shown in Fig. 6–19. Any additional members added to this basic element would be redundant in supporting the force **P**. A *simple space truss* can be built from this basic tetrahedral element by adding three additional members and a joint, and continuing in this manner to form a system of multiconnected tetrahedrons.

Assumptions for Design.   The members of a space truss may be treated as two-force members provided the external loading is applied at the joints and the joints consist of ball-and-socket connections. These assumptions are justified if the welded or bolted connections of the joined members intersect at a common point and the weight of the members can be neglected. In cases where the weight of a member is to be included in the analysis, it is generally satisfactory to apply it as a vertical force, half of its magnitude applied at each end of the member.

Procedure for Analysis

Either the method of joints or the method of sections can be used to determine the forces developed in the members of a simple space truss.

Method of Joints.

If the forces in *all* the members of the truss are to be determined, then the method of joints is most suitable for the analysis. Here it is necessary to apply the three equilibrium equations $\Sigma F_x = 0$, $\Sigma F_y = 0$, $\Sigma F_z = 0$ to the forces acting at each joint. Remember that the solution of many simultaneous equations can be avoided if the force analysis begins at a joint having at least one known force and at most three unknown forces. Also, if the three-dimensional geometry of the force system at the joint is hard to visualize, it is recommended that a Cartesian vector analysis be used for the solution.

Method of Sections.

If only a *few* member forces are to be determined, the method of sections can be used. When an imaginary section is passed through a truss and the truss is separated into two parts, the force system acting on one of the segments must satisfy the *six* equilibrium equations: $\Sigma F_x = 0$, $\Sigma F_y = 0$, $\Sigma F_z = 0$, $\Sigma M_x = 0$, $\Sigma M_y = 0$, $\Sigma M_z = 0$ (Eqs. 5–6). By proper choice of the section and axes for summing forces and moments, many of the unknown member forces in a space truss can be computed *directly*, using a single equilibrium equation.

EXAMPLE 6.8

Determine the forces acting in the members of the space truss shown in Fig. 6–20a. Indicate whether the members are in tension or compression.

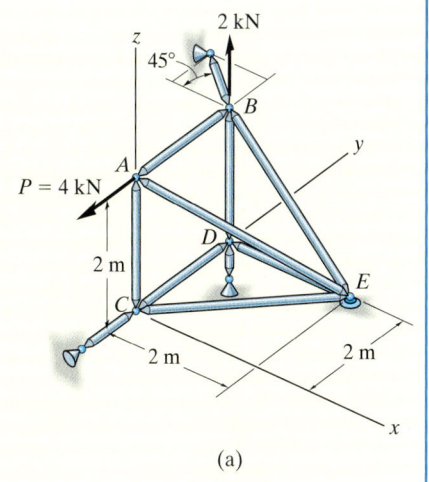

(a)

SOLUTION

Since there are one known force and three unknown forces acting at joint A, the force analysis of the truss will begin at this joint.

Joint A. (Fig. 6–20b). Expressing each force acting on the free-body diagram of joint A as a Cartesian vector, we have

$$\mathbf{P} = \{-4\mathbf{j}\} \text{ kN}, \qquad \mathbf{F}_{AB} = F_{AB}\mathbf{j}, \quad \mathbf{F}_{AC} = -F_{AC}\mathbf{k},$$

$$\mathbf{F}_{AE} = F_{AE}\left(\frac{\mathbf{r}_{AE}}{r_{AE}}\right) = F_{AE}(0.577\mathbf{i} + 0.577\mathbf{j} - 0.577\mathbf{k})$$

For equilibrium,

$$\Sigma\mathbf{F} = \mathbf{0}; \qquad \mathbf{P} + \mathbf{F}_{AB} + \mathbf{F}_{AC} + \mathbf{F}_{AE} = \mathbf{0}$$

$$-4\mathbf{j} + F_{AB}\mathbf{j} - F_{AC}\mathbf{k} + 0.577F_{AE}\mathbf{i} + 0.577F_{AE}\mathbf{j} - 0.577F_{AE}\mathbf{k} = \mathbf{0}$$

$$\Sigma F_x = 0; \qquad\qquad 0.577F_{AE} = 0$$

$$\Sigma F_y = 0; \qquad -4 + F_{AB} + 0.577F_{AE} = 0$$

$$\Sigma F_z = 0; \qquad\qquad -F_{AC} - 0.577F_{AE} = 0$$

$$F_{AC} = F_{AE} = 0 \qquad\qquad \textit{Ans.}$$

$$F_{AB} = 4 \text{ kN} \quad (\text{T}) \qquad\qquad \textit{Ans.}$$

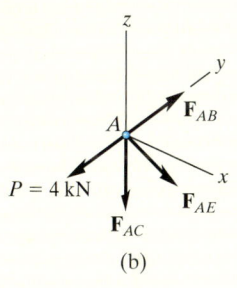

(b)

Since F_{AB} is known, joint B can be analyzed next.

Joint B. (Fig. 6–20c).

$$\Sigma F_x = 0; \qquad -R_B \cos 45° + 0.707F_{BE} = 0$$

$$\Sigma F_y = 0; \qquad\qquad -4 + R_B \sin 45° = 0$$

$$\Sigma F_z = 0; \qquad 2 + F_{BD} - 0.707F_{BE} = 0$$

$$R_B = F_{BE} = 5.66 \text{ kN} \quad (\text{T}), \qquad F_{BD} = 2 \text{ kN} \quad (\text{C}) \qquad \textit{Ans.}$$

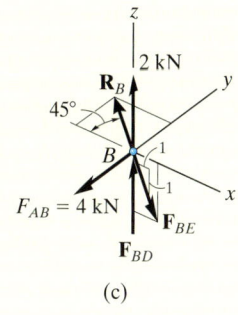

(c)

Fig. 6–20

The *scalar* equations of equilibrium may also be applied directly to the forces acting on the free-body diagrams of joints D and C since the force components are easily determined. Show that

$$F_{DE} = F_{DC} = F_{CE} = 0 \qquad\qquad \textit{Ans.}$$

PROBLEMS

All problem solutions must include FBDs.

6–50. Two space trusses are used to equally support the uniform 50-kg sign. Determine the force developed in members *AB, AC,* and *BC* of truss *ABCD* and state if the members are in tension or compression. Horizontal short links support the truss at joints *B* and *D* and there is a ball-and-socket joint at *C.*

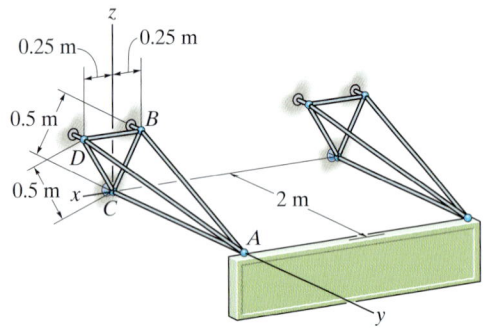

Prob. 6–50

6–51. Determine the force in each member of the space truss and state if the members are in tension or compression. *Hint:* The support reaction at *E* acts along member *EB.* Why?

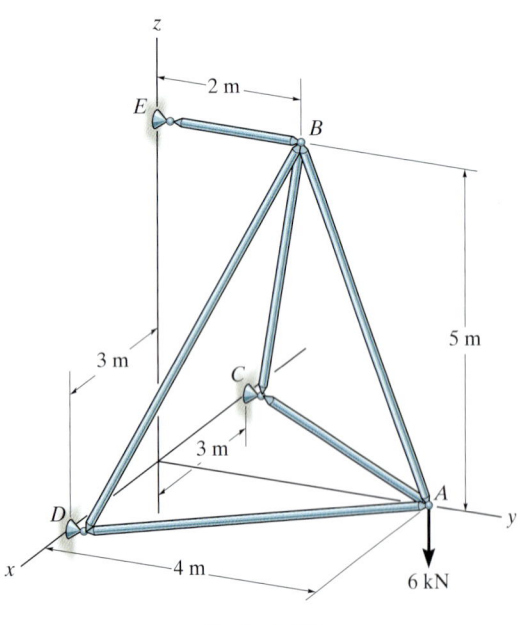

Prob. 6–51

***6–52.** Determine the force in each member of the space truss and state if the members are in tension or compression. The truss is supported by rollers at *A, B,* and *C.*

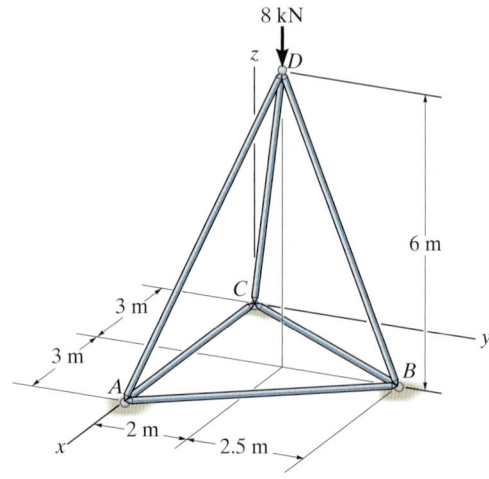

Prob. 6–52

6–53. The space truss supports a force $\mathbf{F} = [300\mathbf{i} + 400\mathbf{j} - 500\mathbf{k}]$ N. Determine the force in each member, and state if the members are in tension or compression.

6–54. The space truss supports a force $\mathbf{F} = [-400\mathbf{i} + 500\mathbf{j} + 600\mathbf{k}]$ N. Determine the force in each member, and state if the members are in tension or compression.

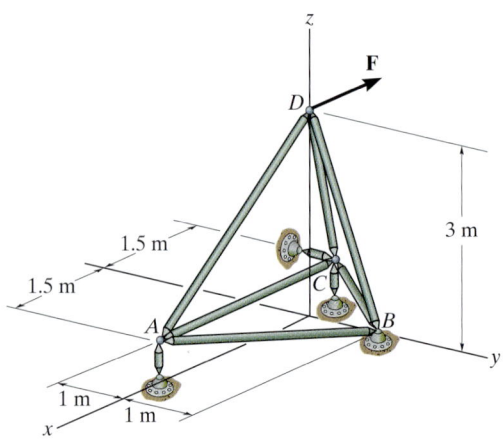

Probs. 6–53/54

6–55. Determine the force in each member of the space truss and state if the members are in tension or compression. The truss is supported by ball-and-socket joints at C, D, E, and G.

6–57. Determine the force in members BE, BC, BF, and CE of the space truss, and state if the members are in tension or compression.

6–58. Determine the force in members AF, AB, AD, ED, FD, and BD of the space truss, and state if the members are in tension or compression.

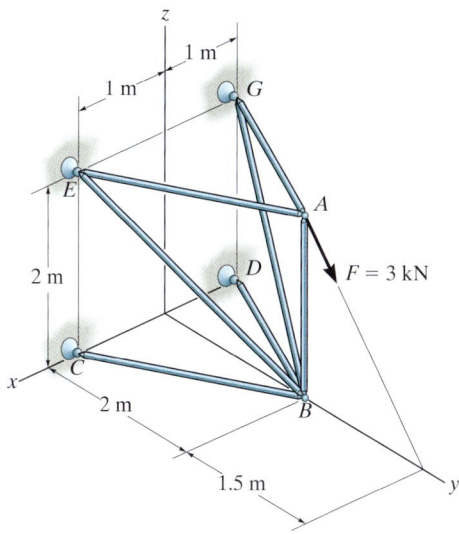

Prob. 6–55

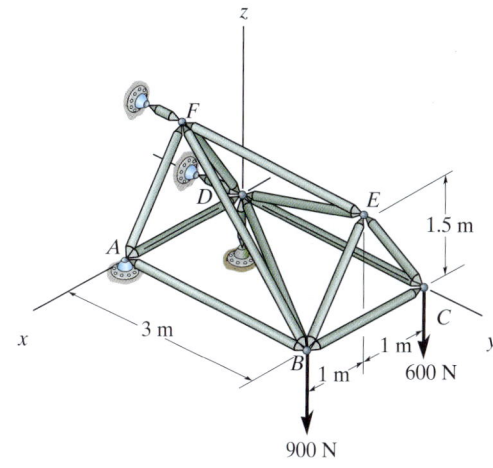

Probs. 6–57/58

6–59. If the truss supports a force of $F = 200$ N, determine the force in each member and state if the members are in tension or compression.

***6–60.** If each member of the space truss can support a maximum force of 600 N in compression and 800 N in tension, determine the greatest force F the truss can support.

***6–56.** The space truss is used to support vertical forces at joints B, C, and D. Determine the force in each member and state if the members are in tension or compression. There is a roller at E, and A and F are ball-and-socket joints.

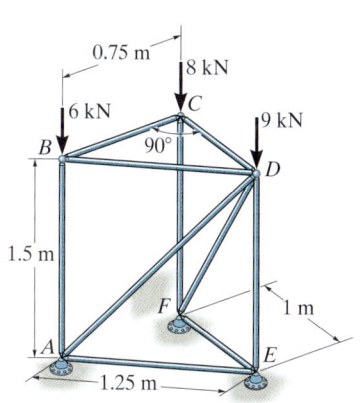

Prob. 6–56

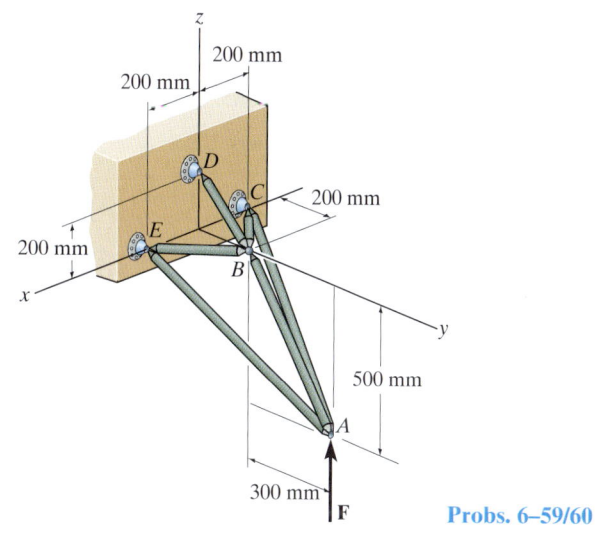

Probs. 6–59/60

6.6   Frames and Machines

Frames and machines are two types of structures which are often composed of pin-connected *multiforce members*, i.e., members that are subjected to more than two forces. *Frames* are used to support loads, whereas *machines* contain moving parts and are designed to transmit and alter the effect of forces. Provided a frame or machine contains no more supports or members than are necessary to prevent its collapse, the forces acting at the joints and supports can be determined by applying the equations of equilibrium to each of its members. Once these forces are obtained, it is then possible to *design* the size of the members, connections, and supports using the theory of mechanics of materials and an appropriate engineering design code.

Free-Body Diagrams.

In order to determine the forces acting at the joints and supports of a frame or machine, the structure must be disassembled and the free-body diagrams of its parts must be drawn. The following important points *must* be observed:

- Isolate each part by drawing its *outlined shape*. Then show all the forces and/or couple moments that act on the part. Make sure to *label* or *identify* each known and unknown force and couple moment with reference to an established *x, y* coordinate system. Also, indicate any dimensions used for taking moments. Most often the equations of equilibrium are easier to apply if the forces are represented by their rectangular components. As usual, the sense of an unknown force or couple moment can be assumed.

- Identify all the two-force members in the structure and represent their free-body diagrams as having two equal but opposite collinear forces acting at their points of application. (See Sec. 5.4.) By recognizing the two-force members, we can avoid solving an unnecessary number of equilibrium equations.

- Forces common to *any* two *contacting* members act with equal magnitudes but opposite sense on the respective members. If the two members are treated as a *"system" of connected members*, then these forces are *"internal"* and are *not shown* on the *free-body diagram of the system*; however, if the free-body diagram of *each member* is drawn, the forces are *"external"* and *must* be shown as equal in magnitude and opposite in direction on each of the two free-body diagrams.

The following examples graphically illustrate how to draw the free-body diagrams of a dismembered frame or machine. In all cases, the weight of the members is neglected.

This large crane is a typical example of a framework.

Common tools such as these pliers act as simple machines. Here the applied force on the handles creates a much larger force at the jaws.

EXAMPLE 6.9

For the frame shown in Fig. 6–21a, draw the free-body diagram of (a) each member, (b) the pins at B and A, and (c) the two members connected together.

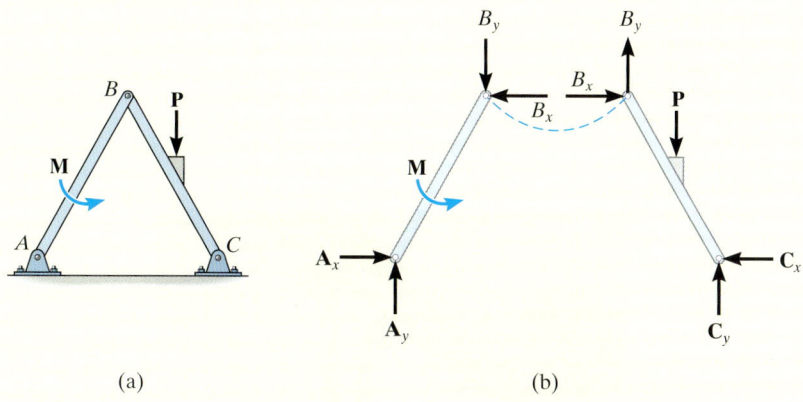

(a)   (b)

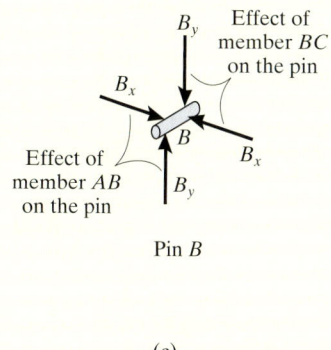

Pin B

(c)

SOLUTION

Part (a). By inspection, members BA and BC are *not* two-force members. Instead, as shown on the free-body diagrams, Fig. 6–21b, BC is subjected to a force from each of the pins at B and C and the external force **P**. Likewise, AB is subjected to a force from each of the pins at A and B and the external couple moment **M**. The pin forces are represented by their x and y components.

Part (b). The pin at B is subjected to only *two forces*, i.e., the force of member BC and the force of member AB. For *equilibrium* these forces (or their respective components) must be equal but opposite, Fig. 6–21c. Realize that Newton's third law is applied between the pin and its connected members, i.e., the effect of the pin on the two members, Fig. 6–21b, and the equal but opposite effect of the two members on the pin, Fig. 6–21c. In the same manner, there are three forces on pin A, Fig. 6–21d, caused by the force components of member AB and each of the two pin leaves.

Part (c). The free-body diagram of both members connected together, yet removed from the supporting pins at A and C, is shown in Fig. 6–21e. The force components $\mathbf{B}_x$ and $\mathbf{B}_y$ are *not shown* on this diagram since they are *internal* forces (Fig. 6–21b) and therefore cancel out. Also, to be consistent when later applying the equilibrium equations, the unknown force components at A and C must act in the *same sense* as those shown in Fig. 6–21b.

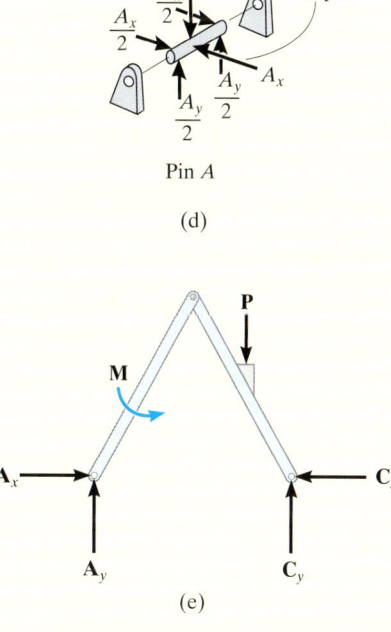

Pin A

(d)

(e)

Fig. 6–21

EXAMPLE | 6.10

A constant tension in the conveyor belt is maintained by using the device shown in Fig. 6–22a. Draw the free-body diagrams of the frame and the cylinder (or pulley) that the belt surrounds. The suspended block has a weight of W.

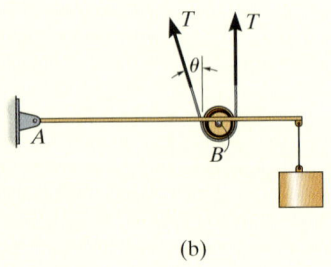

(b)

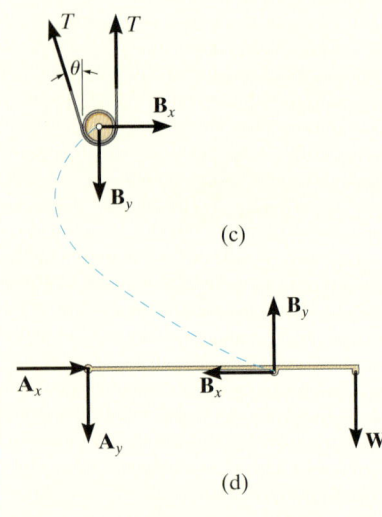

(c)

(d)

(a)

Fig. 6–22

SOLUTION

The idealized model of the device is shown in Fig. 6–22b. Here the angle θ is assumed to be known. From this model, the free-body diagrams of the pulley and frame are shown in Figs. 6–22c and 6–22d, respectively. Note that the force components $\mathbf{B}_x$ and $\mathbf{B}_y$ that the pin at B exerts on the pulley must be equal but opposite to the ones acting on the frame. See Fig. 6–21c of Example 6.9.

EXAMPLE | 6.11

For the frame shown in Fig. 6–23a, draw the free-body diagrams of (a) the entire frame including the pulleys and cords, (b) the frame without the pulleys and cords, and (c) each of the pulleys.

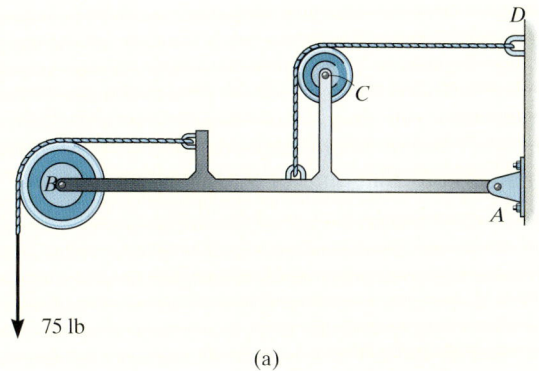

(a)

SOLUTION

Part (a).   When the entire frame including the pulleys and cords is considered, the interactions at the points where the pulleys and cords are connected to the frame become pairs of *internal* forces which cancel each other and therefore are not shown on the free-body diagram, Fig. 6–23b.

Part (b).   When the cords and pulleys are removed, their effect *on the frame* must be shown, Fig. 6–23c.

Part (c).   The force components B_x, B_y, C_x, C_y of the pins on the pulleys, Fig. 6–23d, are equal but opposite to the force components exerted by the pins on the frame, Fig. 6–23c. See Example 6.9.

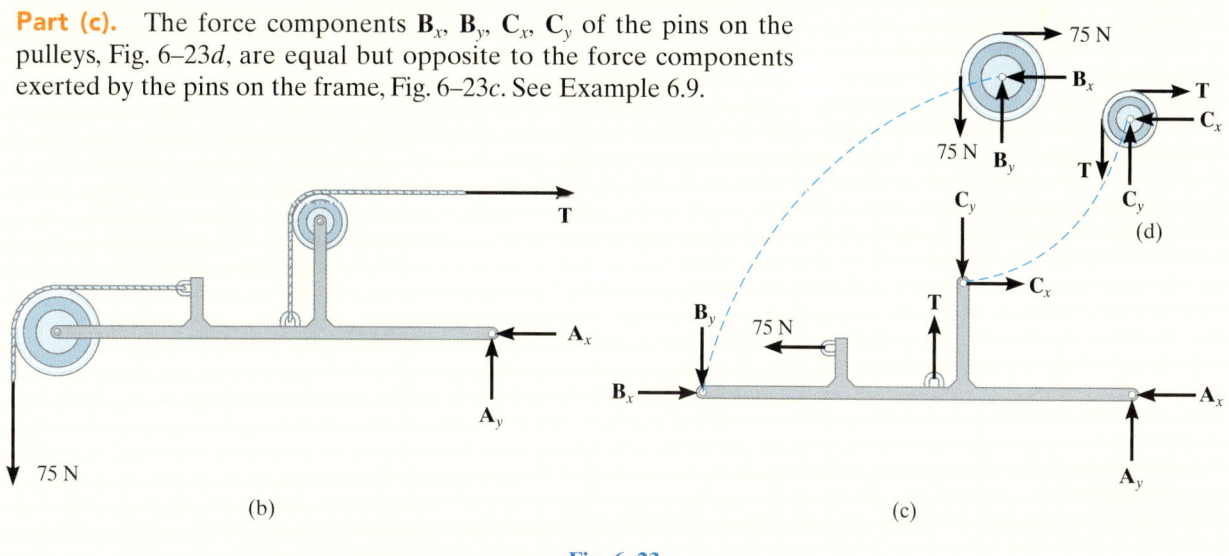

(b)

(c)

(d)

Fig. 6–23

EXAMPLE | 6.12

(a)

Fig. 6–24

Draw the free-body diagrams of the members of the backhoe, shown in the photo, Fig. 6–24a. The bucket and its contents have a weight W.

SOLUTION

The idealized model of the assembly is shown in Fig. 6–24b. By inspection, members AB, BC, BE, and HI are all two-force members since they are pin connected at their end points and no other forces act on them. The free-body diagrams of the bucket and the stick are shown in Fig. 6–24c. Note that pin C is subjected to only two forces, whereas the pin at B is subjected to three forces, Fig. 6–24d. The free-body diagram of the entire assembly is shown in Fig. 6–24e.

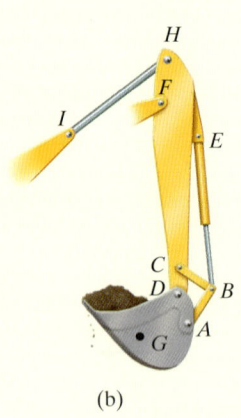

(b)

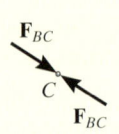

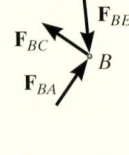

(d)

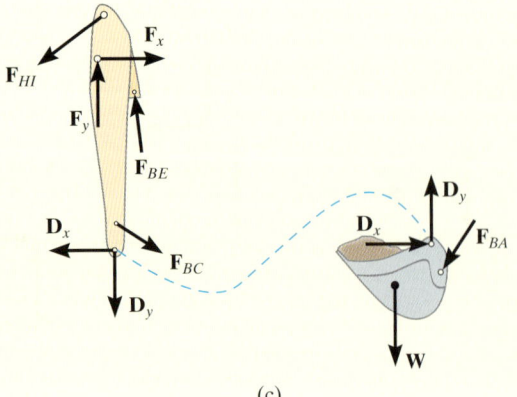

(c)

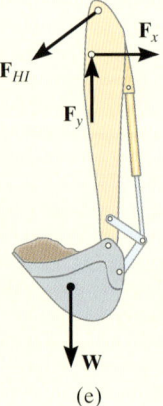

(e)

EXAMPLE 6.13

Draw the free-body diagram of each part of the smooth piston and link mechanism used to crush recycled cans, Fig. 6–25a.

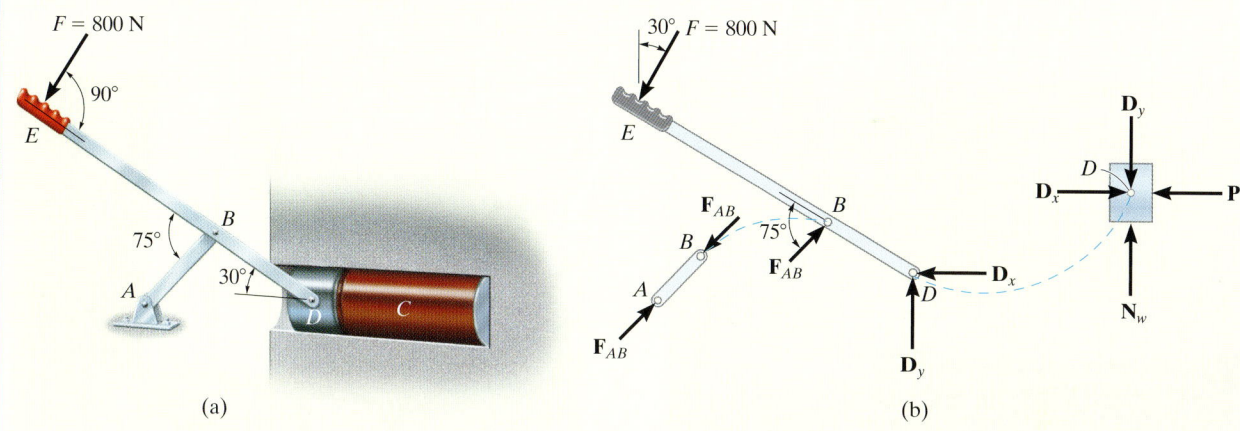

(a)

(b)

Fig. 6–25

SOLUTION

By inspection, member AB is a two-force member. The free-body diagrams of the three parts are shown in Fig. 6–25b. Since the pins at B and D *connect only two parts together*, the forces there are shown as equal but opposite on the separate free-body diagrams of their connected members. In particular, four components of force act on the piston: D_x and D_y represent the effect of the pin (or lever EBD), N_w is the *resultant force* of the wall support, and P is the resultant compressive force caused by the can C. The directional sense of each of the unknown forces is assumed, and the correct sense will be established after the equations of equilibrium are applied.

NOTE: A free-body diagram of the entire assembly is shown in Fig. 6–25c. Here the forces between the components are internal and are not shown on the free-body diagram.

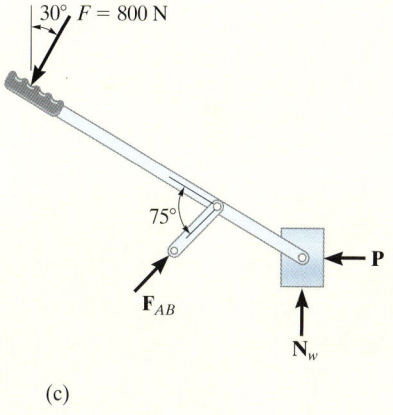

(c)

Before proceeding, it is highly recommended that you cover the solutions to the previous examples and attempt to draw the requested free-body diagrams. When doing so, make sure the work is neat and that all the forces and couple moments are properly labeled. When finished, challenge yourself and solve the following four problems.

CONCEPTUAL PROBLEMS

P6–1. Draw the free-body diagrams of each of the crane boom segments *AB*, *BC*, and *BD*. Only the weights of *AB* and *BC* are significant. Assume *A* and *B* are pins.

P6–3. Draw the free-body diagrams of the boom *ABCDF* and the stick *FGH* of the bucket lift. Neglect the weights of the members. The bucket weighs *W*. The two–force members are *BI*, *CE*, *DE* and *GE*. Assume all indicated points of connection are pins.

P6–1

P6–3

P6–2. Draw the free-body diagrams of the boom *ABCD* and the stick *EDFGH* of the backhoe. The weights of these two members are significant. Neglect the weights of all the other members, and assume all indicated points of connection are pins.

P6–4. To operate the can crusher one pushes down on the lever arm *ABC* which rotates about the fixed pin at *B*. This moves the side links *CD* downward, which causes the guide plate *E* to also move downward and thereby crush the can. Draw the free-body diagrams of the lever, side link, and guide plate. Make up some reasonable numbers and do an equilibrium analysis to show how much an applied vertical force at the handle is magnified when it is transmitted to the can. Assume all points of connection are pins and the guides for the plate are smooth.

P6–2

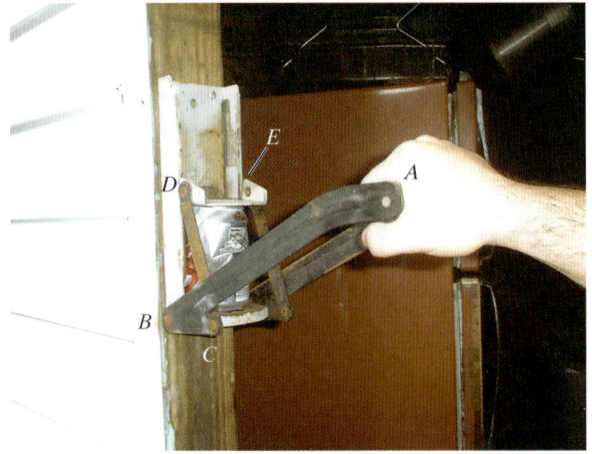

P6–4

Procedure for Analysis

The joint reactions on frames or machines (structures) composed of multiforce members can be determined using the following procedure.

Free-Body Diagram.

- Draw the free-body diagram of the entire frame or machine, a portion of it, or each of its members. The choice should be made so that it leads to the most direct solution of the problem.

- When the free-body diagram of a group of members of a frame or machine is drawn, the forces between the connected parts of this group are internal forces and are not shown on the free-body diagram of the group.

- Forces common to two members which are in contact act with equal magnitude but opposite sense on the respective free-body diagrams of the members.

- Two-force members, regardless of their shape, have equal but opposite collinear forces acting at the ends of the member.

- In many cases it is possible to tell by inspection the proper sense of the unknown forces acting on a member; however, if this seems difficult, the sense can be assumed.

- Remember that a couple moment is a free vector and can act at any point on the free-body diagram. Also, a force is a sliding vector and can act at any point along its line of action.

Equations of Equilibrium.

- Count the number of unknowns and compare it to the total number of equilibrium equations that are available. In two dimensions, there are three equilibrium equations that can be written for each member.

- Sum moments about a point that lies at the intersection of the lines of action of as many of the unknown forces as possible.

- If the solution of a force or couple moment magnitude is found to be negative, it means the sense of the force is the reverse of that shown on the free-body diagram.

EXAMPLE 6.14

Determine the tension in the cables and also the force **P** required to support the 600-N force using the frictionless pulley system shown in Fig. 6–26a.

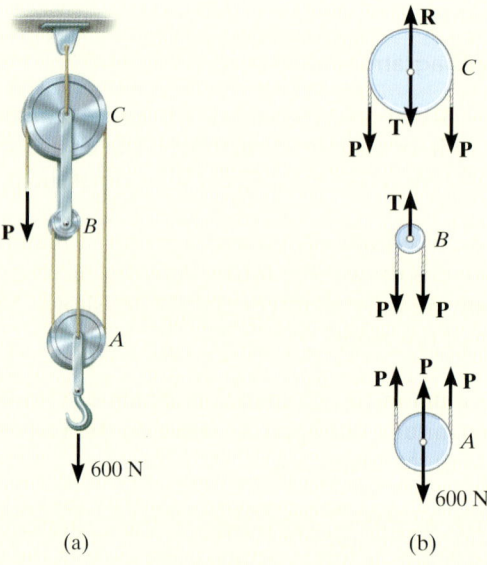

(a)   (b)

Fig. 6–26

SOLUTION

Free-Body Diagram.   A free-body diagram of each pulley *including* its pin and a portion of the contacting cable is shown in Fig. 6–26b. Since the cable is *continuous*, it has a *constant tension P* acting throughout its length. The link connection between pulleys *B* and *C* is a two-force member, and therefore it has an unknown tension *T* acting on it. Notice that the *principle of action, equal but opposite reaction* must be carefully observed for forces **P** and **T** when the *separate* free-body diagrams are drawn.

Equations of Equilibrium.   The three unknowns are obtained as follows:

Pulley A

$$+\uparrow\Sigma F_y = 0; \qquad 3P - 600 \text{ N} = 0 \qquad P = 200 \text{ N} \qquad \textit{Ans.}$$

Pulley B

$$+\uparrow\Sigma F_y = 0; \qquad T - 2P = 0 \qquad T = 400 \text{ N} \qquad \textit{Ans.}$$

Pulley C

$$+\uparrow\Sigma F_y = 0; \qquad R - 2P - T = 0 \qquad R = 800 \text{ N} \qquad \textit{Ans.}$$

EXAMPLE 6.15

A 500-kg elevator car in Fig. 6–27a is being hoisted by motor A using the pulley system shown. If the car is traveling with a constant speed, determine the force developed in the two cables. Neglect the mass of the cable and pulleys.

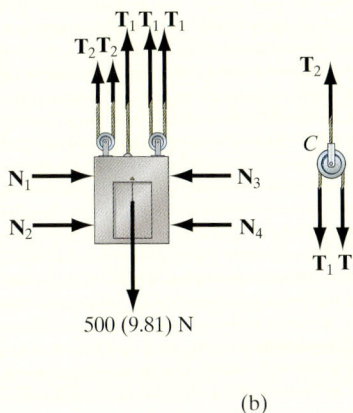

(b)

(a)

Fig. 6–27

SOLUTION

Free-Body Diagram.   We can solve this problem using the free-body diagrams of the elevator car and pulley C, Fig. 6–27b. The tensile forces developed in the cables are denoted as T_1 and T_2.

Equations of Equilibrium.   For pulley C,

$$+\uparrow \Sigma F_y = 0; \quad T_2 - 2T_1 = 0 \quad \text{or} \quad T_2 = 2T_1 \qquad (1)$$

For the elevator car,

$$+\uparrow \Sigma F_y = 0; \quad 3T_1 + 2T_2 - 500(9.81)\ \text{N} = 0 \qquad (2)$$

Substituting Eq. (1) into Eq. (2) yields

$$3T_1 + 2(2T_1) - 500(9.81)\ \text{N} = 0$$

$$T_1 = 700.71\ \text{N} = 701\ \text{N} \qquad \textit{Ans.}$$

Substituting this result into Eq. (1),

$$T_2 = 2(700.71)\ \text{N} = 1401\ \text{N} = 1.40\ \text{kN} \qquad \textit{Ans.}$$

6

EXAMPLE | 6.16

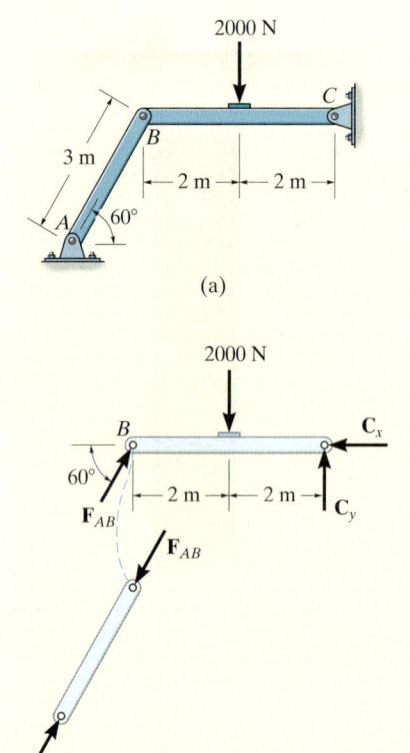

(a)

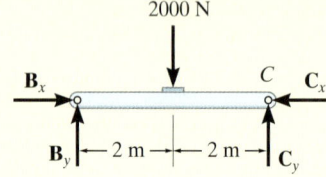

(b)

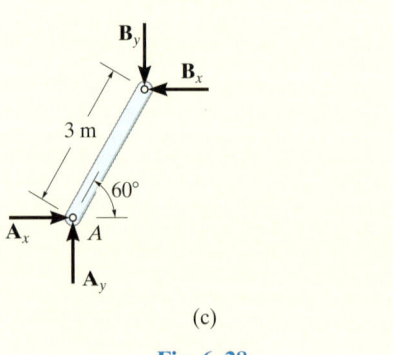

(c)

Fig. 6–28

Determine the horizontal and vertical components of force which the pin at C exerts on member BC of the frame in Fig. 6–28a.

SOLUTION I

Free-Body Diagrams.   By inspection it can be seen that AB is a two-force member. The free-body diagrams are shown in Fig. 6–28b.

Equations of Equilibrium.   The *three unknowns* can be determined by applying the three equations of equilibrium to member CB.

$$\zeta + \Sigma M_C = 0; \quad 2000 \text{ N}(2 \text{ m}) - (F_{AB} \sin 60°)(4 \text{ m}) = 0 \quad F_{AB} = 1154.7 \text{ N}$$

$$\xrightarrow{+} \Sigma F_x = 0; \quad 1154.7 \cos 60° \text{ N} - C_x = 0 \quad C_x = 577 \text{ N} \qquad Ans.$$

$$+\uparrow \Sigma F_y = 0; \quad 1154.7 \sin 60° \text{ N} - 2000 \text{ N} + C_y = 0$$

$$C_y = 1000 \text{ N} \qquad Ans.$$

SOLUTION II

Free-Body Diagrams.   If one does not recognize that AB is a two-force member, then more work is involved in solving this problem. The free-body diagrams are shown in Fig. 6–28c.

Equations of Equilibrium.   The *six unknowns* are determined by applying the three equations of equilibrium to each member.

Member AB

$$\zeta + \Sigma M_A = 0; \quad B_x(3 \sin 60° \text{ m}) - B_y(3 \cos 60° \text{ m}) = 0 \qquad (1)$$

$$\xrightarrow{+} \Sigma F_x = 0; \quad A_x - B_x = 0 \qquad (2)$$

$$+\uparrow \Sigma F_y = 0; \quad A_y - B_y = 0 \qquad (3)$$

Member BC

$$\zeta + \Sigma M_C = 0; \quad 2000 \text{ N}(2 \text{ m}) - B_y(4 \text{ m}) = 0 \qquad (4)$$

$$\xrightarrow{+} \Sigma F_x = 0; \quad B_x - C_x = 0 \qquad (5)$$

$$+\uparrow \Sigma F_y = 0; \quad B_y - 2000 \text{ N} + C_y = 0 \qquad (6)$$

The results for C_x and C_y can be determined by solving these equations in the following sequence: 4, 1, 5, then 6. The results are

$$B_y = 1000 \text{ N}$$

$$B_x = 577 \text{ N}$$

$$C_x = 577 \text{ N} \qquad Ans.$$

$$C_y = 1000 \text{ N} \qquad Ans.$$

By comparison, Solution I is simpler since the requirement that F_{AB} in Fig. 6–28b be equal, opposite, and collinear at the ends of member AB automatically satisfies Eqs. 1, 2, and 3 above and therefore eliminates the need to write these equations. *As a result, save yourself some time and effort by always identifying the two-force members before starting the analysis!*

EXAMPLE 6.17

The compound beam shown in Fig. 6–29a is pin connected at B. Determine the components of reaction at its supports. Neglect its weight and thickness.

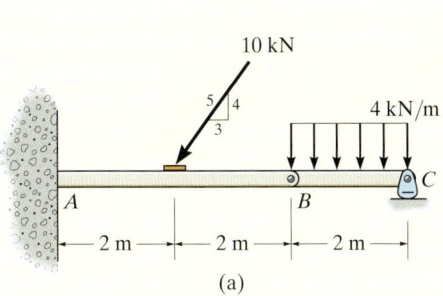

(a)

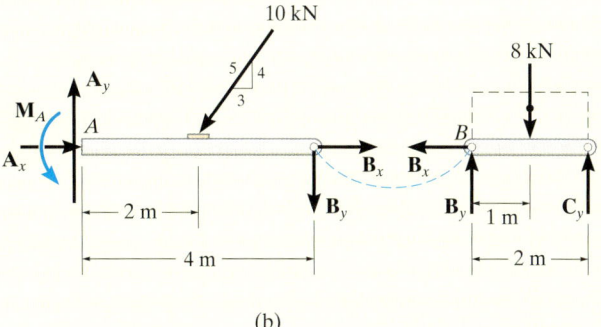

(b)

Fig. 6–29

SOLUTION

Free-Body Diagrams. By inspection, if we consider a free-body diagram of the *entire beam ABC*, there will be three unknown reactions at A and one at C. These four unknowns cannot all be obtained from the three available equations of equilibrium, and so for the solution it will become necessary to dismember the beam into its two segments, as shown in Fig. 6–29b.

Equations of Equilibrium. The six unknowns are determined as follows:

Segment BC

$$\xrightleftharpoons{} \Sigma F_x = 0; \qquad\qquad\qquad\qquad B_x = 0$$

$$\zeta + \Sigma M_B = 0; \qquad -8 \text{ kN}(1 \text{ m}) + C_y(2 \text{ m}) = 0$$

$$+\uparrow \Sigma F_y = 0; \qquad\qquad B_y - 8 \text{ kN} + C_y = 0$$

Segment AB

$$\xrightarrow{} \Sigma F_x = 0; \qquad A_x - (10 \text{ kN})\left(\tfrac{3}{5}\right) + B_x = 0$$

$$\zeta + \Sigma M_A = 0; \qquad M_A - (10 \text{ kN})\left(\tfrac{4}{5}\right)(2 \text{ m}) - B_y(4 \text{ m}) = 0$$

$$+\uparrow \Sigma F_y = 0; \qquad A_y - (10 \text{ kN})\left(\tfrac{4}{5}\right) - B_y = 0$$

Solving each of these equations successively, using previously calculated results, we obtain

$A_x = 6$ kN	$A_y = 12$ kN	$M_A = 32$ kN · m	*Ans.*
$B_x = 0$	$B_y = 4$ kN		
$C_y = 4$ kN			*Ans.*

EXAMPLE | **6.18**

The two planks in Fig. 6–30a are connected together by cable BC and a smooth spacer DE. Determine the reactions at the smooth supports A and F, and also find the force developed in the cable and spacer.

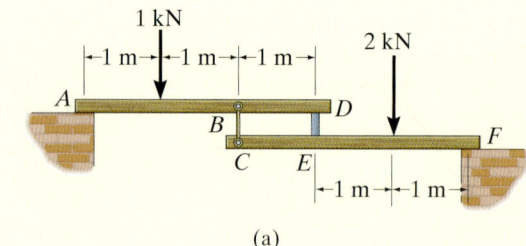

(a)

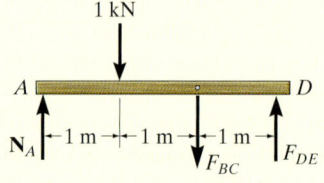   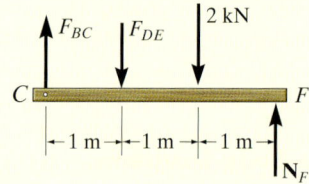

(b)

Fig. 6–30

SOLUTION

Free-Body Diagrams. The free-body diagram of each plank is shown in Fig. 6–30b. It is important to apply Newton's third law to the interaction forces F_{BC} and F_{DE} as shown.

Equations of Equilibrium. For plank AD,

$$\zeta + \Sigma M_A = 0; \qquad F_{DE}(3 \text{ m}) - F_{BC}(2 \text{ m}) - 1 \text{ kN} (1 \text{ m}) = 0$$

For plank CF,

$$\zeta + \Sigma M_F = 0; \qquad F_{DE}(2 \text{ m}) - F_{BC}(3 \text{ m}) + 2 \text{ kN} (1 \text{ m}) = 0$$

Solving simultaneously,

$$F_{DE} = 1.40 \text{ kN} \quad F_{BC} = 1.60 \text{ kN} \qquad\qquad Ans.$$

Using these results, for plank AD,

$$+\uparrow \Sigma F_y = 0; \qquad N_A + 1.40 \text{ kN} - 1.60 \text{ kN} - 1 \text{ kN} = 0$$

$$N_A = 1.20 \text{ kN} \qquad\qquad Ans.$$

And for plank CF,

$$+\uparrow \Sigma F_y = 0; \qquad N_F + 1.60 \text{ kN} - 1.40 \text{ kN} - 2 \text{ kN} = 0$$

$$N_F = 1.80 \text{ kN} \qquad\qquad Ans.$$

NOTE: Draw the free-body diagram of the system of *both* planks and apply $\Sigma M_A = 0$ to determine N_F. Then use the free-body diagram of CEF to determine F_{DE} and F_{BC}.

EXAMPLE 6.19

The 75-kg man in Fig. 6–31a attempts to lift the 40-kg uniform beam off the roller support at B. Determine the tension developed in the cable attached to B and the normal reaction of the man on the beam when this is about to occur.

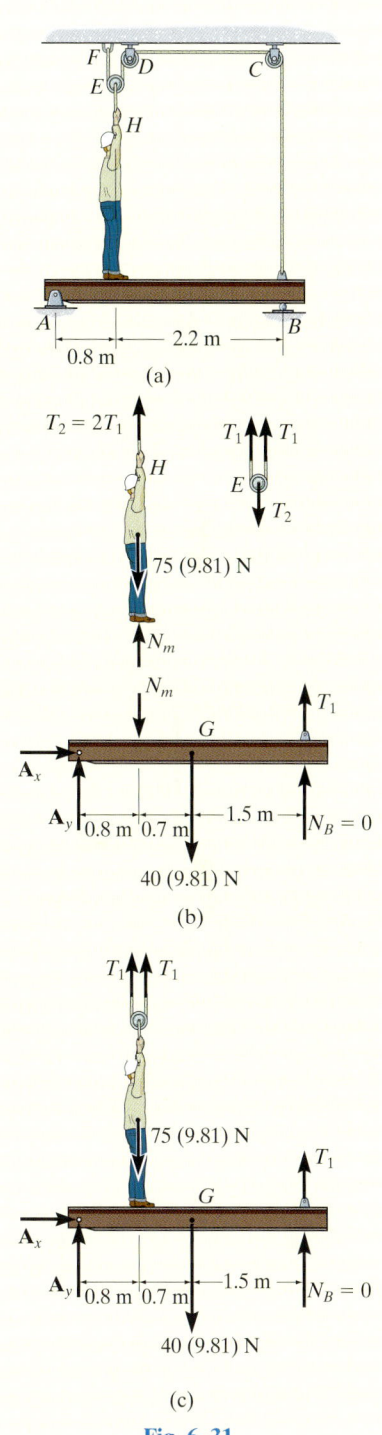

SOLUTION

Free-Body Diagrams. The tensile force in the cable will be denoted as T_1. The free-body diagrams of the pulley E, the man, and the beam are shown in Fig. 6–31b. Since the man must lift the beam off the roller B then $N_B = 0$. When drawing each of these diagrams, it is very important to apply Newton's third law.

Equations of Equilibrium. Using the free-body diagram of pulley E,

$$+\uparrow\Sigma F_y = 0; \qquad 2T_1 - T_2 = 0 \quad \text{or} \quad T_2 = 2T_1 \tag{1}$$

Referring to the free-body diagram of the man using this result,

$$+\uparrow\Sigma F_y = 0 \qquad N_m + 2T_1 - 75(9.81)\text{ N} = 0 \tag{2}$$

Summing moments about point A on the beam,

$$\zeta+\Sigma M_A = 0; \ T_1(3\text{ m}) - N_m(0.8\text{ m}) - [40(9.81)\text{ N}](1.5\text{ m}) = 0 \tag{3}$$

Solving Eqs. 2 and 3 simultaneously for T_1 and N_m, then using Eq. (1) for T_2, we obtain

$$T_1 = 256\text{ N} \qquad N_m = 224\text{ N} \qquad T_2 = 512\text{ N} \qquad \textit{Ans.}$$

SOLUTION II

A direct solution for T_1 can be obtained by considering the beam, the man, and pulley E as a *single system*. The free-body diagram is shown in Fig. 6–31c. Thus,

$$\zeta+\Sigma M_A = 0; \qquad 2T_1(0.8\text{ m}) - [75(9.81)\text{ N}](0.8\text{ m})$$
$$- [40(9.81)\text{ N}](1.5\text{ m}) + T_1(3\text{ m}) = 0$$
$$T_1 = 256\text{ N} \qquad \textit{Ans.}$$

With this result Eqs. 1 and 2 can then be used to find N_m and T_2.

Fig. 6–31

EXAMPLE 6.20

The smooth disk shown in Fig. 6–32a is pinned at D and has a weight of 20 lb. Neglecting the weights of the other members, determine the horizontal and vertical components of reaction at pins B and D.

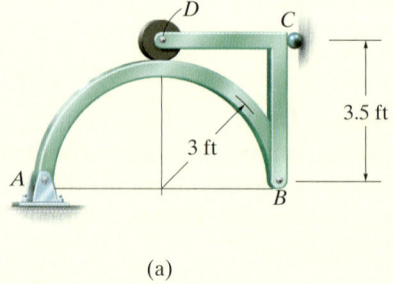

(a)

SOLUTION

Free-Body Diagrams.   The free-body diagrams of the entire frame and each of its members are shown in Fig. 6–32b.

Equations of Equilibrium.   The eight unknowns can of course be obtained by applying the eight equilibrium equations to each member—three to member AB, three to member BCD, and two to the disk. (Moment equilibrium is automatically satisfied for the disk.) If this is done, however, all the results can be obtained only from a simultaneous solution of some of the equations. (Try it and find out.) To avoid this situation, it is best first to determine the three support reactions on the *entire* frame; then, using these results, the remaining five equilibrium equations can be applied to two other parts in order to solve successively for the other unknowns.

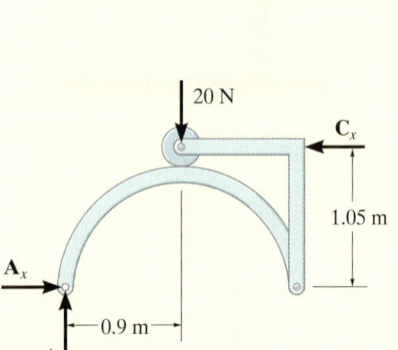

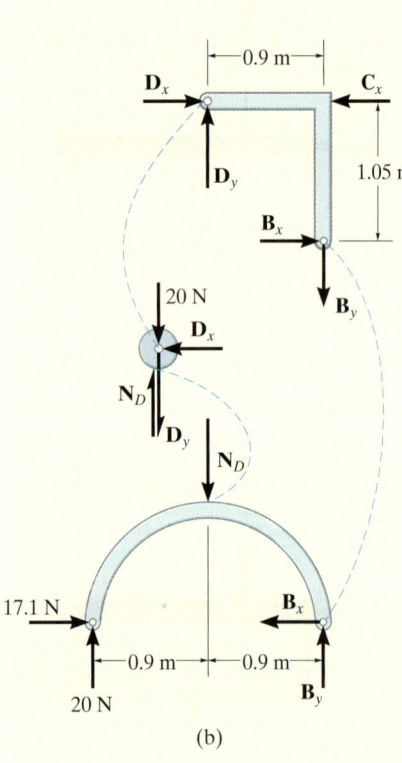

(b)

Fig. 6–32

Entire Frame

$\zeta +\Sigma M_A = 0;$ $-20 \text{ N} (0.9 \text{ m}) + C_x(1.05 \text{ m}) = 0$ $C_x = 17.1 \text{ N}$

$\xrightarrow{+} \Sigma F_x = 0;$ $A_x - 17.1 \text{ N} = 0$ $A_x = 17.1 \text{ N}$

$+\uparrow \Sigma F_y = 0;$ $A_y - 20 \text{ N} = 0$ $A_y = 20 \text{ N}$

Member AB

$\xrightarrow{+} \Sigma F_x = 0;$ $17.1 \text{ N} - B_x = 0$ $B_x = 17.1 \text{ N}$ *Ans.*

$\zeta +\Sigma M_B = 0;$ $-20 \text{ N} (1.8 \text{ m}) + N_D(0.9 \text{ m}) = 0$ $N_D = 40 \text{ N}$

$+\uparrow \Sigma F_y = 0;$ $20 \text{ N} - 40 \text{ N} + B_y = 0$ $B_y = 20 \text{ N}$ *Ans.*

Disk

$\xrightarrow{+} \Sigma F_x = 0;$ $D_x = 0$ *Ans.*

$+\uparrow \Sigma F_y = 0;$ $40 \text{ N} - 20 \text{ N} - D_y = 0$ $D_y = 20 \text{ N}$ *Ans.*

EXAMPLE 6.21

The frame in Fig. 6–33a supports the 50-kg cylinder. Determine the horizontal and vertical components of reaction at A and the force at C.

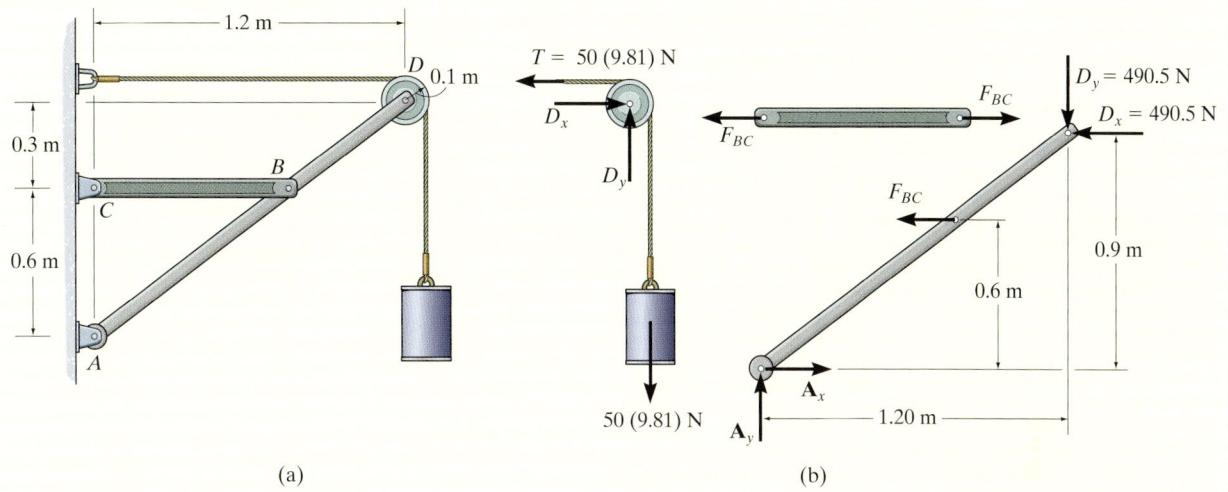

Fig. 6–33

SOLUTION

Free-Body Diagrams. The free-body diagram of pulley D, along with the cylinder and a portion of the cord (a system), is shown in Fig. 6–33b. Member BC is a two-force member as indicated by its free-body diagram. The free-body diagram of member ABD is also shown.

Equations of Equilibrium. We will begin by analyzing the equilibrium of the pulley. The moment equation of equilibrium is automatically satisfied with $T = 50(9.81)$ N, and so

$$\xrightarrow{+} \Sigma F_x = 0; \qquad D_x - 50(9.81)\ \text{N} = 0 \quad D_x = 490.5\ \text{N}$$

$$+\uparrow \Sigma F_y = 0; \qquad D_y - 50(9.81)\ \text{N} = 0 \quad D_y = 490.5\ \text{N} \qquad \textit{Ans.}$$

Using these results, F_{BC} can be determined by summing moments about point A on member ABD.

$$\zeta + \Sigma M_A = 0;\ F_{BC}(0.6\ \text{m}) + 490.5\ \text{N}(0.9\ \text{m}) - 490.5\ \text{N}(1.20\ \text{m}) = 0$$

$$F_{BC} = 245.25\ \text{N} \qquad \textit{Ans.}$$

Now A_x and A_y can be determined by summing forces.

$$\xrightarrow{+} \Sigma F_x = 0; \quad A_x - 245.25\ \text{N} - 490.5\ \text{N} = 0 \quad A_x = 736\ \text{N} \qquad \textit{Ans.}$$

$$+\uparrow \Sigma F_y = 0; \qquad\qquad A_y - 490.5\ \text{N} = 0 \quad A_y = 490.5\ \text{N} \qquad \textit{Ans.}$$

EXAMPLE 6.22

Determine the force the pins at A and B exert on the two-member frame shown in Fig. 6–34a.

SOLUTION I

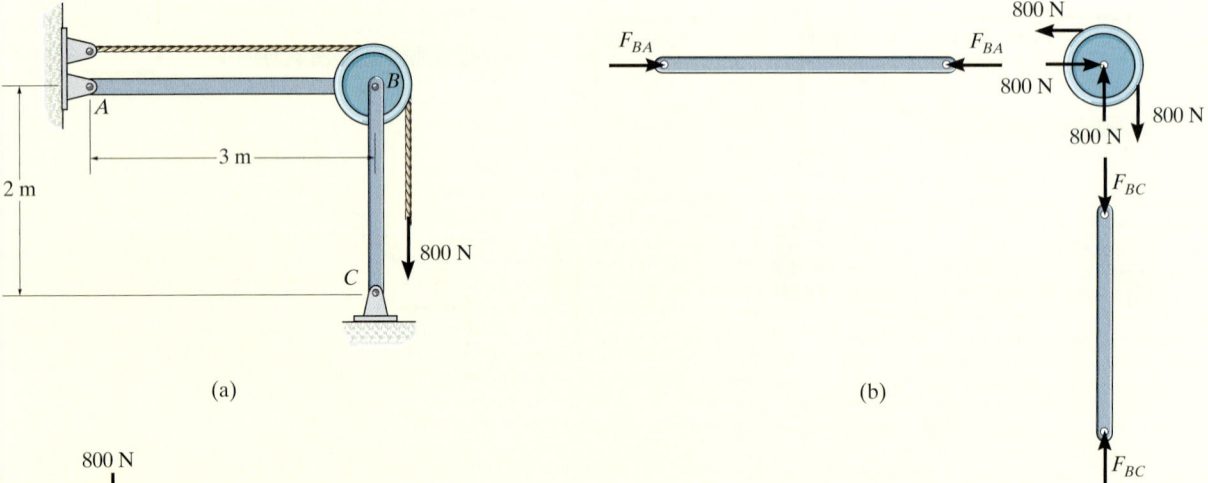

(a)   (b)

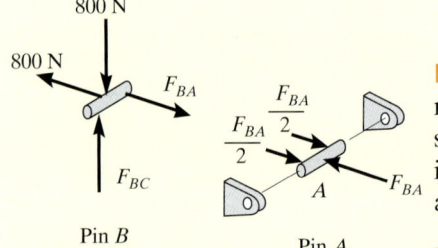

Pin B Pin A

(c)   (d)

Free-Body Diagrams. By inspection AB and BC are two-force members. Their free-body diagrams, along with that of the pulley, are shown in Fig. 6–34b. In order to solve this problem we must also include the free-body diagram of the pin at B because this pin connects all *three members* together, Fig. 6–34c.

Equations of Equilibrium: Apply the equations of force equilibrium to pin B.

$$\xrightarrow{+} \Sigma F_x = 0; \quad F_{BA} - 800\ \text{N} = 0; \quad F_{BA} = 800\ \text{N} \quad \textit{Ans.}$$

$$+\uparrow \Sigma F_y = 0; \quad F_{BC} - 800\ \text{N} = 0; \quad F_{BC} = 800\ \text{N} \quad \textit{Ans.}$$

NOTE: The free-body diagram of the pin at A, Fig. 6–34d, indicates how the force F_{AB} is balanced by the force $(F_{AB}/2)$ exerted on the pin by each of the two pin leaves.

SOLUTION II

Free-Body Diagram. If we realize that AB and BC are two-force members, then the free-body diagram of the *entire frame* produces an easier solution, Fig. 6–34e. The force equations of equilibrium are the same as those above. Note that moment equilibrium will be satisfied, regardless of the radius of the pulley.

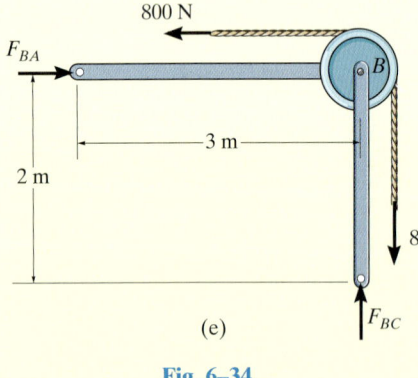

(e)

Fig. 6–34

FUNDAMENTAL PROBLEMS

All problem solutions must include FBDs.

F6–13. Determine the force *P* needed to hold the 60-N weight in equilibrium.

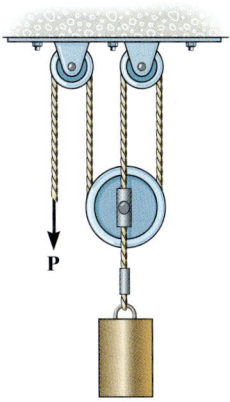

F6–13

F6–14. Determine the horizontal and vertical components of reaction at pin *C*.

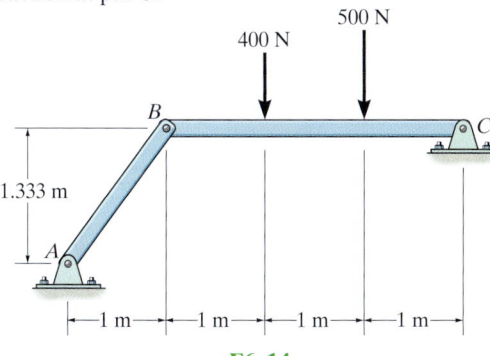

F6–14

F6–15. If a 100-N force is applied to the handles of the pliers, determine the clamping force exerted on the smooth pipe *B* and the magnitude of the resultant force that one of the members exerts on pin *A*.

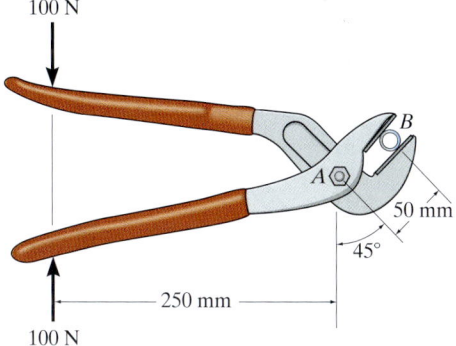

F6–15

F6–16. Determine the horizontal and vertical components of reaction at pin *C*.

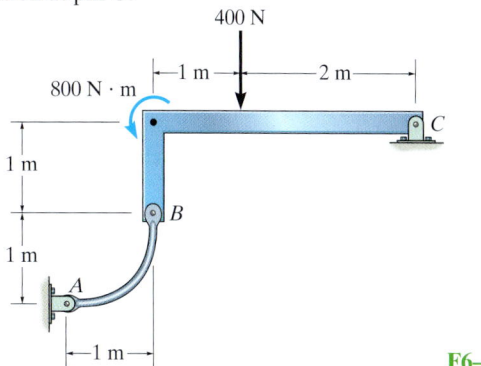

F6–16

F6–17. Determine the normal force that the 100-N plate *A* exerts on the 30-N plate *B*.

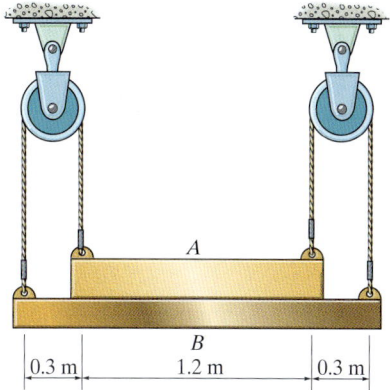

F6–17

F6–18. Determine the force *P* needed to lift the load. Also, determine the proper placement *x* of the hook for equilibrium. Neglect the weight of the beam.

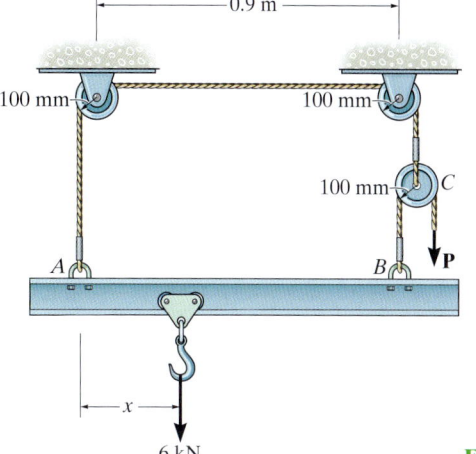

F6–18

6

F6–19. Determine the components of reaction at *A* and *B*.

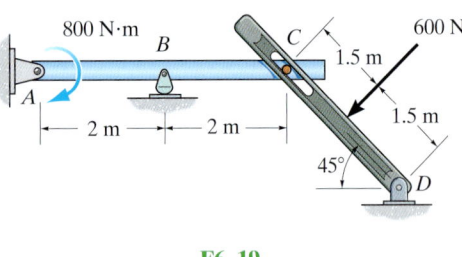

F6–19

F6–20. Determine the components of reaction at *D*.

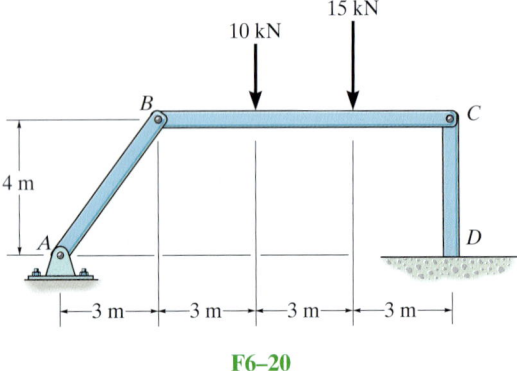

F6–20

F6–21. Determine the components of reaction at *A* and *C*.

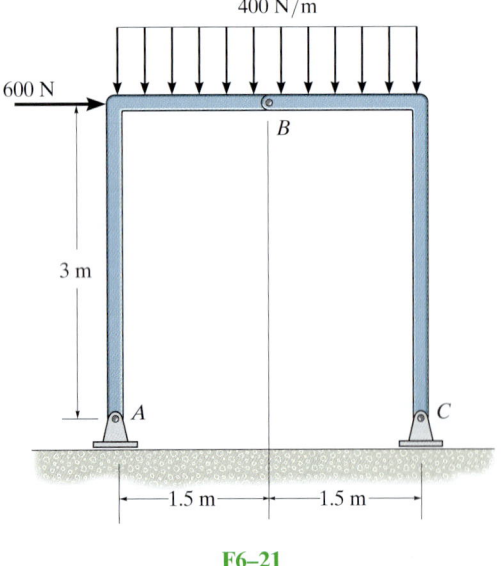

F6–21

F6–22. Determine the components of reaction at *C*.

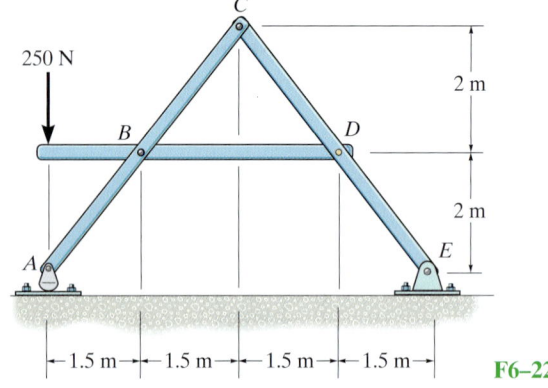

F6–22

F6–23. Determine the components of reaction at *E*.

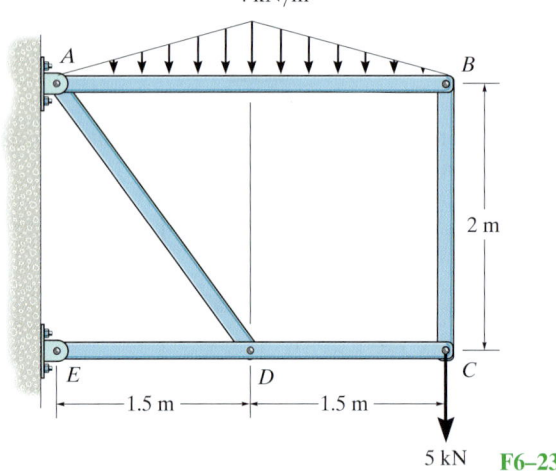

F6–23

F6–24. Determine the components of reaction at *D* and the components of reaction the pin at *A* exerts on member *BA*.

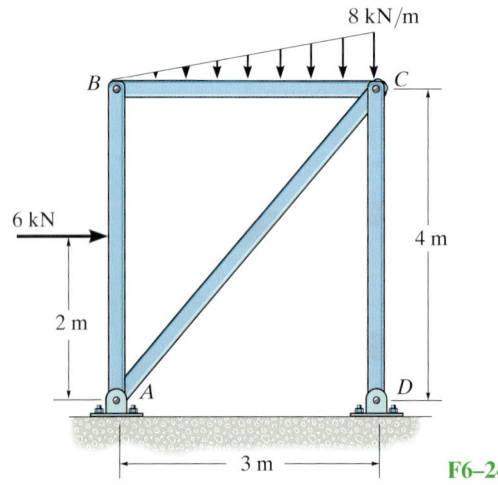

F6–24

PROBLEMS

All problem solutions must include FBDs.

6–61. Determine the force **P** required to hold the 50-kg mass in equilibrium.

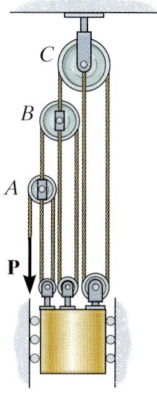

Prob. 6–61

6–62. Determine the force **P** required to hold the 150-kg crate in equilibrium. The two cables are connected to the bottom of the hanger.

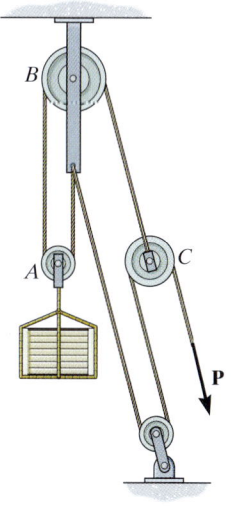

Prob. 6–62

6–63. The principles of a *differential chain block* are indicated schematically in the figure. Determine the magnitude of force **P** needed to support the 800-N force. Also, find the distance *x* where the cable must be attached to bar *AB* so the bar remains horizontal. All pulleys have a radius of 60 mm.

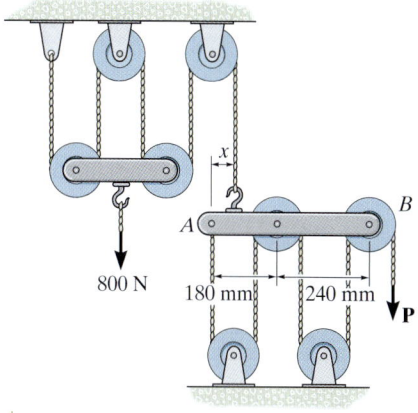

Prob. 6–63

*6–64. Determine the force *P* needed to support the 20-kg mass using the *Spanish Burton rig*. Also, what are the reactions at the supporting hooks *A*, *B*, and *C*?

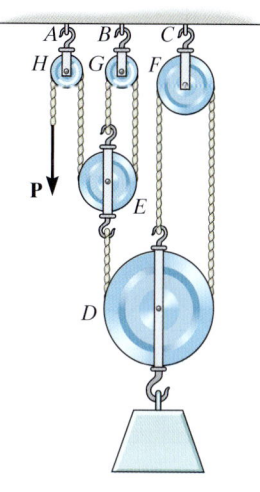

Prob. 6–64

6–65. Determine the greatest force P that can be applied to the frame if the largest force resultant acting at A can have a magnitude of 5 kN.

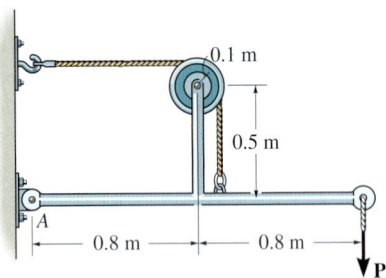

Prob. 6–65

6–66. Determine the horizontal and vertical components of force that the pins at A, B, and C exert on their connecting members.

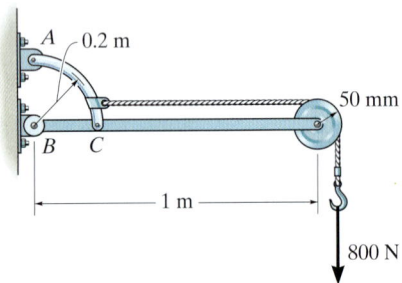

Prob. 6–66

6–67. The double link grip is used to lift the beam. If the beam weighs 4 kN, determine the horizontal and vertical components of force acting on the pin at A and the horizontal and vertical components of force that the flange of the beam exerts on the jaw at B.

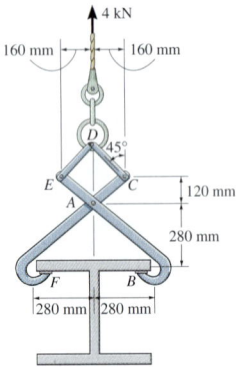

Prob. 6–67

*6–68. Determine the greatest force P that can be applied to the frame if the largest force resultant acting at A can have a magnitude of 2 kN.

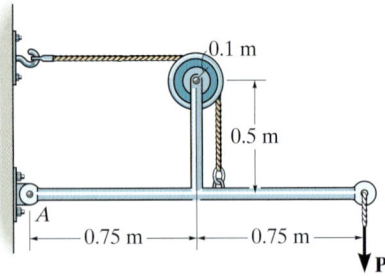

Prob. 6–68

6–69. Determine the horizontal and vertical components of force that pins A and C exert on the frame.

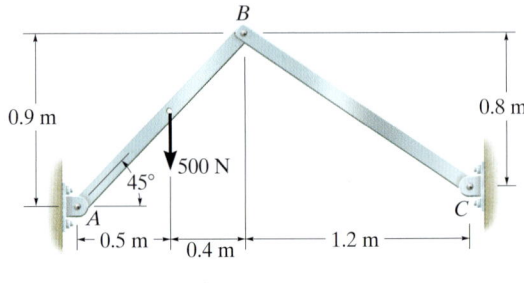

Prob. 6–69

6–70. The toggle clamp is subjected to a force $\mathbf{F}$ at the handle. Determine the vertical clamping force acting at E.

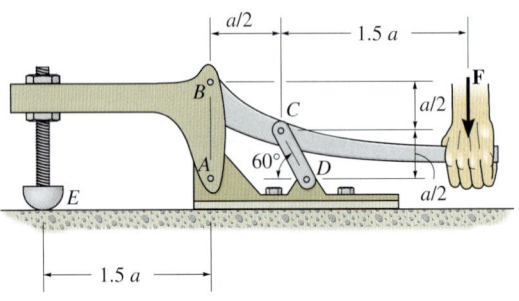

Prob. 6–70

6–71. Determine the support reactions at *A*, *C*, and *E* on the compound beam which is pin connected at *B* and *D*.

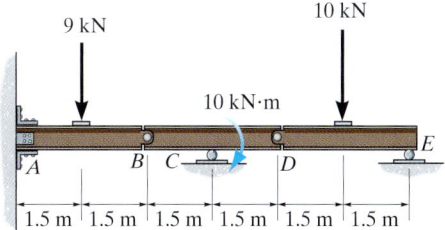

9 kN

10 kN

10 kN·m

A   *B* *C*   *D*   *E*

1.5 m 1.5 m 1.5 m 1.5 m 1.5 m 1.5 m

Prob. 6–71

***6–72.** Determine the horizontal and vertical components of force at pins *A*, *B*, and *C*, and the reactions at the fixed support *D* of the three-member frame.

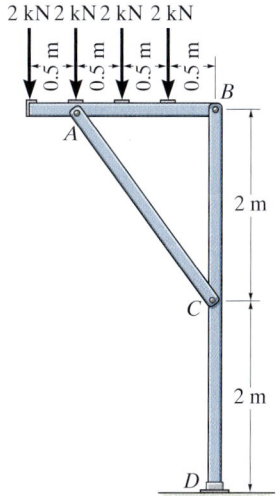

2 kN 2 kN 2 kN 2 kN

0.5 m 0.5 m 0.5 m 0.5 m

A   *B*

2 m

C

2 m

D

Prob. 6–72

6–73. The compound beam is fixed at *A* and supported by a rocker at *B* and *C*. There are hinges (pins) at *D* and *E*. Determine the reactions at the supports.

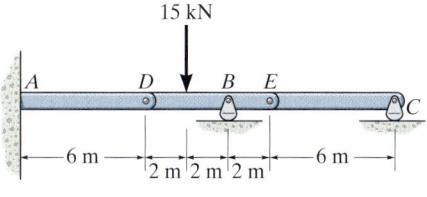

15 kN

A   *D*   *B* *E*   *C*

6 m   2 m 2 m 2 m   6 m

Prob. 6–73

6–74. The platform scale consists of a combination of third and first class levers so that the load on one lever becomes the effort that moves the next lever. Through this arrangement, a small weight can balance a massive object. If *x* = 450 mm, determine the required mass of the counterweight *S* required to balance a 90-kg load, *L*.

6–75. The platform scale consists of a combination of third and first class levers so that the load on one lever becomes the effort that moves the next lever. Through this arrangement, a small weight can balance a massive object. If *x* = 450 mm and, the mass of the counterweight *S* is 2 kg, determine the mass of the load *L* required to maintain the balance.

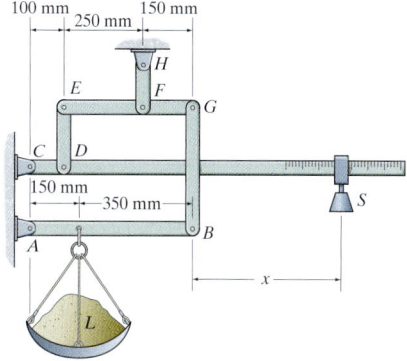

100 mm   150 mm
250 mm

E   *F*
H
G

C *D*

150 mm

350 mm

A   *B*

S

x

L

Probs. 6–74/75

***6–76.** Determine the horizontal and vertical components of force which the pins at *A*, *B*, and *C* exert on member *ABC* of the frame.

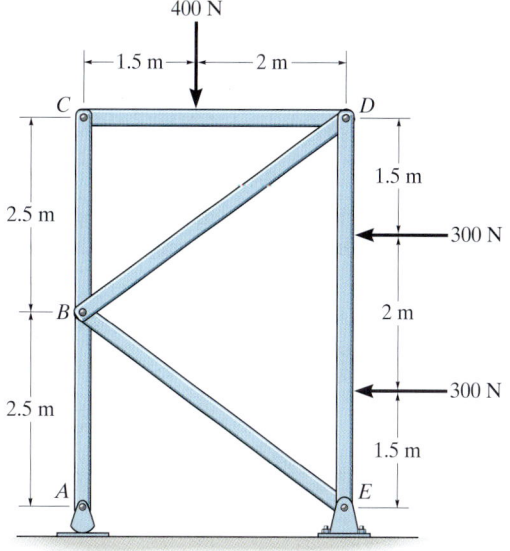

400 N

1.5 m   2 m

C   *D*

1.5 m

2.5 m

300 N

B

2 m

2.5 m

300 N

1.5 m

A   *E*

Prob. 6–76

6–77. Determine the required mass of the suspended cylinder if the tension in the chain wrapped around the freely turning gear is to be 2 kN. Also, what is the magnitude of the resultant force on pin A?

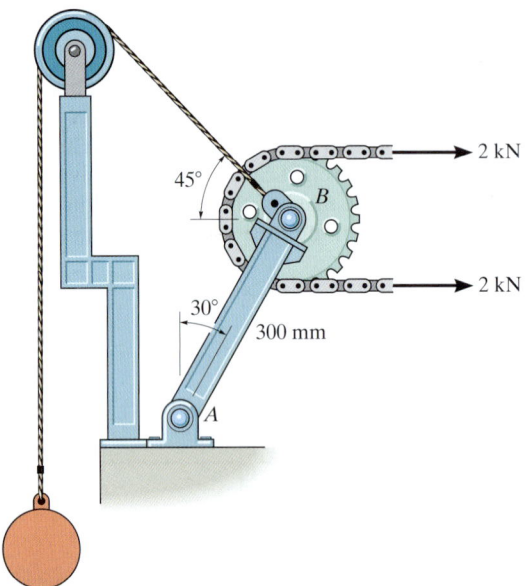

Prob. 6–77

6–78. Determine the reactions on the collar at A and the pin at C. The collar fits over a smooth rod, and rod AB is fixed connected to the collar.

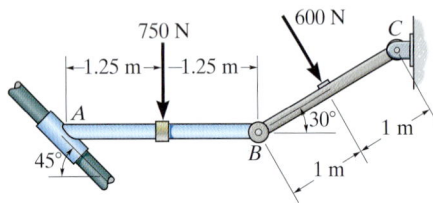

Prob. 6–78

6–79. The toggle clamp is subjected to a force $\mathbf{F}$ at the handle. Determine the vertical clamping force acting at E.

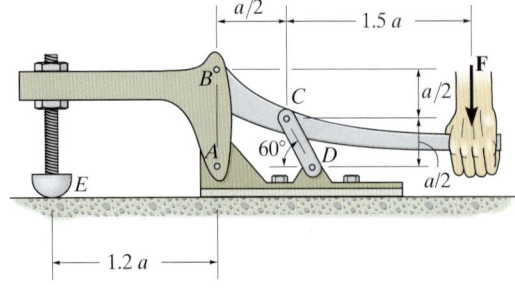

Prob. 6–79

***6–80.** Determine the force that the jaws J of the metal cutters exert on the smooth cable C if 100-N forces are applied to the handles. The jaws are pinned at E and A, and D and B. There is also a pin at F.

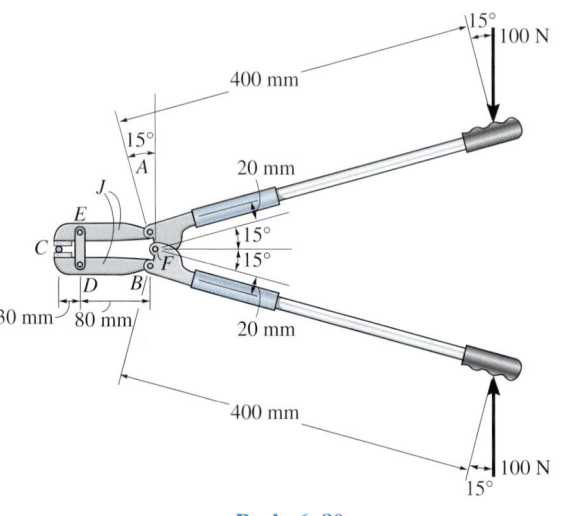

Prob. 6–80

6–81. The engine hoist is used to support the 200-kg engine. Determine the force acting in the hydraulic cylinder AB, the horizontal and vertical components of force at the pin C, and the reactions at the fixed connection D.

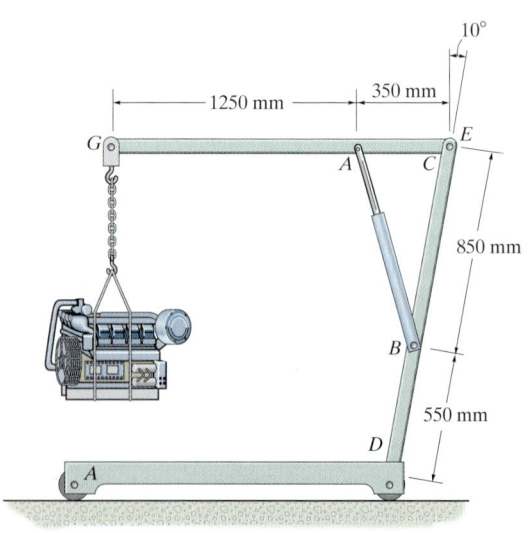

Prob. 6–81

6–82.   Determine the horizontal and vertical components of force that the pins at A, B, and C exert on the frame. The cylinder has a mass of 80 kg. The pulley has a radius of 0.1 m.

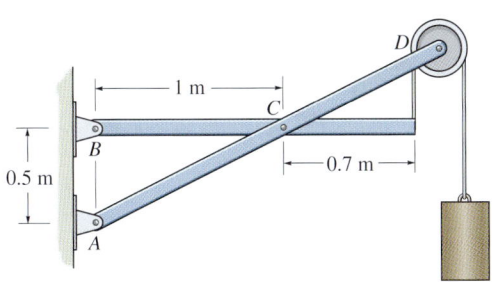

Prob. 6–82

6–83.   Determine the horizontal and vertical components of force that pins A and C exert on the frame.

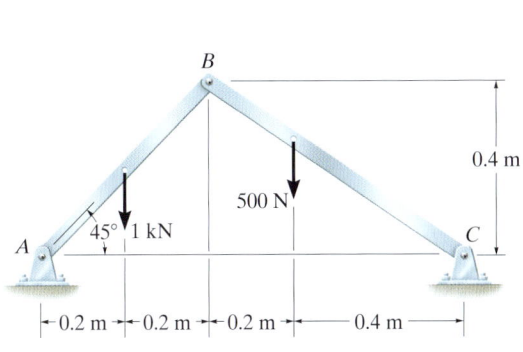

Prob. 6–83

*6–84.   The compound beam is fixed supported at A and supported by rockers at B and C. If there are hinges (pins) at D and E, determine the reactions at the supports A, B, and C.

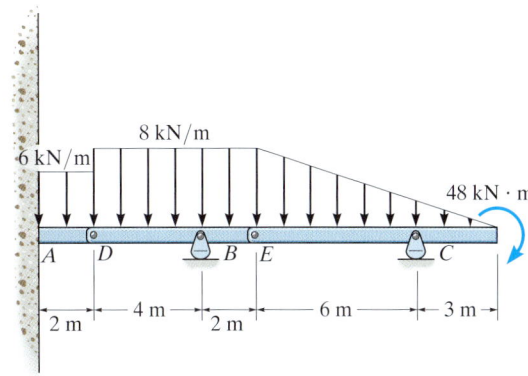

Prob. 6–84

6–85.   The pruner multiplies blade-cutting power with the compound leverage mechanism. If a 20-N force is applied to the handles, determine the cutting force generated at A. Assume that the contact surface at A is smooth.

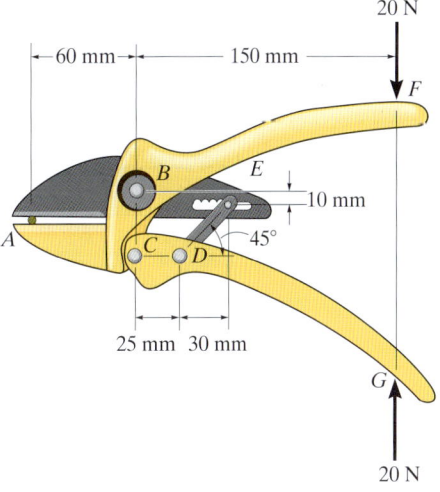

Prob. 6–85

6–86. The pipe cutter is clamped around the pipe P. If the wheel at A exerts a normal force of $F_A = 80$ N on the pipe, determine the normal forces of wheels B and C on the pipe. Also compute the pin reaction on the wheel at C. The three wheels each have a radius of 7 mm and the pipe has an outer radius of 10 mm.

*6–88. Show that the weight W_1 of the counterweight at H required for equilibrium is $W_1 = (b/a)W$, and so it is independent of the placement of the load W on the platform.

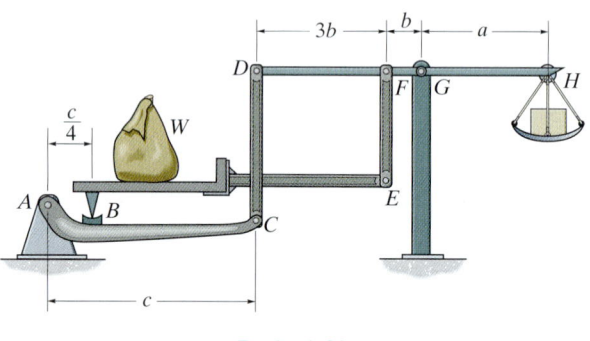

Prob. 6–88

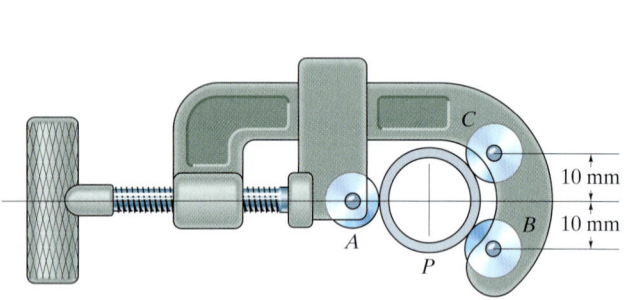

Prob. 6–86

6–87. The link is used to hold the rod in place. Determine the required axial force on the screw at E if the largest force to be exerted on the rod at B, C, or D is to be 100 N. Also, find the magnitude of the force reaction at pin A. Assume all surfaces of contact are smooth.

6–89. The derrick is pin connected to the pivot at A. Determine the largest mass that can be supported by the derrick if the maximum force that can be sustained by the pin at A is 18 kN.

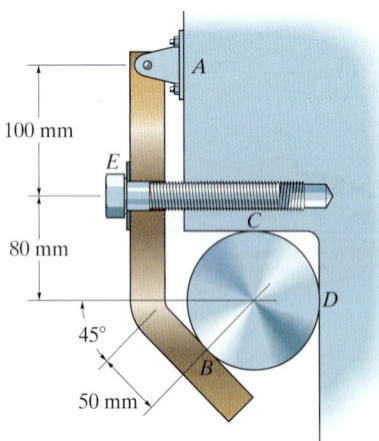

Prob. 6–87

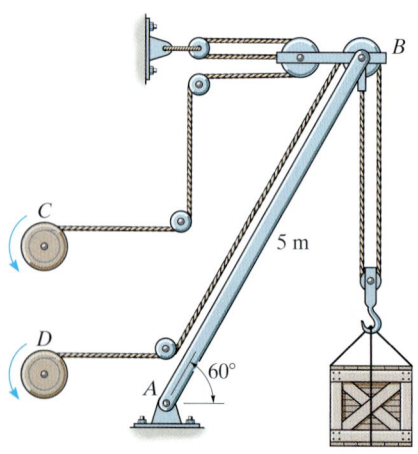

Prob. 6–89

6–90. Determine the force that the jaws J of the metal cutters exert on the smooth cable C if 200-N forces are applied to the handles. The jaws are pinned at E and A, and D and B. There is also a pin at F.

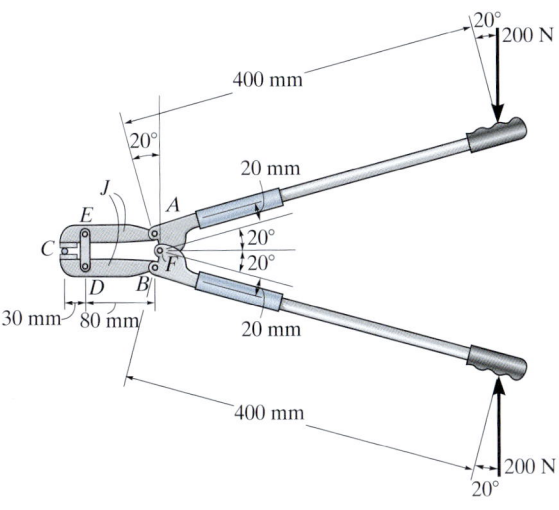

Prob. 6–90

*6–92.** The scissors lift consists of *two* sets of cross members and *two* hydraulic cylinders, DE, symmetrically located on *each side* of the platform. The platform has a uniform mass of 60 kg, with a center of gravity at G_1. The load of 85 kg, with center of gravity at G_2, is centrally located between each side of the platform. Determine the force in each of the hydraulic cylinders for equilibrium. Rollers are located at B and D.

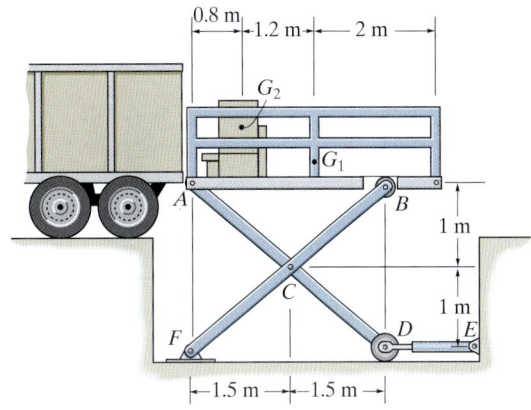

Prob. 6–92

6–91. The compound beam is pin supported at B and supported by rockers at A and C. There is a hinge (pin) at D. Determine the reactions at the supports.

6–93. The two disks each have a mass of 20 kg and are attached at their centers by an elastic cord that has a stiffness of $k = 2 \text{ kN/m}$. Determine the stretch of the cord when the system is in equilibrium and the angle θ of the cord.

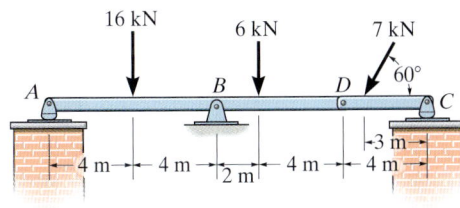

Prob. 6–91

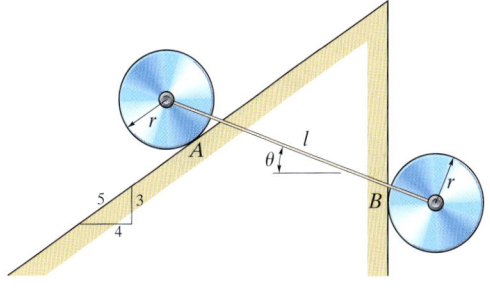

Prob. 6–93

6–94. A man having a weight of 875 N (≈ 87.5 kg) attempts to hold himself using one of the two methods shown. Determine the total force he must exert on bar *AB* in each case and the normal reaction he exerts on the platform at *C*. Neglect the weight of the platform.

6–95. A man having a weight of 875 N (≈ 87.5 kg) attempts to hold himself using one of the two methods shown. Determine the total force he must exert on bar *AB* in each case and the normal reaction he exerts on the platform at *C*. The platform has a weight of 150 N (≈ 15 kg).

6–97. Operation of exhaust and intake valves in an automobile engine consists of the cam *C*, push rod *DE*, rocker arm *EFG* which is pinned at *F*, and a spring and valve, *V*. If the compression in the spring is 20 mm when the valve is open as shown, determine the normal force acting on the cam lobe at *C*. Assume the cam and bearings at *H*, *I*, and *J* are smooth. The spring has a stiffness of 300 N/m.

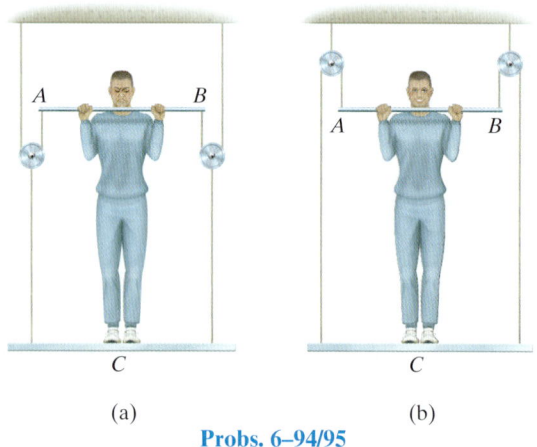

(a)　　(b)

Probs. 6–94/95

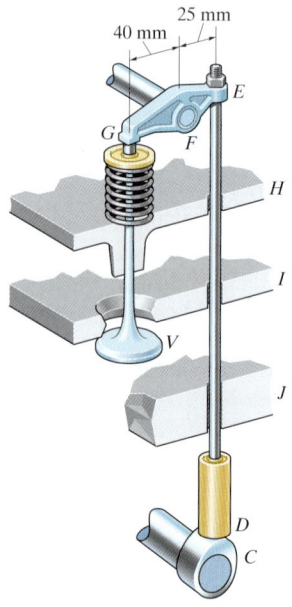

Prob. 6–97

*****6–96.** The double link grip is used to lift the beam. If the beam weighs 8 kN, determine the horizontal and vertical components of force acting on the pin at *A* and the horizontal and vertical components of force that the flange of the beam exerts on the jaw at *B*.

6–98. Determine the horizontal and vertical components of force that the pins at *A*, *B*, and *C* exert on their connecting members.

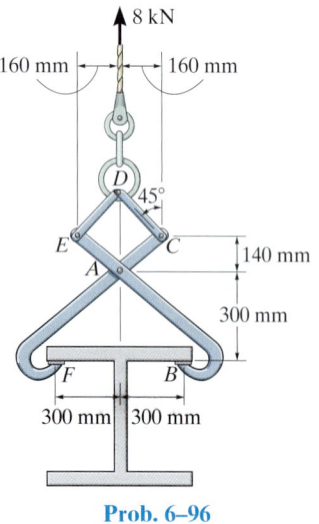

Prob. 6–96

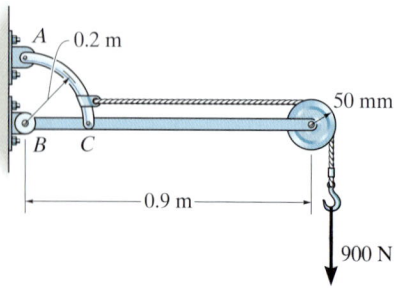

Prob. 6–98

6–99. If a clamping force of 300 N is required at A, determine the amount of force **F** that must be applied to the handle of the toggle clamp.

*6–100. If a force of $F = 350$ N is applied to the handle of the toggle clamp, determine the resulting clamping force at A.

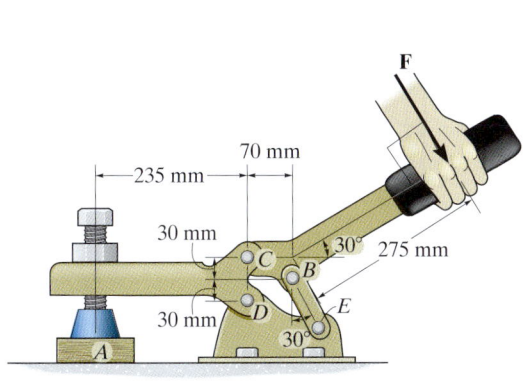

Probs. 6–99/100

6–102. The tractor boom supports the uniform mass of 600 kg in the bucket which has a center of mass at G. Determine the force in each hydraulic cylinder AB and CD and the resultant force at pins E and F. The load is supported equally on each side of the tractor by a similar mechanism.

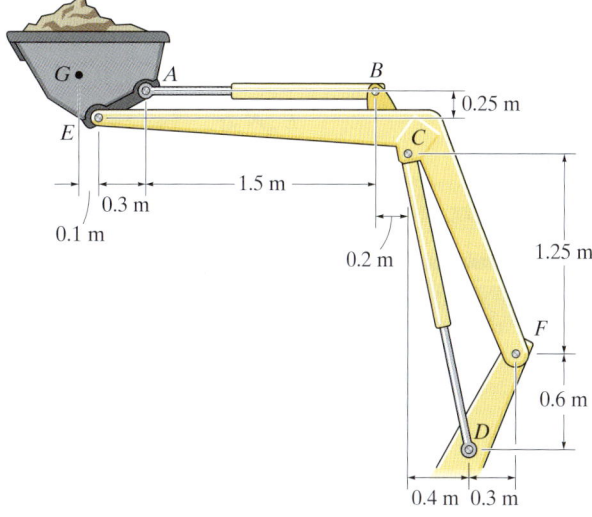

Prob. 6–102

6–101. The skid steer loader has a mass of 1.18 Mg, and in the position shown the center of mass is at G_1. If there is a 300-kg stone in the bucket, with center of mass at G_2, determine the reactions of each pair of wheels A and B on the ground and the force in the hydraulic cylinder CD and at the pin E. There is a similar linkage on each side of the loader.

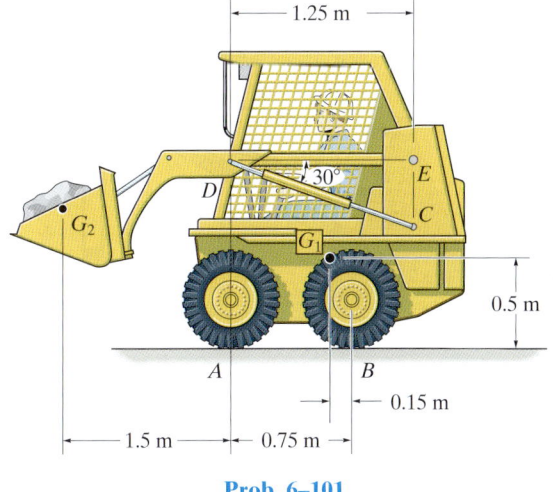

Prob. 6–101

6–103. The nail cutter consists of the handle and the two cutting blades. Assuming the blades are pin connected at B and the surface at D is smooth, determine the normal force on the fingernail when a force of 5 N is applied to the handles as shown. The pin AC slides through a smooth hole at A and is attached to the bottom member at C.

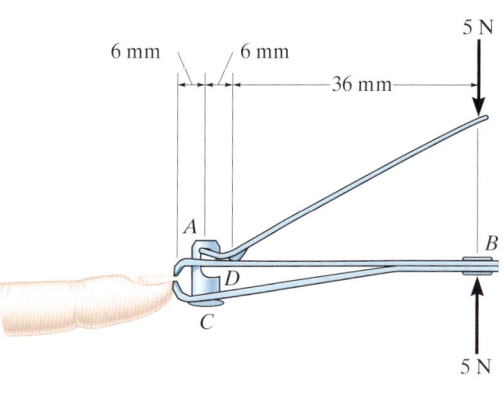

Prob. 6–103

*6–104. Determine the force created in the hydraulic cylinders *EF* and *AD* in order to hold the shovel in equilibrium. The shovel load has a mass of 1.25 Mg and a center of gravity at *G*. All joints are pin connected.

6–106. If $P = 75$ N, determine the force F that the toggle clamp exerts on the wooden block.

6–107. If the wooden block exerts a force of $F = 600$ N on the toggle clamp, determine the force P applied to the handle.

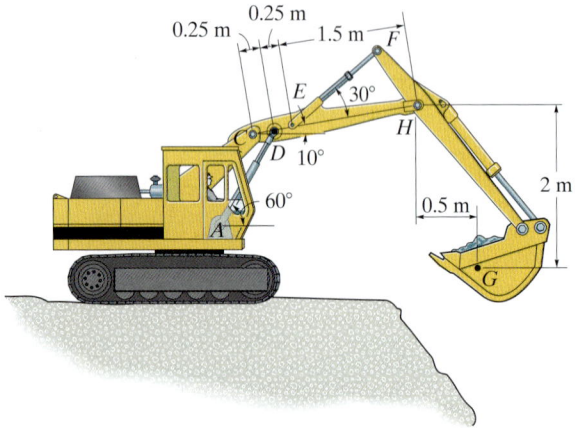

Prob. 6–104

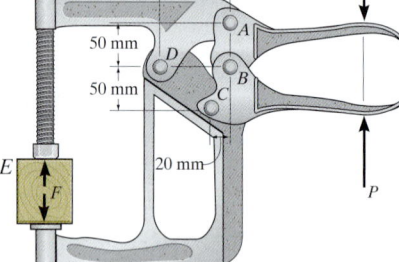

Probs. 6–106/107

6–105. The hoist supports the 125-kg engine. Determine the force in member *DB* and in the hydraulic cylinder *H* of member *FB*.

*6–108. The pillar crane is subjected to the load having a mass of 500 kg. Determine the force developed in the tie rod *AB* and the horizontal and vertical reactions at the pin support *C* when the boom is tied in the position shown.

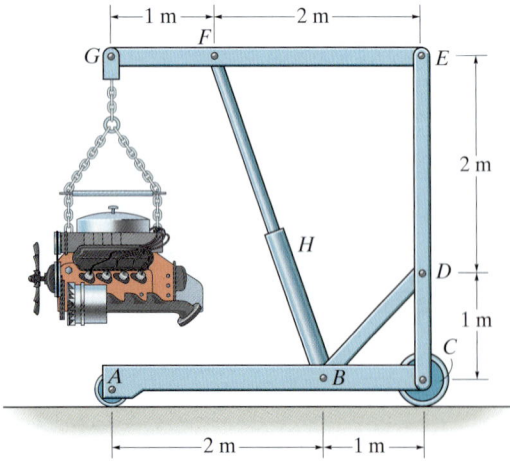

Prob. 6–105

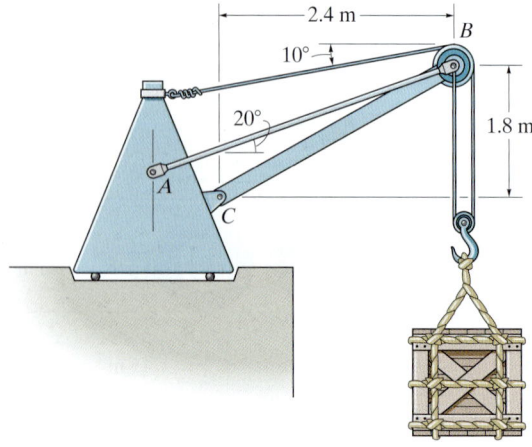

Prob. 6–108

6–109. The symmetric coil tong supports the coil which has a mass of 800 kg and center of mass at *G*. Determine the horizontal and vertical components of force the linkage exerts on plate *DEIJH* at points *D* and *E*. The coil exerts only vertical reactions at *K* and *L*.

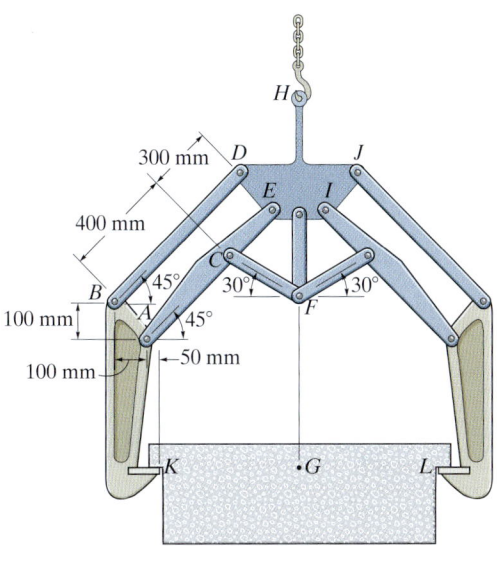

Prob. 6–109

6–110. A 300-kg counterweight, with center of mass at *G*, is mounted on the pitman crank *AB* of the oil-pumping unit. If a force of *F* = 5 kN is to be developed in the fixed cable attached to the end of the walking beam *DEF*, determine the torque *M* that must be supplied by the motor.

6–111. A 300-kg counterweight, with center of mass at *G*, is mounted on the pitman crank *AB* of the oil-pumping unit. If the motor supplies a torque of *M* = 2500 N · m, determine the force **F** developed in the fixed cable attached to the end of the walking beam *DEF*

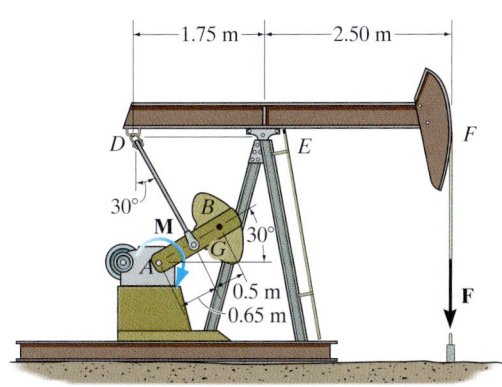

Probs. 6–110/111

*6–112.** The aircraft-hangar door opens and closes slowly by means of a motor which draws in the cable *AB*. If the door is made in two sections (bifold) and each section has a uniform weight *W* and length *L*, determine the force in the cable as a function of the door's position θ. The sections are pin-connected at *C* and *D* and the bottom is attached to a roller that travels along the vertical track.

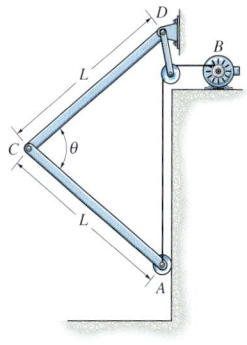

Prob. 6–112

6–113. Determine the horizontal and vertical components of reaction which the pins exert on member *AB* of the frame.

6–114. Determine the horizontal and vertical components of reaction which the pins exert on member *EDC* of the frame.

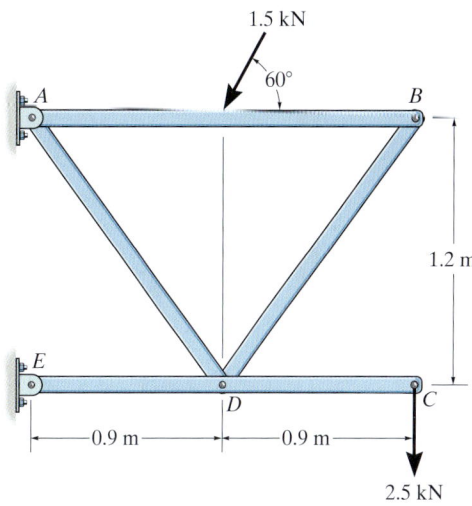

Probs. 6–113/114

6–115. The handle of the sector press is fixed to gear G, which in turn is in mesh with the sector gear C. Note that AB is pinned at its ends to gear C and the underside of the table EF, which is allowed to move vertically due to the smooth guides at E and F. If the gears exert tangential forces between them, determine the compressive force developed on the cylinder S when a vertical force of 40 N is applied to the handle of the press.

6–117. The four-member "A" frame is supported at A and E by smooth collars and at G by a pin. All the other joints are ball-and-sockets. If the pin at G will fail when the resultant force there is 800 N, determine the largest vertical force P that can be supported by the frame. Also, what are the x, y, z force components which member BD exerts on members EDC and ABC? The collars at A and E and the pin at G only exert force components on the frame.

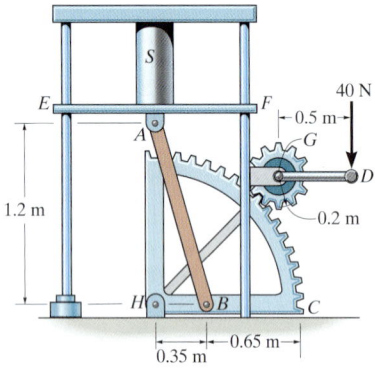

Prob. 6–115

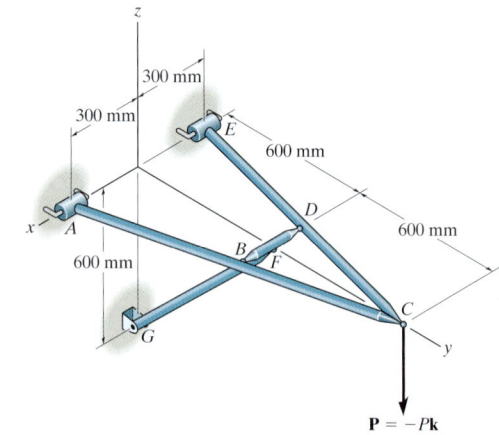

Prob. 6–117

***6–116.** The structure is subjected to the loading shown. Member AD is supported by a cable AB and roller at C and fits through a smooth circular hole at D. Member ED is supported by a roller at D and a pole that fits in a smooth snug circular hole at E. Determine the x, y, z components of reaction at E and the tension in cable AB.

6–118. The structure is subjected to the loadings shown. Member AB is supported by a ball-and-socket at A and smooth collar at B. Member CD is supported by a pin at C. Determine the x, y, z components of reaction at A and C.

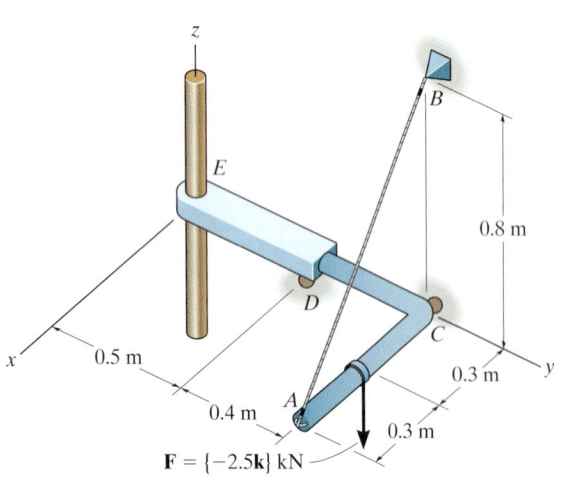

Prob. 6–116

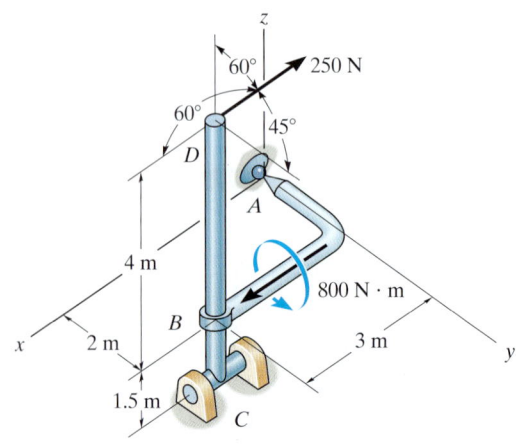

Prob. 6–118

CHAPTER REVIEW

Simple Truss

A simple truss consists of triangular elements connected together by pinned joints. The forces within its members can be determined by assuming the members are all two-force members, connected concurrently at each joint. The members are either in tension or compression, or carry no force.

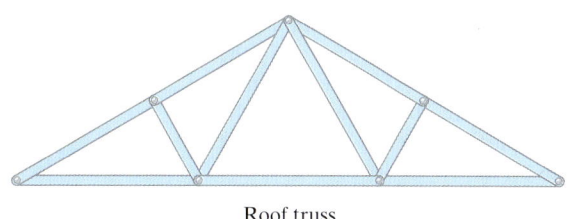

Roof truss

Method of Joints

The method of joints states that if a truss is in equilibrium, then each of its joints is also in equilibrium. For a plane truss, the concurrent force system at each joint must satisfy force equilibrium.

To obtain a numerical solution for the forces in the members, select a joint that has a free-body diagram with at most two unknown forces and one known force. (This may require first finding the reactions at the supports.)

Once a member force is determined, use its value and apply it to an adjacent joint.

Remember that forces that are found to *pull* on the joint are *tensile forces*, and those that *push* on the joint are *compressive forces*.

To avoid a simultaneous solution of two equations, set one of the coordinate axes along the line of action of one of the unknown forces and sum forces perpendicular to this axis. This will allow a direct solution for the other unknown.

The analysis can also be simplified by first identifying all the zero-force members.

$$\Sigma F_x = 0$$

$$\Sigma F_y = 0$$

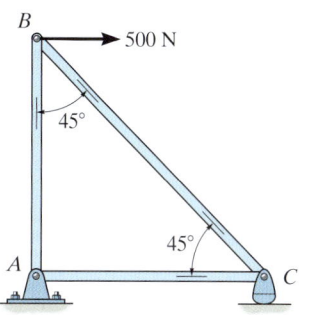

Method of Sections

The method of sections states that if a truss is in equilibrium, then each segment of the truss is also in equilibrium. Pass a section through the truss and the member whose force is to be determined. Then draw the free-body diagram of the sectioned part having the least number of forces on it.

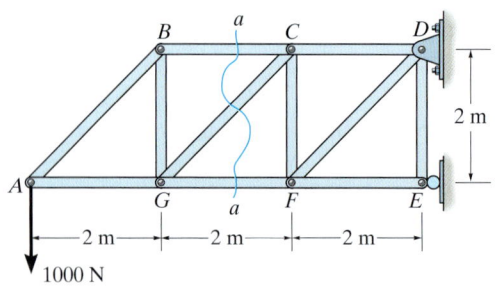

Sectioned members subjected to *pulling* are in *tension*, and those that are subjected to *pushing* are in *compression*.

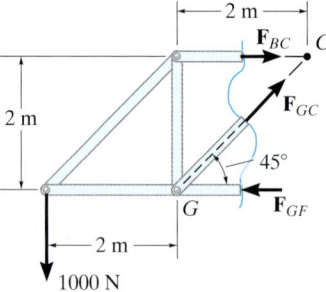

Three equations of equilibrium are available to determine the unknowns.

$$\Sigma F_x = 0$$
$$\Sigma F_y = 0$$
$$\Sigma M_O = 0$$

If possible, sum forces in a direction that is perpendicular to two of the three unknown forces. This will yield a direct solution for the third force.

$$+\uparrow \Sigma F_y = 0$$
$$-1000 \text{ N} + F_{GC} \sin 45° = 0$$
$$F_{GC} = 1.41 \text{ kN (T)}$$

Sum moments about the point where the lines of action of two of the three unknown forces intersect, so that the third unknown force can be determined directly.

$$\zeta + \Sigma M_C = 0$$
$$1000 \text{ N}(4 \text{ m}) - F_{GF} (2 \text{ m}) = 0$$
$$F_{GF} = 2 \text{ kN (C)}$$

Space Truss

A space truss is a three-dimensional truss built from tetrahedral elements, and is analyzed using the same methods as for plane trusses. The joints are assumed to be ball-and-socket connections.

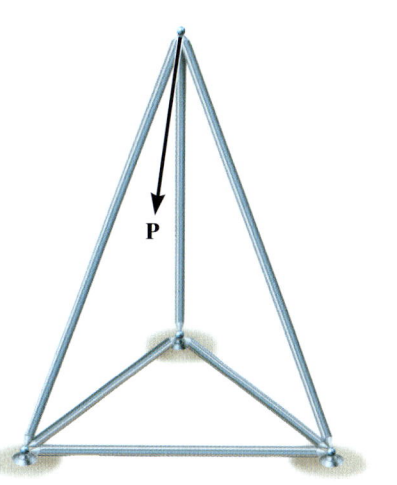

Frames and Machines

Frames and machines are structures that contain one or more multiforce members, that is, members with three or more forces or couples acting on them. Frames are designed to support loads, and machines transmit and alter the effect of forces.

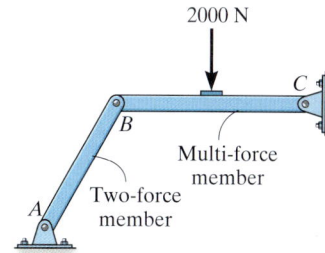

The forces acting at the joints of a frame or machine can be determined by drawing the free-body diagrams of each of its members or parts. The principle of action–reaction should be carefully observed when indicating these forces on the free-body diagram of each adjacent member or pin. For a coplanar force system, there are three equilibrium equations available for each member.

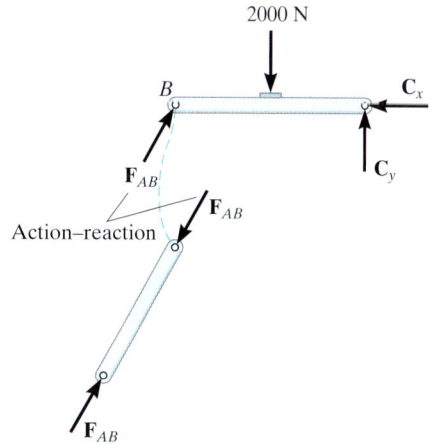

To simplify the analysis, be sure to recognize all two-force members. They have equal but opposite collinear forces at their ends.

REVIEW PROBLEMS

6–119. The tractor boom supports the uniform mass of 500 kg in the bucket which has a center of mass at *G*. Determine the force in each hydraulic cylinder *AB* and *CD* and the resultant force at pins *E* and *F*. The load is supported equally on each side of the tractor by a similar mechanism.

6–121. Determine the horizontal and vertical components of force at pins *A* and *C* of the two-member frame.

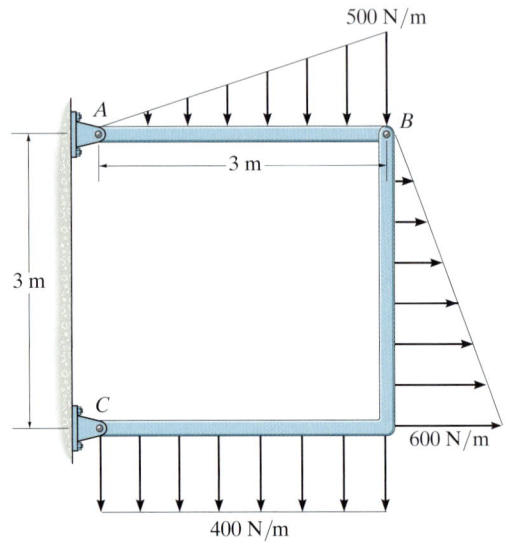

Prob. 6–121

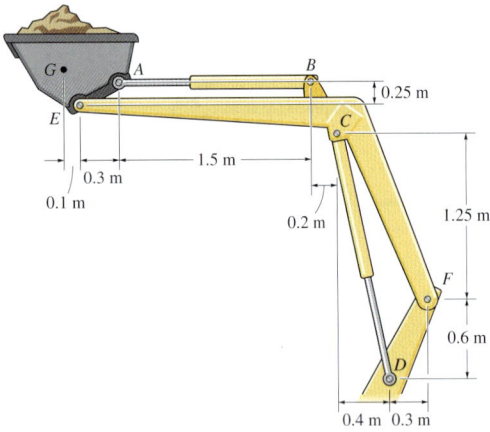

Prob. 6–119

***6–120.** Determine the force in each member of the truss and state if the members are in tension or compression.

6–122. The clamping hooks are used to lift the uniform smooth 500-kg plate. Determine the resultant compressive force that the hook exerts on the plate at *A* and *B*, and the pin reaction at *C*.

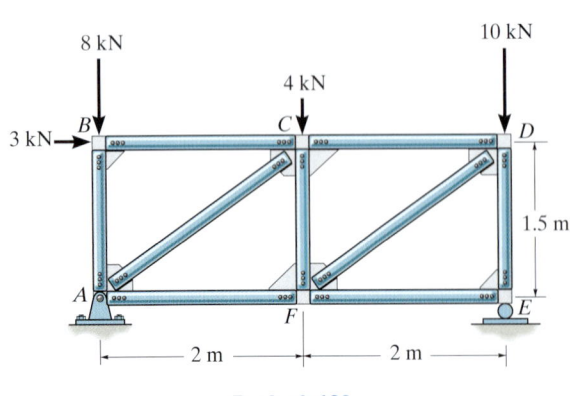

Prob. 6–120

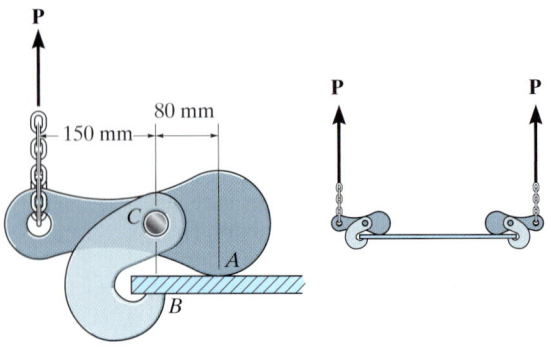

Prob. 6–122

6–123. The spring has an unstretched length of 0.3 m. Determine the mass m of each uniform link if the angle $\theta = 20°$ for equilibrium.

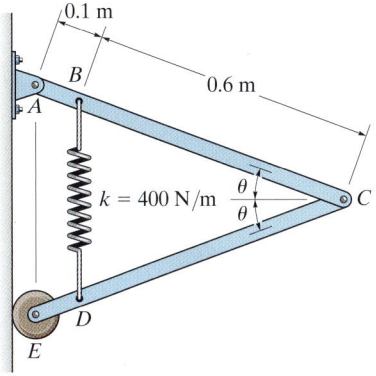

Prob. 6–123

6–125. Determine the horizontal and vertical components of force that pins A and B exert on the two-member frame. Set $F = 500$ N.

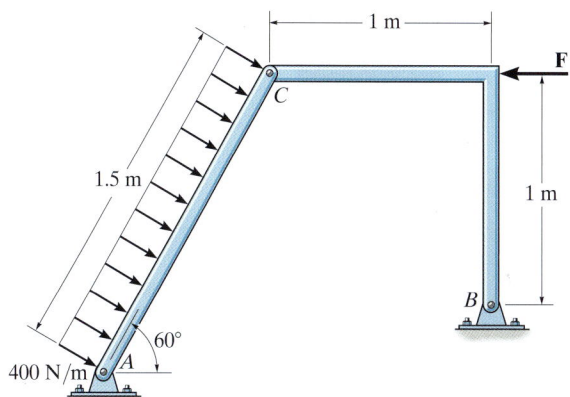

Prob. 6–125

*6–124. Determine the horizontal and vertical components of force that the pins A and B exert on the two-member frame. Set $F = 0$.

6–126. Determine the force in each member of the truss and state if the members are in tension or compression.

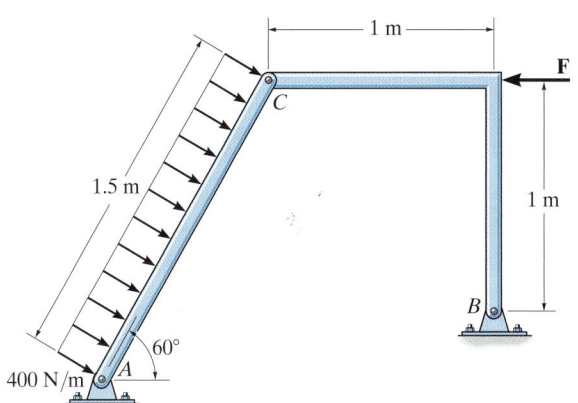

Prob. 6–124

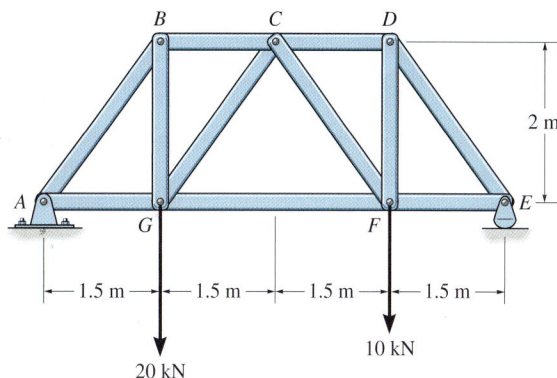

Prob. 6–126

Chapter 7

When external loads are placed upon these beams and columns, the loads within them must be determined if they are to be properly designed. In this chapter we will study how to determine these internal loadings.

Internal Forces

CHAPTER OBJECTIVES

■ To show how to use the method of sections to determine the internal loadings in a member.

■ To generalize this procedure by formulating equations that can be plotted so that they describe the internal shear and moment throughout a member.

■ To analyze the forces and study the geometry of cables supporting a load.

7.1 Internal Loadings Developed in Structural Members

To design a structural or mechanical member it is necessary to know the loading acting within the member in order to be sure the material can resist this loading. Internal loadings can be determined by using the *method of sections*. To illustrate this method, consider the cantilever beam in Fig. 7–1a. If the internal loadings acting on the cross section at point B are to be determined, we must pass an imaginary section a–a perpendicular to the axis of the beam through point B and then separate the beam into two segments. The internal loadings acting at B will then be exposed and become *external* on the free-body diagram of each segment, Fig. 7–1b.

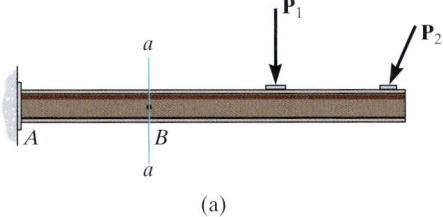

(a)

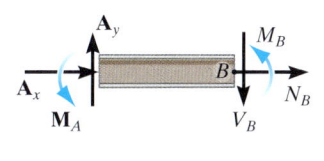

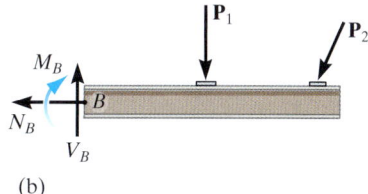

(b)

Fig. 7–1

In each case, the link on the backhoe is a two-force member. In the top photo it is subjected to both bending and an axial load at its center. By making the member straight, as in the bottom photo, then only an axial force acts within the member.

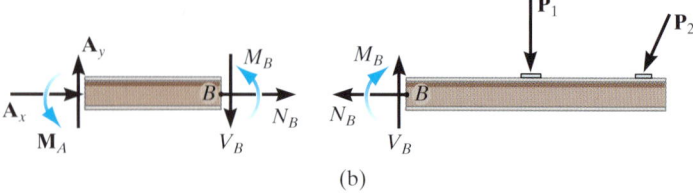

(b)

Fig. 7–1 (Repeated)

The force component $\mathbf{N}_B$ that acts *perpendicular* to the cross section is termed the *normal force*. The force component $\mathbf{V}_B$ that is tangent to the cross section is called the *shear force*, and the couple moment $\mathbf{M}_B$ is referred to as the *bending moment*. The force components prevent the relative translation between the two segments, and the couple moment prevents the relative rotation. According to Newton's third law, these loadings must act in opposite directions on each segment, as shown in Fig. 7–1b. They can be determined by applying the equations of equilibrium to the free-body diagram of either segment. In this case, however, the right segment is the better choice since it does not involve the unknown support reactions at A. A direct solution for $\mathbf{N}_B$ is obtained by applying $\Sigma F_x = 0$, $\mathbf{V}_B$ is obtained from $\Sigma F_y = 0$, and $\mathbf{M}_B$ can be obtained by applying $\Sigma M_B = 0$, since the moments of $\mathbf{N}_B$ and $\mathbf{V}_B$ about B are zero.

In two dimensions, we have shown that three internal loading resultants exist, Fig. 7–2a; however in three dimensions, a general internal force and couple moment resultant will act at the section. The x, y, z components of these loadings are shown in Fig. 7–2b. Here $\mathbf{N}_y$ is the *normal force*, and $\mathbf{V}_x$ and $\mathbf{V}_z$ are *shear force components*. $\mathbf{M}_y$ is a *torsional or twisting moment*, and $\mathbf{M}_x$ and $\mathbf{M}_z$ are *bending moment components*. For most applications, these *resultant loadings* will act at the geometric center or centroid (C) of the section's cross-sectional area. Although the magnitude for each loading generally will be different at various points along the axis of the member, the method of sections can always be used to determine their values.

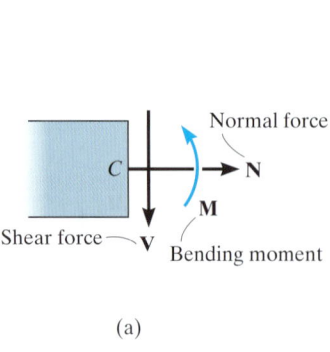

(a)

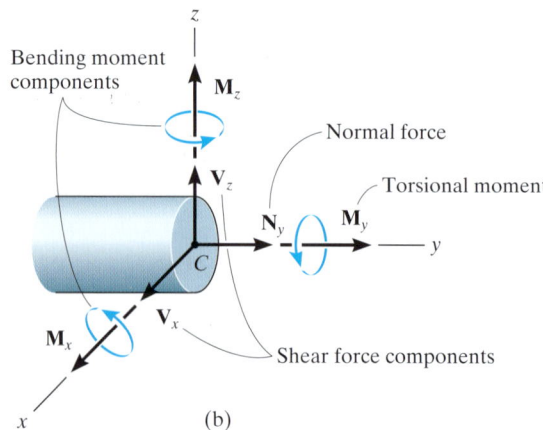

(b)

Fig. 7–2

Sign Convention. For problems in two dimensions engineers generally use a sign convention to report the three internal loadings **N**, **V**, and **M**. Although this sign convention can be arbitrarily assigned, the one that is widely accepted will be used here, Fig. 7–3. The normal force is said to be positive if it creates *tension*, a positive shear force will cause the beam segment on which it acts to rotate clockwise, and a positive bending moment will tend to bend the segment on which it acts in a concave upward manner. Loadings that are opposite to these are considered negative.

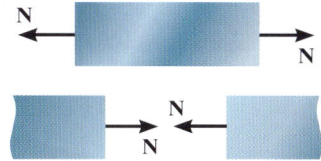

Positive normal force

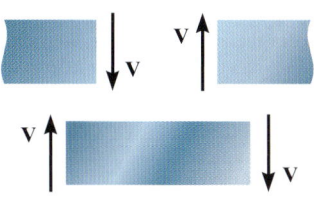

Positive shear

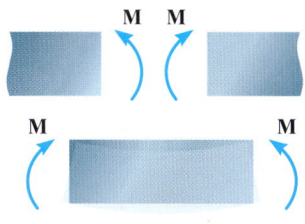

Positive moment

Fig. 7–3

Procedure for Analysis

The method of sections can be used to determine the internal loadings on the cross section of a member using the following procedure.

Support Reactions.

- Before the member is sectioned, it may first be necessary to determine its support reactions. Once obtained, the equilibrium equations can then be used to solve for the internal loadings after the member is sectioned.

Free-Body Diagram.

- It is important to *keep* all distributed loadings, couple moments, and forces acting on the member in their *exact locations*, *then* pass an imaginary section through the member, perpendicular to its axis at the point where the internal loadings are to be determined.

- After the section is made, draw a free-body diagram of the segment that has the least number of loads on it, and indicate the components of the internal force and couple moment resultants at the cross section acting in their positive directions in accordance with the established sign convention.

Equations of Equilibrium.

- Moments should be summed at the section. This way the normal and shear forces at the section are elminated, and we can obtain a direct solution for the moment.

- If the solution of the equilibrium equations yields a negative scalar, the sense of the quantity is opposite to that shown on the free-body diagram.

The designer of this shop crane realized the need for additional reinforcement around the joint in order to prevent severe internal bending of the joint when a large load is suspended from the chain hoist.

Please refer to the Companion Website
for the animation: *Free-Body Diagram
for a Simple Beam*

Determine the normal force, shear force, and bending moment acting just to the left, point B, and just to the right, point C, of the 6-kN force on the beam in Fig. 7–4a.

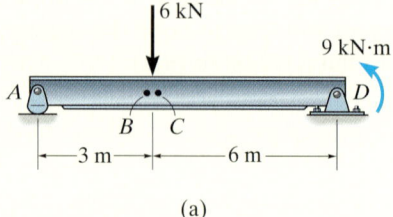

(a)

SOLUTION

Support Reactions.   The free-body diagram of the beam is shown in Fig. 7–4b. When determining the *external reactions*, realize that the 9-kN $\cdot$ m couple moment is a free vector and therefore it can be placed *anywhere* on the free-body diagram of the entire beam. Here we will only determine A_y, since the left segments will be used for the analysis.

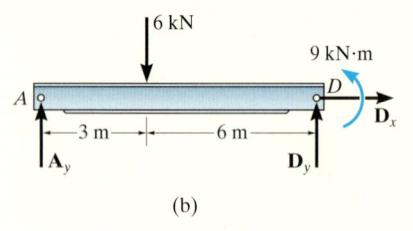

(b)

$$\zeta + \Sigma M_D = 0; \quad 9 \text{ kN} \cdot \text{m} + (6 \text{ kN})(6 \text{ m}) - A_y(9 \text{ m}) = 0$$
$$A_y = 5 \text{ kN}$$

Free-Body Diagrams.   The free-body diagrams of the left segments AB and AC of the beam are shown in Figs. 7–4c and 7–4d. In this case the 9-kN $\cdot$ m couple moment is *not included* on these diagrams since it must be kept in its *original position* until *after* the section is made and the appropriate segment is isolated.

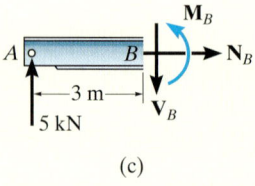

(c)

Equations of Equilibrium.

Segment AB

$$\xrightarrow{+} \Sigma F_x = 0; \qquad\qquad\qquad N_B = 0 \qquad\qquad\qquad \textit{Ans.}$$

$$+\uparrow \Sigma F_y = 0; \qquad\qquad 5 \text{ kN} - V_B = 0 \quad V_B = 5 \text{ kN} \qquad \textit{Ans.}$$

$$\zeta + \Sigma M_B = 0; \quad -(5 \text{ kN})(3 \text{ m}) + M_B = 0 \quad M_B = 15 \text{ kN} \cdot \text{m} \quad \textit{Ans.}$$

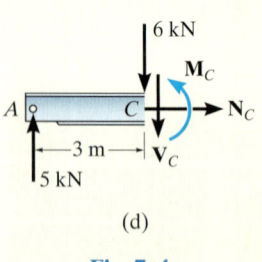

(d)

Fig. 7–4

Segment AC

$$\xrightarrow{+} \Sigma F_x = 0; \qquad\qquad\qquad N_C = 0 \qquad\qquad\qquad \textit{Ans.}$$

$$+\uparrow \Sigma F_y = 0; \quad 5 \text{ kN} - 6 \text{ kN} - V_C = 0 \quad V_C = -1 \text{ kN} \qquad \textit{Ans.}$$

$$\zeta + \Sigma M_C = 0; \quad -(5 \text{ kN})(3 \text{ m}) + M_C = 0 \quad M_C = 15 \text{ kN} \cdot \text{m} \quad \textit{Ans.}$$

NOTE: The negative sign indicates that $\mathbf{V}_C$ acts in the opposite sense to that shown on the free-body diagram. Also, the moment arm for the 5-kN force in both cases is approximately 3 m since B and C are "almost" coincident.

EXAMPLE 7.2

Determine the normal force, shear force, and bending moment at C of the beam in Fig. 7–5a.

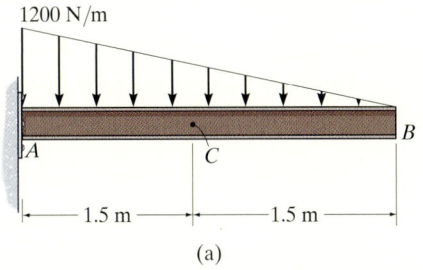

(a)

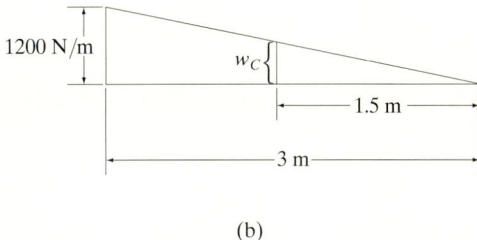

(b)

Fig. 7–5

SOLUTION

Free-Body Diagram. It is not necessary to find the support reactions at A since segment BC of the beam can be used to determine the internal loadings at C. The intensity of the triangular distributed load at C is determined using similar triangles from the geometry shown in Fig. 7–5b, i.e.,

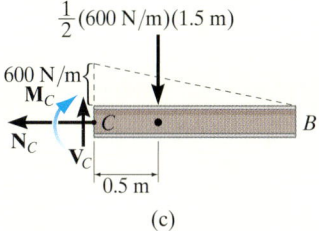

(c)

$$w_C = (1200 \text{ N/m}) \left(\frac{1.5 \text{ m}}{3 \text{ m}} \right) = 600 \text{ N/m}$$

The distributed load acting on segment BC can now be replaced by its resultant force, and its location is indicated on the free-body diagram, Fig. 7–5c.

Equations of Equilibrium.

$$\xrightarrow{+} \Sigma F_x = 0; \qquad\qquad N_C = 0 \qquad\qquad \textit{Ans.}$$

$$+\uparrow \Sigma F_y = 0; \qquad V_C - \tfrac{1}{2}(600 \text{ N/m})(1.5 \text{ m}) = 0$$

$$V_C = 450 \text{ N} \qquad\qquad \textit{Ans.}$$

$$\zeta + \Sigma M_C = 0; \quad -M_C - \tfrac{1}{2}(600 \text{ N/m})(1.5 \text{ m})(0.5 \text{ m}) = 0$$

$$M_C = -225 \text{ N} \qquad\qquad \textit{Ans.}$$

The negative sign indicates that $\mathbf{M}_C$ acts in the opposite sense to that shown on the free-body diagram.

EXAMPLE | 7.3

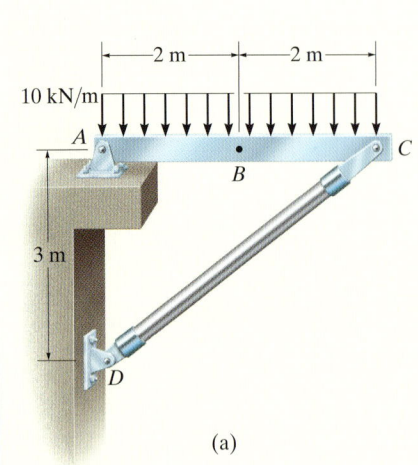

(a)

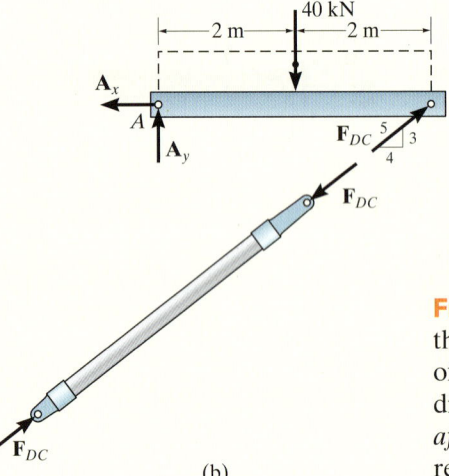

(b)

Fig. 7–6

Determine the normal force, shear force, and bending moment acting at point B of the two-member frame shown in Fig. 7–6a.

SOLUTION

Support Reactions.   A free-body diagram of each member is shown in Fig. 7–6b. Since CD is a two-force member, the equations of equilibrium need to be applied only to member AC.

$$\zeta + \Sigma M_A = 0; \quad -40 \text{ kN } (2 \text{ m}) + \left(\tfrac{3}{5}\right) F_{DC}(4 \text{ m}) = 0 \quad F_{DC} = 33.33 \text{ kN}$$

$$\xrightarrow{+} \Sigma F_x = 0; \quad -A_x + \left(\tfrac{4}{5}\right)(33.33 \text{ kN}) = 0 \quad A_x = 26.67 \text{ kN}$$

$$+\uparrow \Sigma F_y = 0; \quad A_y - 40 \text{ kN } + \left(\tfrac{3}{5}\right)(33.33 \text{ kN}) = 0 \quad A_y = 20 \text{ kN}$$

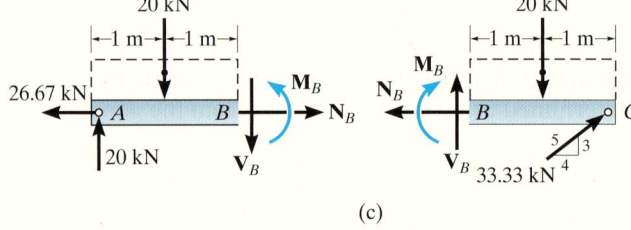

(c)

Free-Body Diagrams.   Passing an imaginary section perpendicular to the axis of member AC through point B yields the free-body diagrams of segments AB and BC shown in Fig. 7–6c. When constructing these diagrams it is important to keep the distributed loading where it is until *after the section is made*. Only then can it be replaced by a single resultant force.

Equations of Equilibrium.   Applying the equations of equilibrium to segment AB, we have

$$\xrightarrow{+} \Sigma F_x = 0; \qquad N_B - 26.67 \text{ kN} = 0 \quad N_B = 26.7 \text{ kN} \qquad \textit{Ans.}$$

$$+\uparrow \Sigma F_y = 0; \qquad 20 \text{ kN} - 20 \text{ kN} - V_B = 0 \quad V_B = 0 \qquad \textit{Ans.}$$

$$\zeta + \Sigma M_B = 0; \qquad M_B - 20 \text{ kN } (2 \text{ m}) + 20 \text{ kN } (1 \text{ m}) = 0$$

$$M_B = 20 \text{ kN} \cdot \text{m} \qquad \textit{Ans.}$$

NOTE: As an exercise, try to obtain these same results using segment BC.

EXAMPLE 7.4

Determine the normal force, shear force, and bending moment acting at point E of the frame loaded as shown in Fig. 7–7a.

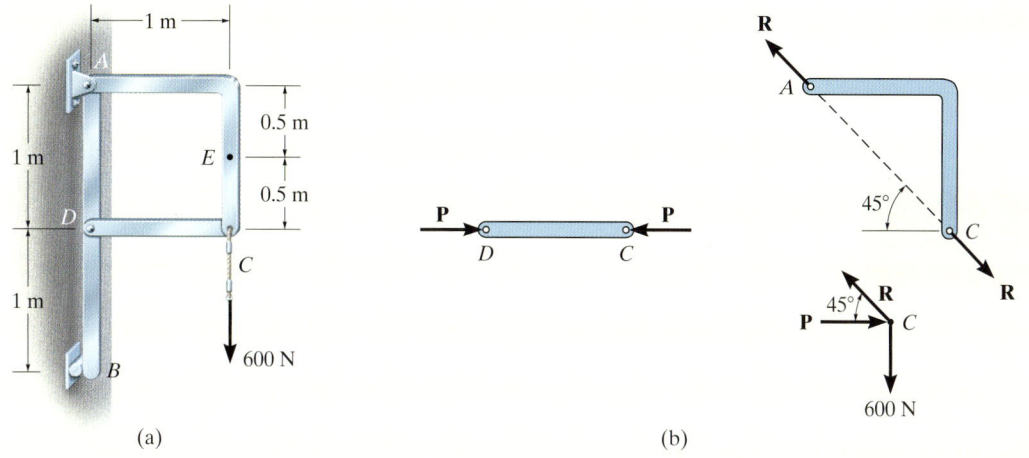

(a)   (b)

SOLUTION

Support Reactions.   By inspection, members AC and CD are two-force members, Fig. 7–7b. In order to determine the internal loadings at E, we must first determine the force **R** acting at the end of member AC. To obtain it, we will analyze the equilibrium of the pin at C.

Summing forces in the vertical direction on the pin, Fig. 7–7b, we have

$$+\uparrow \Sigma F_y = 0; \qquad R \sin 45° - 600 \text{ N} = 0 \qquad R = 848.5 \text{ N}$$

Free-Body Diagram.   The free-body diagram of segment CE is shown in Fig. 7–7c.

Equations of Equilibrium.

$$\xrightarrow{+} \Sigma F_x = 0; \qquad 848.5 \cos 45° \text{ N} - V_E = 0 \qquad V_E = 600 \text{ N} \qquad \textit{Ans.}$$

$$+\uparrow \Sigma F_y = 0; \qquad -848.5 \sin 45° \text{ N} + N_E = 0 \qquad N_E = 600 \text{ N} \qquad \textit{Ans.}$$

$$\zeta + \Sigma M_E = 0; \qquad 848.5 \cos 45° \text{ N}(0.5 \text{ m}) - M_E = 0$$

$$M_E = 300 \text{ N} \cdot \text{m} \qquad \textit{Ans.}$$

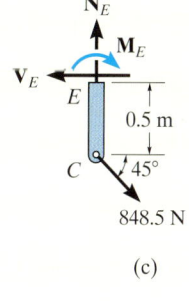

(c)

Fig. 7–7

NOTE: These results indicate a poor design. Member AC should be *straight* (from A to C) so that bending within the member is *eliminated*. If AC were straight then the internal force would only create tension in the member.

EXAMPLE | 7.5

(a)

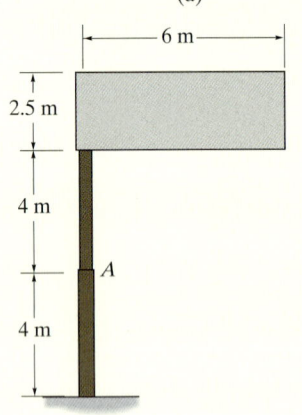

(b)

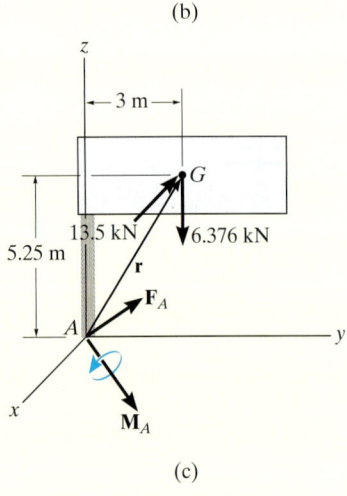

Fig. 7–8

The uniform sign shown in Fig. 7–8a has a mass of 650 kg and is supported on the fixed column. Design codes indicate that the expected maximum uniform wind loading that will occur in the area where it is located is 900 Pa. Determine the internal loadings at A.

SOLUTION

The idealized model for the sign is shown in Fig. 7–8b. Here the necessary dimensions are indicated. We can consider the free-body diagram of a section above point A since it does not involve the support reactions.

Free-Body Diagram.   The sign has a weight of $W = 650(9.81)$ N = 6.376 kN, and the wind creates a resultant force of $F_w = 900$ N/m^2(6 m)(2.5 m) = 13.5 kN, which acts perpendicular to the face of the sign. These loadings are shown on the free-body diagram, Fig. 7–8c.

Equations of Equilibrium.   Since the problem is three dimensional, a vector analysis will be used.

$$\Sigma \mathbf{F} = \mathbf{0}; \qquad\qquad \mathbf{F}_A - 13.5\mathbf{i} - 6.376\mathbf{k} = \mathbf{0}$$

$$\mathbf{F}_A = \{13.5\mathbf{i} + 6.38\mathbf{k}\} \text{ kN} \qquad\qquad Ans.$$

$$\Sigma \mathbf{M}_A = \mathbf{0}; \qquad\qquad \mathbf{M}_A + \mathbf{r} \times (\mathbf{F}_w + \mathbf{W}) = \mathbf{0}$$

$$\mathbf{M}_A + \begin{vmatrix} \mathbf{i} & \mathbf{j} & \mathbf{k} \\ 0 & 3 & 5.25 \\ -13.5 & 0 & -6.376 \end{vmatrix} = \mathbf{0}$$

$$\mathbf{M}_A = \{19.1\mathbf{i} + 70.9\mathbf{j} - 40.5\mathbf{k}\} \text{ kN} \cdot \text{m} \qquad\qquad Ans.$$

NOTE: Here $\mathbf{F}_{A_z} = \{6.38\mathbf{k}\}$ kN represents the normal force, whereas $\mathbf{F}_{A_x} = \{13.5\mathbf{i}\}$ kN is the shear force. Also, the torsional moment is $\mathbf{M}_{A_z} = \{-40.5\mathbf{k}\}$ kN · m, and the bending moment is determined from its components $\mathbf{M}_{A_x} = \{19.1\mathbf{i}\}$ kN · m and $\mathbf{M}_{A_y} = \{70.9\mathbf{j}\}$ kN · m; i.e., $(M_b)_A = \sqrt{(M_A)_x^2 + (M_A)_y^2} = 73.4$ kN · m.

FUNDAMENTAL PROBLEMS

All problem solutions must include FBDs.

F7–1. Determine the normal force, shear force, and moment at point C.

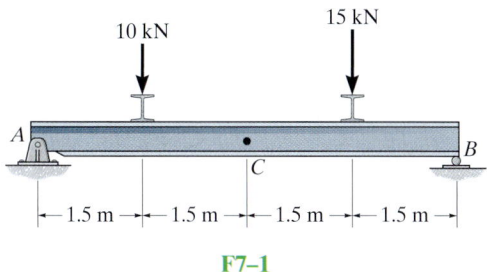

F7–1

F7–2. Determine the normal force, shear force, and moment at point C.

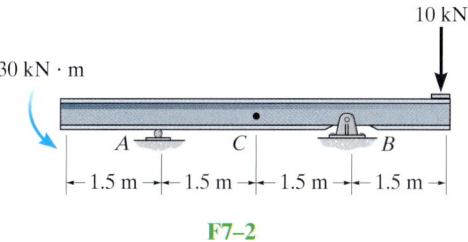

F7–2

F7–3. Determine the normal force, shear force, and moment at point C.

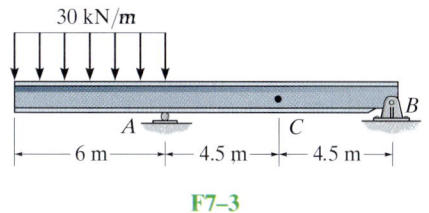

F7–3

F7–4. Determine the normal force, shear force, and moment at point C.

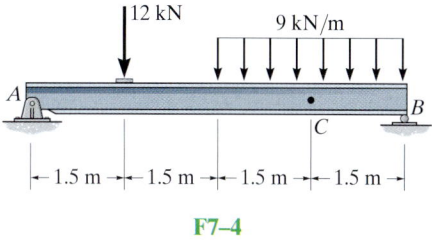

F7–4

F7–5. Determine the normal force, shear force, and moment at point C.

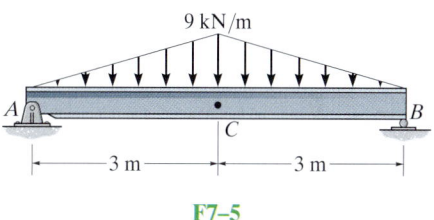

F7–5

F7–6. Determine the normal force, shear force, and moment at point C. Assume A is pinned and B is a roller.

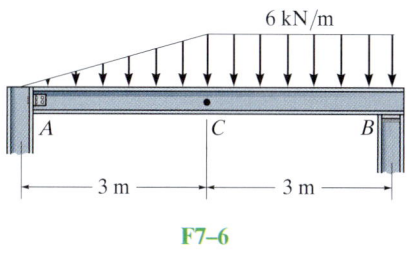

F7–6

7

PROBLEMS

All problem solutions must include FBDs.

7–1. The forces act on the shaft shown. Determine the internal normal force at points A, B, and C.

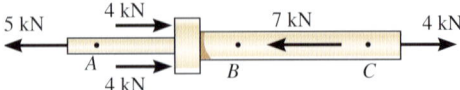

Prob. 7–1

7–2. Three torques act on the shaft. Determine the internal torque at points A, B, C, and D.

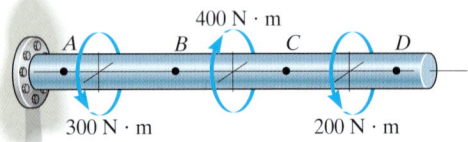

Prob. 7–2

7–3. The strongback or lifting beam is used for materials handling. If the suspended load has a weight of 2 kN and a center of gravity of G, determine the placement d of the padeyes on the top of the beam so that there is no moment developed within the length AB of the beam. The lifting bridle has two legs that are positioned at 45°, as shown.

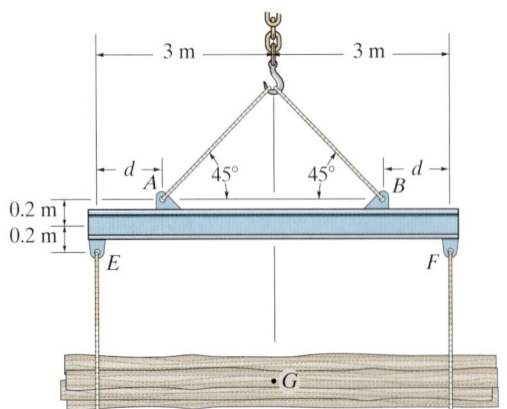

Prob. 7–3

*7–4. Determine the normal force, shear force, and moment at a section passing through point D of the two-member frame.

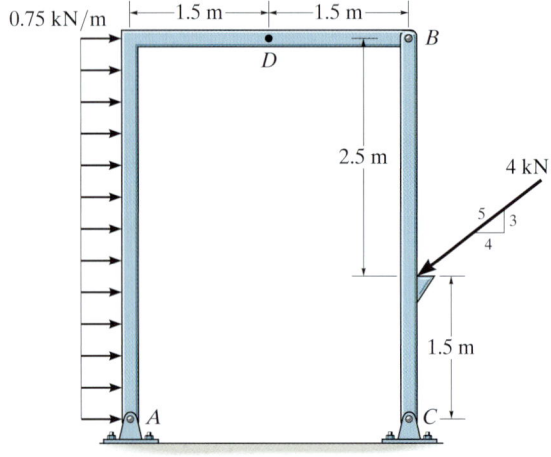

Prob. 7–4

7–5. Determine the internal normal force, shear force, and moment at points A and B in the column.

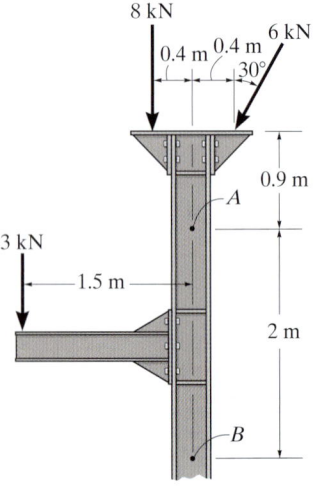

Prob. 7–5

7–6.   Determine the distance a as a fraction of the beam's length L for locating the roller support so that the moment in the beam at B is zero.

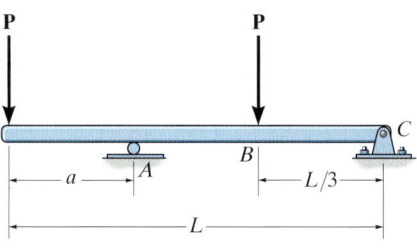

Prob. 7–6

7–7.   Determine the internal normal force, shear force, and moment at points E and D of the compound beam.

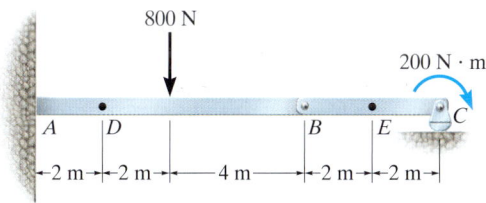

Prob. 7–7

*7–8. Determine the internal normal force, shear force, and moment at point C.

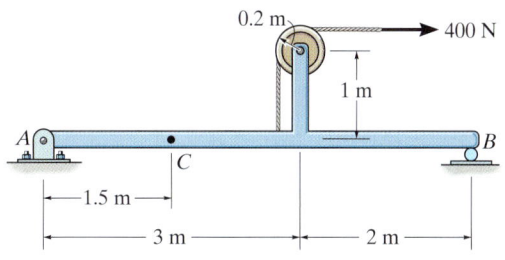

Prob. 7–8

7–9.   Determine the normal force, shear force, and moment at a section passing through point C. Take $P = 8$ kN.

7–10.   The cable will fail when subjected to a tension of 2 kN. Determine the largest vertical load P the frame will support and calculate the internal normal force, shear force, and moment at a section passing through point C for this loading.

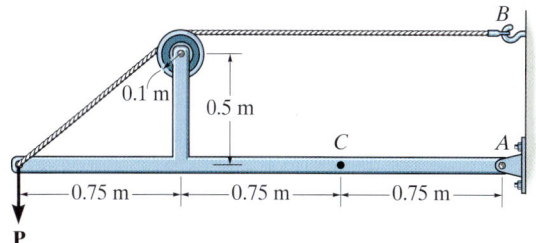

Probs. 7–9/10

7–11.   Determine the internal normal force, shear force, and moment at points D and E in the compound beam. Point E is located just to the left of the 10-kN concentrated load. Assume the support at A is fixed and the connection at B is a pin.

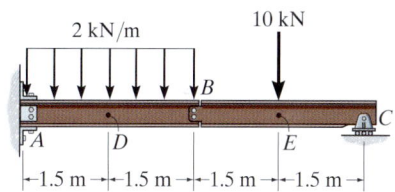

Prob. 7–11

*7–12.   Determine the internal normal force, shear force, and the moment at points C and D.

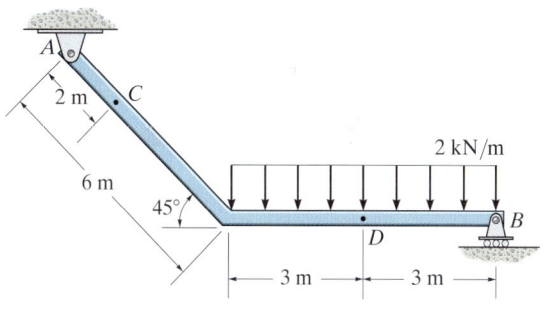

Prob. 7–12

7

7–13. Determine the internal normal force, shear force, and moment at point C in the simply supported beam.

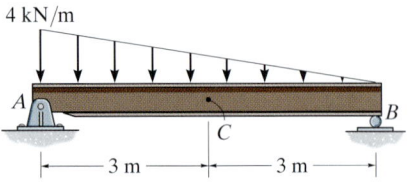

4 kN/m

A C B

3 m   3 m

Prob. 7–13

7–14. Determine the normal force, shear force, and moment at a section passing through point D. Take $w = 150$ N/m.

7–15. The beam AB will fail if the maximum internal moment at D reaches 800 N·m or the normal force in member BC becomes 1500 N. Determine the largest load w it can support.

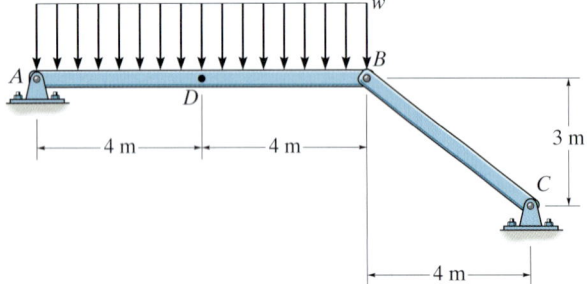

w

A D B

4 m   4 m

3 m

C

4 m

Probs. 7–14/15

***7–16.** Determine the internal normal force, shear force, and moment at point D in the beam.

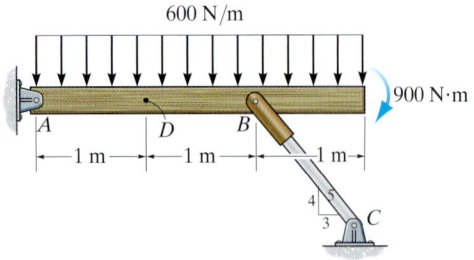

600 N/m

900 N·m

A D B

1 m   1 m   1 m

4
3   C

Prob. 7–16

7–17. Determine the normal force, shear force, and moment at a section passing through point E of the two-member frame.

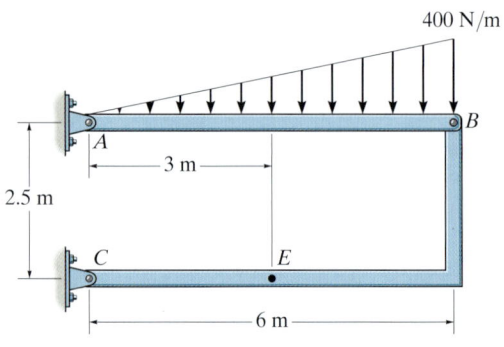

400 N/m

A B

3 m

2.5 m

C E

6 m

Prob. 7–17

7–18. Determine the internal normal force, shear force, and moment at point C in the cantilever beam.

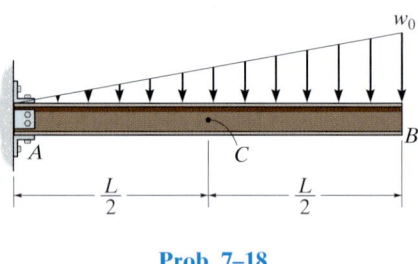

w_0

A C B

$\dfrac{L}{2}$ $\dfrac{L}{2}$

Prob. 7–18

7–19. Determine the internal normal force, shear force, and moment at points E and F in the beam.

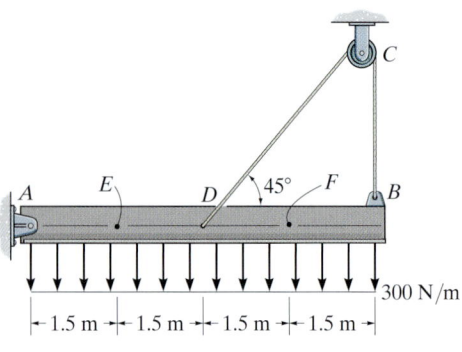

C

A E D 45° F B

300 N/m

1.5 m   1.5 m   1.5 m   1.5 m

Prob. 7–19

*7–20. Determine the internal normal force, shear force, and moment at points C and D in the simply supported beam. Point D is located just to the left of the 5-kN force.

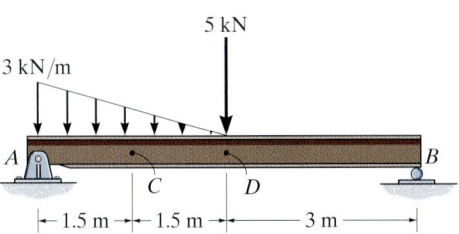

Prob. 7–20

7–21. Determine the internal normal force, shear force, and moment at points D and E in the overhang beam. Point D is located just to the left of the roller support at B, where the couple moment acts.

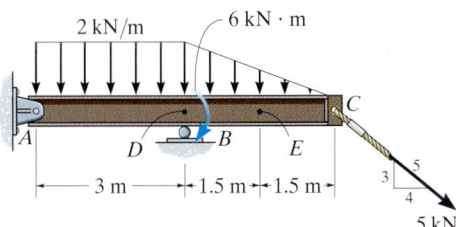

Prob. 7–21

7–22. Determine the internal normal force, shear force, and moment at points E and F in the compound beam. Point F is located just to the left of the 15-kN force and 25-kN · m couple moment.

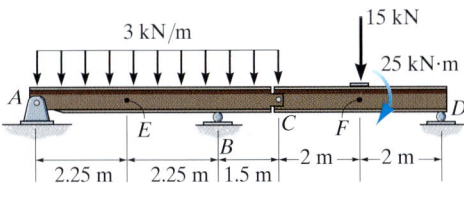

Prob. 7–22

7–23. Determine the internal normal force, shear force, and moment at points D and E in the frame. Point D is located just above the 400-N force.

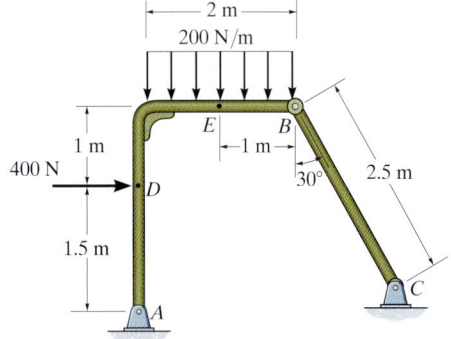

Prob. 7–23

*7–24. Determine the internal normal force, shear force, and bending moment at point C.

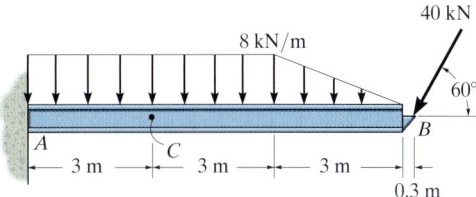

Prob. 7–24

7–25. Determine the internal normal force, shear force, and moment at point C in the double-overhang beam.

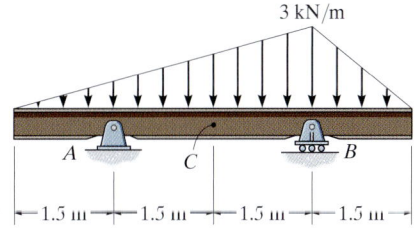

Prob. 7–25

7–26. Determine the ratio of a/b for which the shear force will be zero at the midpoint C of the beam.

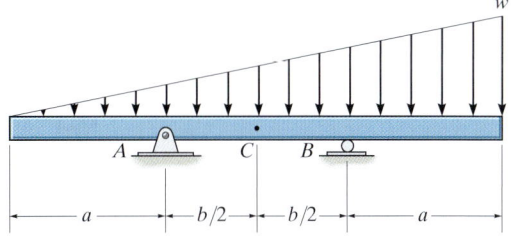

Prob. 7–26

7

7–27. Determine the normal force, shear force, and moment at a section passing through point D of the two-member frame.

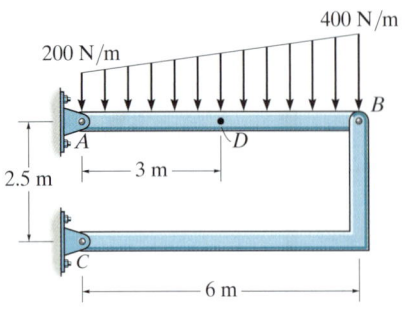

Prob. 7–27

*7–28. Determine the internal normal force, shear force, and moment at point C in the simply supported beam. Point C is located just to the right of the 2.5-kN · m couple moment.

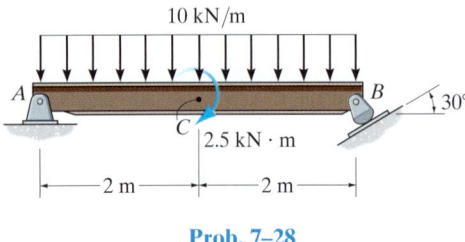

Prob. 7–28

7–29. Determine the internal normal force, shear force, and moment at point D of the two-member frame.

7–30. Determine the internal normal force, shear force, and moment at point E of the two-member frame.

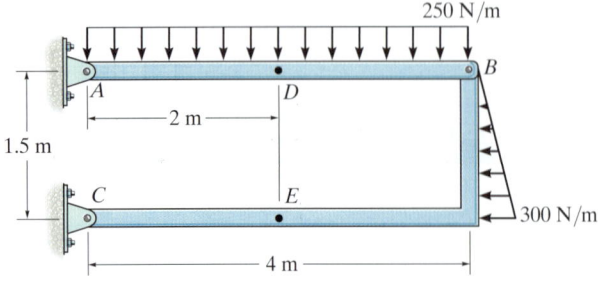

Probs. 7–29/30

7–31. The hook supports the 4-kN load. Determine the internal normal force, shear force, and moment at point A.

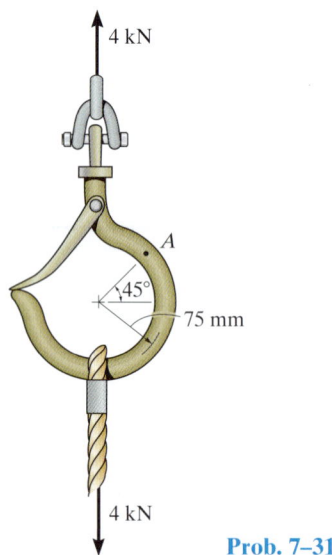

Prob. 7–31

*7–32. Determine the internal normal force, shear force, and moment at points C and D in the simply supported beam. Point D is located just to the left of the 10-kN concentrated load.

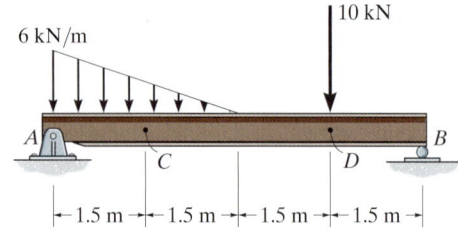

Prob. 7–32

7–33. Determine the distance a in terms of the beam's length L between the symmetrically placed supports A and B so that the internal moment at the center of the beam is zero.

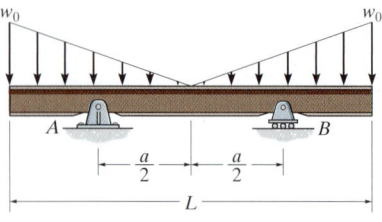

Prob. 7–33

7–34. The beam has a weight w per unit length. Determine the internal normal force, shear force, and moment at point C due to its weight.

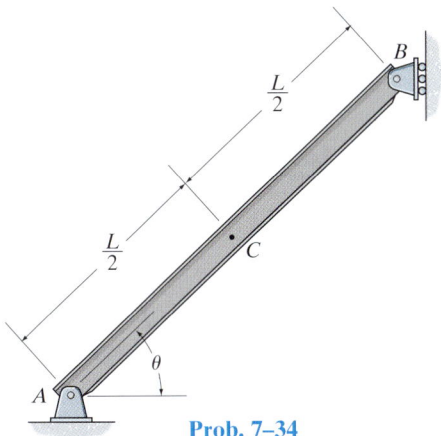

Prob. 7–34

7–35. Determine the internal normal force, shear force, and bending moment at points E and F of the frame.

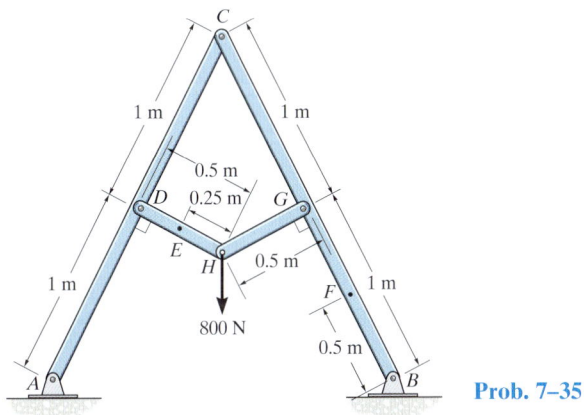

Prob. 7–35

*7–36. Determine the distance a between the supports in terms of the shaft's length L so that the bending moment in the *symmetric* shaft is zero at the shaft's center. The intensity of the distributed load at the center of the shaft is w_0. The supports are journal bearings.

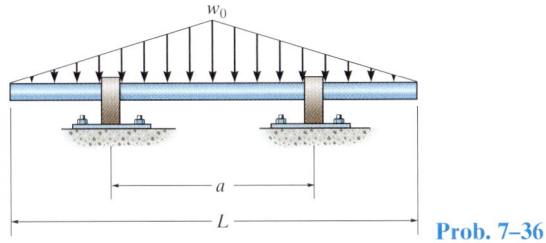

Prob. 7–36

7–37. Determine the internal normal force, shear force, and moment acting at point C. The cooling unit has a total mass of 225 kg with a center of mass at G.

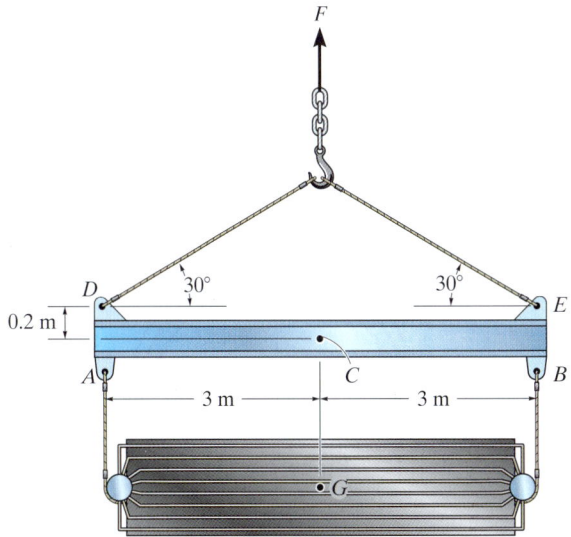

Prob. 7–37

7–38. Determine the internal normal force, shear force, and moment at points D and E of the frame which supports the 100-kg crate. Neglect the size of the smooth peg at C.

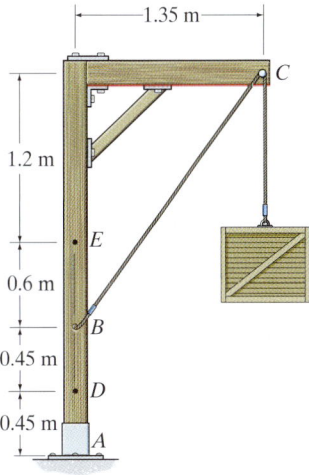

Prob. 7–38

7–39. The semicircular arch is subjected to a uniform distributed load along its axis of w_0 per unit length. Determine the internal normal force, shear force, and moment in the arch at $\theta = 45°$.

*7–40. The semicircular arch is subjected to a uniform distributed load along its axis of w_0 per unit length. Determine the internal normal force, shear force, and moment in the arch at $\theta = 120°$.

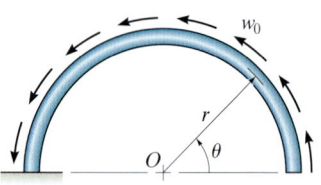

Probs. 7–39/40

7–41. Determine the x, y, z components of internal loading at a section passing through point C in the pipe assembly. Neglect the weight of the pipe. Take $\mathbf{F}_1 = \{350\mathbf{j} - 400\mathbf{k}\}$ kN and $\mathbf{F}_2 = \{150\mathbf{i} - 300\mathbf{k}\}$ kN.

7–42. Determine the x, y, z components of internal loading at a section passing through point C in the pipe assembly. Neglect the weight of the pipe. Take $\mathbf{F}_1 = \{-80\mathbf{i} + 200\mathbf{j} - 300\mathbf{k}\}$ kN and $\mathbf{F}_2 = \{250\mathbf{i} - 150\mathbf{j} - 200\mathbf{k}\}$ kN.

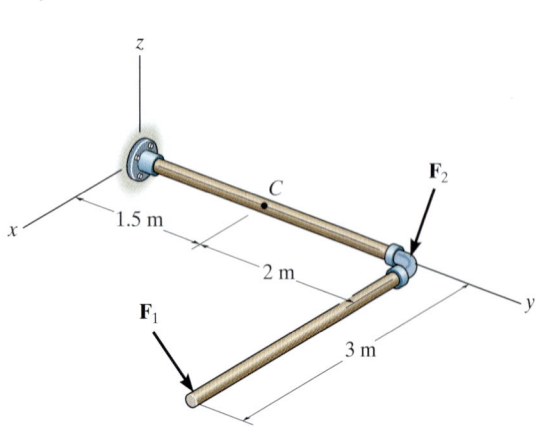

Probs. 7–41/42

7–43. Determine the x, y, z components of internal loading in the rod at point D. $\mathbf{F} = \{7\mathbf{i} - 12\mathbf{j} - 5\mathbf{k}\}$ kN.

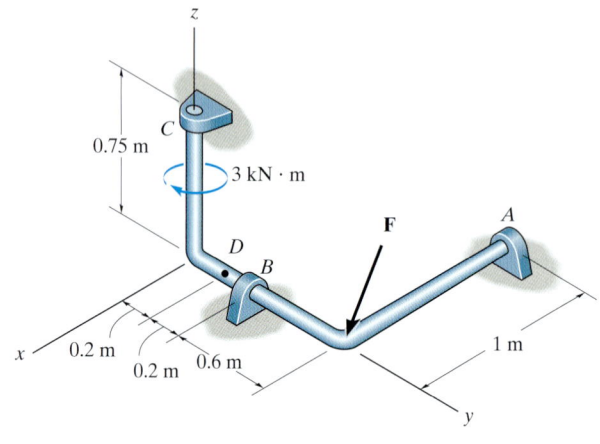

Prob. 7–43

*7–44. Determine the x, y, z components of internal loading in the rod at point E. $\mathbf{F} = \{7\mathbf{i} - 12\mathbf{j} - 5\mathbf{k}\}$ kN.

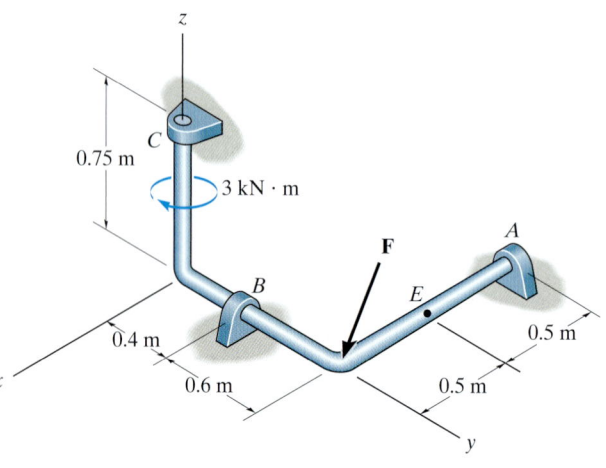

Prob. 7–44

*7.2   Shear and Moment Equations and Diagrams

Beams are structural members designed to support loadings applied perpendicular to their axes. In general, they are long and straight and have a constant cross-sectional area. They are often classified as to how they are supported. For example, a *simply supported beam* is pinned at one end and roller supported at the other, as in Fig. 7–9a, whereas a *cantilevered beam* is fixed at one end and free at the other. The actual design of a beam requires a detailed knowledge of the *variation* of the internal shear force V and bending moment M acting at *each point* along the axis of the beam.*

These *variations* of V and M along the beam's axis can be obtained by using the method of sections discussed in Sec. 7.1. In this case, however, it is necessary to section the beam at an arbitrary distance x from one end and then apply the equations of equilibrium to the segment having the length x. Doing this we can then obtain V and M as functions of x.

In general, the internal shear and bending-moment functions will be discontinuous, or their slopes will be discontinuous, at points where a distributed load changes or where concentrated forces or couple moments are applied. Because of this, these functions must be determined for *each* segment of the beam located between any two discontinuities of loading. For example, segments having lengths x_1, x_2, and x_3 will have to be used to describe the variation of V and M along the length of the beam in Fig. 7–9a. These functions will be valid *only* within regions from 0 to a for x_1, from a to b for x_2, and from b to L for x_3. If the resulting functions of x are plotted, the graphs are termed the *shear diagram* and *bending-moment diagram*, Fig. 7–9b and Fig. 7–9c, respectively.

To save on material and thereby produce an efficient design, these beams, also called girders, have been tapered, since the internal moment in the beam will be larger at the supports, or piers, than at the center of the span.

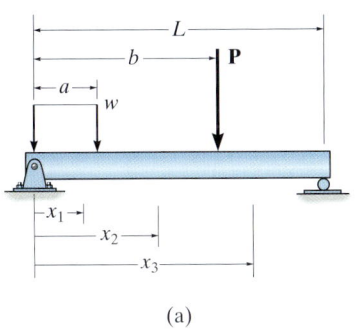

(a)

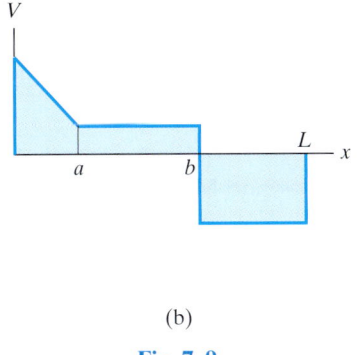

(b)

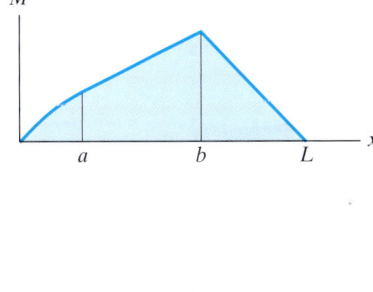

(c)

Fig. 7–9

*The internal normal force is not considered for two reasons. In most cases, the loads applied to a beam act perpendicular to the beam's axis and hence produce only an internal shear force and bending moment. And for design purposes, the beam's resistance to shear, and particularly to bending, is more important than its ability to resist a normal force.

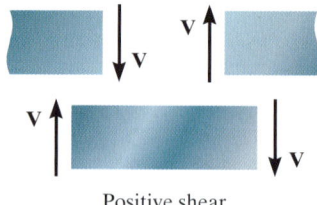

Positive shear

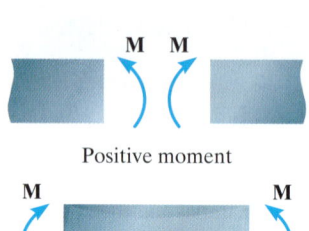

Positive moment

Beam sign convention

Fig. 7–10

Procedure for Analysis

The shear and bending-moment diagrams for a beam can be constructed using the following procedure.

Support Reactions.

- Determine all the reactive forces and couple moments acting on the beam and resolve all the forces into components acting perpendicular and parallel to the beam's axis.

Shear and Moment Functions.

- Specify separate coordinates x having an origin at the beam's left end and extending to regions of the beam *between* concentrated forces and/or couple moments, or where the distributed loading is continuous.

- Section the beam at each distance x and draw the free-body diagram of one of the segments. Be sure **V** and **M** are shown acting in their *positive sense*, in accordance with the sign convention given in Fig. 7–10.

- The shear V is obtained by summing forces perpendicular to the beam's axis.

- The moment M is obtained by summing moments about the sectioned end of the segment.

Shear and Moment Diagrams.

- Plot the shear diagram (V versus x) and the moment diagram (M versus x). If computed values of the functions describing V and M are *positive*, the values are plotted above the x axis, whereas *negative* values are plotted below the x axis.

- Generally, it is convenient to plot the shear and bending-moment diagrams directly below the free-body diagram of the beam.

This extended towing arm must resist both bending and shear loadings throughout its length due to the weight of the vehicle. The variation of these loadings must be known if the arm is to be properly designed.

EXAMPLE | 7.6

Draw the shear and moment diagrams for the shaft shown in Fig. 7–11a. The support at A is a thrust bearing and the support at C is a journal bearing.

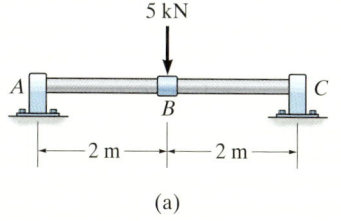

(a)

SOLUTION

Support Reactions.   The support reactions are shown on the shaft's free-body diagram, Fig. 7–11d.

Shear and Moment Functions.   The shaft is sectioned at an arbitrary distance x from point A, extending within the region AB, and the free-body diagram of the left segment is shown in Fig. 7–11b. The unknowns **V** and **M** are assumed to act in the *positive sense* on the right-hand face of the segment according to the established sign convention. Applying the equilibrium equations yields

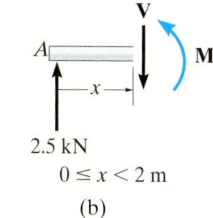

(b)

$$+\uparrow \Sigma F_y = 0; \qquad\qquad V = 2.5 \text{ kN} \qquad\qquad (1)$$
$$\zeta + \Sigma M = 0; \qquad\qquad M = 2.5x \text{ kN} \cdot \text{m} \qquad\qquad (2)$$

A free-body diagram for a left segment of the shaft extending from A a distance x, within the region BC is shown in Fig. 7–11c. As always, **V** and **M** are shown acting in the positive sense. Hence,

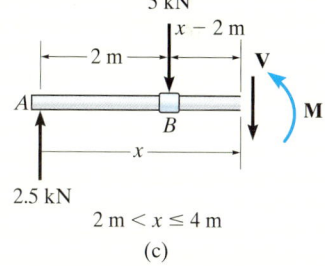

(c)

$$+\uparrow \Sigma F_y = 0; \qquad\qquad 2.5 \text{ kN} - 5 \text{ kN} - V = 0$$
$$V = -2.5 \text{ kN} \qquad\qquad (3)$$
$$\zeta + \Sigma M = 0; \qquad M + 5 \text{ kN}(x - 2 \text{ m}) - 2.5 \text{ kN}(x) = 0$$
$$M = (10 - 2.5x) \text{ kN} \cdot \text{m} \qquad\qquad (4)$$

Shear and Moment Diagrams.   When Eqs. 1 through 4 are plotted within the regions in which they are valid, the shear and moment diagrams shown in Fig. 7–11d are obtained. The shear diagram indicates that the internal shear force is always 2.5 kN (positive) within segment AB. Just to the right of point B, the shear force changes sign and remains at a constant value of −2.5 kN for segment BC. The moment diagram starts at zero, increases linearly to point B at $x - 2$ m, where $M_{max} = 2.5 \text{ kN}(2 \text{ m}) = 5 \text{ kN} \cdot \text{m}$, and thereafter decreases back to zero.

NOTE: It is seen in Fig. 7–11d that the graphs of the shear and moment diagrams are discontinuous where the concentrated force acts, i.e., at points A, B, and C. For this reason, as stated earlier, it is necessary to express both the shear and moment functions separately for regions between concentrated loads. It should be realized, however, that all loading discontinuities are mathematical, arising from the *idealization of a concentrated force and couple moment.* Physically, loads are always applied over a finite area, and if the actual load variation could be accounted for, the shear and moment diagrams would then be continuous over the shaft's entire length.

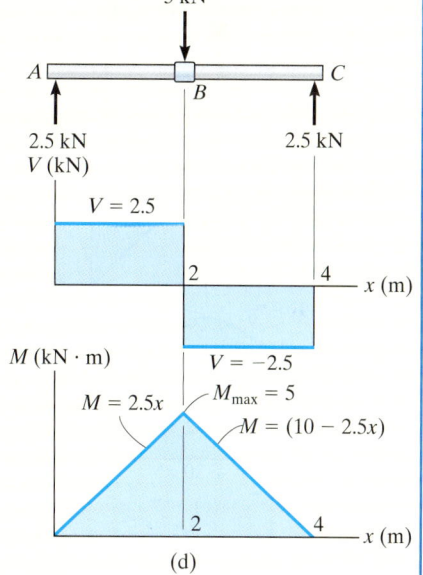

(d)

Fig. 7–11

7

EXAMPLE 7.7

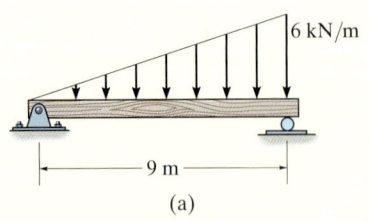

(a)

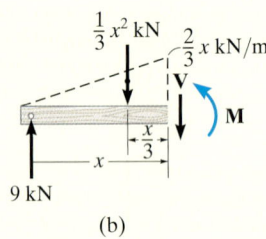

(b)

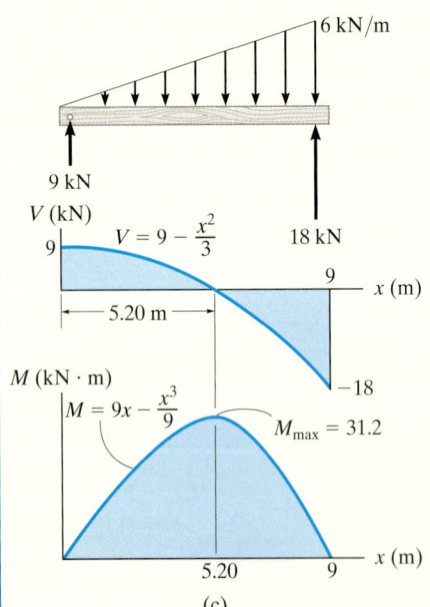

(c)

Fig. 7–12

Draw the shear and moment diagrams for the beam shown in Fig. 7–12a.

SOLUTION

Support Reactions. The support reactions are shown on the beam's free-body diagram, Fig. 7–12c.

Shear and Moment Functions. A free-body diagram for a left segment of the beam having a length x is shown in Fig. 7–12b. Due to proportional triangles, the distributed loading acting at the end of this segment has an intensity of $w/x = 6/9$ or $w = (2/3)x$. It is replaced by a resultant force *after* the segment is isolated as a free-body diagram. The *magnitude* of the resultant force is equal to $\frac{1}{2}(x)\left(\frac{2}{3}x\right) = \frac{1}{3}x^2$. This force *acts through the centroid* of the distributed loading area, a distance $\frac{1}{3}x$ from the right end. Applying the two equations of equilibrium yields

$$+\uparrow \Sigma F_y = 0; \qquad\qquad 9 - \frac{1}{3}x^2 - V = 0$$

$$V = \left(9 - \frac{x^2}{3}\right) \text{kN} \qquad (1)$$

$$\zeta + \Sigma M = 0; \qquad\qquad M + \frac{1}{3}x^2\left(\frac{x}{3}\right) - 9x = 0$$

$$M = \left(9x - \frac{x^3}{9}\right) \text{kN} \cdot \text{m} \qquad (2)$$

Shear and Moment Diagrams. The shear and moment diagrams shown in Fig. 7–12c are obtained by plotting Eqs. 1 and 2.
   The point of *zero shear* can be found using Eq. 1:

$$V = 9 - \frac{x^2}{3} = 0$$

$$x = 5.20 \text{ m}$$

NOTE: It will be shown in Sec. 7.3 that this value of x happens to represent the point on the beam where the *maximum moment* occurs. Using Eq. 2, we have

$$M_{\text{max}} = \left(9(5.20) - \frac{(5.20)^3}{9}\right) \text{kN} \cdot \text{m}$$

$$= 31.2 \text{ kN} \cdot \text{m}$$

FUNDAMENTAL PROBLEMS

F7–7. Determine the shear and moment as a function of x, and then draw the shear and moment diagrams.

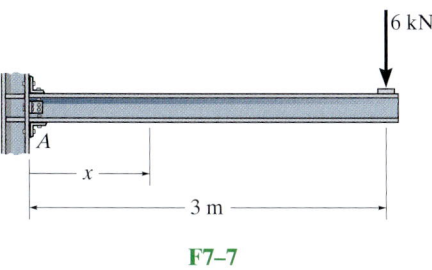

F7–7

F7–8. Determine the shear and moment as a function of x, and then draw the shear and moment diagrams.

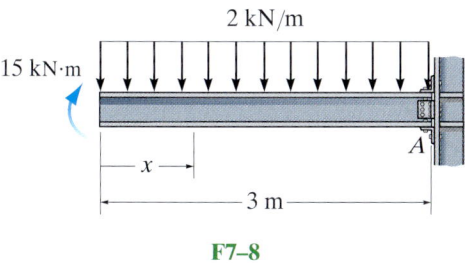

F7–8

F7–9. Determine the shear and moment as a function of x, and then draw the shear and moment diagrams.

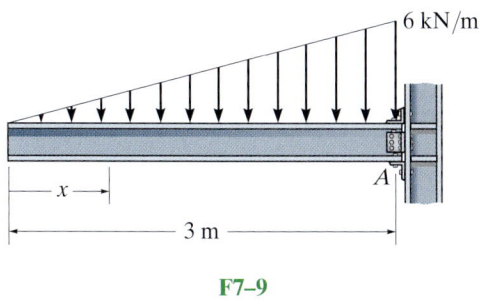

F7–9

F7–10. Determine the shear and moment as a function of x, and then draw the shear and moment diagrams.

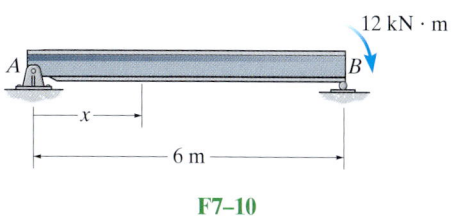

F7–10

F7–11. Determine the shear and moment as a function of x, where $0 \leq x < 3$ m and 3 m $< x \leq 6$ m, and then draw the shear and moment diagrams.

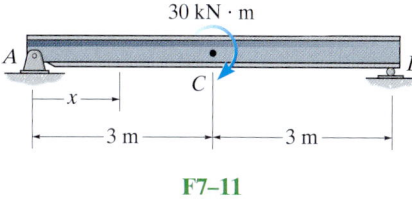

F7–11

F7–12. Determine the shear and moment as a function of x, where $0 \leq x < 3$ m and 3 m $< x \leq 6$ m, and then draw the shear and moment diagrams.

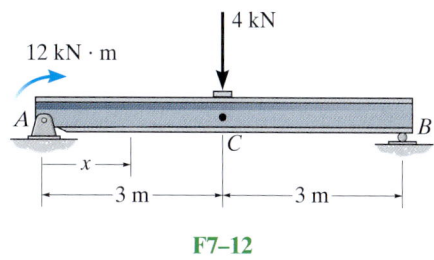

F7–12

7

PROBLEMS

For each of the following problems, establish the x axis with the origin at the left side of the beam, and obtain the internal shear and moment as a function of x. Use these results to plot the shear and moment diagrams.

7–45. Draw the shear and moment diagrams for the overhang beam.

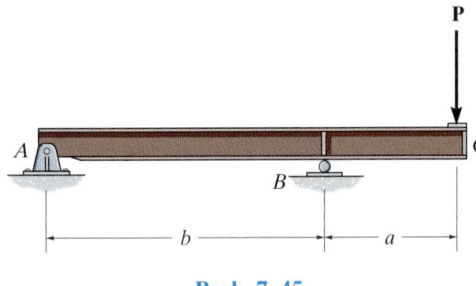

Prob. 7–45

7–46. Draw the shear and moment diagrams for the beam.

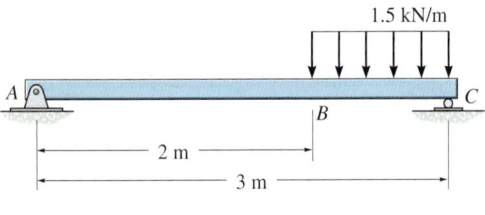

Prob. 7–46

7–47. Draw the shear and moment diagrams for the beam (a) in terms of the parameters shown; (b) set $P = 4$ kN, $a = 1.5$ m, $L = 3.6$ m.

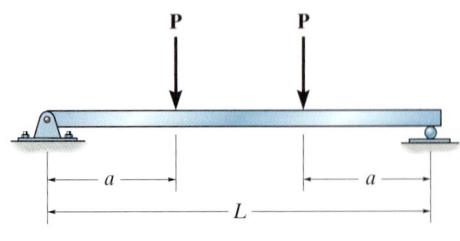

Prob. 7–47

*7–48. Draw the shear and moment diagrams for the beam (a) in terms of the parameters shown; (b) set $M_0 = 500$ N $\cdot$ m, $L = 8$ m.

7–49. If $L = 9$ m, the beam will fail when the maximum shear force is $V_{max} = 5$ kN or the maximum bending moment is $M_{max} = 2$ kN $\cdot$ m. Determine the magnitude M_0 of the largest couple moments it will support.

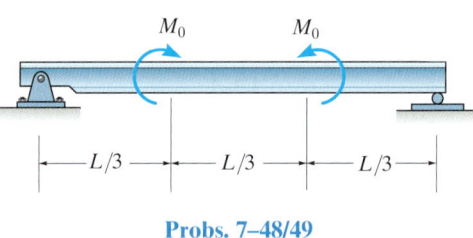

Probs. 7–48/49

7–50. Draw the shear and moment diagrams for the cantilever beam.

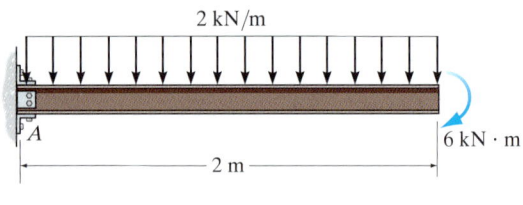

Prob. 7–50

7–51. Draw the shear and moment diagrams for the beam.

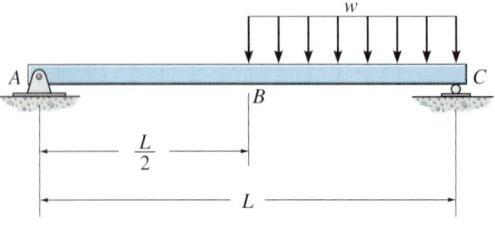

Prob. 7–51

*7–52. Draw the shear and moment diagrams for the beam.

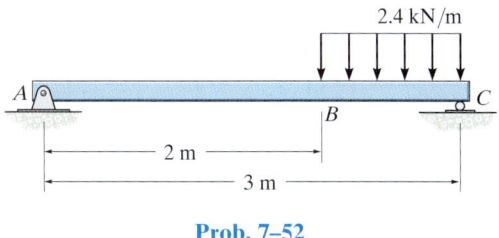

2.4 kN/m

A   B   C
2 m
3 m

Prob. 7–52

7–53. Draw the shear and moment diagrams for the beam.

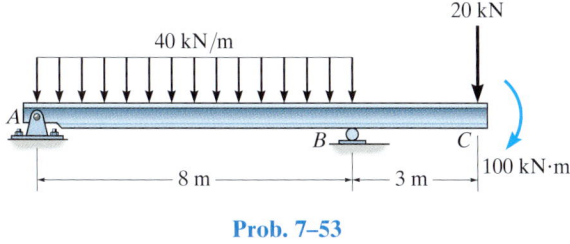

20 kN
40 kN/m
A   B   C
8 m   3 m
100 kN·m

Prob. 7–53

7–54. Draw the shear and moment diagrams for the simply supported beam.

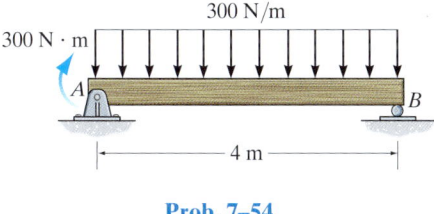

300 N/m
300 N · m
A   B
4 m

Prob. 7–54

7–55. Draw the shear and moment diagrams for the simply supported beam.

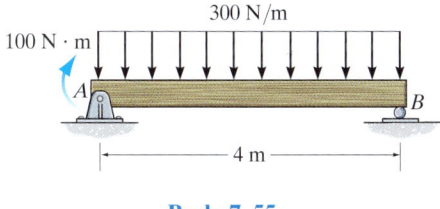

300 N/m
100 N · m
A   B
4 m

Prob. 7–55

*7–56. Draw the shear and bending-moment diagrams for beam ABC. Note that there is a pin at B.

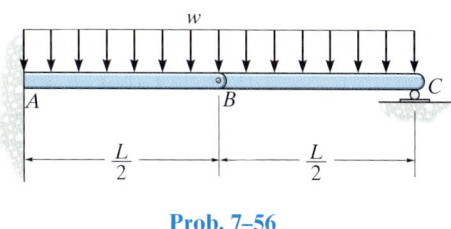

w
A   B   C
$\dfrac{L}{2}$ $\dfrac{L}{2}$

Prob. 7–56

7–57. Draw the shear and moment diagrams for the beam.

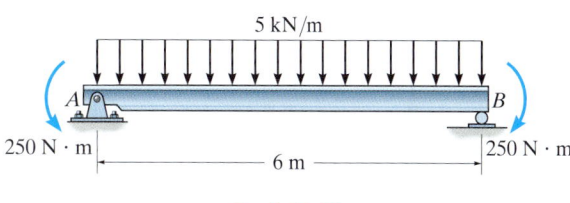

5 kN/m
A   B
250 N · m
6 m
250 N · m

Prob. 7–57

7–58. Draw the shear and moment diagrams for the compound beam. The beam is pin-connected at E and F.

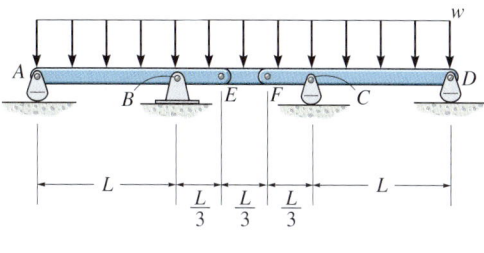

w
A   B   E   F   C   D
L $\dfrac{L}{3}$ $\dfrac{L}{3}$ $\dfrac{L}{3}$ L

Prob. 7–58

7

7–59. Draw the shear and moment diagrams for the beam.

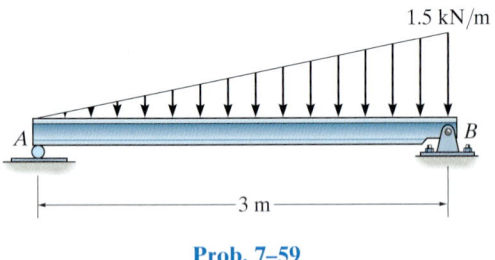

1.5 kN/m

A B

3 m

Prob. 7–59

*7–60. The beam will fail when the maximum internal moment is M_{max}. Determine the position x of the concentrated force **P** and its smallest magnitude that will cause failure.

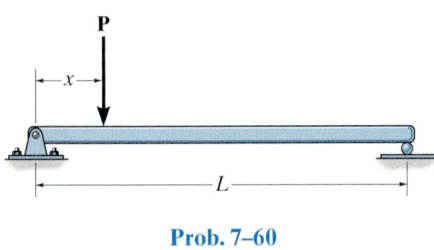

P

x

L

Prob. 7–60

7–61. Draw the shear and moment diagrams for the beam.

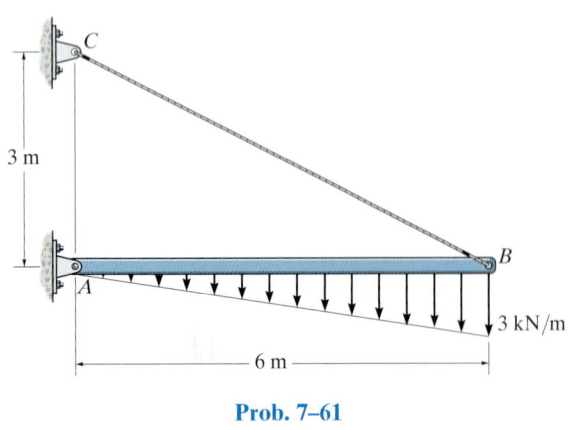

C

3 m

B

A

6 m

3 kN/m

Prob. 7–61

7–62. The cantilevered beam is made of material having a specific weight γ. Determine the shear and moment in the beam as a function of x.

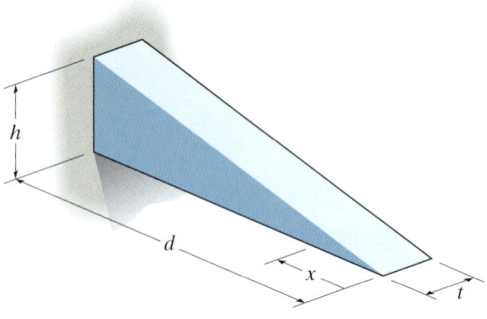

h

d

x

t

Prob. 7–62

7–63. Draw the shear and moment diagrams for the overhang beam.

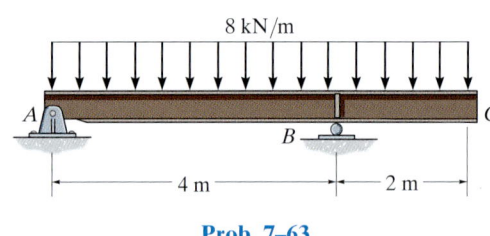

8 kN/m

A B C

4 m   2 m

Prob. 7–63

*7–64. Draw the shear and moment diagrams for the beam.

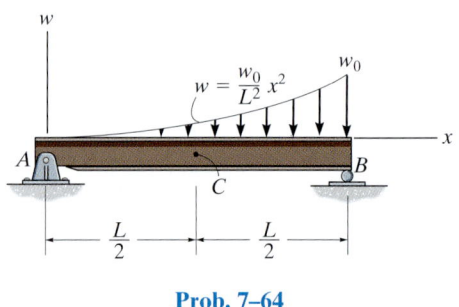

w

$w = \dfrac{w_0}{L^2} x^2$ w_0

A C B x

$\dfrac{L}{2}$ $\dfrac{L}{2}$

Prob. 7–64

7–65. Draw the shear and bending-moment diagrams for the beam.

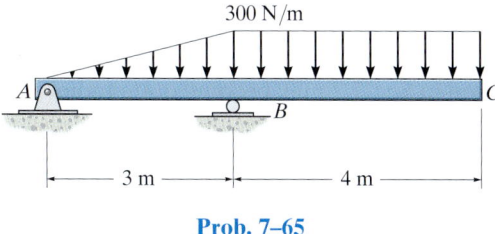

Prob. 7–65

7–66. Draw the shear and moment diagrams for the beam.

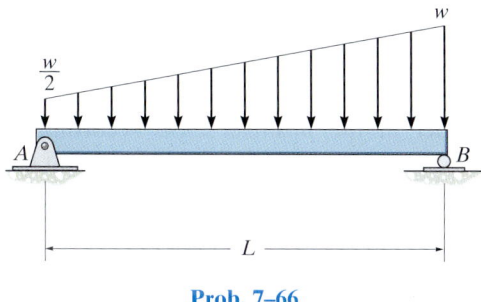

Prob. 7–66

7–67. Determine the internal normal force, shear force, and moment in the curved rod as a function of θ, where $0° \le \theta \le 90°$.

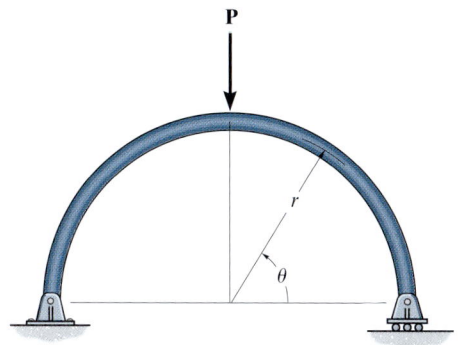

Prob. 7–67

*__7–68.__ Determine the normal force, shear force, and moment in the curved rod as a function of θ.

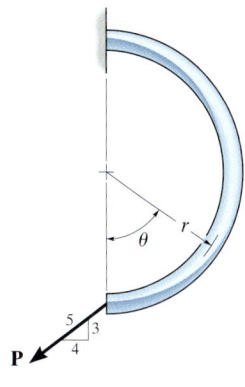

Prob. 7–68

7–69. The quarter circular rod lies in the horizontal plane and supports a vertical force **P** at its end. Determine the magnitudes of the components of the internal shear force, moment, and torque acting in the rod as a function of the angle θ.

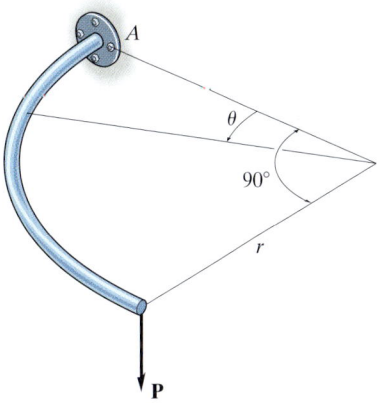

Prob. 7–69

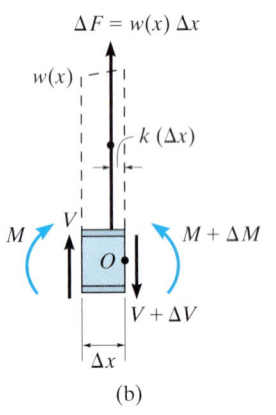

In order to design the beam used to support these power lines, it is important to first draw the shear and moment diagrams for the beam.

*7.3   Relations between Distributed Load, Shear, and Moment

If a beam is subjected to several concentrated forces, couple moments, and distributed loads, the method of constructing the shear and bending-moment diagrams discussed in Sec. 7.2 may become quite tedious. In this section a simpler method for constructing these diagrams is discussed—a method based on differential relations that exist between the load, shear, and bending moment.

Distributed Load.   Consider the beam AD shown in Fig. 7–13a, which is subjected to an arbitrary load $w = w(x)$ and a series of concentrated forces and couple moments. In the following discussion, the *distributed load* will be considered *positive* when the *loading acts upward* as shown. A free-body diagram for a small segment of the beam having a length Δx is chosen at a point x along the beam which is *not* subjected to a concentrated force or couple moment, Fig. 7–13b. Hence any results obtained will not apply at these points of concentrated loading. The internal shear force and bending moment shown on the free-body diagram are assumed to act in the *positive sense* according to the established sign convention. Note that both the shear force and moment acting on the right-hand face must be increased by a small, finite amount in order to keep the segment in equilibrium. The distributed loading has been replaced by a resultant force $\Delta F = w(x)\,\Delta x$ that acts at a fractional distance $k(\Delta x)$ from the right end, where $0 < k < 1$ [for example, if $w(x)$ is *uniform*, $k = \frac{1}{2}$].

Relation Between the Distributed Load and Shear.   If we apply the force equation of equilibrium to the segment, then

$$+\uparrow \Sigma F_y = 0; \qquad V + w(x)\Delta x - (V + \Delta V) = 0$$
$$\Delta V = w(x)\Delta x$$

Dividing by Δx, and letting $\Delta x \to 0$, we get

$$\frac{dV}{dx} = w(x)$$

$$\frac{\text{slope of}}{\text{shear diagram}} = \frac{\text{distributed load}}{\text{intensity}}$$

(7–1)

Fig. 7–13

If we rewrite the above equation in the form $dV = w(x)dx$ and perform an integration between any two points B and C on the beam, we see that

$$\Delta V = \int w(x)\,dx$$

$$\begin{array}{cc} \text{Change in} \\ \text{shear} \end{array} = \begin{array}{cc} \text{Area under} \\ \text{loading curve} \end{array}$$

(7–2)

Relation Between the Shear and Moment.
If we apply the moment equation of equilibrium about point O on the free-body diagram in Fig. 7–13b, we get

$$\zeta + \Sigma M_O = 0; \quad (M + \Delta M) - [w(x)\Delta x]\,k\Delta x - V\Delta x - M = 0$$
$$\Delta M = V\Delta x + k\,w(x)\Delta x^2$$

Dividing both sides of this equation by Δx, and letting $\Delta x \to 0$, yields

$$\frac{dM}{dx} = V$$

$$\begin{array}{cc} \text{Slope of} \\ \text{moment diagram} \end{array} = \text{Shear}$$

(7–3)

In particular, notice that the maximum bending moment $|M|_{\max}$ will occur at the point where the slope $dM/dx = 0$, since this is where the shear is equal to zero.

If Eq. 7–3 is rewritten in the form $dM = \int V\,dx$ and integrated between any two points B and C on the beam, we have

$$\Delta M = \int V\,dx$$

$$\begin{array}{cc} \text{Change in} \\ \text{moment} \end{array} = \begin{array}{cc} \text{Area under} \\ \text{shear diagram} \end{array}$$

(7–4)

As stated previously, the above equations do not apply at points where a *concentrated* force or couple moment acts. These two special cases create *discontinuities* in the shear and moment diagrams, and as a result, each deserves separate treatment.

Force.
A free-body diagram of a small segment of the beam in Fig. 7–13a, taken from under one of the forces, is shown in Fig. 7–14a. Here force equilibrium requires

$$+\uparrow \Sigma F_y = 0; \qquad \Delta V = F$$

(7–5)

Since the *change in shear is positive*, the shear diagram will "jump" *upward when* $\mathbf{F}$ *acts upward* on the beam. Likewise, the jump in shear (ΔV) is downward when $\mathbf{F}$ acts downward.

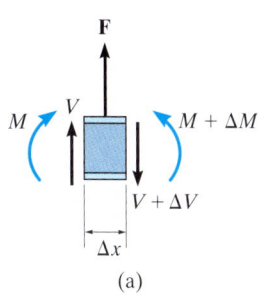

(a)

Fig. 7–14

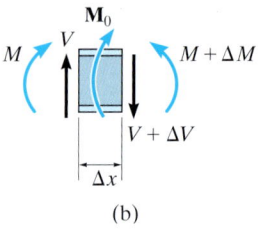

(b)

Fig. 7–14 (cont.)

Couple Moment.

If we remove a segment of the beam in Fig. 7–13a that is located at the couple moment $\mathbf{M}_0$, the free-body diagram shown in Fig. 7–14b results. In this case letting $\Delta x \to 0$, moment equilibrium requires

$$\zeta + \Sigma M = 0; \qquad\qquad \Delta M = M_0 \qquad\qquad (7\text{–}6)$$

Thus, the *change in moment is positive*, or the moment diagram will "jump" *upward if* $\mathbf{M}_0$ *is clockwise.* Likewise, the jump ΔM is downward when $\mathbf{M}_0$ is counterclockwise.

The examples which follow illustrate application of the above equations when used to construct the shear and moment diagrams. After working through these examples, it is recommended that you also go back and solve Examples 7.6 and 7.7 using this method.

This concrete beam is used to support the deck. Its size and the placement of steel reinforcement within it can be determined once the shear and moment diagrams have been established.

Important Points

- The slope of the shear diagram at a point is equal to the intensity of the distributed loading, where positive distributed loading is upward, i.e., $dV/dx = w(x)$.

- The change in the shear ΔV between two points is equal to *the area* under the distributed-loading curve between the points.

- If a concentrated force acts upward on the beam, the shear will jump upward by the same amount.

- The slope of the moment diagram at a point is equal to the shear, i.e., $dM/dx = V$.

- The change in the moment ΔM between two points is equal to the *area* under the shear diagram between the two points.

- If a *clock*wise couple moment acts on the beam, the shear will not be affected; however, the moment diagram will jump *upward* by the amount of the moment.

- Points of *zero shear* represent points of *maximum or minimum moment* since $dM/dx = 0$.

- Because two integrations of $w = w(x)$ are involved to first determine the change in shear, $\Delta V = \int w(x)\,dx$, then to determine the change in moment, $\Delta M = \int V\,dx$, then if the loading curve $w = w(x)$ is a polynomial of degree n, $V = V(x)$ will be a curve of degree $n + 1$, and $M = M(x)$ will be a curve of degree $n + 2$.

EXAMPLE 7.8

Draw the shear and moment diagrams for the cantilever beam in Fig. 7–15a.

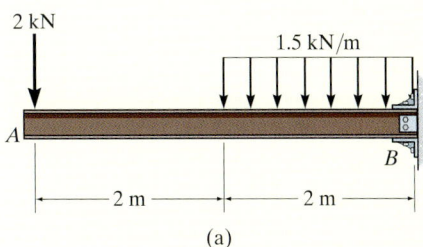

(a)

Fig. 7–15

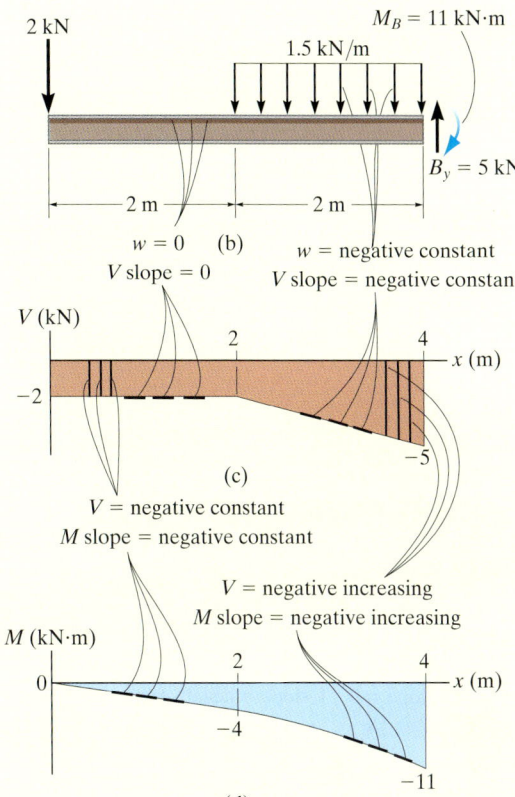

SOLUTION

The support reactions at the fixed support B are shown in Fig. 7–15b.

Shear Diagram. The shear at end A is -2 kN. This value is plotted at $x = 0$, Fig. 7–15c. Notice how the shear diagram is constructed by following the slopes defined by the loading w. The shear at $x = 4$ m is -5 kN, the reaction on the beam. This value can be verified by finding the area under the distributed loading; i.e.,

$$V|_{x=4\,m} = V|_{x=2\,m} + \Delta V = -2\text{ kN} - (1.5\text{ kN/m})(2\text{ m}) = -5\text{ kN}$$

Moment Diagram. The moment of zero at $x = 0$ is plotted in Fig. 7–15d. Construction of the moment diagram is based on knowing that its slope is equal to the shear at each point. The change of moment from $x = 0$ to $x = 2$ m is determined from the area under the shear diagram. Hence, the moment at $x = 2$ m is

$$M|_{x=2\,m} = M|_{x=0} + \Delta M = 0 + [-2\text{ kN}(2\text{ m})] = -4\text{ kN} \cdot \text{m}$$

This same value can be determined from the method of sections, Fig. 7–15e.

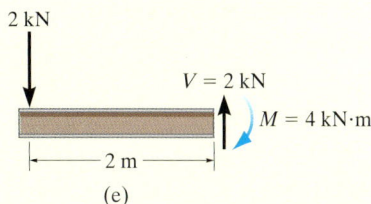

(e)

7

EXAMPLE | 7.9

Draw the shear and moment diagrams for the overhang beam in Fig. 7–16a.

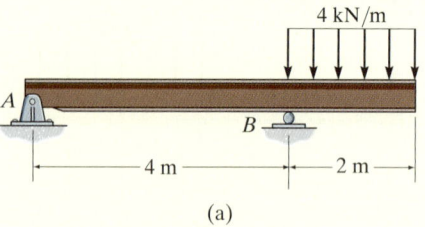

4 kN/m

A

B

4 m

2 m

(a)

Fig. 7–16

SOLUTION

The support reactions are shown in Fig. 7–16b.

Shear Diagram.   The shear of −2 kN at end A of the beam is plotted at $x = 0$, Fig. 7–16c. The slopes are determined from the loading and from this the shear diagram is constructed, as indicated in the figure. In particular, notice the positive jump of 10 kN at $x = 4$ m due to the force B_y, as indicated in the figure.

Moment Diagram.   The moment of zero at $x = 0$ is plotted, Fig. 7–16d, then following the behavior of the slope found from the shear diagram, the moment diagram is constructed. The moment at $x = 4$ m is found from the area under the shear diagram.

$$M\big|_{x=4\,m} = M\big|_{x=0} + \Delta M = 0 + [-2\ kN(4\ m)] = -8\ kN \cdot m$$

We can also obtain this value by using the method of sections, as shown in Fig. 7–16e.

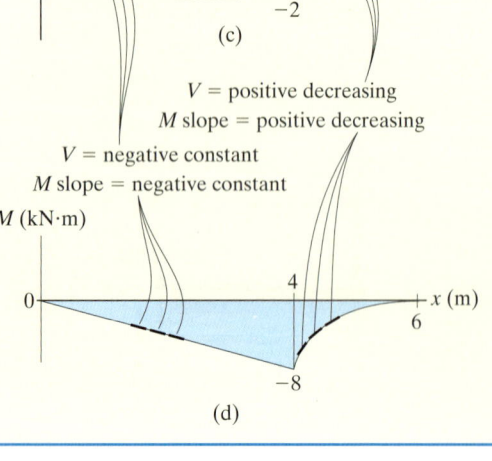

4 kN/m

A

4 m

B

2 m

$A_y = 2$ kN

$B_y = 10$ kN

(b)

$w = 0$ $w = $ negative constant
V slope $= 0$ V slope $= $ negative constant

V (kN)

8

0

4

6

x (m)

−2

(c)

$V = $ positive decreasing
M slope $= $ positive decreasing

$V = $ negative constant
M slope $= $ negative constant

M (kN·m)

0

4

6

x (m)

−8

(d)

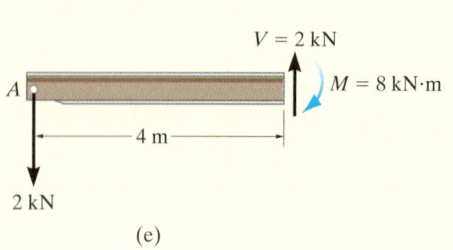

$V = 2$ kN

A

4 m

$M = 8$ kN·m

2 kN

(e)

EXAMPLE 7.10

The shaft in Fig. 7–17a is supported by a thrust bearing at A and a journal bearing at B. Draw the shear and moment diagrams.

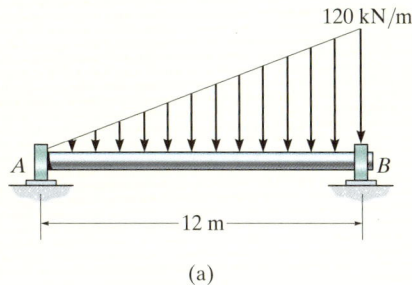

120 kN/m

A

B

12 m

(a)

Fig. 7–17

SOLUTION

The support reactions are shown in Fig. 7–17b.

Shear Diagram.   As shown in Fig. 7–17c, the shear at $x=0$ is +240. Following the slope defined by the loading, the shear diagram is constructed, where at B its value is −480 kN. Since the shear changes sign, the point where $V=0$ must be located. To do this we will use the method of sections. The free-body diagram of the left segment of the shaft, sectioned at an arbitrary position x within the region $0 \leq x < 12$ m, is shown in Fig. 7–17e. Notice that the intensity of the distributed load at x is $w=10x$, which has been found by proportional triangles, i.e., $120/12 = w/x$.

   Thus, for $V=0$,

$$+\uparrow \Sigma F_y = 0; \qquad 240 \text{ kN} - \tfrac{1}{2}(10x)x = 0$$

$$x = 6.93 \text{ m}$$

Moment Diagram.   The moment diagram starts at 0 since there is no moment at A, then it is constructed based on the slope as determined from the shear diagram. The maximum moment occurs at $x=6.93$ m, where the shear is equal to zero, since $dM/dx = V = 0$, Fig. 7–17e,

$$\zeta +\Sigma M = 0;$$
$$M_{max} + \tfrac{1}{2}[(10)(6.93)]\, 6.93\left(\tfrac{1}{3}(6.93)\right) - 240(6.93) = 0$$

$$M_{max} = 1109 \text{ kN} \cdot \text{m}$$

Finally, notice how integration, first of the loading w which is linear, produces a shear diagram which is parabolic, and then a moment diagram which is cubic.

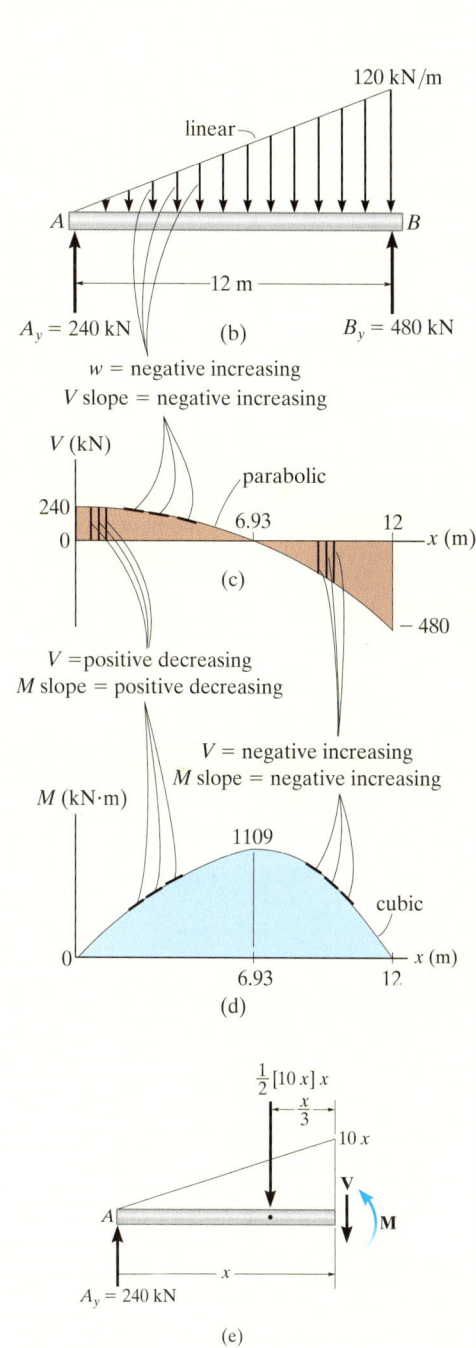

120 kN/m

linear

A

B

12 m

$A_y = 240$ kN   (b) $B_y = 480$ kN

w = negative increasing
V slope = negative increasing

V (kN)

parabolic

240

6.93   12
0   x (m)

(c)

−480

V =positive decreasing
M slope = positive decreasing

V = negative increasing
M slope = negative increasing

M (kN·m)

1109

cubic

0   x (m)
6.93   12

(d)

$\tfrac{1}{2}[10\,x]\,x$

$\tfrac{x}{3}$

10 x

A
                    V
                    M

x

$A_y = 240$ kN

(e)

7

FUNDAMENTAL PROBLEMS

F7–13. Draw the shear and moment diagrams for the beam.

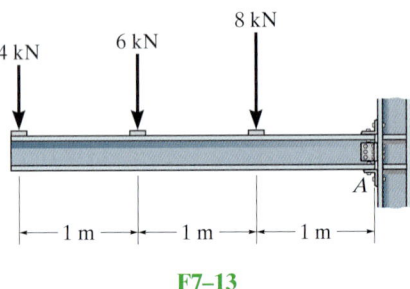

F7–13

F7–14. Draw the shear and moment diagrams for the beam.

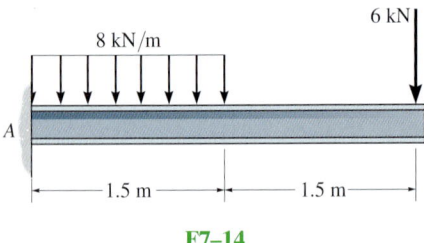

F7–14

F7–15. Draw the shear and moment diagrams for the beam.

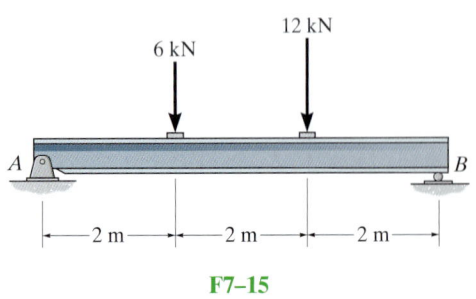

F7–15

F7–16. Draw the shear and moment diagrams for the beam.

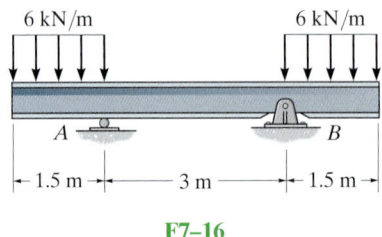

F7–16

F7–17. Draw the shear and moment diagrams for the beam.

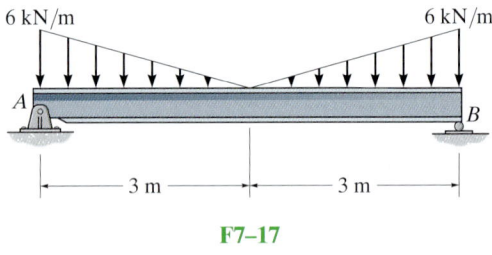

F7–17

F7–18. Draw the shear and moment diagrams for the beam.

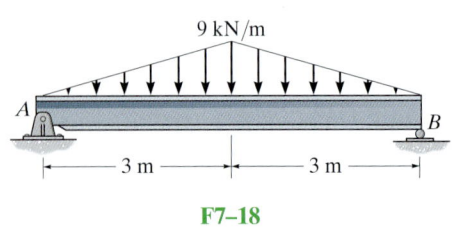

F7–18

PROBLEMS

7–70. Draw the shear and moment diagrams for the beam.

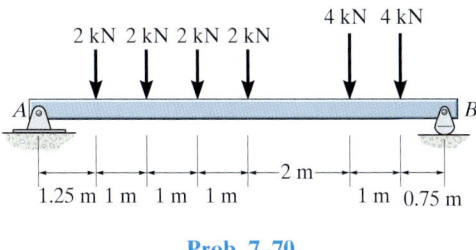

Prob. 7–70

7–71. Draw the shear and moment diagrams for the beam.

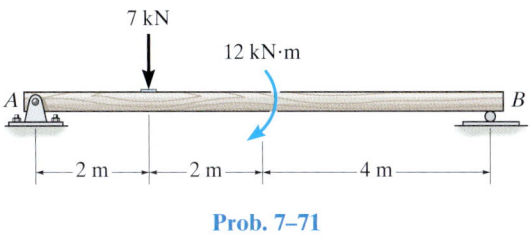

Prob. 7–71

*7–72. Draw the shear and moment diagrams for the beam.

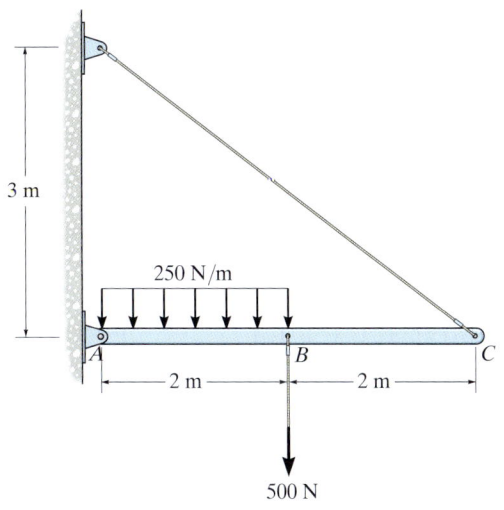

Prob. 7–72

7–72. Draw the shear and moment diagrams for the beam.

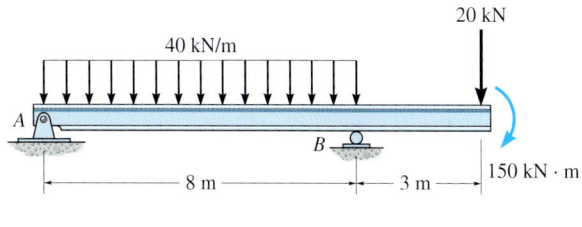

Prob. 7–73

7–74. Draw the shear and moment diagrams for the simply-supported beam.

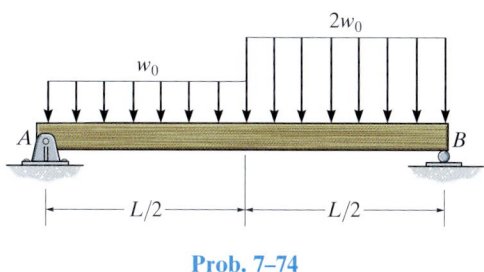

Prob. 7–74

7–75. Draw the shear and moment diagrams for the beam. The support at *A* offers no resistance to vertical load.

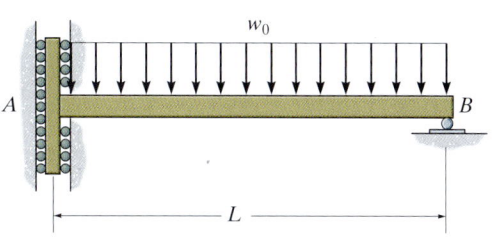

Prob. 7–75

7

*7–76. Draw the shear and moment diagrams for the beam.

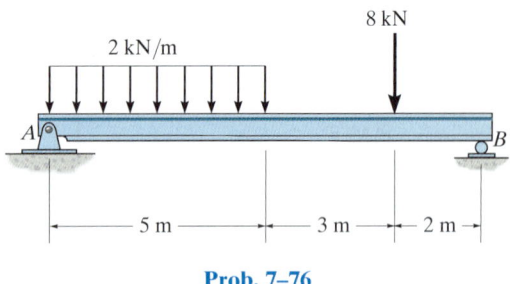

Prob. 7–76

7–77. The shaft is supported by a thrust bearing at *A* and a journal bearing at *B*. Draw the shear and moment diagrams for the shaft.

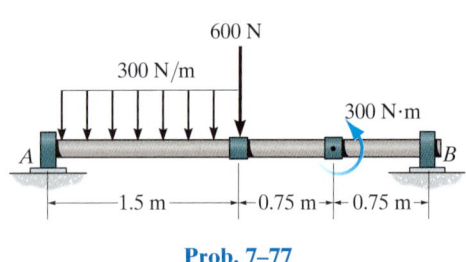

Prob. 7–77

7–78. Draw the shear and moment diagrams for the shaft. The support at *A* is a thrust bearing and at *B* it is a journal bearing.

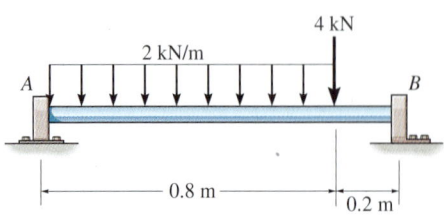

Prob. 7–78

7–79. Draw the shear and moment diagrams for the beam.

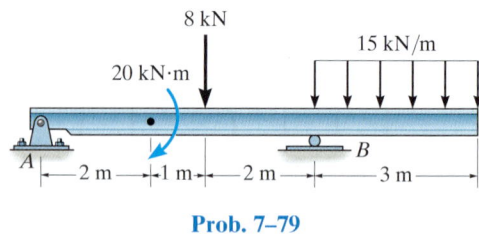

Prob. 7–79

*7–80. Draw the shear and moment diagrams for the compound supported beam.

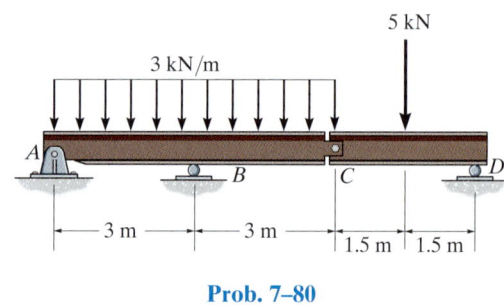

Prob. 7–80

7–81. Draw the shear and moment diagrams for the beam.

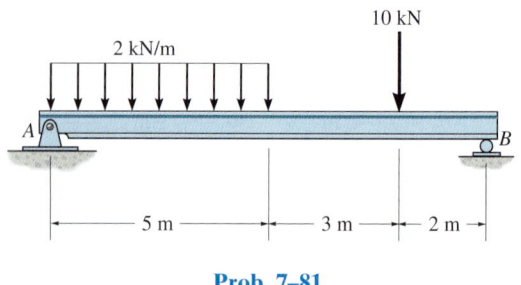

Prob. 7–81

7–82. Draw the shear and moment diagrams for the overhang beam.

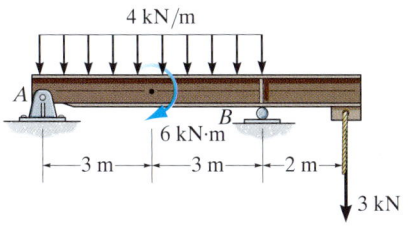

4 kN/m

A

B

6 kN·m

├─ 3 m ─┼─ 3 m ─┼─ 2 m ─┤

3 kN

Prob. 7–82

7–83. Draw the shear and moment diagrams for the shaft. The support at *A* is a journal bearing and at *B* it is a thrust bearing.

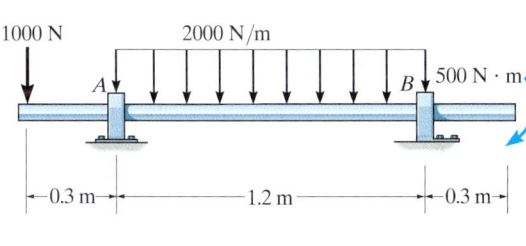

1000 N

2000 N/m

A

B   500 N · m

├─0.3 m─┼──── 1.2 m ────┼─0.3 m─┤

Prob. 7–83

*****7–84.** Draw the shear and moment diagrams for the beam.

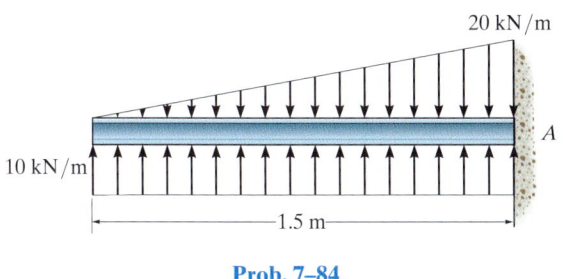

20 kN/m

A

10 kN/m

├──── 1.5 m ────┤

Prob. 7–84

7–85. Draw the shear and moment diagrams for the beam.

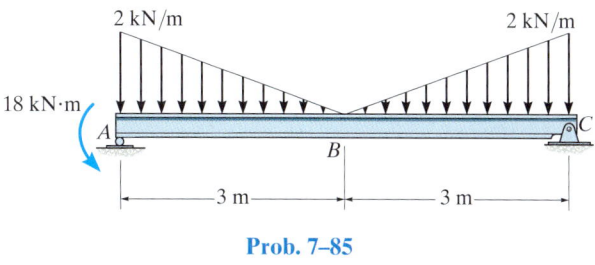

2 kN/m   2 kN/m

18 kN·m

A

B

C

├──── 3 m ────┼──── 3 m ────┤

Prob. 7–85

7–86. Draw the shear and moment diagrams for the beam.

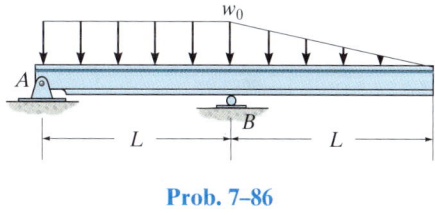

w_0

A

B

├──── L ────┼──── L ────┤

Prob. 7–86

7–87. Draw the shear and moment diagrams for the beam.

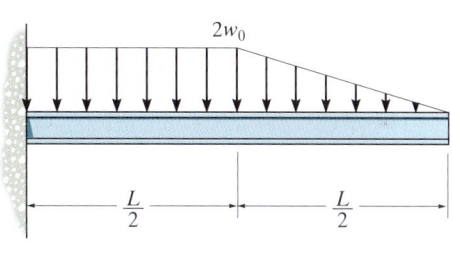

$2w_0$

├── $\frac{L}{2}$ ──┼── $\frac{L}{2}$ ──┤

Prob. 7–87

7

*7–88. Draw the shear and moment diagrams for the beam.

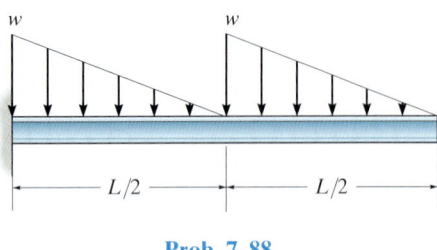

Prob. 7–88

7–89. The shaft is supported by a smooth thrust bearing at *A* and a smooth journal bearing at *B*. Draw the shear and moment diagrams for the shaft.

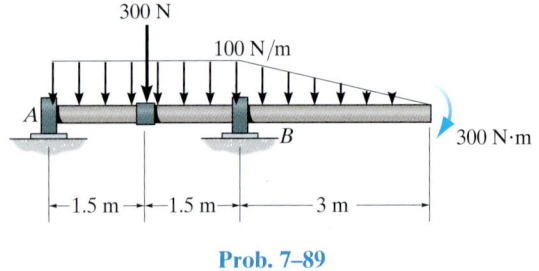

Prob. 7–89

7–90. Draw the shear and moment diagrams for the overhang beam.

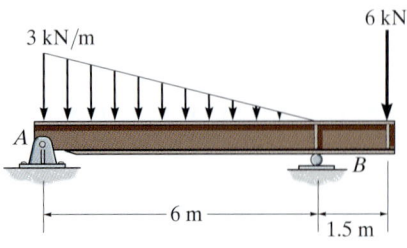

Prob. 7–90

7–91. Draw the shear and moment diagrams for the beam.

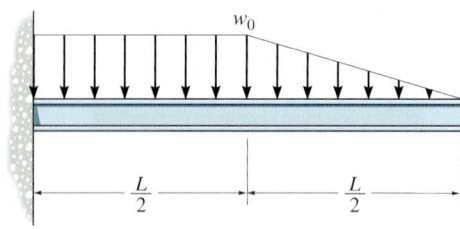

Prob. 7–91

*7–92. The beam consists of three segments pin connected at *B* and *E*. Draw the shear and moment diagrams for the beam.

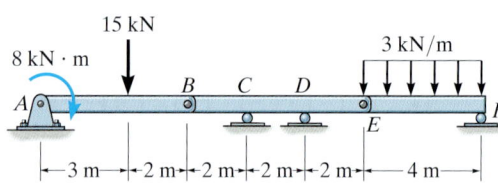

Prob. 7–92

7–93. Draw the shear and moment diagrams for the beam.

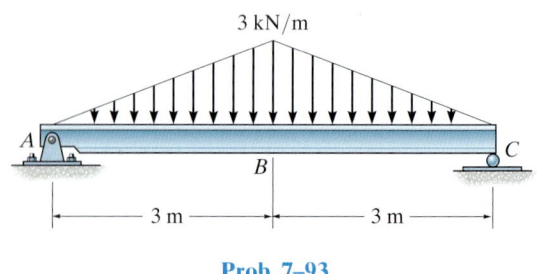

Prob. 7–93

*7.4   Cables

Flexible cables and chains combine strength with lightness and often are used in structures for support and to transmit loads from one member to another. When used to support suspension bridges and trolley wheels, cables form the main load-carrying element of the structure. In the force analysis of such systems, the weight of the cable itself may be neglected because it is often small compared to the load it carries. On the other hand, when cables are used as transmission lines and guys for radio antennas and derricks, the cable weight may become important and must be included in the structural analysis.

Each of the cable segments remains approximately straight as they support the weight of these traffic lights.

Three cases will be considered in the analysis that follows. In each case we will make the assumption that the cable is *perfectly flexible* and *inextensible*. Due to its flexibility, the cable offers no resistance to bending, and therefore, the tensile force acting in the cable is always tangent to the cable at points along its length. Being inextensible, the cable has a constant length both before and after the load is applied. As a result, once the load is applied, the geometry of the cable remains unchanged, and the cable or a segment of it can be treated as a rigid body.

Cable Subjected to Concentrated Loads.

When a cable of negligible weight supports several concentrated loads, the cable takes the form of several straight-line segments, each of which is subjected to a constant tensile force. Consider, for example, the cable shown in Fig. 7–18, where the distances $h, L_1, L_2,$ and L_3 and the loads P_1 and P_2 are known. The problem here is to determine the *nine unknowns* consisting of the tension in each of the *three* segments, the *four* components of reaction at A and B, and the *two* sags y_C and y_D at points C and D. For the solution we can write *two* equations of force equilibrium at each of points A, B, C, and D. This results in a total of *eight equations.** To complete the solution, we need to know something about the geometry of the cable in order to obtain the necessary ninth equation. For example, if the cable's total *length L* is specified, then the Pythagorean theorem can be used to relate each of the three segmental lengths, written in terms of $h, y_C, y_D, L_1, L_2,$ and L_3, to the total length L. Unfortunately, this type of problem cannot be solved easily by hand. Another possibility, however, is to specify one of the sags, either y_C or y_D, instead of the cable length. By doing this, the equilibrium equations are then sufficient for obtaining the unknown forces and the remaining sag. Once the sag at each point of loading is obtained, the length of the cable can then be determined by trigonometry. The following example illustrates a procedure for performing the equilibrium analysis for a problem of this type.

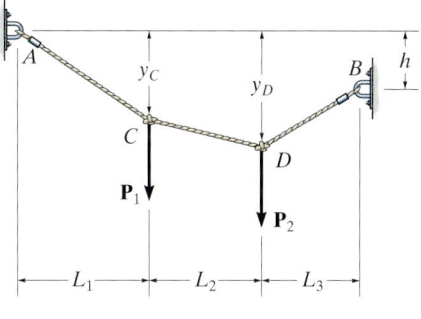

Fig. 7–18

---

*As will be shown in the following example, the eight equilibrium equations *also* can be written for the entire cable, or any part thereof. But *no more* than *eight* independent equations are available.

EXAMPLE   7.11

Determine the tension in each segment of the cable shown in Fig. 7–19a.

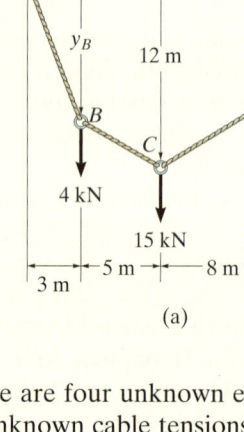

(a)

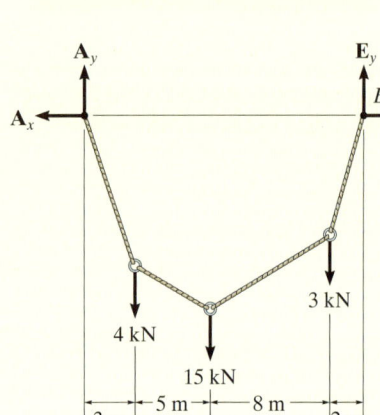

(b)

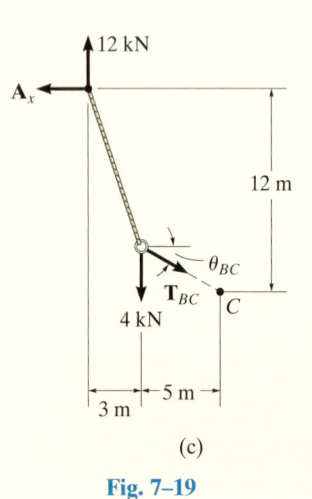

(c)

Fig. 7–19

SOLUTION

By inspection, there are four unknown external reactions (A_x, A_y, E_x, and E_y) and four unknown cable tensions, one in each cable segment. These eight unknowns along with the two unknown sags y_B and y_D can be determined from *ten* available equilibrium equations. One method is to apply the force equations of equilibrium ($\Sigma F_x = 0$, $\Sigma F_y = 0$) to each of the five points A through E. Here, however, we will take a more direct approach.

Consider the free-body diagram for the entire cable, Fig. 7–19b. Thus,

$$\xrightarrow{+} \Sigma F_x = 0; \qquad\qquad -A_x + E_x = 0$$
$$\curvearrowleft +\Sigma M_E = 0;$$

$$-A_y(18\text{ m}) + 4\text{ kN}(15\text{ m}) + 15\text{ kN}(10\text{ m}) + 3\text{ kN}(2\text{ m}) = 0$$
$$A_y = 12\text{ kN}$$

$$+\uparrow \Sigma F_y = 0; \qquad 12\text{ kN} - 4\text{ kN} - 15\text{ kN} - 3\text{ kN} + E_y = 0$$
$$E_y = 10\text{ kN}$$

Since the sag $y_C = 12$ m is known, we will now consider the leftmost section, which cuts cable BC, Fig. 7–19c.

$$\curvearrowleft +\Sigma M_C = 0;\quad A_x(12\text{ m}) - 12\text{ kN}(8\text{ m}) + 4\text{ kN}(5\text{ m}) = 0$$
$$A_x = E_x = 6.33\text{ kN}$$

$$\xrightarrow{+} \Sigma F_x = 0; \qquad T_{BC}\cos\theta_{BC} - 6.33\text{ kN} = 0$$
$$+\uparrow \Sigma F_y = 0;\quad 12\text{ kN} - 4\text{ kN} - T_{BC}\sin\theta_{BC} = 0$$

Thus,

$$\theta_{BC} = 51.6°$$
$$T_{BC} = 10.2\text{ kN} \qquad\qquad \textit{Ans.}$$

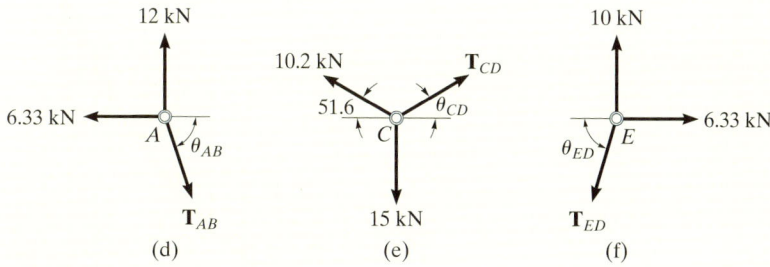

Fig. 7–19 (cont.)

Proceeding now to analyze the equilibrium of points A, C, and E in sequence, we have

Point A. (Fig. 7–19d).

$$\xrightarrow{+} \Sigma F_x = 0; \qquad T_{AB} \cos \theta_{AB} - 6.33 \text{ kN} = 0$$

$$+\uparrow \Sigma F_y = 0; \qquad -T_{AB} \sin \theta_{AB} + 12 \text{ kN} = 0$$

$$\theta_{AB} = 62.2°$$

$$T_{AB} = 13.6 \text{ kN} \qquad Ans.$$

Point C. (Fig. 7–19e).

$$\xrightarrow{+} \Sigma F_x = 0; \qquad T_{CD} \cos \theta_{CD} - 10.2 \cos 51.6° \text{ kN} = 0$$

$$+\uparrow \Sigma F_y = 0; \qquad T_{CD} \sin \theta_{CD} + 10.2 \sin 51.6° \text{ kN} - 15 \text{ kN} = 0$$

$$\theta_{CD} = 47.9°$$

$$T_{CD} = 9.44 \text{ kN} \qquad Ans.$$

Point E. (Fig. 7–19f).

$$\xrightarrow{+} \Sigma F_x = 0; \qquad 6.33 \text{ kN} - T_{ED} \cos \theta_{ED} = 0$$

$$+\uparrow \Sigma F_y = 0; \qquad 10 \text{ kN} - T_{ED} \sin \theta_{ED} = 0$$

$$\theta_{ED} = 57.7°$$

$$T_{ED} = 11.8 \text{ kN} \qquad Ans.$$

NOTE: By comparison, the maximum cable tension is in segment AB since this segment has the greatest slope (θ) and it is required that for any cable segment the horizontal component $T \cos \theta = A_x = E_x$ (a constant). Also, since the slope angles that the cable segments make with the horizontal have now been determined, it is possible to determine the sags y_B and y_D, Fig. 7–19a, using trigonometry.

The cable and suspenders are used to support the uniform load of a gas pipe which crosses the river.

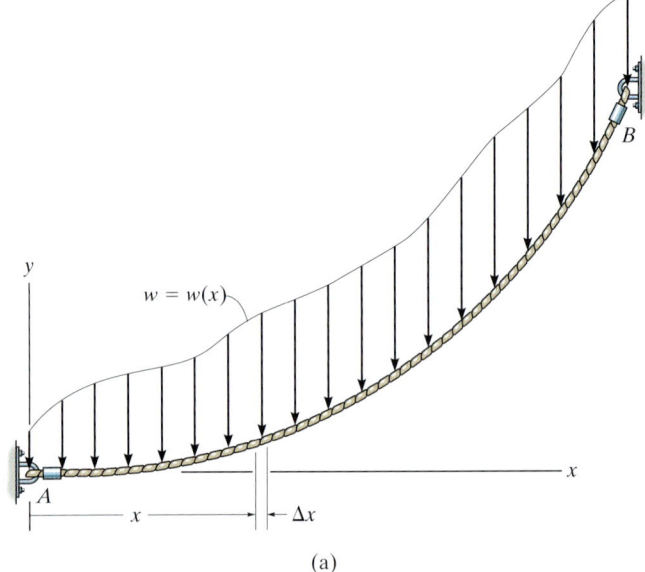

(a)

Fig. 7–20

Cable Subjected to a Distributed Load.

Let us now consider the weightless cable shown in Fig. 7–20a, which is subjected to a distributed loading $w = w(x)$ that is *measured in the x direction*. The free-body diagram of a small segment of the cable having a length Δs is shown in Fig. 7–20b. Since the tensile force changes in both magnitude and direction along the cable's length, we will denote this change on the free-body diagram by ΔT. Finally, the distributed load is represented by its resultant force $w(x)(\Delta x)$, which acts at a fractional distance $k(\Delta x)$ from point O, where $0 < k < 1$. Applying the equations of equilibrium, we have

$$\xrightarrow{+} \Sigma F_x = 0; \qquad\qquad -T\cos\theta + (T + \Delta T)\cos(\theta + \Delta\theta) = 0$$

$$+\uparrow \Sigma F_y = 0; \qquad -T\sin\theta - w(x)(\Delta x) + (T + \Delta T)\sin(\theta + \Delta\theta) = 0$$

$$\zeta + \Sigma M_O = 0; \qquad w(x)(\Delta x)k(\Delta x) - T\cos\theta\,\Delta y + T\sin\theta\,\Delta x = 0$$

Dividing each of these equations by Δx and taking the limit as $\Delta x \to 0$, and therefore $\Delta y \to 0$, $\Delta\theta \to 0$, and $\Delta T \to 0$, we obtain

$$\frac{d(T\cos\theta)}{dx} = 0 \qquad\qquad (7\text{--}7)$$

$$\frac{d(T\sin\theta)}{dx} - w(x) = 0 \qquad\qquad (7\text{--}8)$$

$$\frac{dy}{dx} = \tan\theta \qquad\qquad (7\text{--}9)$$

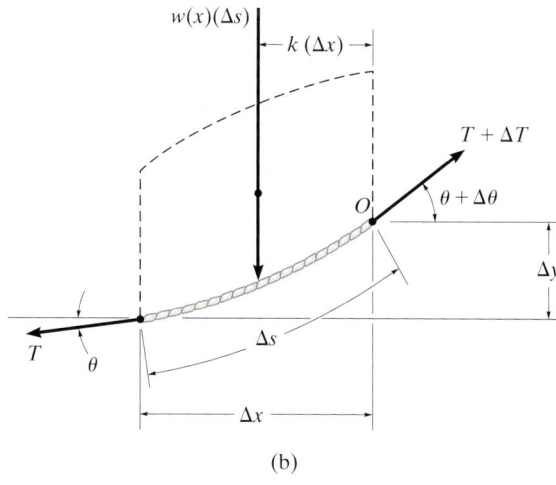

(b)

Fig. 7–20 (cont.)

Integrating Eq. 7–7, we have

$$T \cos \theta = \text{constant} = F_H \qquad (7\text{–}10)$$

where F_H represents the horizontal component of tensile force at *any point* along the cable.

Integrating Eq. 7–8 gives

$$T \sin \theta = \int w(x)\, dx \qquad (7\text{–}11)$$

Dividing Eq. 7–11 by Eq. 7–10 eliminates T. Then, using Eq. 7–9, we can obtain the slope of the cable.

$$\tan \theta = \frac{dy}{dx} = \frac{1}{F_H} \int w(x)\, dx$$

Performing a second integration yields

$$y = \frac{1}{F_H} \int \left(\int w(x)\, dx \right) dx \qquad (7\text{–}12)$$

This equation is used to determine the curve for the cable, $y = f(x)$. The horizontal force component F_H and the additional two constants, say C_1 and C_2, resulting from the integration are determined by applying the boundary conditions for the curve.

The cables of the suspension bridge exert very large forces on the tower and the foundation block which have to be accounted for in their design.

EXAMPLE | 7.12

The cable of a suspension bridge supports half of the uniform road surface between the two towers at A and B, Fig. 7–21a. If this distributed loading is w_0, determine the maximum force developed in the cable and the cable's required length. The span length L and sag h are known.

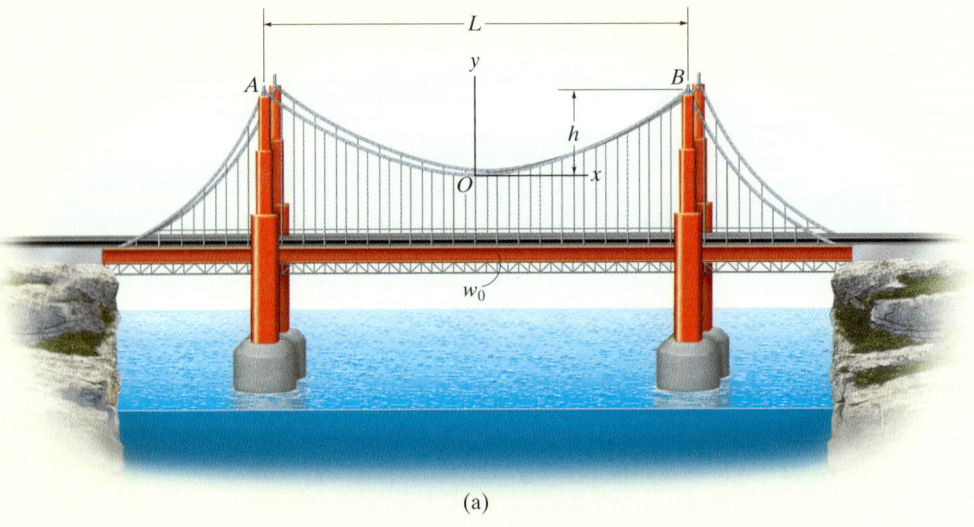

(a)

Fig. 7–21

SOLUTION

We can determine the unknowns in the problem by first finding the equation of the curve that defines the shape of the cable using Eq. 7–12. For reasons of symmetry, the origin of coordinates has been placed at the cable's center. Noting that $w(x) = w_0$, we have

$$y = \frac{1}{F_H} \int \left(\int w_0 \, dx \right) dx$$

Performing the two integrations gives

$$y = \frac{1}{F_H} \left(\frac{w_0 x^2}{2} + C_1 x + C_2 \right) \qquad (1)$$

The constants of integration may be determined using the boundary conditions $y = 0$ at $x = 0$ and $dy/dx = 0$ at $x = 0$. Substituting into Eq. 1 and its derivative yields $C_1 = C_2 = 0$. The equation of the curve then becomes

$$y = \frac{w_0}{2F_H} x^2 \qquad (2)$$

This is the equation of a *parabola*. The constant F_H may be obtained using the boundary condition $y = h$ at $x = L/2$. Thus,

$$F_H = \frac{w_0 L^2}{8h} \tag{3}$$

Therefore, Eq. 2 becomes

$$y = \frac{4h}{L^2} x^2 \tag{4}$$

Since F_H is known, the tension in the cable may now be determined using Eq. 7–10, written as $T = F_H/\cos\theta$. For $0 \leq \theta < \pi/2$, the maximum tension will occur when θ is *maximum*, i.e., at point B, Fig. 7–21a. From Eq. 2, the slope at this point is

$$\left.\frac{dy}{dx}\right|_{x=L/2} = \tan\theta_{max} = \left.\frac{w_0}{F_H} x\right|_{x=L/2}$$

or

$$\theta_{max} = \tan^{-1}\left(\frac{w_0 L}{2F_H}\right) \tag{5}$$

Therefore,

$$T_{max} = \frac{F_H}{\cos(\theta_{max})} \tag{6}$$

Using the triangular relationship shown in Fig. 7–21b, which is based on Eq. 5, Eq. 6 may be written as

$$T_{max} = \frac{\sqrt{4F_H^2 + w_0^2 L^2}}{2}$$

Substituting Eq. 3 into the above equation yields

$$T_{max} = \frac{w_0 L}{2}\sqrt{1 + \left(\frac{L}{4h}\right)^2} \qquad \textit{Ans.}$$

For a differential segment of cable length ds, we can write

$$ds = \sqrt{(dx)^2 + (dy)^2} = \sqrt{1 + \left(\frac{dy}{dx}\right)^2}\, dx$$

Hence, the total length of the cable can be determined by integration. Using Eq. 4, we have

$$\mathcal{L} = \int ds = 2\int_0^{L/2}\sqrt{1 + \left(\frac{8h}{L^2}x\right)^2}\, dx \tag{7}$$

Integrating yields

$$\mathcal{L} = \frac{L}{2}\left[\sqrt{1 + \left(\frac{4h}{L}\right)^2} + \frac{L}{4h}\sinh^{-1}\left(\frac{4h}{L}\right)\right] \qquad \textit{Ans.}$$

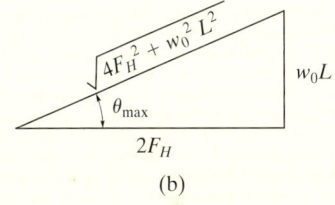

(b)

Fig. 7–21 (cont.)

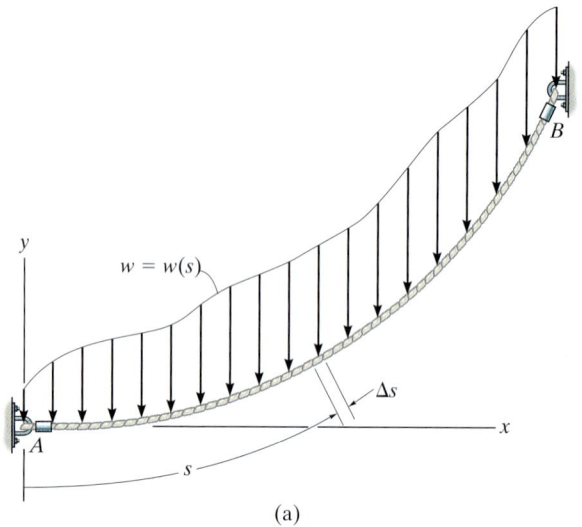

$w = w(s)$

Δs

s

(a)

Fig. 7–22

Cable Subjected to Its Own Weight.

When the weight of a cable becomes important in the force analysis, the loading function along the cable will be a function of the arc length s rather than the projected length x. To analyze this problem, we will consider a generalized loading function $w = w(s)$ acting along the cable as shown in Fig. 7–22a. The free-body diagram for a small segment Δs of the cable is shown in Fig. 7–22b. Applying the equilibrium equations to the force system on this diagram, one obtains relationships identical to those given by Eqs. 7–7 through 7–9, but with s replacing x in Eqs. 7–7 and 7–8. Therefore, we can show that

$$T \cos \theta = F_H$$

$$T \sin \theta = \int w(s) \, ds \tag{7–13}$$

$$\frac{dy}{dx} = \frac{1}{F_H} \int w(s) \, ds \tag{7–14}$$

To perform a direct integration of Eq. 7–14, it is necessary to replace dy/dx by ds/dx. Since

$$ds = \sqrt{dx^2 + dy^2}$$

then

$$\frac{dy}{dx} = \sqrt{\left(\frac{ds}{dx}\right)^2 - 1}$$

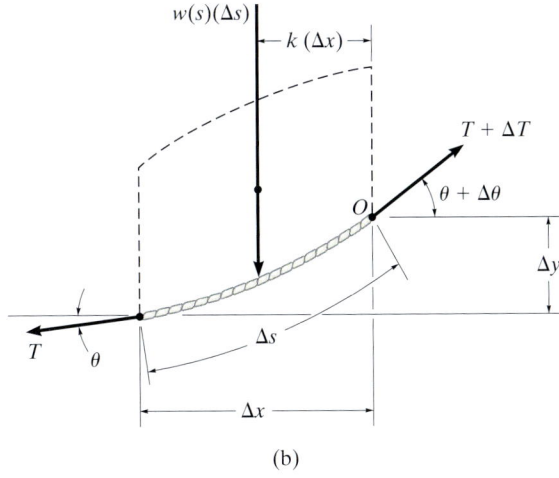

(b)

Fig. 7–22 (cont.)

Therefore,

$$\frac{ds}{dx} = \left[1 + \frac{1}{F_H^2} \left(\int w(s)\, ds \right)^2 \right]^{1/2}$$

Separating the variables and integrating we obtain

$$x = \int \frac{ds}{\left[1 + \dfrac{1}{F_H^2} \left(\displaystyle\int w(s)\, ds \right)^2 \right]^{1/2}} \qquad (7\text{--}15)$$

The two constants of integration, say C_1 and C_2, are found using the boundary conditions for the curve.

Electrical transmission towers must be designed to support the weights of the suspended power lines. The weight and length of the cables can be determined since they each form a catenary curve.

EXAMPLE | 7.13

Determine the deflection curve, the length, and the maximum tension in the uniform cable shown in Fig. 7–23. The cable has a weight per unit length of $w_0 = 5 \text{ N/m}$.

SOLUTION

For reasons of symmetry, the origin of coordinates is located at the center of the cable. The deflection curve is expressed as $y = f(x)$. We can determine it by first applying Eq. 7–15, where $w(s) = w_0$.

$$x = \int \frac{ds}{\left[1 + (1/F_H^2)\left(\int w_0\, ds\right)^2\right]^{1/2}}$$

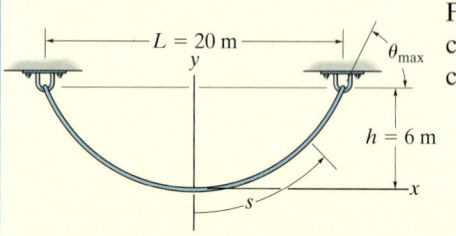

Fig. 7–23

Integrating the term under the integral sign in the denominator, we have

$$x = \int \frac{ds}{[1 + (1/F_H^2)(w_0 s + C_1)^2]^{1/2}}$$

Substituting $u = (1/F_H)(w_0 s + C_1)$ so that $du = (w_0/F_H)\, ds$, a second integration yields

$$x = \frac{F_H}{w_0}(\sinh^{-1} u + C_2)$$

or

$$x = \frac{F_H}{w_0}\left\{ \sinh^{-1}\left[\frac{1}{F_H}(w_0 s + C_1)\right] + C_2 \right\} \qquad (1)$$

To evaluate the constants note that, from Eq. 7–14,

$$\frac{dy}{dx} = \frac{1}{F_H}\int w_0\, ds \quad \text{or} \quad \frac{dy}{dx} = \frac{1}{F_H}(w_0 s + C_1)$$

Since $dy/dx = 0$ at $s = 0$, then $C_1 = 0$. Thus,

$$\frac{dy}{dx} = \frac{w_0 s}{F_H} \qquad (2)$$

The constant C_2 may be evaluated by using the condition $s = 0$ at $x = 0$ in Eq. 1, in which case $C_2 = 0$. To obtain the deflection curve, solve for s in Eq. 1, which yields

$$s = \frac{F_H}{w_0}\sinh\left(\frac{w_0}{F_H}x\right) \qquad (3)$$

Now substitute into Eq. 2, in which case

$$\frac{dy}{dx} = \sinh\left(\frac{w_0}{F_H}x\right)$$

Hence,

$$y = \frac{F_H}{w_0}\cosh\left(\frac{w_0}{F_H}x\right) + C_3$$

If the boundary condition $y = 0$ at $x = 0$ is applied, the constant $C_3 = -F_H/w_0$, and therefore the deflection curve becomes

$$y = \frac{F_H}{w_0}\left[\cosh\left(\frac{w_0}{F_H}x\right) - 1\right] \tag{4}$$

This equation defines the shape of a *catenary curve*. The constant F_H is obtained by using the boundary condition that $y = h$ at $x = L/2$, in which case

$$h = \frac{F_H}{w_0}\left[\cosh\left(\frac{w_0 L}{2F_H}\right) - 1\right] \tag{5}$$

Since $w_0 = 5$ N/m, $h = 6$ m, and $L = 20$ m, Eqs. 4 and 5 become

$$y = \frac{F_H}{5\text{ N/m}}\left[\cosh\left(\frac{5\text{ N/m}}{F_H}x\right) - 1\right] \tag{6}$$

$$6\text{ m} = \frac{F_H}{5\text{ N/m}}\left[\cosh\left(\frac{50\text{ N}}{F_H}\right) - 1\right] \tag{7}$$

Equation 7 can be solved for F_H by using a trial-and-error procedure. The result is

$$F_H = 45.9\text{ N}$$

and therefore the deflection curve, Eq. 6, becomes

$$y = 9.19[\cosh(0.109x) - 1]\text{ m} \qquad \text{Ans.}$$

Using Eq. 3, with $x = 10$ m, the half-length of the cable is

$$\frac{\mathscr{L}}{2} = \frac{45.9\text{ N}}{5\text{ N/m}}\sinh\left[\frac{5\text{ N/m}}{45.9\text{ N}}(10\text{ m})\right] = 12.1\text{ m}$$

Hence,

$$\mathscr{L} = 24.2\text{ m} \qquad \text{Ans.}$$

Since $T = F_H/\cos\theta$, the maximum tension occurs when θ is maximum, i.e., at $s = \mathscr{L}/2 = 12.1$ m. Using Eq. 2 yields

$$\left.\frac{dy}{dx}\right|_{s=12.1\text{ m}} = \tan\theta_{max} = \frac{5\text{ N/m}(12.1\text{ m})}{45.9\text{ N}} = 1.32$$

$$\theta_{max} = 52.8°$$

And so,

$$T_{max} = \frac{F_H}{\cos\theta_{max}} = \frac{45.9\text{ N}}{\cos 52.8°} = 75.9\text{ N} \qquad \text{Ans.}$$

PROBLEMS

Neglect the weight of the cable in the following problems, unless specified.

7–94. Determine the tension in each segment of the cable and the cable's total length. Set $P = 400$ N.

7–95. If each cable segment can support a maximum tension of 375 N, determine the largest load P that can be applied.

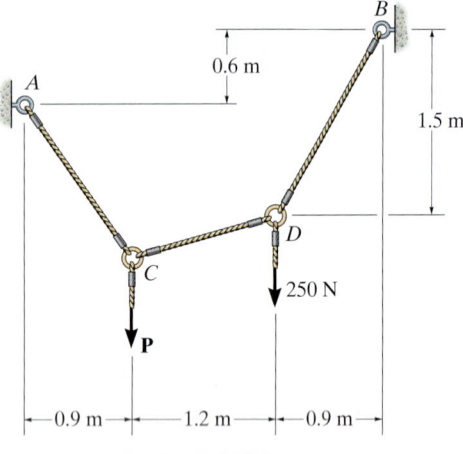

Probs. 7–94/95

***7–96.** Cable $ABCD$ supports the 10-kg lamp E and the 15-kg lamp F. Determine the maximum tension in the cable and the sag of point B.

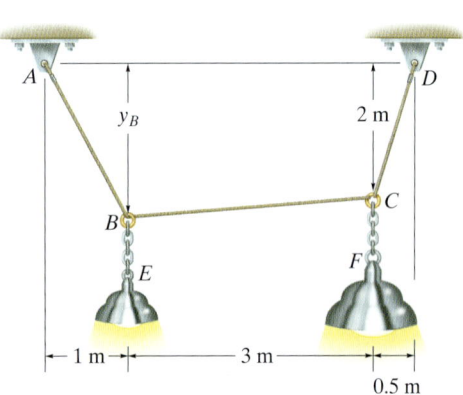

Prob. 7–96

7–97. The cable supports the three loads shown. Determine the sags y_B and y_D of points B and D. Take $P_1 = 2000$ N, $P_2 = 1250$ N.

7–98. The cable supports the three loads shown. Determine the magnitude of $\mathbf{P}_1$ if $P_2 = 1500$ N and $y_B = 2.4$ m. Also find the sag y_D.

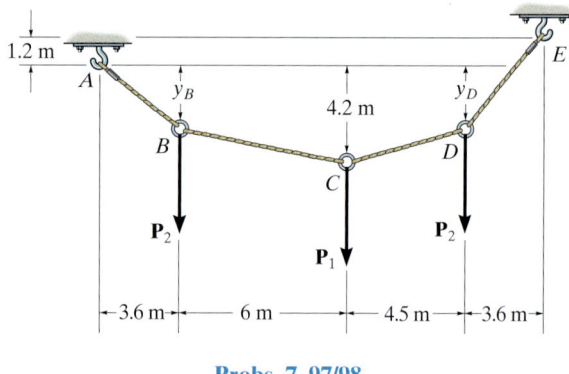

Probs. 7–97/98

7–99. If cylinders E and F have a mass of 20 kg and 40 kg, respectively, determine the tension developed in each cable and the sag y_C.

***7–100.** If cylinder E has a mass of 20 kg and each cable segment can sustain a maximum tension of 400 N, determine the largest mass of cylinder F that can be supported. Also, what is the sag y_C?

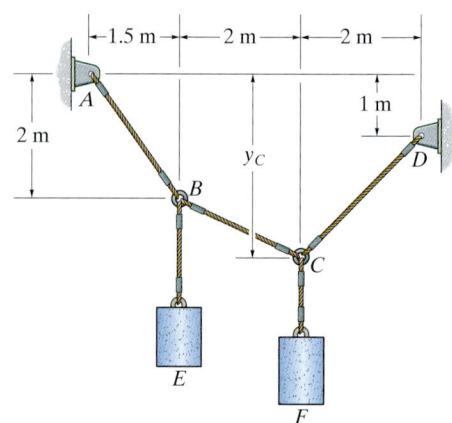

Probs. 7–99/100

7–101. If $h = 5$ m, determine the maximum tension developed in the chain and its length. The chain has a mass per unit length of 8 kg/m.

*7–104. The cable AB is subjected to a uniform loading of 200 N/m. If the weight of the cable is neglected and the slope angles at points A and B are 30° and 60°, respectively, determine the curve that defines the cable shape and the maximum tension developed in the cable.

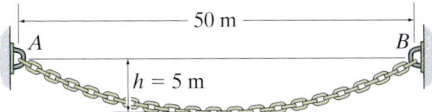

Prob. 7–101

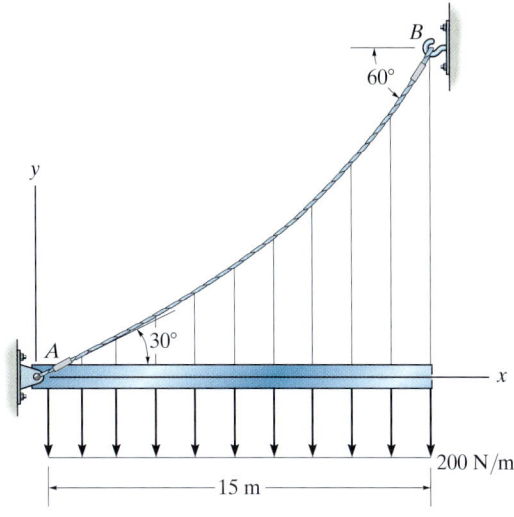

Prob. 7–104

7–102. The cable supports the uniform distributed load of $w_0 = 12$ kN/m. Determine the tension in the cable at each support A and B.

7–103. Determine the maximum uniform distributed load w_0 the cable can support if the maximum tension the cable can sustain is 20 kN.

7–105. The bridge deck has a weight per unit length of 80 kN/m. It is supported on each side by a cable. Determine the tension in each cable at the piers A and B.

7–106. If each of the two side cables that support the bridge deck can sustain a maximum tension of 50 MN, determine the allowable uniform distributed load w_0 caused by the weight of the bridge deck.

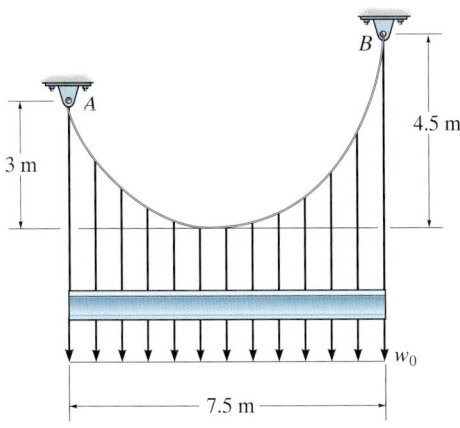

Probs. 7–102/103

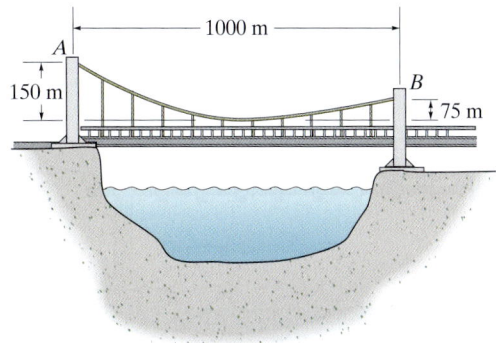

Probs. 7–105/106

7–107. Cylinders C and D are attached to the end of the cable. If D has a mass of 600 kg, determine the required mass of C, the maximum sag h of the cable, and the length of the cable between the pulleys A and B. The beam has a mass per unit length of 50 kg/m.

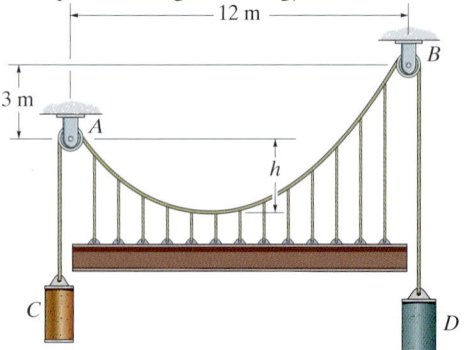

Prob. 7–107

*__7–108.__ The cable will break when the maximum tension reachs $T_{max} = 10$ kN. Determine the minimum sag h if it supports the uniform distributed load of $w = 600$ N/m.

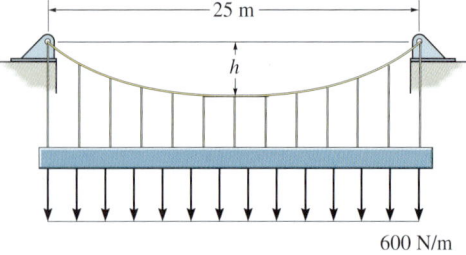

Prob. 7–108

7–109. If the pipe has a mass per unit length of 1500 kg/m, determine the maximum tension developed in the cable.

7–110. If the pipe has a mass per unit length of 1500 kg/m, determine the minimum tension developed in the cable.

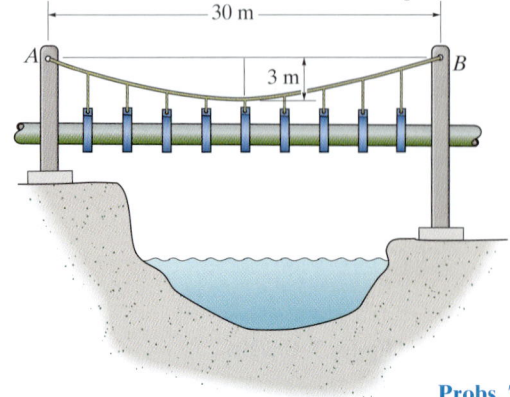

Probs. 7–109/110

7–111. If the slope of the cable at support A is zero, determine the deflection curve $y = f(x)$ of the cable and the maximum tension developed in the cable.

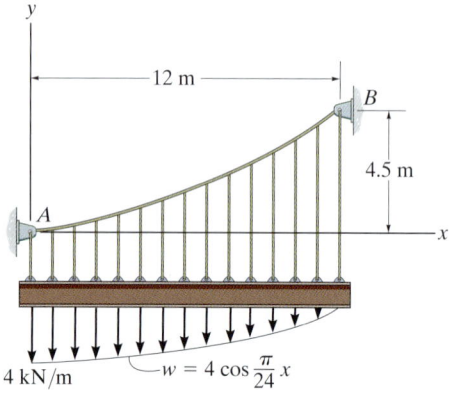

Prob. 7–111

*__7–112.__ Determine the maximum tension developed in the cable if it is subjected to a uniform load of 600 N/m.

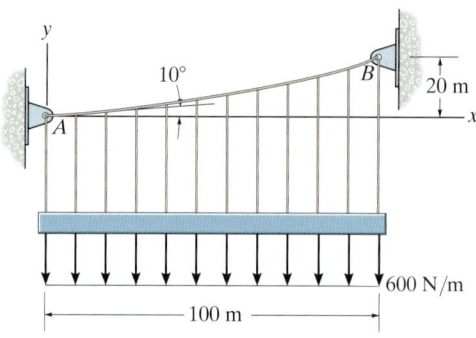

Prob. 7–112

7–113. Determine the maximum uniform distributed loading w_0 N/m that the cable can support if it is capable of sustaining a maximum tension of 60 kN.

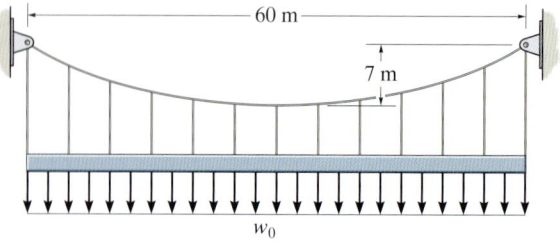

Prob. 7–113

7–114.   A cable has a weight of 30 N/m ($\approx$ 3 kg/m) and is supported at points that are 25 m apart and at the same elevation. If it has a length of 26 m, determine the sag.

7–115.   A wire has a weight of 2 N/m ($\approx$ 0.2 kg/m). If it can span 10 m and has a sag of 1.2 m, determine the length of the wire. The ends of the wire are supported from the same elevation.

*7–116.   The 10 kg/m cable is suspended between the supports A and B. If the cable can sustain a maximum tension of 1.5 kN and the maximum sag is 3 m, determine the maximum distance L between the supports

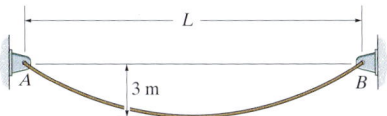

Prob. 7–116

7–117.   Show that the deflection curve of the cable discussed in Example 7.13 reduces to Eq. 4 in Example 7.12 when the *hyperbolic cosine function* is expanded in terms of a series and only the first two terms are retained. (The answer indicates that the *catenary* may be replaced by a *parabola* in the analysis of problems in which the sag is small. In this case, the cable weight is assumed to be uniformly distributed along the horizontal.)

7–118.   If the horizontal towing force is $T = 20$ kN and the chain has a mass per unit length of 15 kg/m, determine the maximum sag h. Neglect the buoyancy effect of the water on the chain. The boats are stationary.

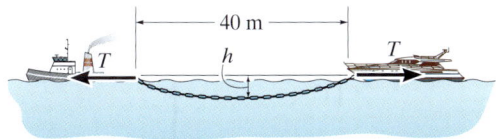

Prob. 7–118

7–119.   The cable has a mass of 0.5 kg/m and is 25 m long. Determine the vertical and horizontal components of force it exerts on the top of the tower.

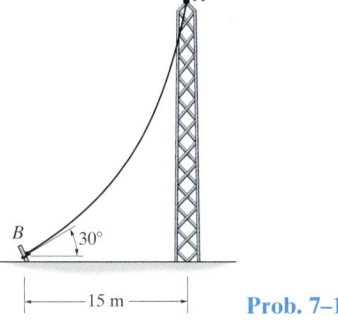

Prob. 7–119

*7–120.   A 50-m cable is suspended between two points a distance of 15 m apart and at the same elevation. If the minimum tension in the cable is 200 kN, determine the total weight of the cable and the maximum tension developed in the cable.

7–121.   The 80-m-long chain is fixed at its ends and hoisted at its midpoint B using a crane. If the chain has a weight of 0.5 kN/m, determine the minimum height h of the hook in order to lift the chain *completely* off the ground. What is the horizontal force at pin A or C when the chain is in this position? *Hint:* When h is a minimum, the slope at A and C is zero.

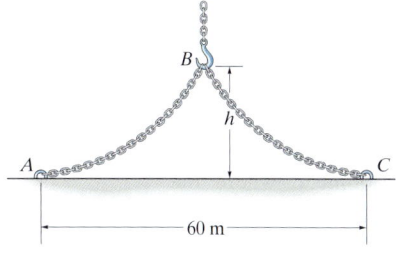

Prob. 7–121

7–122.   A uniform cord is suspended between two points having the same elevation. Determine the sag-to-span ratio so that the maximum tension in the cord equals the cord's total weight.

7–123.   A cable having a weight per unit length of 0.1 kN/m is suspended between supports A and B. Determine the equation of the catenary curve of the cable and the cable's length.

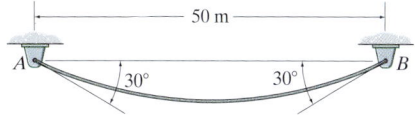

Prob. 7–123

*7–124.   A fiber optic cable is suspended over the poles so that the angle at the supports is $\theta = 22°$. Determine the minimum tension in the cable and the sag. The cable has a mass of 0.9 kg/m and the supports are at the same elevation.

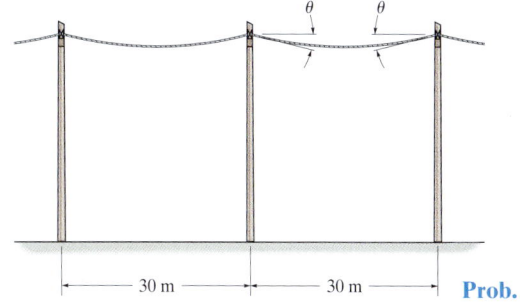

Prob. 7–124

CHAPTER REVIEW

Internal Loadings

If a coplanar force system acts on a member, then in general a resultant internal *normal force* **N**, *shear force* **V**, and *bending moment* **M** will act at any cross section along the member. For two-dimensional problems the positive directions of these loadings are shown in the figure.

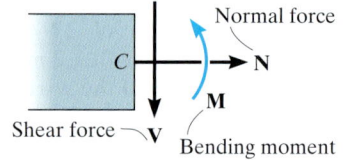

The resultant internal normal force, shear force, and bending moment are determined using the method of sections. To find them, the member is sectioned at the point C where the internal loadings are to be determined. A free-body diagram of one of the sectioned parts is then drawn and the internal loadings are shown in their positive directions.

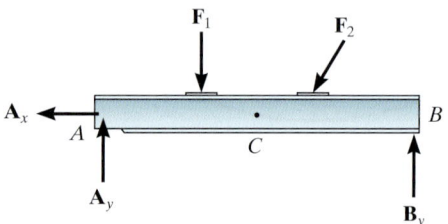

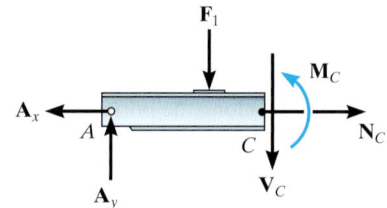

The resultant normal force is determined by summing forces normal to the cross section. The resultant shear force is found by summing forces tangent to the cross section, and the resultant bending moment is found by summing moments about the geometric center or centroid of the cross-sectional area.

$$\Sigma F_x = 0$$
$$\Sigma F_y = 0$$
$$\Sigma M_C = 0$$

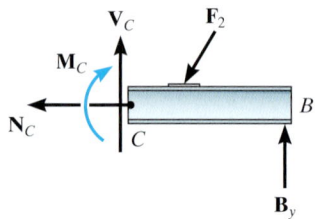

If the member is subjected to a three-dimensional loading, then, in general, a *torsional moment* will also act on the cross section. It can be determined by summing moments about an axis that is perpendicular to the cross section and passes through its centroid.

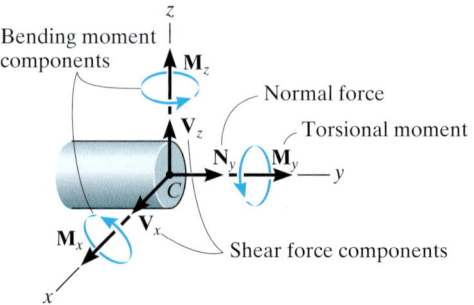

Shear and Moment Diagrams

To construct the shear and moment diagrams for a member, it is necessary to section the member at an arbitrary point, located a distance x from the left end.

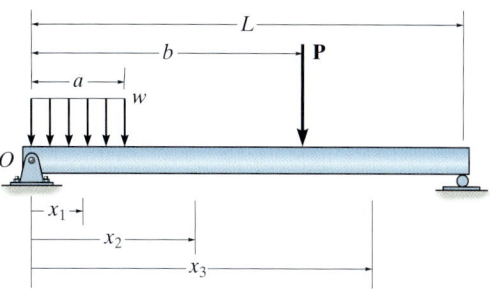

If the external loading consists of changes in the distributed load, or a series of concentrated forces and couple moments act on the member, then different expressions for V and M must be determined within regions between any load discontinuities.

The unknown shear and moment are indicated on the cross section in the positive direction according to the established sign convention, and then the internal shear and moment are determined as functions of x.

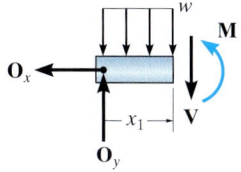

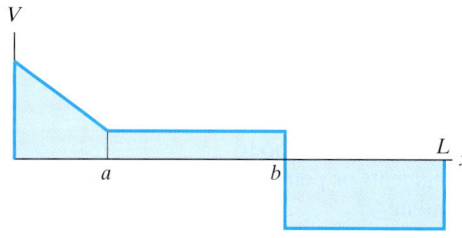

Each of the functions of the shear and moment is then plotted to create the shear and moment diagrams.

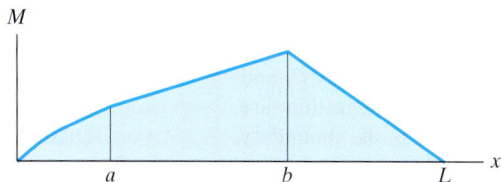

Relations between Shear and Moment

It is possible to plot the shear and moment diagrams quickly by using differential relationships that exist between the distributed loading w, V and M.

The slope of the shear diagram is equal to the distributed loading at any point. The slope is positive if the distributed load acts upward, and vice-versa.

$$\frac{dV}{dx} = w$$

The slope of the moment diagram is equal to the shear at any point. The slope is positive if the shear is positive, or vice-versa.

$$\frac{dM}{dx} = V$$

The change in shear between any two points is equal to the area under the distributed loading between the points.

$$\Delta V = \int w\,dx$$

The change in the moment is equal to the area under the shear diagram between the points.

$$\Delta M = \int V\,dx$$

Cables

When a flexible and inextensible cable is subjected to a series of concentrated forces, then the analysis of the cable can be performed by using the equations of equilibrium applied to free-body diagrams of either segments or points of application of the loading.

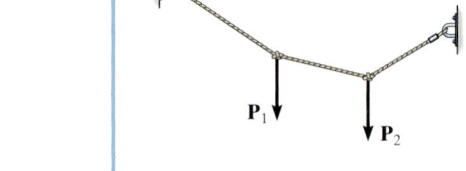

P_1 P_2

If external distributed loads or the weight of the cable are to be considered, then the shape of the cable must be determined by first analyzing the forces on a differential segment of the cable and then integrating this result. The two constants, say C_1 and C_2, resulting from the integration are determined by applying the boundary conditions for the cable.

$$y = \frac{1}{F_H} \int \left(\int w(x)\,dx \right) dx$$

Distributed load

$$x = \int \frac{ds}{\left[1 + \frac{1}{F_H^2} \left(\int w(s)\,ds \right)^2 \right]^{1/2}}$$

Cable weight

REVIEW PROBLEMS

7–125. The beam is supported by a pin at C and a rod AB. Determine the internal normal force, shear force, and moment at point D.

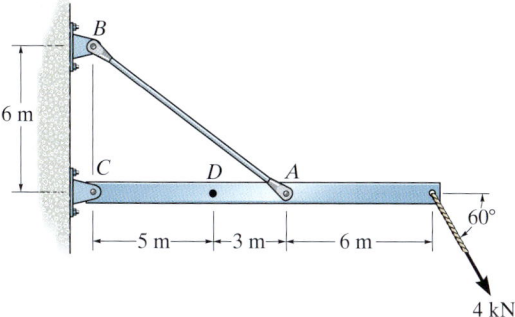

Prob. 7–125

7–126. Draw the shear and moment diagrams for the beam.

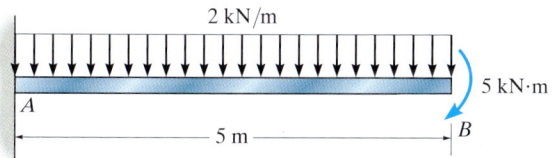

Prob. 7–126

7–127. Determine the distance a between the supports in terms of the beam's length L so that the moment in the *symmetric* beam is zero at the beam's center.

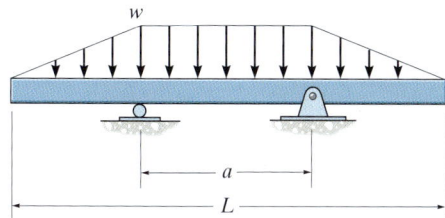

Prob. 7–127

7–128. The stacker crane supports a 1.5-Mg boat with the center of mass at G. Determine the internal normal force, shear force, and moment at point D in the girder. The trolley is free to roll along the girder rail and is located at the position shown. Only vertical reactions occur at A and B.

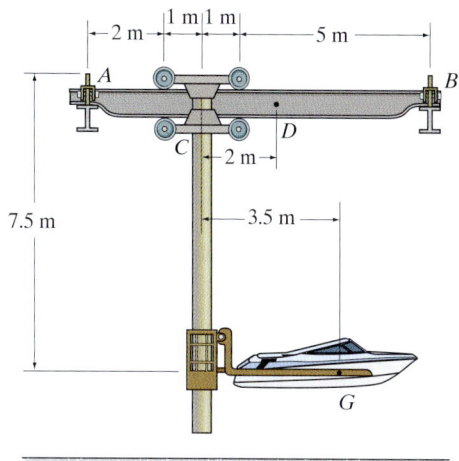

Prob. 7–128

7–129. The yacht is anchored with a chain that has a total length of 40 m and a mass per unit length of 18 kg/m, and the tension in the chain at A is 7 kN. Determine the length of chain l_d which is lying at the bottom of the sea. What is the distance d? Assume that buoyancy effects of the water on the chain are negligible. *Hint:* Establish the origin of the coordinate system at B as shown in order to find the chain length BA.

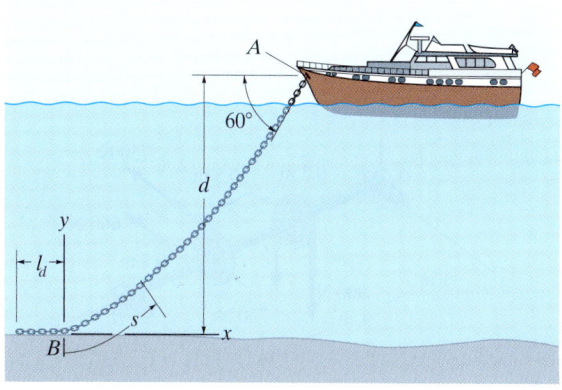

Prob. 7–129

7

7–130. Draw the shear and moment diagrams for the beam ABC.

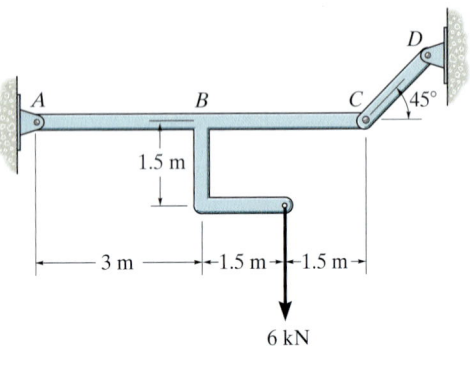

6 kN

Prob. 7–130

***7–132.** A chain is suspended between points at the same elevation and spaced a distance of 60 m apart. If it has a weight per unit length of 0.5 kN/m and the sag is 3 m, determine the maximum tension in the chain.

7–133. Draw the shear and moment diagrams for the beam.

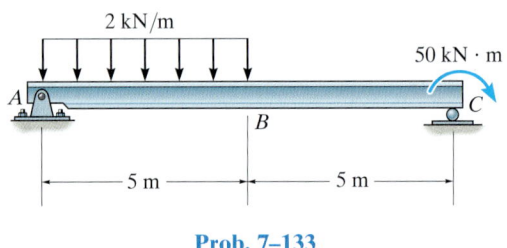

Prob. 7–133

7–131. The shaft is supported by a thrust bearing at A and a journal bearing at B. Determine the x, y, z components of internal loading at point C.

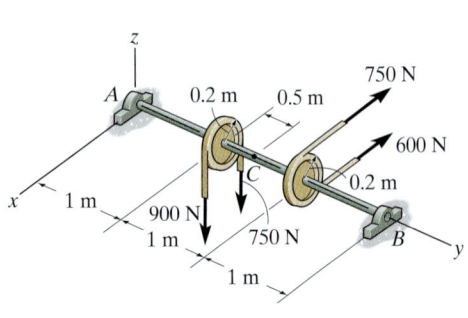

Prob. 7–131

7–134. Determine the normal force, shear force, and moment at points B and C of the beam.

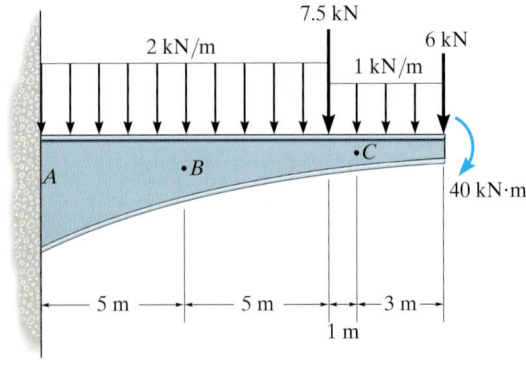

Prob. 7–134

7–135. Draw the shear and moment diagrams for the beam.

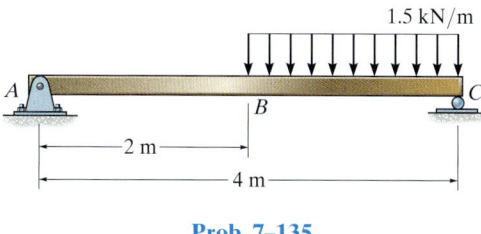

Prob. 7–135

*7–136. If the 45-m-long cable has a mass per unit length of 5 kg/m, determine the equation of the catenary curve of the cable and the maximum tension developed in the cable.

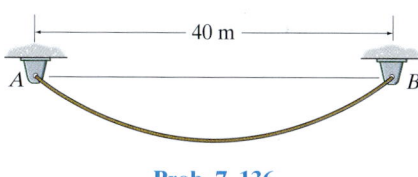

Prob. 7–136

7–137. The traveling crane consists of a 5-m-long beam having a uniform mass per unit length of 20 kg/m. The chain hoist and its supported load exert a force of 8 kN on the beam when $x = 2$ m. Draw the shear and moment diagrams for the beam. The guide wheels at the ends A and B exert only vertical reactions on the beam. Neglect the size of the trolley at C.

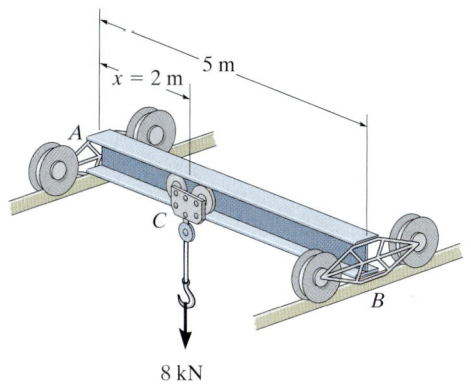

Prob. 7–137

7–138. Draw the shear and moment diagrams for the beam.

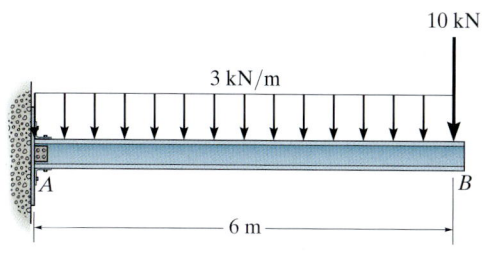

Prob. 7–138

7–139. The suspender bar supports the 300-kg engine. Draw the shear and moment diagrams for the bar.

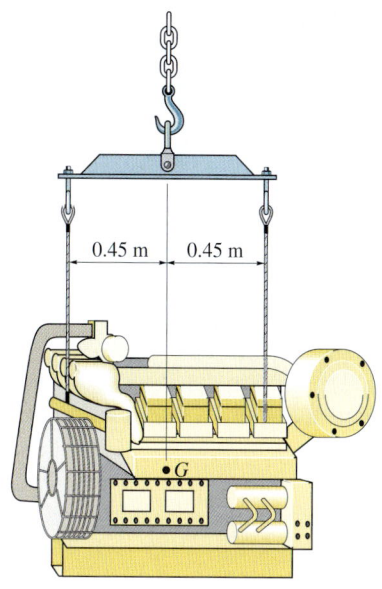

Prob. 7–139

Chapter **8**

The effective design of each brake on this railroad wheel requires that it resist the frictional forces developed between it and the wheel. In this chapter we will study dry friction, and show how to analyze friction forces for various engineering applications.

Friction

CHAPTER OBJECTIVES

■ To introduce the concept of dry friction and show how to analyze the equilibrium of rigid bodies subjected to this force.

■ To present specific applications of frictional force analysis on wedges, screws, belts, and bearings.

■ To investigate the concept of rolling resistance.

8.1   Characteristics of Dry Friction

Friction is a force that resists the movement of two contacting surfaces that slide relative to one another. This force always acts *tangent* to the surface at the points of contact and is directed so as to oppose the possible or existing motion between the surfaces.

In this chapter, we will study the effects of *dry friction*, which is sometimes called *Coulomb friction* since its characteristics were studied extensively by C. A. Coulomb in 1781. Dry friction occurs between the contacting surfaces of bodies when there is no lubricating fluid.*

The heat generated by the abrasive action of friction can be noticed when using this grinder to sharpen a metal blade.

*Another type of friction, called fluid friction, is studied in fluid mechanics.

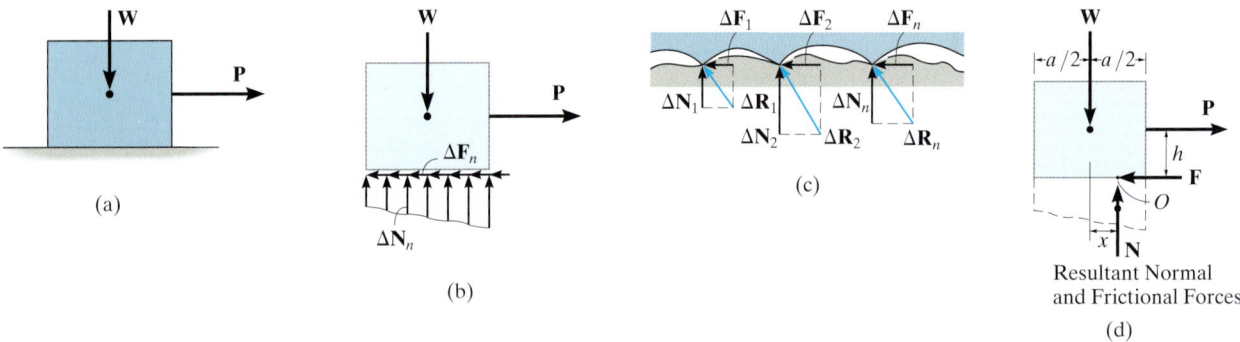

Fig. 8–1

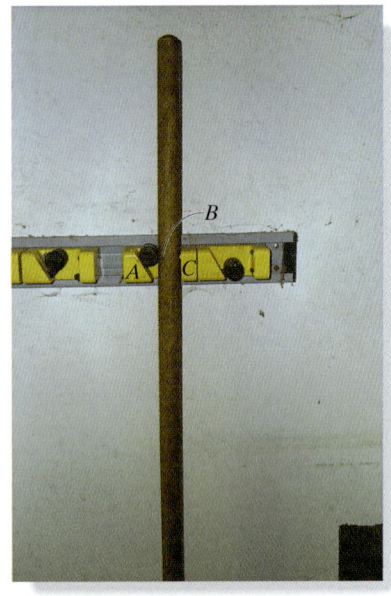

Regardless of the weight of the rake or shovel that is suspended, the device has been designed so that the small roller holds the handle in equilibrium due to frictional forces that develop at the points of contact, A, B, C.

Theory of Dry Friction.

The theory of dry friction can be explained by considering the effects caused by pulling horizontally on a block of uniform weight $\mathbf{W}$ which is resting on a rough horizontal surface that is *nonrigid or deformable,* Fig. 8–1a. The upper portion of the block, however, can be considered rigid. As shown on the free-body diagram of the block, Fig. 8–1b, the floor exerts an uneven *distribution* of both *normal force* $\Delta\mathbf{N}_n$ and *frictional force* $\Delta\mathbf{F}_n$ along the contacting surface. For equilibrium, the normal forces must act *upward* to balance the block's weight $\mathbf{W}$, and the frictional forces act to the left to prevent the applied force $\mathbf{P}$ from moving the block to the right. Close examination of the contacting surfaces between the floor and block reveals how these frictional and normal forces develop, Fig. 8–1c. It can be seen that many microscopic irregularities exist between the two surfaces and, as a result, reactive forces $\Delta\mathbf{R}_n$ are developed at each point of contact.* As shown, each reactive force contributes both a frictional component $\Delta\mathbf{F}_n$ and a normal component $\Delta\mathbf{N}_n$.

Equilibrium.

The effect of the *distributed* normal and frictional loadings is indicated by their *resultants* $\mathbf{N}$ and $\mathbf{F}$ on the free-body diagram, Fig. 8–1d. Notice that $\mathbf{N}$ acts a distance x to the right of the line of action of $\mathbf{W}$, Fig. 8–1d. This location, which coincides with the centroid or geometric center of the normal force distribution in Fig. 8–1b, is necessary in order to balance the "tipping effect" caused by $\mathbf{P}$. For example, if $\mathbf{P}$ is applied at a height h from the surface, Fig. 8–1d, then moment equilibrium about point O is satisfied if $Wx = Ph$ or $x = Ph/W$.

---

*Besides mechanical interactions as explained here, which is referred to as a classical approach, a detailed treatment of the nature of frictional forces must also include the effects of temperature, density, cleanliness, and atomic or molecular attraction between the contacting surfaces. See J. Krim, *Scientific American,* October, 1996.

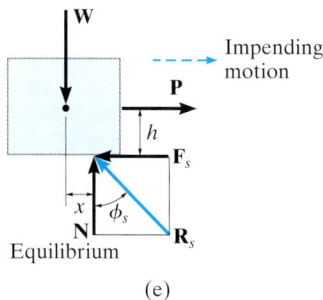

Fig. 8–1 (cont.)

Impending Motion.   In cases where the surfaces of contact are rather "slippery," the frictional force **F** may *not* be great enough to balance **P**, and consequently the block will tend to slip. In other words, as *P* is slowly increased, *F* correspondingly increases until it attains a certain *maximum value F_s*, called the *limiting static frictional force*, Fig. 8–1e. When this value is reached, the block is in *unstable equilibrium* since any further increase in *P* will cause the block to move. Experimentally, it has been determined that this limiting static frictional force F_s is *directly proportional* to the resultant normal force *N*. Expressed mathematically,

$$F_s = \mu_s N \qquad (8\text{–}1)$$

where the constant of proportionality, μ_s (mu "sub" s), is called the *coefficient of static friction*.

Thus, when the block is on the *verge of sliding*, the normal force **N** and frictional force $\mathbf{F}_s$ combine to create a resultant $\mathbf{R}_s$, Fig. 8–1e. The angle ϕ_s (phi "sub" s) that $\mathbf{R}_s$ makes with **N** is called the *angle of static friction*. From the figure,

$$\phi_s = \tan^{-1}\left(\frac{F_s}{N}\right) = \tan^{-1}\left(\frac{\mu_s N}{N}\right) = \tan^{-1}\mu_s$$

Typical values for μ_s are given in Table 8–1. Note that these values can vary since experimental testing was done under variable conditions of roughness and cleanliness of the contacting surfaces. For applications, therefore, it is important that both caution and judgment be exercised when selecting a coefficient of friction for a given set of conditions. When a more accurate calculation of F_s is required, the coefficient of friction should be determined directly by an experiment that involves the two materials to be used.

Table 8–1 Typical Values for μ_s	
Contact Materials	Coefficient of Static Friction (μ_s)
Metal on ice	0.03–0.05
Wood on wood	0.30–0.70
Leather on wood	0.20–0.50
Leather on metal	0.30–0.60
Aluminum on aluminum	1.10–1.70

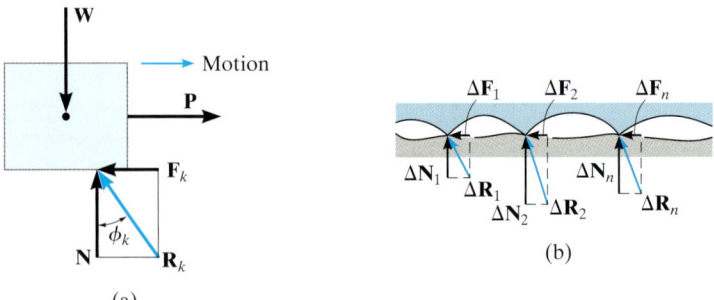

Fig. 8–2

Motion.   If the magnitude of **P** acting on the block is increased so that it becomes slightly greater than F_s, the frictional force at the contacting surface will drop to a smaller value F_k, called the *kinetic frictional force*. The block will begin to slide with increasing speed, Fig. 8–2a. As this occurs, the block will "ride" on top of these peaks at the points of contact, as shown in Fig. 8–2b. The continued breakdown of the surface is the dominant mechanism creating kinetic friction.

Experiments with sliding blocks indicate that the magnitude of the kinetic friction force is directly proportional to the magnitude of the resultant normal force, expressed mathematically as

$$F_k = \mu_k N \qquad (8\text{–}2)$$

Here the constant of proportionality, μ_k, is called the *coefficient of kinetic friction*. Typical values for μ_k are approximately 25 percent *smaller* than those listed in Table 8–1 for μ_s.

As shown in Fig. 8–2a, in this case, the resultant force at the surface of contact, $\mathbf{R}_k$, has a line of action defined by ϕ_k. This angle is referred to as the *angle of kinetic friction*, where

$$\phi_k = \tan^{-1}\left(\frac{F_k}{N}\right) = \tan^{-1}\left(\frac{\mu_k N}{N}\right) = \tan^{-1}\mu_k$$

By comparison, $\phi_s \geq \phi_k$.

The above effects regarding friction can be summarized by referring to the graph in Fig. 8–3, which shows the variation of the frictional force F versus the applied load P. Here the frictional force is categorized in three different ways:

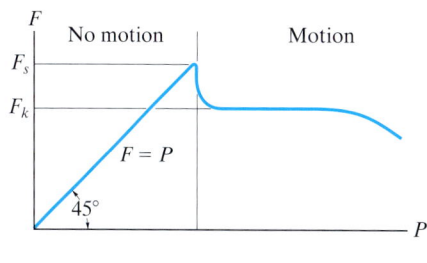

Fig. 8–3

- F is a *static frictional force* if equilibrium is maintained.

- F is a *limiting static frictional force* F_s when it reaches a maximum value needed to maintain equilibrium.

- F is a *kinetic frictional force* F_k when sliding occurs at the contacting surface.

Notice also from the graph that for very large values of P or for high speeds, aerodynamic effects will cause F_k and likewise μ_k to begin to decrease.

Characteristics of Dry Friction.   As a result of *experiments* that pertain to the foregoing discussion, we can state the following rules which apply to bodies subjected to dry friction.

- The frictional force acts *tangent* to the contacting surfaces in a direction *opposed* to the *motion* or tendency for motion of one surface relative to another.

- The maximum static frictional force F_s that can be developed is independent of the area of contact, provided the normal pressure is not very low nor great enough to severely deform or crush the contacting surfaces of the bodies.

- The maximum static frictional force is generally greater than the kinetic frictional force for any two surfaces of contact. However, if one of the bodies is moving with a *very low velocity* over the surface of another, F_k becomes approximately equal to F_s, i.e., $\mu_s \approx \mu_k$.

- When *slipping* at the surface of contact is *about to occur*, the maximum static frictional force is proportional to the normal force, such that $F_s = \mu_s N$.

- When *slipping* at the surface of contact is *occurring*, the kinetic frictional force is proportional to the normal force, such that $F_k = \mu_k N$.

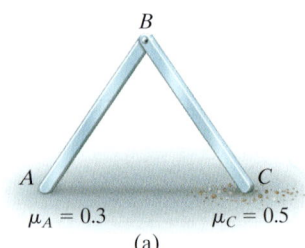

$\mu_A = 0.3$ $\mu_C = 0.5$

(a)

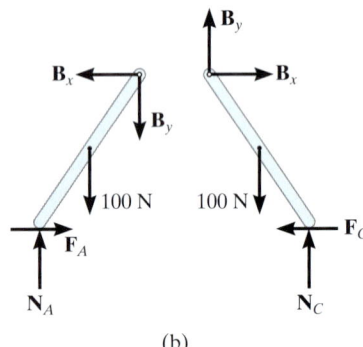

(b)

Fig. 8–4

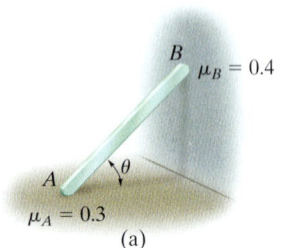

$\mu_B = 0.4$

$\mu_A = 0.3$

(a)

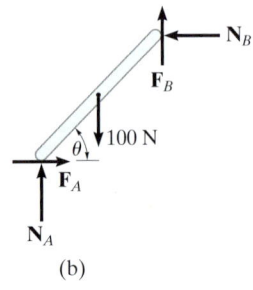

(b)

Fig. 8–5

8.2   Problems Involving Dry Friction

If a rigid body is in equilibrium when it is subjected to a system of forces that includes the effect of friction, the force system must satisfy not only the equations of equilibrium but *also* the laws that govern the frictional forces.

Types of Friction Problems.
In general, there are three types of mechanics problems involving dry friction. They can easily be classified once free-body diagrams are drawn and the total number of unknowns are identified and compared with the total number of available equilibrium equations.

No Apparent Impending Motion.
Problems in this category are strictly equilibrium problems, which require the number of unknowns to be *equal* to the number of available equilibrium equations. Once the frictional forces are determined from the solution, however, their numerical values must be checked to be sure they satisfy the inequality $F \le \mu_s N$; otherwise, slipping will occur and the body will not remain in equilibrium. A problem of this type is shown in Fig. 8–4a. Here we must determine the frictional forces at A and C to check if the equilibrium position of the two-member frame can be maintained. If the bars are uniform and have known weights of 100 N each, then the free-body diagrams are as shown in Fig. 8–4b. There are six unknown force components which can be determined *strictly* from the six equilibrium equations (three for each member). Once F_A, N_A, F_C, and N_C are determined, then the bars will remain in equilibrium provided $F_A \le 0.3N_A$ and $F_C \le 0.5N_C$ are satisfied.

Impending Motion at All Points of Contact.
In this case the total number of unknowns will *equal* the total number of available equilibrium equations *plus* the total number of available frictional equations, $F = \mu N$. When *motion is impending* at the points of contact, then $F_s = \mu_s N$; whereas if the body is *slipping*, then $F_k = \mu_k N$. For example, consider the problem of finding the smallest angle θ at which the 100-N bar in Fig. 8–5a can be placed against the wall without slipping. The free-body diagram is shown in Fig. 8–5b. Here the *five* unknowns are determined from the *three* equilibrium equations and *two* static frictional equations which apply at *both* points of contact, so that $F_A = 0.3N_A$ and $F_B = 0.4N_B$.

Impending Motion at Some Points of Contact.   Here the number of unknowns will be *less* than the number of available equilibrium equations plus the number of available frictional equations or conditional equations for tipping. As a result, several possibilities for motion or impending motion will exist and the problem will involve a determination of the kind of motion which actually occurs. For example, consider the two-member frame in Fig. 8–6a. In this problem we wish to determine the horizontal force P needed to cause movement. If each member has a weight of 100 N, then the free-body diagrams are as shown in Fig. 8–6b. There are *seven* unknowns. For a unique solution we must satisfy the *six* equilibrium equations (three for each member) and only *one* of two possible static frictional equations. This means that as P increases it will either cause slipping at A and no slipping at C, so that $F_A = 0.3 N_A$ and $F_C \le 0.5 N_C$; or slipping occurs at C and no slipping at A, in which case $F_C = 0.5 N_C$ and $F_A \le 0.3 N_A$. The actual situation can be determined by calculating P for each case and then choosing the case for which P is *smaller*. If in both cases the *same value* for P is calculated, which would be highly improbable, then slipping at both points occurs simultaneously; i.e., the *seven unknowns* would satisfy *eight equations*.

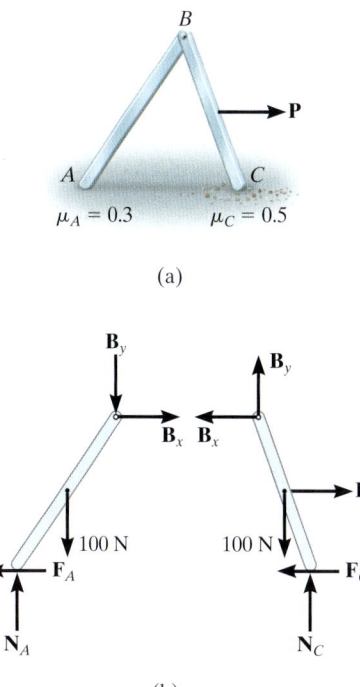

(a)

(b)

Fig. 8–6

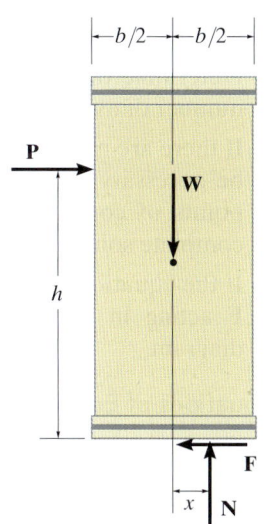

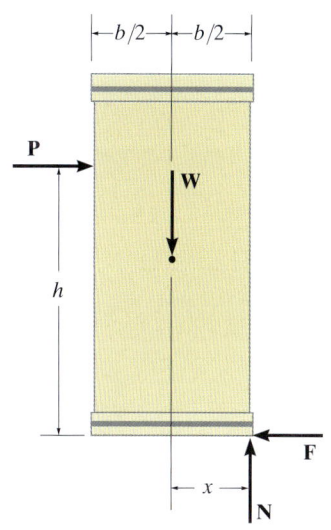

Consider pushing on the uniform crate that has a weight W and sits on the rough surface. As shown on the first free-body diagram, if the magnitude of **P** is small, the crate will remain in equilibrium. As P increases the crate will either be on the verge of slipping on the surface ($F = \mu_s N$), or if the surface is very rough (large μ_s) then the resultant normal force will shift to the corner, $x = b/2$, as shown on the second free-body diagram. At this point the crate will begin to tip over. The crate also has a greater chance of tipping if **P** is applied at a greater height h above the surface, or if its width b is smaller.

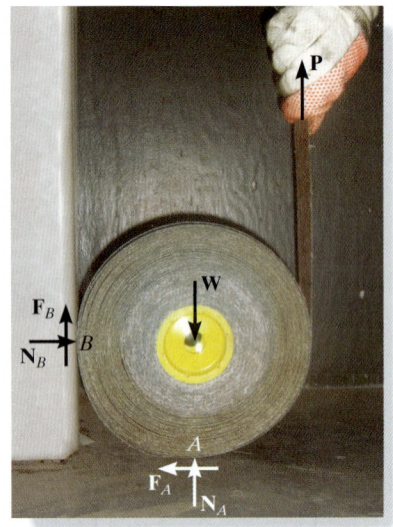

The applied vertical force **P** on this roll must be large enough to overcome the resistance of friction at the contacting surfaces A and B in order to cause rotation.

Equilibrium Versus Frictional Equations.

Whenever we solve a problem such as the one in Fig. 8–4, where the friction force F is to be an "equilibrium force" and satisfies the inequality $F < \mu_s N$, then we can assume the sense of direction of F on the free-body diagram. The correct sense is made known *after* solving the equations of equilibrium for F. If F is a negative scalar the sense of **F** is the reverse of that which was assumed. This convenience of *assuming* the sense of **F** is possible because the equilibrium equations equate to zero the *components of vectors* acting in the *same direction*. However, in cases where the frictional equation $F = \mu N$ is used in the solution of a problem, as in the case shown in Fig. 8–5, then the convenience of *assuming* the sense of **F** is *lost*, since the frictional equation relates only the *magnitudes* of two *perpendicular* vectors. Consequently, **F** *must always* be shown acting with its *correct sense* on the free-body diagram, *whenever* the frictional equation is used for the solution of a problem.

Procedure for Analysis

Equilibrium problems involving dry friction can be solved using the following procedure.

Free-Body Diagrams.

- Draw the necessary free-body diagrams, and unless it is stated in the problem that impending motion or slipping occurs, *always* show the frictional forces as unknowns (i.e., *do not assume $F = \mu N$*).

- Determine the number of unknowns and compare this with the number of available equilibrium equations.

- If there are more unknowns than equations of equilibrium, it will be necessary to apply the frictional equation at some, if not all, points of contact to obtain the extra equations needed for a complete solution.

- If the equation $F = \mu N$ is to be used, it will be necessary to show **F** acting in the correct sense of direction on the free-body diagram.

Equations of Equilibrium and Friction.

- Apply the equations of equilibrium and the necessary frictional equations (or conditional equations if tipping is possible) and solve for the unknowns.

- If the problem involves a three-dimensional force system such that it becomes difficult to obtain the force components or the necessary moment arms, apply the equations of equilibrium using Cartesian vectors.

EXAMPLE 8.1

The uniform crate shown in Fig. 8–7a has a mass of 20 kg. If a force $P = 80$ N is applied to the crate, determine if it remains in equilibrium. The coefficient of static friction is $\mu_s = 0.3$.

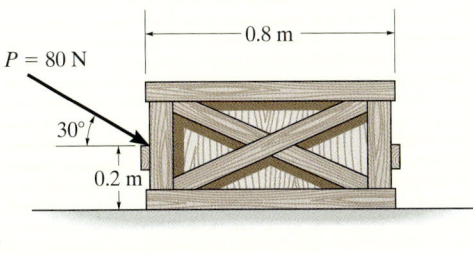

(a)

Fig. 8–7

SOLUTION

Free-Body Diagram.   As shown in Fig. 8–7b, the *resultant* normal force $\mathbf{N}_C$ must act a distance x from the crate's center line in order to counteract the tipping effect caused by **P**. There are *three unknowns*, F, N_C, and x, which can be determined strictly from the *three* equations of equilibrium.

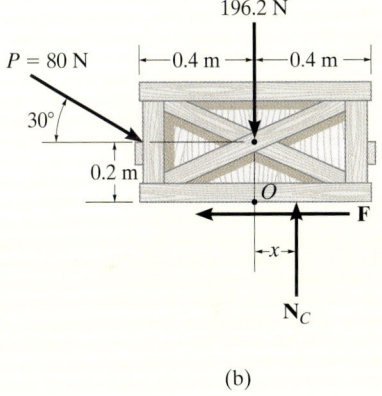

(b)

Equations of Equilibrium.

$$\xrightarrow{+} \Sigma F_x = 0; \qquad\qquad 80 \cos 30° \text{ N} - F = 0$$

$$+\uparrow \Sigma F_y = 0; \qquad -80 \sin 30° \text{ N} + N_C - 196.2 \text{ N} = 0$$

$$\zeta + \Sigma M_O = 0; \quad 80 \sin 30° \text{ N}(0.4 \text{ m}) - 80 \cos 30° \text{ N}(0.2 \text{ m}) + N_C(x) = 0$$

Solving,

$$F = 69.3 \text{ N}$$
$$N_C = 236.2 \text{ N}$$
$$x = -0.00908 \text{ m} = -9.08 \text{ mm}$$

Since x is negative it indicates the *resultant* normal force acts (slightly) to the *left* of the crate's center line. No tipping will occur since $x < 0.4$ m. Also, the *maximum* frictional force which can be developed at the surface of contact is $F_{max} = \mu_s N_C = 0.3(236.2 \text{ N}) = 70.9$ N. Since $F = 69.3$ N < 70.9 N, the crate will *not slip*, although it is very close to doing so.

EXAMPLE | 8.2

(a)

It is observed that when the bed of the dump truck is raised to an angle of $\theta = 25°$ the vending machines will begin to slide off the bed, Fig. 8–8a. Determine the static coefficient of friction between a vending machine and the surface of the truckbed.

SOLUTION

An idealized model of a vending machine resting on the truckbed is shown in Fig. 8–8b. The dimensions have been measured and the center of gravity has been located. We will assume that the vending machine weighs W.

Free-Body Diagram. As shown in Fig. 8–8c, the dimension x is used to locate the position of the resultant normal force **N**. There are four unknowns, $N, F, \mu_s,$ and x.

Equations of Equilibrium.

$$+\searrow\Sigma F_x = 0; \qquad\qquad\qquad\qquad W\sin 25° - F = 0 \qquad (1)$$

$$+\nearrow\Sigma F_y = 0; \qquad\qquad\qquad\qquad N - W\cos 25° = 0 \qquad (2)$$

$$\zeta+\Sigma M_O = 0; \quad -W\sin 25°(0.75\ \text{m}) + W\cos 25°(x) = 0 \qquad (3)$$

Since slipping impends at $\theta = 25°$, using Eqs. 1 and 2, we have

$$F_s = \mu_s N; \qquad W\sin 25° = \mu_s(W\cos 25°)$$

$$\mu_s = \tan 25° = 0.466 \qquad\qquad\qquad Ans.$$

The angle of $\theta = 25°$ is referred to as the *angle of repose*, and by comparison, it is equal to the angle of static friction, $\theta = \phi_s$. Notice from the calculation that θ is independent of the weight of the vending machine, and so knowing θ provides a convenient method for determining the coefficient of static friction.

NOTE: From Eq. 3, we find $x = 0.350\ \text{m}$. Since $0.350\ \text{m} < 0.45\ \text{m}$, indeed the vending machine will slip before it can tip as observed in Fig. 8–8a.

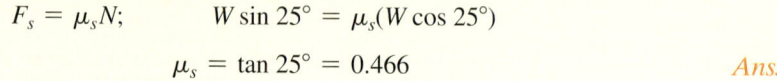

(b)

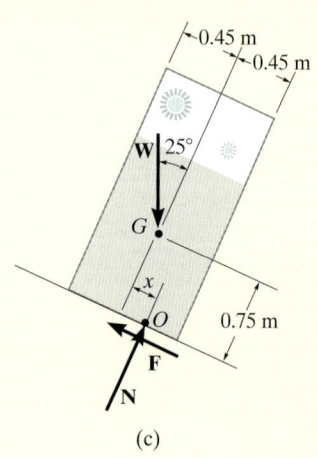

(c)

Fig. 8–8

8

EXAMPLE | 8.3

The uniform 10-kg ladder in Fig. 8–9a rests against the smooth wall at
B, and the end A rests on the rough horizontal plane for which the
coefficient of static friction is $\mu_s = 0.3$. Determine the angle of
inclination θ of the ladder and the normal reaction at B if the ladder is
on the verge of slipping.

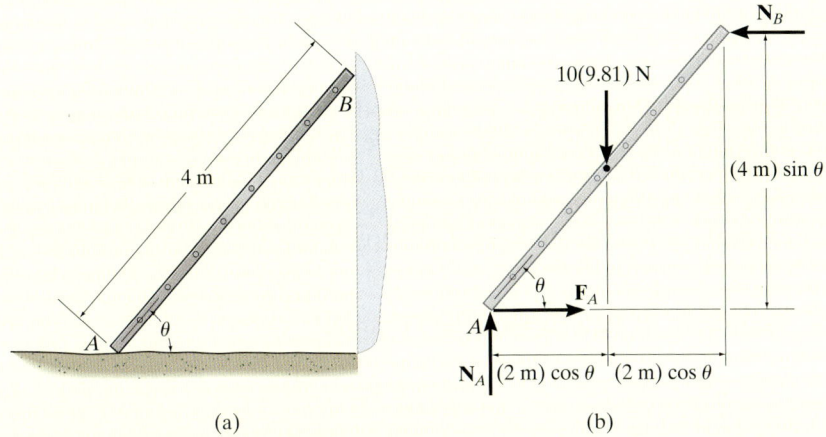

(a)   (b)

Fig. 8–9

SOLUTION

Free-Body Diagram.   As shown on the free-body diagram, Fig. 8–9b,
the frictional force $\mathbf{F}_A$ must act to the right since impending motion at A
is to the left.

Equations of Equilibrium and Friction.   Since the ladder is on the
verge of slipping, then $F_A = \mu_s N_A = 0.3 N_A$. By inspection, N_A can be
obtained directly.

$$+\uparrow \Sigma F_y = 0; \qquad\qquad N_A - 10(9.81)\text{ N} = 0 \qquad\qquad N_A = 98.1\text{ N}$$

Using this result, $F_A = 0.3(98.1\text{ N}) = 29.43\text{ N}$. Now N_B can be found.

$$\xrightarrow{+} \Sigma F_x = 0; \qquad\qquad 29.43\text{ N} - N_B = 0$$

$$N_B = 29.43\text{ N} = 29.4\text{ N} \qquad\qquad \textit{Ans.}$$

Finally, the angle θ can be determined by summing moments about
point A.

$$\zeta + \Sigma M_A = 0; \qquad (29.43\text{ N})(4\text{ m})\sin\theta - [10(9.81)\text{ N}](2\text{ m})\cos\theta = 0$$

$$\frac{\sin\theta}{\cos\theta} = \tan\theta = 1.6667$$

$$\theta = 59.04° = 59.0° \qquad\qquad \textit{Ans.}$$

EXAMPLE 8.4

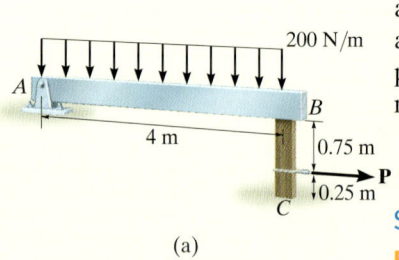

(a)

Beam AB is subjected to a uniform load of 200 N/m and is supported at B by post BC, Fig. 8–10a. If the coefficients of static friction at B and C are $\mu_B = 0.2$ and $\mu_C = 0.5$, determine the force **P** needed to pull the post out from under the beam. Neglect the weight of the members and the thickness of the beam.

SOLUTION

Free-Body Diagrams. The free-body diagram of the beam is shown in Fig. 8–10b. Applying $\Sigma M_A = 0$, we obtain $N_B = 400$ N. This result is shown on the free-body diagram of the post, Fig. 8–10c. Referring to this member, the *four* unknowns F_B, P, F_C, and N_C are determined from the *three* equations of equilibrium and *one* frictional equation applied either at B or C.

Equations of Equilibrium and Friction.

$$\xrightarrow{+}\ \Sigma F_x = 0; \qquad\qquad P - F_B - F_C = 0 \qquad\qquad (1)$$
$$+\uparrow \Sigma F_y = 0; \qquad\qquad N_C - 400 \text{ N} = 0 \qquad\qquad (2)$$
$$\zeta + \Sigma M_C = 0; \qquad -P(0.25 \text{ m}) + F_B(1 \text{ m}) = 0 \qquad (3)$$

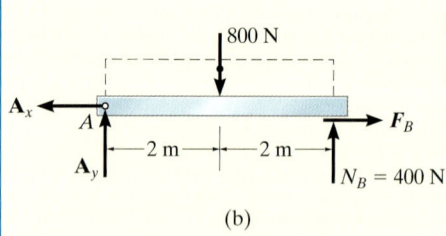

(b)

(Post Slips at B and Rotates about C.)   This requires $F_C \le \mu_C N_C$ and

$$F_B = \mu_B N_B; \qquad\qquad F_B = 0.2(400 \text{ N}) = 80 \text{ N}$$

Using this result and solving Eqs. 1 through 3, we obtain

$$P = 320 \text{ N}$$
$$F_C = 240 \text{ N}$$
$$N_C = 400 \text{ N}$$

Since $F_C = 240$ N $> \mu_C N_C = 0.5(400$ N$) = 200$ N,   slipping   at C occurs. Thus the other case of movement must be investigated.

(Post Slips at C and Rotates about B.)   Here $F_B \le \mu_B N_B$ and

$$F_C = \mu_C N_C; \qquad\qquad F_C = 0.5 N_C \qquad\qquad (4)$$

Solving Eqs. 1 through 4 yields

$$P = 267 \text{ N} \qquad\qquad Ans.$$
$$N_C = 400 \text{ N}$$
$$F_C = 200 \text{ N}$$
$$F_B = 66.7 \text{ N}$$

(c)

Fig. 8–10

Obviously, this case occurs first since it requires a *smaller* value for P.

EXAMPLE 8.5

Blocks A and B have a mass of 3 kg and 9 kg, respectively, and are connected to the weightless links shown in Fig. 8–11a. Determine the largest vertical force $\mathbf{P}$ that can be applied at the pin C without causing any movement. The coefficient of static friction between the blocks and the contacting surfaces is $\mu_s = 0.3$.

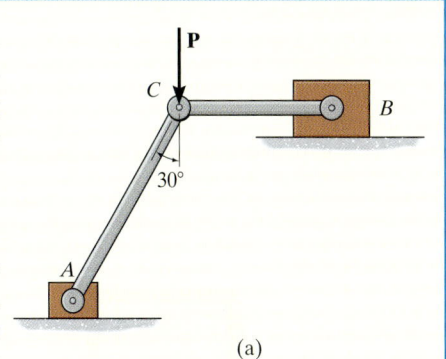

(a)

SOLUTION

Free-Body Diagram. The links are two-force members and so the free-body diagrams of pin C and blocks A and B are shown in Fig. 8–11b. Since the horizontal component of $\mathbf{F}_{AC}$ tends to move block A to the left, $\mathbf{F}_A$ must act to the right. Similarly, $\mathbf{F}_B$ must act to the left to oppose the tendency of motion of block B to the right, caused by $\mathbf{F}_{BC}$. There are seven unknowns and six available force equilibrium equations, two for the pin and two for each block, so that *only one* frictional equation is needed.

Equations of Equilibrium and Friction. The force in links AC and BC can be related to P by considering the equilibrium of pin C.

$$+\uparrow \Sigma F_y = 0; \qquad F_{AC}\cos 30° - P = 0; \qquad F_{AC} = 1.155P$$
$$\xrightarrow{+} \Sigma F_x = 0; \qquad 1.155P \sin 30° - F_{BC} = 0; \qquad F_{BC} = 0.5774P$$

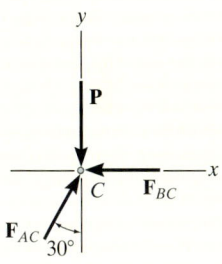

Using the result for F_{AC}, for block A,

$$\xrightarrow{+} \Sigma F_x = 0; \quad F_A - 1.155P \sin 30° = 0; \quad F_A = 0.5774P \qquad (1)$$
$$+\uparrow \Sigma F_y = 0; \quad N_A - 1.155P \cos 30° - 3(9.81\text{ N}) = 0;$$
$$N_A = P + 29.43\text{ N} \qquad (2)$$

Using the result for F_{BC}, for block B,

$$\xrightarrow{+} \Sigma F_x = 0; \quad (0.5774P) - F_B = 0; \qquad F_B = 0.5774P \qquad (3)$$
$$+\uparrow \Sigma F_y = 0; \quad N_B - 9(9.81)\text{ N} = 0; \qquad N_B = 88.29\text{ N}$$

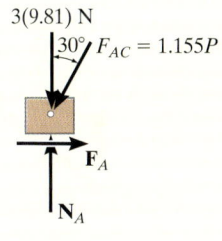

Movement of the system may be caused by the initial slipping of *either* block A or block B. If we assume that block A slips first, then

$$F_A = \mu_s N_A = 0.3 N_A \qquad (4)$$

Substituting Eqs. 1 and 2 into Eq. 4,

$$0.5774P = 0.3(P + 29.43)$$
$$P = 31.8\text{ N} \qquad \qquad \textit{Ans.}$$

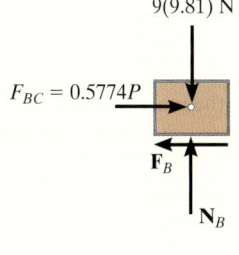

Substituting this result into Eq. 3, we obtain $F_B = 18.4$ N. Since the maximum static frictional force at B is $(F_B)_{max} = \mu_s N_B = 0.3(88.29\text{ N}) = 26.5\text{ N} > F_B$, block B will not slip. Thus, the above assumption is correct. Notice that if the inequality were not satisfied, we would have to assume slipping of block B and then solve for P.

(b)

Fig. 8–11

8

FUNDAMENTAL PROBLEMS

All problem solutions must include FBDs.

F8–1. If $P = 200$ N, determine the friction developed between the 50-kg crate and the ground. The coefficient of static friction between the crate and the ground is $\mu_s = 0.3$.

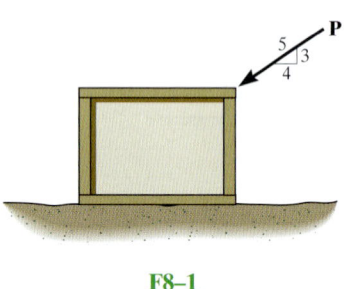

F8–1

F8–2. Determine the minimum force P to prevent the 30-kg rod AB from sliding. The contact surface at B is smooth, whereas the coefficient of static friction between the rod and the wall at A is $\mu_s = 0.2$.

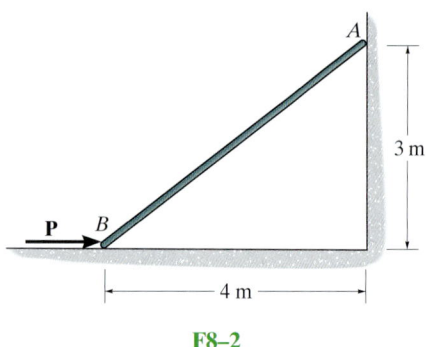

F8–2

F8–3. Determine the maximum force P that can be applied without causing the two 50-kg crates to move. The coefficient of static friction between each crate and the ground is $\mu_s = 0.25$.

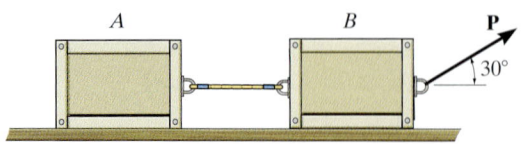

F8–3

F8–4. If the coefficient of static friction at contact points A and B is $\mu_s = 0.3$, determine the maximum force P that can be applied without causing the 100-kg spool to move.

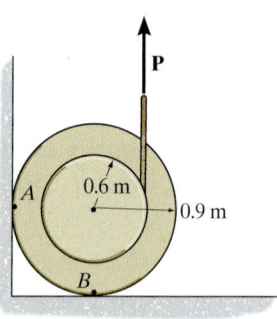

F8–4

F8–5. Determine the maximum force P that can be applied without causing movement of the 100-kg crate that has a center of gravity at G. The coefficient of static friction at the floor is $\mu_s = 0.4$.

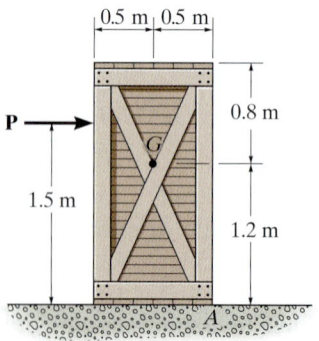

F8–5

F8–6. Determine the minimum coefficient of static friction between the uniform 50-kg spool and the wall so that the spool does not slip.

F8–8. If the coefficient of static friction at all contacting surfaces is μ_s, determine the inclination θ at which the identical blocks, each of weight W, begin to slide.

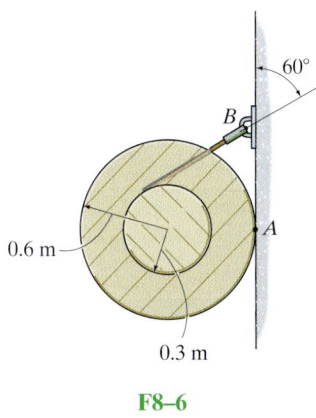

F8–6

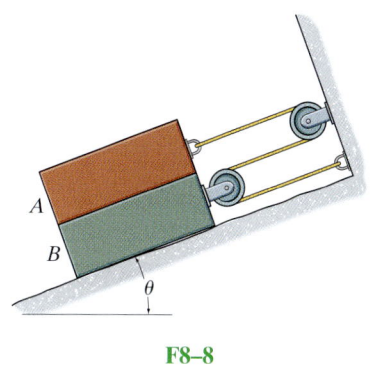

F8–8

F8–7. Blocks A, B, and C have weights of 50 N, 25 N, and 15 N, respectively. Determine the smallest horizontal force P that will cause impending motion. The coefficient of static friction between A and B is $\mu_s = 0.3$, between B and C, $\mu_s' = 0.4$, and between block C and the ground, $\mu_s'' = 0.35$.

F8–9. Blocks A and B have a mass of 7 kg and 10 kg, respectively. Using the coefficients of static friction indicated, determine the largest force P which can be applied to the cord without causing motion. There are pulleys at C and D.

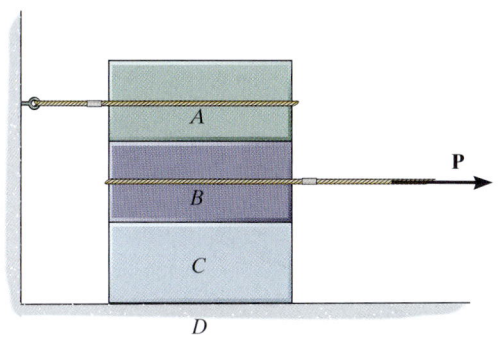

F8–7

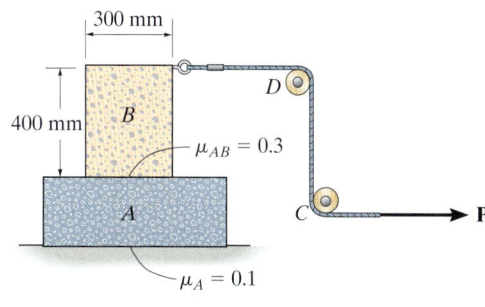

F8–9

8

PROBLEMS

All problem solutions must include FBDs.

8–1. The mine car and its contents have a total mass of 6 Mg and a center of gravity at G. If the coefficient of static friction between the wheels and the tracks is $\mu_s = 0.4$ when the wheels are locked, find the normal force acting on the front wheels at B and the rear wheels at A when the brakes at both A and B are locked. Does the car move?

8–3. If the coefficient of static friction at A is $\mu_s = 0.4$ and the collar at B is smooth so it only exerts a horizontal force on the pipe, determine the minimum distance x so that the bracket can support the cylinder of any mass without slipping. Neglect the mass of the bracket.

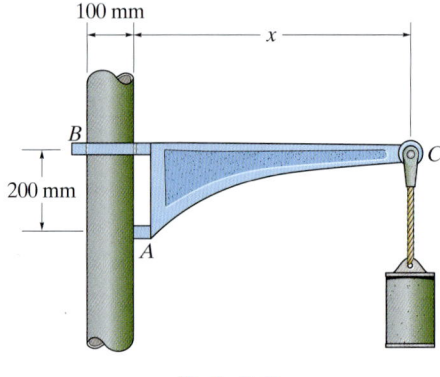

Prob. 8–3

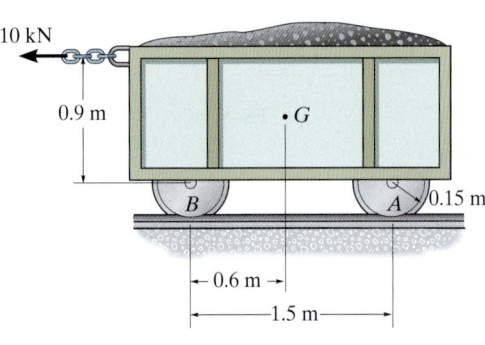

Prob. 8–1

8–2. Determine the maximum force P the connection can support so that no slipping occurs between the plates. There are four bolts used for the connection and each is tightened so that it is subjected to a tension of 4 kN. The coefficient of static friction between the plates is $\mu_s = 0.4$.

***8–4.** If the coefficient of static friction at contacting surface between blocks A and B is μs, and that between block B and bottom is $2\mu_s$, determine the inclination θ at which the identical blocks, each of weight W, begin to slide.

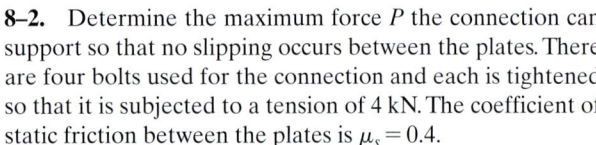

Prob. 8–2

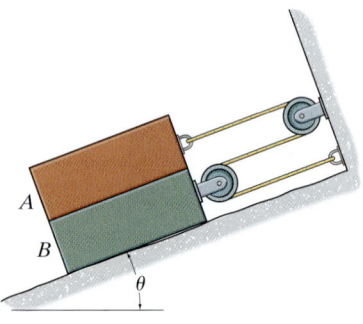

Prob. 8–4

8–5. The coefficients of static and kinetic friction between the drum and brake bar are $\mu_s = 0.4$ and $\mu_k = 0.3$, respectively. If $M = 50$ N·m and $P = 85$ N determine the horizontal and vertical components of reaction at the pin O. Neglect the weight and thickness of the brake. The drum has a mass of 25 kg.

8–7. The block brake consists of a pin-connected lever and friction block at B. The coefficient of static friction between the wheel and the lever is $\mu_s = 0.3$, and a torque of 5 N·m is applied to the wheel. Determine if the brake can hold the wheel stationary when the force applied to the lever is (a) $P = 30$ N, (b) $P = 70$ N.

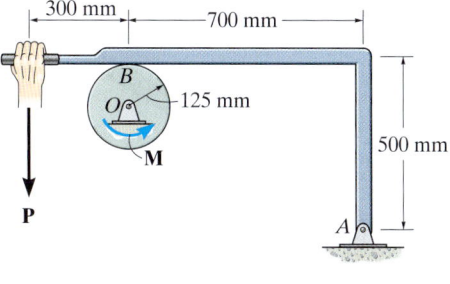

Prob. 8–5

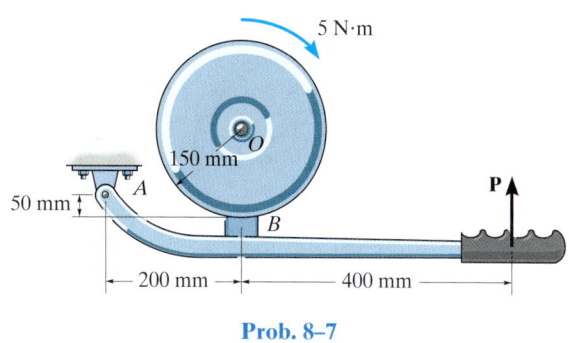

Prob. 8–7

8–6. The coefficient of static friction between the drum and brake bar is $\mu_s = 0.4$. If the moment $M = 35$ N·m, determine the smallest force P that needs to be applied to the brake bar in order to prevent the drum from rotating. Also determine the corresponding horizontal and vertical components of reaction at pin O. Neglect the weight and thickness of the brake bar. The drum has a mass of 25 kg.

*8–8. The block brake consists of a pin-connected lever and friction block at B. The coefficient of static friction between the wheel and the lever is $\mu_s = 0.3$, and a torque of 5 N·m is applied to the wheel. Determine if the brake can hold the wheel stationary when the force applied to the lever is (a) $P = 30$ N, (b) $P = 70$ N.

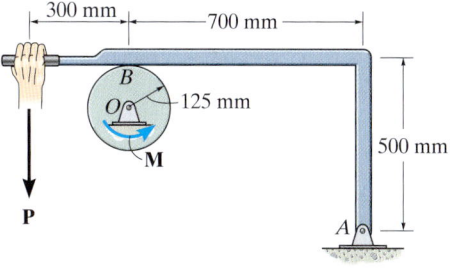

Prob. 8–6

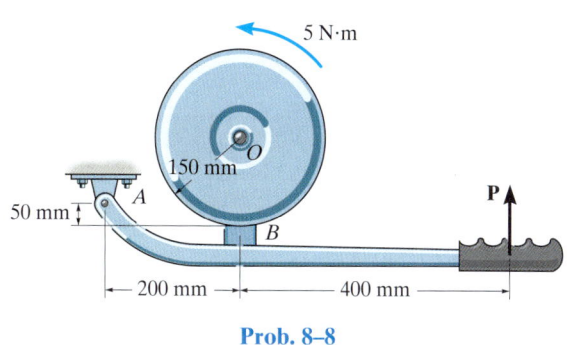

Prob. 8–8

8–9.   The uniform hoop of weight W is suspended from the peg at A and a horizontal force P is slowly applied at B. If the hoop begins to slip at A when $\theta = 30°$, determine the coefficient of static friction between the hoop and the peg.

8–10.   The uniform hoop of weight W is suspended from the peg at A and a horizontal force P is slowly applied at B. If the coefficient of static friction between the hoop and peg is $\mu_s = 0.2$, determine if it is possible for the angle $\theta = 30°$ before the hoop begins to slip.

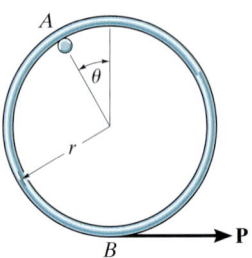

Probs. 8–9/10

8–11.   The fork lift has a weight of 12 kN and a center of gravity at G. If the rear wheels are powered, whereas the front wheels are free to roll, determine the maximum number of 150-kg crates the fork lift can push forward. The coefficient of static friction between the wheels and the ground is $\mu_s = 0.4$, and between each crate and the ground $\mu_s' = 0.35$.

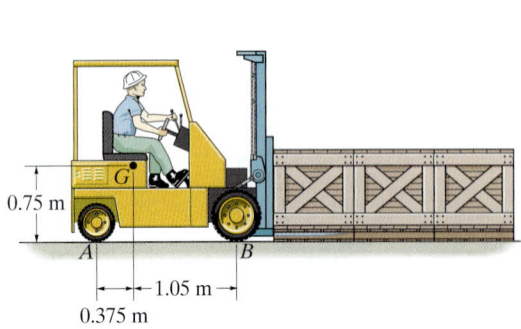

Prob. 8–11

*__*8–12.__*   If a torque of $M = 300$ N·m is applied to the flywheel, determine the force that must be developed in the hydraulic cylinder CD to prevent the flywheel from rotating. The coefficient of static friction between the friction pad at B and the flywheel is $\mu_s = 0.4$.

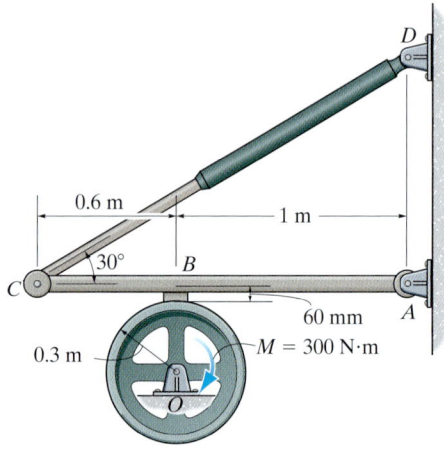

Prob. 8–12

8–13.   The cam is subjected to a couple moment of 5 N·m Determine the minimum force P that should be applied to the follower in order to hold the cam in the position shown. the coefficient of static friction between the cam and the follower is $\mu = 0.4$. The guide at A is smooth.

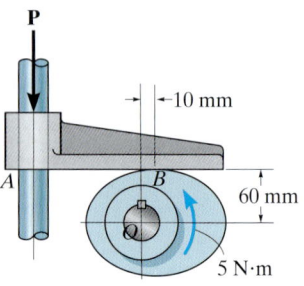

Prob. 8–13

8–14. A 35-kg disk rests on an inclined surface for which $\mu_s = 0.2$. Determine the maximum vertical force **P** that may be applied to link AB without causing the disk to slip at C.

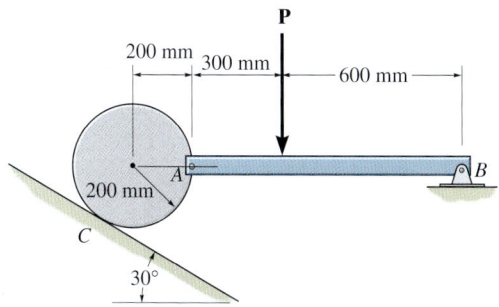

Prob. 8–14

8–15. The car has a mass of 1.6 Mg and center of mass at G. If the coefficient of static friction between the shoulder of the road and the tires is $\mu_s = 0.4$, determine the greatest slope θ the shoulder can have without causing the car to slip or tip over if the car travels along the shoulder at constant velocity.

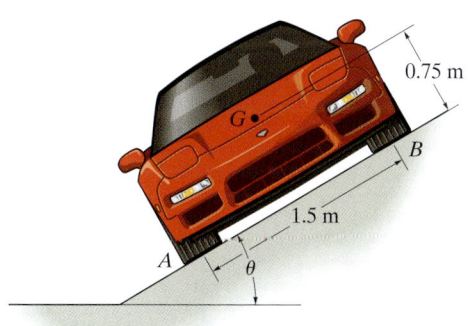

Prob. 8–15

*****8–16.** If the coefficient of static friction between the collars A and B and the rod is $\mu_s = 0.6$, determine the maximum angle θ for the system to remain in equilibrium, regardless of the weight of cylinder D. Links AC and BC have negligible weight and are connected together at C by a pin.

8–17. If $\theta = 15°$, determine the minimum coefficient of static friction between the collars A and B and the rod required for the system to remain in equilibrium, regardless of the weight of cylinder D. Links AC and BC have negligible weight and are connected together at C by a pin.

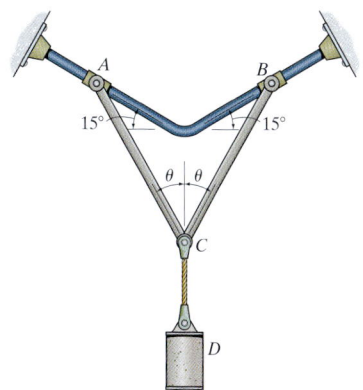

Probs. 8–16/17

8–18. The 5-kg cylinder is suspended from two equal-length cords. The end of each cord is attached to a ring of negligible mass that passes along a horizontal shaft. If the rings can be separated by the greatest distance $d = 400$ mm and still support the cylinder, determine the coefficient of static friction between each ring and the shaft.

8–19. The 5-kg cylinder is suspended from two equal-length cords. The end of each cord is attached to a ring of negligible mass that passes along a horizontal shaft. If the coefficient of static friction between each ring and the shaft is $\mu_s = 0.5$, determine the greatest distance d by which the rings can be separated and still support the cylinder.

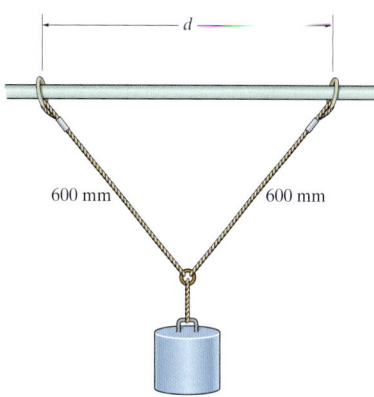

Probs. 8–18/19

*8–20. The pipe is hoisted using the tongs. If the coefficient of static friction at A and B is μ_s, determine the smallest dimension b so that any pipe of inner diameter d can be lifted.

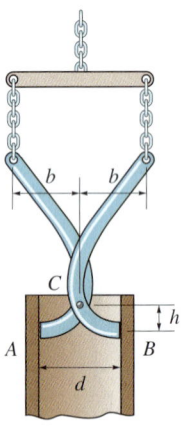

Prob. 8–20

8–21. The uniform pole has a weight W and length L. Its end B is tied to a supporting cord, and end A is placed against the wall, for which the coefficient of static friction is μ_s. Determine the largest angle θ at which the pole can be placed without slipping.

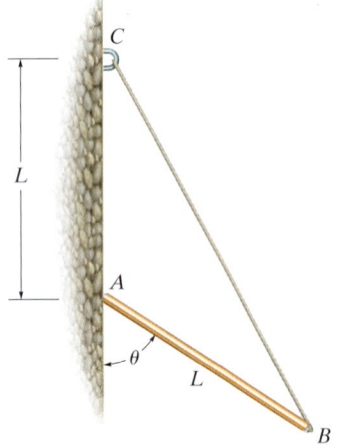

Prob. 8–21

8–22. If the clamping force is $F = 200$ N and each board has a mass of 2 kg, determine the maximum number of boards the clamp can support. The coefficient of static friction between the boards is $\mu_s = 0.3$, and the coefficient of static friction between the boards and the clamp is $\mu_s' = 0.45$.

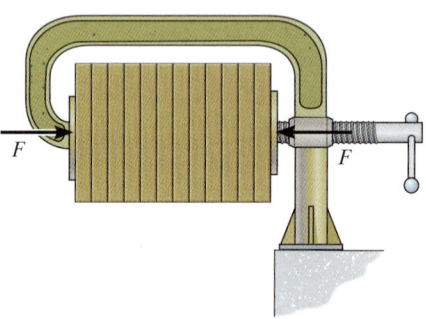

Prob. 8–22

8–23. A 35-kg disk rests on an inclined surface for which $\mu_s = 0.3$. Determine the maximum vertical force $\mathbf{P}$ that may be applied to bar AB without causing the disk to slip at C. Neglect the mass of the bar.

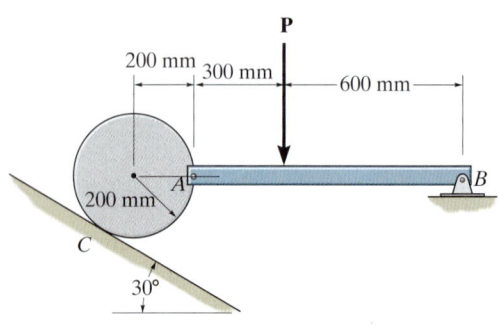

Prob. 8–23

*8–24.   The coefficient of static friction between the shoes at A and B of the tongs and the pallet is $\mu_s' = 0.5$, and between the pallet and the floor $\mu_s = 0.4$. If a horizontal towing force of $P = 300$ N is applied to the tongs, determine the largest mass that can be towed.

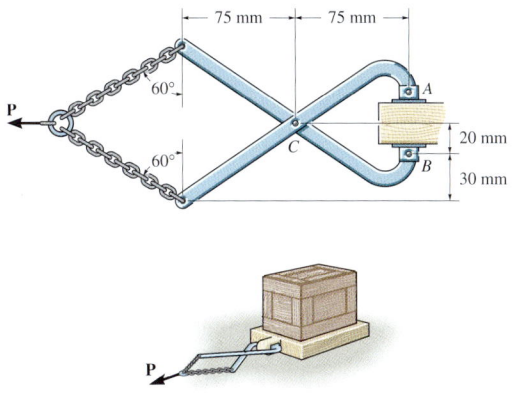

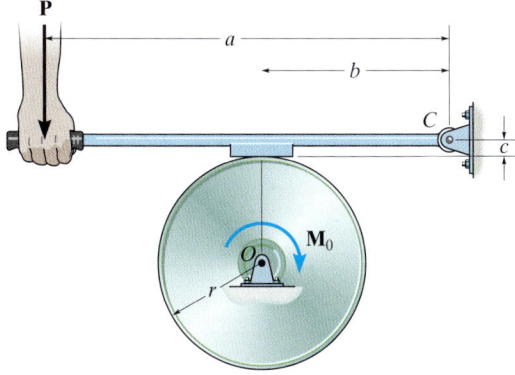

Prob. 8–24

8–25.   The block brake is used to stop the wheel from rotating when the wheel is subjected to a couple moment M_0. If the coefficient of static friction between the wheel and the block is μ_s, determine the smallest force P that should be applied.

8–26.   Show that the brake in Prob. 8–25 is self locking, i.e., $P \le 0$, provided $b/c \le \mu_s$.

8–27.   Solve Prob. 8–25 if the couple moment M_0 is applied counterclockwise.

*8–28.   The 10-kg cylinder is suspended from two equal-length cords. The end of each cord is attached to a ring of negligible mass, which passes along a horizontal shaft. If the coefficient of static friction between each ring and the shaft is $\mu_s = 0.5$, determine the greatest distance d by which the rings can be separated and still support the cylinder.

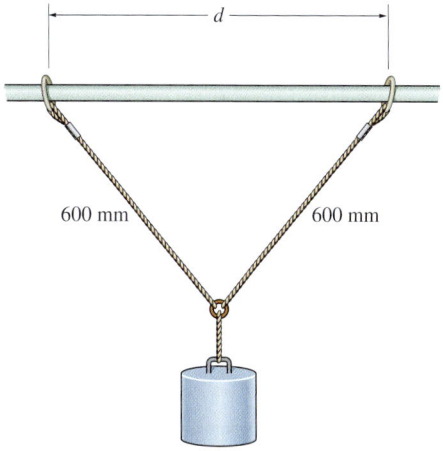

Prob. 8–28

8–29.   The friction pawl is pinned at A and rests against the wheel at B. It allows freedom of movement when the wheel is rotating counterclockwise about C. Clockwise rotation is prevented due to friction of the pawl which tends to bind the wheel. If $(\mu_s)_B = 0.6$, determine the design angle θ which will prevent clockwise motion for any value of applied moment M. Hint: Neglect the weight of the pawl so that it becomes a two-force member.

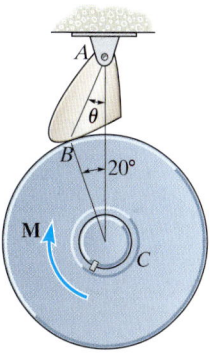

Probs. 8–25/26/27

Prob. 8–29

8–30. If $\theta = 30°$, determine the minimum coefficient of static friction at A and B so that equilibrium of the supporting frame is maintained regardless of the mass of the cylinder C. Neglect the mass of the rods.

8–31. If the coefficient of static friction at A and B is $\mu_s = 0.6$, determine the maximum angle θ so that the frame remains in equilibrium, regardless of the mass of the cylinder. Neglect the mass of the rods.

8–34. The coefficient of static friction between the 150-kg crate and the ground is $\mu_s = 0.3$, while the coefficient of static friction between the 80-kg man's shoes and the ground is $\mu_s' = 0.4$. Determine if the man can move the crate.

8–35. If the coefficient of static friction between the crate and the ground in Prob. 8–34 is $\mu_s = 0.3$, determine the minimum coefficient of static friction between the man's shoes and the ground so that the man can move the crate.

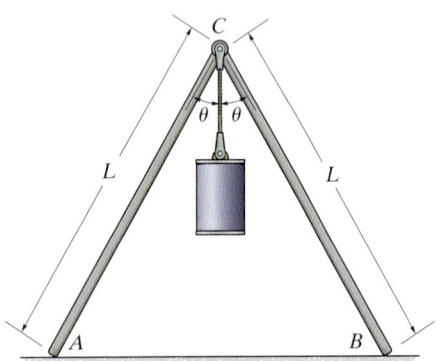

Probs. 8–30/31

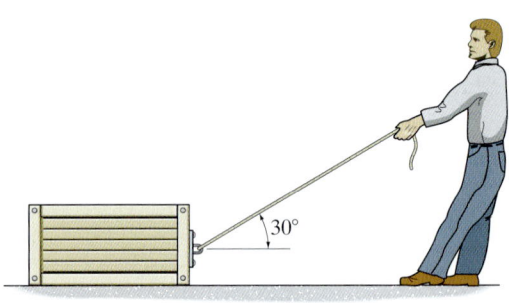

Probs. 8–34/35

*8–32.** The semicylinder of mass m and radius r lies on the rough inclined plane for which $\phi = 10°$ and the coefficient of static friction is $\mu_s = 0.3$. Determine if the semicylinder slides down the plane, and if not, find the angle of tip θ of its base AB.

8–33. The semicylinder of mass m and radius r lies on the rough inclined plane. If the inclination $\phi = 15°$, determine the smallest coefficient of static friction which will prevent the semicylinder from slipping.

*8–36.** The rod has a weight W and rests against the floor and wall for which the coefficients of static friction are μ_A and μ_B, respectively. Determine the smallest value of θ for which the rod will not move.

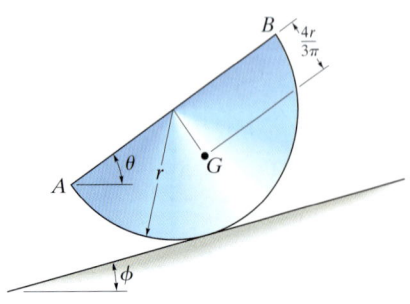

Probs. 8–32/33

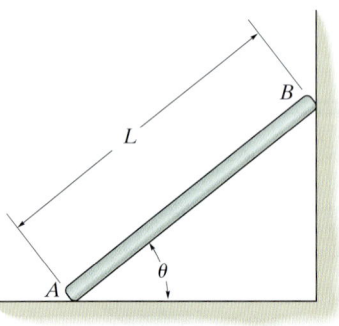

Prob. 8–36

8–37. Determine the magnitude of force **P** needed to start towing the 40-kg crate. Also determine the location of the resultant normal force acting on the crate, measured from point A. Take $\mu_s = 0.3$.

8–38. Determine the friction force on the 40-kg crate, and the resultant normal force and its position x, measured from point A, if the force $P = 300$ N. Take $\mu_s = 0.5$ and $\mu_k = 0.2$.

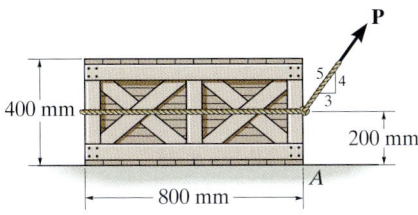

Probs. 8–37/38

***8–40.** The spool of wire having a mass of 150 kg rests on the ground at A and against the wall at B. Determine the force P required to begin pulling the wire horizontally off the spool. The coefficient of static friction between the spool and its points of contact is $\mu_s = 0.25$.

8–41. The spool of wire having a mass of 150 kg rests on the ground at A and against the wall at B. Determine the forces acting on the spool at A and B if $P = 800$ N. The coefficient of static friction between the spool and the ground at point A is $\mu_s = 0.35$. The wall at B is smooth.

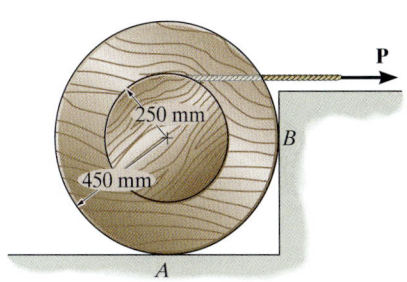

Probs. 8–40/41

8–39. Determine the smallest force the man must exert on the rope in order to move the 80-kg crate. Also, what is the angle θ at this moment? The coefficient of static friction between the crate and the floor is $\mu_s = 0.3$.

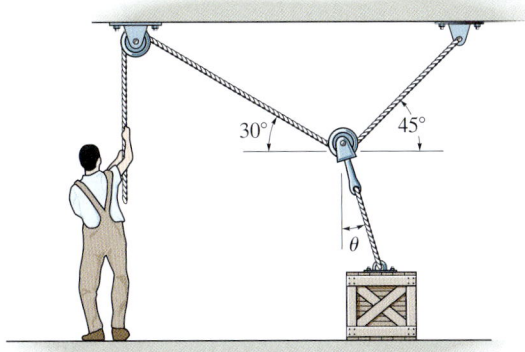

Prob. 8–39

8–42. The friction hook is made from a fixed frame and a cylinder of negligible weight. A piece of paper is placed between the wall and the cylinder. If $\theta = 20°$, determine the smallest coefficient of static friction μ at all points of contact so that any weight W of paper p can be held.

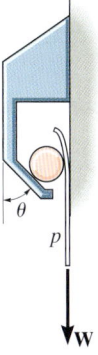

Prob. 8–42

8–43. The uniform rod has a mass of 10 kg and rests on the inside of the smooth ring at B and on the ground at A. If the rod is on the verge of slipping, determine the coefficient of static friction between the rod and the ground.

8–45. Car A has a mass of 1.4 Mg and mass center at G. If car B exerts a horizontal force on A of 2 kN, determine if this force is great enough to move car A. The coefficients of static and kinetic friction between the tires and the road are $\mu_s = 0.5$ and $\mu_k = 0.35$. Assume B's bumper is smooth.

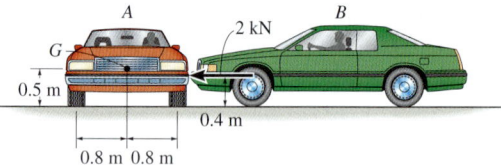

Prob. 8–45

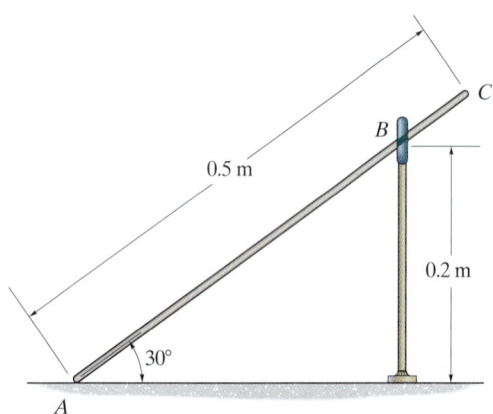

Prob. 8–43

8–46. The beam AB has a negligible mass and thickness and is subjected to a triangular distributed loading. It is supported at one end by a pin and at the other end by a post having a mass of 50 kg and negligible thickness. Determine the minimum force P needed to move the post. The coefficients of static friction at B and C are $\mu_B = 0.4$ and $\mu_C = 0.2$, respectively.

8–47. The beam AB has a negligible mass and thickness and is subjected to a triangular distributed loading. It is supported at one end by a pin and at the other end by a post having a mass of 50 kg and negligible thickness. Determine the two coefficients of static friction at B and at C so that when the magnitude of the applied force is increased to $P = 150$ N, the post slips at both B and C simultaneously.

*8–44. The rings A and C each weigh W and rest on the rod, which has a coefficient of static friction of μ_s. If the suspended ring at B has a weight of $2W$, determine the largest distance d between A and C so that no motion occurs. Neglect the weight of the wire. The wire is smooth and has a total length of l.

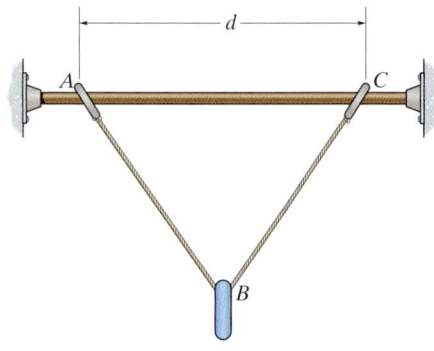

Prob. 8–44

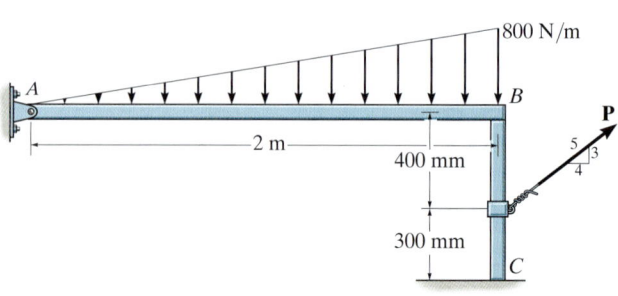

Probs. 8–46/47

*8–48. The beam AB has a negligible mass and thickness and is subjected to a force of 200 N. It is supported at one end by a pin and at the other end by a spool having a mass of 40 kg. If a cable is wrapped around the inner core of the spool, determine the minimum cable force P needed to move the spool. The coefficients of static friction at B and D are $\mu_B = 0.4$ and $\mu_D = 0.2$, respectively.

8–51. The block of weight W is being pulled up the inclined plane of slope α using a force $\mathbf{P}$. If $\mathbf{P}$ acts at the angle ϕ as shown, show that for slipping to occur, $P = W \sin(\alpha + \theta)/\cos(\phi - \theta)$, where θ is the angle of friction; $\theta = \tan^{-1}\mu$.

*8–52. Determine the angle ϕ at which $\mathbf{P}$ should act on the block so that the magnitude of $\mathbf{P}$ is as small as possible to begin pulling the block up the incline. What is the corresponding value of P? The block weighs W and the slope α is known.

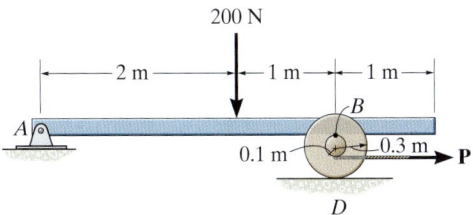

Prob. 8–48

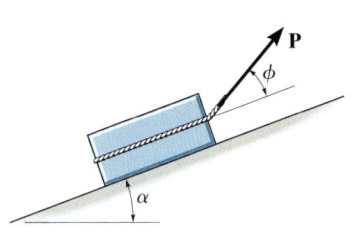

Probs. 8–51/52

8–49. The 45-kg disk rests on the surface for which the coefficient of static friction is $\mu_A = 0.2$. Determine the largest couple moment M that can be applied to the bar without causing motion.

8–50. The 45-kg disk rests on the surface for which the coefficient of static friction is $\mu_A = 0.15$. If $M = 50\ \text{N} \cdot \text{m}$, determine the friction force at A.

8–53. The homogeneous semicylinder has a mass m and mass center at G. Determine the largest angle θ of the inclined plane upon which it rests so that it does not slip down the plane. The coefficient of static friction between the plane and the cylinder is $\mu_s = 0.3$. Also, what is the angle ϕ for this case?

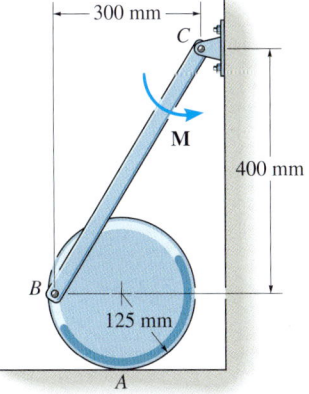

Probs. 8–49/50

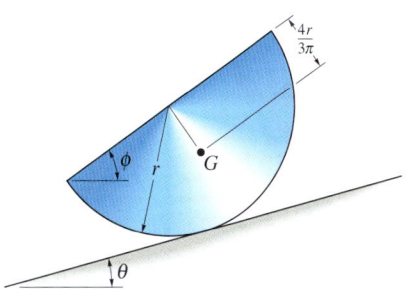

Prob. 8–53

8–54. The uniform beam has a weight W and length $4a$. It rests on the fixed rails at A and B. If the coefficient of static friction at the rails is μ_s, determine the horizontal force P, applied perpendicular to the face of the beam, which will cause the beam to move.

*8–56. The uniform 6-kg slender rod rests on the top center of the 3-kg block. If the coefficients of static friction at the points of contact are $\mu_A = 0.4$, $\mu_B = 0.6$, and $\mu_C = 0.3$, determine the largest couple moment M which can be applied to the rod without causing motion of the rod.

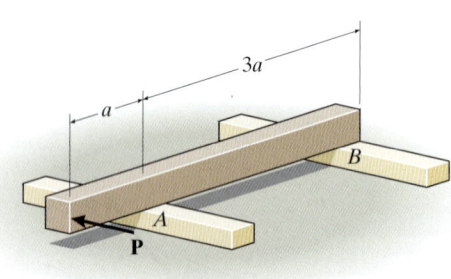

Prob. 8–54

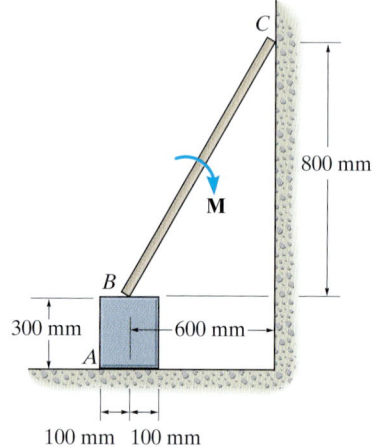

Prob. 8–56

8–55. Determine the greatest angle θ so that the ladder does not slip when it supports the 75-kg man in the position shown. The surface is rather slippery, where the coefficient of static friction at A and B is $\mu_s = 0.3$.

8–57. The disk has a weight W and lies on a plane that has a coefficient of static friction μ. Determine the maximum height h to which the plane can be lifted without causing the disk to slip.

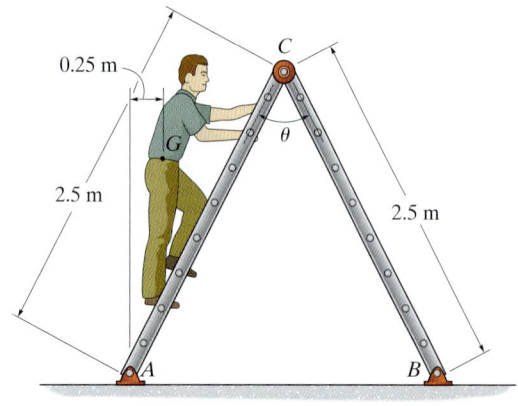

Prob. 8–55

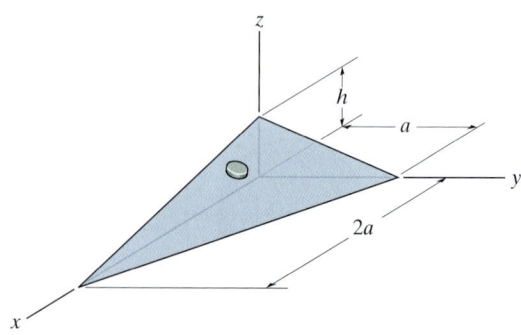

Prob. 8–57

CONCEPTUAL PROBLEMS

P8–1. Is it more effective to move the load forward at constant velocity with the boom fully extended as shown, or should the boom be fully retracted? Power is supplied to the rear wheels. The front wheels are free to roll. Do an equilibrium analysis to explain your answer.

P8–1

P8–2. The lug nut on the free-turning wheel is to be removed using the wrench. Which is the most effective way to apply force to the wrench? Also, why is it best to keep the car tire on the ground rather than first jacking it up? Explain your answers with an equilibrium analysis.

P8–2

P8–3. The rope is used to tow the refrigerator. Is it best to pull slightly up on the rope as shown, pull horizontally, or pull somewhat downwards? Also, is it best to attach the rope at a high position as shown, or at a lower position? Do an equilibrium analysis to explain your answer.

P8–4. The rope is used to tow the refrigerator. In order to prevent yourself from slipping while towing, is it best to pull up as shown, pull horizontally, or pull downwards on the rope? Do an equilibrium analysis to explain your answer.

P8–3/4

P8–5. Is it easier to tow the load by applying a force along the tow bar when it is in an almost horizontal position as shown, or is it better to pull on the bar when it has a steeper slope? Do an equilibrium analysis to explain your answer.

P8–5

8

8.3 Wedges

A *wedge* is a simple machine that is often used to transform an applied force into much larger forces, directed at approximately right angles to the applied force. Wedges also can be used to slightly move or adjust heavy loads.

Consider, for example, the wedge shown in Fig. 8–12a, which is used to *lift* the block by applying a force to the wedge. Free-body diagrams of the block and wedge are shown in Fig. 8–12b. Here we have excluded the weight of the wedge since it is usually *small* compared to the weight **W** of the block. Also, note that the frictional forces $\mathbf{F}_1$ and $\mathbf{F}_2$ must oppose the motion of the wedge. Likewise, the frictional force $\mathbf{F}_3$ of the wall on the block must act downward so as to oppose the block's upward motion. The locations of the resultant normal forces are not important in the force analysis since neither the block nor wedge will "tip." Hence the moment equilibrium equations will not be considered. There are seven unknowns, consisting of the applied force **P**, needed to cause motion of the wedge, and six normal and frictional forces. The seven available equations consist of four force equilibrium equations, $\Sigma F_x = 0$, $\Sigma F_y = 0$ applied to the wedge and block, and three frictional equations, $F = \mu N$, applied at each surface of contact.

If the block is to be *lowered*, then the frictional forces will all act in a sense opposite to that shown in Fig. 8–12b. Provided the coefficient of friction is very *small* or the wedge angle θ is *large*, then the applied force **P** must act to the right to hold the block. Otherwise, **P** may have a reverse sense of direction in order to *pull* on the wedge to remove it. If **P** is *not applied* and friction forces hold the block in place, then the wedge is referred to as *self-locking*.

Please refer to the Companion Website for the animation: *Free-Body Diagram for a Load and a Wedge*

Wedges are often used to adjust the elevation of structural or mechanical parts. Also, they provide stability for objects such as this pipe.

8

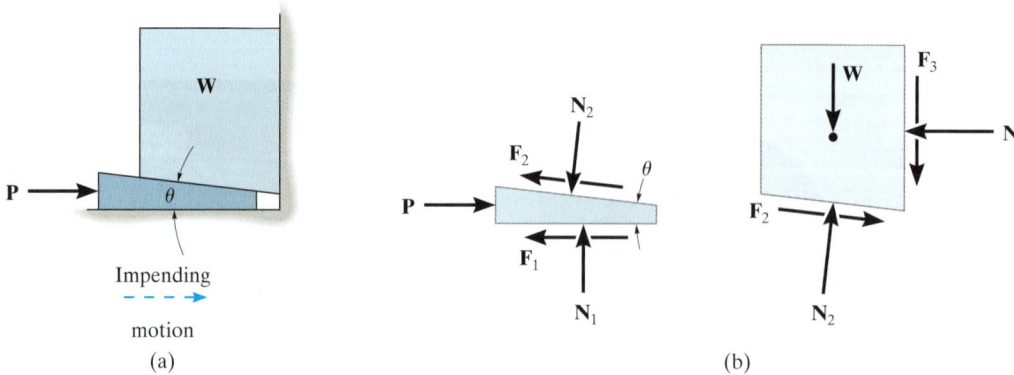

(a)

(b)

Fig. 8–12

EXAMPLE 8.6

The uniform stone in Fig. 8–13a has a mass of 500 kg and is held in the horizontal position using a wedge at B. If the coefficient of static friction is $\mu_s = 0.3$ at the surfaces of contact, determine the minimum force P needed to remove the wedge. Assume that the stone does not slip at A.

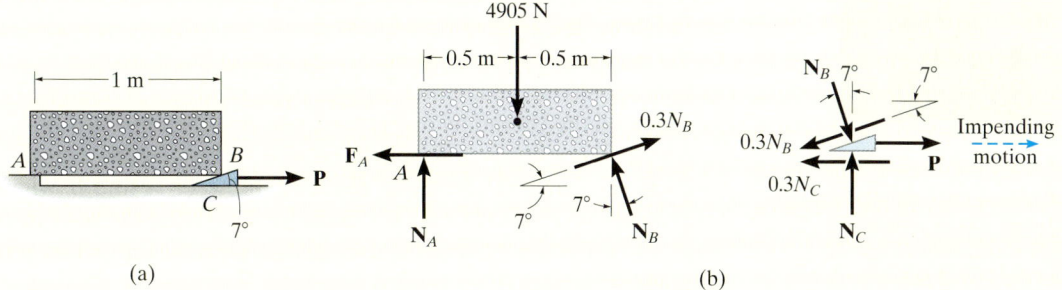

(a)

(b)

Fig. 8–13

SOLUTION

The minimum force P requires $F = \mu_s N$ at the surfaces of contact with the wedge. The free-body diagrams of the stone and wedge are shown in Fig. 8–13b. On the wedge the friction force opposes the impending motion, and on the stone at A, $F_A \leq \mu_s N_A$, since slipping does not occur there. There are five unknowns. Three equilibrium equations for the stone and two for the wedge are available for solution. From the free-body diagram of the stone,

$$\zeta + \Sigma M_A = 0; \quad -4905 \text{ N}(0.5 \text{ m}) + (N_B \cos 7° \text{ N})(1 \text{ m})$$
$$+ (0.3N_B \sin 7° \text{ N})(1 \text{ m}) = 0$$
$$N_B = 2383.1 \text{ N}$$

Using this result for the wedge, we have

$$+\uparrow \Sigma F_y = 0; \quad N_C - 2383.1 \cos 7° \text{ N} - 0.3(2383.1 \sin 7° \text{ N}) = 0$$
$$N_C = 2452.5 \text{ N}$$

$$\xrightarrow{+} \Sigma F_x = 0; \quad 2383.1 \sin 7° \text{ N} - 0.3(2383.1 \cos 7° \text{ N}) +$$
$$P - 0.3(2452.5 \text{ N}) = 0$$
$$P = 1154.9 \text{ N} = 1.15 \text{ kN} \qquad \textit{Ans.}$$

NOTE: Since P is positive, indeed the wedge must be pulled out. If P were zero, the wedge would remain in place (self-locking) and the frictional forces developed at B and C would satisfy $F_B < \mu_s N_B$ and $F_C < \mu_s N_C$.

Square-threaded screws find applications on valves, jacks, and vises, where particularly large forces must be developed along the axis of the screw.

8.4   Frictional Forces on Screws

In most cases, screws are used as fasteners; however, in many types of machines they are incorporated to transmit power or motion from one part of the machine to another. A *square-threaded screw* is commonly used for the latter purpose, especially when large forces are applied along its axis. In this section, we will analyze the forces acting on square-threaded screws. The analysis of other types of screws, such as the V-thread, is based on these same principles.

For analysis, a square-threaded screw, as in Fig. 8–14, can be considered a cylinder having an inclined square ridge or *thread* wrapped around it. If we unwind the thread by one revolution, as shown in Fig. 8–14b, the slope or the *lead angle* θ is determined from $\theta = \tan^{-1}(l/2\pi r)$. Here l and $2\pi r$ are the vertical and horizontal distances between A and B, where r is the mean radius of the thread. The distance l is called the *lead* of the screw and it is equivalent to the distance the screw advances when it turns one revolution.

Upward Impending Motion.
Let us now consider the case of the square-threaded screw jack in Fig. 8–15 that is subjected to upward impending motion caused by the applied torsional moment ***M**. A free-body diagram of the *entire unraveled thread h* in contact with the jack can be represented as a block as shown in Fig. 8–16a. The force **W** is the vertical force acting on the thread or the axial force applied to the shaft, Fig. 8–15, and M/r is the resultant horizontal force produced by the couple moment M about the axis of the shaft. The reaction **R** of the groove on the thread has both frictional and normal components, where $F = \mu_s N$. The angle of static friction is $\phi_s = \tan^{-1}(F/N) = \tan^{-1}\mu_s$. Applying the force equations of equilibrium along the horizontal and vertical axes, we have

$$\xrightarrow{+}\Sigma F_x = 0; \qquad M/r - R\sin(\theta + \phi_s) = 0$$

$$+\uparrow\Sigma F_y = 0; \qquad R\cos(\theta + \phi_s) - W = 0$$

Eliminating R from these equations, we obtain

$$\boxed{M = rW\tan(\theta + \phi_s)} \qquad (8\text{–}3)$$

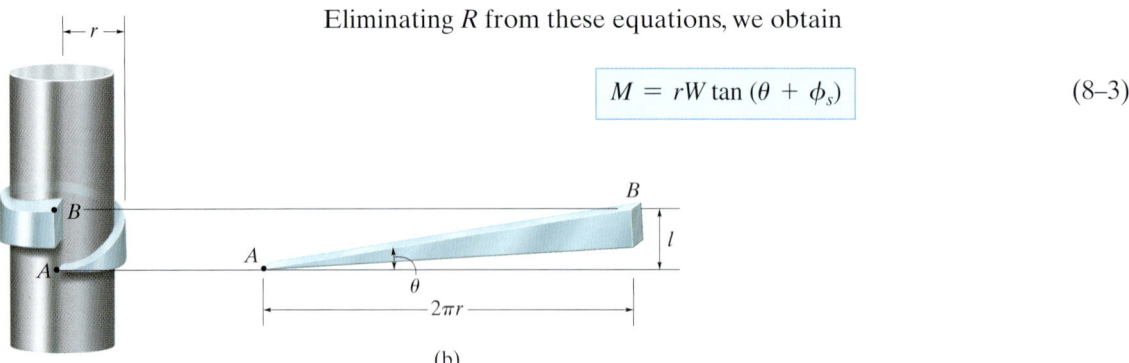

(a)

(b)

*For applications, **M** is developed by applying a horizontal force **P** at a right angle to the end of a lever that would be fixed to the screw.

Fig. 8–14

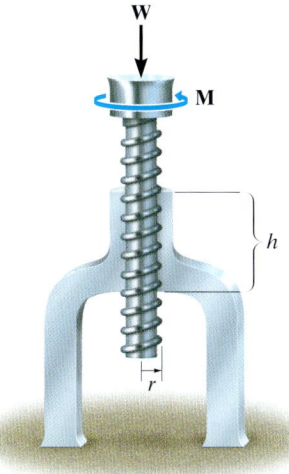

Fig. 8–15

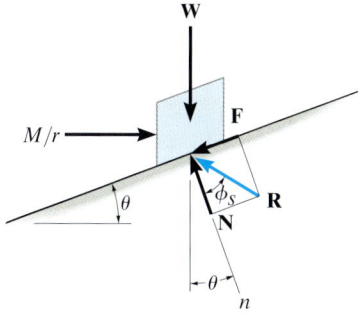

Upward screw motion
(a)

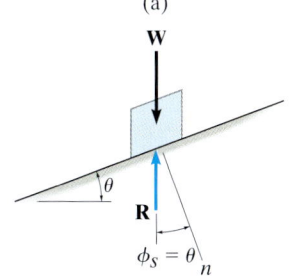

Self-locking screw ($\theta = \phi_S$)
(on the verge of rotating downward)
(b)

Self-Locking Screw.

A screw is said to be *self-locking* if it remains in place under any axial load **W** when the moment **M** is removed. For this to occur, the direction of the frictional force must be reversed so that **R** acts on the other side of **N**. Here the angle of static friction ϕ_s becomes greater than or equal to θ, Fig. 8–16*d*. If $\phi_s = \theta$, Fig. 8–16*b*, then **R** will act vertically to balance **W**, and the screw will be on the verge of winding downward.

Downward Impending Motion, $(\theta > \phi_s)$.

If the screw is not self-locking, it is necessary to apply a moment **M**′ to prevent the screw from winding downward. Here, a horizontal force M'/r is required to push against the thread to prevent it from sliding down the plane, Fig. 8–16*c*. Using the same procedure as before, the magnitude of the moment **M**′ required to prevent this unwinding is

$$M' = rW \tan (\theta - \phi_s) \qquad (8\text{–}4)$$

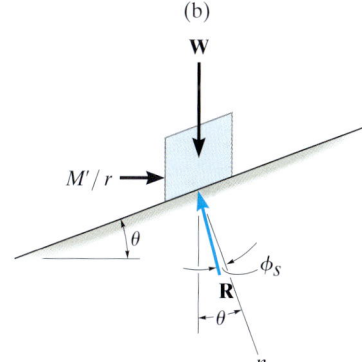

Downward screw motion ($\theta > \phi_S$)
(c)

Downward Impending Motion, $(\phi_s > \theta)$.

If a screw is self-locking, a couple moment **M**″ must be applied to the screw in the opposite direction to wind the screw downward ($\phi_s > \theta$). This causes a reverse horizontal force M''/r that pushes the thread down as indicated in Fig. 8–16*d*. In this case, we obtain

$$M'' = rW \tan (\phi_s - \theta) \qquad (8\text{–}5)$$

If *motion of the screw* occurs, Eqs. 8–3, 8–4, and 8–5 can be applied by simply replacing ϕ_s with ϕ_k.

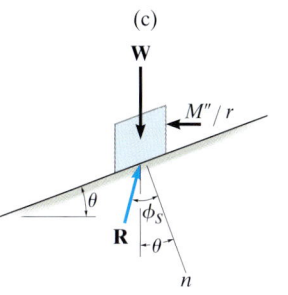

Downward screw motion ($\theta < \phi_S$)
(d)

Fig. 8–16

EXAMPLE   8.7

The turnbuckle shown in Fig. 8–17 has a square thread with a mean radius of 5 mm and a lead of 2 mm. If the coefficient of static friction between the screw and the turnbuckle is $\mu_s = 0.25$, determine the moment **M** that must be applied to draw the end screws closer together.

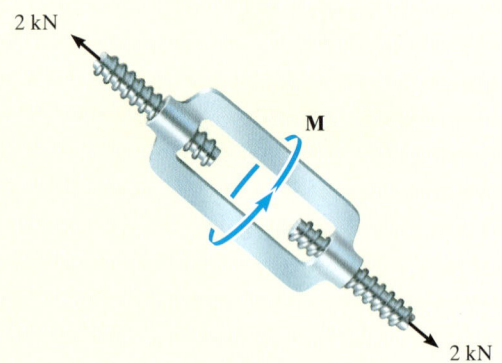

Fig. 8–17

SOLUTION

The moment can be obtained by applying Eq. 8–3. Since friction at *two screws* must be overcome, this requires

$$M = 2[rW \tan(\theta + \phi_s)] \qquad (1)$$

Here $W = 2000$ N, $\phi_s = \tan^{-1}\mu_s = \tan^{-1}(0.25) = 14.04°$, $r = 5$ mm, and $\theta = \tan^{-1}(l/2\pi r) = \tan^{-1}(2 \text{ mm}/[2\pi(5 \text{ mm})]) = 3.64°$. Substituting these values into Eq. 1 and solving gives

$$M = 2[(2000 \text{ N})(5 \text{ mm}) \tan(14.04° + 3.64°)]$$

$$= 6374.7 \text{ N} \cdot \text{mm} = 6.37 \text{ N} \cdot \text{m} \qquad \textit{Ans.}$$

NOTE: When the moment is *removed*, the turnbuckle will be self-locking; i.e., it will not unscrew since $\phi_s > \theta$.

PROBLEMS

8–58. Determine the largest angle θ that will cause the wedge to be self-locking regardless of the magnitude of horizontal force P applied to the blocks. The coefficient of static friction between the wedge and the blocks is $\mu_s = 0.3$. Neglect the weight of the wedge.

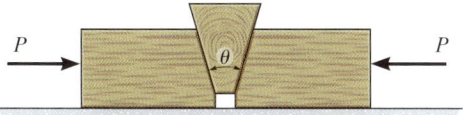

Prob. 8–58

8–59. If the beam AD is loaded as shown, determine the horizontal force P which must be applied to the wedge in order to remove it from under the beam. The coefficients of static friction at the wedge's top and bottom surfaces are $\mu_{CA} = 0.25$ and $\mu_{CB} = 0.35$, respectively. If $P = 0$, is the wedge self-locking? Neglect the weight and size of the wedge and the thickness of the beam.

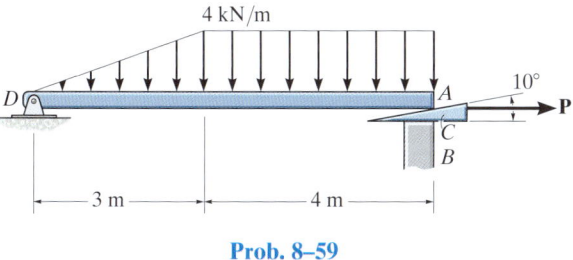

Prob. 8–59

*8–60. The wedge has a negligible weight and a coefficient of static friction $\mu_s = 0.35$ with all contacting surfaces. Determine the largest angle θ so that it is "self-locking." This requires no slipping for any magnitude of the force P applied to the joint.

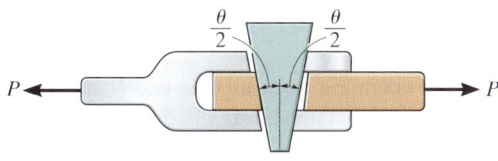

Prob. 8–60

8–61. If the spring is compressed 60 mm and the coefficient of static friction between the tapered stub S and the slider A is $\mu_{SA} = 0.5$, determine the horizontal force $\mathbf{P}$ needed to move the slider forward. The stub is free to move without friction within the fixed collar C. The coefficient of static friction between A and surface B is $\mu_{AB} = 0.4$. Neglect the weights of the slider and stub.

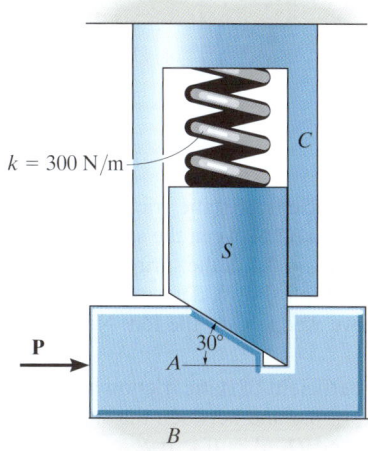

Prob. 8–61

8–62. If $P = 250$ N, determine the required minimum compression in the spring so that the wedge will not move to the right. Neglect the weight of A and B. The coefficient of static friction for all contacting surfaces is $\mu_s = 0.35$. Neglect friction at the rollers.

8–63. Determine the minimum applied force $\mathbf{P}$ required to move wedge A to the right. The spring is compressed a distance of 175 mm. Neglect the weight of A and B. The coefficient of static friction for all contacting surfaces is $\mu_s = 0.35$. Neglect friction at the rollers.

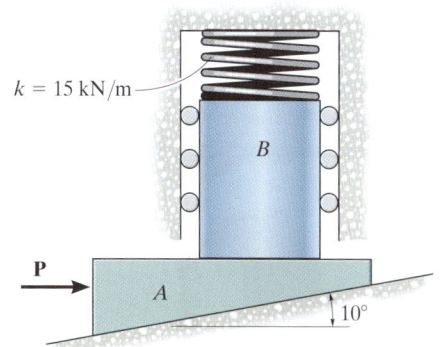

Probs. 8–62/63

*8–64. The wedge blocks are used to hold the specimen in a tension testing machine. Determine the design angle θ of the wedges so that the specimen will not slip regardless of the applied load. The coefficients of static friction are $\mu_A = 0.1$ at A and $\mu_B = 0.6$ at B. Neglect the weight of the blocks.

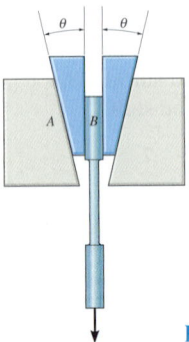

Prob. 8–64

8–65. The coefficient of static friction between wedges B and C is $\mu_s = 0.6$ and between the surfaces of contact B and A and C and D, $\mu'_s = 0.4$. If the spring is compressed 200 mm when in the position shown, determine the smallest force P needed to move wedge C to the left. Neglect the weight of the wedges.

8–66. The coefficient of static friction between the wedges B and C is $\mu_s = 0.6$ and between the surfaces of contact B and A and C and D, $\mu'_s = 0.4$. If $P = 50$ N, determine the smallest allowable compression of the spring without causing wedge C to move to the left. Neglect the weight of the wedges.

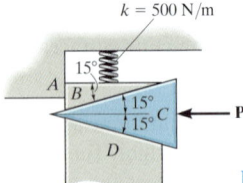

Probs. 8–65/66

8–67. The vise is used to grip the pipe. If a horizontal force of 100 N is applied perpendicular to the end of the 250 mm handle, determine the compressive force F developed in the pipe. The square threads have a mean diameter of 37.5 mm and a lead of 5 mm. How much force must be applied perpendicular to the handle to loosen the vise? Take $\mu_s = 0.3$.

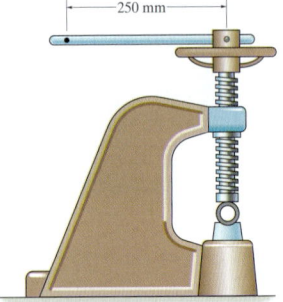

Prob. 8–67

*8–68. Determine the couple forces F that must be applied to the handle of the machinist's vise in order to create a compressive force of 400 N in the block. Neglect friction at the bearing A. The guide at B is smooth so that the axial force on the screw is 400 N. The single square-threaded screw has a mean radius of 6 mm and a lead of 8 mm, and the coefficient of static friction is $\mu_s = 0.27$.

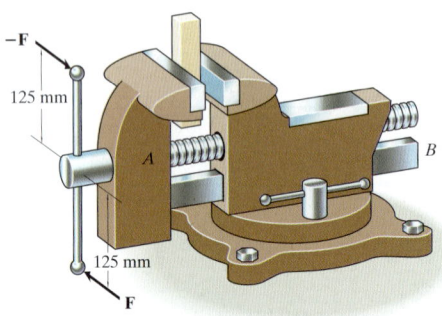

Prob. 8–68

8–69. The column is used to support the upper floor. If a force $F = 80$ N is applied perpendicular to the handle to tighten the screw, determine the compressive force in the column. The square-threaded screw on the jack has a coefficient of static friction of $\mu_s = 0.4$, mean diameter of 25 mm, and a lead of 3 mm.

8–70. If the force $\mathbf{F}$ is removed from the handle of the jack in Prob. 8–69, determine if the screw is self-locking.

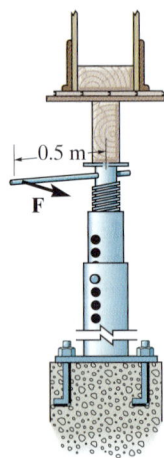

Probs. 8–69/70

8–71. If the clamping force at G is 900 N, determine the horizontal force $\mathbf{F}$ that must be applied perpendicular to the handle of the lever at E. The mean diameter and lead of both single square-threaded screws at C and D are 25 mm and 5 mm, respectively. The coefficient of static friction is $\mu_s = 0.3$.

*8–72. If a horizontal force of $F = 50$ N is applied perpendicular to the handle of the lever at E, determine the clamping force developed at G. The mean diameter and lead of the single square-threaded screw at C and D are 25 mm and 5 mm, respectively. The coefficient of static friction is $\mu_s = 0.3$.

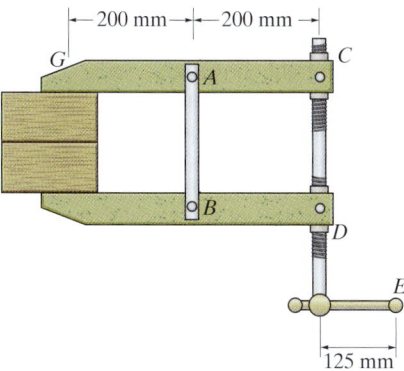

Probs. 8–71/72

8–73. A turnbuckle, similar to that shown in Fig. 8–17, is used to tension member AB of the truss. The coefficient of the static friction between the square threaded screws and the turnbuckle is $\mu_s = 0.5$. The screws have a mean radius of 6 mm and a lead of 3 mm. If a torque of $M = 10$ N · m is applied to the turnbuckle, to draw the screws closer together, determine the force in each member of the truss. No external forces act on the truss.

8–74. A turnbuckle, similar to that shown in Fig. 8–17, is used to tension member AB of the truss. The coefficient of the static friction between the square-threaded screws and the turnbuckle is $\mu_s = 0.5$. The screws have a mean radius of 6 mm and a lead of 3 mm. Determine the torque M which must be applied to the turnbuckle to draw the screws closer together, so that the compressive force of 500 N is developed in member BC.

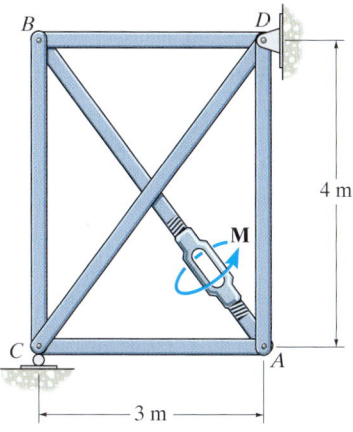

Probs. 8–73/74

8–75. The shaft has a square-threaded screw with a lead of 8 mm and a mean radius of 15 mm. If it is in contact with a plate gear having a mean radius of 30 mm, determine the resisting torque **M** on the plate gear which can be overcome if a torque of 7 N · m is applied to the shaft. The coefficient of static friction at the screw is $\mu_B = 0.2$. Neglect friction of the bearings located at A and B.

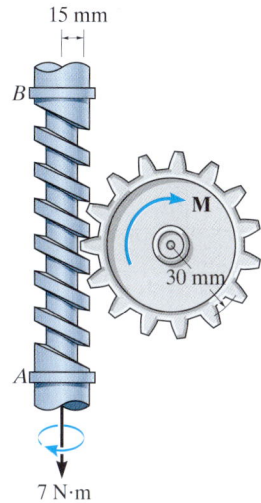

Prob. 8–75

*8–76. The square threaded screw of the clamp has a mean diameter of 14 mm and a lead of 6 mm. If $\mu_s = 0.2$ for the threads, and the torque applied to the handle is 1.5 N · m, determine the compressive force **F** on the block.

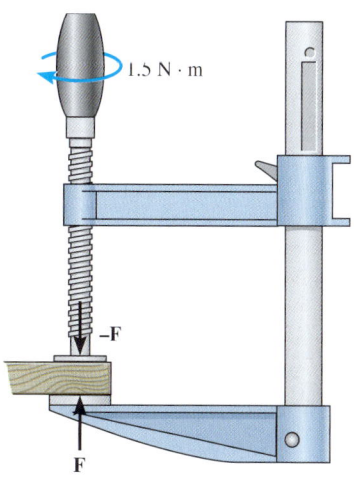

Prob. 8–76

8–77. The fixture clamp consist of a square-threaded screw having a coefficient of static friction of $\mu_s = 0.3$, mean diameter of 3 mm, and a lead of 1 mm. The five points indicated are pin connections. Determine the clamping force at the smooth blocks D and E when a torque of $M = 0.08$ N · m is applied to the handle of the screw.

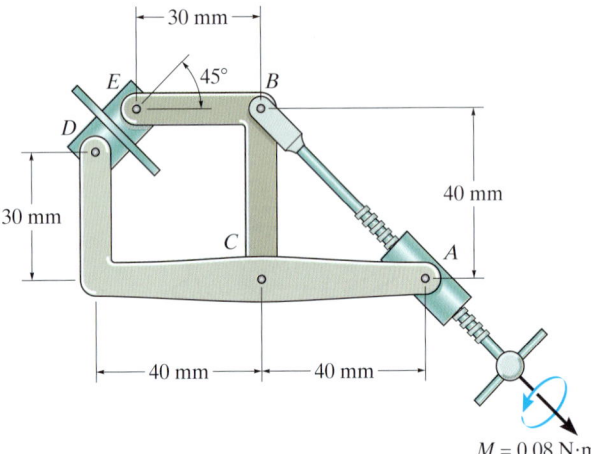

Prob. 8–77

8–78. The braking mechanism consists of two pinned arms and a square-threaded screw with left- and right-hand threads. Thus when turned, the screw draws the two arms together. If the lead of the screw is 4 mm, the mean diameter 12 mm, and the coefficient of static friction is $\mu_s = 0.35$, determine the tension in the screw when a torque of 5 N · m is applied to tighten the screw. If the coefficient of static friction between the brake pads A and B and the circular shaft is $\mu'_s = 0.5$, determine the maximum torque M the brake can resist.

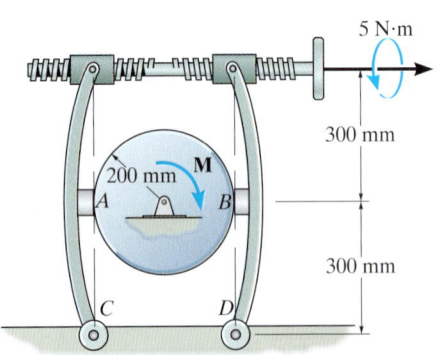

Prob. 8–78

8–79. If a horizontal force of $P = 100$ N is applied perpendicular to the handle of the lever at A, determine the compressive force **F** exerted on the material. Each single square-threaded screw has a mean diameter of 25 mm and a lead of 7.5 mm. The coefficient of static friction at all contacting surfaces of the wedges is $\mu_s = 0.2$, and the coefficient of static friction at the screw is $\mu'_s = 0.15$.

*****8–80.** Determine the horizontal force **P** that must be applied perpendicular to the handle of the lever at A in order to develop a compressive force of 12 kN on the material. Each single square-threaded screw has a mean diameter of 25 mm and a lead of 7.5 mm. The coefficient of static friction at all contacting surfaces of the wedges is $\mu_s = 0.2$, and the coefficient of static friction at the screw is $\mu'_s = 0.15$.

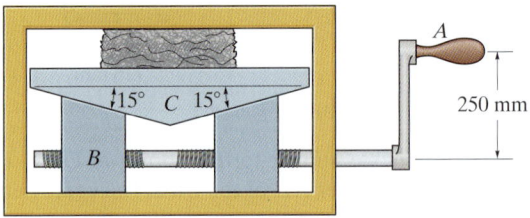

Probs. 8–79/80

8–81. Determine the clamping force on the board A if the screw of the "C" clamp is tightened with a twist of $M = 8$ N · m. The single square-threaded screw has a mean radius of 10 mm, a lead of 3 mm, and the coefficient of static friction is $\mu_s = 0.35$.

8–82. If the required clamping force at the board A is to be 50 N, determine the torque M that must be applied to the handle of the "C" clamp to tighten it down. The single square-threaded screw has a mean radius of 10 mm, a lead of 3 mm, and the coefficient of static friction is $\mu_s = 0.35$.

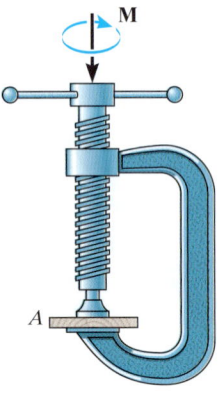

Probs. 8–81/82

8.5 Frictional Forces on Flat Belts

Whenever belt drives or band brakes are designed, it is necessary to determine the frictional forces developed between the belt and its contacting surface. In this section we will analyze the frictional forces acting on a flat belt, although the analysis of other types of belts, such as the V-belt, is based on similar principles.

Consider the flat belt shown in Fig. 8–18a, which passes over a fixed curved surface. The total angle of belt to surface contact in radians is β, and the coefficient of friction between the two surfaces is μ. We wish to determine the tension T_2 in the belt, which is needed to pull the belt counterclockwise over the surface, and thereby overcome both the frictional forces at the surface of contact and the tension T_1 in the other end of the belt. Obviously, $T_2 > T_1$.

Frictional Analysis. A free-body diagram of the belt segment in contact with the surface is shown in Fig. 8–18b. As shown, the normal and frictional forces, acting at different points along the belt, will vary both in magnitude and direction. Due to this *unknown* distribution, the analysis of the problem will first require a study of the forces acting on a differential element of the belt.

A free-body diagram of an element having a length ds is shown in Fig. 8–18c. Assuming either impending motion or motion of the belt, the magnitude of the frictional force $dF = \mu \, dN$. This force opposes the sliding motion of the belt, and so it will increase the magnitude of the tensile force acting in the belt by dT. Applying the two force equations of equilibrium, we have

$$\searrow + \Sigma F_x = 0; \qquad T\cos\left(\frac{d\theta}{2}\right) + \mu \, dN - (T + dT)\cos\left(\frac{d\theta}{2}\right) = 0$$

$$+\nearrow \Sigma F_y = 0; \qquad dN - (T + dT)\sin\left(\frac{d\theta}{2}\right) - T\sin\left(\frac{d\theta}{2}\right) = 0$$

Since $d\theta$ is of *infinitesimal size*, $\sin(d\theta/2) = d\theta/2$ and $\cos(d\theta/2) = 1$. Also, the *product* of the two infinitesimals dT and $d\theta/2$ may be neglected when compared to infinitesimals of the first order. As a result, these two equations become

$$\mu \, dN = dT$$

and

$$dN = T \, d\theta$$

Eliminating dN yields

$$\frac{dT}{T} = \mu \, d\theta$$

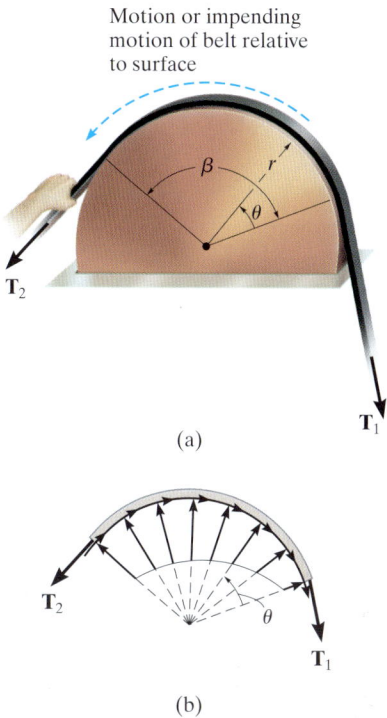

(a)

(b)

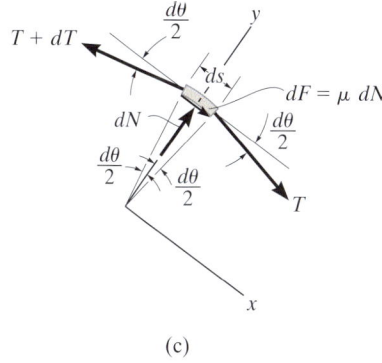

(c)

Fig. 8–18

8

Motion or impending
motion of belt relative
to surface

$\mathbf{T}_2$

β r

θ

$\mathbf{T}_1$

(a)

Fig. 8–18 (Repeated)

Flat or V-belts are often used to transmit the torque developed by a motor to a wheel attached to a pump, fan or blower.

Integrating this equation between all the points of contact that the belt makes with the drum, and noting that $T = T_1$ at $\theta = 0$ and $T = T_2$ at $\theta = \beta$, yields

$$\int_{T_1}^{T_2} \frac{dT}{T} = \mu \int_0^{\beta} d\theta$$

$$\ln \frac{T_2}{T_1} = \mu\beta$$

Solving for T_2, we obtain

$$\boxed{T_2 = T_1 e^{\mu\beta}} \tag{8–6}$$

where

T_2, T_1 = belt tensions; T_1 opposes the direction of motion (or impending motion) of the belt measured relative to the surface, while T_2 acts in the direction of the relative belt motion (or impending motion); because of friction, $T_2 > T_1$

μ = coefficient of static or kinetic friction between the belt and the surface of contact

β = angle of belt to surface contact, measured in radians

$e = 2.718\ldots$, base of the natural logarithm

Note that T_2 is *independent* of the *radius* of the drum, and instead it is a function of the angle of belt to surface contact, β. As a result, this equation is valid for flat belts passing over any curved contacting surface.

EXAMPLE 8.8

The maximum tension that can be developed in the cord shown in Fig. 8–19a is 500 N. If the pulley at A is free to rotate and the coefficient of static friction at the fixed drums B and C is $\mu_s = 0.25$, determine the largest mass of the cylinder that can be lifted by the cord.

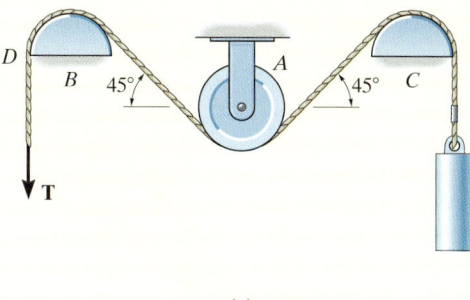

(a)

SOLUTION

Lifting the cylinder, which has a weight $W = mg$, causes the cord to move counterclockwise over the drums at B and C; hence, the maximum tension T_2 in the cord occurs at D. Thus, $F = T_2 = 500$ N. A section of the cord passing over the drum at B is shown in Fig. 8–19b. Since $180° = \pi$ rad the angle of contact between the drum and the cord is $\beta = (135°/180°)\pi = 3\pi/4$ rad. Using Eq. 8–6, we have

$$T_2 = T_1 e^{\mu_s \beta}; \qquad 500 \text{ N} = T_1 e^{0.25[(3/4)\pi]}$$

Hence,

$$T_1 = \frac{500 \text{ N}}{e^{0.25[(3/4)\pi]}} = \frac{500 \text{ N}}{1.80} = 277.4 \text{ N}$$

Since the pulley at A is free to rotate, equilibrium requires that the tension in the cord remains the *same* on both sides of the pulley.

The section of the cord passing over the drum at C is shown in Fig. 8–19c. The weight $W < 277.4$ N. Why? Applying Eq. 8–6, we obtain

$$T_2 = T_1 e^{\mu_s \beta}; \qquad 277.4 \text{ N} = W e^{0.25[(3/4)\pi]}$$

$$W = 153.9 \text{ N}$$

so that

$$m = \frac{W}{g} = \frac{153.9 \text{ N}}{9.81 \text{ m/s}^2}$$

$$= 15.7 \text{ kg} \qquad \qquad \textit{Ans.}$$

Impending
motion

135° B

500 N

T_1

(b)

Impending
motion
135°

C

277.4 N

$W = mg$

(c)

Fig. 8–19

PROBLEMS

8–83. A cylinder having a mass of 250 kg is to be supported by the cord that wraps over the pipe. Determine the smallest vertical force **F** needed to support the load if the cord passes (a) once over the pipe, $\beta = 180°$, and (b) two times over the pipe, $\beta = 540°$. Take $\mu_s = 0.2$.

*8–84. A cylinder having a mass of 250 kg is to be supported by the cord that wraps over the pipe. Determine the largest vertical force **F** that can be applied to the cord without moving the cylinder. The cord passes (a) once over the pipe, $\beta = 180°$, and (b) two times over the pipe, $\beta = 540°$. Take $\mu_s = 0.2$.

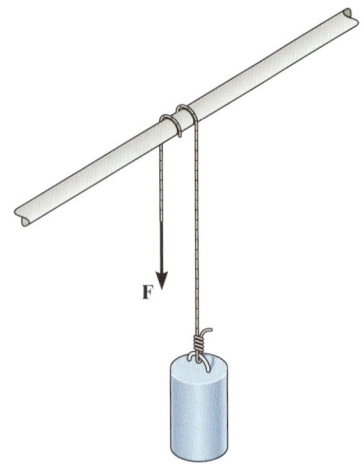

Probs. 8–83/84

8–85. The cord supporting the 6-kg cylinder passes around three pegs, A, B, C, where $\mu_s = 0.2$. Determine the range of values for the magnitude of the horizontal force **P** for which the cylinder will not move up or down.

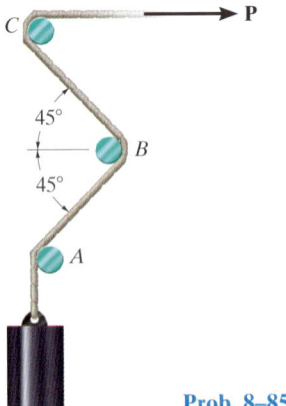

Prob. 8–85

8–86. A force of $P = 25$ N is just sufficient to prevent the 20-kg cylinder from descending. Determine the required force **P** to begin lifting the cylinder. The rope passes over a rough peg with two and half turns.

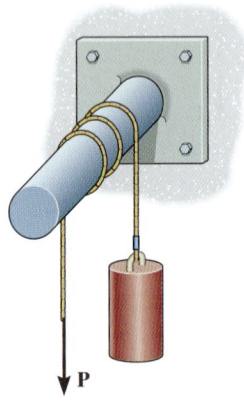

Prob. 8–86

8–87. The 20-kg cylinder A and 50-kg cylinder B are connected together using a rope that passes around a rough peg two and a half turns. If the cylinders are on the verge of moving, determine the coefficient of static friction between the rope and the peg.

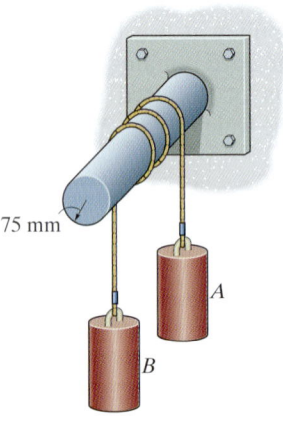

Prob. 8–87

*8–88. The uniform bar AB is supported by a rope that passes over a frictionless pulley at C and a fixed peg at D. If the coefficient of static friction between the rope and the peg is $\mu_D = 0.3$, determine the smallest distance x from the end of the bar at which a 20-N force may be placed and not cause the bar to move.

8–90. The smooth beam is being hoisted using a rope that is wrapped around the beam and passes through a ring at A as shown. If the end of the rope is subjected to a tension **T** and the coefficient of static friction between the rope and ring is $\mu_s = 0.3$, determine the smallest angle of θ for equilibrium.

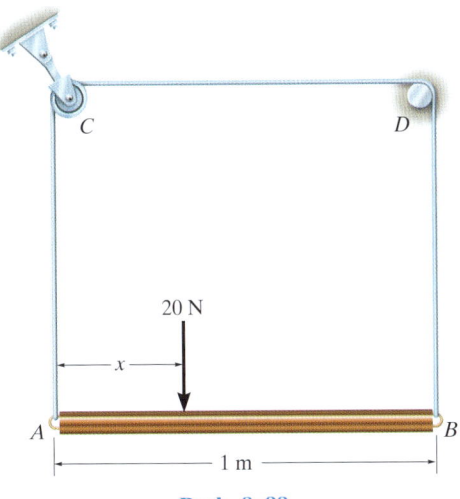

20 N

Prob. 8–88

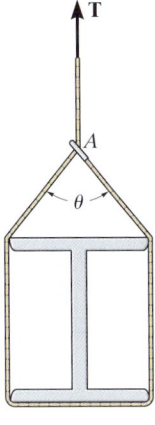

Prob. 8–90

8–89. The truck, which has a mass of 3.4 Mg, is to be lowered down the slope by a rope that is wrapped around a tree. If the wheels are free to roll and the man at A can resist a pull of 300 N, determine the minimum number of turns the rope should be wrapped around the tree to lower the truck at a constant speed. The coefficient of kinetic friction between the tree and rope is $\mu_k = 0.3$.

8–91. A cable is attached to the 20-kg plate B, passes over a fixed peg at C, and is attached to the block at A. Using the coefficients of static friction shown, determine the smallest mass of block A so that it will prevent sliding motion of B down the plane.

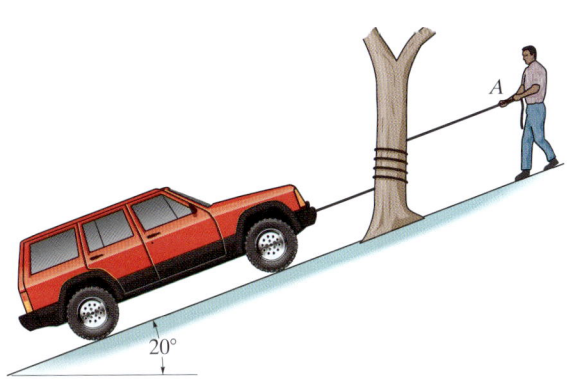

Prob. 8–89

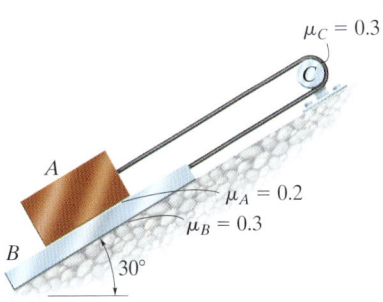

Prob. 8–91

8

*8–92. Determine the smallest lever force P needed to prevent the wheel from rotating if it is subjected to a torque of $M = 250\ \text{N} \cdot \text{m}$. The coefficient of static friction between the belt and the wheel is $\mu_s = 0.3$. The wheel is pin-connected at its center, B.

8–93. Determine the torque M that can be resisted by the band brake if a force of $P = 30\ \text{N}$ is applied to the handle of the lever. The coefficient of static friction between the belt and the wheel is $\mu_s = 0.3$. The wheel is pin-connected at its center, B.

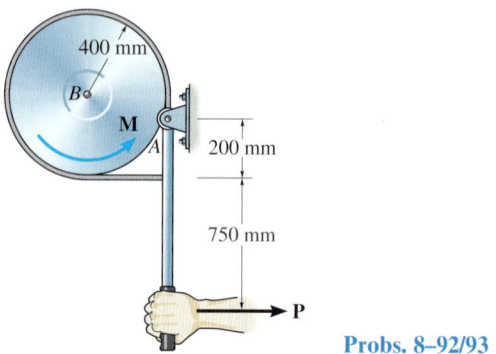

Probs. 8–92/93

8–94. Block A has a mass of 50 kg and rests on surface for which $\mu_s = 0.25$. If the coefficient of static friction between the cord and the fixed peg at C is $\mu_s' = 0.3$, determine the greatest mass of the suspended cylinder D without causing motion.

8–95. Block A rests on the surface for which $\mu_s = 0.25$. If the mass of the suspended cylinder D is 4 kg, determine the smallest mass of block A so that it does not slip or tip. The coefficient of static friction between the cord and the fixed peg at C is $\mu_s' = 0.3$.

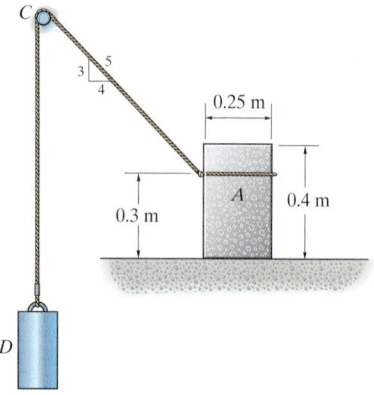

Probs. 8–94/95

*8–96. The 20-kg motor has a center of gravity at G and is pin-connected at C to maintain a tension in the drive belt. Determine the smallest counterclockwise twist or torque M that must be supplied by the motor to turn the disk B if wheel A locks and causes the belt to slip over the disk. No slipping occurs at A. The coefficient of static friction between the belt and the disk is $\mu_s = 0.3$.

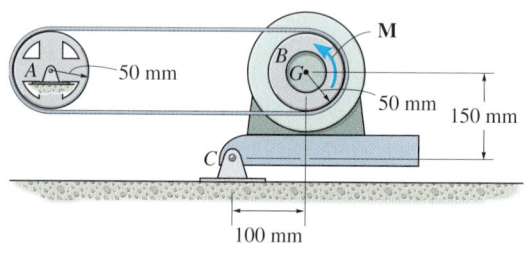

Prob. 8–96

8–97. Determine the smallest force P required to lift the 40-kg crate. The coefficient of static friction between the cable and each peg is $\mu_s = 0.1$.

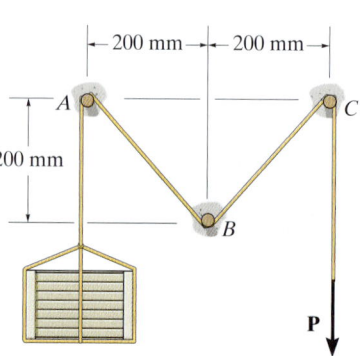

Prob. 8–97

8–98. Show that the frictional relationship between the belt tensions, the coefficient of friction μ, and the angular contacts α and β for the V-belt is $T_2 = T_1 e^{\mu\beta/\sin(\alpha/2)}$.

***8–100.** A 10-kg cylinder D, which is attached to a small pulley B, is placed on the cord as shown. Determine the largest angle θ so that the cord does not slip over the peg at C. The cylinder at E also has a mass of 10 kg, and the coefficient of static friction between the cord and the peg is $\mu_s = 0.1$.

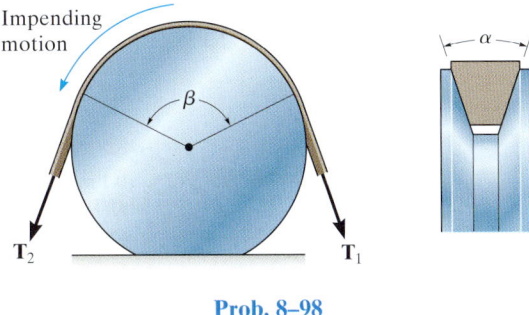

Prob. 8–98

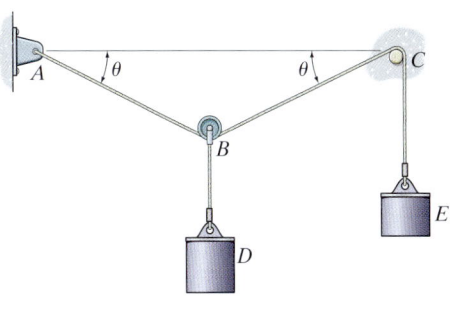

Prob. 8–100

8–99. If a force of $P = 200$ N is applied to the handle of the bell crank, determine the maximum torque **M** that can be resisted so that the flywheel does not rotate clockwise. The coefficient of static friction between the brake band and the rim of the wheel is $\mu_s = 0.3$.

8–101. A V-belt is used to connect the hub A of the motor to wheel B. If the belt can withstand a maximum tension of 1200 N, determine the largest mass of cylinder C that can be lifted and the corresponding torque **M** that must be supplied to A. The coefficient of static friction between the hub and the belt is $\mu_s = 0.3$, and between the wheel and the belt is $\mu_s' = 0.20$. *Hint*: See Prob. 8–98.

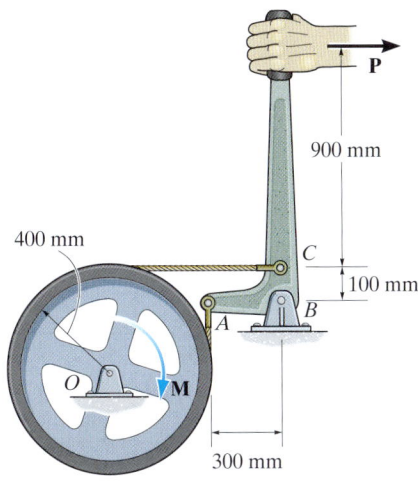

Prob. 8–99

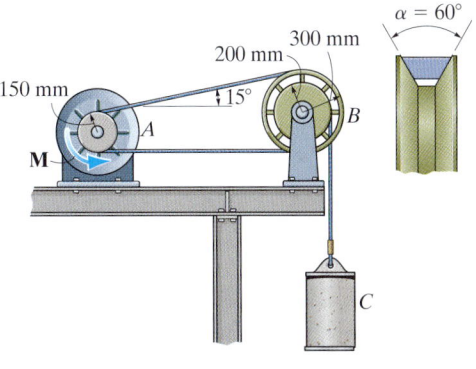

Prob. 8–101

8–102. The 20-kg motor has a center of gravity at G and is pin-connected at C to maintain a tension in the drive belt. Determine the smallest counterclockwise twist or torque $\mathbf{M}$ that must be supplied by the motor to turn the disk B if wheel A locks and causes the belt to slip over the disk. No slipping occurs at A. The coefficient of static friction between the belt and the disk is $\mu_s = 0.35$.

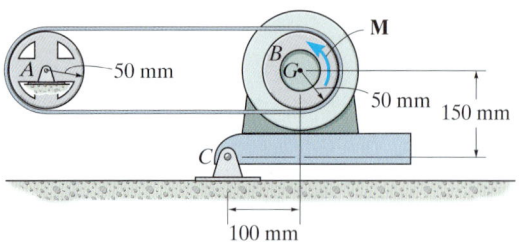

Prob. 8–102

8–103. Blocks A and B have a mass of 100 kg and 150 kg, respectively. If the coefficient of static friction between A and B and between B and C is $\mu_s = 0.25$ and between the ropes and the pegs D and E $\mu'_s = 0.5$ determine the smallest force F needed to cause motion of block B if $P = 30$ N.

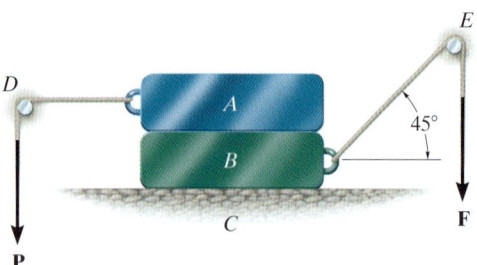

Prob. 8–103

*8–104. Determine the minimum coefficient of static friction μ_s between the cable and the peg and the placement d of the 3-kN force for the uniform 100-kg beam to maintain equilibrium.

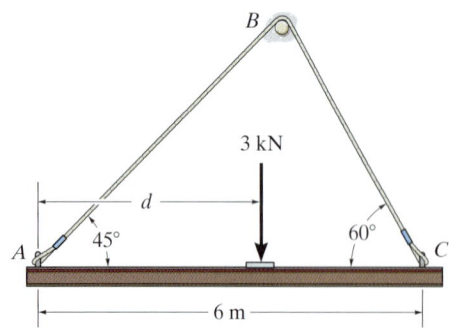

Prob. 8–104

8–105. A conveyer belt is used to transfer granular material and the frictional resistance on the top of the belt is $F = 500$ N. Determine the smallest stretch of the spring attached to the moveable axle of the idle pulley B so that the belt does not slip at the drive pulley A when the torque $\mathbf{M}$ is applied. What minimum torque $\mathbf{M}$ is required to keep the belt moving? The coefficient of static friction between the belt and the wheel at A is $\mu_s = 0.2$.

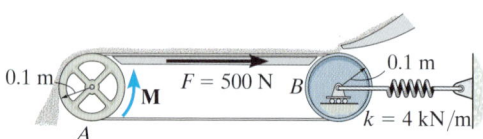

Prob. 8–105

8–106. The belt on the portable dryer wraps around the drum D, idler pulley A, and motor pulley B. If the motor can develop a maximum torque of $M = 0.80$ N · m, determine the smallest spring tension required to hold the belt from slipping. The coefficient of static friction between the belt and the drum and motor pulley is $\mu_s = 0.3$.

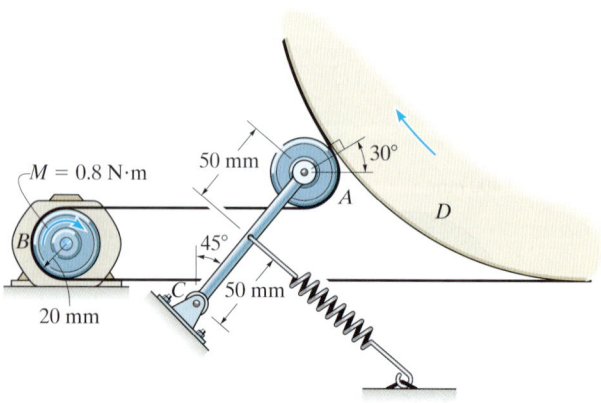

Prob. 8–106

*8.6   Frictional Forces on Collar Bearings, Pivot Bearings, and Disks

Pivot and *collar bearings* are commonly used in machines to support an *axial load* on a rotating shaft. Typical examples are shown in Fig. 8–20. Provided these bearings are not lubricated, or are only partially lubricated, the laws of dry friction may be applied to determine the moment needed to turn the shaft when it supports an axial force.

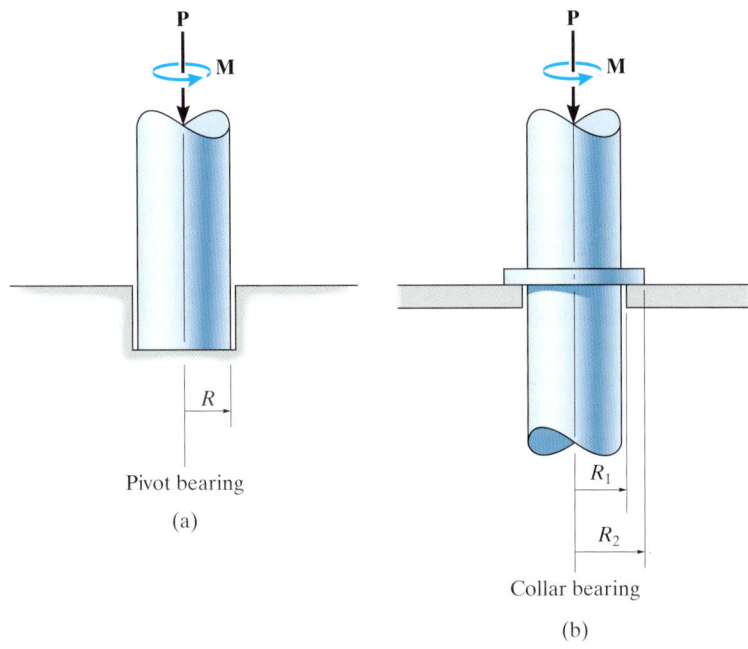

Pivot bearing

(a)

R_1

R_2

Collar bearing

(b)

Fig. 8–20

Frictional Analysis.   The collar bearing on the shaft shown in Fig. 8–21 is subjected to an axial force **P** and has a total bearing or contact area $\pi(R_2^2 - R_1^2)$. Provided the bearing is new and evenly supported, then the normal pressure p on the bearing will be *uniformly distributed* over this area. Since $\Sigma F_z = 0$, then p, measured as a force per unit area, is $p = P/\pi(R_2^2 - R_1^2)$.

The moment needed to cause impending rotation of the shaft can be determined from moment equilibrium about the z axis. A differential area element $dA = (r\,d\theta)(dr)$, shown in Fig. 8–21, is subjected to both a normal force $dN = p\,dA$ and an associated frictional force,

$$dF = \mu_s\,dN = \mu_s p\,dA = \frac{\mu_s P}{\pi(R_2^2 - R_1^2)}\,dA$$

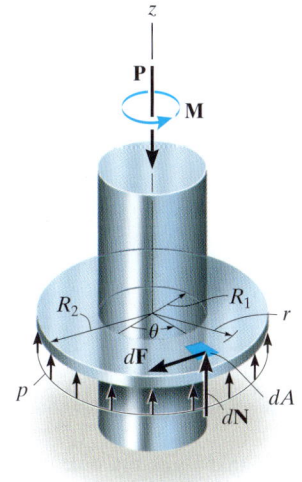

Fig. 8–21

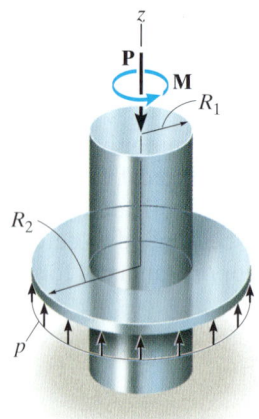

Fig. 8–21 (Repeated)

The normal force does not create a moment about the z axis of the shaft; however, the frictional force does; namely, $dM = r \, dF$. Integration is needed to compute the applied moment **M** needed to overcome all the frictional forces. Therefore, for impending rotational motion,

$$\Sigma M_z = 0; \qquad M - \int_A r \, dF = 0$$

Substituting for dF and dA and integrating over the entire bearing area yields

$$M = \int_{R_1}^{R_2} \int_0^{2\pi} r \left[\frac{\mu_s P}{\pi(R_2^2 - R_1^2)} \right] (r \, d\theta \, dr) = \frac{\mu_s P}{\pi(R_2^2 - R_1^2)} \int_{R_1}^{R_2} r^2 \, dr \int_0^{2\pi} d\theta$$

or

$$M = \frac{2}{3} \mu_s P \left(\frac{R_2^3 - R_1^3}{R_2^2 - R_1^2} \right) \qquad (8\text{–}7)$$

The moment developed at the end of the shaft, when it is *rotating* at constant speed, can be found by substituting μ_k for μ_s in Eq. 8–7.

In the case of a pivot bearing, Fig. 8–20a, then $R_2 = R$ and $R_1 = 0$, and Eq. 8–7 reduces to

$$M = \frac{2}{3} \mu_s P R \qquad (8\text{–}8)$$

Remember that Eqs. 8–7 and 8–8 apply only for bearing surfaces subjected to *constant pressure*. If the pressure is not uniform, a variation of the pressure as a function of the bearing area must be determined before integrating to obtain the moment. The following example illustrates this concept.

The motor that turns the disk of this sanding machine develops a torque that must overcome the frictional forces acting on the disk.

EXAMPLE | 8.9

The uniform bar shown in Fig. 8–22a has a weight of 20 N ($\approx$2 kg). If it is assumed that the normal pressure acting at the contacting surface varies linearly along the length of the bar as shown, determine the couple moment **M** required to rotate the bar. Assume that the bar's width is negligible in comparison to its length. The coefficient of static friction is equal to $\mu_s = 0.3$.

SOLUTION

A free-body diagram of the bar is shown in Fig. 8–22b. The intensity w_0 of the distributed load at the center ($x = 0$) is determined from vertical force equilibrium, Fig. 8–22a.

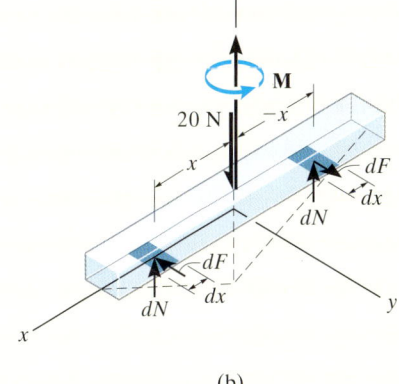

(a)

$$+\uparrow\Sigma F_z = 0; \quad -20\text{ N} + 2\left[\frac{1}{2}(0.5\text{ m})w_0\right] = 0 \qquad {}_0w = 40\text{ N/m}$$

Since $w = 0$ at $x = 0.5$ m, the distributed load expressed as a function of x is

$$w = (40\text{ N/m})\left(1 - \frac{x}{0.5\text{ m}}\right) = 40 - 80x$$

The magnitude of the normal force acting on a differential segment of area having a length dx is therefore

$$dN = w\,dx = (40 - 80x)dx$$

The magnitude of the frictional force acting on the same element of area is

$$dF = \mu_s\,dN = 0.3(40 - 80x)\,dx$$

Hence, the moment created by this force about the z axis is

$$dM = x\,dF = 0\,3(40 - 80x^2)dx$$

The summation of moments about the z axis of the bar is determined by integration, which yields

$$\Sigma M_z = 0; \quad M - 2\int_0^{0.5\text{ m}}(0.3)(40x - 80x^2)\,dx = 0$$

$$M = 6\left(x^2 - \frac{4x^3}{3}\right)\Big|_0^{0.5}$$

$$M = 0.5\text{ N}\cdot\text{m} \qquad \textit{Ans.}$$

(b)

Fig. 8–22

8.7   Frictional Forces on Journal Bearings

When a shaft or axle is subjected to lateral loads, a *journal bearing* is commonly used for support. Provided the bearing is not lubricated, or is only partially lubricated, a reasonable analysis of the frictional resistance on the bearing can be based on the laws of dry friction.

Frictional Analysis.   A typical journal-bearing support is shown in Fig. 8–23a. As the shaft rotates, the contact point moves up the wall of the bearing to some point A where slipping occurs. If the vertical load acting at the end of the shaft is **P**, then the bearing reactive force **R** acting at A will be equal and opposite to **P**, Fig. 8–23b. The moment needed to maintain constant rotation of the shaft can be found by summing moments about the z axis of the shaft; i.e.,

$$\Sigma M_z = 0; \qquad\qquad M - (R \sin \phi_k)r = 0$$

or

$$M = Rr \sin \phi_k \qquad\qquad (8\text{–}9)$$

where ϕ_k is the angle of kinetic friction defined by $\tan \phi_k = F/N = \mu_k N/N = \mu_k$. In Fig. 8–23c, it is seen that $r \sin \phi_k = r_f$. The dashed circle with radius r_f is called the *friction circle*, and as the shaft rotates, the reaction **R** will always be tangent to it. If the bearing is partially lubricated, μ_k is small, and therefore $\sin \phi_k \approx \tan \phi_k \approx \mu_k$. Under these conditions, a reasonable *approximation* to the moment needed to overcome the frictional resistance becomes

$$M \approx Rr\mu_k \qquad\qquad (8\text{–}10)$$

Notice that to minimize friction the bearing radius r should be as small as possible. In practice, however, this type of journal bearing is not suitable for long service since friction between the shaft and bearing will eventually wear down the surfaces. Instead, designers will incorporate "ball bearings" or "rollers" in journal bearings to minimize frictional losses.

Unwinding the cable from this spool requires overcoming friction from the supporting shaft.

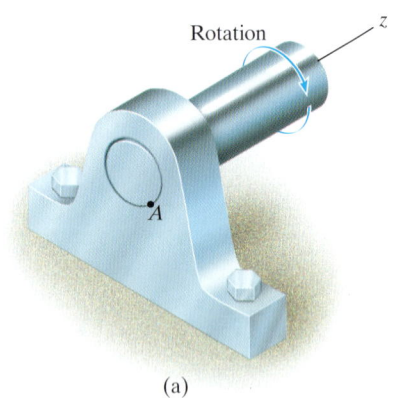

(a)

Fig. 8–23

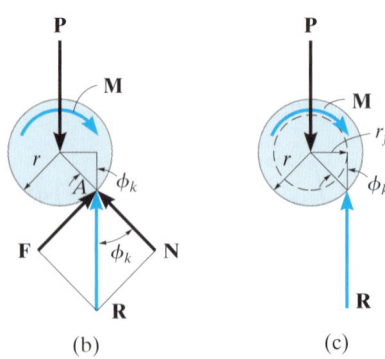

(b)   (c)

EXAMPLE 8.10

The 100-mm-diameter pulley shown in Fig. 8–24a fits loosely on a 10-mm-diameter shaft for which the coefficient of static friction is $\mu_s = 0.4$. Determine the minimum tension T in the belt needed to (a) raise the 100-kg block and (b) lower the block. Assume that no slipping occurs between the belt and pulley and neglect the weight of the pulley.

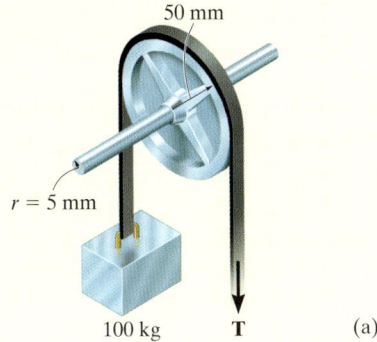

50 mm

$r = 5$ mm

100 kg   **T**   (a)

SOLUTION

Part (a). A free-body diagram of the pulley is shown in Fig. 8–24b. When the pulley is subjected to belt tensions of 981 N each, it makes contact with the shaft at point P_1. As the tension T is *increased*, the contact point will move around the shaft to point P_2 before motion impends. From the figure, the friction circle has a radius $r_f = r \sin \phi_s$. Using the simplification that $\sin \phi_s \approx \tan \phi_s \approx \mu_s$ then $r_f \approx r\mu_s = (5 \text{ mm})(0.4) = 2$ mm, so that summing moments about P_2 gives

$\zeta + \Sigma M_{P_2} = 0;$ $981 \text{ N}(52 \text{ mm}) - T(48 \text{ mm}) = 0$
$$T = 1063 \text{ N} = 1.06 \text{ kN} \qquad Ans.$$

If a more exact analysis is used, then $\phi_s = \tan^{-1} 0.4 = 21.8°$. Thus, the radius of the friction circle would be $r_f = r \sin \phi_s = 5 \sin 21.8° = 1.86$ mm. Therefore,

$\zeta + \Sigma M_{P_2} = 0;$
$981 \text{ N}(50 \text{ mm} + 1.86 \text{ mm}) - T(50 \text{ mm} - 1.86 \text{ mm}) = 0$
$$T = 1057 \text{ N} = 1.06 \text{ kN} \qquad Ans.$$

Part (b). When the block is lowered, the resultant force **R** acting on the shaft passes through point as shown in Fig. 8–24c. Summing moments about this point yields

$\zeta + \Sigma M_{P_3} = 0;$ $981 \text{ N}(48 \text{ mm}) - T(52 \text{ mm}) = 0$
$$T = 906 \text{ N} \qquad Ans.$$

NOTE: Using the approximate analysis, the difference between raising and lowering the block is thus 157 N.

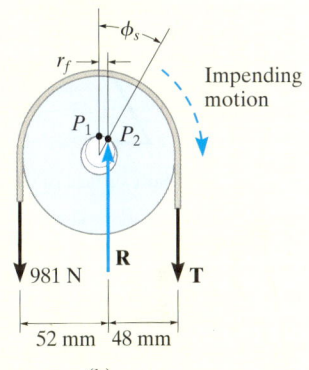

ϕ_s

r_f

Impending motion

P_1 P_2

R

981 N   **T**

52 mm 48 mm

(b)

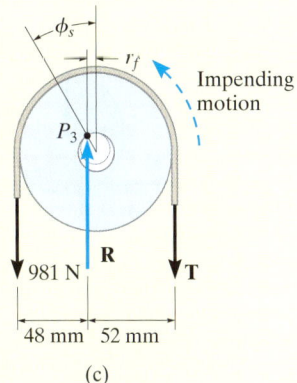

ϕ_s

r_f

Impending motion

P_3

R

981 N   **T**

48 mm 52 mm

(c)

Fig. 8–24

8

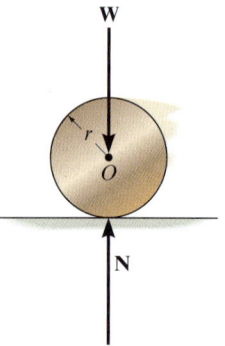

Rigid surface of contact

(a)

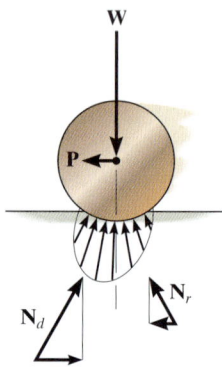

Soft surface of contact

(b)

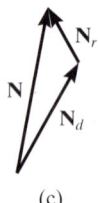

(c)

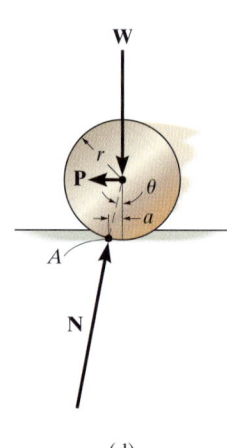

(d)

Fig. 8–25

*8.8   Rolling Resistance

When a *rigid* cylinder rolls at constant velocity along a *rigid* surface, the normal force exerted by the surface on the cylinder acts perpendicular to the tangent at the point of contact, as shown in Fig. 8–25a. Actually, however, no materials are perfectly rigid, and therefore the reaction of the surface on the cylinder consists of a distribution of normal pressure. For example, consider the cylinder to be made of a very hard material, and the surface on which it rolls to be relatively soft. Due to its weight, the cylinder compresses the surface underneath it, Fig. 8–25b. As the cylinder rolls, the surface material in front of the cylinder *retards* the motion since it is being *deformed*, whereas the material in the rear is *restored* from the deformed state and therefore tends to *push* the cylinder forward. The normal pressures acting on the cylinder in this manner are represented in Fig. 8–25b by their resultant forces $\mathbf{N}_d$ and $\mathbf{N}_r$. The magnitude of the force of *deformation*, $\mathbf{N}_d$, and its horizontal component is *always greater* than that of *restoration*, $\mathbf{N}_r$, and consequently a horizontal driving force $\mathbf{P}$ must be applied to the cylinder to maintain the motion. Fig. 8–25b.*

Rolling resistance is caused primarily by this effect, although it is also, to a lesser degree, the result of surface adhesion and relative microsliding between the surfaces of contact. Because the actual force $\mathbf{P}$ needed to overcome these effects is difficult to determine, a simplified method will be developed here to explain one way engineers have analyzed this phenomenon. To do this, we will consider the resultant of the *entire* normal pressure, $\mathbf{N} = \mathbf{N}_d + \mathbf{N}_r$, acting on the cylinder, Fig. 8–25c. As shown in Fig. 8–25d, this force acts at an angle θ with the vertical. To keep the cylinder in equilibrium, i.e., rolling at a constant rate, it is necessary that $\mathbf{N}$ be *concurrent* with the driving force $\mathbf{P}$ and the weight $\mathbf{W}$. Summing moments about point A gives $Wa = P(r \cos \theta)$. Since the deformations are generally very small in relation to the cylinder's radius, $\cos \theta \approx 1$; hence,

$$Wa \approx Pr$$

or

$$\boxed{P \approx \frac{Wa}{r}} \qquad (8\text{–}11)$$

The distance a is termed the *coefficient of rolling resistance,* which has the dimension of length. For instance, $a \approx 0.5$ mm for a wheel rolling on a rail, both of which are made of mild steel. For hardened steel ball

---

*Actually, the deformation force $\mathbf{N}_d$ causes *energy* to be stored in the material as its magnitude is increased, whereas the restoration force $\mathbf{N}_r$, as its magnitude is decreased, allows some of this energy to be released. The remaining energy is *lost* since it is used to heat up the surface, and if the cylinder's weight is very large, it accounts for permanent deformation of the surface. Work must be done by the horizontal force $\mathbf{P}$ to make up for this loss.

bearings on steel, $a \approx 0.1$ mm. Experimentally, though, this factor is difficult to measure, since it depends on such parameters as the rate of rotation of the cylinder, the elastic properties of the contacting surfaces, and the surface finish. For this reason, little reliance is placed on the data for determining a. The analysis presented here does, however, indicate why a heavy load (W) offers greater resistance to motion (P) than a light load under the same conditions. Furthermore, since Wa/r is generally very small compared to $\mu_k W$, the force needed to *roll* a cylinder over the surface will be much less than that needed to *slide* it across the surface. It is for this reason that a roller or ball bearings are often used to minimize the frictional resistance between moving parts.

Rolling resistance of railroad wheels on the rails is small since steel is very stiff. By comparison, the rolling resistance of the wheels of a tractor in a wet field is very large.

EXAMPLE 8.11

A 10-kg steel wheel shown in Fig. 8–26a has a radius of 100 mm and rests on an inclined plane made of soft wood. If θ is increased so that the wheel begins to roll down the incline with constant velocity when $\theta = 1.2°$, determine the coefficient of rolling resistance.

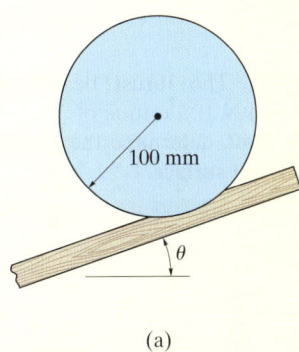

(a)

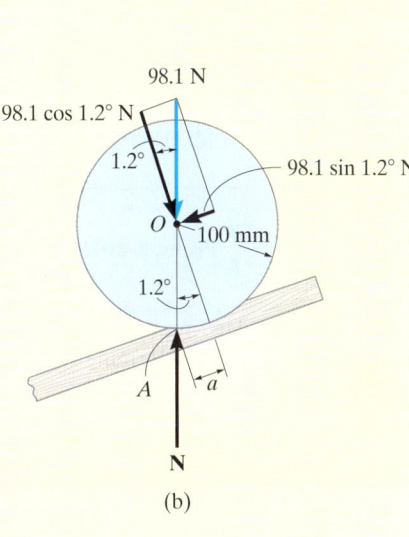

(b)

Fig. 8–26

SOLUTION
As shown on the free-body diagram, Fig. 8–26b, when the wheel has impending motion, the normal reaction **N** acts at point A defined by the dimension a. Resolving the weight into components parallel and perpendicular to the incline, and summing moments about point A, yields

$$\zeta + \Sigma M_A = 0;$$

$$-(98.1 \cos 1.2° \text{ N})(a) + (98.1 \sin 1.2° \text{ N})(100 \cos 1.2° \text{ mm}) = 0$$

Solving, we obtain

$$a = 2.09 \text{ mm} \qquad \qquad \textit{Ans.}$$

8

PROBLEMS

8–107. The collar bearing uniformly supports an axial force of $P = 2000$ N. If the coefficient of static friction is $\mu_s = 0.3$, determine the torque M required to overcome friction.

*8–108. The collar bearing uniformly supports an axial force of $P = 2000$ N. If a torque of $M = 3.6$ N · m is applied to the shaft and causes it to rotate at constant velocity, determine the coefficient of kinetic friction at the surface of contact.

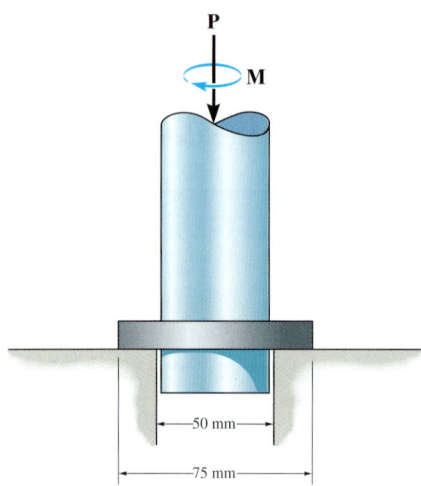

Probs. 8–107/108

8–109. The annular ring bearing is subjected to a thrust of 4000 N. If $\mu_s = 0.35$, determine the torque M that must be applied to overcome friction.

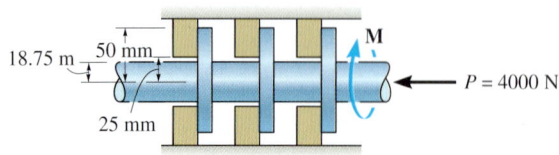

Prob. 8–109

8–110. The shaft is supported by a thrust bearing A and a journal bearing B. Determine the torque M required to rotate the shaft at constant angular velocity. The coefficient of kinetic friction at the thrust bearing is $\mu_k = 0.2$. Neglect friction at B.

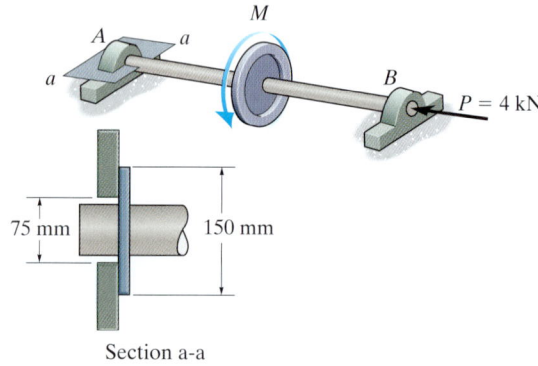

Section a-a

Prob. 8–110

8–111. The thrust bearing supports an axial load of $P = 6$ kN. If a torque of $M = 150$ N · m is required to rotate the shaft, determine the coefficient of static friction at the constant surface.

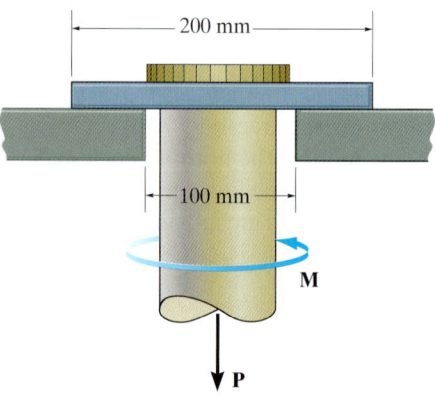

Prob. 8–111

*8–112. Assuming that the variation of pressure at the bottom of the pivot bearing is defined as $p = p_0(R_2/r)$, determine the torque M needed to overcome friction if the shaft is subjected to an axial force **P**. The coefficient of static friction is μ_s. For the solution, it is necessary to determine p_0 in terms of P and the bearing dimensions R_1 and R_2.

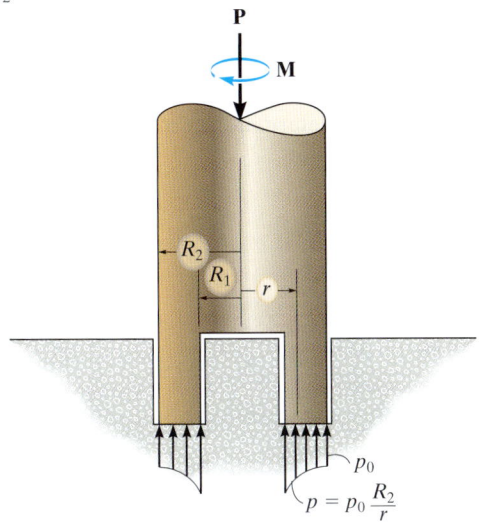

Prob. 8–112

8–113. The plate clutch consists of a flat plate A that slides over the rotating shaft S. The shaft is fixed to the driving plate gear B. If the gear C, which is in mesh with B, is subjected to a torque of $M = 0.8\ \text{N}\cdot\text{m}$, determine the smallest force P, that must be applied via the control arm, to stop the rotation. The coefficient of static friction between the plates A and D is $\mu_s = 0.4$. Assume the bearing pressure between A and D to be uniform.

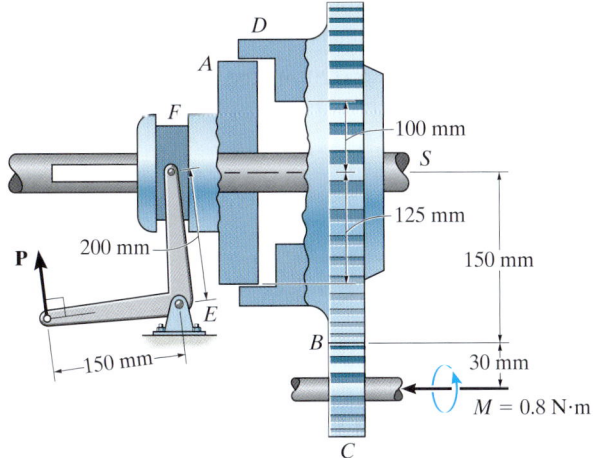

Prob. 8–113

8–114. The conical bearing is subjected to a constant pressure distribution at its surface of contact. If the coefficient of static friction is μ_s, determine the torque M required to overcome friction if the shaft supports an axial force **P**.

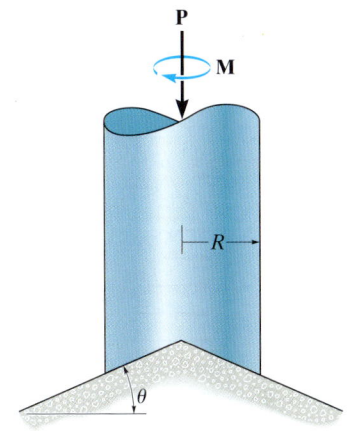

Prob. 8–114

8–115. The pivot bearing is subjected to a pressure distribution at its surface of contact which varies as shown. If the coefficient of static friction is μ, determine the torque M required to overcome friction if the shaft supports an axial force **P**.

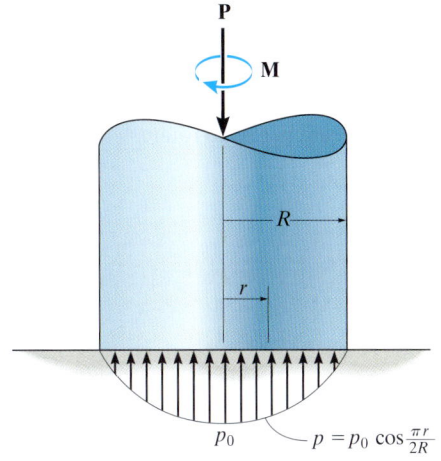

Prob. 8–115

*8–116. A 200-mm-diameter post is driven 3 m into sand for which $\mu_s = 0.3$. If the normal pressure acting *completely around the post* varies linearly with depth as shown, determine the frictional torque **M** that must be overcome to rotate the post.

8–118. The connecting rod is attached to the piston by a 20-mm-diameter pin at B and to the crank shaft by a 50-mm-diameter bearing A. If the piston is moving downwards, and the coefficient of static friction at these points is $\mu_s = 0.2$, determine the radius of the friction circle at each connection.

8–119. The connecting rod is attached to the piston by a 20-mm-diameter pin at B and to the crank shaft by a 50-mm-diameter bearing A. If the piston is moving upwards, and the coefficient of static friction at these points is $\mu_s = 0.3$, determine the radius of the friction circle at each connection.

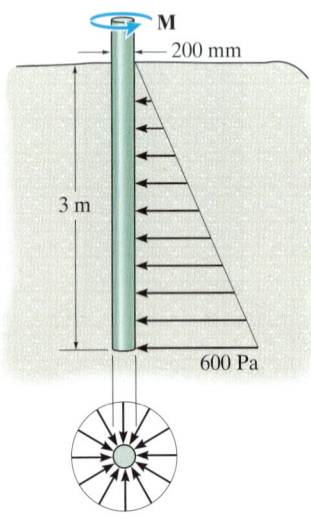

Prob. 8–116

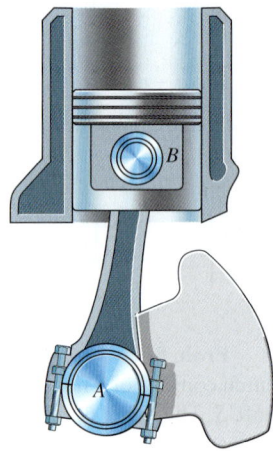

Probs. 8–118/119

8–117. A beam having a uniform weight W rests on the rough horizontal surface having a coefficient of static friction μ_s. If the horizontal force **P** is applied perpendicular to the beam's length, determine the location d of the point O about which the beam begins to rotate.

*8–120. The 5-kg pulley has a diameter of 240 mm and the axle has a diameter of 40 mm. If the coefficient of kinetic friction between the axle and the pulley is $\mu_k = 0.15$, determine the vertical force P on the rope required to lift the 80-kg block at constant velocity.

8–121. Solve Prob. 8–120 if the force **P** is applied horizontally to the right.

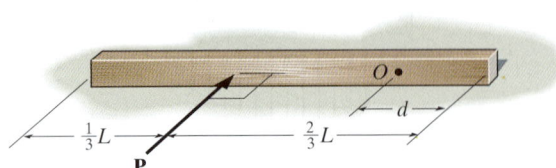

Prob. 8–117

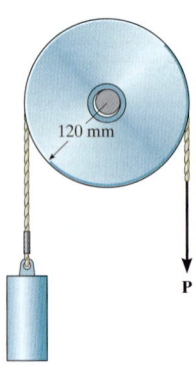

Probs. 8–120/121

8–122. The collar fits *loosely* around a fixed shaft that has a radius of 50 mm. If the coefficient of kinetic friction between the shaft and the collar is $\mu_k = 0.3$, determine the force P on the horizontal segment of the belt so that the collar rotates *counterclockwise* with a constant angular velocity. Assume that the belt does not slip on the collar; rather, the collar slips on the shaft. Neglect the weight and thickness of the belt and collar. The radius, measured from the center of the collar to the mean thickness of the belt, is 56.25 mm.

8–123. The collar fits *loosely* around a fixed shaft that has a radius of 50 mm. If the coefficient of kinetic friction between the shaft and the collar is $\mu_k = 0.3$, determine the force P on the horizontal segment of the belt so that the collar rotates *clockwise* with a constant angular velocity. Assume that the belt does not slip on the collar; rather, the collar slips on the shaft. Neglect the weight and thickness of the belt and collar. The radius, measured from the center of the collar to the mean thickness of the belt, is 56.25 mm.

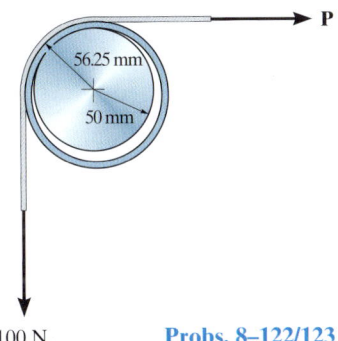

Probs. 8–122/123

*8–124.** A pulley having a diameter of 80 mm and mass of 1.25 kg is supported loosely on a shaft having a diameter of 20 mm. Determine the torque M that must be applied to the pulley to cause it to rotate with constant motion. The coefficient of kinetic friction between the shaft and pulley is $\mu_k = 0.4$. Also calculate the angle θ which the normal force at the point of contact makes with the horizontal. The shaft itself cannot rotate.

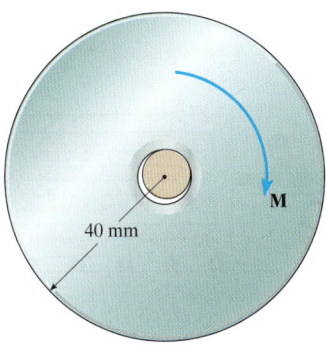

Prob. 8–124

8–125. The 5-kg skateboard rolls down the 5° slope at constant speed. If the coefficient of kinetic friction between the 12.5 mm diameter axles and the wheels is $\mu_k = 0.3$, determine the radius of the wheels. Neglect rolling resistance of the wheels on the surface. The center of mass for the skateboard is at G.

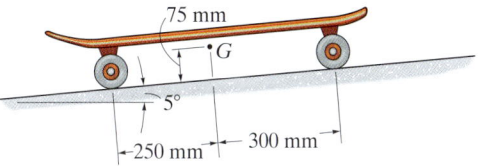

Prob. 8–125

8–126. Determine the force P required to overcome rolling resistance and pull the 50-kg roller up the inclined plane with constant velocity. The coefficient of rolling resistance is $a = 15$ mm.

8–127. Determine the force P required to overcome rolling resistance and support the 50-kg roller if it rolls down the inclined plane with constant velocity. The coefficient of rolling resistance is $a = 15$ mm.

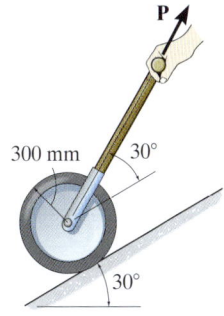

Probs. 8–126/127

*8-128. The lawn roller has a mass of 80 kg. If the arm BA is held at an angle of 30° from the horizontal and the coefficient of rolling resistance for the roller is 25 mm, determine the force P needed to push the roller at constant speed. Neglect friction developed at the axle, A, and assume that the resultant force **P** acting on the handle is applied along arm BA.

Prob. 8–128

8-129. The 1.4-Mg machine is to be moved over a level surface using a series of rollers for which the coefficient of rolling resistance is 0.5 mm at the ground and 0.2 mm at the bottom surface of the machine. Determine the appropriate diameter of the rollers so that the machine can be pushed forward with a horizontal force of $P = 250$ N. *Hint:* Use the result of Prob. 8–131.

Prob. 8–129

8-130. The hand cart has wheels with a diameter of 80 mm. If a crate having a mass of 500 kg is placed on the cart so that each wheel carries an equal load, determine the horizontal force P that must be applied to the handle to overcome the rolling resistance. The coefficient of rolling resistance is 2 mm. Neglect the mass of the cart.

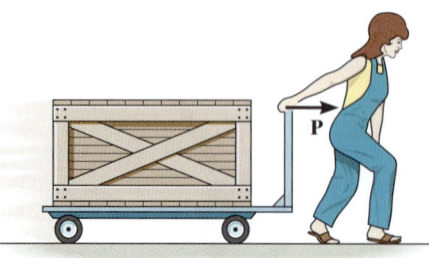

Prob. 8–130

8-131. The cylinder is subjected to a load that has a weight W. If the coefficients of rolling resistance for the cylinder's top and bottom surfaces are a_A and a_B, respectively, show that a horizontal force having a magnitude of $P = [W(a_A + a_B)]/2r$ is required to move the load and thereby roll the cylinder forward. Neglect the weight of the cylinder.

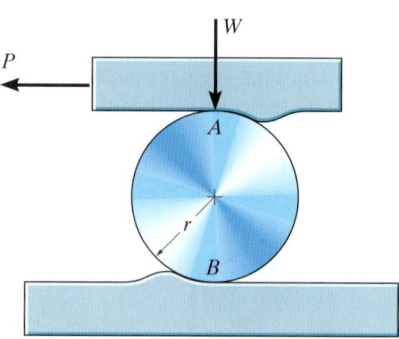

Prob. 8–131

*8-132. A large crate having a mass of 200 kg is moved along the floor using a series of 150-mm-diameter rollers for which the coefficient of rolling resistance is 3 mm at the ground and 7 mm at the bottom surface of the crate. Determine the horizontal force **P** needed to push the crate forward at a constant speed. *Hint:* Use the result of Prob. 8–131.

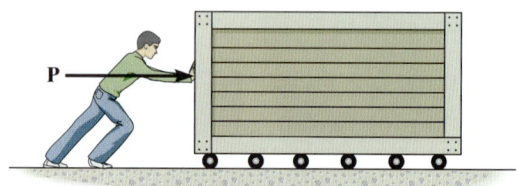

Prob. 8–132

CHAPTER REVIEW

Dry Friction

Frictional forces exist between two rough surfaces of contact. These forces act on a body so as to oppose its motion or tendency of motion.

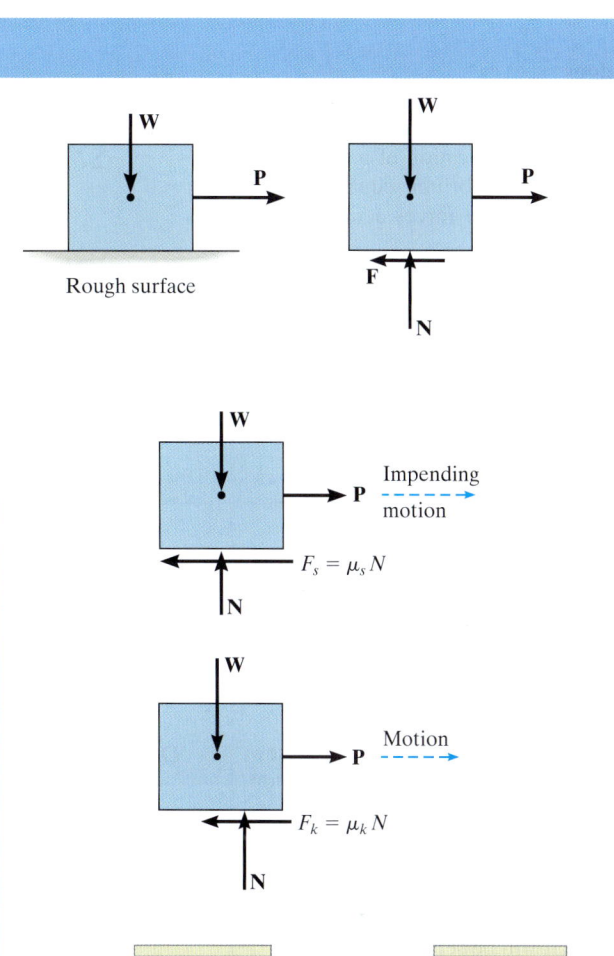

Rough surface

A static frictional force approaches a maximum value of $F_s = \mu_s N$, where μ_s is the *coefficient of static friction*. In this case, motion between the contacting surfaces is *impending*.

If slipping occurs, then the friction force remains essentially constant and equal to $F_k = \mu_k N$. Here μ_k is the *coefficient of kinetic friction*.

The solution of a problem involving friction requires first drawing the free-body diagram of the body. If the unknowns cannot be determined strictly from the equations of equilibrium, and the possibility of slipping occurs, then the friction equation should be applied at the appropriate points of contact in order to complete the solution.

It may also be possible for slender objects, like crates, to tip over, and this situation should also be investigated.

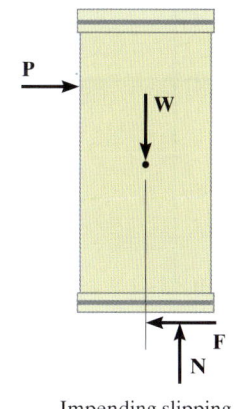

Impending slipping
$F = \mu_s N$

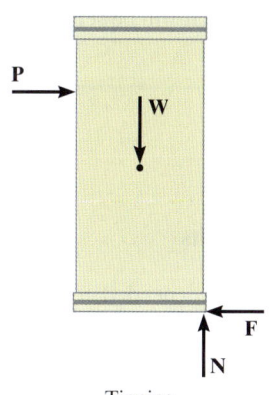

Tipping

Wedges

Wedges are inclined planes used to increase the application of a force. The two force equilibrium equations are used to relate the forces acting on the wedge.

An applied force **P** must push on the wedge to move it to the right.

If the coefficients of friction between the surfaces are large enough, then **P** can be removed, and the wedge will be self-locking and remain in place.

$$\Sigma F_x = 0$$
$$\Sigma F_y = 0$$

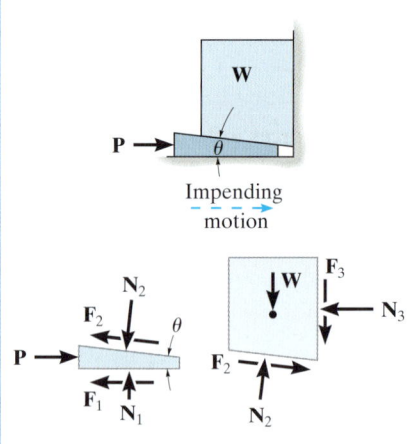

Screws

Square-threaded screws are used to move heavy loads. They represent an inclined plane, wrapped around a cylinder.

The moment needed to turn a screw depends upon the coefficient of friction and the screw's lead angle θ.

If the coefficient of friction between the surfaces is large enough, then the screw will support the load without tending to turn, i.e., it will be self-locking.

$$M = rW \tan(\theta + \phi_s)$$
Upward Impending Screw Motion

$$M' = rW \tan(\theta - \phi_s)$$
Downward Impending Screw Motion
$$\theta > \phi_s$$

$$M'' = rW \tan(\phi_s - \theta)$$
Downward Screw Motion
$$\phi_s > \theta$$

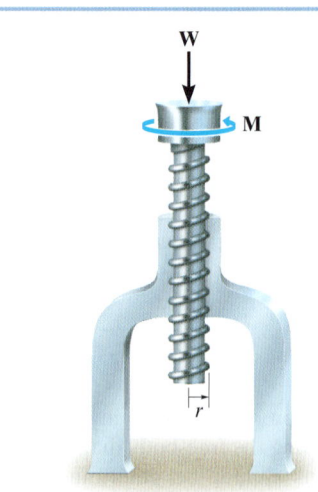

Flat Belts

The force needed to move a flat belt over a rough curved surface depends only on the angle of belt contact, β, and the coefficient of friction.

$$T_2 = T_1 e^{\mu\beta}$$

$$T_2 > T_1$$

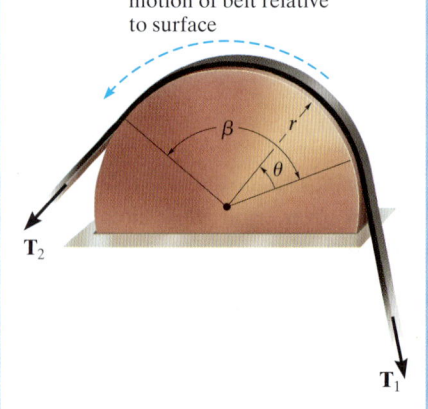

Collar Bearings and Disks

The frictional analysis of a collar bearing or disk requires looking at a differential element of the contact area. The normal force acting on this element is determined from force equilibrium along the shaft, and the moment needed to turn the shaft at a constant rate is determined from moment equilibrium about the shaft's axis.

If the pressure on the surface of a collar bearing is uniform, then integration gives the result shown.

$$M = \frac{2}{3}\mu_s P\left(\frac{R_2^3 - R_1^3}{R_2^2 - R_1^2}\right)$$

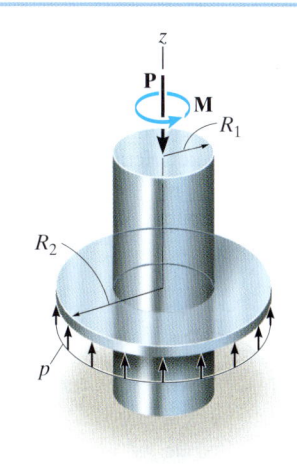

Journal Bearings

When a moment is applied to a shaft in a nonlubricated or partially lubricated journal bearing, the shaft will tend to roll up the side of the bearing until slipping occurs. This defines the radius of a friction circle, and from it the moment needed to turn the shaft can be determined.

$$M = Rr \sin \phi_k$$

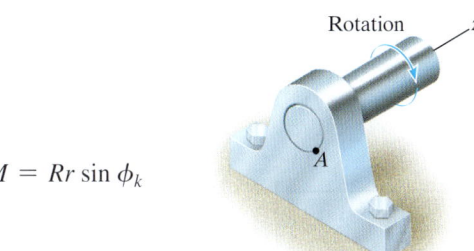
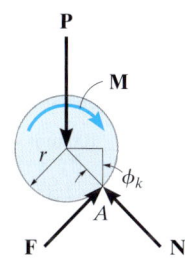

Rolling Resistance

The resistance of a wheel to rolling over a surface is caused by localized *deformation* of the two materials in contact. This causes the resultant normal force acting on the rolling body to be inclined so that it provides a component that acts in the opposite direction of the applied force **P** causing the motion. This effect is characterized using the *coefficient of rolling resistance, a,* which is determined from experiment.

$$P \approx \frac{Wa}{r}$$

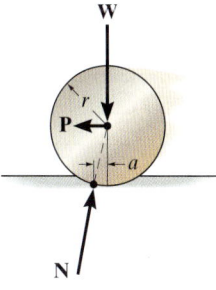

REVIEW PROBLEMS

8–133. Each of the cylinders has a mass of 50 kg. If the coefficients of static friction at the points of contact are $\mu_A = 0.5$, $\mu_B = 0.5$, $\mu_C = 0.5$, and $\mu_D = 0.6$, determine the smallest couple moment M needed to rotate cylinder E.

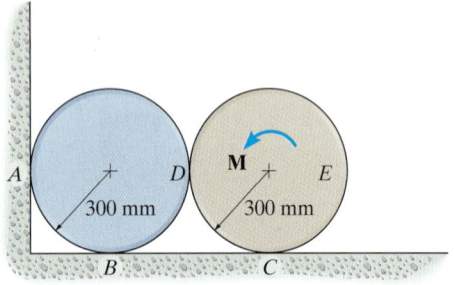

Prob. 8–133

8–134. The clamp is used to tighten the connection between two concrete drain pipes. Determine the least coefficient of static friction at A or B so that the clamp does not slip regardless of the force in the shaft CD.

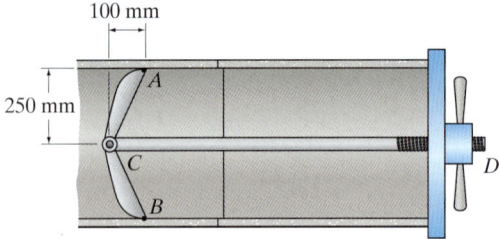

Prob. 8–134

8–135. If $P = 900$ N is applied to the handle of the bell crank, determine the maximum torque M the cone clutch can transmit. The coefficient of static friction at the contacting surface is $\mu_s = 0.3$.

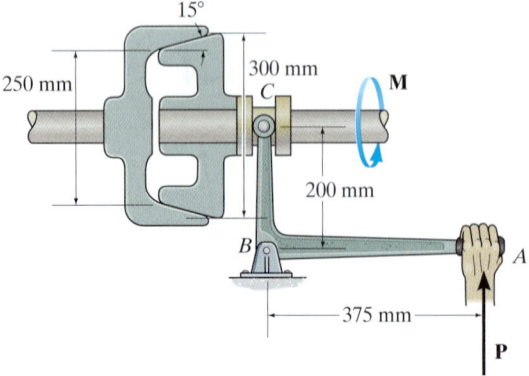

Prob. 8–135

***8–136.** The lawn roller has a mass of 80 kg. If the arm BA is held at an angle of 30° from the horizontal and the coefficient of rolling resistance for the roller is 25 mm, determine the force P needed to push the roller at constant speed. Neglect friction developed at the axle, A, and assume that the resultant force $\mathbf{P}$ acting on the handle is applied along arm BA.

Prob. 8–136

8–137. Two blocks A and B, each having a mass of 6 kg, are connected by the linkage shown. If the coefficients of static friction at the contacting surfaces are $\mu_A = 0.2$ and $\mu_B = 0.8$, determine the largest vertical force P that may be applied to pin C without causing the blocks to slip. Neglect the weight of the links.

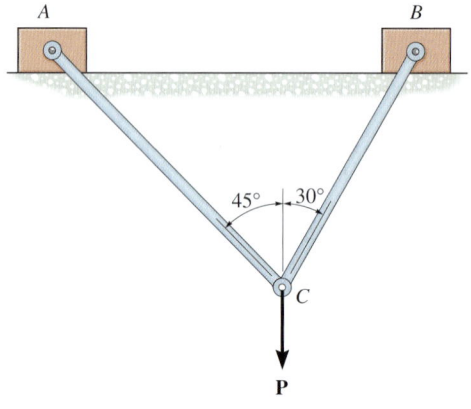

Prob. 8–137

8–138. The uniform 60-kg crate C rests uniformly on a 10-kg dolly D. If the front casters of the dolly at A are locked to prevent rolling while the casters at B are free to roll, determine the maximum force **P** that may be applied without causing motion of the crate. The coefficient of static friction between the casters and the floor is $\mu_f = 0.35$ and between the dolly and the crate, $\mu_d = 0.5$.

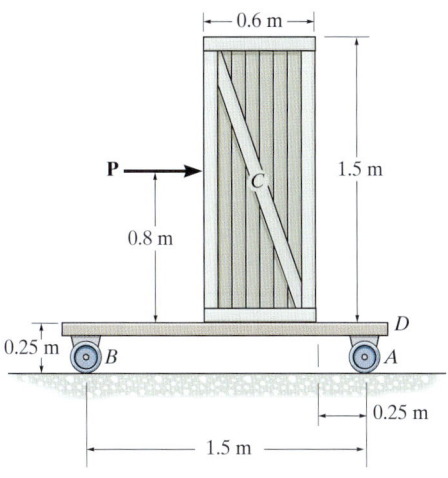

Prob. 8–138

8–139. The 3-Mg four-wheel-drive truck (SUV) has a center of mass at G. Determine the maximum mass of the log that can be towed by the truck. The coefficient of static friction between the log and the ground is $\mu_s = 0.8$, and the coefficient of static friction between the wheels of the truck and the ground is $\mu_s' = 0.4$. Assume that the engine of the truck is powerful enough to generate a torque that will cause all the wheels to slip.

*8–140. A 3-Mg front-wheel-drive truck (SUV) has a center of mass at G. Determine the maximum mass of the log that can be towed by the truck. The coefficient of static friction between the log and the ground is $\mu_s = 0.8$, and the coefficient of static friction between the front wheels of the truck and the ground is $\mu_s' = 0.4$. The rear wheels are free to roll. Assume that the engine of the truck is powerful enough to generate a torque that will cause the front wheels to slip.

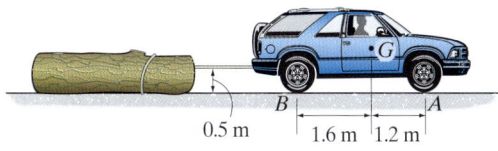

Probs. 8–139/140

8–141. A roofer, having a mass of 70 kg, walks slowly in an upright position down along the surface of a dome that has a radius of curvature of $r = 20$ m. If the coefficient of static friction between his shoes and the dome is $\mu_s = 0.7$, determine the angle θ at which he first begins to slip.

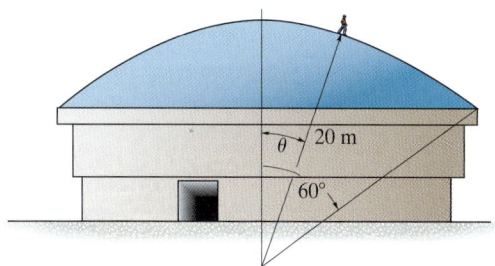

Prob. 8–141

8–142. Determine the minimum horizontal force P required to hold the crate from sliding down the plane. The crate has a mass of 50 kg and the coefficient of static friction between the crate and the plane is $\mu_s = 0.25$.

8–143. Determine the minimum force P required to push the crate up the plane. The crate has a mass of 50 kg and the coefficient of static friction between the crate and the plane is $\mu_s = 0.25$.

*8–144. A horizontal force of $P = 100$ N is just sufficient to hold the crate from sliding down the plane, and a horizontal force of $P = 350$ N is required to just push the crate up the plane. Determine the coefficient of static friction between the plane and the crate, and find the mass ot the crate.

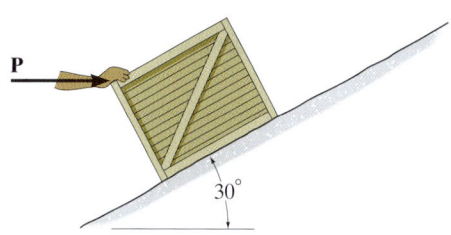

Probs. 8–142/143/144

Chapter 9

When a pressure tank of any shape is designed, it is important to be able to determine its center of gravity, calculate its volume and surface area, and determine the forces of the liquids they contain. All of these topics will be covered in this chapter.

Center of Gravity and Centroid

Video Solutions are available for selected questions in this chapter.

9.1 Center of Gravity, Center of Mass, and the Centroid of a Body

In this section we will first show how to locate the center of gravity for a body, and then we will show that the center of mass and the centroid of a body can be developed using this same method.

Center of Gravity. A body is composed of an infinite number of particles of differential size, and so if the body is located within a gravitational field, then each of these particles will have a weight dW, Fig. 9–1a. These weights will form an approximately parallel force system, and the resultant of this system is the total weight of the body, which passes through a single point called the *center of gravity, G*, Fig. 9–1b.*

---

*This is true as long as the gravity field is assumed to have the same magnitude and direction everywhere. That assumption is appropriate for most engineering applications, since gravity does not vary appreciably between, for instance, the bottom and the top of a building.

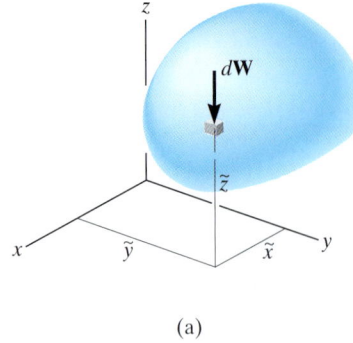

(a)

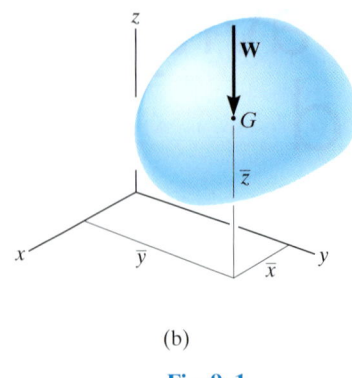

(b)

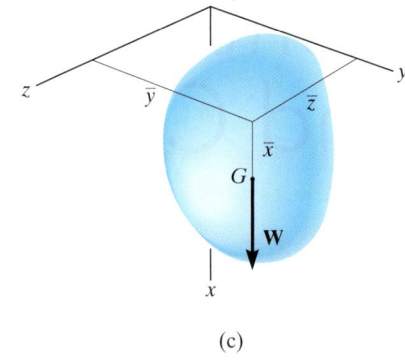

(c)

Fig. 9–1

Using the methods outlined in Sec. 4.8, the weight of the body is the sum of the weights of all of its particles, that is

$$+\downarrow F_R = \Sigma F_z; \qquad\qquad W = \int dW$$

The location of the center of gravity, measured from the y axis, is determined by equating the moment of W about the y axis, Fig. 9–1b, to the sum of the moments of the weights of the particles about this same axis. If dW is located at point $(\tilde{x}, \tilde{y}, \tilde{z})$, Fig. 9–1$a$, then

$$(M_R)_y = \Sigma M_y; \qquad\qquad \bar{x}W = \int \tilde{x}\,dW$$

Similarly, if moments are summed about the x axis,

$$(M_R)_x = \Sigma M_x; \qquad\qquad \bar{y}W = \int \tilde{y}\,dW$$

Finally, imagine that the body is fixed within the coordinate system and this system is rotated 90° about the y axis, Fig. 9–1c. Then the sum of the moments about the y axis gives

$$(M_R)_y = \Sigma M_y; \qquad\qquad \bar{z}W = \int \tilde{z}\,dW$$

Therefore, the location of the center of gravity G with respect to the $x, y,$ z axes becomes

$$\bar{x} = \frac{\int \tilde{x}\,dW}{\int dW} \qquad \bar{y} = \frac{\int \tilde{y}\,dW}{\int dW} \qquad \bar{z} = \frac{\int \tilde{z}\,dW}{\int dW} \qquad (9\text{–}1)$$

Here

$\bar{x}, \bar{y}, \bar{z}$ are the coordinates of the center of gravity G, Fig. 9–1b.
$\tilde{x}, \tilde{y}, \tilde{z}$ are the coordinates of each particle in the body, Fig. 9–1a.

Center of Mass of a Body.   In order to study the *dynamic response* or accelerated motion of a body, it becomes important to locate the body's center of mass C_m, Fig. 9–2. This location can be determined by substituting $dW = g\,dm$ into Eqs. 9–1. Since g is constant, it cancels out, and so

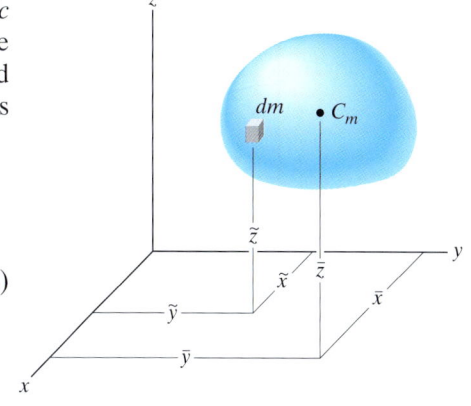

$$\bar{x} = \frac{\displaystyle\int \tilde{x}\,dm}{\displaystyle\int dm} \qquad \bar{y} = \frac{\displaystyle\int \tilde{y}\,dm}{\displaystyle\int dm} \qquad \bar{z} = \frac{\displaystyle\int \tilde{z}\,dm}{\displaystyle\int dm} \qquad (9\text{–}2)$$

Centroid of a Volume.   If the body in Fig. 9–3 is made from a homogeneous material, then its density ρ (rho) will be constant. Therefore, a differential element of volume dV has a mass $dm = \rho\,dV$. Substituting this into Eqs. 9–2 and canceling out ρ, we obtain formulas that locate the *centroid C* or geometric center of the body; namely

Fig. 9–2

$$\bar{x} = \frac{\displaystyle\int_V \tilde{x}\,dV}{\displaystyle\int_V dV} \qquad \bar{y} = \frac{\displaystyle\int_V \tilde{y}\,dV}{\displaystyle\int_V dV} \qquad \bar{z} = \frac{\displaystyle\int_V \tilde{z}\,dV}{\displaystyle\int_V dV} \qquad (9\text{–}3)$$

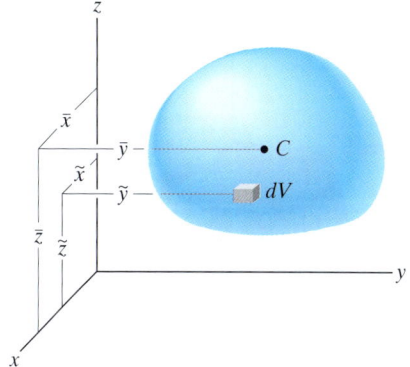

These equations represent a balance of the moments of the volume of the body. Therefore, if the volume possesses two planes of symmetry, then its centroid must lie along the line of intersection of these two planes. For example, the cone in Fig. 9–4 has a centroid that lies on the y axis so that $\bar{x} = \bar{z} = 0$. The location $\bar{y}$ can be found using a single integration by choosing a differential element represented by a *thin disk* having a thickness dy and radius $r = z$. Its volume is $dV = \pi r^2\,dy = \pi z^2\,dy$ and its centroid is at $\tilde{x} = 0, \tilde{y} = y, \tilde{z} = 0$.

Fig. 9–3

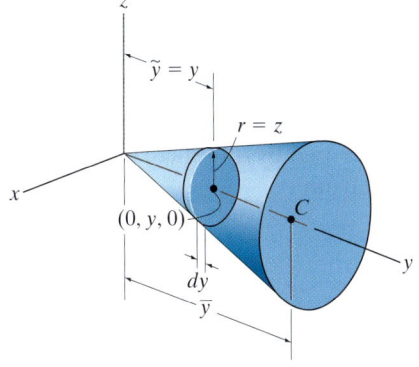

Fig. 9–4

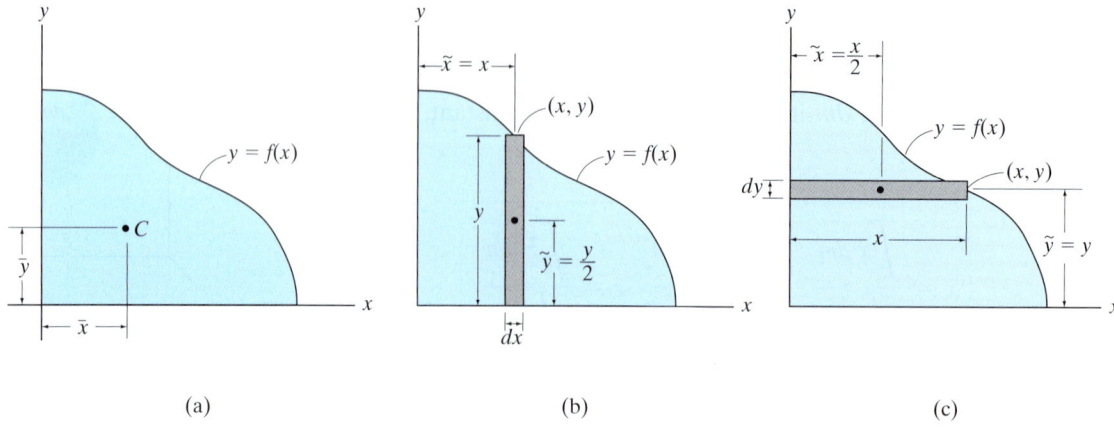

(a)   (b)   (c)

Fig. 9–5

Integration must be used to determine the location of the center of gravity of this goal post due to the curvature of the supporting member.

Centroid of an Area. If an area lies in the x–y plane and is bounded by the curve $y = f(x)$, as shown in Fig. 9–5a, then its centroid will be in this plane and can be determined from integrals similar to Eqs. 9–3, namely,

$$\bar{x} = \frac{\displaystyle\int_A \tilde{x}\, dA}{\displaystyle\int_A dA} \qquad \bar{y} = \frac{\displaystyle\int_A \tilde{y}\, dA}{\displaystyle\int_A dA} \qquad (9\text{–}4)$$

These integrals can be evaluated by performing a *single integration* if we use a *rectangular strip* for the differential area element. For example, if a vertical strip is used, Fig. 9–5b, the area of the element is $dA = y\, dx$, and its centroid is located at $\tilde{x} = x$ and $\tilde{y} = y/2$. If we consider a horizontal strip, Fig. 9–5c, then $dA = x\, dy$, and its centroid is located at $\tilde{x} = x/2$ and $\tilde{y} = y$.

Centroid of a Line. If a line segment (or rod) lies within the x–y plane and it can be described by a thin curve $y = f(x)$, Fig. 9–6a, then its centroid is determined from

$$\bar{x} = \frac{\displaystyle\int_L \tilde{x}\, dL}{\displaystyle\int_L dL} \qquad \bar{y} = \frac{\displaystyle\int_L \tilde{y}\, dL}{\displaystyle\int_L dL} \qquad (9\text{–}5)$$

Here, the length of the differential element is given by the Pythagorean theorem, $dL = \sqrt{(dx)^2 + (dy)^2}$, which can also be written in the form

$$dL = \sqrt{\left(\frac{dx}{dx}\right)^2 dx^2 + \left(\frac{dy}{dx}\right)^2 dx^2}$$

$$= \left(\sqrt{1 + \left(\frac{dy}{dx}\right)^2}\right) dx$$

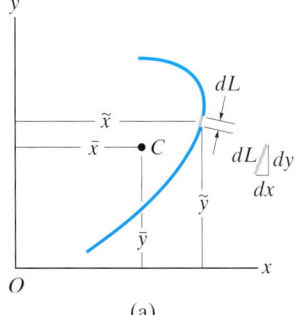

(a)

or

$$dL = \sqrt{\left(\frac{dx}{dy}\right)^2 dy^2 + \left(\frac{dy}{dy}\right)^2 dy^2}$$

$$= \left(\sqrt{\left(\frac{dx}{dy}\right)^2 + 1}\right) dy$$

Either one of these expressions can be used; however, for application, the one that will result in a simpler integration should be selected. For example, consider the rod in Fig. 9–6b, defined by $y = 2x^2$. The length of the element is $dL = \sqrt{1 + (dy/dx)^2}\,dx$, and since $dy/dx = 4x$, then $dL = \sqrt{1 + (4x)^2}\,dx$. The centroid for this element is located at $\tilde{x} = x$ and $\tilde{y} = y$.

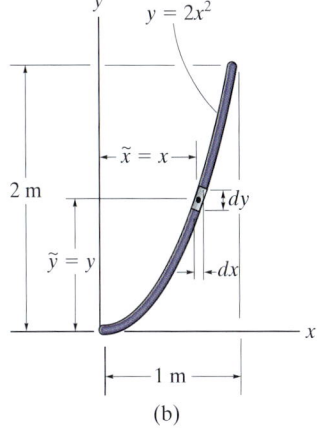

(b)

Fig. 9–6

Important Points

- The centroid represents the geometric center of a body. This point coincides with the center of mass or the center of gravity only if the material composing the body is uniform or homogeneous.

- Formulas used to locate the center of gravity or the centroid simply represent a balance between the sum of moments of all the parts of the system and the moment of the "resultant" for the system.

- In some cases the centroid is located at a point that is not on the object, as in the case of a ring, where the centroid is at its center. Also, this point will lie on any axis of symmetry for the body, Fig. 9–7.

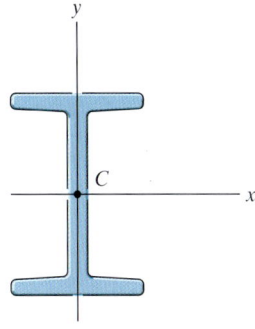

Fig. 9–7

Procedure for Analysis

The center of gravity or centroid of an object or shape can be determined by single integrations using the following procedure.

Differential Element.

- Select an appropriate coordinate system, specify the coordinate axes, and then choose a differential element for integration.

- For lines the element is represented by a differential line segment of length dL.

- For areas the element is generally a rectangle of area dA, having a finite length and differential width.

- For volumes the element can be a circular disk of volume dV, having a finite radius and differential thickness.

- Locate the element so that it touches the arbitrary point (x, y, z) on the curve that defines the boundary of the shape.

Size and Moment Arms.

- Express the length dL, area dA, or volume dV of the element in terms of the coordinates describing the curve.

- Express the moment arms $\tilde{x}$, $\tilde{y}$, $\tilde{z}$ for the centroid or center of gravity of the element in terms of the coordinates describing the curve.

Integrations.

- Substitute the formulations for $\tilde{x}$, $\tilde{y}$, $\tilde{z}$ and dL, dA, or dV into the appropriate equations (Eqs. 9–1 through 9–5).

- Express the function in the integrand in terms of the *same variable as the differential thickness of the element.*

- The limits of the integral are defined from the two extreme locations of the element's differential thickness, so that when the elements are "summed" or the integration performed, the entire region is covered.

EXAMPLE 9.1

Locate the centroid of the rod bent into the shape of a parabolic arc as shown in Fig. 9–8.

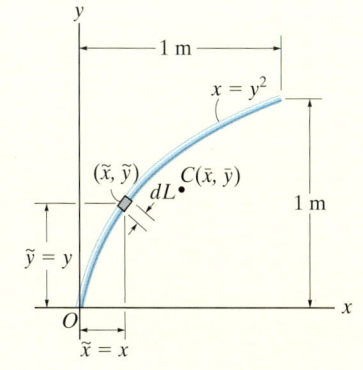

Fig. 9–8

SOLUTION

Differential Element.   The differential element is shown in Fig. 9–8. It is located on the curve at the *arbitrary point* (x, y).

Area and Moment Arms.   The differential element of length dL can be expressed in terms of the differentials dx and dy using the Pythagorean theorem.

$$dL = \sqrt{(dx)^2 + (dy)^2} = \sqrt{\left(\frac{dx}{dy}\right)^2 + 1}\, dy$$

Since $x = y^2$, then $dx/dy = 2y$. Therefore, expressing dL in terms of y and dy, we have

$$dL = \sqrt{(2y)^2 + 1}\, dy$$

As shown in Fig. 9–8, the centroid of the element is located at $\tilde{x} = x$, $\tilde{y} = y$.

Integrations.   Applying Eq. 9–5 and using the integration formula to evaluate the integrals, we get

$$\bar{x} = \frac{\displaystyle\int_L \tilde{x}\, dL}{\displaystyle\int_L dL} = \frac{\displaystyle\int_0^{1\,m} x\sqrt{4y^2 + 1}\, dy}{\displaystyle\int_0^{1\,m} \sqrt{4y^2 + 1}\, dy} = \frac{\displaystyle\int_0^{1\,m} y^2\sqrt{4y^2 + 1}\, dy}{\displaystyle\int_0^{1\,m} \sqrt{4y^2 + 1}\, dy}$$

$$= \frac{0.6063}{1.479} = 0.410 \text{ m} \qquad\qquad Ans.$$

$$\bar{y} = \frac{\displaystyle\int_L \tilde{y}\, dL}{\displaystyle\int_L dL} = \frac{\displaystyle\int_0^{1\,m} y\sqrt{4y^2 + 1}\, dy}{\displaystyle\int_0^{1\,m} \sqrt{4y^2 + 1}\, dy} = \frac{0.8484}{1.479} = 0.574 \text{ m} \qquad Ans.$$

NOTE: These results for C seem reasonable when they are plotted on Fig. 9–8.

EXAMPLE   9.2

Locate the centroid of the circular wire segment shown in Fig. 9–9.

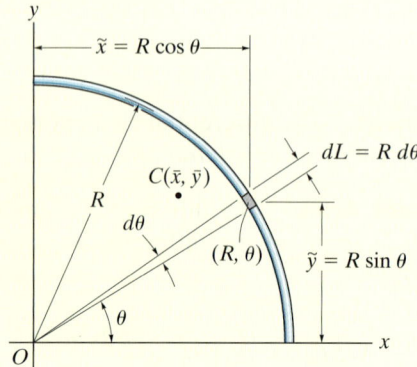

Fig. 9–9

SOLUTION

Polar coordinates will be used to solve this problem since the arc is circular.

Differential Element. A differential circular arc is selected as shown in the figure. This element lies on the curve at (R, θ).

Length and Moment Arm. The length of the differential element is $dL = R\, d\theta$, and its centroid is located at $\tilde{x} = R \cos \theta$ and $\tilde{y} = R \sin \theta$.

Integrations. Applying Eqs. 9–5 and integrating with respect to θ, we obtain

$$\bar{x} = \frac{\displaystyle\int_L \tilde{x}\, dL}{\displaystyle\int_L dL} = \frac{\displaystyle\int_0^{\pi/2} (R \cos \theta) R\, d\theta}{\displaystyle\int_0^{\pi/2} R\, d\theta} = \frac{R^2 \displaystyle\int_0^{\pi/2} \cos \theta\, d\theta}{R \displaystyle\int_0^{\pi/2} d\theta} = \frac{2R}{\pi} \quad \textit{Ans.}$$

$$\bar{y} = \frac{\displaystyle\int_L \tilde{y}\, dL}{\displaystyle\int_L dL} = \frac{\displaystyle\int_0^{\pi/2} (R \sin \theta) R\, d\theta}{\displaystyle\int_0^{\pi/2} R\, d\theta} = \frac{R^2 \displaystyle\int_0^{\pi/2} \sin \theta\, d\theta}{R \displaystyle\int_0^{\pi/2} d\theta} = \frac{2R}{\pi} \quad \textit{Ans.}$$

NOTE: As expected, the two coordinates are numerically the same due to the symmetry of the wire.

EXAMPLE 9.3

Determine the distance $\bar{y}$ measured from the x axis to the centroid of the area of the triangle shown in Fig. 9–10.

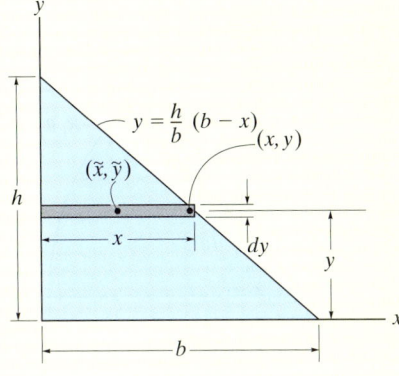

$$y = \frac{h}{b}(b - x)$$

Fig. 9–10

SOLUTION

Differential Element.   Consider a rectangular element having a thickness dy, and located in an arbitrary position so that it intersects the boundary at (x, y), Fig. 9–10.

Area and Moment Arms.   The area of the element is $dA = x\,dy$ $= \dfrac{b}{h}(h - y)\,dy$, and its centroid is located a distance $\tilde{y} = y$ from the x axis.

Integration.   Applying the second of Eqs. 9–4 and integrating with respect to y yields

$$\bar{y} = \frac{\displaystyle\int_A \tilde{y}\,dA}{\displaystyle\int_A dA} = \frac{\displaystyle\int_0^h y\left[\frac{b}{h}(h - y)\,dy\right]}{\displaystyle\int_0^h \frac{b}{h}(h - y)\,dy} = \frac{\frac{1}{6}bh^2}{\frac{1}{2}bh}$$

$$= \frac{h}{3} \qquad\qquad\qquad Ans.$$

NOTE: This result is valid for any shape of triangle. It states that the centroid is located at one-third the height, measured from the base of the triangle.

EXAMPLE | **9.4**

Locate the centroid for the area of a quarter circle shown in Fig. 9–11.

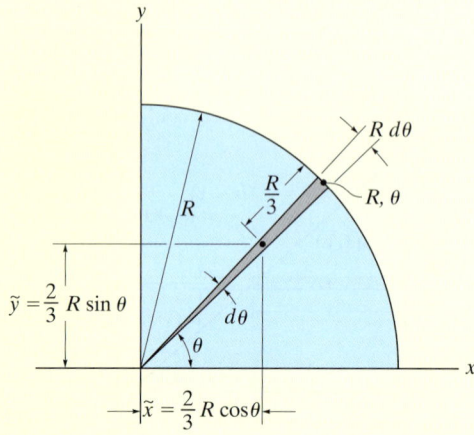

Fig. 9–11

SOLUTION

Differential Element. Polar coordinates will be used, since the boundary is circular. We choose the element in the shape of a *triangle*, Fig. 9–11. (Actually the shape is a circular sector; however, neglecting higher-order differentials, the element becomes triangular.) The element intersects the curve at point (R, θ).

Area and Moment Arms. The area of the element is

$$dA = \tfrac{1}{2}(R)(R\,d\theta) = \frac{R^2}{2}\,d\theta$$

and using the results of Example 9.3, the centroid of the (triangular) element is located at $\tilde{x} = \tfrac{2}{3}R\cos\theta$, $\tilde{y} = \tfrac{2}{3}R\sin\theta$.

Integrations. Applying Eqs. 9–4 and integrating with respect to θ, we obtain

$$\bar{x} = \frac{\displaystyle\int_A \tilde{x}\,dA}{\displaystyle\int_A dA} = \frac{\displaystyle\int_0^{\pi/2}\left(\tfrac{2}{3}R\cos\theta\right)\dfrac{R^2}{2}\,d\theta}{\displaystyle\int_0^{\pi/2}\dfrac{R^2}{2}\,d\theta} = \frac{\left(\tfrac{2}{3}R\right)\displaystyle\int_0^{\pi/2}\cos\theta\,d\theta}{\displaystyle\int_0^{\pi/2}d\theta} = \frac{4R}{3\pi} \qquad Ans.$$

$$\bar{y} = \frac{\displaystyle\int_A \tilde{y}\,dA}{\displaystyle\int_A dA} = \frac{\displaystyle\int_0^{\pi/2}\left(\tfrac{2}{3}R\sin\theta\right)\dfrac{R^2}{2}\,d\theta}{\displaystyle\int_0^{\pi/2}\dfrac{R^2}{2}\,d\theta} = \frac{\left(\tfrac{2}{3}R\right)\displaystyle\int_0^{\pi/2}\sin\theta\,d\theta}{\displaystyle\int_0^{\pi/2}d\theta} = \frac{4R}{3\pi} \qquad Ans.$$

EXAMPLE | 9.5

Locate the centroid of the area shown in Fig. 9–12a.

SOLUTION I

Differential Element. A differential element of thickness dx is shown in Fig. 9–12a. The element intersects the curve at the *arbitrary point* (x, y), and so it has a height y.

Area and Moment Arms. The area of the element is $dA = y\,dx$, and its centroid is located at $\tilde{x} = x$, $\tilde{y} = y/2$.

Integrations. Applying Eqs. 9–4 and integrating with respect to x yields

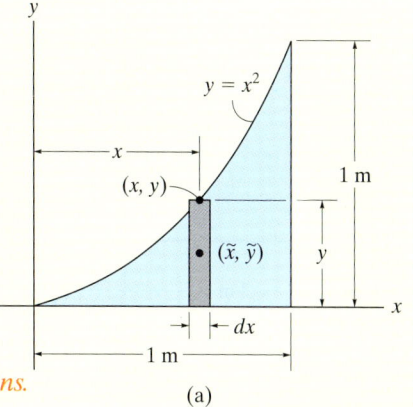

(a)

$$\bar{x} = \frac{\int_A \tilde{x}\,dA}{\int_A dA} = \frac{\int_0^{1\,m} xy\,dx}{\int_0^{1\,m} y\,dx} = \frac{\int_0^{1\,m} x^3\,dx}{\int_0^{1\,m} x^2\,dx} = \frac{0.250}{0.333} = 0.75\ m \qquad Ans.$$

$$\bar{y} = \frac{\int_A \tilde{y}\,dA}{\int_A dA} = \frac{\int_0^{1\,m} (y/2)y\,dx}{\int_0^{1\,m} y\,dx} = \frac{\int_0^{1\,m} (x^2/2)x^2\,dx}{\int_0^{1\,m} x^2\,dx} = \frac{0.100}{0.333} = 0.3\ m \quad Ans.$$

SOLUTION II

Differential Element. The differential element of thickness dy is shown in Fig. 9–12b. The element intersects the curve at the *arbitrary point* (x, y), and so it has a length $(1 - x)$.

Area and Moment Arms. The area of the element is $dA = (1 - x)\,dy$, and its centroid is located at

$$\tilde{x} = x + \left(\frac{1 - x}{2}\right) = \frac{1 + x}{2}, \quad \tilde{y} = y$$

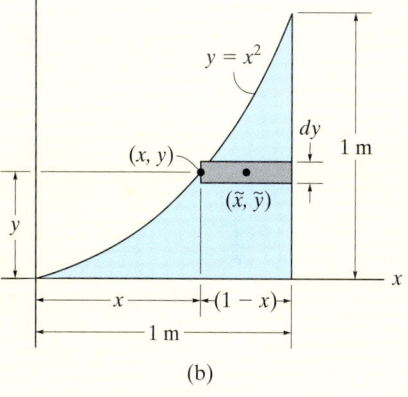

(b)

Fig. 9–12

Integrations. Applying Eqs. 9–4 and integrating with respect to y, we obtain

$$\bar{x} = \frac{\int_A \tilde{x}\,dA}{\int_A dA} = \frac{\int_0^{1\,m} [(1 + x)/2](1 - x)\,dy}{\int_0^{1\,m} (1 - x)\,dy} = \frac{\frac{1}{2}\int_0^{1\,m} (1 - y)\,dy}{\int_0^{1\,m} (1 - \sqrt{y})\,dy} = \frac{0.250}{0.333} = 0.75\ m \quad Ans.$$

$$\bar{y} = \frac{\int_A \tilde{y}\,dA}{\int_A dA} = \frac{\int_0^{1\,m} y(1 - x)\,dy}{\int_0^{1\,m} (1 - x)\,dy} = \frac{\int_0^{1\,m} (y - y^{3/2})\,dy}{\int_0^{1\,m} (1 - \sqrt{y})\,dy} = \frac{0.100}{0.333} = 0.3\ m \qquad Ans.$$

NOTE: Plot these results and notice that they seem reasonable. Also, for this problem, elements of thickness dx offer a simpler solution.

EXAMPLE | 9.6

Locate the centroid of the semi-elliptical area shown in Fig. 9–13a.

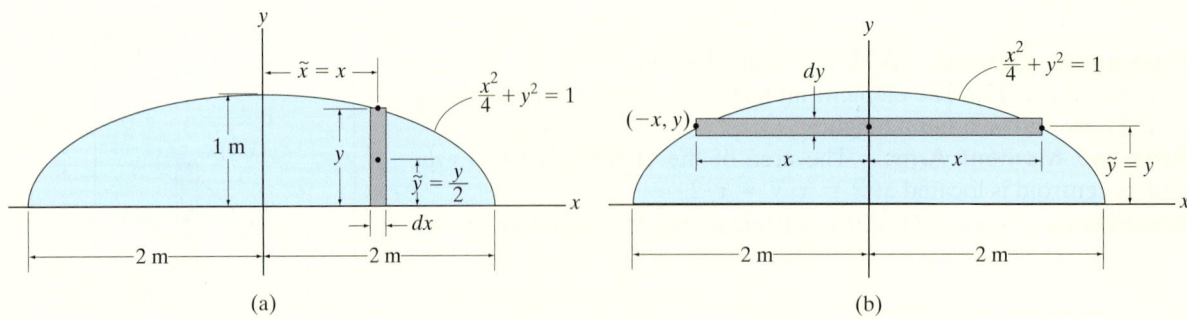

(a)   (b)

Fig. 9–13

SOLUTION I

Differential Element.   The rectangular differential element parallel to the y axis shown shaded in Fig. 9–13a will be considered. This element has a thickness of dx and a height of y.

Area and Moment Arms.   Thus, the area is $dA = y\,dx$, and its centroid is located at $\tilde{x} = x$ and $\tilde{y} = y/2$.

Integration.   Since the area is symmetrical about the y axis,

$$\bar{x} = 0 \qquad\qquad Ans.$$

Applying the second of Eqs. 9–4 with $y = \sqrt{1 - \dfrac{x^2}{4}}$, we have

$$\bar{y} = \frac{\displaystyle\int_A \tilde{y}\,dA}{\displaystyle\int_A dA} = \frac{\displaystyle\int_{-2\,m}^{2\,m} \frac{y}{2}(y\,dx)}{\displaystyle\int_{-2\,m}^{2\,m} y\,dx} = \frac{\dfrac{1}{2}\displaystyle\int_{-2\,m}^{2\,m}\left(1 - \frac{x^2}{4}\right)dx}{\displaystyle\int_{-2\,m}^{2\,m}\sqrt{1 - \frac{x^2}{4}}\,dx} = \frac{4/3}{\pi} = 0.424\ m \qquad Ans.$$

SOLUTION II

Differential Element.   The shaded rectangular differential element of thickness dy and width $2x$, parallel to the x axis, will be considered, Fig. 9–13b.

Area and Moment Arms.   The area is $dA = 2x\,dy$, and its centroid is at $\tilde{x} = 0$ and $\tilde{y} = y$.

Integration.   Applying the second of Eqs. 9–4, with $x = 2\sqrt{1 - y^2}$, we have

$$\bar{y} = \frac{\displaystyle\int_A \tilde{y}\,dA}{\displaystyle\int_A dA} = \frac{\displaystyle\int_0^{1\,m} y(2x\,dy)}{\displaystyle\int_0^{1\,m} 2x\,dy} = \frac{\displaystyle\int_0^{1\,m} 4y\sqrt{1 - y^2}\,dy}{\displaystyle\int_0^{1\,m} 4\sqrt{1 - y^2}\,dy} = \frac{4/3}{\pi}\ m = 0.424\ m \qquad Ans.$$

EXAMPLE | 9.7

Locate the $\bar{y}$ centroid for the paraboloid of revolution, shown in Fig. 9–14.

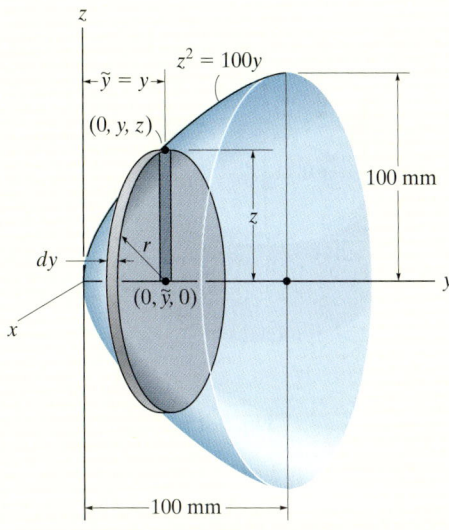

Fig. 9–14

SOLUTION

Differential Element.   An element having the shape of a *thin disk* is chosen. This element has a thickness dy, it intersects the generating curve at the *arbitrary point* $(0, y, z)$, and so its radius is $r = z$.

Volume and Moment Arm.   The volume of the element is $dV = (\pi z^2)\, dy$, and its centroid is located at $\tilde{y} = y$.

Integration.   Applying the second of Eqs. 9–3 and integrating with respect to y yields.

$$\bar{y} = \frac{\displaystyle\int_V \tilde{y}\, dV}{\displaystyle\int_V dV} = \frac{\displaystyle\int_0^{100\text{ mm}} y(\pi z^2)\, dy}{\displaystyle\int_0^{100\text{ mm}} (\pi z^2)\, dy} = \frac{100\pi \displaystyle\int_0^{100\text{ mm}} y^2\, dy}{100\pi \displaystyle\int_0^{100\text{ mm}} y\, dy} = 66.7 \text{ mm} \quad \textit{Ans.}$$

EXAMPLE | 9.8

Determine the location of the center of mass of the cylinder shown in Fig. 9–15 if its density varies directly with the distance from its base, i.e., $\rho = 200z \text{ kg/m}^3$.

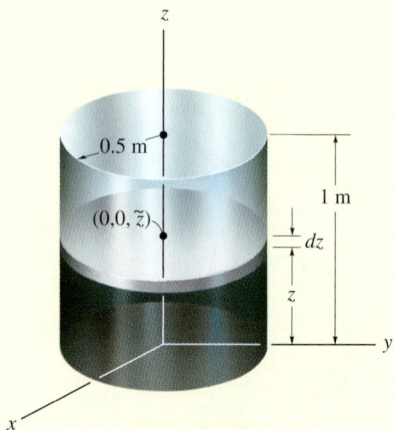

Fig. 9–15

SOLUTION

For reasons of material symmetry,

$$\bar{x} = \bar{y} = 0 \qquad \qquad \textit{Ans.}$$

Differential Element. A disk element of radius 0.5 m and thickness dz is chosen for integration, Fig. 9–15, since the *density of the entire element is constant* for a given value of z. The element is located along the z axis at the *arbitrary point* $(0, 0, z)$.

Volume and Moment Arm. The volume of the element is $dV = \pi(0.5)^2 \, dz$, and its centroid is located at $\tilde{z} = z$.

Integrations. Using the third of Eqs. 9–2 with $dm = \rho \, dV$ and integrating with respect to z, noting that $\rho = 200z$, we have

$$\bar{z} = \frac{\displaystyle\int_V \tilde{z}\rho \, dV}{\displaystyle\int_V \rho \, dV} = \frac{\displaystyle\int_0^{1\text{ m}} z(200z)\left[\pi(0.5)^2 \, dz\right]}{\displaystyle\int_0^{1\text{ m}} (200z)\pi(0.5)^2 \, dz}$$

$$= \frac{\displaystyle\int_0^{1\text{ m}} z^2 \, dz}{\displaystyle\int_0^{1\text{ m}} z \, dz} = 0.667 \text{ m} \qquad \textit{Ans.}$$

FUNDAMENTAL PROBLEMS

F9–1.   Determine the centroid $(\bar{x}, \bar{y})$ of the shaded area.

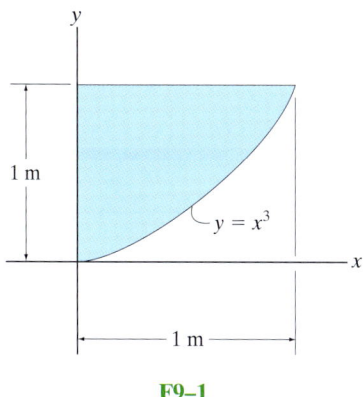

F9–1

F9–2.   Determine the centroid $(\bar{x}, \bar{y})$ of the shaded area.

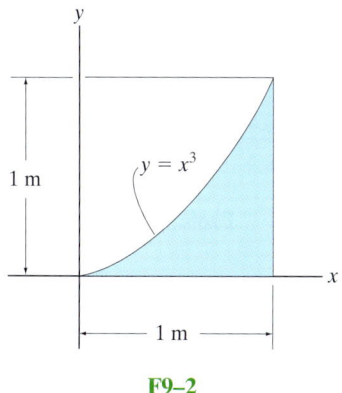

F9–2

F9–3.   Determine the centroid $\bar{y}$ of the shaded area.

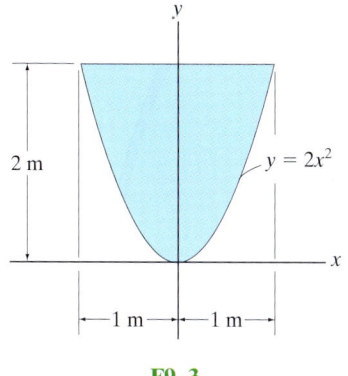

F9–3

F9–4.   Locate the center mass $\bar{x}$ of the straight rod if its mass per unit length is given by $m = m_0(1 + x^2/L^2)$.

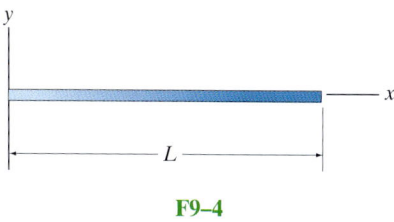

F9–4

F9–5.   Locate the centroid $\bar{y}$ of the homogeneous solid formed by revolving the shaded area about the y axis.

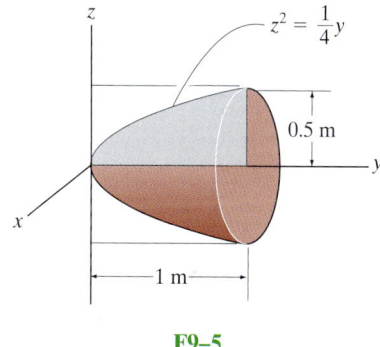

F9–5

F9–6.   Locate the centroid $\bar{z}$ of the homogeneous solid formed by revolving the shaded area about the z axis.

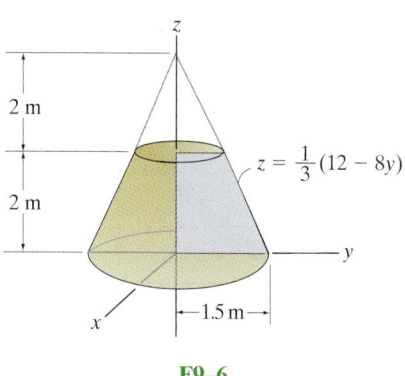

F9–6

PROBLEMS

9–1. Locate the center of mass of the homogeneous rod bent into the shape of a circular arc.

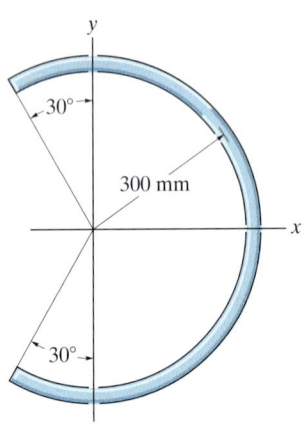

Prob. 9–1

9–2. Determine the distance $\bar{x}$ to the center of mass of the homogeneous rod bent into the shape shown. If the rod has a mass per unit length of 0.5 kg/m, determine the reactions at the fixed support O.

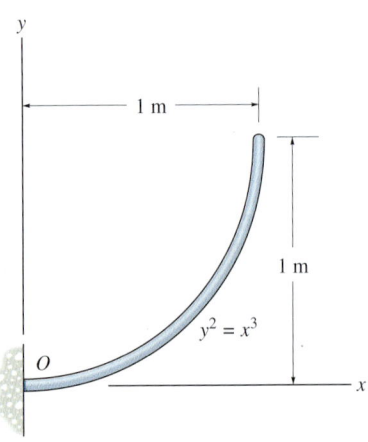

Prob. 9–2

9–3. Determine the mass and the location of the center of mass $\bar{x}$ of the rod if its mass per unit length is $m = m_0(1 + x/L)$.

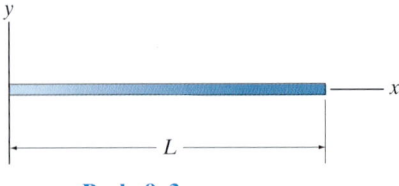

Prob. 9–3

*9–4. Locate the centroid of the parabolic area.

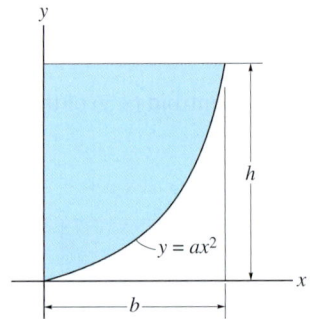

Prob. 9–4

9–5. Locate the centroid $(\bar{x}, \bar{y})$ of the uniform rod. Evaluate the integrals using a numerical method.

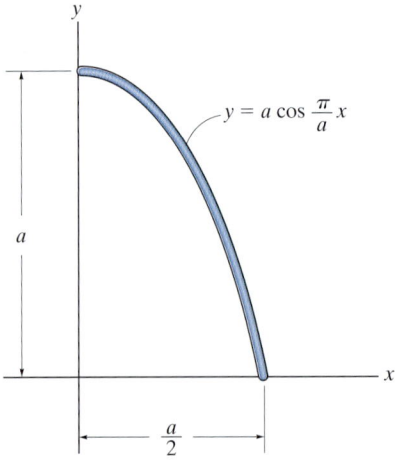

Prob. 9–5

9–6. Locate the centroid $\bar{y}$ of the area.

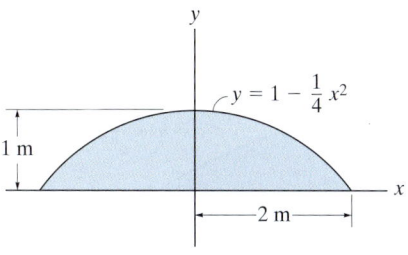

Prob. 9–6

9–7. Locate the centroid $\bar{x}$ of the parabolic area.

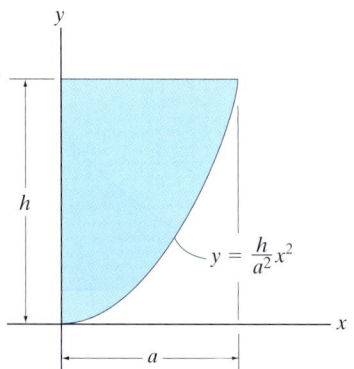

Prob. 9–7

*9–8. Locate the centroid $\bar{y}$ of the parabolic area.

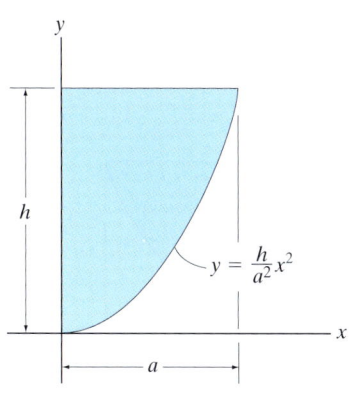

Prob. 9–8

9–9. Locate the centroid $\bar{x}$ of the area.

9–10. Locate the centroid $\bar{y}$ of the area.

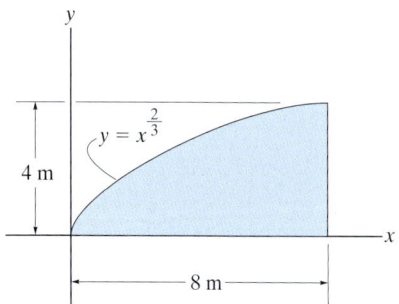

Probs. 9–9/10

9–11. Locate the centroid $\bar{x}$ of the area.

*9–12. Locate the centroid $\bar{y}$ of the area.

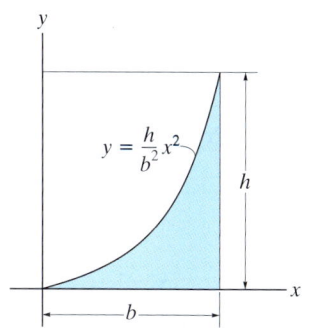

Probs. 9–11/12

9–13. Locate the centroid $\bar{x}$ of the area.

9–14. Locate the centroid $\bar{y}$ of the area.

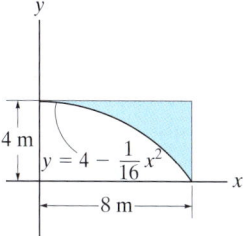

Probs. 9–13/14

9

9–15. Locate the centroid $\bar{x}$ of the area.

*9–16. Locate the centroid $\bar{y}$ of the area.

9–18. Locate the centroid $\bar{x}$ of the area.

9–19. Locate the centroid $\bar{y}$ of the area.

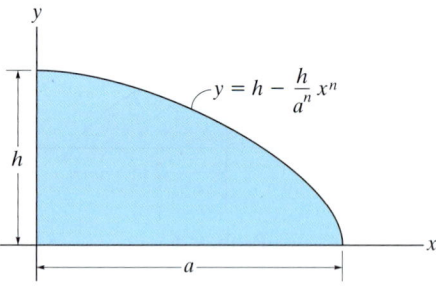

Probs. 9–18/19

*9–20. Locate the centroid $\bar{y}$ of the shaded area.

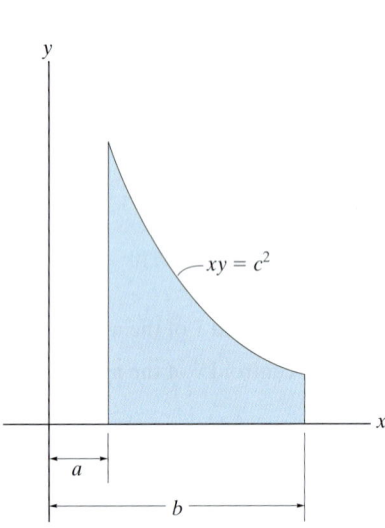

Probs. 9–15/16

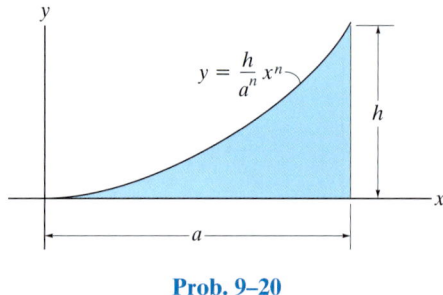

Prob. 9–20

9–17. Locate the centroid $\bar{x}$ of the shaded area.

9–21. Locate the centroid $\bar{x}$ of the area.

9–22. Locate the centroid $\bar{y}$ of the area.

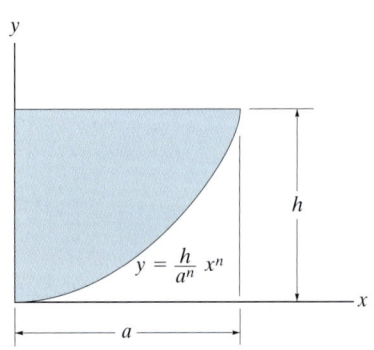

Prob. 9–17

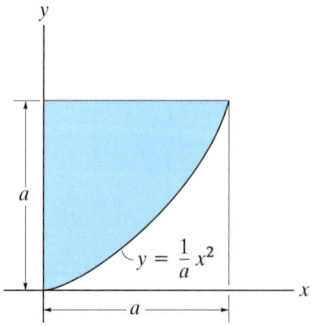

Probs. 9–21/22

9–23. Locate the centroid $\bar{x}$ of the quarter elliptical area.

*9–24. Locate the centroid $\bar{y}$ of the quarter elliptical area.

9–26. Locate the centroid $\bar{x}$ of the area.

9–27. Locate the centroid $\bar{y}$ of the area.

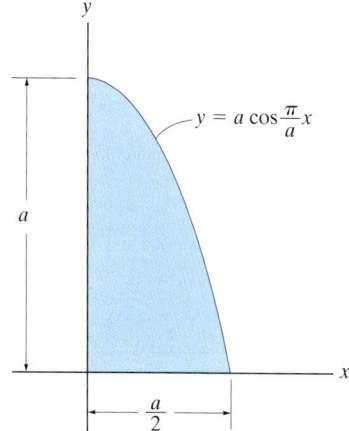

Probs. 9–26/27

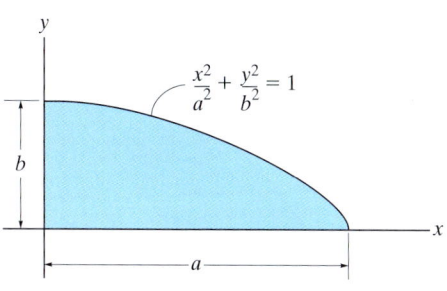

Probs. 9–23/24

*9–28. Locate the centroid $\bar{x}$ of the area.

9–29. Locate the centroid $\bar{y}$ of the area.

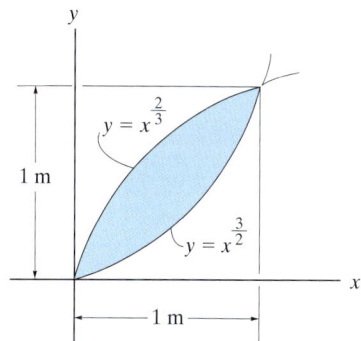

Probs. 9–28/29

9–25. Locate the centroid $\bar{y}$ of the shaded area. Solve the problem by evaluating the integrals using Simpson's rule.

9–30. Locate the centroid $\bar{x}$ of the area.

9–31. Locate the centroid $\bar{y}$ of the area.

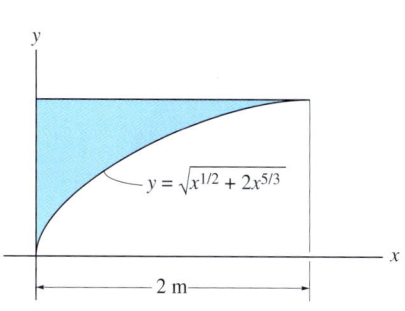

Prob. 9–25

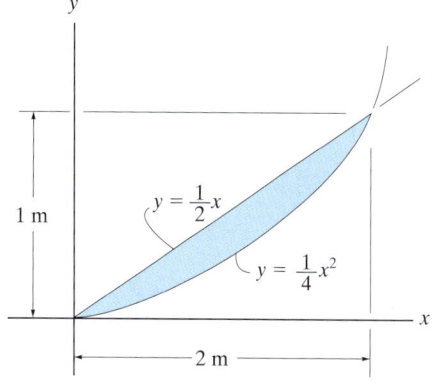

Probs. 9–30/31

*9–32. Locate the centroid $\bar{x}$ of the area.

9–33. Locate the centroid $\bar{y}$ of the area.

9–35. Locate the centroid $\bar{x}$ of the shaded area.

*9–36. Locate the centroid $\bar{y}$ of the shaded area.

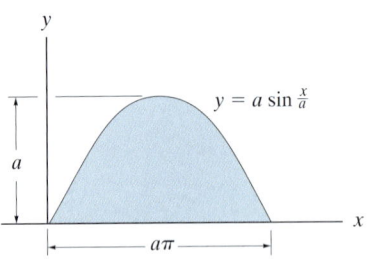

Probs. 9–32/33

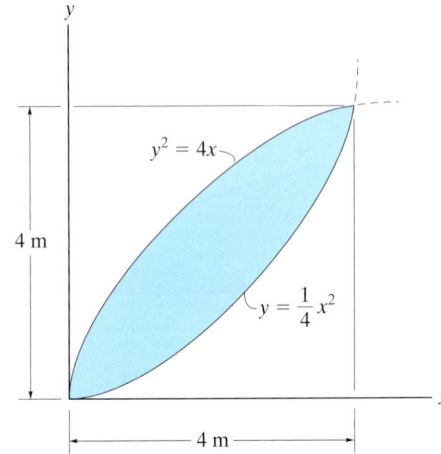

Probs. 9–35/36

9–34. The steel plate is 0.3 m thick and has a density of 7850 kg/m^3. Determine the location of its center of mass. Also find the reactions at the pin and roller support.

9–37. If the density at any point in the quarter circular plate is defined by $\rho = \rho_0 xy$, where ρ_0 is a constant, determine the mass and locate the center of mass $(\bar{x}, \bar{y})$ of the plate. The plate has a thickness t.

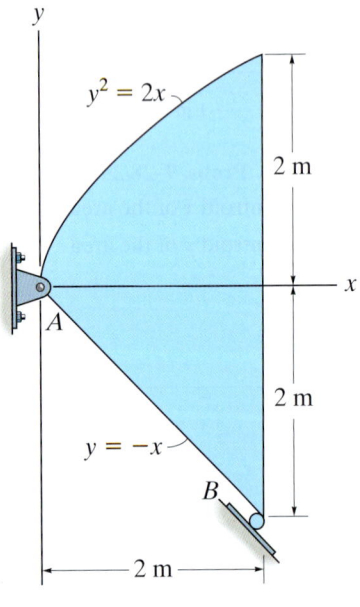

Prob. 9–34

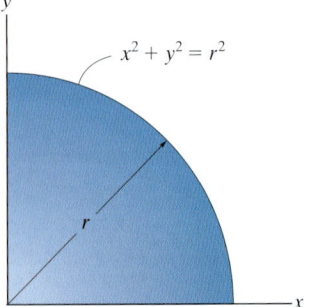

Prob. 9–37

9–38.   Determine the location $\bar{r}$ of the centroid C of the cardioid, $r = a(1 - \cos\theta)$.

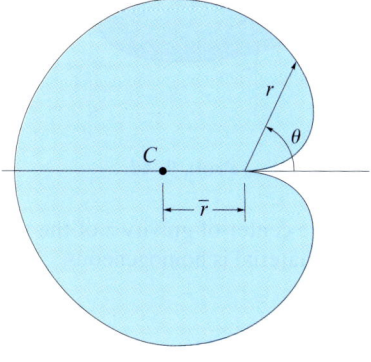

Prob. 9–38

9–39.   Locate the centroid $\bar{y}$ of the paraboloid.

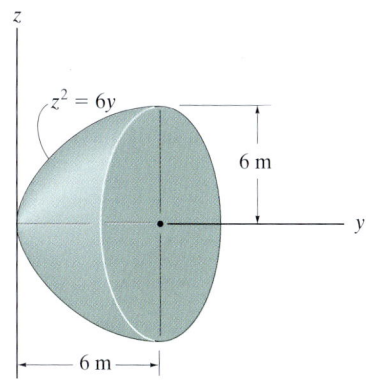

Prob. 9–39

*9–40.   Locate the center of gravity of the volume. The material is homogeneous.

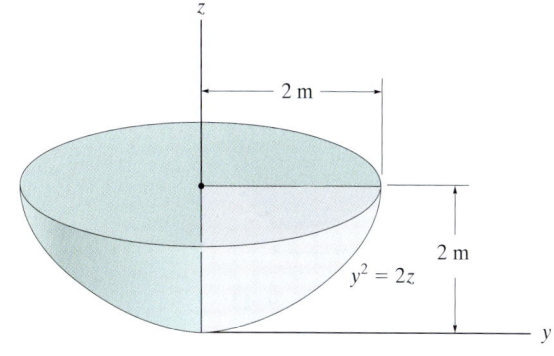

Prob. 9–40

9–41.   Locate the centroid $\bar{z}$ of the hemisphere.

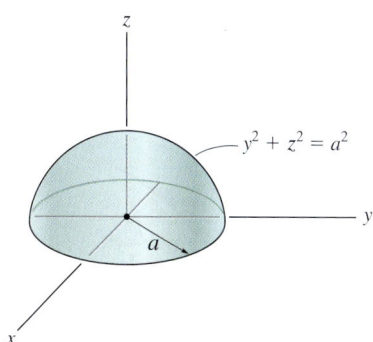

Prob. 9–41

9–42. Locate the centroid $\bar{y}$ of the paraboloid.

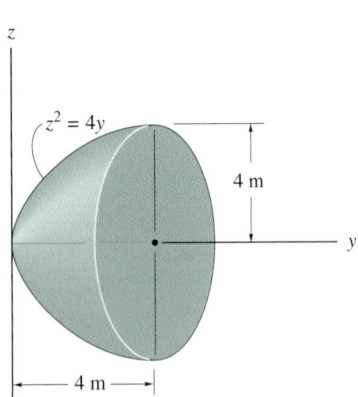

Prob. 9–42

9–43. Locate the centroid of the solid.

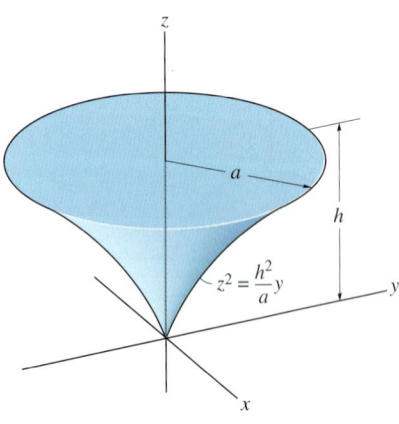

Prob. 9–43

9–44. Locate the centroid of the ellipsoid of revolution.

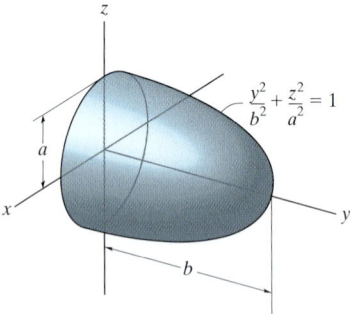

Prob. 9–44

9–45. Locate the center of gravity $\bar{z}$ of the frustum of the paraboloid. The material is homogeneous.

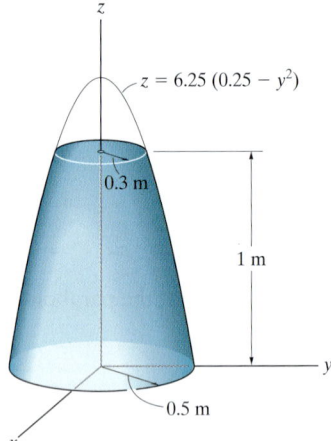

Prob. 9–45

9–46. The hemisphere of radius r is made from a stack of very thin plates such that the density varies with height, $\rho = kz$, where k is a constant. Determine its mass and the distance $\bar{z}$ to the center of mass G.

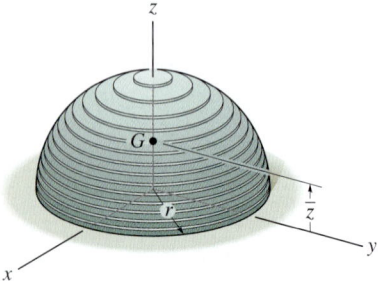

Prob. 9–46

9–47. Locate the centroid of the quarter-cone.

9–49. The king's chamber of the Great Pyramid of Giza is located at its centroid. Assuming the pyramid to be a solid, prove that this point is at $\bar{z} = \frac{1}{4}h$, *Suggestion:* Use a rectangular differential plate element having a thickness dz and area $(2x)(2y)$.

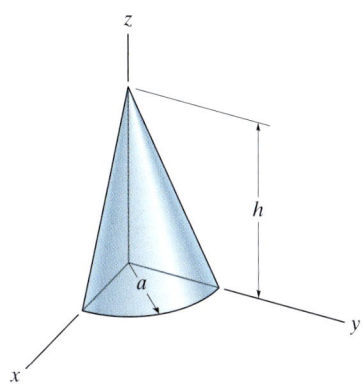

Prob. 9–47

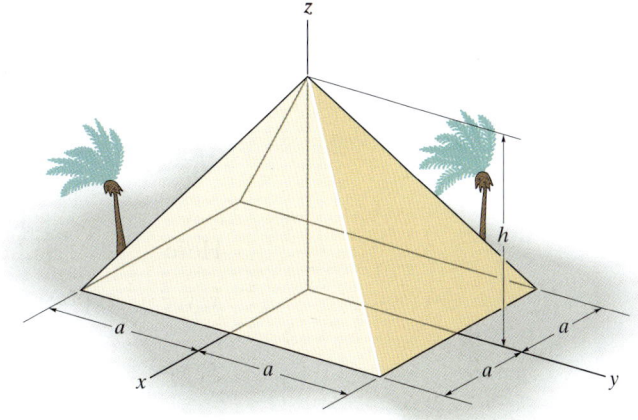

Prob. 9–49

9–50. Determine the location $\bar{z}$ of the centroid for the tetrahedron. *Suggestion:* Use a triangular "plate" element parallel to the x–y plane and of thickness dz.

*9–48. Locate the centroid $\bar{z}$ of the frustum of the right-circular cone.

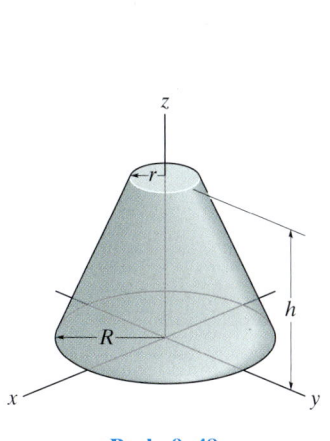

Prob. 9–48

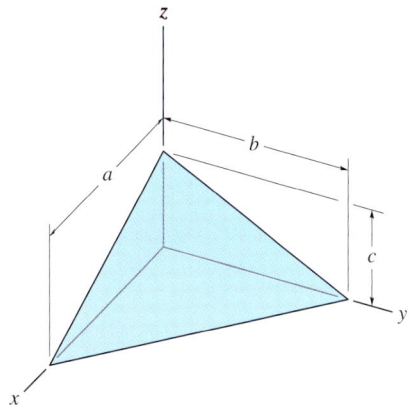

Prob. 9–50

9.2   Composite Bodies

A *composite body* consists of a series of connected "simpler" shaped bodies, which may be rectangular, triangular, semicircular, etc. Such a body can often be sectioned or divided into its composite parts and, provided the *weight* and location of the center of gravity of each of these parts are known, we can then eliminate the need for integration to determine the center of gravity for the entire body. The method for doing this follows the same procedure outlined in Sec. 9.1. Formulas analogous to Eqs. 9–1 result; however, rather than account for an infinite number of differential weights, we have instead a finite number of weights. Therefore,

$$\bar{x} = \frac{\Sigma \tilde{x} W}{\Sigma W} \quad \bar{y} = \frac{\Sigma \tilde{y} W}{\Sigma W} \quad \bar{z} = \frac{\Sigma \tilde{z} W}{\Sigma W} \qquad (9\text{--}6)$$

Here

$\bar{x}, \bar{y}, \bar{z}$ represent the coordinates of the center of gravity G of the composite body.

$\tilde{x}, \tilde{y}, \tilde{z}$ represent the coordinates of the center of gravity of each composite part of the body.

ΣW is the sum of the weights of all the composite parts of the body, or simply the total weight of the body.

When the body has a *constant density or specific weight*, the center of gravity *coincides* with the centroid of the body. The centroid for composite lines, areas, and volumes can be found using relations analogous to Eqs. 9–6; however, the W's are replaced by L's, A's, and V's, respectively. Centroids for common shapes of lines, areas, shells, and volumes that often make up a composite body are given in the table on the inside back cover.

In order to determine the force required to tip over this concrete barrier it is first necessary to determine the location of its center of gravity G. Due to symmetry, G will lie on the vertical axis of symmetry.

Procedure for Analysis

The location of the center of gravity of a body or the centroid of a composite geometrical object represented by a line, area, or volume can be determined using the following procedure.

Composite Parts.

- Using a sketch, divide the body or object into a finite number of composite parts that have simpler shapes.
- If a composite body has a *hole*, or a geometric region having no material, then consider the composite body without the hole and consider the hole as an *additional* composite part having *negative* weight or size.

Moment Arms.

- Establish the coordinate axes on the sketch and determine the coordinates $\tilde{x}$, $\tilde{y}$, $\tilde{z}$ of the center of gravity or centroid of each part.

Summations.

- Determine $\bar{x}$, $\bar{y}$, $\bar{z}$ by applying the center of gravity equations, Eqs. 9–6, or the analogous centroid equations.
- If an object is *symmetrical* about an axis, the centroid of the object lies on this axis.

If desired, the calculations can be arranged in tabular form, as indicated in the following three examples.

The center of gravity of this water tank can be determined by dividing it into composite parts and applying Eqs. 9–6.

EXAMPLE 9.9

Locate the centroid of the wire shown in Fig. 9–16a.

SOLUTION

Composite Parts.   The wire is divided into three segments as shown in Fig. 9–16b.

Moment Arms.   The location of the centroid for each segment is determined and indicated in the figure. In particular, the centroid of segment ① is determined either by integration or by using the table on the inside back cover.

Summations.   For convenience, the calculations can be tabulated as follows:

Segment	L (mm)	$\tilde{x}$ (mm)	$\tilde{y}$ (mm)	$\tilde{z}$ (mm)	$\tilde{x}L$ (mm²)	$\tilde{y}L$ (mm²)	$\tilde{z}L$ (mm²)
1	$\pi(60) = 188.5$	60	−38.2	0	11 310	−7200	0
2	40	0	20	0	0	800	0
3	20	0	40	−10	0	800	−200
	$\Sigma L = 248.5$				$\Sigma \tilde{x}L = 11\ 310$	$\Sigma \tilde{y}L = -5600$	$\Sigma \tilde{z}L = -200$

Thus,

$$\bar{x} = \frac{\Sigma \tilde{x}L}{\Sigma L} = \frac{11\ 310}{248.5} = 45.5 \text{ mm} \qquad \textit{Ans.}$$

$$\bar{y} = \frac{\Sigma \tilde{y}L}{\Sigma L} = \frac{-5600}{248.5} = -22.5 \text{ mm} \qquad \textit{Ans.}$$

$$\bar{z} = \frac{\Sigma \tilde{z}L}{\Sigma L} = \frac{-200}{248.5} = -0.805 \text{ mm} \qquad \textit{Ans.}$$

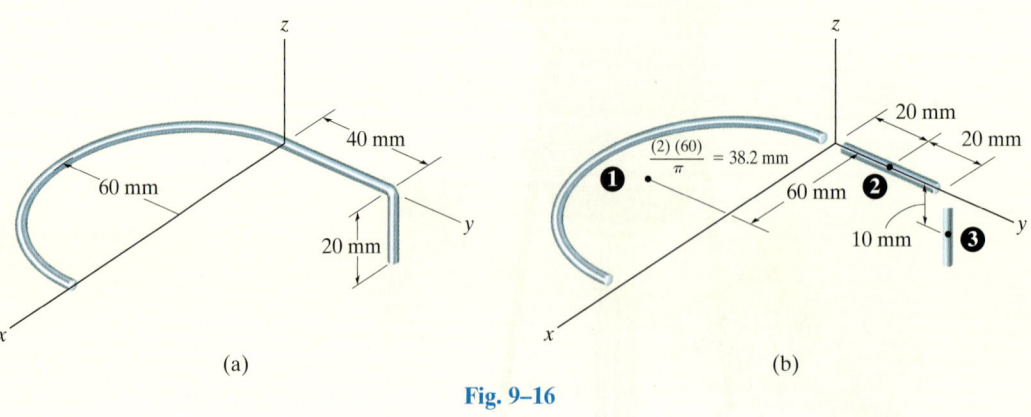

(a)   (b)

Fig. 9–16

EXAMPLE | 9.10

Locate the centroid of the plate area shown in Fig. 9–17a.

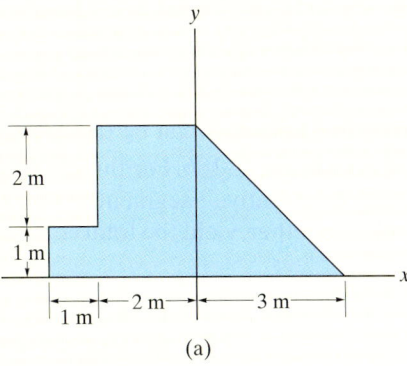

(a)

Fig. 9–17

SOLUTION

Composite Parts. The plate is divided into three segments as shown in Fig. 9–17b. Here the area of the small rectangle ③ is considered "negative" since it must be subtracted from the larger one ②.

Moment Arms. The centroid of each segment is located as indicated in the figure. Note that the $\tilde{x}$ coordinates of ② and ③ are *negative*.

Summations. Taking the data from Fig. 9–17b, the calculations are tabulated as follows:

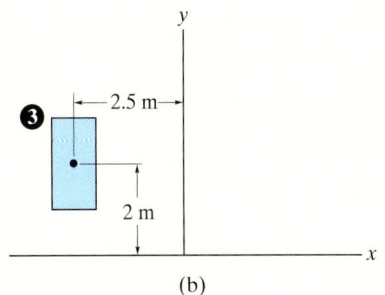

Segment	A (m²)	$\tilde{x}$ (m)	$\tilde{y}$ (m)	$\tilde{x}A$ (m³)	$\tilde{y}A$ (m³)
1	$\frac{1}{2}(3)(3) = 4.5$	1	1	4.5	4.5
2	$(3)(3) = 9$	−1.5	1.5	−13.5	13.5
3	$-(2)(1) = -2$	−2.5	2	5	−4
	$\Sigma A = 11.5$			$\Sigma \tilde{x}A = -4$	$\Sigma \tilde{y}A = 14$

Thus,

$$\bar{x} = \frac{\Sigma \tilde{x}A}{\Sigma A} = \frac{-4}{11.5} = -0.348 \text{ m} \qquad Ans.$$

$$\bar{y} = \frac{\Sigma \tilde{y}A}{\Sigma A} = \frac{14}{11.5} = 1.22 \text{ m} \qquad Ans.$$

(b)

NOTE: If these results are plotted in Fig. 9–17a, the location of point C seems reasonable.

EXAMPLE | 9.11

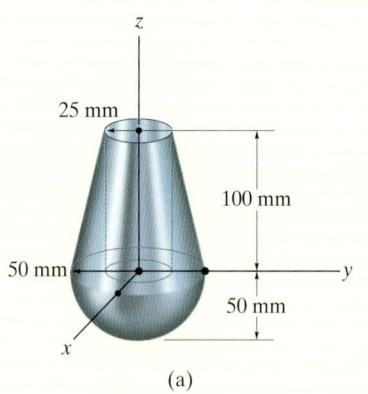

(a)

Fig. 9–18

Locate the center of mass of the assembly shown in Fig. 9–18a. The conical frustum has a density of $\rho_c = 8 \text{ Mg/m}^3$, and the hemisphere has a density of $\rho_h = 4 \text{ Mg/m}^3$. There is a 25-mm-radius cylindrical hole in the center of the frustum.

SOLUTION

Composite Parts.   The assembly can be thought of as consisting of four segments as shown in Fig. 9–18b. For the calculations, ③ and ④ must be considered as "negative" segments in order that the four segments, when added together, yield the total composite shape shown in Fig. 9–18a.

Moment Arm.   Using the table on the inside back cover, the computations for the centroid $\tilde{z}$ of each piece are shown in the figure.

Summations.   Because of *symmetry*, note that

$$\bar{x} = \bar{y} = 0 \qquad\qquad Ans.$$

Since $W = mg$, and g is constant, the third of Eqs. 9–6 becomes $\bar{z} = \Sigma \tilde{z}m/\Sigma m$. The mass of each piece can be computed from $m = \rho V$ and used for the calculations. Also, $1 \text{ Mg/m}^3 = 10^{-6} \text{ kg/mm}^3$, so that

Segment	m (kg)	$\tilde{z}$ (mm)	$\tilde{z}m$ (kg · mm)
1	$8(10^{-6})\left(\frac{1}{3}\right)\pi(50)^2(200) = 4.189$	50	209.440
2	$4(10^{-6})\left(\frac{2}{3}\right)\pi(50)^3 = 1.047$	-18.75	-19.635
3	$-8(10^{-6})\left(\frac{1}{3}\right)\pi(25)^2(100) = -0.524$	$100 + 25 = 125$	-65.450
4	$-8(10^{-6})\pi(25)^2(100) = -1.571$	50	-78.540
	$\Sigma m = 3.142$		$\Sigma \tilde{z}m = 45.815$

Thus, $$\tilde{z} = \frac{\Sigma \tilde{z}m}{\Sigma m} = \frac{45.815}{3.142} = 14.6 \text{ mm} \qquad Ans.$$

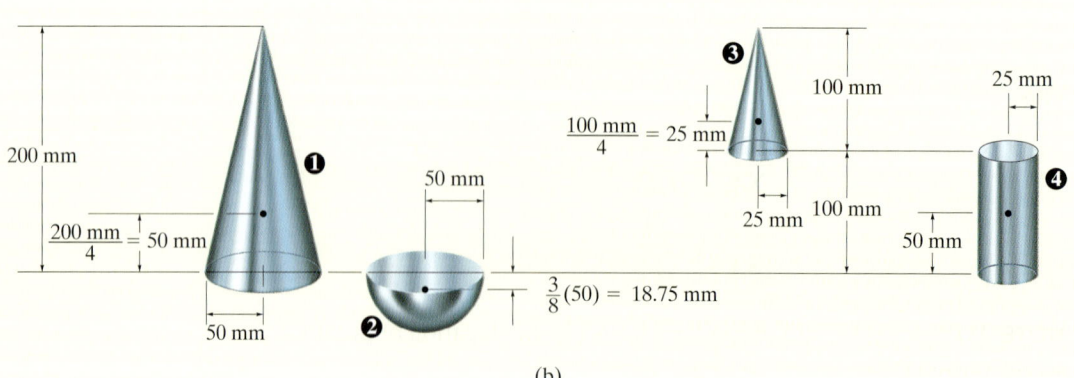

(b)

FUNDAMENTAL PROBLEMS

F9–7. Locate the centroid $(\bar{x}, \bar{y}, \bar{z})$ of the wire bent in the shape shown.

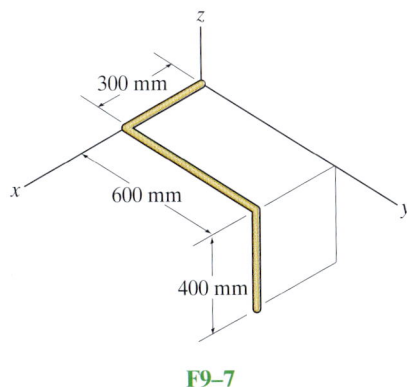

F9–7

F9–8. Locate the centroid $\bar{y}$ of the beam's cross-sectional area.

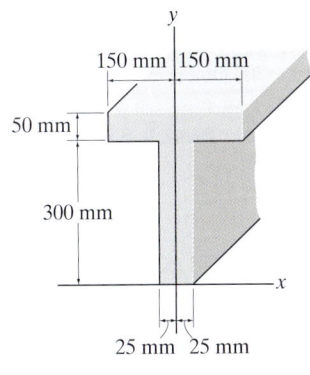

F9–8

F9–9. Locate the centroid $\bar{y}$ of the beam's cross-sectional area.

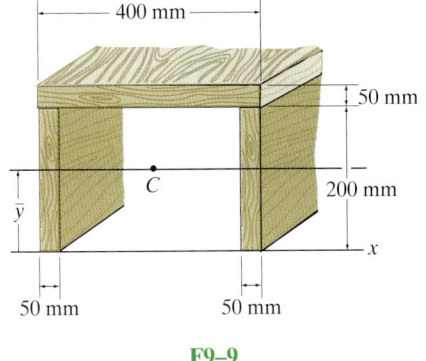

F9–9

F9–10. Locate the centroid $(\bar{x}, \bar{y})$ of the cross-sectional area.

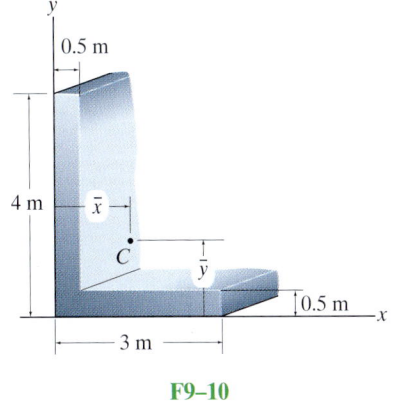

F9–10

F9–11. Locate the center of mass $(\bar{x}, \bar{y}, \bar{z})$ of the homogeneous solid block.

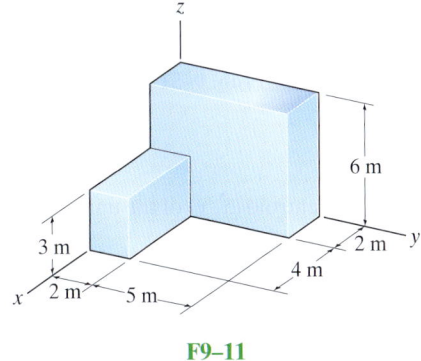

F9–11

F9–12. Determine the center of mass $(\bar{x}, \bar{y}, \bar{z})$ of the homogeneous solid block.

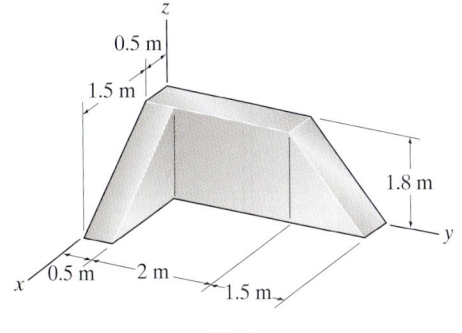

F9–12

9

PROBLEMS

9–51. The truss is made from five members, each having a length of 4 m and a mass of 7 kg/m. If the mass of the gusset plates at the joints and the thickness of the members can be neglected, determine the distance d to where the hoisting cable must be attached, so that the truss does not tip (rotate) when it is lifted.

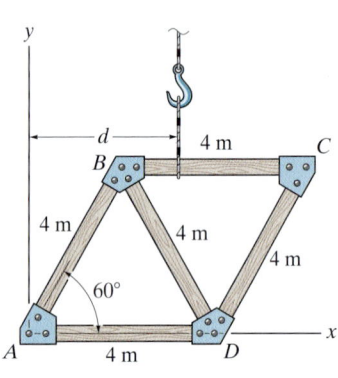

Prob. 9–51

*9–52. Locate the centroid $(\bar{x}, \bar{y})$ of the uniform wire bent in the shape shown.

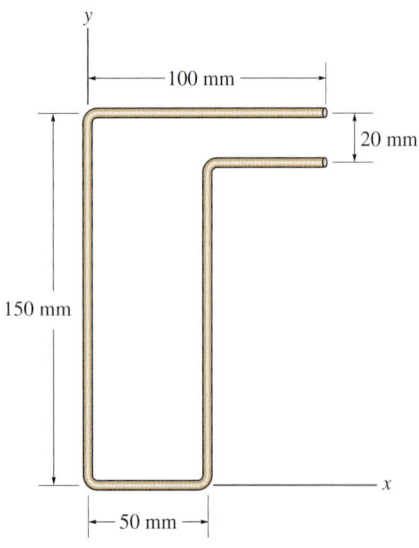

Prob. 9–52

9–53. Locate the centroid $(\bar{x}, \bar{y})$ of the cross section. All the dimensions are measured to the centerline thickness of each thin segment.

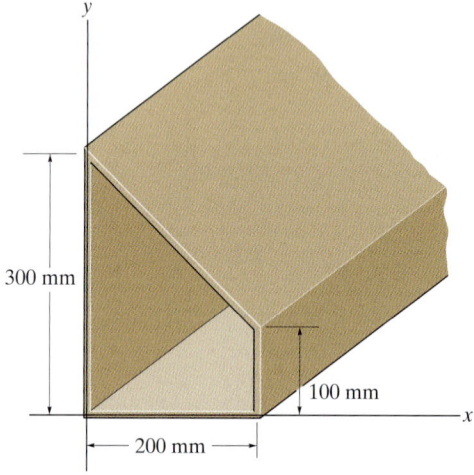

Prob. 9–53

9–54. Locate the centroid $(\bar{x}, \bar{y})$ of the metal cross section. Neglect the thickness of the material and slight bends at the corners.

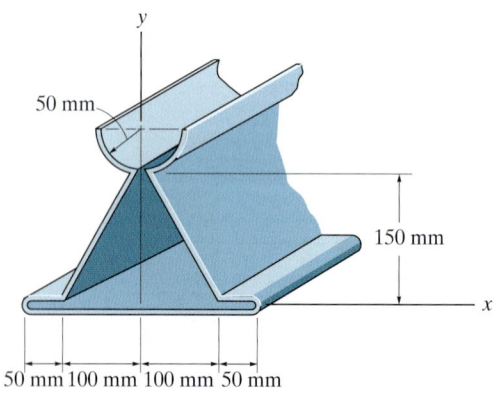

Prob. 9–54

9–55. Each of the three members of the frame has a mass per unit length of 6 kg/m. Locate the position $(\bar{x}, \bar{y})$ of the center of gravity. Neglect the size of the pins at the joints and the thickness of the members. Also, calculate the reactions at the pin A and roller E.

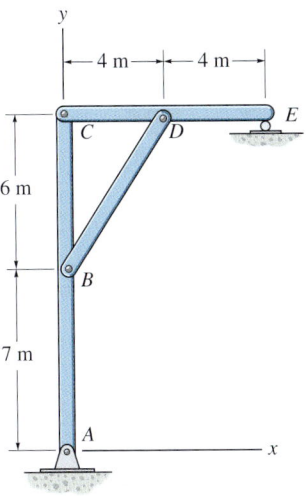

Prob. 9–55

*9–56.** The steel and aluminum plate assembly is bolted together and fastened to the wall. Each plate has a constant width in the z direction of 200 mm and thickness of 20 mm. If the density of A and B is $\rho_s = 7.85$ Mg/m³, and for C, $\rho_{al} = 2.71$ Mg/m³, determine the location $\bar{x}$ of the center of mass. Neglect the size of the bolts.

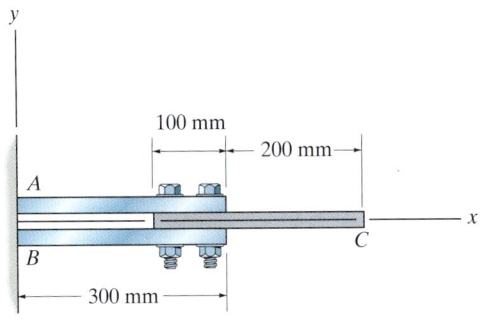

Prob. 9–56

9–57. To determine the location of the center of gravity of the automobile it is first placed in a *level position*, with the two wheels on one side resting on the scale platform P. In this position the scale records a reading of W_1. Then, one side is elevated to a convenient height c as shown. The new reading on the scale is W_2. If the automobile has a total weight of W, determine the location of its center of gravity $G(\bar{x}, \bar{y})$.

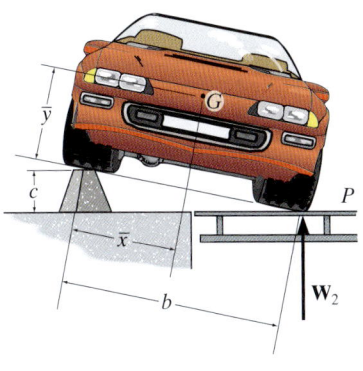

Prob. 9–57

9–58. Determine the location $\bar{y}$ of the centroidal axis $\bar{x}-\bar{x}$ of the beam's cross-sectional area. Neglect the size of the corner welds at A and B for the calculation.

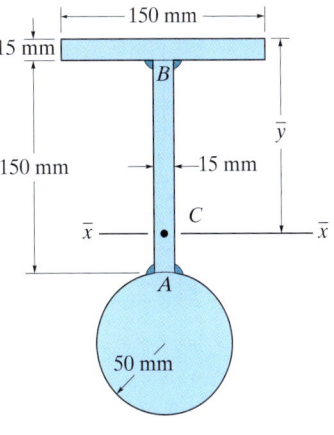

Prob. 9–58

9–59. Locate the centroid $\bar{y}$ of the cross-sectional area of the built-up beam.

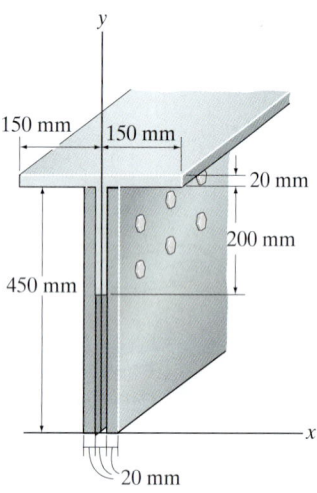

Prob. 9–59

*9–60.** Locate the centroid $\bar{y}$ of the cross-sectional area of the built-up beam.

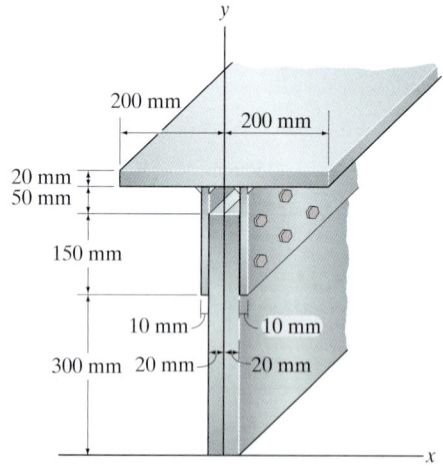

Prob. 9–60

9–61. Locate the centroid $(\bar{x}, \bar{y})$ of the member's cross-sectional area.

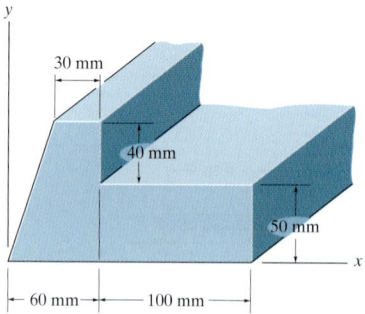

Prob. 9–61

9–62. Locate the centroid $\bar{y}$ of the bulb-tee cross section.

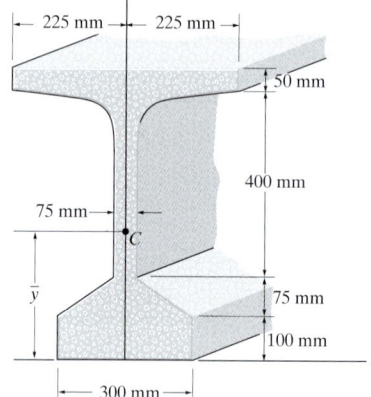

Prob. 9–62

9–63. The gravity wall is made of concrete. Determine the location $(\bar{x}, \bar{y})$ of the center of gravity G for the wall.

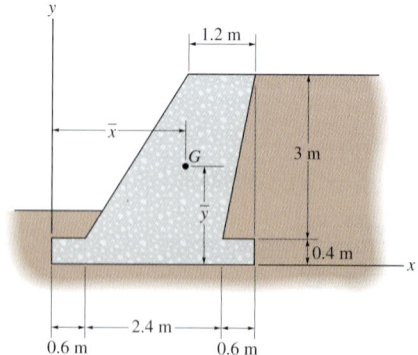

Prob. 9–63

*9–64. Locate the centroid $\bar{y}$ of the concrete beam having the tapered cross section shown.

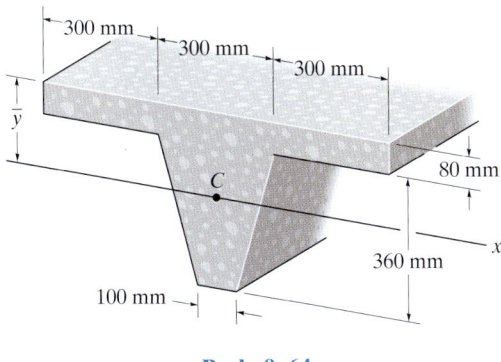

Prob. 9–64

9–65. Locate the center of gravity $G(\bar{x}, \bar{y})$ of the streetlight. Neglect the thickness of each segment. The mass per unit length of each segment is as follows: $\rho_{AB} = 12$ kg/m, $\rho_{BC} = 8$ kg/m, $\rho_{CD} = 5$ kg/m, and $\rho_{DE} = 2$ kg/m.

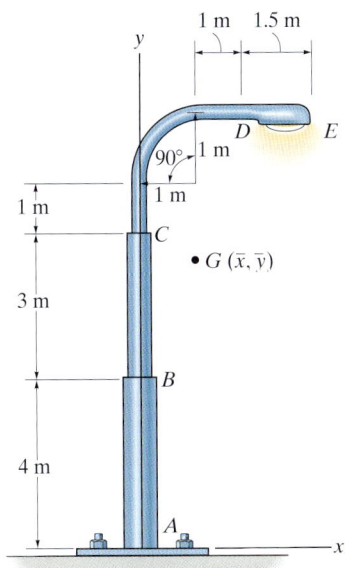

Prob. 9–65

9–66. Locate the centroid $\bar{y}$ of the cross-sectional area of the beam constructed from a channel and a plate. Assume all corners are square and neglect the size of the weld at A.

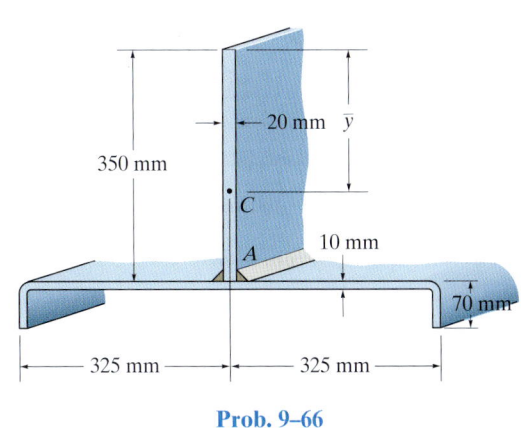

Prob. 9–66

9–67. An aluminum strut has a cross section referred to as a deep hat. Locate the centroid $\bar{y}$ of its area. Each segment has a thickness of 10 mm.

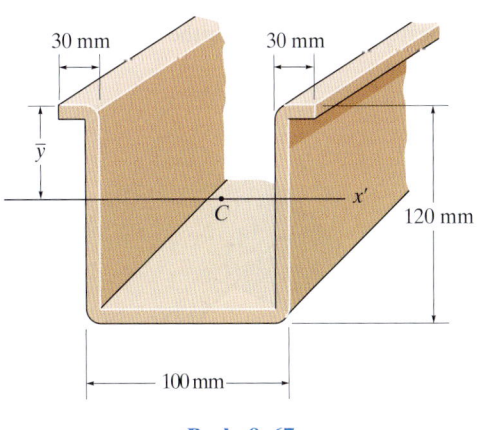

Prob. 9–67

*9–68.   Locate the centroid $\bar{y}$ of the beam's cross-sectional area.

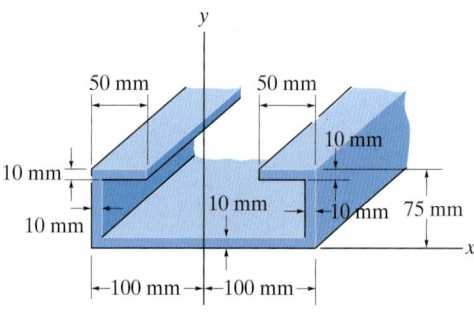

Prob. 9–68

9–69.   Determine the location $\bar{x}$ of the centroid C of the shaded area that is part of a circle having a radius r.

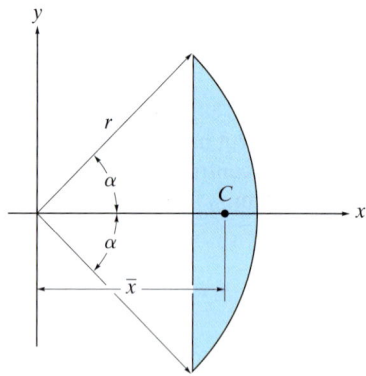

Prob. 9–69

9–70.   Locate the centroid $\bar{y}$ for the cross-sectional area of the angle.

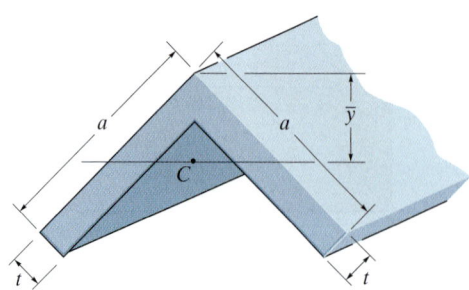

Prob. 9–70

9–71.   Determine the location $\bar{y}$ of the centroid of the beam's cross-sectional area. Neglect the size of the corner welds at A and B for the calculation.

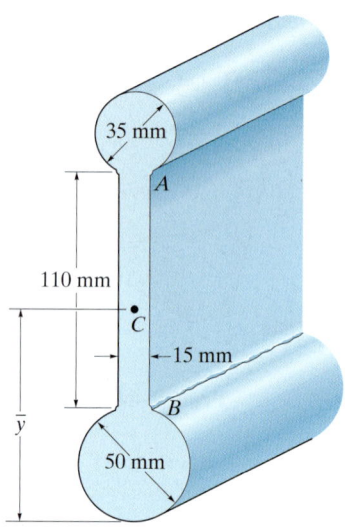

Prob. 9–71

*9–72. Uniform blocks having a length L and mass m are stacked one on top of the other, with each block overhanging the other by a distance d, as shown. If the blocks are glued together, so that they will not topple over, determine the location $\bar{x}$ of the center of mass of a pile of n blocks.

9–73.   Uniform blocks having a length L and mass m are stacked one on top of the other, with each block overhanging the other by a distance d, as shown. Show that the maximum number of blocks which can be stacked in this manner is $n < L/d$.

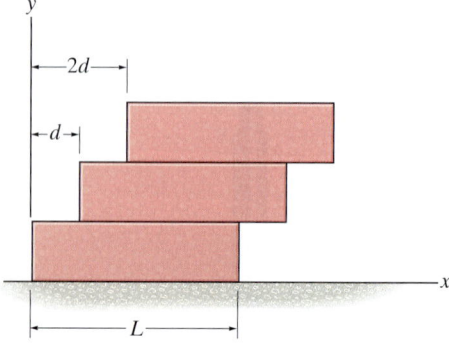

Probs. 9–72/73

9–74. The assembly is made from a steel hemisphere, $\rho_{st} = 7.80$ Mg/m³, and an aluminum cylinder, $\rho_{al} = 2.70$ Mg/m³. Determine the mass center of the assembly if the height of the cylinder is $h = 200$ mm.

9–75. The assembly is made from a steel hemisphere, $\rho_{st} = 7.80$ Mg/m³, and an aluminum cylinder, $\rho_{al} = 2.70$ Mg/m³. Determine the height h of the cylinder so that the mass center of the assembly is located at $\bar{z} = 160$ mm.

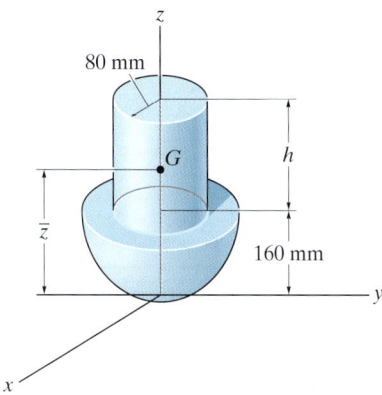

Probs. 9–74/75

*9–76. Locate the centroid $\bar{y}$ for the beam's cross-sectional area.

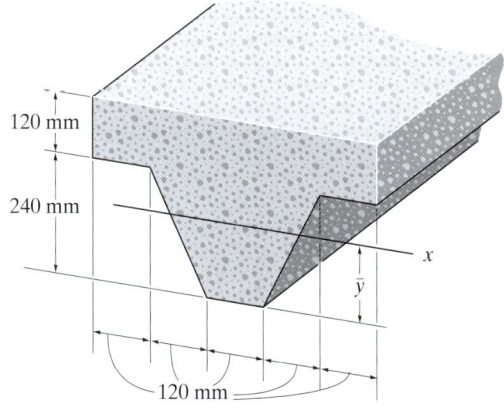

Prob. 9–76

9–77. Locate the centroid $\bar{y}$ for the strut's cross-sectional area.

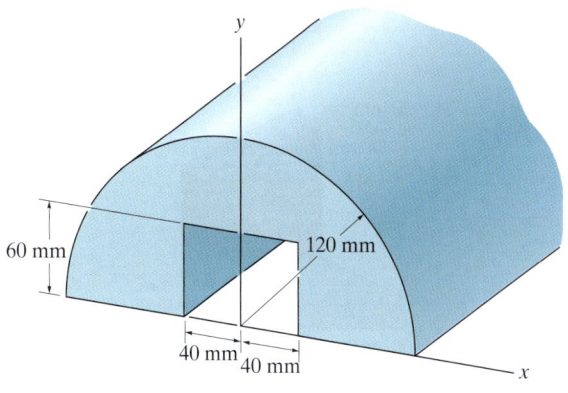

Prob. 9–77

9–78. An aluminum strut has a cross section referred to as a deep hat. Locate the centroid $\bar{y}$ of its area. Each segment has a thickness of 10 mm.

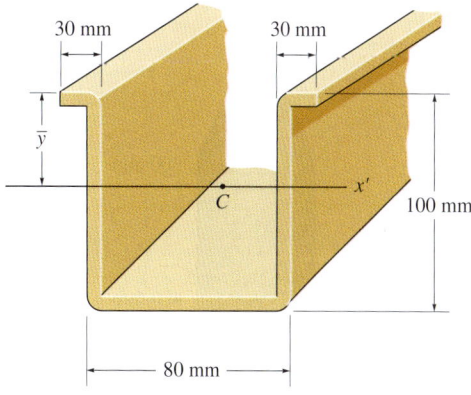

Prob. 9–78

9–79. Locate the center of mass of the block. Materials 1, 2, and 3 have densities of 2.70 Mg/m³, 5.70 Mg/m³, and 7.80 Mg/m³, respectively.

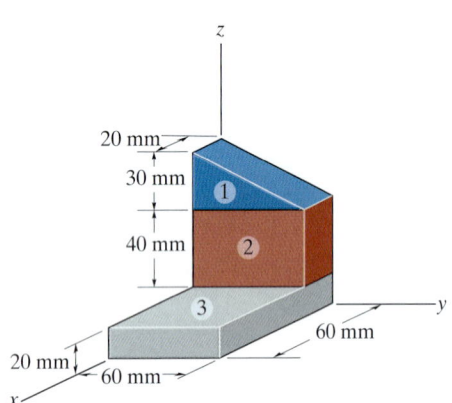

20 mm
30 mm
40 mm
①
②
③
60 mm
20 mm
60 mm

Prob. 9–79

*9–80.** Locate the centroid $\bar{z}$ of the homogenous solid formed by boring a hemispherical hole into the cylinder that is capped with a cone.

9–81. Locate the center of mass $\bar{z}$ of the solid formed by boring a hemispherical hole into a cylinder that is capped with a cone. The cone and cylinder are made of materials having densities of 7.80 Mg/m³ and 2.70 Mg/m³, respectively.

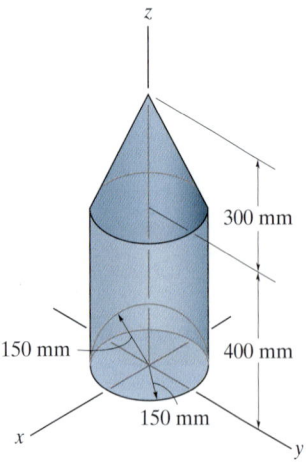

300 mm
150 mm
400 mm
150 mm

Probs. 9–80/81

9–82. Determine the distance h to which a 100-mm-diameter hole must be bored into the base of the cone so that the center of mass of the resulting shape is located at $\bar{z} = 115$ mm. The material has a density of 8 Mg/m³.

9–83. Determine the distance $\bar{z}$ to the centroid of the shape that consists of a cone with a hole of height $h = 50$ mm bored into its base.

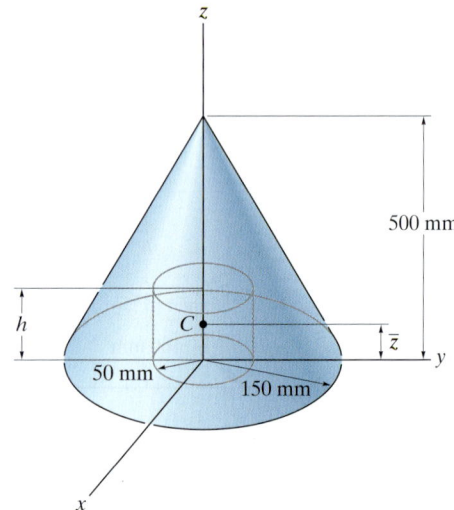

500 mm
h
C
$\bar{z}$
50 mm
150 mm

Probs. 9–82/83

*9–84.** Determine the distance $\bar{x}$ to the centroid of the solid which consists of a cylinder with a hole of length $h = 50$ mm bored into its base.

9–85. Determine the distance h to which a hole must be bored into the cylinder so that the center of mass of the assembly is located at $\bar{x} = 64$ mm. The material has a density of 8 Mg/m³.

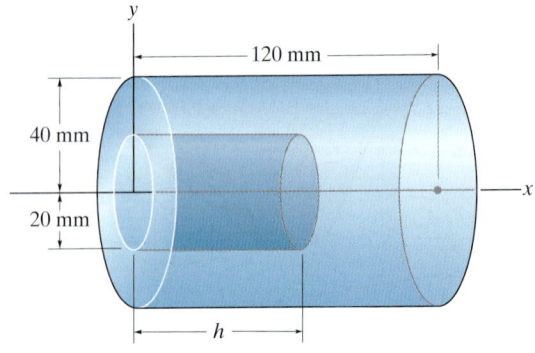

120 mm
40 mm
20 mm
h

Probs. 9–84/85

9–86. Locate the center of mass $\bar{z}$ of the assembly. The assembly consists of a cylindrical center core, A, having a density of 7.90 Mg/m³, and a cylindrical outer part, B, and a cone cap, C, each having a density of 2.70 Mg/m³.

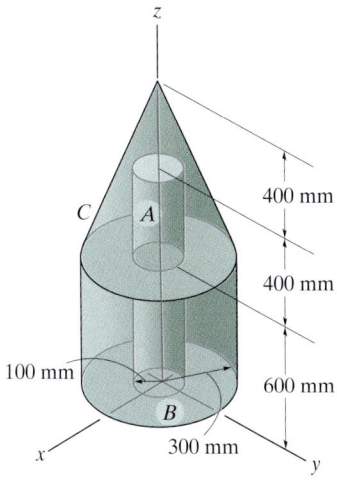

Prob. 9–86

9–87. Locate the center of mass $\bar{z}$ of the assembly. The material has a density of $\rho = 3$ Mg/m³. There is a 30-mm diameter hole bored through the center.

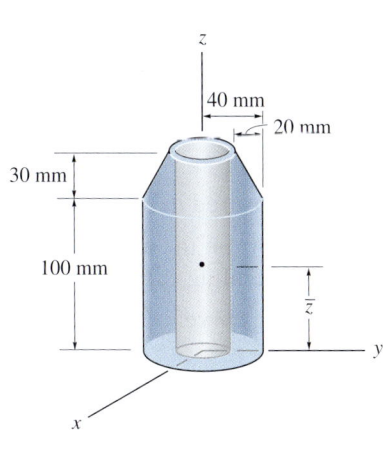
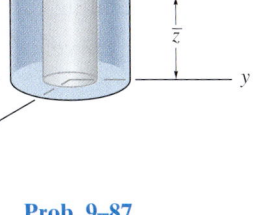

Prob. 9–87

*9–88.** A hole having a radius r is to be drilled in the center of the homogeneous block. Determine the depth h of the hole so that the center of gravity G is as low as possible.

Prob. 9–88

9–89. Locate the center of mass $\bar{z}$ of the assembly. The cylinder and the cone are made from materials having densities of 5 Mg/m³ and 9 Mg/m³, respectively.

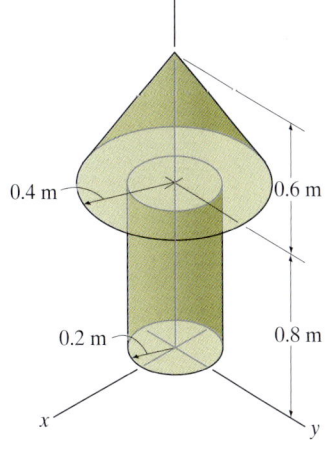

Prob. 9–89

*9.3   Theorems of Pappus and Guldinus

The two *theorems of Pappus and Guldinus* are used to find the surface area and volume of any body of revolution. They were first developed by Pappus of Alexandria during the fourth century A.D. and then restated at a later time by the Swiss mathematician Paul Guldin or Guldinus (1577–1643).

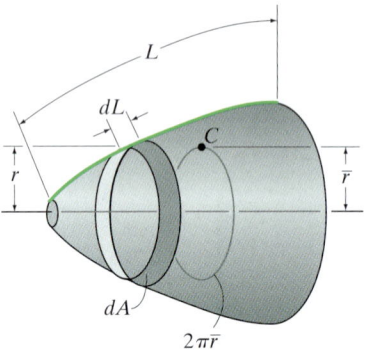

Fig. 9–19

The amount of roofing material used on this storage building can be estimated by using the first theorem of Pappus and Guldinus to determine its surface area.

Surface Area.   If we revolve a *plane curve* about an axis that does not intersect the curve we will generate a *surface area of revolution*. For example, the surface area in Fig. 9–19 is formed by revolving the curve of length L about the horizontal axis. To determine this surface area, we will first consider the differential line element of length dL. If this element is revolved 2π radians about the axis, a ring having a surface area of $dA = 2\pi r \, dL$ will be generated. Thus, the surface area of the entire body is $A = 2\pi \int r \, dL$. Since $\int r \, dL = \bar{r} L$ (Eq. 9–5), then $A = 2\pi \bar{r} L$. If the curve is revolved only through an angle θ (radians), then

$$A = \theta \bar{r} L \qquad (9\text{–}7)$$

where

A = surface area of revolution

θ = angle of revolution measured in radians, $\theta \leq 2\pi$

$\bar{r}$ = perpendicular distance from the axis of revolution to the centroid of the generating curve

L = length of the generating curve

Therefore the first theorem of Pappus and Guldinus states that *the area of a surface of revolution equals the product of the length of the generating curve and the distance traveled by the centroid of the curve in generating the surface area.*

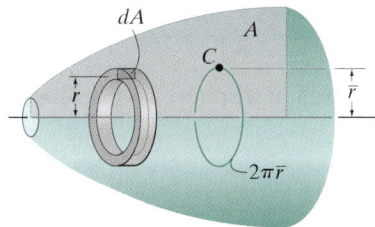

Fig. 9–20

Volume.   A *volume* can be generated by revolving a *plane area* about an axis that does not intersect the area. For example, if we revolve the shaded area A in Fig. 9–20 about the horizontal axis, it generates the volume shown. This volume can be determined by first revolving the differential element of area dA 2π radians about the axis, so that a ring having the volume $dV = 2\pi r\, dA$ is generated. The entire volume is then $V = 2\pi \int r\, dA$. However, $\int r\, dA = \bar{r}\, A$, Eq. 9–4, so that $V = 2\pi \bar{r} A$. If the area is only revolved through an angle θ (radians), then

$$V = \theta \bar{r} A \qquad\qquad (9\text{–}8)$$

where

$V = $ volume of revolution

$\theta = $ angle of revolution measured in radians, $\theta \leq 2\pi$

$\bar{r} = $ perpendicular distance from the axis of revolution to the centroid of the generating area

$A = $ generating area

Therefore the second theorem of Pappus and Guldinus states that *the volume of a body of revolution equals the product of the generating area and the distance traveled by the centroid of the area in generating the volume.*

Composite Shapes.   We may also apply the above two theorems to lines or areas that are composed of a series of composite parts. In this case the total surface area or volume generated is the addition of the surface areas or volumes generated by each of the composite parts. If the perpendicular distance from the axis of revolution to the centroid of each composite part is $\tilde{r}$, then

$$A = \theta \Sigma (\tilde{r} L) \qquad\qquad (9\text{–}9)$$

and

$$V = \theta \Sigma (\tilde{r} A) \qquad\qquad (9\text{–}10)$$

Application of the above theorems is illustrated numerically in the following examples.

The volume of fertilizer contained within this silo can be determined using the second theorem of Pappus and Guldinus.

EXAMPLE 9.12

Show that the surface area of a sphere is $A = 4\pi R^2$ and its volume is $V = \frac{4}{3}\pi R^3$.

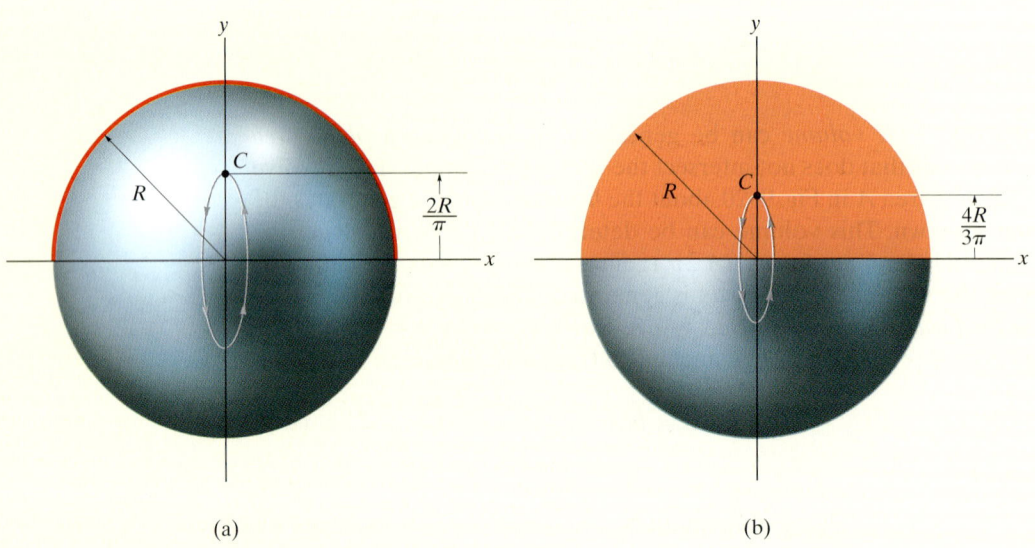

(a)   (b)

Fig. 9–21

SOLUTION

Surface Area. The surface area of the sphere in Fig. 9–21a is generated by revolving a semicircular *arc* about the x axis. Using the table on the inside back cover, it is seen that the centroid of this arc is located at a distance $\bar{r} = 2R/\pi$ from the axis of revolution (x axis). Since the centroid moves through an angle of $\theta = 2\pi$ rad to generate the sphere, then applying Eq. 9–7 we have

$$A = \theta \bar{r} L; \qquad A = 2\pi \left(\frac{2R}{\pi}\right)\pi R = 4\pi R^2 \qquad Ans.$$

Volume. The volume of the sphere is generated by revolving the semicircular *area* in Fig. 9–21b about the x axis. Using the table on the inside back cover to locate the centroid of the area, i.e., $\bar{r} = 4R/3\pi$, and applying Eq. 9–8, we have

$$V = \theta \bar{r} A; \qquad V = 2\pi \left(\frac{4R}{3\pi}\right)\left(\frac{1}{2}\pi R^2\right) = \frac{4}{3}\pi R^3 \qquad Ans.$$

EXAMPLE 9.13

Determine the surface area and volume of the full solid in Fig. 9–22a.

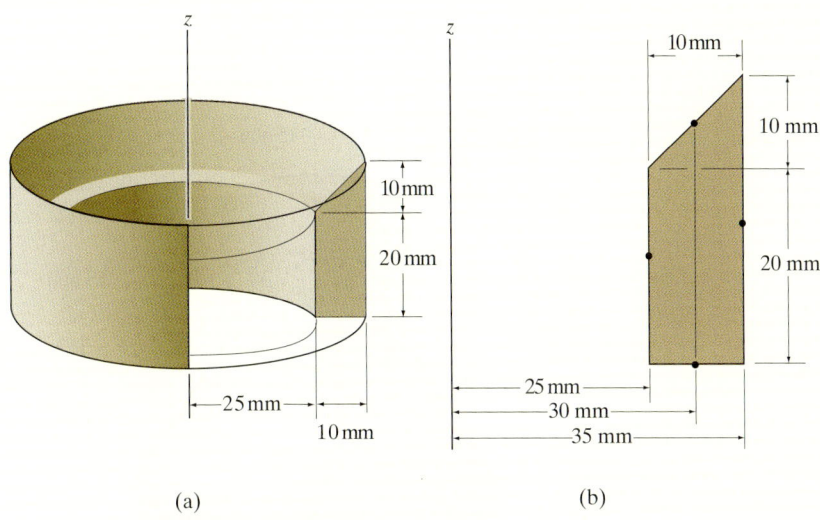

(a)   (b)

Fig. 9–22

SOLUTION

Surface Area.   The surface area is generated by revolving the four line segments shown in Fig. 9–22b 2π radians about the z axis. The distances from the centroid of each segment to the z axis are also shown in the figure. Applying Eq. 9–7 yields

$$A = 2\pi\Sigma \bar{r}L = 2\pi[(25 \text{ mm})(20 \text{ mm}) + (30 \text{ mm})\left(\sqrt{(10 \text{ mm})^2 + (10 \text{ mm})^2}\right)$$
$$+ (35 \text{ mm})(30 \text{ mm}) + (30 \text{ mm})(10 \text{ mm})]$$
$$= 14\,290 \text{ mm}^2 \qquad\qquad Ans.$$

Volume.   The volume of the solid is generated by revolving the two area segments shown in Fig. 9–22c 2π radians about the z axis. The distances from the centroid of each segment to the z axis are also shown in the figure. Applying Eq. 9–10, we have

$$V = 2\pi\Sigma \bar{r}A = 2\pi\left\{(31.667 \text{ mm})\left[\frac{1}{2}(10 \text{ mm})(10 \text{ mm})\right] \right.$$
$$\left. + (30 \text{ mm})[(20 \text{ mm})(10 \text{ mm})]\right\}$$
$$= 47\,648 \text{ mm}^3 \qquad\qquad Ans.$$

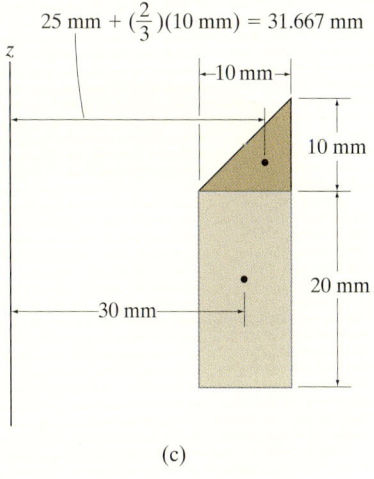

(c)

FUNDAMENTAL PROBLEMS

F9–13. Determine the surface area and volume of the solid formed by revolving the shaded area 360° about the z axis.

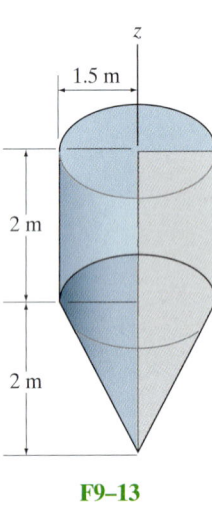

F9–13

F9–14. Determine the surface area and volume of the solid formed by revolving the shaded area 360° about the z axis.

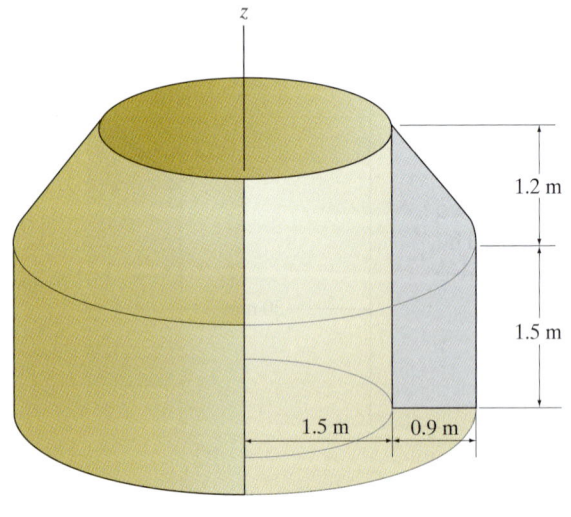

F9–14

F9–15. Determine the surface area and volume of the solid formed by revolving the shaded area 360° about the z axis.

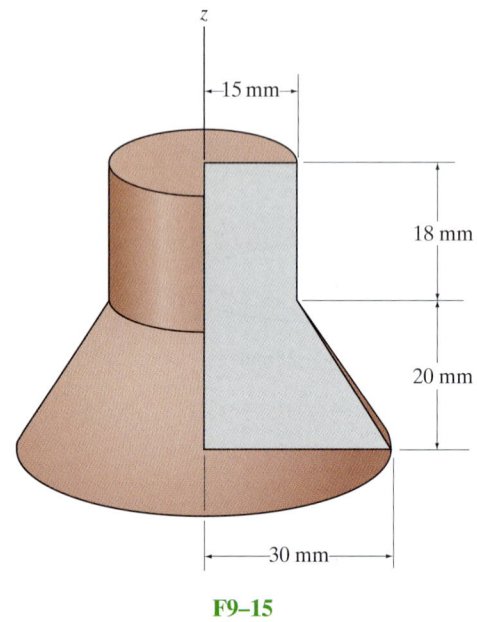

F9–15

F9–16. Determine the surface area and volume of the solid formed by revolving the shaded area 360° about the z axis.

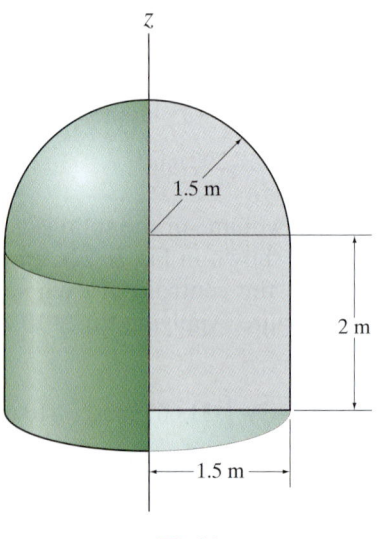

F9–16

PROBLEMS

9–90. The water tank AB has a hemispherical top and is fabricated from thin steel plate. Determine the volume within the tank.

9–91. The water tank AB has a hemispherical roof and is fabricated from thin steel plate. If a liter of paint can cover 3 m² of the tank's surface, determine how many liters are required to coat the surface of the tank from A to B.

9–94. The *rim* of a flywheel has the cross section A–A shown. Determine the volume of material needed for its construction.

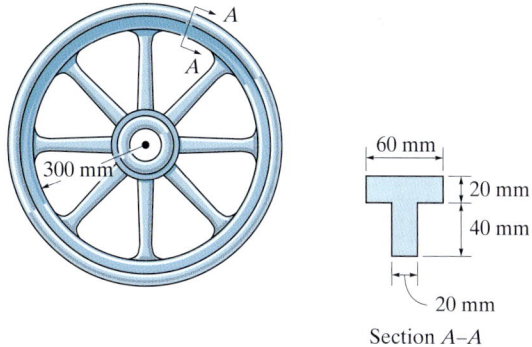

Section A–A

Prob. 9–94

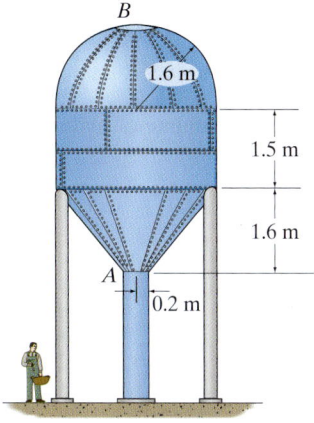

Probs. 9–90/91

*9–92. Determine the outside surface area of the hopper.

9–93. The hopper is filled to its top with coal. Determine the volume of coal if the voids (air space) are 30 percent of the volume of the hopper.

9–95. Determine the surface area of the tank, which consists of a cylinder and hemispherical cap.

*9–96. Determine the volume of the tank, which consists of a cylinder and hemispherical cap.

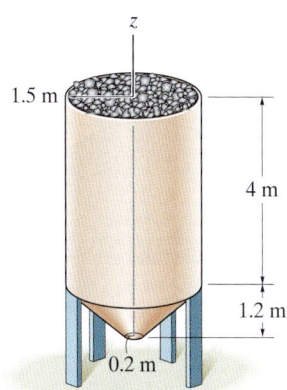

Probs. 9–92/93

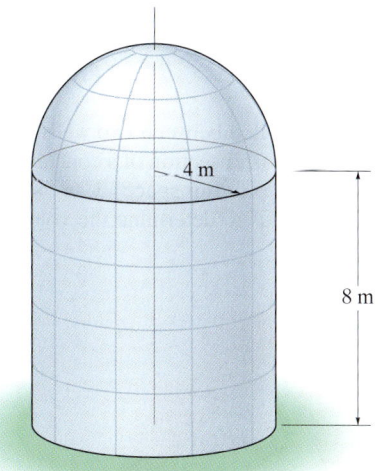

Probs. 9–95/96

9–97. The process tank is used to store liquids during manufacturing. Estimate the outside surface area of the tank. The tank has a flat top and the plates from which the tank is made have negligible thickness.

9–98. The process tank is used to store liquids during manufacturing. Estimate the volume of the tank. The tank has a flat top and the plates from which the tank is made have negligible thickness.

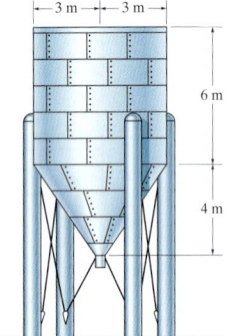

Probs. 9–97/98

9–99. The hopper is filled to its top with coal. Estimate the volume of coal if the voids (air space) are 35 percent of the volume of the hopper.

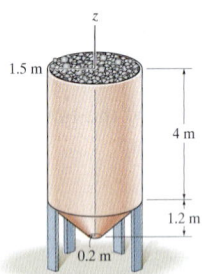

Prob. 9–99

***9–100.** Sand is piled between two walls as shown. Assume the pile to be a quarter section of a cone and that 26 percent of this volume is voids (air space). Use the second theorem of Pappus–Guldinus to determine the volume of sand.

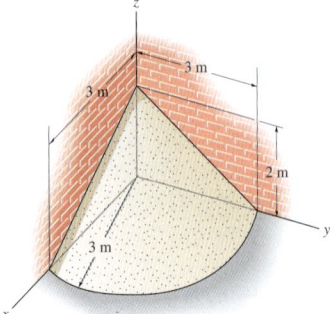

Prob. 9–100

9–101. Determine the surface area and the volume of the ring formed by rotating the square about the vertical axis.

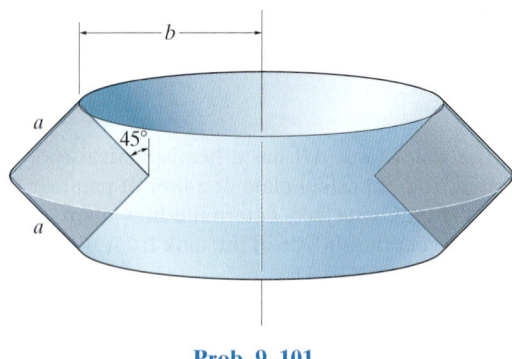

Prob. 9–101

9–102. Determine the volume of material needed to make the casting.

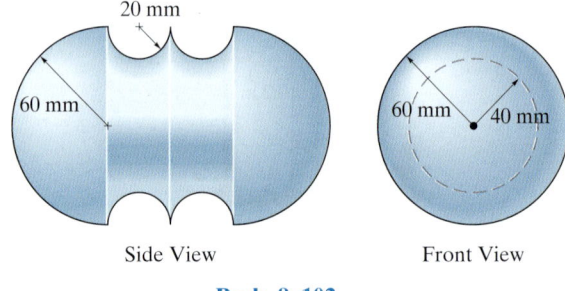

Side View   Front View

Prob. 9–102

9–103. Determine the surface area from A to B of the tank.

***9–104.** Determine the volume within the thin-walled tank from A to B.

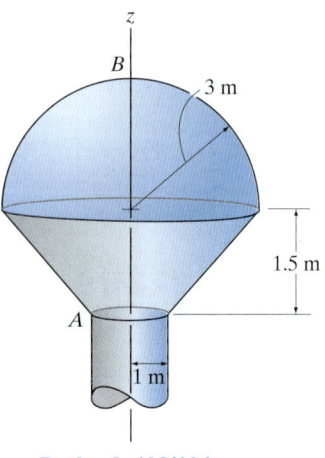

Probs. 9–103/104

9–105. A steel wheel has a diameter of 840 mm and a cross section as shown in the figure. Determine the total mass of the wheel if $\rho = 5$ Mg/m³.

9–107. Using integration, determine both the area and the centroidal distance $\bar{x}$ of the shaded area. Then, using the second theorem of Pappus–Guldinus, determine the volume of the solid generated by revolving the area about the y axis.

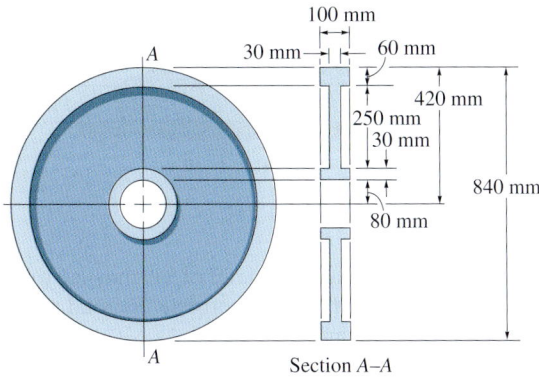

Prob. 9–105

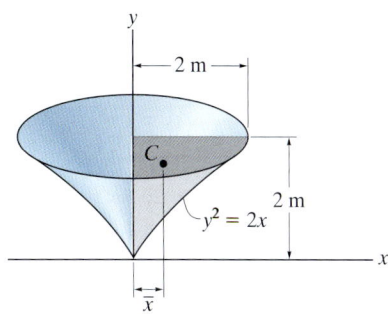

Prob. 9–107

9–106. Determine the volume of an ellipsoid formed by revolving the shaded area about the x axis using the second theorem of Pappus–Guldinus. The area and centroid y of the shaded area should first be obtained by using integration.

*9–108.** Determine the height h to which liquid should be poured into the conical cup so that it contacts half the surface area on the inside of the cup.

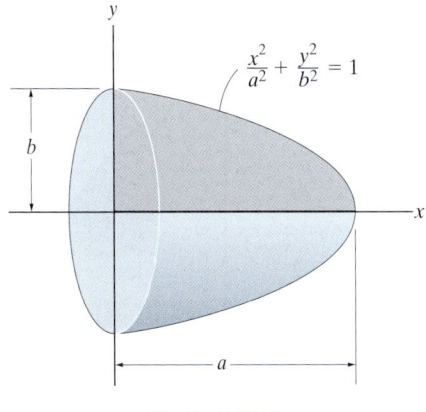

Prob. 9–106

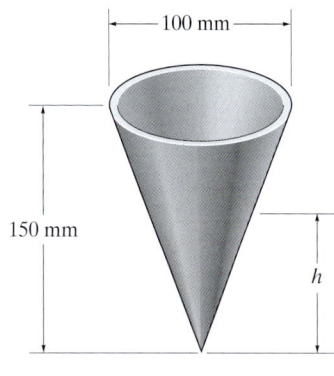

Prob. 9–108

9–109. Determine the interior surface area of the brake piston. It consists of a full circular part. Its cross section is shown in the figure.

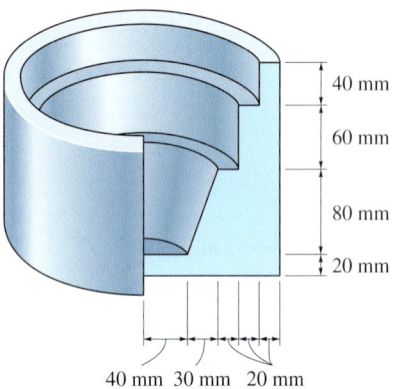

40 mm

60 mm

80 mm

20 mm

40 mm 30 mm 20 mm

Prob. 9–109

9–110. Determine the surface area and the volume of the conical solid.

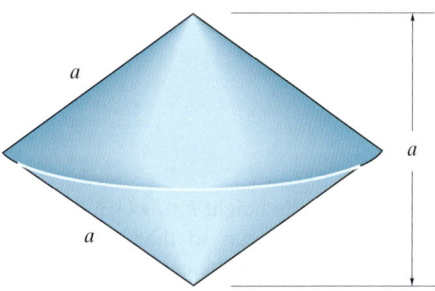

a

a

a

Prob. 9–110

9–111. Determine the height h to which liquid should be poured into the cup so that it contacts half the surface area on the inside of the cup. Neglect the cup's thickness for the calculation.

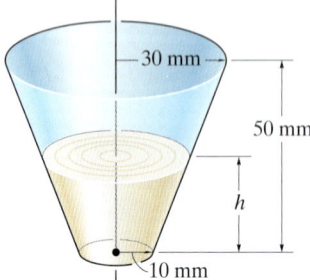

30 mm

50 mm

h

10 mm

Prob. 9–111

***9–112.** The water tank has a paraboloid-shaped roof. If one liter of paint can cover 3 m² of the tank, determine the number of liters required to coat the roof.

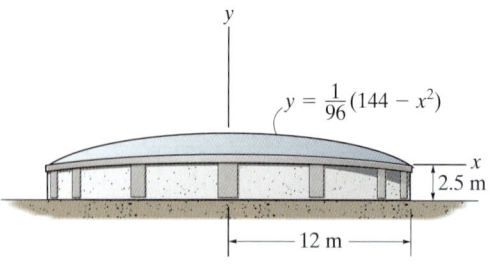

y

$y = \frac{1}{96}(144 - x^2)$

x

2.5 m

12 m

Prob. 9–112

9–113. A steel wheel has a diameter of 840 mm and a cross section as shown in the figure. Determine the total mass of the wheel if $\rho = 5 \text{ Mg/m}^3$.

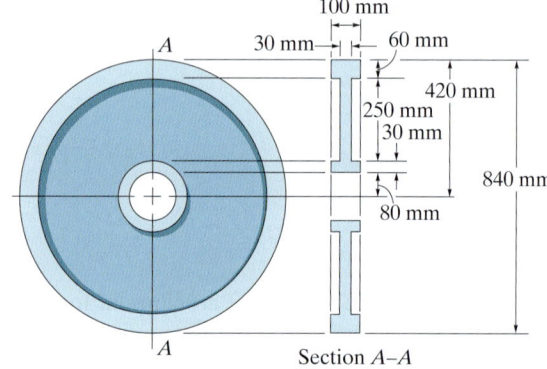

100 mm

A 30 mm   60 mm

420 mm

250 mm

30 mm

840 mm

80 mm

A Section A–A

Prob. 9–113

9–114. Determine the surface area of the roof of the structure if it is formed by rotating the parabola about the y axis.

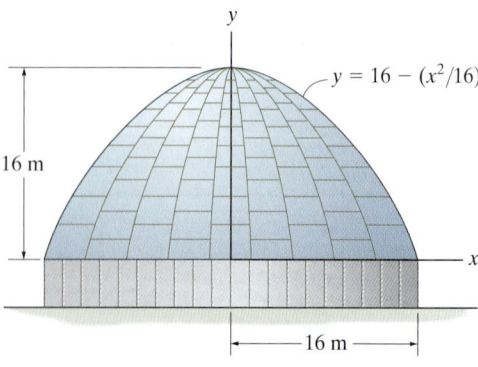

y

$y = 16 - (x^2/16)$

16 m

x

16 m

Prob. 9–114

*9.4   Resultant of a General Distributed Loading

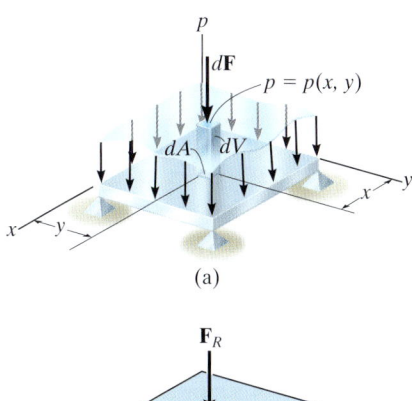

(a)

In Sec. 4.9, we discussed the method used to simplify a two-dimensional distributed loading to a single resultant force acting at a specific point. In this section we will generalize this method to include flat surfaces that have an arbitrary shape and are subjected to a variable load distribution. Consider, for example, the flat plate shown in Fig. 9–23a, which is subjected to the loading defined by $p = p(x, y)$ Pa, where 1 Pa (pascal) = $1\ \text{N/m}^2$. Knowing this function, we can determine the resultant force $\mathbf{F}_R$ acting on the plate and its location $(\bar{x}, \bar{y})$, Fig. 9–23b.

(b)

Fig. 9–23

Magnitude of Resultant Force.   The force $d\mathbf{F}$ acting on the differential area $dA\ \text{m}^2$ of the plate, located at the arbitrary point (x, y), has a magnitude of $dF = [p(x, y)\ \text{N/m}^2](dA\ \text{m}^2) = [p(x, y)\ dA]\ \text{N}$. Notice that $p(x, y)\ dA = dV$, the colored differential *volume element* shown in Fig. 9–23a. The *magnitude* of $\mathbf{F}_R$ is the sum of the differential forces acting over the plate's *entire surface area A*. Thus:

$$F_R = \Sigma F; \qquad F_R = \int_A p(x, y)\ dA = \int_V dV = V \qquad (9\text{–}11)$$

This result indicates that the *magnitude of the resultant force is equal to the total volume under the distributed-loading diagram.*

The resultant of a wind loading that is distributed on the front or side walls of this building must be calculated using integration in order to design the framework that holds the building together.

Location of Resultant Force.   The location $(\bar{x}, \bar{y})$ of $\mathbf{F}_R$ is determined by setting the moments of $\mathbf{F}_R$ equal to the moments of all the differential forces $d\mathbf{F}$ about the respective y and x axes: From Figs. 9–23a and 9–23b, using Eq. 9–11, this results in

$$\bar{x} = \frac{\displaystyle\int_A x p(x, y)\ dA}{\displaystyle\int_A p(x, y)\ dA} = \frac{\displaystyle\int_V x\ dV}{\displaystyle\int_V dV} \qquad \bar{y} = \frac{\displaystyle\int_A y p(x, y)\ dA}{\displaystyle\int_A p(x, y)\ dA} = \frac{\displaystyle\int_V y\ dV}{\displaystyle\int_V dV} \qquad (9\text{–}12)$$

Hence, the *line of action of the resultant force passes through the geometric center or centroid of the volume under the distributed-loading diagram.*

*9.5   Fluid Pressure

According to Pascal's law, a fluid at rest creates a pressure p at a point that is the *same* in *all* directions. The magnitude of p, measured as a force per unit area, depends on the specific weight γ or mass density ρ of the fluid and the depth z of the point from the fluid surface.* The relationship can be expressed mathematically as

$$p = \gamma z = \rho g z \qquad (9\text{--}13)$$

where g is the acceleration due to gravity. This equation is valid only for fluids that are assumed *incompressible*, as in the case of most liquids. Gases are compressible fluids, and since their density changes significantly with both pressure and temperature, Eq. 9–13 cannot be used.

To illustrate how Eq. 9–13 is applied, consider the submerged plate shown in Fig. 9–24. Three points on the plate have been specified. Since point B is at depth z_1 from the liquid surface, the *pressure* at this point has a magnitude $p_1 = \gamma z_1$. Likewise, points C and D are both at depth z_2; hence, $p_2 = \gamma z_2$. In all cases, the pressure acts *normal* to the surface area dA located at the specified point.

Using Eq. 9–13 and the results of Sec. 9.4, it is possible to determine the resultant force caused by a liquid and specify its location on the surface of a submerged plate. Three different shapes of plates will now be considered.

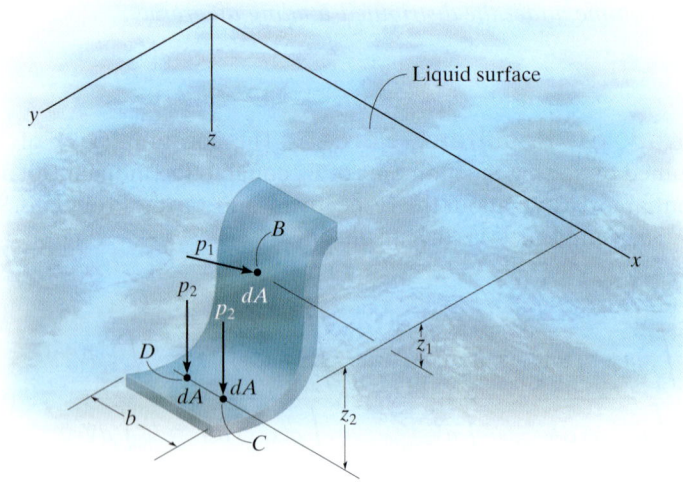

Fig. 9–24

*In particular, for water $\gamma = \rho g = 9810 \text{ N/m}^3$ since $\rho = 1000 \text{ kg/m}^3$ and $g = 9.81 \text{ m/s}^2$.

Flat Plate of Constant Width.

A flat rectangular plate of constant width, which is submerged in a liquid having a specific weight γ, is shown in Fig. 9–25a. Since pressure varies linearly with depth, Eq. 9–13, the distribution of pressure over the plate's surface is represented by a trapezoidal volume having an intensity of $p_1 = \gamma z_1$ at depth z_1 and $p_2 = \gamma z_2$ at depth z_2. As noted in Sec. 9.4, the magnitude of the *resultant force* $\mathbf{F}_R$ is equal to the *volume* of this loading diagram and $\mathbf{F}_R$ has a *line of action* that passes through the volume's centroid C. Hence, $\mathbf{F}_R$ does *not* act at the centroid of the plate; rather, it acts at point P, called the *center of pressure*.

Since the plate has a *constant width*, the loading distribution may also be viewed in two dimensions, Fig. 9–25b. Here the loading intensity is measured as force/length and varies linearly from $w_1 = bp_1 = b\gamma z_1$ to $w_2 = bp_2 = b\gamma z_2$. The magnitude of $\mathbf{F}_R$ in this case equals the trapezoidal *area*, and $\mathbf{F}_R$ has a *line of action* that passes through the area's *centroid C*. For numerical applications, the area and location of the centroid for a trapezoid are tabulated on the inside back cover.

The walls of the tank must be designed to support the pressure loading of the liquid that is contained within it.

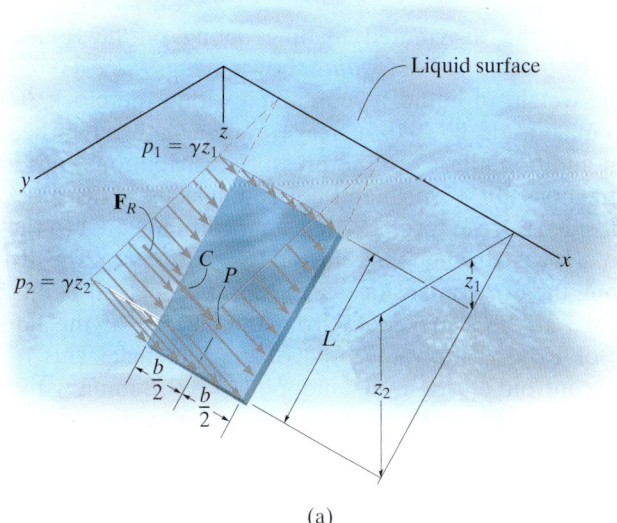

(a)

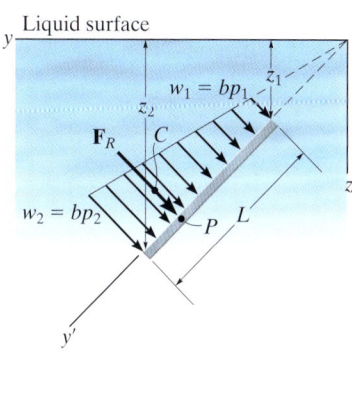

(b)

Fig. 9–25

9

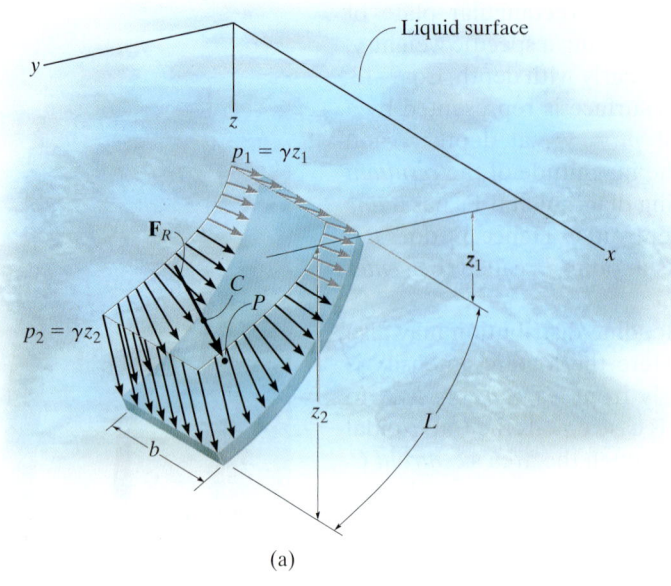

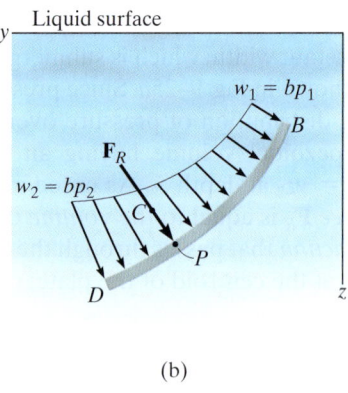

(b)

(a)

Fig. 9–26

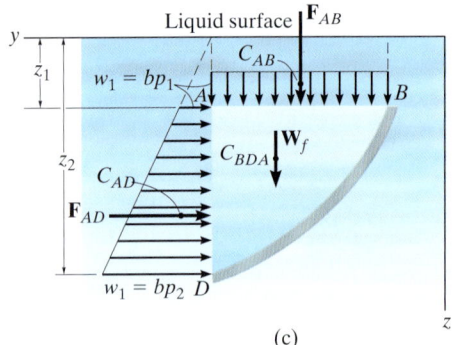

(c)

Curved Plate of Constant Width.

When a submerged plate of constant width is curved, the pressure acting normal to the plate continually changes both its magnitude and direction, and therefore calculation of the magnitude of $\mathbf{F}_R$ and its location P is more difficult than for a flat plate. Three- and two-dimensional views of the loading distribution are shown in Figs. 9–26a and 9–26b, respectively. Although integration can be used to solve this problem, a simpler method exists. This method requires separate calculations for the horizontal and vertical *components* of $\mathbf{F}_R$.

For example, the distributed loading acting on the plate can be represented by the *equivalent loading* shown in Fig. 9–26c. Here the plate supports the weight of liquid W_f contained within the block BDA. This force has a magnitude $W_f = (\gamma b)(\text{area}_{BDA})$ and acts through the centroid of BDA. In addition, there are the pressure distributions caused by the liquid acting along the vertical and horizontal sides of the block. Along the vertical side AD, the force $\mathbf{F}_{AD}$ has a magnitude equal to the area of the trapezoid. It acts through the centroid C_{AD} of this area. The distributed loading along the horizontal side AB is *constant* since all points lying in this plane are at the same depth from the surface of the liquid. The magnitude of $\mathbf{F}_{AB}$ is simply the area of the rectangle. This force acts through the centroid C_{AB} or at the midpoint of AB. Summing these three forces yields $\mathbf{F}_R = \Sigma\mathbf{F} = \mathbf{F}_{AD} + \mathbf{F}_{AB} + \mathbf{W}_f$. Finally, the location of the center of pressure P on the plate is determined by applying $M_R = \Sigma M$, which states that the moment of the resultant force about a convenient reference point such as D or B, in Fig. 9–26b, is equal to the sum of the moments of the three forces in Fig. 9–26c about this same point.

Flat Plate of Variable Width. The pressure distribution acting on the surface of a submerged plate having a variable width is shown in Fig. 9–27. If we consider the force $d\mathbf{F}$ acting on the differential area strip dA, parallel to the x axis, then its magnitude is $dF = p \, dA$. Since the depth of dA is z, the pressure on the element is $p = \gamma z$. Therefore, $dF = (\gamma z)dA$ and so the resultant force becomes

$$F_R = \int dF = \gamma \int z \, dA$$

If the depth to the centroid C' of the area is $\bar{z}$, Fig. 9–27, then, $\int z \, dA = \bar{z}A$. Substituting, we have

$$F_R = \gamma \bar{z} A \qquad (9\text{–}14)$$

The resultant force of the water pressure and its location on the elliptical back plate of this tank truck must be determined by integration.

In other words, *the magnitude of the resultant force acting on any flat plate is equal to the product of the area A of the plate and the pressure $p = \gamma \bar{z}$ at the depth of the area's centroid C'.* As discussed in Sec. 9.4, this force is also equivalent to the volume under the pressure distribution. Realize that its line of action passes through the centroid C of this *volume* and intersects the plate at the center of pressure P, Fig. 9–27. Notice that the location of C' does not coincide with the location of P.

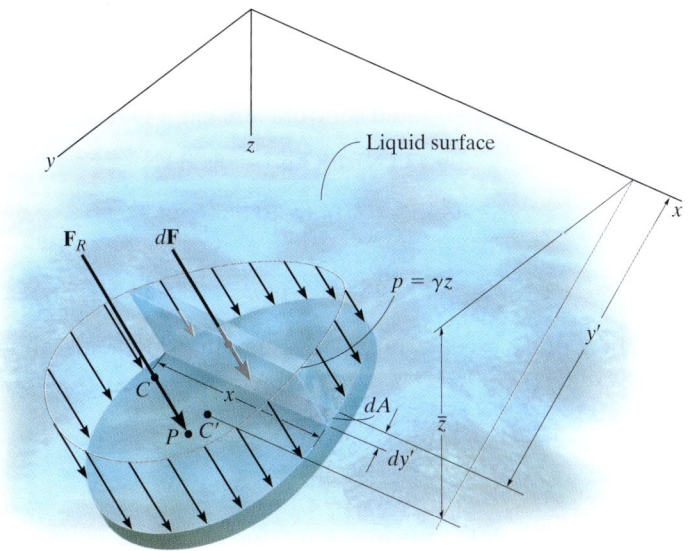

Fig. 9–27

EXAMPLE | 9.14

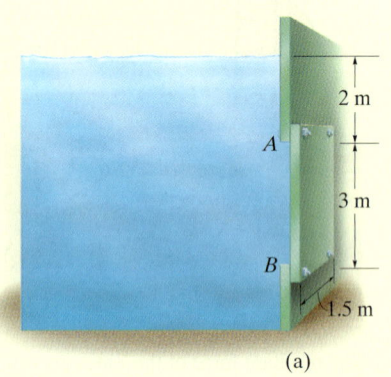

(a)

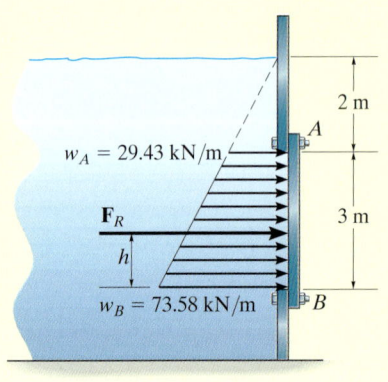

(b)

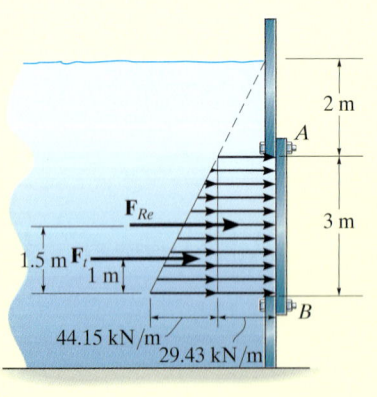

(c)

Fig. 9–28

Determine the magnitude and location of the resultant hydrostatic force acting on the submerged rectangular plate AB shown in Fig. 9–28a. The plate has a width of 1.5 m; $\rho_w = 1000 \text{ kg/m}^3$.

SOLUTION I

The water pressures at depths A and B are

$$p_A = \rho_w g z_A = (1000 \text{ kg/m}^3)(9.81 \text{ m/s}^2)(2 \text{ m}) = 19.62 \text{ kPa}$$

$$p_B = \rho_w g z_B = (1000 \text{ kg/m}^3)(9.81 \text{ m/s}^2)(5 \text{ m}) = 49.05 \text{ kPa}$$

Since the plate has a constant width, the pressure loading can be viewed in two dimensions as shown in Fig. 9–28b. The intensities of the load at A and B are

$$w_A = b p_A = (1.5 \text{ m})(19.62 \text{ kPa}) = 29.43 \text{ kN/m}$$

$$w_B = b p_B = (1.5 \text{ m})(49.05 \text{ kPa}) = 73.58 \text{ kN/m}$$

From the table on the inside back cover, the magnitude of the resultant force $\mathbf{F}_R$ created by this distributed load is

$$F_R = \text{area of a trapezoid} = \tfrac{1}{2}(3)(29.4 + 73.6) = 154.5 \text{ kN} \qquad \textit{Ans.}$$

This force acts through the centroid of this area,

$$h = \frac{1}{3}\left(\frac{2(29.43) + 73.58}{29.43 + 73.58}\right)(3) = 1.29 \text{ m} \qquad \textit{Ans.}$$

measured upward from B, Fig. 9–31b.

SOLUTION II

The same results can be obtained by considering two components of $\mathbf{F}_R$, defined by the triangle and rectangle shown in Fig. 9–28c. Each force acts through its associated centroid and has a magnitude of

$$F_{Re} = (29.43 \text{ kN/m})(3 \text{ m}) = 88.3 \text{ kN}$$

$$F_t = \tfrac{1}{2}(44.15 \text{ kN/m})(3 \text{ m}) = 66.2 \text{ kN}$$

Hence,

$$F_R = F_{Re} + F_t = 88.3 + 66.2 = 154.5 \text{ kN} \qquad \textit{Ans.}$$

The location of $\mathbf{F}_R$ is determined by summing moments about B, Figs. 9–28b and c, i.e.,

$$\zeta + (M_R)_B = \Sigma M_B; \quad (154.5)h = 88.3(1.5) + 66.2(1)$$

$$h = 1.29 \text{ m} \qquad \textit{Ans.}$$

NOTE: Using Eq. 9–14, the resultant force can be calculated as $F_R = \gamma \bar{z} A = (9810 \text{ N/m}^3)(3.5 \text{ m})(3 \text{ m})(1.5 \text{ m}) = 154.5 \text{ kN}$.

EXAMPLE 9.15

Determine the magnitude of the resultant hydrostatic force acting on the surface of a seawall shaped in the form of a parabola as shown in Fig. 9–29a. The wall is 5 m long; $\rho_w = 1020$ kg/m³.

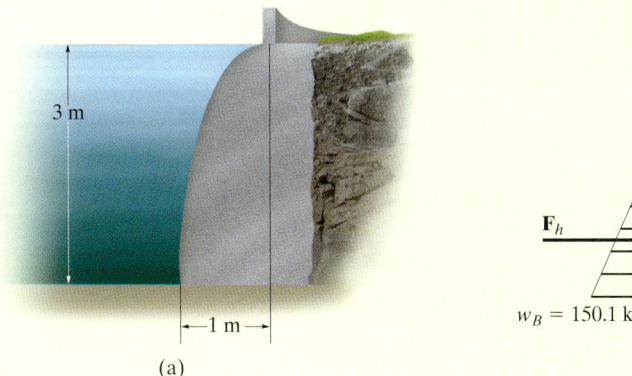

(a)   (b)

Fig. 9–29

SOLUTION

The horizontal and vertical components of the resultant force will be calculated, Fig. 9–29b. Since

$$p_B = \rho_w g z_B = (1020 \text{ kg/m}^3)(9.81 \text{ m/s}^2)(3 \text{ m}) = 30.02 \text{ kPa}$$

then

$$w_B = b p_B = 5 \text{ m}(30.02 \text{ kPa}) = 150.1 \text{ kN/m}$$

Thus,

$$F_h = \tfrac{1}{2}(3 \text{ m})(150.1 \text{ kN/m}) = 225.1 \text{ kN}$$

The area of the parabolic section ABC can be determined using the formula for a parabolic area $A = \tfrac{1}{3}ab$. Hence, the weight of water within this 5-m-long region is

$$F_v = (\rho_w g b)(\text{area}_{ABC})$$
$$= (1020 \text{ kg/m}^3)(9.81 \text{ m/s}^2)(5 \text{ m})\left[\tfrac{1}{3}(1 \text{ m})(3 \text{ m})\right] = 50.0 \text{ kN}$$

The resultant force is therefore

$$F_R = \sqrt{F_h^2 + F_v^2} = \sqrt{(225.1 \text{ kN})^2 + (50.0 \text{ kN})^2}$$
$$= 231 \text{ kN} \qquad\qquad\qquad Ans.$$

9

EXAMPLE | 9.16

Determine the magnitude and location of the resultant force acting on the triangular end plates of the water trough shown in Fig. 9–30a; $\rho_w = 1000 \text{ kg/m}^3$.

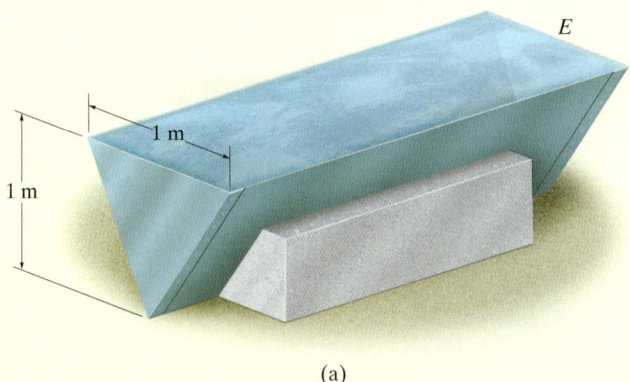

(a)

SOLUTION

The pressure distribution acting on the end plate E is shown in Fig. 9–30b. The magnitude of the resultant force is equal to the volume of this loading distribution. We will solve the problem by integration. Choosing the differential volume element shown in the figure, we have

$$dF = dV = p \, dA = \rho_w g z (2x \, dz) = 19\ 620 z x \, dz$$

The equation of line AB is

$$x = 0.5(1 - z)$$

Hence, substituting and integrating with respect to z from $z = 0$ to $z = 1$ m yields

$$F = V = \int_V dV = \int_0^{1\,\text{m}} (19\ 620) z [0.5(1 - z)] \, dz$$

$$= 9810 \int_0^{1\,\text{m}} (z - z^2) \, dz = 1635 \text{ N} = 1.64 \text{ kN} \qquad Ans.$$

This resultant passes through the *centroid of the volume*. Because of symmetry,

$$\bar{x} = 0 \qquad Ans.$$

Since $\tilde{z} = z$ for the volume element, then

$$\bar{z} = \frac{\displaystyle\int_V \tilde{z} \, dV}{\displaystyle\int_V dV} = \frac{\displaystyle\int_0^{1\,\text{m}} z(19\ 620) z [0.5(1 - z)] \, dz}{1635} = \frac{9810 \displaystyle\int_0^{1\,\text{m}} (z^2 - z^3) \, dz}{1635}$$

$$= 0.5 \text{ m} \qquad Ans.$$

NOTE: We can also determine the resultant force by applying Eq. 9–14, $F_R = \gamma \bar{z} A = \left(9810 \text{ N/m}^3\right)\left(\tfrac{1}{3}\right)(1 \text{ m})\left[\tfrac{1}{2}(1 \text{ m})(1 \text{ m})\right] = 1.64 \text{ kN}$.

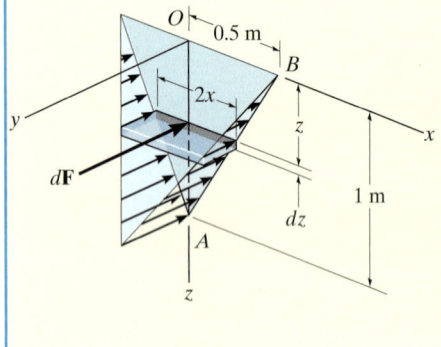

(b)

Fig. 9–30

FUNDAMENTAL PROBLEMS

F9–17.   Determine the magnitude of the hydrostatic force acting per meter length of the wall. Water has a density of $\rho = 1 \text{ Mg/m}^3$.

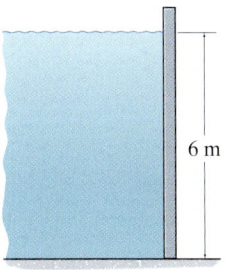

F9–17

F9–18.   Determine the magnitude of the hydrostatic force acting on gate AB, which has a width of 1 m. The specific weight of water is $\gamma = 9.81 \text{ kN/m}^3$.

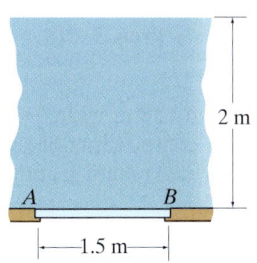

F9–18

F9–19.   Determine the magnitude of the hydrostatic force acting on gate AB, which has a width of 1.5 m. Water has a density of $\rho = 1 \text{ Mg/m}^3$.

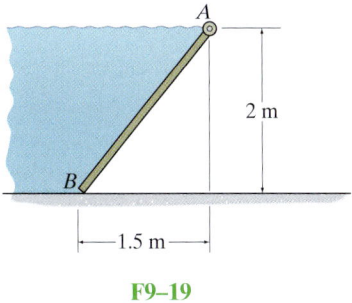

F9–19

F9–20.   Determine the magnitude of the hydrostatic force acting on gate AB, which has a width of 2 m. Water has a density of $\rho = 1 \text{ Mg/m}^3$.

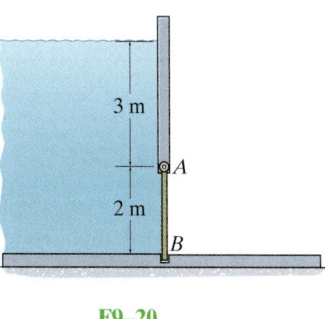

F9–20

F9–21.   Determine the magnitude of the hydrostatic force acting on gate AB, which has a width of 0.6 m. The specific weight of water is $\gamma = 9.81 \text{ kN/m}^3$.

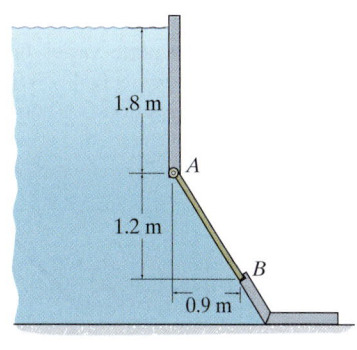

F9–21

9

PROBLEMS

9–115. The concrete "gravity" dam is held in place by its own weight. If the density of concrete is $\rho_c = 2.5$ Mg/m³, and water has a density of $\rho_w = 1.0$ Mg/m³, determine the smallest dimension d that will prevent the dam from overturning about its end A.

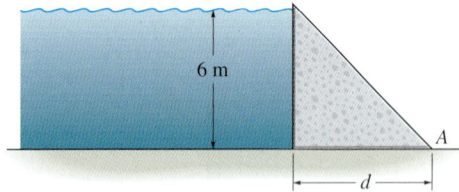

Prob. 9–115

*9–116. The loading acting on a square plate is represented by a parabolic pressure distribution. Determine the magnitude of the resultant force and the coordinates $(\bar{x}, \bar{y})$ of the point where the line of action of the force intersects the plate. Also, what are the reactions at the rollers B and C and the ball-and-socket joint A? Neglect the weight of the plate.

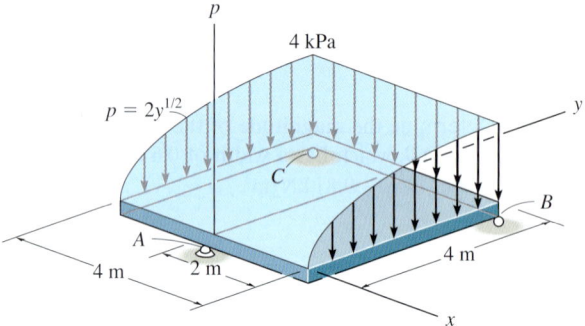

Prob. 9–116

9–117. The load over the plate varies linearly along the sides of the plate such that $p = \frac{2}{3}\,[x(4-y)]$ kPa. Determine the resultant force and its position $(\bar{x}, \bar{y})$ on the plate.

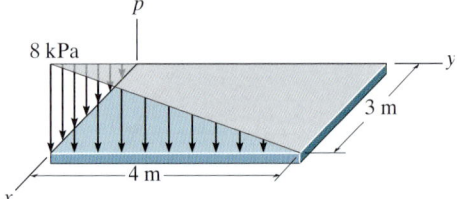

Prob. 9–117

9–118. The rectangular plate is subjected to a distributed load over its *entire surface*. The load is defined by the expression $p = p_0 \sin(\pi x/a)\sin(\pi y/b)$, where p_0 represents the pressure acting at the center of the plate. Determine the magnitude and location of the resultant force acting on the plate.

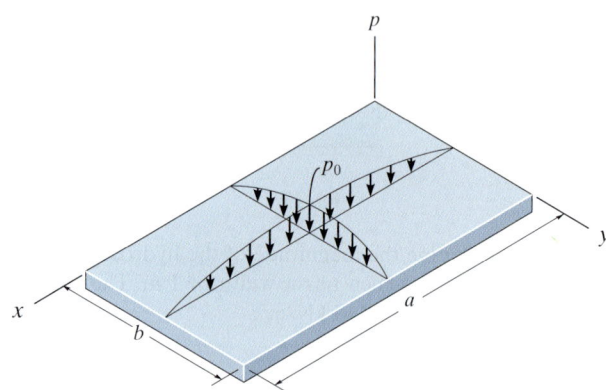

Prob. 9–118

9–119. The gate AB is 8 m wide. Determine the horizontal and vertical components of force acting on the pin at B and the vertical reaction at the smooth support A. $\rho_w = 1.0$ Mg/m³.

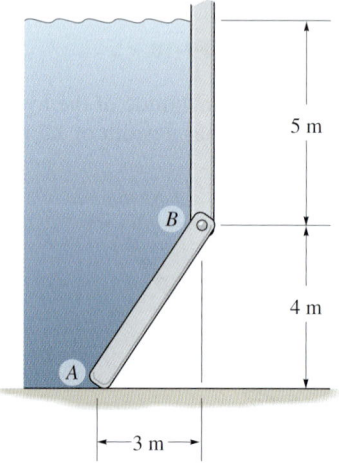

Prob. 9–119

*9–120. The tank is filled with water to a depth of $d = 4$ m. Determine the resultant force the water exerts on side A and side B of the tank. If oil instead of water is placed in the tank, to what depth d should it reach so that it creates the same resultant forces? $\rho_o = 900$ kg/m^3 and $\rho_w = 1000$ kg/m^3.

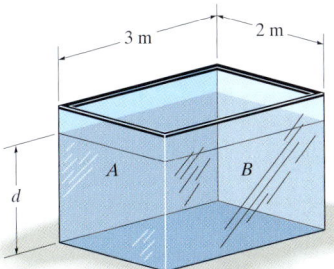

Prob. 9–120

9–121. When the tide water A subsides, the tide gate automatically swings open to drain the marsh B. For the condition of high tide shown, determine the horizontal reactions developed at the hinge C and stop block D. The length of the gate is 6 m and its height is 4 m. $\rho_w = 1.0$ Mg/m^3.

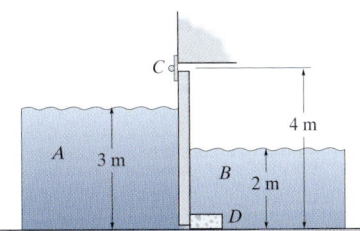

Prob. 9–121

9–122. The symmetric concrete "gravity" dam is held in place by its own weight. If the density of concrete is $\rho_c = 2.5$ Mg/m^3, and water has a density of $\rho_w = 1.0$ Mg/m^3, determine the smallest distance d at its base that will prevent the dam from overturning about its end A. The dam has a width of 8 m.

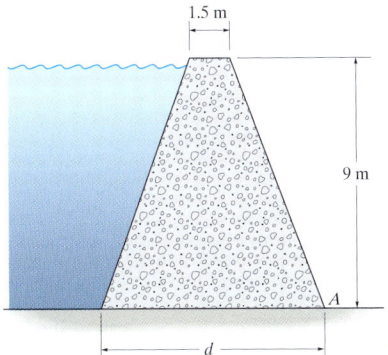

Prob. 9–122

9–123. Determine the magnitude of the resultant hydrostatic force acting on the dam and its location, measured from the top surface of the water. The width of the dam is 8 m; $\rho_w = 1.0$ Mg/m^3.

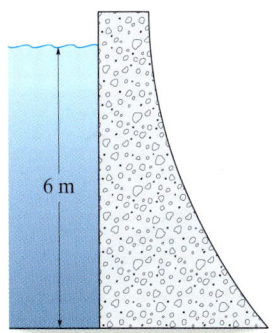

Prob. 9–123

*9–124. The concrete dam in the shape of a quarter circle. Determine the magnitude of the resultant hydrostatic force that acts on the dam per meter of length. The density of water is $\rho_w = 1$ Mg/m^3.

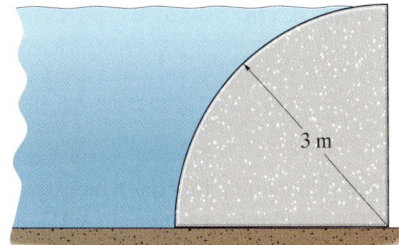

Prob. 9–124

9–125. The storage tank contains oil having a specific weight of $\gamma_o = 9$ kN/m^3. If the tank is 6 m wide, calculate the resultant force acting on the inclined side BC of the tank, caused by the oil, and specify its location along BC, measured from B. Also compute the total resultant force acting on the bottom of the tank.

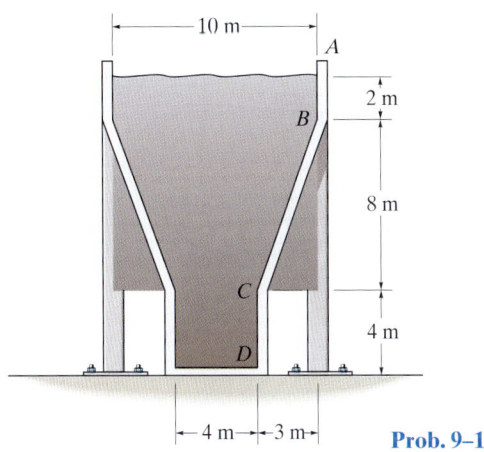

Prob. 9–125

9–126. The tank is filled to the top ($y = 0.5$ m) with water having a density of $\rho_w = 1.0$ Mg/m^3. Determine the resultant force of the water pressure acting on the flat end plate C of the tank, and its location, measured from the top of the tank.

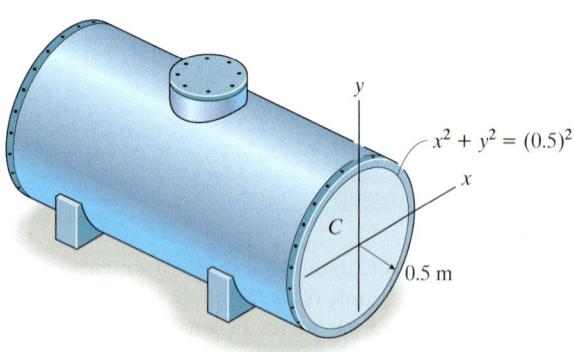

Prob. 9–126

9–127. The tank is filled with a liquid that has a density of 900 kg/m^3. Determine the resultant force that it exerts on the elliptical end plate, and the location of the center of pressure, measured from the x axis.

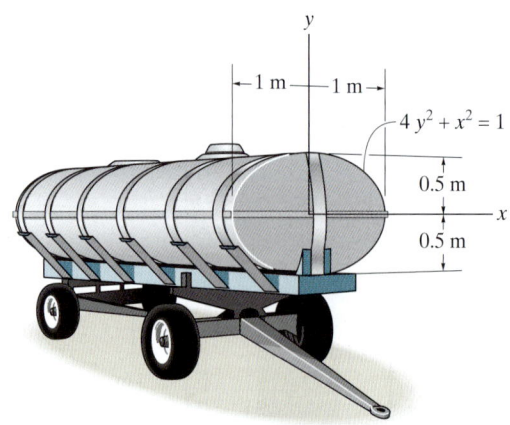

Prob. 9–127

*9–128. The underwater tunnel in the aquatic center is fabricated from a transparent polycarbonate material formed in the shape of a parabola. Determine the magnitude of the hydrostatic force that acts per meter length along the surface AB of the tunnel. The density of the water is $\rho_w = 1000$ kg/m^3.

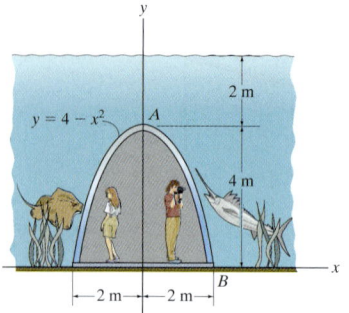

Prob. 9–128

9–129. The semicircular tunnel passes under a river that is 9 m deep. Determine the vertical resultant hydrostatic force acting per meter of length along the length of the tunnel. The tunnel is 6 m wide; $\rho_w = 1.0$ Mg/m^3.

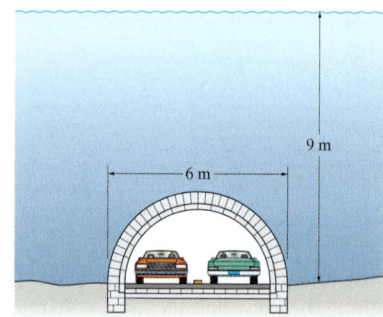

Prob. 9–129

9–130. The arched surface AB is shaped in the form of a quarter circle. If it is 8 m long, determine the horizontal and vertical components of the resultant force caused by the water acting on the surface; $\rho_w = 1.0$ Mg/m^3.

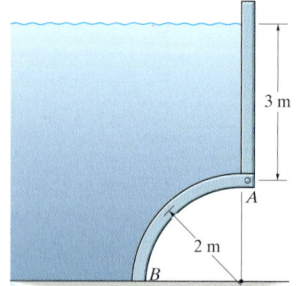

Prob. 9–130

CHAPTER REVIEW

Center of Gravity and Centroid

The *center of gravity G* represents a point where the weight of the body can be considered concentrated. The distance from an axis to this point can be determined from a balance of moments, which requires that the moment of the weight of all the particles of the body about this axis must equal the moment of the entire weight of the body about the axis.

$$\bar{x} = \frac{\int \tilde{x}\,dW}{\int dW}$$

$$\bar{y} = \frac{\int \tilde{y}\,dW}{\int dW}$$

$$\bar{z} = \frac{\int \tilde{z}\,dW}{\int dW}$$

The center of mass will coincide with the center of gravity provided the acceleration of gravity is constant.

$$\bar{x} = \frac{\int_L \tilde{x}\,dL}{\int_L dL} \qquad \bar{y} = \frac{\int_L \tilde{y}\,dL}{\int_L dL} \qquad \bar{z} = \frac{\int_L \tilde{z}\,dL}{\int_L dL}$$

The *centroid* is the location of the geometric center for the body. It is determined in a similar manner, using a moment balance of geometric elements such as line, area, or volume segments. For bodies having a continuous shape, moments are summed (integrated) using differential elements.

$$\bar{x} = \frac{\int_A \tilde{x}\,dA}{\int_A dA} \qquad \bar{y} = \frac{\int_A \tilde{y}\,dA}{\int_A dA} \qquad \bar{z} = \frac{\int_A \tilde{z}\,dA}{\int_A dA}$$

$$\bar{x} = \frac{\int_V \tilde{x}\,dV}{\int_V dV} \qquad \bar{y} = \frac{\int_V \tilde{y}\,dV}{\int_A dV} \qquad \bar{z} = \frac{\int_V \tilde{z}\,dV}{\int_V dV}$$

The center of mass will coincide with the centroid provided the material is homogeneous, i.e., the density of the material is the same throughout. The centroid will always lie on an axis of symmetry.

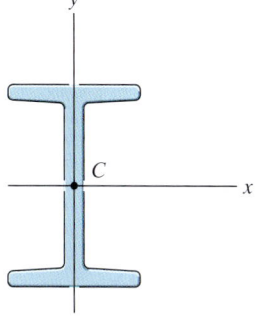

Composite Body

If the body is a composite of several shapes, each having a known location for its center of gravity or centroid, then the location of the center of gravity or centroid of the body can be determined from a discrete summation using its composite parts.

$$\bar{x} = \frac{\Sigma \tilde{x} W}{\Sigma W}$$

$$\bar{y} = \frac{\Sigma \tilde{y} W}{\Sigma W}$$

$$\bar{z} = \frac{\Sigma \tilde{z} W}{\Sigma W}$$

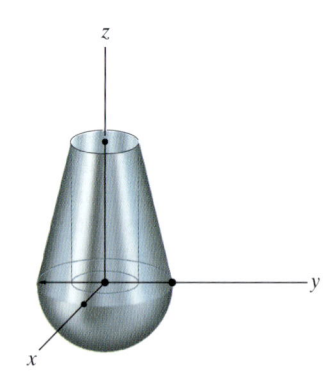

Theorems of Pappus and Guldinus

The theorems of Pappus and Guldinus can be used to determine the surface area and volume of a body of revolution.

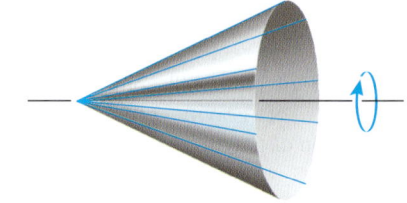

The *surface area* equals the product of the length of the generating curve and the distance traveled by the centroid of the curve needed to generate the area.

$$A = \theta \bar{r} L$$

The *volume* of the body equals the product of the generating area and the distance traveled by the centroid of this area needed to generate the volume.

$$V = \theta \bar{r} A$$

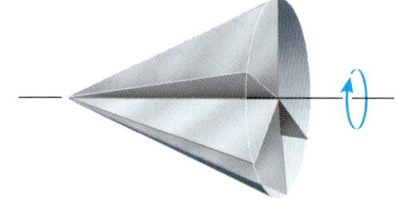

General Distributed Loading

The magnitude of the resultant force is equal to the total volume under the distributed-loading diagram. The line of action of the resultant force passes through the geometric center or centroid of this volume.

$$F_R = \int_A p(x, y)\, dA = \int_V dV$$

$$\bar{x} = \frac{\displaystyle\int_V x\, dV}{\displaystyle\int_V dV}$$

$$\bar{y} = \frac{\displaystyle\int_V y\, dV}{\displaystyle\int_V dV}$$

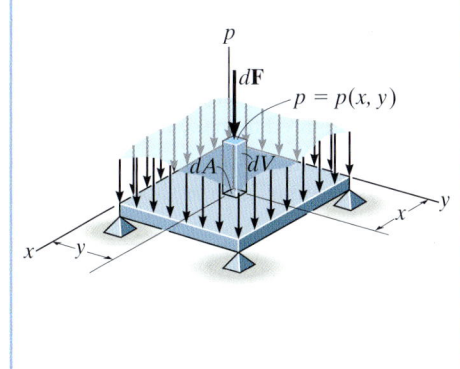

Fluid Pressure

The pressure developed by a liquid at a point on a submerged surface depends upon the depth of the point and the density of the liquid in accordance with Pascal's law, $p = \rho g h = \gamma h$. This pressure will create a *linear distribution* of loading on a flat vertical or inclined surface.

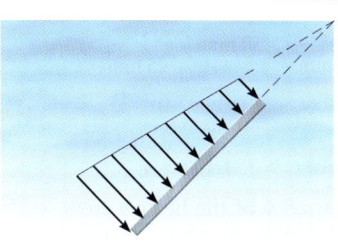

If the surface is horizontal, then the loading will be *uniform*.

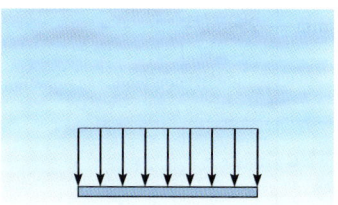

In any case, the resultants of these loadings can be determined by finding the volume under the loading curve or using $F_R = \gamma \bar{z} A$, where $\bar{z}$ is the depth to the centroid of the plate's area. The line of action of the resultant force passes through the centroid of the volume of the loading diagram and acts at a point P on the plate called the center of pressure.

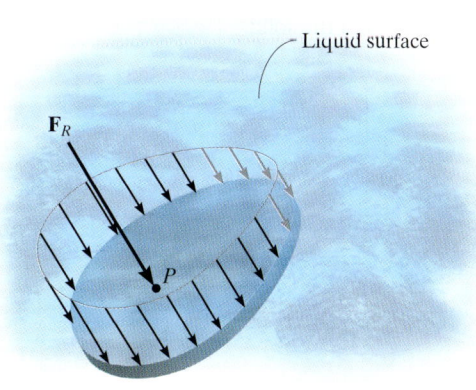

REVIEW PROBLEMS

9–131. A circular V-belt has an inner radius of 600 mm and a cross-sectional area as shown. Determine the volume of material required to make the belt.

*9–132. A circular V-belt has an inner radius of 600 mm and a cross-sectional area as shown. Determine the surface area of the belt.

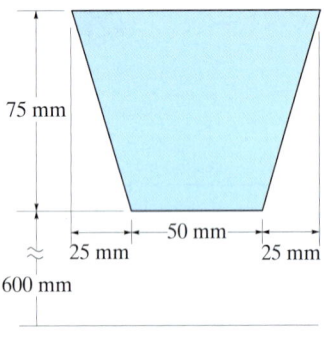

Probs. 9–131/132

9–133. Locate the centroid $\bar{y}$ of the beam's cross-sectional area.

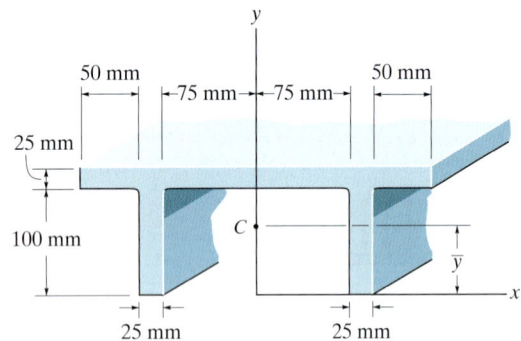

Prob. 9–133

9–134. Locate the centroid of the solid.

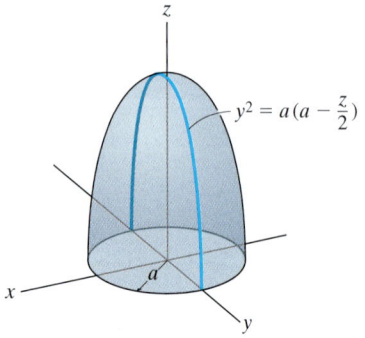

Prob. 9–134

9–135. Locate the centroid $\bar{x}$ of the triangular area.

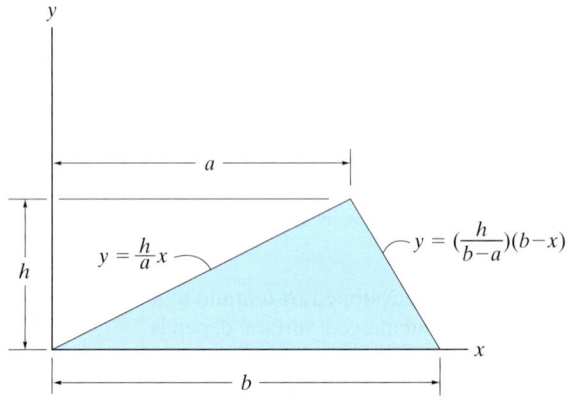

Prob. 9–135

*9–136. Locate the centroid $\bar{y}$ of the bulb-tee cross section.

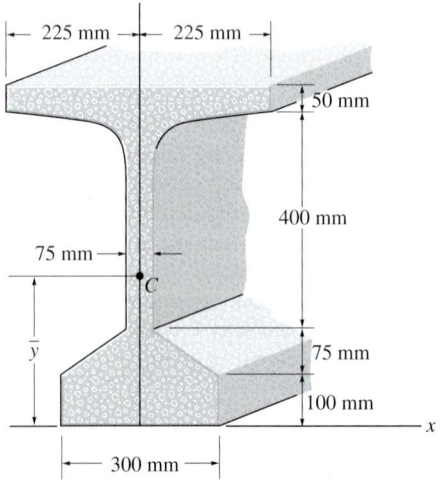

Prob. 9–136

9–137. Determine the location $(\bar{x}, \bar{y})$ of the center of mass of the turbine and compressor assembly. The mass and the center of mass of each of the various components are indicated below.

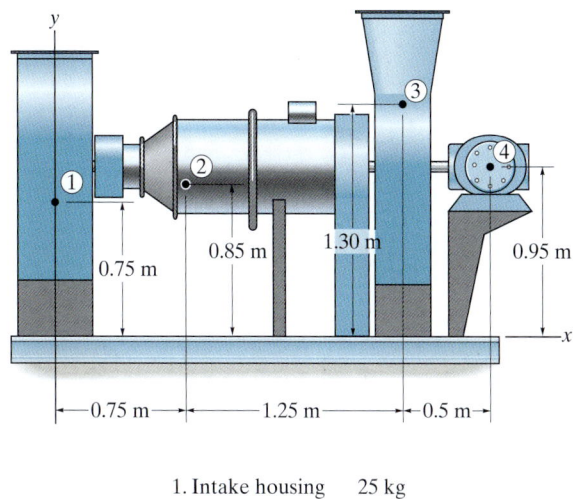

1. Intake housing	25 kg
2. Turbine	80 kg
3. Exhaust housing	30 kg
4. Compressor	105 kg

Prob. 9–137

9–138. Each of the three homogeneous plates welded to the rod has a density of 6 Mg/m³ and a thickness of 10 mm. Determine the length l of plate C and the angle of placement, θ, so that the center of mass of the assembly lies on the y axis. Plates A and B lie in the x–y and z–y planes, respectively.

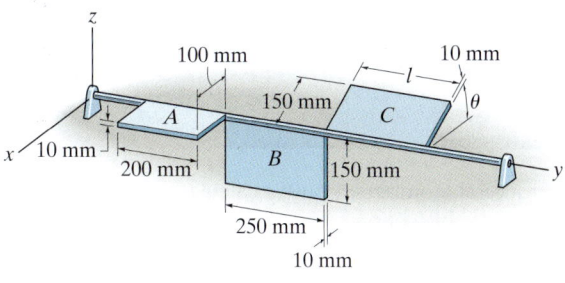

Prob. 9–138

9–139. The gate AB is 5 m wide. Determine the horizontal and vertical components of force acting on the pin at B and the vertical reaction at the smooth support A; $\rho_w = 1.0 \text{ Mg/m}^3$.

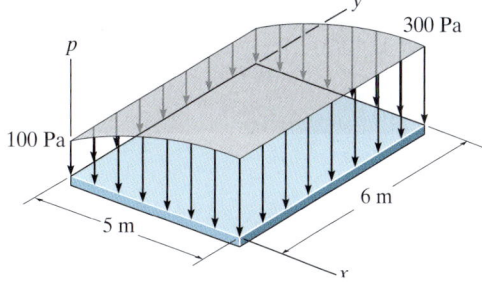

Prob. 9–139

***9–140.** The pressure loading on the plate is described by the function $p = \{-240/(x + 1) + 340\}$ Pa. Determine the magnitude of the resultant force and coordinates of the point where the line of action of the force intersects the plate.

Prob. 9–140

9

Chapter 10

The design of these structural members requires calculation of their cross-sectional moment of inertia. In this chapter we will discuss how this is done.

Moments of Inertia

CHAPTER OBJECTIVES

■ To develop a method for determining the moment of inertia for
an area.

■ To introduce the product of inertia and show how to determine
the maximum and minimum moments of inertia for an area.

■ To discuss the mass moment of inertia.

10.1 Definition of Moments of Inertia for Areas

Whenever a distributed loading acts perpendicular to an area and its
intensity varies linearly, the computation of the moment of the loading
distribution about an axis will involve a quantity called the *moment of
inertia of the area*. For example, consider the plate in Fig. 10–1, which is
subjected to a fluid pressure p. As discussed in Sec. 9.5, this pressure p varies
linearly with depth, such that $p = \gamma y$, where γ is the specific weight of the
fluid. Thus, the force acting on the differential area dA of the plate is
$dF = p\, dA = (\gamma y)dA$. The moment of this force about the x axis is therefore
$dM = y\, dF = \gamma y^2 dA$, and so integrating dM over the entire area of the plate
yields $M = \gamma \int y^2 dA$. The integral $\int y^2 dA$ is called the *moment of inertia* I_x
of the area about the x axis. Integrals of this form often arise in formulas
used in fluid mechanics, mechanics of materials, structural mechanics, and
mechanical design, and so the engineer needs to be familiar with the
methods used for their computation.

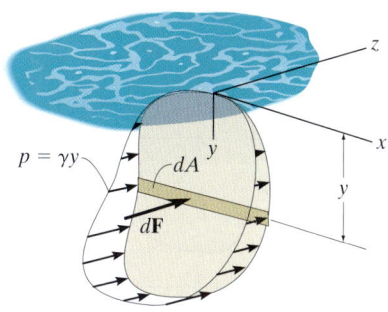

$p = \gamma y$

Fig. 10–1

Please refer to the Companion Website
for the animation: *Moment of Inertia*

Moment of Inertia.   By definition, the moments of inertia of a differential area dA about the x and y axes are $dI_x = y^2 \, dA$ and $dI_y = x^2 \, dA$, respectively, Fig. 10–2. For the entire area A the *moments of inertia* are determined by integration; i.e.,

$$
\begin{aligned}
I_x &= \int_A y^2 \, dA \\[2mm]
I_y &= \int_A x^2 \, dA
\end{aligned}
\tag{10–1}
$$

We can also formulate this quantity for dA about the "pole" O or z axis, Fig. 10–2. This is referred to as the *polar moment of inertia*. It is defined as $dJ_O = r^2 \, dA$, where r is the perpendicular distance from the pole (z axis) to the element dA. For the entire area the *polar moment of inertia* is

$$
J_O = \int_A r^2 \, dA = I_x + I_y
\tag{10–2}
$$

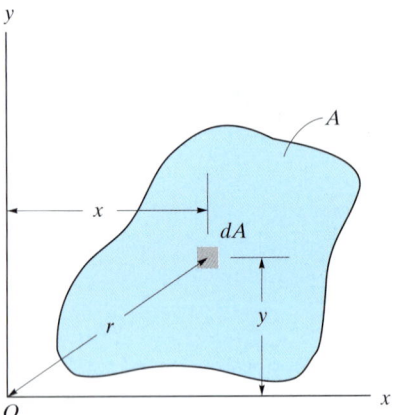

Fig. 10–2

This relation between J_O and I_x, I_y is possible since $r^2 = x^2 + y^2$, Fig. 10–2.

From the above formulations it is seen that I_x, I_y, and J_O will *always* be *positive* since they involve the product of distance squared and area. Furthermore, the units for moment of inertia involve length raised to the fourth power, e.g., m^4, mm^4.

10.2   Parallel-Axis Theorem for an Area

The *parallel-axis theorem* can be used to find the moment of inertia of an area about *any axis* that is parallel to an axis passing through the centroid and about which the moment of inertia is known. To develop this theorem, we will consider finding the moment of inertia of the shaded area shown in Fig. 10–3 about the x axis. To start, we choose a differential element dA located at an arbitrary distance y' from the *centroidal x' axis*. If the distance between the parallel x and x' axes is d_y, then the moment of inertia of dA about the x axis is $dI_x = (y' + d_y)^2 \, dA$. For the entire area,

$$
\begin{aligned}
I_x &= \int_A (y' + d_y)^2 \, dA \\[2mm]
&= \int_A y'^2 \, dA + 2d_y \int_A y' \, dA + d_y^2 \int_A dA
\end{aligned}
$$

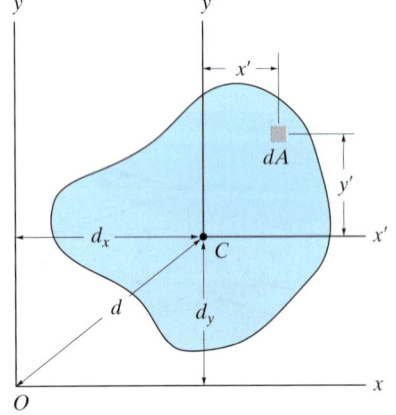

Fig. 10–3

10

The first integral represents the moment of inertia of the area about the centroidal axis, $\bar{I}_{x'}$. The second integral is zero since the x' axis passes through the area's centroid C; i.e., $\int y' \, dA = \bar{y}' \int dA = 0$ since $\bar{y}' = 0$. Since the third integral represents the total area A, the final result is therefore

$$I_x = \bar{I}_{x'} + Ad_y^2 \tag{10–3}$$

A similar expression can be written for I_y; i.e.,

$$I_y = \bar{I}_{y'} + Ad_x^2 \tag{10–4}$$

And finally, for the polar moment of inertia, since $\bar{J}_C = \bar{I}_{x'} + \bar{I}_{y'}$ and $d^2 = d_x^2 + d_y^2$, we have

$$J_O = \bar{J}_C + Ad^2 \tag{10–5}$$

The form of each of these three equations states that *the moment of inertia for an area about an axis is equal to its moment of inertia about a parallel axis passing through the area's centroid plus the product of the area and the square of the perpendicular distance between the axes.*

In order to predict the strength and deflection of this beam, it is necessary to calculate the moment of inertia of the beam's cross-sectional area.

10.3 Radius of Gyration of an Area

The *radius of gyration* of an area about an axis has units of length and is a quantity that is often used for the design of columns in structural mechanics. Provided the areas and moments of inertia are *known*, the radii of gyration are determined from the formulas

$$k_x = \sqrt{\frac{I_x}{A}}$$

$$k_y = \sqrt{\frac{I_y}{A}} \tag{10–6}$$

$$k_O = \sqrt{\frac{J_O}{A}}$$

The form of these equations is easily remembered since it is similar to that for finding the moment of inertia for a differential area about an axis. For example, $I_x = k_x^2 A$; whereas for a differential area, $dI_x = y^2 \, dA$.

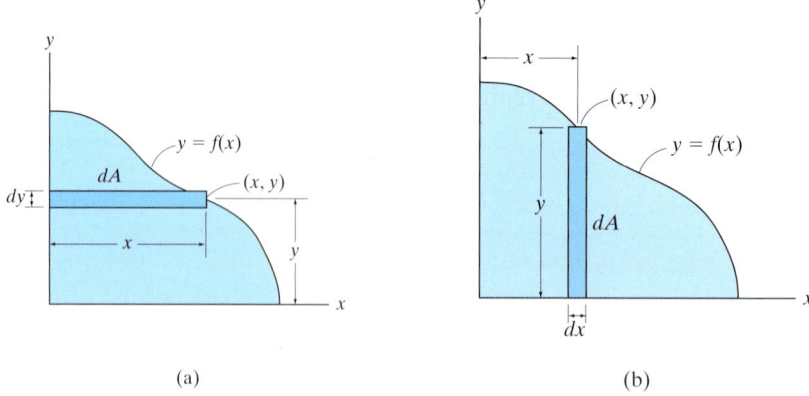

(a)   (b)

Fig. 10–4

Procedure for Analysis

In most cases the moment of inertia can be determined using a single integration. The following procedure shows two ways in which this can be done.

- If the curve defining the boundary of the area is expressed as $y = f(x)$, then select a rectangular differential element such that it has a finite length and differential width.

- The element should be located so that it intersects the curve at the *arbitrary point* (x, y).

Case 1.

- Orient the element so that its length is *parallel* to the axis about which the moment of inertia is computed. This situation occurs when the rectangular element shown in Fig. 10–4a is used to determine I_x for the area. Here the entire element is at a distance y from the x axis since it has a thickness dy. Thus $I_x = \int y^2 \, dA$. To find I_y, the element is oriented as shown in Fig. 10–4b. This element lies at the *same* distance x from the y axis so that $I_y = \int x^2 \, dA$.

Case 2.

- The length of the element can be oriented *perpendicular* to the axis about which the moment of inertia is computed; however, Eq. 10–1 *does not apply* since all points on the element will *not* lie at the same moment-arm distance from the axis. For example, if the rectangular element in Fig. 10–4a is used to determine I_y, it will first be necessary to calculate the moment of inertia of the *element* about an axis parallel to the y axis that passes through the element's centroid, and then determine the moment of inertia of the *element* about the y axis using the parallel-axis theorem. Integration of this result will yield I_y. See Examples 10.2 and 10.3.

EXAMPLE | 10.1

Determine the moment of inertia for the rectangular area shown in Fig. 10–5 with respect to (a) the centroidal x' axis, (b) the axis x_b passing through the base of the rectangle, and (c) the pole or z' axis perpendicular to the $x'–y'$ plane and passing through the centroid C.

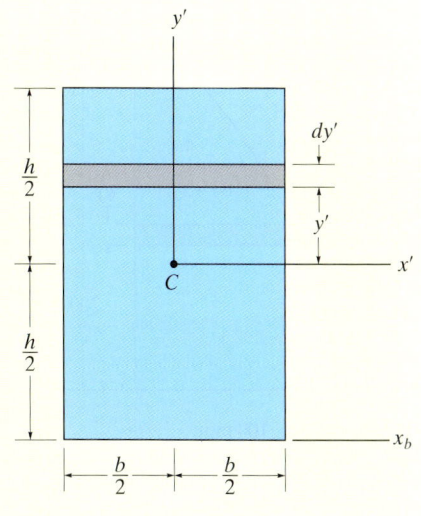

Fig. 10–5

SOLUTION (CASE 1)

Part (a). The differential element shown in Fig. 10–5 is chosen for integration. Because of its location and orientation, the *entire element* is at a distance y' from the x' axis. Here it is necessary to integrate from $y' = -h/2$ to $y' = h/2$. Since $dA = b\,dy'$, then

$$\bar{I}_{x'} = \int_A y'^2\, dA = \int_{-h/2}^{h/2} y'^2(b\,dy') = b\int_{-h/2}^{h/2} y'^2\, dy'$$

$$\bar{I}_{x'} = \frac{1}{12}bh^3 \qquad\qquad \textit{Ans.}$$

Part (b). The moment of inertia about an axis passing through the base of the rectangle can be obtained by using the above result of part (a) and applying the parallel-axis theorem, Eq. 10–3.

$$I_{x_b} = \bar{I}_{x'} + Ad_y^2$$

$$= \frac{1}{12}bh^3 + bh\left(\frac{h}{2}\right)^2 = \frac{1}{3}bh^3 \qquad\qquad \textit{Ans.}$$

Part (c). To obtain the polar moment of inertia about point C, we must first obtain $\bar{I}_{y'}$, which may be found by interchanging the dimensions b and h in the result of part (a), i.e.,

$$\bar{I}_{y'} = \frac{1}{12}hb^3$$

Using Eq. 10–2, the polar moment of inertia about C is therefore

$$\bar{J}_C = \bar{I}_{x'} + \bar{I}_{y'} = \frac{1}{12}bh(h^2 + b^2) \qquad\qquad \textit{Ans.}$$

10

EXAMPLE | 10.2

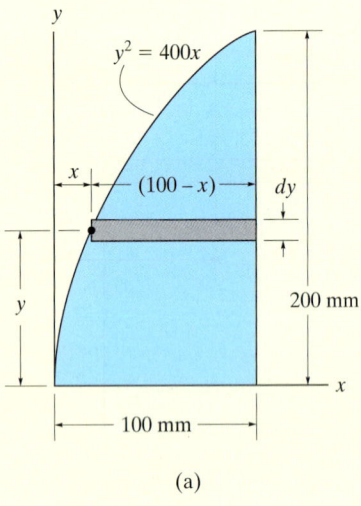

(a)

Determine the moment of inertia for the shaded area shown in Fig. 10–6a about the x axis.

SOLUTION I (CASE 1)

A differential element of area that is *parallel* to the x axis, as shown in Fig. 10–6a, is chosen for integration. Since this element has a thickness dy and intersects the curve at the *arbitrary point* (x, y), its area is $dA = (100 - x)\, dy$. Furthermore, the element lies at the same distance y from the x axis. Hence, integrating with respect to y, from $y = 0$ to $y = 200$ mm, yields

$$I_x = \int_A y^2\, dA = \int_0^{200\text{ mm}} y^2(100 - x)\, dy$$

$$= \int_0^{200\text{ mm}} y^2\left(100 - \frac{y^2}{400}\right) dy = \int_0^{200\text{ mm}} \left(100y^2 - \frac{y^4}{400}\right) dy$$

$$= 107(10^6)\text{ mm}^4 \qquad\qquad\qquad\qquad \textit{Ans.}$$

SOLUTION II (CASE 2)

A differential element *parallel* to the y axis, as shown in Fig. 10–6b, is chosen for integration. It intersects the curve at the *arbitrary point* (x, y). In this case, all points of the element do *not* lie at the same distance from the x axis, and therefore the parallel-axis theorem must be used to determine the *moment of inertia of the element* with respect to this axis. For a rectangle having a base b and height h, the moment of inertia about its centroidal axis has been determined in part (a) of Example 10.1. There it was found that $\bar{I}_{x'} = \frac{1}{12}bh^3$. For the differential element shown in Fig. 10–6b, $b = dx$ and $h = y$, and thus $d\bar{I}_{x'} = \frac{1}{12}dx\, y^3$. Since the centroid of the element is $\tilde{y} = y/2$ from the x axis, the moment of inertia of the element about this axis is

$$dI_x = d\bar{I}_{x'} + dA\, \tilde{y}^2 = \frac{1}{12}dx\, y^3 + y\, dx\left(\frac{y}{2}\right)^2 = \frac{1}{3}y^3\, dx$$

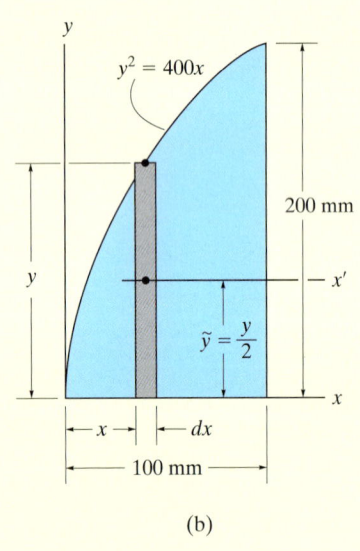

(b)

Fig. 10–6

(This result can also be concluded from part (b) of Example 10.1.) Integrating with respect to x, from $x = 0$ to $x = 100$ mm, yields

$$I_x = \int dI_x = \int_0^{100\text{ mm}} \frac{1}{3}y^3\, dx = \int_0^{100\text{ mm}} \frac{1}{3}(400x)^{3/2}\, dx$$

$$= 107(10^6)\text{ mm}^4 \qquad\qquad\qquad\qquad \textit{Ans.}$$

EXAMPLE 10.3

Determine the moment of inertia with respect to the x axis for the circular area shown in Fig. 10–7a.

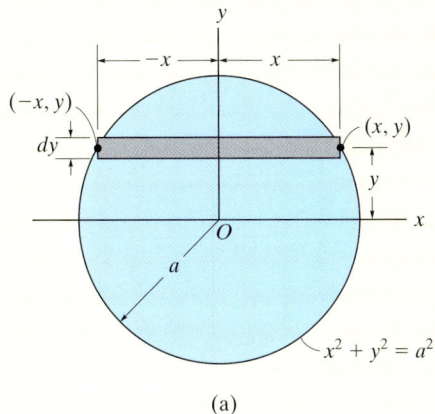

(a)

SOLUTION I (CASE 1)

Using the differential element shown in Fig. 10–7a, since $dA = 2x\, dy$, we have

$$I_x = \int_A y^2\, dA = \int_A y^2(2x)\, dy$$

$$= \int_{-a}^{a} y^2\left(2\sqrt{a^2 - y^2}\right) dy = \frac{\pi a^4}{4} \qquad \textit{Ans.}$$

SOLUTION II (CASE 2)

When the differential element shown in Fig. 10–7b is chosen, the centroid for the element happens to lie on the x axis, and since $\bar{I}_{x'} = \frac{1}{12}bh^3$ for a rectangle, we have

$$dI_x = \frac{1}{12}dx(2y)^3$$

$$= \frac{2}{3}y^3\, dx$$

Integrating with respect to x yields

$$I_x = \int_{-a}^{a} \frac{2}{3}(a^2 - x^2)^{3/2}\, dx = \frac{\pi a^4}{4} \qquad \textit{Ans.}$$

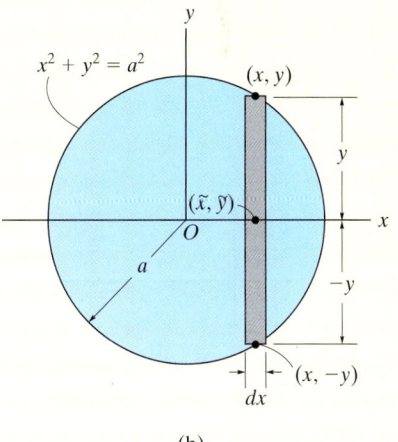

(b)

Fig. 10–7

NOTE: By comparison, Solution I requires much less computation. Therefore, if an integral using a particular element appears difficult to evaluate, try solving the problem using an element oriented in the other direction.

10

FUNDAMENTAL PROBLEMS

F10–1. Determine the moment of inertia of the shaded area about the x axis.

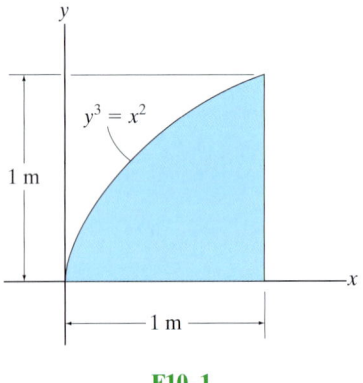

F10–1

F10–2. Determine the moment of inertia of the shaded area about the x axis.

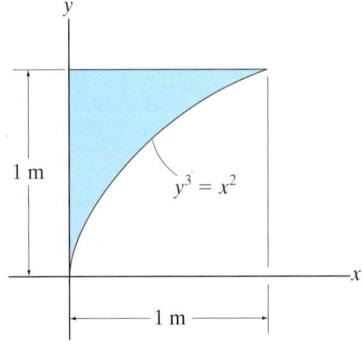

F10–2

F10–3. Determine the moment of inertia of the shaded area about the y axis.

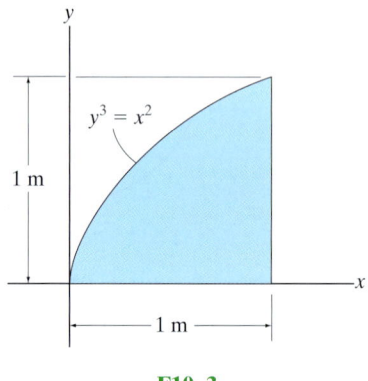

F10–3

F10–4. Determine the moment of inertia of the shaded area about the y axis.

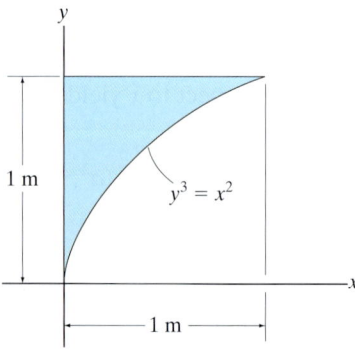

F10–4

PROBLEMS

10–1. Determine the moment of inertia of the shaded area about the *x* axis.

10–2. Determine the moment of inertia of the shaded area about the *y* axis.

10–5. Determine the moment of inertia of the area about the *x* axis.

10–6. Determine the moment of inertia of the area about the *y* axis.

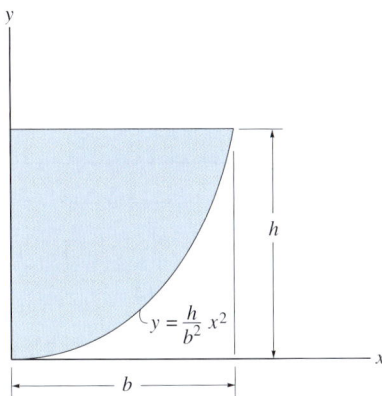

Probs. 10–1/2

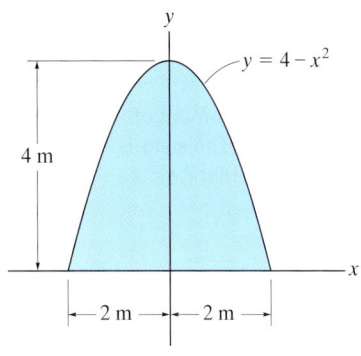

Probs. 10–5/6

10–3. Determine the moment of inertia of the area about the *x* axis.

10–4. Determine the moment of inertia of the area about the *y* axis.

10–7. Determine the moment of inertia of the area about the *x* axis.

10–8. Determine the moment of inertia of the area about the *y* axis.

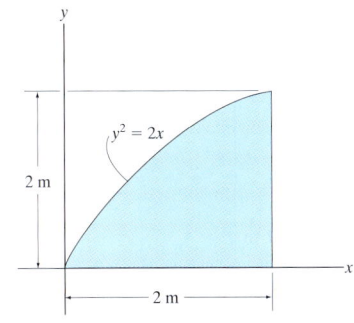

Probs. 10–3/4

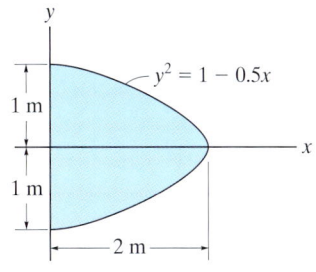

Probs. 10–7/8

10

10–9. Determine the moment of inertia of the area about the x axis.

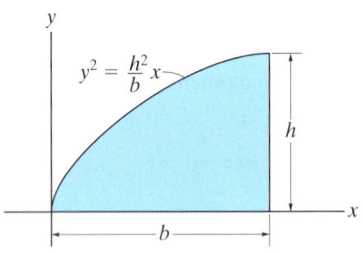

$$y^2 = \frac{h^2}{b}x$$

Prob. 10–9

10–10. Determine the moment of inertia for the thin strip of area about the x axis. The strip is oriented at an angle θ from the x axis. Assume that $t \ll l$.

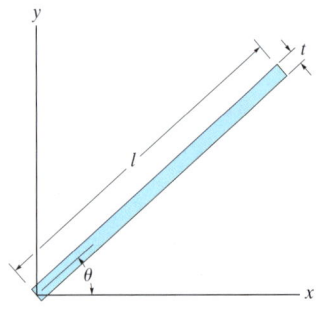

Prob. 10–10

10–11. Determine the moment of inertia of the area about the x axis.

***10–12.** Determine the moment of inertia of the area about the y axis.

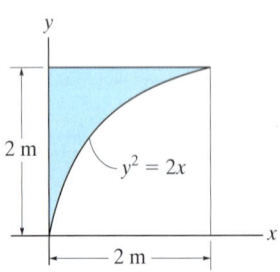

$$y^2 = 2x$$

Probs. 10–11/12

10–13. Determine the moment of inertia for the shaded area about the x axis.

10–14. Determine the moment of inertia for the shaded area about the y axis.

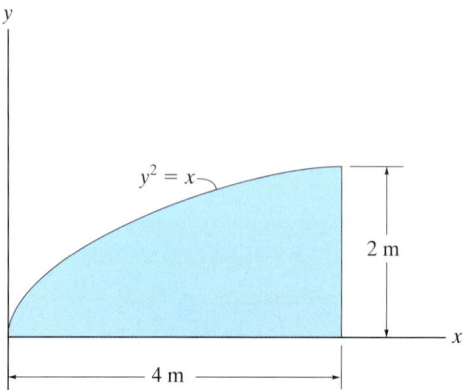

$$y^2 = x$$

Probs. 10–13/14

10–15. Determine the moment of inertia of the shaded area about the y axis. Use Simpson's rule to evaluate the integral.

***10–16.** Determine the moment of inertia of the shaded area about the x axis. Use Simpson's rule to evaluate the integral.

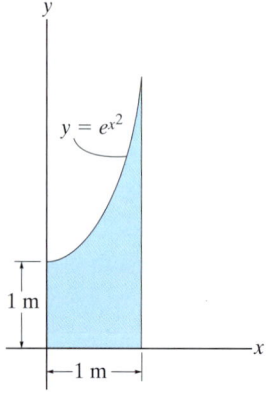

$$y = e^{x^2}$$

Probs. 10–15/16

10–17. Determine the moment of inertia of the shaded area about the x axis.

10–18. Determine the moment of inertia of the shaded area about the y axis.

10–21. Determine the moment of inertia of the shaded area about the x axis.

10–22. Determine the moment of inertia of the shaded area about the y axis.

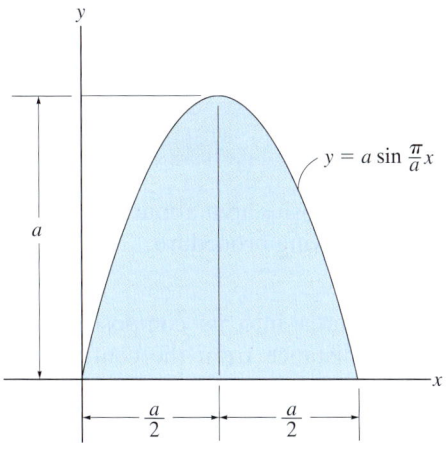

Probs. 10–17/18

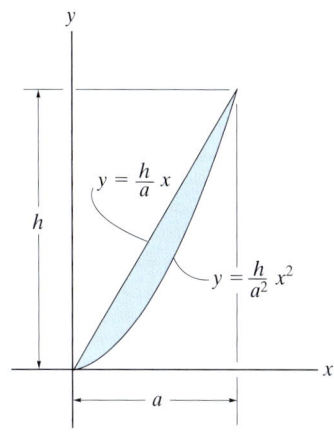

Probs. 10–21/22

10–19. Determine the moment of inertia of the shaded area about the x axis.

*__10–20.__ Determine the moment of inertia of the shaded area about the y axis.

10–23. Determine the moment of inertia of the shaded area about the x axis.

*__10–24.__ Determine the moment of inertia of the shaded area about the y axis.

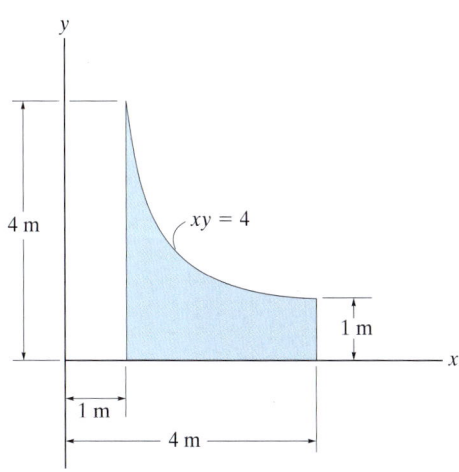

Probs. 10–19/20

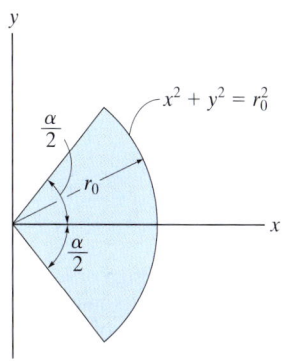

Probs. 10–23/24

10.4 Moments of Inertia for Composite Areas

A composite area consists of a series of connected "simpler" parts or shapes, such as rectangles, triangles, and circles. Provided the moment of inertia of each of these parts is known or can be determined about a common axis, then the moment of inertia for the composite area about this axis equals the *algebraic sum* of the moments of inertia of all its parts.

Procedure for Analysis

The moment of inertia for a composite area about a reference axis can be determined using the following procedure.

Composite Parts.

- Using a sketch, divide the area into its composite parts and indicate the perpendicular distance from the centroid of each part to the reference axis.

Parallel-Axis Theorem.

- If the centroidal axis for each part does not coincide with the reference axis, the parallel-axis theorem, $I = \bar{I} + Ad^2$, should be used to determine the moment of inertia of the part about the reference axis. For the calculation of $\bar{I}$, use the table on the inside back cover.

Summation.

- The moment of inertia of the entire area about the reference axis is determined by summing the results of its composite parts about this axis.

- If a composite part has a "hole", its moment of inertia is found by "subtracting" the moment of inertia of the hole from the moment of inertia of the entire part including the hole.

For design or analysis of this T-beam, engineers must be able to locate the centroid of its cross-sectional area, and then find the moment of inertia of this area about the centroidal axis.

EXAMPLE | 10.4

Determine the moment of inertia of the area shown in Fig. 10–8a about the x axis.

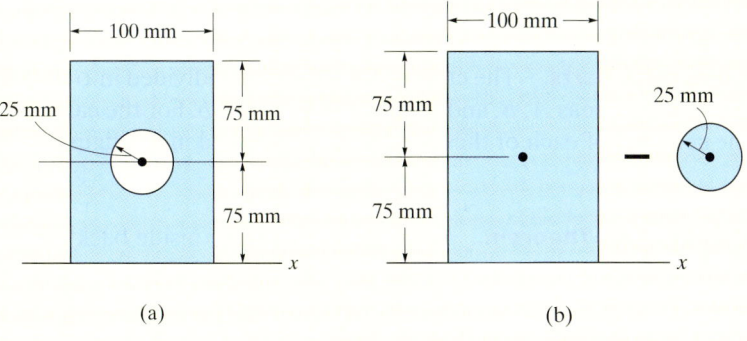

Fig. 10–8

SOLUTION

Composite Parts.   The area can be obtained by *subtracting* the circle from the rectangle shown in Fig. 10–8b. The centroid of each area is located in the figure.

Parallel-Axis Theorem.   The moments of inertia about the x axis are determined using the parallel-axis theorem and the geometric properties formulae for circular and rectangular areas $I_x = \frac{1}{4}\pi r^4$; $I_x = \frac{1}{12}bh^3$, found on the inside back cover.

Circle

$$I_x = \bar{I}_{x'} + Ad_y^2$$

$$= \frac{1}{4}\pi(25)^4 + \pi(25)^2(75)^2 = 11.4(10^6)\ \text{mm}^4$$

Rectangle

$$I_x = \bar{I}_{x'} + Ad_y^2$$

$$= \frac{1}{12}(100)(150)^3 + (100)(150)(75)^2 = 112.5(10^6)\ \text{mm}^4$$

Summation.   The moment of inertia for the area is therefore

$$I_x = -11.4(10^6) + 112.5(10^6)$$

$$= 101(10^6)\ \text{mm}^4 \qquad\qquad \textit{Ans.}$$

10

EXAMPLE   10.5

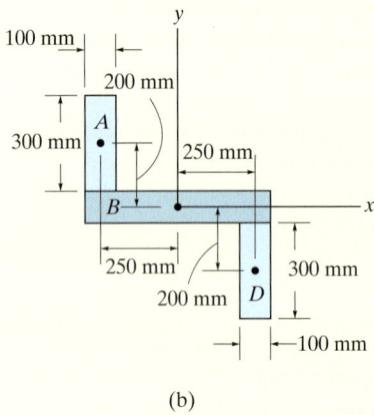

(a)

Fig. 10–9

(b)

Determine the moments of inertia for the cross-sectional area of the member shown in Fig. 10–9a about the x and y centroidal axes.

SOLUTION

Composite Parts.   The cross section can be subdivided into the three rectangular areas A, B, and D shown in Fig. 10–9b. For the calculation, the centroid of each of these rectangles is located in the figure.

Parallel-Axis Theorem.   From the table on the inside back cover, or Example 10.1, the moment of inertia of a rectangle about its centroidal axis is $\bar{I} = \frac{1}{12}bh^3$. Hence, using the parallel-axis theorem for rectangles A and D, the calculations are as follows:

Rectangles A and D

$$I_x = \bar{I}_{x'} + Ad_y^2 = \frac{1}{12}(100)(300)^3 + (100)(300)(200)^2$$

$$= 1.425(10^9)\ \text{mm}^4$$

$$I_y = \bar{I}_{y'} + Ad_x^2 = \frac{1}{12}(300)(100)^3 + (100)(300)(250)^2$$

$$= 1.90(10^9)\ \text{mm}^4$$

Rectangle B

$$I_x = \frac{1}{12}(600)(100)^3 = 0.05(10^9)\ \text{mm}^4$$

$$I_y = \frac{1}{12}(100)(600)^3 = 1.80(10^9)\ \text{mm}^4$$

Summation.   The moments of inertia for the entire cross section are thus

$$I_x = 2[1.425(10^9)] + 0.05(10^9)$$

$$= 2.90(10^9)\ \text{mm}^4 \qquad \qquad Ans.$$

$$I_y = 2[1.90(10^9)] + 1.80(10^9)$$

$$= 5.60(10^9)\ \text{mm}^4 \qquad \qquad Ans.$$

10

FUNDAMENTAL PROBLEMS

F10–5. Determine the moment of inertia of the beam's cross-sectional area about the centroidal x and y axes.

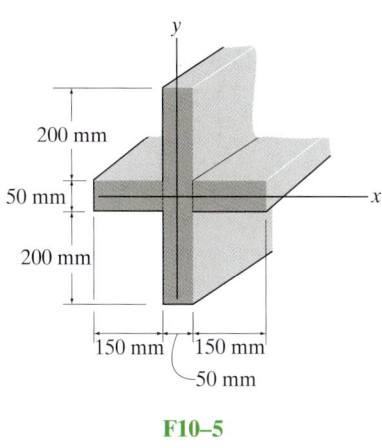

F10–5

F10–7. Determine the moment of inertia of the cross-sectional area of the channel with respect to the y axis.

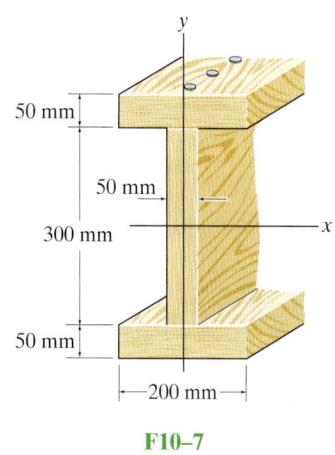

F10–7

F10–6. Determine the moment of inertia of the beam's cross-sectional area about the centroidal x and y axes.

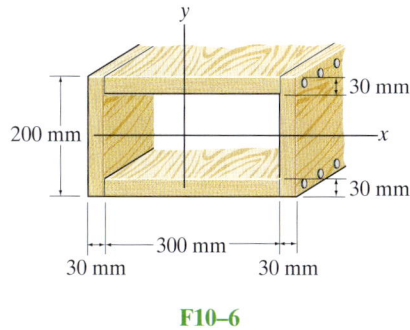

F10–6

F10–8. Determine the moment of inertia of the cross-sectional area of the T-beam with respect to the x' axis passing through the centroid of the cross section.

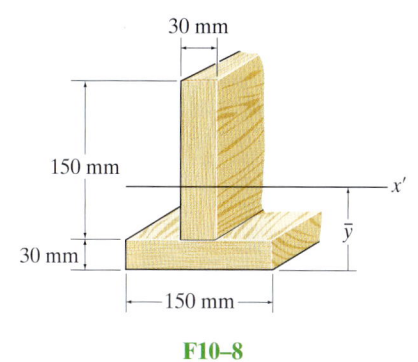

F10–8

10

PROBLEMS

10–25. Determine the moment of inertia of the composite area about the x axis.

10–26. Determine the moment of inertia of the composite area about the y axis.

*10–28. Determine the moment of inertia of the beam's cross-sectional area about the x axis.

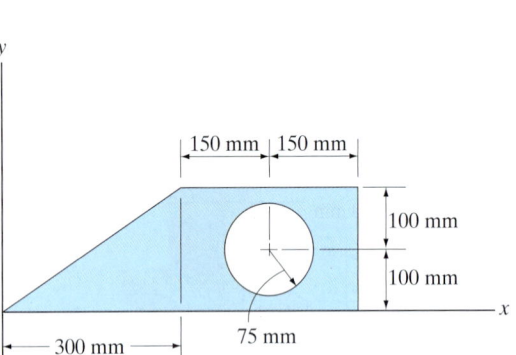

Probs. 10–25/26

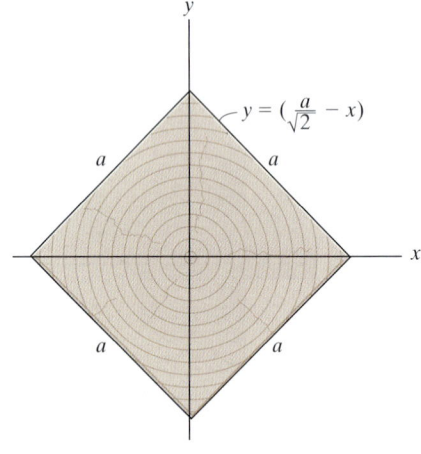

Prob. 10–28

10–27. Determine the radius of gyration k_x for the column's cross-sectional area.

10–29. Locate the centroid $\bar{y}$ of the cross section and determine the moment of inertia of the section about the x' axis.

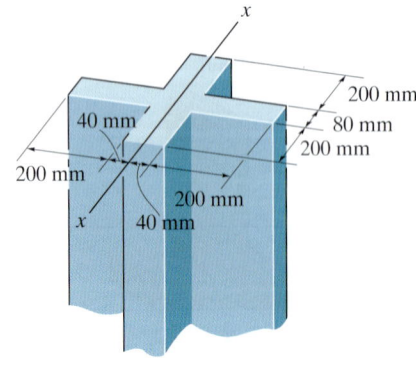

Prob. 10–27

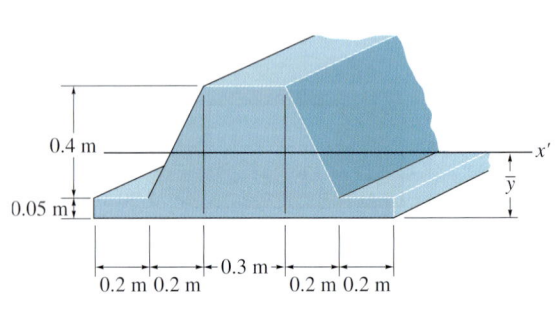

Prob. 10–29

10–30. Determine the distance $\bar{x}$ to the centroid of the beam's cross-sectional area, then find the moment of inertia about the y' axis.

10–31. Determine the moment of inertia of the beam's cross-sectional area about the x' axis.

10–34. Determine the moment of inertia of the beam's cross-sectional area about the y axis.

10–35. Determine $\bar{y}$, which locates the centroidal axis x' for the cross-sectional area of the T-beam, and then find the moment of inertia about the x' axis.

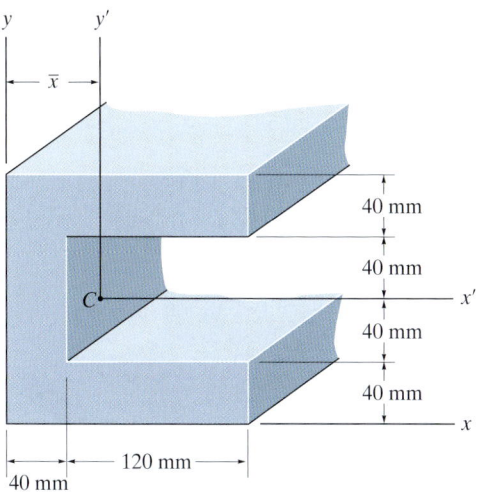

Probs. 10–30/31

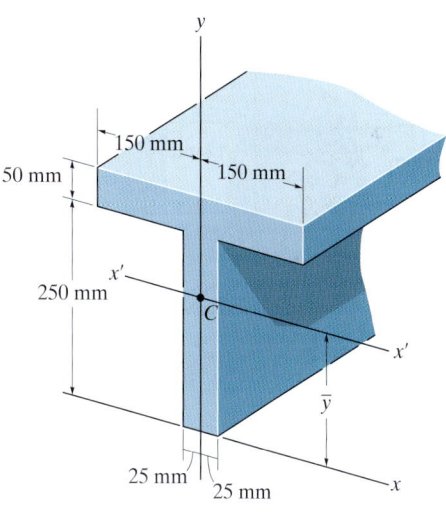

Probs. 10–34/35

*__10–32.__ Determine the moment of inertia of the shaded area about the x axis.

10–33. Determine the moment of inertia of the shaded area about the y axis.

*__10–36.__ Determine the moment of inertia of the beam's cross-sectional area about the x axis.

10–37. Determine the moment of inertia of the beam's cross-sectional area about the y axis.

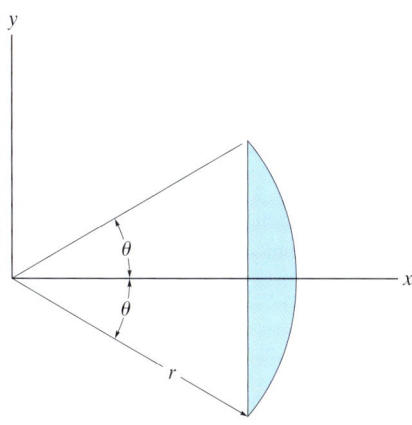

Probs. 10–32/33

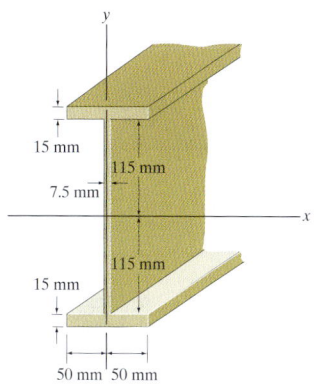

Probs. 10–36/37

10–38. Determine the moment of inertia of the beam's cross-sectional area about the *x* axis.

10–39. Determine the moment of inertia of the beam's cross-sectional area about the *y* axis.

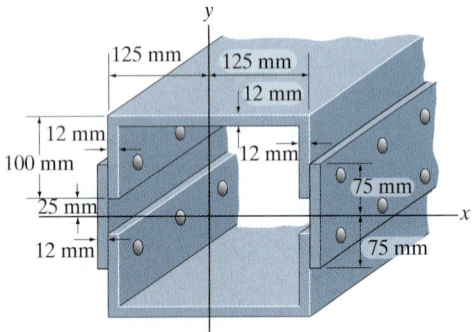

Probs. 10–38/39

*10–40. Determine the moment of inertia of the cross-sectional area about the *x* axis.

10–41. Locate the centroid $\bar{x}$ of the beam's cross-sectional area, and then determine the moment of inertia of the area about the centroidal *y'* axis.

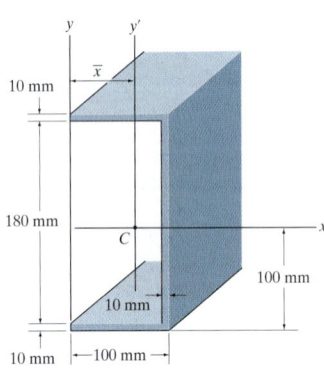

Probs. 10–40/41

10–42. Determine the moment of inertia of the beam's cross-sectional area about the *x* axis.

10–43. Determine the moment of inertia of the beam's cross-sectional area about the *y* axis.

*10–44. Determine the distance $\bar{y}$ to the centroid *C* of the beam's cross-sectional area and then compute the moment of inertia $\bar{I}_{x'}$ about the *x'* axis.

10–45. Determine the distance $\bar{x}$ to the centroid *C* of the beam's cross-sectional area and then compute the moment of inertia $\bar{I}_{y'}$ about the *y'* axis.

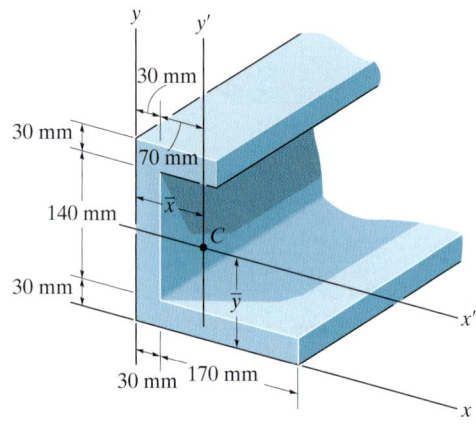

Probs. 10–42/43/44/45

10–46. Determine the distance $\bar{y}$ to the centroid for the beam's cross-sectional area; then determine the moment of inertia about the *x'* axis.

10–47. Determine the moment of inertia of the beam's cross-sectional area about the *y* axis.

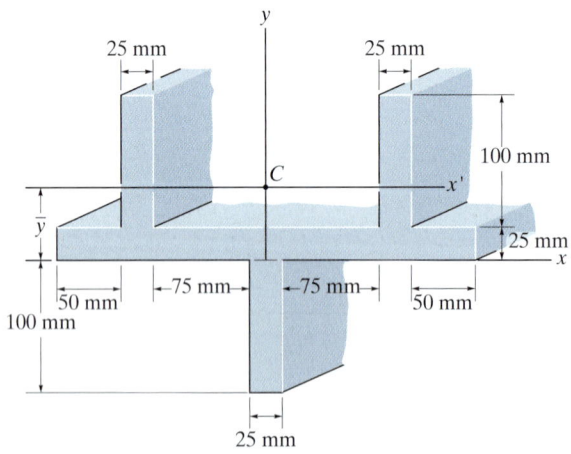

Probs. 10–46/47

*10–48. Determine the beam's moment of inertia I_x about the centroidal x axis.

10–49. Determine the beam's moment of inertia I_y about the centroidal y axis.

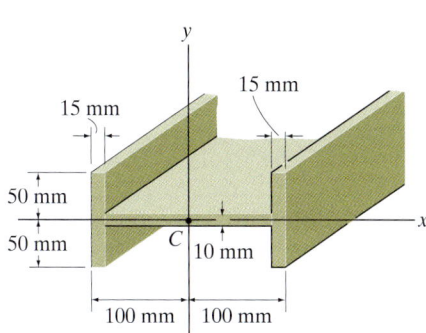

Probs. 10–48/49

10–51. Determine the moment of inertia of the beam's cross-sectional area with respect to the x' centroidal axis. Neglect the size of all the rivet heads, R, for the calculation. Handbook values for the area, moment of inertia, and location of the centroid C of one of the angles are listed in the figure.

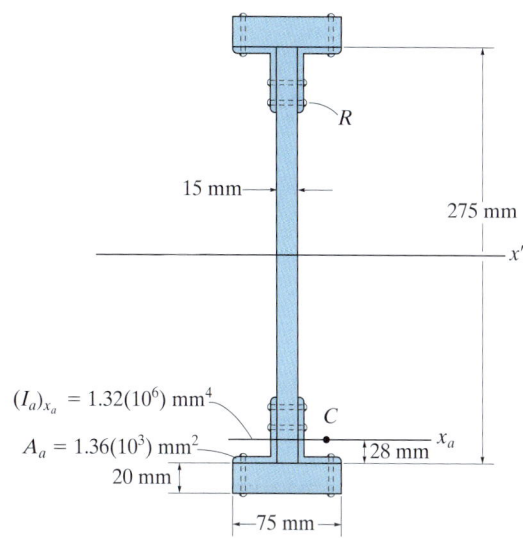

$(I_a)_{x_a} = 1.32(10^6)$ mm^4

$A_a = 1.36(10^3)$ mm^2

Prob. 10–51

10–50. Locate the centroid $\bar{y}$ of the cross section and determine the moment of inertia of the section about the x' axis.

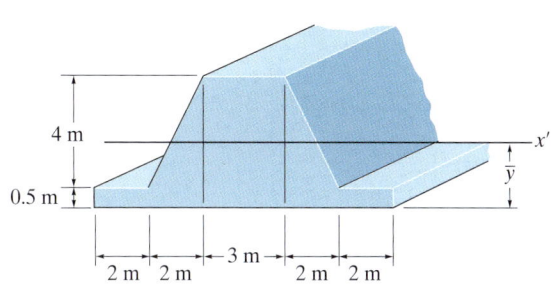

Prob. 10–50

*10–52. Determine the moment of inertia of the parallelogram about the x' axis, which passes through the centroid C of the area.

10–53. Determine the moment of inertia of the parallelogram about the y' axis, which passes through the centroid C of the area.

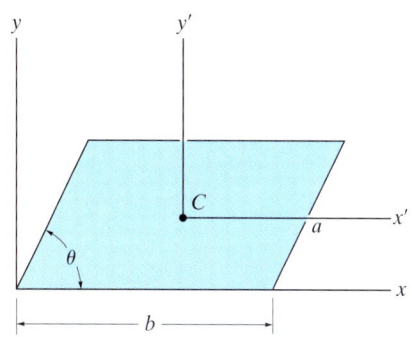

Probs. 10–52/53

10

Fig. 10–10

*10.5   Product of Inertia for an Area

It will be shown in the next section that the property of an area, called the product of inertia, is required in order to determine the *maximum* and *minimum* moments of inertia for the area. These maximum and minimum values are important properties needed for designing structural and mechanical members such as beams, columns, and shafts.

The *product of inertia* of the area in Fig. 10–10 with respect to the x and y axes is defined as

$$I_{xy} = \int_A xy \, dA \qquad (10\text{–}7)$$

If the element of area chosen has a differential size in two directions, as shown in Fig. 10–10, a double integration must be performed to evaluate I_{xy}. Most often, however, it is easier to choose an element having a differential size or thickness in only one direction in which case the evaluation requires only a single integration (see Example 10.6).

Like the moment of inertia, the product of inertia has units of length raised to the fourth power, e.g., m^4, mm^4. However, since x or y may be negative, the product of inertia may either be positive, negative, or zero, depending on the location and orientation of the coordinate axes. For example, the product of inertia I_{xy} for an area will be *zero* if either the x or y axis is an axis of *symmetry* for the area, as in Fig. 10–11. Here every element dA located at point (x, y) has a corresponding element dA located at $(x, -y)$. Since the products of inertia for these elements are, respectively, $xy \, dA$ and $-xy \, dA$, the algebraic sum or integration of all the elements that are chosen in this way will cancel each other. Consequently, the product of inertia for the total area becomes zero. It also follows from the definition of I_{xy} that the "sign" of this quantity depends on the quadrant where the area is located. As shown in Fig. 10–12, if the area is rotated from one quadrant to another, the sign of I_{xy} will change.

The effectiveness of this beam to resist bending can be determined once its moments of inertia and its product of inertia are known.

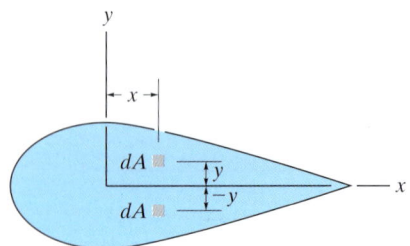

Fig. 10–11

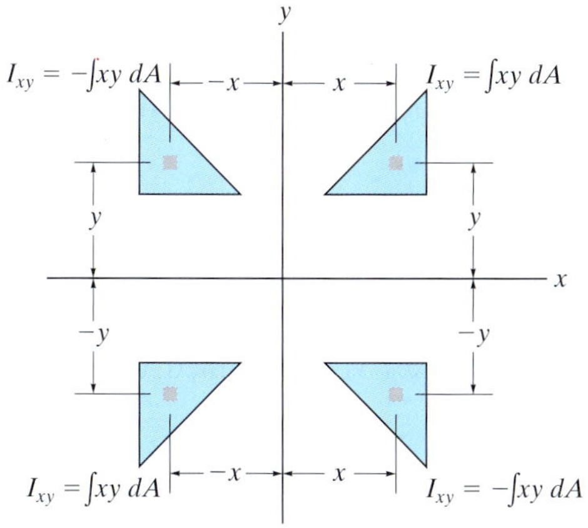

$$I_{xy} = -\int xy \, dA \qquad\qquad\qquad I_{xy} = \int xy \, dA$$

$$I_{xy} = \int xy \, dA \qquad\qquad\qquad I_{xy} = -\int xy \, dA$$

Fig. 10–12

Parallel-Axis Theorem.

Consider the shaded area shown in Fig. 10–13, where x' and y' represent a set of axes passing through the *centroid* of the area, and x and y represent a corresponding set of parallel axes. Since the product of inertia of dA with respect to the x and y axes is $dI_{xy} = (x' + d_x)(y' + d_y) \, dA$, then for the entire area,

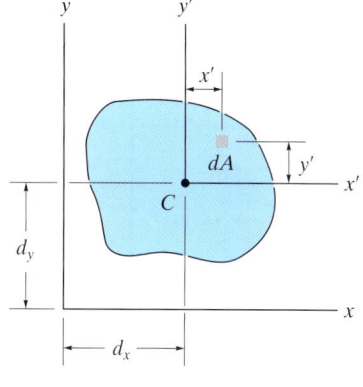

$$I_{xy} = \int_A (x' + d_x)(y' + d_y) \, dA$$

$$= \int_A x'y' \, dA + d_x \int_A y' \, dA + d_y \int_A x' \, dA + d_x d_y \int_A dA$$

Fig. 10–13

The first term on the right represents the product of inertia for the area with respect to the centroidal axes, $\bar{I}_{x'y'}$. The integrals in the second and third terms are zero since the moments of the area are taken about the centroidal axis. Realizing that the fourth integral represents the entire area A, the parallel-axis theorem for the product of inertia becomes

$$I_{xy} = \bar{I}_{x'y'} + A d_x d_y \qquad\qquad (10\text{–}8)$$

It is important that the *algebraic signs* for d_x and d_y be maintained when applying this equation.

10

EXAMPLE | 10.6

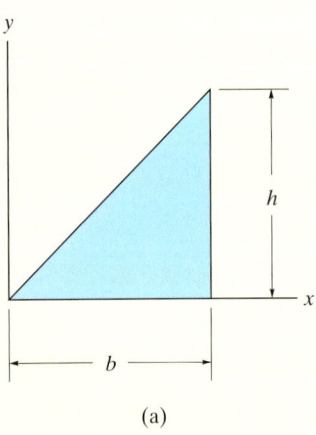

(a)

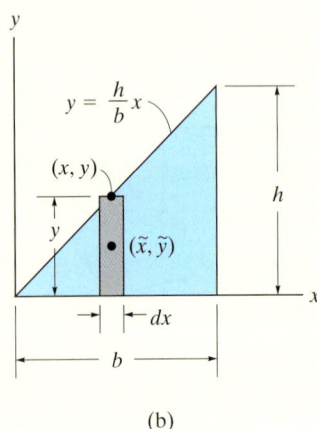

(b)

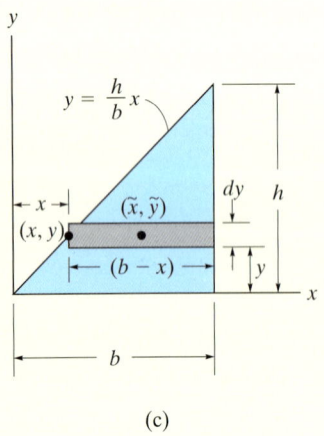

(c)

Fig. 10–14

Determine the product of inertia I_{xy} for the triangle shown in Fig. 10–14a.

SOLUTION I

A differential element that has a thickness dx, as shown in Fig. 10–14b, has an area $dA = y\,dx$. The product of inertia of this element with respect to the x and y axes is determined using the parallel-axis theorem.

$$dI_{xy} = d\bar{I}_{x'y'} + dA\,\tilde{x}\,\tilde{y}$$

where $\tilde{x}$ and $\tilde{y}$ locate the *centroid* of the element or the origin of the x', y' axes. (See Fig. 10–13.) Since $d\bar{I}_{x'y'} = 0$, due to symmetry, and $\tilde{x} = x$, $\tilde{y} = y/2$, then

$$dI_{xy} = 0 + (y\,dx)x\left(\frac{y}{2}\right) = \left(\frac{h}{b}x\,dx\right)x\left(\frac{h}{2b}x\right)$$

$$= \frac{h^2}{2b^2}x^3\,dx$$

Integrating with respect to x from $x = 0$ to $x = b$ yields

$$I_{xy} = \frac{h^2}{2b^2}\int_0^b x^3\,dx = \frac{b^2h^2}{8} \qquad \textit{Ans.}$$

SOLUTION II

The differential element that has a thickness dy, as shown in Fig. 10–14c, can also be used. Its area is $dA = (b - x)\,dy$. The *centroid* is located at point $\tilde{x} = x + (b - x)/2 = (b + x)/2$, $\tilde{y} = y$, so the product of inertia of the element becomes

$$dI_{xy} = d\bar{I}_{x'y'} + dA\,\tilde{x}\,\tilde{y}$$

$$= 0 + (b - x)\,dy\left(\frac{b + x}{2}\right)y$$

$$= \left(b - \frac{b}{h}y\right)dy\left[\frac{b + (b/h)y}{2}\right]y = \frac{1}{2}y\left(b^2 - \frac{b^2}{h^2}y^2\right)dy$$

Integrating with respect to y from $y = 0$ to $y = h$ yields

$$I_{xy} = \frac{1}{2}\int_0^h y\left(b^2 - \frac{b^2}{h^2}y^2\right)dy = \frac{b^2h^2}{8} \qquad \textit{Ans.}$$

EXAMPLE 10.7

Determine the product of inertia for the cross-sectional area of the member shown in Fig. 10–15a, about the x and y centroidal axes.

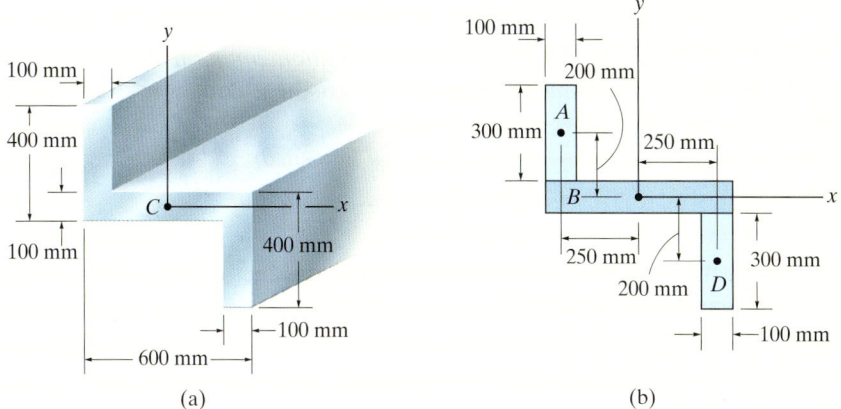

(a)   (b)

Fig. 10–15

SOLUTION

As in Example 10.5, the cross section can be subdivided into three composite rectangular areas A, B, and D, Fig. 10–15b. The coordinates for the centroid of each of these rectangles are shown in the figure. Due to symmetry, the product of inertia of *each rectangle* is *zero* about a set of x', y' axes that passes through the centroid of each rectangle. Using the parallel-axis theorem, we have

Rectangle A

$$I_{xy} = \bar{I}_{x'y'} + A d_x d_y$$
$$= 0 + (300)(100)(-250)(200) = -1.50(10^9) \text{ mm}^4$$

Rectangle B

$$I_{xy} = \bar{I}_{x'y'} + A d_x d_y$$
$$= 0 + 0 = 0$$

Rectangle D

$$I_{xy} = \bar{I}_{x'y'} + A d_x d_y$$
$$= 0 + (300)(100)(250)(-200) = -1.50(10^9) \text{ mm}^4$$

The product of inertia for the entire cross section is therefore

$$I_{xy} = -1.50(10^9) + 0 - 1.50(10^9) = -3.00(10^9) \text{ mm}^4 \quad Ans.$$

NOTE: This negative result is due to the fact that rectangles A and D have centroids located with negative x and negative y coordinates, respectively.

*10.6   Moments of Inertia for an Area about Inclined Axes

In structural and mechanical design, it is sometimes necessary to calculate the moments and product of inertia I_u, I_v, and I_{uv} for an area with respect to a set of inclined u and v axes when the values for θ, I_x, I_y, and I_{xy} are *known*. To do this we will use *transformation equations* which relate the x, y and u, v coordinates. From Fig. 10–16, these equations are

$$u = x \cos \theta + y \sin \theta$$

$$v = y \cos \theta - x \sin \theta$$

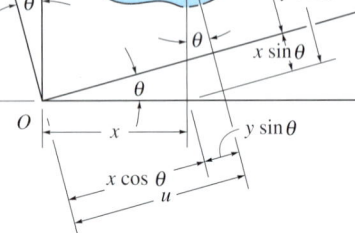

Fig. 10–16

With these equations, the moments and product of inertia of dA about the u and v axes become

$$dI_u = v^2\, dA = (y \cos \theta - x \sin \theta)^2\, dA$$

$$dI_v = u^2\, dA = (x \cos \theta + y \sin \theta)^2\, dA$$

$$dI_{uv} = uv\, dA = (x \cos \theta + y \sin \theta)(y \cos \theta - x \sin \theta)\, dA$$

Expanding each expression and integrating, realizing that $I_x = \int y^2\, dA$, $I_y = \int x^2\, dA$, and $I_{xy} = \int xy\, dA$, we obtain

$$I_u = I_x \cos^2 \theta + I_y \sin^2 \theta - 2I_{xy} \sin \theta \cos \theta$$

$$I_v = I_x \sin^2 \theta + I_y \cos^2 \theta + 2I_{xy} \sin \theta \cos \theta$$

$$I_{uv} = I_x \sin \theta \cos \theta - I_y \sin \theta \cos \theta + I_{xy}(\cos^2 \theta - \sin^2 \theta)$$

Using the trigonometric identities $\sin 2\theta = 2 \sin \theta \cos \theta$ and $\cos 2\theta = \cos^2 \theta - \sin^2 \theta$ we can simplify the above expressions, in which case

$$I_u = \frac{I_x + I_y}{2} + \frac{I_x - I_y}{2} \cos 2\theta - I_{xy} \sin 2\theta$$

$$I_v = \frac{I_x + I_y}{2} - \frac{I_x - I_y}{2} \cos 2\theta + I_{xy} \sin 2\theta \qquad (10\text{–}9)$$

$$I_{uv} = \frac{I_x - I_y}{2} \sin 2\theta + I_{xy} \cos 2\theta$$

Notice that if the first and second equations are added together, we can show that the polar moment of inertia about the z axis passing through point O is, as expected, *independent* of the orientation of the u and v axes; i.e.,

$$J_O = I_u + I_v = I_x + I_y$$

Principal Moments of Inertia.   Equations 10–9 show that I_u, I_v, and I_{uv} depend on the angle of inclination, θ, of the u, v axes. We will now determine the orientation of these axes about which the moments of inertia for the area are maximum and minimum. This particular set of axes is called the *principal axes* of the area, and the corresponding moments of inertia with respect to these axes are called the *principal moments of inertia*. In general, there is a set of principal axes for every chosen origin O. However, for structural and mechanical design, the origin O is located at the centroid of the area.

The angle which defines the orientation of the principal axes can be found by differentiating the first of Eqs. 10–9 with respect to θ and setting the result equal to zero. Thus,

$$\frac{dI_u}{d\theta} = -2\left(\frac{I_x - I_y}{2}\right)\sin 2\theta - 2I_{xy}\cos 2\theta = 0$$

Therefore, at $\theta = \theta_p$,

$$\tan 2\theta_p = \frac{-I_{xy}}{(I_x - I_y)/2} \qquad (10\text{–}10)$$

The two roots θ_{p_1} and θ_{p_2} of this equation are 90° apart, and so they each specify the inclination of one of the principal axes. In order to substitute them into Eq. 10–9, we must first find the sine and cosine of $2\theta_{p_1}$ and $2\theta_{p_2}$. This can be done using these ratios from the triangles shown in Fig. 10–17, which are based on Eq. 10–10.

Substituting each of the sine and cosine ratios into the first or second of Eqs. 10–9 and simplifying, we obtain

$$I_{\substack{max \\ min}} = \frac{I_x + I_y}{2} \pm \sqrt{\left(\frac{I_x - I_y}{2}\right)^2 + I_{xy}^2} \qquad (10\text{–}11)$$

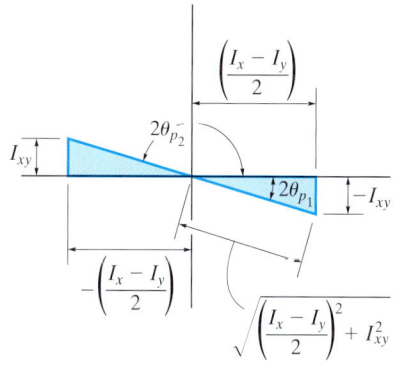

Fig. 10–17

10

Depending on the sign chosen, this result gives the maximum or minimum moment of inertia for the area. Furthermore, if the above trigonometric relations for θ_{p_1} and θ_{p_2} are substituted into the third of Eqs. 10–9, it can be shown that $I_{uv} = 0$; that is, the *product of inertia with respect to the principal axes is zero*. Since it was indicated in Sec. 10.6 that the product of inertia is zero with respect to any symmetrical axis, it therefore follows that *any symmetrical axis represents a principal axis of inertia for the area*.

EXAMPLE | 10.8

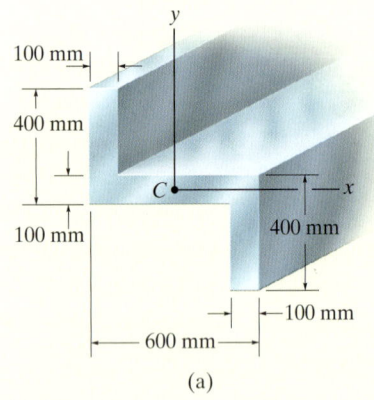

(a)

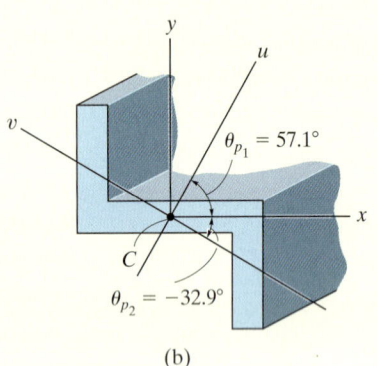

(b)

Fig. 10–18

Determine the principal moments of inertia and the orientation of the principal axes for the cross-sectional area of the member shown in Fig. 10–18a with respect to an axis passing through the centroid.

SOLUTION

The moments and product of inertia of the cross section with respect to the x, y axes have been determined in Examples 10.5 and 10.7. The results are

$$I_x = 2.90(10^9) \text{ mm}^4 \quad I_y = 5.60(10^9) \text{ mm}^4 \quad I_{xy} = -3.00(10^9) \text{ mm}^4$$

Using Eq. 10–10, the angles of inclination of the principal axes u and v are

$$\tan 2\theta_p = \frac{-I_{xy}}{(I_x - I_y)/2} = \frac{-[-3.00(10^9)]}{[2.90(10^9) - 5.60(10^9)]/2} = -2.22$$

$$2\theta_p = -65.8° \text{ and } 114.2°$$

Thus, by inspection of Fig. 10–18b,

$$\theta_{p_2} = -32.9° \quad \text{and} \quad \theta_{p_1} = 57.1° \qquad \textit{Ans.}$$

The principal moments of inertia with respect to these axes are determined from Eq. 10–11. Hence,

$$I_{\min}^{\max} = \frac{I_x + I_y}{2} \pm \sqrt{\left(\frac{I_x - I_y}{2}\right)^2 + I_{xy}^2}$$

$$= \frac{2.90(10^9) + 5.60(10^9)}{2}$$

$$\pm \sqrt{\left[\frac{2.90(10^9) - 5.60(10^9)}{2}\right]^2 + [-3.00(10^9)]^2}$$

$$I_{\min}^{\max} = 4.25(10^9) \pm 3.29(10^9)$$

or

$$I_{\max} = 7.54(10^9) \text{ mm}^4 \quad I_{\min} = 0.960(10^9) \text{ mm}^4 \qquad \textit{Ans.}$$

NOTE: The maximum moment of inertia, $I_{\max} = 7.54(10^9) \text{ mm}^4$, occurs with respect to the u axis since *by inspection* most of the cross-sectional area is farthest away from this axis. Or, stated in another manner, $I_{\max}$ occurs about the u axis since this axis is located within $\pm 45°$ of the y axis, which has the larger value of I ($I_y > I_x$). Also, this can be concluded by substituting the data with $\theta = 57.1°$ into the first of Eqs. 10–9 and solving for I_u.

*10.7   Mohr's Circle for Moments of Inertia

Equations 10–9 to 10–11 have a graphical solution that is convenient to use and generally easy to remember. Squaring the first and third of Eqs. 10–9 and adding, it is found that

$$\left(I_u - \frac{I_x + I_y}{2}\right)^2 + I_{uv}^2 = \left(\frac{I_x - I_y}{2}\right)^2 + I_{xy}^2$$

Here I_x, I_y, and I_{xy} are *known constants*. Thus, the above equation may be written in compact form as

$$(I_u - a)^2 + I_{uv}^2 = R^2$$

When this equation is plotted on a set of axes that represent the respective moment of inertia and the product of inertia, as shown in Fig. 10–19, the resulting graph represents a *circle* of radius

$$R = \sqrt{\left(\frac{I_x - I_y}{2}\right)^2 + I_{xy}^2}$$

and having its center located at point $(a, 0)$, where $a = (I_x + I_y)/2$. The circle so constructed is called *Mohr's circle*, named after the German engineer Otto Mohr (1835–1918).

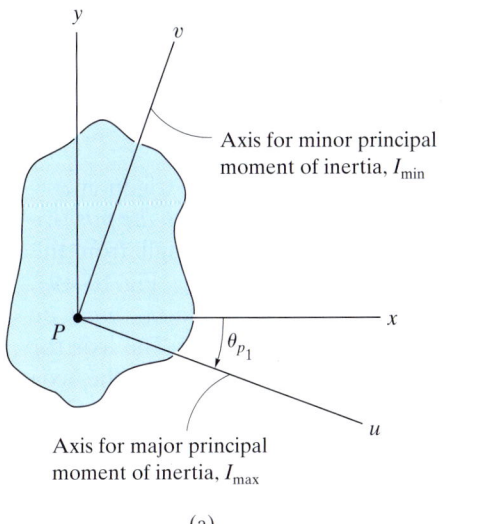

(a)

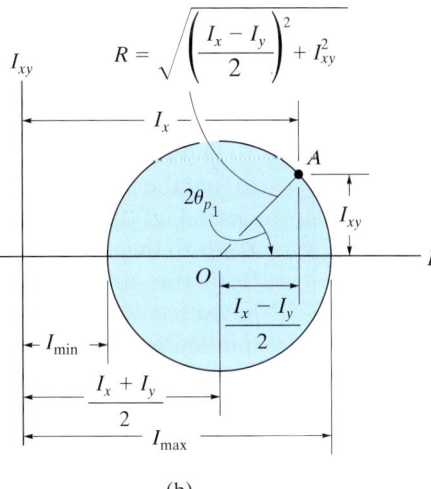

(b)

Fig. 10–19

Procedure for Analysis

The main purpose in using Mohr's circle here is to have a convenient means for finding the principal moments of inertia for an area. The following procedure provides a method for doing this.

Determine I_x, I_y, and I_{xy}.

• Establish the x, y axes and determine I_x, I_y, and I_{xy}, Fig. 10–19a.

Construct the Circle.

• Construct a rectangular coordinate system such that the abscissa represents the moment of inertia I, and the ordinate represents the product of inertia I_{xy}, Fig. 10–19b.

• Determine the center of the circle, O, which is located at a distance $(I_x + I_y)/2$ from the origin, and plot the reference point A having coordinates (I_x, I_{xy}). Remember, I_x is always positive, whereas I_{xy} can be either positive or negative.

• Connect the reference point A with the center of the circle and determine the distance OA by trigonometry. This distance represents the radius of the circle, Fig. 10–19b. Finally, draw the circle.

Principal Moments of Inertia.

• The points where the circle intersects the I axis give the values of the principal moments of inertia I_{min} and I_{max}. Notice that, as expected, the *product of inertia will be zero at these points*, Fig. 10–19b.

Principal Axes.

• To find the orientation of the major principal axis, use trigonometry to find the angle $2\theta_{p_1}$, *measured from the radius OA to the positive I axis*, Fig. 10–19b. This angle represents *twice* the angle from the x axis to the axis of maximum moment of inertia I_{max}, Fig. 10–19a. Both the angle on the circle, $2\theta_{p_1}$, and the angle θ_{p_1} *must be measured in the same sense*, as shown in Fig. 10–19. The axis for minimum moment of inertia I_{min} is perpendicular to the axis for I_{max}.

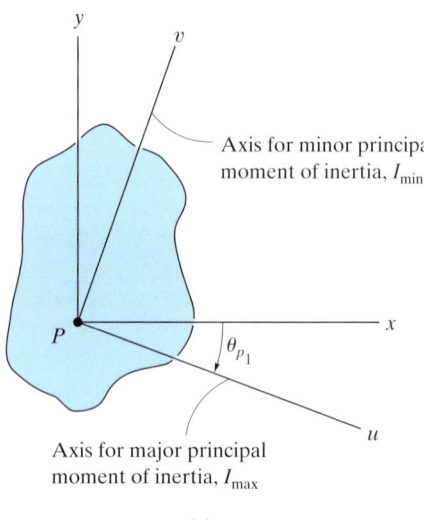

(a)

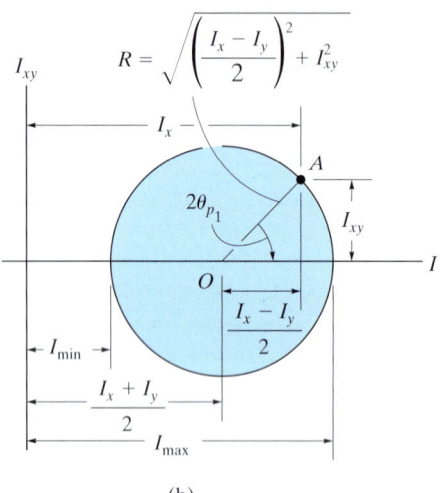

$$R = \sqrt{\left(\frac{I_x - I_y}{2}\right)^2 + I_{xy}^2}$$

(b)

Fig. 10–19 (Repeated)

Using trigonometry, the above procedure can be verified to be in accordance with the equations developed in Sec. 10.6.

EXAMPLE 10.9

Using Mohr's circle, determine the principal moments of inertia and the orientation of the major principal axes for the cross-sectional area of the member shown in Fig. 10–20a, with respect to an axis passing through the centroid.

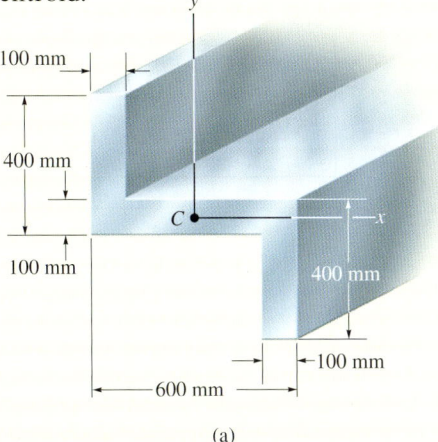

(a)

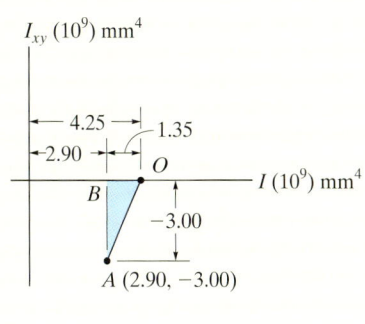

(b)

SOLUTION

Determine I_x, I_y, I_{xy}. The moments and product of inertia have been determined in Examples 10.5 and 10.7 with respect to the x, y axes shown in Fig. 10–20a. The results are $I_x = 2.90(10^9)$ mm^4, $I_y = 5.60(10^9)$ mm^4, and $I_{xy} = -3.00(10^9)$ mm^4.

Construct the Circle. The I and I_{xy} axes are shown in Fig. 10–20b. The center of the circle, O, lies at a distance $(I_x + I_y)/2 = (2.90 + 5.60)/2 = 4.25$ from the origin. When the reference point $A(I_x, I_{xy})$ or $A(2.90, -3.00)$ is connected to point O, the radius OA is determined from the triangle OBA using the Pythagorean theorem.

$$OA = \sqrt{(1.35)^2 + (-3.00)^2} = 3.29$$

The circle is constructed in Fig. 10–20c.

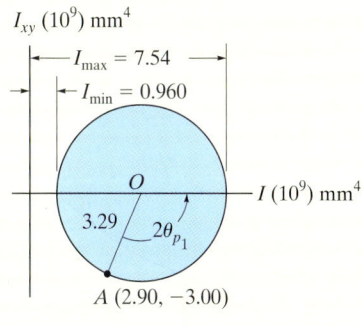

(c)

Principal Moments of Inertia. The circle intersects the I axis at points $(7.54, 0)$ and $(0.960, 0)$. Hence,

$$I_{max} = (4.25 + 3.29)10^9 = 7.54(10^9) \text{ mm}^4 \qquad Ans.$$
$$I_{min} = (4.25 - 3.29)10^9 = 0.960(10^9) \text{ mm}^4 \qquad Ans.$$

Principal Axes. As shown in Fig. 10–20c, the angle $2\theta_{p_1}$ is determined from the circle by measuring counterclockwise from OA to the direction of the *positive I* axis. Hence,

$$2\theta_{p_1} = 180° - \sin^{-1}\left(\frac{|BA|}{|OA|}\right) = 180° - \sin^{-1}\left(\frac{3.00}{3.29}\right) = 114.2°$$

The principal axis for $I_{max} = 7.54(10^9)$ mm^4 is therefore oriented at an angle $\theta_{p_1} = 57.1°$, measured *counterclockwise*, from the *positive x* axis to the *positive u* axis. The v axis is perpendicular to this axis. The results are shown in Fig. 10–20d.

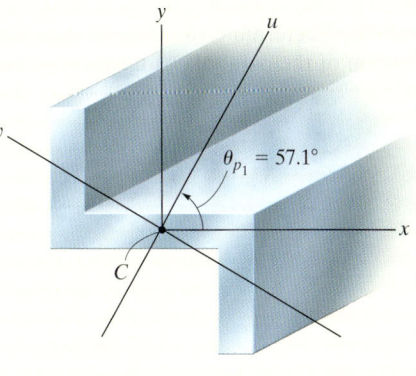

(d)

Fig. 10–20

PROBLEMS

10–54. Determine the product of inertia of the thin strip of area with respect to the x and y axes. The strip is oriented at an angle θ from the x axis. Assume that $t \ll l$.

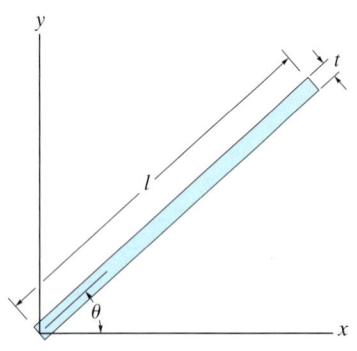

Prob. 10–54

10–55. Determine the product of inertia of the shaded area with respect to the x and y axes.

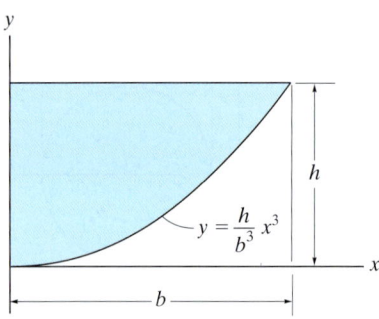

Prob. 10–55

*10–56. Determine the product of inertia of the shaded portion of the parabola with respect to the x and y axes.

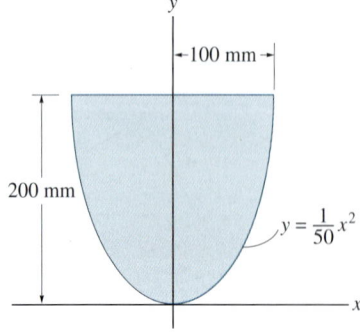

Prob. 10–56

10–57. Determine the product of inertia of the shaded area with respect to the x and y axes.

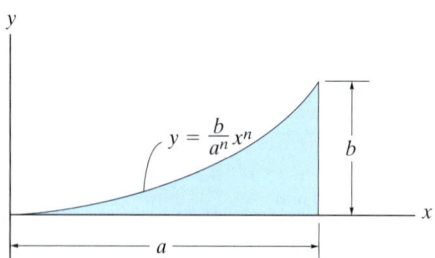

Prob. 10–57

10–58. Determine the product of inertia of the shaded area with respect to the x and y axes, and then use the parallel-axis theorem to find the product of inertia of the area with respect to the centroidal x' and y' axes.

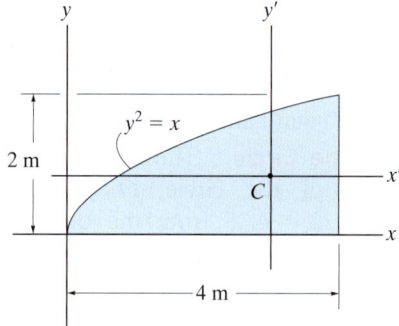

Prob. 10–58

10–59. Determine the product of inertia of the shaded area with respect to the x and y axes. Use Simpson's rule to evaluate the integral.

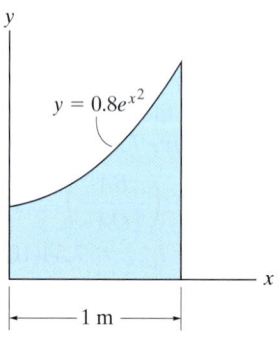

Prob. 10–59

*10–60. Determine the product of inertia of the shaded area with respect to the x and y axes.

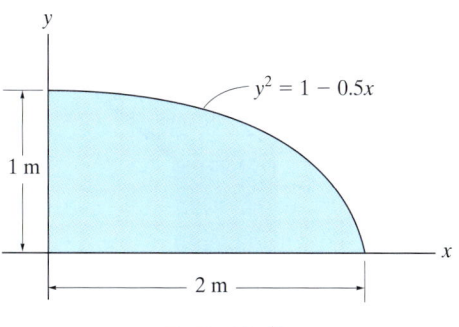

$y^2 = 1 - 0.5x$

1 m

2 m

Prob. 10–60

10–61. Determine the product of inertia of the parallelogram with respect to the x and y axes.

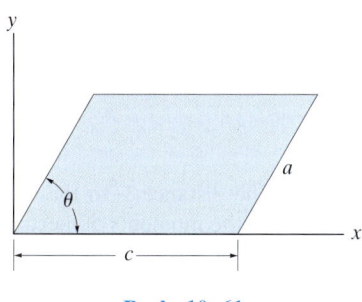

θ

a

c

Prob. 10–61

10–62. Determine the product of inertia of the parabolic area with respect to the x and y axes.

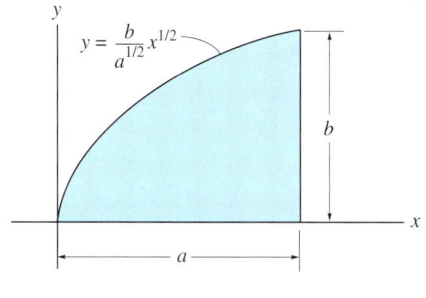

$y = \dfrac{b}{a^{1/2}}x^{1/2}$

b

a

Prob. 10–62

10–63. Determine the product of inertia for the beam's cross-sectional area with respect to the u and v axes.

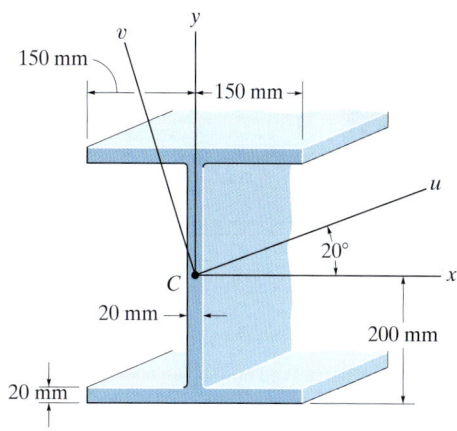

150 mm

150 mm

v

y

20 mm

20 mm

C

20°

u

x

200 mm

Prob. 10–63

*10–64. Determine the product of inertia of the beam's cross-sectional area with respect to the x and y axes that have their origin located at the centroid C.

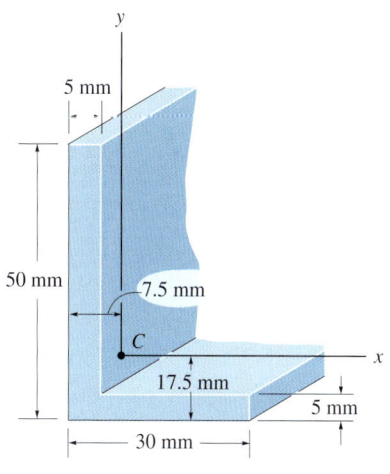

5 mm

50 mm

7.5 mm

C

17.5 mm

5 mm

30 mm

Prob. 10–64

10

10–65. Determine the product of inertia for the beam's cross-sectional area with respect to the x and y axes that have their origin located at the centroid C.

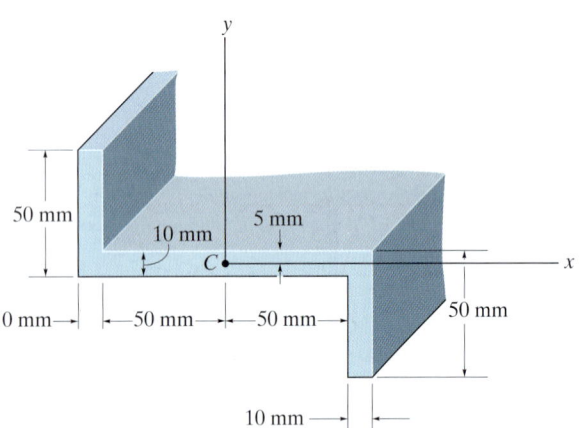

Prob. 10–65

10–66. Determine the product of inertia of the cross-sectional area with respect to the x and y axes.

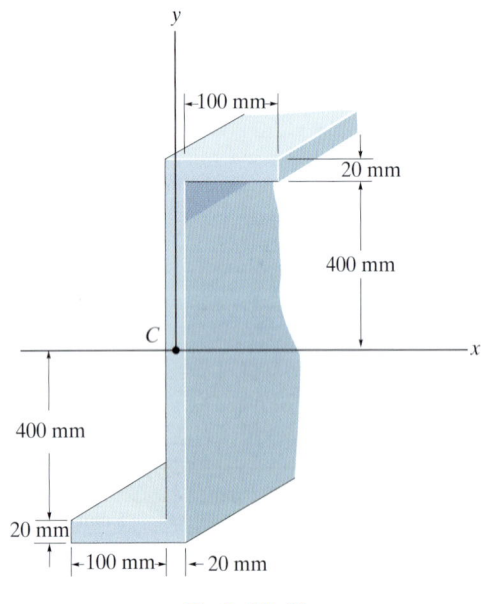

Prob. 10–66

10–67. Determine the product of inertia of the beam's cross-sectional area with respect to the x and y axes.

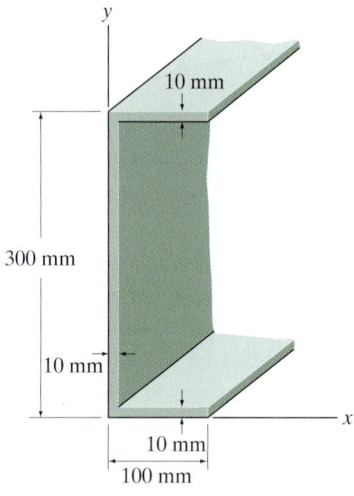

Prob. 10–67

10–68. Determine the distance $\bar{y}$ to the centroid of the area and then calculate the moments of inertia I_u and I_v of the channel's cross-sectional area. The u and v axes have their origin at the centroid C. For the calculation, assume all corners to be square.

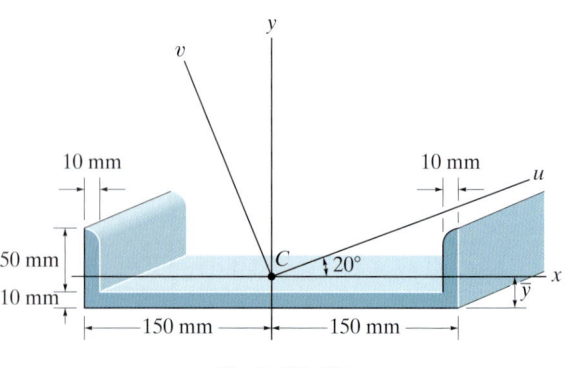

Prob. 10–68

10–69. Determine the moments of inertia I_u and I_v of the shaded area.

*10–72. Locate the centroid $\bar{y}$ of the beam's cross-sectional area and then determine the moments of inertia and the product of inertia of this area with respect to the u and v axes.

10–73. Solve Prob. 10–72 using Mohr's circle.

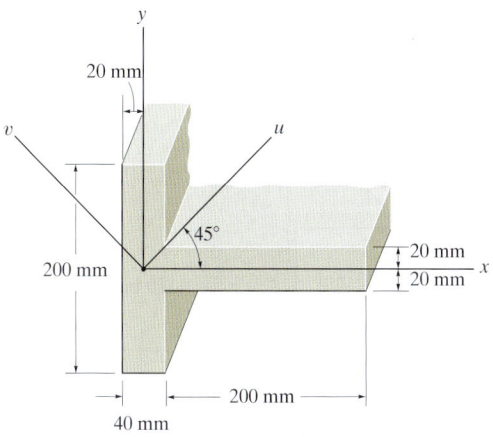

Prob. 10–69

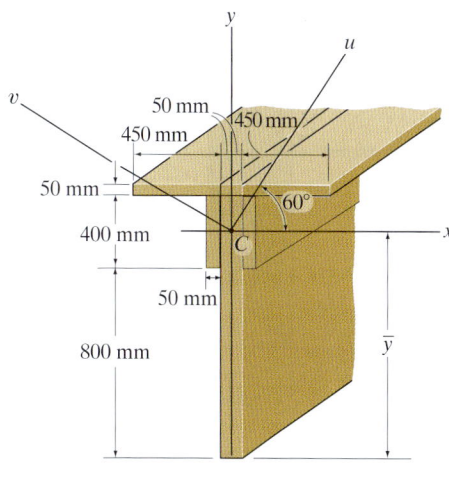

Probs. 10–72/73

10–70. Determine the moments of inertia and the product of inertia of the beam's cross sectional area with respect to the u and v axes.

10–71. Solve Prob. 10–70 using Mohr's circle. *Hint:* Once the circle is established, rotate $2\theta = 60°$ counterclockwise from the reference OA, then find the coordinates of the points that define the diameter of the circle.

10–74. Locate the centroid $\bar{y}$ of the beam's cross-sectional area and then determine the moments of inertia of this area and the product of inertia with respect to the u and v axes. The axes have their origin at the centroid C.

10–75. Solve Prob. 10–74 using Mohr's circle.

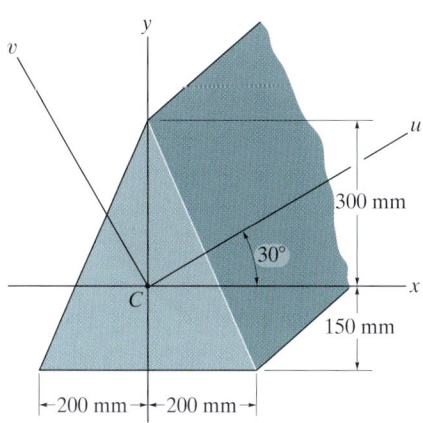

Probs. 10–70/71

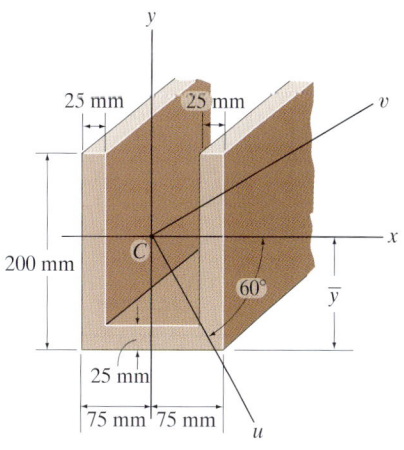

Probs. 10–74/75

*10–76.   Locate the centroid $\bar{x}$ of the beam's cross-sectional area and then determine the moments of inertia and the product of inertia of this area with respect to the u and v axes. The axes have their origin at the centroid C.

10–77.   Solve Prob. 10–76 using Mohr's circle.

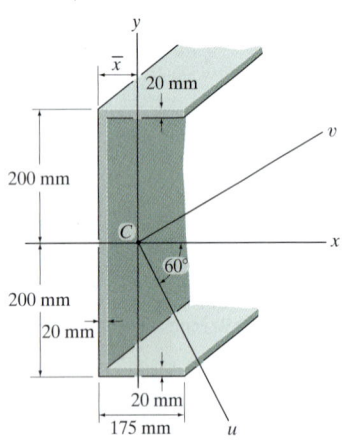

Probs. 10–76/77

10–78.   Determine the principal moments of inertia for the angle's cross-sectional area with respect to a set of principal axes that have their origin located at the centroid C. Use the equation developed in Section 10.7. For the calculation, assume all corners to be square.

10–79.   Solve Prob. 10–78 using Mohr's circle.

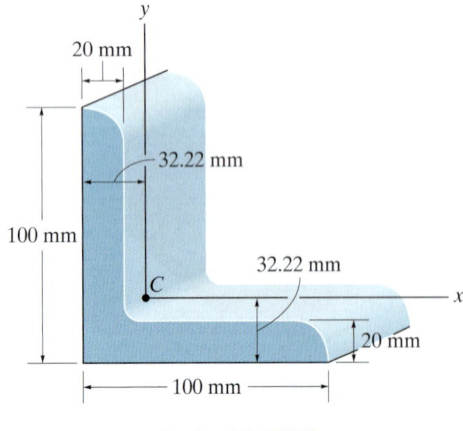

Probs. 10–78/79

*10–80.   Determine the orientation of the principal axes, which have their origin at centroid C of the beam's cross-sectional area. Also, find the principal moments of inertia.

10–81.   Solve Prob. 10–80 using Mohr's circle.

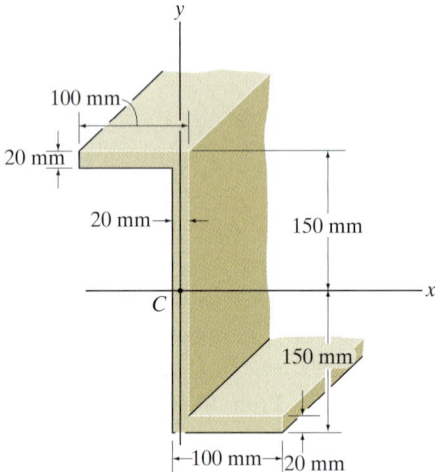

Probs. 10–80/81

10–82.   Locate the centroid $\bar{y}$ of the beam's cross-sectional area and then determine the moments of inertia of this area and the product of inertia with respect to the u and v axes. The axes have their origin at the centroid C.

10–83.   Solve Prob. 10–82 using Mohr's circle.

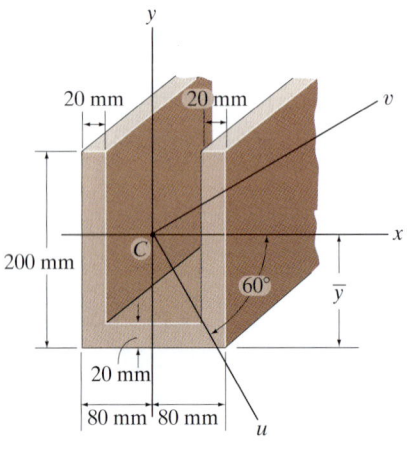

Probs. 10–82/83

10.8   Mass Moment of Inertia

The mass moment of inertia of a body is a measure of the body's resistance to angular acceleration. Since it is used in dynamics to study rotational motion, methods for its calculation will now be discussed.*

Consider the rigid body shown in Fig. 10–21. We define the *mass moment of inertia* of the body about the z axis as

$$I = \int_m r^2 \, dm \qquad (10\text{–}12)$$

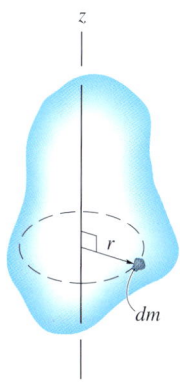

Fig. 10–21

Here r is the perpendicular distance from the axis to the arbitrary element dm. Since the formulation involves r, the value of I is *unique* for each axis about which it is computed. The axis which is generally chosen, however, passes through the body's mass center G. Common units used for its measurement are $kg \cdot m^2$.

If the body consists of material having a density ρ, then $dm = \rho \, dV$, Fig. 10–22a. Substituting this into Eq. 10–12, the body's moment of inertia is then computed using *volume elements* for integration; i.e.

$$I = \int_V r^2 \rho \, dV \qquad (10\text{–}13)$$

For most applications, ρ will be a *constant*, and so this term may be factored out of the integral, and the integration is then purely a function of geometry.

$$I = \rho \int_V r^2 \, dV \qquad (10\text{–}14)$$

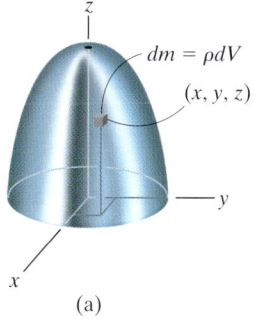

(a)

Fig. 10–22

*Another property of the body, which measures the symmetry of the body's mass with respect to a coordinate system, is the mass product of inertia. This property most often applies to the three-dimensional motion of a body and is discussed in *Engineering Mechanics: Dynamics* (Chapter 21).

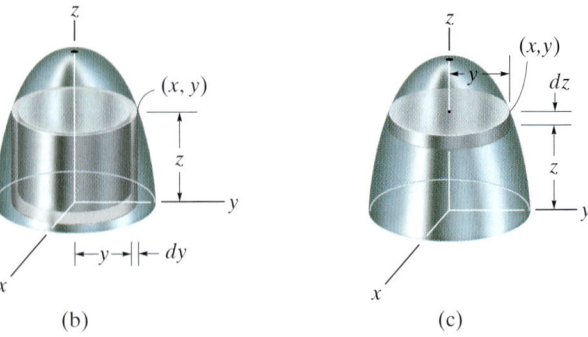

Fig. 10–22 (cont'd)

Procedure for Analysis

If a body is symmetrical with respect to an axis, as in Fig. 10–22, then its mass moment of inertia about the axis can be determined by using a single integration. Shell and disk elements are used for this purpose.

Shell Element.

- If a *shell element* having a height z, radius y, and thickness dy is chosen for integration, Fig. 10–22b, then its volume is $dV = (2\pi y)(z)\,dy$.

- This element can be used in Eq. 10–13 or 10–14 for determining the moment of inertia I_z of the body about the z axis since the *entire element*, due to its "thinness," lies at the *same* perpendicular distance $r = y$ from the z axis (see Example 10.10).

Disk Element.

- If a disk element having a radius y and a thickness dz is chosen for integration, Fig. 10–22c, then its volume is $dV = (\pi y^2)\,dz$.

- In this case the element is *finite* in the radial direction, and consequently its points *do not* all lie at the *same radial distance r* from the z axis. As a result, Eqs. 10–13 or 10–14 *cannot* be used to determine I_z. Instead, to perform the integration using this element, it is first necessary to determine the moment of inertia *of the element* about the z axis and then integrate this result (see Example 10.11).

EXAMPLE 10.10

Determine the mass moment of inertia of the cylinder shown in Fig. 10–23a about the z axis. The density of the material, ρ, is constant.

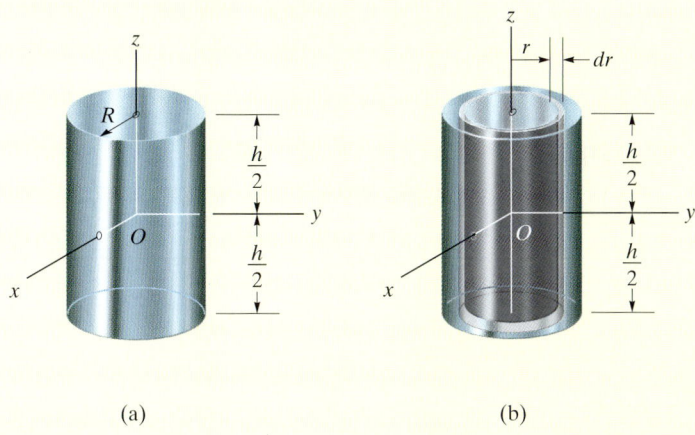

(a)   (b)

Fig. 10–23

SOLUTION

Shell Element.   This problem will be solved using the *shell element* in Fig. 10–23b and thus only a single integration is required. The volume of the element is $dV = (2\pi r)(h)\,dr$, and so its mass is $dm = \rho\,dV = \rho(2\pi hr\,dr)$. Since the *entire element* lies at the same distance r from the z axis, the moment of inertia *of the element* is

$$dI_z = r^2\,dm = \rho 2\pi hr^3\,dr$$

Integrating over the entire cylinder yields

$$I_z = \int_m r^2\,dm = \rho 2\pi h \int_0^R r^3\,dr = \frac{\rho\pi}{2}R^4 h$$

Since the mass of the cylinder is

$$m = \int_m dm = \rho 2\pi h \int_0^R r\,dr = \rho\pi hR^2$$

then

$$I_z = \frac{1}{2}mR^2 \qquad\qquad \textit{Ans.}$$

10

EXAMPLE | 10.11

A solid is formed by revolving the shaded area shown in Fig. 10–24a about the y axis. If the density of the material is 2 Mg/m³, determine the mass moment of inertia about the y axis.

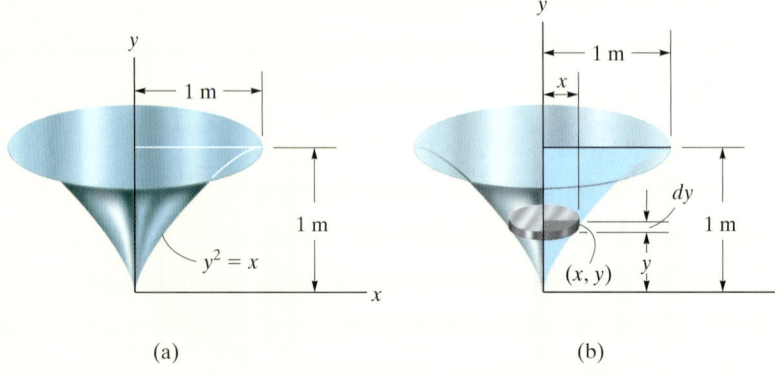

Fig. 10–24

SOLUTION

Disk Element.   The moment of inertia will be determined using this *disk element*, as shown in Fig. 10–24b. Here the element intersects the curve at the arbitrary point (x, y) and has a mass

$$dm = \rho \, dV = \rho(\pi x^2) \, dy$$

Although all points on the element are *not* located at the same distance from the y axis, it is still possible to determine the moment of inertia dI_y *of the element* about the y axis. In the previous example it was shown that the moment of inertia of a homogeneous cylinder about its longitudinal axis is $I = \frac{1}{2}mR^2$, where m and R are the mass and radius of the cylinder. Since the height of the cylinder is not involved in this formula, we can also use this result for a disk. Thus, for the disk element in Fig. 10–24b, we have

$$dI_y = \frac{1}{2}(dm)x^2 = \frac{1}{2}[\rho(\pi x^2) \, dy]x^2$$

Substituting $x = y^2$, $\rho = 2$ Mg/m³, and integrating with respect to y, from $y = 0$ to $y = 1$ ft, yields the moment of inertia for the entire solid.

$$I_y = \frac{2\pi}{2} \int_0^{1\,m} x^4 \, dy = \frac{2\pi}{2} \int_0^{1\,m} y^8 \, dy = 0.349 \text{ Mg} \cdot \text{m}^2 = 349 \text{ kg} \cdot \text{m}^2 \quad \textit{Ans.}$$

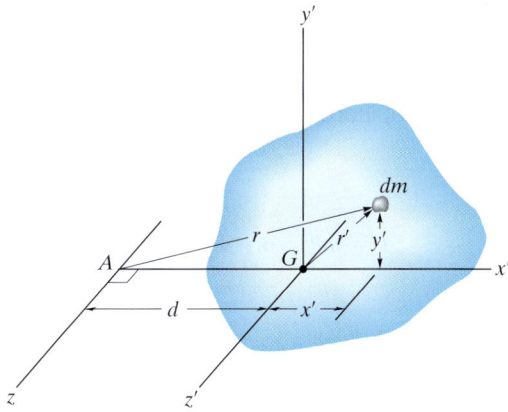

Fig. 10–25

Parallel-Axis Theorem. If the moment of inertia of the body about an axis passing through the body's mass center is known, then the moment of inertia about any other *parallel axis* can be determined by using the *parallel-axis theorem.* To derive this theorem, consider the body shown in Fig. 10–25. The z' axis passes through the mass center G, whereas the corresponding *parallel z axis* lies at a constant distance d away. Selecting the differential element of mass dm, which is located at point (x', y'), and using the Pythagorean theorem, $r^2 = (d + x')^2 + y'^2$, the moment of inertia of the body about the z axis is

$$I = \int_m r^2 \, dm = \int_m [(d + x')^2 + y'^2] \, dm$$

$$= \int_m (x'^2 + y'^2) \, dm + 2d \int_m x' \, dm + d^2 \int_m dm$$

Since $r'^2 = x'^2 + y'^2$, the first integral represents I_G. The second integral is equal to *zero*, since the z' axis passes through the body's mass center, i.e., $\int x' \, dm = \bar{x} \int dm = 0$ since $\bar{x} = 0$. Finally, the third integral is the total mass m of the body. Hence, the moment of inertia about the z axis becomes

$$\boxed{I = I_G + md^2} \tag{10–15}$$

where

I_G = moment of inertia about the z' axis passing through the mass center G

m = mass of the body

d = distance between the parallel axes

10

Radius of Gyration.   Occasionally, the moment of inertia of a body about a specified axis is reported in handbooks using the *radius of gyration, k*. This value has units of length, and when it and the body's mass m are known, the moment of inertia can be determined from the equation

$$I = mk^2 \quad \text{or} \quad k = \sqrt{\frac{I}{m}} \qquad (10\text{--}16)$$

Note the *similarity* between the definition of k in this formula and r in the equation $dI = r^2\, dm$, which defines the moment of inertia of a differential element of mass dm of the body about an axis.

Composite Bodies.   If a body is constructed from a number of simple shapes such as disks, spheres, and rods, the moment of inertia of the body about any axis z can be determined by adding algebraically the moments of inertia of all the composite shapes calculated about the same axis. Algebraic addition is necessary since a composite part must be considered as a negative quantity if it has already been included within another part—as in the case of a "hole" subtracted from a solid plate. Also, the parallel-axis theorem is needed for the calculations if the center of mass of each composite part does not lie on the z axis. For calculations, a table of some simple shapes is given on the inside back cover.

This flywheel, which operates a metal cutter, has a large moment of inertia about its center. Once it begins rotating it is difficult to stop it and therefore a uniform motion can be effectively transferred to the cutting blade.

EXAMPLE 10.12

If the plate shown in Fig. 10–26a has a density of 8000 kg/m³ and a thickness of 10 mm, determine its mass moment of inertia about an axis perpendicular to the page and passing through the pin at O.

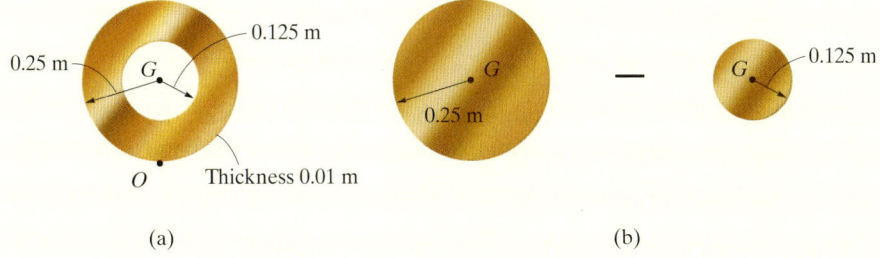

(a)   (b)

Fig. 10–26

SOLUTION

The plate consists of two composite parts, the 250-mm-radius disk *minus* a 125-mm-radius disk, Fig. 10–26b. The moment of inertia about O can be determined by finding the moment of inertia of each of these parts about O and then *algebraically* adding the results. The calculations are performed by using the parallel-axis theorem in conjunction with the mass moment of inertia formula for a circular disk, $I_G = \frac{1}{2}mr^2$, as found on the inside back cover.

Disk. The moment of inertia of a disk about an axis perpendicular to the plane of the disk and passing through G is $I_G = \frac{1}{2}mr^2$. The mass center of both disks is 0.25 m from point O. Thus,

$$m_d = \rho_d V_d = 8000 \text{ kg/m}^3 \, [\pi(0.25 \text{ m})^2(0.01 \text{ m})] = 15.71 \text{ kg}$$

$$(I_O)_d = \frac{1}{2}m_d r_d^2 + m_d d^2$$

$$= \frac{1}{2}(15.71 \text{ kg})(0.25 \text{ m})^2 + (15.71 \text{ kg})(0.25 \text{ m})^2$$

$$= 1.473 \text{ kg} \cdot \text{m}^2$$

Hole. For the smaller disk (hole), we have

$$m_h = \rho_h V_h = 8000 \text{ kg/m}^3 \, [\pi(0.125 \text{ m})^2(0.01 \text{ m})] = 3.93 \text{ kg}$$

$$(I_O)_h = \frac{1}{2}m_h r_h^2 + m_h d^2$$

$$= \frac{1}{2}(3.93 \text{ kg})(0.125 \text{ m})^2 + (3.93 \text{ kg})(0.25 \text{ m})^2$$

$$= 0.276 \text{ kg} \cdot \text{m}^2$$

The moment of inertia of the plate about the pin is therefore

$$I_O = (I_O)_d - (I_O)_h$$

$$= 1.473 \text{ kg} \cdot \text{m}^2 - 0.276 \text{ kg} \cdot \text{m}^2$$

$$= 1.20 \text{ kg} \cdot \text{m}^2 \qquad\qquad Ans.$$

10

EXAMPLE | 10.13

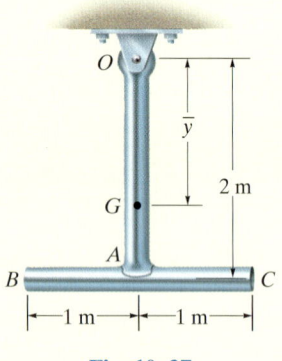

Fig. 10–27

The pendulum in Fig. 10–27 consists of two thin rods each having a mass of 9 kg. Determine the pendulum's mass moment of inertia about an axis passing through (a) the pin at O, and (b) the mass center G of the pendulum.

SOLUTION

Part (a).   Using the table on the inside back cover, the moment of inertia of rod OA about an axis perpendicular to the page and passing through the end point O of the rod is $I_O = \frac{1}{3}ml^2$. Hence,

$$(I_{OA})_O = \frac{1}{3}ml^2 = \frac{1}{3}(9 \text{ kg})(2 \text{ m})^2 = 12 \text{ kg} \cdot \text{m}^2$$

Realize that this same value may be determined using $I_G = \frac{1}{12}ml^2$ and the parallel-axis theorem; i.e.,

$$(I_{OA})_O = \frac{1}{12}ml^2 + md^2 = \frac{1}{12}(9 \text{ kg})(2 \text{ m})^2 + 9 \text{ kg}(1 \text{ m})^2$$

$$= 12 \text{ kg} \cdot \text{m}^2$$

For rod BC we have

$$(I_{BC})_O = \frac{1}{12}ml^2 + md^2 = \frac{1}{12}(9 \text{ kg})(2 \text{ m})^2 + 9 \text{ kg}(2 \text{ m})^2$$

$$= 39 \text{ kg} \cdot \text{m}^2$$

The moment of inertia of the pendulum about O is therefore

$$I_O = 12 + 39 = 51 \text{ kg} \cdot \text{m}^2 \qquad \textit{Ans.}$$

Part (b).   The mass center G will be located relative to the pin at O. Assuming this distance to be $\bar{y}$, Fig. 10–27, and using the formula for determining the mass center, we have

$$\bar{y} = \frac{\Sigma \tilde{y}m}{\Sigma m} = \frac{1(9) + 2(9)}{(9) + (9)} = 1.50 \text{ m}$$

The moment of inertia I_G may be computed in the same manner as I_O, which requires successive applications of the parallel-axis theorem in order to transfer the moments of inertia of rods OA and BC to G. A more direct solution, however, involves applying the parallel-axis theorem using the result for I_O determined above; i.e.,

$$I_O = I_G + md^2; \qquad 51 \text{ kg} \cdot \text{m}^2 = I_G + (18 \text{ kg})(1.50 \text{ m})^2$$

$$I_G = 10.5 \text{ kg} \cdot \text{m}^2 \qquad \textit{Ans.}$$

PROBLEMS

*10–84. Determine the moment of inertia of the thin ring about the z axis. The ring has a mass m.

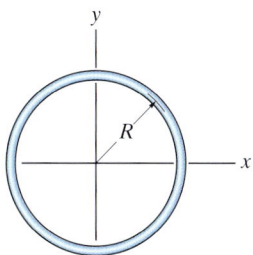

Prob. 10–84

10–85. Determine the moment of inertia of the semi-ellipsoid with respect to the x axis and express the result in terms of the mass m of the semiellipsoid. The material has a constant density ρ.

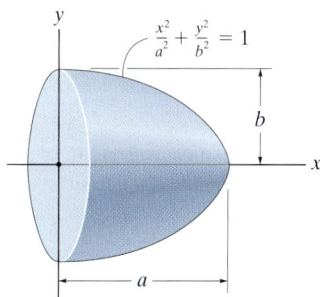

$$\frac{x^2}{a^2} + \frac{y^2}{b^2} = 1$$

Prob. 10–85

10–86. Determine the moment of inertia I_x of the right circular cone and express the result in terms of the total mass m of the cone. The cone has a constant density ρ.

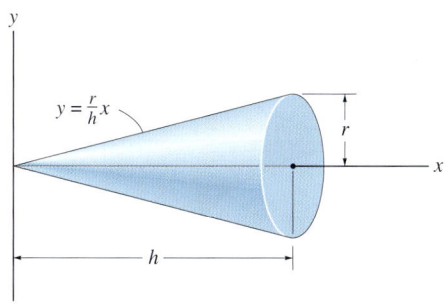

$$y = \frac{r}{h}x$$

Prob. 10–86

10–87. Determine the radius of gyration k_x of the paraboloid. The density of the material is $\rho = 5 \text{ Mg/m}^3$.

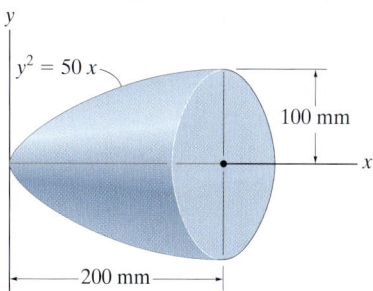

$y^2 = 50\,x$

100 mm

200 mm

Prob. 10–87

*10–88. Determine the moment of inertia of the ellipsoid with respect to the x axis and express the result in terms of the mass m of the ellipsoid. The material has a constant density ρ.

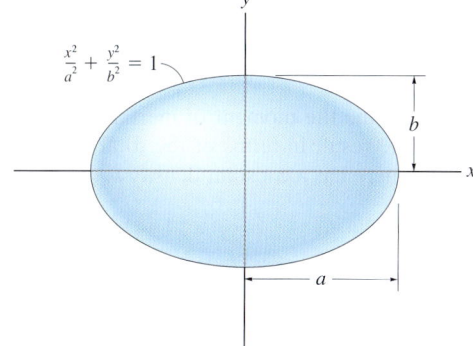

$$\frac{x^2}{a^2} + \frac{y^2}{b^2} = 1$$

Prob. 10–88

10–89. Determine the moment of inertia of the homogenous triangular prism with respect to the y axis. Express the result in terms of the mass m of the prism. *Hint:* For integration, use thin plate elements parallel to the x–y plane having a thickness of dz.

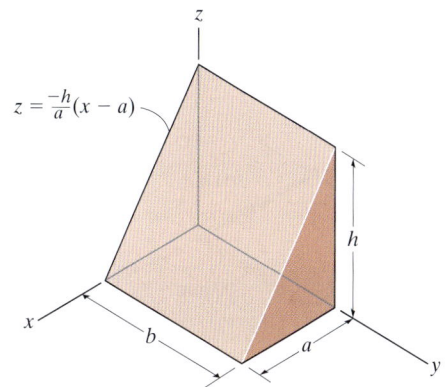

$$z = \frac{-h}{a}(x - a)$$

Prob. 10–89

10

10–90. Determine the mass moment of inertia I_z of the solid formed by revolving the shaded area around the z axis. The density of the materials is ρ. Express the result in terms of the mass m of the solid.

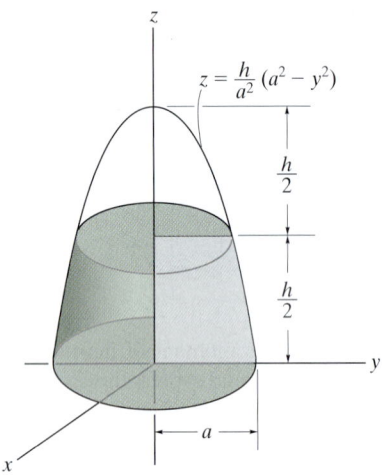

$$z = \frac{h}{a^2}(a^2 - y^2)$$

$\frac{h}{2}$

$\frac{h}{2}$

a

Prob. 10–90

10–91. Determine the moment of inertia I_x of the sphere and express the result in terms of the total mass m of the sphere. The sphere has a constant density ρ.

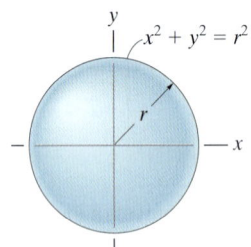

$$x^2 + y^2 = r^2$$

r

Prob. 10–91

*10–92. Determine the mass moment of inertia of the 2-kg bent rod about the z axis.

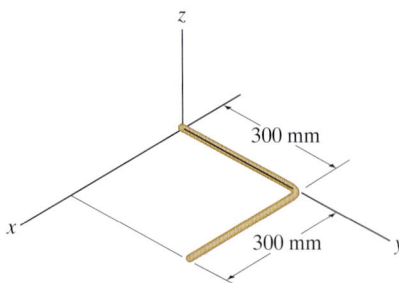

300 mm

300 mm

Prob. 10–92

10–93. Determine the mass moment of inertia I_y of the solid formed by revolving the shaded area around the y axis. The density of the material is ρ. Express the result in terms of the mass m of the solid.

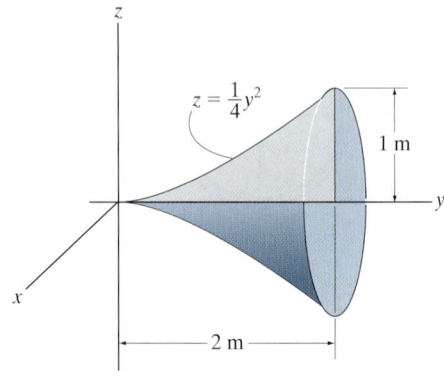

$$z = \frac{1}{4}y^2$$

1 m

2 m

Prob. 10–93

10–94. Determine the mass moment of inertia I_y of the solid formed by revolving the shaded area around the y axis. The total mass of the solid is 1500 kg.

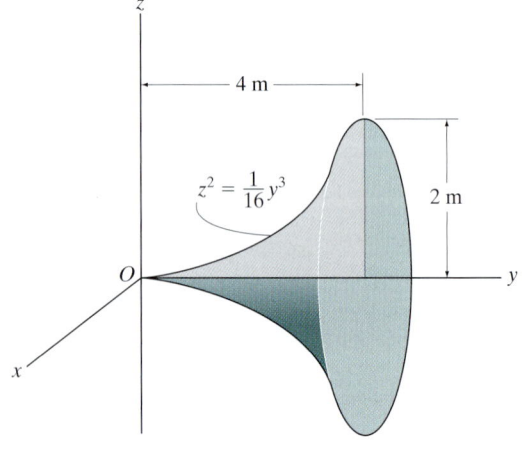

4 m

$$z^2 = \frac{1}{16}y^3$$

2 m

O

Prob. 10–94

10

10–95. Determine the moment of inertia I_z of the frustrum of the cone which has a conical depression. The material has a density of 200 kg/m³.

10–97. The pendulum consists of the 3-kg slender rod and the 5-kg thin plate. Determine the location $\bar{y}$ of the center of mass G of the pendulum; then find the mass moment of inertia of the pendulum about an axis perpendicular to the page and passing through G.

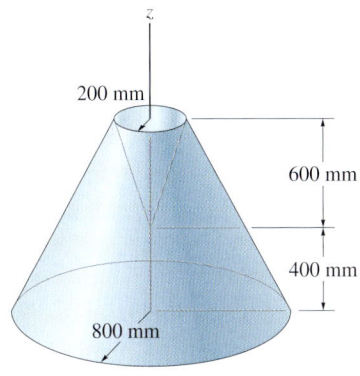

Prob. 10–95

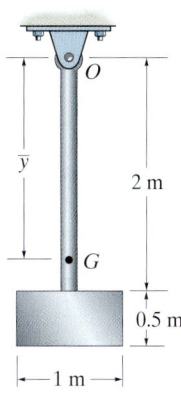

Prob. 10–97

*10–96.** The pendulum consists of a disk having a mass of 6 kg and slender rods AB and DC which have a mass of 2 kg/m. Determine the length L of DC so that the center of the mass is at the bearing O. What is the moment of inertia of the assembly about an axis perpendicular to the page and passing through point O?

10–98. Determine the location $\bar{y}$ of the center of mass G of the assembly and then calculate the moment of inertia about an axis perpendicular to the page and passing through G. The block has a mass of 3 kg and the mass of the semicylinder is 5 kg.

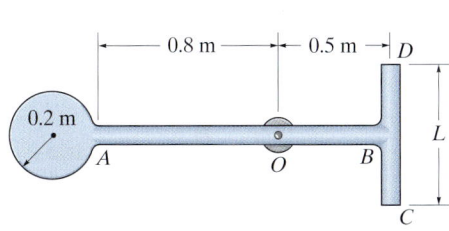

Prob. 10–96

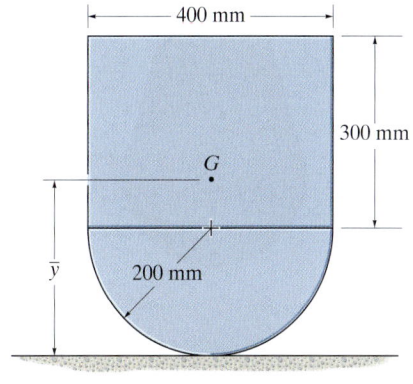

Prob. 10–98

10

10–99.   If the large ring, small ring and each of the spokes weigh 500 N, 75 N, and 100 N, respectively, determine the mass moment of inertia of the wheel about an axis perpendicular to the page and passing through point A.

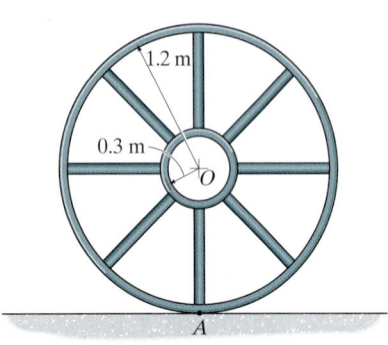

Prob. 10–99

***10–100.**   Determine the mass moment of inertia of the assembly about the z axis. The density of the material is 7.85 Mg/m³.

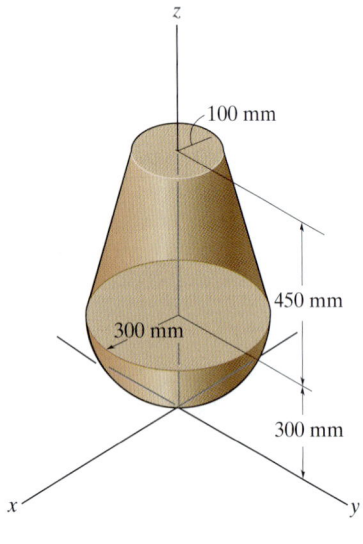

Prob. 10–100

10–101.   Determine the moment of inertia I_z of the frustum of the cone which has a conical depression. The material has a density of 200 kg/m³.

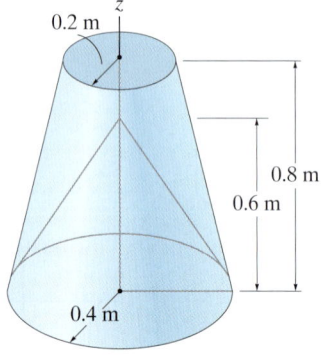

Prob. 10–101

10–102.   The pendulum consists of a plate having a weight of 60 kg and a slender rod having a weight of 20 kg. Determine the radius of gyration of the pendulum about an axis perpendicular to the page and passing through point O.

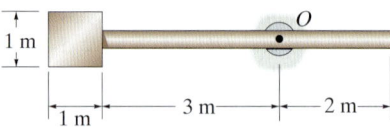

Prob. 10–102

10–103.   The slender rods have a mass of 3 kg/m. Determine the moment of inertia of the assembly about an axis perpendicular to the page and passing through point A.

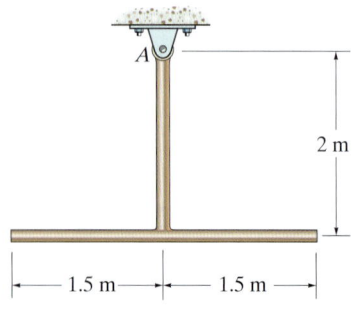

Prob. 10–103

10

*10–104.** Determine the moment of inertia I_z of the frustrum of the cone which has a conical depression. The material has a density of 2000 kg/m^3.

10–106. The thin plate has a mass per unit area of 10 kg/m^2. Determine its mass moment of inertia about the y axis.

10–107. The thin plate has a mass per unit area of 10 kg/m^2. Determine its mass moment of inertia about the z axis.

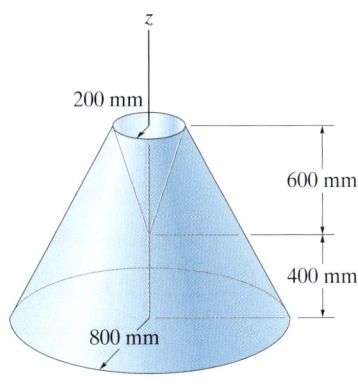

Prob. 10–104

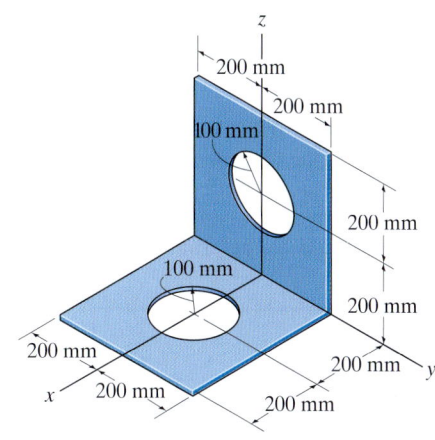

Probs. 10–106/107

10–105. Determine the moment of inertia of the wire triangle about an axis perpendicular to the page and passing through point O. Also, locate the mass center G and determine the moment of inertia about an axis perpendicular to the page and passing through point G. The wire has a mass of 0.3 kg/m. Neglect the size of the ring at O.

*10–108.** Determine the mass moment of inertia of the overhung crank about the x axis. The material is steel having a density of $\rho = 7.85$ Mg/m^3.

10–109. Determine the mass moment of inertia of the overhung crank about the x' axis. The material is steel having a density of $\rho = 7.85$ Mg/m^3.

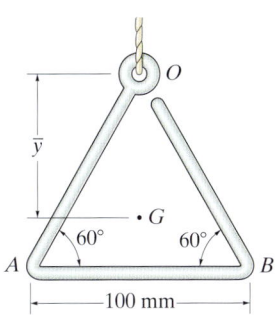

Prob. 10–105

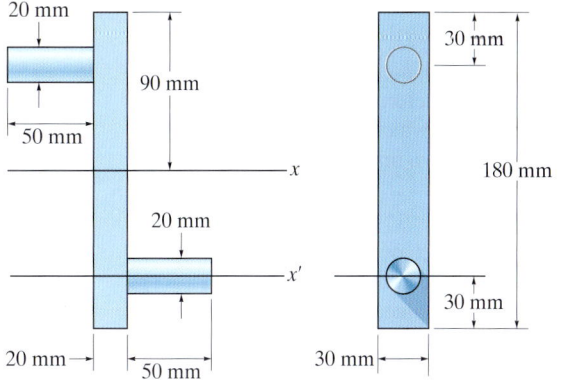

Probs. 10–108/109

CHAPTER REVIEW

Area Moment of Inertia

The *area moment of inertia* represents the second moment of the area about an axis. It is frequently used in formulas related to the strength and stability of structural members or mechanical elements.

If the area shape is irregular but can be described mathematically, then a differential element must be selected and integration over the entire area must be performed to determine the moment of inertia.

$$I_x = \int_A y^2 \, dA$$

$$I_y = \int_A x^2 \, dA$$

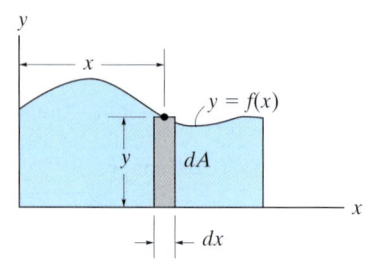

Parallel-Axis Theorem

If the moment of inertia for an area is known about a centroidal axis, then its moment of inertia about a parallel axis can be determined using the parallel-axis theorem.

$$I = \bar{I} + Ad^2$$

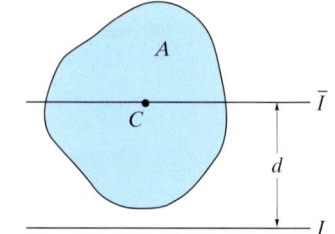

Composite Area

If an area is a composite of common shapes, as found on the inside back cover, then its moment of inertia is equal to the algebraic sum of the moments of inertia of each of its parts.

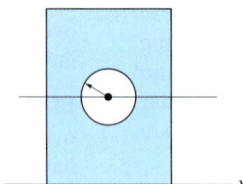
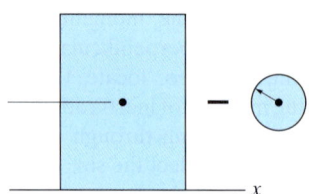

Product of Inertia

The *product of inertia* of an area is used in formulas to determine the orientation of an axis about which the moment of inertia for the area is a maximum or minimum.

If the product of inertia for an area is known with respect to its centroidal x', y' axes, then its value can be determined with respect to any x, y axes using the parallel-axis theorem for the product of inertia.

$$I_{xy} = \int_A xy \, dA$$

$$I_{xy} = \bar{I}_{x'y'} + A d_x d_y$$

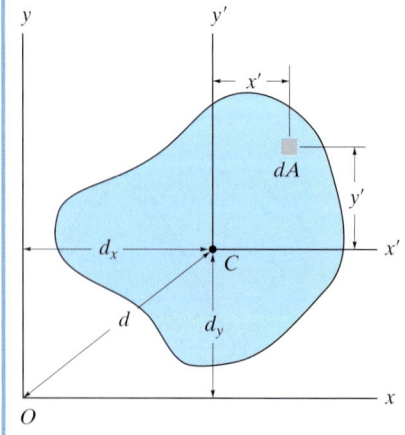

Principal Moments of Inertia

Provided the moments of inertia, I_x and I_y, and the product of inertia, I_{xy}, are known, then the transformation formulas, or Mohr's circle, can be used to determine the maximum and minimum or *principal moments of inertia* for the area, as well as finding the orientation of the principal axes of inertia.

$$I_{\substack{max \\ min}} = \frac{I_x + I_y}{2} \pm \sqrt{\left(\frac{I_x - I_y}{2}\right)^2 + I_{xy}^2}$$

$$\tan 2\theta_p = \frac{-I_{xy}}{(I_x - I_y)/2}$$

Mass Moment of Inertia

The *mass moment of inertia* is a property of a body that measures its resistance to a change in its rotation. It is defined as the "second moment" of the mass elements of the body about an axis.

$$I = \int_m r^2 \, dm$$

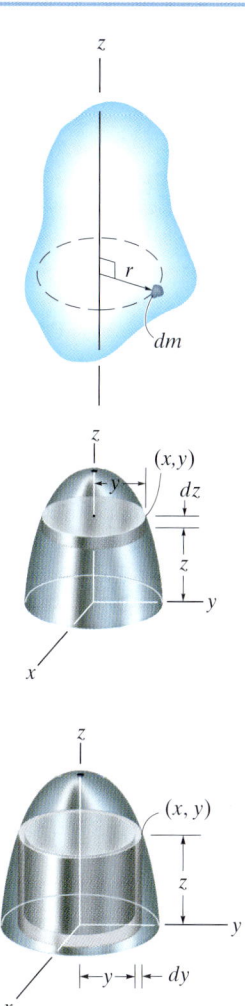

For homogeneous bodies having axial symmetry, the mass moment of inertia can be determined by a single integration, using a disk or shell element.

$$I = \rho \int_V r^2 \, dV$$

The mass moment of inertia of a composite body is determined by using tabular values of its composite shapes, found on the inside back cover, along with the parallel-axis theorem.

$$I = I_G + md^2$$

10

REVIEW PROBLEMS

10–110. Determine the radius of gyration k_x for the column's cross-sectional area.

*10–112. Determine the product of inertia of the shaded area with respect to the x and y axes.

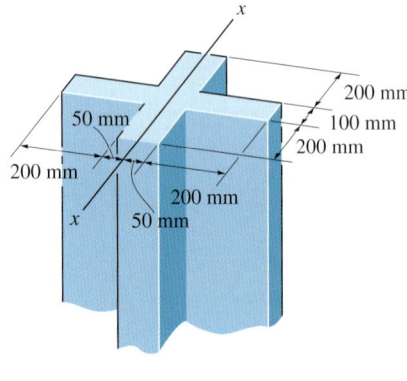

Prob. 10–110

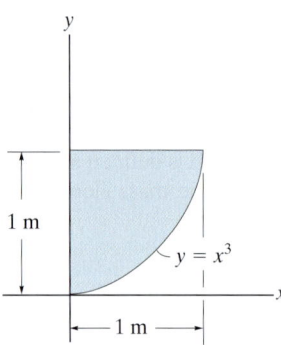

Prob. 10–112

10–111. Determine the area moment of inertia of the area about the x axis. Then, using the parallel-axis theorem, find the area moment of inertia about the x' axis that passes through the centroid C of the area. $\bar{y} = 120$ mm.

10–113. Determine the area moment of inertia of the triangular area about (a) the x axis, and (b) the centroidal x' axis.

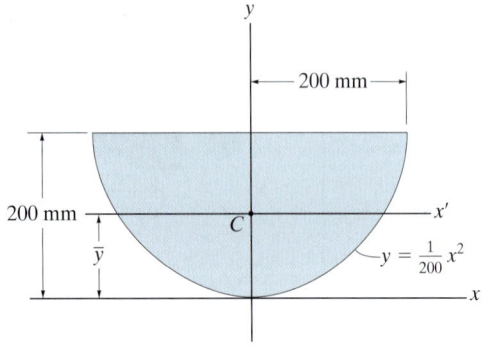

Prob. 10–111

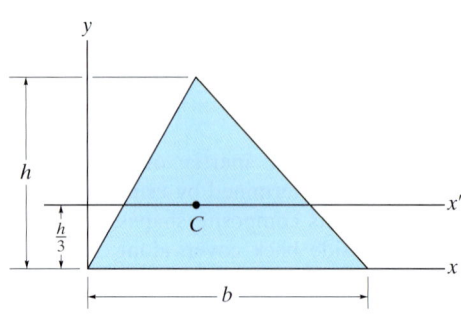

Prob. 10–113

10–114. Determine the mass moment of inertia I_x of the body and express the result in terms of the total mass m of the body. The density is constant.

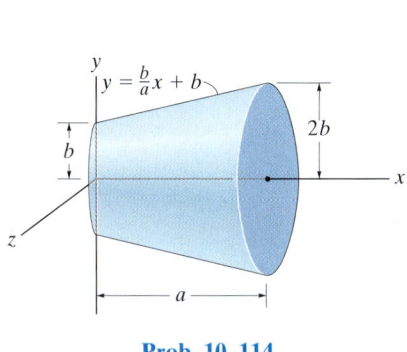

Prob. 10–114

10–115. Determine the moment of inertia for the shaded area about the x axis.

*****10–116.** Determine the moment of inertia for the shaded area about the y axis.

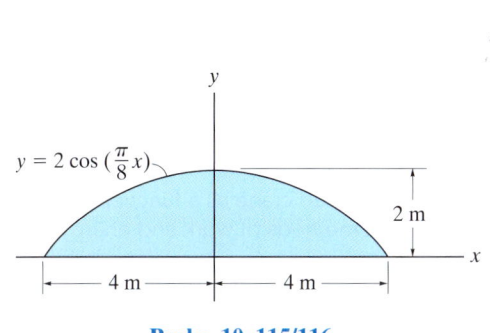

Probs. 10–115/116

10–117. Determine the area moments of inertia I_u and I_v and the product of inertia I_{uv} for the semicircular area.

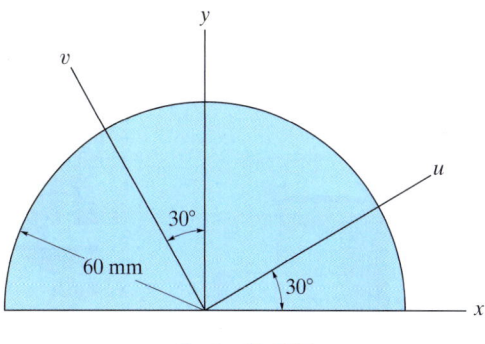

Prob. 10–117

10–118. Determine the area moment of inertia of the beam's cross-sectional area about the x axis which passes through the centroid C.

10–119. Determine the area moment of inertia of the beam's cross-sectional area about the y axis which passes through the centroid C.

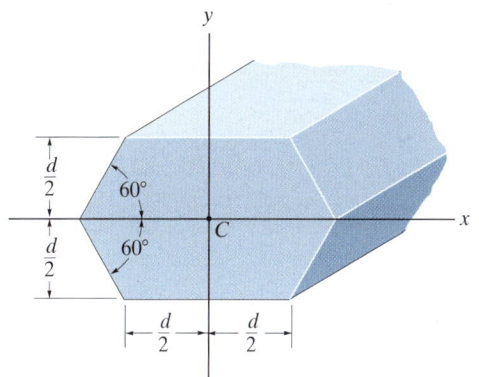

Probs. 10–118/119

10

Chapter 11

Equilibrium and stability of this articulated crane as a function of its position can be determined using the methods of work and energy, which are explained in this chapter.

Virtual Work

Video Solutions are available for selected questions in this chapter.

11.1   Definition of Work

The *principle of virtual work* was proposed by the Swiss mathematician Jean Bernoulli in the eighteenth century. It provides an alternative method for solving problems involving the equilibrium of a particle, a rigid body, or a system of connected rigid bodies. Before we discuss this principle, however, we must first define the work produced by a force and by a couple moment.

Work of a Force.   A force does work when it undergoes a displacement in the direction of its line of action. Consider, for example, the force $\mathbf{F}$ in Fig. 11–1a that undergoes a differential displacement $d\mathbf{r}$. If θ is the angle between the force and the displacement, then the component of $\mathbf{F}$ in

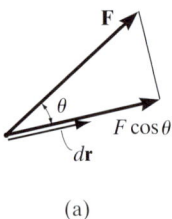

(a)

the direction of the displacement is $F \cos \theta$. And so the work produced by **F** is

$$dU = F \, dr \cos \theta$$

Notice that this expression is also the product of the force F and the component of displacement in the direction of the force, $dr \cos \theta$, Fig. 11–1*b*. If we use the definition of the dot product (Eq. 2–11) the work can also be written as

$$dU = \mathbf{F} \cdot d\mathbf{r}$$

(b)

Fig. 11–1

As the above equations indicate, work is a *scalar*, and like other scalar quantities, it has a magnitude that can either be *positive* or *negative*.

In the SI system, the unit of work is a *joule* (J), which is the work produced by a 1-N force that displaces through a distance of 1 m in the direction of the force ($1 \, \text{J} = 1 \, \text{N} \cdot \text{m}$).

The moment of a force has this same combination of units; however, the concepts of moment and work are in no way related. A moment is a vector quantity, whereas work is a scalar.

Work of a Couple Moment.

The rotation of a couple moment also produces work. Consider the rigid body in Fig. 11–2, which is acted upon by the couple forces **F** and **–F** that produce a couple moment **M** having a magnitude $M = Fr$. When the body undergoes the differential displacement shown, points A and B move $d\mathbf{r}_A$ and $d\mathbf{r}_B$ to their final positions A' and B', respectively. Since $d\mathbf{r}_B = d\mathbf{r}_A + d\mathbf{r}'$, this movement can be thought of as a *translation* $d\mathbf{r}_A$, where A and B move to A' and B'', and a *rotation* about A', where the body rotates through the angle $d\theta$ about A. The couple forces do no work during the translation $d\mathbf{r}_A$ because each force undergoes the same amount of displacement in opposite directions, thus canceling out the work. During rotation, however, **F** is displaced $dr' = r \, d\theta$, and so it does work $dU = F \, dr' = F r \, d\theta$. Since $M = Fr$, the work of the couple moment **M** is therefore

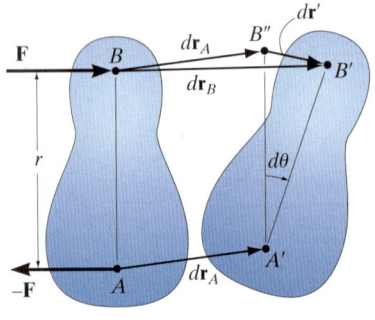

Fig. 11–2

$$dU = M d\theta$$

If **M** and $d\theta$ have the same sense, the work is *positive*; however, if they have the opposite sense, the work will be *negative*.

11

Virtual Work.   The definitions of the work of a force and a couple have been presented in terms of *actual movements* expressed by differential displacements having magnitudes of *dr* and *dθ*. Consider now an *imaginary* or *virtual movement* of a body in static equilibrium, which indicates a displacement or rotation that is *assumed* and *does not actually exist*. These movements are first-order differential quantities and will be denoted by the symbols δr and $\delta \theta$ (delta *r* and delta *θ*), respectively. The *virtual work* done by a force having a virtual displacement δr is

$$\delta U = F \cos \theta \, \delta r \qquad (11\text{--}1)$$

Similarly, when a couple undergoes a virtual rotation $\delta \theta$ in the plane of the couple forces, the *virtual work* is

$$\delta U = M \, \delta \theta \qquad (11\text{--}2)$$

11.2   Principle of Virtual Work

The principle of virtual work states that if a body is in equilibrium, then the algebraic sum of the virtual work done by all the forces and couple moments acting on the body is zero for any virtual displacement of the body. Thus,

$$\delta U = 0 \qquad (11\text{--}3)$$

For example, consider the free-body diagram of the particle (ball) that rests on the floor, Fig. 11–3. If we "imagine" the ball to be displaced downwards a virtual amount δy, then the weight does positive virtual work, $W \, \delta y$, and the normal force does negative virtual work, $-N \, \delta y$. For equilibrium the total virtual work must be zero, so that $\delta U = W \, \delta y - N \, \delta y = (W - N) \, \delta y = 0$. Since $\delta y \neq 0$, then $N = W$ as required by applying $\Sigma F_y = 0$.

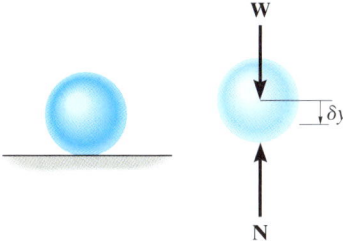

Fig. 11–3

11

In a similar manner, we can also apply the virtual-work equation $\delta U = 0$ to a rigid body subjected to a coplanar force system. Here, separate virtual translations in the x and y directions, and a virtual rotation about an axis perpendicular to the x–y plane that passes through an arbitrary point O, will correspond to the three equilibrium equations, $\Sigma F_x = 0$, $\Sigma F_y = 0$, and $\Sigma M_O = 0$. When writing these equations, it is *not necessary* to include the work done by the *internal forces* acting within the body since a rigid body *does not deform* when subjected to an external loading, and furthermore, when the body moves through a virtual displacement, the internal forces occur in equal but opposite collinear pairs, so that the corresponding work done by each pair of forces will cancel.

To demonstrate an application, consider the simply supported beam in Fig. 11–4a. When the beam is given a virtual rotation $\delta\theta$ about point B, Fig. 11–4b, the only forces that do work are $\mathbf{P}$ and $\mathbf{A}_y$. Since $\delta y = l\,\delta\theta$ and $\delta y' = (l/2)\,\delta\theta$, the virtual work equation for this case is $\delta U = A_y(l\,\delta\theta) - P(l/2)\,\delta\theta = (A_y l - Pl/2)\,\delta\theta = 0$. Since $\delta\theta \neq 0$, then $A_y = P/2$. Excluding $\delta\theta$, notice that the terms in parentheses actually represent the application of $\Sigma M_B = 0$.

As seen from the above two examples, no added advantage is gained by solving particle and rigid-body equilibrium problems using the principle of virtual work. This is because for each application of the virtual-work equation, the virtual displacement, common to every term, factors out, leaving an equation that could have been obtained in a more *direct manner* by simply applying an equation of equilibrium.

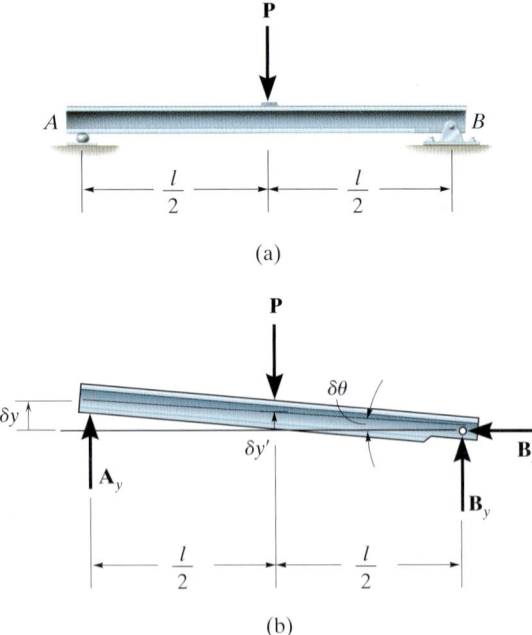

(a)

(b)

Fig. 11–4

11.3   Principle of Virtual Work for a System of Connected Rigid Bodies

The method of virtual work is particularly effective for solving equilibrium problems that involve a system of several *connected* rigid bodies, such as the ones shown in Fig. 11–5.

Each of these systems is said to have only one degree of freedom since the arrangement of the links can be completely specified using only one coordinate θ. In other words, with this single coordinate and the length of the members, we can locate the position of the forces **F** and **P**.

In this text, we will only consider the application of the principle of virtual work to systems containing one degree of freedom.* Because they are less complicated, they will serve as a way to approach the solution of more complex problems involving systems with many degrees of freedom. The procedure for solving problems involving a system of frictionless connected rigid bodies follows.

Please refer to the Companion Website for the animation: *Principle of Virtual Work*

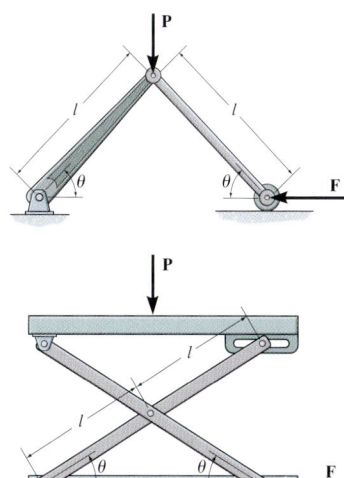

Fig. 11–5

Important Points

- A force does work when it moves through a displacement in the direction of the force. A couple moment does work when it moves through a collinear rotation. Specifically, positive work is done when the force or couple moment and its displacement have the same sense of direction.

- The principle of virtual work is generally used to determine the equilibrium configuration for a system of multiple connected members.

- A virtual displacement is imaginary; i.e., it does not really happen. It is a differential displacement that is given in the positive direction of a position coordinate.

- Forces or couple moments that do not virtually displace do no virtual work.

This scissors lift has one degree of freedom. Without the need for dismembering the mechanism, the force in the hydraulic cylinder *AB* required to provide the lift can be determined *directly* by using the principle of virtual work.

---

*This method of applying the principle of virtual work is sometimes called the *method of virtual displacements* because a virtual displacement is applied, resulting in the calculation of a real force. Although it is not used here, we can also apply the principle of virtual work as a *method of virtual forces*. This method is often used to apply a virtual force and then determine the displacements of points on deformable bodies. See R. C. Hibbeler, *Mechanics of Materials*, 8th edition, Pearson/Prentice Hall, 2011.

Procedure for Analysis

Free-Body Diagram.

- Draw the free-body diagram of the entire system of connected bodies and define the *coordinate q*.

- Sketch the "deflected position" of the system on the free-body diagram when the system undergoes a *positive virtual* displacement δq.

Virtual Displacements.

- Indicate *position coordinates s*, each measured from a *fixed point* on the free-body diagram. These coordinates are directed to the forces that do work.

- Each of these coordinate axes should be *parallel* to the line of action of the force to which it is directed, so that the virtual work along the coordinate axis can be calculated.

- Relate each of the position coordinates s to the coordinate q; then *differentiate* these expressions in order to express each virtual displacement δs in terms of δq.

Virtual-Work Equation.

- Write the *virtual-work equation* for the system assuming that, whether possible or not, each position coordinate s undergoes a *positive* virtual displacement δs. If a force or couple moment is in the same direction as the positive virtual displacement, the work is positive. Otherwise, it is negative.

- Express the work of *each* force and couple moment in the equation in terms of δq.

- Factor out this common displacement from all the terms, and solve for the unknown force, couple moment, or equilibrium position q.

11

EXAMPLE 11.1

Determine the angle θ for equilibrium of the two-member linkage shown in Fig. 11–6a. Each member has a mass of 10 kg.

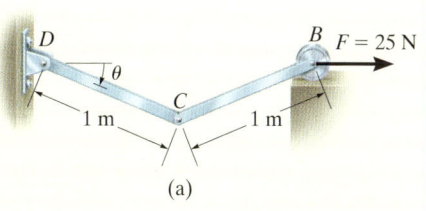

(a)

SOLUTION

Free-Body Diagram. The system has only one degree of freedom since the location of both links can be specified by the single coordinate, $(q =)\, \theta$. As shown on the free-body diagram in Fig. 11–6b, when θ has a *positive* (clockwise) virtual rotation $\delta\theta$, only the force **F** and the two 98.1-N weights do work. (The reactive forces D_x and D_y are fixed, and B_y does not displace along its line of action.)

Virtual Displacements. If the origin of coordinates is established at the *fixed* pin support D, then the position of **F** and **W** can be specified by the *position coordinates* x_B and y_w. In order to determine the work, note that, as required, these coordinates are parallel to the lines of action of their associated forces. Expressing these position coordinates in terms of θ and taking the derivatives yields

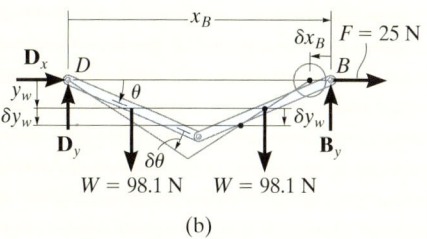

(b)

Fig. 11–6

$$x_B = 2(1 \cos \theta)\ \text{m} \quad \delta x_B = -2 \sin \theta\ \delta\theta\ \text{m} \quad (1)$$
$$y_w = \tfrac{1}{2}(1 \sin \theta)\ \text{m} \quad \delta y_w = 0.5 \cos \theta\ \delta\theta\ \text{m} \quad (2)$$

It is seen by the *signs* of these equations, and indicated in Fig. 11–6b, that an *increase* in θ (i.e., $\delta\theta$) causes a *decrease* in x_B and an *increase* in y_w.

Virtual-Work Equation. If the virtual displacements δx_B and δy_w were *both positive*, then the forces **W** and **F** would do positive work since the forces and their corresponding displacements would have the same sense. Hence, the virtual-work equation for the displacement $\delta\theta$ is

$$\delta U = 0; \qquad W\,\delta y_w + W\,\delta y_w + F\,\delta x_B = 0 \qquad (3)$$

Substituting Eqs. 1 and 2 into Eq. 3 in order to relate the virtual displacements to the common virtual displacement $\delta\theta$ yields

$$98.1(0.5 \cos \theta\ \delta\theta) + 98.1(0.5 \cos \theta\ \delta\theta) + 25(-2 \sin \theta\ \delta\theta) = 0$$

Notice that the "negative work" done by **F** (force in the opposite sense to displacement) has actually been *accounted for* in the above equation by the "negative sign" of Eq. 1. Factoring out the *common displacement* $\delta\theta$ and solving for θ, noting that $\delta\theta \neq 0$, yields

$$(98.1 \cos \theta - 50 \sin \theta)\,\delta\theta = 0$$

$$\theta = \tan^{-1}\frac{98.1}{50} = 63.0° \qquad \textit{Ans.}$$

NOTE: If this problem had been solved using the equations of equilibrium, it would be necessary to dismember the links and apply three scalar equations to *each* link. The principle of virtual work, by means of calculus, has eliminated this task so that the answer is obtained directly.

11

EXAMPLE | 11.2

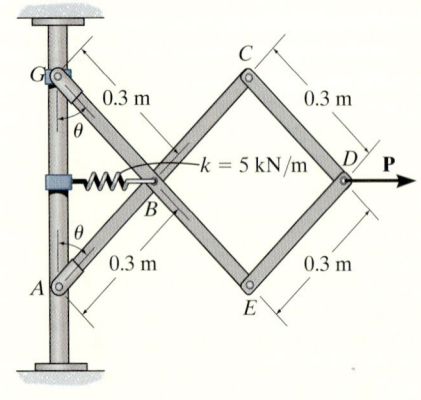

(a)

Fig. 11–7

Determine the required force P in Fig. 11–7a needed to maintain equilibrium of the scissors linkage when $\theta = 60°$. The spring is unstretched when $\theta = 30°$. Neglect the mass of the links.

SOLUTION

Free-Body Diagram. Only $\mathbf{F}_s$ and $\mathbf{P}$ do work when θ undergoes a *positive* virtual displacement $\delta\theta$, Fig. 11–7b. For the arbitrary position θ, the spring is stretched $(0.3 \text{ m}) \sin\theta - (0.3 \text{ m}) \sin 30°$, so that

$$F_s = ks = 5000 \text{ N/m} [(0.3 \text{ m}) \sin\theta - (0.3 \text{ m}) \sin 30°]$$

$$= (1500 \sin\theta - 750) \text{ N}$$

Virtual Displacements. The position coordinates, x_B and x_D, measured from the *fixed point A*, are used to locate $\mathbf{F}_s$ and $\mathbf{P}$. These coordinates are parallel to the line of action of their corresponding forces. Expressing x_B and x_D in terms of the angle θ using trigonometry,

$$x_B = (0.3 \text{ m}) \sin\theta$$

$$x_D = 3[(0.3 \text{ m}) \sin\theta] = (0.9 \text{ m}) \sin\theta$$

Differentiating, we obtain the virtual displacements of points B and D.

$$\delta x_B = 0.3 \cos\theta \, \delta\theta \qquad (1)$$

$$\delta x_D = 0.9 \cos\theta \, \delta\theta \qquad (2)$$

Virtual-Work Equation. Force $\mathbf{P}$ does positive work since it acts in the positive sense of its virtual displacement. The spring force $\mathbf{F}_s$ does negative work since it acts opposite to its positive virtual displacement. Thus, the virtual-work equation becomes

$$\delta U = 0; \qquad -F_s \, \delta x_B + P \delta x_D = 0$$

$$-[1500 \sin\theta - 750] (0.3 \cos\theta \, \delta\theta) + P (0.9 \cos\theta \, \delta\theta) = 0$$

$$[0.9P + 225 - 450 \sin\theta] \cos\theta \, \delta\theta = 0$$

Since $\cos\theta \, \delta\theta \neq 0$, then this equation requires

$$P = 500 \sin\theta - 250$$

When $\theta = 60°$,

$$P = 500 \sin 60° - 250 = 183 \text{ N} \qquad \textit{Ans.}$$

EXAMPLE | 11.3

If the box in Fig. 11–8a has a mass of 10 kg, determine the couple moment M needed to maintain equilibrium when $\theta = 60°$. Neglect the mass of the members.

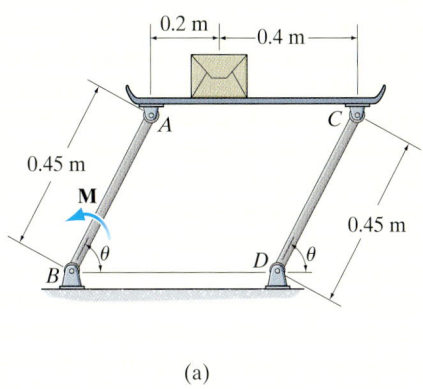

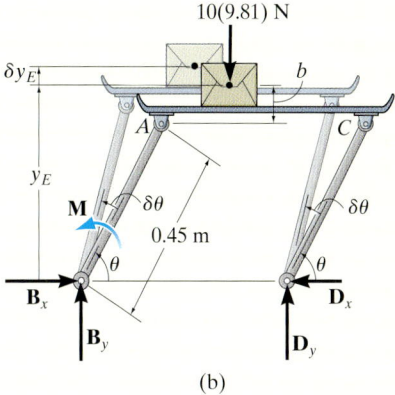

(a)   (b)

Fig. 11–8

SOLUTION

Free-Body Diagram.   When θ undergoes a positive virtual displacement $\delta\theta$, only the couple moment $\mathbf{M}$ and the weight of the box do work, Fig. 11–8b.

Virtual Displacements.   The position coordinate y_E, measured from the *fixed point B*, locates the weight, 10(9.81) N. Here,

$$y_E = (0.45 \text{ m}) \sin \theta + b$$

where b is a constant distance. Differentiating this equation, we obtain

$$\delta y_E = 0.45 \text{ m} \cos \theta \; \delta\theta \qquad (1)$$

Virtual-Work Equation.   The virtual-work equation becomes

$$\delta U = 0; \qquad\qquad M \, \delta\theta - [10(9.81) \text{ N}]\delta y_E = 0$$

Substituting Eq. 1 into this equation

$$M \, \delta\theta - 10(9.81) \text{ N}(0.45 \text{ m} \cos \theta \; \delta\theta) = 0$$
$$\delta\theta(M - 44.145 \cos \theta) = 0$$

Since $\delta\theta \neq 0$, then

$$M - 44.145 \cos \theta = 0$$

Since it is required that $\theta = 60°$, then

$$M = 44.145 \cos 60° = 22.1 \text{ N} \cdot \text{m} \qquad\qquad Ans.$$

11

EXAMPLE 11.4

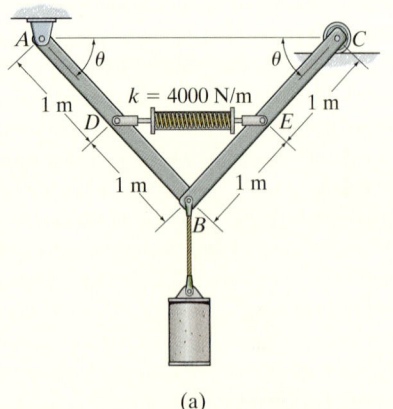

(a)

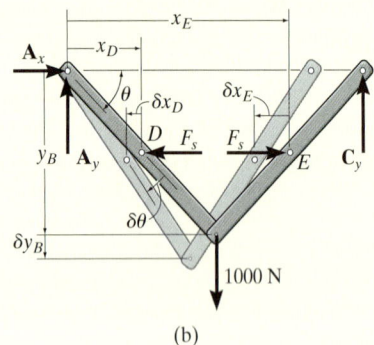

(b)

Fig. 11–9

The mechanism in Fig. 11–9a supports the 50-lb cylinder. Determine the angle θ for equilibrium if the spring has an unstretched length of 2 m when $\theta = 0°$. Neglect the mass of the members.

SOLUTION

Free-Body Diagram.   When the mechanism undergoes a positive virtual displacement $\delta\theta$, Fig. 11–9b, only $\mathbf{F}_s$ and the 1000-N force do work. Since the final length of the spring is $2(1 \text{ m} \cos \theta)$, then

$$F_s = ks = (4000 \text{ N/m})(2 \text{ m} - 2 \text{ m} \cos \theta) = (8000 - 8000 \cos \theta) \text{ N}$$

Virtual Displacements.   The position coordinates x_D and x_E are established from the *fixed point* A to locate $\mathbf{F}_s$ at D and at E. The coordinate y_B, also measured from A, specifies the position of the 50-lb force at B. The coordinates can be expressed in terms of θ using trigonometry.

$$x_D = (1 \text{ m}) \cos \theta$$

$$x_E = 3[(1 \text{ m}) \cos \theta] = (3 \text{ m}) \cos \theta$$

$$y_B = (2 \text{ m}) \sin \theta$$

Differentiating, we obtain the virtual displacements of points D, E, and B as

$$\delta x_D = -1 \sin \theta \, \delta\theta \tag{1}$$

$$\delta x_E = -3 \sin \theta \, \delta\theta \tag{2}$$

$$\delta y_B = 2 \cos \theta \, \delta\theta \tag{3}$$

Virtual-Work Equation.   The virtual-work equation is written as if all virtual displacements are positive, thus

$$\delta U = 0; \qquad F_s \, \delta x_E + 1000 \, \delta y_B - F_s \, \delta x_D = 0$$

$$(8000 - 8000 \cos \theta)(-3 \sin \theta \, \delta\theta) + 1000(2 \cos \theta \, \delta\theta)$$

$$- (8000 - 8000 \cos \theta)(-1 \sin\theta \, \delta\theta) = 0$$

$$\delta\theta(1600 \sin \theta \cos \theta - 1600 \sin \theta + 2000 \cos \theta) = 0$$

Since $\delta\theta \neq 0$, then

$$1600 \sin \theta \cos \theta - 1600 \sin \theta + 2000 \cos \theta = 0$$

Solving by trial and error,

$$\theta = 34.9° \qquad\qquad Ans.$$

11

FUNDAMENTAL PROBLEMS

F11–1. Determine the required magnitude of force **P** to maintain equilibrium of the linkage at $\theta = 60°$. Each link has a mass of 20 kg.

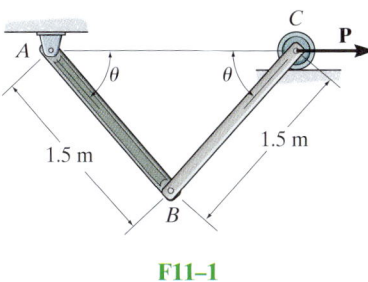

F11–1

F11–2. Determine the magnitude of force **P** required to hold the 50-kg smooth rod in equilibrium at $\theta = 60°$.

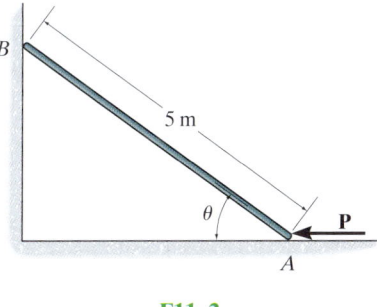

F11–2

F11–3. The linkage is subjected to a force of $P = 2$ kN. Determine the angle θ for equilibrium. The spring is unstretched when $\theta = 0°$. Neglect the mass of the links.

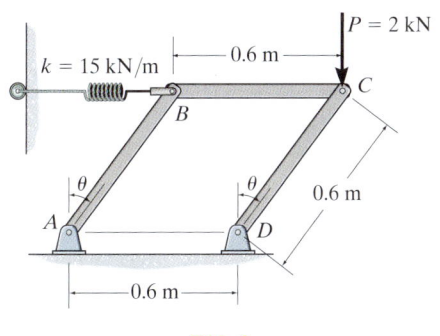

F11–3

F11–4. The linkage is subjected to a force of $P = 6$ kN. Determine the angle θ for equilibrium. The spring is unstretched at $\theta = 60°$. Neglect the mass of the links.

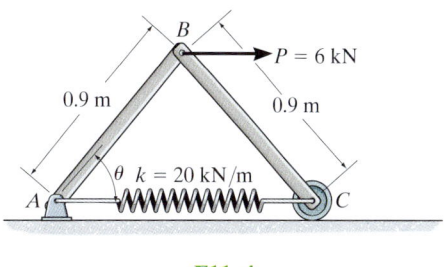

F11–4

F11–5. Determine the angle θ where the 50-kg bar is in equilibrium. The spring is unstretched at $\theta = 60°$.

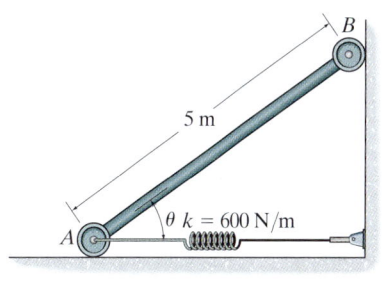

F11–5

F11–6. The scissors linkage is subjected to a force of $P = 150$ N. Determine the angle θ for equilibrium. The spring is unstretched at $\theta = 0°$. Neglect the mass of the links

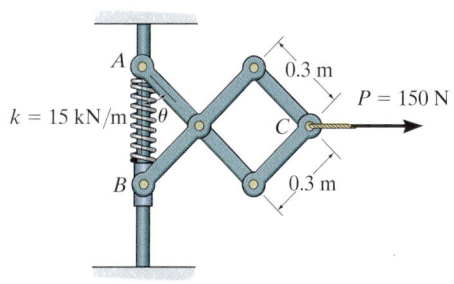

F11–6

11

PROBLEMS

11–1. The scissors jack supports a load **P**. Determine the axial force in the screw necessary for equilibrium when the jack is in the position θ. Each of the four links has a length $2L$ and is pin-connected at its center. Points B and D can move horizontally.

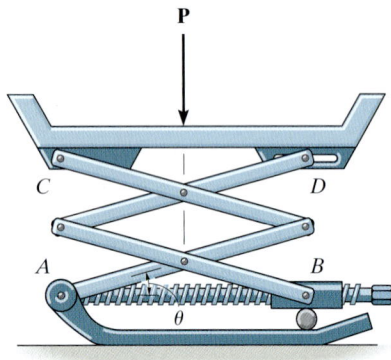

Prob. 11–1

11–2. Determine the force **F** acting on the cord which is required to maintain equilibrium of the horizontal 10-kg bar AB. *Hint:* Express the total *constant vertical length l* of the cord in terms of position coordinates s_1 and s_2. The derivative of this equation yields a relationship between δ_1 and δ_2.

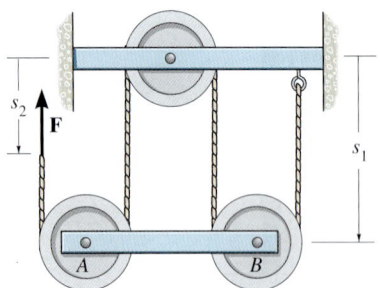

Prob. 11–2

11–3. The machine shown is used for forming metal plates. It consists of two toggles ABC and DEF, which are operated by hydraulic cylinder BE. The toggles push the moveable bar FC forward, pressing the plate p into the cavity. If the force which the plate exerts on the head is $P = 8$ kN, determine the force **F** in the hydraulic cylinder when $\theta = 30°$.

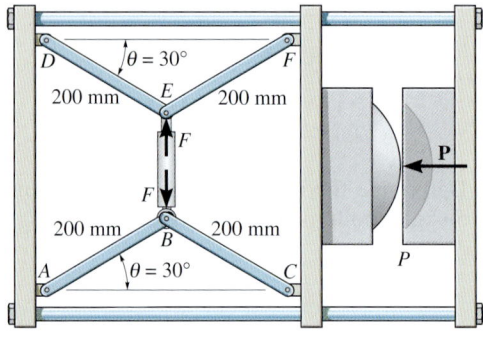

Prob. 11–3

*11–4.** The spring has an unstretched length of 0.3 m. Determine the angle θ for equilibrium if the uniform links each have a mass of 5 kg.

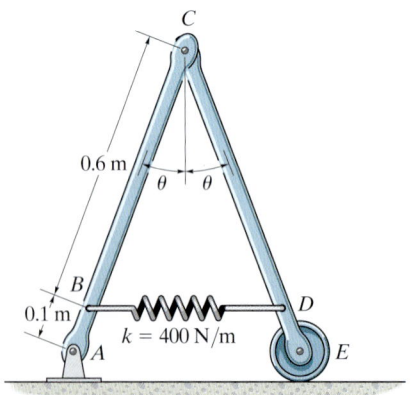

Prob. 11–4

11–5. Each member of the pin-connected mechanism has a mass of 8 kg. If the spring is unstretched when $\theta = 0°$, determine the angle θ for equilibrium. Set $k = 2500$ N/m and $M = 50$ N·m.

11–6. Each member of the pin-connected mechanism has a mass of 8 kg. If the spring is unstretched when $\theta = 0°$, determine the required stiffness k so that the mechanism is in equilibrium when $\theta = 30°$. Set $M = 0$.

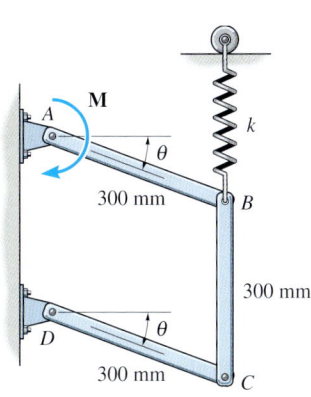

Probs. 11–5/6

11–7. The scissors jack supports a load **P**. Determine the axial force in the screw necessary for equilibrium when the jack is in the position θ. Each of the four links has a length L and is pin-connected at its center. Points B and D can move horizontally.

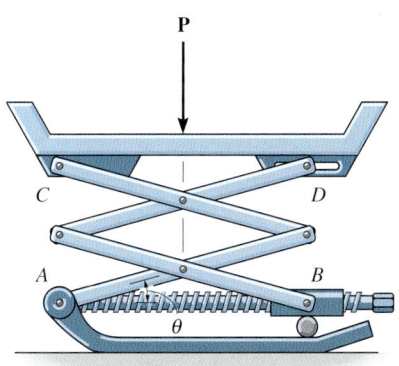

Prob. 11–7

*11–8.** The toggle joint is subjected to the load **P**. Determine the compressive force F it creates on the cylinder at A as a function of θ.

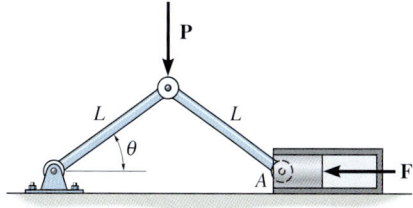

Prob. 11–8

11–9. Determine the force F needed to lift the block having a mass of 50 kg. *Hint:* Note that the coordinates s_A and s_B can be related to the *constant* vertical length l of the cord.

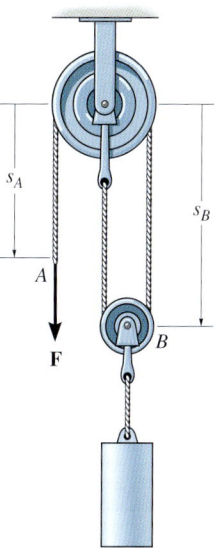

Prob. 11–9

11

11–10. The thin rod of weight W rest against the smooth wall and floor. Determine the magnitude of force **P** needed to hold it in equilibrium for a given angle θ.

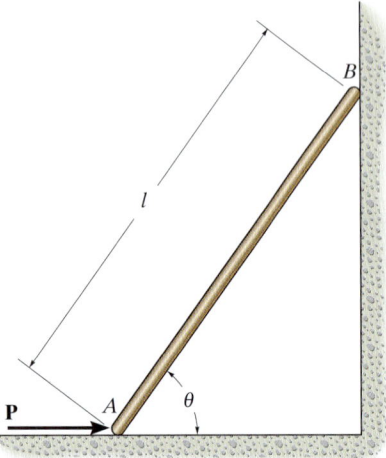

Prob. 11–10

11–11. Determine the force **F** acting on the cord which is required to maintain equilibrium of the horizontal 20-kg bar AB. *Hint:* Express the total *constant vertical length l* of the cord in terms of position coordinates s_1 and s_2. The derivative of this equation yields a relationship between δ_1 and δ_2.

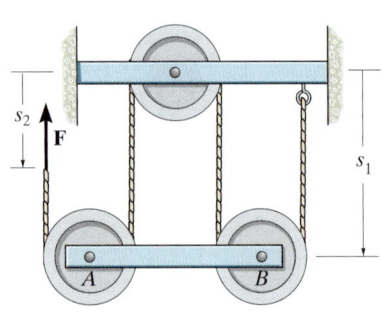

Prob. 11–11

****11–12.** The assembly is used for exercise. It consist of four pin-connected bars, each of length L, and a spring of stiffness k and unstretched length a ($<2L$). If horizontal forces **P** and $-$**P** are applied to the handles so that θ is slowly decreased, determine the angle θ at which the magnitude of **P** becomes a maximum.

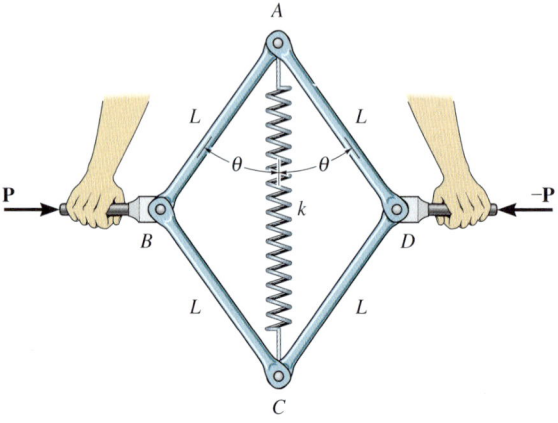

Prob. 11–12

11–13. Each member of the pin-connected mechanism has a mass of 8 kg. If the spring is unstretched when $\theta = 0°$, determine the angle θ for equilibrium. Set $k = 3000 \text{ N/m}$ and $M = 100 \text{ N} \cdot \text{m}$.

11–14. Each member of the pin-connected mechanism has a mass of 8 kg. If the spring is unstretched when $\theta = 0°$, determine the required stiffness k so that the mechanism is in equilibrium when $\theta = 25°$. Set **M** $= 0$.

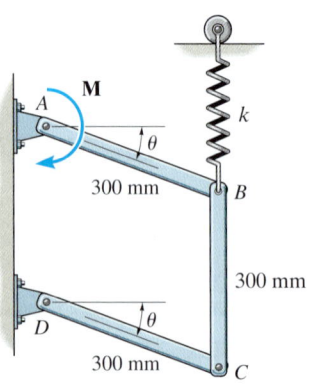

Probs. 11–13/14

11–15. The service window at a fast-food restaurant consists of glass doors that open and close automatically using a motor which supplies a torque **M** to each door. The far end, A and B, move along the horizontal guides. If a food tray becomes stuck between the doors as shown, determine the horizontal force the doors exert on the tray at the position θ.

11–17. If the spring has a torsional stiffness of $k = 300$ N·m/rad and it is unstretched when $\theta = 90°$, determine the angle θ when the frame is in equilibrium.

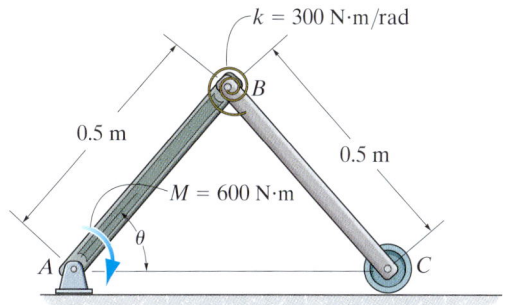

Prob. 11–17

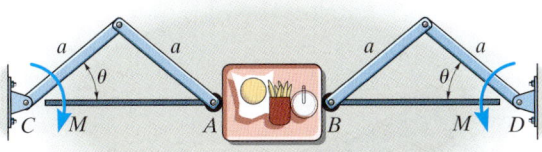

Prob. 11–15

11–18. A 5-kg uniform serving table is supported on each side by pairs of two identical links, AB and CD, and springs CE. If the bowl has a mass of 1 kg, determine the angle θ where the table is in equilibrium. The springs each have a stiffness of $k = 400$ N/m and are unstretched when $\theta = 90°$. Neglect the mass of the links.

11–19. A 5-kg uniform serving table is supported on each side by two pairs of identical links, AB and CD, and springs CE. If the bowl has a mass of 1 kg and is in equilibrium when $\theta = 60°$, determine the stiffness k of each spring. The springs are unstretched when $\theta = 90°$. Neglect the mass of the links.

*11–16. If the spring is unstretched when $\theta = 30°$, the mass of the cylinder is 25 kg, and the mechanism is in equilibrium when $\theta = 45°$, determine the stiffness k of the spring. Rod AB slides freely through the collar at A. Neglect the mass of the rods.

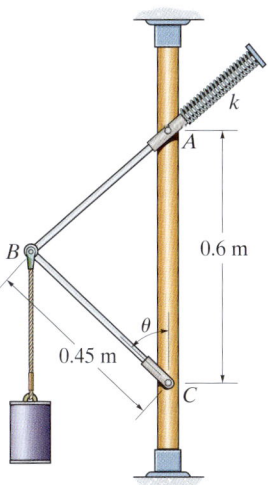

Prob. 11–16

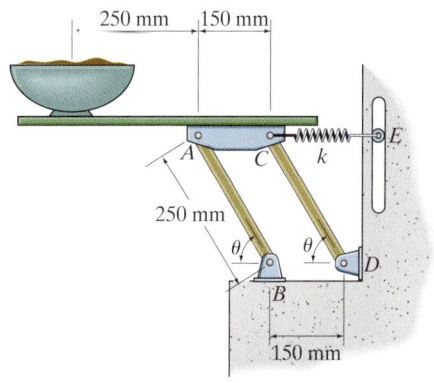

Probs. 11–18/19

11

*11–20. The "Nuremberg scissors" is subjected to a horizontal force of $P = 600$ N. Determine the angle θ for equilibrium. The spring has a stiffness of $k = 15$ kN/m and is unstretched when $\theta = 15°$.

11–21. The "Nuremberg scissors" is subjected to a horizontal force of $P = 600$ N. Determine the stiffness k of the spring for equilibrium when $\theta = 60°$. The spring is unstretched when $\theta = 15°$.

11–23. The crankshaft is subjected to a torque of $M = 50$ N·m. Determine the horizontal compressive force F applied to the piston for equilibrium when $\theta = 60°$.

*11–24. The crankshaft is subjected to a torque of $M = 50$ N·m. Determine the horizontal compressive force F and plot the result of F (ordinate) versus θ (abscissa) for $0° \le \theta \le 90°$.

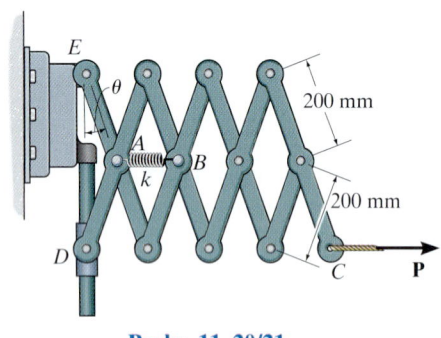

Probs. 11–20/21

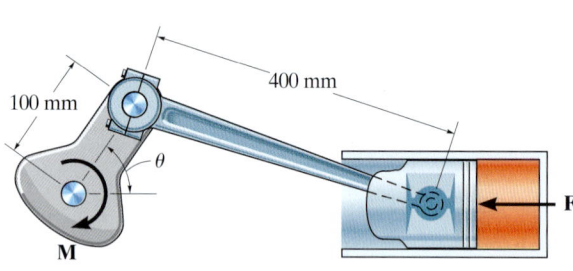

Probs. 11–23/24

11–22. The dumpster has a weight W and a center of gravity at G. Determine the force in the hydraulic cylinder needed to hold it in the general position θ.

11–25. Rods AB and BC have centers of mass located at their midpoints. If all contacting surfaces are smooth and BC has a mass of 100 kg, determine the appropriate mass of AB required for equilibrium.

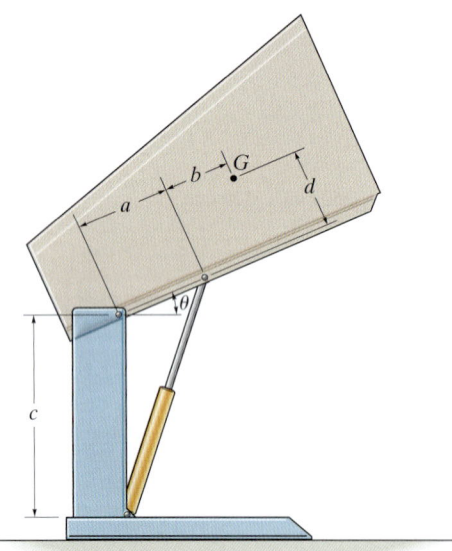

Prob. 11–22

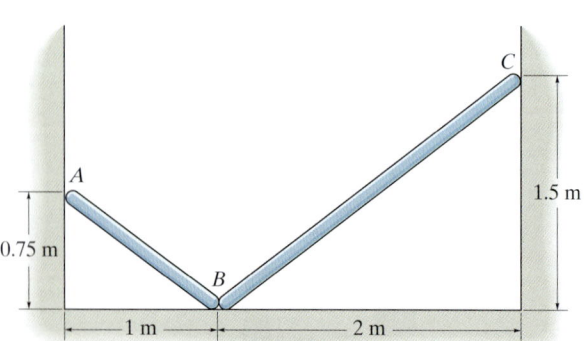

Prob. 11–25

11

*11.4   Conservative Forces

If the work of a force only depends upon its initial and final positions, and is *independent* of the path it travels, then the force is referred to as a *conservative force*. The weight of a body and the force of a spring are two examples of conservative forces.

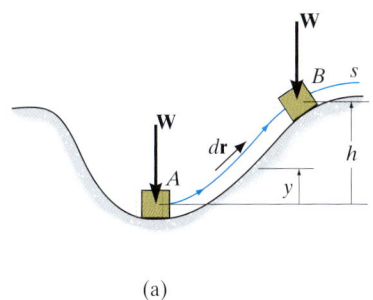

Weight.   Consider a block of weight **W** that travels along the path in Fig. 11–10*a*. When it is displaced up the path by an amount *d***r**, then the work is $dU = \mathbf{W} \cdot d\mathbf{r}$ or $dU = -W(dr \cos \theta) = -W dy$, as shown in Fig. 11–10*b*. In this case, the work is *negative* since **W** acts in the opposite sense of *dy*. Thus, if the block moves from *A* to *B*, through the vertical displacement *h*, the work is

(a)

$$U = -\int_0^h W\, dy = -Wh$$

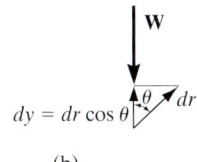

$dy = dr \cos \theta$

(b)

Fig. 11–10

The weight of a body is therefore a conservative force, since the work done by the weight depends only on the *vertical displacement* of the body, and is independent of the path along which the body travels.

Spring Force.   Now consider the linearly elastic spring in Fig. 11–11, which undergoes a displacement *ds*. The work done by the spring force on the block is $dU = -F_s\, ds = -ks\, ds$. The work is *negative* because $\mathbf{F}_s$ acts in the opposite sense to that of *ds*. Thus, the work of $\mathbf{F}_s$ when the block is displaced from $s = s_1$ to $s = s_2$ is

$$U = -\int_{s_1}^{s_2} ks\, ds = -\left(\tfrac{1}{2} ks_2^2 - \tfrac{1}{2} ks_1^2 \right)$$

Here the work depends only on the spring's initial and final positions, s_1 and s_2, measured from the spring's unstretched position. Since this result is independent of the path taken by the block as it moves, then a spring force is also a *conservative force*.

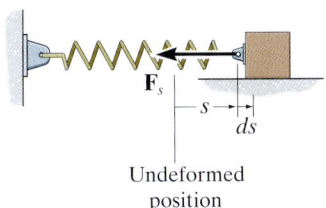

Undeformed
position

Fig. 11–11

11

Friction.   In contrast to a conservative force, consider the force of *friction* exerted on a sliding body by a fixed surface. The work done by the frictional force depends on the path; the longer the path, the greater the work. Consequently, frictional forces are *nonconservative*, and most of the work done by them is dissipated from the body in the form of heat.

*11.5   Potential Energy

When a conservative force acts on a body, it gives the body the capacity to do work. This capacity, measured as *potential energy*, depends on the location of the body relative to a fixed reference position or datum.

Gravitational Potential Energy.   If a body is located a distance y *above* a fixed horizontal reference or datum as in Fig. 11–12, the weight of the body has *positive* gravitational potential energy V_g since **W** has the capacity of doing positive work when the body is moved back down to the datum. Likewise, if the body is located a distance y *below* the datum, V_g is *negative* since the weight does negative work when the body is moved back up to the datum. At the datum, $V_g = 0$.

Measuring y as *positive upward*, the gravitational potential energy of the body's weight **W** is therefore

$$V_g = Wy \tag{11-4}$$

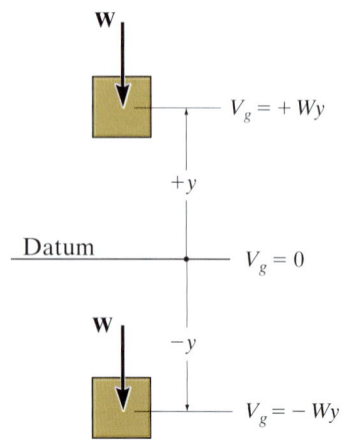

Fig. 11–12

Elastic Potential Energy.   When a spring is either elongated or compressed by an amount s from its unstretched position (the datum), the energy stored in the spring is called *elastic potential energy*. It is determined from

$$V_e = \tfrac{1}{2} ks^2 \tag{11-5}$$

This energy is always a positive quantity since the spring force acting on the attached body does *positive* work on the body as the force returns the body to the spring's unstretched position, Fig. 11–13.

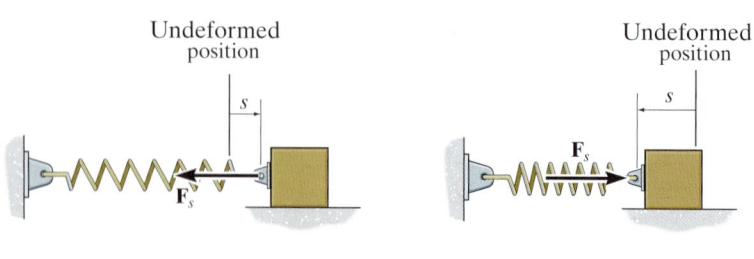

$$V_e = +\tfrac{1}{2} ks^2$$

Fig. 11–13

Potential Function.

In the general case, if a body is subjected to *both* gravitational and elastic forces, the *potential energy or potential function V* of the body can be expressed as the algebraic sum

$$V = V_g + V_e \qquad (11\text{--}6)$$

where measurement of V depends on the location of the body with respect to a selected datum in accordance with Eqs. 11–4 and 11–5.

In particular, if a *system* of frictionless connected rigid bodies has a *single degree of freedom*, such that its vertical distance from the datum is defined by the coordinate q, then the potential function for the system can be expressed as $V = V(q)$. The work done by all the weight and spring forces acting on the system in moving it from q_1 to q_2, is measured by the *difference* in V; i.e.,

$$U_{1-2} = V(q_1) - V(q_2) \qquad (11\text{--}7)$$

For example, the potential function for a system consisting of a block of weight **W** supported by a spring, as in Fig. 11–14, can be expressed in terms of the coordinate $(q =\ )\ y$, measured from a fixed datum located at the unstretched length of the spring. Here

$$V = V_g + V_e$$
$$= -Wy + \tfrac{1}{2}ky^2 \qquad (11\text{--}8)$$

If the block moves from y_1 to y_2, then applying Eq. 11–7 the work of **W** and $\mathbf{F}_s$ is

$$U_{1-2} = V(y_1) - V(y_2) = -W(y_1 - y_2) + \tfrac{1}{2}ky_1^2 - \tfrac{1}{2}ky_2^2$$

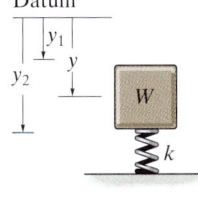

(a)

Fig. 11–14

*11.6   Potential-Energy Criterion for Equilibrium

If a frictionless connected system has one degree of freedom, and its position is defined by the coordinate q, then if it displaces from q to $q + dq$, Eq. 11–7 becomes

$$dU = V(q) - V(q + dq)$$

or

$$dU = -dV$$

If the system is in equilibrium and undergoes a *virtual displacement* δq, rather than an actual displacement dq, then the above equation becomes $\delta U = -\delta V$. However, the principle of virtual work requires that $\delta U = 0$, and therefore, $\delta V = 0$, and so we can write $\delta V = (dV/dq)\delta q = 0$. Since $\delta q \neq 0$, this expression becomes

$$\boxed{\frac{dV}{dq} = 0} \tag{11–9}$$

Hence, *when a frictionless connected system of rigid bodies is in equilibrium, the first derivative of its potential function is zero.* For example, using Eq. 11–8 we can determine the equilibrium position for the spring and block in Fig. 11–14a. We have

$$\frac{dV}{dy} = -W + ky = 0$$

Hence, the equilibrium position $y = y_{eq}$ is

$$y_{eq} = \frac{W}{k}$$

Of course, this *same result* can be obtained by applying $\Sigma F_y = 0$ to the forces acting on the free-body diagram of the block, Fig. 11–14b.

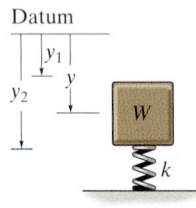

Datum

y_1

y

y_2

W

k

(a)

W

$F_s = ky_{eq}$

(b)

Fig. 11–14 (cont'd)

*11.7   Stability of Equilibrium Configuration

The potential function V of a system can also be used to investigate the stability of the equilibrium configuration, which is classified as *stable*, *neutral*, or *unstable*.

Stable Equilibrium.   A system is said to be *stable* if a system has a tendency to return to its original position when a small displacement is given to the system. The potential energy of the system in this case is at its *minimum*. A simple example is shown in Fig. 11–15a. When the disk is given a small displacement, its center of gravity G will always move (rotate) back to its equilibrium position, which is at the *lowest point* of its path. This is where the potential energy of the disk is at its *minimum*.

Neutral Equilibrium.   A system is said to be in *neutral equilibrium* if the system still remains in equilibrium when the system is given a small displacement away from its original position. In this case, the potential energy of the system is *constant*. Neutral equilibrium is shown in Fig. 11–15b, where a disk is pinned at G. Each time the disk is rotated, a new equilibrium position is established and the potential energy remains unchanged.

Unstable Equilibrium.   A system is said to be *unstable* if it has a tendency to be *displaced further away* from its original equilibrium position when it is given a small displacement. The potential energy of the system in this case is a *maximum*. An unstable equilibrium position of the disk is shown in Fig. 11–15c. Here the disk will rotate away from its equilibrium position when its center of gravity is slightly displaced. At this *highest point*, its potential energy is at a *maximum*.

The counterweight at A balances the weight of the deck B of this simple lift bridge. By applying the method of potential energy we can study the stability of the structure for various equilibrium positions of the deck.

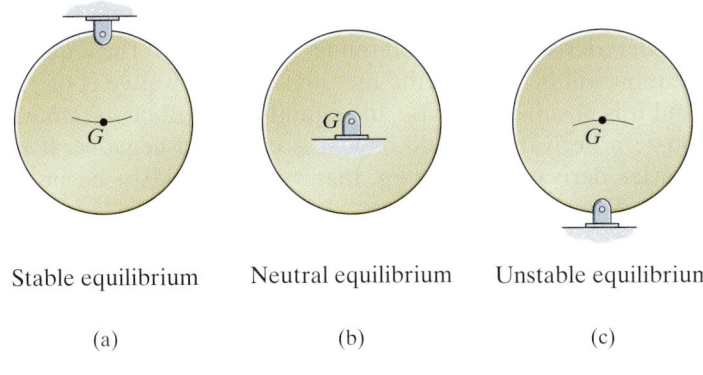

Stable equilibrium	Neutral equilibrium	Unstable equilibrium
(a)	(b)	(c)

Fig. 11–15

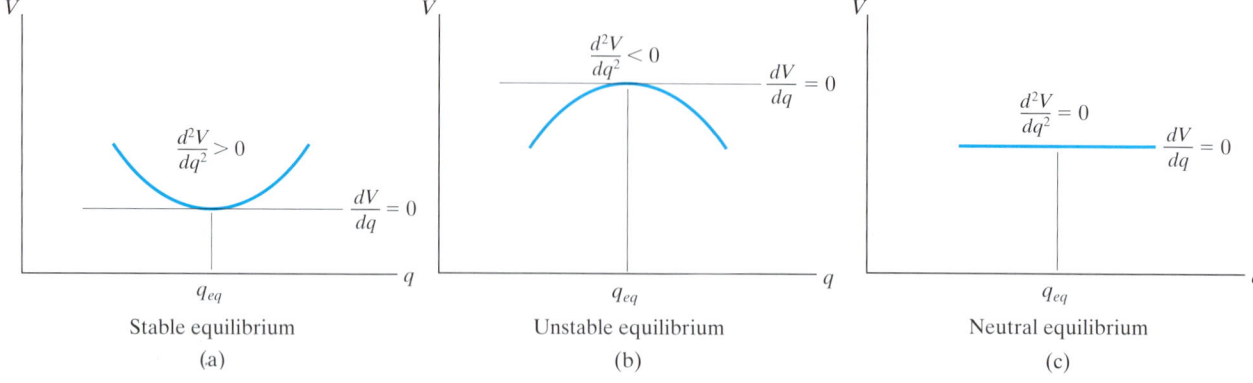

Fig. 11–16

One-Degree-of-Freedom System.

If a system has only one degree of freedom, and its position is defined by the coordinate q, then the potential function V for the system in terms of q can be plotted, Fig. 11–16. Provided the system is in *equilibrium*, then dV/dq, which represents the slope of this function, must be equal to zero. An investigation of stability at the equilibrium configuration therefore requires that the second derivative of the potential function be evaluated.

If d^2V/dq^2 is greater than zero, Fig. 11–16a, the potential energy of the system will be a *minimum*. This indicates that the equilibrium configuration is *stable*. Thus,

$$\frac{dV}{dq} = 0, \qquad \frac{d^2V}{dq^2} > 0 \qquad \text{stable equilibrium} \qquad (11\text{–}10)$$

If d^2V/dq^2 is less than zero, Fig. 11–16b, the potential energy of the system will be a *maximum*. This indicates an *unstable* equilibrium configuration. Thus,

$$\frac{dV}{dq} = 0, \qquad \frac{d^2V}{dq^2} < 0 \qquad \text{unstable equilibrium} \qquad (11\text{–}11)$$

Finally, if d^2V/dq^2 is equal to zero, it will be necessary to investigate the higher order derivatives to determine the stability. The equilibrium configuration will be *stable* if the first non-zero derivative is of an *even* order and it is *positive*. Likewise, the equilibrium will be *unstable* if this first non-zero derivative is odd or if it is even and negative. If all the higher order derivatives are *zero*, the system is said to be in *neutral equilibrium*, Fig. 11–16c. Thus,

$$\frac{dV}{dq} = \frac{d^2V}{dq^2} = \frac{d^3V}{dq^3} = \cdots = 0 \qquad \text{neutral equilibrium} \quad (11\text{–}12)$$

This condition occurs only if the potential-energy function for the system is constant at or around the neighborhood of q_{eq}.

During high winds and when going around a curve, these sugar-cane trucks can become unstable and tip over since their center of gravity is high off the road when they are fully loaded.

Procedure for Analysis

Using potential-energy methods, the equilibrium positions and the stability of a body or a system of connected bodies having a single degree of freedom can be obtained by applying the following procedure.

Potential Function.

- Sketch the system so that it is in the *arbitrary position* specified by the coordinate q.

- Establish a horizontal *datum* through a *fixed point** and express the gravitational potential energy V_g in terms of the weight W of each member and its vertical distance y from the datum, $V_g = Wy$.

- Express the elastic potential energy V_e of the system in terms of the stretch or compression, s, of any connecting spring, $V_e = \frac{1}{2}ks^2$.

- Formulate the potential function $V = V_g + V_e$ and express the *position coordinates* y and s in terms of the single coordinate q.

Equilibrium Position.

- The equilibrium position of the system is determined by taking the first derivative of V and setting it equal to zero, $dV/dq = 0$.

Stability.

- Stability at the equilibrium position is determined by evaluating the second or higher-order derivatives of V.

- If the second derivative is greater than zero, the system is stable; if all derivatives are equal to zero, the system is in neutral equilibrium; and if the second derivative is less than zero, the system is unstable.

11

*The location of the datum is *arbitrary*, since only the *changes* or differentials of V are required for investigation of the equilibrium position and its stability.

EXAMPLE 11.5

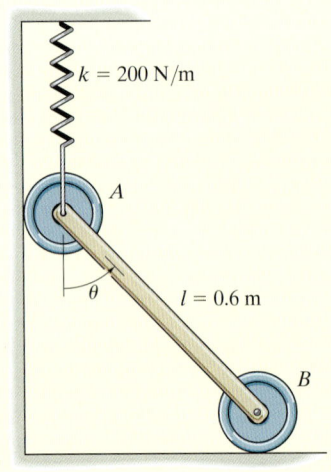

(a)

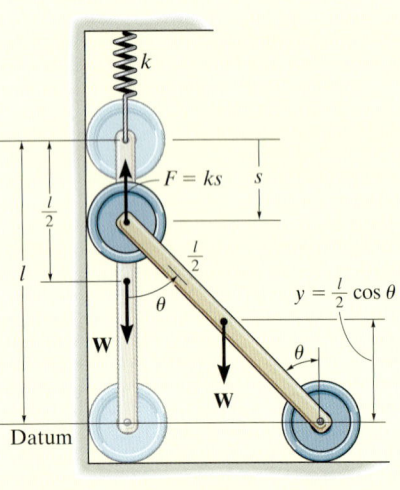

Datum

(b)

Fig. 11–17

The uniform link shown in Fig. 11–17a has a mass of 10 kg. If the spring is unstretched when $\theta = 0°$, determine the angle θ for equilibrium and investigate the stability at the equilibrium position.

SOLUTION

Potential Function. The datum is established at the bottom of the link, Fig. 11–17b. When the link is located in the arbitrary position θ, the spring increases its potential energy by stretching and the weight decreases its potential energy. Hence,

$$V = V_e + V_g = \frac{1}{2}ks^2 + Wy$$

Since $l = s + l \cos \theta$ or $s = l(1 - \cos \theta)$, and $y = (l/2) \cos \theta$, then

$$V = \frac{1}{2}kl^2(1 - \cos \theta)^2 + W\left(\frac{l}{2} \cos \theta\right)$$

Equilibrium Position. The first derivative of V is

$$\frac{dV}{d\theta} = kl^2(1 - \cos \theta) \sin \theta - \frac{Wl}{2} \sin \theta = 0$$

or

$$l\left[kl(1 - \cos \theta) - \frac{W}{2}\right] \sin \theta = 0$$

This equation is satisfied provided

$$\sin \theta = 0 \qquad \theta = 0° \qquad \qquad Ans.$$

or

$$\theta = \cos^{-1}\left(1 - \frac{W}{2kl}\right) = \cos^{-1}\left[1 - \frac{10(9.81)}{2(200)(0.6)}\right] = 53.8° \quad Ans.$$

Stability. The second derivative of V is

$$\frac{d^2V}{d\theta^2} = kl^2(1 - \cos \theta) \cos \theta + kl^2 \sin \theta \sin \theta - \frac{Wl}{2} \cos \theta$$

$$= kl^2(\cos \theta - \cos 2\theta) - \frac{Wl}{2} \cos \theta$$

Substituting values for the constants, with $\theta = 0°$ and $\theta = 53.8°$, yields

$$\left.\frac{d^2V}{d\theta^2}\right|_{\theta=0°} = 200(0.6)^2(\cos 0° - \cos 0°) - \frac{10(9.81)(0.6)}{2} \cos 0°$$

$$= -29.4 < 0 \qquad \text{(unstable equilibrium at } \theta = 0°) \qquad Ans.$$

$$\left.\frac{d^2V}{d\theta^2}\right|_{\theta=53.8°} = 200(0.6)^2(\cos 53.8° - \cos 107.6°) - \frac{10(9.81)(0.6)}{2} \cos 53.8°$$

$$= 46.9 > 0 \qquad \text{(stable equilibrium at } \theta = 53.8°) \qquad Ans.$$

EXAMPLE 11.6

If the spring AD in Fig. 11–18a has a stiffness of 18 kN/m and is unstretched when $\theta = 60°$, determine the angle θ for equilibrium. The load has a mass of 1.5 Mg. Investigate the stability at the equilibrium position.

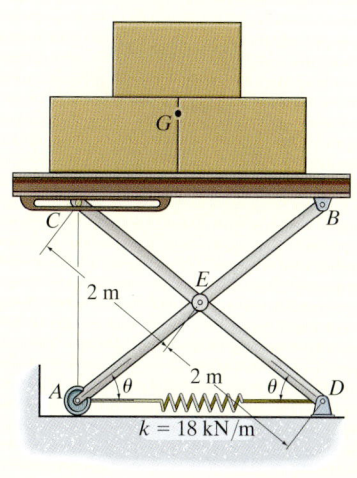

(a)

SOLUTION

Potential Energy.   The gravitational potential energy for the load with respect to the fixed datum, shown in Fig. 11–18b, is

$$V_g = mgy = 1500(9.81) \text{ N}[(4 \text{ m}) \sin \theta + h] = 58\,860 \sin \theta + 14\,715h$$

where h is a constant distance. From the geometry of the system, the elongation of the spring when the load is on the platform is
$s = (4 \text{ m}) \cos \theta - (4 \text{ m}) \cos 60° = (4 \text{ m}) \cos \theta - 2 \text{ m}$.
   Thus, the elastic potential energy of the system is

$$V_e = \tfrac{1}{2}ks^2 = \tfrac{1}{2}(18\,000 \text{ N/m})(4 \text{ m} \cos \theta - 2 \text{ m})^2 = 9000(4 \cos \theta - 2)^2$$

The potential energy function for the system is therefore

$$V = V_g + V_e = 58\,860 \sin \theta + 14\,715h + 9000(4 \cos \theta - 2)^2 \quad (1)$$

Equilibrium.   When the system is in equilibrium,

$$\frac{dV}{d\theta} = 58\,860 \cos \theta + 18\,000(4 \cos \theta - 2)(-4 \sin \theta) = 0$$

$$58\,860 \cos \theta - 288\,000 \sin \theta \cos \theta + 144\,000 \sin \theta = 0$$

Since $\sin 2\theta = 2 \sin \theta \cos \theta$,

$$58\,860 \cos \theta - 144\,000 \sin 2\theta + 144\,000 \sin \theta = 0$$

Solving by trial and error,

$$\theta = 28.18° \text{ and } \theta = 45.51° \qquad \textit{Ans.}$$

Stability.   Taking the second derivative of Eq. 1,

$$\frac{d^2V}{d\theta^2} = -58\,860 \sin \theta - 288\,000 \cos 2\theta + 144\,000 \cos \theta$$

Substituting $\theta = 28.18°$ yields

$$\frac{d^2V}{d\theta^2} = -60\,402 < 0 \qquad \text{Unstable} \qquad \textit{Ans.}$$

And for $\theta = 45.51°$,

$$\frac{d^2V}{d\theta^2} = 64\,073 > 0 \qquad \text{Stable} \qquad \textit{Ans.}$$

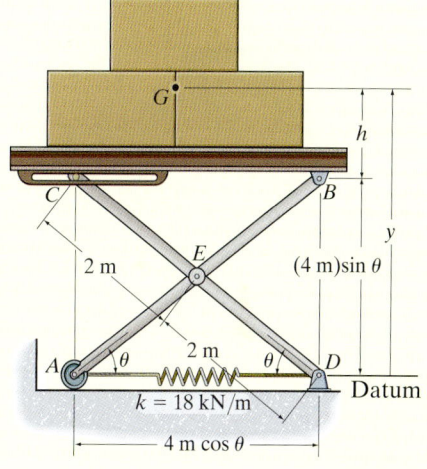

(b)

Fig. 11–18

11

EXAMPLE 11.7

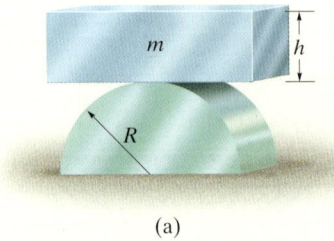

(a)

The uniform block having a mass m rests on the top surface of the half cylinder, Fig. 11–19a. Show that this is a condition of unstable equilibrium if $h > 2R$.

SOLUTION

Potential Function. The datum is established at the base of the cylinder, Fig. 11–19b. If the block is displaced by an amount θ from the equilibrium position, the potential function is

$$V = V_e + V_g$$

$$= 0 + mgy$$

From Fig. 11–19b,

$$y = \left(R + \frac{h}{2} \right) \cos \theta + R\theta \sin \theta$$

Thus,

$$V = mg \left[\left(R + \frac{h}{2} \right) \cos \theta + R\theta \sin \theta \right]$$

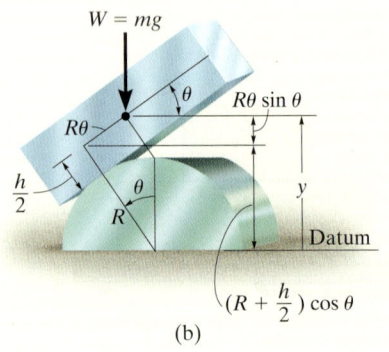

(b)

Fig. 11–19

Equilibrium Position.

$$\frac{dV}{d\theta} = mg \left[-\left(R + \frac{h}{2} \right) \sin \theta + R \sin \theta + R\theta \cos \theta \right] = 0$$

$$= mg \left(-\frac{h}{2} \sin \theta + R\theta \cos \theta \right) = 0$$

Note that $\theta = 0°$ satisfies this equation.

Stability. Taking the second derivative of V yields

$$\frac{d^2V}{d\theta^2} = mg \left(-\frac{h}{2} \cos \theta + R \cos \theta - R\theta \sin \theta \right)$$

At $\theta = 0°$,

$$\left. \frac{d^2V}{d\theta^2} \right|_{\theta=0°} = -mg \left(\frac{h}{2} - R \right)$$

Since all the constants are positive, the block is in unstable equilibrium provided $h > 2R$, because then $d^2V/d\theta^2 < 0$.

PROBLEMS

11–26. If the potential energy for a conservative one-degree-of-freedom system is expressed by the relation $V = (3y^3 + 2y^2 - 4y + 50)$ J, where y is given in meters, determine the equilibrium positions and investigate the stability at each position.

11–27. If the potential function for a conservative one-degree-of-freedom system is $V = (8x^3 - 2x^2 - 10)$ J, where x is given in meters, determine the positions for equilibrium and investigate the stability at each of these positions.

*11–28. If the potential function for a conservative one-degree-of-freedom system is $V = (12 \sin 2\theta + 15 \cos \theta)$ J, where $0° < \theta < 180°$, determine the positions for equilibrium and investigate the stability at each of these positions.

11–29. If the potential energy for a conservative one-degree-of-freedom system is expressed by the relation $V = (24 \sin \theta + 10 \cos 2\theta)$ J, $0° \le \theta \le 90°$, determine the equilibrium positions and investigate the stability at each position.

11–30. If the potential function for a conservative one-degree-of-freedom system is $V = (10 \cos 2\theta + 25 \sin \theta)$ J, where $0° < \theta < 180°$, determine the positions for equilibrium and investigate the stability at each of these positions.

11–31. Determine the angle θ for equilibrium and investigate the stability of the mechanism in this position. The spring has a stiffness of $k = 1.5$ kN/m and is unstretched when $\theta = 90°$. The block A has a mass of 40 kg. Neglect the mass of the links.

*11–32. The spring of the scale has an unstretched length of a. Determine the angle θ for equilibrium when a weight W is supported on the platform. Neglect the weight of the members. What value W would be required to keep the scale in neutral equilibrium when $\theta = 0°$?

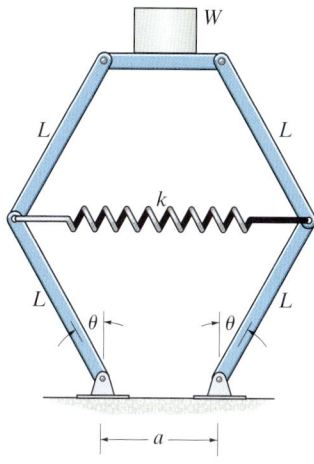

Prob. 11–32

11–33. The cup has a hemispherical bottom and a mass m. Determine the position h of the center of mass G so that the cup is in neutral equilibrium.

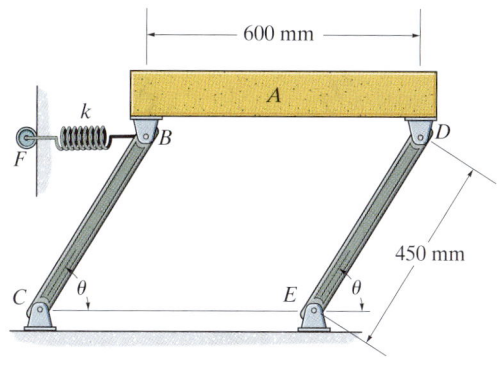

Prob. 11–31

Prob. 11–33

11

11–34. The homogeneous cone has a conical cavity cut into it as shown. Determine the depth of d of the cavity in terms of h so that the cone balances on the pivot and remains in neutral equilibrium.

*11–36.** Determine the angle θ for equilibrium and investigate the stability at this position. The bars each have a mass of 3 kg and the suspended block D has a mass of 7 kg. Cord DC has a total length of 1 m.

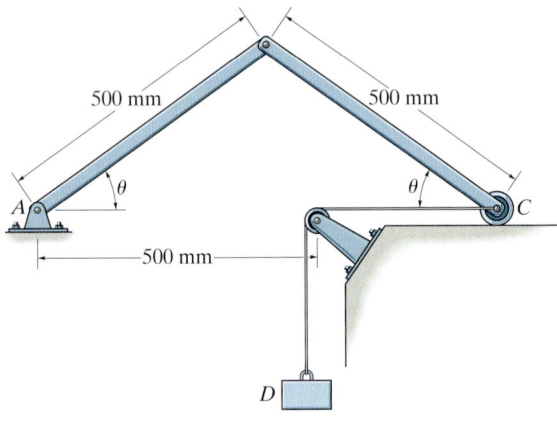

Prob. 11–36

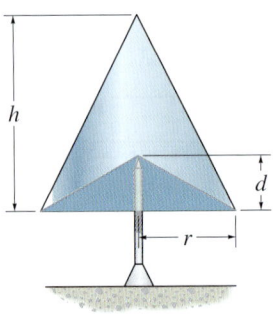

Prob. 11–34

11–35. The uniform beam has a mass of 200 kg. If the contacting surfaces are smooth, determine the angle θ for equilibrium and investigate the stability of the beam when it is in this position. The spring has an unstretched length of 0.5 m.

11–37. Each of the two springs has an unstretched length of 500 mm. Determine the mass m of the cylinder when it is held in the equilibrium position shown, i.e., $y = 1$ m.

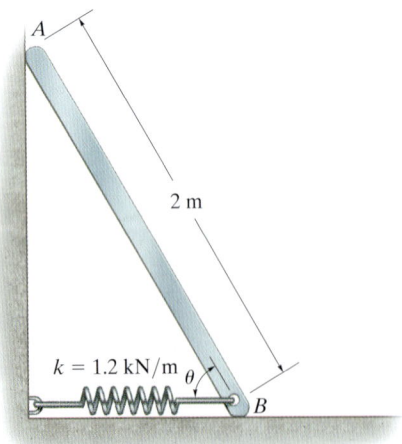

Prob. 11–35

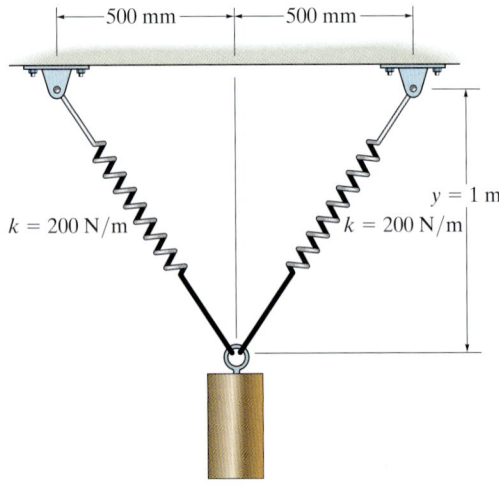

Prob. 11–37

11–38. A homogeneous block rests on top of the cylindrical surface. Derive the relationship between the radius of the cylinder, r, and the dimension of the block, b, for stable equilibrium. *Hint*: Establish the potential energy function for a small angle θ, i.e., approximate $\sin \theta \approx 0$, and $\cos \theta \approx 1 - \theta^2/2$.

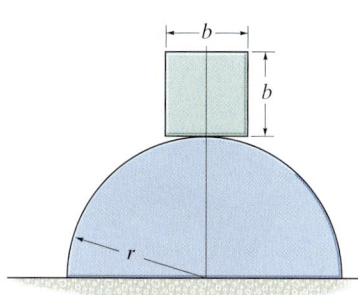

Prob. 11–38

11–39. The uniform rod AB has a mass of 80 kg. If spring DC is unstretched when $\theta = 90°$, determine the angle θ for equilibrium and investigate the stability at the equilibrium position. The spring always acts in the horizontal position due to the roller guide at D.

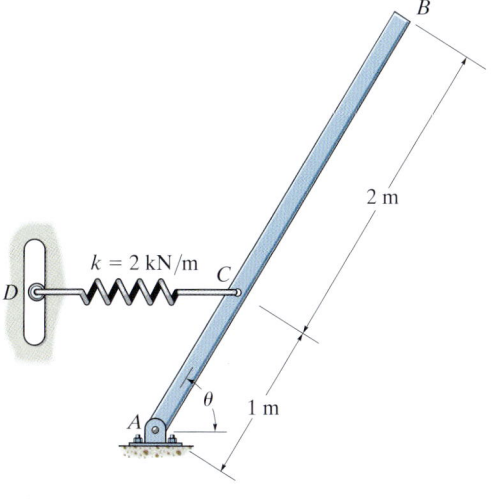

Prob. 11–39

*11–40.** A spring with a torsional stiffness k is attached to the pin at B. It is unstretched when the rod assembly is in the vertical position. Determine the weight W of the block that results in neutral equilibrium. *Hint:* Establish the potential energy function for a small angle θ, i.e., approximate $\sin \theta \approx 0$, and $\cos \theta \approx 1 - \theta^2/2$.

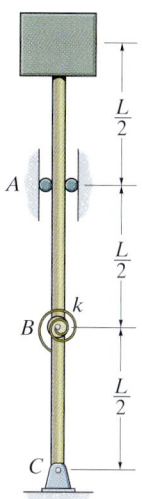

Prob. 11–40

11–41. Each bar has a mass per length of m_0. Determine the angles θ and ϕ at which they are suspended in equilibrium. The contact at A is smooth, and both are pin connected at B.

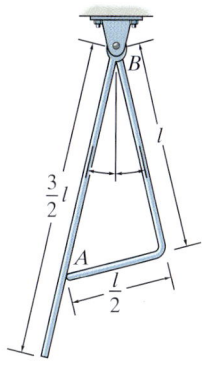

Prob. 11–41

11

11–42. The small postal scale consists of a counterweight W_1, connected to the members having negligible weight. Determine the weight W_2 that is on the pan in terms of the angles θ and ϕ and the dimensions shown. All members are pin connected.

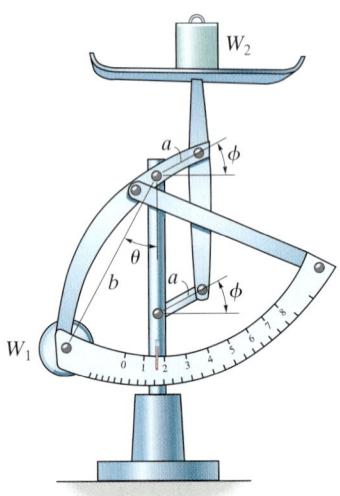

Prob. 11–42

11–43. If the spring has a torsional stiffness $k = 300\ \text{N} \cdot \text{m/rad}$ and is unwound when $\theta = 90°$, determine the angle for equilibrium if the sphere has a mass of 20 kg. Investigate the stability at this position. Collar C can slide freely along the vertical guide. Neglect the weight of the rods and collar C.

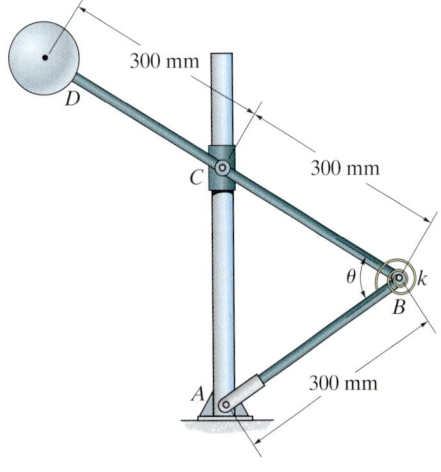

Prob. 11–43

*11–44. The truck has a mass of 20 Mg and a mass center at G. Determine the steepest grade θ along which it can park without overturning and investigate the stability in this position.

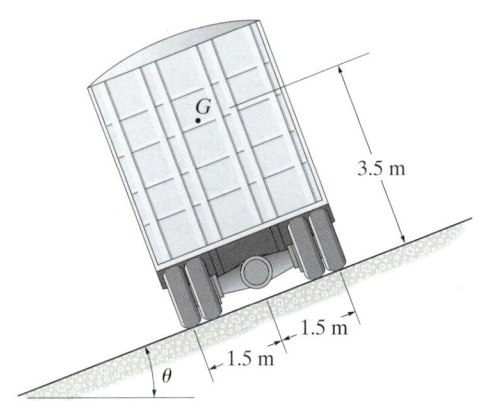

Prob. 11–44

11–45. The cylinder is made of two materials such that it has a mass of m and a center of gravity at point G. Show that when G lies above the centroid C of the cylinder, the equilibrium is unstable.

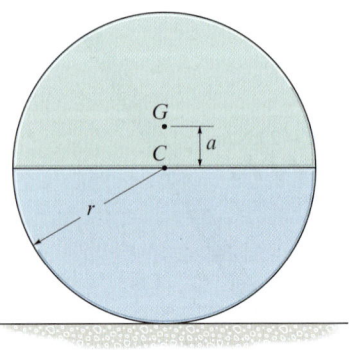

Prob. 11–45

11–46. If each of the three links of the mechanism has a weight W, determine the angle θ for equilibrium. The spring, which always remains vertical, is unstretched when $\theta = 0°$.

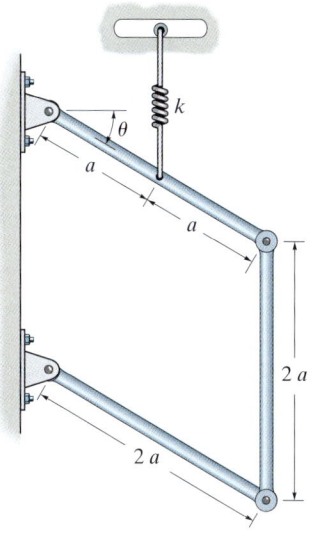

Prob. 11–46

11–47. If the uniform rod OA has a mass of 12 kg, determine the mass m that will hold the rod in equilibrium when $\theta = 30°$. Point C is coincident with B when OA is horizontal. Neglect the size of the pulley at B.

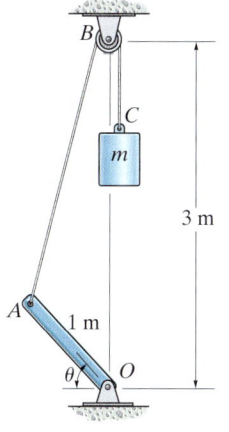

Prob. 11–47

11–48. The triangular block of weight W rests on the smooth corners which are a distance a apart. If the block has three equal sides of length d, determine the angle θ for equilibrium.

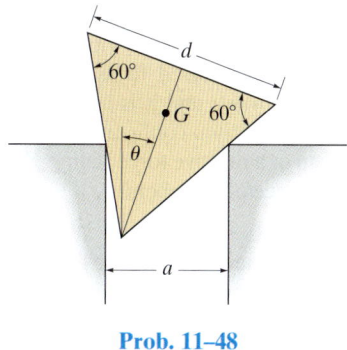

Prob. 11–48

11–49. Two uniform bars, each having a weight W, are pin connected at their ends. If they are placed over a smooth cylindrical surface, show that the angle θ for equilibrium must satisfy the equation $\cos\theta/\sin^3\theta = a/2r$.

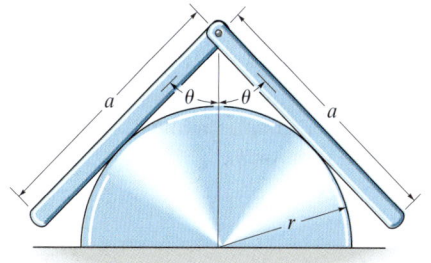

Prob. 11–49

11

CHAPTER REVIEW

Principle of Virtual Work

The forces on a body will do *virtual work* when the body undergoes an *imaginary* differential displacement or rotation.

δy, $\delta y'$ –virtual displacements

$\delta\theta$ –virtual rotation

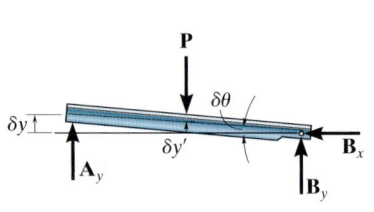

For equilibrium, the sum of the virtual work done by all the forces acting on the body must be equal to zero for any virtual displacement. This is referred to as the *principle of virtual work*, and it is useful for finding the equilibrium configuration for a mechanism or a reactive force acting on a series of connected members.

$$\delta U = 0$$

If the system of connected members has one degree of freedom, then its position can be specified by one independent coordinate, such as θ.

To apply the principle of virtual work, it is first necessary to use *position coordinates* to locate all the forces and moments on the mechanism that will do work when the mechanism undergoes a virtual movement $\delta\theta$.

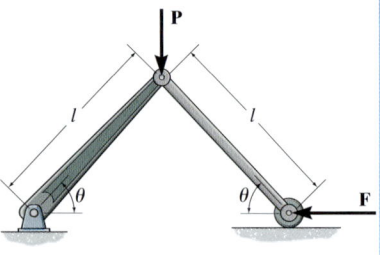

The coordinates are related to the independent coordinate θ and then these expressions are differentiated in order to relate the *virtual* coordinate displacements to the virtual displacement $\delta\theta$.

Finally, the equation of virtual work is written for the mechanism in terms of the common virtual displacement $\delta\theta$, and then it is set equal to zero. By factoring $\delta\theta$ out of the equation, it is then possible to determine either the unknown force or couple moment, or the equilibrium position θ.

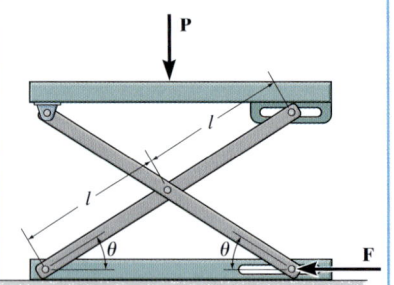

11

Potential-Energy Criterion for Equilibrium

When a system is subjected only to conservative forces, such as weight and spring forces, then the equilibrium configuration can be determined using the *potential-energy function V* for the system.

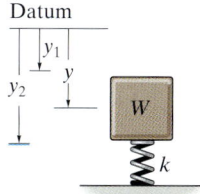

Datum

$$V = V_g + V_e = -Wy + \tfrac{1}{2} ky^2$$

The potential-energy function is established by expressing the weight and spring potential energy for the system in terms of the independent coordinate q.

Once the potential-energy function is formulated, its first derivative is set equal to zero. The solution yields the equilibrium position q_{eq} for the system.

$$\frac{dV}{dq} = 0$$

The stability of the system can be investigated by taking the second derivative of V.

$$\frac{dV}{dq} = 0, \quad \frac{d^2V}{dq^2} > 0 \quad \text{stable equilibrium}$$

$$\frac{dV}{dq} = 0, \quad \frac{d^2V}{dq^2} < 0 \quad \text{unstable equilibrium}$$

$$\frac{dV}{dq} = \frac{d^2V}{dq^2} = \frac{d^3V}{dq^3} = \cdots = 0 \quad \text{neutral equilibrium}$$

11

REVIEW PROBLEMS

11–50. The piston C moves vertically between the two smooth walls. If the spring has a stiffness of $k = 1.5$ kN/m and is unstretched when $\theta = 0°$, determine the couple **M** that must be applied to link AB to hold the mechanism in equilibrium; $\theta = 30°$.

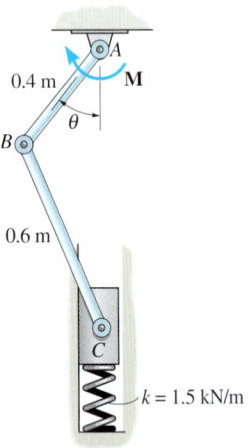

0.4 m   **M**

θ

B

0.6 m

C

$k = 1.5$ kN/m

Prob. 11–50

11–51. The uniform bar AB weighs 50 N (≈ 5 kg) If the attached spring is unstretched when $\theta = 90°$, use the method of virtual work and determine the angle θ for equilibrium. Note that the spring always remains in the vertical position due to the roller guide.

11–52. Solve Prob. 11-51 using the principle of potential energy. Investigate the stability of the bar when it is in the equilibrium position.

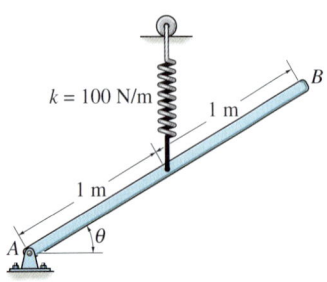

$k = 100$ N/m B

1 m

1 m

A θ

Probs. 11–51/52

11–53. The uniform right circular cone having a mass m is suspended from the cord as shown. Determine the angle θ at which it hangs from the wall for equilibrium. Is the cone in stable equilibrium?

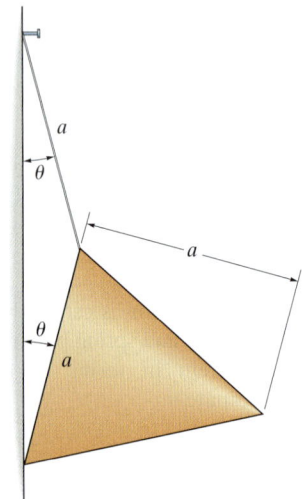

a

θ

a

θ

a

Prob. 11–53

11–54. The assembly shown consists of a semicylinder and a rectangular block. If the block weighs 40 N and the semicylinder weighs 10 N, investigate the stability when the assembly is resting in the equilibrium position. Set $h = 100$ mm.

11–55. The 10-N (≈ 1-kg) semicylinder supports the block which has a specific weight of $\gamma = 13.5$ kN/m³. Determine the height h of the block which will produce neutral equilibrium in the position shown.

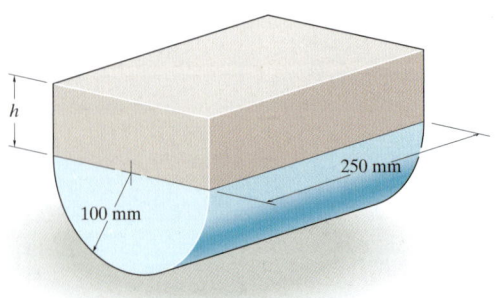

h

250 mm

100 mm

Probs. 11–54/55

*11–56. If the potential energy for a conservative one-degree-of-freedom system is expressed by the relation $V = (4x^3 - x^2 - 3x + 10)$ J, where x is given in meters, determine the equilibrium positions and investigate the stability at each position.

11–57. Determine the weight of block G required to balance the differential lever when the 100-N ($\approx$ 10-kg) load F is placed on the pan. The lever is in balance when the load and block are not on the lever. Take $x = 300$ mm.

11–58. If load F weighs 100 N and block G weighs 10 N, determine its position x for equilibrium of the differential lever. The lever is in balance when the load and block are not on the lever.

*11–60. A 5-kg uniform serving table is supported on each side by pairs of two identical links, AB and CD, and springs CE. If the bowl has a mass of 1 kg, determine the angle θ where the table is in equilibrium. The springs each have a stiffness of $k = 200$ N/m and are unstretched when $\theta = 90°$. Neglect the mass of the links.

11–61. A 5-kg uniform serving table is supported on each side by two pairs of identical links, AB and CD, and springs CE. If the bowl has a mass of 1 kg and is in equilibrium when $\theta = 45°$, determine the stiffness k of each spring. The springs are unstretched when $\theta = 90°$. Neglect the mass of the links.

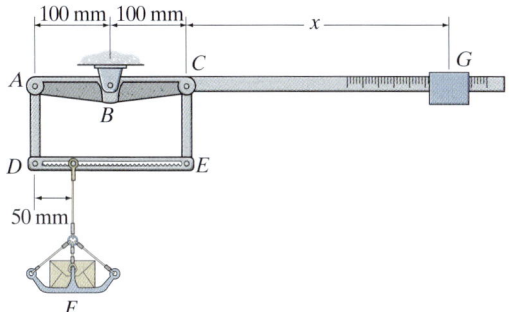

Probs. 11–57/58

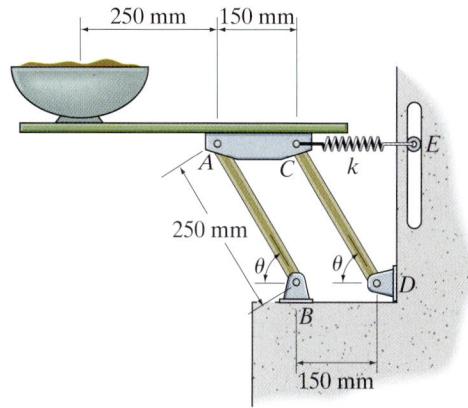

Probs. 11–60/61

11–59. A force $\mathbf{P}$ is applied to the end of the lever. Determine the horizontal force F on the piston for equilibrium.

11–62. Rods AB and BC have centers of mass located at their midpoints. If all contacting surfaces are smooth and BC has a mass of 150 kg, determine the appropriate mass of AB required for equilibrium.

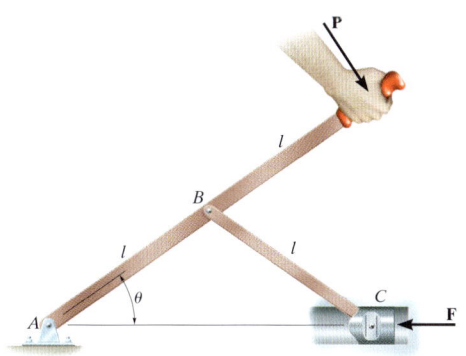

Prob. 11–59

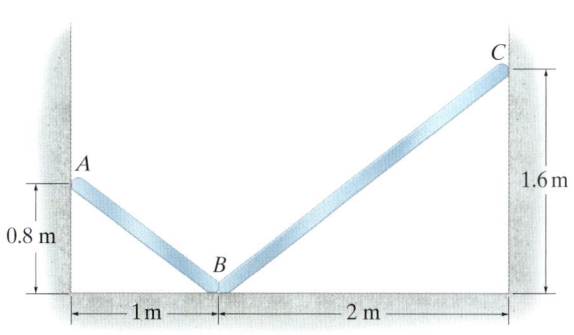

Prob. 11–62

11

Mathematical Review and Expressions

Geometry and Trigonometry Review

The angles θ in Fig. A–1 are equal between the transverse and two parallel lines.

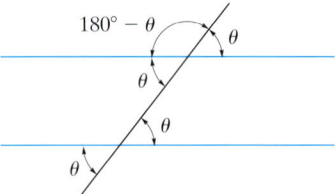

Fig. A–1

For a line and its normal, the angles θ in Fig. A–2 are equal.

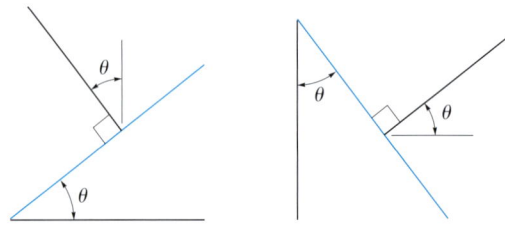

Fig. A–2

For the circle in Fig. A–3 $s = \theta r$, so that when $\theta = 360° = 2\pi$ rad then the circumference is $s = 2\pi r$. Also, since $180° = \pi$ rad, then θ (rad) $= (\pi/180°)\theta°$. The area of the circle is $A = \pi r^2$.

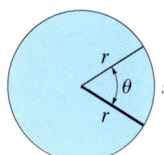

Fig. A–3

The sides of a similar triangle can be obtained by proportion as in Fig. A–4, where $\dfrac{a}{A} = \dfrac{b}{B} = \dfrac{c}{C}$.

For the right triangle in Fig. A–5, the Pythagorean theorem is

$$h = \sqrt{(o)^2 + (a)^2}$$

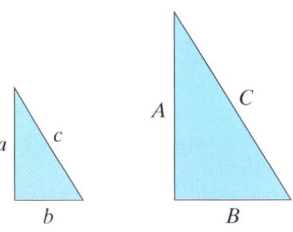

Fig. A–4

The trigonometric functions are

$$\sin \theta = \frac{o}{h}$$

$$\cos \theta = \frac{a}{h}$$

$$\tan \theta = \frac{o}{a}$$

This is easily remembered as "soh, cah, toa", i.e., the sine is the opposite over the hypotenuse, etc. The other trigonometric functions follow from this.

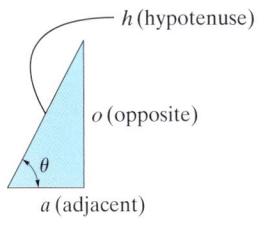

Fig. A–5

$$\csc \theta = \frac{1}{\sin \theta} = \frac{h}{o}$$

$$\sec \theta = \frac{1}{\cos \theta} = \frac{h}{a}$$

$$\cot \theta = \frac{1}{\tan \theta} = \frac{a}{o}$$

Trigonometric Identities

$$\sin^2 \theta + \cos^2 \theta = 1$$

$$\sin(\theta \pm \phi) = \sin \theta \cos \phi \pm \cos \theta \sin \phi$$

$$\sin 2\theta = 2 \sin \theta \cos \theta$$

$$\cos(\theta \pm \phi) = \cos \theta \cos \phi \mp \sin \theta \sin \phi$$

$$\cos 2\theta = \cos^2 \theta - \sin^2 \theta$$

$$\cos \theta = \pm \sqrt{\frac{1 + \cos 2\theta}{2}}, \quad \sin \theta = \pm \sqrt{\frac{1 - \cos 2\theta}{2}}$$

$$\tan \theta = \frac{\sin \theta}{\cos \theta}$$

$$1 + \tan^2 \theta = \sec^2 \theta \qquad 1 + \cot^2 \theta = \csc^2 \theta$$

Quadratic Formula

If $ax^2 + bx + c = 0$, then $x = \dfrac{-b \pm \sqrt{b^2 - 4ac}}{2a}$

Hyperbolic Functions

$$\sinh x = \frac{e^x - e^{-x}}{2},$$

$$\cosh x = \frac{e^x + e^{-x}}{2},$$

$$\tanh x = \frac{\sinh x}{\cosh x}$$

Power-Series Expansions

$$\sin x = x - \frac{x^3}{3!} + \cdots, \quad \cos x = 1 - \frac{x^2}{2!} + \cdots$$

$$\sinh x = x + \frac{x^3}{3!} + \cdots, \quad \cosh x = 1 + \frac{x^2}{2!} + \cdots$$

Derivatives

$$\frac{d}{dx}(u^n) = nu^{n-1}\frac{du}{dx} \qquad \frac{d}{dx}(\sin u) = \cos u \frac{du}{dx}$$

$$\frac{d}{dx}(uv) = u\frac{dv}{dx} + v\frac{du}{dx} \qquad \frac{d}{dx}(\cos u) = -\sin u \frac{du}{dx}$$

$$\frac{d}{dx}\left(\frac{u}{v}\right) = \frac{v\dfrac{du}{dx} - u\dfrac{dv}{dx}}{v^2} \qquad \frac{d}{dx}(\tan u) = \sec^2 u \frac{du}{dx}$$

$$\frac{d}{dx}(\cot u) = -\csc^2 u \frac{du}{dx} \qquad \frac{d}{dx}(\sinh u) = \cosh u \frac{du}{dx}$$

$$\frac{d}{dx}(\sec u) = \tan u \sec u \frac{du}{dx} \qquad \frac{d}{dx}(\cosh u) = \sinh u \frac{du}{dx}$$

$$\frac{d}{dx}(\csc u) = -\csc u \cot u \frac{du}{dx}$$

A

Integrals

$$\int x^n \, dx = \frac{x^{n+1}}{n+1} + C, n \neq -1$$

$$\int \frac{dx}{a+bx} = \frac{1}{b}\ln(a+bx) + C$$

$$\int \frac{dx}{a+bx^2} = \frac{1}{2\sqrt{-ab}}\ln\left[\frac{a+x\sqrt{-ab}}{a-x\sqrt{-ab}}\right] + C,$$
$$ab < 0$$

$$\int \frac{x \, dx}{a+bx^2} = \frac{1}{2b}\ln(bx^2+a) + C$$

$$\int \frac{x^2 \, dx}{a+bx^2} = \frac{x}{b} - \frac{a}{b\sqrt{ab}}\tan^{-1}\frac{x\sqrt{ab}}{a} + C, ab > 0$$

$$\int \sqrt{a+bx} \, dx = \frac{2}{3b}\sqrt{(a+bx)^3} + C$$

$$\int x\sqrt{a+bx} \, dx = \frac{-2(2a-3bx)\sqrt{(a+bx)^3}}{15b^2} + C$$

$$\int x^2\sqrt{a+bx} \, dx =$$
$$\frac{2(8a^2-12abx+15b^2x^2)\sqrt{(a+bx)^3}}{105b^3} + C$$

$$\int \sqrt{a^2-x^2} \, dx = \frac{1}{2}\left[x\sqrt{a^2-x^2} + a^2\sin^{-1}\frac{x}{a}\right] + C,$$
$$a > 0$$

$$\int x\sqrt{a^2-x^2} \, dx = -\frac{1}{3}\sqrt{(a^2-x^2)^3} + C$$

$$\int x^2\sqrt{a^2-x^2} \, dx = -\frac{x}{4}\sqrt{(a^2-x^2)^3}$$
$$+ \frac{a^2}{8}\left(x\sqrt{a^2-x^2} + a^2\sin^{-1}\frac{x}{a}\right) + C, a > 0$$

$$\int \sqrt{x^2 \pm a^2} \, dx =$$
$$\frac{1}{2}\left[x\sqrt{x^2 \pm a^2} \pm a^2\ln\left(x + \sqrt{x^2 \pm a^2}\right)\right] + C$$

$$\int x\sqrt{x^2 \pm a^2} \, dx = \frac{1}{3}\sqrt{(x^2 \pm a^2)^3} + C$$

$$\int x^2\sqrt{x^2 \pm a^2} \, dx = \frac{x}{4}\sqrt{(x^2 \pm a^2)^3}$$
$$\mp \frac{a^2}{8}x\sqrt{x^2 \pm a^2} - \frac{a^4}{8}\ln\left(x + \sqrt{x^2 \pm a^2}\right) + C$$

$$\int \frac{dx}{\sqrt{a+bx}} = \frac{2\sqrt{a+bx}}{b} + C$$

$$\int \frac{x \, dx}{\sqrt{x^2 \pm a^2}} = \sqrt{x^2 \pm a^2} + C$$

$$\int \frac{dx}{\sqrt{a+bx+cx^2}} = \frac{1}{\sqrt{c}}\ln\left[\sqrt{a+bx+cx^2} + \right.$$
$$\left. x\sqrt{c} + \frac{b}{2\sqrt{c}}\right] + C, c > 0$$

$$= \frac{1}{\sqrt{-c}}\sin^{-1}\left(\frac{-2cx-b}{\sqrt{b^2-4ac}}\right) + C, c < 0$$

$$\int \sin x \, dx = -\cos x + C$$

$$\int \cos x \, dx = \sin x + C$$

$$\int x\cos(ax) \, dx = \frac{1}{a^2}\cos(ax) + \frac{x}{a}\sin(ax) + C$$

$$\int x^2\cos(ax) \, dx = \frac{2x}{a^2}\cos(ax) + \frac{a^2x^2-2}{a^3}\sin(ax) + C$$

$$\int e^{ax} \, dx = \frac{1}{a}e^{ax} + C$$

$$\int xe^{ax} \, dx = \frac{e^{ax}}{a^2}(ax-1) + C$$

$$\int \sinh x \, dx = \cosh x + C$$

$$\int \cosh x \, dx = \sinh x + C$$

Fundamental Problems
Partial Solutions And Answers

Chapter 2

F2–1.
$$F_R = \sqrt{(2 \text{ kN})^2 + (6 \text{ kN})^2 - 2(2 \text{ kN})(6 \text{ kN}) \cos 105°}$$
$$= 6.798 \text{ kN} = 6.80 \text{ kN} \qquad \qquad \qquad \textit{Ans.}$$
$$\frac{\sin \phi}{6 \text{ kN}} = \frac{\sin 105°}{6.798 \text{ kN}}, \quad \phi = 58.49°$$
$$\theta = 45° + \phi = 45° + 58.49° = 103° \qquad \textit{Ans.}$$

F2–2. $F_R = \sqrt{200^2 + 500^2 - 2(200)(500) \cos 140°}$
$$= 666 \text{ N} \qquad \qquad \qquad \qquad \textit{Ans.}$$

F2–3. $F_R = \sqrt{600^2 + 800^2 - 2(600)(800) \cos 60°}$
$$= 721.11 \text{ N} = 721 \text{ N} \qquad \qquad \textit{Ans.}$$
$$\frac{\sin \alpha}{800} = \frac{\sin 60°}{721.11}; \quad \alpha = 73.90°$$
$$\phi = \alpha - 30° = 73.90° - 30° = 43.9° \qquad \textit{Ans.}$$

F2–4. $\dfrac{F_u}{\sin 45°} = \dfrac{30}{\sin 105°}; \qquad F_u = 22.0 \text{ N} \qquad \textit{Ans.}$
$$\dfrac{F_v}{\sin 30°} = \dfrac{30}{\sin 105°}; \qquad F_v = 15.5 \text{ N} \qquad \textit{Ans.}$$

F2–5. $\dfrac{F_{AB}}{\sin 105°} = \dfrac{450}{\sin 30°}$
$$F_{AB} = 869 \text{ N} \qquad \qquad \qquad \textit{Ans.}$$
$$\dfrac{F_{AC}}{\sin 45°} = \dfrac{450}{\sin 30°}$$
$$F_{AC} = 636 \text{ N} \qquad \qquad \qquad \textit{Ans.}$$

F2–6. $\dfrac{F}{\sin 30°} = \dfrac{6}{\sin 105°} \qquad F = 3.11 \text{ kN} \qquad \textit{Ans.}$
$$\dfrac{F_v}{\sin 45°} = \dfrac{6}{\sin 105°} \qquad F_v = 4.39 \text{ kN} \qquad \textit{Ans.}$$

F2–7. $(F_1)_x = 0 \quad (F_1)_y = 300 \text{ N} \qquad \textit{Ans.}$
$$(F_2)_x = -(450 \text{ N}) \cos 45° = -318 \text{ N} \qquad \textit{Ans.}$$
$$(F_2)_y = (450 \text{ N}) \sin 45° = 318 \text{ N} \qquad \textit{Ans.}$$
$$(F_3)_x = \left(\tfrac{3}{5}\right) 600 \text{ N} = 360 \text{ N} \qquad \textit{Ans.}$$
$$(F_3)_y = \left(\tfrac{4}{5}\right) 600 \text{ N} = 480 \text{ N} \qquad \textit{Ans.}$$

F2–8. $F_{Rx} = 300 + 400 \cos 30° - 250\left(\tfrac{4}{5}\right) = 446.4 \text{ N}$
$$F_{Ry} = 400 \sin 30° + 250\left(\tfrac{3}{5}\right) = 350 \text{ N}$$
$$F_R = \sqrt{(446.4)^2 + 350^2} = 567 \text{ N} \qquad \textit{Ans.}$$
$$\theta = \tan^{-1} \tfrac{350}{446.4} = 38.1° \qquad \textit{Ans.}$$

F2–9.
$$\xrightarrow{+}(F_R)_x = \Sigma F_x;$$
$$(F_R)_x = -(700 \text{ N}) \cos 30° + 0 + \left(\tfrac{3}{5}\right)(600 \text{ N})$$
$$= -246.22 \text{ N}$$
$$+\uparrow(F_R)_y = \Sigma F_y;$$
$$(F_R)_y = -(700 \text{ N}) \sin 30° - 400 \text{ N} - \left(\tfrac{4}{5}\right)(600 \text{ N})$$
$$= -1230 \text{ N}$$
$$F_R = \sqrt{(246.22 \text{ N})^2 + (1230 \text{ N})^2} = 1254 \text{ N} \qquad \textit{Ans.}$$
$$\phi = \tan^{-1} \left(\tfrac{1230 \text{ N}}{246.22 \text{ N}}\right) = 78.68°$$
$$\theta = 180° + \phi = 180° + 78.68° = 259° \qquad \textit{Ans.}$$

F2–10. $\xrightarrow{+}(F_R)_x = \Sigma F_x;$
$$750 \text{ N} = F \cos \theta + \left(\tfrac{5}{13}\right)(325 \text{ N}) + (600 \text{ N}) \cos 45°$$
$$+\uparrow(F_R)_y = \Sigma F_y;$$
$$0 = F \sin \theta + \left(\tfrac{12}{13}\right)(325 \text{ N}) - (600 \text{ N}) \sin 45°$$
$$\tan \theta = 0.6190 \quad \theta = 31.76° = 31.8° \qquad \textit{Ans.}$$
$$F = 236 \text{ N} \qquad \qquad \qquad \qquad \textit{Ans.}$$

F2–11. $\xrightarrow{+}(F_R)_x = \Sigma F_x;$
$$(80 \text{ N}) \cos 45° = F \cos \theta + 50 \text{ N} - \left(\tfrac{3}{5}\right) 90 \text{ N}$$
$$+\uparrow(F_R)_y = \Sigma F_y;$$
$$-(80 \text{ N}) \sin 45° = F \sin \theta - \left(\tfrac{4}{5}\right)(90 \text{ N})$$
$$\tan \theta = 0.2547 \quad \theta = 14.29° = 14.3° \qquad \textit{Ans.}$$
$$F = 62.5 \text{ N} \qquad \qquad \qquad \qquad \textit{Ans.}$$

F2–12. $(F_R)_x = 15\left(\tfrac{4}{5}\right) + 0 + 15\left(\tfrac{4}{5}\right) = 24 \text{ kN} \rightarrow$
$$(F_R)_y = 15\left(\tfrac{3}{5}\right) + 20 - 15\left(\tfrac{3}{5}\right) = 20 \text{ kN} \uparrow$$
$$F_R = 31.2 \text{ kN} \qquad \qquad \qquad \textit{Ans.}$$
$$\theta = 39.8° \qquad \qquad \qquad \qquad \textit{Ans.}$$

F2–13. $F_x = 75 \cos 30° \sin 45° = 45.93 \text{ N}$
$$F_y = 75 \cos 30° \cos 45° = 45.93 \text{ N}$$
$$F_z = -75 \sin 30° = -37.5 \text{ N}$$
$$\alpha = \cos^{-1} \left(\tfrac{45.93}{75}\right) = 52.2° \qquad \textit{Ans.}$$
$$\beta = \cos^{-1} \left(\tfrac{45.93}{75}\right) = 52.2° \qquad \textit{Ans.}$$
$$\gamma = \cos^{-1} \left(\tfrac{-37.5}{75}\right) = 120° \qquad \textit{Ans.}$$

F2–14. $\cos \beta = \sqrt{1 - \cos^2 120° - \cos^2 60°} = \pm 0.7071$

Require $\beta = 135°$.

$\mathbf{F} = F\mathbf{u}_F = (500 \text{ N})(-0.5\mathbf{i} - 0.7071\mathbf{j} + 0.5\mathbf{k})$
$= \{-250\mathbf{i} - 354\mathbf{j} + 250\mathbf{k}\} \text{ N}$ *Ans.*

F2–15. $\cos^2\alpha + \cos^2 135° + \cos^2 120° = 1$

$\alpha = 60°$

$\mathbf{F} = F\mathbf{u}_F = (500 \text{ N})(0.5\mathbf{i} - 0.7071\mathbf{j} - 0.5\mathbf{k})$
$= \{250\mathbf{i} - 354\mathbf{j} - 250\mathbf{k}\} \text{ N}$ *Ans.*

F2–16. $F_z = (50 \text{ N}) \sin 45° = 35.36 \text{ N}$
$F' = (50 \text{ N}) \cos 45° = 35.36 \text{ N}$
$F_x = \left(\frac{3}{5}\right)(35.36 \text{ N}) = 21.21 \text{ N}$
$F_y = \left(\frac{4}{5}\right)(35.36 \text{ N}) = 28.28 \text{ N}$
$\mathbf{F} = \{-21.2\mathbf{i} + 28.3\mathbf{j} + 35.4\mathbf{k}\} \text{ N}$ *Ans.*

F2–17. $F_z = (750 \text{ N}) \sin 45° = 530.33 \text{ N}$
$F' = (750 \text{ N}) \cos 45° = 530.33 \text{ N}$
$F_x = (530.33 \text{ N}) \cos 60° = 265.2 \text{ N}$
$F_y = (530.33 \text{ N}) \sin 60° = 459.3 \text{ N}$
$\mathbf{F}_2 = \{265\mathbf{i} - 459\mathbf{j} + 530\mathbf{k}\} \text{ N}$ *Ans.*

F2–18. $\mathbf{F}_1 = \left(\frac{4}{5}\right)(500 \text{ N}) \mathbf{j} + \left(\frac{3}{5}\right)(500 \text{ N})\mathbf{k}$
$= \{400\mathbf{j} + 300\mathbf{k}\} \text{ N}$
$\mathbf{F}_2 = [(800 \text{ N}) \cos 45°] \cos 30° \, \mathbf{i}$
$+ [(800 \text{ N}) \cos 45°] \sin 30°\mathbf{j}$
$+ (800 \text{ N}) \sin 45° \,(-\mathbf{k})$
$= \{489.90\mathbf{i} + 282.84\mathbf{j} - 565.69\mathbf{k}\} \text{ N}$
$\mathbf{F}_R = \mathbf{F}_1 + \mathbf{F}_2 = \{490\mathbf{i} + 683\mathbf{j} - 266\mathbf{k}\} \text{ N}$ *Ans.*

F2–19. $\mathbf{r}_{AB} = \{-6\mathbf{i} + 6\mathbf{j} + 3\mathbf{k}\} \text{ m}$ *Ans.*
$r_{AB} = \sqrt{(-6 \text{ m})^2 + (6 \text{ m})^2 + (3 \text{ m})^2} = 9 \text{ m}$ *Ans.*
$\alpha = 132°, \quad \beta = 48.2°, \quad \gamma = 70.5°$ *Ans.*

F2–20. $\mathbf{r}_{AB} = \{-4\mathbf{i} + 2\mathbf{j} + 4\mathbf{k}\} \text{ m}$ *Ans.*
$r_{AB} = \sqrt{(-4 \text{ m})^2 + (2 \text{ m})^2 + (4 \text{ m})^2} = 6 \text{ m}$ *Ans.*
$\alpha = \cos^{-1}\left(\frac{-4 \text{ m}}{6 \text{ m}}\right) = 131.8°$
$\theta = 180° - 131.8° = 48.2°$ *Ans.*

F2–21. $\mathbf{r}_{AB} = \{2\mathbf{i} + 3\mathbf{j} - 6\mathbf{k}\} \text{ m}$
$\mathbf{F}_{AB} = F_{AB}\mathbf{u}_{AB}$
$= (630 \text{ N})\left(\frac{2}{7}\mathbf{i} + \frac{3}{7}\mathbf{j} - \frac{6}{7}\mathbf{k}\right)$
$= \{180\mathbf{i} + 270\mathbf{j} - 540\mathbf{k}\} \text{ N}$ *Ans.*

F2–22. $\mathbf{F} = F\mathbf{u}_{AB} = 900\text{N}\left(-\frac{4}{9}\mathbf{i} + \frac{7}{9}\mathbf{j} - \frac{4}{9}\mathbf{k}\right)$
$= \{-400\mathbf{i} + 700\mathbf{j} - 400\mathbf{k}\} \text{ N}$ *Ans.*

F2–23. $\mathbf{F}_B = F_B\mathbf{u}_B$
$= (840 \text{ N})\left(\frac{3}{7}\mathbf{i} - \frac{2}{7}\mathbf{j} - \frac{6}{7}\mathbf{k}\right)$
$= \{360\mathbf{i} - 240\mathbf{j} - 720\mathbf{k}\} \text{ N}$
$\mathbf{F}_C = F_C\mathbf{u}_C$
$= (420 \text{ N})\left(\frac{2}{7}\mathbf{i} + \frac{3}{7}\mathbf{j} - \frac{6}{7}\mathbf{k}\right)$
$= \{120\mathbf{i} + 180\mathbf{j} - 360\mathbf{k}\} \text{ N}$
$F_R = \sqrt{(480 \text{ N})^2 + (-60 \text{ N})^2 + (-1080 \text{ N})^2}$
$= 1.18 \text{ kN}$ *Ans.*

F2–24. $\mathbf{F}_B = F_B\mathbf{u}_B$
$= (600 \text{ N})\left(-\frac{1}{3}\mathbf{i} + \frac{2}{3}\mathbf{j} - \frac{2}{3}\mathbf{k}\right)$
$= \{-200\mathbf{i} + 400\mathbf{j} - 400\mathbf{k}\} \text{ N}$
$\mathbf{F}_C = F_C\mathbf{u}_C$
$= (490 \text{ N})\left(-\frac{6}{7}\mathbf{i} + \frac{3}{7}\mathbf{j} - \frac{2}{7}\mathbf{k}\right)$
$= \{-420\mathbf{i} + 210\mathbf{j} - 140\mathbf{k}\} \text{ N}$
$\mathbf{F}_R = \mathbf{F}_B + \mathbf{F}_C = \{-620\mathbf{i} + 610\mathbf{j} - 540\mathbf{k}\} \text{ N}$ *Ans.*

F2–25. $\mathbf{u}_{AO} = -\frac{1}{3}\mathbf{i} + \frac{2}{3}\mathbf{j} - \frac{2}{3}\mathbf{k}$
$\mathbf{u}_F = -0.5345\mathbf{i} + 0.8018\mathbf{j} + 0.2673\mathbf{k}$
$\theta = \cos^{-1}(\mathbf{u}_{AO} \cdot \mathbf{u}_F) = 57.7°$ *Ans.*

F2–26. $\mathbf{u}_{AB} = -\frac{3}{5}\mathbf{j} + \frac{4}{5}\mathbf{k}$
$\mathbf{u}_F = \frac{4}{5}\mathbf{i} - \frac{3}{5}\mathbf{j}$
$\theta = \cos^{-1}(\mathbf{u}_{AB} \cdot \mathbf{u}_F) = 68.9°$ *Ans.*

F2–27. $\mathbf{u}_{OA} = \frac{12}{13}\mathbf{i} + \frac{5}{13}\mathbf{j}$
$\mathbf{u}_{OA} \cdot \mathbf{j} = u_{OA}(1) \cos \theta$
$\cos \theta = \frac{5}{13}; \quad \theta = 67.4°$ *Ans.*

F2–28. $\mathbf{u}_{OA} = \frac{12}{13}\mathbf{i} + \frac{5}{13}\mathbf{j}$
$\mathbf{F} = F\mathbf{u}_F = [650\mathbf{j}] \text{ N}$
$F_{OA} = \mathbf{F} \cdot \mathbf{u}_{OA} = 250 \text{ N}$
$\mathbf{F}_{OA} = F_{OA}\,\mathbf{u}_{OA} = \{231\mathbf{i} + 96.2\mathbf{j}\} \text{ N}$ *Ans.*

F2–29.
$$\mathbf{F} = (400\text{ N})\frac{\{4\,\mathbf{i} + 1\,\mathbf{j} - 6\,\mathbf{k}\}\text{ m}}{\sqrt{(4\text{ m})^2 + (1\text{ m})^2 + (-6\text{ m})^2}}$$
$$= \{219.78\mathbf{i} + 54.94\mathbf{j} - 329.67\mathbf{k}\}\text{ N}$$
$$\mathbf{u}_{AO} = \frac{\{-4\,\mathbf{j} - 6\,\mathbf{k}\}\text{ m}}{\sqrt{(-4\text{ m})^2 + (-6\text{ m})^2}}$$
$$= -0.5547\mathbf{j} - 0.8321\mathbf{k}$$
$$(F_{AO})_{\text{proj}} = \mathbf{F} \cdot \mathbf{u}_{AO} = 244\text{ N} \qquad\qquad Ans.$$

F2–30.
$$\mathbf{F} = [(-600\text{ N})\cos 60°]\sin 30°\,\mathbf{i}$$
$$+ [(600\text{ N})\cos 60°]\cos 30°\,\mathbf{j}$$
$$+ [(600\text{ N})\sin 60°]\,\mathbf{k}$$
$$= \{-150\mathbf{i} + 259.81\mathbf{j} + 519.62\mathbf{k}\}\text{ N}$$
$$\mathbf{u}_A = -\tfrac{2}{3}\mathbf{i} + \tfrac{2}{3}\mathbf{j} + \tfrac{1}{3}\mathbf{k}$$
$$(F_A)_{\text{par}} = \mathbf{F} \cdot \mathbf{u}_A = 446.41\text{ N} = 446\text{ N} \qquad Ans.$$
$$(F_A)_{\text{per}} = \sqrt{(600\text{ N})^2 - (446.41\text{ N})^2}$$
$$= 401\text{ N} \qquad\qquad Ans.$$

F2–31.
$$\mathbf{F} = 56\text{ N}\,(\tfrac{3}{7}\mathbf{i} - \tfrac{6}{7}\mathbf{j} + \tfrac{2}{7}\mathbf{k})$$
$$= \{24\mathbf{i} - 48\mathbf{j} + 16\mathbf{k}\}\text{ N}$$
$$(F_{AO})_{\parallel} = \mathbf{F} \cdot \mathbf{u}_{AO} = (24\mathbf{i} - 48\mathbf{j} + 16\mathbf{k}) \cdot (\tfrac{3}{7}\mathbf{i} - \tfrac{6}{7}\mathbf{j} - \tfrac{2}{7}\mathbf{k})$$
$$= 46.86\text{ N} = 46.9\text{ N} \qquad Ans.$$
$$(F_{AO})_{\perp} = \sqrt{F^2 - (F_{AO})_{\parallel}} = \sqrt{(56)^2 - (46.86)^2}$$
$$= 30.7\text{ N} \qquad\qquad Ans.$$

Chapter 3

F3–1. $\xrightarrow{+} \Sigma F_x = 0;\ \tfrac{4}{5}F_{AC} - F_{AB}\cos 30° = 0$
$$+\uparrow \Sigma F_y = 0;\ \tfrac{3}{5}F_{AC} + F_{AB}\sin 30° - 550 = 0$$
$$F_{AB} = 478\text{ N} \qquad\qquad Ans.$$
$$F_{AC} = 518\text{ N} \qquad\qquad Ans.$$

F3–2. $+\uparrow \Sigma F_y = 0;\ -2(7.5)\sin \theta + 3.5 = 0$
$$\theta = 13.5°$$
$$L_{ABC} = 2\left(\frac{1.5\text{ m}}{\cos 13.5°}\right) = 3.085\text{ m} \qquad Ans.$$

F3–3. $\xrightarrow{+} \Sigma F_x = 0;\ T\cos \theta - T\cos \phi = 0$
$$\phi = \theta$$
$$+\uparrow \Sigma F_y = 0;\ 2T\sin \theta - 49.05\text{ N} = 0$$
$$\theta = \tan^{-1}\left(\frac{0.15\text{ m}}{0.2\text{ m}}\right) = 36.87°$$
$$T = 40.9\text{ N} \qquad\qquad Ans.$$

F3–4. $+\nearrow \Sigma F_x = 0;\ \tfrac{4}{5}(F_{sp}) - 5(9.81)\sin 45° = 0$
$$F_{sp} = 43.35\text{ N}$$
$$F_{sp} = k(l - l_0);\ 43.35 = 200(0.5 - l_0)$$
$$l_0 = 0.283\text{ m} \qquad\qquad Ans.$$

F3–5. $+\uparrow \Sigma F_y = 0;\ (392.4\text{ N})\sin 30° - m_A(9.81) = 0$
$$m_A = 20\text{ kg} \qquad\qquad Ans.$$

F3–6. $+\uparrow \Sigma F_y = 0;\ T_{AB}\sin 15° - 10(9.81)\text{ N} = 0$
$$T_{AB} = 379.03\text{ N} = 379\text{ N} \qquad Ans.$$
$$\xrightarrow{+} \Sigma F_x = 0;\ T_{BC} - 379.03\text{ N}\cos 15° = 0$$
$$T_{BC} = 366.11\text{ N} = 366\text{ N} \qquad Ans.$$
$$\xrightarrow{+} \Sigma F_x = 0;\ T_{CD}\cos \theta - 366.11\text{ N} = 0$$
$$+\uparrow \Sigma F_y = 0;\ T_{CD}\sin \theta - 15(9.81)\text{ N} = 0$$
$$T_{CD} = 395\text{ N} \qquad\qquad Ans.$$
$$\theta = 21.9° \qquad\qquad Ans.$$

F3–7. $\Sigma F_x = 0;\ \left[\left(\tfrac{3}{5}\right)F_3\right]\left(\tfrac{3}{5}\right) + 600\text{ N} - F_2 = 0 \quad (1)$
$$\Sigma F_y = 0;\ \left(\tfrac{4}{5}\right)F_1 - \left[\left(\tfrac{3}{5}\right)F_3\right]\left(\tfrac{4}{5}\right) = 0 \quad (2)$$
$$\Sigma F_z = 0;\ \left(\tfrac{4}{5}\right)F_3 + \left(\tfrac{3}{5}\right)F_1 - 900\text{ N} = 0 \quad (3)$$
$$F_3 = 776\text{ N} \qquad\qquad Ans.$$
$$F_1 = 466\text{ N} \qquad\qquad Ans.$$
$$F_2 = 879\text{ N} \qquad\qquad Ans.$$

F3–8. $\Sigma F_z = 0;\ F_{AD}\left(\tfrac{4}{5}\right) - 900 = 0$
$$F_{AD} = 1125\text{ N} = 1.125\text{ kN} \qquad Ans.$$
$$\Sigma F_y = 0;\ F_{AC}\left(\tfrac{4}{5}\right) - 1125\left(\tfrac{3}{5}\right) = 0$$
$$F_{AC} = 843.75\text{ N} = 844\text{ N} \qquad Ans.$$
$$\Sigma F_x = 0;\ F_{AB} - 843.75\left(\tfrac{3}{5}\right) = 0$$
$$F_{AB} = 506.25\text{ N} = 506\text{ N} \qquad Ans.$$

F3–9. $\mathbf{F}_{AD} = F_{AD}\left(\dfrac{\mathbf{r}_{AD}}{r_{AD}}\right) = \tfrac{1}{3}F_{AD}\mathbf{i} - \tfrac{2}{3}F_{AD}\mathbf{j} + \tfrac{2}{3}F_{AD}\mathbf{k}$
$$\Sigma F_z = 0;\qquad \tfrac{2}{3}F_{AD} - 600 = 0$$
$$F_{AD} = 900\text{ N} \qquad\qquad Ans.$$
$$\Sigma F_y = 0;\qquad F_{AB}\cos 30° - \tfrac{2}{3}(900) = 0$$
$$F_{AB} = 692.82\text{ N} = 693\text{ N} \qquad Ans.$$
$$\Sigma F_x = 0;\qquad \tfrac{1}{3}(900) + 692.82\sin 30° - F_{AC} = 0$$
$$F_{AC} = 646.41\text{ N} = 646\text{ N} \qquad Ans.$$

F3–10. $\mathbf{F}_{AC} = F_{AC}\{-\cos 60°\sin 30°\,\mathbf{i}$
$$+ \cos 60°\cos 30°\,\mathbf{j} + \sin 60°\,\mathbf{k}\}$$
$$= -0.25F_{AC}\,\mathbf{i} + 0.4330F_{AC}\mathbf{j} + 0.8660F_{AC}\,\mathbf{k}$$
$$\mathbf{F}_{AD} = F_{AD}\{\cos 120°\,\mathbf{i} + \cos 120°\,\mathbf{j} + \cos 45°\,\mathbf{k}\}$$
$$= -0.5F_{AD}\,\mathbf{i} - 0.5F_{AD}\,\mathbf{j} + 0.7071F_{AD}\,\mathbf{k}$$
$$\Sigma F_y = 0;\ 0.4330F_{AC} - 0.5F_{AD} = 0$$
$$\Sigma F_z = 0;\ 0.8660F_{AC} + 0.7071F_{AD} - 300 = 0$$
$$F_{AD} = 175.74\text{ N} = 176\text{ N} \qquad Ans.$$
$$F_{AC} = 202.92\text{ N} = 203\text{ N} \qquad Ans.$$
$$\Sigma F_x = 0;\ F_{AB} - 0.25(202.92) - 0.5(175.74) = 0$$
$$F_{AB} = 138.60\text{ N} = 139\text{ N} \qquad Ans.$$

F3–11. $\mathbf{F}_B = F_B\left(\dfrac{\mathbf{r}_{AB}}{r_{AB}}\right)$

$= F_B\left[\dfrac{\{-3\mathbf{i} + 1.5\mathbf{j} + 1\mathbf{k}\}\ \text{m}}{\sqrt{(-3\ \text{m})^2 + (1.5\ \text{m})^2 + (1\ \text{m})^2}}\right]$

$= -\tfrac{6}{7}F_B\mathbf{i} + \tfrac{3}{7}F_B\mathbf{j} + \tfrac{2}{7}F_B\mathbf{k}$

$\mathbf{F}_C = F_C\left(\dfrac{\mathbf{r}_{AC}}{r_{AC}}\right)$

$= F_C\left[\dfrac{\{-3\mathbf{i} - 1\mathbf{j} + 1.5\mathbf{k}\}\ \text{m}}{\sqrt{(-3\ \text{m})^2 + (-1\ \text{m})^2 + (1.5\ \text{m})^2}}\right]$

$= -\tfrac{6}{7}F_C\mathbf{i} - \tfrac{2}{7}F_C\mathbf{j} + \tfrac{3}{7}F_C\mathbf{k}$

$\mathbf{F}_D = F_D\mathbf{i}$

$\mathbf{W} = \{-150\mathbf{k}\}\ \text{N}$

$\Sigma F_x = 0;\ -\tfrac{6}{7}F_B - \tfrac{6}{7}F_C + F_D = 0$　(1)

$\Sigma F_y = 0;\ \tfrac{3}{7}F_B - \tfrac{2}{7}F_C = 0$　(2)

$\Sigma F_z = 0;\ \tfrac{2}{7}F_B + \tfrac{3}{7}F_C - 150 = 0$　(3)

$F_B = 162\ \text{N}$　*Ans.*

$F_C = 1.5(162\ \text{N}) = 242\ \text{N}$　*Ans.*

$F_D = 346.15\ \text{N} = 346\ \text{N}$　*Ans.*

Chapter 4

F4–1. $\circlearrowleft +M_O = -\left(\tfrac{4}{5}\right)(100\ \text{N})(2\ \text{m}) - \left(\tfrac{3}{5}\right)(100\ \text{N})(5\ \text{m})$

$= -460\ \text{N}\cdot\text{m} = 460\ \text{N}\cdot\text{m}\ \circlearrowright$　*Ans.*

F4–2. $\circlearrowleft +M_O = [(300\ \text{N})\sin 30°][0.4\ \text{m} + (0.3\ \text{m})\cos 45°]$

$-\ [(300\ \text{N})\cos 30°][(0.3\ \text{m})\sin 45°]$

$= 36.7\ \text{N}\cdot\text{m}$　*Ans.*

F4–3. $\circlearrowleft +M_O = (60\ \text{kN})(4\ \text{m} + (3\ \text{m})\cos 45° - 1\ \text{m})$

$= 307\ \text{kN}\cdot\text{m}$　*Ans.*

F4–4. $\circlearrowright +M_O = 50\sin 60°\ (0.1 + 0.2\cos 45° + 0.1)$

$-\ 50\cos 60°(0.2\sin 45°)$

$= 11.2\ \text{N}\cdot\text{m}$　*Ans*

F4–5. $\circlearrowleft +M_O = 60\sin 50°\ (2.5) + 60\cos 50°\ (0.25)$

$= 124.5\ \text{kN}\cdot\text{m}$　*Ans.*

F4–6. $\circlearrowleft +M_O = 500\sin 45°\ (3 + 3\cos 45°)$

$-\ 500\cos 45°\ (3\sin 45°)$

$= 1.06\ \text{kN}\cdot\text{m}$　*Ans.*

F4–7. $\circlearrowleft +(M_R)_O = \Sigma Fd;$

$(M_R)_O = -(600\ \text{N})(1\ \text{m})$

$+\ (500\ \text{N})[3\ \text{m} + (2.5\ \text{m})\cos 45°]$

$-\ (300\text{N})[(2.5\ \text{m})\sin 45°]$

$= 1254\ \text{N}\cdot\text{m} = 1.25\ \text{kN}\cdot\text{m}$　*Ans.*

F4–8. $\circlearrowleft +(M_R)_O = \Sigma Fd;$

$(M_R)_O = \left[\left(\tfrac{3}{5}\right)500\ \text{N}\right](0.425\ \text{m})$

$-\ \left[\left(\tfrac{4}{5}\right)500\ \text{N}\right](0.25\ \text{m})$

$-\ [(600\ \text{N})\cos 60°](0.25\ \text{m})$

$-\ [(600\ \text{N})\sin 60°](0.425\ \text{m})$

$= -268\ \text{N}\cdot\text{m} = 268\ \text{N}\cdot\text{m}\ \circlearrowright$　*Ans.*

F4–9. $\circlearrowleft +(M_R)_O = \Sigma Fd;$

$(M_R)_O = (30\cos 30°\ \text{kN})(3\ \text{m} + 3\sin 30°\ \text{m})$

$-\ (30\sin 30°\ \text{kN})(3\cos 30°\ \text{m})$

$+\ (20\ \text{kN})(3\cos 30°\ \text{m})$

$= 129.9\ \text{kN}\cdot\text{m}$　*Ans.*

F4–10. $\mathbf{F} = F\mathbf{u}_{AB} = 500\ \text{N}\left(\tfrac{4}{5}\mathbf{i} - \tfrac{3}{5}\mathbf{j}\right) = \{400\mathbf{i} - 300\mathbf{j}\}\ \text{N}$

$\mathbf{M}_O = \mathbf{r}_{OA} \times \mathbf{F} = \{3\mathbf{j}\}\ \text{m} \times \{400\mathbf{i} - 300\mathbf{j}\}\ \text{N}$

$= \{-1200\mathbf{k}\}\ \text{N}\cdot\text{m}$　*Ans.*

or

$\mathbf{M}_O = \mathbf{r}_{OB} \times \mathbf{F} = \{4\mathbf{i}\}\ \text{m} \times \{400\mathbf{i} - 300\mathbf{j}\}\ \text{N}$

$= \{-1200\mathbf{k}\}\ \text{N}\cdot\text{m}$　*Ans.*

F4–11. $\mathbf{F} = F\mathbf{u}_{BC}$

$= 120\ \text{kN}\left[\dfrac{\{4\mathbf{i} - 4\mathbf{j} - 2\mathbf{k}\}\ \text{m}}{\sqrt{(4\ \text{m})^2 + (-4\ \text{m})^2 + (-2\ \text{m})^2}}\right]$

$= \{80\mathbf{i} - 80\mathbf{j} - 40\mathbf{k}\}\ \text{kN}$

$\mathbf{M}_O = \mathbf{r}_C \times \mathbf{F} = \begin{vmatrix} \mathbf{i} & \mathbf{j} & \mathbf{k} \\ 5 & 0 & 0 \\ 80 & -80 & -40 \end{vmatrix}$

$= \{200\mathbf{j} - 400\mathbf{k}\}\ \text{kN}\cdot\text{m}$　*Ans.*

or

$\mathbf{M}_O = \mathbf{r}_D \times \mathbf{F} = \begin{vmatrix} \mathbf{i} & \mathbf{j} & \mathbf{k} \\ 1 & 4 & 2 \\ 80 & -80 & -40 \end{vmatrix}$

$= \{200\mathbf{j} - 400\mathbf{k}\}\ \text{kN}\cdot\text{m}$　*Ans.*

F4–12. $\mathbf{F}_R = \mathbf{F}_1 + \mathbf{F}_2$

$= \{(100 - 200)\mathbf{i} + (-120 + 250)\mathbf{j}$

$+\ (75 + 100)\mathbf{k}\}\ \text{kN}$

$= \{-100\mathbf{i} + 130\mathbf{j} + 175\mathbf{k}\}\ \text{kN}$

$(\mathbf{M}_R)_O = \mathbf{r}_A \times \mathbf{F}_R = \begin{vmatrix} \mathbf{i} & \mathbf{j} & \mathbf{k} \\ 4 & 5 & 3 \\ -100 & 130 & 175 \end{vmatrix}$

$= \{485\mathbf{i} - 1000\mathbf{j} + 1020\mathbf{k}\}\ \text{kN}\cdot\text{m}$　*Ans.*

F4–13. $M_x = \mathbf{i} \cdot (\mathbf{r}_{OB} \times \mathbf{F}) = \begin{vmatrix} 1 & 0 & 0 \\ 0.3 & 0.4 & -0.2 \\ 300 & -200 & 150 \end{vmatrix}$

$= 20 \text{ N} \cdot \text{m}$ *Ans.*

F4–14. $\mathbf{u}_{OA} = \dfrac{\mathbf{r}_A}{r_A} = \dfrac{\{0.3\mathbf{i} + 0.4\mathbf{j}\} \text{ m}}{\sqrt{(0.3 \text{ m})^2 + (0.4 \text{ m})^2}} = 0.6\,\mathbf{i} + 0.8\,\mathbf{j}$

$M_{OA} = \mathbf{u}_{OA} \cdot (\mathbf{r}_{AB} \times \mathbf{F}) = \begin{vmatrix} 0.6 & 0.8 & 0 \\ 0 & 0 & -0.2 \\ 300 & -200 & 150 \end{vmatrix}$

$= -72 \text{ N} \cdot \text{m}$ *Ans.*

$|M_{OA}| = 72 \text{ N} \cdot \text{m}$

F4–15. Scalar Analysis

The magnitudes of the force components are

$F_x = |200 \cos 120°| = 100 \text{ N}$

$F_y = 200 \cos 60° = 100 \text{ N}$

$F_z = 200 \cos 45° = 141.42 \text{ N}$

$M_x = -F_y(z) + F_z(y)$

$= -(100 \text{ N})(0.25 \text{ m}) + (141.42 \text{ N})(0.3 \text{ m})$

$= 17.4 \text{ N} \cdot \text{m}$ *Ans.*

Vector Analysis

$M_x = \begin{vmatrix} 1 & 0 & 0 \\ 0 & 0.3 & 0.25 \\ -100 & 100 & 141.42 \end{vmatrix} = 17.4 \text{ N} \cdot \text{m}$ *Ans.*

F4–16. $M_y = \mathbf{j} \cdot (\mathbf{r}_A \times \mathbf{F}) = \begin{vmatrix} 0 & 1 & 0 \\ -3 & -4 & 2 \\ 30 & -20 & 50 \end{vmatrix}$

$= 210 \text{ N} \cdot \text{m}$ *Ans.*

F4–17. $\mathbf{u}_{AB} = \dfrac{\mathbf{r}_{AB}}{r_{AB}} = \dfrac{\{-4\mathbf{i} + 3\mathbf{j}\} \text{ m}}{\sqrt{(-4 \text{ m})^2 + (3 \text{ m})^2}} = -0.8\mathbf{i} + 0.6\mathbf{j}$

$M_{AB} = \mathbf{u}_{AB} \cdot (\mathbf{r}_{AC} \times \mathbf{F})$

$= \begin{vmatrix} -0.8 & 0.6 & 0 \\ 0 & 0 & 2 \\ 50 & -40 & 20 \end{vmatrix} = -4 \text{ kN} \cdot \text{m}$

$\mathbf{M}_{AB} = M_{AB}\mathbf{u}_{AB} = \{3.2\mathbf{i} - 2.4\mathbf{j}\} \text{ kN} \cdot \text{m}$ *Ans.*

F4–18. Scalar Analysis

The magnitudes of the force components are

$F_x = \left(\frac{3}{5}\right)\left[\frac{4}{5}(500)\right] = 240 \text{ N}$

$F_y = \frac{4}{5}\left[\frac{4}{5}(500)\right] = 320 \text{ N}$

$F_z = \frac{3}{5}(500) = 300 \text{ N}$

$M_x = -F_y(z) + F_z(y)$

$= -320(3) + 300(2) = -360 \text{ N} \cdot \text{m}$ *Ans.*

$M_y = -F_x(z) - F_z(x)$

$= -240(3) - 300(-2) = -120 \text{ N} \cdot \text{m}$ *Ans.*

$M_z = F_x(y) - F_y(x)$

$= 240(2) - 320(-2) = -160 \text{ N} \cdot \text{m}$ *Ans.*

Vector Analysis

$\mathbf{F} = \{-240\mathbf{i} + 320\mathbf{j} + 300\mathbf{k}\} \text{ N}$

$\mathbf{r}_{OA} = \{-2\mathbf{i} + 2\mathbf{j} + 3\mathbf{k}\} \text{ m}$

$M_x = \mathbf{i} \cdot (\mathbf{r}_{OA} \times \mathbf{F}) = -360 \text{ N} \cdot \text{m}$

$M_y = \mathbf{j} \cdot (\mathbf{r}_{OA} \times \mathbf{F}) = -120 \text{ N} \cdot \text{m}$

$M_z = \mathbf{k} \cdot (\mathbf{r}_{OA} \times \mathbf{F}) = -160 \text{ N} \cdot \text{m}$

F4–19. $\curvearrowleft +M_{C_R} = \Sigma M_A = 400(3) - 400(5) + 300(5)$

$+ 200(0.2) = 740 \text{ N} \cdot \text{m}$ *Ans.*

Also,

$\curvearrowleft +M_{C_R} = 300(5) - 400(2) + 200(0.2)$

$= 740 \text{ N} \cdot \text{m}$ *Ans.*

F4–20. $\curvearrowright +M_{C_R} = 300(0.4) + 200(0.4) + 150(0.4)$

$= 260 \text{ N} \cdot \text{m}$ *Ans.*

F4–21. $\curvearrowright +(M_B)_R = \Sigma M_B$

$-1.5 \text{ kN} \cdot \text{m} = (2 \text{ kN})(0.3 \text{ m}) - F(0.9 \text{ m})$

$F = 2.33 \text{ kN}$ *Ans.*

F4–22. $\curvearrowright +M_C = 10\left(\frac{3}{5}\right)(2) - 10\left(\frac{4}{5}\right)(4) = -20 \text{ kN} \cdot \text{m}$

$= 20 \text{ kN} \cdot \text{m} \curvearrowleft$ *Ans.*

F4–23. $\mathbf{u}_1 = \dfrac{\mathbf{r}_1}{r_1} = \dfrac{\{-2\mathbf{i} + 2\mathbf{j} + 3.5\mathbf{k}\} \text{ m}}{\sqrt{(-2 \text{ m})^2 + (2 \text{ m})^2 + (3.5 \text{ m})^2}}$

$= -\frac{2}{4.5}\mathbf{i} + \frac{2}{4.5}\mathbf{j} + \frac{3.5}{4.5}\mathbf{k}$

$\mathbf{u}_2 = -\mathbf{k}$

$\mathbf{u}_3 = \frac{1.5}{2.5}\mathbf{i} - \frac{2}{2.5}\mathbf{j}$

$(\mathbf{M}_c)_1 = (M_c)_1 \mathbf{u}_1$

$= (450 \text{ kN} \cdot \text{m})\left(\frac{2}{4.5}\,\mathbf{i} + \frac{2}{4.5}\mathbf{j} + \frac{3.5}{4.5}\mathbf{k}\right)$

$= \{-200\mathbf{i} + 200\mathbf{j} + 350\mathbf{k}\} \text{ kN} \cdot \text{m}$

$(\mathbf{M}_c)_2 = (M_c)_2 \mathbf{u}_2 = (250 \text{ kN} \cdot \text{m})(-\mathbf{k})$

$= \{-250\mathbf{k}\} \text{ kN} \cdot \text{m}$

$(\mathbf{M}_c)_3 = (M_c)_3\,\mathbf{u}_3 = (300 \text{ kN} \cdot \text{m})\left(\frac{1.5}{2.5}\,\mathbf{i} - \frac{2}{2.5}\mathbf{j}\right)$

$= \{180\mathbf{i} - 240\mathbf{j}\} \text{ kN} \cdot \text{m}$

$(\mathbf{M}_c)_R = \Sigma \mathbf{M}_c;$

$(\mathbf{M}_c)_R = \{-20\mathbf{i} - 40\mathbf{j} + 100\mathbf{k}\} \text{ kN} \cdot \text{m}$ *Ans.*

F4–24. $\mathbf{F}_B = \left(\frac{4}{5}\right)(450 \text{ N})\mathbf{j} - \left(\frac{3}{5}\right)(450 \text{ N})\mathbf{k}$

$\qquad = \{360\mathbf{j} - 270\mathbf{k}\}$ N

$\qquad \mathbf{M}_c = \mathbf{r}_{AB} \times \mathbf{F}_B = \begin{vmatrix} \mathbf{i} & \mathbf{j} & \mathbf{k} \\ 0.4 & 0 & 0 \\ 0 & 360 & -270 \end{vmatrix}$

$\qquad = \{108\mathbf{j} + 144\mathbf{k}\}$ N $\cdot$ m $\qquad$ *Ans.*

Also,

$\qquad \mathbf{M}_c = (\mathbf{r}_A \times \mathbf{F}_A) + (\mathbf{r}_B \times \mathbf{F}_B)$

$\qquad = \begin{vmatrix} \mathbf{i} & \mathbf{j} & \mathbf{k} \\ 0 & 0 & 0.3 \\ 0 & -360 & 270 \end{vmatrix} + \begin{vmatrix} \mathbf{i} & \mathbf{j} & \mathbf{k} \\ 0.4 & 0 & 0.3 \\ 0 & 360 & -270 \end{vmatrix}$

$\qquad = \{108\mathbf{j} + 144\mathbf{k}\}$ N $\cdot$ m $\qquad$ *Ans.*

F4–25. $\overset{+}{\leftarrow} F_{Rx} = \Sigma F_x; \quad F_{Rx} = 200 - \frac{3}{5}(100) = 140$ N

$\qquad + \downarrow F_{Ry} = \Sigma F_y; \quad F_{Ry} = 150 - \frac{4}{5}(100) = 70$ N

$\qquad\qquad F_R = \sqrt{140^2 + 70^2} = 157$ N $\qquad$ *Ans.*

$\qquad\qquad \theta = \tan^{-1}\left(\frac{70}{140}\right) = 26.6° \quad \nearrow$ $\qquad$ *Ans.*

$\qquad \zeta + M_{A_R} = \Sigma M_A;$

$\qquad\qquad M_{A_R} = \frac{3}{5}(100)(0.4) - \frac{4}{5}(100)(0.6) + 150(0.3)$

$\qquad\qquad M_{R_A} = 21.0$ N $\cdot$ m $\qquad$ *Ans.*

F4–26. $\overset{+}{\rightarrow} F_{Rx} = \Sigma F_x; \quad F_{Rx} = \frac{4}{5}(50) = 40$ N

$\qquad + \downarrow F_{Ry} = \Sigma F_y; \quad F_{Ry} = 40 + 30 + \frac{3}{5}(50)$

$\qquad\qquad = 100$ N

$\qquad\qquad F_R = \sqrt{(40)^2 + (100)^2} = 108$ N $\qquad$ *Ans.*

$\qquad\qquad \theta = \tan^{-1}\left(\frac{100}{40}\right) = 68.2° \quad \searrow$ $\qquad$ *Ans.*

$\qquad \zeta + M_{A_R} = \Sigma M_A;$

$\qquad\qquad M_{A_R} = 30(3) + \frac{3}{5}(50)(6) + 200$

$\qquad\qquad = 470$ N $\cdot$ m $\qquad$ *Ans.*

F4–27. $\overset{+}{\rightarrow} (F_R)_x = \Sigma F_x;$

$\qquad\qquad (F_R)_x = 900 \sin 30° = 450$ N $\rightarrow$

$\qquad + \uparrow (F_R)_y = \Sigma F_y;$

$\qquad\qquad (F_R)_y = -900 \cos 30° - 300$

$\qquad\qquad = -1079.42$ N $= 1079.42$ N $\downarrow$

$\qquad\qquad F_R = \sqrt{450^2 + 1079.42^2}$

$\qquad\qquad = 1169.47$ N $= 1.17$ kN $\qquad$ *Ans.*

$\qquad\qquad \theta = \tan^{-1}\left(\frac{1079.42}{450}\right) = 67.4° \quad \searrow$ $\qquad$ *Ans.*

$\qquad \zeta + (M_R)_A = \Sigma M_A;$

$\qquad\qquad (M_R)_A = 300 - 900 \cos 30° (0.75) - 300(2.25)$

$\qquad\qquad = -959.57$ N $\cdot$ m

$\qquad\qquad = 960$ N $\cdot$ m $\circlearrowright$ $\qquad$ *Ans.*

F4–28. $\overset{+}{\rightarrow} (F_R)_x = \Sigma F_x;$

$\qquad\qquad (F_R)_x = 150\left(\frac{3}{5}\right) + 50 - 100\left(\frac{4}{5}\right) = 60$ N$\rightarrow$

$\qquad + \uparrow (F_R)_y = \Sigma F_y;$

$\qquad\qquad (F_R)_y = -150\left(\frac{4}{5}\right) - 100\left(\frac{3}{5}\right)$

$\qquad\qquad = -180$ N $= 180$ N $\downarrow$

$\qquad\qquad F_R = \sqrt{60^2 + 180^2} = 189.74$ N $= 190$ N $\qquad$ *Ans.*

$\qquad\qquad \theta = \tan^{-1}\left(\frac{180}{60}\right) = 71.6° \quad \searrow$ $\qquad$ *Ans.*

$\qquad \zeta + (M_R)_A = \Sigma M_A;$

$\qquad\qquad (M_R)_A = 100\left(\frac{4}{5}\right)(0.1) - 100\left(\frac{3}{5}\right)(0.6) - 150\left(\frac{4}{5}\right)(0.3)$

$\qquad\qquad = -64.0 = 64.0$ N $\cdot$ m $\circlearrowright$ $\qquad$ *Ans.*

F4–29. $\mathbf{F}_R = \Sigma\mathbf{F};$

$\qquad\qquad \mathbf{F}_R = \mathbf{F}_1 + \mathbf{F}_2$

$\qquad\qquad = (-300\mathbf{i} + 150\mathbf{j} + 200\mathbf{k}) + (-450\mathbf{k})$

$\qquad\qquad = \{-300\mathbf{i} + 150\mathbf{j} - 250\mathbf{k}\}$ N $\qquad$ *Ans.*

$\qquad\qquad \mathbf{r}_{OA} = (2 - 0)\mathbf{j} = \{2\mathbf{j}\}$ m

$\qquad\qquad \mathbf{r}_{OB} = (-1.5 - 0)\mathbf{i} + (2 - 0)\mathbf{j} + (1 - 0)\mathbf{k}$

$\qquad\qquad = \{-1.5\mathbf{i} + 2\mathbf{j} + 1\mathbf{k}\}$ m

$\qquad (\mathbf{M}_R)_O = \Sigma\mathbf{M};$

$\qquad (\mathbf{M}_R)_O = \mathbf{r}_{OB} \times \mathbf{F}_1 + \mathbf{r}_{OA} \times \mathbf{F}_2$

$\qquad = \begin{vmatrix} \mathbf{i} & \mathbf{j} & \mathbf{k} \\ -1.5 & 2 & 1 \\ -300 & 150 & 200 \end{vmatrix} + \begin{vmatrix} \mathbf{i} & \mathbf{j} & \mathbf{k} \\ 0 & 2 & 0 \\ 0 & 0 & -450 \end{vmatrix}$

$\qquad = \{-650\mathbf{i} + 375\mathbf{k}\}$ N $\cdot$ m $\qquad$ *Ans.*

F4–30. $\mathbf{F}_1 = \{-100\mathbf{j}\}$ N

$\qquad \mathbf{F}_2 = (200 \text{ N})\left[\dfrac{\{-0.4\mathbf{i} - 0.3\mathbf{k}\} \text{ m}}{\sqrt{(-0.4 \text{ m})^2 + (-0.3 \text{ m})^2}}\right]$

$\qquad = \{-160\mathbf{i} - 120\mathbf{k}\}$ N

$\qquad \mathbf{M}_c = \{-75\mathbf{i}\}$ N $\cdot$ m

$\qquad \mathbf{F}_R = \{-160\mathbf{i} - 100\mathbf{j} - 120\mathbf{k}\}$ N $\qquad$ *Ans.*

$\qquad (\mathbf{M}_R)_O = (0.3\mathbf{k}) \times (-100\mathbf{j})$

$\qquad + \begin{vmatrix} \mathbf{i} & \mathbf{j} & \mathbf{k} \\ 0 & 0.5 & 0.3 \\ -160 & 0 & -120 \end{vmatrix} + (-75\mathbf{i})$

$\qquad = \{-105\mathbf{i} - 48\mathbf{j} + 80\mathbf{k}\}$ N $\cdot$ m $\qquad$ *Ans.*

F4–31. $+ \downarrow F_R = \Sigma F_y; \quad F_R = 500 + 250 + 500$

$\qquad\qquad = 1250$ kN $\qquad$ *Ans.*

$\qquad \zeta + F_R x = \Sigma M_O;$

$\qquad 1250(x) = 500(3) + 250(6) + 500(9)$

$\qquad\qquad x = 6$ m $\qquad$ *Ans.*

F4–32. $\xrightarrow{+} (F_R)_x = \Sigma F_x;$

$(F_R)_x = 100\left(\frac{3}{5}\right) + 50 \sin 30° = 85 \text{ kN} \rightarrow$

$+\uparrow (F_R)_y = \Sigma F_y;$

$(F_R)_y = 200 + 50 \cos 30° - 100\left(\frac{4}{5}\right)$

$= 163.30 \text{ kN}\uparrow$

$F_R = \sqrt{85^2 + 163.30^2} = 184 \text{ kN}$

$\theta = \tan^{-1}\left(\frac{163.30}{85}\right) = 62.5°$ ⊿   *Ans.*

$\zeta +(M_R)_A = \Sigma M_A;$

$163.30(d) = 200(3) - 100\left(\frac{4}{5}\right)(6) + 50 \cos 30°(9)$

$d = 3.12 \text{ m}$ *Ans.*

F4–33. $\xrightarrow{+} (F_R)_x = \Sigma F_x;$

$(F_R)_x = 15\left(\frac{4}{5}\right) = 12 \text{ kN} \rightarrow$

$+\uparrow (F_R)_y = \Sigma F_y;$

$(F_R)_y = -20 + 15\left(\frac{3}{5}\right) = -11 \text{ kN} = 11 \text{ kN}\downarrow$

$F_R = \sqrt{12^2 + 11^2} = 16.3 \text{ kN}$ *Ans.*

$\theta = \tan^{-1}\left(\frac{11}{12}\right) = 42.5°$ ⦧   *Ans.*

$\zeta +(M_R)_A = \Sigma M_A;$

$-11(d) = -20(2) - 15\left(\frac{4}{5}\right)(2) + 15\left(\frac{3}{5}\right)(6)$

$d = 0.909 \text{ m}$ *Ans.*

F4–34. $\xrightarrow{+} (F_R)_x = \Sigma F_x;$

$(F_R)_x = \left(\frac{3}{5}\right) 5 \text{ kN} - 8 \text{ kN}$

$= -5 \text{ kN} = 5 \text{ kN} \leftarrow$

$+\uparrow (F_R)_y = \Sigma F_y;$

$(F_R)_y = -6 \text{ kN} - \left(\frac{4}{5}\right) 5 \text{ kN}$

$= -10 \text{ kN} = 10 \text{ kN}\downarrow$

$F_R = \sqrt{5^2 + 10^2} = 11.2 \text{ kN}$ *Ans.*

$\theta = \tan^{-1}\left(\frac{10 \text{ kN}}{5 \text{ kN}}\right) = 63.4°$ ⦨   *Ans.*

$\zeta +(M_R)_A = \Sigma M_A;$

$5 \text{ kN}(d) = 8 \text{ kN}(3 \text{ m}) - 6 \text{ kN}(0.5 \text{ m})$

$- \left[\left(\frac{4}{5}\right)5 \text{ kN}\right](2 \text{ m})$

$- \left[\left(\frac{3}{5}\right)5 \text{ kN}\right](4 \text{ m})$

$d = 0.2 \text{ m}$ *Ans.*

F4–35. $+\downarrow F_R = \Sigma F_z;$ $F_R = 400 + 500 - 100$

$= 800 \text{ N}$ *Ans.*

$M_{Rx} = \Sigma M_x;\ -800y = -400(4) - 500(4)$

$y = 4.50 \text{ m}$ *Ans.*

$M_{Ry} = \Sigma M_y;\ \ 800x = 500(4) - 100(3)$

$x = 2.125 \text{ m}$ *Ans.*

F4–36. $+\downarrow F_R = \Sigma F_z;$

$F_R = 200 + 200 + 100 + 100$

$= 600 \text{ N}$ *Ans.*

$\zeta +M_{Rx} = \Sigma M_x;$

$-600y = 200(1) + 200(1) + 100(3) - 100(3)$

$y = -0.667 \text{ m}$ *Ans.*

$\complement +M_{Ry} = \Sigma M_y;$

$600x = 100(3) + 100(3) + 200(2) - 200(3)$

$x = 0.667 \text{ m}$ *Ans.*

F4–37. $+\uparrow F_R = \Sigma F_y;$

$-F_R = -6(1.5) - 9(3) - 3(1.5)$

$F_R = 40.5 \text{ kN}\downarrow$ *Ans.*

$\zeta +(M_R)_A = \Sigma M_A;$

$-40.5(d) = 6(1.5)(0.75)$

$- 9(3)(1.5) - 3(1.5)(3.75)$

$d = 1.25 \text{ m}$ *Ans.*

F4–38. $F_R = \frac{1}{2}(3)(150) + 4(150) = 825 \text{ kN}$ *Ans.*

$\complement +M_{A_R} = \Sigma M_A;$

$825d = \left[\frac{1}{2}(3)(150)\right](2) + [4(150)](5)$

$d = 4.18 \text{ m}$ *Ans.*

F4–39. $+\uparrow F_R = \Sigma F_y;$

$-F_R = -\frac{1}{2}(6)(3) - \frac{1}{2}(6)(6)$

$F_R = 27 \text{ kN}\downarrow$ *Ans.*

$\zeta +(M_R)_A = \Sigma M_A;$

$-27(d) = \frac{1}{2}(6)(3)(1) - \frac{1}{2}(6)(6)(2)$

$d = 1 \text{ m}$ *Ans.*

F4–40. $+\downarrow F_R = \Sigma F_y;$

$F_R = \frac{1}{2}(50)(6) + 150(6) + 500$

$= 1550 \text{ kN}$ *Ans.*

$\complement +M_{A_R} = \Sigma M_A;$

$1550d = \left[\frac{1}{2}(50)(6)\right](4) + [150(6)](3) + 500(9)$

$d = 5.03 \text{ m}$ *Ans.*

F4–41. $+\uparrow F_R = \Sigma F_y;$

$-F_R = -\frac{1}{2}(3)(4.5) - 3(6)$

$F_R = 24.75 \text{ kN}\downarrow$ *Ans.*

$\zeta +(M_R)_A = \Sigma M_A;$

$-24.75(d) = -\frac{1}{2}(3)(4.5)(1.5) - 3(6)(3)$

$d = 2.59 \text{ m}$ *Ans.*

F4–42. $F_R = \int w(x)\, dx = \int_0^4 2.5x^3\, dx = 160$ N

$\curvearrowleft + M_{A_R} = \Sigma M_A;$

$$\bar{x} = \frac{\int x w(x)\, dx}{\int w(x)\, dx} = \frac{\int_0^4 2.5x^4\, dx}{160} = 3.20 \text{ m } Ans.$$

Chapter 5

F5–1. $\xrightarrow{+} \Sigma F_x = 0; \quad -A_x + 500\left(\tfrac{3}{5}\right) = 0$

$A_x = 300$ kN   *Ans.*

$\curvearrowleft + \Sigma M_A = 0; \quad B_y(10) - 500\left(\tfrac{4}{5}\right)(5) - 600 = 0$

$B_y = 260$ kN   *Ans.*

$+\uparrow \Sigma F_y = 0; \quad A_y + 260 - 500\left(\tfrac{4}{5}\right) = 0$

$A_y = 140$ kN   *Ans.*

F5–2. $\curvearrowleft + \Sigma M_A = 0;$

$F_{CD} \sin 45°(1.5 \text{ m}) - 4 \text{ kN}(3 \text{ m}) = 0$

$F_{CD} = 11.31$ kN $= 11.3$ kN   *Ans.*

$\xrightarrow{+} \Sigma F_x = 0; \quad A_x + (11.31 \text{ kN}) \cos 45° = 0$

$A_x = -8$ kN $= 8$ kN $\leftarrow$ *Ans.*

$+\uparrow \Sigma F_y = 0;$

$A_y + (11.31 \text{ kN}) \sin 45° - 4 \text{ kN} = 0$

$A_y = -4$ kN $= 4$ kN $\downarrow$ *Ans.*

F5–3. $\curvearrowleft + \Sigma M_A = 0;$

$N_B[6 \text{ m} + (6 \text{ m}) \cos 45°]$

$- 10 \text{ kN}[2 \text{ m} + (6 \text{ m}) \cos 45°]$

$- 5 \text{ kN}(4 \text{ m}) = 0$

$N_B = 8.047$ kN $= 8.05$ kN   *Ans.*

$\xrightarrow{+} \Sigma F_x = 0;$

$(5 \text{ kN}) \cos 45° - A_x = 0$

$A_x = 3.54$ kN   *Ans.*

$+\uparrow \Sigma F_y = 0;$

$A_y + 8.047 \text{ kN} - (5 \text{ kN}) \sin 45° - 10 \text{ kN} = 0$

$A_y = 5.49$ kN   *Ans.*

F5–4. $\xrightarrow{+} \Sigma F_x = 0; \quad -A_x + 400 \cos 30° = 0$

$A_x = 346$ N   *Ans.*

$+\uparrow \Sigma F_y = 0;$

$A_y - 200 - 200 - 200 - 400 \sin 30° = 0$

$A_y = 800$ N   *Ans.*

$\curvearrowleft + \Sigma M_A = 0;$

$M_A - 200(2.5) - 200(3.5) - 200(4.5)$

$- 400 \sin 30°(4.5) - 400 \cos 30°(3 \sin 60°) = 0$

$M_A = 3.90$ kN $\cdot$ m   *Ans.*

F5–5. $\curvearrowleft + \Sigma M_A = 0;$

$N_C(0.7 \text{ m}) - [25(9.81) \text{ N}] (0.5 \text{ m}) \cos 30° = 0$

$N_C = 151.71$ N $= 152$ N   *Ans.*

$\xrightarrow{+} \Sigma F_x = 0;$

$T_{AB} \cos 15° - (151.71 \text{ N}) \cos 60° = 0$

$T_{AB} = 78.53$ N $= 78.5$ N   *Ans.*

$+\uparrow \Sigma F_y = 0;$

$F_A + (78.53 \text{ N}) \sin 15°$

$+ (151.71 \text{ N}) \sin 60° - 25(9.81) \text{ N} = 0$

$F_A = 93.5$ N   *Ans.*

F5–6. $\xrightarrow{+} \Sigma F_x = 0;$

$N_C \sin 30° - (250 \text{ N}) \sin 60° = 0$

$N_C = 433.0$ N $= 433$ N   *Ans.*

$\curvearrowleft + \Sigma M_B = 0;$

$-N_A \sin 30°(0.15 \text{ m}) - 433.0 \text{ N}(0.2 \text{ m})$

$+ [(250 \text{ N}) \cos 30°](0.6 \text{ m}) = 0$

$N_A = 577.4$ N $= 577$ N   *Ans.*

$+\uparrow \Sigma F_y = 0;$

$N_B - 577.4 \text{ N} + (433.0 \text{ N}) \cos 30°$

$- (250 \text{ N}) \cos 60° = 0$

$N_B = 327$ N   *Ans.*

F5–7. $\Sigma F_z = 0;$

$T_A + T_B + T_C - 20 - 50 = 0$

$\Sigma M_x = 0;$

$T_A(3) + T_C(3) - 50(1.5) - 20(3) = 0$

$\Sigma M_y = 0;$

$-T_B(4) - T_C(4) + 50(2) + 20(2) = 0$

$T_A = 35$ kN, $T_B = 25$ kN, $T_C = 10$ kN   *Ans.*

F5–8. $\Sigma M_y = 0;$

$600 \text{ N}(0.2 \text{ m}) + 900 \text{ N}(0.6 \text{ m}) - F_A(1 \text{ m}) = 0$

$F_A = 660$ N   *Ans.*

$\Sigma M_x = 0;$

$D_z(0.8 \text{ m}) - 600 \text{ N}(0.5 \text{ m}) - 900 \text{ N}(0.1 \text{ m}) = 0$

$D_z = 487.5$ N   *Ans.*

$\Sigma F_x = 0; \qquad D_x = 0$ *Ans.*

$\Sigma F_y = 0; \qquad D_y = 0$ *Ans.*

$\Sigma F_z = 0;$

$T_{BC} + 660 \text{ N} + 487.5 \text{ N} - 900 \text{ N} - 600 \text{ N} = 0$

$T_{BC} = 352.5$ N   *Ans.*

F5–9. $\Sigma F_y = 0$; $400 \text{ N} + C_y = 0$;

$\qquad\qquad\qquad C_y = -400 \text{ N}$ *Ans.*

$\Sigma M_y = 0$; $-C_x (0.4 \text{ m}) - 600 \text{ N} (0.6 \text{ m}) = 0$

$\qquad\qquad\qquad C_x = -900 \text{ N}$ *Ans.*

$\Sigma M_x = 0$; $B_z (0.6 \text{ m}) + 600 \text{ N} (1.2 \text{ m})$

$\qquad\qquad + (-400 \text{ N})(0.4 \text{ m}) = 0$

$\qquad\qquad\qquad B_z = -933.3 \text{ N}$ *Ans.*

$\Sigma M_z = 0$;

$-B_x (0.6 \text{ m}) - (-900 \text{ N})(1.2 \text{ m})$

$\qquad\qquad + (-400 \text{ N})(0.6 \text{ m}) = 0$

$\qquad\qquad\qquad B_x = 1400 \text{ N}$ *Ans.*

$\Sigma F_x = 0$; $1400 \text{ N} + (-900 \text{ N}) + A_x = 0$

$\qquad\qquad\qquad A_x = -500 \text{ N}$ *Ans.*

$\Sigma F_z = 0$; $A_z - 933.3 \text{ N} + 600 \text{ N} = 0$

$\qquad\qquad\qquad A_z = 333.3 \text{ N}$ *Ans.*

F5–10. $\Sigma F_x = 0$; $B_x = 0$ *Ans.*

$\Sigma M_z = 0$;

$C_y (0.4 \text{ m} + 0.6 \text{ m}) = 0$ $C_y = 0$ *Ans.*

$\Sigma F_y = 0$; $A_y + 0 = 0$ $A_y = 0$ *Ans.*

$\Sigma M_x = 0$; $C_z (0.6 \text{ m} + 0.6 \text{ m}) + B_z (0.6 \text{ m})$

$\qquad\qquad - 450 \text{ N}(0.6 \text{ m} + 0.6 \text{ m}) = 0$

$1.2C_z + 0.6B_z - 540 = 0$

$\Sigma M_y = 0$; $-C_z (0.6 \text{ m} + 0.4 \text{ m})$

$\qquad\qquad - B_z (0.6 \text{ m}) + 450 \text{ N}(0.6 \text{ m}) = 0$

$\qquad\qquad -C_z - 0.6B_z + 270 = 0$

$C_z = 1350 \text{ N}$ $B_z = -1800 \text{ N}$ *Ans.*

$\Sigma F_z = 0$;

$A_z + 1350 \text{ N} + (-1800 \text{ N}) - 450 \text{ N} = 0$

$A_z = 900 \text{ N}$ *Ans.*

F5–11. $\Sigma F_y = 0$; $A_y = 0$ *Ans.*

$\Sigma M_x = 0$; $-9(3) + F_{CE}(3) = 0$

$\qquad\qquad F_{CE} = 9 \text{ kN}$ *Ans.*

$\Sigma M_z = 0$; $F_{CF}(3) - 6(3) = 0$

$\qquad\qquad F_{CF} = 6 \text{ kN}$ *Ans.*

$\Sigma M_y = 0$; $9(4) - A_z(4) - 6(1.5) = 0$

$\qquad\qquad A_z = 6.75 \text{ kN}$ *Ans.*

$\Sigma F_x = 0$; $A_x + 6 - 6 = 0$ $A_x = 0$ *Ans.*

$\Sigma F_z = 0$; $F_{DB} + 9 - 9 + 6.75 = 0$

$\qquad\qquad F_{DB} = -6.75 \text{ kN}$ *Ans.*

F5–12. $\Sigma F_x = 0$; $A_x = 0$ *Ans.*

$\Sigma F_y = 0$; $A_y = 0$ *Ans.*

$\Sigma F_z = 0$; $A_z + F_{BC} - 80 = 0$

$\Sigma M_x = 0$; $(M_A)_x + 0.6F_{BC} - 80(0.6) = 0$

$\Sigma M_y = 0$; $0.3F_{BC} - 80(0.15) = 0$ $F_{BC} = 40 \text{ N}$ *Ans.*

$\Sigma M_z = 0$; $(M_A)_z = 0$ *Ans.*

$\qquad A_z = 40 \text{ N}$ $(M_A)_x = 24 \text{ N} \cdot \text{m}$ *Ans.*

Chapter 6

F6–1.   *Joint A.*

$+\uparrow\Sigma F_y = 0$; $225 \text{ kN} - F_{AD} \sin 45° = 0$

$F_{AD} = 318.20 \text{ kN} = 318 \text{ kN (C)}$ *Ans.*

$\xrightarrow{+}\Sigma F_x = 0$; $F_{AB} - (318.20 \text{ kN}) \cos 45° = 0$

$F_{AB} = 225 \text{ kN (T)}$ *Ans.*

Joint B.

$\xrightarrow{+}\Sigma F_x = 0$; $F_{BC} - 225 \text{ kN} = 0$

$F_{BC} = 225 \text{ kN (T)}$ *Ans.*

$+\uparrow\Sigma F_y = 0$; $F_{BD} = 0$ *Ans.*

Joint D.

$\xrightarrow{+}\Sigma F_x = 0$;

$F_{CD} \cos 45° + (318.20 \text{ kN}) \cos 45° - 450 \text{ kN} = 0$

$F_{CD} = 318.20 \text{ kN} = 318 \text{ kN (T)}$ *Ans.*

F6–2.   *Joint D.*

$+\uparrow\Sigma F_y = 0$; $\frac{3}{5} F_{CD} - 300 = 0$;

$F_{CD} = 500 \text{ kN (T)}$ *Ans.*

$\xrightarrow{+} \Sigma F_x = 0$; $-F_{AD} + \frac{4}{5}(500) = 0$

$F_{AD} = 400 \text{ kN (C)}$ *Ans.*

$F_{BC} = 500 \text{ kN (T)}$, $F_{AC} = F_{AB} = 0$ *Ans.*

F6–3. $A_x = 0$, $A_y = C_y = 400 \text{ kN}$

Joint A.

$+\uparrow\Sigma F_y = 0$; $-\frac{3}{5} F_{AE} + 400 = 0$

$F_{AE} = 667 \text{ kN (C)}$ *Ans.*

Joint C.

$+\uparrow\Sigma F_y = 0$; $-F_{DC} + 400 = 0$;

$F_{DC} = 400 \text{ kN (C)}$ *Ans.*

F6–4. *Joint C.*

$+\uparrow\Sigma F_y = 0;$ $2F\cos 30° - P = 0$

$F_{AC} = F_{BC} = F = \dfrac{P}{2\cos 30°} = 0.5774P$ (C)

Joint B.

$\xrightarrow{+}\Sigma F_x = 0;$ $0.5774P\cos 60° - F_{AB} = 0$

$F_{AB} = 0.2887P$ (T)

$F_{AB} = 0.2887P = 2$ kN

$P = 6.928$ kN

$F_{AC} = F_{BC} = 0.5774P = 1.5$ kN

$P = 2.598$ kN

The *smaller value* of P is chosen,

$P = 2.598$ kN $= 2.60$ kN   *Ans.*

F6–5. $F_{CB} = 0$ *Ans.*

$F_{CD} = 0$ *Ans.*

$F_{AE} = 0$ *Ans.*

$F_{DE} = 0$ *Ans.*

F6–6. *Joint C.*

$+\uparrow\Sigma F_y = 0;$ 259.81 kN $- F_{CD}\sin 30° = 0$

$F_{CD} = 519.62$ kN $= 520$ kN (C)   *Ans.*

$\xrightarrow{+}\Sigma F_x = 0;$ $(519.62$ kN$)\cos 30° - F_{BC} = 0$

$F_{BC} = 450$ kN (T)   *Ans.*

Joint D.

$+\nearrow\Sigma F_{y'} = 0;$ $F_{BD}\cos 30° = 0$ $F_{BD} = 0$ *Ans.*

$+\searrow\Sigma F_{x'} = 0;$ $F_{DE} - 519.62$ kN $= 0$

$F_{DE} = 519.62$ kN $= 520$ kN (C)   *Ans.*

Joint B.

$\uparrow\Sigma F_y = 0;$ $F_{BE}\sin\phi = 0$ $F_{BE} = 0$ *Ans.*

$\xrightarrow{+}\Sigma F_x = 0;$ 450 kN $- F_{AB} = 0$

$F_{AB} = 450$ kN (T)   *Ans.*

Joint A.

$+\uparrow\Sigma F_y = 0;$ 340.19 kN $- F_{AE} = 0$

$F_{AE} = 340$ kN (C)   *Ans.*

F6–7. $+\uparrow\Sigma F_y = 0;$ $F_{CF}\sin 45° - 600 - 800 = 0$

$F_{CF} = 1980$ N (T)   *Ans.*

$\zeta+\Sigma M_C = 0;$ $F_{FE}(1) - 800(1) = 0$

$F_{FE} = 800$ N (T)   *Ans.*

$\zeta+\Sigma M_F = 0;$ $F_{BC}(1) - 600(1) - 800(2) = 0$

$F_{BC} = 2200$ N (C)   *Ans.*

F6–8. $\zeta+\Sigma M_A = 0;$ $G_y(12\text{ m}) - 20\text{ kN}(2\text{ m})$

$- 30\text{ kN}(4\text{ m}) - 40\text{ kN}(6\text{ m}) = 0$

$G_y = 33.33$ kN

$+\uparrow\Sigma F_y = 0;$ $F_{KC} + 33.33$ kN $- 40$ kN $= 0$

$F_{KC} = 6.67$ kN (C)   *Ans.*

$\zeta+ \Sigma M_K = 0;$

33.33 kN$(8\text{ m}) - 40$ kN$(2\text{ m}) - F_{CD}(3\text{ m}) = 0$

$F_{CD} = 62.22$ kN $= 62.2$ kN (T)   *Ans.*

$\xrightarrow{+}\Sigma F_x = 0;$ $F_{LK} - 62.22$ kN $= 0$

$F_{LK} = 62.2$ kN (C)   *Ans.*

F6–9. From the geometry of the truss,

$\phi = \tan^{-1}(3\text{ m}/2\text{ m}) = 56.31°.$

$\zeta+\Sigma M_K = 0;$

33.33 kN$(8\text{ m}) - 40$ kN$(2\text{ m}) - F_{CD}(3\text{ m}) = 0$

$F_{CD} = 62.2$ kN (T)   *Ans.*

$\zeta+\Sigma M_D = 0;$ 33.33 kN$(6\text{ m}) - F_{KJ}(3\text{ m}) = 0$

$F_{KJ} = 66.7$ kN (C)   *Ans.*

$+\uparrow\Sigma F_y = 0;$

33.33 kN $- 40$ kN $+ F_{KD}\sin 56.31° = 0$

$F_{KD} = 8.01$ kN (T)   *Ans.*

F6–10. From the geometry of the truss,

$\tan\phi = \dfrac{(3\text{ m})\tan 30°}{1\text{ m}} = 1.732$ $\phi = 60°$

$\zeta+\Sigma M_C = 0;$

$F_{EF}\sin 30°(2\text{ m}) + 300\text{ N}(2\text{ m}) = 0$

$F_{EF} = -600$ N $= 600$ N (C)   *Ans.*

$\zeta+\Sigma M_D = 0;$

$300\text{ N}(2\text{ m}) - F_{CF}\sin 60° (2\text{ m}) = 0$

$F_{CF} = 346.41$ N $= 346$ N (T)   *Ans.*

$\zeta+\Sigma M_F = 0;$

$300\text{ N}(3\text{ m}) - 300\text{ N}(1\text{ m}) - F_{BC}(3\text{ m})\tan 30° = 0$

$F_{BC} = 346.41$ N $= 346$ N (T)   *Ans.*

F6–11. From the geometry of the truss,

$\theta = \tan^{-1}(1\text{ m}/2\text{ m}) = 26.57°$

$\phi = \tan^{-1}(3\text{ m}/2\text{ m}) = 56.31°.$

The location of O can be found using similar triangles.

$$\frac{1\text{ m}}{2\text{ m}} = \frac{2\text{ m}}{2\text{ m} + x}$$

$$4\text{ m} = 2\text{ m} + x$$

$$x = 2\text{ m}$$

$\zeta + \Sigma M_G = 0;$

$26.25 \text{ kN}(4 \text{ m}) - 15 \text{ kN}(2 \text{ m}) - F_{CD}(3 \text{ m}) = 0$

$\qquad F_{CD} = 25 \text{ kN (T)} \qquad Ans.$

$\zeta + \Sigma M_D = 0;$

$26.25 \text{ kN}(2 \text{ m}) - F_{GF} \cos 26.57°(2 \text{ m}) = 0$

$\qquad F_{GF} = 29.3 \text{ kN (C)} \qquad Ans.$

$\zeta + \Sigma M_O = 0; \quad 15 \text{ kN}(4 \text{ m}) - 26.25 \text{ kN}(2 \text{ m})$

$\qquad\qquad - F_{GD} \sin 56.31°(4 \text{ m}) = 0$

$\qquad F_{GD} = 2.253 \text{ kN} = 2.25 \text{ kN (T)} \qquad Ans.$

F6–12. $\quad \zeta + \Sigma M_H = 0;$

$F_{DC}(4 \text{ m}) + 12 \text{ kN}(3 \text{ m}) - 16 \text{ kN}(7 \text{ m}) = 0$

$\qquad F_{DC} = 19 \text{ kN (C)} \qquad Ans.$

$\zeta + \Sigma M_D = 0;$

$12 \text{ kN}(7 \text{ m}) - 16 \text{ kN}(3 \text{ m}) - F_{HI}(4 \text{ m}) = 0$

$\qquad F_{HI} = 9 \text{ kN (C)} \qquad Ans.$

$\zeta + \Sigma M_C = 0; \quad F_{JI} \cos 45°(4 \text{ m}) + 12 \text{ kN}(7 \text{ m})$

$\qquad\qquad - 9 \text{ kN}(4 \text{ m}) - 16 \text{ kN}(3 \text{ m}) = 0$

$\qquad F_{JI} = 0 \qquad Ans.$

F6–13. $\quad +\uparrow\Sigma F_y = 0; \quad 3P - 60 = 0$

$\qquad P = 20 \text{ N} \qquad Ans.$

F6–14. $\quad \zeta + \Sigma M_C = 0;$

$-\left(\frac{4}{5}\right)(F_{AB})(3) + 400(2) + 500(1) = 0$

$\qquad F_{AB} = 541.67 \text{ N}$

$\xrightarrow{+}\Sigma F_x = 0; -C_x + \frac{3}{5}(541.67) = 0$

$\qquad C_x = 325 \text{ N} \qquad Ans.$

$+\uparrow\Sigma F_y = 0; C_y + \frac{4}{5}(541.67) - 400 - 500 = 0$

$\qquad C_y = 467 \text{ N} \qquad Ans.$

F6–15. $\quad \zeta + \Sigma M_A = 0; 100 \text{ N}(250 \text{ mm}) - N_B(50 \text{ mm}) = 0$

$\qquad N_B = 500 \text{ N} \qquad Ans.$

$\xrightarrow{+}\Sigma F_x = 0; \quad (500 \text{ N}) \sin 45° - A_x = 0$

$\qquad A_x = 353.55 \text{ N}$

$+\uparrow\Sigma F_y = 0; A_y - 100 \text{ N} - (500 \text{ N}) \cos 45° = 0$

$\qquad A_y = 453.55 \text{ N}$

$F_A = \sqrt{(353.55 \text{ N})^2 + (453.55 \text{ N})^2}$

$\qquad = 575 \text{ N} \qquad Ans.$

F6–16. $\quad \zeta + \Sigma M_C = 0;$

$F_{AB} \cos 45°(1) - F_{AB} \sin 45°(3)$

$\qquad\qquad\qquad + 800 + 400(2) = 0$

$\qquad F_{AB} = 1131.37 \text{ N}$

$\xrightarrow{+}\Sigma F_x = 0; -C_x + 1131.37 \cos 45° = 0$

$\qquad C_x = 800 \text{ N} \qquad Ans.$

$+\uparrow\Sigma F_y = 0; -C_y + 1131.37 \sin 45° - 400 = 0$

$\qquad C_y = 400 \text{ N} \qquad Ans.$

F6–17. Plate A:

$+\uparrow\Sigma F_y = 0; \quad 2T + N_{AB} - 100 = 0$

Plate B:

$+\uparrow\Sigma F_y = 0; \quad 2T - N_{AB} - 30 = 0$

$\qquad T = 32.5 \text{ N}, N_{AB} = 35 \text{ N} \qquad Ans.$

F6–18. Pulley C:

$+\uparrow\Sigma F_y = 0; \quad T - 2P = 0; T = 2P$

Beam:

$+\uparrow\Sigma F_y = 0; \quad 2P + P - 6 = 0$

$\qquad P = 2 \text{ kN} \qquad Ans.$

$\zeta + \Sigma M_A = 0; \quad 2(1) - 6(x) = 0$

$\qquad x = 0.333 \text{ m} \qquad Ans.$

F6–19. Member CD

$\zeta + \Sigma M_D = 0; \quad 600(1.5) - N_C(3) = 0$

$\qquad N_C = 300 \text{ N}$

Member ABC

$\zeta + \Sigma M_A = 0; \quad -800 + B_y(2) - (300 \sin 45°) 4 = 0$

$\qquad B_y = 824.26 = 824 \text{ N} \qquad Ans.$

$\xrightarrow{+}\Sigma F_x = 0; \quad A_x - 300 \cos 45° = 0;$

$\qquad A_x = 212 \text{ N} \qquad Ans.$

$+\uparrow\Sigma F_y = 0; \quad -A_y + 824.26 - 300 \sin 45° = 0;$

$\qquad A_y = 612 \text{ N} \qquad Ans.$

F6–20. AB is a two-force member.

Member BC

$\zeta + \Sigma M_c = 0; \quad 15(3) + 10(6) - F_{BC}\left(\frac{4}{5}\right)(9) = 0$

$\qquad F_{BC} = 14.58 \text{ kN}$

$\xrightarrow{+}\Sigma F_x = 0; \quad (14.58)\left(\frac{3}{5}\right) - C_x = 0;$

$\qquad C_x = 8.75 \text{ kN}$

$+\uparrow\Sigma F_y = 0; \quad (14.58)\left(\frac{4}{5}\right) - 10 - 15 + C_y = 0;$

$\qquad C_y = 13.3 \text{ kN}$

Member CD

$\xrightarrow{+}\Sigma F_x = 0; \qquad 8.75 - D_x = 0; \quad D_x = 8.75 \text{ kN} \qquad Ans.$

$+\uparrow\Sigma F_y = 0; \qquad -13.3 + D_y = 0; \quad D_y = 13.3 \text{ kN} \qquad Ans.$

$\zeta + \Sigma M_D = 0; \quad -8.75(4) + M_D = 0; \quad M_D = 35 \text{ kN} \cdot \text{m} \qquad Ans.$

F6–21. Entire frame

$\zeta +\Sigma M_A = 0; \quad -600(3) - [400(3)](1.5) + C_y(3) = 0$

$C_y = 1200$ N   *Ans.*

$+\uparrow \Sigma F_y = 0; \quad A_y - 400(3) + 1200 = 0$

$A_y = 0$ *Ans.*

$\xrightarrow{+}\Sigma F_x = 0; \quad 600 - A_x - C_x = 0$

Member AB

$\zeta +\Sigma M_B = 0; \quad 400(1.5)(0.75) - A_x(3) = 0$

$A_x = 150$ N   *Ans.*

$C_x = 450$ N   *Ans.*

These same results can be obtained by considering members AB and BC.

F6–22. Entire frame

$\zeta +\Sigma M_E = 0; \quad 250(6) - A_y(6) = 0$

$A_y = 250$ N

$\xrightarrow{+}\Sigma F_x = 0; \quad E_x = 0$

$+\uparrow \Sigma F_y = 0; \quad 250 - 250 + E_y = 0; \quad E_y = 0$

Member BD

$\zeta +\Sigma M_D = 0; \quad 250(4.5) - B_y(3) = 0;$

$B_y = 375$ N

Member ABC

$\zeta +\Sigma M_C = 0; \quad -250(3) + 375(1.5) + B_x(2) = 0$

$B_x = 93.75$ N

$\xrightarrow{+}\Sigma F_x = 0; \quad C_x - B_x = 0; \quad C_x = 93.75$ N   *Ans.*

$+\uparrow \Sigma F_y = 0; \quad 250 - 375 + C_y = 0; \quad C_y = 125$ N   *Ans.*

F6–23. *AD, CB* are two-force members.

Member AB

$\zeta +\Sigma M_A = 0; \quad -[\frac{1}{2}(3)(4)](1.5) + B_y(3) = 0$

$B_y = 3$ kN

Since BC is a two-force member $C_y = B_y = 3$ kN and $C_x = 0 \ (\Sigma M_B = 0)$.

Member EDC

$\zeta +\Sigma M_E = 0; \quad F_{DA}(\frac{4}{5})(1.5) - 5(3) - 3(3) = 0$

$F_{DA} = 20$ kN

$\xrightarrow{+}\Sigma F_x = 0; \quad E_x - 20(\frac{3}{5}) = 0; \quad E_x = 12$ kN   *Ans.*

$+\uparrow \Sigma F_y = 0; \quad -E_y + 20(\frac{4}{5}) - 5 - 3 = 0;$

$E_y = 8$ kN   *Ans.*

F6–24. *AC* and *DC* are two-force members.

Member BC

$\zeta +\Sigma M_C = 0; \quad [\frac{1}{2}(3)(8)](1) - B_y(3) = 0$

$B_y = 4$ kN

Member BA

$\zeta +\Sigma M_B = 0; \quad 6(2) - A_x(4) = 0$

$A_x = 3$ kN   *Ans.*

$+\uparrow \Sigma F_y = 0; \quad -4 \text{ kN} + A_y = 0; \quad A_y = 4 \text{ kN } Ans.$

Entire Frame

$\zeta +\Sigma M_A = 0; \quad -6(2) - [\frac{1}{2}(3)(8)](2) + D_y(3) = 0$

$D_y = 12$ kN   *Ans.*

Since DC is a two-force member $(\Sigma M_C = 0)$ then

$D_x = 0$ *Ans.*

Chapter 7

F7–1. $\zeta +\Sigma M_A = 0; \quad B_y(6) - 10(1.5) - 15(4.5) = 0$

$B_y = 13.75$ kN

$\xrightarrow{+}\Sigma F_x = 0; \quad N_C = 0$ *Ans.*

$+\uparrow \Sigma F_y = 0; \quad V_C + 13.75 - 15 = 0$

$V_C = 1.25$ kN   *Ans.*

$\zeta +\Sigma M_C = 0; \quad 13.75(3) - 15(1.5) - M_C = 0$

$M_C = 18.75$ kN·m   *Ans.*

F7–2. $\zeta +\Sigma M_B = 0; \quad 30 - 10(1.5) - A_y(3) = 0$

$A_y = 5$ kN

$\xrightarrow{+}\Sigma F_x = 0; \quad N_C = 0$ *Ans.*

$+\uparrow \Sigma F_y = 0; \quad 5 - V_C = 0$

$V_C = 5$ kN   *Ans.*

$\zeta +\Sigma M_C = 0; \quad M_C + 30 - 5(1.5) = 0$

$M_C = -22.5$ kN·m   *Ans.*

F7–3. $\xrightarrow{+}\Sigma F_x = 0; \quad B_x = 0$

$\zeta +\Sigma M_A = 0; \quad 30(6)(3) - B_y(9) = 0$

$B_y = 60$ kN

$\xrightarrow{+}\Sigma F_x = 0; \quad N_C = 0$ *Ans.*

$\zeta +\uparrow \Sigma F_y = 0; \quad V_C - 60 = 0$

$V_C = 60$ kN   *Ans.*

$\zeta +\Sigma M_C = 0; \quad -M_C - 60(4.5) = 0$

$M_C = -270$ kN·m   *Ans.*

F7–4. $\zeta +\Sigma M_A = 0; \quad B_y(6) - 12(1.5) - 9(3)(4.5) = 0$

$B_y = 23.25$ kN

$\xrightarrow{+}\Sigma F_x = 0; \quad N_C = 0$ *Ans.*

$+\uparrow \Sigma F_y = 0; \quad V_C + 23.25 - 9(1.5) = 0$

$V_C = -9.75$ kN   *Ans.*

$\zeta +\Sigma M_C = 0;$

$23.25(1.5) - 9(1.5)(0.75) - M_C = 0$

$M_C = 24.75$ kN·m   *Ans.*

F7–5. $\zeta + \Sigma M_A = 0;$ $B_y(6) - \frac{1}{2}(9)(6)(3) = 0$
$$B_y = 13.5 \text{ kN}$$

$\stackrel{+}{\rightarrow} \Sigma F_x = 0;$ $N_C = 0$ *Ans.*

$+ \uparrow \Sigma F_y = 0;$ $V_C + 13.5 - \frac{1}{2}(9)(3) = 0$
$$V_C = 0$$ *Ans.*

$\zeta + \Sigma M_C = 0;$ $13.5(3) - \frac{1}{2}(9)(3)(1) - M_C = 0$
$$M_C = 27 \text{ kN} \cdot \text{m}$$ *Ans.*

F7–6. $\zeta + \Sigma M_A = 0;$
$$B_y(6) - \frac{1}{2}(6)(3)(2) - 6(3)(4.5) = 0$$
$$B_y = 16.5 \text{ kN}$$

$\stackrel{+}{\rightarrow} \Sigma F_x = 0;$ $N_C = 0$ *Ans.*

$+ \uparrow \Sigma F_y = 0;$ $V_C + 16.5 - 6(3) = 0$
$$V_C = 1.50 \text{ kN}$$ *Ans.*

$\zeta + \Sigma M_C = 0;$ $16.5(3) - 6(3)(1.5) - M_C = 0$
$$M_C = 22.5 \text{ kN} \cdot \text{m}$$ *Ans.*

F7–7. $+ \uparrow \Sigma F_y = 0;$ $6 - V = 0$ $V = 6 \text{ kN}$
$\zeta + \Sigma M_O = 0;$ $M + 18 - 6x = 0$
$$M = (6x - 18) \text{ kN} \cdot \text{m}$$

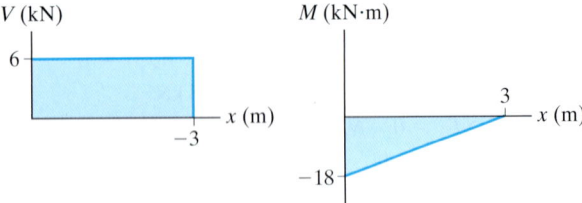

Fig. F7–7

F7–8. $+ \uparrow \Sigma F_y = 0;$ $- V - 2x = 0$
$$V = (-2x) \text{ kN}$$
$\zeta + \Sigma M_O = 0;$ $M + 2x\left(\frac{x}{2}\right) - 15 = 0$
$$M = (15 - x^2) \text{ kN} \cdot \text{m}$$

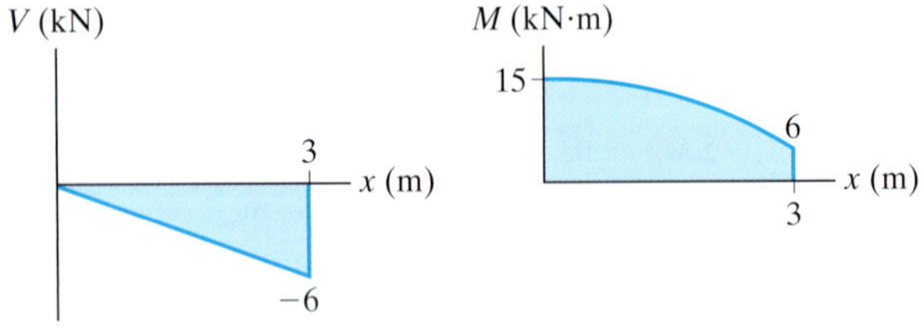

Fig. F7–8

F7–9. $+ \uparrow \Sigma F_y = 0;$ $- V - \frac{1}{2}(2x)(x) = 0$
$$V = -(x^2) \text{ kN}$$
$\zeta + \Sigma M_O = 0;$ $M + \frac{1}{2}(2x)(x)(\frac{x}{3}) = 0$
$$M = -(\frac{1}{3}x^3) \text{ kN} \cdot \text{m}$$

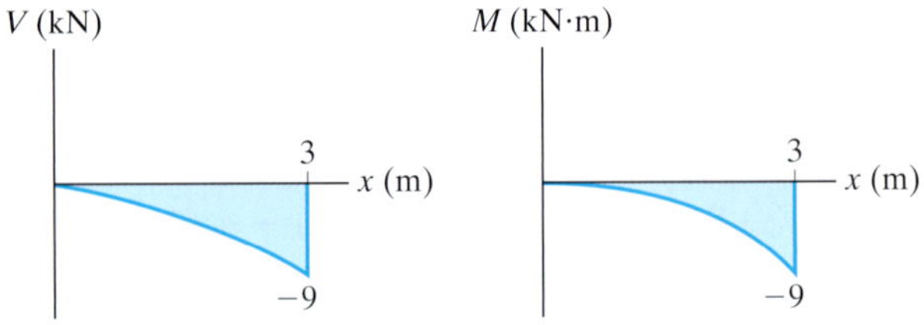

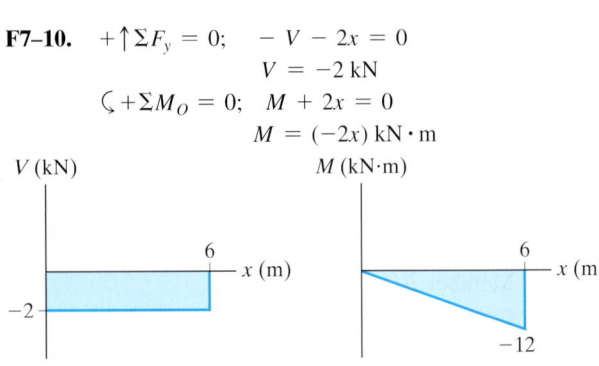

Fig. F7–9

F7–10. $+ \uparrow \Sigma F_y = 0;$ $- V - 2x = 0$
$$V = -2 \text{ kN}$$
$\zeta + \Sigma M_O = 0;$ $M + 2x = 0$
$$M = (-2x) \text{ kN} \cdot \text{m}$$

Fig. F7–10

F7–11.   Region $3 \text{ m} \leq x < 3 \text{ m}$
$+ \uparrow \Sigma F_y = 0;$ $- V - 5 = 0$ $V = -5 \text{ kN}$
$\zeta + \Sigma M_O = 0;$ $M + 5x = 0$
$$M = (-5x) \text{ kN} \cdot \text{m}$$
Region $0 < x \leq 6 \text{ m}$
$+ \uparrow \Sigma F_y = 0;$ $V + 5 = 0$ $V = -5 \text{ kN}$
$\zeta + \Sigma M_O = 0;$ $5(6 - x) - M = 0$
$$M = \left(5(6 - x)\right) \text{ kN} \cdot \text{m}$$

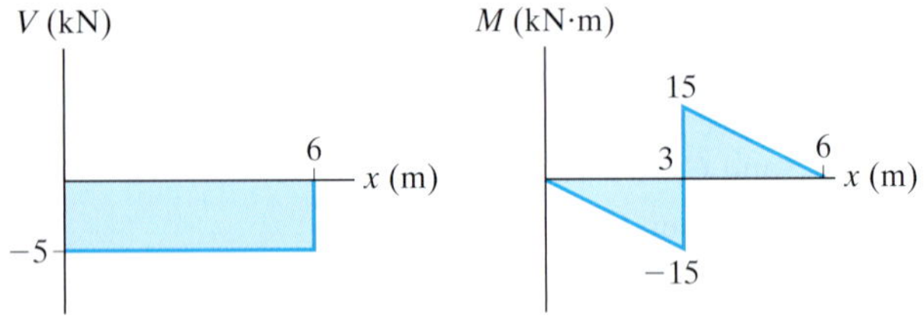

Fig. F7–11

F7–12.   Region $0 \le x < 3$ m
$$+\uparrow \Sigma F_y = 0; \quad V = 0$$
$$\zeta + \Sigma M_O = 0; \quad M - 12 = 0$$
$$M = 12 \text{ kN} \cdot \text{m}$$
Region 3 m $< x \le 6$ m
$$+\uparrow \Sigma F_y = 0; \quad V + 4 = 0 \quad V = -4 \text{ kN}$$
$$\zeta + \Sigma M_O = 0; \quad 4(6 - x) - M = 0$$
$$M = \big(4(6 - x)\big) \text{ kN} \cdot \text{m}$$

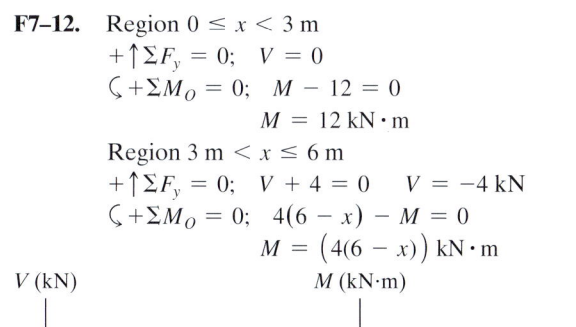

Fig. F7–12

F7–13.

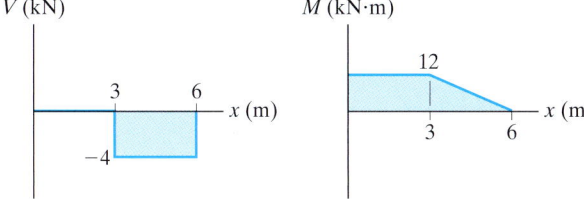

Fig. F7–13

F7–14.

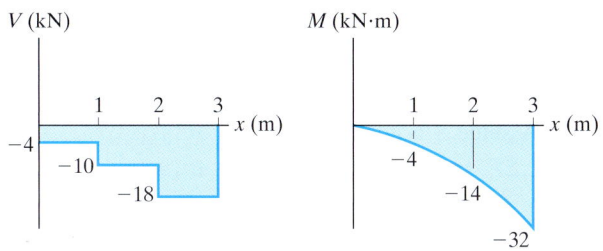

Fig. F7–14

F7–15.

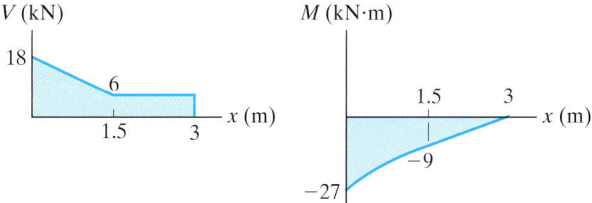

Fig. F7–15

F7–16.

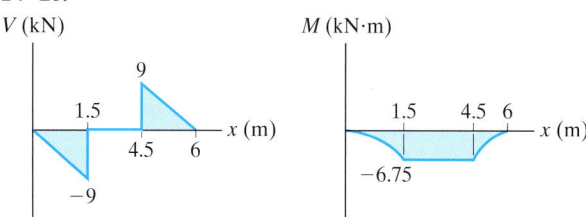

Fig. F7–16

F7–17.

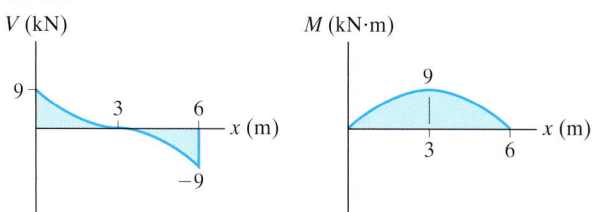

Fig. F7–17

F7–18.

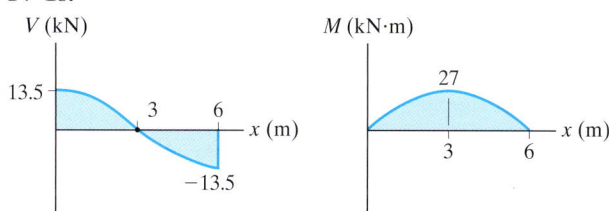

Fig. F7–18

Chapter 8

F8–1. $+\uparrow \Sigma F_y = 0; \quad N - 50(9.81) - 200\big(\frac{3}{5}\big) = 0$
$$N = 610.5 \text{ N}$$
$$\xrightarrow{+} \Sigma F_x = 0; \quad F - 200\big(\frac{4}{5}\big) = 0$$
$$F = 160 \text{ N}$$
$F < F_{max} = \mu_s N = 0.3(610.5) = 183.15$ N,
therefore $F = 160$ N   *Ans.*

F8–2. $\zeta + \Sigma M_B = 0;$
$$N_A(3) + 0.2N_A(4) - 30(9.81)(2) = 0$$
$$N_A = 154.89 \text{ N}$$
$$\xrightarrow{+} \Sigma F_x = 0; \quad P - 154.89 = 0$$
$$P = 154.89 \text{ N} = 155 \text{ N} \qquad \textit{Ans.}$$

F8–3.   Crate A

$$+\uparrow \Sigma F_y = 0; \quad N_A - 50(9.81) = 0$$
$$N_A = 490.5 \text{ N}$$
$$\xrightarrow{+} \Sigma F_x = 0; \quad T - 0.25(490.5) = 0$$
$$T = 122.62 \text{ N}$$

Crate B

$$+\uparrow \Sigma F_y = 0; \quad N_B + P \sin 30° - 50(9.81) = 0$$
$$N_B = 490.5 - 0.5P$$
$$\xrightarrow{+} \Sigma F_x = 0;$$
$$P \cos 30° - 0.25(490.5 - 0.5\,P) - 122.62 = 0$$
$$P = 247 \text{ N} \qquad\qquad \textit{Ans.}$$

F8–4. $\xrightarrow{+} \Sigma F_x = 0; \quad N_A - 0.3N_B = 0$

$$+\uparrow \Sigma F_y = 0;$$
$$N_B + 0.3N_A + P - 100(9.81) = 0$$
$$\zeta + \Sigma M_O = 0;$$
$$P(0.6) - 0.3N_B(0.9) - 0.3\,N_A(0.9) = 0$$
$$N_A = 175.70 \text{ N} \qquad N_B = 585.67 \text{ N}$$
$$P = 343 \text{ N} \qquad\qquad \textit{Ans.}$$

F8–5.   If slipping occurs:

$$+\uparrow \Sigma F_y = 0; \quad N_c - 100(9.81)\text{ N} = 0; \quad N_c = 981 \text{ N}$$
$$\xrightarrow{} \Sigma F_x = 0; \quad P - 0.4(981) = 0; \; P = 392.4 \text{ N}$$

If tipping occurs:

$$\zeta + \Sigma M_A = 0; \quad -P(1.5) + 981(0.5) = 0$$
$$P = 327 \text{ N} \qquad\qquad \textit{Ans.}$$

F8–6.

$$\zeta + \Sigma M_A = 0; \quad 490.5\,(0.6) - T \cos 60°(0.3 \cos 60° + 0.6)$$
$$- T \sin 60°\,(0.3 \sin 60°) = 0$$
$$T = 490.5 \text{ N}$$
$$\xrightarrow{+} \Sigma F_x = 0; \quad 490.5 \sin 60° - N_A = 0; \quad N_A = 424.8 \text{ N}$$
$$+\uparrow \Sigma F_y = 0; \quad \mu_s(424.8) + 490.5 \cos 60° - 490.5 = 0$$
$$\mu_s = 0.577 \qquad\qquad \textit{Ans.}$$

F8–7. A will not move. Assume B is about to slip on C and A, and C is stationary.

$$\xrightarrow{+} \Sigma F_x = 0; \quad P - 0.3\,(50) - 0.4\,(75); \quad P = 45 \text{ N}$$

Assume C is about to slip and B does not slip on C, but is about to slip at A.

$$\xrightarrow{+} \Sigma F_x = 0; \quad P - 0.3\,(50) - 0.35\,(90) = 0$$
$$P = 46.5 \text{ N} > 45 \text{ N}$$
$$P = 45 \text{ N} \qquad\qquad \textit{Ans.}$$

F8–8. A is about to move down the plane and B moves upward.

Block A

$$+\nwarrow \Sigma F_y = 0; \quad N = W \cos \theta$$
$$+\nearrow \Sigma F_x = 0; \quad T + \mu_s(W \cos \theta) - W \sin \theta = 0$$
$$T = W \sin \theta - \mu_s W \cos \theta \qquad\qquad (1)$$

Block B

$$+\nwarrow \Sigma F_y = 0; \quad N' = 2W \cos \theta$$
$$+\nearrow \Sigma F_x = 0; \quad 2T - \mu_s W \cos \theta - \mu_s(2W \cos \theta)$$
$$- W \sin \theta = 0$$

Using Eq.(1);

$$\theta = \tan^{-1} 5\mu_s \qquad\qquad \textit{Ans.}$$

F8–9.   Assume B is about to slip on A, $F_B = 0.3\,N_B$.

$$\xrightarrow{+} \Sigma F_x = 0; \quad P - 0.3\,(10)\,(9.81) = 0$$
$$P = 29.4 \text{ N}$$

Assume B is about to slip on A, $x = 0$.

$$\zeta + \Sigma M_O = 0; \quad 10(9.81)\,(0.15) - P(0.4) = 0$$
$$P = 36.8 \text{ N}$$

Assume A is about to slip, $F_A = 0.1\,N_A$.

$$\xrightarrow{+} \Sigma F_x = 0 \quad P - 0.1\big[7(9.81) + 10(9.81)\big] = 0$$
$$P = 16.7 \text{ N}$$

Choose the smallest result. $P = 16.7 \text{ N} \qquad \textit{Ans.}$

Chapter 9

F9–1.

$$\bar{x} = \frac{\displaystyle\int_A \tilde{x}\, dA}{\displaystyle\int_A dA} = \frac{\dfrac{1}{2}\displaystyle\int_0^{1\text{ m}} y^{2/3}\, dy}{\displaystyle\int_0^{1\text{ m}} y^{1/3} dy} = 0.4 \text{ m} \qquad \textit{Ans.}$$

$$\bar{y} = \frac{\displaystyle\int_A \tilde{y}\, dA}{\displaystyle\int_A dA} = \frac{\displaystyle\int_0^{1\text{ m}} y^{4/3}\, dy}{\displaystyle\int_0^{1\text{ m}} y^{1/3} dy} = 0.571 \text{ m} \qquad \textit{Ans.}$$

F9–2.

$$\bar{x} = \frac{\displaystyle\int_A \tilde{x}\, dA}{\displaystyle\int_A dA} = \frac{\displaystyle\int_0^{1\text{ m}} x(x^3\, dx)}{\displaystyle\int_0^{1\text{ m}} x^3\, dx}$$

$$= 0.8 \text{ m} \qquad\qquad \textit{Ans.}$$

$$\bar{y} = \frac{\int_A \tilde{y}\, dA}{\int_A dA} = \frac{\int_0^{1\,m} \frac{1}{2} x^3 (x^3\, dx)}{\int_0^{1\,m} x^3\, dx}$$

$$= 0.286\ m \qquad\qquad Ans.$$

F9–3. $$\bar{y} = \frac{\int_A \tilde{y}\, dA}{\int_A dA} = \frac{\int_0^{2\,m} y\left(2\left(\frac{y^{1/2}}{\sqrt{2}}\right)\right) dy}{\int_0^{2\,m} 2\left(\frac{y^{1/2}}{\sqrt{2}}\right) dy}$$

$$= 1.2\ m \qquad\qquad Ans.$$

F9–4. $$\bar{x} = \frac{\int_m \tilde{x}\, dm}{\int_m dm} = \frac{\int_0^L x\left[m_0\left(1 + \frac{x^2}{L^2}\right) dx\right]}{\int_0^L m_0\left(1 + \frac{x^2}{L^2}\right) dx}$$

$$= \frac{9}{16} L \qquad\qquad Ans.$$

F9–5. $$\bar{y} = \frac{\int_V \tilde{y}\, dV}{\int_V dV} = \frac{\int_0^{1\,m} y\left(\frac{\pi}{4} y dy\right)}{\int_0^{1\,m} \frac{\pi}{4} y\, dy}$$

$$= 0.667\ m \qquad\qquad Ans.$$

F9–6. $$\bar{z} = \frac{\int_V \tilde{z}\, dV}{\int_V dV} = \frac{\int_0^{2\,ft} z\left[\frac{9\pi}{64}(4-z)^2\, dz\right]}{\int_0^{2\,ft} \frac{9\pi}{64}(4-z)^2\, dz}$$

$$= 0.786\ m \qquad\qquad Ans.$$

F9–7. $$\bar{x} = \frac{\Sigma \tilde{x} L}{\Sigma L} = \frac{150(300) + 300(600) + 300(400)}{300 + 600 + 400}$$

$$= 265\ mm \qquad\qquad Ans.$$

$$\bar{y} = \frac{\Sigma \tilde{y} L}{\Sigma L} = \frac{0(300) + 300(600) + 600(400)}{300 + 600 + 400}$$

$$= 323\ mm \qquad\qquad Ans.$$

$$\bar{z} = \frac{\Sigma \tilde{z} L}{\Sigma L} = \frac{0(300) + 0(600) + (-200)(400)}{300 + 600 + 400}$$

$$= -61.5\ mm \qquad\qquad Ans.$$

F9–8. $$\bar{y} = \frac{\Sigma \tilde{y} A}{\Sigma A} = \frac{150[300(50)] + 325[50(300)]}{300(50) + 50(300)}$$

$$= 237.5\ mm \qquad\qquad Ans.$$

F9–9. $$\bar{y} = \frac{\Sigma \tilde{y} A}{\Sigma A} = \frac{100[2(200)(50)] + 225[50(400)]}{2(200)(50) + 50(400)}$$

$$= 162.5\ mm \qquad\qquad Ans.$$

F9–10. $$\bar{x} = \frac{\Sigma \tilde{x} A}{\Sigma A} = \frac{0.25[4(0.5)] + 1.75[0.5(2.5)]}{4(0.5) + 0.5(2.5)}$$

$$= 0.827\ m \qquad\qquad Ans.$$

$$\bar{y} = \frac{\Sigma \tilde{y} A}{\Sigma A} = \frac{2[4(0.5)] + 0.25[(0.5)(2.5)]}{4(0.5) + (0.5)(2.5)}$$

$$= 1.33\ m \qquad\qquad Ans.$$

F9–11. $$\bar{x} = \frac{\Sigma \tilde{x} V}{\Sigma V} = \frac{1[2(7)(6)] + 4[4(2)(3)]}{2(7)(6) + 4(2)(3)}$$

$$= 1.67\ m \qquad\qquad Ans.$$

$$\bar{y} = \frac{\Sigma \tilde{y} V}{\Sigma V} = \frac{3.5[2(7)(6)] + 1[4(2)(3)]}{2(7)(6) + 4(2)(3)}$$

$$= 2.94\ m \qquad\qquad Ans.$$

$$\bar{z} = \frac{\Sigma \tilde{z} V}{\Sigma V} = \frac{3[2(7)(6)] + 1.5[4(2)(3)]}{2(7)(6) + 4(2)(3)}$$

$$= 2.67\ m \qquad\qquad Ans.$$

F9–12. $$\bar{x} = \frac{\Sigma \tilde{x} V}{\Sigma V}$$

$$= \frac{0.25[0.5(2.5)(1.8)] + 0.25\left[\frac{1}{2}(1.5)(1.8)(0.5)\right] + (1.0)\left[\frac{1}{2}(1.5)(1.8)(0.5)\right]}{0.5(2.5)(1.8) + \frac{1}{2}(1.5)(1.8)(0.5) + \frac{1}{2}(1.5)(1.8)(0.5)}$$

$$= 0.391\ m \qquad\qquad Ans.$$

$$\bar{y} = \frac{\Sigma \tilde{y} V}{\Sigma V} = \frac{5.00625}{3.6} = 1.39\ m \qquad Ans.$$

$$\bar{z} = \frac{\Sigma \tilde{z} V}{\Sigma V} = \frac{2.835}{3.6} = 0.7875\ m \qquad Ans.$$

F9–13. $$A = 2\pi \Sigma \tilde{r} L$$
$$= 2\pi\left[0.75(1.5) + 1.5(2) + 0.75\sqrt{(1.5)^2 + (2)^2}\right]$$
$$= 37.7\ m^2 \qquad\qquad Ans.$$
$$V = 2\pi \Sigma \tilde{r} A$$
$$= 2\pi\left[0.75(1.5)(2) + 0.5\left(\frac{1}{2}\right)(1.5)(2)\right]$$
$$= 18.8\ m^3 \qquad\qquad Ans.$$

F9–14. $$A = 2\pi \Sigma \tilde{r} L$$
$$= 2\pi\left[1.95\sqrt{(0.9)^2 + (1.2)^2} + 2.4(1.5) + 1.95(0.9) + 1.5(2.7)\right]$$
$$= 77.5\ m^2 \qquad\qquad Ans.$$

$V = 2\pi\Sigma\tilde{r}A$

$= 2\pi\left[1.8\left(\frac{1}{2}\right)(0.9)(1.2) + 1.95(0.9)(1.5)\right]$

$= 22.6 \text{ m}^3$ *Ans.*

F9–15. $A = 2\pi\Sigma\tilde{r}L$

$= 2\pi\left[7.5(15) + 15(18) + 22.5\sqrt{15^2 + 20^2} + 15(30)\right]$

$= 8765 \text{ mm}^2$ *Ans.*

$V = 2\pi\Sigma\tilde{r}A$

$= 2\pi\left[7.5(15)(38) + 20\left(\frac{1}{2}\right)(15)(20)\right]$

$= 45\,710 \text{ mm}^3$ *Ans.*

F9–16. $A = 2\pi\Sigma\tilde{r}L$

$= 2\pi\left[\frac{2(1.5)}{\pi}\left(\frac{\pi(1.5)}{2}\right) + 1.5(2) + 0.75(1.5)\right]$

$= 40.1 \text{ m}^2$ *Ans.*

$V = 2\pi\Sigma\tilde{r}A$

$= 2\pi\left[\frac{4(1.5)}{3\pi}\left(\frac{\pi(1.5^2)}{4}\right) + 0.75(1.5)(2)\right]$

$= 21.2 \text{ m}^3$ *Ans.*

F9–17. $w_b = \rho_w ghb = 1000(9.81)(6)(1)$

$= 58.86 \text{ kN/m}$

$F_R = \frac{1}{2}(58.76)(6) = 176.58 \text{ kN} = 177 \text{ kN}$ *Ans.*

F9–18. $w_b = \gamma_w\, hb = 9.81\,(2)(1) = 19.62 \text{ kN/m}$

$F_R = 19.62(1.5) = 29.43 \text{ kN}$ *Ans.*

F9–19. $w_b = \rho_w gh_B b = 1000(9.81)(2)(1.5)$

$= 29.43 \text{ kN/m}$

$F_R = \frac{1}{2}(29.43)\left(\sqrt{(1.5)^2 + (2)^2}\right)$

$= 36.8 \text{ kN}$ *Ans.*

F9–20. $w_A = \rho_w gh_A b = 1000(9.81)(3)(2)$

$= 58.86 \text{ kN/m}$

$w_B = \rho_w gh_B b = 1000(9.81)(5)(2)$

$= 98.1 \text{ kN/m}$

$F_R = \frac{1}{2}(58.86 + 98.1)(2) = 157 \text{ kN}$ *Ans.*

F9–21. $w_A = \gamma_w h_A\, b = 9.81(1.8)(0.6) = 10.59 \text{ kN/m}$

$w_B = \gamma_w h_B\, b = 9.81(3.0)(0.6) = 17.66 \text{ kN/m}$

$F_R = \frac{1}{2}(10.59 + 17.66)\left(\sqrt{(0.9)^2 + (1.2)^2}\right)$

$= 21.2 \text{ kN}$ *Ans.*

Chapter 10

F10–1.

$$I_x = \int_A y^2\, dA = \int_0^{1\,\text{m}} y^2\left[\left(1 - y^{3/2}\right)dy\right] = 0.111 \text{ m}^4 \quad Ans.$$

F10–2.

$$I_x = \int_A y^2\, dA = \int_0^{1\,\text{m}} y^2\left(y^{3/2}\, dy\right) = 0.222 \text{ m}^4 \quad Ans.$$

F10–3.

$$I_y = \int_A x^2\, dA = \int_0^{1\,\text{m}} x^2\left(x^{2/3}\right)dx = 0.273 \text{ m}^4 \quad Ans.$$

F10–4.

$$I_y = \int_A x^2\, dA = \int_0^{1\,\text{m}} x^2\left[(1 - x^{2/3})\, dx\right] = 0.0606 \text{ m}^4 \quad Ans.$$

F10–5. $I_x = \left[\frac{1}{12}(50)\left(450^3\right) + 0\right] + \left[\frac{1}{12}(300)\left(50^3\right) + 0\right]$

$= 383\left(10^6\right) \text{ mm}^4$ *Ans.*

$I_y = \left[\frac{1}{12}(450)\left(50^3\right) + 0\right]$

$+ 2\left[\frac{1}{12}(50)\left(150^3\right) + (150)(50)(100)^2\right]$

$= 183\left(10^6\right) \text{ mm}^4$ *Ans.*

F10–6. $I_x = \frac{1}{12}(360)\left(200^3\right) - \frac{1}{12}(300)\left(140^3\right)$

$= 171\left(10^6\right) \text{ mm}^4$ *Ans.*

$I_y = \frac{1}{12}(200)\left(360^3\right) - \frac{1}{12}(140)\left(300^3\right)$

$= 463\left(10^6\right) \text{ mm}^4$ *Ans.*

F10–7. $I_y = 2\left[\frac{1}{12}(50)\left(200^3\right) + 0\right]$

$+ \left[\frac{1}{12}(300)\left(50^3\right) + 0\right]$

$= 69.8\,(10^6) \text{ mm}^4$ *Ans.*

F10–8.

$$\bar{y} = \frac{\Sigma\tilde{y}\,A}{\Sigma A} = \frac{15(150)(30) + 105(30)(150)}{150(30) + 30(150)} = 60 \text{ mm}$$

$$\bar{I}_{x'} = \Sigma(\bar{I} + Ad^2)$$

$= \left[\frac{1}{12}(150)(30)^3 + (150)(30)(60 - 15)^2\right]$

$+ \left[\frac{1}{12}(30)(150)^3 + 30(150)(105 - 60)^2\right]$

$= 27.0\,(10^6) \text{ mm}^4$ *Ans.*

Chapter 11

F11–1. $y_G = 0.75 \sin \theta$ $\delta y_G = 0.75 \cos \theta \, \delta\theta$

$x_C = 2(1.5) \cos \theta$ $\delta x_C = -3 \sin \theta \, \delta\theta$

$\delta U = 0; \quad 2W\delta y_G + P\delta x_C = 0$

$\quad (294.3 \cos \theta - 3P \sin \theta)\delta\theta = 0$

$P = 98.1 \cot \theta \big|_{\theta=60°} = 56.6 \text{ N}$ *Ans.*

F11–2. $x_A = 5 \cos \theta$ $\delta x_A = -5 \sin \theta \, \delta\theta$

$y_G = 2.5 \sin \theta$ $\delta y_G = 2.5 \cos \theta \, \delta\theta$

$\delta U = 0; \quad\quad -P\delta x_A + (-W\delta y_G) = 0$

$\quad (5P \sin \theta - 1226.25 \cos \theta)\delta\theta = 0$

$P = 245.25 \cot \theta \big|_{\theta=60°} = 142 \text{ N}$ *Ans.*

F11–3. $x_B = 0.6 \sin \theta$ $\delta x_B = 0.6 \cos \theta \, \delta\theta$

$y_C = 0.6 \cos \theta$ $\delta y_C = -0.6 \sin \theta \, \delta\theta$

$\delta U = 0; \quad\quad -F_{sp}\delta x_B + (-P\delta y_C) = 0$

$\quad -9(10^3) \sin \theta \, (0.6 \cos \theta \, \delta\theta)$

$\quad - 2000(-0.6 \sin \theta \, \delta\theta) = 0$

$\sin \theta = 0 \quad\quad \theta = 0°$ *Ans.*

$\quad\quad -5400 \cos \theta + 1200 = 0$

$\theta = 77.16° = 77.2°$ *Ans.*

F11–4. $x_B = 0.9 \cos \theta$ $\delta x_B = -0.9 \sin \theta \, \delta\theta$

$x_C = 2(0.9 \cos \theta)$ $\delta x_C = -1.8 \sin \theta \, \delta\theta$

$\delta U = 0; \quad P\delta x_B + \left(-F_{sp} \, \delta x_C\right) = 0$

$6(10^3)(-0.9 \sin \theta \, \delta\theta)$

$\quad - 36(10^3)(\cos \theta - 0.5)(-1.8 \sin \theta \, \delta\theta) = 0$

$\sin \theta \, (64\,800 \cos \theta - 37\,800)\delta\theta = 0$

$\sin \theta = 0 \quad\quad \theta = 0°$ *Ans.*

$\quad 64\,800 \cos \theta - 37\,800 = 0$

$\quad\quad \theta = 54.31° = 54.3°$ *Ans.*

F11–5. $y_G = 2.5 \sin \theta$ $\delta y_G = 2.5 \cos \theta \, \delta\theta$

$x_A = 5 \cos \theta$ $\delta x_C = -5 \sin \theta \, \delta\theta$

$\delta U = 0; \quad\quad \left(-F_{sp}\delta x_A\right) - W\delta y_G = 0$

$\quad (15\,000 \sin \theta \cos \theta - 7500 \sin \theta$

$\quad\quad\quad - 1226.25 \cos \theta)\delta\theta = 0$

$\theta = 56.33° = 56.3°$ *Ans.*

or $\theta = 9.545° = 9.55°$ *Ans.*

F11–6. $F_{sp} = 15\,000(0.6 - 0.6 \cos \theta)$

$x_C = 3[0.3 \sin \theta]$ $\delta x_C = 0.9 \cos \theta \, \delta\theta$

$y_B = 2[0.3 \cos \theta]$ $\delta y_B = -0.6 \sin \theta \, \delta\theta$

$\delta U = 0; \quad\quad P\delta x_C + F_{sp}\delta y_B = 0$

$\quad (135 \cos \theta - 5400 \sin \theta + 5400 \sin \theta \cos \theta)\delta\theta = 0$

$\theta = 20.9°$ *Ans.*

Answers to Selected Problems

Chapter 1

1–1.   **a.**   4.66 m
       b.   55.6 s
       c.   4.56 kN
       d.   2.77 Mg

1–2.   **a.**   GN/s
       b.   Gg/N
       c.   GN/(kg · s)

1–3.   **a.**   km/s
       b.   mm
       c.   Gs/kg
       d.   mm · N

1–5.   **a.**   Gg/s
       b.   kN/m
       c.   kN/(kg · s)

1–6.   **a.**   MN/s
       b.   kg/N
       c.   GN/(kg · s)

1–7.   **a.**   3.53 Gg
       b.   34.6 MN
       c.   5.68 MN
       d.   3.53 Gg

1–9.   **a.**   5.10 g
       b.   25.5 Mg
       c.   8.15 Gg

1–10. **a.** $0.04\ MN^2$
       b. $25\ \mu m^2$
       c. $0.064\ km^3$

1–11. **a.** $0.185\ Mg^2$
       b. $4\ \mu g^2$
       c. $0.0122\ km^3$

1–13. **a.**   8.653 s
       b.   8.368 kN
       c.   893 g

1–14. 26.9 μm · kg/N

1–15. F = 10.0 nN
       $W_1 = 78.5$ N
       $W_2 = 118$ N

Chapter 2

2–1. $F_R = 867$ N
       $\phi = 108°$

2–2. $F_R = 497$ N
       $\phi = 155°$

2–3.   F = 960 N
       $\theta = 45.2°$

2–5. $F_{1u} = 205$ N
       $F_{1v} = 160$ N

2–6. $F_{2u} = 376$ N
       $F_{2v} = 482$ N

2–7. $F_R = 3.92$ kN
       $\theta = 78.6°$

2–9. $F_{1u} = 129$ N
       $F_{1v} = 183$ N

2–10. $F_{2u} = 77.6$ N
       $F_{2v} = 150$ N

2–11. $F_A = 774$ N
       $F_B = 346$ N

2–13. $F_x = -125$ N
       $F_{y'} = 317$ N

2–14. $F_{x'} = -183$ N
       $F_y = 344$ N

2–15. $F_R = 10.8$ kN
       $\phi = 3.16°$

2–17. $F_A = 3.66$ kN
       $F_B = 7.07$ kN

2–18. $F_A = 8.66$ kN
       $F_B = 5.00$ kN
       $\theta = 60°$

2–19. $F_R = 19.2$ N
       $\theta = 2.37°$ ⦣

2–21. $\theta = 75.5°$

2–22. $\phi = \dfrac{\theta}{2}$

$$F_R = 2F \cos\left(\frac{\theta}{2}\right)$$

2–23. $\theta = 36.9°$
       $F_R = 920$ N

2–26. $F_A = 439$ N
       $F_B = 311$ N
       $\theta = 60°$

2–27. $F_A = 520$ N
       $F_B = 300$ N

2–29. $F_R = 4.01$ kN
       $\phi = 16.2°$

2–30. $\theta = 90°$
       $F_B = 1$ kN
       $F_R = 1.73$ kN

2–31. $F_R = 400$ N
       $\theta = 60°$

2–33. $F_R = 546$ N
$\theta = 253°$

2–34. $\mathbf{F}_1 = \{200\mathbf{i} + 346\mathbf{j}\}$ N
$\mathbf{F}_2 = \{177\mathbf{i} - 177\mathbf{j}\}$ N

2–35. $F_R = 413$ N
$\theta = 24.2°$

2–37. $F_R = 1.96$ kN
$\theta = 4.12°$

2–38. $F = 11.3$ kN

2–39. $\mathbf{F}_1 = \{400\mathbf{i} + 693\mathbf{j}\}$ N
$\mathbf{F}_2 = \{-424\mathbf{i} + 424\mathbf{j}\}$ N
$\mathbf{F}_3 = \{600\mathbf{i} - 250\mathbf{j}\}$ N

2–41. $F_R = 25.1$ kN
$\theta = 184.5°$

2–42. $\theta = 68.6°$
$F_B = 960$ N

2–43. $F_R = 839$ N
$\theta = 14.8°$

2–45. $F_1 = 143$ N
$F_R = 91.9$ N

2–46. $\phi = 29.2°$, $F_1 = 528$ N

2–47. $F_R = \sqrt{F_1^2 + F_2^2 + 2F_1F_2\cos\phi}$
$\tan\theta = \dfrac{F_1\sin\phi}{F_2 + F_1\cos\phi}$

2–49. $\phi = 42.4°$
$F_1 = 731$ N

2–50. $\theta = 29.1°$
$F_1 = 275$ N

2–51. $F_R = 1.03$ kN
$\theta = 87.9°$

2–53. $F = 2.03$ kN
$F_R = 7.87$ kN

2–54. $\theta = 37.0°$
$F_1 = 889$ N

2–55. $F_R = 717$ N
$\phi = 37.1°$

2–57. $F_R = 1.23$ kN
$\theta = 6.08°$

2–58. $\phi = 10.9°$
$F_1 = 474$ N

2–59. $F_R = \sqrt{(0,5F_1 + 300)^2 + (0.866F_1 - 240)^2}$
$F_R = 380$ N
$F_1 = 57.8$ N

2–61. $\mathbf{F}_1 = \{-159\mathbf{i} + 276\mathbf{j} + 318\mathbf{k}\}$ N
$\mathbf{F}_2 = \{424\mathbf{i} + 300\mathbf{j} - 300\mathbf{k}\}$ N
$\gamma = 120°$

2–62. $\mathbf{F}_R = \{265.16\mathbf{i} + 575.5\mathbf{j} + 18.20\mathbf{k}\}$ N
$F_R = 634$ N
$\alpha = 65.3°$
$\beta = 24.8°$
$\gamma = 88.4°$

2–63. $F_x = 40$ N
$F_y = 40$ N
$F_z = 56.6$ N

2–65. $F_R = 50$ N
$\alpha = 59.1°$
$\beta = 52.2°$
$\gamma = 53.1°$

2–66. $F_1 = 87.7$ N
$\alpha_1 = 46.9°$
$\beta_1 = 125°$
$\gamma_1 = 62.9°$
$F_2 = 98.6$ N
$\alpha_2 = 114°$
$\beta_2 = 150°$
$\gamma_2 = 72.3°$

2–67. $\mathbf{F}_1 = \{-2\mathbf{j}\}$ N
$\mathbf{F}_2 = \{2.5\mathbf{i} + 3.54\mathbf{j} + 2.5\mathbf{k}\}$ N

2–69. $F_R = 733$ N
$\alpha = 53.5°$
$\beta = 65.3°$
$\gamma = 133°$

2–70. $F_R = 369$ N
$\alpha = 19.5°$
$\beta = 78.3°$
$\gamma = 105°$

2–71. $F_R = 754$ N
$\alpha = 121°$
$\beta = 52.5°$
$\gamma = 53.1°$

2–73. $F_3 = 428$ N
$\alpha = 88.3°$
$\beta = 20.6°$
$\gamma = 69.5°$

2–74. $F_3 = 250$ N
$\alpha = 87.0°$
$\beta = 142°$
$\gamma = 53.1°$

2–75. $\alpha = 46.1°$
$\beta = 114°$
$\gamma = 53.1°$

2–77. $\mathbf{F}_1 = \{225\mathbf{j} + 268\mathbf{k}\}$ N
$\mathbf{F}_2 = \{70.7\mathbf{i} + 50.0\mathbf{j} - 50.0\mathbf{k}\}$ N
$\mathbf{F}_3 = \{125\mathbf{i} - 177\mathbf{j} + 125\mathbf{k}\}$ N

$F_R = 407$ N
$\alpha_x = 61.3°$
$\beta_y = 76.0°$
$\gamma_z = 32.5°$

2–78. $F_3 = 166$ N
$\alpha = 97.5°$
$\beta = 63.7°$
$\gamma = 27.5°$

2–79. $\alpha_{F1} = 36.9°$
$\beta_{F1} = 90.0°$
$\gamma_{F1} = 53.1°$
$\alpha_R = 69.3°$
$\beta_R = 52.2°$
$\gamma_R = 45.0°$

2–81. $\alpha_1 = 90°$
$\beta_1 = 53.1°$
$\gamma_1 = 66.4°$

2–82. $F_2 = 363$ N
$\alpha_2 = 15.8°$
$\beta_2 = 104°$
$\gamma_2 = 82.6°$

2–83. $F_R = 180$ N
$\alpha = 147°$
$\beta = 119°$
$\gamma = 75.0°$

2–85. $F = 2.02$ kN
$F_y = 0.523$ kN

2–86. $\gamma_{AB} = 467$ mm

2–87. $r_{AD} = 1.50$ m
$r_{BD} = 1.50$ m
$r_{CD} = 1.73$ m

2–89. $x = 5.06$ m
$y = 3.61$ m
$z = 6.51$ m

2–90. $\mathbf{F}_B = \{-400\mathbf{i} - 200\mathbf{j} + 400\mathbf{k}\}$ N
$\mathbf{F}_C = \{-200\mathbf{i} + 200\mathbf{j} + 350\mathbf{k}\}$ N

2–91. $F_R = 960$ N
$\alpha = 129°$
$\beta = 90°$
$\gamma = 38.7°$

2–93. $F_R = 1.17$ kN
$\alpha = 68.0°$
$\beta = 96.8°$
$\gamma = 157°$

2–94. $F_R = 1.49$ kN
$\alpha = 79.8°$
$\beta = 88.4°$
$\gamma = 169.7°$

2–95. $d = 6.71$ km

2–97. $F_R = 3.59$ kN
$x = 1.25$ m
$z = 2.20$ m

2–98. $\mathbf{F}_A = \{285\mathbf{j} - 93.0\mathbf{k}\}$ N
$\mathbf{F}_C = \{159\mathbf{i} + 183\mathbf{j} - 59.7\mathbf{k}\}$ N

2–99. $F_R = 316$ N
$\alpha = 60.1°$
$\beta = 74.6°$
$\gamma = 146°$

2–101. $\mathbf{r}_A = \{14.9\mathbf{i} + 7.5\mathbf{j} - 18.6\mathbf{k}\}$ m

2–102. $\mathbf{F}_{EA} = \{12\mathbf{i} - 8\mathbf{j} - 24\mathbf{k}\}$ kN
$\mathbf{F}_{EB} = \{12\mathbf{i} + 8\mathbf{j} - 24\mathbf{k}\}$ kN
$\mathbf{F}_{EC} = \{-12\mathbf{i} + 8\mathbf{j} - 24\mathbf{k}\}$ kN
$\mathbf{F}_{ED} = \{-12\mathbf{i} - 8\mathbf{j} - 24\mathbf{k}\}$ kN
$\mathbf{F}_R = \{-96\mathbf{k}\}$ kN

2–103. $x = 3.82$ m, $y = 2.12$ m, $z = 1.88$ m

2–105. $F_R = 1.50$ kN
$\alpha = 77.6°$
$\beta = 90.6°$
$\gamma = 168°$

2–106. $\mathbf{F}_A = \{144.12\mathbf{i} - 83.21\mathbf{j} - 249.62\mathbf{k}\}$ N
$\mathbf{F}_B = \{-144.12\mathbf{i} - 83.21\mathbf{j} - 249.62\mathbf{k}\}$ N
$\mathbf{F}_C = \{166.41\mathbf{j} - 249.62\mathbf{k}\}$ N
$F_R = 748.85$ N
$\alpha = 90°$
$\beta = 90°$
$\gamma = 180°$

2–107. $F = 260.4$ N

2–109. $F_C = 442$ N; $F_B = 318$ N; $F_D = 866$ N

2–110. $|\mathbf{r}_{AB}| = 567.2$ m

2–111. $\mathbf{F}_A = \{-1.46\mathbf{i} + 5.82\mathbf{k}\}$ kN
$\mathbf{F}_B = \{0.970\mathbf{i} - 1.68\mathbf{j} + 7.76\mathbf{k}\}$ kN
$\mathbf{F}_C = \{0.857\mathbf{i} + 0.857\mathbf{j} + 4.85\mathbf{k}\}$ kN
$\alpha = 88.8°$
$\beta = 92.6°$
$\gamma = 2.81°$

2–113. $\theta = 82.0°$

2–114. $\theta = 74.2°$

2–115. $r_{BC} = 5.39$ m

2–117. $r_1 \cdot u_2 = 2.99$ m, $r_2 \cdot u_1 = 1.99$ m

2–118. $\text{Proj}\,F = 0.667$ kN

2–119. $\theta = 64.6°$

2–121. $F_u = 71.6$ N

2–122. $(F_{AB})_z = 3$ kN

2–123. $(F_{AC})_z = 2.846$ kN

2–125. $(F_{BC})_{\text{paral}} = 245$ N
$(F_{BC})_{\text{per}} = 316$ N

2–126. $\theta = 70.5°$

2–127. $\phi = 65.8°$

2–129. $(\mathbf{F}_1)_{OA} = 18.5$ N
$(\mathbf{F}_2)_{OA} = 21.3$ N

2–130. $\theta = 132°$

2–131. $\theta = 74.4°$
$\theta = 55.4°$

2–133. $\theta = 97.3°$
2–134. $\theta = 127.8°$
2–135. $F_{AB} = 31.1$ N
2–137. $F_{\parallel} = 82.4$ N
$F_{\perp} = 594$ N
2–138. $\mathbf{F} = -300 \sin 30° \sin 30°\mathbf{i} + 300 \cos 30°\mathbf{j} +$
$300 \sin 30° \cos 30°\mathbf{k}$
$F_x = 75$ N
$F_y = 260$ N
2–139. $\mathbf{F} = (-300 \sin 30° \sin 30°\mathbf{i} + 300 \cos 30°\mathbf{j} +$
$300 \sin 30° \cos 30°\mathbf{k})$
$F_{OA} = 242$ N
2–141. $F_{1x} = 141$ N
$F_{1y} = 141$ N
$F_{2x} = -130$ N
$F_{2y} = 75$ N
2–142. $F_R = 217$ N
$\theta = 87.0°$
2–143. $F_u = 186$ N
$F_v = 98.7$ N
2–145. $F_R = 25.1$ kN
$\theta = 185°$
2–146. $\mathrm{F}_{proj} = 48.0$ N
2–147. $\mathrm{F}_R = 178$ N
$\theta = 85.2°$
2–149. $\theta = 53.5°$
$F_{AB} = 621$ N

Chapter 3
3–1. $F_2 = 9.45$ kN
$F_1 = 3.49$ kN
3–2. $\theta = 15.36° \; F_1 = 4.91$ kN
3–3. $F_{AC} = F_{AB} = F = \{2.45 \csc \theta\}$ kN
$l = 2.80$ m
3–5. $T = 13.3$ kN, $F = 10.2$ kN
3–6. $T = 14.3$ kN, $\theta = 36.3°$
3–7. $F_{BC} = 2.99$ kN, $F_{AB} = 3.78$ kN
3–9. $F_{AB} = 239$ N, $F_{AC} = 243$ N
3–10. $F_{CA} = 80.0$ N, $F_{CB} = 90.4$ N
3–11. $\theta = 64.3°$
$F_{CB} = 85.2$ N
$F_{CA} = 42.6$ N
3–13. $F = 95.0$ N
3–14. $d = 0.997$ m
3–15. $F_{BD} = 171$ N
$F_{BC} = 145$ N
3–17. $m = 2.37$ kg
3–18. $x_{AC} = 0.793$ m
$x_{AB} = 0.467$ m
3–19. $m = 8.56$ kg
3–21. $d = 1.56$ m

3–22. $F_A = F_D = 14.9$ N
$F_C = F_B = 40.8$ N
3–23. $P = 147$ N
3–25. $F_{BC} = 2.90$ kN
$y = 841$ mm
3–26. $F_1 = 1.83$ kN
$F_2 = 9.60$ kN
3–27. $\theta = 4.69°$
$F_1 = 4.31$ kN
3–29. $m = 20.4$ kg
3–30. $\tan\theta = dy/dx = 5.0 \,(0.4)$
$m_B = 3.58$ kg
$N = 19.7$ N
3–31. $F_{CD} = 359$ N
$F_{BD} = 440$ N
$F_{AB} = 622$ N
$F_{BC} = 228$ N
3–33. $y = 6.59$ m
3–34. $F = 1.13$ mN
3–35. $(3.5 - x)/\cos \phi + x/\cos\phi = 5$
$x = 1.38$ m, $T = 687$ N
3–37. $\theta = 15°$
$F_{AB} = 98.1$ N
3–38. $F_{AC} = F_{AB} = (2.45 \csc\theta)$ kN
$l = 1.72$ m
3–39. $T_{AB} = 340$ N
$T_{AE} = 170$ N
$T_{BD} = 490$ N
$T_{BC} = 562$ N
3–41. $F_{AB} = 98.6$ N, $F_{AC} = 267$ N
3–42. $d = 2.42$ m
3–43. $P = 1.61$ kN
$\alpha = 136°$
$\beta = 128°$
$\gamma = 72°$
3–45. $F_1 = 800$ N
$F_2 = 147$ N
$F_3 = 564$ N
3–46. $F_{DA} = 107.7$ N
$F_{DB} = 69.9$ N
$F_{DC} = 88.1$ N
3–47. $S_{OB} = 327$ mm
$S_{OA} = 218$ mm
3–49. $F = 843$ N
3–50. $F_{AO} = 319$ N
$F_{AB} = 110$ N
$F_{AC} = 85.8$ N
3–51. $W = 138$ N
3–53. $F_{AB} = 520$ N
$F_{AC} = 260$ N
$F_{AD} = 260$ N
$d = 3.61$ m

3–54. $F_{AB} = 1.21$ kN, $F_{AC} = 606$ N
$F_{AD} = 750$ N
3–55. $F_{AB} = 1.31$ kN
$F_{AC} = 763$ N
$F_{AD} = 708.5$ N
3–57. $W = 55.77$ N
3–58. $F_{AB} = 6.848$ kN
$F_{AC} = 3.721$ kN
$F_{AD} = 8.518$ kN
3–59. $F_{AB} = 7.337$ kN
$F_{AC} = 4.568$ kN
$F_{AD} = 7.098$ kN
3–61. $F_{AB} = 831$ N
$F_{AC} = 35.6$ N
$F_{AD} = 415$ N
3–62. $m = 90.3$ kg
3–63. $F_{AB} = F_{AC} = F_{AD} = 426$ N
3–65. $F_B = 858$ N
$F_C = 0$
$F_D = 858$ N
3–66. $m = 2.62$ Mg
3–67. $F_{AB} = 2.52$ kN
$F_{CB} = 2.52$ kN
$F_{BD} = 3.64$ kN
3–69. $\theta = 23.24°$, $F_1 = 2.63$ kN
3–70. $F_2 = 9.26$ kN, $F_1 = 1.91$ kN
3–71. Yes, Romeo can climb up the rope.
Yes, Romeo and Juliet can climb down.
3–73. $F_{AD} = 1.20$ kN
$F_{AC} = 0.40$ kN
$F_{AB} = 0.80$ kN
3–74. $F_{AB} = 873.4$ N
$l = 0.703$ m
3–75. $w = \dfrac{2W}{L-y}\sqrt{(L-y)^2 - d^2}$
3–77. $F_1 = 400$ N
$F_2 = 280$ N
$F_3 = 357$ N

Chapter 4

4–5. $M_P = 3.15$ kN · m (*Counterclockwise*)
4–6. $M_A = \{1.18 \cos\theta(7.5 + x)\}$ kN · m (*Clockwise*)
The maximum moment at A occurs when $\theta = 0°$
and $x = 5$ m.
$\zeta + (M_A)_{max} = 14.7$ kN · m (*Clockwise*)
4–7. $\zeta + (M_{F1})_A = 433$ N · m (*Clockwise*)
$\zeta + (M_{F2})_A = 1.30$ kN · m (*Clockwise*)
$\zeta + (M_{F3})_A = 800$ N · m (*Clockwise*)
4–9. $\zeta + M_0 = 3.57$ kN · m $\circlearrowright$
4–10. $\zeta + M_P = 3.15$ kN · m $\circlearrowright$
4–11. $(M_R)_A = 2.08$ kN · m (*Counterclockwise*)

4–13. **a.** $\zeta + M_A = 73.9$ N · m
b. $F_C = 82.2$ N
4–14. $\theta_{max} = 37.9°$, $M_{Amax} = 79.812$ N · m
$\theta_{min} = 128°$, $M_{Amin} = 0$ N · m
4–15. $(M_R)_A = 2.09$ N · m (*Clockwise*)
4–17. $\zeta + M_A = 0.418$ N · m (*Counterclockwise*)
$\zeta + M_B = 4.92$ N · m (*Clockwise*)
4–18. $(M_R)_A = 76.0$ kN · m (*Counterclockwise*)
4–19. $M_C = 4.97$ Mg
4–21. $\theta = 28.6°$
4–22. $M_A = 13.0$ N · m
$F = 35.2$ N
4–23. $\zeta + (M_0)_{max} = 80$ kN · m $\circlearrowright$
$x = 24.0$ m
4–25. $M_A = -2.532$ kN · m (*Clockwise*)
4–26. $F_3 = 1.593$ kN
4–27. $M_A = -100$ N · m (*Clockwise*)
4–29. $F = 239$ N
4–30. $\mathbf{M}_A = \{-16.0\mathbf{i} - 32.1\mathbf{k}\}$ N · m
4–31. $\mathbf{M}_O = \{-720\mathbf{i} + 120\mathbf{j} - 660\mathbf{k}\}$ N · m
4–33. $\mathbf{M}_O = [1080\mathbf{i} + 720\mathbf{j}]$ N · m
4–34. $\mathbf{M}_O = [-720\mathbf{i} + 720\mathbf{j}]$ N · m
4–35. $M_{max} = 90°$, $M_{min} = 0°$, $180°$
4–37. $\mathbf{M}_B = \{-37.6\mathbf{i} + 90.7\mathbf{j} - 155\mathbf{k}\}$ N · m
4–38. $\mathbf{M}_A = \{1.56\mathbf{i} - 0.750\mathbf{j} - 1.00\mathbf{k}\}$ kN · m
4–39. $\mathbf{M}_B = \{1.25\mathbf{i} - 0.937\mathbf{j} - 1.95\mathbf{k}\}$ kN · m
4–41. $\mathbf{M}_B = \{10.6\mathbf{i} + 13.1\mathbf{j} + 29.2\mathbf{k}\}$ N · m
4–42. $\mathbf{M}_O = \{373\mathbf{i} - 99.9\mathbf{j} + 173\mathbf{k}\}$ N · m
4–43. $\mathbf{M}_O = \{-84\mathbf{i} - 8\mathbf{j} - 39\mathbf{k}\}$ kN · m
4–45. $y = 2$m
$z = 1$ m
4–46. $y = 1$ m
$z = 3$ m
$d = 1.15$ m
4–47. $(M_{AB})_1 = 72.0$ N · m
$(M_{AB})_2 = 0$
$(M_{AB})_3 = 0$
4–49. $\mathbf{M}_{OD} = \{11.3\mathbf{i} + 11.3\mathbf{j} + 5.67\mathbf{k}\}$ N · m
4–50. $M_{CA} = 226$ N · m
4–51. $M_X = 137$ N · m
4–53. $M_{xmax} = 22.5$ N · m Adequate!
4–54. $M_x = -21.7$ N · m
4–55. $F = 139$ N
4–57. $\mathbf{M}_x = \{77-1\mathbf{i}\}$ N · m
$\mathbf{M}_z = \{-91.9\mathbf{k}\}$ N · m
4–58. $M_x = 73.0$ N · m
4–59. $F = 771$ N
4–61. $F = 20.2$ N
4–62. **a.** $M_x = 15.0$ kN · m, $M_y = 4.00$ kN · m,
$M_z = 36.0$ kN · m
b. $M_x = 15.0$ kN · m, $M_y = 4.00$ kN · m,
$M_z = 36.0$ kN · m

4–63. $\mathbf{M}_{AC} = \{11.5\mathbf{i} + 8.64\mathbf{j}\}$ kN · m

4–65. $F_B = 192$ N

$F_A = 236$ N

4–66. $M_x = 0.211$ N · m

4–67. $F = 133$ N

$P = 800$ N

4–69. $F = 625$ N

4–70. $F = 111$ N

4–71. $\theta = 56.1°$

4–73. $M_C = 22.5$ N · m ↺

4–74. $F = 83.3$ N

4–75. $N = 26.0$ N

4–77. $P = 70.7$ N

4–78. $P = 49.5$ N

4–79. $|\mathbf{M}| = 45.1$ N · m

4–81. $(M_C)_R = 5.20$ kN · m (*Clockwise*)

4–82. $F = 14.2$ kN

4–83. $\mathbf{M}_C = \{-5\mathbf{i} + 8.75\mathbf{j}\}$ N · m

4–85. $\mathbf{M}_R = \{52.0\mathbf{i} + 50\mathbf{j} + 30\mathbf{k}\}$ N · m

$M_R = 78.1$ N · m

$\alpha = 48.3°$

$\beta = 50.2°$

$\gamma = 67.4°$

4–86. $M_R = 59.9$ N · m

$\alpha = 99.0°$

$\beta = 106°$

$\gamma = 18.3°$

4–87. $P = 200$ N

4–89. $\mathbf{M}_R = \{-12.1\mathbf{i} - 10.0\mathbf{j} - 17.3\mathbf{k}\}$ N · m

4–90. $d = 342$ mm

4–91. $M_c = 40.8$ N · m

$\alpha = 11.3°$

$\beta = 101°$

$\gamma = 90°$

4–93. $\mathbf{M}_c = \{7.01\mathbf{i} + 42.1\mathbf{j}\}$ N · m

4–94. $F = 35.1$ N

4–95. $(M_C)_R = 71.9$ N · m

$\alpha = 44.2°$

$\beta = 131°$

$\gamma = 103°$

4–97. $F_R = 2.10$ kN

$\theta = 81.6°$ ∠

$M_O = 10.6$ kN · m ↺

4–98. $F_R = 2.10$kN

$\theta = 81.6°$ ∠

$M_P = 16.8$ kN · m ↺

4–99. $F_R = 5.93$ kN

$\theta = 77.8°$ ↘

$M_{RA} = 34.8$ kN · m (*Clockwise*)

4–101. $|F_R| = 294$ N

4–102. $F = 1302$ N , $\theta = 84.5°$ ∠

$x = 7.36$ m

4–103. $F = 1302$ N

$\theta = 84.5°$

$x = 1.36$ m (to the right)

4–105. $F_R = 8.27$ kN

$\theta = 69.9°$ ↘

$(M_R)_A = 9.77$ kN · m (*Clockwise*)

4–106. $F_R = 650$ N

$\theta = 72.0°$ ↘

$(M_R)_A = 431$ N · m (*Counterclockwise*)

4–107. $\mathbf{F}_R = \{270\mathbf{k}\}$ N

$\mathbf{M}_{RO} = \{-2.22\mathbf{i}\}$ N · m

4–109. $\mathbf{F}_R = \{400\mathbf{i} + 300\mathbf{j} - 650\mathbf{k}\}$ N

$\mathbf{M}_{RA} = \{-3100\mathbf{i} + 4800\mathbf{j}\}$ N · m

4–110. $\mathbf{F}_R = \{-40\mathbf{j} - 40\mathbf{k}\}$ N

$\mathbf{M}_{RA} = \{-12\mathbf{j} + 12\mathbf{k}\}$ N · m

4–111. $\mathbf{F}_R = \{-28.3\mathbf{j} - 68.3\mathbf{k}\}$ N

$\mathbf{M}_{RA} = \{-20.5\mathbf{j} + 8.49\mathbf{k}\}$ N · m

4–113. $\mathbf{F}_v = \{187.5\mathbf{i} - 324.76\mathbf{j}\}$N

$M_O = -100.481$ N · m

4–114. $\mathbf{F}_v = \{187.5\mathbf{i} - 324.76\mathbf{j}\}$N

$M_P = 736.58$ N · m (*Counterclockwise*)

4–115. $F = 4.427$ kN

$\theta = 71.565°$ (*Counterclockwise*)

$d = 3.524$ m

4–117. $F = 1302$ N

$\theta = 84.5°$ ↗

$x = 8.51$ m

4–118. $F = 1302$ N

$\theta = 84.5°$ ↗

$x = 2.52$ m (to the right)

4–119. $F_R = 4.50$ kN

$d = 2.22$ m

4–121. $F_R = 991$ N

$\theta = 63.0°$ ↗

$x = 2.64$ m

4–122. $\mathbf{F}_R = \{0.077\mathbf{i} + 2.188\mathbf{j}\}$ kN

$M_O = 6.92$ kN · m

4–123. $F_R = 2.19$ kN

$M_P = 0.358$ kN · m

4–125. $F_R = 542$ N

$\theta = 10.6°$ ↘

$d = 2.17$ m

4–126. $F_R = 50.2$ kN

$\theta = 84.3°$

$d = 4.79$ m

4–127. $F_C = 600$ N, $F_D = 500$ N

4–129. $\mathbf{F}_R = \{8\mathbf{i} + 6\mathbf{j} + 8\mathbf{k}\}$ kN

$\mathbf{M}_R = \{-10\mathbf{i} - 30\mathbf{j} - 20\mathbf{k}\}$ kN · m

4–130. $F_R = 140$ kN ↓

$y = 7.14$ m

$x = 5.71$ m

4–131. $F_R = 140$kN, $x = 6.43$ m, $y = 7.29$ m

4–133. $F_A = 30$ kN
$F_B = 20$ kN
$F_R = 190°$ kN

4–134. $\mathbf{F}_R = \{141\mathbf{i} + 100\mathbf{j} + 159\mathbf{k}\}$ N
$\mathbf{M}_{RO} = \{122\mathbf{i} - 183\mathbf{k}\}$ N · m

4–135. $F_R = -10$ kN
$x = 1.00$ m
$z = 1.40$ m

4–137. $F_R = 990$ N
$M_R = 3.07$ kN · m
$x = 1.16$ m
$y = 2.06$ m

4–138. $F_{RO} = 70$ N ↓
$x = 0.107$ m

4–139. $F_R = 6.75$ kN ↓
$\bar{x} = 2.5$ m

4–141. $F_R = 0$
$M_{RA} = 3.75$ kN · m

4–142. $F_R = 90$ kN ↓
$M_{RO} = 338$ kN · m ↻

4–143. $F_R = 0.525$ kN ↑
$d = 0.171$ m

4–145. $F_R = 60$ kN
$x = 3.89$ m

4–146. $F_R = \dfrac{1}{2}w_0 L$ ↓
$\bar{x} = \dfrac{5}{12}L$

4–147. $F_R = 11.25$ kN
$d = 4.05$ m

4–149. $a = 1.539$ m
$b = 5.625$ m

4–150. $F_R = 51.0$ kN ↓
$M_{RO} = 914$ kN · m (*Clockwise*)

4–151. $F_R = 51.0$ kN ↓
$d = 17.9$ m

4–153. $F_R = 3.46$ kN
$\theta = 47.5°$ ↘
$M_{RB} = 5.04$ kN · m ↻

4–154. $F_R = 1.35$ kN
$\theta = 42.0°$ ↗
$y = 0.1$ m

4–155. $F_R = 1.35$ kN
$x = 0.556$ m

4–157. $F_R = 36$ kN ↓
$\bar{x} = 2.17$ m

4–158. $F_R = 3$ kN
$M_O = 2.25$ kN · m

4–159. $F_R = 107$ kN ←
$h = 1.60$ m

4–161. $F_R = 14.9$ kN ↓
$\bar{x} = 2.27$ m

4–162. $F_R = 14.9$ kN
$\bar{x} = 2.27$ m

4–163. $\mathbf{M}_{CR} = \{63.6\mathbf{i} - 170\mathbf{j} + 264\mathbf{k}\}$ N · m

4–165. $\mathbf{M}_O = \{1.06\mathbf{i} + 1.06\mathbf{j} - 4.03\mathbf{k}\}$ N · m
$\alpha = 75.7°$
$\beta = 75.7°$
$\gamma = 160°$

4–166. $\mathbf{M}_{RP} = \{-26\mathbf{i} + 357\mathbf{j} + 127\mathbf{k}\}$ N · m

4–167. $M_O = 28.1$ N · m (*Clockwise*)
$\theta = 88.6°$
$(M_O)_{max} = 32.0$ N · m (*Clockwise*)

4–169. $\mathbf{M}_O = \{440\mathbf{i} + 220\mathbf{j} + 990\mathbf{k}\}$ N · m

4–170. $\mathbf{M}_A = \{-59.7\mathbf{i} - 159\mathbf{k}\}$ N · m

4–171. $M_{a-a} = 59.7$ N · m

4–173. $M_z = 15.5$ N · m

Chapter 5

5–10. $N_B = 3.46$ kN
$A_x = 1.73$ kN
$A_y = 1.00$ kN

5–11. $B_y = 586$ N
$F_A = 413$ N

5–13. $B_y = 736$ N
$A_x = 920$ N
$B_x = 920$ N

5–14. $F_{BC} = 3.92$ kN
$A_x = 3.13$ kN
$A_y = 950$ N

5–15. $F_B = 10.5$ N
$A_x = 42.0$ N
$A_y = 10.5$ N

5–17. $F_B = 105$ N

5–18. $N_A = 105$ kN
$B_x = 97.4$ kN
$B_y = 269$ kN

5–19. $B_x = 989$ N
$A_x = 989$ N
$B_y = 186$ N

5–21. $w_1 = 413$ kN/m
$w_2 = 407$ kN/m

5–22. $k = 1.33$ kN/m
$A_x = 398$ N
$A_y = 300$ N

5–23. $\theta = 23.1°$
$A_x = 353$ N
$A_y = 300$ N

5–25. $M_A = 10.6$ N · m

5–26. $N_B = 2.11$ N
$F_A = 2.81$ N

5–27. $F_B = 3.68$ kN, $x_A = 53.2$ mm, $x_B = 50.5$ mm

5–29. $\alpha = 1.02°$

5–30. $k_B = 2.50$ kN/m

5–31. $F_{AB} = 38.3$ kN
$C_x = 30.7$ kN
$C_y = 3.37$ kN

5–33. $F_{BD} = 628$ N
$C_x = 432$ N
$C_y = 68.2$ N

5–34. $\theta = a\cos\left(\dfrac{L + \sqrt{L^2 + 128r^2}}{16r}\right)$

5–35. $l = 0.67664$ m
$\theta = 2.1332$ N
$F_s = 2.1332$ N
$C_x = 432$ N
$N_B = 1.89$ N
$F_A = 2.52$ N

5–37. $F_{CB} = 782$ N
$A_x = 625$ N
$A_y = 681$ N

5–38. $F_2 = 724$ N
$F_1 = 1.45$ kN
$F_A = 1.75$ kN

5–39. $N_C = 213$ N
$A_x = 105$ N
$A_y = 118$ N

5–41. $T = \dfrac{W\cos\theta}{2\sin(\phi-\theta)}$

$A_x = \dfrac{W\cos\phi\cos\theta}{2\sin(\phi-\theta)}$

$A_x = \dfrac{W(\sin\phi\cos\theta - 2\cos\phi\sin\theta)}{2\sin(\phi-\theta)}$

5–42. $F = 10$ mN

5–43. $k = 250$ N/m

5–45. $F_B = 6.38$ N
$A_x = 3.19$ N
$A_y = 2.48$ N

5–46. $d = \dfrac{3a}{4}$

5–47. $\Delta_A = 0.1067$ m
$\Delta_B = 0.05333$ m

5–49. $P = 71.3$ N
$N = 446$ N

5–50. $\theta = \tan^{-1}\dfrac{b}{a}$

5–51. $F_B = 0.3\,P$
$F_C = 0.6\,P$
$x_C = 0.6\,P/k$

5–53. $T = \dfrac{W\cos\theta}{2\sin(\phi-\theta)}$

$A_x = \dfrac{W\cos\phi\cos\theta}{2\sin(\phi-\theta)}$

$A_y = \dfrac{W(\sin\phi\cos\theta - 2\cos\phi\sin\theta)}{2\sin(\phi-\theta)}$

5–54. $d = \dfrac{a}{\cos^3\theta}$

5–55. $N_A = 346$ N
$N_B = 693$ N
$a = 0.650$ m

5–57. $F = 5.20$ kN
$N_A = 17.3$ kN
$N_B = 24.9$ kN

5–58. $\theta = 63.4°$
$T = 29.2$ kN

5–59. $a = \sqrt{(4r^2 l)^{\frac{2}{3}} - 4r^2}$

5–61. $h = \sqrt{\dfrac{s^2 - l^2}{3}}$

5–62. $T = 1.84$ kN
$F = 6.18$ kN

5–63. $R_D = 113.1$ kN
$R_E = 113.1$ kN
$R_F = 68.7$ kN

5–65. $N_C = 289$ N
$N_A = 213$ N
$N_B = 332$ N

5–66. $T_{BC} = T_{BD} = 17$ kN
$A_x = 0$
$A_y = 11.3$ kN
$A_z = -15.7$ kN

5–67. $T_B = 16.7$ kN
$A_x = 0$
$A_y = 5.00$ kN
$A_z = 16.7$ kN

5–69. $C_y = -318$ N
$C_z = 500$ N
$B_x = 239$ N
$B_z = -273$ N
$A_x = -239$ N
$A_z = 90.9$ N

5–70. $d = \dfrac{r}{2}\left(1 + \dfrac{W}{P}\right)$

5–71. $N_B = 1000$ N
$(M_A)_z = 0$
$A_y = 0$
$(M_A)_y = -560$ N · m
$A_z = 400$ N

5–73. $T_{CD} = 3$ kN
$T_{EF} = 2.25$ kN
$T_{AB} = 0.75$ kN

5–74. $x = 0.667$ m
$y = 0.667$ m

5–75. $A_y = 0$
$T = 1.23$ kN
$B_x = -433$ N
$B_z = 1.42$ kN
$A_x = 867$ N
$A_z = 711$ N

5–77. $P = 0.5\ W$

5–78. $C_y = 450$ N, $C_z = 250$ N
$B_x = 25$ N, $B_z = 1.125$ N
$A_x = 475$ N, $A_z = 125$ N

5–79. $F_{AC} = F_{BC} = 6.13$ kN
$F_{DE} = 19.62$ kN

5–81. $F_{BD} = 294$ N
$F_{BC} = 589$ N
$A_x = 0$
$A_y = 589$ N
$A_z = 490.5$ N

5–82. $T_{BD} = T_{CD} = 116.7$ N
$A_x = 66.7$ N
$A_y = 0$
$A_z = 100$ N

5–83. $T = 58.0$ N, $C_z = 87.0$ N, $C_y = 28.8$ N, $D_x = 0$
$D_y = 79.2$ N, $D_z = 58.0$ N

5–85. $F = 354$ N

5–86. $N_A = 8.00$ kN, $B_x = 5.20$ kN, $B_y = 5.00$ kN

5–87. $N_B = 400$ N, $F_A = 721$ N

5–89. $\theta = \tan^{-1}\left(\dfrac{1}{2}\cot\psi - \dfrac{1}{2}\cot\phi\right)$

5–90. $A_x = 35.1$ N, $A_y = 343$ N, $N_B = 343$ N

5–91. $A_x = 0$
$A_y = -200$ N, $A_z = 150$ N, $(M_A)_x = 100$ N · m
$(M_A)_y = 0$, $(M_A)_z = 500$ N · m

Chapter 6

6–1. $F_{AB} = 60$ kN (T), $F_{BD} = 40$ kN (C)
$F_{AD} = 84.9$ kN (C), $F_{DC} = 141$ kN (T)
$F_{BC} = 60$ kN (T), $F_{DE} = 160$ kN (C)

6–2. $F_{AD} = 113$ kN (C)
$F_{AB} = 80$ kN (T)
$F_{BD} = 0$
$F_{BC} = 80$ kN (T)
$F_{DC} = 113$ kN (T)
$F_{DE} = 160$ kN (C)

6–3. $F_{CD} = 5.21$ kN(C)
$F_{CB} = 4.17$ kN (T), $F_{AD} = 1.46$ kN (C)
$F_{AB} = 4.17$ kN (T), $F_{BD} = 4$ kN (T)

6–5. $F_{DC} = 400$ N (C)
$F_{DA} = 300$ N (C)
$F_{BA} = 250$ N (T)
$F_{BC} = 200$ N (T)
$F_{CA} = 283$ N (C)

6–6. $F_{DE} = 1.00$ kN (C)
$F_{DC} = 800$ N (T)
$F_{CE} = 900$ N(C)
$F_{CB} = 800$ N (T)
$F_{EB} = 750$ N (T)
$F_{EA} = 1.75$ kN (C)

6–7. $F_{JD} = 33.3$ kN (T)
$F_{AL} = F_{GH} = F_{LK} = F_{HI} = 28.3$ kN (C)
$F_{AB} = F_{GF} = F_{BC} = F_{FE} = F_{CD} = F_{ED} = 20$ kN (T)
$F_{BL} = F_{FH} = F_{LC} = F_{HE} = 0$
$F_{CK} = F_{EI} = 10$ kN (T)
$F_{KJ} = F_{IJ} = 23.6$ kN (C)
$F_{KD} = F_{ID} = 7.45$ kN (C)

6–9. $F_{AB} = 7.5$ kN (T), $F_{AE} = 4.5$ kN (C)
$F_{ED} = 4.5$ kN (C), $F_{EB} = 8$ kN (T)
$F_{BD} = 19.8$ kN (C), $F_{BC} = 18.5$ kN (T)

6–10. $F_{AB} = 196$ N (T), $F_{AE} = 118$ N (C)
$F_{ED} = 118$ N (C), $F_{EB} = 216$ N (T)
$F_{BD} = 1.04$ kN (C), $F_{BC} = 857$ N (T)

6–11. $F_{DE} = 16.3$ kN (C), $F_{DC} = 8.40$ kN (T)
$F_{EA} = 8.85$ kN (C), $F_{EC} = 6.20$ kN (C)
$F_{CF} = 8.77$ kN (T), $F_{CB} = 2.20$ kN (T)
$F_{BA} = 3.11$ kN (T), $F_{BF} = 6.20$ kN (C)
$F_{FA} = 6.20$ kN (T)

6–13. $F_{GB} = 30$ kN (T)
$F_{AF} = 20$ kN (C), $F_{AB} = 22.4$ kN (C)
$F_{BF} = 20$ kN (T), $F_{BC} = 20$ kN (T)
$F_{FC} = 28.3$ kN (C), $F_{FE} = 0$
$F_{ED} = 0$, $F_{EC} = 20.0$ kN(T), $F_{DC} = 0$

6–14. $F_{CB} = 3.00$ kN (T)
$F_{CD} = 2.60$ kN (C)
$F_{DE} = 2.60$ kN (C)
$F_{DB} = 2.00$ kN (T)
$F_{BE} = 2.00$ kN (C)
$F_{BA} = 5.00$ kN (T)

6–15. $F_{CB} = 8.00$ kN (T)
$F_{CD} = 6.93$ kN (C)
$F_{DE} = 6.93$ kN (C)
$F_{DB} = 4.00$ kN (T)
$F_{BE} = 4.00$ kN (C)
$F_{BA} = 12.0$ kN (T)

6–17. Maximum force:
$F_{DC} = F_{CB} = F_{CE} = F_{BE} = F_{BA} = 1.1547P$
$P = 5.20$ kN

6–18. $F_{CD} = 3.61$ kN (C), $F_{CB} = 3$ kN (T)
$F_{BA} = 3$ kN (T), $F_{BD} = 3$ kN (C)
$F_{DA} = 2.70$ kN (T), $F_{DE} = 6.31$ kN (C)

6–19. $F_{CD} = 467$ N (C), $F_{CB} = 389$ N (T)
$F_{BA} = 389$ N (T), $F_{BD} = 314$ N (C)
$F_{DE} = 1.20$ kN (C), $F_{DA} = 736$ N (T)

6–21. $P = 1.25$ kN

6–22. $F_{FE} = 0.667\ P$ (T), $F_{FD} = 1.67\ P$ (T)
$F_{AB} = 0.471\ P$ (C), $F_{AE} = 1.67\ P$ (T)
$F_{AC} = 1.49\ P$ (C), $F_{BF} = 1.41\ P$ (T),
$F_{BD} = 1.49\ P$ (C), $F_{EC} = 1.41\ P$ (T),
$F_{CD} = 0.471\ P$ (C)

6–23. $F_{CD} = 0.577\ P$ (C), $F_{DB} = 0.289\ P$ (T)
$F_{CE} = 0.577\ P$ (T), $F_{BC} = 0.577\ P$ (C)
$F_{BE} = 0.577\ P$ (T), $F_{AB} = 0.577\ P$ (C)
$F_{AE} = 0.577\ P$ (T)

6–25. $F_{BA} = P\ \csc 2\theta$ (C), $F_{BC} = P\ \cot 2\theta$ (C)
$F_{CA} = (\cot\theta\cos\theta - \sin\theta + 2\cos\theta)\ P$ (T)
$F_{CD} = (\cot 2\theta + 1)\ P$ (C)
$F_{DA} = (\cot 2\theta + 1)(\cos 2\theta)\ (P)$ (C)

6–26. Maximum force: $F_{CA} = 2.732\ P$ (T),
$F_{CD} = 1.577\ P$ (C), $P_{max} = 732$ N

6–27. $F_{GF} = 8.08$ kN (T)
$F_{BC} = 7.70$ kN (C)
$F_{CG} = 0.770$ kN (C)

6–29. $F_{LD} = 0$, $F_{LK} = -112.5$ kN (C)
$F_{CD} = 112.5$ kN(T), $F_{KD} = -50$ kN (C)

6–30. $F_{JI} = -75$ kN (C)
$F_{JE} = -50$ kN (C)
$F_{DE} = 75$ kN (T)

6–31. $F_{CB} = 5$ kN (T), $F_{BE} = 21.2$ kN (T)
$F_{EF} = -25$ kN (C)

6–33. $F_{BC} = -3.25$ kN (C)
$F_{CH} = 1.923$ kN (T)

6–34. $F_{CD} = -1.917$ kN (C), $F_{GF} = 1.533$ kN (T)

6–35. $F_{BC} = 10.4$ kN (C), $F_{HG} = 9.15$ kN (T),
$F_{HC} = 2.24$ kN (T)

6–37. $F_{KJ} = 3.07$ kN (C), $F_{CD} = 3.07$ kN (T)
$F_{ND} = 0.167$ kN (T), $F_{NJ} = 0.167$ kN (C)

6–38. $F_{JI} = 2.13$ kN (C)
$F_{DE} = 2.13$ kN (T)

6–39. $F_{IC} = 5.62$ kN (C), $F_{CG} = 9.00$ kN (T)

6–41. $F_{CB} = 3.60$ kN (T), $F_{GC} = 1.80$ kN (C)
$F_{FG} = 4.02$ kN (C)

6–42. $F_{ML} = 38.4$ kN (T)
$F_{DE} = -37.1$ kN (C)
$F_{DL} = -3.8$ kN (C)

6–43. $F_{EF} = -37.1$ kN (C)
$F_{EL} = 6$ kN (T)

6–45. $F_{AB} = P$ (T), $F_{EF} = P$ (C), $F_{BF} = 1.41\ P$ (C)

6–46. Method of Joints: By inspection, members *BN,*
NC, DO, OC, HJ, LE and *G* are zero–force
members.
$F_{CD} = 5.625$ kN (T), $F_{CM} = 2.00$ kN (T)

6–47. Method of Joints: By inspection, members *BN,*
NC, DO, OC HJ, LE and *JG* are zero–force
members.
$F_{EF} = 7.88$ kN (T), $F_{LK} = 9.25$ kN (C)
$F_{ED} = 1.94$ kN (T)

6–49. $F_{BG} = \{-600\ \csc\ \theta\}$ N
$F_{BC} = -200L$ N
$F_{HG} = 400L$ N

6–50. $F_{AB} = 584$ N (T)
$F_{AC} = -1133$ N (C)
$F_{BC} = -142$ N (C)

6–51. $F_{AB} = 6.46$ kN (T), $F_{AC} = F_{AD} = 1.50$ kN (C)
$F_{BC} = F_{BD} = 3.70$ kN (C), $F_{BE} = 4.80$ kN (T)

6–53. $F_{BD} = 896$ N (C), $F_{DC} = 554$ N (T)
$F_{DA} = 146$ N (C), $F_{AB} = 52.1$ N (T)
$F_{AC} = 31.25$ N (T), $F_{CB} = 448$ N (C)

6–54. $F_{DB} = 474$N (C), $F_{DC} = 146$ N (T)
$F_{DA} = 1.08$ kN (T), $F_{AB} = 385$ N (C)
$F_{AC} = 231$ N (C), $F_{CB} = 281$ N (T)

6–55. $F_{BC} = F_{BD} = 1.34$ kN (C), $F_{AB} = 2.4$ kN (C)
$F_{AG} = F_{AE} = 1.01$ kN (T), $F_{BG} = 1.80$ kN (T)
$F_{BE} = 1.80$ kN (T)

6–57. $F_{CE} = 721$ N (T), $F_{BC} = 400$ N (C)
$F_{BE} = 0$, $F_{BF} = 2.10$ kN (T)

6–58. $F_{AB} = 1.50$ kN (C), $F_{AF} = 1.08$ kN (C)
$F_{AD} = 600$ N (T), $F_{FD} = 0$, $F_{ED} = 1.40$ kN (C)
$F_{BD} = 361$ N (C)

6–59. $F_{AE} = F_{AC} = 220$ N (T)
$F_{AB} = 583$ N (C)
$F_{BD} = 707$ N (C)
$F_{BE} = F_{BC} = 141.4$ N (T)

6–61. $P = 18.9$ N

6–62. $P = 368$ N

6–63. $P = 40.0$ N, $x = 240$ mm

6–65. $P = 2.24$ kn

6–66. $A_x = 4.20$ kN, $B_x = 4.20$ kN, $A_y = 4.00$ kN
$B_y = 3.20$ kN, $C_x = 3.40$ kN, $C_y = 4.00$ kN

6–67. $B_x = 4.00$ kN
$B_y = 2.00$ kN
$A_x = 6.00$ kN
$A_y = 0$

6–69. $C_x = 167$ N, $C_y = 111$ N
$A_x = 167$ N, $A_y = 389$ N

6–70. $3.64\ F$

6–71. $N_E = 5$ kN
$D_x = 0$
$N_C = 16.7$ kN
$A_x = 0$
$A_y = 2.67$ kN
$M_A = 21.5$ kN · m

6–73. $C_y = 0$
$B_y = 7.50$ kN
$M_A = 45.0$ kN · m
$A_y = -7.50$ kN
$A_x = 0$

6–74. $m = 1.71$ kg

6–75. $m = 106$ kg

6–77. $m = 366$ kg, $F_A = 2.93$ kN

6–78. $N_A = 3.67$ kN, $M_A = 5.55$ kN · m
$C_x = 2.89$ kN, $C_y = 1.32$ kN

6–79. $F_E = 4.55\ F$

6–81. $F_{AB} = 9.23$ kN, $C_x = 2.17$ kN, $C_y = 7.01$ kN
$D_x = 0$, $D_y = 1.96$ kN, $M_D = 2.66$ kN · m

6–82. $A_x = 2.98$ kN, $A_y = 235$ N
$B_x = 2.98$ kN, $B_y = 549$ N
$C_x = 2.98$ kN, $C_y = 1.33$ kN

6–83. $A_x = 500$ N, $A_y = 1000$ N
$C_x = 500$ N, $C_y = 500$ N

6–85. $F_A = 130$ N

6–86. $N_B = N_C = 49.5$ N

6–87. $R_E = 177$ N, $R_A = 128$ N

6–89. $m = 1.11$ Mg

6–90. $F_C = 28.8$ kN

6–91. $A_y = 3.09$ kN, $C_y = 1.52$ kN
$B_x = 3.5$ kN, $B_y = 23.5$ kN

6–93. $N_A = 490.5$ N, $N_B = 294.3$ N, $T = 353.70$ N,
$\theta = 33.7°$, $x = 177$ mm

6–94. a. $F = 875$ N, $N_C = 1750$ N
b. $F = 437.5$ N, $N_C = 437.5$ N

6–95. a. $F = 1025$ N, $N_C = 1900$ N
b. $F = 512.5$ N, $N_C = 362.5$ N

6–97. $T = 9.60$ N

6–98. $B_x = 4.28$ kN, $B_y = 3.15$ kN, $A_x = 4.28$ kN
$A_y = 4.05$ kN, $C_x = 3.38$ kN, $C_y = 4.05$ kN

6–99. $F = 370$ N

6–101. $N_A = 11.1$ kN, $N_B = 3.37$ kN, $F_{CD} = 6.5$ kN
$F_E = 5.88$ kN

6–102. $F_{AB} = 1.18$ kN, $F_E = 3.17$ kN, $F_{CD} = 19.6$ kN
$F_F = 16.8$ kN

6–103. $F_N = 26.25$ N

6–105. $F_{DB} = 2.601$ kN, $F_{FB} = 1.94$ kN

6–106. $F = 562.5$ N

6–107. $P = 80$ N

6–109. $E_x = 6.79$ kN, $E_y = 1.55$ kN
$D_x = 981$ N, $D_y = 981$ N

6–110. $M = 2.43$ kN · m

6–111. $F = 5.07$ kN

6–113. $A_x = 0.2629$ kN, $A_y = 0.6495$ kN
$B_x = 0.4871$ kN, $B_y = 0.6495$ kN

6–114. $E_x = 4.7243$ kN, $E_y = 2.5$ kN
$D_x = 4.7243$ kN, $D_y = 5$ kN

6–115. $F_s = 286$ N

6–117. $P = B_z = D_z = B_y = D_y = 283$ N
$B_x = D_x = 0$

6–118. $A_x = -172.3$ N, $A_y = -114.8$ N, $A_z = 0$ N
$C_x = 47.3$ N, $C_y = -61.9$ N, $C_z = -125$ N
$M_{Cy} = -429$ N · m, $M_{Cz} = 0$ N · m

6–119. $F_{AB} = 981$ N, $F_E = 2.64$ kN
$F_{CD} = 16.3$ kN, $F_F = 14.0$ kN

6–121. $A_y = 250$ N, $A_x = 1.40$ kN, $C_x = 500$ N,
$C_y = 1.70$ kN

6–122. $N_A = 4.60$ kN
$C_y = 7.05$ kN
$N_B = 7.05$ kN

6–123. $m = 3.86$ kg

6–125. $A_x = 117$ N, $A_y = 397$ N
$B_x = 97.4$ N, $B_y = 97.4$ N

6–126. $F_{AB} = 21.9$ kN (C), $F_{BG} = 17.5$ kN (T)
$F_{AG} = 13.1$ kN (T), $F_{BC} = 13.1$ kN (C)
$F_{GC} = 3.12$ kN (T), $F_{GF} = 11.2$ kN (T)
$F_{CF} = 3.12$ kN (C), $F_{CD} = 9.38$ kN (C)
$F_{DE} = 15.6$ kN (C), $F_{DF} = 12.5$ kN (T)
$F_{EF} = 9.38$ kN (T)

Chapter 7

7–1. $N_A = 5$ kN, $N_C = 4$ kN, $N_B = 3$ kN

7–2. $T_A = 100$ N · m, $T_B = 200$ N · m
$T_C = 200$ N · m, $T_D = 0$

7–3. $d = 0.200$ m

7–5. $V_A = 3$ kN, $N_A = 13.2$ kN, $M_A = 3.82$ kN · m
$V_B = 3$ kN, $N_B = 16.2$ kN, $M_B = 14.3$ kN · m

7–6. $a = \dfrac{L}{3}$

7–7. $N_E = 0$, $V_E = -50$ N, $M_E = -100$ N · m
$N_D = 0$, $V_D = 750$ N, $M_D = -1300$ N · m

7–9. $N_C = -30$ kN, $V_C = -8$ kN, $M_C = 6$ kN · m

7–10. $P = 0.533$ kN, $N_C = -2$ kN, $V_C = -0.533$ kN
$M_C = 0.400$ kN · m

7–11. $M_E = 7.5$ kN · m, $N_E = 0$, $V_E = 5$ kN
$M_D = -9.75$ kN · m, $N_D = 0$, $V_D = 8$ kN

7–13. $N_C = 0$, $V_C = -1$ kN, $M_C = 9$ kN · m

7–14. $N_D = -800$ N, $V_D = 0$, $M_D = 1.20$ kN · m

7–15. $w = 100$ N/m

7–17. $N_E = -1.92$ kN, $V_E = 800$ N, $M_E = 2.40$ kN · m

7–18. $N_C = 0$
$V_C = \dfrac{3w_0 L}{8}$

$M_C = -\dfrac{5}{48} w_0 L^2$

7–19. $N_E = 470$ N, $V_E = 215$ N
$M_E = 660$ N · m, $N_F = 0$
$V_F = -215$ N, $M_F = 660$ N · m

7–21. $N_D = 4$ kN, $V_D = -9$ kN, $M_D = -18$ kN $\cdot$ m
$N_E = 4$ kN, $V_E = 3.75$ kN, $M_E = -4.875$ kN $\cdot$ m

7–22. $N_E = 0$, $V_E = -1.17$ kN, $M_E = 4.97$ kN $\cdot$ m
$N_F = 0$, $V_F = 1.25$ kN, $M_F = 2.5$ kN $\cdot$ m

7–23. $V_D = 168$ N, $N_D = -110$ N, $M_D = 348$ N $\cdot$ m
$N_E = -168$ N, $V_E = -90.4$ N, $M_E = 190$ N $\cdot$ m

7–25. $N_C = 0$, $V_C = 0$, $M_C = 1.5$ kN $\cdot$ m

7–26. $A_y = \left(\dfrac{w}{6b}\right)(2a + b)(a - b)$, $\dfrac{a}{b} = \dfrac{1}{4}$

7–27. $N_D = 2.40$ kN, $V_D = 50$ N, $M_D = 1.35$ kN $\cdot$ m

7–29. $N_D = 1.26$ kN, $V_D = 0$, $M_D = 500$ N $\cdot$ m

7–30. $N_E = -1.48$ kN
$V_E = 500$ N
$M_E = 1000$ N $\cdot$ m

7–31. $N_A = V_A = 2.83$ kN, $M_A = 212$ N $\cdot$ m

7–33. $a = \dfrac{2}{3}L$

7–34. $N_C = -\dfrac{wL}{2}\csc\theta$, $V_C = 0$, $M_C = \dfrac{wL^2}{8}\cos\theta$

7–35. $V_E = 0$, $N_E = 894$ N, $M_E = 0$, $V_F = 447$ N,
$N_F = -224$ N, $M_F = 224$ N $\cdot$ m

7–37. $N_C = -1.91$ kN, $V_C = 0$, $M_C = 382$ N $\cdot$ m

7–38. $N_D = 981$ N, $V_D = 0$, $M_D = 1.32$ kN $\cdot$ m
$N_E = 1.77$ kN, $V_E = 589$ N, $M_E = 1.68$ kN $\cdot$ m

7–39. $V = -0.293\,rw_0$, $N = -0.707\,rw_0$
$M = -0.0783\,r^2 w_0$

7–41. $C_x = -150$ kN, $C_y = -350$ kN, $C_z = 700$ kN
$M_{Cx} = 1.40$ MN $\cdot$ m, $M_{Cy} = -1.20$ MN $\cdot$ m,
$M_{Cz} = -750$ kN $\cdot$ m

7–42. $N_C = -170$ kN, $(V_C)_y = -50$ kN
$(V_C)_z = 500$ kN
$(M_C)_x = 1$ MN $\cdot$ m
$(M_C)_y = 900$ kN $\cdot$ m
$(M_C)_z = -260$ kN $\cdot$ m

7–43. $N_{Dy} = -65$ kN, $V_{Dx} = 116$ kN, $V_{Dz} = 0$
$M_{Dx} = 49.2$ kN $\cdot$ m
$M_{Dy} = 87.0$ kN $\cdot$ m
$M_{Dz} = 26.2$ kN $\cdot$ m

7–45. For $0 \le x < b$, $V = -\dfrac{Pa}{b}$, $M = -\dfrac{Pa}{b}x$.
For $b < x \le a + b$, $V = P$, $M = -P(a + b - x)$.

7–46. For $0 \le x < 2$ m, $V = 0.25$ kN, $M = (0.25x)$ kN $\cdot$ m.
For 2 m $< x \le 3$ m, $V = (3.25 - 1.5x)$ kN.
$M = (-0.75x^2 + 3.25x - 3.0)$ kN $\cdot$ m.

7–47. **a.** For $0 \le x < a$, $V = P$, $M = Px$.
For $a < x < L - a$, $V = 0$, $M = Pa$.
For $L - a < x \le L$, $V = -P$.
b. For $0 \le x < 1.5$ m, $V = 4$ kN, $M = (4x)$ kN $\cdot$ m.
For 1.5 m $< x < 2.1$ m, $V = 0$, $M = 6$ kN $\cdot$ m.
For 2.1 m $< x \le 3.6$ m, $V = -4$ kN,
$M = (14.4 - 4x)$ kN $\cdot$ m.

7–49. $M_{max} = 2$ kN $\cdot$ m

7–50. $V = (4 - 2x)$ kN
$M = (-x^2 + 4x - 10)$ kN $\cdot$ m

7–51. For $0 \le x < \dfrac{L}{2}$,
$$V = \dfrac{wL}{8},$$
$$M = \dfrac{wL}{8}x.$$
For $\dfrac{L}{2} < x \le L$,
$$V = \dfrac{w}{8}(5L - 8x),$$
$$M = \dfrac{w}{8}(-L^2 + 5Lx - 4x^2).$$

7–53. For $0 \le x < 8$m, $V = 140 - 40x$ kN,
$M = (140x - 20x^2)$.
For 8m $< x \le 11$m, $V = 20$ kN,
$M = (20x - 320)$ kN $\cdot$ m.

7–54. $V = (525 - 300x)$ kN
$M = (-150x^2 + 525x + 300)$ kN $\cdot$ m

7–57. $V = 5(3 - x)$
$M = (15x - 2.5x^2 - 0.25)$ kN $\cdot$ m

7–58. For $0 \le x < L$, $V = \dfrac{w}{18}(7L - 18x)$,
$$M = \dfrac{w}{18}\left(7Lx - 9x^2\right).$$
For $L \le x < 2L$, $V = \dfrac{w}{2}(3L - 2x)$,
$$M = \dfrac{w}{18}(27Lx - 20L^2 - 9x^2).$$
For $2L < x \le 3L$, $V = \dfrac{w}{18}(47L - 18x)$,
$$M = \dfrac{w}{18}(47Lx - 9x^2 - 60L^2).$$

7–59. $x = 1.732$ m
$M_{max} = 0.866$

7–61. $V = \left\{3.00 - \dfrac{x^2}{4}\right\}$ kN
$M = \left\{3.00x - \dfrac{x^3}{12}\right\}$ kN $\cdot$ m

7–62. $V = \dfrac{\gamma ht}{2d}x^2$, $M = -\dfrac{\gamma ht}{6d}x^3$

7–63. For $0 \le x < 5$ m, $V = 2.5 - 2x$, $M = 2.5x - x^2$.
For 5 m $< x \le 10$ m, $V = -7.5$, $M = -7.5x - 25$.

7–65. For $0 \le x < 3$ m, $V = \{-650 - 50.0x^2\}$ N,
$M = \{-650x - 16.7x^3\}$ N $\cdot$ m.
For 3 m $< x \le 7$ m, $V = \{2100 - 300x\}$ N,
$M = \{-150(7 - x)^2\}$ N $\cdot$ m.

7–66. $V = \dfrac{w}{12L}\left(4L^2 - 6Lx - 3x^2\right)$,
$M = \dfrac{w}{12L}\left(4L^2 x - 3Lx^2 - x^3\right)$, $M_{max} = 0.0940wL^2$

7–67. $N = \dfrac{P}{2}\cos\theta$

$V = \dfrac{P}{2}\sin\theta$

$M = \dfrac{Pr}{2}(1 - \cos\theta)$

7–69. $V = |P|$

$M = |Pr\cos\theta|$

$T = |Pr(1 - \sin\theta)|$

7–94. $T_{BD} = 390.9$ N, $T_{AC} = 378.4$ kN
$T_{CD} = 218.4$ N, $l = 4.674$ m

7–95. $P = 360$ N

7–97. $y_B = 2.60$ m, $y_D = 2.11$ m

7–98. $P_1 = 3288.1$ N, $y_D = 1.933$ m

7–99. $T_{AB} = 413$ N
$T_{BC} = 282$ N
$T_{CD} = 358$ N
$y_C = 3.08$ m

7–101. $T_{max} = 5.36$ kN, $L = 51.3$ m

7–102. $T_A = 46.4$ kN
$T_B = 54.5$ kN

7–103. $w_o = 4.40$ kN/m

7–105. $T_B = 48.7$ MN, $T_A = 51.4$ MN

7–106. $w_o = 77.8$ kN/m

7–107. $m_C = 478$ kg
$h = 0.827$ m
$L = 13.2$ m

7–109. $T_{max} = 594$ kN

7–110. $T_{min} = 552$ kN

7–111. $y = 4.5\left(1 - \cos\dfrac{\pi}{24}x\right)$ m

$T_{max} = 60.2$ kN

7–113. $w_o = 0.846$ kN/m

7–114. $h = 3.104$ m

7–115. $L = 10.39$ m

7–118. $h = 1.47$ m

7–119. $(F_V)_A = 165$ N, $(F_H)_A = 73.9$ N

7–121. $F_A = F_C = 11.1$ kN, $h = 23.5$ m

7–122. $\dfrac{h}{L} = 0.141$

7–123. $y = 45.512\{\cosh(0.0219722x) - 1\}$ m
$L = 52.55$ m

7–125. $N_D = -6.08$ kN
$V_D = -2.6$ kN, $M_D = -12.99$ kN · m

7–126. $V = 10 - 2x$
$M = 10x - x^2 - 30$

7–127. $a = 0.366$ L

7–129. $l_d = 5.67$ m, $d = 19.8$ m

7–130. For $0 \le x < 3$ m, $V = 1.50$ kN,
$M = \{1.50\,x\}$ kN · m.
For 3 m $< x \le 6$ m, $V = -4.50$ kN,
$M = \{27.0 - 4.50x\}$ kN · m.

7–131. $(V_C)_x = 450$ N, $N_C = 0$, $(V_C)_z = -550$ N
$(M_C)_x = -825$ N · m, $T_C = 30$ N · m,
$(M_C)_z = 675$ N · m

7–133. For $0 \le x < 5$ m, $V = (2.5 - 2x)$ kN,
$M = (2.5x - x^2)$ kN · m.
For 5 m $< x \le 10$ m, $V = -7.5$ kN,
$M = (25 - 7.5x)$ kN · m.

7–134. $N_C = 0$, $V_C = 9.0$ kN, $M_C = -62.5$ kN · m,
$N_B = 0$, $V_B = 27.5$ kN, $M_B = -184.5$ kN · m

7–135. $V_1(x) = A_y \dfrac{1}{\text{kN}}$

$M_1(x) = A_y x \dfrac{1}{\text{kN} \cdot \text{m}}$

$V_2(x) = \left[A_y - w(x - a)\right]\dfrac{1}{\text{kN}}$

$M_2(x) = \left[A_y x - w\dfrac{(x-a)^2}{2}\right]\dfrac{1}{\text{kN} \cdot \text{m}}$

7–137. For $0 \le x = 2$ m, $V = \{5.29 - 0.196x\}$ kN,
$M = \{5.29x - 0.0981x^2\}$ kN · m.
For 2 m $< x \le 5$ m, $V = \{-0.196x - 2.71\}$ kN,
$M = \{16.0 - 2.71x - 0.0981x^2\}$ kN · m.

7–139. For $0 \le x < 0.45$ m, $V = -1.47$ kN,
$M = (-1.47x)$ kN · m.
For 0.45 m $\le x \le 0.9$ m, $V = 1.47$ kN,
$M = (1.47x - 1.324)$ kN · m.

Chapter 8

8–1. $N_A = 16.5$ kN,
$N_B = 42.3$ kN, the **mine car does not move**.

8–2. $P = 12.8$ kN

8–3. $x = 0.5$ m

8–5. $O_x = 46.4$ N, $O_y = 400$ N

8–6. $O_x = 280$ N, $O_y = 945$ N

8–7. **a.** No
b. Yes

8–9. $\mu = 0.27$

8–10. It is not possible since
$\mu = 0.2 < 0.27$.

8–11. $n = 6$

8–13. $P = 147$ N

8–14. $P = 182$ N, $N_C = 606.6$ N

8–15. $\theta = 21.8°$

8–17. $\mu_s = 0.577$

8–18. $\mu = 0.354$

8–19. $d = 537$ mm

8–21. $\theta = 2\tan^{-1}\mu s$

8–22. Slip at clamp, n = 9.17, slip between end boards;
n = 8.12, *use* n = 8.

8–23. $P = 371.4$ N

8–25. $P = \dfrac{M_O(b - \mu_s c)}{\mu_s ra}$

8–26. $P = \dfrac{M_O(b + \mu_s c)}{\mu_s r a}$

$P < 0$ if $(b - \mu_s c) < 0$ i.e. if $\dfrac{b}{c} < \mu_s$

8–27. $P = \dfrac{M_O(b + \mu_s c)}{\mu_s r a}$

8–29. $\theta = 11.0°$
8–30. $\mu_s = 0.577$
8–31. $\theta = 31.0°$
8–33. $\mu_s = 0.268$
8–34. Thus, **he can move the crate**.
8–35. $\mu_{s'} = 0.376$
8–37. $P = 140$ N, $x_A = 523.5$ mm
8–38. $F_C = 30.5$ N, $N_C = 152.3$ N
$x_1 = 0.79$ m
8–39. $\theta = 7.50°$, $T = 452$ N
8–41. $F_A = 0.44$ kN, $N_A = 1.47$ kN, $N_B = 1.24$ kN
8–42. $\mu = 0.176$
8–43. $\mu_s = 0.509$
8–45. Can A will not move.
8–46. $P = 355$ N
8–47. $\mu_C = 0.0734$, $\mu_B = 0.0964$
8–49. $M = 77.3$ N · m
8–50. $F_A = 71.4$ N
8–53. $\phi = 42.6°$
8–54. $P = \dfrac{1}{2}\mu_s W$
8–55. $\theta = 33.4°$
8–57. $\mathbf{N} = (-a\mathbf{j} + h\mathbf{k}) \times (2a\mathbf{i} - a\mathbf{j})$, $n = \mathbf{N}/N$,
$h = \dfrac{2}{\sqrt{5}}a\mu$
8–58. $\theta = 33.4°$
8–59. $P = 5.53$ kN
Since a force P (> 0) is required to pull out the wedge, **the wedge will be self–locking when** $P = 0$.
8–61. $P = 34.5$ N
8–62. $x = 18.3$ mm
8–63. $P = 2.39$ kN
8–65. $P = 304$ N
8–66. $x = 32.9$ mm
8–67. $F = 3.84$ kN, $P = 73.3$ kN
8–69. $W = 7.19$kN
8–70. Since $\phi s > \theta p$, **the screw is self–locking.**
8–71. $F = 66.7$ N
8–73. $F_{AB} = 1.38$ kN (T), $F_{BD} = 828$ N (C)
$F_{BC} = 1.10$ kN (C), $F_{AC} = 828$ N (C)
$F_{AD} = 1.10$ kN (C), $F_{CD} = 1.38$ kN (T)
8–74. $M = 4.53$N · m
8–75. $M = 48.3$ N · m
8–77. $F_E = 72.7$ N
$F_D = 72.7$ N

8–78. $P = 880$ N
$M = 352$ N · m
8–79. $T = 4.02$ kN, $F = 11.6$ kN
8–81. $P = 1.98$ kN
8–82. $M = 0.202$ N · m
8–83. **a.** $F = 1.31$ kN
b. $F = 372$ N
8–85. 15.9 N $< P < 217.4$ N
8–86. $P = 1.54$ kN
8–87. $\mu_s = 0.0583$
8–89. Approx. 2 turns $(695°)$
8–90. $\theta = 99.2°$
8–91. $m_A = 2.22$ kg
8–93. $M = 177$ N · m
8–94. $m_D = 25.6$ kg
8–95. $m_A = 7.82$ kg
8–97. $P = 736$ N
8–99. $M = 187$ N · m
8–101. $m_C = 136$ kg, $M = 134$ N · m
8–102. $M = 3.93$ N · m
8–103. $F = 2.49$ kN
8–105. $M = 50.0$ N · m, $x = 286$ mm
8–106. $F_s = 85.4$ N
8–107. $M = 19$ N · m
8–109. $M = 54.4$ N · m
8–110. $M = 46.7$ N · m
8–111. $\mu_s = 0.321$
8–113. $P = 118$ N
8–114. $M = \dfrac{2\mu_s PR}{3\cos\theta}$
8–115. $M = 0.521$ PμR
8–117. $d = 0.140$ L
8–118. $(\gamma f)_A = 5$ mm, $(\gamma f)_B = 2$ mm
8–119. $(\gamma f)_A = 7.5$ mm, $(\gamma f)_B = 3$ mm
8–121. $P = 814$ N
8–122. $P = 68.97$ N
8–123. $P = 145.0$ N
8–125. $r = 20.6$ mm
8–126. $P = 299$ N
8–127. $P = 266$ N
8–129. $d = 38.5$ mm
8–130. $P = 245$ N
8–133. $M = 90.6$ N·m
8–134. $\mu_s = 0.4$
8–135. $M = 270$ N·m
8–137. $P = 40.2$ N
8–138. Check if crate slips on dolly, if crate tips, or if dolly tips. $P = 196$ N.
8–139. $m_I = 1500$ kg
8–141. $\theta = 35.0°$
8–142. $P = 140$ N
8–143. $P = 474$ N

Chapter 9

9–1. $\bar{x} = 124\,\text{mm}$

$\bar{y} = 0$

9–2. $\bar{x} = 0.546\,\text{m}, O_x = 0, O_y = 7.06\,\text{N}$

$M_O = 3.85\,\text{N} \cdot \text{m}$

9–3. $m = \dfrac{3}{2} m_o L, \ \bar{x} = \dfrac{5}{9}L$

9–5. $\bar{x} = 0.299a$

$\bar{y} = 0.537a$

9–6. $\bar{y} = \dfrac{2}{5}\text{m}$

9–7. $\bar{x} = \dfrac{3}{8}a$

9–9. $\bar{x} = 5\,\text{m}$

9–10. $\bar{y} = 1.43\,\text{m}$

9–11. $\bar{x} = \dfrac{3b}{4}$

9–13. $\bar{x} = 6\,\text{m}$

9–14. $\bar{y} = 2.8\,\text{m}$

9–15. $\bar{x} = \dfrac{b - a}{\ln\dfrac{b}{a}}$

9–17. $\bar{x} = \dfrac{n+1}{2(n+2)}a$

9–18. $\bar{x} = \dfrac{(n+1)}{2(n+2)}a$

9–19. $\bar{y} = \dfrac{hn}{2n+1}$

9–21. $\bar{x} = \dfrac{3a}{8}$

9–22. $\bar{y} = \dfrac{3a}{5}$

9–23. $\bar{x} = \dfrac{4a}{3\pi}$

9–25. $\bar{x} = 0.649\,\text{m}$

9–26. $\bar{x} = \left(\dfrac{\pi - 2}{2\pi}\right)a$

9–27. $\bar{y} = \dfrac{\pi}{8}a$

9–29. $\bar{y} = \dfrac{25}{56}\text{m}$

9–30. $\bar{x} = 1\,\text{m}$

9–31. $\bar{y} = 0.4\,\text{m}$

9–33. $\bar{y} = \dfrac{\pi a}{8}$

9–34. $\bar{x} = 1.26\,\text{m}$

$\bar{y} = 0.143\,\text{m}$

$N_B = 47.9\,\text{kN}$

$A_x = 33.9\,\text{kN}$

$A_y = 73.9\,\text{kN}$

9–35. $x_c = 1.80\,\text{m}$

9–37. $dm = \rho\,dV = \rho_0 xy\,(ty\,dx), m = \dfrac{1}{8}\rho_0 r^4 t$

$\bar{x} = \dfrac{8}{15}r$

$\bar{y} = \dfrac{8}{15}r$

9–38. $\bar{r} = 0.833a$

9–39. $\bar{y} = 4\,\text{m}$

9–41. $\bar{z} = \dfrac{3}{8}a$

9–42. $\bar{y} = 2.67\,\text{m}$

9–43. $\bar{z} = \dfrac{5h}{6}$

9–45. $z_C = 0.422\,\text{m}$

9–46. $\bar{z} = \dfrac{8}{15}r$

9–47. $\bar{z} = \dfrac{h}{4}, \bar{x} = \bar{y} = \dfrac{a}{\pi}$

9–50. $\bar{z} = \dfrac{c}{4}$

9–51. $d = 3\,\text{m}$

9–53. $\bar{x} = 77.3\,\text{mm}$

$\bar{y} = 121\,\text{mm}$

9–54. $\bar{x} = 0, \bar{y} = 58.3\,\text{mm}$

9–55. $\bar{x} = 1.65\,\text{m}, \bar{y} = 9.25\,\text{m}$

$A_x = 0, A_y = 1.32\,\text{kN}$

$E_y = 342\,\text{N}$

9–57. $\bar{x} = \dfrac{W_1}{W}b$

$\bar{y} = \dfrac{b(W_2 - W_1)\sqrt{b^2 - c^2}}{cW}$

9–58. $\bar{y} = 154\,\text{mm}$

9–59. $\bar{y} = 293\,\text{mm}$

9–61. $\bar{x} = 77.2\,\text{mm}, \bar{y} = 31.7\,\text{mm}$

9–62. $\bar{y} = 291\,\text{mm}$

9–63. $\bar{x} = 2.22\,\text{m}, \bar{y} = 1.41\,\text{m}$

9–65. $x_c = 0.2\,\text{m}, y_c = 4.365\,\text{m}$

9–66. $\bar{y} = 272\,\text{mm}$

9–67. $\bar{y} = 66.1\,\text{mm}$

9–69. $\bar{x}=\dfrac{\frac{2}{3}r\sin^3\alpha}{\alpha-\dfrac{\sin 2\alpha}{2}}$

9–70. $\bar{y}=\dfrac{\sqrt{2}(a^2+at-t^2)}{2(2a-t)}$

9–71. $\bar{y}=85.9\,\text{mm}$

9–73. $n\le\dfrac{L}{d}$

9–74. $\bar{z}=122\,\text{mm}$

9–75. $h=385\,\text{mm}$

9–77. $y_c=56.6\,\text{m}$

9–78. $\bar{y}=53.0\,\text{mm}$

9–79. $\bar{x}=22.7\,\text{mm}$

$\bar{y}=29.5\,\text{mm}$

$\bar{z}=22.6\,\text{mm}$

9–81. $\bar{z}=359\,\text{mm}$

9–82. $h=323\,\text{mm}$

9–83. $\bar{z}=128\,\text{mm}$

9–85. $h=48\,\text{mm}$

9–86. $\bar{z}=463\,\text{mm}$

9–87. $\bar{z}=58.1\,\text{mm}$

9–89. $\bar{z}=754\,\text{mm}$

9–90. $V=25.5\,\text{m}^3$

9–91. 14.4 liters

9–93. $V=22.1\,\text{m}^3$

9–94. $V=4.25\,(10^6)\,\text{mm}^3$

9–95. $A=302\,\text{m}^2$

9–97. $A=188\,\text{m}^2$

9–98. $V=207\,\text{m}^3$

9–99. $V_c=20.5\,\text{m}^3$

9–101. $A=8\pi ba,\ V=2\pi ba^2$

9–102. $A=1.403\,(10^6)\,\text{mm}^3$

9–103. $A=88.0\,\text{m}^2$

9–105. $m=138\,\text{kg}$

9–106. $V=\dfrac{2}{3}\pi ab^2$

9–107. $A=1.33\,\text{m}^2$

$\bar{x}=0.6\,\text{m}$

$V=5.03\,\text{m}^3$

9–109. $A=119\,(10^3)\,\text{mm}^2$

9–110. $A=\sqrt{3}\pi a^2$

$V=\dfrac{\pi}{4}a^3$

9–111. $h=29.9\,\text{mm}$

9–113. $m=138\,\text{kg}$

9–114. $A=1365\,\text{m}^2$

9–115. $d=2.68\,\text{m}$

9–117. $F_R=\dfrac{2}{3}(xdx)[(4-y)dy],\ F_R=24.0\,\text{kN}$

$\bar{x}=2.00\,\text{m},\ \bar{y}=1.33\,\text{m}$

9–118. $F_R=\dfrac{4ab}{\pi^2}p_0$

$\bar{x}=\dfrac{a}{2}\quad \bar{y}=\dfrac{b}{2}$

9–119. $A_y=2.51\,\text{MN}$
$B_x=2.20\,\text{MN}$
$B_y=859\,\text{kN}$

9–121. $D_x=101\,\text{kN}$
$C_x=46.6\,\text{kN}$

9–122. $d=3.65\,\text{m}$

9–123. $F=1.41\,\text{MN},\ h=4\,\text{m}$

9–125. $F_R=2.77\,\text{MN}$
$h=5.22\,\text{m},\ F_{bot}=3.02\,\text{MN}$

9–126. $F=3.85\,\text{kN}$
$d'=0.625\,\text{m}$

9–127. $F_R=6.93\,\text{kN},\ \bar{y}=-0.125\,\text{m}$

9–129. $F=391\,\text{kN/m}$

9–130. $F_x=628\,\text{kN}$
$F_y=538\,\text{kN}$

9–131. $V=22.7\,(10)^{-3}\,\text{m}^3$

9–133. $\bar{y}=87.5\,\text{mm}$

9–134. $\bar{x}=\bar{y}=0,\ \bar{z}=\dfrac{2}{3}a$

9–135. $x_C=\dfrac{a+b}{3}$

9–137. $x_C=1.594\,\text{m},\ y_C=0.940\,\text{m}$

9–138. $l=265\,\text{mm}$
$\theta=70.4°$

9–139. $A_y=1.77\,\text{MN},\ B_x=1.57\,\text{MN},\ B_y=594\,\text{kN}$

Chapter 10

10–1. $I_x=\dfrac{2}{7}bh^3$

10–2. $I_y=\dfrac{2}{15}hb^3$

10–3. $I_x=2.13\,\text{m}^4$

10–5. $I_x=39.0\,\text{m}^4$

10–6. $I_y=8.53\,\text{m}^4$

10–7. $I_x=0.533\,\text{m}^4$

10–9. $I_x=\dfrac{2}{15}bh^3$

10–10. $I_x=\dfrac{1}{3}tl^3\sin^2(\theta)$

10–11. $I_x=3.20\,\text{m}^4$

10–13. $I_x=4.27\,\text{m}^4$

10–14. $I_y=36.7\,\text{m}^4$

10–15. $I_y=0.628\,\text{m}^4$

10–17. $I_x=\dfrac{4a^4}{9\pi}$

10–18. $I_y = \left(\dfrac{\pi^2 - 4}{\pi^3}\right)a^4$

10–19. $I_x = 10 \text{ m}^4$

10–21. $I_x = \dfrac{ah^3}{28}$

10–22. $I_y = \dfrac{a^3 h}{20}$

10–23. $dA = (r \, d\theta)dr, \; y = r\sin\theta, \; I_x = \dfrac{r_0^4}{8}(\alpha - \sin\alpha)$

10–25. $I_x = 798 \, (10^6) \text{ mm}^4$

10–26. $I_y = 10.3 \, (10^9) \text{ mm}^4$

10–27. $k_x = 103.5 \text{ mm}$

10–29. $\overline{y} = 0.181 \text{ m}, \; I_x = 4.23(10^{-3}) \text{ m}^4$

10–30. $\overline{x} = 68.0 \text{ mm}, \; I_{y'} = 36.9(10^6) \text{ mm}^4$

10–31. $I_{x'} = 49.5(10^6) \text{ mm}^4$

10–33. $I_y = \dfrac{r^4}{4}(\theta + \dfrac{1}{2}\sin 2\theta - 2\sin\theta\cos^3\theta)$

10–34. $I_y = 115(10^6) \text{ mm}^4$

10–35. $\overline{y} = 207 \text{ mm}$

$\overline{I}_{x'} = 222(10^6) \text{ mm}^4$

10–37. $I_y = 2.51(10^6) \text{ mm}^4$

10–38. $I_x = 115(10^6) \text{ mm}^4$

10–39. $I_y = 153(10^6) \text{ mm}^4$

10–41. $\overline{x} = 71.32 \text{ mm}$

$I_y = 3.60(10^6) \text{ mm}^4$

10–42. $I_x = 154(10^6) \text{ mm}^4$

10–43. $I_y = 91.3(10^6) \text{ mm}^4$

10–45. $\overline{x} = 61.6 \text{ mm}, \; \overline{I}_{y'} = 41.2(10^6) \text{ mm}^4$

10–46. $\overline{y} = 22.5 \text{ mm}, \; I_{x'} = 34.4(10^6) \text{ mm}^4$

10–47. $I_y = 122(10^6) \text{ mm}^4$

10–49. $I_y = 29.8(10^6) \text{ mm}^4$

10–50. $\overline{y} = 1.81 \text{ m}, \; \overline{I}_x = 42.33 \text{ m}^4$

10–51. $\overline{I}_{x'} = 162(10^6) \text{ mm}^4$

10–53. $\overline{I}_{y'} = \dfrac{ab\sin\theta}{12}(b^2 + a^2\cos^2\theta)$

10–54. $I_{xy} = \dfrac{1}{6}l^3 t \sin 2\theta$

10–55. $I_{xy} = \dfrac{3}{16}b^2 h^2$

10–57. $I_{xy} = \dfrac{a^2 b^2}{4(n+1)}$

10–58. $\overline{I}_{x'y'} = 1.07 \text{ m}^4$

10–59. $I_{xy} = 0.511 \text{ m}^4$

10–61. $I_{xy} = \dfrac{a^2 c \sin^2\theta}{12}(4a\cos\theta + 3c)$

10–62. $I_{xy} = \dfrac{1}{6}a^2 b^2$

10–63. $I_{uv} = 135(10)^6 \text{ mm}^4$

10–65. $I_{xy} = -1.10(10^6) \text{ mm}^4$

10–66. $I_{xy} = 98.4(10^6) \text{ mm}^4$

10–67. $I_{xy} = 17.1 \, (10^6) \text{ mm}^4$

10–69. $I_u = 85.3 \, (10^6) \text{ mm}^4$

$I_v = 85.3 \, (10^6) \text{ mm}^4$

10–70. $I_u = 909 \, (10^6) \text{ mm}^4$

$I_v = 703 \, (10^6) \text{ mm}^4$

$I_{uv} = 179 \, (10^6) \text{ mm}^4$

10–71. $I_u = 909 \, (10^6) \text{ mm}^4$

$I_v = 703 \, (10^6) \text{ mm}^4$

$I_{uv} = 179 \, (10^6) \text{ mm}^4$

10–73. $\overline{y} = 82.5 \text{ mm}$

$I_u = 109 \, (10^8) \text{ mm}^4$

$I_v = 238 \, (10^8) \text{ mm}^4$

$I_{uv} = 111 \, (10^8) \text{ mm}^4$

10–74. $\overline{y} = 82.5 \text{ mm}$

$I_u = 43.4 \, (10^6) \text{ mm}^4$

$I_v = 47.0 \, (10^6) \text{ mm}^4$

$I_{uv} = -3.08 \, (10^6) \text{ mm}^4$

10–75. $\overline{y} = 82.5 \text{ mm}$

$I_u = 43.4 \, (10^6) \text{ mm}^4$

$I_v = 47.0 \, (10^6) \text{ mm}^4$

$I_{uv} = -3.08 \, (10^6) \text{ mm}^4$

10–77. $\overline{x} = 48.2 \text{ mm}$

$I_u = 112 \, (10^6) \text{ mm}^4$

$I_v = 258 \, (10^6) \text{ mm}^4$

$I_{uv} = -126 \, (10^6) \text{ mm}^4$

10–78. $I_{max} = 4.92 \, (10^6) \text{ mm}^4, \; I_{min} = 1.36 \, (10^6) \text{ mm}^4$

10–79. $I_{max} = 4.92 \, (10^6) \text{ mm}^4, \; I_{min} = 1.36 \, (10^6) \text{ mm}^4$

10–81. $\theta = 12.3° \; (Counterclockwise)$

$\theta = 77.7° \; (Clockwise)$

10–82. $\overline{y} = 79.23 \text{ mm}$

$I_u = 42.2 \, (10^6) \text{ mm}^4$

$I_v = 41.9 \, (10^6) \text{ mm}^4$

$I_{uv} = 0.28 \, (10^6) \text{ mm}^4$

10–83. $\overline{y} = 79.23 \text{ mm}$

$I_u = 42.19 \, (10^6) \text{ mm}^4$

$I_v = 41.86 \, (10^6) \text{ mm}^4$

$I_{uv} = 0.28 \, (10^6) \text{ mm}^4$

10–85. $I_x = \dfrac{2}{5}mb^2$

10–86. $I_x = \dfrac{3}{10}mr^2$

10–87. $k_x = 57.7 \text{ mm}$

10–89. $I_y = \dfrac{m}{6}(a^2 + h^2)$

10–90. $I_z = \dfrac{7}{18}ma^2$

10–91. $I_x = \dfrac{2}{5}mr^2$

10–93. $I_y = \dfrac{5}{18}m$

10–94. $I_y = 1.71(10^3)$ kg · m²

10–95. $I_z = 34.2$ kg · m²

10–97. $\bar{y} = 1.78$ m, $I_G = 4.45$ kg · m²

10–98. $\bar{y} = 203$ mm, $I_G = 0.230$ kg · m²

10–99. $I_A = 327.3$ kg · m²

10–101. $I_z = 1.53$ kg · m²

10–102. $k_O = 3.15$ m

10–103. $I_A = 50.75$ kg · m²

10–105. $I_O = 0.450\,(10^{-3})$ kg · m²
$\bar{y} = 57.7$ mm
$I_G = 0.150\,(10^{-3})$ kg · m²

10–106. $I_y = 0.144$ kg · m²

10–107. $I_z = 0.113$ kg · m²

10–109. $I_{z'} = 7.19$ kg · m²

10–110. $k_x = 109$ mm

10–111. $I_x = 914\,(10^6)$ mm⁴
$\bar{I}_x = 146(10^6)$ mm⁴

10–113. a. $I_x = \dfrac{bh^3}{12}$

b. $\bar{I}_x = \dfrac{bh^3}{36}$

10–114. $I_x = \dfrac{93}{70}mb^2$

10–115. $I_x = 9.05$ m⁴

10–117. $I_u = 5.09\,(10^6)$ mm⁴, $I_v = 5.09\,(10^6)$ mm⁴, $I_{uv} = 0$

10–118. $I_y = 0.0954\,d^4$

10–119. $I_y = 0.187d^4$

Chapter 11
11–1. $F = 2P\cot\theta$

11–2. $F = 24.5$ N

11–3. $F = 4.62$ kN

11–5. $\theta = 27.4°$ or $72.7°$

11–6. $k = 1.05$ kN/m

11–7. $F = 2P\cot\theta$

11–9. $F = 245.25$ N

11–10. $P = \dfrac{W}{2}\cot\theta$

11–11. $F = 49.1$ N

11–13. $\theta = 42.7°$, $\theta = 54.8°$

11–14. $k = 1.24$ kN/m

11–15. $F = \dfrac{M}{2a\sin\theta}$

11–17. $M_{sp} = 300[2(\pi/2 - \theta)]$, $\theta = 61.4°$

11–18. $\theta = 90°$, $\theta = 17.1°$

11–19. $k = 136$ N/m

11–21. $k = 9.88$ kN/m

11–22. $F = \left(\dfrac{W(a+b-d\tan\theta)}{ac}\right)\sqrt{a^2 + c^2 + 2ac\sin\theta}$

11–23. $F = 512$ N, $\delta_x = -0.09769\,\delta\theta$

11–25. $m = 100$ kg

11–26. $y = -925$ m Unstable
$y = 0.481$ m

11–27. $x = 0$ Unstable
$x = 0.167$ m Stable

11–29. $\theta = 36.9°$ Unstable
$\theta = 90°$ Stable

11–30. $\theta = 38.7°$ Unstable
$\theta = 90°$ Stable
$\theta = 141°$ Unstable

11–31. $\theta = 35.5°$ Unstable
$\theta = 90°$ Stable

11–33. $h = 0$

11–34. $d = \dfrac{h}{3}$

11–35. $\theta = 36.4°$ Unstable
$\theta = 62.3°$ Stable

11–37. $M = 22.5$ kg

11–38. $b < 2r$

11–39. $\theta_1 = 36.1°$ Unstable
$\theta_2 = 90°$ Stable

11–41. $\theta = 9.18°$, $\phi = 17.38°$

11–42. $W_2 = W_1\left(\dfrac{b}{a}\right)\dfrac{\sin\theta}{\cos\phi}$

11–43. $\theta = 76.8°$ Stable

11–46. $\theta = \sin^{-1}\left(\dfrac{4W}{ka}\right)$, $\theta = 90°$

11–47. $m = 5.29$ kg

11–49. $V = 2W\,(r\csc\theta - (a/2)\cos\theta)$

11–50. $M = 42.5$ N·m

11–51. $\theta = 90°$, $\theta = 30°$

11–53. $\theta = 9.46°$ Stable

11–54. $\theta = 0$ Unstable

11–55. $h = 35.46$ mm

11–57. $W_G = 12.5$ N

11–58. $x = 400$ mm

11–59. $F = P\csc(\theta)$

11–61. $k = 166$ N/m

11–62. $m_{AB} = 150$ kg

Index

Fundamental Equations of Statics

Cartesian Vector

$$\mathbf{A} = A_x\mathbf{i} + A_y\mathbf{j} + A_z\mathbf{k}$$

Magnitude

$$A = \sqrt{A_x^2 + A_y^2 + A_z^2}$$

Directions

$$\mathbf{u}_A = \frac{\mathbf{A}}{A} = \frac{A_x}{A}\mathbf{i} + \frac{A_y}{A}\mathbf{j} + \frac{A_z}{A}\mathbf{k}$$
$$= \cos\alpha\mathbf{i} + \cos\beta\mathbf{j} + \cos\gamma\mathbf{k}$$
$$\cos^2\alpha + \cos^2\beta + \cos^2\gamma = 1$$

Dot Product

$$\mathbf{A} \cdot \mathbf{B} = AB\cos\theta$$
$$= A_xB_x + A_yB_y + A_zB_z$$

Cross Product

$$\mathbf{C} = \mathbf{A} \times \mathbf{B} = \begin{vmatrix} \mathbf{i} & \mathbf{j} & \mathbf{k} \\ A_x & A_y & A_z \\ B_x & B_y & B_z \end{vmatrix}$$

Cartesian Position Vector

$$\mathbf{r} = (x_2 - x_1)\mathbf{i} + (y_2 - y_1)\mathbf{j} + (z_2 - z_1)\mathbf{k}$$

Cartesian Force Vector

$$\mathbf{F} = F\mathbf{u} = F\left(\frac{\mathbf{r}}{r}\right)$$

Moment of a Force

$$M_o = Fd$$
$$\mathbf{M}_o = \mathbf{r} \times \mathbf{F} = \begin{vmatrix} \mathbf{i} & \mathbf{j} & \mathbf{k} \\ r_x & r_y & r_z \\ F_x & F_y & F_z \end{vmatrix}$$

Moment of a Force About a Specified Axis

$$M_a = \mathbf{u} \cdot \mathbf{r} \times \mathbf{F} = \begin{vmatrix} u_x & u_y & u_z \\ r_x & r_y & r_z \\ F_x & F_y & F_z \end{vmatrix}$$

Simplification of a Force and Couple System

$$\mathbf{F}_R = \Sigma\mathbf{F}$$
$$(\mathbf{M}_R)_O = \Sigma\mathbf{M} + \Sigma\mathbf{M}_O$$

Equilibrium

Particle

$$\Sigma F_x = 0, \Sigma F_y = 0, \Sigma F_z = 0$$

Rigid Body-Two Dimensions

$$\Sigma F_x = 0, \Sigma F_y = 0, \Sigma M_O = 0$$

Rigid Body-Three Dimensions

$$\Sigma F_x = 0, \Sigma F_y = 0, \Sigma F_z = 0$$
$$\Sigma M_{x'} = 0, \Sigma M_{y'} = 0, \Sigma M_{z'} = 0$$

Friction

Static (maximum) $F_s = \mu_s N$

Kinetic $F_k = \mu_k N$

Center of Gravity

Particles or Discrete Parts

$$\bar{r} = \frac{\Sigma\tilde{r}W}{\Sigma W}$$

Body

$$\bar{r} = \frac{\int \tilde{r}\, dW}{\int dW}$$

Area and Mass Moments of Inertia

$$I = \int r^2\, dA \qquad I = \int r^2\, dm$$

Parallel-Axis Theorem

$$I = \bar{I} + Ad^2 \qquad I = \bar{I} + md^2$$

Radius of Gyration

$$k = \sqrt{\frac{I}{A}} \qquad k = \sqrt{\frac{I}{m}}$$

Virtual Work

$$\delta U = 0$$

SI Prefixes

Multiple	Exponential Form	Prefix	SI Symbol
1 000 000 000	10^9	giga	G
1 000 000	10^6	mega	M
1 000	10^3	kilo	k

Submultiple			
0.001	10^{-3}	milli	m
0.000 001	10^{-6}	micro	μ
0.000 000 001	10^{-9}	nano	n

Conversion Factors (FPS) to (SI)

Quantity	Unit of Measurement (FPS)	Equals	Unit of Measurement (SI)
Force	lb		4.448 N
Mass	slug		14.59 kg
Length	ft		0.3048 m

Conversion Factors (FPS)

1 ft = 12 in. (inches)
1 mi. (mile) = 5280 ft
1 kip (kilopound) = 1000 lb
1 ton = 2000 lb

Geometric Properties of Line and Area Elements

Centroid Location	Centroid Location	Area Moment of Inertia

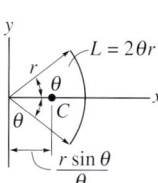

$L = 2\theta r$

Circular arc segment

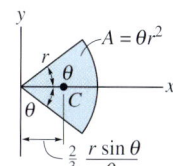

$A = \theta r^2$

$\frac{2}{3}\frac{r \sin \theta}{\theta}$

Circular sector area

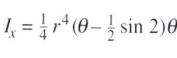

$$I_x = \tfrac{1}{4} r^4 (\theta - \tfrac{1}{2}\sin 2)\theta$$

$$I_y = \tfrac{1}{4} r^4 (\theta + \tfrac{1}{2}\sin 2)\theta$$

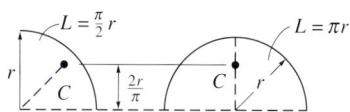

$L = \frac{\pi}{2} r$ $L = \pi r$

$\frac{2r}{\pi}$

Quarter and semicircle arcs

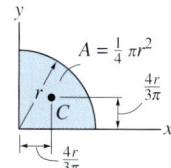

$A = \tfrac{1}{4}\pi r^2$

$\frac{4r}{3\pi}$

$\frac{4r}{3\pi}$

Quarter circle area

$$I_x = \tfrac{1}{16}\pi r^4$$

$$I_y = \tfrac{1}{16}\pi r^4$$

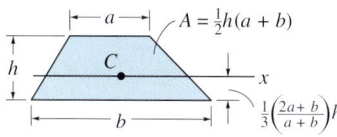

$A = \tfrac{1}{2}h(a + b)$

$\frac{1}{3}\left(\frac{2a+b}{a+b}\right)h$

Trapezoidal area

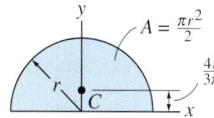

$A = \frac{\pi r^2}{2}$

$\frac{4r}{3\pi}$

Semicircular area

$$I_x = \tfrac{1}{8}\pi r^4$$

$$I_y = \tfrac{1}{8}\pi r^4$$

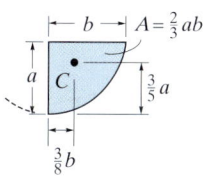

$A = \tfrac{2}{3}ab$

$\tfrac{3}{5}a$

$\tfrac{3}{8}b$

Semiparabolic area

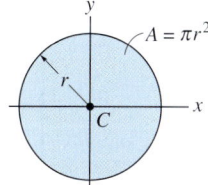

$A = \pi r^2$

Circular area

$$I_x = \tfrac{1}{4}\pi r^4$$

$$I_y = \tfrac{1}{4}\pi r^4$$

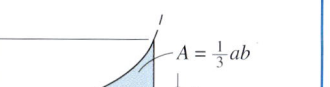

$A = \tfrac{1}{3}ab$

$\tfrac{3}{10}b$

$\tfrac{3}{4}a$

Exparabolic area

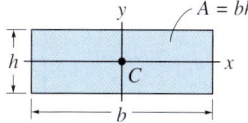

$A = bh$

Rectangular area

$$I_x = \tfrac{1}{12}bh^3$$

$$I_y = \tfrac{1}{12}hb^3$$

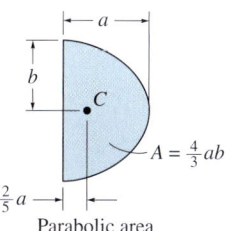

$A = \tfrac{4}{3}ab$

$\tfrac{2}{5}a$

Parabolic area

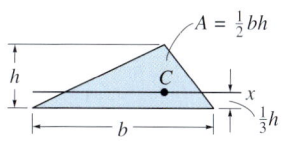

$A = \tfrac{1}{2}bh$

$\tfrac{1}{3}h$

Triangular area

$$I_x = \tfrac{1}{36}bh^3$$

Center of Gravity and Mass Moment of Inertia of Homogeneous Solids

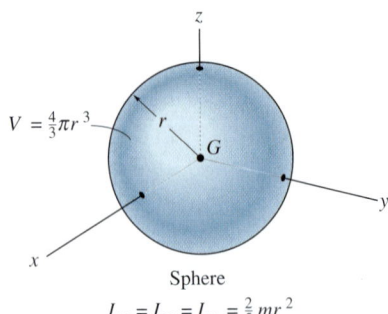

$V = \frac{4}{3}\pi r^3$

Sphere

$I_{xx} = I_{yy} = I_{zz} = \frac{2}{5}mr^2$

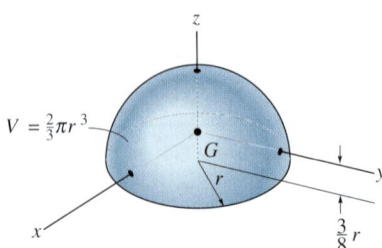

$V = \frac{2}{3}\pi r^3$

Hemisphere

$I_{xx} = I_{yy} = 0.259mr^2 \quad I_{zz} = \frac{2}{5}mr^2$

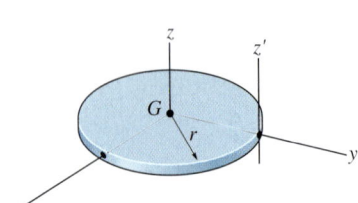

Thin Circular disk

$I_{xx} = I_{yy} = \frac{1}{4}mr^2 \quad I_{zz} = \frac{1}{2}mr^2 \quad I_{z'z'} = \frac{3}{2}mr^2$

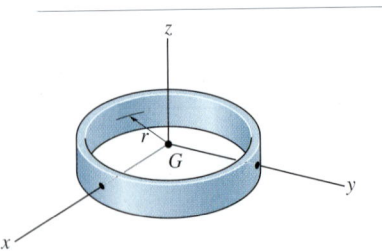

Thin ring

$I_{xx} = I_{yy} = \frac{1}{2}mr^2 \quad I_{zz} = mr^2$

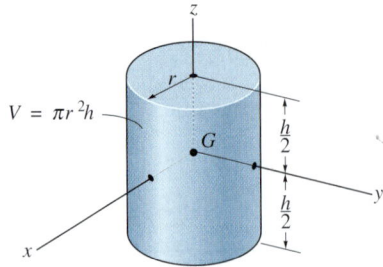

$V = \pi r^2 h$

Cylinder

$I_{xx} = I_{yy} = \frac{1}{12}m(3r^2 + h^2) \quad I_{zz} = \frac{1}{2}mr^2$

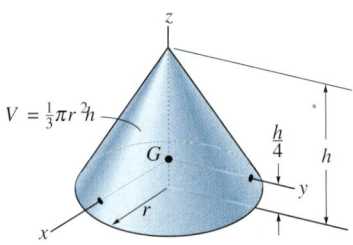

$V = \frac{1}{3}\pi r^2 h$

Cone

$I_{xx} = I_{yy} = \frac{3}{80}m(4r^2 + h^2) \quad I_{zz} = \frac{3}{10}mr^2$

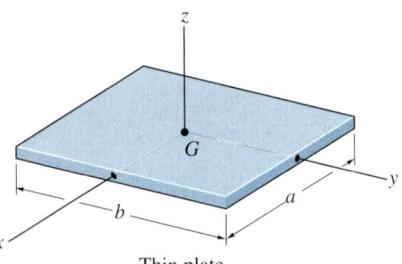

Thin plate

$I_{xx} = \frac{1}{12}mb^2 \quad I_{yy} = \frac{1}{12}ma^2 \quad I_{zz} = \frac{1}{12}m(a^2 + b^2)$

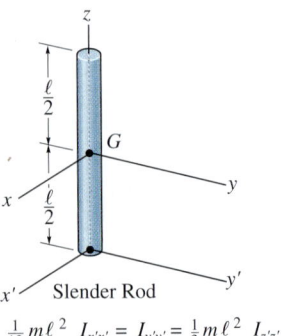

Slender Rod

$I_{xx} = I_{yy} = \frac{1}{12}m\ell^2 \quad I_{x'x'} = I_{y'y'} = \frac{1}{3}m\ell^2 \quad I_{z'z'} = 0$